Introduction to Operations Research

McGRAW-HILL SERIES
IN INDUSTRIAL ENGINEERING AND MANAGEMENT SCIENCE

Barnes: *Statistical Analysis for Engineers and Scientists: A Computer-Based Approach*
Bedworth, Henderson, and Wolfe: *Computer-Integrated Design and Manufacturing*
Black: *The Design of the Factory with a Future*
Blank: *Statistical Procedures for Engineering, Management, and Science*
Bridger: *Introduction to Ergonomics*
Denton: *Safety Management: Improving Performance*
Dervitsiotis: *Operations Management*
Grant and Leavenworth: *Statistical Quality Control*
Hicks: *Industrial Engineering and Management: A New Perspective*
Hillier and Lieberman: *Introduction to Mathematical Programming*
Hillier and Lieberman: *Introduction to Operations Research*
Huchingson: *New Horizons for Human Factors in Design*
Juran and Gryna: *Quality Planning and Analysis: From Product Development through Use*
Khoshnevis: *Discrete Systems Simulation*
Kolarik: *Creating Quality: Concepts, Systems, Strategies, and Tools*
Law and Kelton: *Simulation Modeling and Analysis*
Lehrer: *White-Collar Productivity*
Moen, Nolan, and Provost: *Improving Quality through Planned Experimentation*
Nelson: *Stochastic Modeling: Analysis and Simulation*
Niebel, Draper, and Wysk: *Modern Manufacturing Process Engineering*
Oliver: *Decision Analysis and Forecasting*
Pegden: *Introduction to Simulation Using SIMAN*
Polk: *Methods Analysis and Work Measurement*
Riggs and West: *Engineering Economics*
Taguchi, Elsayed, and Hsiang: *Quality Engineering in Production Systems*
Wu and Coppins: *Linear Programming and Extensions*

Introduction to Operations Research

Sixth Edition

Frederick S. Hillier
Professor of Operations Research
Stanford University

Gerald J. Lieberman
Professor Emeritus of Operations Research
and Statistics
Stanford University

McGraw-Hill, Inc.
New York St. Louis San Francisco Auckland Bogotá Caracas
Lisbon London Madrid Mexico City Milan Montreal
New Delhi San Juan Singapore Sydney Tokyo Toronto

INTRODUCTION TO OPERATIONS RESEARCH

This book is printed on acid-free paper.

1 2 3 4 5 6 7 8 9 0 DOH DOH 9 0 9 8 7 6 5

P/N 028956-5

This book was set in Times Roman by York Graphic Services, Inc.
The editors were Eric M. Munson and Scott Amerman;
the production supervisor was Elizabeth J. Strange.
The cover was designed by Carla Bauer.
R. R. Donnelley & Sons Company was printer and binder.

Library of Congress Cataloging-in-Publication Data

Hillier, Frederick S.
Introduction to operations research / Frederick S. Hillier, Gerald J. Lieberman.—6th ed.
p. cm.
Includes index.
1. Operations research. I. Lieberman, Gerald J. II. Title.
T57.6.H53 1995
658.4′032—dc20 94-48290

About the Authors

Frederick S. Hillier was born and raised in Aberdeen, Washington, where he was an award winner in statewide high school contests in essay writing, mathematics, debate, and music. As an undergraduate at Stanford University, he ranked first in his engineering class, won the McKinsey Prize for technical writing, and won the Hamilton Award for combining excellence in engineering with notable achievements in the humanities and social sciences. Upon his graduation with a B.S. degree in Industrial Engineering, he was awarded three national fellowships (National Science Foundation, Tau Beta Pi, and Danforth) for graduate study at Stanford with specialization in operations research. After receiving his Ph.D. degree, he joined the faculty of Stanford University, where he is now Professor of Operations Research.

Dr. Hillier's research has extended into a variety of areas, including integer programming, queueing theory and its application, statistical quality control, and the application of operations research to the design of production systems and to capital budgeting. He has published widely, and his seminal papers have been selected for republication in books of selected readings at least ten times. He was the first-prize winner of a research contest on ''Capital Budgeting of Interrelated Projects'' sponsored by The Institute of Management Sciences and the U.S. Office of Naval Research. He also has served as Treasurer of the Operations Research Society of America, Vice President for Meetings of The Institute of Management Sciences (TIMS), and Co-General Chairman of the 1989 TIMS International Meeting in Osaka, Japan.

In addition to *Introduction to Operations Research* and the two companion volumes, *Introduction to Mathematical Programming* and *Introduction to Stochastic Models in Operations Research,* his books are *The Evaluation of Risky Interrelated Investments* (North-Holland, 1969) and *Queueing Tables and Graphs* (Elsevier North-Holland, 1981, co-authored by O. S. Yu, with D. M. Avis, L. D. Fossett, F. D. Lo, and M. I. Reiman).

Gerald J. Lieberman is Professor Emeritus of Operations Research and Statistics at Stanford University. He has served as Vice Provost and Dean of Graduate Studies and Research, as well as Provost, at Stanford, and was the founding chair of the Department of Operations Research. He is both an engineer (having received an undergraduate degree in mechanical engineering from Cooper Union) and an operations research statistician (with an A. M. from Columbia University in mathematical statistics, and a Ph.D. from Stanford University in statistics).

His research interests have been in the stochastic areas of operations research, often at the interface of applied probability and statistics. He has published extensively in the areas of reliability and quality control, and in the modeling of complex systems, including their optimal design, when resources are limited.

Dr. Lieberman's professional honors include being elected to the National Academy of Engineering, receiving the Shewhart Medal of the American Society for Quality Control, receiving the Cuthbertson Award for exceptional service to Stanford University, and serving as a fellow at the Center for Advanced Study in the Behavioral Sciences. He also served as president of The Institute of Management Sciences.

In addition to *Introduction to Operations Research* and the two companion volumes, *Introduction to Mathematical Programming* and *Introduction to Stochastic Models in Operations Research,* his books are *Handbook of Industrial Statistics* (Prentice-Hall, 1955, co-authored by A. H. Bowker), *Tables of the Non-Central t-Distribution* (Stanford University Press, 1957, co-authored by G. J. Resnikoff), *Tables of the Hypergeometric Probability Distribution* (Stanford University Press, 1961, co-authored by D. Owen), and *Engineering Statistics,* Second Edition (Prentice-Hall, 1972, co-authored by A. H. Bowker).

To Our Parents

Contents

Preface

Over the past 27 years, we have been deeply gratified by the widespread response to our first five editions. At the outset, we never dreamed that we now would have had the privilege of helping to introduce several hundred thousand students around the world to our field. It has been a heavy responsibility, but we have enjoyed the challenge of meeting the needs of new generations of students. As always, our goal for the current edition has been to help define the modern approach to teaching operations research effectively at an introductory level.

One key element of a modern approach now is the use of the computer. We believe that our OR Courseware accompanying the book is helping to usher in a new era for making effective use of the computer in a modern introductory course. This innovative tutorial software—featuring demonstration examples, interactive routines, and automatic routines—has been expanded, improved (much larger problems now can be solved), and fully integrated with the text and problems (including a guided tour at the end of Chap. 1 and documentation in Appendix 1). Many new routines have been added, including some for the network analysis and simulation chapters, so now *every* chapter after the first two introductory chapters has useful routines for aiding the learning process.

The software now has 17 demonstration examples, thereby supplementing the book with 17 additional examples for those who need them without adding many more pages for those who do not. Furthermore, these demos vividly demonstrate the evolution of an algorithm in ways that cannot be duplicated on the printed page. A new enhancement is the ability to backtrack in each demo to enable quickly referring back to a preceding screen.

The interactive routines also are a key tutorial feature of the software. Each one enables the student to interactively execute one of the algorithms of operations research, making the needed decision at each step while the computer does the needed arithmetic. To get the student started properly, the computer also points out any mistake made on the first iteration (where possible). By enabling the student to focus on concepts rather than mindless number crunching when doing homework to learn an algorithm, we have found that these interactive routines make the learning process *far* more efficient and effective as well as more stimulating. Our students say that they cannot imagine having to do this homework by hand instead.

Perhaps the biggest improvement in the software over the preceding edition has been in the automatic routines. Several new ones have been added, and they have been made considerably more powerful to enable solving essentially any textbook problem or problem encountered in a student project in an introductory course. For example, the automatic simplex method routine now can handle up to 50 functional constraints and

50 decision variables (not counting slack, surplus, and artificial variables), whereas it previously was limited to only 6 functional constraints and 10 variables of all kinds. The output also has been redesigned to resemble that in popular software packages such as LINDO in order to better demonstrate what can be done with commercial software. An open option then is to next introduce the students to a commercial software package as a supplement to our tutorial software. With this option, whenever a problem indicates that an automatic routine from the OR Courseware should be used, the corresponding routine (when available) from the commercial software package can be used instead.

One key change for this edition has been the addition of more than 250 new problems. Some of these have been adapted from previous operations research examinations given by the Society of Actuaries, which has provided a rich and diverse source of interesting problems. Included are some sets of true–false questions to enable students to test their understanding of key concepts. All problems now are organized and numbered by section at the end of each chapter. The problems also are coded to indicate when to use a helpful routine in the OR Courseware.

The new problems include six larger *case problems* that have been added at the ends of Chaps. 3, 4, 6, 12, 15, and 21. In contrast to the usual textbook problems, these case problems develop relatively elaborate and realistic problem scenarios and then require relatively challenging and comprehensive analyses with substantial use of the computer (using automatic routines in the OR Courseware). Therefore, they are suitable for student projects, working either individually or in teams, and can then lead to class discussion of the analysis.

Another key change has been the addition of some dramatic case studies of real applications. We begin in Chap. 1 with a tabular summary of 15 award-winning studies and the great impact that they had on their organizations. We then expand on some of these studies throughout Chap. 2 to illustrate the operations research approach. Problems at the end of Chap. 2 involve further investigation of these studies, which can provide the basis for class discussion. Section 3.5 then presents three case studies in substantial detail, including the factors that contributed to the success of the studies. These case studies should excite students about the importance and relevance of operations research.

Another important enhancement for the current edition has been a considerable increase in the number of formulation examples, including especially some in the first linear programming chapter (Chap. 3) and the integer programming chapter (Chap. 12). In fact, the latter chapter includes a new section devoted entirely to interesting formulation examples.

Other additions include a greatly expanded section on recent algorithmic developments in integer programming and a new section on utility theory (from an applied viewpoint).

As always, we have provided enough material to give the instructor some flexibility in picking and choosing what to cover. However, we do not believe that it is appropriate for an introductory textbook to be encyclopedic in length and coverage. Therefore, with all the additions described above, we felt that it was important to make some difficult choices of old material to eliminate. In addition to trimming the verbiage in a number of places, we have deleted the following material that was in the fifth edition: the chapter on formulating linear programming models (except for moving the goal programming section, one formulation example, and most of the formulation problems to other chapters), the chapter on reliability, and several sections, including those on the transshipment problem (now treated briefly as a special case of the minimum cost flow problem), multidivisional problems, and evaluating travel time for designing queueing systems, along with the appendix on simultaneous linear equations and tables for the chi-square and Poisson distributions.

Another option available for instructors who want to focus on linear programming is to take advantage of the other Hillier-Lieberman book published by McGraw-Hill: *Introduction to Mathematical Programming* (2d ed., 1995). This book is virtually identical to Chaps. 1 to 13 (plus appendixes) of this book with the one exception that Chap. 8 here (The Transportation and Assignment Problems) has been replaced by a much longer Chap. 8 (Special Types of Linear Programming Problems) that also includes sections on the transshipment problem, multidivisional problems, the decomposition principle, multitime period problems, multidivisional multitime period problems, stochastic programming, and chance-constrained programming.

Outside the changes, additions, and deletions already described for this book, the organization remains almost the same as in the fifth edition. One small change is that the section on the dual simplex method (formerly Sec. 9.2, but now Sec. 7.1) has been moved next to the related chapter on duality theory and sensitivity analysis (Chap. 6). Another is that the chapter on the transportation and assignment problems (now Chap. 8) has been moved next to the chapter on network analysis (Chap. 9), with increased emphasis on the network representations of these problems.

Every chapter has received significant revision and updating, ranging from modest refining to extensive rewriting. Some areas receiving a major revision include the chapters on the simplex method (giving more emphasis to geometric insight), decision analysis, and Markov decision processes, as well as the sections on primal-dual forms and continuous-time Markov chains.

The overall thrust of all the revision efforts has been to build upon the strengths of previous editions while thoroughly updating the material and integrating the software to fully meet the needs of students preparing for a career in the twenty-first century. We think that the net effect has been to make this edition even more of a ''student's book''—clear, interesting, and well-organized with lots of helpful examples and illustrations, good motivation and perspective, easy-to-find important material, and enjoyable homework, and without too much notation, terminology, and dense mathematics. We believe and trust that the numerous instructors who have used previous editions will agree that this is the best edition yet.

The prerequisites for a course using this book can be relatively modest. As with previous editions, the mathematics has been kept at a relatively elementary level. Most of Chaps. 1 to 13 (introduction, linear programming, and mathematical programming) require no mathematics beyond high school algebra. Calculus is used only in Chap. 13 (Nonlinear Programming) and in one example in Chap. 10 (Dynamic Programming). Matrix notation is used in Chap. 5 (The Theory of the Simplex Method), Chap. 6 (Duality Theory and Sensitivity Analysis), Sec. 7.4 (An Interior-Point Algorithm), and Chap. 13, but the only background needed for this is presented in Appendix 4. For Chaps. 14 to 21 (probabilistic models), a previous introduction to probability theory is assumed, and calculus is used in a few places. In general terms, the mathematical maturity that a student achieves through taking an elementary calculus course is useful throughout Chaps. 14 to 21 and for the more advanced material in the preceding chapters.

The content of the book is aimed largely at the upper-division undergraduate level (including well-prepared sophomores) and at first-year (master's level) graduate students. Because of the book's great flexibility, there are many ways to package the material into a course. Chapters 1 and 2 give an introduction to the subject of operations research. Chapters 3 to 13 (on linear programming and on mathematical programming) may essentially be covered independently of Chaps. 14 to 21 (on probabilistic models), and vice versa. Furthermore, the individual chapters among Chaps. 3 to 13 are almost independent, except that they all use basic material presented in Chap. 3 and perhaps in Chap. 4. Chapter 6 and Sec. 7.2 also draw upon Chap. 5. Sections 7.1 and 7.2 use parts of Chap. 6. Section 9.6 assumes an acquaintance with the problem formulations in Secs. 8.1, and 8.3, while prior exposure to Secs. 7.3 and 8.2 is helpful (but not essential) in Sec. 9.7. Within Chaps. 14 to 21, there is considerable flexibility of coverage, although some integration of the material is available.

An elementary survey course covering linear programming, mathematical programming, and some probabilistic models can be presented in a quarter (40 hours) or semester by selectively drawing from material throughout the book. For example, a good survey of the field can be obtained from Chaps. 1, 2, 3, 4, 15, 17, 18, 20, and 21, along with parts of Chaps. 9, 10, 12, and 13. A more extensive elementary survey course can be completed in two quarters (60 to 80 hours) by excluding just a few chapters, for example, Chaps. 7, 11, and 19. Chapters 1 to 8 (and perhaps part of Chap. 9) form an excellent basis for a (one-quarter) course in linear programming. The material in Chaps. 9 to 13 covers topics for another (one-quarter) course in other deterministic models. Finally, the material in Chaps. 14 to 21 covers the probabilistic (stochastic) models of operations research suitable for presentation in a (one-quarter) course. In fact, these latter three courses (the material in the entire text) can be viewed as a basic one-year sequence in the techniques of operations research, forming the core of a master's degree program. Each course outlined has been presented at either the undergraduate or graduate level at Stanford University, and this text has been used in the manner suggested.

Again, as in previous editions, we thank our wives, Ann and Helen, for their encouragement and support during the long process of preparing this sixth edition. Our children, David, John, and Mark Hillier, Janet Lieberman Argyres, and Joanne, Michael, and Diana Lieberman have literally grown up with the book and our periodic hibernations to prepare a new edition. Now, most of them have used the book as a text in their own college courses, given considerable advice, and even (in the case of Mark Hillier) become a full-fledged collaborator. It is a joy to see them and (we trust) the book reach maturity together.

Acknowledgments

We are deeply indebted to many people for their part in making this revision possible. However, we must particularly single out two individuals, Professors James G. Morris (University of Wisconsin) and Bruce Pollack-Johnson (Villanova University), who expended an enormous effort to provide a page-by-page critique of the fifth edition. Their advice has been extremely helpful, and we are only sorry that we could not give it full justice in this one revision cycle.

Others who made helpful comments on how to improve the fifth edition (and its software) are literally too numerous to mention. However, we would like to acknowledge particularly the advice and support of Stephen Argyres, David Chang, Jeffery Cochran, Richard Cottle, George Dantzig, B. Curtis Eaves, Peter Glynn, Herbert Hackney, Dwight Hillier, Mark Hillier, Donald Iglehart, Michael Munson, Walter Murray, Charles H. Reilly, Sheldon Ross, Shirley Ross, Michael Saunders, Siegfried Schaible, John Tuttle, Robert and Linda VanderPloeg, Arthur Veinott, Jr., and Morris Weigelt. We also thank many dozens of Stanford students who gave us helpful written suggestions at our request.

Ann Hillier devoted numerous long days and nights to sitting with a Macintosh, doing word processing and constructing many figures and tables. She was a vital member of the team.

We feel very fortunate to have had the services of Mark Hillier to develop the software accompanying the book. Born the same year as the first edition, Mark now is a faculty member in the Management Science Department at the University of Washington.

We also would like to thank the Society of Actuaries for its permission to adapt problems from its previously released operations research examinations. (The first author is serving as a consultant to ACT, the American College Testing Program, for the development of these examinations.)

It was a real pleasure working with McGraw-Hill's thoroughly professional editorial and production staff, including Eric Munson, Scott Amerman, and Elizabeth Strange.

FREDERICK S. HILLIER
GERALD J. LIEBERMAN

Introduction to Operations Research

1

Introduction

1.1 The Origins of Operations Research

Since the advent of the industrial revolution, the world has seen a remarkable growth in the size and complexity of organizations. The artisans' small shops of an earlier era have evolved into the billion-dollar corporations of today. An integral part of this revolutionary change has been a tremendous increase in the division of labor and segmentation of management responsibilities in these organizations. The results have been spectacular. However, along with its blessings, this increasing specialization has created new problems, problems that are still occurring in many organizations. One problem is a tendency for the many components of an organization to grow into relatively autonomous empires with their own goals and value systems, thereby losing sight of how their activities and objectives mesh with those of the overall organization. What is best for one component frequently is detrimental to another, so the components may end up working at cross purposes. A related problem is that as the complexity and specialization in an organization increase, it becomes more and more difficult to allocate the available resources to the various activities in a way that is most effective for the organization as a whole. These kinds of problems and the need to find a better way to solve them provided the environment for the emergence of **operations research** (commonly referred to as **OR**).

The roots of OR can be traced back many decades, when early attempts were made to use a scientific approach in the management of organizations. However, the beginning of the activity called *operations research* has generally been attributed to the military services early in World War II. Because of the war effort, there was an urgent need to allocate scarce resources to the various military operations and to the activities within each operation in an effective manner. Therefore, the British and then the U.S. military management called upon a large number of scientists to apply a scientific approach to dealing with this and other strategic and tactical problems. In effect, they were asked to do *research on* (military) *operations*. These teams of scientists were the first OR teams. By developing effective methods of using the new tool of radar, these teams were instrumental in winning the Air Battle of Britain. Through their research on how to better manage convoy and antisubmarine operations, they also played a major role in winning the Battle of the North Atlantic. Similar efforts assisted the Island Campaign in the Pacific.

When the war ended, the success of OR in the war effort spurred interest in applying OR outside the military as well. As the industrial boom following the war was running its course, the problems caused by the increasing complexity and specialization in organizations were again coming to the forefront. It was becoming apparent to a growing number of people, including business consultants who had served on or with the OR teams during the war, that these were basically the same problems that had been faced by the military but in a different context. By the early 1950s, these individuals had introduced the use of OR to a variety of organizations in business, industry, and government. The rapid spread of OR soon followed.

At least two other factors that played a key role in the rapid growth of OR during this period can be identified. One was the substantial progress that was made early in improving the techniques to OR. After the war, many of the scientists who had participated on OR teams or who had heard about this work were motivated to pursue research relevant to the field; important advancements in the state of the art resulted. A prime example is the *simplex method* for solving linear programming problems, developed by George Dantzig in 1947. Many of the standard tools of OR, such as linear programming, dynamic programming, queueing theory, and inventory theory, were relatively well developed before the end of the 1950s.

A second factor that gave great impetus to the growth of the field was the onslaught of the *computer revolution*. A large amount of computation is usually required to deal most effectively with the complex problems typically considered by OR. Doing this by hand would often be out of the question. Therefore, the development of electronic digital computers, with their ability to perform arithmetic calculations thousands or even millions of times faster than a human being can, was a tremendous boon to OR. A further boost came in the 1980s with the development of increasingly powerful personal computers accompanied by good software packages for doing OR. This brought the use of OR within the easy reach of much larger numbers of people. Today, literally millions of individuals have ready access to OR software. Consequently, a whole range of computers from mainframes to laptops now are being routinely used to solve OR problems.

1.2 The Nature of Operations Research

As its name implies, operations research involves "research on operations." Thus, operations research is applied to problems that concern how to conduct and coordinate the *operations* (i.e., the *activities*) within an organization. The nature of the organization is essentially immaterial, and, in fact, OR has been applied extensively in such diverse areas as manufacturing, transportation, construction, telecommunications, financial planning, health care, the military, and public services, to name just a few. Therefore, the breadth of application is unusually wide.

The *research* part of the name means that operations research uses an approach that resembles the way research is conducted in established scientific fields. To a considerable extent, the *scientific method* is used to investigate the problem of concern. (In fact, the term *management science* sometimes is used as a synonym for operations research.) In particular, the process begins by carefully observing and formulating the problem, including gathering all relevant data. The next step is to construct a scientific (typically mathematical) model that attempts to abstract the essence of the real problem. It is then hypothesized that this model is a sufficiently precise representation of the essential features of the situation that the conclusions (solutions) obtained from the model are also valid for the real problem. Next, suitable experiments are conducted to test this hypothesis, modify it as needed, and eventually verify some form of the hypothesis. (This step is frequently referred to as *model validation.*) Thus, in a certain sense, operations research involves creative scientific research into the fundamental properties of operations. However, there is more to it than this. Specifically, OR is also concerned with the practical management of the organization. Therefore, to be successful, OR must also provide positive, understandable conclusions to the decision maker(s) when they are needed.

Still another characteristic of OR is its broad viewpoint. As implied in the preceding section, OR adopts an organizational point of view. Thus, it attempts to resolve the conflicts of interest among the components of the organization in a way that is best for the organization as a whole. This does not imply that the study of each problem must give explicit consideration to all aspects of the organization; rather, the objectives being sought must be consistent with those of the overall organization.

An additional characteristic is that OR frequently attempts to find a *best* solution (referred to as an *optimal* solution) for the problem under consideration. (We say *a* best instead of *the* best solution because there may be multiple solutions tied as best.) Rather than simply improving the status quo, the goal is to identify a best possible course of action. Although it must be interpreted carefully in terms of the practical needs of management, this "search for optimality" is an important theme in OR.

All these characteristics lead quite naturally to still another one. It is evident that no single individual should be expected to be an expert on all the many aspects of OR work or the problems typically considered; this would require a group of individuals having diverse backgrounds and skills. Therefore, when a full-fledged OR study of a new problem is undertaken, it is usually necessary to use a *team approach.* Such an OR team typically needs to include individuals who collectively are highly trained in mathematics, statistics and probability theory, economics, business administration, computer science, engineering and the physical sciences, the behavioral sciences, and the special techniques of OR. The team also needs to have the necessary experience and variety of skills to give appropriate consideration to the many ramifications of the problem throughout the organization.

1.3 The Impact of Operations Research

Operations research has had an impressive impact on improving the efficiency of numerous organizations around the world. In the process, OR has made a significant contribution to increasing the productivity of the economies of various countries. There now are more than 30 member countries in the International Federation of Operational Research Societies (IFORS), with each country having a national operations research society.

It appears that the impact of OR will continue to grow. For example, upon entering the 1990s, the U.S. Bureau of Labor Statistics predicted that OR will be the third-fastest-growing career area for U.S. college graduates from 1990 to 2005. It also predicted that 100,000 people will be employed as operations research analysts in the United States by the year 2005.

To give you a better notion of the wide applicability of OR, we list some actual award-winning applications in Table 1.1. Note the diversity of organizations and applications in the first two columns. The curious reader can find a complete article describing each application in the January-February issue of *Interfaces* for the year cited in the third column of the table. The fourth column lists the chapters in *this* book that describe the kinds of OR techniques that were used in the application. (Note that many of the applications combine a variety of techniques.) The last column indicates that these applications typically resulted in annual savings in the millions (or even tens of millions) of dollars. Furthermore, additional benefits not recorded in the table (e.g., improved service to customers and better managerial control) sometimes were considered to be even more important than these financial benefits. (You will have an opportunity to investigate these less tangible benefits further in Prob. 1.3-1.)

Although most routine OR studies provide considerably more modest benefits than these award-winning applications, the figures in the rightmost column of Table 1.1 do accurately reflect the dramatic impact that large, well-designed OR studies occasionally can have.

We will briefly describe some of these applications in the next chapter, and then we present two in greater detail as case studies in Sec. 3.5.

1.4 Algorithms and OR Courseware

An important part of this book is the presentation of the major **algorithms** (iterative solution procedures) of OR for solving certain types of problems. Some of these algorithms are amazingly efficient and are routinely used on problems involving hundreds or thousands of variables. Outside the classroom, the algorithms normally are executed on computers because of the relatively extensive numerical calculations involved.

Table 1.1 **Some Applications of Operations Research**

Organization	Nature of Application	Year of Publication*	Related Chapters†	Annual Savings
The Netherlands Rijkswaterstatt	Develop national water management policy, including mix of new facilities, operating procedures, and pricing.	1985	2–8, 13, 21	$15 million
Monsanto Corp.	Optimize production operations in chemical plants to meet production targets with minimum cost.	1985	2, 12	$2 million
Weyerhauser Co.	Optimize how trees are cut into wood products to maximize their yield.	1986	2, 10	$15 million
Eletrobras/CEPAL, Brazil	Optimally allocate hydro and thermal resources in the national electrical generating system.	1986	10	$43 million
United Airlines	Schedule shift work at reservation offices and airports to meet customer needs with minimum cost.	1986	2–9, 12, 15, 16, 18	$6 million
Citgo Petroleum Corp.	Optimize refinery operations and the supply, distribution, and marketing of products.	1987	2–9, 18	$70 million
SANTOS, Ltd., Australia	Optimize capital investments for producing natural gas over a 25-year period.	1987	2–6, 13, 21	$3 million
San Francisco Police Department	Optimally schedule and deploy police patrol officers with a computerized system.	1989	2–4, 12, 18	$11 million
Electric Power Research Institute	Manage oil and coal inventories for electric utilities to balance inventory costs and risk of shortages.	1989	17, 21	$59 million
Texaco, Inc.	Optimally blend available ingredients into gasoline products to meet quality and sales requirements.	1989	2, 13	$30 million
IBM	Integrate a national network of spare-parts inventories to improve service support.	1990	2, 17, 21	$20 million + $250 million less inventory
Yellow Freight System, Inc.	Optimize the design of a national trucking network and the routing of shipments.	1992	2, 9, 13, 18, 21	$17.3 million
U.S. Military Airlift Command	Quickly coordinate aircraft, crews, cargo, and passengers to run the Operation Desert Storm airlift.	1992	10	Victory
American Airlines	Design a system of fare structures, overbooking, and coordinating flights to increase revenues.	1992	2, 10, 12, 17, 18	$500 million more revenue
New Haven Health Dept.	Design an effective needle exchange program to combat the spread of HIV/AIDS.	1993	2	33% less HIV/AIDS

* Pertains to January-February issues of *Interfaces* in which complete articles can be found describing the application.

† Refers to chapters in this book that describe the kinds of OR techniques used in the application.

To aid the student in learning these algorithms, personal *tutorial software* (entitled OR Courseware) is packaged with most versions of the book. Separate diskettes are available for either an IBM (or IBM-compatible) personal computer with a graphics card or a Macintosh. (If the type of diskette accompanying this book is not the one you want, the Instructor's Manual, available to the teacher, contains diskettes of both types that can be copied.) Although it is possible to comprehend the material in this book without software, we highly recommend using this software as a fully integrated supplement to the book.

Three types of routines are included in the software. One is *demonstration examples* that display and explain the algorithms in action. These ''demos'' supplement the examples in the book.

For doing homework problems, the second type of routine—*interactive execution of algorithms*—commonly will be used. The computer does all the routine calculations while the student focuses on learning and executing the logic of the algorithm.

A third type available is routines for *automatic execution of algorithms.* This type will be used to test the formulation of models (unreasonable output indicates a mistake in the formulation) and to perform subsequent analysis, much as practitioners do with the output of production codes. (Most OR software packages contain only routines of this type.)

At the end of almost every chapter, we list the specific routines that are relevant for that chapter and its problems. Next to each problem where helpful routines are available, we indicate the types of routines.

Problem 1.4-1 below gives you a preliminary guided tour through the software.

PROBLEMS

1.3-1. Select one of the applications listed in Table 1.1. Read the article describing the application in the January-February issue of *Interfaces* for the year given in the third column of the table. Write a one-page description of the benefits (including nonfinancial benefits) that resulted from this application of OR.

1.4-1. Follow the instructions below, and then turn in the printout you obtain.

A Guided Tour through the OR Courseware

(1) Refer to Appendix 1 for documentation on the software, including instructions for getting started.
(2) Choose the program MathProg by following the instructions in Appendix 1. (The other program, ProbMod, is similar.) Read the initial screens giving a general introduction to the software.
(3) Select the Area menu. Note the choices available and then choose *General Linear Programming.*
(4) Select the Option menu. Note that different options are available for some procedures in this area. (You do not need to choose any particular option for this guided tour.)
(5) Select the Procedure menu. Choose *Enter or Revise a General Linear Programming Model.* [This procedure must precede the others in this menu. The current unavailability of the others is indicated by their being either dimmed (Macintosh) or unnumbered (IBM-compatibles).]
(6) Select the Help menu and choose *Introduction to Current Procedure.* After reading the general introduction to the procedure on the screen, press Return or Enter to return to the procedure. You soon will use this procedure to enter the model found in the next step.

(7) Select the Demo menu and choose *Graphical Method.* Press Enter or Return once to go to the second screen. Note the mathematical model displayed at the bottom of the screen. You will enter this model in the next step. Press Esc to exit from the demo and return to the procedure for entering such a model.
(8) To enter this model, press the following number keys (*always* followed by the Enter key) in turn:

(*a*) Number of decision variables: Press 2, Enter (or Return).
(*b*) Number of functional constraints: Press 2, Enter.
(*c*) Max: Press Enter (this accepts the default value).
(*d*) First coefficient of x_1: Press 20, Enter.
(*e*) First coefficient of x_2: Press 10, Enter.
(*f*) Second coefficient of x_1: Press 1, Enter.
(*g*) Second coefficient of x_2: Press $-$ (minus sign), 1, Enter.
(*h*) First $\leq$: Press Enter.
(*i*) Right side of first inequality: Press 1, Enter.
(*j*) Third coefficient of x_1: Press 3, Enter.
(*k*) Third coefficient of x_2: Press 1, Enter.
(*l*) Second $\leq$: Press Enter.
(*m*) Right side of second inequality: Press 7, Enter.

If you make a mistake, you can move the selection rectangle to that spot and reenter the desired number or symbol. You may also go to the Help menu at any point for specific help. Try this now by going to the Help menu to choose *Specific Help on Current Step,* read how to move the selection rectangle, then move the selection rectangle back to *Number of Functional Constraints* [part (*b*)] and reenter 2.

When you are done with this step, repeat step 7 to verify that the model you have just entered is the one found there, and then go to step 9.
(9) Select the Procedure menu. Choose *Solve Automatically by the Simplex Method* to solve the model you entered in step 8.
(10) To print this solution, check that the computer is connected to a printer and that the printer is turned on. Then select the File menu, choose *Print,* and press Enter.
(11) Select the File menu and choose *Quit.* (Whenever using the software, this step is available at *any* time to quit the program when desired.)

2

Overview of the Operations Research Modeling Approach

The bulk of this book is devoted to the mathematical methods of operations research (OR). This is quite appropriate because these quantitative techniques form the main part of what is known about OR. However, it does not imply that practical OR studies are primarily mathematical exercises. As a matter of fact, the mathematical analysis often represents only a relatively small part of the total effort required. The purpose of this chapter is to place things into better perspective by describing all the major phases of a typical OR study.

One way of summarizing the usual (overlapping) phases of an OR study is the following:

1. Define the problem of interest and gather relevant data.
2. Formulate a mathematical model to represent the problem.
3. Develop a computer-based procedure for deriving solutions to the problem from the model.
4. Test the model and refine it as needed.
5. Prepare for the ongoing application of the model as prescribed by management.
6. Implement.

Each of these phases will be discussed in turn in the following sections.

Most of the award-winning OR studies introduced in Table 1.1 provide excellent examples of how to execute these phases well. We will intersperse snippets from these examples throughout the chapter, with references to invite your further reading.

2.1 Defining the Problem and Gathering Data

In contrast to textbook examples, most practical problems encountered by OR teams are initially described to them in a vague, imprecise way. Therefore, the first order of business is to study the relevant system and develop a well-defined statement of the problem to be considered. This includes determining such things as the appropriate objectives, constraints on what can be done, interrelationships between the area to be studied and other areas of the organization, possible alternative courses of action, time limits for making a decision, and so on. This process of problem definition is a crucial one because it greatly affects how relevant the conclusions of the study will be. It is difficult to extract a ''right'' answer from the ''wrong'' problem!

The first thing to recognize is that an OR team is normally working in an *advisory capacity*. The team members are not just given a problem and told to solve it however they see fit. Instead, they are advising management (often one key decision maker). The team performs a detailed technical analysis of the problem and then presents recommendations to management. Frequently, the report to management will identify a number of alternatives that are particularly attractive under different assumptions or over a different range of values of some policy parameter that can be evaluated only by management (e.g., the trade-off between *cost* and *benefits*). Management evaluates the study and its recommendations, takes into account a variety of intangible factors, and makes the final decision based on its best judgment. Consequently, it is vital for the OR team to get on the same wavelength as management, including identifying the ''right'' problem from management's viewpoint, and to build the support of management for the course that the study is taking.

Ascertaining the *appropriate objectives* is a very important aspect of problem definition. To do this, it is necessary first to identify the member (or members) of management who actually will be making the decisions concerning the system under study and then to probe into this individual's thinking regarding the pertinent objectives. (Involving the decision maker from the outset also is essential to build her or his support for the implementation of the study.)

By its nature, OR is concerned with the welfare of the *entire organization* rather than that of only certain of its components. An OR study seeks solutions that are optimal for the overall organization rather than suboptimal solutions that are best for only one component. Therefore, the objectives that are formulated ideally should be those of the entire organization. However, this is not always convenient. Many problems primarily concern only a portion of the organization, so the analysis would become unwieldy if the stated objectives were too general and if explicit consideration were given to all side effects on the rest of the organization. Instead, the objectives used in the study should be as specific as they can be while still encompassing the main goals of the decision maker and maintaining a reasonable degree of consistency with the higher-level objectives of the organization.

For profit-making organizations, one possible approach to circumventing the problem of suboptimization is to use *long-run profit maximization* (considering the time value of money) as the sole objective. The adjective *long-run* indicates that this objective provides the flexibility to consider activities that do not translate into profits *immediately* (e.g., research and development projects) but need to do so *eventually* in order to be worthwhile. This approach has considerable merit. This objective is specific enough to be used conveniently, and yet it seems to be broad enough to encompass the basic goal of profit-making organizations. In fact, some people believe that all other legitimate objectives can be translated into this one.

However, in actual practice, many profit-making organizations do not use this approach. A number of studies of U.S. corporations have found that management tends to adopt the goal of *satisfactory profits,* combined with *other objectives,* instead of focusing on long-run profit maximization. (In fact, inadequate consideration of long-run profits sometimes is cited as a major reason why U.S. industry may be losing its competitive edge over other leading countries.) Typically, some of these *other* objectives might be to maintain stable profits, increase (or maintain) one's share of the market, provide for product diversification, maintain stable prices, improve worker morale, maintain family control of the business, and increase company prestige. Fulfilling these objectives might achieve long-run profit maximization, but the relationship may be sufficiently obscure that it may not be convenient to incorporate them all into this one objective.

Furthermore, there are additional considerations involving social responsibilities that are distinct from the profit motive. The five parties generally affected by a business firm located in a single country are (1) the *owners* (stockholders, etc.), who desire profits (dividends, stock appreciation, and so on); (2) the *employees,* who desire steady employment at reasonable wages; (3) the *customers,* who desire a reliable product at a reasonable price; (4) the *suppliers,* who desire integrity and a reasonable selling price for their goods; and (5) the *government* and hence the *nation,* which desire payment of fair taxes and consideration of the national interest. All five parties make essential contributions to the firm, and the firm should not be viewed as the exclusive servant of any one party for the exploitation of others. By the same token, international corporations acquire additional obligations to follow socially responsible practices. Therefore, while granting that management's prime responsibility is to make profits (which ultimately benefits all five parties), we note that its broader social responsibilities also must be recognized.

OR teams typically spend a surprisingly large amount of time *gathering relevant data* about the problem. Much data usually are needed both to gain an accurate understanding of the problem and to provide the needed input for the mathematical model being formulated in the next phase of study. Frequently, much of the needed data will not be available when the study begins, either because the information never has been kept or because what was kept is outdated or in the wrong form. Therefore, it often is necessary to install a new computer-based *management information system* to collect the necessary data on an ongoing basis and in the needed form. The OR team normally needs to enlist the assistance of various other key individuals in the organization to track down all the vital data. Even with this effort, much of the data may be quite "soft," i.e., rough estimates based only on educated guesses. Typically, an OR team will spend considerable time trying to improve the precision of the data and then will make do with the best that can be obtained.

Examples: An OR study done for the **San Francisco Police Department**[1] resulted in the development of a computerized system for optimally scheduling and deploying

[1] P. E. Taylor and S. J. Huxley, "A Break from Tradition for the San Francisco Police: Patrol Officer Scheduling Using an Optimization-Based Decision Support System," *Interfaces,* **19**(1): 4–24, Jan.-Feb. 1989. See especially pp. 4–11.

police patrol officers. The new system provided annual savings of $11 million, an annual $3 million increase in traffic citation revenues, and a 20 percent improvement in response times. In assessing the *appropriate objectives* for this study, three fundamental objectives were identified:

1. Maintain a high level of citizen safety.
2. Maintain a high level of officer morale.
3. Minimize the cost of operations.

To satisfy the first objective, the police department and city government jointly established a desired level of protection. The mathematical model then imposed the requirement that this level of protection be achieved. Similarly, the model imposed the requirement of balancing the workload equitably among officers in order to work toward the second objective. Finally, the third objective was incorporated by adopting the long-term goal of minimizing the number of officers needed to meet the first two objectives.

The **Health Department of New Haven, Connecticut** used an OR team[1] to design an effective needle exchange program to combat the spread of the virus that causes AIDS (HIV), and succeeded in reducing the HIV infection rate among program clients by 33 percent. The key part of this study was an innovative *data collection program* to obtain the needed input for mathematical models of HIV transmission. This program involved complete tracking of *each* needle (and syringe), including the identity, location, and date for each person receiving the needle and each person returning the needle during an exchange, as well as testing whether the returned needle was HIV-positive or HIV-negative.

An OR study done for the **Citgo Petroleum Corporation**[2] optimized both refinery operations and the supply, distribution, and marketing of its products, thereby achieving a profit improvement of approximately $70 million per year. *Data collection* also played a key role in this study. The OR team held data requirement meetings with top Citgo management to ensure the eventual and continual quality of data. A state-of-the-art management database system was developed and installed on a mainframe computer. In cases where needed data did not exist, LOTUS 1-2-3 screens were created to help operations personnel input the data, and then the data from the personal computers (PCs) were uploaded to the mainframe computer. Before data was inputted to the mathematical model, a preloader program was used to check for data errors and inconsistencies. Initially, the preloader generated a log of error messages 1 inch thick! Eventually, the number of error and warning messages (indicating bad or questionable numbers) was reduced to less than 10 for each new run.

We will describe the overall Citgo study in much more detail in Sec. 3.5.

2.2 Formulating a Mathematical Model

After the decision maker's problem is defined, the next phase is to reformulate this problem in a form that is convenient for analysis. The conventional OR approach for

[1] E. H. Kaplan and E. O'Keefe, "Let the Needles Do the Talking! Evaluating the New Haven Needle Exchange," *Interfaces,* **23**(1): 7–26, Jan.-Feb. 1993. See especially pp. 12–14.

[2] D. Klingman, N. Phillips, D. Steiger, R. Wirth, and W. Young, "The Challenges and Success Factors in Implementing an Integrated Products Planning System for Citgo," *Interfaces,* **16**(3): 1–19, May-June 1986. See especially pp. 11–14. Also see D. Klingman, N. Phillips, D. Steiger, and W. Young, "The Successful Deployment of Management Science throughout Citgo Petroleum Corporation," *Interfaces,* **17**(1): 4–25, Jan.-Feb. 1987. See especially pp. 13–15.

doing this is to construct a mathematical model that represents the essence of the problem. Before discussing how to formulate such a model, we first explore the nature of models in general and of mathematical models in particular.

Models, or idealized representations, are an integral part of everyday life. Common examples include model airplanes, portraits, globes, and so on. Similarly, models play an important role in science and business, as illustrated by models of the atom, models of genetic structure, mathematical equations describing physical laws of motion or chemical reactions, graphs, organizational charts, and industrial accounting systems. Such models are invaluable for abstracting the essence of the subject of inquiry, showing interrelationships, and facilitating analysis.

Mathematical models are also idealized representations, but they are expressed in terms of mathematical symbols and expressions. Such laws of physics as $F = ma$ and $E = mc^2$ are familiar examples. Similarly, the mathematical model of a business problem is the system of equations and related mathematical expressions that describe the essence of the problem. Thus, if there are n related quantifiable decisions to be made, they are represented as **decision variables** (say, $x_1, x_2, \ldots, x_n$) whose respective values are to be determined. The appropriate measure of performance (e.g., profit) is then expressed as a mathematical function of these decision variables (for example, $P = 3x_1 + 2x_2 + \cdots + 5x_n$). This function is called the **objective function**. Any restrictions on the values that can be assigned to these decision variables are also expressed mathematically, typically by means of inequalities or equations (for example, $x_1 + 3x_1x_2 + 2x_2 \leq 10$). Such mathematical expressions for the restrictions often are called **constraints**. The constants (namely, the coefficients and right-hand sides) in the constraints and the objective function are called the **parameters** of the model. The mathematical model might then say that the problem is to choose the values of the decision variables so as to maximize the objective function, subject to the specified constraints. Such a model, and minor variations of it, typifies the models used in OR.

Determining the appropriate values to assign to the parameter of the model (one value per parameter) is both a critical and a challenging part of the model-building process. In contrast to textbook problems where the numbers are given to you, determining parameter values for real problems requires *gathering relevant data.* As discussed in the preceding section, gathering accurate data frequently is difficult. Therefore, the value assigned to a parameter often is, of necessity, only a rough estimate. Because of the uncertainty about the true value of the parameter, it is important to analyze how the solution derived from the model would change (if at all) if the value assigned to the parameter were changed to other plausible values. This process is referred to as **sensitivity analysis**, as discussed further in the next section (and much of Chap. 6).

Although we refer to "the" mathematical model of a business problem, real problems normally don't have just a single "right" model. Section 2.4 will describe how the process of testing a model typically leads to a succession of models that provide better and better representations of the problem. It is even possible that two or more completely different types of models may be developed to help analyze the same problem.

You will see numerous examples of mathematical models throughout the remainder of this book. One particularly important type that is studied in the next several chapters is the **linear programming model**, where the mathematical functions appearing in both the objective function and the constraints are all linear functions. In the next chapter, specific linear programming models are constructed to fit such diverse problems as determining (1) the mix of products that maximizes profit, (2) the design of radiation therapy that effectively attacks a tumor while minimizing the damage to nearby healthy tissue, (3) the allocation of acreage to crops that maximizes total net return, and (4) the combination of pollution abatement methods that achieves air quality standards at minimum cost.

Mathematical models have many advantages over a verbal description of the problem. One obvious advantage is that a mathematical model describes a problem much more concisely. This tends to make the overall structure of the problem more comprehensible, and it helps to reveal important cause-and-effect relationships. In this way, it indicates more clearly what additional data are relevant to the analysis. It also facilitates dealing with the problem in its entirety and considering all its interrelationships simultaneously. Finally, a mathematical model forms a bridge to the use of high-powered mathematical techniques and computers to analyze the problem. Indeed, packaged software for both microcomputers and mainframe computers has become widely available for many mathematical models.

However, there are pitfalls to be avoided when you use mathematical models. Such a model is necessarily an abstract idealization of the problem, so approximations and simplifying assumptions generally are required if the model is to be *tractable* (capable of being solved). Therefore, care must be taken to ensure that the model remains a valid representation of the problem. The proper criterion for judging the validity of a model is whether the model predicts the relative effects of the alternative courses of action with sufficient accuracy to permit a sound decision. Consequently, it is not necessary to include unimportant details or factors that have approximately the same effect for all the alternative courses of action considered. It is not even necessary that the absolute magnitude of the measure of performance be approximately correct for the various alternatives, provided that their relative values (i.e., the differences between their values) are sufficiently precise. Thus all that is required is that there be a high *correlation* between the prediction by the model and what would actually happen in the real world. To ascertain whether this requirement is satisfied, it is important to do considerable *testing* and consequent modifying of the model, which will be the subject of Sec. 2.4. Although this testing phase is placed later in the chapter, much of this *model validation* work actually is conducted during the model-building phase of the study to help guide the construction of the mathematical model.

In developing the model, a good approach is to begin with a very simple version and then move in evolutionary fashion toward more elaborate models that more nearly reflect the complexity of the real problem. This process of *model enrichment* continues only as long as the model remains tractable. The basic trade-off under constant consideration is between the *precision* and the *tractability* of the model. (See Selected Reference 5 for a detailed description of this process.)

A crucial step in formulating an OR model is the construction of the objective function. This requires developing a quantitative measure of performance relative to each of the decision maker's ultimate objectives that were identified while the problem was being defined. If there are multiple objectives, their respective measures commonly are then transformed and combined into a composite measure, called the **overall measure of performance**. This overall measure might be something tangible (e.g., profit) corresponding to a higher goal of the organization, or it might be abstract (e.g., utility). In the latter case, the task of developing this measure tends to be a complex one requiring a careful comparison of the objectives and their relative importance. After the overall measure of performance is developed, the objective function is then obtained by expressing this measure as a mathematical function of the decision variables. Alternatively, there also are methods for explicitly considering multiple objectives simultaneously, and one of these (goal programming) is discussed in Chap. 7.

Examples: An OR study done for **Monsanto Corp.**[1] was concerned with optimizing production operations in Monsanto's chemical plants to minimize the cost of meeting the target for the amount of a certain chemical product (maleic anhydride) to be produced in a given month. The decisions to be made are the dial setting for each of the catalytic reactors used to produce this product, where the setting determines both the amount produced and the cost of operating the reactor. The form of the resulting mathematical model is as follows:

Choose the values of the *decision variables* R_{ij}
$(i = 1, 2, \ldots, r; j = 1, 2, \ldots, s)$
so as to

$$\text{Minimize} \quad \sum_{i=1}^{r} \sum_{j=1}^{s} c_{ij} R_{ij},$$

subject to

$$\sum_{i=1}^{r} \sum_{j=1}^{s} p_{ij} R_{ij} \geq T$$

$$\sum_{j=1}^{s} R_{ij} = 1, \qquad \text{for } i = 1, 2, \ldots, r$$

$$R_{ij} = 0 \text{ or } 1,$$

where $R_{ij} = \begin{cases} 1 & \text{if reactor } i \text{ is operated at setting } j \\ 0 & \text{otherwise} \end{cases}$

c_{ij} = cost for reactor i at setting j
p_{ij} = production of reactor i at setting j
T = production target
r = number of reactors
s = number of settings (including off position)

The *objective function* for this model is $\Sigma \Sigma c_{ij} R_{ij}$. The *constraints* are given in the three lines below the objective function. The *parameters* are c_{ij}, p_{ij}, and T. For Monsanto's application, this model has over 1,000 *decision variables* R_{ij} (that is, $rs >$ 1,000). Its use led to annual savings of approximately $2 million.

The Netherlands government agency responsible for water control and public works, the **Rijkswaterstatt**, commissioned a major OR study[2] to guide the development of a new national water management policy. The new policy saved hundreds of millions of dollars in investment expenditures and reduced agricultural damage by about $15 million per year, while decreasing thermal and algae pollution. Rather than formulating *one* mathematical model, this OR study developed a comprehensive, integrated system of 50 models! Furthermore, for some of the models, both simple and complex versions were developed. The simple version was used to gain basic insights,

[1] R. F. Boykin, "Optimizing Chemical Production at Monsanto," *Interfaces,* **15**(1): 88–95, Jan.-Feb. 1985. See especially pp. 92–93.

[2] B. F. Goeller and the PAWN team: "Planning the Netherlands' Water Resources," *Interfaces,* **15**(1): 3–33, Jan.-Feb. 1985. See especially pp. 7–18.

including trade-off analyses. The complex version then was used in the final rounds of the analysis or whenever greater accuracy or more detailed outputs were desired. The overall OR study directly involved over 125 person-years of effort (more than one-third in data gathering), created several dozen computer programs, and structured an enormous amount of data.

2.3 Deriving Solutions from the Model

After a mathematical model is formulated for the problem under consideration, the next phase in an OR study is to develop a procedure (usually a computer-based procedure) for deriving solutions to the problem from this model. You might think that this must be the major part of the study, but actually it is not in most cases. Sometimes, in fact, it is a relatively simple step, in which one of the standard **algorithms** (iterative solution procedures) of OR is applied on a computer by using one of a number of readily available software packages. For experienced OR practitioners, finding a solution is the fun part, whereas the real work comes in the preceding and following steps, including the *post-optimality analysis,* discussed later in this section.

Since much of this book is devoted to the subject of how to obtain solutions for various important types of mathematical models, little needs to be said about it here. However, we do need to discuss the nature of such solutions.

A common theme in OR is the search for an **optimal**, or best, **solution**. Indeed, many procedures have been developed, and are presented in this book, for finding such solutions for certain kinds of problems. However, it needs to be recognized that these solutions are optimal only with respect to the model being used. Since the model necessarily is an idealized rather than an exact representation of the real problem, there cannot be any utopian guarantee that the optimal solution for the model will prove to be the best possible solution that could have been implemented for the real problem. There just are too many imponderables and uncertainties associated with real problems. However, if the model is well formulated and tested, the resulting solution should tend to be a good approximation to an ideal course of action for the real problem. Therefore, rather than be deluded into demanding the impossible, you should make the test of the practical success of an OR study hinge on whether it provides a better guide for action than can be obtained by other means.

Eminent management scientist and Nobel Laureate in economics Herbert Simon points out that **satisficing** is much more prevalent than optimizing in actual practice. In coining the term *satisficing* as a combination of the words *satisfactory* and *optimizing,* Simon is describing the tendency of managers to seek a solution that is ''good enough'' for the problem at hand. Rather than trying to develop an overall measure of performance to optimally reconcile conflicts between various desirable objectives (including well-established criteria for judging the performance of different segments of the organization), a more pragmatic approach may be used. Goals may be set to establish minimum satisfactory levels of performance in various areas, based perhaps on past levels of performance or on what the competition is achieving. If a solution is found that enables all these goals to be met, it is likely to be adopted without further ado. Such is the nature of satisficing.

The distinction between optimizing and satisficing reflects the difference between theory and the realities frequently faced in trying to implement that theory in practice. In the words of one of England's OR leaders Samuel Eilon, "optimizing is the science of the ultimate; satisficing is the art of the feasible."[1]

OR teams attempt to bring as much of the "science of the ultimate" as possible to the decision-making process. However, the successful team does so in full recognition of the overriding need of the decision maker to obtain a satisfactory guide for action in a reasonable time. Therefore, the goal of an OR study should be to conduct the study in an optimal manner, regardless of whether this involves finding an optimal solution to the model. Thus, in addition to pursuing the science of the ultimate, the team should consider the cost of the study and the disadvantages of delaying its completion and then attempt to maximize the net benefits resulting from the study. In recognition of this concept, OR teams occasionally use only **heuristic procedures** (i.e., intuitively designed procedures that do not guarantee an optimal solution) to find a good **suboptimal solution**. This is most often the case when the time or cost required to find an optimal solution for an adequate model of the problem would be very large.

The discussion thus far has implied that an OR study seeks to find only *one* solution, which may or may not be required to be optimal. In fact, this usually is not the case. An optimal solution for the original model may be far from ideal for the real problem, so additional analysis is needed. Therefore, **post-optimality analysis** (analysis done after an optimal solution is found) is a very important part of most OR studies.

In part, post-optimality analysis involves conducting **sensitivity analysis** to determine which parameters of the model are most critical (the "sensitive parameters") in determining the solution. A common definition of *sensitive parameter* (used throughout this book) is the following.

> For a mathematical model with specified values for all its parameters, the model's **sensitive parameters** are the parameters whose values cannot be changed without changing the optimal solution.

Identifying the sensitive parameters is important, because this identifies the parameters whose value must be assigned with special care to avoid distorting the output of the model.

The value assigned to a parameter commonly is just an *estimate* of some quantity (e.g., unit profit) whose exact value will become known only after the solution has been implemented. Therefore, after the sensitive parameters are identified, special attention is given to estimating each one more closely, or at least its range of likely values. One then seeks a solution that remains a particularly good one for all the various combinations of likely values of the sensitive parameters.

If the solution is implemented on an ongoing basis, any later change in the value of a sensitive parameter immediately signals a need to change the solution.

In some cases, certain parameters of the model represent policy decisions (e.g., resource allocations). If so, there frequently is some flexibility in the values assigned to these parameters. Perhaps some can be increased by decreasing others. Post-optimality analysis includes the investigation of such trade-offs.

In conjunction with the study phase discussed in the next section (testing the model), post-optimality analysis also involves obtaining a sequence of solutions that comprises a series of improving approximations to the ideal course of action. Thus the apparent weaknesses in the initial solution are used to suggest improvements in the

[1] S. Eilon, "Goals and Constraints in Decision-making," *Operational Research Quarterly*, **23:** 3–15, 1972—address given at the 1971 annual conference of the Canadian Operational Research Society.

model, its input data, and perhaps the solution procedure. A new solution is then obtained, and the cycle is repeated. This process continues until the improvements in the succeeding solutions become too small to warrant continuation. Even then, a number of alternative solutions (perhaps solutions that are optimal for one of several plausible versions of the model and its input data) may be presented to management for the final selection. As suggested in Sec. 2.1, this presentation of alternative solutions would normally be done whenever the final choice among these alternatives should be based on considerations that are best left to the judgment of management.

Example: Consider again the **Rijkswaterstatt** OR study of national water management policy for the Netherlands, introduced at the end of the preceding section. This study did not conclude by recommending just a single solution. Instead, a number of attractive alternatives were identified, analyzed, and compared. The final choice was left to the Dutch political process, culminating with approval by Parliament. *Sensitivity analysis* played a major role in this study. For example, certain parameters of the models represented environmental standards. Sensitivity analysis included assessing the impact on water management problems if the values of these parameters were changed from the current environmental standards to other reasonable values. Sensitivity analysis also was used to assess the impact of changing the assumptions of the models, e.g., the assumption on the effect of future international treaties on the amount of pollution entering the Netherlands. A variety of *scenarios* (e.g., an extremely dry year and an extremely wet year) also were analyzed, with appropriate probabilities assigned.

2.4 Testing the Model

Developing a large mathematical model is analogous in some ways to developing a large computer program. When the first version of the computer program is completed, it inevitably contains many bugs. The program must be thoroughly tested to try to find and correct as many bugs as possible. Eventually, after a long succession of improved programs, the programmer (or programming team) concludes that the current program now is generally giving reasonably valid results. Although some minor bugs undoubtedly remain hidden in the program (and may never be detected), the major bugs have been sufficiently eliminated that the program now can be reliably used.

Similarly, the first version of a large mathematical model inevitably contains many flaws. Some relevant factors or interrelationships undoubtedly have not been incorporated into the model, and some parameters undoubtedly have not been estimated correctly. This is inevitable, given the difficulty of communicating and understanding all the aspects and subtleties of a complex operational problem as well as the difficulty of collecting reliable data. Therefore, before you use the model, it must be thoroughly tested to try to identify and correct as many flaws as possible. Eventually, after a long succession of improved models, the OR team concludes that the current model now is giving reasonably valid results. Although some minor flaws undoubtedly remain hidden in the model (and may never be detected), the major flaws have been sufficiently eliminated that the model now can be reliably used.

This process of testing and improving a model to increase its validity is commonly referred to as **model validation**.

It is difficult to describe how model validation is done, because the process depends greatly on the nature of the problem being considered and the model being used. However, we make a few general comments, and then we give some examples. (See Selected Reference 1 for a detailed discussion.)

Since the OR team may spend months developing all the detailed pieces of the model, it is easy to ''lose the forest for the trees.'' Therefore, after the details (''the trees'') of the initial version of the model are completed, a good way to begin model validation is to take a fresh look at the overall model (''the forest'') to check for obvious errors or oversights. The group doing this review preferably should include at least one individual who did not participate in the formulation of the model. Reexamining the definition of the problem and comparing it with the model may help to reveal mistakes. It is also useful to make sure that all the mathematical expressions are *dimensionally consistent* in the units used. Additional insight into the validity of the model can sometimes be obtained by varying the values of the parameters and/or the decision variables and checking to see whether the output from the model behaves in a plausible manner. This is often especially revealing when the parameters or variables are assigned extreme values near their maxima or minima.

A more systematic approach to testing the model is to use a **retrospective test**. When it is applicable, this test involves using historical data to reconstruct the past and then determining how well the model and the resulting solution would have performed if they had been used. Comparing the effectiveness of this hypothetical performance with what actually happened then indicates whether using this model tends to yield a significant improvement over current practice. It may also indicate areas where the model has shortcomings and requires modifications. Furthermore, by using alternative solutions from the model and estimating their hypothetical historical performances, considerable evidence can be gathered regarding how well the model predicts the relative effects of alternative courses of actions.

On the other hand, a disadvantage of retrospective testing is that it uses the same data that guided the formulation of the model. The crucial question is whether the past is truly representative of the future. It it is not, then the model might perform quite differently in the future than it would have in the past.

To circumvent this disadvantage of retrospective testing, it is sometimes useful to continue the status quo temporarily. This provides new data that were not available when the model was constructed. These data are then used in the same ways as those described here to evaluate the model.

Documenting the process used for model validation is important. This helps to increase confidence in the model for subsequent users. Furthermore, if concerns arise in the future about the model, this documentation will be helpful in diagnosing where problems may lie.

Examples: Consider once again the **Rijkswaterstatt** OR study of national water management policy for the Netherlands, discussed at the end of Secs. 2.2 and 2.3. The process of model validation in this case had three main parts. First, the OR team checked the general behavior of the models by checking whether the results from each model moved in reasonable ways when changes were made in the values of the model parameters. Second, retrospective testing was done. Third, a careful technical review of the models, methodology, and results was conducted by individuals unaffiliated with the project, including Dutch experts. This process led to a number of important new insights and improvements in the models.

Many new insights also were gleaned during the model validation phase of the OR study for the **Citgo Petroleum Corp.**, discussed at the end of Sec. 2.1. In this case, the model of refinery operations was tested by collecting the actual inputs and outputs of the refinery for a series of months, using these inputs to fix the model inputs, and then comparing the model outputs with the actual refinery outputs. The process of properly calibrating and recalibrating the model was a lengthy one, but ultimately led to routine use of the model to provide critical decision information. As already mentioned in Sec. 2.1, the validation and correction of input data for the models also played an important role in this study.

Our next example concerns an OR study done for **IBM**[1] to integrate its national network of spare-parts inventories to improve service support for IBM's customers. This study resulted in a new inventory system that improved customer service while reducing the value of IBM's inventories by over $250 million and saving an additional $20 million per year through improved operational efficiency. A particularly interesting aspect of the model validation phase of this study was the way that *future users* of the inventory system were incorporated into the testing process. Because these future users (IBM managers in functional areas responsible for implementation of the inventory system) were skeptical about the system being developed, representatives were appointed to a *user team* to serve as advisers to the OR team. After a preliminary version of the new system had been developed (based on a multiechelon inventory model), a *preimplementation test* of the system was conducted. Extensive feedback from the user team led to major improvements in the proposed system.

2.5 Preparing to Apply the Model

What happens after the testing phase has been completed and an acceptable model has been developed? If the model is to be used repeatedly, the next step is to install a well-documented *system* for applying the model as prescribed by management. This system will include the model, solution procedure (including post-optimality analysis), and operating procedures for implementation. Then, even as personnel changes, the system can be called on at regular intervals to provide a specific numerical solution.

This system usually is *computer-based.* In fact, a considerable number of computer programs often need to be used and integrated. *Databases* and *management information systems* may provide up-to-date input for the model each time it is used, in which case interface programs are needed. After a solution procedure (another program) is applied to the model, additional computer programs may trigger the implementation of the results automatically. In other cases, an *interactive* computer-based system called a **decision support system** is installed to help managers use data and models to support (rather than replace) their decision making as needed. Another program may generate *managerial reports* (in the language of management) that interpret the output of the model and its implications for application.

In major OR studies, several months (or longer) may be required to develop, test, and install this computer system. Part of this effort involves developing and imple-

[1] M. Cohen, P. V. Kamesam, P. Kleindorfer, H. Lee, and A. Tekerian, "Optimizer: IBM's Multi-Echelon Inventory System for Managing Service Logistics," *Interfaces,* **20**(1): 65–82, Jan.-Feb. 1990. See especially pp. 73–76.

menting a process for maintaining the system throughout its future use. As conditions change over time, this process should modify the computer system (including the model) accordingly.

Examples: The **IBM** OR study introduced at the end of Sec. 2.4 provides a good example of a particularly large computer system for applying a model. The system developed, called *Optimizer,* provides optimal control of service levels and spare-parts inventories throughout IBM's U.S. parts distribution network, which includes two central automated warehouses, dozens of field distribution centers and parts stations, and many thousands of outside locations. The parts inventory maintained in this network is valued in the billions of dollars. Optimizer consists of four major modules. A forecasting system module contains a few programs for estimating the failure rates of individual types of parts. A data delivery system module consists of approximately 100 programs that process over 15 gigabytes of data to provide the input for the model. A decision system module then solves the model on a weekly basis to optimize control of the inventories. The fourth module includes six programs that integrate Optimizer into IBM's Parts Inventory Management System (PIMS). PIMS is a sophisticated information and control system that contains millions of lines of code.

Our next example also involves a large computer system for applying a model to control operations over a national network. This system, called *SYSNET,* was developed as the result of an OR study done for **Yellow Freight System, Inc.**[1] Yellow Freight annually handles over 15 million shipments by motor carrier over a network of 630 terminals throughout the United States. SYSNET is used to optimize both the routing of shipments and the design of the network. Because SYSNET requires extensive information about freight flows and forecasts, transportation and handling costs, and so on, a major part of the OR study involved integrating SYSNET into the corporate management information system. This integration enabled periodic updating of all the input for the model. The implementation of SYSNET resulted in annual savings of approximately $17.3 million as well as improved service to customers.

Our next example illustrates a *decision support system.* A system of this type was developed for **Texaco**[2] to help plan and schedule its blending operations at its various refineries. Called *OMEGA* (Optimization Method for the Estimation of Gasoline Attributes), it is an *interactive* system based on a nonlinear optimization model that is implemented on both personal computers and larger computers. Input data can be entered either manually or by interfacing with refinery databases. The user has considerable flexibility in choosing an objective function and constraints to fit the current situation as well as in asking a series of *what-if questions* (i.e., questions about *what* would happen *if* the assumed conditions change). OMEGA is maintained centrally by Texaco's information technology department, which enables constant updating to reflect new government regulations, other business changes, and changes in refinery operations. The implementation of OMEGA is credited with annual savings of more than $30 million as well as improved planning, quality control, and marketing information.

[1] J. W. Braklow, W. W. Graham, S. M. Hassler, K. E. Peck, and W. B. Powell, "Interactive Optimization Improves Service and Performance for Yellow Freight System," *Interfaces,* **22**(1): 147–172, Jan.-Feb. 1992. See especially p. 163.

[2] C. W. DeWitt, L. S. Lasdon, A. D. Waren, D. A. Brenner, and S. A. Melhem, "OMEGA: An Improved Gasoline Blending System for Texaco," *Interfaces,* **19**(1): 85–101, Jan.-Feb. 1989. See especially pp. 93–95.

2.6 Implementation

After a system is developed for applying the model, the last phase of an OR study is to implement this system as prescribed by management. This phase is a critical one because it is here, and only here, that the benefits of the study are reaped. Therefore, it is important for the OR team to participate in launching this phase, both to make sure that model solutions are accurately translated to an operating procedure and to rectify any flaws in the solutions that are then uncovered.

The success of the implementation phase depends a great deal upon the support of both top management and operating management. The OR team is much more likely to gain this support if it has kept management well informed and encouraged management's active guidance throughout the course of the study. Good communications help to ensure that the study accomplishes what management wanted and so deserves implementation. They also give management a greater sense of ownership of the study, which encourages their support for implementation.

The implementation phase involves several steps. First, the OR team gives operating management a careful explanation of the new system to be adopted and how it relates to operating realities. Next, these two parties share the responsibility for developing the procedures required to put this system into operation. Operating management then sees that a detailed indoctrination is given to the personnel involved, and the new course of action is initiated. If successful, the new system may be used for years to come. With this in mind, the OR team monitors the initial experience with the course of action taken and seeks to identify any modifications that should be made in the future.

Upon culmination of a study, it is appropriate for the OR team to *document* its methodology clearly and accurately enough that the work is *reproducible. Replicability* should be part of the professional ethical code of the operations researcher. This condition is especially crucial when controversial public policy issues are being studied.

Examples: This last point about *documenting* an OR study is illustrated by the **Rijkswaterstatt** study of national water management policy for the Netherlands, discussed at the end of Secs. 2.2, 2.3, and 2.4. Management wanted unusually thorough and extensive documentation, both to support the new policy and to use in training new analysts or in performing new studies. Requiring several years to complete, this documentation aggregated 4,000 single-spaced pages and 21 volumes!

Our next example concerns the **IBM** OR study discussed at the end of Secs. 2.4 and 2.5. Careful planning was required to implement the complex Optimizer system for controlling IBM's national network of spare-parts inventories. Three factors proved to be especially important in achieving a successful implementation. As discussed in Sec. 2.4, the first was the inclusion of a *user team* (consisting of operational managers) as advisers to the OR team throughout the study. By the time of the implementation phase, these operational managers had a strong sense of ownership and so had become ardent supporters for installing Optimizer in their functional areas. A second success factor was a very extensive *user acceptance test* whereby users could identify problem areas that needed rectifying prior to full implementation. The third key was that the new system was *phased in gradually,* with careful testing at each phase, so the major bugs could be eliminated before the system went live nationally.

Our final example concerns **Yellow Freight's** SYSNET system for routing shipments over a national network, as described at the end of the preceding section. In this case, there were four key elements to the implementation process. The first was selling the concept to upper management. This was successfully done by validating the accuracy of the cost model and then holding *interactive sessions* for upper management that demonstrated the effectiveness of the system. The second element was the development of an implementation strategy for gradually phasing in the new system while identifying and eliminating its flaws. The third was the working closely with operational managers to install the system properly, provide the needed support tools, train the personnel who will use the system, and convince them of the usefulness of the system. The final key element was the provision to management of incentives and enforcement for the effective implementation of the system.

2.7 Conclusions

Although the remainder of this book focuses primarily on *constructing* and *solving* mathematical models, in this chapter we have tried to emphasize that this constitutes only a portion of the overall process involved in conducting a typical OR study. The other phases described here also are very important to the success of the study. Try to keep in perspective the role of the model and the solution procedure in the overall process, as you move through the subsequent chapters. Then, after you gain a deeper understanding of mathematical models, we suggest that you plan to return to review this chapter again in order to further sharpen this perspective.

OR is closely intertwined with the use of computers. Until recent years, these generally were mainframe computers, but now microcomputers and workstations are being widely used.

In concluding this discussion of the major phases of an OR study, note that there are many exceptions to the ''rules'' prescribed in this chapter. By its very nature, OR requires considerable ingenuity and innovation, so it is impossible to write down any standard procedure that should always be followed by OR teams. Rather, the preceding description may be viewed as a model that roughly represents how successful OR studies are conducted.

SELECTED REFERENCES

1. Gass, S. I., ''Decision-Aiding Models: Validation, Assessment, and Related Issues for Policy Analysis,'' *Operations Research,* **31:** 603–631, 1983.
2. Gass, S. I., ''Model World: Danger, Beware the User as Modeler,'' *Interfaces,* **20**(3): 60–64, May-June 1990.
3. Hall, R. W., ''What's So Scientific about MS/OR?'' *Interfaces,* **15**(2): 40–45, March-April 1985.
4. Miser, H. J., ''The Easy Chair: Observation and Experimentation,'' *Interfaces,* **19**(5): 23–30, Sept.-Oct. 1989.
5. Morris, W. T., ''On the Art of Modeling,'' *Management Science,* **13:** B707–717, 1967.
6. Simon, H. A., ''Prediction and Prescription in Systems Modeling,'' *Operations Research,* **38:** 7–14, 1990.
7. Tilanus, C. B., O. B. DeGans, and J. K. Lenstra (eds.), *Quantitative Methods in Management: Case Studies of Failures and Successes.* Wiley, New York, 1986.
8. Williams, H. P., *Model Building in Mathematical Programming,* 3d ed., Wiley, New York, 1990.

2.1-1. Read the article footnoted in Sec. 2.1 that describes an OR study done for the San Francisco Police Department.

(*a*) Summarize the background that led to undertaking this study.
(*b*) Define part of the problem being addressed by identifying the six directives for the scheduling system to be developed.
(*c*) Describe how the needed data were gathered.
(*d*) List the various tangible and intangible benefits that resulted from the study.

2.1-2. Read the article footnoted in Sec. 2.1 that describes an OR study done for the Health Department of New Haven, Connecticut.

(*a*) Summarize the background that led to undertaking this study.
(*b*) Outline the system developed to track and test each needle and syringe in order to gather the needed data.
(*c*) Summarize the initial results from this tracking and testing system.
(*d*) Describe the impact and potential impact of this study on public policy.

2.2-1. Read the article footnoted in Sec. 2.2 that describes an OR study done for the Rijkswaterstatt of the Netherlands. (Focus especially on pp. 3–20 and 30–32.)

(*a*) Summarize the background that led to undertaking this study.
(*b*) Summarize the purpose of each of the five mathematical models described on pp. 10–18.
(*c*) Summarize the "impact measures" (measures of performance) for comparing policies that are described on pp. 6–7 of this article.
(*d*) List the various tangible and intangible benefits that resulted from the study.

2.2-2. Read Selected Reference 3.

(*a*) Identify the author's example of a model in the natural sciences and of a model in OR.
(*b*) Describe the author's viewpoint about how basic precepts of using models to do research in the natural sciences can also be used to guide *research on operations* (OR).

2.3-1. Refer to Selected Reference 3.

(*a*) Describe the author's viewpoint about whether the sole goal in using a model should be to find its optimal solution.
(*b*) Summarize the author's viewpoint about the complementary roles of modeling, evaluating information from the model, and then applying the decision maker's judgment when deciding on a course of action.

2.4-1. Refer to pp. 18–20 of the article footnoted in Sec. 2.2 that describes an OR study done for the Rijkswaterstatt of the Netherlands. Describe an important lesson that was gained from model validation in this study.

2.4-2. Read Selected Reference 4. Summarize the author's viewpoint about the roles of observation and experimentation in the model validation process.

2.4-3. Read pp. 603–617 of Selected Reference 1.

(*a*) What does the author say about whether a model can be completely validated?
(*b*) Summarize the distinctions made between *model validity, data validity, logical/mathematical validity, predictive validity, operational validity,* and *dynamic validity.*
(*c*) Describe the role of *sensitivity analysis* in testing the *operational validity* of a model.
(*d*) What does the author say about whether there is a validation methodology that is appropriate for all models?
(*e*) Cite the page in the article that lists basic validation steps.

2.5-1. Read the article footnoted in Sec. 2.5 that describes an OR study done for Texaco.
(*a*) Summarize the background that led to undertaking this study.
(*b*) Briefly describe the user interface with the decision support system OMEGA that was developed as a result of this study.
(*c*) OMEGA is constantly being updated and extended to reflect changes in the operating environment. Briefly describe the various kinds of changes involved.
(*d*) Summarize how OMEGA is used.
(*e*) List the various tangible and intangible benefits that resulted from the study.

2.5-2. Refer to the article footnoted in Sec. 2.5 that describes an OR study done for Yellow Freight System, Inc.
(*a*) Referring to pp. 147–149 of this article, summarize the background that led to undertaking this study.
(*b*) Referring to p. 150, briefly describe the computer system SYSNET that was developed as a result of this study. Also summarize the applications of SYSNET.
(*c*) Referring to pp. 162–163, describe why the *interactive* aspects of SYSNET proved important.
(*d*) Referring to p. 163, summarize the outputs from SYSNET.
(*e*) Referring to pp. 168–172, summarize the various benefits that have resulted from using SYSNET.

2.6-1. Refer to pp. 163–167 of the article footnoted in Sec. 2.5 that describes an OR study done for Yellow Freight System, Inc., and the resulting computer system SYSNET.
(*a*) Briefly describe how the OR team gained the support of upper management for implementing SYSNET.
(*b*) Briefly describe the implementation strategy that was developed.
(*c*) Briefly describe the field implementation.
(*d*) Briefly describe how management incentives and enforcement were used in implementing SYSNET.

2.6-2. Read the article footnoted in Sec. 2.4 that describes an OR study done for IBM and the resulting computer system Optimizer.
(*a*) Summarize the background that led to undertaking this study.
(*b*) List the complicating factors that the OR team members faced when they started developing a model and a solution algorithm.
(*c*) Briefly describe the preimplementation test of Optimizer.
(*d*) Briefly describe the field implementation test.
(*e*) Briefly describe national implementation.
(*f*) List the various tangible and intangible benefits that resulted from the study.

2.7-1. Read Selected Reference 2. The author describes 13 detailed phases of any OR study that develops and applies a computer-based model, whereas this chapter describes six broader phases. For each of these broader phases, list the detailed phases that fall partially or primarily within the broader phase.

3

Introduction to Linear Programming

The development of linear programming has been ranked among the most important scientific advances of the mid-20th century, and we must agree with this assessment. Its impact since just 1950 has been extraordinary. Today it is a standard tool that has saved many thousands or millions of dollars for most companies or businesses of even moderate size in the various industrialized countries of the world; and its use in other sectors of society has been spreading rapidly. A major proportion of all scientific computation on computers is devoted to the use of linear programming. Dozens of textbooks have been written about linear programming, and *published* articles describing important applications now number in the hundreds.

What is the nature of this remarkable tool, and what kinds of problems does it address? You will gain insight into this topic as you work through subsequent examples. However, a verbal summary may help provide perspective. Briefly, the most common type of application involves the general problem of allocating *limited resources* among *competing activities* in a best possible (i.e., *optimal*) way. More precisely, this problem involves selecting the level of certain activities that compete for scarce resources that are necessary to perform those activities. The choice of activity levels then dictates how much of each resource will be consumed by each activity. The variety of situations to which this description applies is diverse, indeed, ranging from the allocation of production facilities to products to the allocation of national resources to domestic needs, from portfolio selection to the selection of shipping patterns, from agricultural planning to the design of radiation therapy, and so on. However, the one common ingredient in each of these situations is the necessity for allocating resources to activities by choosing the levels of those activities.

Linear programming uses a mathematical model to describe the problem of concern. The adjective *linear* means that all the mathematical functions in this model are required to be *linear functions.* The word *programming* does not refer here to computer programming; rather, it is essentially a synonym for *planning.* Thus linear programming involves the *planning of activities* to obtain an optimal result, i.e., a result that reaches the specified goal best (according to the mathematical model) among all feasible alternatives.

Although allocating resources to activities is the most common type of application, linear programming has numerous other important applications as well. In fact, *any* problem whose mathematical model fits the very general format for the linear programming model is a linear programming problem. Furthermore, a remarkably efficient solution procedure, called the **simplex method**, is available for solving linear programming problems of even enormous size. These are some of the reasons for the tremendous impact of linear programming in recent decades.

Because of its great importance, we devote this and the next six chapters specifically to linear programming. After this chapter introduces the general features of linear programming, Chaps. 4 and 5 focus on the simplex method. Chapter 6 discusses the further analysis of linear programming problems *after* the simplex method has been initially applied. Chapter 7 presents several widely used extensions of the simplex method and introduces an *interior-point algorithm* that sometimes can be used to solve even larger linear programming problems than the simplex method can handle. Chapter 8 considers some special types of linear programming problems whose importance warrants individual study.

You also can look forward to seeing applications of linear programming to other areas of operations research (OR) in several later chapters.

We begin this chapter by developing a miniature prototype example of a linear programming problem. This example is small enough to be solved graphically in a straightforward way. We then present the general *linear programming model* and its basic assumptions. The chapter concludes with some additional examples of linear programming applications, including three case studies.

3.1 Prototype Example

The WYNDOR GLASS CO. produces high-quality glass products, including windows and glass doors. It has three plants. Aluminum frames and hardware are made in Plant 1, wood frames are made in Plant 2, and Plant 3 produces the glass and assembles the products.

Because of declining earnings, top management has decided to revamp the company's product line. Unprofitable products are being discontinued, releasing production capacity to launch two new products having large sales potential:

Product 1: An 8-foot glass door with aluminum framing
Product 2: A 4×6 foot double-hung wood-framed window

Product 1 requires some of the production capacity in Plants 1 and 3, but none in Plant 2. Product 2 needs only Plants 2 and 3. The marketing division has concluded that the company could sell as much of either product as could be produced by these plants. However, because both products would be competing for the same production capacity in Plant 3, it is not clear which *mix* of the two products would be *most profitable*. Therefore, an OR team has been formed to study this question.

The OR team began by having discussions with upper management to identify management's objectives for the study. These discussions led to developing the following definition of the problem:

> Determine what the *production rates* should be for the two products in order to *maximize their total profit,* subject to the restrictions imposed by the limited production capacities available in the three plants. (Each product will be produced in batches of 20, so the *production rate* is defined as the number of batches produced per week.) *Any* combination of production rates that satisfies these restrictions is permitted, including producing none of one product and as much as possible of the other.

The OR team also identified the data that needed to be gathered:

1. Number of hours of production time available per week in each plant for these new products. (Most of the time in these plants already is committed to current products.)
2. Number of hours of production time used in each plant for each batch produced of each new product.
3. Profit per batch produced of each new product. (*Profit per batch produced* was chosen as an appropriate measure after the team concluded that the incremental profit from each additional batch produced would be roughly *constant* regardless of the total number of batches produced. Because no substantial costs will be incurred to initiate the production and marketing of these new products, the total profit from each one is approximately this *profit per batch produced* times *the number of batches produced.*)

Obtaining reasonable estimates of these quantities required enlisting the help of key personnel in various units of the company. Staff in the manufacturing division provided the data in the first category above. Developing estimates for the second category of data required some analysis by the manufacturing engineers involved in designing the production processes for the new products. By analyzing cost data from these same engineers and the marketing division, along with a pricing decision from the marketing division, the accounting department developed estimates for the third category.

Table 3.1 summarizes the data gathered.

The OR team immediately recognized that this was a linear programming problem of the classic **product mix** type, and the team next undertook the formulation of the corresponding mathematical model.

FORMULATION AS A LINEAR PROGRAMMING PROBLEM: To formulate the mathematical (linear programming) model for this problem, let

x_1 = number of batches of product 1 produced per week

Table 3.1 Data for the Wyndor Glass Co. Problem

	Production Time per Batch, Hours		
	Product		
Plant	1	2	Production Time Available per Week, Hours
1	1	0	4
2	0	2	12
3	3	2	18
Profit per batch	\$3,000	\$5,000	

x_2 = number of batches of product 2 produced per week

Z = total profit per week (in thousands of dollars) from producing these two products

Thus x_1 and x_2 are the *decision variables* for the model. Using the bottom row of Table 3.1, we obtain

$$Z = 3x_1 + 5x_2.$$

The objective is to choose the values of x_1 and x_2 so as to *maximize* $Z = 3x_1 + 5x_2$, subject to the restrictions imposed on their values by the limited production capacities available in the three plants. Table 3.1 indicates that each batch of product 1 produced per week uses 1 hour of production time per week in Plant 1, whereas only 4 hours per week is available. This restriction is expressed mathematically by the inequality $x_1 \leq 4$. Similarly, Plant 2 imposes the restriction that $2x_2 \leq 12$. The number of hours of production time used per week in Plant 3 by choosing x_1 and x_2 as the new products' production rates would be $3x_1 + 2x_2$. Therefore, the mathematical statement of the Plant 3 restriction is $3x_1 + 2x_2 \leq 18$. Finally, since production rates cannot be negative, it is necessary to restrict the decision variables to be nonnegative: $x_1 \geq 0$ and $x_2 \geq 0$.

To summarize, in the mathematical language of linear programming, the problem is to choose values of x_1 and x_2 so as to

$$\text{Maximize} \qquad Z = 3x_1 + 5x_2,$$

subject to the restrictions

$$\begin{aligned} x_1 \qquad &\leq 4 \\ 2x_2 &\leq 12 \\ 3x_1 + 2x_2 &\leq 18 \end{aligned}$$

and

$$x_1 \geq 0, \qquad x_2 \geq 0.$$

(Notice how the layout of the coefficients of x_1 and x_2 in this linear programming model essentially duplicates the information summarized in Table 3.1.)

Graphical Solution: This very small problem has only two decision variables and therefore only two dimensions, so a graphical procedure can be used to solve it.

This procedure involves constructing a two-dimensional graph with x_1 and x_2 as the axes. The first step is to identify the values of (x_1, x_2) that are permitted by the restrictions. This is done by drawing each line that borders the range of permissible values for one restriction. To begin, note that the nonnegativity restrictions $x_1 \geq 0$ and $x_2 \geq 0$ require (x_1, x_2) to lie on the *positive* side of the axes (including actually *on* either axis), i.e., in the first quadrant. Next, observe that the restriction $x_1 \leq 4$ means that (x_1, x_2) cannot lie to the right of the line $x_1 = 4$. These results are shown in Fig. 3.1, where the shaded area contains the only values of (x_1, x_2) that are still allowed.

In a similar fashion, the restriction $2x_2 \leq 12$ (or, equivalently, $x_2 \leq 6$) implies that the line $2x_2 = 12$ should be added to the boundary of the permissible region. The final restriction, $3x_1 + 2x_2 \leq 18$, requires plotting the points (x_1, x_2) such that $3x_1 + 2x_2 = 18$ (another line) to complete the boundary. (Note that the points such that $3x_1 + 2x_2 \leq 18$ are those that lie either underneath or on the line $3x_1 + 2x_2 = 18$, so this is the limiting line above which points do not satisfy the inequality.) The resulting region of permissible values of (x_1, x_2), called the **feasible region**, is shown in Fig. 3.2. (The demo called *Graphical Method* in your OR Courseware provides a more detailed example of constructing a feasible region.)

The final step is to pick out the point in this feasible region that maximizes the value of $Z = 3x_1 + 5x_2$. To discover how to perform this step efficiently, begin by trial and error. Try, for example, $Z = 10 = 3x_1 + 5x_2$ to see if there are in the permissible region any values of (x_1, x_2) that yield a value of Z as large as 10. By drawing the line $3x_1 + 5x_2 = 10$ (see Fig. 3.3), you can see that there are many points on this line that lie within the region. Having gained perspective by trying this arbitrarily chosen value of $Z = 10$, you should next try a larger arbitrary value of Z, say, $Z = 20 = 3x_1 + 5x_2$. Again, Fig. 3.3 reveals that a segment of the line $3x_1 + 5x_2 = 20$ lies within the region, so that the maximum permissible value of Z must be at least 20.

Now notice in Fig. 3.3 that the two lines just constructed are parallel. This is no coincidence, since *any* line constructed in this way has the form $Z = 3x_1 + 5x_2$ for the chosen value of Z, which implies that $5x_2 = -3x_1 + Z$ or, equivalently,

$$x_2 = -\tfrac{3}{5}x_1 + \tfrac{1}{5}Z$$

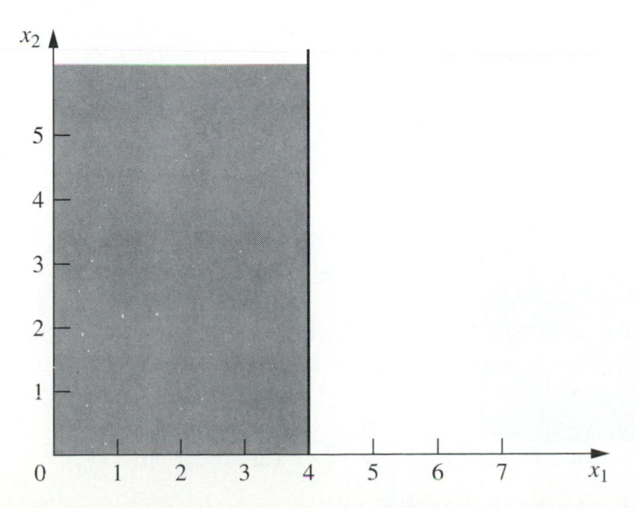

Figure 3.1 Shaded area shows values of (x_1, x_2) allowed by $x_1 \geq 0$, $x_2 \geq 0$, $x_1 \leq 4$.

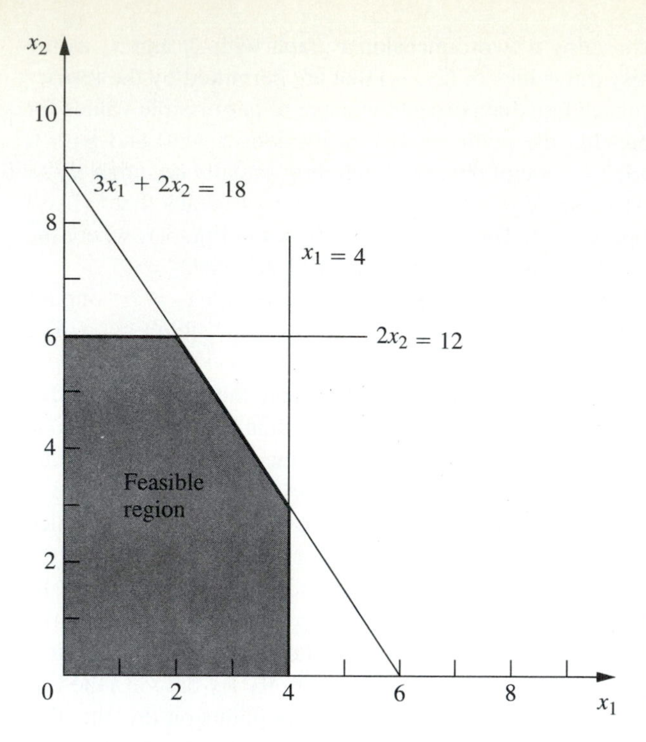

Figure 3.2 Shaded area shows the set of permissible values of (x_1, x_2), called the feasible region.

This last equation, called the **slope-intercept form** of the objective function, demonstrates that the *slope* of the line is $-\frac{3}{5}$ (since each unit increase in x_1 changes x_2 by $-\frac{3}{5}$), whereas the *intercept* of the line with the x_2 axis is $\frac{1}{5}Z$ (since $x_2 = \frac{1}{5}Z$ when $x_1 = 0$). The fact that the slope is fixed at $-\frac{3}{5}$ means that *all* lines constructed in this way are parallel.

Again, comparing the $10 = 3x_1 + 5x_2$ and $20 = 3x_1 + 5x_2$ lines in Fig. 3.3, we note that the line giving a larger value of Z ($Z = 20$) is farther up and away from the

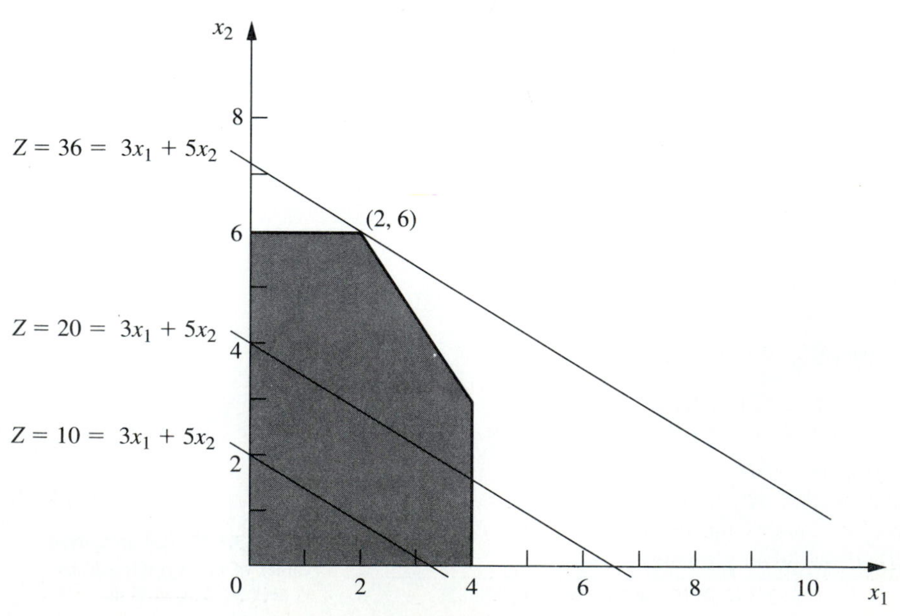

Figure 3.3 The value of (x_1, x_2) that maximizes $3x_1 + 5x_2$ is (2, 6).

origin than the other line ($Z = 10$). This fact also is implied by the slope-intercept form of the objective function, which indicates that the intercept with the x_1 axis ($\frac{1}{3}Z$) increases when the value chosen for Z is increased.

These observations imply that our trial-and-error procedure for constructing lines in Fig. 3.3 involves nothing more than drawing a family of parallel lines containing at least one point in the feasible region and selecting the line that corresponds to the largest value of Z. Figure 3.3 shows that this line passes through the point (2, 6), indicating that the **optimal solution** is $x_1 = 2$ and $x_2 = 6$. The equation of this line is $3x_1 + 5x_2 = 3(2) + 5(6) = 36 = Z$, indicating that the optimal value of Z is $Z = 36$. The point (2, 6) lies at the intersection of the two lines $2x_2 = 12$ and $3x_1 + 2x_2 = 18$, shown in Fig. 3.2, so that this point can be calculated algebraically as the simultaneous solution of these two equations.

Having seen the trial-and-error procedure for finding the optimal point (2, 6), you now can streamline this approach for other problems. Rather than draw several parallel lines, it is sufficient to form a single line with a ruler to establish the slope. Then move the ruler with fixed slope through the feasible region in the direction of improving Z. (When the objective is to *minimize* Z, move the ruler in the direction that *decreases* Z.) Stop moving the ruler at the last instant that it still passes through a point in this region. This point is the desired *optimal solution.*

This procedure often is referred to as the **graphical method** for linear programming. It can be used to solve any linear programming problem with two decision variables. With some difficulty, it is possible to extend the method to three decision variables but not more than three. (The next chapter will focus on the *simplex method* for solving larger problems.)

CONCLUSIONS: The OR team used this approach to find that the optimal solution is $x_1 = 2, x_2 = 6$, with $Z = 36$. This solution indicates that the Wyndor Glass Co. should produce products 1 and 2 at the rate of 2 batches per week and 6 batches per week, respectively, with a resulting total profit of $36,000 per week. No other mix of the two products would be so profitable—*according to the model.*

However, we emphasized in Chap. 2 that well-conducted OR studies do not simply find *one* solution for the *initial* model formulated and then stop. All six phases described in Chap. 2 are important, including thorough testing of the model (see Sec. 2.4) and post-optimality analysis (see Sec. 2.3).

In full recognition of these practical realities, the OR team now is ready to evaluate the validity of the model more critically (to be continued in Sec. 3.3) and to perform sensitivity analysis on the effect of the estimates in Table 3.1 being different because of inaccurate estimation, changes of circumstances, etc. (to be continued in Sec. 6.7).

Continuing the Learning Process with Your OR Courseware

This is the first of many points in the book where you are likely to find it helpful to use your *OR Courseware* (packaged in the back of most versions of this book). This software includes a complete demonstration example of the *graphical method* introduced in this section. Like the many other demonstration examples accompanying other sections of the book, this computer demonstration highlights concepts that are difficult to convey on the printed page.

To view this demonstration example, call the MathProg program, choose *General Linear Programming* under the Area menu, and then choose *Graphical Method* under the Demo menu. You may refer to Appendix 1 for documentation on the software, including instructions for getting started.

Another feature of your OR Courseware is a collection of routines for *interactively* executing the various solution procedures presented throughout the book. One such routine is for the *graphical method.* Like the others, this routine performs the detailed work (such as drawing graphs) while you make the decisions step by step (e.g., deciding where a line should be located), thereby enabling you to focus on concepts rather than getting bogged down in time-consuming details. Therefore, you probably will want to use this routine for your homework on this section. When you finish a problem, you can print out everything you have done for your homework by choosing the print command under the File menu.

To use this routine, choose *General Linear Programming* under the Area menu. Then choose *Enter or Revise a General Linear Programming Model* under the Procedure menu to enter your model. Finally, choose *Solve Interactively by the Graphical Method* under the Procedure menu.

Another routine, Solve Automatically by the Simplex Method (under the Procedure menu), also should be helpful for the upcoming sections of this chapter. When you formulate a linear programming model with more than two decision variables (so the graphical method cannot be used), this routine enables you to still find an optimal solution immediately. Doing so also is helpful for *model validation,* since finding a *nonsensical* optimal solution signals that you have made a mistake in formulating the model.

3.2 The Linear Programming Model

The Wyndor Glass Co. problem is intended to illustrate a typical linear programming problem (miniature version). However, linear programming is too versatile to be completely characterized by a single example. In this section we discuss the general characteristics of linear programming problems, including the various legitimate forms of the mathematical model for linear programming.

Let us begin with some basic terminology and notation. The first column of Table 3.2 summarizes the components of the Wyndor Glass Co. problem. The second column then introduces more general terms for these same components that will fit many linear programming problems. The key terms are *resources* and *activities,* where m denotes the number of different kinds of resources that can be used and n denotes the number of activities being considered. Some typical resources are money and particular kinds of machines, equipment, vehicles, and personnel. Examples of activities include investing in particular projects, advertising in particular media, and shipping goods from a particular source to a particular destination. In any application of linear programming, all the activities may be of one general kind (such as any one of these three examples), and then the individual activities would be particular alternatives within this general category.

As described in the introduction to this chapter, the most common type of application of linear programming involves allocating resources to activities. The amount available of each resource is limited, so a careful allocation of resources to activities must be made. Determining this allocation involves choosing the *levels* of the activities that achieve the best possible value of the *overall measure of performance.*

Table 3.2 **Common Terminology for Linear Programming**

Prototype Example	General Problem
Production capacities of plants 3 plants	Resources m resources
Production of products 2 products Production rate of product j, x_j	Activities n activities Level of activity j, x_j
Profit Z	Overall measure of performance Z

Certain symbols are commonly used to denote the various components of a linear programming model. These symbols are listed below, along with their interpretation for the general problem of allocating resources to activities.

Z = value of overall measure of performance.

x_j = level of activity j (for $j = 1, 2, \ldots, n$).

c_j = increase in Z that would result from each unit increase in level of activity j.

b_i = amount of resource i that is available for allocation to activities (for $i = 1, 2, \ldots, m$).

a_{ij} = amount of resource i consumed by each unit of activity j.

The model poses the problem in terms of making decisions about the levels of the activities, so $x_1, x_2, \ldots, x_n$ are called the **decision variables.** As summarized in Table 3.3, the values of c_j, b_i, and a_{ij} (for $i = 1, 2, \ldots, m$ and $j = 1, 2, \ldots, n$) are the *input constants* for the model. The c_j, b_i, and a_{ij} are also referred to as the **parameters** of the model.

Notice the correspondence between Table 3.3 and Table 3.1.

A Standard Form of the Model

Proceeding as for the Wyndor Glass Co. problem, we can now formulate the mathematical model for this general problem of allocating resources to activities. In particular,

Table 3.3 **Data Needed for a Linear Programming Model Involving the Allocation of Resources to Activities**

	Resource Usage per Unit of Activity				
	Activity				
Resource	1	2	...	n	Amount of Resource Available
1	a_{11}	a_{12}	...	a_{1n}	b_1
2	a_{21}	a_{22}	...	a_{2n}	b_2
. . .	...	...	...	...	. . .
m	a_{m1}	a_{m2}	...	a_{mn}	b_m
Contribution to Z per unit of activity	c_1	c_2	...	c_n	

this model is to select the values for $x_1, x_2, \ldots, x_n$ so as to

$$\text{Maximize} \quad Z = c_1 x_1 + c_2 x_2 + \cdots + c_n x_n,$$

subject to the restrictions

$$\begin{aligned} a_{11}x_1 + a_{12}x_2 + \cdots + a_{1n}x_n &\leq b_1 \\ a_{21}x_1 + a_{22}x_2 + \cdots + a_{2n}x_n &\leq b_2 \\ &\vdots \\ a_{m1}x_1 + a_{m2}x_2 + \cdots + a_{mn}x_n &\leq b_m, \end{aligned}$$

and

$$x_1 \geq 0, \quad x_2 \geq 0, \quad \ldots, \quad x_n \geq 0.$$

We call this *our standard form*[1] for the linear programming problem. Any situation whose mathematical formulation fits this model is a linear programming problem.

Notice that the model for the Wyndor Glass Co. problem fits our standard form, with $m = 3$ and $n = 2$.

Common terminology for the linear programming model can now be summarized. The function being maximized, $c_1x_1 + c_2x_2 + \cdots + c_nx_n$, is called the **objective function**. The restrictions normally are referred to as **constraints.** The first m constraints (those with a *function* of all the variables $a_{i1}x_1 + a_{i2}x_2 + \cdots + a_{in}x_n$ on the left-hand side) are sometimes called **functional constraints**. Similarly, the $x_j \geq 0$ restrictions are called **nonnegativity constraints** (or **nonnegativity conditions**).

Other Forms

We now hasten to add that the preceding model does not actually fit the natural form of some linear programming problems. The other *legitimate forms* are the following:

1. Minimizing rather than maximizing the objective function:

$$\text{Minimize} \quad Z = c_1x_1 + c_2x_2 + \cdots + c_nx_n.$$

2. Some functional constraints with a greater-than-or-equal-to inequality:

$$a_{i1}x_1 + a_{i2}x_2 + \cdots + a_{in}x_n \geq b_i \qquad \text{for some values of } i.$$

3. Some functional constraints in equation form:

$$a_{i1}x_1 + a_{i2}x_2 + \cdots + a_{in}x_n = b_i \qquad \text{for some values of } i.$$

4. Deleting the nonnegativity constraints for some decision variables:

$$x_j \text{ unrestricted in sign} \qquad \text{for some values of } j.$$

Any problem that mixes some of or all these forms with the remaining parts of the preceding model is still a linear programming problem. Our interpretation of the words *allocating limited resources among competing activities* may no longer apply very well, if at all; but regardless of the interpretation or context, all that is required is that the mathematical statement of the problem fit the allowable forms.

[1] This is called *our* standard form rather than *the* standard form because some textbooks adopt other forms.

Terminology for Solutions of the Model

You may be used to having the term *solution* mean the final answer to a problem, but the convention in linear programming (and its extensions) is quite different. Here, *any* specification of values for the decision variables ($x_1, x_2, \ldots, x_n$) is called a **solution**, regardless of whether it is a desirable or even an allowable choice. Different types of solutions are then identified by using an appropriate adjective.

> A **feasible solution** is a solution for which *all* the constraints are *satisfied.*
> An **infeasible solution** is a solution for which *at least one* constraint is *violated.*

In the example, the points (2, 3) and (4, 1) in Fig. 3.2 are *feasible solutions,* while the points (−1, 3) and (4, 4) are *infeasible solutions.*

> The **feasible region** is the collection of all feasible solutions.

The feasible region in the example is the entire shaded area in Fig. 3.2.

It is possible for a problem to have **no feasible solutions**. This would have happened in the example if the new products had been required to return a net profit of at least $50,000 per week to justify discontinuing part of the current product line. The corresponding constraint, $3x_1 + 5x_2 \geq 50$, would eliminate the entire feasible region, so no mix of new products would be superior to the status quo. This case is illustrated in Fig. 3.4.

Given that there are feasible solutions, the goal of linear programming is to find a best feasible solution, as measured by the value of the objective function in the model.

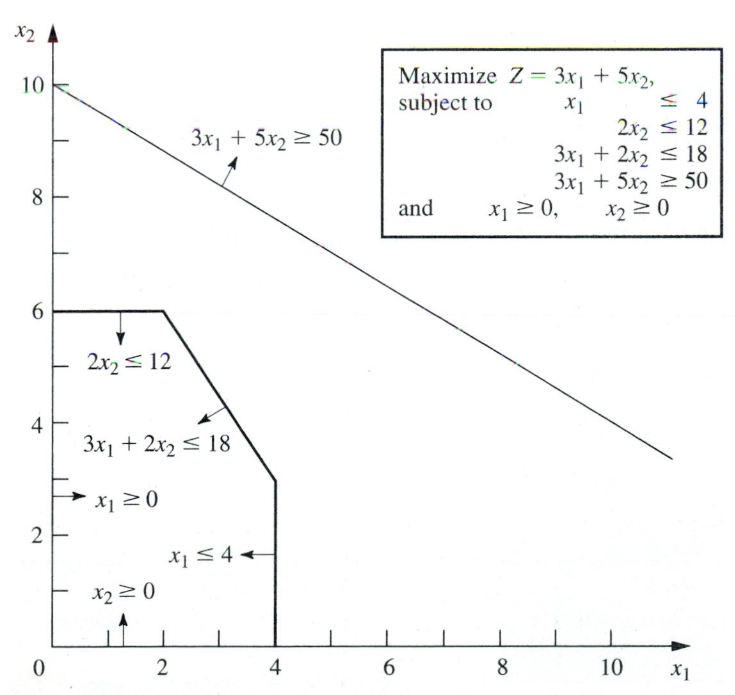

Figure 3.4 The Wyndor Glass Co. problem would have no feasible solutions if the constraint $3x_1 + 5x_2 \geq 50$ were added to the problem.

An **optimal solution** is a feasible solution that has the *most favorable value* of the objective function.

The **most favorable value** is the *largest value* if the objective function is to be *maximized,* whereas it is the *smallest value* if the objective function is to be *minimized.*

Most problems will have just one optimal solution. However, it is possible to have more than one. This would occur in the example if the *profit per batch produced* of product 2 were changed to \$2,000. This changes the objective function to $Z = 3x_1 + 2x_2$, so that all the points on the line segment connecting (2, 6) and (4, 3) would be optimal. This case is illustrated in Fig. 3.5. As in this case, *any* problem having **multiple optimal solutions** will have an *infinite* number of them, each with the same optimal value of the objective function.

Another possibility is that a problem has **no optimal solutions**. This occurs only if (1) it has no feasible solutions or (2) the constraints do not prevent improving the value of the objective function (Z) indefinitely in the favorable direction (positive or negative). For example, the latter case would result if the last two functional constraints were mistakenly deleted in the example. This case is illustrated in Fig. 3.6.

We next introduce a special type of feasible solution that plays the key role when the simplex method searches for an optimal solution.

A **corner-point feasible (CPF) solution** is a solution that lies at a corner of the feasible region.

Figure 3.7 highlights the five CPF solutions for the example.

Sections 4.1 and 5.1 will delve into the various useful properties of CPF solutions for problems of any size, including the following relationship with optimal solutions.

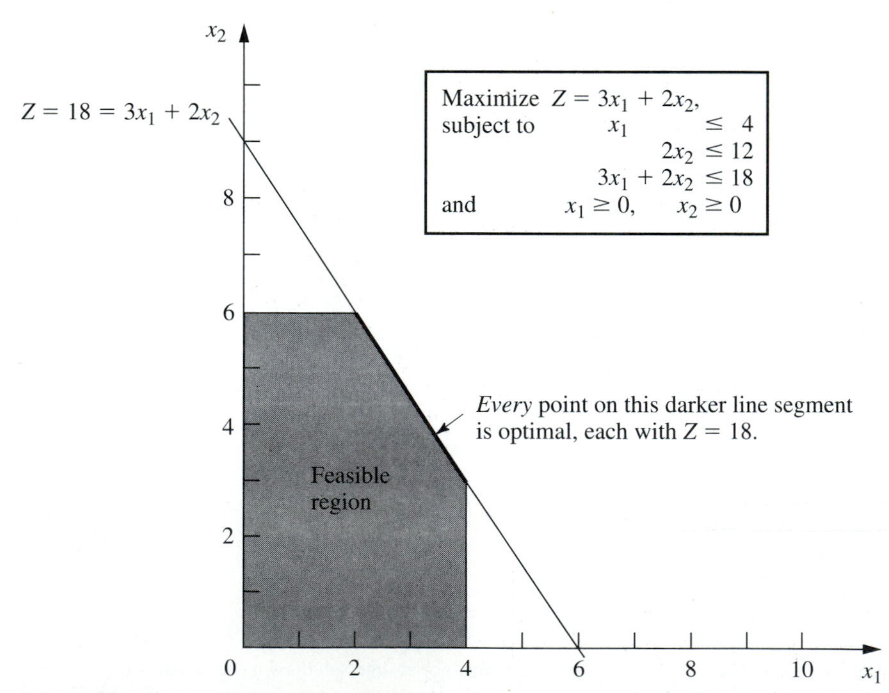

Figure 3.5 The Wyndor Glass Co. problem would have multiple optimal solutions if the objective function were changed to $Z = 3x_1 + 2x_2$.

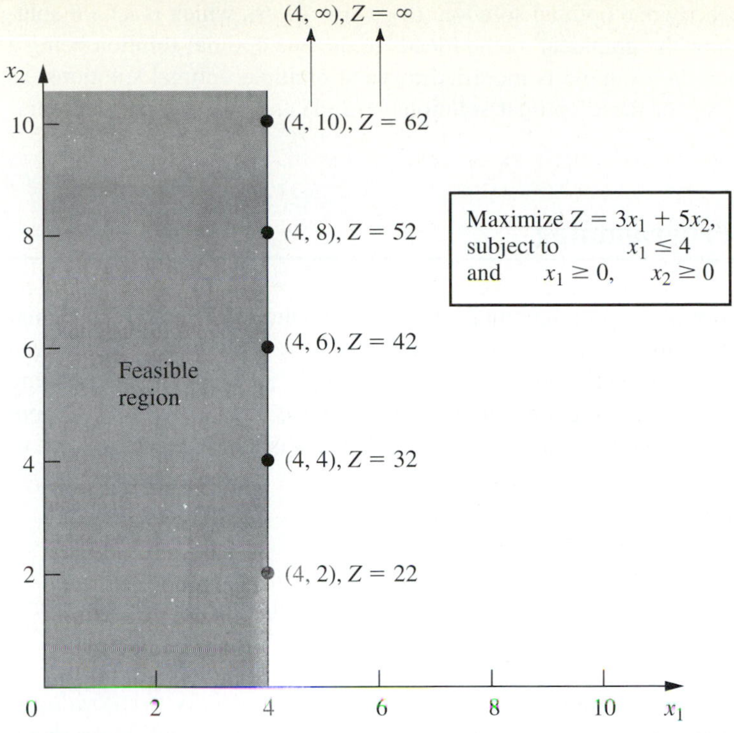

Figure 3.6 The Wyndor Glass Co. problem would have no optimal solutions if the only functional constraint were $x_1 \leq 4$, because x_2 then could be increased indefinitely in the feasible region without ever reaching the maximum value of $Z = 3x_1 + 5x_2$.

Relationship between optimal solutions and CPF solutions: Consider any linear programming problem with feasible solutions and a bounded feasible region. The problem must possess CPF solutions and at least one optimal solution. Furthermore, the best CPF solution *must* be an optimal solution. Thus, if a problem has exactly one optimal solution, it *must* be a CPF solution. If the problem has multiple optimal solutions, at least two *must* be CPF solutions.

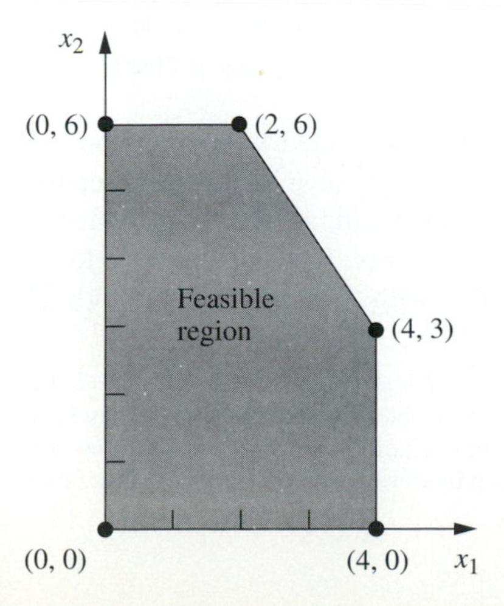

Figure 3.7 The five dots are the five CPF solutions for the Wyndor Glass Co. problem.

The example has exactly one optimal solution, $(x_1, x_2) = (2, 6)$, which is a CPF solution. (Think about how the graphical method leads to the one optimal solution being a CPF solution.) When the example is modified to yield multiple optimal solutions, as shown in Fig. 3.5, two of these optimal solutions—(2, 6) and (4, 3)—are CPF solutions.

3.3 Assumptions of Linear Programming

All the assumptions of linear programming actually are implicit in the model formulation given in Sec. 3.2. However, it is good to highlight these assumptions so you can more easily evaluate how well linear programming applies to any given problem. Furthermore, we still need to see why the OR team for the Wyndor Glass Co. concluded that a linear programming formulation provided a satisfactory representation of the problem.

Proportionality

Proportionality is an assumption about both the objective function and the functional constraints, as summarized below.

> **Proportionality assumption:** The contribution of each activity to the *value of the objective function Z* is *proportional* to the *level of the activity* x_j, as represented by the $c_j x_j$ term in the objective function. Similarly, the contribution of each activity to the *left-hand side of each functional constraint* is *proportional* to the *level of the activity* x_j, as represented by the $a_{ij} x_j$ term in the constraint. Consequently, this assumption rules out any exponent other than 1 for any variable in any term of any function (whether the objective function or the function on the left-hand side of a functional constraint) in a linear programming model.[1]

To illustrate this assumption, consider the first term ($3x_1$) in the objective function ($Z = 3x_1 + 5x_2$) for the Wyndor Glass Co. problem. This term represents the profit generated per week (in thousands of dollars) by producing product 1 at the rate of x_1 batches per week. The *proportionality satisfied* column of Table 3.4 shows the case that was assumed in Sec. 3.1, namely, that this profit is indeed proportional to x_1 so that $3x_1$ is the appropriate term for the objective function. By contrast, the next three columns show different hypothetical cases where the proportionality assumption would be violated.

Case 1 would arise if there were *start-up costs* associated with initiating the production of product 1. For example, there might be costs involved with setting up the production facilities. There might also be costs associated with arranging the distribution of the new product. Because these are one-time costs, they would need to be amortized on a per-week basis to be commensurable with Z (profit in thousands of

[1] When the function includes any *cross-product terms,* proportionality should be interpreted to mean that *changes* in the function value are proportional to *changes* in each variable (x_j) individually, given any fixed values for all the other variables. Therefore, a cross-product term satisfies proportionality as long as each variable in the term has an exponent of 1. (However, any cross-product term violates the *additivity assumption,* discussed next.)

Table 3.4 **Examples of Satisfying or Violating Proportionality**

	Profit from Product 1 ($000 per Week)			
		Proportionality Violated		
x_1	Proportionality Satisfied	Case 1	Case 2	Case 3
0	0	0	0	0
1	3	2	3	3
2	6	5	7	5
3	9	8	12	6
4	12	11	18	6

dollars per week). Suppose that this amortization were done and that the total start-up cost amounted to reducing Z by 1, but that the profit without considering the start-up cost would be $3x_1$. This would mean that the contribution from product 1 to Z should be $3x_1 - 1$ for $x_1 > 0$, whereas the contribution would be $3x_1 = 0$ when $x_1 = 0$ (no start-up cost). This profit function,[1] which is given by the solid curve in Fig. 3.8, certainly is *not* proportional to x_1.

At first glance, it might appear that *Case 2* in Table 3.4 is quite similar to Case 1. However, Case 2 actually arises in a very different way. There no longer is a start-up cost, and the profit from the first unit of product 1 per week is indeed 3, as originally

[1] If the contribution from product 1 to Z were $3x_1 - 1$ for *all* $x_1 \geq 0$, including $x_1 = 0$, then the fixed constant, -1, could be deleted from the objective function without changing the optimal solution and proportionality would be restored. However, this "fix" does not work here because the -1 constant does not apply when $x_1 = 0$.

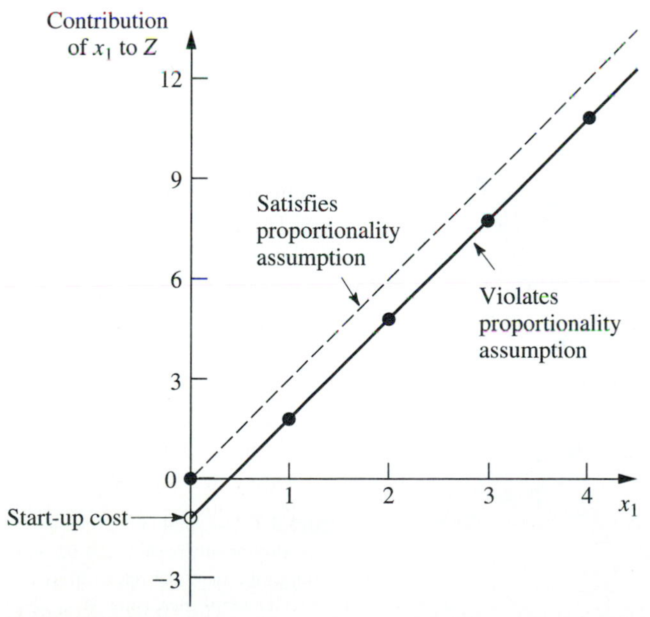

Figure 3.8 The solid curve violates the proportionality assumption because of the start-up cost that is incurred when x_1 is increased from zero. The values at the dots are given by the *Case 1* column of Table 3.4.

assumed. However, there now is an *increasing marginal return;* i.e., the *slope* of the *profit function* for product 1 (see the solid curve in Fig. 3.9) keeps increasing as x_1 is increased. This violation of proportionality might occur because of economies of scale that can sometimes be achieved at higher levels of production, e.g., through the use of more efficient high-volume machinery, longer production runs, quantity discounts for large purchases of raw materials, and the learning-curve effect whereby workers become more efficient as they gain experience with a particular mode of production. As the incremental cost goes down, the incremental profit will go up (assuming constant marginal revenue).

The reverse of Case 2 is *Case 3,* where there is a *decreasing marginal return.* In this case, the *slope* of the *profit function* for product 1 (given by the solid curve in Fig. 3.10) keeps decreasing as x_1 is increased. This violation of proportionality might occur because the *marketing costs* need to go up more than proportionally to attain increases in the level of sales. For example, it might be possible to sell product 1 at the rate of 1 per week ($x_1 = 1$) with no advertising, whereas attaining sales to sustain a production rate of $x_1 = 2$ might require a moderate amount of advertising, $x_1 = 3$ might necessitate an extensive advertising campaign, and $x_1 = 4$ might require also lowering the price.

All three cases are hypothetical examples of ways in which the proportionality assumption could be violated. What is the actual situation? The actual profit from producing product 1 (or any other product) is derived from the sales revenue minus various direct and indirect costs. Inevitably, some of these cost components are not strictly proportional to the production rate, perhaps for one of the reasons illustrated above. However, the real question is whether, after all the components of profit have been accumulated, proportionality is a reasonable approximation for practical modeling purposes. For the Wyndor Glass Co. problem, the OR team checked both the objective function and the functional constraints. The conclusion was that proportionality could indeed be assumed without serious distortion.

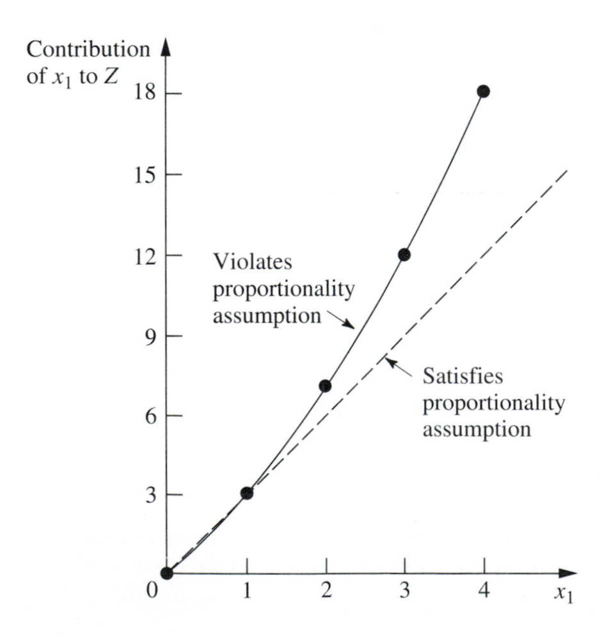

Figure 3.9 The solid curve violates the proportionality assumption because its slope (the marginal return from product 1) keeps increasing as x_1 is increased. The values at the dots are given by the *Case 2* column of Table 3.4.

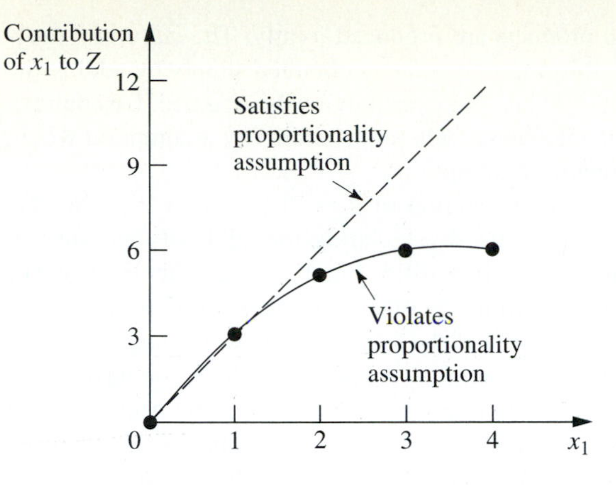

Figure 3.10 The solid curve violates the proportionality assumption because its slope (the marginal return from product 1) keeps decreasing as x_1 is increased. The values at the dots are given by the *Case 3* column in Table 3.4.

For other problems, what happens when the proportionality assumption does not hold even as a reasonable approximation? In most cases, this means you must use *nonlinear programming* instead (presented in Chap. 13). However, we do point out in Sec. 13.8 that a certain important kind of nonproportionality can still be handled by linear programming by reformulating the problem appropriately. Furthermore, if the assumption is violated only because of start-up costs, there is an extension of linear programming (*mixed integer programming*) that can be used, as discussed in Sec. 12.2 (the fixed-charge problem).

Additivity

Although the proportionality assumption rules out exponents other than one, it does not prohibit *cross-product terms* (terms involving the product of two or more variables). The additivity assumption does rule out this latter possibility, as summarized below.

> **Additivity assumption:** *Every* function in a linear programming model (whether the objective function or the function on the left-hand side of a functional constraint) is the *sum* of the *individual contributions* of the respective activities.

To make this definition more concrete and clarify why we need to worry about this assumption, let us look at some examples. Table 3.5 shows some possible cases for the objective function for the Wyndor Glass Co. problem. In each case, the *individual contributions* from the products are just as assumed in Sec. 3.1, namely, $3x_1$ for product 1 and $5x_2$ for product 2. The difference lies in the last row, which gives the

Table 3.5 **Examples of Satisfying or Violating Additivity for the Objective Function**

	Value of Z		
		Additivity Violated	
(x_1, x_2)	Additivity Satisfied	Case 1	Case 2
(1, 0)	3	3	3
(0, 1)	5	5	5
(1, 1)	8	9	7

function value for Z when the two products are produced jointly. The *additivity satisfied* column shows the case where this *function value* is obtained simply by adding the first two rows $(3 + 5 = 8)$, so that $Z = 3x_1 + 5x_2$ as previously assumed. By contrast, the next two columns show hypothetical cases where the additivity assumption would be violated (but not the proportionality assumption).

Case 1 corresponds to an objective function of $Z = 3x_1 + 5x_2 + x_1x_2$, so that $Z = 3 + 5 + 1 = 9$ for $(x_1, x_2) = (1, 1)$, thereby violating the additivity assumption that $Z = 3 + 5$. (The proportionality assumption still is satisfied since after the value of one variable is fixed, the increment in Z from the other variable is proportional to the value of that variable.) This case would arise if the two products were *complementary* in some way that *increases* profit. For example, suppose that a major advertising campaign would be required to market either new product produced by itself, but that the same single campaign can effectively promote both products if the decision is made to produce both. Because a major cost is saved for the second product, their joint profit is somewhat more than the *sum* of their individual profits when each is produced by itself.

Case 2 also violates the additivity assumption because of the extra term in its objective function, $Z = 3x_1 + 5x_2 - x_1x_2$, so that $Z = 3 + 5 - 1 = 7$ for $(x_1, x_2) = (1, 1)$. As the reverse of the first case, Case 2 would arise if the two products were *competitive* in some way that *decreased* their joint profit. For example, suppose that both products need to use the same machinery and equipment. If either product were produced by itself, this machinery and equipment would be dedicated to this one use. However, producing both products would require switching the production processes back and forth, with substantial time and cost involved in temporarily shutting down the production of one product and setting up for the other. Because of this major extra cost, their joint profit is somewhat less than the *sum* of their individual profits when each is produced by itself.

The same kinds of interaction between activities can affect the additivity of the constraint functions. For example, consider the third functional constraint of the Wyndor Glass Co. problem: $3x_1 + 2x_2 \le 18$. (This is the only constraint involving both products.) This constraint concerns the production capacity of Plant 3, where 18 hours of production time per week is available for the two new products, and the function on the left-hand side $(3x_1 + 2x_2)$ represents the number of hours of production time per week that would be used by these products. The *additivity satisfied* column of Table 3.6 shows this case as is, whereas the next two columns display cases where the function has an extra cross-product term that violates additivity. For all three columns, the *individual contributions* from the products toward using the capacity of Plant 3 are just as assumed previously, namely, $3x_1$ for product 1 and $2x_2$ for product 2, or $3(2) = 6$ for $x_1 = 2$ and $2(3) = 6$ for $x_2 = 3$. As was true for Table 3.5, the difference lies in the last

Table 3.6 **Examples of Satisfying or Violating Additivity for a Functional Constraint**

	Amount of Resource Used		
		Additivity Violated	
(x_1, x_2)	Additivity Satisfied	Case 3	Case 4
(2, 0)	6	6	6
(0, 3)	6	6	6
(2, 3)	12	15	10.8

row, which now gives the *total function value* for production time used when the two products are produced jointly.

For Case 3 (see Table 3.6), the production time used by the two products is given by the function $3x_1 + 2x_2 + 0.5x_1x_2$, so the *total function value* is $6 + 6 + 3 = 15$ when $(x_1, x_2) = (2, 3)$, which violates the additivity assumption that the value is just $6 + 6 = 12$. This case can arise in exactly the same way as described for Case 2; namely, extra time is wasted switching the production processes back and forth between the two products. The extra cross-product term $(0.5x_1x_2)$ would give the production time wasted in this way. (Note that wasting time switching between products leads to a positive cross-product term here, where the total function is measuring production time used, whereas it led to a negative cross-product term for Case 2 because the total function there measures profit.)

For Case 4, the function for production time used is $3x_1 + 2x_2 - 0.1x_1^2x_2$, so the *function value* for $(x_1, x_2) = (2, 3)$ is $6 + 6 - 1.2 = 10.8$. This case could arise in the following way. As in Case 3, suppose that the two products require the same type of machinery and equipment. But suppose now that the time required to switch from one product to the other would be relatively small. Because each product goes through a sequence of production operations, individual production facilities normally dedicated to that product would incur occasional idle periods. During these otherwise idle periods, these facilities can be used by the other product. Consequently, the total production time used when the two products are produced jointly would be less than the *sum* of the production times used by the individual products when each is produced by itself.

After analyzing the possible kinds of interaction between the two products illustrated by these four cases, the OR team concluded that none played a major role in the actual Wyndor Glass Co. problem. Therefore, the additivity assumption was adopted as a reasonable approximation.

For other problems, if additivity is not a reasonable assumption, so that some of or all the mathematical functions of the model need to be *nonlinear* (because of the cross-product terms), you definitely enter the realm of nonlinear programming (Chap. 13).

Divisibility

Our next assumption concerns the values allowed for the decision variables.

> **Divisibility assumption:** Decision variables in a linear programming model are allowed to have *any* values, including *noninteger* values, that satisfy the functional and nonnegativity constraints. Thus, these variables are *not* restricted to just integer values. Since each decision variable represents the level of some activity, it is being assumed that the activities can be run at *fractional levels.*

For the Wyndor Glass Co. problem, the decision variables represent production rates (the number of batches of a product produced per week). Since these production rates can have *any* fractional values within the feasible region, the divisibility assumption does hold.

In certain situations, the divisibility assumption does not hold because some of or all the decision variables must be restricted to *integer values.* Mathematical models with this restriction are called *integer programming* models, and they are discussed in Chap. 12.

Certainty

Our last assumption concerns the *parameters* of the model, namely, the coefficients in the objective function c_j, the coefficients in the functional constraints a_{ij}, and the right-hand sides of the functional constraints b_i.

Certainty assumption: The value assigned to each parameter of a linear programming model is assumed to be a *known constant.*

In real applications, the certainty assumption is seldom satisfied precisely. Linear programming models usually are formulated to select some future course of action. Therefore, the parameter values used would be based on a prediction of future conditions, which inevitably introduces some degree of uncertainty.

For this reason it is usually important to conduct **sensitivity analysis** after a solution is found that is optimal under the assumed parameter values. As discussed in Sec. 2.3, one purpose is to identify the *sensitive* parameters (those whose value cannot be changed without changing the optimal solution), since any later change in the value of a sensitive parameter immediately signals a need to change the solution being used.

Sensitivity analysis plays an important role in the analysis of the Wyndor Glass Co. problem, as you will see in Sec. 6.7. However, it is necessary to acquire some more background before we finish that story.

Occasionally, the degree of uncertainty in the parameters is too great to be amenable to sensitivity analysis. In this case, it is necessary to treat the parameters explicitly as *random variables.* Formulations of this kind have been developed, but they are beyond the scope of this book.

The Assumptions in Perspective

We emphasized in Sec. 2.2 that a mathematical model is intended to be only an idealized representation of the real problem. Approximations and simplifying assumptions generally are required in order for the model to be tractable. Adding too much detail and precision can make the model too unwieldy for useful analysis of the problem. All that is really needed is that there be a reasonably high correlation between the prediction of the model and what would actually happen in the real problem.

This advice certainly is applicable to linear programming. It is very common in real applications of linear programming that almost *none* of the four assumptions hold completely. Except perhaps for the *divisibility assumption,* minor disparities are to be expected. This is especially true for the *certainty assumption,* so sensitivity analysis normally is a must to compensate for the violation of this assumption.

However, it is important for the OR team to examine the four assumptions for the problem under study and to analyze just how large the disparities are. If any of the assumptions are violated in a major way, then a number of useful alternative models are available, as presented in later chapters of the book. A disadvantage of these other models is that the algorithms available for solving them are not nearly as powerful as those for linear programming, but this gap has been closing in some cases. For some applications, the powerful linear programming approach is used for the initial analysis, and then a more complicated model is used to refine this analysis.

As you work through the examples in the next section, you will find it good practice to analyze how well each of the four assumptions of linear programming applies.

3.4 Additional Examples

The Wyndor Glass Co. problem is a prototype example of linear programming in several respects: It involves allocating limited resources among competing activities, its model fits our standard form, and its context is the traditional one of improved business planning. However, the applicability of linear programming is much wider. In this section we begin broadening our horizons. As you study the following examples, note that it is their underlying mathematical model rather than their context that characterizes them as linear programming problems. Then give some thought to how the same mathematical model could arise in many other contexts by merely changing the names of the activities and so forth.

These examples are scaled-down versions of actual applications (including two that are included in the case studies presented in the next section).

Design of Radiation Therapy

MARY has just been diagnosed as having a cancer at a fairly advanced stage. Specifically, she has a large malignant tumor in the bladder area (a "whole bladder lesion").

Mary is to receive the most advanced medical care available to give her every possible chance for survival. This care will include extensive *radiation therapy.*

Radiation therapy involves using an external beam treatment machine to pass ionizing radiation through the patient's body, damaging both cancerous and healthy tissues. Normally, several beams are precisely administered from different angles in a two-dimensional plane. Due to attenuation, each beam delivers more radiation to the tissue near the entry point than to the tissue near the exit point. Scatter also causes some delivery of radiation to tissue outside the direct path of the beam. Because tumor cells are typically microscopically interspersed among healthy cells, the radiation dosage throughout the tumor region must be large enough to kill the malignant cells, which are slightly more radiosensitive, yet small enough to spare the healthy cells. At the same time, the aggregate dose to critical tissues must not exceed established tolerance levels, in order to prevent complications that can be more serious than the disease itself. For the same reason, the total dose to the entire healthy anatomy must be minimized.

Because of the need to carefully balance all these factors, the design of radiation therapy is a very delicate process. The goal of the design is to select the combination of beams to be used, and the intensity of each one, to generate the best possible dose distribution. (The dose strength at any point in the body is measured in units called *kilorads.*) Once the treatment design has been developed, it is administered in many installments, spread over several weeks.

In Mary's case, the size and location of her tumor make the design of her treatment an even more delicate process than usual. Figure 3.11 shows a diagram of a cross section of the tumor viewed from above, as well as nearby critical tissues to avoid. These tissues include critical organs (e.g., the rectum) as well as bony structures (e.g., the femurs and pelvis) that will attenuate the radiation. Also shown are the entry point and direction for the only two beams that can be used with any modicum of safety in this case. (Actually, we are simplifying the example at this point, because normally dozens of possible beams must be considered.)

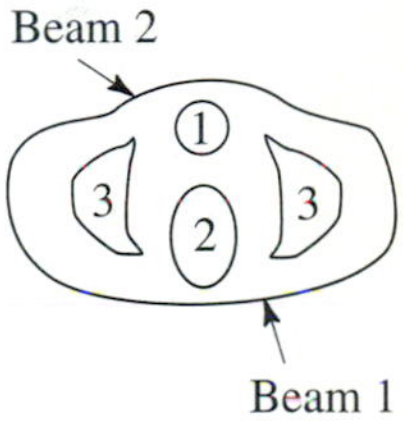

For any proposed beam of given intensity, the analysis of what the resulting radiation absorption by various parts of the body would be requires a complicated

Beam 2

Beam 1

1. Bladder and tumor
2. Rectum, coccyx, etc.
3. Femur, part of pelvis, etc.

Figure 3.11 Cross-section of Mary's tumor (viewed from above), nearby critical tissues, and the radiation beams being used.

process. In brief, based on careful anatomical analysis, the energy distribution within the two-dimensional cross section of the tissue can be plotted on an isodose map, where the contour lines represent the dose strength as a percentage of the dose strength at the entry point. A fine grid then is placed over the isodose map. By summing the radiation absorbed in the squares containing each type of tissue, the average dose that is absorbed by the tumor, healthy anatomy, and critical tissues can be calculated. With more than one beam (administered sequentially), the radiation absorption is additive.

After thorough analysis of this type, the medical team has carefully estimated the data needed to design Mary's treatment, as summarized in Table 3.7. The first column lists the areas of the body that must be considered, and then the next two columns give the fraction of the radiation dose at the entry point for each beam that is absorbed by the respective areas on average. For example, if the dose level at the entry point for beam 1 is 1 kilorad, then an average of 0.4 kilorad will be absorbed by the entire healthy anatomy in the two-dimensional plane, an average of 0.3 kilorad will be absorbed by nearby critical tissues, an average of 0.5 kilorad will be absorbed by the various parts of the tumor, and 0.6 kilorad will be absorbed by the center of the tumor. The last column gives the restrictions on the total dosage from both beams that is absorbed on average by the respective areas of the body. In particular, the average dosage absorption for the healthy anatomy must be *as small as possible,* the critical tissues must *not exceed* 2.7 kilorads, the average over the entire tumor must *equal* 6 kilorads, and the center of the tumor must be *at least* 6 kilorads.

Formulation as a Linear Programming Problem: The two decision variables x_1 and x_2 represent the dose (in kilorads) at the entry point for beam 1 and beam 2, respectively. Because the total dosage reaching the healthy anatomy is to be minimized, let Z denote this quantity. The data from Table 3.7 can then be used directly to formulate the following linear programming model.[1]

$$\text{Minimize} \quad Z = 0.4x_1 + 0.5x_2,$$

subject to

$$0.3x_1 + 0.1x_2 \leq 2.7$$
$$0.5x_1 + 0.5x_2 = 6$$
$$0.6x_1 + 0.4x_2 \geq 6$$

[1] Actually, Table 3.7 simplifies the real situation, so the real model would be somewhat more complicated than this one and would have dozens of variables and constraints. For details about the general situation, see D. Sonderman and P. G. Abrahamson, "Radiotherapy Treatment Design Using Mathematical Programming Models," *Operations Research,* **33**:705–725, 1985, and its ref. 1.

Table 3.7 **Data for the Design of Mary's Radiation Therapy**

Area	*Fraction of Entry Dose Absorbed by Area (Average)*		Restriction on Total Average Dosage, Kilorads
	Beam 1	Beam 2	
Healthy anatomy	0.4	0.5	Minimize
Critical tissues	0.3	0.1	≤ 2.7
Tumor region	0.5	0.5	$= 6$
Center of tumor	0.6	0.4	≥ 6

and

$$x_1 \geq 0, \qquad x_2 \geq 0.$$

Notice the differences between this model and the one in Sec. 3.1 for the Wyndor Glass Co. problem. The latter model involved *maximizing Z*, and all the functional constraints were in $\leq$ form. This new model does not fit this same standard form, but it does incorporate three other *legitimate* forms described in Sec. 3.2, namely, *minimizing Z*, functional constraints in $=$ form, and functional constraints in $\geq$ form.

However, both models have only two variables, so this new problem also can be solved by the *graphical method* illustrated in Sec. 3.1. Figure 3.12 shows the graphical solution. The *feasible region* consists of just the dark line segment between (6, 6) and (7.5, 4.5), because the points on this segment are the only ones that simultaneously satisfy all the constraints. (Note that the equality constraint limits the feasible region to the line containing this line segment, and then the other two functional constraints

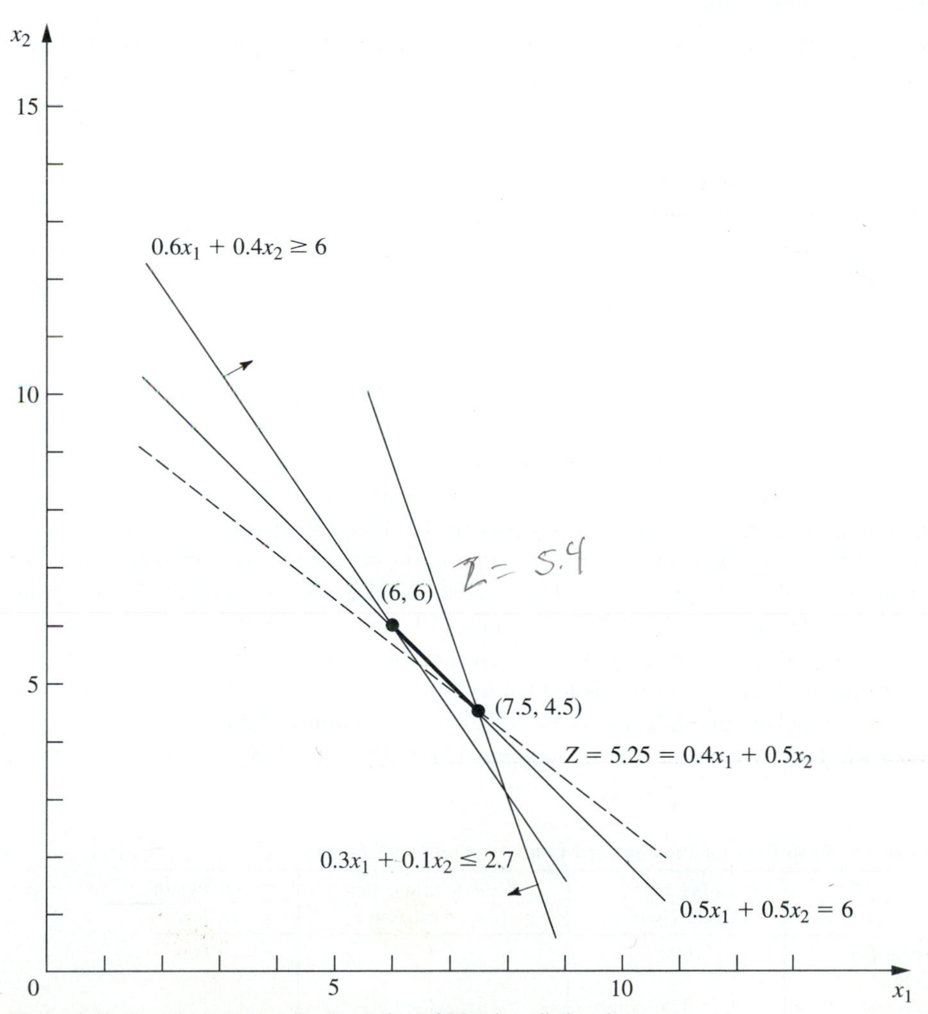

Figure 3.12 Graphical solution for the design of Mary's radiation therapy.

Table 3.8 **Resource Data for the Southern Confederation of Kibbutzim**

Kibbutz	Usable Land (Acres)	Water Allocation (Acre Feet)
1	400	600
2	600	800
3	300	375

determine the two endpoints of the line segment.) The dashed line is the objective function line that passes through the optimal solution $(x_1, x_2) = (7.5, 4.5)$ with $Z = 5.25$. This solution is optimal rather than the point (6, 6) because *decreasing* Z (for positive values of Z) pushes the objective function line toward the origin (where $Z = 0$). And $Z = 5.25$ for (7.5, 4.5) is less than $Z = 5.4$ for (6, 6).

Thus the optimal design is to use a total dose at the entry point of 7.5 kilorads for beam 1 and 4.5 kilorads for beam 2.

Regional Planning

The SOUTHERN CONFEDERATION OF KIBBUTZIM is a group of three kibbutzim (communal farming communities) in Israel. Overall planning for this group is done in its Coordinating Technical Office. This office currently is planning agricultural production for the coming year.

The agricultural output of each kibbutz is limited by both the amount of available irrigable land and the quantity of water allocated for irrigation by the Water Commissioner (a national government official). These data are given in Table 3.8.

The crops suited for this region include sugar beets, cotton, and sorghum, and these are the three being considered for the upcoming season. These crops differ primarily in their expected net return per acre and their consumption of water. In addition, the Ministry of Agriculture has set a maximum quota for the total acreage that can be devoted to each of these crops by the Southern Confederation of Kibbutzim, as shown in Table 3.9.

Because of the limited water available for irrigation, the Southern Confederation of Kibbutzim will not be able to use all its irrigable land for planting crops in the upcoming season. To ensure equity between the three kibbutzim, it has been agreed that every kibbutz will plant the same proportion of its available irrigable land. For example, if kibbutz 1 plants 200 of its available 400 acres, then kibbutz 2 must plant 300 of its 600 acres, while kibbutz 3 plants 150 acres of its 300 acres. However, any combination of the crops may be grown at any of the kibbutzim. The job facing the Coordinating Technical Office is to plan how many acres to devote to each crop at the respective kibbutzim while satisfying the given restrictions. The objective is to maximize the total net return to the Southern Confederation of Kibbutzim as a whole.

Table 3.9 **Crop Data for the Southern Confederation of Kibbutzim**

Crop	Maximum Quota (Acres)	Water Consumption (Acre Feet/Acre)	Net Return ($/Acre)
Sugar beets	600	3	1,000
Cotton	500	2	750
Sorghum	325	1	250

#3

Table 3.10 **Decision Variables for the Southern Confederation of Kibbutzim Problem**

	Allocation (Acres)		
	Kibbutz		
Crop	1	2	3
Sugar beets	x_1	x_2	x_3
Cotton	x_4	x_5	x_6
Sorghum	x_7	x_8	x_9

Formulation as a Linear Programming Problem: The quantities to be decided upon are the number of acres to devote to each of the three crops at each of the three kibbutzim. The decision variables x_j $(j = 1, 2, \ldots, 9)$ represent these nine quantities, as shown in Table 3.10.

Since the measure of effectiveness Z is the total net return, the resulting linear programming model for this problem is

Maximize $\quad Z = 1{,}000(x_1 + x_2 + x_3) + 750(x_4 + x_5 + x_6) + 250(x_7 + x_8 + x_9),$

subject to the following constraints:

1. Usable land for each kibbutz:

$$x_1 + x_4 + x_7 \le 400$$
$$x_2 + x_5 + x_8 \le 600$$
$$x_3 + x_6 + x_9 \le 300$$

2. Water allocation for each kibbutz:

$$3x_1 + 2x_4 + x_7 \le 600$$
$$3x_2 + 2x_5 + x_8 \le 800$$
$$3x_3 + 2x_6 + x_9 \le 375$$

3. Total acreage for each crop:

$$x_1 + x_2 + x_3 \le 600$$
$$x_4 + x_5 + x_6 \le 500$$
$$x_7 + x_8 + x_9 \le 325$$

4. Equal proportion of land planted:

$$\frac{x_1 + x_4 + x_7}{400} = \frac{x_2 + x_5 + x_8}{600}$$
$$\frac{x_2 + x_5 + x_8}{600} = \frac{x_3 + x_6 + x_9}{300}$$
$$\frac{x_3 + x_6 + x_9}{300} = \frac{x_1 + x_4 + x_7}{400}$$

5. Nonnegativity:

$$x_j \ge 0, \qquad \text{for } j = 1, 2, \ldots, 9.$$

Table 3.11 **Optimal Solution for the Southern Confederation of Kibbutzim Problem**

	Best Allocation (Acres)		
	Kibbutz		
Crop	1	2	3
Sugar beets	$133\frac{1}{3}$	100	25
Cotton	100	250	150
Sorghum	0	0	0

This completes the model, except that the equality constraints are not yet in an appropriate form for a linear programming model because some of the variables are on the right-hand side. Hence their final form[1] is

$$3(x_1 + x_4 + x_7) - 2(x_2 + x_5 + x_8) = 0$$

$$(x_2 + x_5 + x_8) - 2(x_3 + x_6 + x_9) = 0$$

$$4(x_3 + x_6 + x_9) - 3(x_1 + x_4 + x_7) = 0$$

The Coordinating Technical Office formulated this model and then applied the simplex method (developed in the next chapter) to find an optimal solution

$$(x_1, x_2, x_3, x_4, x_5, x_6, x_7, x_8, x_9) = (133\tfrac{1}{3}, 100, 25, 100, 250, 150, 0, 0, 0),$$

as shown in Table 3.11. The resulting optimal value of the objective function is $Z = 633{,}333\frac{1}{3}$, that is, a total net return of $633,333.33.

Controlling Air Pollution

The NORI & LEETS CO., one of the major producers of steel in its part of the world, is located in the city of Steeltown and is the only large employer there. Steeltown has grown and prospered along with the company, which now employs nearly 50,000 residents. Therefore, the attitude of the townspeople always has been, "What's good for Nori & Leets is good for the town." However, this attitude is now changing; uncontrolled air pollution from the company's furnaces is ruining the appearance of the city and endangering the health of its residents.

A recent stockholders' revolt resulted in the election of a new enlightened board of directors for the company. These directors are determined to follow socially responsible policies, and they have been discussing with Steeltown city officials and citizens' groups what to do about the air pollution problem. Together they have worked out stringent air quality standards for the Steeltown airshed.

The three main types of pollutants in this airshed are particulate matter, sulfur oxides, and hydrocarbons. The new standards require that the company reduce its annual emission of these pollutants by the amounts shown in Table 3.12. The board of directors has instructed management to have the engineering staff determine how to achieve these reductions in the most economical way.

The steelworks has two primary sources of pollution, namely, the blast furnaces for making pig iron and the open-hearth furnaces for changing iron into steel. In both

[1] Actually, any one of these equations is redundant and can be deleted if desired. Also, because of these equations, any two of the land constraints also could be deleted. (Can you see why?)

Table 3.12 **Clean Air Standards for the Nori & Leets Co.**

Pollutant	Required Reduction in Annual Emission Rate (Million Pounds)
Particulates	60
Sulfur oxides	150
Hydrocarbons	125

cases the engineers have decided that the most effective types of abatement methods are (1) increasing the height of the smokestacks,[1] (2) using filter devices (including gas traps) in the smokestacks, and (3) including cleaner, high-grade materials among the fuels for the furnaces. Each of these methods has a technological limit on how heavily it can be used (e.g., a maximum feasible increase in the height of the smokestacks), but there also is considerable flexibility for using the method at a fraction of its technological limit.

Table 3.13 shows how much emission (in millions of pounds per year) can be eliminated from each type of furnace by fully using any abatement method to its technological limit. For purposes of analysis, it is assumed that each method also can be used less fully to achieve any fraction of the emission-rate reductions shown in this table. Furthermore, the fractions can be different for blast furnaces and for open-hearth furnaces. For either type of furnace, the emission reduction achieved by each method is not substantially affected by whether the other methods also are used.

After these data were developed, it became clear that no single method by itself could achieve all the required reductions. On the other hand, combining all three methods at full capacity on both types of furnaces (which would be prohibitively expensive if the company's products are to remain competitively priced) is much more than adequate. Therefore, the engineers concluded that they would have to use some combination of the methods, perhaps with fractional capacities, based upon the relative costs. Furthermore, because of the differences between the blast and the open-hearth furnaces, the two types probably should not use the same combination.

An analysis was conducted to estimate the total annual cost that would be incurred by each abatement method. A method's annual cost includes increased operating and maintenance expenses as well as reduced revenue due to any loss in the efficiency

[1] Subsequent to this study, this particular abatement method has become a controversial one. Because its effect is to reduce ground-level pollution by spreading emissions over a greater distance, environmental groups contend that this creates more acid rain by keeping sulfur oxides in the air longer. Consequently, the U.S. Environmental Protection Agency adopted new rules in 1985 to remove incentives for using tall smokestacks.

Table 3.13 **Reduction in Emission Rate (in Millions of Pounds per Year) from the Maximum Feasible Use of an Abatement Method for Nori & Leets Co.**

	Taller Smokestacks		*Filters*		*Better Fuels*	
Pollutant	Blast Furnaces	Open-Hearth Furnaces	Blast Furnaces	Open-Hearth Furnaces	Blast Furnaces	Open-Hearth Furnaces
Particulates	12	9	25	20	17	13
Sulfur oxides	35	42	18	31	56	49
Hydrocarbons	37	53	28	24	29	20

Table 3.14 **Total Annual Cost from the Maximum Feasible Use of an Abatement Method for Nori & Leets Co. ($ Millions)**

Abatement Method	Blast Furnaces	Open-Hearth Furnaces
Taller smokestacks	8	10
Filters	7	6
Better fuels	11	9

of the production process caused by using the method. The other major cost is the *start-up cost* (the initial capital outlay) required to install the method. To make this one-time cost commensurable with the ongoing annual costs, the time value of money was used to calculate the annual expenditure (over the expected life of the method) that would be equivalent in value to this start-up cost.

This analysis led to the total annual cost estimates (in millions of dollars) given in Table 3.14 for using the methods at their full abatement capacities. It also was determined that the cost of a method being used at a lower level is roughly proportional to the fraction of the abatement capacity given in Table 3.13 that is achieved. Thus, for any given fraction achieved, the total annual cost would be roughly that fraction of the corresponding quantity in Table 3.14.

The stage now was set to develop the general framework of the company's plan for pollution abatement. This plan specifies which types of abatement methods will be used and at what fractions of their abatement capacities for (1) the blast furnaces and (2) the open-hearth furnaces. Because of the combinatorial nature of the problem of finding a plan that satisfies the requirements with the smallest possible cost, an OR team was formed to solve the problem. The team adopted a linear programming approach, formulating the model summarized next.

Formulation as a Linear Programming Problem: This problem has six decision variables x_j, $j = 1, 2, \ldots, 6$, each representing the use of one of the three abatement methods for one of the two types of furnaces, expressed as a *fraction of the abatement capacity* (so x_j cannot exceed 1). The ordering of these variables is shown in Table 3.15. Because the objective is to minimize total cost while satisfying the emission reduction requirements, the data in Tables 3.12, 3.13, and 3.14 yield the following model:

$$\text{Minimize} \quad Z = 8x_1 + 10x_2 + 7x_3 + 6x_4 + 11x_5 + 9x_6,$$

subject to the following constraints:

1. Emission reduction:

$$12x_1 + 9x_2 + 25x_3 + 20x_4 + 17x_5 + 13x_6 \geq 60$$
$$35x_1 + 42x_2 + 18x_3 + 31x_4 + 56x_5 + 49x_6 \geq 150$$
$$37x_1 + 53x_2 + 28x_3 + 24x_4 + 29x_5 + 20x_6 \geq 125$$

2. Technological limit:

$$x_j \leq 1, \qquad \text{for } j = 1, 2, \ldots, 6$$

3. Nonnegativity:

$$x_j \geq 0, \qquad \text{for } j = 1, 2, \ldots, 6.$$

Table 3.15 **Decision Variables (Fraction of the Maximum Feasible Use of an Abatement Method) for Nori & Leets Co.**

Abatement Method	Blast Furnaces	Open-Hearth Furnaces
Taller smokestacks	x_1	x_2
Filters	x_3	x_4
Better fuels	x_5	x_6

The OR team used this model[1] to find a minimum-cost plan

$$(x_1, x_2, x_3, x_4, x_5, x_6) = (1, 0.623, 0.343, 1, 0.048, 1),$$

with $Z = 32.16$ (total annual cost of \$32.16 million). Sensitivity analysis then was conducted to explore the effect of making possible adjustments in the air standards given in Table 3.12, as well as to check on the effect of any inaccuracies in the cost data given in Table 3.14. (This story is continued in Case Problem CP6-1 at the end of Chap. 6.) Next came detailed planning and managerial review. Soon after, this program for controlling air pollution was fully implemented by the company, and the citizens of Steeltown breathed deep (cleaner) sighs of relief.

Reclaiming Solid Wastes

The SAVE-IT COMPANY operates a reclamation center that collects four types of solid waste materials and treats them so that they can be amalgamated into a salable product. (Treating and amalgamating are separate processes.) Three different grades of this product can be made (see the first column of Table 3.16), depending upon the mix of the materials used. Although there is some flexibility in the mix for each grade, quality standards may specify the minimum or maximum amount allowed for the proportion of a material in the product grade. (This proportion is the weight of the material expressed as a percentage of the total weight for the product grade.) For each of the two higher grades, a fixed percentage is specified for one of the materials. These specifications are given in Table 3.16 along with the cost of amalgamation and the selling price for each grade.

The reclamation center collects its solid waste materials from regular sources and so is normally able to maintain a steady rate for treating them. Table 3.17 gives the

[1] An equivalent formulation can express each decision variable in natural units for its abatement method; for example, x_1 and x_2 could represent the number of *feet* that the heights of the smokestacks are increased.

Table 3.16 **Product Data for Save-It Co.**

Grade	Specification	Amalgamation Cost per Pound (\$)	Selling Price per Pound (\$)
A	Material 1: Not more than 30% of total Material 2: Not less than 40% of total Material 3: Not more than 50% of total Material 4: Exactly 20% of total	3.00	8.50
B	Material 1: Not more than 50% of total Material 2: Not less than 10% of total Material 4: Exactly 10% of total	2.50	7.00
C	Material 1: Not more than 70% of total	2.00	5.50

Table 3.17 **Solid Waste Materials Data for the Save-It Co.**

Material	Pounds per Week Available	Treatment Cost per Pound ($)	Additional Restrictions
1	3,000	3.00	1. For each material, at least half of the pounds per week available should be collected and treated.
2	2,000	6.00	2. $30,000 per week should be used to treat these materials.
3	4,000	4.00	
4	1,000	5.00	

quantities available for collection and treatment each week, as well as the cost of treatment, for each type of material.

The Save-It Co. is solely owned by Green Earth, an organization devoted to dealing with environmental issues, so Save-It's profits are used to help support Green Earth's activities. Green Earth has raised contributions and grants, amounting to $30,000 per week, to be used exclusively to cover the entire treatment cost for the solid waste materials. The board of directors of Green Earth has instructed the management of Save-It to divide this money among the materials in such a way that *at least half* of the amount available of each material is actually collected and treated. These additional restrictions are listed in Table 3.17.

Within the restrictions specified in Tables 3.16 and 3.17, management wants to determine the *amount* of each product grade to produce *and* the exact *mix* of materials to be used for each grade so as to maximize the total weekly profit (total sales income minus total amalgamation and treatment cost).

Formulation as a Linear Programming Problem: Before attempting to construct a linear programming model, we must give careful consideration to the proper definition of the decision variables. Although this definition is often obvious, it sometimes becomes the crux of the entire formulation. After clearly identifying what information is really desired and the most convenient form for conveying this information by means of decision variables, we can develop the objective function and the constraints on the values of these decision variables.

In this particular problem, the decisions to be made are well defined, but the appropriate means of conveying this information may require some thought. (Try it and see if you first obtain the following *inappropriate* choice of decision variables.)

Because one set of decisions is the *amount* of each product grade to produce, it would seem natural to define one set of decision variables accordingly. Proceeding tentatively along this line, we define

$$y_i = \text{number of pounds of product grade } i \text{ produced per week} \qquad (i = A, B, C).$$

The other set of decisions is the *mix* of materials for each product grade. This mix is identified by the proportion of each material in the product grade, which would suggest defining the other set of decision variables as

$$z_{ij} = \text{proportion of material } j \text{ in product grade } i \qquad (i = A, B, C; j = 1, 2, 3, 4).$$

However, Table 3.17 gives both the treatment cost and the availability of the materials by *quantity* (pounds) rather than *proportion,* so it is this *quantity* information that needs to be recorded in some of the constraints. For material j ($j = 1, 2, 3, 4$),

$$\text{Number of pounds of material } j \text{ used per week} = z_{Aj}y_A + z_{Bj}y_B + z_{Cj}y_C.$$

For example, since Table 3.17 indicates that 3,000 pounds of material 1 is available per week, one constraint in the model would be

$$z_{A1}y_A + z_{B1}y_B + z_{C1}y_C \leq 3{,}000.$$

Unfortunately, this is not a legitimate linear programming constraint. The expression on the left-hand side is *not* a linear function because it involves products of variables. Therefore, a linear programming model cannot be constructed with these decision variables.

Fortunately, there is another way of defining the decision variables that will fit the linear programming format. (Do you see how to do it?) It is accomplished by merely replacing each *product* of the old decision variables by a single variable! In other words, define

$$x_{ij} = z_{ij}y_i \qquad (\text{for } i = A, B, C;\ j = 1, 2, 3, 4)$$
$$= \text{number of pounds of material } j \text{ allocated to product grade } i \text{ per week,}$$

and then we let the x_{ij} be the decision variables. Combining the x_{ij} in different ways yields the following quantities needed in the model (for $i = A, B, C$; $j = 1, 2, 3, 4$).

$$x_{i1} + x_{i2} + x_{i3} + x_{i4} = \text{number of pounds of product grade } i \text{ produced per week.}$$

$$x_{Aj} + x_{Bj} + x_{Cj} = \text{number of pounds of material } j \text{ used per week.}$$

$$\frac{x_{ij}}{x_{i1} + x_{i2} + x_{i3} + x_{i4}} = \text{proportion of material } j \text{ in product grade } i.$$

The fact that this last expression is a *nonlinear* function does not cause a complication. For example, consider the first specification for product grade A in Table 3.16 (the proportion of material 1 should not exceed 30 percent). This restriction gives the nonlinear constraint

$$\frac{x_{A1}}{x_{A1} + x_{A2} + x_{A3} + x_{A4}} \leq 0.3.$$

However, multiplying through both sides of this inequality by the denominator yields an *equivalent* constraint

$$x_{A1} \leq 0.3(x_{A1} + x_{A2} + x_{A3} + x_{A4}),$$

so

$$0.7x_{A1} - 0.3x_{A2} - 0.3x_{A3} - 0.3x_{A4} \leq 0,$$

which is a legitimate linear programming constraint.

With this adjustment, the three quantities given above lead directly to all the functional constraints of the model. The objective function is based on management's objective of maximizing total weekly profit (total sales income *minus* total amalgamation cost) from the three product grades. The cost for treating the solid waste materials is not included in calculating the profit, because this expenditure is covered by the contributions and grants of \$30,000 per week for this purpose. Thus, for each product grade, the profit per pound is obtained by subtracting the amalgamation cost given in the third column of Table 3.16 from the selling price in the fourth column. These *differences* provide the coefficients for the objective function.

Therefore, the complete linear programming model is

$$\text{Maximize} \quad Z = 5.5(x_{A1} + x_{A2} + x_{A3} + x_{A4}) + 4.5(x_{B1} + x_{B2} + x_{B3} + x_{B4}) + 3.5(x_{C1} + x_{C2} + x_{C3} + x_{C4}),$$

subject to the following constraints:

1. Mixture specifications (second column of Table 3.16):

$$x_{A1} \le 0.3(x_{A1} + x_{A2} + x_{A3} + x_{A4}) \qquad \text{(grade } A\text{, material 1)}$$
$$x_{A2} \ge 0.4(x_{A1} + x_{A2} + x_{A3} + x_{A4}) \qquad \text{(grade } A\text{, material 2)}$$
$$x_{A3} \le 0.5(x_{A1} + x_{A2} + x_{A3} + x_{A4}) \qquad \text{(grade } A\text{, material 3)}$$
$$x_{A4} = 0.2(x_{A1} + x_{A2} + x_{A3} + x_{A4}) \qquad \text{(grade } A\text{, material 4).}$$

$$x_{B1} \le 0.5(x_{B1} + x_{B2} + x_{B3} + x_{B4}) \qquad \text{(grade } B\text{, material 1)}$$
$$x_{B2} \ge 0.1(x_{B1} + x_{B2} + x_{B3} + x_{B4}) \qquad \text{(grade } B\text{, material 2)}$$
$$x_{B4} = 0.1(x_{B1} + x_{B2} + x_{B3} + x_{B4}) \qquad \text{(grade } B\text{, material 4).}$$

$$x_{C1} \le 0.7(x_{C1} + x_{C2} + x_{C3} + x_{C4}) \qquad \text{(grade } C\text{, material 1).}$$

2. Availability of materials (second column of Table 3.17):

$$x_{A1} + x_{B1} + x_{C1} \le 3{,}000 \qquad \text{(material 1)}$$
$$x_{A2} + x_{B2} + x_{C2} \le 2{,}000 \qquad \text{(material 2)}$$
$$x_{A3} + x_{B3} + x_{C3} \le 4{,}000 \qquad \text{(material 3)}$$
$$x_{A4} + x_{B4} + x_{C4} \le 1{,}000 \qquad \text{(material 4).}$$

3. Restrictions on amounts treated (right side of Table 3.17):

$$x_{A1} + x_{B1} + x_{C1} \ge 1{,}500 \qquad \text{(material 1)}$$
$$x_{A2} + x_{B2} + x_{C2} \ge 1{,}000 \qquad \text{(material 2)}$$
$$x_{A3} + x_{B3} + x_{C3} \ge 2{,}000 \qquad \text{(material 3)}$$
$$x_{A4} + x_{B4} + x_{C4} \ge 500 \qquad \text{(material 4).}$$

4. Restriction on treatment cost (right side of Table 3.17):

$$3(x_{A1} + x_{B1} + x_{C1}) + 6(x_{A2} + x_{B2} + x_{C2}) + 4(x_{A3} + x_{B3} + x_{C3}) + 5(x_{A4} + x_{B4} + x_{C4}) = 30{,}000.$$

5. Nonnegativity constraints:

$$x_{A1} \ge 0, \qquad x_{A2} \ge 0, \qquad \ldots, \qquad x_{C4} \ge 0.$$

This formulation completes the model, except that the constraints for the mixture specifications need to be rewritten in the proper form for a linear programming model by bringing all variables to the left-hand side and combining terms, as follows:

Mixture specifications:

$$0.7x_{A1} - 0.3x_{A2} - 0.3x_{A3} - 0.3x_{A4} \le 0 \qquad \text{(grade } A\text{, material 1)}$$
$$-0.4x_{A1} + 0.6x_{A2} - 0.4x_{A3} - 0.4x_{A4} \ge 0 \qquad \text{(grade } A\text{, material 2)}$$

$$-0.5x_{A1} - 0.5x_{A2} + 0.5x_{A3} - 0.5x_{A4} \leq 0 \quad \text{(grade } A\text{, material 3)}$$
$$-0.2x_{A1} - 0.2x_{A2} - 0.2x_{A3} + 0.8x_{A4} = 0 \quad \text{(grade } A\text{, material 4).}$$

$$0.5x_{B1} - 0.5x_{B2} - 0.5x_{B3} - 0.5x_{B4} \leq 0 \quad \text{(grade } B\text{, material 1)}$$
$$-0.1x_{B1} + 0.9x_{B2} - 0.1x_{B3} - 0.1x_{B4} \geq 0 \quad \text{(grade } B\text{, material 2)}$$
$$-0.1x_{B1} - 0.1x_{B2} - 0.1x_{B3} + 0.9x_{B4} = 0 \quad \text{(grade } B\text{, material 4).}$$

$$0.3x_{C1} - 0.7x_{C2} - 0.7x_{C3} - 0.7x_{C4} \leq 0 \quad \text{(grade } C\text{, material 1).}$$

An optimal solution for this model is shown in Table 3.18, and then these x_{ij} values are used to calculate the other quantities of interest given in the table. The resulting optimal value of the objective function is $Z = 35{,}108.90$ (a total weekly profit of $35,108.90).

The Save-It Co. problem is an example of a **blending problem**. The objective for a blending problem is to find the best blend of ingredients into final products to meet certain specifications. Some of the earliest applications of linear programming were for *gasoline blending,* where petroleum ingredients were blended to obtain various grades of gasoline. The award-winning OR study at Texaco discussed at the end of Sec. 2.5 dealt with gasoline blending (although Texaco used a *nonlinear* programming model). Other blending problems involve such final products as steel, fertilizer, and animal feed.

Case Study?

Personnel Scheduling

UNION AIRWAYS is adding more flights to and from its hub airport, and so it needs to hire additional customer service agents. However, it is not clear just how many more should be hired. Management recognizes the need for cost control while also consistently providing a satisfactory level of service to customers. Therefore, an OR team is studying how to schedule the agents to provide satisfactory service with the smallest personnel cost.

Based on the new schedule of flights, an analysis has been made of the *minimum* number of customer service agents that need to be on duty at different times of the day to provide a satisfactory level of service. The rightmost column of Table 3.19 shows the number of agents needed for the time periods given in the first column. The other entries in this table reflect one of the provisions in the company's current contract with

Table 3.18 **Optimal Solution for the Save-It Co. Problem**

	Pounds Used per Week				
	Material				Number of Pounds
Grade	1	2	3	4	Produced per Week
A	412.3 (19.2%)	859.6 (40%)	447.4 (20.8%)	429.8 (20%)	2149
B	2587.7 (50%)	517.5 (10%)	1552.6 (30%)	517.5 (10%)	5175
C	0	0	0	0	0
Total	3000	1377	2000	947	

Table 3.19 **Data for the Union Airways Personnel Scheduling Problem**

Time Period	Time Periods Covered — Shift 1	2	3	4	5	Minimum Number of Agents Needed
6:00 A.M. to 8:00 A.M.	✔					48
8:00 A.M. to 10:00 A.M.	✔	✔				79
10:00 A.M. to 12:00 A.M.	✔	✔				65
12:00 A.M. to 2:00 P.M.	✔	✔	✔			87
2:00 P.M. to 4:00 P.M.		✔	✔			64
4:00 P.M. to 6:00 P.M.			✔	✔		73
6:00 P.M. to 8:00 P.M.			✔	✔		82
8:00 P.M. to 10:00 P.M.				✔		43
10:00 P.M. to 12:00 P.M.				✔	✔	52
12:00 P.M. to 6:00 A.M.					✔	15
Daily cost per agent	$170	$160	$175	$180	$195	

the union that represents the customer service agents. The provision is that each agent work an 8-hour shift 5 days per week, and the authorized shifts are

Shift 1: 6:00 A.M. to 2:00 P.M.
Shift 2: 8:00 A.M. to 4:00 P.M.
Shift 3: 12:00 A.M. (noon) to 8:00 P.M.
Shift 4: 4:00 P.M. to 12:00 P.M. (midnight)
Shift 5: 10:00 P.M. to 6:00 A.M.

Checkmarks in the main body of Table 3.19 show the hours covered by the respective shifts. Because some shifts are less desirable than others, the wages specified in the contract differ by shift. For each shift, the daily compensation (including benefits) for each agent is shown in the bottom row. The problem is to determine how many agents should be assigned to the respective shifts each day to minimize the *total* personnel cost for agents, based on this bottom row, while meeting (or surpassing) the service requirements given in the rightmost column.

Formulation as a Linear Programming Problem: Linear programming problems always involve finding the best *mix of activity levels.* The key to formulating this particular problem is to recognize the nature of the activities.

Activities correspond to shifts, where the *level* of each activity is the number of agents assigned to that shift. Thus this problem involves finding the *best mix of shift sizes.* Since the decision variables always are the levels of the activities, the five decision variables here are

$$x_j = \text{number of agents assigned to shift } j, \qquad \text{for } j = 1, 2, 3, 4, 5.$$

The main restrictions on the values of these decision variables are that the number of agents working during each time period must satisfy the minimum requirement given in the rightmost column of Table 3.19. For example, for 2:00 P.M. to 4:00 P.M., the total number of agents assigned to the shifts that cover this time period (shifts 2 and 3) must be at least 64, so

$$x_2 + x_3 \geq 64$$

is the functional constraint for this time period.

Because the objective is to minimize the total cost of the agents assigned to the five shifts, the coefficients in the objective function are given by the last row of Table 3.19.

Therefore, the complete linear programming model is

$$\text{Minimize} \quad Z = 170x_1 + 160x_2 + 175x_3 + 180x_4 + 195x_5,$$

subject to

$$
\begin{array}{llll}
x_1 & & \geq 48 & \text{(6–8 A.M.)} \\
x_1 + x_2 & & \geq 79 & \text{(8–10 A.M.)} \\
x_1 + x_2 & & \geq 65 & \text{(10–12 A.M.)} \\
x_1 + x_2 + x_3 & & \geq 87 & \text{(12 A.M.–2 P.M.)} \\
\quad x_2 + x_3 & & \geq 64 & \text{(2–4 P.M.)} \\
\qquad x_3 + x_4 & & \geq 73 & \text{(4–6 P.M.)} \\
\qquad x_3 + x_4 & & \geq 82 & \text{(6–8 P.M.)} \\
\qquad\quad x_4 & & \geq 43 & \text{(8–10 P.M.)} \\
\qquad\quad x_4 + x_5 & & \geq 52 & \text{(10–12 P.M.)} \\
\qquad\qquad x_5 & & \geq 15 & \text{(12 P.M.–6 A.M.)}
\end{array}
$$

and

$$x_j \geq 0, \qquad \text{for } j = 1, 2, 3, 4, 5.$$

With a keen eye, you might have noticed that the third constraint, $x_1 + x_2 \geq 65$, actually is not necessary because the second constraint, $x_1 + x_2 \geq 79$, ensures that $x_1 + x_2$ will be larger than 65. Thus, $x_1 + x_2 \geq 65$ is a *redundant* constraint that can be deleted. Similarly, the sixth constraint, $x_3 + x_4 \geq 73$, also is a *redundant* constraint because the seventh constraint is $x_3 + x_4 \geq 82$. (In fact, three of the nonnegativity constraints—$x_1 \geq 0$, $x_4 \geq 0$, $x_5 \geq 0$—also are redundant constraints because of the first, eighth, and tenth functional constraints: $x_1 \geq 48$, $x_4 \geq 43$, and $x_5 \geq 15$. However, no computational advantage is gained by deleting these three nonnegativity constraints.)

The optimal solution for this model is $(x_1, x_2, x_3, x_4, x_5) = (48, 31, 39, 43, 15)$. This yields $Z = 30{,}610$, that is, a total daily personnel cost of \$30,610.

This problem is an example where the divisibility assumption of linear programming actually is not satisfied. The number of agents assigned to each shift needs to be an integer. Strictly speaking, the model should have an additional constraint for each decision variable specifying that the variable must have an integer value. Adding these constraints would convert the linear programming model to an integer programming model (the topic of Chap. 12).

Without these constraints, the optimal solution given above turned out to have integer values anyway, so no harm was done by not including the constraints. (The form of the functional constraints made this outcome a likely one.) If some of the variables had turned out to be noninteger, the easiest approach would have been to *round up* to integer values. (Rounding up is feasible for this example because all the functional constraints are in $\geq$ form with nonnegative coefficients.) Rounding up does not ensure obtaining an optimal solution for the integer programming model, but the error introduced by rounding up such large numbers would be negligible for most practical situations. Alternatively, integer programming techniques described in Chap. 12 could be used to solve exactly for an optimal solution with integer values.

Section 3.5 includes a case study of how United Airlines used linear programming to develop a personnel scheduling system on a vastly larger scale than this example.

Distributing Goods through a Distribution Network

THE PROBLEM: The DISTRIBUTION UNLIMITED CO. will be producing the same new product at two different factories, and then the product must be shipped to two warehouses, where either factory can supply either warehouse. The distribution network available for shipping this product is shown in Fig. 3.13, where F1 and F2 are the two factories, W1 and W2 are the two warehouses, and DC is a distribution center. The amounts to be shipped from F1 and F2 are shown to their left, and the amounts to be received at W1 and W2 are shown to their right. Each arrow represents a feasible shipping lane. Thus F1 can ship directly to W1 and has three possible routes (F1 → DC → W2, F1 → F2 → DC → W2, and F1 → W1 → W2) for shipping to W2. Factory F2 has just one route to W2 (F2 → DC → W2) and one to W1 (F2 → DC → W2 → W1). The cost per unit shipped through each shipping lane is shown next to the arrow. Also shown next to F1 → F2 and DC → W2 are the maximum amounts that can be shipped through these lanes. The other lanes have sufficient shipping capacity to handle everything these factories can send.

The decision to be made concerns how much to ship through each shipping lane. The objective is to minimize the total shipping cost.

FORMULATION AS A LINEAR PROGRAMMING PROBLEM: With seven shipping lanes, we need seven decision variables ($x_{\text{F1-F2}}, x_{\text{F1-DC}}, x_{\text{F1-W1}}, x_{\text{F2-DC}}, x_{\text{DC-W2}}, x_{\text{W1-W2}}, x_{\text{W2-W1}}$) to represent the amounts shipped through the respective lanes.

There are several restrictions on the values of these variables. In addition to the usual nonnegativity constraints, there are two *upper-bound constraints,* $x_{\text{F1-F2}} \leq 10$ and $x_{\text{DC-W2}} \leq 80$, imposed by the limited shipping capacities for the two lanes, F1 → F2 and DC → W2. All the other restrictions arise from five *net flow constraints,* one for each of the five locations. These constraints have the following form.
Net flow constraint for each location:

$$\text{Amount shipped out} - \text{Amount shipped in} = \text{Required amount}.$$

As indicated in Fig. 3.13, these required amounts are 50 for F1, 40 for F2, −30 for W1, and −60 for W2.

What is the required amount for DC? All the units produced at the factories are ultimately needed at the warehouses, so any units shipped from the factories to the distribution center should be forwarded to the warehouses. Therefore, the total amount shipped from the distribution center to the warehouses should *equal* the total amount shipped from the factories to the distribution center. In other words, the *difference* of these two shipping amounts (the required amount for the net flow constraint) should be *zero.*

Since the objective is to minimize the total shipping cost, the coefficients for the objective function come directly from the unit shipping costs given in Fig. 3.13. There-

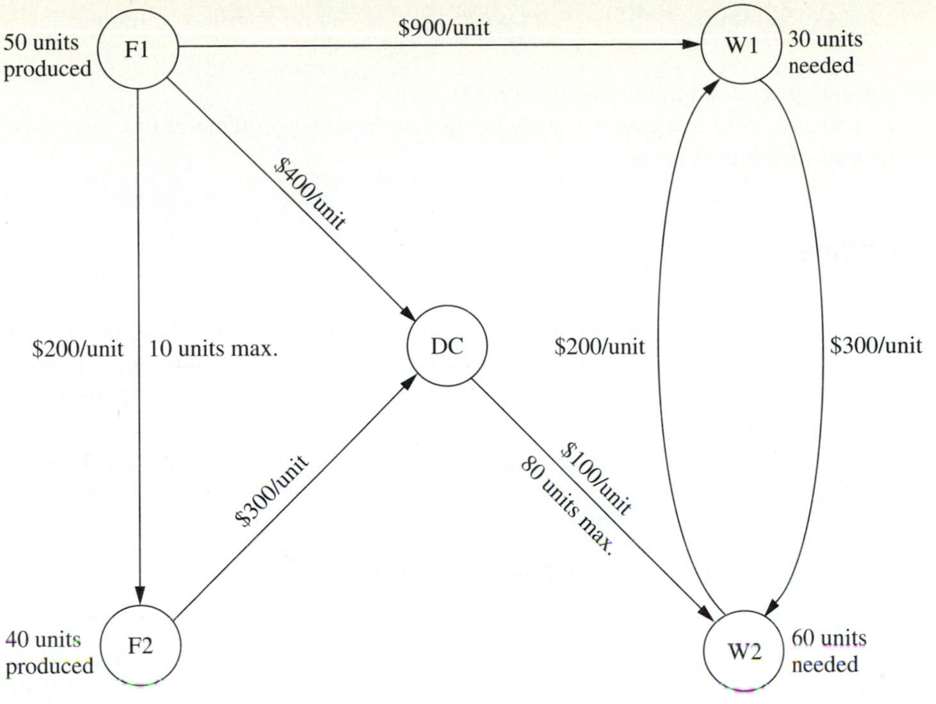

Figure 3.13 The distribution network for Distribution Unlimited Co.

fore, by using money units of hundreds of dollars in this objective function, the complete linear programming model is

$$\text{Minimize} \quad Z = 2x_{\text{F1-F2}} + 4x_{\text{F1-DC}} + 9x_{\text{F1-W1}} + 3x_{\text{F2-DC}} + x_{\text{DC-W2}} + 3x_{\text{W1-W2}} + 2x_{\text{W2-W1}},$$

subject to the following constraints:

1. Net flow constraints:

$$x_{\text{F1-F2}} + x_{\text{F1-DC}} + x_{\text{F1-W1}} = 50 \quad \text{(factory 1)}$$

$$-x_{\text{F1-F2}} + x_{\text{F2-DC}} = 40 \quad \text{(factory 2)}$$

$$-x_{\text{F1-DC}} - x_{\text{F2-DC}} + x_{\text{DC-W2}} = 0 \quad \text{(distribution center)}$$

$$-x_{\text{F1-W1}} + x_{\text{W1-W2}} - x_{\text{W2-W1}} = -30 \quad \text{(warehouse 1)}$$

$$-x_{\text{DC-W2}} - x_{\text{W1-W2}} + x_{\text{W2-W1}} = -60 \quad \text{(warehouse 2)}$$

2. Upper-bound constraints:

$$x_{\text{F1-F2}} \le 10, \qquad x_{\text{DC-W2}} \le 80$$

3. Nonnegativity constraints:

$$x_{\text{F1-F2}} \ge 0, \quad x_{\text{F1-DC}} \ge 0, \quad x_{\text{F1-W1}} \ge 0, \quad x_{\text{F2-DC}} \ge 0, \quad x_{\text{DC-W2}} \ge 0,$$
$$x_{\text{W1-W2}} \ge 0, \quad x_{\text{W2-W1}} \ge 0.$$

You will see this problem again in Sec. 9.6, where we focus on linear programming problems of this type (called the *minimum cost flow problem*). In Sec. 9.7, we will solve for its optimal solution:

$$x_{F1\text{-}F2} = 0, \quad x_{F1\text{-}DC} = 40, \quad x_{F1\text{-}W1} = 10, \quad x_{F2\text{-}DC} = 40, \quad x_{DC\text{-}W2} = 80,$$
$$x_{W1\text{-}W2} = 0, \quad x_{W2\text{-}W1} = 20.$$

The resulting total shipping cost is $49,000.

You also will see a case study involving a *much* larger problem of this same type at the end of the next section.

3.5 Some Case Studies

To give you a better perspective about the great impact that linear programming can have, we now present three case studies of real applications. The first one will bear some strong similarities to the Wyndor Glass Co. problem, but on a realistic scale. Similarly, the second and third cases are the realistic versions of the last two examples presented in the preceding section (the Union Airways and Distribution Unlimited examples).

Choosing the Product Mix at Ponderosa Industrial[1]

PONDEROSA INDUSTRIAL is a plywood manufacturer based in Anhuac, Chihuahua, which supplies 25 percent of the plywood in Mexico. Like any plywood manufacturer, Ponderosa's many products are differentiated by thickness and by the quality of the wood. The plywood market in Mexico is competitive, so the market establishes the prices of the products. The prices can fluctuate considerably from month to month, and there may be great differences between the products in their price movements from even one month to the next. As a result, each product's contribution to Ponderosa's total profit is continually varying, and in different ways for different products.

Because of its pronounced effect on profits, a critical issue facing management is the choice of *product mix*—how much to produce of each product—on a monthly basis. This choice is a very complex one, since it must also take into account the current amounts available of various resources needed to produce the products. The most important resources are logs in four quality categories and production capacities for both the pressing operation and the polishing operation.

Since 1980, linear programming has been used on a monthly basis to guide the product-mix decision. The linear programming model has an objective of maximizing the total profit from all products. The model's constraints include the various resource constraints as well as other relevant restrictions such as the minimum amount of a product that must be provided to regular customers and the maximum amount that can be sold. (To aid planning for the procurement of raw materials, the model also considers the impact of the product-mix decision for the upcoming month on production in the following month.) The model has 90 decision variables and 45 functional constraints.

This model is used each month to find the product mix for the upcoming month that will be optimal if the estimated values of the various parameters of the model prove to be accurate. However, since some of the parameter values can change quickly (e.g., the unit profits of the products), *sensitivity analysis* is done to determine the effect if the estimated values turn out to be inaccurate. The results indicate when adjustments

[1] A. Roy, E. E. DeFalomir, and L. Lasdon, "An Optimization-Based Decision Support System for a Product Mix Problem," *Interfaces,* **12**(2):26–33, April 1982.

in the product mix should be made (if time permits) as unanticipated market changes occur that affect the price (and so the unit profit) of certain products.

One key decision each month concerns the number of logs to purchase in each of the four quality categories. The amounts available for the upcoming month's production actually are parameters of the model. Therefore, after the purchase decision is made and then the corresponding optimal product mix is determined, *post-optimality analysis* is conducted to investigate the effect of adjusting the purchase decision. For example, it is very easy with linear programming to check what the impact on total profit would be if a quick purchase were to be made of additional logs in a certain quality category to enable increasing production for the upcoming month.

Ponderosa's linear programming system is *interactive,* so management receives an immediate response to its ''what-if'' questions about the impact of encountering parameter values that differ from those in the original model. What if a quick purchase of logs of a certain kind were made? What if product prices were to fluctuate in a certain way? A variety of such scenarios can be investigated. Management has effectively used this power to reach better decisions than the ''optimal'' product mix from the original model.

The impact of linear programming at Ponderosa has been reported to be ''tremendous.'' It has led to a dramatic shift in the types of plywood products emphasized by the company. The improved product-mix decisions are credited with increasing the overall profitability of the company by 20 percent. Other contributions of linear programming include better utilization of raw material, capital equipment, and personnel.

Two factors helped make this application of linear programming so successful. One factor is that a *natural-language* financial planning system was interfaced with the codes for finding an optimal solution for the linear programming model. Using natural language rather than mathematical symbols to display the components of the linear programming model and its output made the process understandable and meaningful for the managers making the product-mix decisions. Reporting to management in the language of managers is necessary for the successful application of linear programming.

The other factor was that the linear programming system used was *interactive.* As mentioned earlier, after an optimal solution is obtained for one version of the model, this feature enabled managers to ask a variety of what-if questions and receive immediate responses. Better decisions frequently were reached by exploring other plausible scenarios, and this process also gave managers more confidence that their decision would perform well under most foreseeable circumstances.

In any application, this ability to respond quickly to management's needs and queries through post-optimality analysis (whether interactive or not) is a vital part of a linear programming study.

Personnel Scheduling at United Airlines[1]

Despite unprecedented industry competition in 1983 and 1984, UNITED AIRLINES managed to achieve substantial growth with service to 48 new airports. In 1984, it became the only airline with service to cities in all 50 states. Its 1984 operating profit reached $564 million, with revenues of $6.2 billion, an increase of 6 percent over 1983 while costs grew by less than 2 percent.

[1] T. J. Holloran and J. E. Bryn, ''United Airlines Station Manpower Planning System,'' *Interfaces,* **16**(1): 39–50, Jan.–Feb. 1986.

Cost control is essential to competing successfully in the airline industry. In 1982, upper management of United Airlines initiated an OR study of its personnel scheduling as part of the cost control measures associated with the airline's 1983–1984 expansion. The goal was to schedule personnel at the airline's reservations offices and airports so as to minimize the cost of providing the necessary service to customers.

At the time, United Airlines employed over 4,000 reservations sales representatives and support personnel at its 11 reservations offices and about 1,000 customer service agents at its 10 largest airports. Some were part-time, working shifts from 2 to 8 hours; most were full-time, working 8- or 10-hour shifts. Shifts start at several different times. Each reservations office is open (by telephone) 24 hours per day, as is each of the major airports. However, the number of employees needed at each location to provide the required level of service varies greatly during the 24-hour day and may fluctuate considerably from one half-hour to the next.

Trying to design the work schedules for all the employees at a given location to meet these service requirements most efficiently is a nightmare of combinatorial considerations. Once an employee begins working, he or she will be there continuously for the entire shift (2 to 10 hours, depending on the employee), *except* for either a meal break or short rest breaks every 2 hours. Given the *minimum* number of employees needed on duty for *each* half-hour interval over a 24-hour day (where these requirements change from day to day over a 7-day week), *how many* employees of *each shift length* should begin work at *what start time* over *each* 24-hour day of a 7-day week? Fortunately, linear programming thrives on such combinatorial nightmares.

Actually, several OR techniques described in this book are used in the computerized planning system developed to attack this problem. Both *forecasting* (Chap. 18) and *queueing theory* (Chaps. 15 and 16) are used to determine the minimum number of employees needed on duty for each half-hour interval. *Integer programming* (Chap. 12) is used to determine the times of day at which shifts will be allowed to start. However, the core of the planning system is *linear programming,* which does all the actual scheduling to provide the needed service with the smallest possible labor cost. A complete work schedule is developed for the first full week of a month, and then it is reused for the remainder of the month. This process is repeated each month to reflect changing conditions.

Although the details about the linear programming model have not been published, it is clear that the basic approach used is the one illustrated by the Union Airways example of personnel scheduling in Sec. 3.4. The objective function being minimized represents the total personnel cost for the location being scheduled. The main functional constraints require that the number of employees on duty during each time period not fall below minimum acceptable levels.

However, the Union Airways example only has five decision variables. By contrast, the United Airlines model for some locations has over 20,000 decision variables! The difference is that a real application must consider myriad important details that can be ignored in a textbook example. For example, the United Airlines model takes into account such things as the meal and break assignment times for each employee scheduled, differences in shift lengths for different employees, and days off over a weekly schedule, among other scheduling details.

This application of linear programming is reported to have had "an overwhelming impact not only on United management and members of the manpower planning group, but also for many who had never before heard of management science (OR) or mathematical modeling.'' It earned rave reviews from upper management, operating managers, and affected employees alike. For example, one manager described the scheduling system as

> Magical, . . . just as the [customer] lines begin to build, someone shows up for work, and just as you begin to think you're overstaffed, people start going home.[1]

In more tangible terms, this application is credited with saving United Airlines more than $6 million *annually* in just direct salary and benefit costs. Other benefits include improved customer service and reduced need for support staff.

One factor that helped make this application of linear programming so successful was ''the support of operational managers and their staffs.'' This was a lesson learned by experience, because the OR team initially failed to establish a good line of communication with the operating managers, who then resisted the team's initial recommendations. The team leaders describe their mistake as follows:

> The cardinal rule for earning the trust and respect of operating managers and support staffs—''getting them involved in the development process''—had been violated.[2]

The team then worked much more closely with the operating managers—with outstanding results.

Planning Supply, Distribution, and Marketing at Citgo Petroleum Corporation[3]

CITGO PETROLEUM CORPORATION specializes in refining and marketing petroleum. It has annual sales of several billion dollars, which ranks it among the 150 largest industrial companies in the United States.

After several years of financial losses, Citgo was acquired in 1983 by Southland Corporation, the owner of the 7-Eleven convenience store chain (whose sales include 2 billion gallons of quality motor fuels annually). To turn Citgo's financial losses around, Southland created a task force composed of Southland personnel, Citgo personnel, and outside consultants. An eminent OR consultant was appointed director of the task force to report directly to both the president of Citgo and the chairman of the board of Southland.

During 1984 and 1985, this task force applied various OR techniques (as well as information systems technologies) throughout the corporation. It is reported that these OR applications ''have changed the way Citgo does business and resulted in approximately $70 million per year profit improvement.''[4]

The two most important applications are both *linear programming (LP) systems* that provide management with powerful planning support. One, called the *refinery LP system,* led to great improvements in refinery yield, substantial reductions in the cost of labor, and other important cost savings. This system contributed approximately $50 million to profit improvement in 1985. (See the end of Sec. 2.4 for discussion of the key role that model validation played in the development of this system.)

[1] Holloran and Bryn, ''United Airlines Station Manpower Planning System,'' p. 49.

[2] Ibid, p. 47.

[3] See the references cited in the last footnote of Sec. 2.1.

[4] Klingman et al., ''The Successful Deployment of Management Science,'' p. 4.

However, we will focus here on the other linear programming system, called the Supply, Distribution, and Marketing Modeling System (or just the *SDM system*). The SDM system is particularly interesting because it is based on a special kind of linear programming model that uses *networks,* just like the model for the Distribution Unlimited example presented at the end of Sec. 3.4. The model for the SDM system provides a representation of Citgo's entire marketing and distribution network.

Citgo owns or leases 36 product storage terminals which are supplied through five distribution center terminals via a distribution network of pipelines, tankers, and barges. Citgo also sells product from over 350 exchange terminals that are shared with other petroleum marketers. To supply its customers, product may be acquired by Citgo from its refinery in Lake Charles, Louisiana, or from spot purchases on one of five major spot markets, product exchanges, and trades with other industry refiners. These product acquisition decisions are made daily. However, the time from such a decision until the product reaches the intended customers can be as long as 11 weeks. Therefore, the linear programming model uses an 11-week planning horizon.

The SDM system is used to coordinate the supply, distribution, and marketing of each of Citgo's major products (four grades of motor fuel and no. 2 fuel oil) throughout the United States. Management uses the system to make decisions such as where to sell, what price to charge, where to buy or trade, how much to buy or trade, how much to hold in inventory, and how much to ship by each mode of transportation. Linear programming guides these decisions and when to implement them so as to minimize total cost or maximize total profit. The SDM system also is used in ''what-if'' sessions, where management asks what-if questions about scenarios that differ from those assumed in the original model.

The linear programming model in the SDM system has the same form as the model for the Distribution Unlimited example presented at the end of Sec. 3.4. In fact, both models fit an important special kind of linear programming problem, called the *minimum cost flow problem,* that will be discussed in Sec. 9.6. The main functional constraints for such models are *equality constraints,* where each one prescribes what the net flow of goods out of a specific location must be.

The Distribution Unlimited model has just seven decision variables and five equality constraints. By contrast, the Citgo model for each major product has about 15,000 decision variables and 3,000 equality constraints!

At the end of Sec. 2.1, we described the important role that *data collection* and *data verification* played in developing the Citgo models. With such huge models, a massive amount of data must be gathered to determine all the parameter values. A state-of-the-art management database system was developed for this purpose. Before data are input to a model, a preloader program was used to check for data errors and inconsistencies. The importance of doing so was brought forcefully home to the task force when the initial run of the preloader program generated a log of error messages an inch thick! It was clear that the data collection process needed to be thoroughly debugged to help ensure the validity of the models.

The SDM linear programming system greatly improved the efficiency of Citgo's supply, distribution, and marketing operations, enabling a huge reduction in product inventory with no drop in service levels. In particular, the value of petroleum products held in inventory was reduced by \$116.5 million. This huge reduction in capital tied up in carrying inventory resulted in saving about \$14 million annually in interest expenses for borrowed capital, adding \$14 million to Citgo's annual profits. Improvements in coordination, pricing, and purchasing decisions were estimated to add at least another

$2.5 million to annual profits. Many *indirect* benefits also were attributed to this application of linear programming, including improved data, better pricing strategies, and elimination of unnecessary product terminals, as well as improved communication and coordination between supply, distribution, marketing, and refinery groups.

Some of the factors that contributed to the success of this OR study are the same as for the two preceding case studies. Like Ponderosa Industrial, one factor was the development of output reports in the language of managers to really meet their needs. These output reports are designed to be easy for managers to understand and use, and the reports address the issues that are important to management. Also as with Ponderosa, another factor was that management was able to respond quickly to the dynamics of the industry by using the linear programming system extensively in what-if sessions. As in so many applications of linear programming, post-optimality analysis proved more important than the initial optimal solution obtained for one version of the model.

Much as in the United Airlines application, another factor was the enthusiastic support of operational managers during the development and implementation of this linear programming system.

However, the most important factor was the unlimited support provided to the task force by top management, ranging right up to the chief executive officer and the chairman of the board of Citgo's parent company, Southland Corporation. As mentioned earlier, the director of the task force (an eminent OR consultant) reported directly to both the president of Citgo and the chairman of the board of Southland. This backing by top management included strong organizational and financial support.

The organizational support took a variety of forms. One example is the creation and staffing of the position of senior vice-president of operations coordination to evaluate and coordinate recommendations based on the models which spanned organizational boundaries.

When discussing both this linear programming system and other OR applications implemented by the task force, team members described the financial support of top management as follows:

> The total cost of the systems implemented, $20 million to $30 million, was the greatest obstacle to this project. However, because of the information explosion in the petroleum industry, top management realized that numerous information systems were essential to gather, store, and analyze data. The incremental cost of adding management science (OR) technologies to these computers and systems was small, in fact very small in light of the enormous benefits they provided.[1]

3.6 Conclusions

Linear programming is a powerful technique for dealing with the problem of allocating limited resources among competing activities as well as other problems having a similar mathematical formulation. It has become a standard tool of great importance for numerous business and industrial organizations. Furthermore, almost any social organization is concerned with allocating resources in some context, and there is a growing recognition of the extremely wide applicability of this technique.

However, not all problems of allocating limited resources can be formulated to fit a linear programming model, even as a reasonable approximation. When one or more

[1] Klingman et al., "The Successful Deployment of Management Science," p. 21.

of the assumptions of linear programming is violated seriously, it may then be possible to apply another mathematical programming model instead, e.g., the models of integer programming (Chap. 12) or nonlinear programming (Chap. 13).

SELECTED REFERENCES

1. Anderson, D. R., D. J. Sweeney, and T. A. Williams, *An Introduction to Management Science,* 7th ed., West, St. Paul, Minn., 1994, chaps. 2, 4.
2. Gass, S., *An Illustrated Guide to Linear Programming,* Dover Publications, New York, 1990.
3. Williams, H. P., *Model Building in Mathematical Programming,* 3d ed., Wiley, New York, 1990.

RELEVANT ROUTINES IN YOUR OR COURSEWARE

A Demonstration Example: *Graphical Method*

Interactive Routines: *Enter or Revise a General Linear Programming Model*
Solve Interactively by the Graphical Method

An Automatic Routine: *Solve Automatically by the Simplex Method*

See the end of Sec. 3.1 for descriptions of these routines. To access them, call the MathProg program and then choose *General Linear Programming* under the Area menu. See Appendix 1 for documentation of the software.

PROBLEMS[1]

To the left of each of the following problems (or their parts), we have inserted a D (for demo), I (for interactive routine), or A (for automatic routine) whenever a corresponding routine listed above can be helpful. An asterisk on the I or A indicates that this routine definitely should be used (unless your instructor gives you contrary instructions). An asterisk on the problem number indicates that at least a partial answer is given in the back of the book.

D, I* **3.1-1.*** Use the graphical method to solve the problem:

$$\text{Maximize} \quad Z = 2x_1 + x_2,$$

subject to

$$x_2 \leq 10$$
$$2x_1 + 5x_2 \leq 60$$
$$x_1 + x_2 \leq 18$$
$$3x_1 + x_2 \leq 44$$

and

$$x_1 \geq 0, \quad x_2 \geq 0.$$

D, I* **3.1-2.** Use the graphical method to solve the problem:

$$\text{Maximize} \quad Z = 10x_1 + 20x_2,$$

[1] Problems 3.1-4 to 3.1-10, 3.2-1, 3.2-2, 3.4-5, 3.4-6, and 3.4-10 have been adapted, with permission, from previous operations research examinations given by the Society of Actuaries.

subject to

$$x_1 + 2x_2 \le 15$$
$$x_1 + x_2 \le 12$$
$$5x_1 + 3x_2 \le 45$$

and

$$x_1 \ge 0, \qquad x_2 \ge 0.$$

3.1-3.* A manufacturing firm has discontinued the production of a certain unprofitable product line. This act created considerable excess production capacity. Management is considering devoting this excess capacity to one or more of three products; call them products 1, 2, and 3. The available capacity on the machines that might limit output is summarized in the following table:

Machine Type	Available Time (in Machine Hours per Week)
Milling machine	500
Lathe	350
Grinder	150

The number of machine hours required for each unit of the respective products is as follows:

Productivity Coefficient (in Machine Hours per Unit)

Machine Type	Product 1	Product 2	Product 3
Milling machine	9	3	5
Lathe	5	4	0
Grinder	3	0	2

The sales department indicates that the sales potential for products 1 and 2 exceeds the maximum production rate and that the sales potential for product 3 is 20 units per week. The unit profit would be $50, $20, and $25, respectively, on products 1, 2, and 3. The objective is to determine how much of each product the firm should produce to maximize profit.

(*a*) Formulate a linear programming model for this problem.

A* (*b*) Use a computer to solve this model by the simplex method.

3.1-4. A television manufacturing company has to decide on the number of 27- and 20-inch sets to be produced at one of its factories. Market research indicates that at most 40 of the 27-inch sets and 10 of the 20-inch sets can be sold per month. The maximum number of work hours available is 500 per month. A 27-inch set requires 20 work hours, and a 20-inch set requires 10 work hours. Each 27-inch set sold produces a profit of $120, and each 20-inch set produces a profit of $80. A wholesaler has agreed to purchase all the television sets produced if the numbers do not exceed the maxima indicated by the market research.

(*a*) Formulate a linear programming model for this problem.

D, I* (*b*) Solve this model graphically.

A* (*c*) Use a computer to solve this model by the simplex method.

3.1-5. A company produces two products requiring resources P and Q. Management wants to determine how many units of each product to produce so as to maximize profit. For each unit of product 1, 1 unit of resource P and 2 units of resource Q are required. For each unit of product 2, 3 units of resource P and 2 units of resource Q are required. The company has 200 units of resource P and 300 units of resource Q. Each unit of product 1 gives a profit of $1, and each unit of product 2, up to 60 units, gives a profit of $2. Any excess over 60 units of product 2 brings no profit, so such an excess has been ruled out.

(*a*) Formulate a linear programming model for this problem.

D, I* (*b*) Solve this model graphically. What is the resulting total profit?

A* (*c*) Use a computer to solve this model by the simplex method.

3.1-6. An insurance company is introducing two new product lines: special risk insurance and mortgages. The expected profit is 5 per unit on special risk insurance and 2 per unit on mortgages.

Management wishes to establish sales quotas for the new product lines to maximize total expected profit. The work requirements are as follows:

	Work Hours per Unit		
Department	Special Risk	Mortgage	Work Hours Available
Underwriting	3	2	2400
Administration	0	1	800
Claims	2	0	1200

(*a*) Formulate a linear programming model for this problem.

D, I* (*b*) Use the graphical method to solve this model.

3.1-7. You are the production manager for a manufacturer of three types of spare parts for automobiles. The manufacture of each part requires processing on each of two machines, with the following processing time (in hours):

	Part		
Machine	A	B	C
1	0.02	0.03	0.05
2	0.05	0.02	0.04

Each machine is available 40 hours per month. Each part manufactured will yield a unit profit as follows:

	Part		
	A	B	C
Profit	50	40	30

You want to determine the mix of spare parts to produce in order to maximize total profit.

(*a*) Formulate a linear programming model for this problem.

A* (*b*) Use a computer to solve this model by the simplex method.

D, I **3.1-8.** Consider the following problem, where the value of c_1 has not yet been ascertained.

$$\text{Maximize} \quad Z = c_1x_1 + x_2,$$

subject to

$$x_1 + x_2 \le 6$$
$$x_1 + 2x_2 \le 10$$

and

$$x_1 \ge 0, \quad x_2 \ge 0.$$

Use graphical analysis to determine the optimal solution(s) for (x_1, x_2) for the various possible values of c_1 $(-\infty < c_1 < \infty)$.

D, I **3.1-9.** Consider the following problem, where the value of k has not yet been ascertained.

$$\text{Maximize} \quad Z = x_1 + 2x_2,$$

subject to

$$-x_1 + x_2 \le 2$$
$$x_2 \le 3$$
$$kx_1 + x_2 \le 2k + 3 \qquad \text{where } k \ge 0$$

and

$$x_1 \ge 0, \quad x_2 \ge 0.$$

The solution currently being used is $x_1 = 2$, $x_2 = 3$. Use graphical analysis to determine the values of k such that this solution actually is optimal.

D, I **3.1-10.** Consider the following problem, where the values of c_1 and c_2 have not yet been ascertained.

$$\text{Maximize} \quad Z = c_1x_1 + c_2x_2,$$

subject to

$$2x_1 + x_2 \le 11$$
$$-x_1 + 2x_2 \le 2$$

and

$$x_1 \ge 0, \quad x_2 \ge 0.$$

Use graphical analysis to determine the optimal solution(s) for (x_1, x_2) for the various possible values of c_1 and c_2.

3.2-1. The following table summarizes the key facts about two products, A and B, and the resources, Q, R, and S, required to produce them.

	Resource Use per Unit Produced		
Resource	Product A	Product B	Amount of Resource Available
Q	2	1	2
R	1	2	2
S	3	3	4
Profit/unit	3	2	

All the assumptions of linear programming hold.

(*a*) Formulate a linear programming model for this problem.

D, I* (*b*) Solve this model graphically.

A* (*c*) Use a computer to solve this model by the simplex method.

3.2-2. The shaded area in the following graph represents the feasible region of a linear programming problem whose objective function is to be maximized. Label each of the following statements as true or false, and then justify your answer based on the graphical method. In each case, give an example of an objective function that illustrates your answer.

(*a*) If (3, 3) produces a larger value of the objective function than (0, 2) and (6, 3), then (3, 3) must be an optimal solution.

(*b*) If (3, 3) is an optimal solution and multiple optimal solutions exist, then either (0, 2) or (6, 3) must also be an optimal solution.

(*c*) The point (0, 0) cannot be an optimal solution.

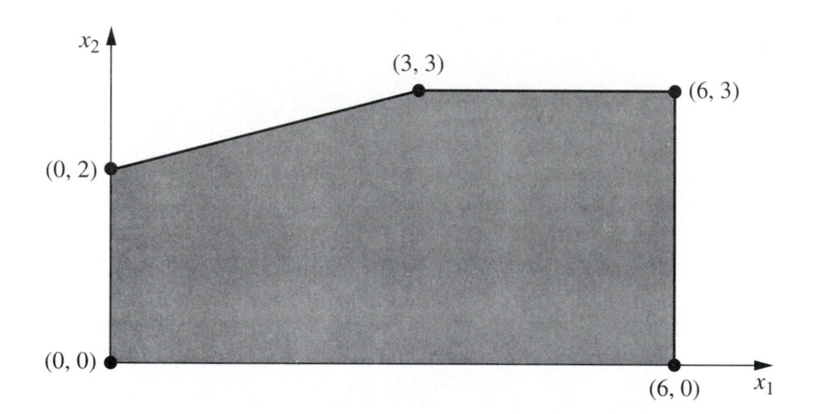

3.2-3.* Suppose you have just inherited $6,000 and you want to invest it. Upon hearing this news, two different friends have offered you an opportunity to become a partner in two different entrepreneurial ventures, one planned by each friend. In both cases, this investment would involve expending some of your time next summer as well as putting up cash. Becoming a *full* partner in the first friend's venture would require an investment of $5,000 and 400 hours, and your estimated profit (ignoring the value of your time) would be $4,500. The corresponding figures for the second friend's venture are $4,000 and 500 hours, with an estimated profit to you of $4,500. However, both friends are flexible and would allow you to come in at any fraction of a full partnership you would like; your share of the profit would be proportional to this fraction.

Because you were looking for an interesting summer job anyway (maximum of 600 hours), you have decided to participate in one friend's or both friends' ventures in whichever combination would maximize your total estimated profit. You now need to solve the problem of finding the best combination.

(*a*) Describe the analogy between this problem and the Wyndor Glass Co. problem discussed in Sec. 3.1. Then construct and fill in a table like Table 3.3 for this problem, identifying both the activities and the resources.

(*b*) Formulate a linear programming model for this problem.

D, I* (*c*) Solve this model graphically. What is your total estimated profit?

3.3-1. Reconsider Prob. 3.2-3. Indicate why each of the four assumptions of linear programming (Sec. 3.3) appears to be reasonably satisfied for this problem. Is one assumption more doubtful than the others? If so, what should be done to take this into account?

3.3-2. Consider a problem with two decision variables, x_1 and x_2, which represent the levels of activities 1 and 2, respectively. For each variable, the permissible values are 0, 1, and 2, where the feasible combinations of these values for the two variables are determined from a variety of constraints. The objective is to maximize a certain measure of performance denoted by Z. The values of Z for the possibly feasible values of (x_1, x_2) are estimated to be those given in the following table:

	x_2		
x_1	0	1	2
0	0	4	8
1	3	8	13
2	6	12	18

Based on this information, indicate whether this problem completely satisfies each of the four assumptions of linear programming. Justify your answers.

3.4-1.* For each of the four assumptions of linear programming discussed in Sec. 3.3, write a one-paragraph analysis of how well you feel it applies to each of the following examples, given in Sec. 3.4.

(*a*) Design of radiation therapy (Mary)
(*b*) Regional planning (Southern Confederation of Kibbutzim)
(*c*) Controlling air pollution (Nori & Leets Co.)

3.4-2. For each of the four assumptions of linear programming discussed in Sec. 3.3, write a one-paragraph analysis of how well it applies to each of the following examples, given in Sec. 3.4.

(*a*) Reclaiming solid wastes (Save-It Co.)
(*b*) Personnel scheduling (Union Airways)
(*c*) Distributing goods through a distribution network (Distribution Unlimited Co.)

D, I* **3.4-3.*** Use the graphical method to solve this problem:

$$\text{Minimize} \quad Z = 15x_1 + 20x_2,$$

subject to

$$x_1 + 2x_2 \geq 10$$
$$2x_1 - 3x_2 \leq 6$$
$$x_1 + x_2 \geq 6$$

and

$$x_1 \geq 0, \qquad x_2 \geq 0.$$

D, I* **3.4-4.** Use the graphical method to solve this problem:

$$\text{Minimize} \quad Z = 3x_1 + 2x_2,$$

subject to

$$x_1 + 2x_2 \leq 12$$
$$2x_1 + 3x_2 = 12$$
$$2x_1 + x_2 \geq 8$$

and

$$x_1 \geq 0, \qquad x_2 \geq 0.$$

D, I **3.4-5.** Consider the following problem, where the value of c_1 has not yet been ascertained.

$$\text{Maximize} \quad Z = c_1x_1 + 2x_2,$$

subject to

$$4x_1 + x_2 \leq 12$$
$$x_1 - x_2 \geq 2$$

and

$$x_1 \geq 0, \quad x_2 \geq 0.$$

Use graphical analysis to determine the optimal solution(s) for (x_1, x_2) for the various possible values of c_1.

3.4-6. You are given the following nutritional and cost information regarding steak and potatoes:

Ingredient	*Grams of Ingredient per Serving*		Daily Requirement (Grams)
	Steak	Potatoes	
Carbohydrates	5	15	≥ 50
Protein	20	5	≥ 40
Fat	15	2	≤ 60
Cost per serving	$4	$2	

You wish to determine the number of daily servings (it may be fractional) of steak and potatoes that will meet these requirements at a minimum cost.

(*a*) Formulate a linear programming model for this problem.

D, I* (*b*) Solve this model graphically.

A* (*c*) Use a computer to solve this model by the simplex method.

3.4-7. A farmer is raising pigs for market and wishes to determine the quantities of the available types of feed (corn, tankage, and alfalfa) that should be given to each pig. Since pigs will eat any mix of these feed types, the objective is to determine which mix will meet certain nutritional requirements at a *minimum cost.* The number of units of each type of basic nutritional ingredient contained within 1 kilogram of each feed type is given in the following table, along with the daily nutritional requirements and feed costs:

Nutritional Ingredient	Kilogram of Corn	Kilogram of Tankage	Kilogram of Alfalfa	Minimum Daily Requirement
Carbohydrates	90	20	40	200
Protein	30	80	60	180
Vitamins	10	20	60	150
Cost (¢)	84	72	60	

(*a*) Formulate a linear programming model for this problem.

A* (*b*) Solve this model by the simplex method.

3.4-8.* A certain corporation has three branch plants with excess production capacity. Fortunately, the corporation has a new product ready to begin production, and all three plants have this capability, so some of the excess capacity can be used in this way. This product can be made in three sizes—large, medium, and small—that yield a net unit profit of $420, $360, and $300, respectively. Plants 1, 2, and 3 have the excess capacity to produce 750, 900, and 450 units per day of this product, respectively, regardless of the size or combination of sizes involved.

The amount of available in-process storage space also imposes a limitation on the production rates of the new product. Plants 1, 2, and 3 have 13,000, 12,000, and 5,000 square feet, respectively, of in-process storage space available for a day's production of this product. Each unit of the large, medium, and small sizes produced per day requires 20, 15, and 12 square feet, respectively.

Sales forecasts indicate that if available, 900, 1,200, and 750 units of the large, medium, and small sizes, respectively, would be sold per day.

At each plant, some employees will need to be laid off unless most of the plant's excess production capacity can be used to produce the new product. To avoid layoffs if possible, management has decided that the plants should use the same percentage of their excess capacity to produce the new product.

Management wishes to know how much of each size should be produced by each plant to maximize profit.

(*a*) Formulate a linear programming model for this problem.

A* (*b*) Solve this model by the simplex method.

3.4-9. A cargo plane has three compartments for storing cargo: front, center, and back. These compartments have capacity limits on both *weight* and *space,* as summarized below:

Compartment	Weight Capacity (Tons)	Space Capacity (Cubic Feet)
Front	12	7,000
Center	18	9,000
Back	10	5,000

Furthermore, the weight of the cargo in the respective compartments must be the same proportion of that compartment's weight capacity to maintain the balance of the airplane.

The following four cargoes have been offered for shipment on an upcoming flight as space is available:

Cargo	Weight (Tons)	Volume (Cubic Feet/Ton)	Profit ($/Ton)
1	20	500	320
2	16	700	400
3	25	600	360
4	13	400	290

Any portion of these cargoes can be accepted. The objective is to determine how much (if any) of each cargo should be accepted and how to distribute each among the compartments to maximize the total profit for the flight.

(*a*) Formulate a linear programming model for this problem.

A* (*b*) Solve this model by the simplex method.

3.4-10. You are an investment manager for a small company. You may purchase only three kinds of assets, each of which costs 100 (in money units of thousands of dollars) per unit. Fractional units may be purchased. The assets produce income 5, 10, and 20 years from now, and that income is needed to cover minimum cash flow requirements in those years, as shown in the following table.

	Income per Unit of Asset			
Years	Asset 1	Asset 2	Asset 3	Minimum Cash Flow Required
5	200	100	50	4000
10	50	50	100	0
20	0	150	200	3000

You wish to determine the mix of investments in these assets that will cover the cash flow requirements while minimizing the total amount invested.

(*a*) Formulate a linear programming model for this problem.

A* (*b*) Solve this model by the simplex method.

3.4-11.* An investor has money-making activities *A* and *B* available at the beginning of each of the next 5 years (call them years 1 to 5). Each dollar invested in *A* at the beginning of a year returns $1.40 (a profit of $0.40) two years later (in time for immediate reinvestment). Each dollar invested in *B* at the beginning of a year returns $1.70 three years later.

In addition, money-making activities *C* and *D* will each be available at one time in the future. Each dollar invested in *C* at the beginning of year 2 returns $1.90 at the end of year 5. Each dollar invested in *D* at the beginning of year 5 returns $1.30 at the end of year 5.

The investor begins with $60,000 and wishes to know which investment plan maximizes the amount of money that can be accumulated by the beginning of year 6.

(*a*) Formulate a linear programming model for this problem.

A* (*b*) Solve this model by the simplex method.

3.4-12. A company desires to blend a new alloy of 40 percent tin, 35 percent zinc, and 25 percent lead from several available alloys having the following properties:

	Alloy				
Property	1	2	3	4	5
Percentage of tin	60	25	45	20	50
Percentage of zinc	10	15	45	50	40
Percentage of lead	30	60	10	30	10
Cost ($/pound)	22	20	25	24	27

The objective is to determine the proportions of these alloys that should be blended to produce the new alloy at a minimum cost.

(*a*) Formulate a linear programming model for this problem.

A* (*b*) Solve this model by the simplex method.

3.4-13. At the beginning of the fall semester, the director of the computer facility of a certain university is confronted with the problem of assigning different working hours to her operators. Because all the operators are currently enrolled in the university, they are available to work only a limited number of hours each day.

There are six operators (four men and two women). They all have different wage rates because of differences in their experience with computers and in their programming ability. The following table shows their wage rates, along with the maximum number of hours that each can work each day.

		Maximum Hours of Availability				
Operator	Wage Rate ($/Hour)	Mon.	Tue.	Wed.	Thurs.	Fri.
K. C.	10.00	6	0	6	0	6
D. H.	10.10	0	6	0	6	0
H. B.	9.90	4	8	4	0	4
S. C.	9.80	5	5	5	0	5
K. S.	10.80	3	0	3	8	0
N. K.	11.30	0	0	0	6	2

Each operator is guaranteed a certain minimum number of hours per week that will maintain an adequate knowledge of the operation. This level is set arbitrarily at 8 hours per week for the male operators and 7 hours per week for the female operators (K. S. and N. K.).

The computer facility is to be open for operation from 8 A.M. to 10 P.M. Monday through Friday with exactly one operator on duty during these hours. On Saturdays and Sundays, the computer is to be operated by other staff.

Because of a tight budget, the director has to minimize cost. She wishes to determine the number of hours she should assign to each operator on each day.

(*a*) Formulate a linear programming model for this problem.

A* (*b*) Solve this model by the simplex method.

3.4-14. A company needs to lease warehouse storage space over the next 5 months. Just how much space will be required in each of these months is known. However, since these space requirements are quite different, it may be most economical to lease only the amount needed each month on a month-by-month basis. On the other hand, the additional cost for leasing space for additional months is much less than for the first month, so it may be less expensive to lease the maximum amount needed for the entire 5 months. Another option is the intermediate approach of changing the total amount of space leased (by adding a new lease and/or having an old lease expire) at least once but not every month.

The space requirement (in thousands of square feet) and the leasing costs (in hundreds of dollars) for the various leasing periods are as follows:

Month	Required Space (000 Square Feet)	Leasing Period (Months)	Cost per 1,000 Square Feet Leased ($00)
1	30	1	650
2	20	2	1,000
3	40	3	1,350
4	10	4	1,600
5	50	5	1,900

The objective is to minimize the total leasing cost for meeting the space requirements.

(*a*) Formulate a linear programming model for this problem.

A* (*b*) Solve this model by the simplex method.

3.4-15. A large paper manufacturing company has 10 paper mills and a large number (say, 1,000) of customers to be supplied. It uses three alternative types of machines and four types of raw materials to make five different types of paper. Therefore, the company needs to develop a detailed production distribution plan on a monthly basis, with an objective of minimizing the total cost of producing and distributing the paper during the month. Specifically, it is necessary to determine jointly the amount of each type of paper to be made at each paper mill on each type of machine *and* the amount of each type of paper to be shipped from each paper mill to each customer.

The relevant data can be expressed symbolically as follows:

D_{jk} = number of units of paper type k demanded by customer j,

r_{klm} = number of units of raw material m needed to produce 1 unit of paper type k on machine type l,

R_{im} = number of units of raw material m available at paper mill i,

c_{kl} = number of capacity units of machine type l that will produce 1 unit of paper type k,

C_{il} = number of capacity units of machine type l available at paper mill i,

P_{ikl} = production cost for each unit of paper type k produced on machine type l at paper mill i,

T_{ijk} = transportation cost for each unit of paper type k shipped from paper mill i to customer j.

Using these symbols, formulate a linear programming model for this problem.

3.5-1. Read the article footnoted in Sec. 3.5 that describes the first case study presented in that section: "Choosing the Product Mix at Ponderosa Industrial."

(*a*) Describe the two factors which, according to the article, often hinder the use of optimization models by managers.

(*b*) Section 3.5 indicates without elaboration that using linear programming at Ponderosa "led to a dramatic shift in the types of plywood products emphasized by the company." Identify this shift.

(*c*) With the success of this application, management then was eager to use optimization for other problems as well. Identify these other problems.

(*d*) Photocopy the two pages of appendixes that give the mathematical formulation of the problem and the structure of the linear programming model.

3.5-2. Read the article footnoted in Sec. 3.5 that describes the second case study presented in that section: "Personnel Scheduling at United Airlines."

(*a*) Describe how United Airlines prepared shift schedules at airports and reservations offices prior to this OR study.

(*b*) When this study began, the problem definition phase defined five specific project requirements. Identify these project requirements.

(*c*) At the end of the presentation of the corresponding example in Sec. 3.4 (personnel scheduling at Union Airways), we pointed out that the divisibility assumption does not hold for this kind of application. An integer solution is needed, but linear programming may provide an optimal solution that is noninteger. How does United Airlines deal with this problem?

(*d*) Describe the flexibility built into the scheduling system to satisfy the group culture at each office. Why was this flexibility needed?

(*e*) Briefly describe the tangible and intangible benefits that resulted from the study.

3.5-3. Read the 1986 article footnoted in Sec. 2.1 that describes the third case study presented in Sec. 3.5: "Planning Supply, Distribution, and Marketing at Citgo Petroleum Corporation."

(*a*) What happened during the years preceding this OR study that made it vastly more important to control the amount of capital tied up in inventory?
(*b*) What geographical area is spanned by Citgo's distribution network of pipelines, tankers, and barges? Where does Citgo market its products?
(*c*) What time periods are included in the model?
(*d*) Which computer did Citgo use to solve the model? What were typical run times?
(*e*) Who are the four types of model users? How does each one use the model?
(*f*) List the major types of reports generated by the SDM system.
(*g*) What were the major implementation challenges for this study?
(*h*) List the direct and indirect benefits that were realized from this study.

CASE PROBLEM

CP3-1. The school board of a certain city has made the decision to close one of its middle schools (sixth, seventh, and eighth grades) at the end of this school year and reassign all next year's middle-school students to the three remaining middle schools. The school district provides busing for all middle-school students who must travel more than approximately 1 mile, so the school board wants a plan for reassigning the students that will minimize the total busing cost. The annual cost per student of busing from each of the six residential areas of the city to each of the schools is shown in the following table (along with other basic data for next year), where 0 indicates that busing is not needed and a dash indicates an infeasible assignment.

Area	No. of Students	Percentage in 6th Grade	Percentage in 7th Grade	Percentage in 8th Grade	*Busing Cost per Student ($)*		
					School 1	School 2	School 3
1	450	32	38	30	300	0	700
2	600	37	28	35	—	400	500
3	550	30	32	38	600	300	200
4	350	28	40	32	200	500	—
5	500	39	34	27	0	—	400
6	450	34	28	38	500	300	0
				School capacity:	900	1,100	1,000

The school board also has imposed the restriction that each grade must constitute between 30 and 35 percent of each school's population. The above table shows the percentage of each area's middle-school population for next year that falls into each of the three grades. The school attendance zone boundaries can be drawn so as to split any given area among more than one school, but assume that the percentages shown in the table will continue to hold for any partial assignment of an area to a school.

You have been hired as an OR consultant to assist the school board in determining how many students in each area should be assigned to each school.

(*a*) Formulate a linear programming model for this problem.
(*b*) For each of the four assumptions of linear programming discussed in Sec. 3.3, analyze how well it applies to this problem.
A* (*c*) Use a computer to solve the model formulated in part (*a*) by the simplex method.

After seeing your results from part (*c*), the school board expresses concern about all the splitting of residential areas among multiple schools. The board members indicate that they "would like to keep each neighborhood together."

(*d*) Adjust the results from part (*c*) as well as you can to enable each area to be assigned to just one school. How much does this increase the total busing cost? (This line of analysis will be pursued more rigorously in the case problem of Chap. 12.)

The school board is considering eliminating some busing to reduce costs. Option 1 is to eliminate busing only for students traveling 1 to 1.5 miles, where the cost per student is given in the table as $200. Option 2 is to also eliminate busing for students traveling 1.5 to 2 miles, where the estimated cost per student is $300.

A* (*e*) Revise the model from part (*a*) to fit Option 1, and solve. Compare these results with those from part (*c*), including the reduction in total busing cost.

A* (*f*) Repeat part (*e*) for Option 2.

The school board now needs to choose among the three alternative busing plans (the current one or Option 1 or Option 2). One important factor is busing costs. However, the school board also wants to place equal weight on a second factor—the inconvenience and safety problems caused by forcing students to travel by foot or bicycle a substantial distance (more than 1 mile, and especially more than 1.5 miles). Therefore, the board members want to choose a plan that provides the best trade-off between these two factors.

(*g*) Use your results from parts (*c*), (*e*), and (*f*) to summarize the key information related to these two factors that the school board needs to make this decision.

(*h*) Which decision do you think should be made? Why?

4

Solving Linear Programming Problems: The Simplex Method

We now are ready to begin studying the *simplex method,* a general procedure for solving linear programming problems. Developed by George Dantzig in 1947, it has proved to be a remarkably efficient method that is used routinely to solve huge problems on today's computers. Except for its use on tiny problems, this method is always executed on a computer, and sophisticated software packages are widely available. Extensions and variations of the simplex method also are used to perform *post-optimality analysis* (including sensitivity analysis) on the model.

This chapter describes and illustrates the main features of the simplex method. The first section introduces its general nature, including its geometric interpretation. The following three sections then develop the procedure for solving any linear programming model that is in our standard form (maximization, all functional constraints in $\leq$ form, and nonnegativity constraints on all variables) and has only *nonnegative* right-hand sides b_i in the functional constraints. Certain details on resolving ties are deferred to Sec. 4.5, and Sec. 4.6 describes how to adapt the simplex method to other model forms. Next we discuss post-optimality analysis (Sec. 4.7), and then we conclude the chapter with a description of the computer implementation of the simplex method (Sec. 4.8).

4.1 The Essence of the Simplex Method

The simplex method is an *algebraic* procedure. However, its underlying concepts are *geometric.* Understanding these geometric concepts provides a strong intuitive feeling for how the simplex method operates and what makes it so efficient. Therefore, before delving into algebraic details, we focus in this section on the big picture from a geometric viewpoint.

To illustrate the general geometric concepts, we shall use the Wyndor Glass Co. example presented in Sec. 3.1. (Sections 4.2 and 4.3 use the *algebra* of the simplex method to solve this same example.) Section 5.1 will elaborate further on these geometric concepts for larger problems.

To refresh your memory, the model and graph for this example are repeated in Fig. 4.1. The five constraint boundaries and their points of intersection are highlighted in this figure because they are the keys to the analysis. Here, each **constraint boundary** is a line that forms the boundary of what is permitted by the corresponding constraint. The points of intersection are the **corner-point solutions** of the problem. The five that lie on the corners of the *feasible region*—(0, 0), (0, 6), (2, 6), (4, 3), and (4, 0)—are the *corner-point feasible solutions* (**CPF solutions**). [The other three—(0, 9), (4, 6), and (6, 0)—are called *corner-point infeasible solutions*.]

In this example, each corner-point solution lies at the intersection of *two* constraint boundaries. (For a linear programming problem with n decision variables, each

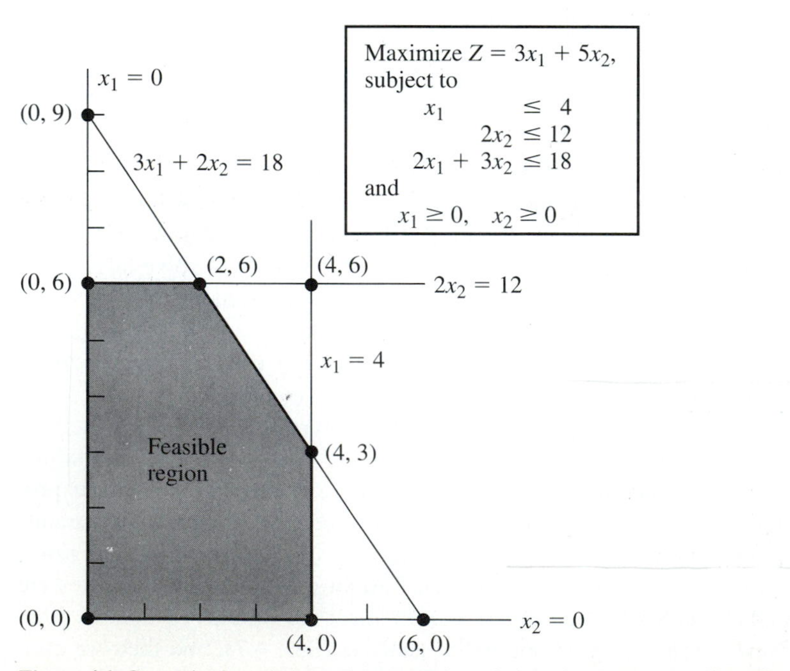

Figure 4.1 Constraint boundaries and corner-point solutions for the Wyndor Glass Co. problem.

of its corner-point solutions lies at the intersection of n constraint boundaries.[1]) Certain pairs of the CPF solutions in Fig. 4.1 share a constraint boundary, and other pairs do not. It will be important to distinguish between these cases by using the following general definitions.

For any linear programming problem with n decision variables, two CPF solutions are **adjacent** to each other if they share $n - 1$ constraint boundaries. The two adjacent CPF solutions are connected by a line segment that lies on these same shared constraint boundaries. Such a line segment is referred to as an **edge** of the feasible region.

Since $n = 2$ in the example, two of its CPF solutions are adjacent if they share *one* constraint boundary; for example, (0, 0) and (0, 6) are adjacent because they share the $x_1 = 0$ constraint boundary. The feasible region in Fig. 4.1 has five edges, consisting of the five line segments forming the boundary of this region. Note that two edges emanate from each CPF solution. Thus, each CPF solution has two adjacent CPF solutions (each lying at the other end of one of the two edges), as enumerated in Table 4.1. (In each row of this table, the CPF solution in the first column is adjacent to each of the two CPF solutions in the second column, but the two CPF solutions in the second column are *not* adjacent to each other.)

One reason for our interest in adjacent CPF solutions is the following general property about such solutions, which provides a very useful way of checking whether a CPF solution is an optimal solution.

Optimality test: Consider any linear programming problem that possesses at least one optimal solution. If a CPF solution has no *adjacent* CPF solutions that are *better* (as measured by Z), then it *must* be an *optimal* solution.

Thus, for the example, (2, 6) must be optimal simply because its $Z = 36$ is larger than $Z = 30$ for (0, 6) and $Z = 27$ for (4, 3). (We will delve further into why this property holds in Sec. 5.1.) This optimality test is the one used by the simplex method for determining when an optimal solution has been reached.

Now we are ready to apply the simplex method to the example.

Solving the Example

Here is an outline of what the simplex method does (from a geometric viewpoint) to solve the Wyndor Glass Co. problem. At each step, first the conclusion is stated and then the reason is given in parentheses. (Refer to Fig. 4.1 for a visualization.)

Table 4.1 **Adjacent CPF Solutions for Each CPF Solution of the Wyndor Glass Co. Problem**

CPF Solution	Its Adjacent CPF Solutions
(0, 0)	(0, 6) and (4, 0)
(0, 6)	(2, 6) and (0, 0)
(2, 6)	(4, 3) and (0, 6)
(4, 3)	(4, 0) and (2, 6)
(4, 0)	(0, 0) and (4, 3)

[1] Although a corner-point solution is defined in terms of n constraint boundaries whose intersection gives this solution, it also is possible that one or more *additional* constraint boundaries pass through this same point.

Initialization: Choose (0, 0) as the *initial* CPF solution to examine. (This is a convenient choice because no calculations are required to identify this CPF solution.)

Optimality Test: Conclude that (0, 0) is *not* an optimal solution. (Adjacent CPF solutions are better.)

Iteration 1: Move to a better *adjacent* CPF solution, (0, 6), by performing the following three steps.

1. Between the two edges of the feasible region that emanate from (0, 0), choose to move along the edge that leads up the x_2 axis. (With an objective function of $Z = 3x_1 + 5x_2$, moving up the x_2 axis increases Z at a faster rate than moving along the x_1 axis.)
2. Stop at the first new constraint boundary: $2x_2 = 12$. [Moving farther in the direction selected in step 1 leaves the feasible region; e.g., moving to the second new constraint boundary hit when moving in that direction gives (0, 9), which is a corner-point *infeasible* solution.]
3. Solve for the intersection of the new set of constraint boundaries: (0, 6). (The equations for these constraint boundaries, $x_1 = 0$ and $2x_2 = 12$, immediately yield this solution.)

Optimality Test: Conclude that (0, 6) is *not* an optimal solution. (An adjacent CPF solution is better.)

Iteration 2: Move to a better adjacent CPF solution, (2, 6), by performing the following three steps.

1. Between the two edges of the feasible region that emanate from (0, 6), choose to move along the edge that leads to the right. (Moving along this edge increases Z, whereas backtracking to move back down the x_2 axis decreases Z.)
2. Stop at the first new constraint boundary encountered when moving in that direction: $3x_1 + 2x_2 = 12$. (Moving farther in the direction selected in step 1 leaves the feasible region.)
3. Solve for the intersection of the new set of constraint boundaries: (2, 6). (The equations for these constraint boundaries, $x_1 = 0$ and $2x_2 = 12$, immediately yield this solution.)

Optimality Test: Conclude that (2, 6) *is* an optimal solution, so stop. (None of the adjacent CPF solutions are better.)

This sequence of CPF solutions examined is shown in Fig. 4.2, where each circled number identifies which iteration obtained that solution.

Now let us look at the six key solution concepts of the simplex method that provide the rationale behind the above steps. (Keep in mind that these concepts also apply for solving problems with more than two decision variables where a graph like Fig. 4.2 is not available to help quickly find an optimal solution.)

The Key Solution Concepts

The first solution concept is based directly on the relationship between optimal solutions and CPF solutions given at the end of Sec. 3.2.

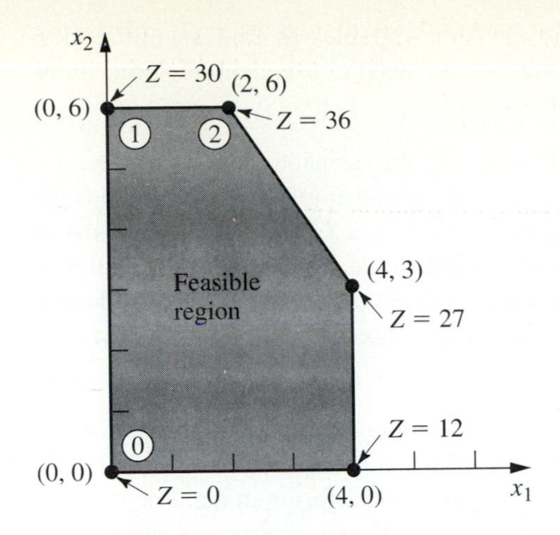

Figure 4.2 This graph shows the sequence of CPF solutions (⓪,①,②) examined by the simplex method for the Wyndor Glass Co. problem. The optimal solution (2, 6) is found after just three solutions are examined.

> **Solution concept 1:** The simplex method focuses solely on CPF solutions. For any problem with at least one optimal solution, finding one requires only finding a best CPF solution.[1]

Since the number of feasible solutions generally is infinite, reducing the number of solutions that need to be examined to a small finite number (just three in Fig. 4.2) is a tremendous simplification.

The next solution concept defines the flow of the simplex method.

> **Solution concept 2:** The simplex method is an *iterative algorithm* (a systematic solution procedure that keeps repeating a fixed series of steps, called an *iteration*, until a desired result has been obtained) with the following structure.

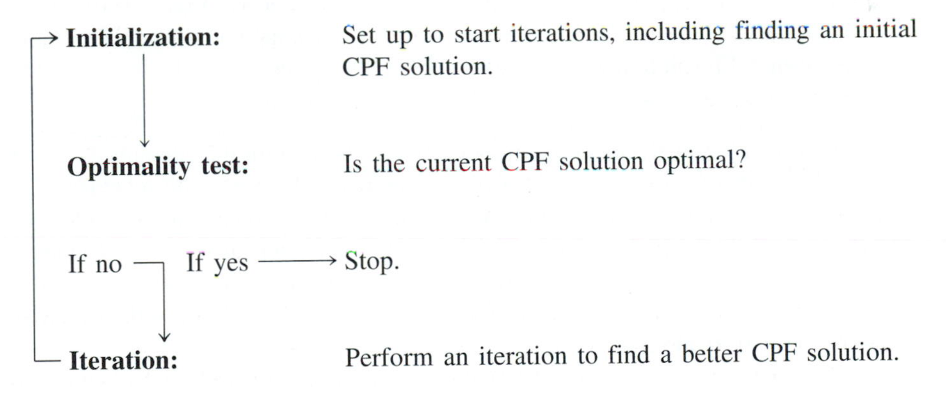

When the example was solved, note how this flow diagram was followed through two iterations until an optimal solution was found.

We next focus on how to get started.

> **Solution concept 3:** Whenever possible, the initialization of the simplex method chooses the *origin* (all decision variables equal to zero) to be the initial CPF

[1] The only restriction is that the problem must possess CPF solutions. This is ensured if the feasible region is bounded.

solution. When there are too many decision variables to find an initial CPF solution graphically, this choice eliminates the need to use algebraic procedures to find and solve for an initial CPF solution.

Choosing the origin commonly is possible when all the decision variables have nonnegativity constraints, because the intersection of these constraint boundaries yields the origin as a corner-point solution. This solution then is a CPF solution *unless* it is *infeasible* because it violates one or more of the functional constraints. If it is infeasible, special procedures described in Sec. 4.6 are needed to find the initial CPF solution.

The next solution concept concerns the choice of a better CPF solution at each iteration.

> **Solution concept 4:** Given a CPF solution, it is much quicker computationally to gather information about its *adjacent* CPF solutions than about other CPF solutions. Therefore, each time the simplex method performs an iteration to move from the current CPF solution to a better one, it *always* chooses a CPF solution that is *adjacent* to the current one. No other CPF solutions are considered. Consequently, the entire path followed to eventually reach an optimal solution is along the *edges* of the feasible region.

The next focus is on which adjacent CPF solution to choose at each iteration.

> **Solution concept 5:** After the current CPF solution is identified, the simplex method examines each of the edges of the feasible region that emanate from this CPF solution. Each of these edges leads to an *adjacent* CPF solution at the other end, but the simplex method does not even take the time to solve for the adjacent CPF solution. Instead, it simply identifies the *rate of improvement in Z* that would be obtained by moving along the edge. Among the edges with a *positive* rate of improvement in *Z*, it then chooses to move along the one with the *largest* rate of improvement in *Z*. The iteration is completed by first solving for the adjacent CPF solution at the other end of this one edge and then relabeling this adjacent CPF solution as the *current* CPF solution for the optimality test and (if needed) the next iteration.

At the first iteration of the example, moving from (0, 0) along the edge on the x_1 axis would give a rate of improvement in Z of 3 (Z increases by 3 per unit increase in x_1), whereas moving along the edge on the x_2 axis would give a rate of improvement in Z of 5 (Z increases by 5 per unit increase in x_2), so the decision is made to move along the latter edge. At the second iteration, the only edge emanating from (0, 6) that would yield a *positive* rate of improvement in Z is the edge leading to (2, 6), so the decision is made to move next along this edge.

The final solution concept clarifies how the optimality test is performed efficiently.

> **Solution concept 6:** Solution concept 5 describes how the simplex method examines each of the edges of the feasible region that emanate from the current CPF solution. This examination of an edge leads to quickly identifying the rate of improvement in Z that would be obtained by moving along the edge toward the adjacent CPF solution at the other end. A *positive* rate of improvement in Z implies that the adjacent CPF solution is *better* than the current CPF solution (since we are assuming maximization), whereas a *negative* rate of improvement in Z implies that the adjacent CPF solution is *worse*. Therefore, the optimality

test consists simply of checking whether *any* of the edges give a *positive* rate of improvement in Z. If *none* do, then the current CPF solution is optimal.

In the example, moving along *either* edge from (2, 6) decreases Z. This fact immediately gives the conclusion that (2, 6) is optimal.

4.2 Setting Up the Simplex Method

The preceding section stressed the geometric concepts that underlie the simplex method. However, this algorithm normally is run on a computer, which can follow only algebraic instructions. Therefore, it is necessary to translate the conceptually geometric procedure just described into a usable algebraic procedure. In this section, we introduce the *algebraic language* of the simplex method and relate it to the concepts of the preceding section.

The algebraic procedure is based on solving systems of equations. Therefore, the first step in setting up the simplex method is to convert the functional *inequality constraints* to equivalent *equality constraints.* (The nonnegativity constraints are left as inequalities because they are treated separately.) This conversion is accomplished by introducing **slack variables.** To illustrate, consider the first functional constraint in the Wyndor Glass Co. example of Sec. 3.1

$$x_1 \le 4.$$

The slack variable for this constraint is defined to be

$$x_3 = 4 - x_1,$$

which is the amount of slack in the left-hand side of the inequality. Thus

$$x_1 + x_3 = 4.$$

Given this equation, $x_1 \le 4$ if and only if $4 - x_1 = x_3 \ge 0$. Therefore, the original constraint $x_1 \le 4$ is entirely *equivalent* to the pair of constraints

$$x_1 + x_3 = 4 \qquad \text{and} \qquad x_3 \ge 0.$$

Upon the introduction of slack variables for the other functional constraints, the original linear programming model for the example (shown below on the left) can now be replaced by the equivalent model (called the *augmented form* of the model) shown below on the right:

Original Form of the Model

Maximize $Z = 3x_1 + 5x_2,$

subject to

$$x_1 \le 4$$
$$2x_2 \le 12$$
$$3x_1 + 2x_2 \le 18$$

and

$$x_1 \ge 0, \qquad x_2 \ge 0.$$

Augmented Form of the Model[1]

Maximize $Z = 3x_1 + 5x_2,$

subject to

$$(1) \quad x_1 + x_3 = 4$$
$$(2) \quad 2x_2 + x_4 = 12$$
$$(3) \quad 3x_1 + 2x_2 + x_5 = 18$$

and

$$x_j \ge 0, \quad \text{for } j = 1, 2, 3, 4, 5.$$

[1] The slack variables are not shown in the objective function because the coefficients there are 0.

Although both forms of the model represent exactly the same problem, the new form is much more convenient for algebraic manipulation and for identification of CPF solutions. We call this the **augmented form** of the problem, because the original form has been *augmented* by some supplementary variables needed to apply the simplex method.

If a slack variable equals 0 in the current solution, then this solution lies on the constraint boundary for the corresponding functional constraint. A value greater than 0 means that the solution lies on the *feasible* side of this constraint boundary, whereas a value less than 0 means that the solution lies on the *infeasible* side of this constraint boundary. A demonstration of these properties is provided by the demonstration in your OR Courseware entitled *Interpretation of the Slack Variables.*

The terminology used in the preceding section (corner-point solutions, etc.) applies to the original form of the problem. We now introduce the corresponding terminology for the augmented form.

> An **augmented solution** is a solution for the original variables (the *decision variables*) that has been augmented by the corresponding values of the *slack variables.*

For example, augmenting the solution (3, 2) in the example yields the augmented solution (3, 2, 1, 8, 5) because the corresponding values of the slack variables are $x_3 = 1$, $x_4 = 8$, and $x_5 = 5$.

> A **basic solution** is an *augmented* corner-point solution.

To illustrate, consider the corner-point infeasible solution (4, 6) in Fig. 4.1. Augmenting it with the resulting values of the slack variables $x_3 = 0$, $x_4 = 0$, and $x_5 = -6$ yields the corresponding basic solution (4, 6, 0, 0, −6).

The fact that corner-point solutions (and so basic solutions) can be either feasible or infeasible implies the following definition:

> A **basic feasible (BF) solution** is an *augmented* CPF solution.

Thus the CPF solution (0, 6) in the example is equivalent to the BF solution (0, 6, 4, 0, 6) for the problem in augmented form.

The only difference between basic solutions and corner-point solutions (or between BF solutions and CPF solutions) is whether the values of the slack variables are included. For any basic solution, the corresponding corner-point solution is obtained simply by deleting the slack variables. Therefore, the geometric and algebraic relationships between these two solutions are very close, as described in Sec. 5.1.

Because the terms *basic solution* and *basic feasible solution* are very important parts of the standard vocabulary of linear programming, we now need to clarify their algebraic properties. For the augmented form of the example, notice that the system of functional constraints has two more variables (5) than equations (3). This fact gives us 2 *degrees of freedom* in solving the system, since any two variables can be chosen to be set equal to any arbitrary value in order to solve the three equations in terms of the remaining three variables (barring redundancies). The simplex method uses zero for this arbitrary value. Thus, two of the variables (called the *nonbasic variables*) are set equal to zero, and then the simultaneous solution of the three equations for the other three variables (called the *basic variables*) is a *basic solution.* These properties are described in the following general definitions.

Set to 0 and Solve

A **basic solution** has the following properties:

1. Each variable is designated as either a nonbasic variable or a basic variable.
2. The number of basic variables equals the number of functional constraints

(now equations). Therefore, the number of nonbasic variables equals the total number of variables *minus* the number of functional constraints.

3. The **nonbasic variables** are set equal to zero.
4. The values of the **basic variables** are obtained as the simultaneous solution of the system of equations (functional constraints in augmented form). (The set of basic variables is often referred to as **the basis**.)
5. If the basic variables satisfy the *nonnegativity constraints,* the basic solution is a **BF solution**.

To illustrate these definitions, consider again the BF solution (0, 6, 4, 0, 6). This solution was obtained before by augmenting the CPF solution (0, 6). However, another way to obtain this same solution is to choose x_1 and x_4 to be the two nonbasic variables, and so the two variables are set equal to zero. The three equations then yield, respectively, $x_3 = 4$, $x_2 = 6$, and $x_5 = 6$ as the solution for the three basic variables, as shown below (with the basic variables in bold type):

$$x_1 = 0 \text{ and } x_4 = 0 \text{ so}$$

$$
\begin{array}{llll}
(1) & x_1 + \mathbf{x_3} = 4 & & \mathbf{x_3} = 4 \\
(2) & 2\mathbf{x_2} + x_4 = 12 & & \mathbf{x_2} = 6 \\
(3) & 3x_1 + 2\mathbf{x_2} + \mathbf{x_5} = 18 & & \mathbf{x_5} = 6
\end{array}
$$

Because all three of these basic variables are nonnegative, this *basic solution* (0, 6, 4, 0, 6) is indeed a *BF solution*.

Just as certain pairs of CPF solutions are *adjacent*, the corresponding pairs of BF solutions also are said to be adjacent. Here is an easy way to tell when two BF solutions are adjacent.

> Two BF solutions are **adjacent** if all but one of their nonbasic variables are the same (so *all but one* of their *basic variables* also are the same, although perhaps with different numerical values).

Consequently, moving from the current BF solution to an adjacent one involves switching one variable from nonbasic to basic and vice versa for one other variable (and then adjusting the values of the basic variables to continue satisfying the system of equations).

To illustrate *adjacent BF solutions,* consider one pair of adjacent CPF solutions in Fig. 4.1: (0, 0) and (0, 6). Their augmented solutions, (0, 0, 4, 12, 18) and (0, 6, 4, 0, 6), automatically are adjacent BF solutions. However, you do not need to look at Fig. 4.1 to draw this conclusion. Another signpost is that their nonbasic variables, (x_1, x_2) and (x_1, x_4), are the same with just the one exception—x_2 has been replaced by x_4. Consequently, moving from (0, 0, 4, 12, 18) to (0, 6, 4, 0, 6) involves switching x_2 from nonbasic to basic and vice versa for x_4.

When we deal with the problem in augmented form, it is convenient to consider and manipulate the objective function equation at the same time as the new constraint equations. Therefore, before we start the simplex method, the problem needs to be rewritten once again in an equivalent way:

Maximize

$$Z,$$

subject to

$$
\begin{array}{llcccccl}
(0) & Z - 3x_1 - 5x_2 & & & & & = & 0 \\
(1) & x_1 & & + x_3 & & & = & 4 \\
(2) & & 2x_2 & & + x_4 & & = & 12 \\
(3) & 3x_1 + 2x_2 & & & & + x_5 & = & 18
\end{array}
$$

and

$$x_j \geq 0, \qquad \text{for } j = 1, 2, \ldots, 5.$$

It is just as if Eq. (0) actually were one of the original constraints; but because it already is in equality form, no slack variable is needed. While adding one more equation, we also have added one more unknown (Z) to the system of equations. Therefore, when using Eqs. (1) to (3) to obtain a basic solution as described above, we use Eq. (0) to solve for Z at the same time.

Somewhat fortuitously, the model for the Wyndor Glass Co. problem fits *our standard form*, and all its functional constraints have nonnegative right-hand sides b_i. If this had not been the case, then additional adjustments would have been needed at this point before the simplex method was applied. These details are deferred to Sec. 4.6, and we now focus on the simplex method itself.

4.3 The Algebra of the Simplex Method

We continue to use the prototype example of Sec. 3.1, as rewritten at the end of Sec. 4.2, for illustrative purposes. To start connecting the geometric and algebraic concepts of the simplex method, we begin by outlining side by side in Table 4.2 how the simplex method solves this example from both a geometric and an algebraic viewpoint. The geometric viewpoint (first presented in Sec. 4.1) is based on the *original form* of the model (no slack variables), so again refer to Fig. 4.1 for a visualization when you examine the second column of the table. Refer to the *augmented form* of the model presented at the end of Sec. 4.2 when you examine the third column of the table.

We now fill in the details for each step of the third column of Table 4.2.

Initialization

The choice of x_1 and x_2 to be the *nonbasic* variables (the variables set equal to zero) for the initial BF solution is based on solution concept 3 in Sec. 4.1. This choice eliminates the work required to solve for the *basic variables* (x_3, x_4, x_5) from the following system of equations (where the basic variables are shown in bold type):

$$
\begin{array}{llcccccl}
 & & & & & & & x_1 = 0 \text{ and } x_2 = 0 \text{ so} \\
(1) & x_1 & & + \mathbf{x_3} & & = 4 & & \mathbf{x_3} = 4 \\
(2) & & 2x_2 & & + \mathbf{x_4} & = 12 & & \mathbf{x_4} = 12 \\
(3) & 3x_1 + 2x_2 & & & + \mathbf{x_5} & = 18 & & \mathbf{x_5} = 18
\end{array}
$$

Thus the **initial BF solution** is (0, 0, 4, 12, 18).

Table 4.2 **Geometric and Algebraic Interpretations of How the Simplex Method Solves the Wyndor Glass Co. Problem**

Method Sequence	Geometric Interpretation	Algebraic Interpretation
Initialization	Choose (0, 0) to be the initial CPF solution.	Choose x_1 and x_2 to be the nonbasic variables (= 0) for the initial BF solution: (0, 0, 4, 12, 18).
Optimality test	Not optimal, because moving along either edge from (0, 0) increases Z.	Not optimal, because increasing either nonbasic variable (x_1 or x_2) increases Z.
Iteration 1		
Step 1	Move up the edge lying on the x_2 axis.	Increase x_2 while adjusting other variable values to satisfy the system of equations.
Step 2	Stop when the first new constraint boundary ($2x_2 = 12$) is reached.	Stop when the first basic variable (x_3, x_4, or x_5) drops to zero (x_4).
Step 3	Find the intersection of the new pair of constraint boundaries: (0, 6) is the new CPF solution.	With x_2 now a basic variable and x_4 now a nonbasic variable, solve the system of equations: (0, 6, 4, 0, 6) is the new BF solution.
Optimality test	Not optimal, because moving along the edge from (0, 6) to the right increases Z.	Not optimal, because increasing one variable (x_1) increases Z.
Iteration 2		
Step 1	Move along this edge to the right.	Increase x_1 while adjusting other variable values to satisfy the system of equations.
Step 2	Stop when the first new constraint boundary ($3x_1 + 2x_2 = 18$) is reached.	Stop when the first basic variable (x_2, x_3, or x_5) drops to zero (x_5).
Step 3	Find the intersection of the new pair of constraint boundaries: (2, 6) is the new CPF solution.	With x_1 now a basic variable and x_5 now a nonbasic variable, solve the system of equations: (2, 6, 2, 0, 0) is the new BF solution.
Optimality test	(2, 6) is optimal, because moving along either edge from (2, 6) decreases Z.	(2, 6, 2, 0, 0) is optimal, because increasing either nonbasic variable (x_4 or x_5) decreases Z.

Notice that this solution can be read immediately because each equation has just one basic variable, which has a coefficient of 1, and this basic variable does not appear in any other equation. You will soon see that when the set of basic variables changes, the simplex method uses an algebraic procedure (Gaussian elimination) to convert the equations to this same convenient form for reading every subsequent BF solution as well. This form is called **proper form from Gaussian elimination**.

Optimality Test

The objective function is

$$Z = 3x_1 + 5x_2,$$

so $Z = 0$ for the initial BF solution. Because none of the basic variables (x_3, x_4, x_5)

have a *nonzero* coefficient in this objective function, the coefficient of each nonbasic variable (x_1, x_2) gives the rate of improvement in Z if that variable were to be increased from zero (while the values of the basic variables are adjusted to continue satisfying the system of equations).[1] These rates of improvement (3 and 5) are *positive*. Therefore, based on solution concept 6 in Sec. 4.1, we conclude that (0, 0, 4, 12, 18) is not optimal.

For each BF solution examined after subsequent iterations, at least one basic variable has a nonzero coefficient in the objective function. Therefore, the optimality test then will use the new Eq. (0) to rewrite the objective function in terms of just the nonbasic variables, as you will see later.

Determining the Direction of Movement (Step 1 of an Iteration)

Increasing one nonbasic variable from zero (while adjusting the values of the basic variables to continue satisfying the system of equations) corresponds to moving along one edge emanating from the current CPF solution. Based on solution concepts 4 and 5 in Sec. 4.1, the choice of which nonbasic variable to increase is made as follows:

$$Z = 3x_1 + 5x_2$$

Increase x_1? Rate of improvement in $Z = 3$.

Increase x_2? Rate of improvement in $Z = 5$.

$5 > 3$, so choose x_2 to increase.

As indicated next, we call x_2 the *entering basic variable* for iteration 1.

> At any iteration of the simplex method, the purpose of step 1 is to choose one *nonbasic variable* to increase from zero (while the values of the basic variables are adjusted to continue satisfying the system of equations). Increasing this nonbasic variable from zero will convert it to a *basic variable* for the next BF solution. Therefore, this variable is called the **entering basic variable** for the current iteration (because it is entering the basis).

Determining Where to Stop (Step 2 of an Iteration)

Step 2 addresses the question of how far to increase the entering basic variable x_2 before stopping. Increasing x_2 increases Z, so we want to go as far as possible without leaving the feasible region. The requirement to satisfy the functional constaints in augmented form (shown below) means that increasing x_2 (while keeping the nonbasic variable $x_1 = 0$) changes the values of some of the basic variables as shown on the right.

$$x_1 = 0, \quad \text{so}$$

$$(1) \quad x_1 + \boldsymbol{x_3} = 4 \qquad \boldsymbol{x_3} = 4$$

$$(2) \quad 2x_2 + \boldsymbol{x_4} = 12 \qquad \boldsymbol{x_4} = 12 - 2x_2$$

$$(3) \quad 3x_1 + 2x_2 + \boldsymbol{x_5} = 18 \qquad \boldsymbol{x_5} = 18 - 2x_2.$$

[1] Note that this interpretation of the coefficients of the x_j variables is based on these variables being on the right-hand side, $Z = 3x_1 + 5x_2$. When these variables are brought to the left-hand side for Eq. (0), $Z - 3x_1 - 5x_2 = 0$, the nonzero coefficients change their signs.

The other requirement for feasibility is that all the variables be *nonnegative.* The nonbasic variables (including the entering basic variable) are nonnegative, but we need to check how far x_2 can be increased without violating the nonnegativity constrains for the basic variables.

$$\mathbf{x_3} = 4 \geq 0 \qquad \Rightarrow \text{no upper bound on } x_2.$$

$$\mathbf{x_4} = 12 - 2x_2 \geq 0 \Rightarrow x_2 \leq \frac{12}{2} = 6 \quad \leftarrow \text{minimum.}$$

$$\mathbf{x_5} = 18 - 2x_2 \geq 0 \Rightarrow x_2 \leq \frac{18}{2} = 9.$$

Thus, x_2 can be increased just to 6, at which point $\mathbf{x_4}$ has dropped to 0. Increasing x_2 beyond 6 would cause x_4 to become negative, which would violate feasibility.

These calculations are referred to as the **minimum ratio test**. Referring to the system of equations, we see that the ratio for each equation is the ratio of the right-hand side to the coefficient of the entering basic variable (except that we ignore any equation where this coefficient is zero or negative since such a coefficient leads to no upper bound on the entering basic variable). The basic variable in the equation with the minimum ratio is the one that drops to zero first as the entering basic variable is increased.

> At any iteration of the simplex method, step 2 uses the *minimum ratio test* to determine which basic variable drops to zero first as the entering basic variable is increased. Decreasing this basic variable to zero will convert it to a *nonbasic variable* for the next BF solution. Therefore, this variable is called the **leaving basic variable** for the current iteration (because it is leaving the basis).

Thus $\mathbf{x_4}$ is the leaving basic variable for iteration 1 of the example.

Solving for the New BF Solution (Step 3 of an Iteration)

Increasing $x_2 = 0$ to $x_2 = 6$ moves us from the *initial* BF solution on the left to the *new* BF solution on the right.

	Initial BF solution	New BF solution
Nonbasic variables:	$x_1 = 0, \quad x_2 = 0$	$x_1 = 0, \quad x_4 = 0$
Basic variables:	$x_3 = 4, \quad x_4 = 12, \quad x_5 = 18$	$x_3 = ?, \quad x_2 = 6, \quad x_5 = ?$

The purpose of step 3 is to convert the system of equations to a more convenient form (proper form from Gaussian elimination) for conducting the optimality test and (if needed) the next iteration with this new BF solution. In the process, this form also will identify the values of x_3 and x_5 for the new solution.

Here again is the complete original system of equations, where the *new* basic variables are shown in bold type (with Z playing the role of the basic variable in the objective function equation):

$$
\begin{array}{ll}
(0) & \mathbf{Z} - 3x_1 - 5\mathbf{x_2} = 0 \\
(1) & x_1 + \mathbf{x_3} = 4 \\
(2) & 2\mathbf{x_2} + x_4 = 12 \\
(3) & 3x_1 + 2\mathbf{x_2} + \mathbf{x_5} = 18.
\end{array}
$$

Thus, x_2 has replaced x_4 as the basic variable in Eq. (2). To solve this system of equations for Z, x_2, x_3, and x_5, we need to perform some **elementary algebraic operations** (multiply or divide an equation by a nonzero constant; add or subtract a multiple of one equation to another equation) to reproduce the current pattern of coefficients of x_4 (0, 0, 1, 0) as the new coefficients of x_2. In particular, divide Eq. (2) by 2 to obtain

$$(2) \qquad x_2 + \tfrac{1}{2}x_4 = 6.$$

Next, add 5 times this new Eq. (2) to Eq. (0), and subtract 2 times this new Eq. (2) from Eq. (3). The resulting complete new system of equations is

$$\begin{array}{llllllll}
(0) & Z - 3x_1 & & & + \tfrac{5}{2}x_4 & & = 30 \\
(1) & x_1 & & + x_3 & & & = 4 \\
(2) & & x_2 & & + \tfrac{1}{2}x_4 & & = 6 \\
(3) & 3x_1 & & & - x_4 & + x_5 & = 6.
\end{array}$$

Since $x_1 = 0$ and $x_4 = 0$, the equations in this form immediately yield the new BF solution, $(x_1, x_2, x_3, x_4, x_5) = (0, 6, 4, 0, 6)$, which yields $Z = 30$.

This procedure for obtaining the simultaneous solution of a system of linear equations is called the *Gauss-Jordan method of elimination,* or **Gaussian elimination** for short.[1] The key concept for this method is the use of elementary algebraic operations to reduce the original system of equations to proper form from Gaussian elimination, where each basic variable has been eliminated from all but one equation (*its* equation) and has a coefficient of +1 in that equation.

Optimality Test for the New BF Solution

The current Eq. (0) gives the value of the objective function in terms of just the current nonbasic variables

$$Z = 30 + 3x_1 - \tfrac{5}{2}x_4.$$

Increasing either of these nonbasic variables from zero (while adjusting the values of the basic variables to continue satisfying the system of equations) would result in moving toward one of the two *adjacent* BF solutions. Because x_1 has a *positive* coefficient, increasing x_1 would lead to an adjacent BF solution that is better than the current BF solution, so the current solution is not optimal.

Iteration 2 and the Resulting Optimal Solution

Since $Z = 30 + 3x_1 - \tfrac{5}{2}x_4$, Z can be increased by increasing x_1, but not x_4. Therefore, step 1 chooses x_1 to be the entering basic variable.

For step 2, the current system of equations yields the following conclusions about how far x_1 can be increased (with $x_4 = 0$):

$$x_3 = 4 - x_1 \geq 0 \quad \Rightarrow x_1 \leq \frac{4}{1} = 4.$$

[1] Actually, there are some technical differences between the Gauss-Jordan method of elimination and Gaussian elimination, but we shall not make this distinction.

$$x_2 = 6 \geq 0 \qquad \Rightarrow \text{ no upper bound on } x_1.$$

$$x_5 = 6 - 3x_1 \geq 0 \Rightarrow x_1 \leq \frac{6}{3} = 2 \quad \leftarrow \text{minimum}.$$

Therefore, the minimum ratio test indicates that x_5 is the leaving basic variable.

For step 3, with x_1 replacing x_5 as a basic variable, we perform elementary algebraic operations on the current system of equations to reproduce the current pattern of coefficients of x_5 (0, 0, 0, 1) as the new coefficients of x_1. This yields the following new system of equations:

$$
\begin{array}{llllll}
(0) & Z & & & + \frac{3}{2}x_4 + x_5 & = 36 \\
(1) & & & x_3 & + \frac{1}{3}x_4 - \frac{1}{3}x_5 & = 2 \\
(2) & & x_2 & & + \frac{1}{2}x_4 & = 6 \\
(3) & & x_1 & & - \frac{1}{3}x_4 + \frac{1}{3}x_5 & = 2.
\end{array}
$$

Therefore, the next BF solution is $(x_1, x_2, x_3, x_4, x_5) = (2, 6, 2, 0, 0)$, yielding $Z = 36$. To apply the optimality test to this new BF solution, we use the current Eq. (0) to express Z in terms of just the current nonbasic variables,

$$Z = 36 - \frac{3}{2}x_4 - x_5.$$

Increasing either x_4 or x_5 would *decrease* Z, so neither adjacent BF solution is as good as the current one. Therefore, based on solution concept 6 in Sec. 4.1, the current BF solution must be optimal.

In terms of the original form of the problem (no slack variables), the optimal solution is $x_1 = 2$, $x_2 = 6$, which yields $Z = 36$.

To see another example of applying the simplex method, we recommend that you now view the demonstration entitled *Simplex Method—Algebraic Form* in your OR Courseware. This vivid demonstration simultaneously displays both the algebra and the geometry of the simplex method as it dynamically evolves step by step. Like the many other demonstration examples accompanying other sections of the book (including the next section), this computer demonstration highlights concepts that are difficult to convey on the printed page.

To further help you learn the simplex method efficiently, your OR Courseware includes a procedure entitled *Solve Interactively by the Simplex Method.* This routine performs nearly all the calculations while you make the decisions step by step, thereby enabling you to focus on concepts rather than get bogged down in a lot of number crunching. Therefore, you probably will want to use this routine for your homework on this section. The software will help you get started by letting you know whenever you make a mistake on the first iteration of a problem.

The next section includes a summary of the simplex method for a more convenient tabular form.

4.4 The Simplex Method in Tabular Form

The algebraic form of the simplex method presented in Sec. 4.3 may be the best one for learning the underlying logic of the algorithm. However, it is not the most convenient form for performing the required calculations. When you need to solve a problem by hand (or interactively with your OR Courseware), we recommend the *tabular form* described in this section.[1]

The tabular form of the simplex method records only the essential information, namely, (1) the coefficients of the variables, (2) the constants on the right-hand sides of the equations, and (3) the basic variable appearing in each equation. This saves writing the symbols for the variables in each of the equations, but what is even more important is the fact that it permits highlighting the numbers involved in arithmetic calculations and recording the computations compactly.

Table 4.3 compares the initial system of equations for the Wyndor Glass Co. problem in algebraic form (on the left) and in tabular form (on the right), where the table on the right is called a *simplex tableau.* The basic variable for each equation is shown in bold type on the left and in the first column of the simplex tableau on the right. [Although only the x_j variables are basic or nonbasic, Z plays the role of the basic variable for Eq. (0).] All variables *not* listed in this *basic variable* column (x_1, x_2) automatically are *nonbasic variables.* After we set $x_1 = 0$, $x_2 = 0$, the *right side* column gives the resulting solution for the basic variables, so that the initial BF solution is $(x_1, x_2, x_3, x_4, x_5) = (0, 0, 4, 12, 18)$ which yields $Z = 0$.

> The *tabular form* of the simplex method uses a **simplex tableau** to compactly display the system of equations yielding the current BF solution. For this solution, each variable in the leftmost column equals the corresponding number in the rightmost column (and variables not listed equal zero). When the optimality test or an iteration is performed, the only relevant numbers are those to the right of the Z column. The term **row** refers to just a row of numbers to the right of the Z column (including the *right side* number), where row i corresponds to Eq. (i).

We summarize the tabular form of the simplex method below and, at the same time, briefly describe its application to the Wyndor Glass Co. problem. Keep in mind that the logic is identical to that for the algebraic form presented in the preceding section. Only the form for displaying both the current system of equations and the subsequent iteration has changed (plus we shall no longer bother to bring variables to the right-hand side of an equation before drawing our conclusions in the optimality test or in steps 1 and 2 of an iteration).

Summary of the Simplex Method (and Iteration 1 for the Example)

INITIALIZATION: Introduce slack variables. Select the *decision variables* to be the *initial nonbasic variables* (set equal to zero) and the *slack variables* to be the *initial*

[1] A form more convenient for automatic execution on a computer is presented in Sec. 5.2.

Table 4.3 **Initial System of Equations for the Wyndor Glass Co. Problem**

(a) Algebraic Form	*(b) Tabular Form*								
			Coefficient of:						
	Basic Variable	Eq.	Z	x_1	x_2	x_3	x_4	x_5	Right Side
(0) $\mathbf{Z} - 3x_1 - 5x_2 = 0$	Z	(0)	1	−3	−5	0	0	0	0
(1) $x_1 + \mathbf{x_3} = 4$	x_3	(1)	0	1	0	1	0	0	4
(2) $2x_2 + \mathbf{x_4} = 12$	x_4	(2)	0	0	2	0	1	0	12
(3) $3x_1 + 2x_2 + \mathbf{x_5} = 18$	x_5	(3)	0	3	2	0	0	1	18

basic variables. (See Sec. 4.6 for the necessary adjustments if the model is not in our standard form—maximization, only $\leq$ functional constraints, and all nonnegativity constraints—or if any b_i values are negative.)

For the Example: This selection yields the initial simplex tableau shown in Table 4.3*b*, so the initial BF solution is (0, 0, 4, 12, 18).

Optimality Test: The current BF solution is optimal if and only if *every* coefficient in row 0 is nonnegative (≥ 0). If it is, stop; otherwise, go to an iteration to obtain the next BF solution, which involves changing one nonbasic variable to a basic variable (step 1) and vice versa (step 2) and then solving for the new solution (step 3).

For the Example: Just as $Z = 3x_1 + 5x_2$ indicates that increasing either x_1 or x_2 will increase Z, so the current BF solution is not optimal, the same conclusion is drawn from the equation $Z - 3x_1 - 5x_2 = 0$. These coefficients of -3 and -5 are shown in row 0 of Table 4.3*b*.

Iteration

Step 1:

Determine the *entering basic variable* by selecting the variable (automatically a nonbasic variable) with the *negative coefficient* having the largest absolute value (i.e., the "most negative" coefficient) in Eq. (0). Put a box around the column below this coefficient, and call this the **pivot column.**

For the Example: The most negative coefficient is -5 for x_2 ($5 > 3$), so x_2 is to be changed to a basic variable. (This change is indicated in Table 4.4 by the box around the x_2 column below -5.)

Step 2:

Determine the leaving basic variable by applying the minimum ratio test.

Minimum Ratio Test

1. Pick out each coefficient in the pivot column that is strictly positive (> 0).
2. Divide each of these coefficients into the *right side* entry for the same row.
3. Identify the row that has the *smallest* of these ratios.
4. The basic variable for that row is the leaving basic variable, so replace that variable by the entering basic variable in the basic variable column of the next simplex tableau.

Put a box around this row and call it the **pivot row.** Also call the number that is in *both* boxes the **pivot number.**

Table 4.4 **Applying the Minimum Ratio Test to Determine the First Leaving Basic Variable for the Wyndor Glass Co. Problem**

Basic Variable	Eq.	Coefficient of:						Right Side	Ratio
		Z	x_1	x_2	x_3	x_4	x_5		
Z	(0)	1	-3	-5	0	0	0	0	
x_3	(1)	0	1	0	1	0	0	4	
x_4	(2)	0	0	2	0	1	0	12	$\rightarrow \frac{12}{2} = 6 \leftarrow$ minimum
x_5	(3)	0	3	2	0	0	1	18	$\rightarrow \frac{18}{2} = 9$

Table 4.5 **Simplex Tableaux for the Wyndor Glass Co. Problem after the First Pivot Row Is Divided by the First Pivot Number**

Iteration	Basic Variable	Eq.	Coefficient of: Z	x_1	x_2	x_3	x_4	x_5	Right Side
0	Z	(0)	1	−3	−5	0	0	0	0
	x_3	(1)	0	1	0	1	0	0	4
	x_4	(2)	0	0	2	0	1	0	12
	x_5	(3)	0	3	2	0	0	1	18
1	Z	(0)	1						
	x_3	(1)	0						
	x_2	(2)	0	0	1	0	$\frac{1}{2}$	0	6
	x_5	(3)	0						

For the Example: The calculations for the minimum ratio test are shown at the right of Table 4.4. Thus, row 2 is the pivot row (see the box around this row in the first simplex tableau of Table 4.5), and x_4 is the leaving basic variable. In the next simplex tableau (see Table 4.5), x_2 replaces x_4 as the basic variable for row 2.

Step 3:

Solve for the *new BF solution* by using **elementary row operations** (multiply or divide a row by a nonzero constant; add or subtract a multiple of one row to another row) to construct a new simplex tableau in proper form from Gaussian elimination below the current one, and then return to the optimality test.

For the Example: Since x_2 is replacing x_4 as a basic variable, we need to reproduce the first tableau's pattern of coefficients in the column of x_4 (0, 0, 1, 0) in the second tableau's column of x_2. To start, divide the pivot row (row 2) by the pivot number (2), which gives the new row 2 shown in Table 4.5. Next, we add to row 0 the product, 5 times the new row 2. Then we subtract from row 3 the product, 2 times the new row 2 (or equivalently, subtract from row 3 the *old* row 2). These calculations yield the new tableau shown in Table 4.6 for iteration 1. Thus, the new BF solution is (0, 6, 4, 0, 6), with $Z = 30$. We next return to the optimality test to check if the new BF solution is optimal. Since the new row 0 still has a negative coefficient (-3 for x_1), the solution is not optimal, and so at least one more iteration is needed.

Table 4.6 **First Two Simplex Tableaux for the Wyndor Glass Co. Problem**

Iteration	Basic Variable	Eq.	Coefficient of: Z	x_1	x_2	x_3	x_4	x_5	Right Side
0	Z	(0)	1	−3	−5	0	0	0	0
	x_3	(1)	0	1	0	1	0	0	4
	x_4	(2)	0	0	2	0	1	0	12
	x_5	(3)	0	3	2	0	0	1	18
1	Z	(0)	1	−3	0	0	$\frac{5}{2}$	0	30
	x_3	(1)	0	1	0	1	0	0	4
	x_2	(2)	0	0	1	0	$\frac{1}{2}$	0	6
	x_5	(3)	0	3	0	0	−1	1	6

Table 4.7 **Steps 1 and 2 of Iteration 2 for the Wyndor Glass Co. Problem**

Iteration	Basic Variable	Eq.	Z	Coefficient of: x_1	x_2	x_3	x_4	x_5	Right Side	Ratio
1	Z	(0)	1	−3	0	0	$\frac{5}{2}$	0	30	
	x_3	(1)	0	1	0	1	0	0	4	$\frac{4}{1} = 4$
	x_2	(2)	0	0	1	0	$\frac{1}{2}$	0	6	
	x_5	(3)	0	3	0	0	−1	1	6	$\frac{6}{3} = 2$ ←minimum

Iteration 2 for the Example and the Resulting Optimal Solution

The second iteration starts anew from the second tableau of Table 4.6 to find the next BF solution. Following the instructions for steps 1 and 2, we find x_1 as the entering basic variable and x_5 as the leaving basic variable, as shown in Table 4.7.

For step 3, we start by dividing the pivot row (row 3) in Table 4.7 by the pivot number (3). Next, we add to row 0 the product, 3 times the new row 3. Then we subtract the new row 3 from row 1.

We now have the set of tableaux shown in Table 4.8. Therefore, the new BF solution is (2, 6, 2, 0, 0), with $Z = 36$. Going to the optimality test, we find that this solution is *optimal* because none of the coefficients in row 0 is negative, so the algorithm is finished. Consequently, the optimal solution for the Wyndor Glass Co. problem (before slack variables are introduced) is $x_1 = 2$, $x_2 = 6$.

Now compare Table 4.8 with the work done in Sec. 4.3 to verify that these two forms of the simplex method really are *equivalent.* Then note how the algebraic form is superior for learning the logic behind the simplex method, but the tabular form organizes the work being done in a considerably more convenient and compact form. We generally use the tabular form from now on.

An additional example of applying the simplex method in tabular form is available to you in the OR Courseware. See the demonstration entitled *Simplex Method—Tabular Form.*

Table 4.8 **Complete Set of Simplex Tableaux for the Wyndor Glass Co. Problem**

Iteration	Basic Variable	Eq.	Z	Coefficient of: x_1	x_2	x_3	x_4	x_5	Right Side
0	Z	(0)	1	−3	−5	0	0	0	0
	x_3	(1)	0	1	0	1	0	0	4
	x_4	(2)	0	0	2	0	1	0	12
	x_5	(3)	0	3	2	0	0	1	18
1	Z	(0)	1	−3	0	0	$\frac{5}{2}$	0	30
	x_3	(1)	0	1	0	1	0	0	4
	x_2	(2)	0	0	1	0	$\frac{1}{2}$	0	6
	x_5	(3)	0	3	0	0	−1	1	6
2	Z	(0)	1	0	0	0	$\frac{3}{2}$	1	36
	x_3	(1)	0	0	0	1	$\frac{1}{3}$	$-\frac{1}{3}$	2
	x_2	(2)	0	0	1	0	$\frac{1}{2}$	0	6
	x_1	(3)	0	1	0	0	$-\frac{1}{3}$	$\frac{1}{3}$	2

4.5 Tie Breaking in the Simplex Method

You may have noticed in the preceding two sections that we never said what to do if the various choice rules of the simplex method do not lead to a clear-cut decision, because of either ties or other similar ambiguities. We discuss these details now.

Tie for the Entering Basic Variable

Step 1 of each iteration chooses the nonbasic variable having the *negative* coefficient with the *largest absolute value* in the current Eq. (0) as the entering basic variable. Now suppose that two or more nonbasic variables are tied for having the largest negative coefficient (in absolute terms). For example, this would occur in the first iteration for the Wyndor Glass Co. problem if its objective function were changed to $Z = 3x_1 + 3x_2$, so that the initial Eq. (0) became $Z - 3x_1 - 3x_2 = 0$. How should this tie be broken?

The answer is that the selection between these contenders may be made *arbitrarily*. The optimal solution will be reached eventually, regardless of the tied variable chosen, and there is no convenient method for predicting in advance which choice will lead there sooner. In this example, the simplex method happens to reach the optimal solution (2, 6) in three iterations with x_1 as the initial entering basic variable, versus two iterations if x_2 is chosen.

Tie for the Leaving Basic Variable—Degeneracy

Now suppose that two or more basic variables tie for being the leaving basic variable in step 2 of an iteration. Does it matter which one is chosen? Theoretically it does, and in a very critical way, because of the following sequence of events that could occur. First, all the tied basic variables reach zero simultaneously as the entering basic variable is increased. Therefore, the one or ones *not* chosen to be the leaving basic variable also will have a value of zero in the new BF solution. (Note that basic variables with a value of *zero* are called **degenerate,** and the same term is applied to the corresponding BF solution.) Second, if one of these degenerate basic variables retains its value of zero until it is chosen at a subsequent iteration to be a leaving basic variable, the corresponding entering basic variable also must remain zero (since it cannot be increased without making the leaving basic variable negative), so the value of Z must remain unchanged. Third, if Z may remain the same rather than increase at each iteration, the simplex method may then go around in a loop, repeating the same sequence of solutions periodically rather than eventually increasing Z toward an optimal solution. In fact, examples have been artificially constructed so that they do become entrapped in just such a perpetual loop.

Fortunately, although a perpetual loop is theoretically possible, it has rarely been known to occur in practical problems. If a loop were to occur, one could always get out of it by changing the choice of the leaving basic variable. Furthermore, special rules[1] have been constructed for breaking ties so that such loops are always avoided. How-

[1] See R. Bland, "New Finite Pivoting Rules for the Simplex Method," *Mathematics of Operations Research,* **2:** 103–107, 1977.

ever, these rules frequently are ignored in actual application, and they will not be repeated here. For your purposes, just break this kind of tie arbitrarily and proceed without worrying about the degenerate basic variables that result.

No Leaving Basic Variable—Unbounded Z

In step 2 of an iteration there is one other possible outcome that we have not yet discussed, namely, that *no* variable qualifies to be the leaving basic variable.[1] This outcome would occur if the entering basic variable could be increased *indefinitely* without giving negative values to *any* of the current basic variables. In tabular form, this means that *every* coefficient in the pivot column (excluding row 0) is either negative or zero.

As illustrated in Table 4.9, this situation arises in the example displayed in Fig. 3.6 on p. 37. In this example, the last two functional constraints of the Wyndor Glass Co. problem have been overlooked and so are not included in the model. Note in Fig. 3.6 how x_2 can be increased indefinitely (thereby increasing Z indefinitely) without ever leaving the feasible region. Then note in Table 4.9 that x_2 is the entering basic variable but the only coefficient in the pivot column is zero. Because the minimum ratio test uses only coefficients that are greater than zero, there is no ratio to provide a leaving basic variable.

The interpretation of a tableau like the one shown in Table 4.9 is that the constraints do not prevent the value of the objective function Z increasing indefinitely, so the simplex method would stop with the message that Z is *unbounded.* Because even linear programming has not discovered a way of making infinite profits, the real message for practical problems is that a mistake has been made! The model probably has been misformulated, either by omitting relevant constraints or by stating them incorrectly. Alternatively, a computational mistake may have occurred.

Multiple Optimal Solutions

We mentioned in Sec. 3.2 (under the definition of **optimal solution**) that a problem can have more than one optimal solution. This fact was illustrated in Fig. 3.5 by changing the objective function in the Wyndor Glass Co. problem to $Z = 3x_1 + 2x_2$, so that every point on the line segment between (2, 6) and (4, 3) is optimal. Thus, all optimal solutions are a *weighted average* of these two optimal CPF solutions

$$(x_1, x_2) = w_1(2, 6) + w_2(4, 3),$$

where the weights w_1 and w_2 are numbers that satisfy the relationships

[1] Note that the analogous case (no *entering* basic variable) cannot occur in step 1 of an iteration, because the optimality test would stop the algorithm first by indicating that an optimal solution had been reached.

Table 4.9 **Initial Simplex Tableau for the Wyndor Glass Co. Problem without the Last Two Functional Constraints**

Basic Variable	Eq.	Coefficient of:				Right Side	Ratio
		Z	x_1	x_2	x_3		
Z	(0)	1	−3	−5	0	0	
x_3	(1)	0	1	0	1	4	None

With $x_1 = 0$ and x_2 increasing, $x_3 = 4 - 1x_1 - 0x_2 = 4 > 0$.

$$w_1 + w_2 = 1 \qquad \text{and} \qquad w_1 \geq 0, \qquad w_2 \geq 0.$$

For example, $w_1 = \frac{1}{3}$ and $w_2 = \frac{2}{3}$ give

$$(x_1, x_2) = \tfrac{1}{3}(2, 6) + \tfrac{2}{3}(4, 3) = (\tfrac{2}{3} + \tfrac{8}{3}, \tfrac{6}{3} + \tfrac{6}{3}) = (\tfrac{10}{3}, 4)$$

as one optimal solution.

In general, any weighted average of two or more solutions (vectors) where the weights are nonnegative and sum to 1 is called a **convex combination** of these solutions. Thus, every optimal solution in the example is a convex combination of (2, 6) and (4, 3).

This example is typical of problems with multiple optimal solutions.

> As indicated at the end of Sec. 3.2, *any* linear programming problem with multiple optimal solutions (and a bounded feasible region) has at least two CPF solutions that are optimal. *Every* optimal solution is a convex combination of these optimal CPF solutions. Consequently, in augmented form, every optimal solution is a convex combination of the optimal BF solutions.

(Problems 4.5-3 and 4.5-4 guide you through the reasoning behind this conclusion.)

The simplex method automatically stops after *one* optimal BF solution is found. However, for many applications of linear programming, there are intangible factors not incorporated into the model that can be used to make meaningful choices between alternative optimal solutions. In such cases, these other optimal solutions should be identified as well. As indicated above, this requires finding all the other optimal BF solutions, and then every optimal solution is a convex combination of the optimal BF solutions.

After the simplex method finds one optimal BF solution, you can detect if there are any others and, if so, find them as follows:

> Whenever a problem has more than one optimal BF solution, at least one of the nonbasic variables has a coefficient of zero in the final row 0, so increasing any such variable will not change the value of Z. Therefore, these other optimal BF solutions can be identified (if desired) by performing additional iterations of the simplex method, each time choosing a nonbasic variable with a zero coefficient as the entering basic variable.[1]

To illustrate, consider again the case just mentioned, where the objective function in the Wyndor Glass Co. problem is changed to $Z = 3x_1 + 2x_2$. The simplex method obtains the first three tableaux shown in Table 4.10 and stops with an optimal BF solution. However, because a nonbasic variable (x_3) then has a zero coefficient in row 0, we perform one more iteration in Table 4.10 to identify the other optimal BF solution. Thus the two optimal BF solutions are (4, 3, 0, 6, 0) and (2, 6, 2, 0, 0), each yielding $Z = 18$. Notice that the last tableau also has a *nonbasic* variable (x_4) with a zero coefficient in row 0. This situation is inevitable because the extra iteration does not change row 0, so this leaving basic variable necessarily retains its zero coefficient. Making x_4 an entering basic variable now would only lead back to the third tableau. (Check this.) Therefore, these two are the only BF solutions that are optimal, and all *other* optimal solutions are a convex combination of these two.

$$(x_1, x_2, x_3, x_4, x_5) = w_1(2, 6, 2, 0, 0) + w_2(4, 3, 0, 6, 0),$$

$$w_1 + w_2 = 1, \qquad w_1 \geq 0, \qquad w_2 \geq 0.$$

[1] If such an iteration has no *leaving* basic variable, this indicates that the feasible region is unbounded and the entering basic variable can be increased indefinitely without changing the value of Z.

Table 4.10 **Complete Set of Simplex Tableaux to Obtain All Optimal BF Solutions for the Wyndor Glass Co. Problem with $c_2 = 2$**

Iteration	Basic Variable	Eq.	Coefficient of: Z	x_1	x_2	x_3	x_4	x_5	Right Side	Solution Optimal?
0	Z	(0)	1	−3	−2	0	0	0	0	No
	x_3	(1)	0	1	0	1	0	0	4	
	x_4	(2)	0	0	2	0	1	0	12	
	x_5	(3)	0	3	2	0	0	1	18	
1	Z	(0)	1	0	−2	3	0	0	12	No
	x_1	(1)	0	1	0	1	0	0	4	
	x_4	(2)	0	0	2	0	1	0	12	
	x_5	(3)	0	0	2	−3	0	1	6	
2	Z	(0)	1	0	0	**0**	0	1	18	Yes
	x_1	(1)	0	1	0	1	0	0	4	
	x_4	(2)	0	0	0	3	1	−1	6	
	x_2	(3)	0	0	1	$-\frac{3}{2}$	0	$\frac{1}{2}$	3	
Extra	Z	(0)	1	0	0	0	**0**	1	18	Yes
	x_1	(1)	0	1	0	0	$-\frac{1}{3}$	$\frac{1}{3}$	2	
	x_3	(2)	0	0	0	1	$\frac{1}{3}$	$-\frac{1}{3}$	2	
	x_2	(3)	0	0	1	0	$\frac{1}{2}$	0	6	

4.6 Adapting to Other Model Forms

Thus far we have presented the details of the simplex method under the assumptions that the problem is in our standard form (maximize Z subject to functional constraints in $\leq$ form and nonnegativity constraints on all variables) and that $b_i \geq 0$ for all $i = 1, 2, \ldots, m$. In this section we point out how to make the adjustments required for other legitimate forms of the linear programming model. You will see that all these adjustments can be made during the initialization, so the rest of the simplex method can then be applied just as you have learned it already.

The only serious problem introduced by the other forms for functional constraints (the $=$ or $\geq$ forms, or having a negative right-hand side) lies in identifying an *initial BF solution.* Before, this initial solution was found very conveniently by letting the slack variables be the initial basic variables, so that each one just equals the *nonnegative* right-hand side of its equation. Now, something else must be done. The standard approach that is used for all these cases is the **artificial-variable technique.** This technique constructs a more convenient *artificial problem* by introducing a dummy variable (called an *artificial variable*) into each constraint that needs one. This new variable is introduced just for the purpose of being the initial basic variable for that equation. The usual nonnegativity constraints are placed on these variables, and the objective function also is modified to impose an exorbitant penalty on their having values larger than zero. The iterations of the simplex method then automatically force the artificial variables to disappear (become zero), one at a time, until they are all gone, after which the *real* problem is solved.

To illustrate the artificial-variable technique, first we consider the case where the only nonstandard form in the problem is the presence of one or more equality constraints.

Equality Constraints

Any equality constraint

$$a_{i1}x_1 + a_{i2}x_2 + \cdots + a_{in}x_n = b_i$$

actually is equivalent to a pair of inequality constraints:

$$a_{i1}x_1 + a_{i2}x_2 + \cdots + a_{in}x_n \le b_i$$

$$a_{i1}x_1 + a_{i2}x_2 + \cdots + a_{in}x_n \ge b_i.$$

However, rather than making this substitution and thereby increasing the number of constraints, it is more convenient to use the artificial-variable technique described next.

Suppose that the Wyndor Glass Co. problem in Sec. 3.1 is modified to *require* that Plant 3 be used at full capacity. The only resulting change in the linear programming model is that the third constraint, $3x_1 + 2x_2 \le 18$, instead becomes an equality constraint

$$3x_1 + 2x_2 = 18,$$

so that the complete model becomes the one shown in the upper right-hand corner of Fig. 4.3 on page 106. This figure also shows in darker ink the feasible region which now consists of *just* the line segment connecting (2, 6) and (4, 3).

After the slack variables still needed for the inequality constraints are introduced, the system of equations for the augmented form of the problem becomes

$$\begin{array}{llr}
(0) & Z - 3x_1 - 5x_2 & = 0 \\
(1) & x_1 \qquad\quad + x_3 & = 4 \\
(2) & 2x_2 \qquad + x_4 & = 12 \\
(3) & 3x_1 + 2x_2 & = 18.
\end{array}$$

Unfortunately, these equations do not have an obvious initial BF solution because there is no longer a slack variable to use as the initial basic variable for Eq. (3). It is necessary to find an initial BF solution to start the simplex method.

This difficulty can be circumvented in the following way.

Construct an **artificial problem** that has the same optimal solution as the real problem by making two modifications of the real problem.

1. Apply the **artificial-variable technique** by introducing a *nonnegative* **artificial variable** (call it $\bar{x}_5$)[1] into Eq. (3), just as if it were a slack variable

$$(3) \qquad 3x_1 + 2x_2 + \bar{x}_5 = 18.$$

[1] We shall always label the artificial variables by putting a bar over them.

2. Assign an *overwhelming penalty* to having $\bar{x}_5 > 0$ by changing the objective function $Z = 3x_1 + 5x_2$ to

$$Z = 3x_1 + 5x_2 - M\bar{x}_5,$$

where M symbolically represents a *huge* positive number. (This method of forcing $\bar{x}_5$ to be $\bar{x}_5 = 0$ in the optimal solution is called the **Big *M* method**.)

Now find the optimal solution for the real problem by applying the simplex method to the artificial problem, starting with the following initial BF solution:

Initial BF Solution

Nonbasic variables: $x_1 = 0, \quad x_2 = 0$

Basic variables: $x_3 = 4, \quad x_4 = 12, \quad \bar{x}_5 = 18.$

Because $\bar{x}_5$ plays the role of the slack variable for the third constraint in the artificial problem, this constraint is equivalent to $3x_1 + 2x_2 \le 18$ (just as for the original Wyndor Glass Co. problem in Sec. 3.1). We show below the resulting artificial problem (before augmenting) next to the real problem.

The Real Problem

Maximize $Z = 3x_1 + 5x_2,$

subject to

$$x_1 \le 4$$
$$2x_2 \le 12$$
$$3x_1 + 2x_2 = 18$$

and

$$x_1 \ge 0, \quad x_2 \ge 0.$$

The Artificial Problem

Define $\bar{x}_5 = 18 - 3x_1 - 2x_2.$

Maximize $Z = 3x_1 + 5x_2 - M\bar{x}_5,$

subject to

$$x_1 \le 4$$
$$2x_2 \le 12$$
$$3x_1 + 2x_2 \le 18$$
$$(\text{so} \quad 3x_1 + 2x_2 + \bar{x}_5 = 18)$$

and

$$x_1 \ge 0, \quad x_2 \ge 0, \quad \bar{x}_5 \ge 0.$$

Therefore, just as in Sec. 3.1, the feasible region for (x_1, x_2) for the artificial problem is the one shown in Fig. 4.4 on page 107. The only portion of this feasible region that coincides with the feasible region for the real problem is where $\bar{x}_5 = 0$ (so $3x_1 + 2x_2 = 18$).

Figure 4.4 also shows the order in which the simplex method examines the CPF solutions (or BF solutions after augmenting), where each circled number identifies which iteration obtained that solution. Note that the simplex method moves counterclockwise here whereas it moved clockwise for the original Wyndor Glass Co. problem (see Fig. 4.2). The reason for this difference is the extra term $-M\bar{x}_5$ in the objective function for the artificial problem.

In particular, the system of equations after the artificial problem is augmented is

$$
\begin{array}{llr}
(0) & Z - 3x_1 - 5x_2 \qquad\qquad + M\bar{\mathbf{x}}_5 = & 0 \\
(1) & x_1 \qquad + \mathbf{x}_3 \qquad\qquad = & 4 \\
(2) & 2x_2 \qquad + \mathbf{x}_4 \qquad = & 12 \\
(3) & 3x_1 + 2x_2 \qquad\qquad + \quad \bar{\mathbf{x}}_5 = & 18
\end{array}
$$

where the initial basic variables $(x_3, x_4, \bar{x}_5)$ are shown in bold type. This system is not yet in proper form from Gaussian elimination because $\bar{x}_5$ has a nonzero coefficient in

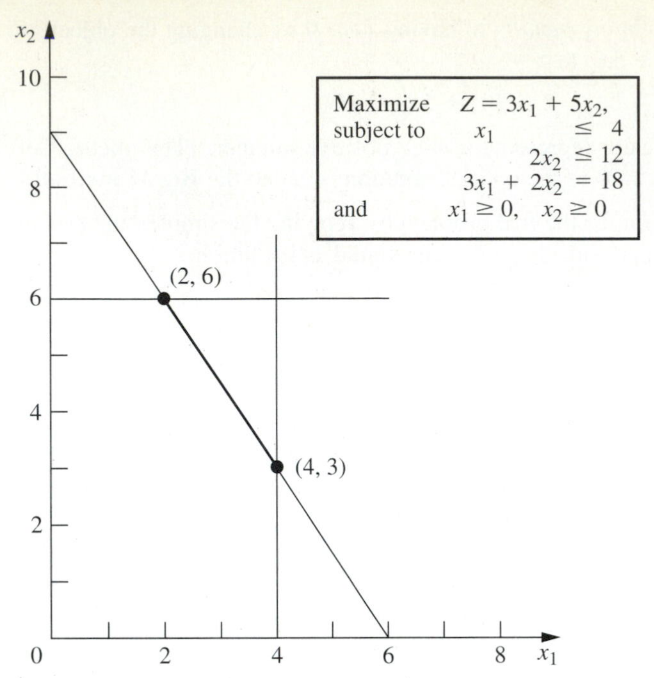

Figure 4.3 When the third functional constraint becomes an equality constraint, the feasible region for the Wyndor Glass Co. problem becomes the line segment between (2, 6) and (4, 3).

Eq. (0). Therefore, before the simplex method can apply the optimality test and find the entering basic variable, we must obtain this form by subtracting from Eq. (0) the product, M times Eq. (3).

$$\begin{array}{lr} & Z - 3x_1 - 5x_2 + M\bar{x}_5 = 0 \\ & -M(3x_1 + 2x_2 + \quad \bar{x}_5 = 18) \\ \hline \text{New (0)} & Z - (3M + 3)x_1 - (2M + 5)x_2 = -18M. \end{array}$$

This new Eq. (0) gives Z in terms of *just* the nonbasic variables (x_1, x_2),

$$Z = -18M + (3M + 3)x_1 + (2M + 5)x_2.$$

Since $3M + 3 > 2M + 5$ (remember that M represents a huge number), increasing x_1 increases Z at a faster rate than increasing x_2 does, so x_1 is chosen as the entering basic variable. This leads to the move from (0, 0) to (4, 0) at iteration 1, shown in Fig. 4.4, thereby increasing Z by $4(3M + 3)$.

The quantities involving M never appear in the system of equations except for Eq. (0), so they need to be taken into account only in the optimality test and when an entering basic variable is determined. One way of dealing with these quantities is to assign some particular (huge) numerical value to M and use the resulting coefficients in Eq. (0) in the usual way. However, this approach may result in significant rounding errors that invalidate the optimality test. Therefore, it is better to do what we have just shown, namely, to express each coefficient in Eq. (0) as a linear function $aM + b$ of the *symbolic* quantity M by separately recording and updating the current numerical value of (1) the *multiplicative* factor a and (2) the *additive* term b. Because M is assumed to be so large that b always is negligible compared with M when $a \neq 0$, the decisions in

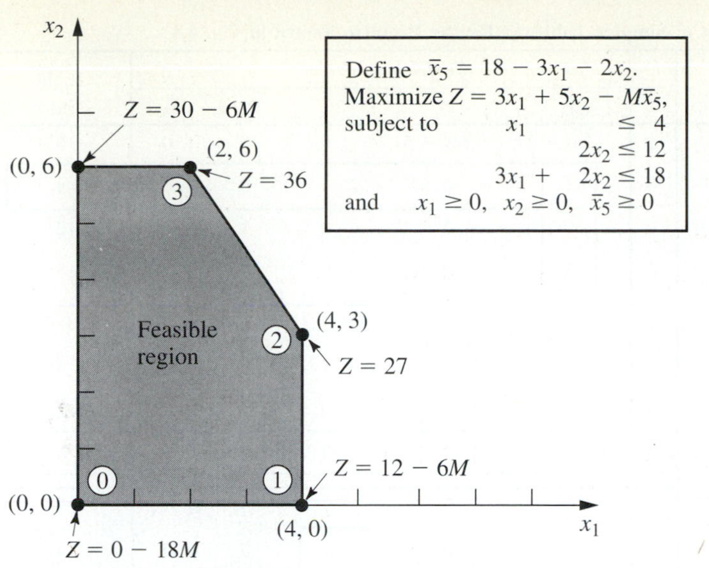

Figure 4.4 This graph shows the feasible region and the sequence of CPF solutions (⓪,①,②,③) examined by the simplex method for the artificial problem that corresponds to the real problem of Fig. 4.3.

the optimality test and the choice of the entering basic variable are made by using just the *multiplicative* factors in the usual way, except for breaking ties with the *additive* factors.

Using this approach on the example yields the simplex tableaux shown in Table 4.11. Note that the artificial variable $\bar{x}_5$ is a *basic variable* ($\bar{x}_5 > 0$) in the first two tableaux and a *nonbasic variable* ($\bar{x}_5 = 0$) in the last two. Therefore, the first two BF solutions for this artificial problem are *infeasible* for the real problem whereas the last two also are BF solutions for the real problem.

This example involved only one equality constraint. If a linear programming model has more than one, each is handled in just the same way. (If the right-hand side is negative, multiply through both sides by -1 first.)

Negative Right-Hand Sides

The technique mentioned in the preceding sentence for dealing with an equality constraint with a negative right-hand side (namely, multiply through both sides by -1) also works for any inequality constraint with a negative right-hand side. Multiplying through both sides of an inequality by -1 also reverses the direction of the inequality; i.e., $\leq$ changes to $\geq$ or vice versa. For example, doing this to the constraint

$$x_1 - x_2 \leq -1 \qquad \text{(that is, } x_1 \leq x_2 - 1\text{)}$$

gives the equivalent constraint

$$-x_1 + x_2 \geq 1 \qquad \text{(that is, } x_2 - 1 \geq x_1\text{)}$$

but now the right-hand side is positive. Having nonnegative right-hand sides for all the functional constraints enables the simplex method to begin, because (after augmenting)

Table 4.11 **Complete Set of Simplex Tableaux for the Problem Shown in Fig. 4.4**

Iteration	Basic Variable	Eq.	Coefficient of: Z	x_1	x_2	x_3	x_4	$\bar{x}_5$	Right Side
0	Z	(0)	1	$-3M-3$	$-2M-5$	0	0	0	$-18M$
	x_3	(1)	0	1	0	1	0	0	4
	x_4	(2)	0	0	2	0	1	0	12
	$\bar{x}_5$	(3)	0	3	2	0	0	1	18
1	Z	(0)	1	0	$-2M-5$	$3M+3$	0	0	$-6M+12$
	x_1	(1)	0	1	0	1	0	0	4
	x_4	(2)	0	0	2	0	1	0	12
	$\bar{x}_5$	(3)	0	0	2	-3	0	1	6
2	Z	(0)	1	0	0	$-\frac{9}{2}$	0	$M+\frac{5}{2}$	27
	x_1	(1)	0	1	0	1	0	0	4
	x_4	(2)	0	0	0	3	1	-1	6
	x_2	(3)	0	0	1	$-\frac{3}{2}$	0	$\frac{1}{2}$	3
Extra	Z	(0)	1	0	0	0	$\frac{3}{2}$	$M+1$	36
	x_1	(1)	0	1	0	0	$-\frac{1}{3}$	$\frac{1}{3}$	2
	x_3	(2)	0	0	0	1	$\frac{1}{3}$	$-\frac{1}{3}$	2
	x_2	(3)	0	0	1	0	$\frac{1}{2}$	0	6

these right-hand sides become the respective values of the *initial basic variables,* which must satisfy nonnegativity constraints.

We next focus on how to augment $\geq$ constraints, such as $-x_1 + x_2 \geq 1$, with the help of the artificial-variable technique.

Functional Constraints in ≥ Form

To illustrate how the artificial-variable technique deals with functional constraints in $\geq$ form, we will use the model for designing Mary's radiation therapy, as presented in Sec. 3.4. For your convenience, this model is repeated below, where we have placed a box around the constraint of special interest here.

Radiation Therapy Example

Minimize $\quad Z = 0.4x_1 + 0.5x_2,$

subject to

$$0.3x_1 + 0.1x_2 \leq 2.7$$
$$0.5x_1 + 0.5x_2 = 6$$
$$\boxed{0.6x_1 + 0.4x_2 \geq 6}$$

and

$$x_1 \geq 0, \qquad x_2 \geq 0.$$

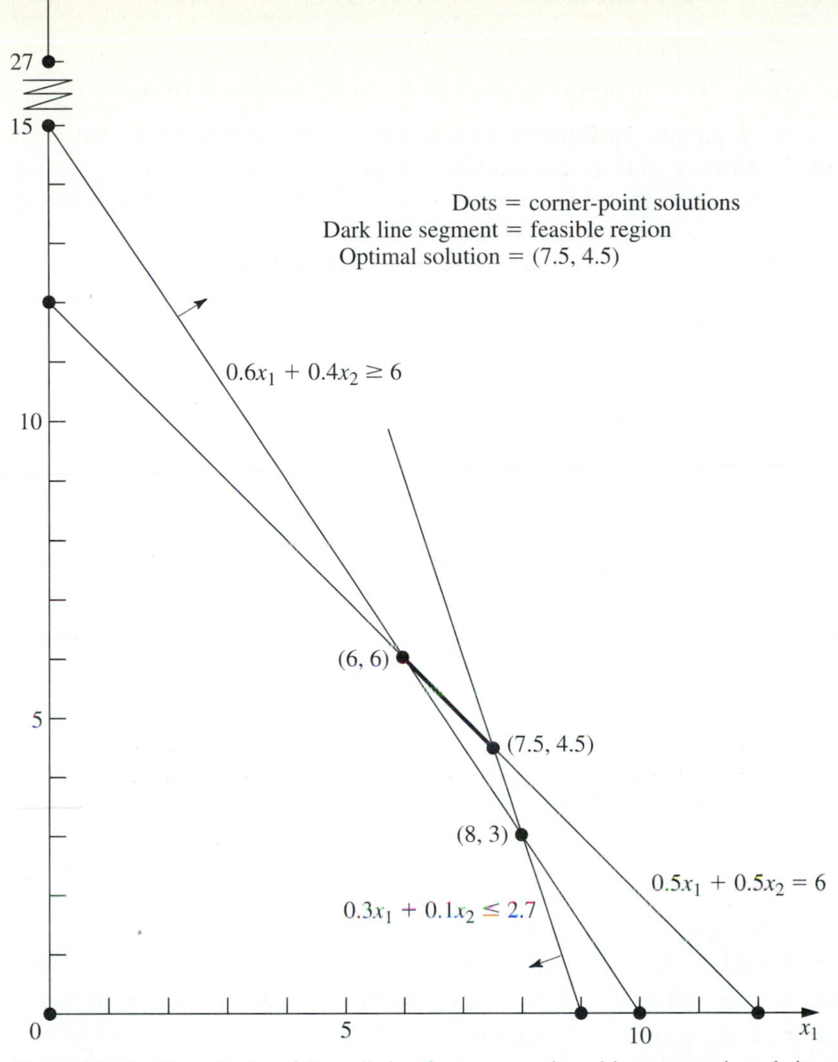

Figure 4.5 Graphical display of the radiation therapy example and its corner-point solutions.

The graphical solution for this example (originally presented in Fig. 3.12) is repeated here in a slightly different form in Fig. 4.5. The three lines in the figure, along with the two axes, constitute the five constraint boundaries of the problem. The dots lying at the intersection of a pair of constraint boundaries are the *corner-point solutions.* The only two corner-point *feasible* solutions are (6, 6) and (7.5, 4.5), and the feasible region is the line segment connecting these two points. The optimal solution is $(x_1, x_2) = (7.5, 4.5)$, with $Z = 5.25$.

We soon will show how the simplex method solves this problem by directly solving the corresponding artificial problem. However, first we must describe how to deal with the third constraint.

Our approach involves introducing *both* a surplus variable x_5 (defined as $x_5 = 0.6x_1 + 0.4x_2 - 6$) and an artificial variable $\bar{x}_6$, as shown next.

$$0.6x_1 + 0.4x_2 \geq 6$$
$$\rightarrow \quad 0.6x_1 + 0.4x_2 - x_5 = 6 \qquad (x_5 \geq 0)$$
$$\rightarrow \quad 0.6x_1 + 0.4x_2 - x_5 + \bar{x}_6 = 6 \qquad (x_5 \geq 0, \bar{x}_6 \geq 0).$$

Here x_5 is called a **surplus variable** because it subtracts the surplus of the left-hand side over the right-hand side to convert the inequality constraint to an equivalent equality constraint. Once this conversion is accomplished, the artificial variable is introduced just as for any equality constraint.

After a slack variable x_3 is introduced into the first constraint, an artificial variable $\bar{x}_4$ is introduced into the second constraint, and the Big M method is applied, the complete artificial problem (in augmented form) is

$$\text{Minimize} \quad Z = 0.4x_1 + 0.5x_2 + M\bar{x}_4 + M\bar{x}_6,$$

subject to

$$0.3x_1 + 0.1x_2 + x_3 = 2.7$$
$$0.5x_1 + 0.5x_2 + \bar{x}_4 = 6$$
$$0.6x_1 + 0.4x_2 - x_5 + \bar{x}_6 = 6$$

and $\quad x_1 \geq 0, \quad x_2 \geq 0, \quad x_3 \geq 0, \quad \bar{x}_4 \geq 0, \quad x_5 \geq 0, \quad \bar{x}_6 \geq 0.$

Note that the coefficients of the artificial variables in the objective function are $+M$, instead of $-M$, because we now are minimizing Z. Thus, even though $\bar{x}_4 > 0$ and/or $\bar{x}_6 > 0$ is possible for a feasible solution for the artificial problem, the huge unit penalty of $+M$ prevents this from occurring in an optimal solution.

As usual, introducing artificial variables enlarges the feasible region. Compare below the original constraints for the real problem with the corresponding constraints on (x_1, x_2) for the artificial problem.

Constraints on (x_1, x_2) *for the Real Problem*	*Constraints on* (x_1, x_2) *for the Artificial Problem*
$0.3x_1 + 0.1x_2 \leq 2.7$	$0.3x_1 + 0.1x_2 \leq 2.7$
$0.5x_1 + 0.5x_2 = 6$	$0.5x_1 + 0.5x_2 \leq 6$ (= holds when $\bar{x}_4 = 0$)
$0.6x_1 + 0.4x_2 \geq 6$	No such constraint (except when $\bar{x}_6 = 0$)
$x_1 \geq 0, \quad x_2 \geq 0$	$x_1 \geq 0, \quad x_2 \geq 0$

Introducing the artificial variable $\bar{x}_4$ to play the role of a slack variable in the second constraint allows values of (x_1, x_2) *below* the $0.5x_1 + 0.5x_2 = 6$ line in Fig. 4.5. Introducing x_5 and $\bar{x}_6$ into the third constraint of the real problem (and moving these variables to the right-hand side) yields the equation

$$0.6x_1 + 0.4x_2 = 6 + x_5 - \bar{x}_6.$$

Because both x_5 and $\bar{x}_6$ are constrained only to be nonnegative, their difference $x_5 - \bar{x}_6$ can be any positive or negative number. Therefore, $0.6x_1 + 0.4x_2$ can have any value, which has the effect of eliminating the third constraint from the artificial problem and allowing points on either side of the $0.6x_1 + 0.4x_2 = 6$ line in Fig. 4.5. (We keep the third constraint in the system of equations only because it will become relevant again later, after the Big M method forces $\bar{x}_6$ to be zero.) Consequently, the feasible region for the artificial problem is the entire polyhedron in Fig. 4.5 whose vertices are (0, 0), (9, 0), (7.5, 4.5), and (0, 12).

Since the origin now is feasible for the artificial problem, the simplex method starts with (0, 0) as the initial CPF solution, i.e., with $(x_1, x_2, x_3, \bar{x}_4, x_5, \bar{x}_6) = (0, 0, 2.7, 6, 0, 6)$ as the initial BF solution. (Making the origin feasible as a convenient starting point for the simplex method is the whole point of creating the artificial problem.) We soon will trace the entire path followed by the simplex method from the origin to the optimal solution for both the artificial and real problems. But, first, how does the simplex method handle *minimization?*

Minimization

One straightforward way of minimizing Z with the simplex method is to exchange the roles of the positive and negative coefficients in row 0 for both the optimality test and step 1 of an iteration. However, rather than changing our instructions for the simplex method for this case, we present the following simple way of converting any minimization problem to an equivalent maximization problem:

$$\textit{Minimizing} \qquad Z = \sum_{j=1}^{n} c_j x_j$$

is equivalent to

$$\textit{maximizing} \qquad -Z = \sum_{j=1}^{n} (-c_j) x_j;$$

i.e., the two formulations yield the same optimal solution(s).

The two formulations are equivalent because the smaller Z is, the larger $-Z$ is, so the solution that gives the *smallest* value of Z in the entire feasible region must also give the *largest* value of $-Z$ in this region.

Therefore, in the radiation therapy example, we make the following change in the formulation:

$$\begin{aligned} &\text{Minimize} && Z = 0.4x_1 + 0.5x_2 \\ \rightarrow \quad &\text{Maximize} && -Z = -0.4x_1 - 0.5x_2. \end{aligned}$$

After artificial variables $\bar{x}_4$ and $\bar{x}_6$ are introduced and then the Big M method is applied, the corresponding conversion is

$$\begin{aligned} &\text{Minimize} && Z = 0.4x_1 + 0.5x_2 + M\bar{x}_4 + M\bar{x}_6 \\ \rightarrow \quad &\text{Maximize} && -Z = -0.4x_1 - 0.5x_2 - M\bar{x}_4 - M\bar{x}_6. \end{aligned}$$

Solving the Radiation Therapy Example

We now are nearly ready to apply the simplex method to the radiation therapy example. By using the maximization form just obtained, the entire system of equations is now

$$\begin{aligned}
&(0) \qquad -Z + 0.4x_1 + 0.5x_2 + M\bar{x}_4 + M\bar{x}_6 = 0 \\
&(1) \qquad 0.3x_1 + 0.1x_2 + \mathbf{x_3} = 2.7 \\
&(2) \qquad 0.5x_1 + 0.5x_2 + \bar{\mathbf{x}}_4 = 6 \\
&(3) \qquad 0.6x_1 + 0.4x_2 - x_5 + \bar{\mathbf{x}}_6 = 6.
\end{aligned}$$

The basic variables $(x_3, \bar{x}_4, \bar{x}_6)$ for the initial BF solution (for this artificial problem) are shown in bold type.

Note that this system of equations is not yet in proper form from Gaussian elimination, as required by the simplex method, since the basic variables $\bar{x}_4$ and $\bar{x}_6$ still need to be algebraically eliminated from Eq. (0). Because $\bar{x}_4$ and $\bar{x}_6$ both have a coefficient of M, Eq. (0) needs to have subtracted from it *both* M times Eq. (2) *and* M times Eq. (3). The calculations for all the coefficients (and the right-hand sides) are summarized below, where the vectors are the relevant rows of the simplex tableau corresponding to the above system of equations.

Row 0:

$$
\begin{array}{rrrrrrr}
[0.4, & 0.5, & 0, & M, & 0, & M, & 0] \\
-M[0.5, & 0.5, & 0, & 1, & 0, & 0, & 6] \\
-M[0.6, & 0.4, & 0, & 0, & -1, & 1, & 6] \\
\hline
\text{New row } 0 = [-1.1M + 0.4, & -0.9M + 0.5, & 0, & 0, & M, & 0, & -12M]
\end{array}
$$

The resulting initial simplex tableau, ready to begin the simplex method, is shown at the top of Table 4.12. Applying the simplex method in just the usual way then yields the sequence of simplex tableaux shown in the rest of Table 4.12. For the optimality test and the selection of the entering basic variable at each iteration, the quantities involving M are treated just as discussed in connection with Table 4.11. Specifically, whenever M is present, only its multiplicative factor is used, unless there is a tie, in which case the tie is broken by using the corresponding additive terms. Just

Table 4.12 **The Big M Method for the Radiation Therapy Example**

Iteration	Basic Variable	Eq.	*Coefficient of:* Z	x_1	x_2	x_3	$\bar{x}_4$	x_5	$\bar{x}_6$	Right Side
0	Z	(0)	−1	$-1.1M + 0.4$	$-0.9M + 0.5$	0	0	M	0	$-12M$
	x_3	(1)	0	0.3	0.1	1	0	0	0	2.7
	$\bar{x}_4$	(2)	0	0.5	0.5	0	1	0	0	6
	$\bar{x}_6$	(3)	0	0.6	0.4	0	0	−1	1	6
1	Z	(0)	−1	0	$-\frac{16}{30}M + \frac{11}{30}$	$\frac{11}{3}M - \frac{4}{3}$	0	M	0	$-2.1M - 3.6$
	x_1	(1)	0	1	$\frac{1}{3}$	$\frac{10}{3}$	0	0	0	9
	$\bar{x}_4$	(2)	0	0	$\frac{1}{3}$	$-\frac{5}{3}$	1	0	0	1.5
	$\bar{x}_6$	(3)	0	0	0.2	−2	0	−1	1	0.6
2	Z	(0)	−1	0	0	$-\frac{5}{3}M + \frac{7}{3}$	0	$-\frac{5}{3}M + \frac{11}{6}$	$\frac{8}{3}M - \frac{11}{6}$	$-0.5M - 4.7$
	x_1	(1)	0	1	0	$\frac{20}{3}$	0	$\frac{5}{3}$	$-\frac{5}{3}$	8
	$\bar{x}_4$	(2)	0	0	0	$\frac{5}{3}$	1	$\frac{5}{3}$	$-\frac{5}{3}$	0.5
	x_2	(3)	0	0	1	−10	0	−5	5	3
3	Z	(0)	−1	0	0	0.5	$M - 1.1$	0	M	−5.25
	x_1	(1)	0	1	0	5	−1	0	0	7.5
	x_5	(2)	0	0	0	1	0.6	1	−1	0.3
	x_2	(3)	0	0	1	−5	3	0	0	4.5

such a tie occurs in the last selection of an entering basic variable (see the next-to-last tableau), where the coefficients of x_3 and x_5 in row 0 both have the same multiplicative factor of $-\frac{5}{3}$. Comparing the additive terms, $\frac{11}{6} < \frac{7}{3}$ leads to choosing x_5 as the entering basic variable.

Note in Table 4.12 the progression of values of the artificial variables $\bar{x}_4$ and $\bar{x}_6$ and of Z. We start with large values, $\bar{x}_4 = 6$ and $\bar{x}_6 = 6$, with $Z = 12M$ $(-Z = -12M)$. The first iteration greatly reduces these values. The Big M method succeeds in driving $\bar{x}_6$ to zero (as a new nonbasic variable) at the second iteration and then in doing the same to $\bar{x}_4$ at the next iteration. With both $\bar{x}_4 = 0$ and $\bar{x}_6 = 0$, the basic solution given in the last tableau is guaranteed to be feasible for the real problem. Since it passes the optimality test, it also is optimal.

Now see what the Big M method has done graphically in Fig. 4.6. The feasible region for the artificial problem initially has four CPF solutions—(0, 0), (9, 0), (0, 12), and (7.5, 4.5)—and then replaces the first three with two new CPF solutions—(8, 3),

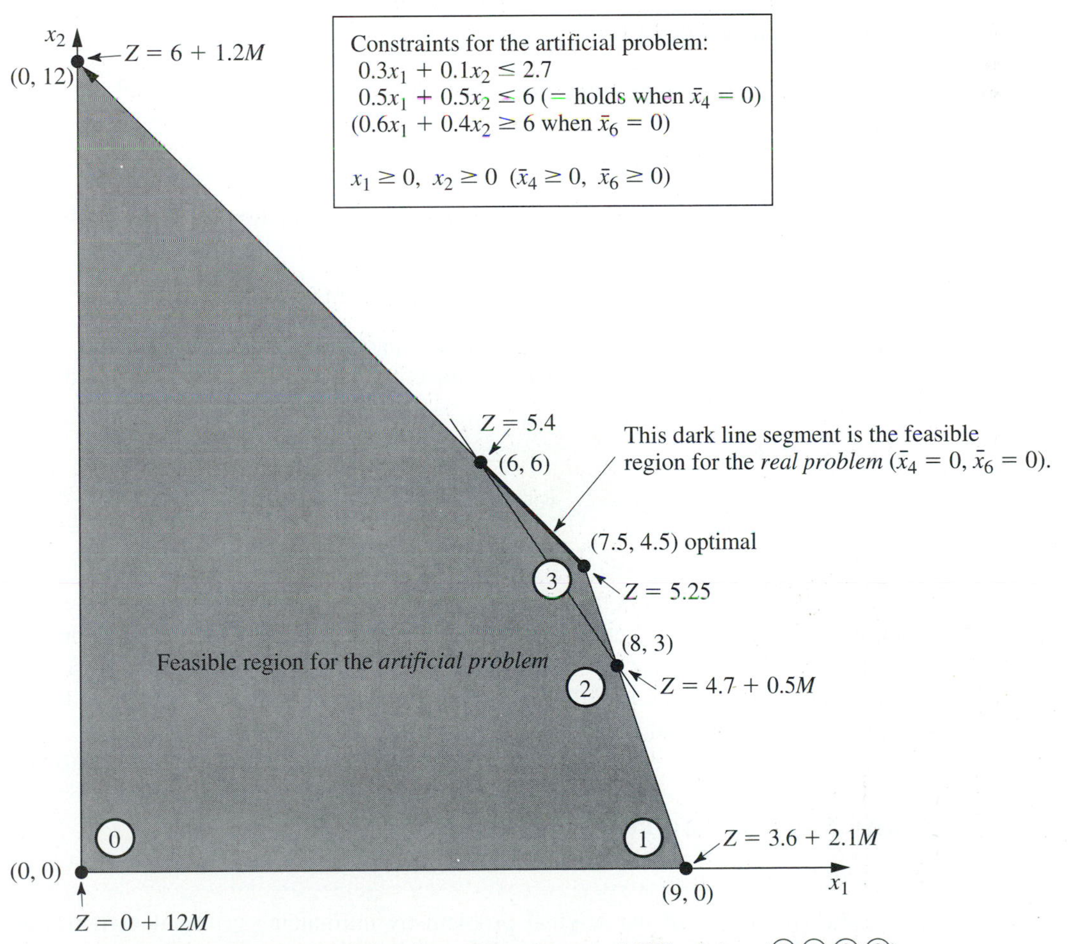

Figure 4.6 This graph shows the feasible region and the sequence of CPF solutions (⓪,①,②,③) examined by the simplex method (with the Big M method) for the artificial problem that corresponds to the real problem of Fig. 4.5.

(6, 6)—after $\bar{x}_6$ decreases to $\bar{x}_6 = 0$ so that $0.6x_1 + 0.4x_2 \geq 6$ becomes an additional constraint. (Note that all four initial CPF solutions, including the origin, actually were corner-point *infeasible* solutions for the real problem shown in Fig. 4.5.) Starting with the origin as the convenient initial CPF solution for the artificial problem, we move around the boundary to three other CPF solutions. The last of these is the first one that also is feasible for the real problem. Fortuitously, this first feasible solution also is optimal, so no additional iterations are needed.

For other problems with artificial variables, it may be necessary to perform additional iterations to reach an optimal solution after the first feasible solution is obtained for the real problem. (This was the case for the example solved in Table 4.11.) Thus the Big M method can be thought of as having two phases. In the first phase, all the artificial variables are driven to zero (because of the penalty of M per unit for being greater than zero) in order to reach an initial BF solution for the *real* problem. In the second phase, all artificial variables are kept at zero (because of this same penalty) while the simplex method generates a sequence of BF solutions leading to an optimal solution. The *two-phase method* described next is a streamlined procedure for performing these two phases directly, without even introducing M explicitly.

Skip?

The Two-Phase Method

For the radiation therapy example just solved in Table 4.12, recall its real objective function

Real problem: Minimize $Z = 0.4x_1 + 0.5x_2$.

However, the Big M method uses the following objective function (or its equivalent in maximization form) throughout the entire procedure:

Big M method: Minimize $Z = 0.4x_1 + 0.5x_2 + M\bar{x}_4 + Mx_6$.

Since the first two coefficients are negligible compared to M, the two-phase method is able to drop M by using the following two objective functions with completely different definitions of Z in turn.

Two-Phase Method:

Phase 1: Minimize $Z = \bar{x}_4 + \bar{x}_6$ (until $\bar{x}_4 = 0$, $\bar{x}_6 = 0$).

Phase 2: Minimize $Z = 0.4x_1 + 0.5x_2$ (with $\bar{x}_4 = 0$, $\bar{x}_6 = 0$).

The phase 1 objective function is obtained by dividing the Big M method objective function by M and then dropping the negligible terms. Since phase 1 concludes by obtaining a BF solution for the real problem (one where $\bar{x}_4 = 0$ and $\bar{x}_6 = 0$), this solution is then used as the *initial* BF solution for applying the simplex method to the real problem (with its real objective function) in phase 2.

Before solving the example in this way, we summarize the general method.

Summary of the Two-Phase Method

Initialization:

Revise the constraints of the original problem by introducing artificial variables as needed to obtain an obvious initial BF solution for the artificial problem.

Phase 1: Use the simplex method to solve the linear programming problem:

Minimize $Z = \Sigma$ artificial variables, subject to revised constraints.

The optimal solution obtained for this problem (with $Z = 0$) will be a BF solution for the real problem.

Phase 2: Drop the artificial variables (they are all zero now anyway).[1] Starting from the BF solution obtained at the end of phase 1, use the simplex method to solve the real problem.

For the example, the problems to be solved by the simplex method in the respective phases are summarized below.

Phase 1 Problem (Radiation Therapy Example):

$$\text{Minimize} \qquad Z = \bar{x}_4 + \bar{x}_6,$$

subject to

$$\begin{aligned} 0.3x_1 + 0.1x_2 + x_3 \qquad\qquad\qquad &= 2.7 \\ 0.5x_1 + 0.5x_2 \qquad + \bar{x}_4 \qquad\quad &= 6 \\ 0.6x_1 + 0.4x_2 \qquad\qquad - x_5 + \bar{x}_6 &= 6 \end{aligned}$$

and

$$x_1 \geq 0, \quad x_2 \geq 0, \quad x_3 \geq 0, \quad \bar{x}_4 \geq 0, \quad x_5 = 0, \quad \bar{x}_6 \geq 0.$$

Phase 2 Problem (Radiation Therapy Example):

$$\text{Minimize} \qquad Z = 0.4x_1 + 0.5x_2,$$

subject to

$$\begin{aligned} 0.3x_1 + 0.1x_2 + x_3 \qquad &= 2.7 \\ 0.5x_1 + 0.5x_2 \qquad\qquad &= 6 \\ 0.6x_1 + 0.4x_2 \qquad - x_5 &= 6 \end{aligned}$$

and

$$x_1 \geq 0, \quad x_2 \geq 0, \quad x_3 \geq 0, \quad x_5 \geq 0.$$

The only differences between these two problems are in the objective function and in the inclusion (phase 1) or exclusion (phase 2) of the artificial variables $\bar{x}_4$ and $\bar{x}_6$. Without the artificial variables, the phase 2 problem does not have an obvious *initial BF solution.* The sole purpose of solving the phase 1 problem is to obtain a BF solution with $\bar{x}_4 = 0$ and $\bar{x}_6 = 0$ so that this solution can be used as the initial BF solution for phase 2.

Table 4.13 shows the result of applying the simplex method to this phase 1 problem. [Row 0 in the initial tableau is obtained by converting Minimize $Z = \bar{x}_4 + \bar{x}_6$ to Maximize $(-Z) = -\bar{x}_4 - \bar{x}_6$ and then using *elementary row operations* to eliminate the basic variables $\bar{x}_4$ and $\bar{x}_6$ from $-Z + \bar{x}_4 + \bar{x}_6 = 0$.] In the next-to-last tableau, there is a tie for the *entering basic variable* between x_3 and x_5, which is broken arbitrarily in favor of x_3. The solution obtained at the end of phase 1, then, is $(x_1, x_2, x_3, \bar{x}_4, x_5, \bar{x}_6) = (6, 6, 0.3, 0, 0, 0)$ or, after $\bar{x}_4$ and $\bar{x}_6$ are dropped, $(x_1, x_2, x_3, x_5) = (6, 6, 0.3, 0)$.

[1] We are skipping over three other possibilities here: (1) artificial variables > 0 (discussed in the next subsection), (2) artificial variables that are degenerate basic variables, and (3) retaining the artificial variables as nonbasic variables in phase 2 (and not allowing them to become basic) as an aid to subsequent postoptimality analysis. Your OR Courseware allows you to explore these possibilities.

Table 4.13 Phase 1 of the Two-Phase Method for the Radiation Therapy Example

Iteration	Basic Variable	Eq.	*Coefficient of:* Z	x_1	x_2	x_3	$\bar{x}_4$	x_5	$\bar{x}_6$	Right Side
0	Z	(0)	−1	−1.1	−0.9	0	0	1	0	−12
	x_3	(1)	0	0.3	0.1	1	0	0	0	2.7
	$\bar{x}_4$	(2)	0	0.5	0.5	0	1	0	0	6
	$\bar{x}_6$	(3)	0	0.6	0.4	0	0	−1	1	6
1	Z	(0)	−1	0	$-\frac{16}{30}$	$\frac{11}{3}$	0	1	0	−2.1
	x_1	(1)	0	1	$\frac{1}{3}$	$\frac{10}{3}$	0	0	0	9
	$\bar{x}_4$	(2)	0	0	$\frac{1}{3}$	$-\frac{5}{3}$	1	0	0	1.5
	$\bar{x}_6$	(3)	0	0	0.2	−2	0	−1	1	0.6
2	Z	(0)	−1	0	0	$-\frac{5}{3}$	0	$-\frac{5}{3}$	$\frac{8}{3}$	−0.5
	x_1	(1)	0	1	0	$\frac{20}{3}$	0	$\frac{5}{3}$	$-\frac{5}{3}$	8
	$\bar{x}_4$	(2)	0	0	0	$\frac{5}{3}$	1	$\frac{5}{3}$	$-\frac{5}{3}$	0.5
	x_2	(3)	0	0	1	−10	0	−5	5	3
3	Z	(0)	−1	0	0	0	1	0	1	0
	x_1	(1)	0	1	0	0	−4	−5	5	6
	x_3	(2)	0	0	0	1	$\frac{3}{5}$	1	−1	0.3
	x_2	(3)	0	0	1	0	6	5	−5	6

As claimed in the summary, this solution from phase 1 is indeed a BF solution for the *real* problem (the phase 2 problem) because it is the solution (after you set $x_5 = 0$) to the system of equations consisting of the three functional constraints for the phase 2 problem. In fact, after the $\bar{x}_4$ and $\bar{x}_6$ columns are deleted, Table 4.13 shows one way of using Gaussian elimination to solve this system of equations by reducing the system to the form displayed in the final tableau.

Table 4.14 shows the preparations for beginning phase 2 after phase 1 is completed. Starting from the final tableau in Table 4.13, we drop the artificial variables ($\bar{x}_4$ and $\bar{x}_6$), substitute the phase 2 objective function ($-Z = 0.4x_1 - 0.5x_2$ in maximization form) into row 0, and then restore the proper form from Gaussian elimination (by algebraically eliminating the basic variables x_1 and x_2 from row 0). Thus, row 0 in the last tableau is obtained by performing the following *elementary row operations* in the next-to-last tableau: from row 0 subtract both the product, 0.4 times row 1, and the product, 0.5 times row 3. Except for the deletion of the two columns, note that rows 1 to 3 never change. The only adjustments occur in row 0 in order to replace the phase 1 objective function by the phase 2 objective function.

The last tableau in Table 4.14 is the initial tableau for applying the simplex method to the phase 2 problem, as shown at the top of Table 4.15. Just one iteration then leads to the optimal solution shown in the second tableau: $(x_1, x_2, x_3, x_5) = (7.5, 4.5, 0, 0.3)$. This solution is the desired optimal solution for the real problem of interest rather than the artificial problem constructed for phase 1.

Table 4.14 Preparing to Begin Phase 2 for the Radiation Therapy Example

	Basic Variable	Eq.	Coefficient of: Z	x_1	x_2	x_3	$\bar{x}_4$	x_5	$\bar{x}_6$	Right Side
Final Phase 1 tableau	Z	(0)	-1	0	0	0	1	0	1	0
	x_1	(1)	0	1	0	0	-4	-5	5	6
	x_3	(2)	0	0	0	1	$\frac{3}{5}$	1	-1	0.3
	x_2	(3)	0	0	1	0	6	5	-5	6
Drop $\bar{x}_4$ and $\bar{x}_6$	Z	(0)	-1	0	0	0		0		0
	x_1	(1)	0	1	0	0		-5		6
	x_3	(2)	0	0	0	1		1		0.3
	x_2	(3)	0	0	1	0		5		6
Substitute phase 2 objective function	Z	(0)	-1	0.4	0.5	0		0		0
	x_1	(1)	0	1	0	0		-5		6
	x_3	(2)	0	0	0	1		1		0.3
	x_2	(3)	0	0	1	0		5		6
Restore proper form from Gaussian elimination	Z	(0)	-1	0	0	0		-0.5		-5.4
	x_1	(1)	0	1	0	0		-5		6
	x_3	(2)	0	0	0	1		1		0.3
	x_2	(3)	0	0	1	0		5		6

Now we see what the two-phase method has done graphically in Fig. 4.7. Starting at the origin, phase 1 examines a total of four CPF solutions for the artificial problem. The first three actually were corner-point infeasible solutions for the real problem shown in Fig. 4.5. The fourth CPF solution, at (6, 6), is the first one that also is feasible for the real problem, so it becomes the initial CPF solution for phase 2. One iteration leads to the optimal CPF solution at (7.5, 4.5).

Table 4.15 Phase 2 of the Two-Phase Method for the Radiation Therapy Example

Iteration	Basic Variable	Eq.	Coefficient of: Z	x_1	x_2	x_3	x_5	Right Side
0	Z	(0)	-1	0	0	0	-0.5	-5.4
	x_1	(1)	0	1	0	0	-5	6
	x_3	(2)	0	0	0	1	1	0.3
	x_2	(3)	0	0	1	0	5	6
1	Z	(0)	-1	0	0	0.5	0	-5.25
	x_1	(1)	0	1	0	5	0	7.5
	x_5	(2)	0	0	0	1	1	0.3
	x_2	(3)	0	0	1	-5	0	4.5

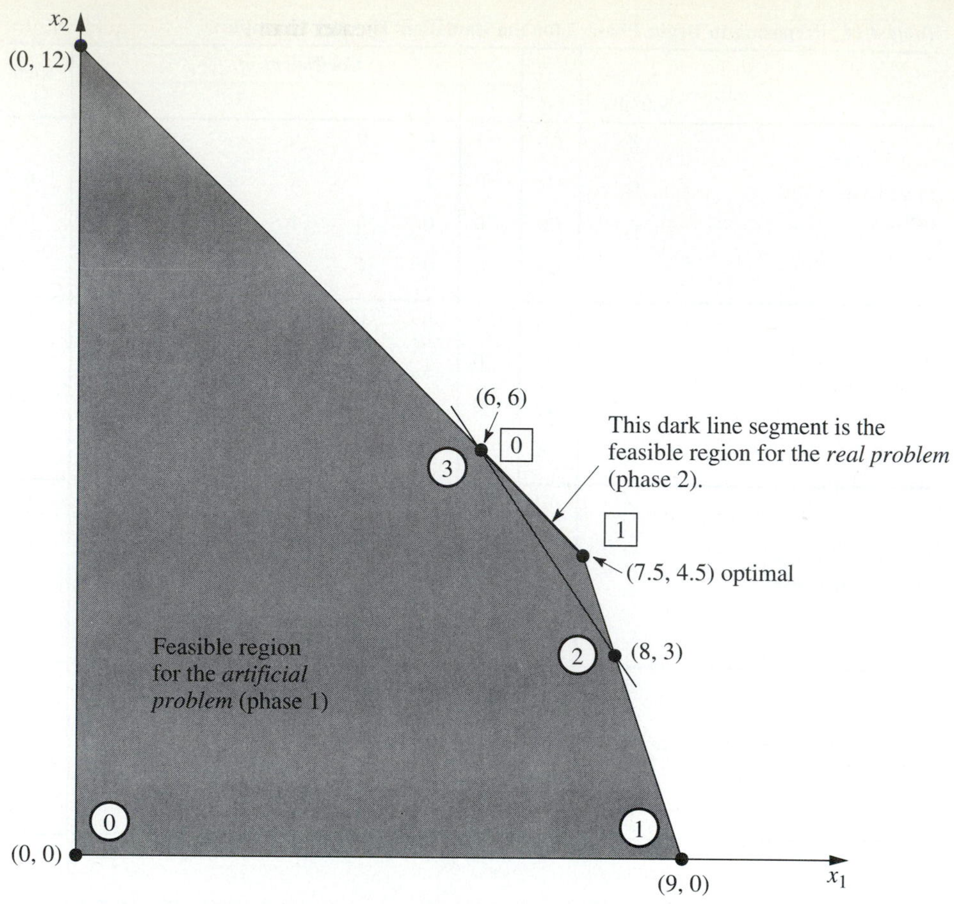

Figure 4.7 This graph shows the sequence of CPF solutions for phase 1 (⓪,①,②,③) and then for phase 2 (0, 1) when the two-phase method is applied to the radiation therapy example.

If the tie for the entering basic variable in the next-to-last tableau of Table 4.13 had been broken in the other way, then phase 1 would have gone directly from (8, 3) to (7.5, 4.5). After (7.5, 4.5) was used to set up the initial simplex tableau for phase 2, the *optimality test* would have revealed that this solution was optimal, so no iterations would be done.

It is interesting to compare the big M and two-phase methods. Begin with their objective functions.

Big M Method:

$$\text{Minimize} \quad Z = 0.4x_1 + 0.5x_2 + M\bar{x}_4 + M\bar{x}_6.$$

Two-Phase Method:

Phase 1: Minimize $Z = \bar{x}_4 + \bar{x}_6.$

Phase 2: Minimize $Z = 0.4x_1 + 0.5x_2.$

Because the $M\bar{x}_4$ and $M\bar{x}_6$ terms dominate the $0.4x_1$ and $0.5x_2$ terms in the objective function for the Big M method, this objective function is essentially equivalent to the phase 1 objective function as long as $\bar{x}_4$ and/or $\bar{x}_6$ is greater than zero. Then, when both $\bar{x}_4 = 0$ and $\bar{x}_6 = 0$, the objective function for the Big M method becomes completely equivalent to the phase 2 objective function.

Because of these virtual equivalencies in objective functions, the Big M and two-phase methods generally have the same sequence of BF solutions. The one possible exception occurs when there is a tie for the entering basic variable in phase 1 of the two-phase method, as happened in the third tableau of Table 4.13. Notice that the first three tableaux of Tables 4.12 and 4.13 are almost identical, with the only difference being that the multiplicative factors of M in Table 4.12 become the sole quantities in the corresponding spots in Table 4.13. Consequently, the additive terms that broke the tie for the entering basic variable in the third tableau of Table 4.12 were not present to break this same tie in Table 4.13. The result for this example was an extra iteration for the two-phase method. Generally, however, the advantage of having the additive factors is minimal.

The two-phase method streamlines the Big M method by using only the multiplicative factors in phase 1 and by dropping the artificial variables in phase 2. (The Big M method could combine the multiplicative and additive factors by assigning an actual huge number to M, but this might create numerical instability problems.) For these reasons, the two-phase method is commonly used in computer codes.

No Feasible Solutions

So far in this section we have been concerned primarily with the fundamental problem of identifying an initial BF solution when an obvious one is not available. You have seen how the artificial-variable technique can be used to construct an artificial problem and obtain an initial BF solution for this artificial problem instead. Use of either the Big M method or the two-phase method then enables the simplex method to begin its pilgrimage toward the BF solutions, and ultimately toward the optimal solution, for the *real* problem.

However, you should be wary of a certain pitfall with this approach. There may be no obvious choice for the initial BF solution for the very good reason that there are no feasible solutions at all! Nevertheless, by constructing an artificial feasible solution, there is nothing to prevent the simplex method from proceeding as usual and ultimately reporting a supposedly optimal solution.

Fortunately, the artificial-variable technique provides the following signpost to indicate when this has happened:

> If the original problem has *no feasible solutions,* then either the Big M method or phase 1 of the two-phase method yields a final solution that has at least one artificial variable *greater* than zero. Otherwise, they *all* equal zero.

To illustrate, let us change the first constraint in the radiation therapy example (see Fig. 4.5) as follows:

$$0.3x_1 + 0.1x_2 \le 2.7 \qquad \rightarrow \qquad 0.3x_1 + 0.1x_2 \le 1.8,$$

so that the problem no longer has any feasible solutions. Applying the Big M method just as before (see Table 4.12) yields the tableaux shown in Table 4.16. (Phase 1 of the two-phase method yields the same tableaux except that each expression involving M is replaced by just the multiplicative factor.) Hence the Big M method normally would be indicating that the optimal solution is (3, 9, 0, 0, 0, 0.6). However, since an artificial

Table 4.16 **The Big M Method for the Revision of the Radiation Therapy Example That Has No Feasible Solutions**

Iteration	Basic Variable	Eq.	Coefficient of: Z	x_1	x_2	x_3	$\bar{x}_4$	x_5	$\bar{x}_6$	Right Side
0	Z	(0)	-1	$-1.1M + 0.4$	$-0.9M + 0.5$	0	0	M	0	$-12M$
	x_3	(1)	0	0.3	0.1	1	0	0	0	1.8
	$\bar{x}_4$	(2)	0	0.5	0.5	0	1	0	0	6
	$\bar{x}_6$	(3)	0	0.6	0.4	0	0	-1	1	6
1	Z	(0)	-1	0	$-\frac{16}{30}M + \frac{11}{30}$	$\frac{11}{3}M - \frac{4}{3}$	0	M	0	$-5.4M - 2.4$
	x_1	(1)	0	1	$\frac{1}{3}$	$\frac{10}{3}$	0	0	0	6
	$\bar{x}_4$	(2)	0	0	$\frac{1}{3}$	$-\frac{5}{3}$	1	0	0	3
	$\bar{x}_6$	(3)	0	0	0.2	-2	0	-1	1	2.4
2	Z	(0)	-1	0	0	$M + 0.5$	$1.6M - 1.1$	M	0	$-0.6M - 5.7$
	x_1	(1)	0	1	0	5	-1	0	0	3
	x_2	(2)	0	0	1	-5	3	0	0	9
	$\bar{x}_6$	(3)	0	0	0	-1	-0.6	-1	1	0.6

variable $\bar{x}_6 = 0.6 > 0$, the real message here is that the problem has no feasible solutions.

Variables Allowed to Be Negative

In most practical problems, negative values for the decision variables would have no physical meaning, so it is necessary to include nonnegativity constraints in the formulations of their linear programming models. However, this is not always the case. To illustrate, suppose that the Wyndor Glass Co. problem is changed so that product 1 already is in production, and the first decision variable x_1 represents the *increase* in its production rate. Therefore, a negative value of x_1 would indicate that product 1 is to be cut back by that amount. Such reductions might be desirable to allow a larger production rate for the new, more profitable product 2, so negative values should be allowed for x_1 in the model.

Since the procedure for determining the *leaving basic variable* requires that all the variables have nonnegativity constraints, any problem containing variables allowed to be negative must be converted to an *equivalent* problem involving only nonnegative variables before the simplex method is applied. Fortunately, this conversion can be done. The modification required for each variable depends upon whether it has a (negative) lower bound on the values allowed. Each of these two cases is now discussed.

Variables with a Bound on the Negative Values Allowed: Consider any decision variable x_j that is allowed to have negative values which satisfy a constraint of the form

$$x_j \geq L_j,$$

where L_j is some negative constant. This constraint can be converted to a nonnegativity constraint by making the change of variables

$$x_j' = x_j - L_j, \qquad \text{so} \qquad x_j' \geq 0.$$

Thus $x_j' + L_j$ would be substituted for x_j throughout the model, so that the redefined decision variable x_j' cannot be negative. (This same technique can be used when L_j is *positive* to convert a functional constraint $x_j \geq L_j$ to a nonnegativity constraint $x_j' \geq 0$.)

To illustrate, suppose that the current production rate for product 1 in the Wyndor Glass Co. problem is 10. With the definition of x_1 just given, the complete model at this point is the same as that given in Sec. 3.1 except that the nonnegativity constraint $x_1 \geq 0$ is replaced by

$$x_1 \geq -10.$$

To obtain the equivalent model needed for the simplex method, this decision variable would be redefined as the *total* production rate of product 1

$$x_j' = x_1 + 10,$$

which yields the changes in the objective function and constraints as shown:

$$\boxed{\begin{array}{l} Z = 3x_1 + 5x_2 \\ x_1 \qquad\qquad \leq \ \ 4 \\ \qquad\quad 2x_2 \leq 12 \\ 3x_1 + 2x_2 \leq 18 \\ x_1 \geq -10, \qquad x_2 \geq 0 \end{array}} \rightarrow \boxed{\begin{array}{l} Z = 3(x_1' - 10) + 5x_2 \\ x_1' - 10 \qquad\quad \leq \ \ 4 \\ \qquad\qquad\quad 2x_2 \leq 12 \\ 3(x_1' - 10) + 2x_2 \leq 18 \\ x_1' - 10 \geq -10, \qquad x_2 \geq 0 \end{array}} \rightarrow \boxed{\begin{array}{l} Z = -30 + 3x_1' + 5x_2 \\ x_1' \qquad\qquad \leq 14 \\ \qquad\quad 2x_2 \leq 12 \\ 3x_1' + 2x_2 \leq 48 \\ x_1' \geq 0, \qquad x_2 \geq 0 \end{array}}$$

Variables with No Bound on the Negative Values Allowed: In the case where x_j does *not* have a lower-bound constraint in the model formulated, another approach is required: x_j is replaced throughout the model by the *difference* of two new *nonnegative* variables

$$x_j = x_j^+ - x_j^-, \qquad \text{where } x_j^+ \geq 0,\ x_j^- \geq 0.$$

Since x_j^+ and x_j^- can have any nonnegative values, this difference $x_j^+ - x_j^-$ can have *any* value (positive or negative), so it is a legitimate substitute for x_j in the model. But after such substitutions, the simplex method can proceed with just nonnegative variables.

The new variables x_j^+ and x_j^- have a simple interpretation. As explained in the next paragraph, each BF solution for the new form of the model necessarily has the property that *either* $x_j^+ = 0$ or $x_j^- = 0$ (or both). Therefore, at the optimal solution obtained by the simplex method (a BF solution),

$$x_j^+ = \begin{cases} x_j & \text{if } x_j \geq 0, \\ 0 & \text{otherwise;} \end{cases}$$

$$x_j^- = \begin{cases} |x_j| & \text{if } x_j \leq 0, \\ 0 & \text{otherwise;} \end{cases}$$

so that x_j^+ represents the positive part of the decision variable x_j and x_j^- its negative part (as suggested by the superscripts).

For example, if $x_j = 10$, the above expressions give $x_j^+ = 10$ and $x_j^- = 0$. This same value of $x_j = x_j^+ - x_j^- = 10$ also would occur with larger values of x_j^+ and x_j^-

such that $x_j^+ = x_j^- + 10$. Plotting these values of x_j^+ and x_j^- on a two-dimensional graph gives a line with an endpoint at $x_j^+ = 10$, $x_j^- = 0$ to avoid violating the nonnegativity constraints. This endpoint is the only corner-point solution on the line. Therefore, only this endpoint can be part of an overall CPF solution or BF solution involving all the variables of the model. This illustrates why each BF solution necessarily has either $x_j^+ = 0$ or $x_j^- = 0$ (or both).

To illustrate the use of the x_j^+ and x_j^-, let us return to the example on the preceding page where x_1 is redefined as the increase over the current production rate of 10 for product 1 in the Wyndor Glass Co. problem.

However, now suppose that the $x_1 \geq -10$ constraint was not included in the original model because it clearly would not change the optimal solution. (In some problems, certain variables do not need explicit lower-bound constraints because the functional constraints already prevent lower values.) Therefore, before the simplex method is applied, x_1 would be replaced by the difference

$$x_1 = x_1^+ - x_1^-, \qquad \text{where } x_1^+ \geq 0,\ x_1^- \geq 0,$$

as shown:

$$\boxed{\begin{array}{ll} \text{Maximize} & Z = 3x_1 + 5x_2, \\ \text{subject to} & \begin{array}{r} x_1 \qquad \leq 4 \\ 2x_2 \leq 12 \\ 3x_1 + 2x_2 \leq 18 \\ x_2 \geq 0 \text{ (only)} \end{array} \end{array}} \rightarrow \boxed{\begin{array}{ll} \text{Maximize} & Z = 3x_1^+ - 3x_1^- + 5x_2, \\ \text{subject to} & \begin{array}{r} x_1^+ - x_1^- \qquad \leq 4 \\ 2x_2 \leq 12 \\ 3x_1^+ - 3x_1^- + 2x_2 \leq 18 \\ x_1^+ \geq 0, \quad x_1^- \geq 0, \quad x_2 \geq 0 \end{array} \end{array}}$$

From a computational viewpoint, this approach has the disadvantage that the new equivalent model to be used has more variables than the original model. In fact, if *all* the original variables lack lower-bound constraints, the new model will have *twice* as many variables. Fortunately, the approach can be modified slightly so that the number of variables is increased by only one, regardless of how many original variables need to be replaced. This modification is done by replacing each such variable x_j by

$$x_j = x_j' - x'', \qquad \text{where } x_j' \geq 0,\ x'' \geq 0,$$

instead, where x'' is the *same* variable for all relevant j. The interpretation of x'' in this case is that $-x''$ is the current value of the *largest* (in absolute terms) negative original variable, so that x_j' is the amount by which x_j exceeds this value. Thus the simplex method now can make some of the x_j' variables larger than zero even when $x'' > 0$.

4.7 Post-Optimality Analysis

We stressed in Secs. 2.3, 2.4, and 2.5 that *post-optimality analysis*—the analysis done *after* an optimal solution is obtained for the initial version of the model—constitutes a very major and very important part of most operations research studies. The fact that post-optimality analysis is very important is particularly true for typical linear programming applications. In this section, we focus on the role of the simplex method in performing this analysis.

Table 4.17 summarizes the typical steps in post-optimality analysis for linear programming studies. The rightmost column identifies some algorithmic techniques

Table 4.17 **Post-Optimality Analysis for Linear Programming**

Task	Purpose	Technique
Model debugging	Find errors and weaknesses in model	Reoptimization
Model validation	Demonstrate validity of final model	See Sec. 2.4
Final managerial decisions on resource allocations (the b_i values)	Make appropriate division of organizational resources between activities under study and other important activities	Shadow prices
Evaluate estimates of model parameters	Determine crucial estimates that may affect optimal solution for further study	Sensitivity analysis
Evaluate trade-offs between model parameters	Determine best trade-off	Parametric linear programming

that involve the simplex method. These techniques are introduced briefly here with the technical details deferred to later chapters.

Reoptimization

After having found an optimal solution for one version of a linear programming model, we frequently must solve again (often many times) for a slightly different version of the model. We nearly always have to solve again several times during the model debugging stage (described in Secs. 2.3 and 2.4), and we usually have to do so a large number of times during the later stages of post-optimality analysis as well.

One approach is simply to reapply the simplex method from scratch for each new version of the model, even though each run may require hundreds or even thousands of iterations for large problems. However, a *much more efficient* approach is to *reoptimize.* Reoptimization involves deducing how changes in the model get carried along to the *final* simplex tableau (as described in Secs. 5.3 and 6.6). This revised tableau and the optimal solution for the prior model are then used as the *initial tableau* and the *initial basic solution* for solving the new model. If this solution is feasible for the new model, then the simplex method is applied in the usual way, starting from this initial BF solution. If the solution is not feasible, a related algorithm called the *dual simplex method* (described in Sec. 7.1) probably can be applied to find the new optimal solution,[1] starting from this initial basic solution.

The big advantage of this **reoptimization technique** over re-solving from scratch is that an optimal solution for the revised model probably is going to be *much* closer to the prior optimal solution than to an initial BF solution constructed in the usual way for the simplex method. Therefore, assuming that the model revisions were modest, only a few iterations should be required to reoptimize instead of the hundreds or thousands that may be required when you start from scratch. In fact, the optimal solutions for the prior and revised models are frequently the same, in which case the reoptimization technique requires only one application of the optimality test and *no* iterations.

[1] The one requirement for using the dual simplex method here is that the *optimality test* is still passed when applied to row 0 of the *revised* final tableau. If not, then still another algorithm called the *primal-dual method* can be used instead.

Shadow Prices

Recall that linear programming problems often can be interpreted as allocating resources to activities. In particular, when the functional constraints are in $\leq$ form, we interpreted the b_i (the right-hand sides) as the amounts of the respective resources being made available for the activities under consideration. In many cases, there may be some latitude in the amounts that will be made available. If so, the b_i values used in the initial (validated) model actually may represent management's *tentative initial decision* on how much of the organization's resources will be provided to the activities considered in the model instead of to other important activities under the purview of management. From this broader perspective, some of the b_i values can be increased in a revised model, but only if a sufficiently strong case can be made to management that this revision would be beneficial.

Consequently, information on the economic contribution of the resources to the measure of performance (Z) for the current study often would be extremely useful. The simplex method provides this information in the form of *shadow prices* for the respective resources.

> The **shadow price** for resource i (denoted by y_i^*) measures the *marginal value* of this resource, i.e., the rate at which Z could be increased by (slightly) increasing the amount of this resource (b_i) being made available.[1] The simplex method identifies this shadow price by y_i^* = coefficient of the ith slack variable in row 0 of the final simplex tableau.

To illustrate, for the Wyndor Glass Co. problem,

Resource i = production capacity of Plant i ($i = 1, 2, 3$) being made available to the two new products under consideration,

b_i = hours of production time per week being made available in Plant i for these new products.

Providing a substantial amount of production time for the new products would require adjusting production times for the current products, so choosing the b_i value is a difficult managerial decision. The tentative initial decision has been

$$b_1 = 4, \qquad b_2 = 12, \qquad b_3 = 18,$$

as reflected in the basic model considered in Sec. 3.1 and in this chapter. However, management now wishes to evaluate the effect of changing any of the b_i values.

The shadow prices for these three resources provide just the information that management needs. The final tableau in Table 4.8 (see p. 99) yields

$$y_1^* = 0 = \text{shadow price for resource 1},$$

$$y_2^* = \frac{3}{2} = \text{shadow price for resource 2},$$

$$y_3^* = 1 = \text{shadow price for resource 3}.$$

With just two decision variables, these numbers can be verified by checking graphically that individually increasing any b_i by 1 indeed would increase the optimal value of Z by y_i^*. For example, Fig. 4.8 demonstrates this increase for resource 2 by reapply-

[1] The increase in b_i must be sufficiently small that the current set of basic variables remains optimal since this rate (marginal value) changes if the set of basic variables changes.

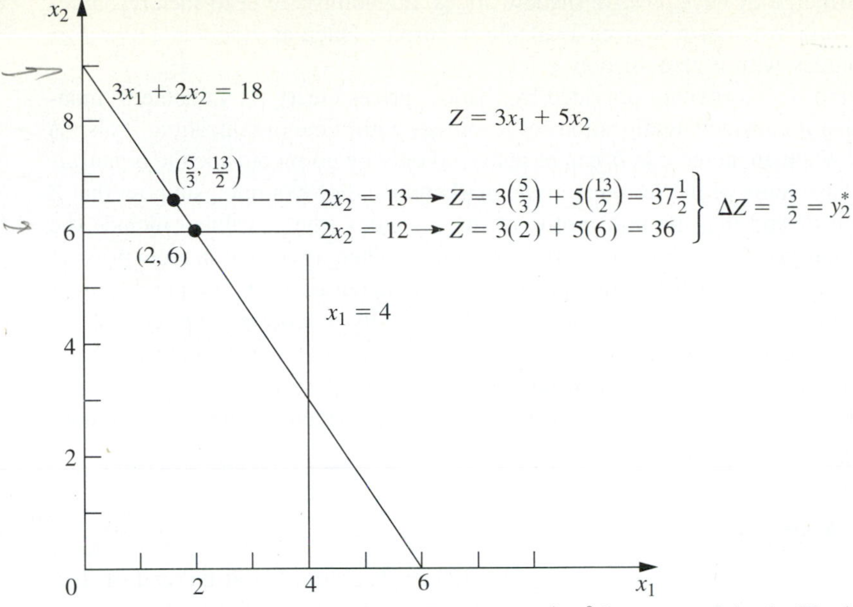

Figure 4.8 This graph shows that the shadow price is $y_2^* = \frac{3}{2}$ for resource 2 for the Wyndor Glass Co. problem. The two dots are the optimal solutions for $b_2 = 12$ or $b_2 = 13$, and plugging these solutions into the objective function reveals that increasing b_2 by 1 increases Z by $y_2^* = \frac{3}{2}$.

ing the graphical method presented in Sec. 3.1. The optimal solution, (2, 6) with $Z = 36$, changes to $(\frac{5}{3}, \frac{13}{2})$ with $Z = 37\frac{1}{2}$ when b_2 is increased by 1 (from 12 to 13), so that

$$y_2^* = \Delta Z = 37\tfrac{1}{2} - 36 = \tfrac{3}{2}.$$

Since Z is expressed in thousands of dollars of profit per week, $y_2^* = \frac{3}{2}$ indicates that adding 1 more hour of production time per week in Plant 2 for these two new products would increase their total profit by $1,500 per week. Should this actually be done? It depends on the marginal profitability of other products currently using this production time. If there is a current product that contributes less than $1,500 of weekly profit per hour of weekly production time in Plant 2, then some shift of production time to the new products would be worthwhile.

We shall continue this story in Sec. 6.7, where the Wyndor OR team uses shadow prices as part of its *sensitivity analysis* of the model.

Figure 4.8 demonstrates that $y_2^* = \frac{3}{2}$ is the rate at which Z could be increased by increasing b_2 slightly. However, it also demonstrates the common phenomenon that this interpretation holds only for a small increase in b_2. Once b_2 is increased beyond 18, the optimal solution stays at (0, 9) with no further increase in Z. (At that point, the set of basic variables in the optimal solution has changed, so a new final simplex tableau will be obtained with new shadow prices, including $y_2^* = 0$.)

Now note in Fig. 4.8 why $y_1^* = 0$. Because the constraint on resource 1, $x_1 \leq 4$, is *not binding* on the optimal solution (2, 6), there is a *surplus* of this resource. Therefore, increasing b_1 beyond 4 cannot yield a new optimal solution with a larger value of Z.

By contrast, the constraints on resources 2 and 3, $2x_2 \leq 12$ and $3x_1 + 2x_2 \leq 18$, are **binding constraints** (constraints that hold with equality at the optimal solution). Because the limited supply of these resources ($b_2 = 12$, $b_3 = 18$) *binds Z from being* increased further, they have *positive* shadow prices. Economists refer to such resources as *scarce goods,* whereas resources available in surplus (such as resource 1) are *free goods* (resources with a zero shadow price).

The kind of information provided by shadow prices clearly is valuable to management when it considers reallocations of resources within the organization. It also is very helpful when an increase in b_i can be achieved only by going outside the organization to purchase more of the resource in the marketplace. For example, suppose that Z represents *profit* and that the unit profits of the activities (the c_j values) include the costs (at regular prices) of all the resources consumed. Then a *positive* shadow price of y_i^* for resource i means that the total profit Z can be increased by y_i^* by purchasing 1 more unit of this resource at its regular price. Alternatively, if a *premium* price must be paid for the resource in the marketplace, then y_i^* represents the *maximum* premium (excess over the regular price) that would be worth paying.[1]

The theoretical foundation for shadow prices is provided by the duality theory described in Chap. 6.

Sensitivity Analysis

When discussing the *certainty assumption* for linear programming at the end of Sec. 3.3, we pointed out that the values used for the model parameters (the a_{ij}, b_i, and c_j identified in Table 3.3) generally are just *estimates* of quantities whose true values will not become known until the linear programming study is implemented at some time in the future. A main purpose of sensitivity analysis is to identify the **sensitive parameters** (i.e., those that cannot be changed without changing the optimal solution). The sensitive parameters are the parameters that will need to be monitored particularly closely as the study is implemented. If it is discovered that the true value of a sensitive parameter differs from its estimated value in the model, this immediately signals a need to change the solution.

How are the sensitive parameters identified? In the case of the b_i, you have just seen that this information is given by the shadow prices provided by the simplex method. In particular, if $y_i^* > 0$, then the optimal solution changes if b_i is changed, so b_i is a sensitive parameter. However, $y_i^* = 0$ implies that the optimal solution is not sensitive to at least small changes in b_i. Consequently, if the value used for b_i is an estimate of the amount of the resource that will be available (rather than a managerial decision), then the b_i values that need to be monitored more closely are those with *positive* shadow prices—especially those with *large* shadow prices.

When there are just two variables, the sensitivity of the various parameters can be analyzed graphically. For example, in Fig. 4.9, $c_1 = 3$ can be changed to any other value from 0 to 7.5 without the optimal solution changing from (2, 6). (The reason is that any value of c_1 within this range keeps the slope of $Z = c_1x_1 + 5x_2$ between the slopes of the lines $2x_2 = 12$ and $3x_1 + 2x_2 = 18$.) Similarly, if $c_2 = 5$ is the only parameter changed, it can have any value greater than 2 without affecting the optimal solution. Hence neither c_1 nor c_2 is a sensitive parameter.

The easiest way to analyze the sensitivity of each of the a_{ij} parameters graphically is to check whether the corresponding constraint is *binding* at the optimal solution. Because $x_1 \leq 4$ is *not* a binding constraint, any sufficiently small change in its coefficients ($a_{11} = 1$, $a_{12} = 0$) is not going to change the optimal solution, so these are

[1] If the unit profits do *not* include the costs of the resources consumed, then y_i^* represents the maximum *total* unit price that would be worth paying to increase b_i.

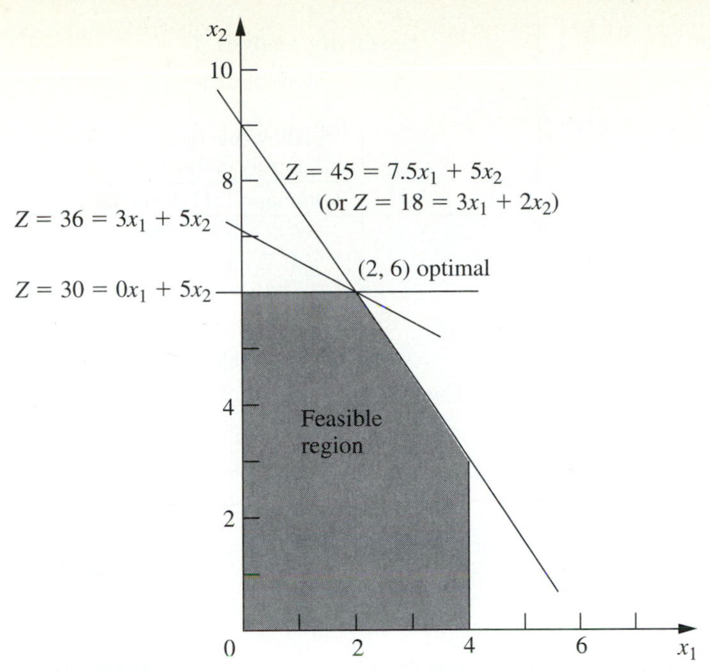

Figure 4.9 This graph demonstrates the sensitivity analysis of c_1 and c_2 for the Wyndor Glass Co. problem. Starting with the original objective function line [where $c_1 = 3$, $c_2 = 5$, and the optimal solution is (2, 6)], the other two lines show the extremes of how much the slope of the objective function line can change and still retain (2, 6) as an optimal solution. Thus, with $c_2 = 5$, the allowable range for c_1 is $0 \leq c_1 \leq 7.5$. With $c_1 = 3$, the allowable range for c_2 is $c_2 \geq 2$.

not sensitive parameters. On the other hand, both $2x_2 \leq 12$ and $3x_1 + 2x_2 \leq 18$ are *binding constraints,* so changing *any* one of their coefficients ($a_{21} = 0$, $a_{22} = 2$, $a_{31} = 3$, $a_{32} = 2$) is going to change the optimal solution, and therefore these are sensitive parameters.

Typically, greater attention is given to performing sensitivity analysis on the b_i and c_j parameters than on the a_{ij} parameters. On real problems with hundreds or thousands of constraints and variables, the effect of changing one a_{ij} value is usually negligible, but changing one b_i or c_j value can have real impact. Furthermore, in many cases, the a_{ij} values are determined by the technology being used (the a_{ij} values are sometimes called *technological coefficients*), so there may be relatively little (or no) uncertainty about their final values. This is fortunate, because there are far more a_{ij} parameters than b_i and c_j parameters for large problems.

For problems with more than two (or possibly three) variables, you cannot analyze the sensitivity of the parameters graphically as was just done for the Wyndor Glass Co. problem. However, you can extract the same kind of information from the simplex method. Getting this information requires using the *fundamental insight* described in Sec. 5.3 to deduce the changes that get carried along to the final simplex tableau as a result of changing the value of a parameter in the original model. The rest of the procedure is described and illustrated in Secs. 6.6 and 6.7.

This procedure normally is incorporated into software packages based on the simplex method so that the output will include basic sensitivity analysis information. Typical is the output from the routine *Solve Automatically by the Simplex Method,* in your OR Courseware, as shown in Fig. 4.10. On the left are the optimal solution, the

Optimal Solution

Value of the objective function: $Z = 36$

Variable	Value
x_1	2
x_2	6

Constraint	Slack or surplus	Shadow price
1	2	0
2	0	1.5
3	0	1

Sensitivity Analysis

Objective function coefficient

Current value	Allowable range to stay optimal	
	Minimum	Maximum
3	0	7.5
5	2	$+\infty$

Right-hand sides

Current value	Allowable range to stay feasible	
	Minimum	Maximum
4	2	$+\infty$
12	6	18
18	12	24

Figure 4.10 The output (including sensitivity analysis information) for the Wyndor Glass Co. problem (as presented in Sec. 3.1) from the automatic routine for the simplex method in the OR Courseware.

resulting slack or surplus in the functional constraints (i.e., the values of the corresponding slack or surplus variables), and the shadow prices associated with these constraints. On the right is the basic sensitivity analysis information for the c_j and b_i parameters. Although it was obtained algebraically, the top table gives the same information as was obtained graphically from Fig. 4.9. When just one b_i value is changed, the bottom table gives the range of values for that b_i over which the originally optimal CPF solution (after adjusting the values of the variables for the change in b_i) will remain feasible. Once again, this information obtained algebraically also can be derived from graphical analysis for this two-variable problem. (See Prob. 4.7-1.) For example, when b_2 is increased from 12 in Fig. 4.8, the originally optimal CPF solution at the intersection of the two constraint boundaries $2x_2 = b_2$ and $3x_1 + 2x_2 = 18$ will remain feasible (including $x_1 \geq 0$) only for $b_2 \leq 18$.

Parametric Linear Programming

Sensitivity analysis involves changing one parameter at a time in the original model to check its effect on the optimal solution. By contrast, **parametric linear programming** (or **parametric programming** for short) involves the systematic study of how the optimal solution changes as *many* of the parameters change *simultaneously* over some range. This study can provide a very useful extension of sensitivity analysis, e.g., to check the effect of "correlated" parameters that change together due to exogenous factors such as the state of the economy. However, a more important application is the investigation of *trade-offs* in parameter values. For example, if the c_j values represent the unit profits of the respective activities, it may be possible to increase some of the c_j values at the expense of decreasing others by an appropriate shifting of personnel and equipment among activities. Similarly, if the b_i values represent the amounts of the respective resources being made available, it may be possible to increase some of the b_i values by agreeing to accept decreases in some of the others. The analysis of such possibilities is discussed and illustrated at the end of Sec. 6.7.

In some applications, the main purpose of the study is to determine the most appropriate trade-off between two basic factors, such as *costs* and *benefits*. The usual approach is to express one of these factors in the objective function (e.g., minimize total cost) and incorporate the other into the constraints (e.g., benefits $\geq$ minimum acceptable level), as was done for the Nori & Leets Co. air pollution problem in Sec. 3.4. Parametric linear programming then enables systematic investigation of what happens when the initial tentative decision on the trade-off (e.g., the minimum acceptable level for the benefits) is changed by improving one factor at the expense of the other.

The algorithmic technique for parametric linear programming is a natural extension of that for sensitivity analysis, so it, too, is based on the simplex method. The procedure is described in Sec. 7.2.

4.8 Computer Implementation

If the electronic computer had never been invented, undoubtedly you would have never heard of linear programming and the simplex method. Even though it is possible to apply the simplex method by hand to solve tiny linear programming problems, the calculations involved are just too tedious to do this on a routine basis. However, the simplex method is ideally suited for execution on a computer. It is the computer revolution that has made possible the widespread application of linear programming in recent decades.

Computer codes for the simplex method now are widely available for essentially all modern computer systems. In fact, major manufacturers of mainframe computers usually supply their customers with a sophisticated software package for mathematical programming that includes the simplex method as well as many of the related procedures described in Chaps. 6 to 9 (including those used for post-optimality analysis). Similar packages also have been developed by independent software development companies and service bureaus.

These production computer codes do not closely follow either the algebraic form or the tabular form of the simplex method presented in Secs. 4.3 and 4.4. These forms can be streamlined considerably for computer implementation. Therefore, the codes use instead a *matrix form* (usually called the *revised simplex method*) that is especially well suited for the computer. This form accomplishes exactly the same things as the algebraic or tabular form, but it does this while computing and storing only the numbers that are actually needed for the current iteration; and then it carries along the essential data in a more compact form. The revised simplex method is described in Sec. 5.2.

The simplex method is used routinely to solve surprisingly large linear programming problems. For example, powerful desktop computers (especially workstations) are solving problems with a *few thousand* functional constraints and *many thousand* decision variables. Mainframe computers are ranging into the *tens of thousands* of functional constraints.[1] For certain *special types* of linear programming problems (such

[1] On problems of this size, the computation time depends greatly upon the linear programming system being used because large savings can be achieved by using special techniques (e.g., *crashing techniques* for quickly finding an advanced initial BF solution). When problems are re-solved periodically after minor updating of the data, much time often is saved by using (or modifying) the last optimal solution to provide the initial basic solution for the new run.

as those described in Chap. 8 and Sec. 9.6), vastly larger problems ranging into the *hundreds of thousands* of functional constraints and *millions* of decision variables now can be solved by *specialized* versions of the simplex method.

Several factors affect how long it will take to solve a linear programming problem by the general simplex method. The most important one is the *number of ordinary functional constraints*. In fact, computation time tends to be roughly proportional to the cube of this number, so that doubling this number may multiply the computation time by a factor of approximately 8. By contrast, the number of variables is a relatively minor factor.[1] Thus doubling the number of variables probably will not even double the computation time. A third factor of some importance is the *density* of the table of constraint coefficients (i.e., the *proportion* of the coefficients that are *not* zero), because this affects the computation time *per iteration*. (For large problems encountered in practice, it is common for the density to be under 5 percent, or even under 1 percent, and this much ''sparcity'' tends to greatly accelerate the simplex method.) One common rule of thumb for the *number of iterations* is that it tends to be roughly twice the number of functional constraints.

One difficulty in dealing with large linear programming problems is the tremendous amount of data involved. For example, a problem with just 1,000 functional constraints and 1,000 decision variables would have 1 million constraint coefficients to be specified! For such problems, clearly it would not be practical to input individual parameter values by hand as you do for small problems with your OR Courseware. Instead, computer programs are needed to access the available databases, transform the data into parameter values, organize the model, etc., as well as generating useful reports of the output.

A key advance in the last decade has been the development of *mathematical programming modeling languages* that slash the time and effort required to prepare these computer programs.

> A mathematical programming **modeling language** is a high-level language for the compact representation of large and complex mathematical programming models (including linear programming models). Thus, a modeling language enables one to transform a conceptual model formulation to an explicit model statement ready to be solved by the appropriate algorithm on the computer of choice. Other *model management tasks* that are expedited by the modeling language include *accessing data*, *transforming data* into model parameters, *modifying* the model, *analyzing solutions* from the model, *producing summary reports* in the vernacular of the decision makers, and *documenting* the model's content.

Prominent among the modeling languages have been GAMS and, more recently, AMPL, as well as LINGO, MathPro, MPL, and others.

With large linear programming problems, it is inevitable that some mistakes and faulty decisions will be made initially in formulating the model and inputting it into the computer. Therefore, as discussed in Sec. 2.4, a thorough process of testing and refining the model (*model validation*) is needed. The usual end product is not a single static model that is solved once by the simplex method. Instead, the OR team and management typically consider a long series of variations on a basic model (sometimes even thousands of variations) to examine different scenarios as part of post-optimality analysis. This entire process is greatly accelerated when it can be carried out *interactively* on

[1] This statement assumes that the revised simplex method described in Sec. 5.2 is being used.

a *desktop computer*. And, with the help of both mathematical programming modeling languages and improving computer technology, this now is becoming common practice.

Until the mid-1980s, linear programming problems were solved almost exclusively on *mainframe computers*. An exciting recent development has been an explosion in the capability of doing linear programming on desktop computers, including microcomputers as well as workstations. There are now dozens of companies around the world marketing microcomputer software based on the simplex method. Most of the packages are for IBM personal computers and IBM-compatibles, but some packages now are available for the Macintosh computer. Many are suitable for commercial applications on problems of substantial size (ranging even into the thousands of functional constraints and decision variables). Solving large problems usually requires additional memory, and the larger versions of some programs require a mathematics coprocessor.

The simplex method occasionally is included as one of many modules in a package designed for a variety of business applications, perhaps in a spreadsheet environment. One example is Microsoft's Excel, which has brought linear programming capabilities to literally millions of individuals who may never have even heard the term previously.

The convenient data entry and editing features of spreadsheets also are very helpful in constructing linear programming models. Many of the current packages are spreadsheet-compatible, and several (for example, VINO and What's Best?) actually perform the optimization within the spreadsheet program.

Some of the linear programming packages include extensions to other areas of mathematical programming such as integer programming (Chap. 12) and nonlinear programming (Chap. 13). Some prominent packages that include such extensions are LINDO, GAMS/MINOS, XPRESS-MP, MPSIII, and CPLEX. All these packages also are available for mainframe computers, as are the powerful linear programming systems from IBM's Optimization Subroutine Library (OSL). In addition, the powerful CPLEX package can be run on supercomputers, including Cray machines.

4.9 The Interior-Point Approach to Solving Linear Programming Problems

The most dramatic new development in operations research (OR) during the 1980s was the discovery of the interior-point approach to solving linear programming problems. This approach appears to have great potential for solving huge linear programming problems that are far beyond the reach of even the simplex method. The approach still has not been fully developed and much research is being done to try to tap its potential.

Some Historical Perspective

In 1984, a young mathematician at AT&T Bell Laboratories, Narendra Karmarkar, created great excitement in the OR community by announcing a new linear programming algorithm that he claimed was many times faster than the simplex method. This dramatic announcement became front-page news in The New York Times and other newspapers.

Then came several years of controversy. Although Karmarkar had provided a theoretical description of his algorithm, the details needed for full computer implementation had been withheld for proprietary reasons. Many other researchers attempted to implement the algorithm to check Karmarkar's claims of vast superiority over the simplex method. Their initial findings fell far short of his claims, but enough progress was made to maintain great interest in this new approach.

Then, in 1988, came another dramatic announcement. AT&T Bell Laboratories was releasing a powerful computer implementation of variants of Karmarkar's algorithm for commercial distribution. Called the AT&T KORBX Linear Programming System, each installation of the entire system (including a dedicated mini-supercomputer) was available initially for just under $9 million.

KORBX proved to be both a scientific success and a commercial failure. It succeeded in solving some very big linear programming problems (many thousand functional constraints) many times faster than the simplex method did. It also solved some huge problems (many tens of thousands of functional constraints) that were too large for the simplex method. However, too few organizations were willing to pay the hefty price for the system. In 1990, this version of KORBX was withdrawn from the market, and AT&T Bell Laboratories soon offered instead a ''portable'' software version (one that can be run on a variety of computers) for approximately $300,000.

Meanwhile, outside AT&T Bell Laboratories, numerous individuals and groups have been conducting both theoretical research and experimental work aimed at developing improved variants of Karmarkar's algorithm. (The number of published research papers triggered by Karmarkar's discovery now is well over a thousand.) A few groups have developed sophisticated computer implementations that rival KORBX. For example, two powerful software packages mentioned at the end of the preceding section, CPLEX and IBM's OSL, added variants of Karmarkar's algorithm during the early 1990s. These variants are available for workstations for well under $10,000.

Such comparisons and prices become obsolete very quickly in this rapidly evolving area. The message of this brief historical account is threefold. First, this approach to solving linear programming problems still is undergoing development. Second, performance is improving rapidly while prices have dropped precipitously. Third, there now has been enough success with this approach to make clear that it will play an important role in the coming years as a complement to the simplex method. We describe this role later in this section.

Now let us look at the key idea behind Karmarkar's algorithm and its variants.

The Key Solution Concept

Although radically different from the simplex method, Karmarkar's algorithm does share a few of the same characteristics. It is an *iterative* algorithm. It gets started by identifying a feasible *trial solution*. At each iteration, it moves from the current trial solution to a better trial solution in the feasible region. It then continues this process until it reaches a trial solution that is (essentially) optimal.

The big difference lies in the nature of these trial solutions. For the simplex method, the trial solutions are *CPF solutions* (or BF solutions after augmenting), so all movement is along edges on the *boundary* of the feasible region. For Karmarkar's algorithm, the trial solutions are **interior points**, i.e., points *inside* the boundary of the feasible region. For this reason, Karmarkar's algorithm and its variants are referred to as **interior-point algorithms**.

Figure 4.11 The curve from (1, 2) to (2, 6) shows a typical path followed by an interior-point algorithm, right through the *interior* of the feasible region for the Wyndor Glass Co. problem.

The key solution concept: Interior-point algorithms shoot through the *interior* of the feasible region toward an optimal solution instead of taking a less direct path around the boundary of the feasible region.

To illustrate, Fig. 4.11 shows the path followed by the interior-point algorithm in your OR Courseware when it is applied to the Wyndor Glass Co. problem, starting from the initial trial solution (1, 2). Note how all the trial solutions (dots) shown on this path are inside the boundary of the feasible region as the path approaches the optimal solution (2, 6). (All the subsequent trial solutions not shown also are inside the boundary of the feasible region.) Contrast this path with the path followed by the simplex method around the boundary of the feasible region from (0, 0) to (0, 6) to (2, 6).

Table 4.18 shows the actual output from your OR Courseware for this problem.[1] (Try it yourself.) Note how the successive trial solutions keep getting closer and closer to the optimal solution, but never literally get there. However, the deviation becomes so infinitesimally small that the final trial solution can be taken to be the optimal solution for all practical purposes.

Section 7.4 presents the details of the specific interior-point algorithm that is implemented in your OR Courseware.

[1] The routine is called *Solve Automatically by the Interior-Point Algorithm*. The Option menu provides two choices for a certain parameter of the algorithm α (defined in Sec. 7.4). The choice used here is the default value of $\alpha = 0.5$.

Table 4.18 **Output of Interior-Point Algorithm in OR Courseware for Wyndor Glass Co. Problem**

Iteration	x_1	x_2	Z
0	1	2	13
1	1.27298	4	23.8189
2	1.37744	5	29.1323
3	1.56291	5.5	32.1887
4	1.80268	5.71816	33.9989
5	1.92134	5.82908	34.9094
6	1.96639	5.90595	35.429
7	1.98385	5.95199	35.7115
8	1.99197	5.97594	35.8556
9	1.99599	5.98796	35.9278
10	1.99799	5.99398	35.9639
11	1.999	5.99699	35.9819
12	1.9995	5.9985	35.991
13	1.99975	5.99925	35.9955
14	1.99987	5.99962	35.9977
15	1.99994	5.99981	35.9989

Comparison with the Simplex Method

One meaningful way of comparing interior-point algorithms with the simplex method is to examine their theoretical properties regarding computational complexity. Karmarkar has proved that the original version of his algorithm is a **polynomial time algorithm**; i.e., the time required to solve *any* linear programming problem can be bounded above by a polynomial function of the size of the problem. Pathological counterexamples have been constructed to demonstrate that the simplex method does not possess this property, so it is an **exponential time algorithm** (i.e., the required time can be bounded above only by an exponential function of the problem size). This difference in *worst-case performance* is noteworthy. However, it tells us nothing about their comparison in average performance on real problems, which is the more crucial issue.

The two basic factors that determine the performance of an algorithm on a real problem are the *average computer time per iteration* and the *number of iterations*. Our next comparisons concern these factors.

Interior-point algorithms are far more complicated than the simplex method. Considerably more extensive computations are required for each iteration to find the next trial solution. Therefore, the computer time per iteration for an interior-point algorithm is many times longer than that for the simplex method.

For fairly small problems, the numbers of iterations needed by an interior-point algorithm and by the simplex method tend to be somewhat comparable. For example, on a problem with 10 functional constraints, roughly 20 iterations would be typical for either kind of algorithm. Consequently, on problems of similar size, the total computer time for an interior-point algorithm will tend to be many times longer than that for the simplex method.

On the other hand, a key advantage of interior-point algorithms is that large problems do not require many more iterations than small problems. For example, a problem with 10,000 functional constraints probably will require well under 100 iterations. Even considering the very substantial computer time per iteration needed for a problem of this size, such a small number of iterations makes the problem quite tractable. By contrast, the simplex method might need 20,000 iterations and so might not finish within a reasonable amount of computer time. Therefore, interior-point algorithms are likely to be faster than the simplex method for such huge problems.

The reason for this very large difference in the number of iterations on huge problems is the difference in the paths followed. At each iteration, the simplex method moves from the current CPF solution to an adjacent CPF solution along an edge on the boundary of the feasible region. Huge problems have an astronomical number of CPF solutions. The path from the initial CPF solution to an optimal solution may be a very circuitous one around the boundary, taking numerous small steps to the next adjacent CPF solution. By contrast, an interior-point algorithm bypasses all this by shooting through the interior of the feasible region toward an optimal solution. Adding more functional constraints adds more constraint boundaries to the feasible region, but has little effect on the number of trial solutions needed on this path through the interior. This makes it possible for interior-point algorithms to solve problems with a huge number of functional constraints.

A final key comparison concerns the ability to perform the various kinds of post-optimality analysis described in Sec. 4.7. The simplex method and its extensions are very well suited to and are widely used for this kind of analysis. Unfortunately, the interior-point approach currently has very limited capability in this area. Given the great importance of post-optimality analysis, this is a crucial drawback of interior-point algorithms. However, we point out next how the simplex method can be combined with the interior-point approach to overcome this drawback.

The Complementary Roles of the Simplex Method and the Interior-Point Approach

Since the interior-point approach still is undergoing development, any predictions about its future role relative to the simplex method are somewhat risky. Nevertheless, we summarize below our current prediction of what will unfold during the course of your career.

We anticipate that the simplex method will continue to be the standard algorithm for the routine use of linear programming. It should continue to be the most efficient algorithm for problems with less than a few hundred functional constraints. However, the interior-point approach should gradually gain widespread use by heavy-duty users of linear programming dealing with relatively large problems. For problems with several hundred functional constraints and a similar or larger number of decision variables, the solution times for the two approaches should tend to be quite comparable. For larger problems, the interior-point approach should tend to be faster than the simplex method in most cases. As the size grows into the tens of thousands of functional constraints, the interior-point approach may be the only one capable of solving the problem.

These generalizations about how the interior-point approach and the simplex method should compare for various problem sizes will not hold across the board. The comparison is affected considerably by the *specific type* of linear programming problem being solved. As time goes on, we should learn much more about how to identify specific types which are better suited for one kind of algorithm.

One of the by-products of the emergence of the interior-point approach has been a major renewal of efforts to improve the efficiency of computer implementations of the simplex method. Impressive progress has been made in recent years, and more lies ahead. At the same time, ongoing research and development of the interior-point approach will further increase its power, and probably at a faster rate than for the simplex method.

Improving computer technology, such as massive parallel processing (a huge number of computer units operating in parallel on different parts of the same problem), also will substantially increase the size of problem that either kind of algorithm can solve. However, it now appears that the interior-point approach has much greater potential to take advantage of parallel processing than the simplex method does.

As discussed earlier, a key disadvantage of the interior-point approach is its very limited capability for performing post-optimality analysis. To overcome this drawback, researchers have been developing procedures for switching over to the simplex method after an interior-point algorithm has finished. Recall that the trial solutions obtained by an interior-point algorithm keep getting closer and closer to an optimal solution (the best CPF solution), but never quite get there. Therefore, a switching procedure requires identifying a CPF solution (or BF solution after augmenting) that is very close to the final trial solution.

For example, by looking at Fig. 4.11, it is easy to see that the final trial solution in Table 4.18 is very near the CPF solution (2, 6). Unfortunately, on problems with thousands of decision variables (so no graph is available), identifying a nearby CPF (or BF) solution is a very challenging and time-consuming task. However, good progress is being made in developing procedures to do this.

Once this nearby BF solution has been found, the optimality test for the simplex method is applied to check whether this actually is the optimal BF solution. If it is not optimal, some iterations of the simplex method are conducted to move from this BF solution to an optimal solution. Generally, only a very few iterations (perhaps one) are needed because the interior-point algorithm has brought us so close to an optimal solution. Therefore, these iterations should be done quite quickly, even on problems that are too huge to be solved from scratch. After an optimal solution is actually reached, the simplex method and its extensions are applied to help perform post-optimality analysis.

Because of the difficulties involved in applying a switching procedure (including the extra computer time), some practitioners will prefer to just use the simplex method from the outset. This makes good sense when you only occasionally encounter problems that are large enough for an interior-point algorithm to be modestly faster (before switching) than the simplex method. This modest speed-up would not justify both the extra computer time for a switching procedure and the high cost of acquiring (and learning to use) a software package based on the interior-point approach. However, for organizations which frequently must deal with extremely large linear programming problems, acquiring a state-of-the-art software package of this kind (including a switching procedure) probably is worthwhile. For sufficiently huge problems, the only available way of solving them may be with such a package.

Applications of huge linear programming models sometimes lead to savings of millions of dollars. Just one such application can pay many times over for a state-of-the-art software package based on the interior-point approach plus switching over to the simplex method at the end.

4.10 Conclusions

The simplex method is an efficient and reliable algorithm for solving linear programming problems. It also provides the basis for performing the various parts of post-optimality analysis very efficiently.

Although it has a useful geometric interpretation, the simplex method is an algebraic procedure. At each iteration, it moves from the current BF solution to a better, adjacent BF solution by choosing both an entering basic variable and a leaving basic variable and then using Gaussian elimination to solve a system of linear equations. When the current solution has no adjacent BF solution that is better, the current solution is optimal and the algorithm stops.

We presented the full algebraic form of the simplex method to convey its logic, and then we streamlined the method to a more convenient tabular form. To set up for starting the simplex method, it is sometimes necessary to use artificial variables to obtain an initial BF solution for an artificial problem. If so, either the Big *M* method or the two-phase method is used to ensure that the simplex method obtains an optimal solution for the real problem.

Microcomputer software packages based on the simplex method now are widely available for dealing with problems with hundreds of functional constraints. Similar packages for workstations make it possible to solve much larger problems (several thousand functional constraints) and then perform post-optimality analysis interactively. Mainframe programs are routinely used to solve and analyze problems ranging up to many thousands of functional constraints and variables.

Interior-point algorithms provide a powerful new tool for solving very large problems.

SELECTED REFERENCES

1. Bazaraa, M. S., J. J. Jarvis, and H. D. Sherali, *Linear Programming and Network Flows,* 2d ed., Wiley, New York, 1990.
2. Bradley, S. P., A. C. Hax, and T. L. Magnanti, *Applied Mathematical Programming,* Addison-Wesley, Reading, MA, 1977.
3. Calvert, J. E., and W. L. Voxman, *Linear Programming,* Harcourt Brace Jovanovich, Orlando, FL, 1989.
4. Schrage, L., *LINDO: An Optimization Modeling System, Text and Software,* 4th ed., Boyd and Fraser, Danvers, MA, 1991.

RELEVANT ROUTINES IN YOUR OR COURSEWARE

Demonstration examples:
Interpretation of the Slack Variables
Simplex Method—Algebraic Form
Simplex Method—Tabular Form

Interactive routines:
Enter or Revise a General Linear Programming Model
Solve Interactively by the Graphical Method
Set Up for the Simplex Method—Interactive Only
Solve Interactively by the Simplex Method

Automatic routines:
Solve Automatically by the Simplex Method
Solve Automatically by the Interior-Point Algorithm

To access these routines, call the MathProg program and then choose *General Linear Programming* under the Area menu. See Appendix 1 for documentation of the software.

PROBLEMS[1]

To the left of each of the following problems (or their parts), we have inserted a D (for Demo), I (for Interactive routine), or A (for Automatic routine) whenever a corresponding routine listed above can be helpful. An asterisk on the I or A indicates that this routine definitely should be used (unless your instructor gives you contrary instructions) and the printout from this routine is all that needs to be turned in to show your work in executing the algorithm. An asterisk on the problem number indicates that at least a partial answer is given in the back of the book.

4.1-1.* Consider the linear programming model (given in the back of the book) that was formulated for Prob. 3.2-3.

(*a*) Use graphical analysis to identify all the *corner-point solutions* for this model. Label each as either feasible or infeasible.
(*b*) Calculate the value of the objective function for each of the CPF solutions. Use this information to identify an optimal solution.
(*c*) Use the solution concepts of the simplex method given in Sec. 4.1 to identify which sequence of CPF solutions might be examined by the simplex method to reach an optimal solution. (*Hint*: There are *two* alternate sequences to be identified for this particular model.)

4.1-2. Repeat Prob. 4.1-1 for the following problem.

$$\text{Maximize} \quad Z = x_1 + 2x_2,$$

subject to

$$x_1 + 3x_2 \leq 8$$
$$x_1 + x_2 \leq 4$$

and

$$x_1 \geq 0, \quad x_2 \geq 0.$$

4.1-3. Repeat Prob. 4.1-1 for the following model.

$$\text{Maximize} \quad Z = 3x_1 + 2x_2,$$

subject to

$$x_1 \leq 4$$
$$x_1 + 3x_2 \leq 15$$
$$2x_1 + x_2 \leq 10$$

and

$$x_1 \geq 0, \quad x_2 \geq 0.$$

4.1-4. Label each of the following statements about linear programming problems as true or false, and then justify your answer.

(*a*) For minimization problems, if the objective function evaluated at a CPF solution is no larger than its value at every adjacent CPF solution, then that solution is optimal.
(*b*) Only CPF solutions can be optimal, so the number of optimal solutions cannot exceed the number of CPF solutions.
(*c*) If multiple optimal solutions exist, then an optimal CPF solution may have an adjacent CPF solution that also is optimal (the same value of Z).

[1] Problems 4.3-5, 4.3-6, 4.4-6 to 4.4-8, 4.5-1, 4.6-10, 4.6-12, 4.7-4, and 4.7-5 have been adapted, with permission, from previous operations research examinations given by the Society of Actuaries.

4.2-1. Reconsider the model in Prob. 4.1-1.

(*a*) Introduce slack variables in order to write the functional constraints in augmented form.

(*b*) For each CPF solution, identify the corresponding BF solution by calculating the values of the slack variables. For each BF solution, use the values of the variables to identify the nonbasic variables and the basic variables.

(*c*) For each BF solution, demonstrate (by plugging in the solution) that, after the nonbasic variables are set equal to zero, this BF solution also is the simultaneous solution of the system of equations obtained in part (*a*).

4.2-2. Reconsider the model in Prob. 4.1-2. Follow the instructions of Prob. 4.2-1 for parts (*a*), (*b*), and (*c*).

(*d*) Repeat part (*b*) for the corner-point infeasible solutions and the corresponding basic infeasible solutions.

(*e*) Repeat part (*c*) for the basic infeasible solutions.

4.2-3. Follow the instructions of Prob. 4.2-1 for the model in Prob. 4.1-3.

D, I*, A **4.3-1.** Use the simplex method (in algebraic form) to solve the model in Prob. 4.1-1.

4.3-2. Reconsider the model in Prob. 4.1-2.

(*a*) Apply the simplex method (in algebraic form) *by hand* to solve this model.

D, I* (*b*) Repeat part (*a*) with the corresponding interactive routine in your OR Courseware.

A* (*c*) Verify the optimal solution you obtained by using the corresponding automatic routine in your OR Courseware.

4.3-3. Follow the instructions of Prob. 4.3-2 for the model in Prob. 4.1-3.

D, I*, A **4.3-4.*** Use the simplex method (in algebraic form) to solve the following problem.

$$\text{Maximize} \quad Z = 4x_1 + 3x_2 + 6x_3,$$

subject to

$$3x_1 + x_2 + 3x_3 \leq 30$$
$$2x_1 + 2x_2 + 3x_3 \leq 40$$

and

$$x_1 \geq 0, \quad x_2 \geq 0, \quad x_3 \geq 0.$$

D, I*, A **4.3-5.** Use the simplex method (in algebraic form) to solve the following problem.

$$\text{Maximize} \quad Z = x_1 + 2x_2 + 4x_3,$$

subject to

$$3x_1 + x_2 + 5x_3 \leq 10$$
$$x_1 + 4x_2 + x_3 \leq 8$$
$$2x_1 + 2x_3 \leq 7$$

and

$$x_1 \geq 0, \quad x_2 \geq 0, \quad x_3 \geq 0.$$

D, I*, A **4.3-6.** Use the simplex method (in algebraic form) to solve the following problem.

$$\text{Maximize} \quad Z = x_1 + 2x_2 + 2x_3,$$

subject to

$$5x_1 + 2x_2 + 3x_3 \leq 15$$
$$x_1 + 4x_2 + 2x_3 \leq 12$$
$$2x_1 + x_3 \leq 8$$

and

$$x_1 \geq 0, \qquad x_2 \geq 0, \qquad x_3 \geq 0.$$

4.3-7. Consider the following problem.

$$\text{Maximize} \qquad Z = 5x_1 + 3x_2 + 4x_3,$$

subject to

$$2x_1 + x_2 + x_3 \leq 20$$
$$3x_1 + x_2 + 2x_3 \leq 30$$

and

$$x_1 \geq 0, \qquad x_2 \geq 0, \qquad x_3 \geq 0.$$

You are given the information that the *nonzero* variables in the optimal solution are x_2 and x_3.

(*a*) Describe how you can use this information to adapt the simplex method to solve this problem in the minimum possible number of iterations (when you start from the usual initial BF solution). Do *not* actually perform any iterations.

(*b*) Use the procedure developed in part (*a*) to solve this problem by hand. (Do *not* use your OR Courseware.)

4.3-8. Consider the following problem.

$$\text{Maximize} \qquad Z = 2x_1 + 4x_2 + 3x_3,$$

subject to

$$x_1 + 3x_2 + 2x_3 \leq 30$$
$$x_1 + x_2 + x_3 \leq 24$$
$$3x_1 + 5x_2 + 3x_3 \leq 60$$

and

$$x_1 \geq 0, \qquad x_2 \geq 0, \qquad x_3 \geq 0.$$

You are given the information that $x_1 > 0$, $x_2 = 0$, and $x_3 > 0$ in the optimal solution.

(*a*) Describe how you can use this information to adapt the simplex method to solve this problem in the minimum possible number of iterations (when you start from the usual initial BF solution). Do *not* actually perform any iterations.

(*b*) Use the procedure developed in part (*a*) to solve this problem by hand. (Do *not* use your OR Courseware.)

4.3-9. Label each of the following statements as true or false, and then justify your answer by referring to specific statements (with page citations) in the chapter.

(*a*) The simplex method's rule for choosing the entering basic variable is used because it always leads to the *best* adjacent BF solution (largest Z).

(*b*) The simplex method's minimum ratio rule for choosing the leaving basic variable is used because making another choice with a larger ratio would yield a basic solution that is not feasible.

(*c*) When the simplex method solves for the next BF solution, elementary algebraic operations are used to eliminate each nonbasic variable from all but one equation (*its* equation) and to give it a coefficient of +1 in that one equation.

D, I*, A **4.4-1.** Repeat Prob. 4.3-1, using the tabular form of the simplex method.

D, I*, A* **4.4-2.** Repeat Prob. 4.3-2, using the tabular form of the simplex method.

D, I*, A* **4.4-3.** Repeat Prob. 4.3-3, using the tabular form of the simplex method.

4.4-4. Consider the following problem.

$$\text{Maximize} \quad Z = 2x_1 + x_2,$$

subject to

$$x_1 + x_2 \leq 40$$
$$4x_1 + x_2 \leq 100$$

and

$$x_1 \geq 0, \quad x_2 \geq 0.$$

D (*a*) Solve this problem graphically by hand. Also identify all the CPF solutions.
D, I* (*b*) Now use your OR Courseware to solve this problem graphically.
D (*c*) Use hand calculations to solve this problem by the simplex method in algebraic form.
D, I*, A (*d*) Now use your OR Courseware to solve this problem interactively by the simplex method in algebraic form.
D (*e*) Use hand calculations to solve this problem by the simplex method in tabular form.
D, I*, A (*f*) Now use your OR Courseware to solve this problem interactively by the simplex method in tabular form.

4.4-5. Repeat Prob. 4.4-4 for the following problem.

$$\text{Maximize} \quad Z = 2x_1 + 3x_2,$$

subject to

$$x_1 + 2x_2 \leq 30$$
$$x_1 + x_2 \leq 20$$

and

$$x_1 \geq 0, \quad x_2 \geq 0.$$

D, I*, A **4.4-6.** Consider the following problem.

$$\text{Maximize} \quad Z = 2x_1 + 4x_2 + 3x_3,$$

subject to

$$3x_1 + 4x_2 + 2x_3 \leq 60$$
$$2x_1 + x_2 + 2x_3 \leq 40$$
$$x_1 + 3x_2 + 2x_3 \leq 80$$

and

$$x_1 \geq 0, \quad x_2 \geq 0, \quad x_3 \geq 0.$$

(*a*) Solve by the simplex method in algebraic form.
(*b*) Solve by the simplex method in tabular form.

D, I*, A **4.4-7.** Consider the following problem.

$$\text{Maximize} \quad Z = 3x_1 + 5x_2 + 6x_3,$$

subject to

$$2x_1 + x_2 + x_3 \leq 4$$
$$x_1 + 2x_2 + x_3 \leq 4$$
$$x_1 + x_2 + 2x_3 \leq 4$$
$${}_1x + {}_2x + {}_3x \leq 3$$

and

$$x_1 \geq 0, \quad x_2 \geq 0, \quad x_3 \geq 0.$$

(*a*) Solve by the simplex method in algebraic form.
(*b*) Solve by the simplex method in tabular form.

D, I*, A **4.4-8.** Consider the following problem.

$$\text{Maximize} \quad Z = 2x_1 - x_2 + x_3,$$

subject to

$$x_1 - x_2 + 3x_3 \leq 4$$
$$2x_1 + x_2 \leq 10$$
$$x_1 - x_2 - x_3 \leq 7$$

and

$$x_1 \geq 0, \quad x_2 \geq 0, \quad x_3 \geq 0.$$

(*a*) Solve by the simplex method in algebraic form.
(*b*) Solve by the simplex method in tabular form.

D, I*, A **4.4-9.** Use the simplex method (in tabular form) to solve the following problem.

$$\text{Maximize} \quad Z = 2x_1 - x_2 + x_3,$$

subject to

$$3x_1 + x_2 + x_3 \leq 6$$
$$x_1 - x_2 + 2x_3 \leq 1$$
$$x_1 + x_2 - x_3 \leq 2$$

and

$$x_1 \geq 0, \quad x_2 \geq 0, \quad x_3 \geq 0.$$

D, I*, A **4.4-10.** Use the simplex method to solve the following problem.

$$\text{Maximize} \quad Z = -x_1 + x_2 + 2x_3,$$

subject to

$$x_1 + 2x_2 - x_3 \leq 20$$
$$-2x_1 + 4x_2 + 2x_3 \leq 60$$
$$2x_1 + 3x_2 + x_3 \leq 50$$

and

$$x_1 \geq 0, \quad x_2 \geq 0, \quad x_3 \geq 0.$$

4.5-1. Consider the following statements about linear programming and the simplex method. Label each statement as true or false, and then justify your answer.

(*a*) In a particular iteration of the simplex method, if there is a tie for which variable should be the leaving basic variable, then the next BF solution must have at least one basic variable equal to zero.
(*b*) If there is no leaving basic variable at some iteration, then the problem has no feasible solutions.
(*c*) If at least one of the basic variables has a coefficient of zero in row 0 of the final tableau, then the problem has multiple optimal solutions.

(*d*) If the problem has multiple optimal solutions, then the problem must have a bounded feasible region.

D, I*, A **4.5-2.** Consider the following problem.

$$\text{Maximize} \quad Z = 5x_1 + x_2 + 3x_3 + 4x_4,$$

subject to

$$x_1 - 2x_2 + 4x_3 + 3x_4 \leq 20$$
$$-4x_1 + 6x_2 + 5x_3 - 4x_4 \leq 40$$
$$2x_1 - 3x_2 + 3x_3 + 8x_4 \leq 50$$

and

$$x_1 \geq 0, \quad x_2 \geq 0, \quad x_3 \geq 0, \quad x_4 \geq 0.$$

Use the simplex method to demonstrate that Z is unbounded.

4.5-3. A basic property of any linear programming problem with a bounded feasible region is that every feasible solution can be expressed as a convex combination of the CPF solutions (perhaps in more than one way). Similarly, for the augmented form of the problem, every feasible solution can be expressed as a convex combination of the BF solutions.

(*a*) Show that *any* convex combination of *any* set of feasible solutions must be a feasible solution (so that any convex combination of CPF solutions must be feasible).

(*b*) Use the result quoted in part (*a*) to show that any convex combination of BF solutions must be a feasible solution.

4.5-4. Using the facts given in Prob. 4.5-3, show that the following statements must be true for any linear programming problem that has a bounded feasible region and multiple optimal solutions:

(*a*) Every convex combination of the optimal BF solutions must be optimal.

(*b*) No other feasible solution can be optimal.

D, I* **4.5-5.** Consider the following problem.

$$\text{Maximize} \quad Z = x_1 + x_2 + x_3 + x_4,$$

subject to

$$x_1 + x_2 \leq 3$$
$$x_3 + x_4 \leq 2$$

and

$$x_j \geq 0, \quad \text{for } j = 1, 2, 3, 4.$$

Use the simplex method to find *all* the optimal BF solutions.

4.6-1.* Consider the following problem.

$$\text{Maximize} \quad Z = 2x_1 + 3x_2,$$

subject to

$$x_1 + 2x_2 \leq 4$$
$$x_1 + x_2 = 3$$

and

$$x_1 \geq 0, \quad x_2 \geq 0.$$

I* (*a*) Solve this problem graphically.

(*b*) Using the Big M method, construct the complete first simplex tableau for the simplex method and identify the corresponding initial (artificial) BF solution. Also identify the initial entering basic variable and the leaving basic variable.

I*, A (*c*) Solve by the simplex method.

4.6-2. Consider the following problem.

$$\text{Maximize} \quad Z = 4x_1 + 2x_2 + 3x_3 + 5x_4,$$

subject to

$$2x_1 + 3x_2 + 4x_3 + 2x_4 = 300$$
$$8x_1 + x_2 + x_3 + 5x_4 = 300$$

and

$$x_j \geq 0, \quad \text{for } j = 1, 2, 3, 4.$$

(*a*) Using the Big M method, construct the complete first simplex tableau for the simplex method and identify the corresponding initial (artificial) BF solution. Also identify the initial entering basic variable and the leaving basic variable.

I*, A (*b*) Solve by the simplex method.

(*c*) Using the two-phase method, construct the complete first simplex tableau for phase 1 and identify the corresponding initial (artificial) BF solution. Also identify the initial entering basic variable and the leaving basic variable.

I* (*d*) Perform phase 1.

(*e*) Construct the complete first simplex tableau for phase 2.

I*, A (*f*) Perform phase 2.

(*g*) Compare the sequence of BF solutions obtained in part (*b*) with that in parts (*d*) and (*f*). Which of these solutions are feasible only for the artificial problem obtained by introducing artificial variables and which are actually feasible for the real problem?

4.6-3. Consider the following problem.

$$\text{Minimize} \quad Z = 3x_1 + 2x_2,$$

subject to

$$2x_1 + x_2 \geq 10$$
$$-3x_1 + 2x_2 \leq 6$$
$$x_1 + x_2 \geq 6$$

and

$$x_1 \geq 0, \quad x_2 \geq 0.$$

I* (*a*) Solve this problem graphically.

(*b*) Using the Big M method, construct the complete first simplex tableau for the simplex method and identify the corresponding initial (artificial) BF solution. Also identify the initial entering basic variable and the leaving basic variable.

I*, A (*c*) Solve by the simplex method.

4.6-4.* Consider the following problem.

$$\text{Minimize} \quad Z = 2x_1 + 3x_2 + x_3,$$

subject to

$$x_1 + 4x_2 + 2x_3 \geq 8$$
$$3x_1 + 2x_2 \qquad \geq 6$$

and

$$x_1 \geq 0, \qquad x_2 \geq 0, \qquad x_3 \geq 0.$$

(*a*) Reformulate this problem to fit our standard form for a linear programming model presented in Sec. 3.2.

I*, A (*b*) Using the Big M method, solve by the simplex method.

I*, A (*c*) Using the two-phase method, solve by the simplex method.

(*d*) Compare the sequence of BF solutions obtained in parts (*b*) and (*c*). Which of these solutions are feasible only for the artificial problem obtained by introducing artificial variables and which are actually feasible for the real problem?

4.6-5. For the Big M method, explain why the simplex method never would choose an artificial variable to be an entering basic variable once all the artificial variables are nonbasic.

4.6-6. Consider the following problem.

$$\text{Maximize} \qquad Z = 2x_1 + 5x_2 + 3x_3,$$

subject to

$$x_1 - 2x_2 + x_3 \geq 20$$
$$2x_1 + 4x_2 + x_3 = 50$$

and

$$x_1 \geq 0, \qquad x_2 \geq 0, \qquad x_3 \geq 0.$$

(*a*) Using the Big M method, construct the complete first simplex tableau for the simplex method and identify the corresponding initial (artificial) BF solution. Also identify the initial entering basic variable and the leaving basic variable.

I*, A (*b*) Solve by the simplex method.

(*c*) Using the two-phase method, construct the complete first simplex tableau for phase 1 and identify the corresponding initial (artificial) BF solution. Also identify the initial entering basic variable and the leaving basic variable.

I* (*d*) Perform phase 1.

(*e*) Construct the complete first simplex tableau for phase 2.

I*, A (*f*) Perform phase 2.

(*g*) Compare the sequence of BF solutions obtained in part (*b*) with that in parts (*d*) and (*f*). Which of these solutions are feasible only for the artificial problem obtained by introducing artificial variables and which are actually feasible for the real problem?

I*, A **4.6-7.** Consider the following problem.

$$\text{Minimize} \qquad Z = 2x_1 + x_2 + 3x_3,$$

subject to

$$5x_1 + 2x_2 + 7x_3 = 420$$
$$3x_1 + 2x_2 + 5x_3 \geq 280$$

and

$$x_1 \geq 0, \qquad x_2 \geq 0, \qquad x_3 \geq 0.$$

Using the two-phase method, solve by the simplex method.

4.6-8*. Consider the following problem.

$$\text{Minimize} \qquad Z = 3x_1 + 2x_2 + 4x_3,$$

subject to

$$2x_1 + x_2 + 3x_3 = 60$$

$$3x_1 + 3x_2 + 5x_3 \geq 120$$

and

$$x_1 \geq 0, \quad x_2 \geq 0, \quad x_3 \geq 0.$$

I*, A (*a*) Using the Big *M* method, solve by the simplex method.

I*, A (*b*) Using the two-phase method, solve by the simplex method.

(*c*) Compare the sequence of BF solutions obtained in parts (*a*) and (*b*). Which of these solutions are feasible only for the artificial problem obtained by introducing artificial variables and which are actually feasible for the real problem?

4.6-9. Consider the following problem.

$$\text{Minimize} \quad Z = 3x_1 + 2x_2 + 7x_3,$$

subject to

$$-x_1 + x_2 \quad = 10$$

$$2x_1 - x_2 + x_3 \geq 10$$

and

$$x_1 \geq 0, \quad x_2 \geq 0, \quad x_3 \geq 0.$$

I*, A (*a*) Using the Big *M* method, solve by the simplex method.

I*, A (*b*) Using the two-phase method, solve by the simplex method.

(*c*) Compare the sequence of BF solutions obtained in parts (*a*) and (*b*). Which of these solutions are feasible only for the artificial problem obtained by introducing artificial variables and which are actually feasible for the real problem?

4.6-10. Consider the following problem.

$$\text{Minimize} \quad Z = 3x_1 + 2x_2 + x_3,$$

subject to

$$x_1 + x_2 \quad = 7$$

$$3x_1 + x_2 + x_3 \geq 10$$

and

$$x_1 \geq 0, \quad x_2 \geq 0, \quad x_3 \geq 0.$$

I*, A (*a*) Using the Big *M* method, solve by the simplex method.

I*, A (*b*) Using the two-phase method, solve by the simplex method.

(*c*) Compare the sequence of BF solutions obtained in parts (*a*) and (*b*). Which of these solutions are feasible only for the artificial problem obtained by introducing artificial variables and which are actually feasible for the real problem?

4.6-11. Label each of the following statements as true or false, and then justify your answer.

(*a*) When a linear programming model has an equality constraint, an artificial variable is introduced into this constraint in order to start the simplex method with an obvious initial basic solution that is feasible for the original model.

(*b*) When an artificial problem is created by introducing artificial variables and using the Big *M* method, if all artificial variables in an optimal solution for the artificial problem are equal to zero, then the real problem has no feasible solutions.

(*c*) The two-phase method is commonly used in practice because it usually requires fewer iterations to reach an optimal solution than the Big *M* method does.

4.6-12. Consider the following problem.

$$\text{Maximize} \quad Z = x_1 + 4x_2 + 2x_3,$$

subject to

$$4x_1 - x_2 + x_3 \le 5$$
$$-x_1 - x_2 + 2x_3 \le 10$$

and

$$x_2 \ge 0, \quad x_3 \ge 0$$

(no nonnegativity constraint for x_1).

(*a*) Reformulate this problem so all variables have nonnegativity constraints.

D, I*, A (*b*) Solve by the simplex method.

4.6-13.* Consider the following problem.

$$\text{Maximize} \quad Z = -x_1 + 4x_2,$$

subject to

$$-3x_1 + x_2 \le 6$$
$$x_1 + 2x_2 \le 4$$
$$x_2 \ge -3$$

(no lower bound constraint for x_1).

I* (*a*) Solve this problem graphically.

(*b*) Reformulate this problem so that it has only two functional constraints and all variables have nonnegativity constraints.

D, I*, A (*c*) Solve by the simplex method.

4.6-14. Consider the following problem.

$$\text{Maximize} \quad Z = -x_1 + 2x_2 + x_3,$$

subject to

$$3x_2 + x_3 \le 120$$
$$x_1 - x_2 - 4x_3 \le 80$$
$$-3x_1 + x_2 + 2x_3 \le 100$$

(no nonnegativity constraints).

(*a*) Reformulate this problem so that all variables have nonnegativity constraints.

D, I*, A (*b*) Solve by the simplex method.

4.6-15. This chapter has described the simplex method as applied to linear programming problems where the objective function is to be maximized. Section 4.6 then described how to convert a minimization problem to an equivalent maximization problem for applying the simplex method. Another option with minimization problems is to make a few modifications in the instructions for the simplex method given in the chapter in order to apply the algorithm directly.

(*a*) Describe what these modifications would need to be.

(*b*) Using the Big *M* method, apply the modified algorithm developed in part (*a*) to solve the following problem directly by hand. (Do not use your OR Courseware.)

$$\text{Minimize} \quad Z = 3x_1 + 8x_2 + 5x_3,$$

subject to

$$3x_2 + 4x_3 \ge 70$$
$$3x_1 + 5x_2 + 2x_3 \ge 70$$

and

$$x_1 \geq 0, \qquad x_2 \geq 0, \qquad x_3 \geq 0.$$

4.6-16. Consider the following problem.

$$\text{Maximize} \quad Z = -2x_1 + x_2 - 4x_3 + 3x_4,$$

subject to

$$x_1 + x_2 + 3x_3 + 2x_4 \leq 4$$
$$x_1 - x_3 + x_4 \geq -1$$
$$2x_1 + x_2 \leq 2$$
$$x_1 + 2x_2 + x_3 + 2x_4 = 2$$

and

$$x_2 \geq 0, \qquad x_3 \geq 0, \qquad x_4 \geq 0$$

(no nonnegativity constraint for x_1).

(*a*) Reformulate this problem to fit our standard form for a linear programming model presented in Sec. 3.2.
(*b*) Using the Big *M* method, construct the complete first simplex tableau for the simplex method and identify the corresponding initial (artificial) BF solution. Also identify the initial entering basic variable and the leaving basic variable.
(*c*) Using the two-phase method, construct row 0 of the first simplex tableau for phase 1.
A* (*d*) Use the automatic routine for the simplex method in your OR Courseware to solve this problem.

I*, A **4.6-17.** Consider the following problem.

$$\text{Maximize} \quad Z = 4x_1 + 5x_2 + 3x_3.$$

subject to

$$x_1 + x_2 + 2x_3 \geq 20$$
$$15x_1 + 6x_2 - 5x_3 \leq 50$$
$$x_1 + 3x_2 + 5x_3 \leq 30$$

and

$$x_1 \geq 0, \qquad x_2 \geq 0, \qquad x_3 \geq 0.$$

Use the simplex method to demonstrate that this problem does not possess any feasible solutions.

4.7-1. Refer to the *allowable range to stay feasible* given in Fig. 4.10 for the respective right-hand sides of the Wyndor Glass Co. problem given in Sec. 3.1. Use graphical analysis to demonstrate that each given allowable range is correct.

4.7-2. Reconsider the model in Prob. 4.1-2. Interpret the right-hand side of the respective functional constraints as the amount available of the respective resources.

I (*a*) Use graphical analysis as in Fig. 4.8 to determine the shadow prices for the respective resources.
I* (*b*) Use graphical analysis to perform sensitivity analysis on this model. In particular, check each parameter of the model to determine whether it is a *sensitive* parameter (a parameter whose value cannot be changed without changing the optimal solution) by examining the graph that identifies the optimal solution.
(*c*) Use graphical analysis as in Fig. 4.9 to determine the allowable range for each c_j value (coefficient of x_j in the objective function) over which the current optimal solution will remain optimal.

(*d*) Changing just one b_i value (the right-hand side of functional constraint i) will shift the corresponding constraint boundary. If the current optimal CPF solution lies on this constraint boundary, this CPF solution also will shift. Use graphical analysis to determine the allowable range for each b_i value over which this CPF solution will remain feasible.

A* (*e*) Verify your answers in parts (*a*), (*c*), and (*d*) by running the automatic routine for the simplex method in your OR Courseware and identifying the section in the output that corresponds to each of these parts.

4.7-3. Repeat Prob. 4.7-2 for the model in Prob. 4.1-3.

4.7-4. You are given the following linear programming problem.

$$\text{Maximize} \quad Z = 4x_1 + 2x_2,$$

subject to

$$\begin{aligned} 2x_1 \qquad\quad &\leq 16 \qquad \text{(resource 1)} \\ x_1 + 3x_2 &\leq 17 \qquad \text{(resource 2)} \\ x_2 &\leq 5 \qquad\ \text{(resource 3)} \end{aligned}$$

and

$$x_1 \geq 0, \qquad x_2 \geq 0.$$

I* (*a*) Solve this problem graphically.

I (*b*) Graphically find the shadow prices for the resources.

(*c*) Determine how many additional units of resource 1 would be needed to increase the optimal value of Z by 15.

4.7-5. Consider the following problem.

$$\text{Maximize} \quad Z = x_1 - 7x_2 + 3x_3,$$

subject to

$$\begin{aligned} 2x_1 + \ x_2 - x_3 &\leq 4 \qquad \text{(resource 1)} \\ 4x_1 - 3x_2 \qquad\ &\leq 2 \qquad \text{(resource 2)} \\ -3x_1 + 2x_2 + x_3 &\leq 3 \qquad \text{(resource 3)} \end{aligned}$$

and

$$x_1 \geq 0, \qquad x_2 \geq 0, \qquad x_3 \geq 0.$$

D, I*, A (*a*) Solve by the simplex method.

(*b*) Identify the shadow prices for the three resources and describe their significance.

4.7-6.* Consider the following problem.

$$\text{Maximize} \quad Z = 2x_1 - 2x_2 + 3x_3,$$

subject to

$$\begin{aligned} -x_1 + x_2 + \ x_3 &\leq \ 4 \qquad \text{(resource 1)} \\ 2x_1 - x_2 + \ x_3 &\leq \ 2 \qquad \text{(resource 2)} \\ x_1 + x_2 + 3x_3 &\leq 12 \qquad \text{(resource 3)} \end{aligned}$$

and

$$x_1 \geq 0, \qquad x_2 \geq 0, \qquad x_3 \geq 0.$$

D, I*, A (*a*) Solve by the simplex method.

(*b*) Identify the shadow prices for the three resources and describe their significance.

4.7-7. Consider the following problem.

$$\text{Maximize} \quad Z = 2x_1 + 4x_2 - x_3,$$

subject to

$$3x_2 - x_3 \le 30 \quad \text{(resource 1)}$$
$$2x_1 - x_2 + x_3 \le 10 \quad \text{(resource 2)}$$
$$4x_1 + 2x_2 - 2x_3 \le 40 \quad \text{(resource 3)}$$

and

$$x_1 \ge 0, \quad x_2 \ge 0, \quad x_3 \ge 0.$$

D, I*, A (*a*) Solve by the simplex method.

(*b*) Identify the shadow prices for the three resources and describe their significance.

4.7-8. Consider the following problem.

$$\text{Maximize} \quad Z = 5x_1 + 4x_2 - x_3 + 3x_4,$$

subject to

$$3x_1 + 2x_2 - 3x_3 + x_4 \le 24 \quad \text{(resource 1)}$$

and

$$x_1 \ge 0, \quad x_2 \ge 0, \quad x_3 \ge 0, \quad x_4 \ge 0.$$

D, I*, A (*a*) Solve by the simplex method.

(*b*) Identify the shadow prices for the two resources and describe their significance.

4.8-1. Consider Prob. 3.4-15. The linear programming model for this problem has more than 5,000 functional constraints and more than 150,000 variables.

(*a*) There are more than 750,000,000 coefficients for these constraints, which creates a storage problem for a computer solution of the model. Given that more than 99 percent of these coefficients are zeros, recommend a way to alleviate this problem.

(*b*) Since the number of nonzero coefficients is well over 100,000, manually inputting these data into the computer would be excessively time-consuming. Given that the number of items of basic raw data is much smaller, recommend a way to alleviate this problem.

A* **4.9-1.** Use the interior-point algorithm in your OR Courseware to solve the model in Prob. 4.1-1. Choose $\alpha = 0.5$ from the Option menu, use $(x_1, x_2) = (0.1, 0.4)$ as the initial trial solution, and run 15 iterations. Draw a graph of the feasible region, and then plot the trajectory of the trial solutions through this feasible region.

A* **4.9-2.** Repeat Prob. 4.9-1 for the model in Prob. 4.1-2.

A* **4.9-3.** Repeat Prob. 4.9-1 for the model in Prob. 4.1-3.

CASE PROBLEM

CP4-1. A farm family owns 640 acres of land. Its members can produce a total of 4,000 person-hours of labor during the winter and spring months and 4,500 person-hours during the summer and fall. If any of these person-hours are not needed, younger members of the family will use them to work on a neighboring farm for $5 per hour during the winter and spring months and $5.50 per hour during the summer and fall.

The farm supports two types of livestock (dairy cows and laying hens) as well as three crops (soybeans, corn, and wheat). (All three are cash crops, but the corn also is a feed crop for the cows and the wheat also is used for chicken feed.) The crops are harvested during the late summer and fall. During the winter months, a decision is made about the mix of livestock and crops for the coming year.

Currently, the family has just completed a particularly successful harvest which has provided an investment fund of $20,000 that can be used to purchase more livestock. (Other money is available for ongoing expenses, including the next planting of crops.) The family currently has 30 cows valued at $35,000 and 2,000 hens valued at $5,000. They wish to keep all this livestock and perhaps purchase more. Each new cow would cost $1,500, and each new hen would cost $3.

Over a year's time, the value of a herd of cows will decrease by about 10 percent and the value of a flock of hens will decrease by about 25 percent due to aging.

Each cow will require 2 acres of land for grazing and 10 person-hours of work per month, while producing a net annual cash income of $850 for the family. The corresponding figures for each hen are as follows: no significant acreage, 0.05 person-hour per month, and an annual net cash income of $4.25. The chicken house can accommodate a maximum of 5,000 hens, and the size of the barn limits the herd to a maximum of 42 cows.

For each acre planted in each of the three crops, the following table gives the number of person-hours of work that will be required during the first and second halves of the year, as well as an estimate of the crop's net value (in either income or savings in purchasing feed for the livestock).

	Soybeans	Corn	Wheat
Winter and spring, person-hours	1.0	0.9	0.6
Summer and fall, person-hours	1.4	1.2	0.7
Net value, $	70	60	40

To provide much of the feed for the livestock, the family wants to plant at least 1 acre of corn for each cow in the coming year's herd and at least 0.05 acre of wheat for each hen in the coming year's flock.

The family wishes to determine how much acreage should be planted in each of the crops and how many cows and hens to have for the coming year to maximize the family's monetary worth at the end of the coming year (the sum of the net income from the livestock for the coming year plus the net value of the crops for the coming year plus what remains from the investment fund plus the value of the livestock at the end of the coming year, minus living expenses of $40,000 for the year).

(*a*) Formulate a linear programming model for this problem.

A* (*b*) Solve this model by the simplex method.

The above estimates of the net value per acre planted in each of the three crops assume good weather conditions. Adverse weather conditions would harm the crops and greatly reduce the resulting value. The scenarios particularly feared by the family are a drought, a flood, an early frost, both a drought and an early frost, and both a flood and an early frost. The estimated net values for the year under these scenarios are shown below.

	Net Value per Acre Planted, $		
Scenario	Soybeans	Corn	Wheat
Drought	−10	−15	0
Flood	15	20	10
Early frost	50	40	30
Drought and early frost	−15	−20	−10
Flood and early frost	10	10	5

A* (*c*) Find an optimal solution under each scenario by making the necessary adjustments to the linear programming model formulated in part (*a*) and then applying the simplex method.

(*d*) For the optimal solution obtained under each of the six scenarios [including the good-weather scenario considered in parts (*a*) and (*b*)], calculate the family's monetary worth at the end of the year if each of the other five scenarios occurred instead. (Note that the computer output already provides what the family's worth would be if the assumed scenario occurred.) In your judgment, which solution provides the best balance between yielding a large monetary worth under good-weather conditions and avoiding an overly small monetary worth under adverse weather conditions?

The family has researched what the weather conditions were in past years as far back as weather records have been kept, and they obtained the following data.

Scenario	Frequency, %
Good weather	40
Drought	20
Flood	10
Early frost	15
Drought and early frost	10
Flood and early frost	5

With these data, the family has decided to use the following approach to making its planting and livestock decisions. Rather than taking the optimistic approach of assuming that good weather conditions will prevail [as done in parts (*a*) and (*b*)], the *average* net value under all weather conditions will be used for each crop (weighting the net values under the various scenarios by the frequencies in the above table). In addition, any solution that would reduce the family's monetary worth by more than $10,000 under any of the scenarios is ruled out.

(*e*) Modify the linear programming model formulated in part (*a*) to fit this new approach.

A* (*f*) Solve this model by the simplex method.

(*g*) Use a shadow price obtained in part (*f*) to analyze whether it would be worthwhile for the family to obtain a bank loan with a 10 percent interest rate to purchase more livestock now beyond what can be obtained with the $20,000 from the investment fund.

(*h*) Use the sensitivity analysis information obtained in part (*f*) to identify how much latitude for error is available in estimating each coefficient in the objective function without changing the optimal solution. Which two coefficients need to be estimated most carefully?

This problem illustrates a kind of situation that is frequently faced by various kinds of organizations. To describe the situation in general terms, an organization faces an uncertain future where any one of a number of scenarios may unfold. Which one will occur depends on conditions that are outside the control of the organization. The organization needs to choose the levels of various activities, but the unit contribution of each activity to the overall measure of performance is greatly affected by which scenario unfolds. Under these circumstances, what is the best mix of activities?

(*i*) Think about specific situations outside of farm management that fit this description. Describe one.

5

The Theory of the Simplex Method

Chapter 4 introduced the basic mechanics of the simplex method. Now we shall delve a little more deeply into this algorithm by examining some of its underlying theory. The first section further develops the general geometric and algebraic properties that form the foundation of the simplex method. We then describe the *matrix form* of the simplex method (called the *revised simplex method*), which streamlines the procedure considerably for computer implementation. Next we present a fundamental insight about a property of the simplex method that enables us to deduce how changes that are made in the original model get carried along to the final simplex tableau. This insight will provide the key to the important topics of Chap. 6 (duality theory and sensitivity analysis).

5.1 Foundations of the Simplex Method

Section 4.1 introduced *corner-point feasible* (*CPF*) *solutions* and the key role they play in the simplex method. These geometric concepts were related to the algebra of the simplex method in Secs. 4.2 and 4.3. However, all this was done in the context of the Wyndor Glass Co. problem, which has only *two decision variables* and so has a straightforward geometric interpretation. How do these concepts generalize to higher dimensions when we deal with larger problems? We address this question in this section.

We begin by introducing some basic terminology for any linear programming problem with n decision variables. While we are doing this, you may find it helpful to refer to Fig. 5.1 (which repeats Fig. 4.1) to interpret these definitions in two dimensions ($n = 2$).

Terminology

It may seem intuitively clear that optimal solutions for any linear programming problem must lie on the boundary of the feasible region, and in fact this is a general property. Because boundary is a geometric concept, our initial definitions clarify how the boundary of the feasible region is identified algebraically.

> The **constraint boundary equation** for any constraint is obtained by replacing its $\leq$, $=$, or $\geq$ sign by an $=$ sign.

Consequently, the form of a constraint boundary equation is $a_{i1}x_1 + a_{i2}x_2 + \cdots + a_{in}x_n = b_i$ for functional constraints and $x_j = 0$ for nonnegativity constraints. Each such equation defines a "flat" geometric shape (called a **hyperplane**) in n-dimensional space, analogous to the line in two-dimensional space and the plane in three-dimensional space. This hyperplane forms the **constraint boundary** for the corresponding constraint. When the constraint has either a $\leq$ or a $\geq$ sign, this *constraint boundary* separates the points that satisfy the constraint (all the points on one side up to

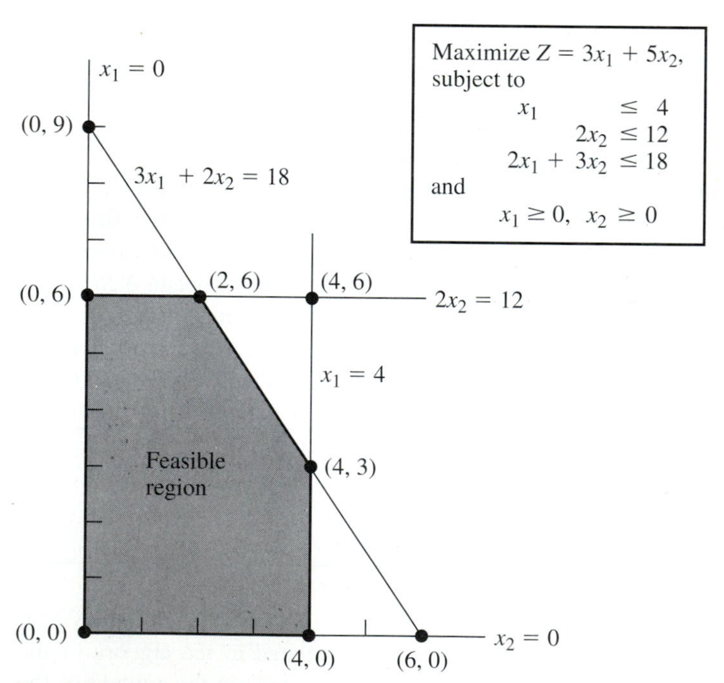

Figure 5.1 Constraint boundaries, constraint boundary equations, and corner-point solutions for the Wyndor Glass Co. problem.

and including the constraint boundary) from the points that violate the constraint (all those on the other side of the constraint boundary). When the constraint has an = sign, only the points on the constraint boundary satisfy the constraint.

For example, the Wyndor Glass Co. problem has five constraints (three functional constraints and two nonnegativity constraints), so it has the five *constraint boundary equations,* shown in Fig. 5.1. Because $n = 2$, the hyperplanes defined by these constraint boundary equations are simply lines. Therefore, the constraint boundaries for the five constraints are the five lines shown in Fig. 5.1.

> The **boundary** of the feasible region contains just those feasible solutions that satisfy one or more of the constraint boundary equations.

Geometrically, any point on the boundary of the feasible region lies on one or more of the hyperplanes defined by the respective constraint boundary equations. Thus, in Fig. 5.1, the boundary consists of the five darker line segments.

Next, we give a general definition of *CPF solution* in n-dimensional space.

> A **corner-point feasible (CPF) solution** is a feasible solution that does not lie on *any* line segment[1] connecting two *other* feasible solutions.

As this definition implies, a feasible solution that *does* lie on a line segment connecting two other feasible solutions is *not* a CPF solution. To illustrate when $n = 2$, consider Fig. 5.1. The point (2, 3) is *not* a CPF solution, because it lies on various such line segments, e.g., the line segment connecting (0, 3) and (4, 3). Similarly, (0, 3) is *not* a CPF solution, because it lies on the line segment connecting (0, 0) and (0, 6). However, (0, 0) *is* a CPF solution, because it is impossible to find two *other* feasible solutions that lie on completely opposite sides of (0, 0). (Try it.)

When the number of decision variables n is greater than 2 or 3, this definition for *CPF solution* is not a very convenient one for identifying such solutions. Therefore, it will prove most helpful to interpret these solutions algebraically. For the Wyndor Glass Co. example, each CPF solution in Fig. 5.1 lies at the intersection of two ($n = 2$) constraint lines; i.e., it is the *simultaneous solution* of a system of two constraint boundary equations. This situation is summarized in Table 5.1, where **defining equations** refer to the constraint boundary equations that yield (define) the indicated CPF solution.

> For any linear programming problem with n decision variables, each CPF solution lies at the intersection of n constraint boundaries; i.e., it is the *simultaneous solution* of a system of n constraint boundary equations.

However, this is not to say that *every* set of n constraint boundary equations chosen from the $n + m$ constraints (n nonnegativity and m functional constraints) yields a CPF solution. In particular, the simultaneous solution of such a system of equations might violate one or more of the other m constraints not chosen, in which case it is a corner-point *infeasible* solution. The example has three such solutions, as summarized in Table 5.2. (Check to see why they are infeasible.)

Furthermore, a system of n constraint boundary equations might have no solution at all. This occurs twice in the example, with the pairs of equations (1) $x_1 = 0$ and $x_1 = 4$ and (2) $x_2 = 0$ and $2x_2 = 12$. Such systems are of no interest to us.

[1] An algebraic expression for a line segment is given in Appendix 2.

Table 5.1 **Defining Equations for Each CPF Solution for the Wyndor Glass Co. Problem**

CPF Solution	Defining Equations
(0, 0)	$x_1 = 0$ $x_2 = 0$
(0, 6)	$x_1 = 0$ $2x_2 = 12$
(2, 6)	$2x_2 = 12$ $3x_1 + 2x_2 = 18$
(4, 3)	$3x_1 + 2x_2 = 18$ $x_1 = 4$
(4, 0)	$x_1 = 4$ $x_2 = 0$

The final possibility (which never occurs in the example) is that a system of n constraint boundary equations has multiple solutions because of redundant equations. You need not be concerned with this case either, because the simplex method circumvents its difficulties.

To summarize for the example, with five constraints and two variables, there are $\binom{5}{2} = 10$ pairs of constraint boundary equations. Five of these pairs became defining equations for CPF solutions (Table 5.1), three became defining equations for corner-point infeasible solutions (Table 5.2), and each of the final two pairs had no solution.

Adjacent CPF Solutions

Section 4.1 introduced adjacent CPF solutions and their role in solving linear programming problems. We now elaborate.

Recall from Chap. 4 that (when we ignore slack, surplus, and artificial variables) each iteration of the simplex method moves from the current CPF solution to an *adjacent* one. What is the *path* followed in this process? What really is meant by *adjacent* CPF solution? First we address these questions from a geometric viewpoint, and then we turn to algebraic interpretations.

Table 5.2 **Defining Equations for Each Corner-Point Infeasible Solution for the Wyndor Glass Co. Problem**

Corner-Point Infeasible Solution	Defining Equations
(0, 9)	$x_1 = 0$ $3x_1 + 2x_2 = 18$
(4, 6)	$2x_2 = 12$ $x_1 = 4$
(6, 0)	$3x_1 + 2x_2 = 18$ $x_2 = 0$

These questions are easy to answer when $n = 2$. In this case, the *boundary* of the feasible region consists of several connected *line segments* forming a *polygon,* as shown in Fig. 5.1 by the five darker line segments. These line segments are the *edges* of the feasible region. Emanating from each CPF solution are *two* such edges leading to an adjacent CPF solution at the other end. (Note in Fig. 5.1 how each CPF solution has two adjacent ones.) The path followed in an iteration is to move along one of these edges from one end to the other. In Fig. 5.1, the first iteration involves moving along the edge from (0, 0) to (0, 6), and then the next iteration moves along the edge from (0, 6) to (2, 6). As Table 5.1 illustrates, each of these moves to an adjacent CPF solution involves just one change in the set of defining equations (constraint boundaries on which the solution lies).

When $n = 3$, the answers are slightly more complicated. To help you visualize what is going on, Fig. 5.2 shows a three-dimensional drawing of a typical feasible region when $n = 3$, where the dots are the CPF solutions. This feasible region is a *polyhedron* rather than the polygon we had with $n = 2$ (Fig. 5.1), because the constraint boundaries now are *planes* rather than lines. The faces of the polyhedron form the *boundary* of the feasible region, where each face is the portion of a constraint boundary that satisfies the other constraints as well. Note that each CPF solution lies at the intersection of three constraint boundaries (sometimes including some of the $x_1 = 0$, $x_2 = 0$, and $x_3 = 0$ constraint boundaries for the nonnegativity constraints), and the solution also satisfies the other constraints. Such intersections that do not satisfy one or more of the other constraints yield corner-point *infeasible* solutions instead.

The darker line segment in Fig. 5.2 depicts the path of the simplex method on a typical iteration. The point (2, 4, 3) is the *current* CPF solution to begin the iteration, and the point (4, 2, 4) will be the new CPF solution at the end of the iteration. The point (2, 4, 3) lies at the intersection of the $x_2 = 4$, $x_1 + x_2 = 6$, and $-x_1 + 2x_3 = 4$ con-

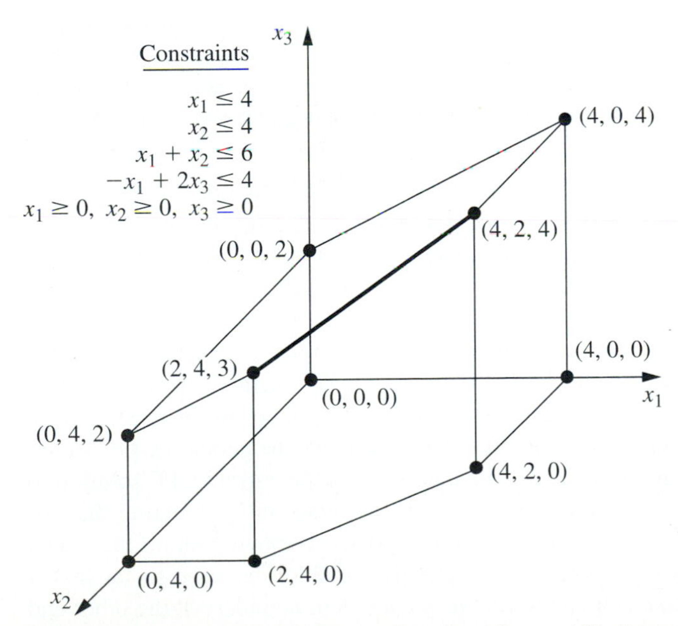

Figure 5.2 Feasible region and CPF solutions for a three-variable linear programming problem.

straint boundaries, so these three equations are the *defining equations* for this CPF solution. If the $x_2 = 4$ defining equation were removed, the intersection of the other two constraint boundaries (planes) would form a line. One segment of this line, shown as the dark line segment from (2, 4, 3) to (4, 2, 4) in Fig. 5.2, lies on the boundary of the feasible region, whereas the rest of the line is infeasible. This line segment is an edge of the feasible region, and its endpoints (2, 4, 3) and (4, 2, 4) are adjacent CPF solutions.

For $n = 3$, all the *edges* of the feasible region are formed in this way as the feasible segment of the line lying at the intersection of two constraint boundaries, and the two endpoints of an edge are *adjacent* CPF solutions. In Fig. 5.2 there are 15 edges of the feasible region, and so there are 15 pairs of adjacent CPF solutions. For the current CPF solution (2, 4, 3), there are three ways to remove one of its three defining equations to obtain an intersection of the other two constraint boundaries, so there are three edges emanating from (2, 4, 3). These edges lead to (4, 2, 4), (0, 4, 2), and (2, 4, 0), so these are the CPF solutions that are adjacent to (2, 4, 3).

For the next iteration, the simplex method chooses one of these three edges, say, the darker line segment in Fig. 5.2, and then moves along this edge away from (2, 4, 3) until it reaches the first new constraint boundary, $x_1 = 4$, at its other endpoint. [We cannot continue farther along this line to the next constraint boundary, $x_2 = 0$, because this leads to a corner-point infeasible solution—(6, 0, 5).] The intersection of this first new constraint boundary with the two constraint boundaries forming the edge yields the *new* CPF solution (4, 2, 4).

When $n > 3$, these same concepts generalize to higher dimensions, except the constraint boundaries now are *hyperplanes* instead of planes. Let us summarize.

> Consider any linear programming problem with n decision variables and a bounded feasible region. A CPF solution lies at the intersection of n constraint boundaries (and satisfies the other constraints as well). An **edge** of the feasible region is a feasible line segment that lies at the intersection of $n - 1$ constraint boundaries, where each endpoint lies on one additional constraint boundary (so that these endpoints are CPF solutions). Two CPF solutions are **adjacent** if the line segment connecting them is an edge of the feasible region. Emanating from each CPF solution are n such edges, each one leading to one of the n adjacent CPF solutions. Each iteration of the simplex method moves from the current CPF solution to an adjacent one by moving along one of these n edges.

When you shift from a geometric viewpoint to an algebraic one, *intersection of constraint boundaries* changes to *simultaneous solution of constraint boundary equations.* The n constraint boundary equations yielding (defining) a CPF solution are its defining equations, where deleting one of these equations yields a line whose feasible segment is an edge of the feasible region.

We next analyze some key properties of CPF solutions and then describe the implications of all these concepts for interpreting the simplex method. However, while the above summary is fresh in your mind, let us give you a preview of its implications. When the simplex method chooses an entering basic variable, the geometric interpretation is that it is choosing one of the edges emanating from the current CPF solution to move along. Increasing this variable from zero (and simultaneously changing the values of the other basic variables accordingly) corresponds to moving along this edge. Having one of the basic variables (the leaving basic variable) decrease so far that it reaches zero corresponds to reaching the first new constraint boundary at the other end of this edge of the feasible region.

Properties of CPF Solutions

We now focus on three key properties of CPF solutions that hold for *any* linear programming problem that has feasible solutions and a bounded feasible region.

Property 1: (*a*) If there is exactly one optimal solution, then it must be a CPF solution. (*b*) If there are multiple optimal solutions (and a bounded feasible region), then at least two must be adjacent CPF solutions.

Property 1 is a rather intuitive one from a geometric viewpoint. First consider Case (*a*), which is illustrated by the Wyndor Glass Co. problem (see Fig. 5.1) where the one optimal solution (2, 6) is indeed a CPF solution. Note that there is nothing special about this example that led to this result. For any problem having just one optimal solution, it always is possible to keep raising the objective function line (hyperplane) until it just touches one point (the optimal solution) at a corner of the feasible region.

We now give an algebraic proof for this case.

Proof of Case (*a*) of Property 1: We set up a *proof by contradiction* by assuming that there is exactly one optimal solution and that it is *not* a CPF solution. We then show below that this assumption leads to a contradiction and so cannot be true. (The solution assumed to be optimal will be denoted by $\mathbf{x}^*$, and its objective function value by Z^*.)

Recall the definition of *CPF solution* (a feasible solution that does not lie on any line segment connecting two other feasible solutions). Since we have assumed that the optimal solution $\mathbf{x}^*$ is not a CPF solution, this implies that there must be two other feasible solutions such that the line segment connecting them contains the optimal solution. Let the vectors $\mathbf{x}'$ and $\mathbf{x}''$ denote these two other feasible solutions, and let Z_1 and Z_2 denote their respective objective function values. Like each other point on the line segment connecting $\mathbf{x}'$ and $\mathbf{x}''$,

$$\mathbf{x}^* = \alpha\mathbf{x}'' + (1 - a)\mathbf{x}'$$

for some value of α such that $0 < \alpha < 1$. Thus

$$Z^* = \alpha Z_2 + (1 - \alpha)Z_1.$$

Since the weights α and $1 - \alpha$ add to 1, the only possibilities for how Z^*, Z_1, and Z_2 compare are (1) $Z^* = Z_1 = Z_2$, (2) $Z_1 < Z^* < Z_2$, and (3) $Z_1 > Z^* > Z_2$. The first possibility implies that $\mathbf{x}'$ and $\mathbf{x}''$ also are optimal, which contradicts the assumption that there is exactly one optimal solution. Both the latter possibilities contradict the assumption that $\mathbf{x}^*$ (not a CPF solution) is optimal. The resulting conclusion is that it is impossible to have a single optimal solution that is not a CPF solution.

Now consider Case (*b*), which was demonstrated in Sec. 3.2 under the definition of *optimal solution* by changing the objective function in the example to $Z = 3x_1 + 2x_2$ (see Fig. 3.5 on page 36). What then happens when you are solving graphically is that the objective function line keeps getting raised until it contains the line segment connecting the two CPF solutions (2, 6) and (4, 3). The same thing would happen in higher dimensions except that an objective function *hyperplane* would keep getting raised until it contained the line segment(s) connecting two (or more) adjacent CPF solutions. As a consequence, *all* optimal solutions can be obtained as weighted averages of optimal CPF solutions. (This situation is described further in Probs. 4.5-3 and 4.5-4.)

The real significance of Property 1 is that it greatly simplifies the search for an optimal solution because now only CPF solutions need to be considered. The magnitude of this simplification is emphasized in Property 2.

Property 2: There are only a *finite* number of CPF solutions.

This property certainly holds in Figs. 5.1 and 5.2, where there are just 5 and 10 CPF solutions, respectively. To see why the number is finite in general, recall that each CPF solution is the simultaneous solution of a system of n out of the $m + n$ constraint boundary equations. The number of different combinations of $m + n$ equations taken n at a time is

$$\binom{m+n}{n} = \frac{(m+n)!}{m!n!},$$

which is a finite number. This number, in turn, in an *upper bound* on the number of CPF solutions. In Fig. 5.1, $m = 3$ and $n = 2$, so there are 10 different systems of two equations, but only half of them yield CPF solutions. In Fig. 5.2, $m = 4$ and $n = 3$, which gives 35 different systems of three equations, but only 10 yield CPF solutions.

Property 2 suggests that, in principle, an optimal solution can be obtained by exhaustive enumeration; i.e., find and compare all the finite number of CPF solutions. Unfortunately, there are finite numbers, and then there are finite numbers that (for all practical purposes) might as well be infinite. For example, a rather small linear programming problem with only $m = 50$ and $n = 50$ would have $100!/(50!)^2 \approx 10^{29}$ systems of equations to be solved! By contrast, the simplex method would need to examine only approximately 100 CPF solutions for a problem of this size. This tremendous savings can be obtained because of the optimality test given in Sec. 4.1 and restated here as Property 3.

Property 3: If a CPF solution has no *adjacent* CPF solutions that are *better* (as measured by Z), then there are no *better* CPF solutions anywhere. Therefore, such a CPF solution is guaranteed to be an *optimal* solution (by Property 1), assuming only that the problem possesses at least one optimal solution (guaranteed if the problem possesses feasible solutions and a bounded feasible region).

To illustrate Property 3, consider Fig. 5.1 for the Wyndor Glass Co. example. For the CPF solution (2, 6), its adjacent CPF solutions are (0, 6) and (4, 3), and neither has a better value of Z than (2, 6) does. This outcome implies that none of the other CPF solutions—(0, 0) and (4, 0)—can be better than (2, 6), so (2, 6) must be optimal.

By contrast, Fig. 5.3 shows a feasible region that can *never* occur for a linear programming problem but that does violate Property 3. The problem shown is identical to the Wyndor Glass Co. example (including the same objective function) *except* for the enlargement of the feasible region to the right of $(\frac{8}{3}, 5)$. Consequently, the adjacent CPF solutions for (2, 6) now are (0, 6) and $(\frac{8}{3}, 5)$, and again neither is better than (2, 6). However, another CPF solution (4, 5) now is better than (2, 6), thereby violating Prop-

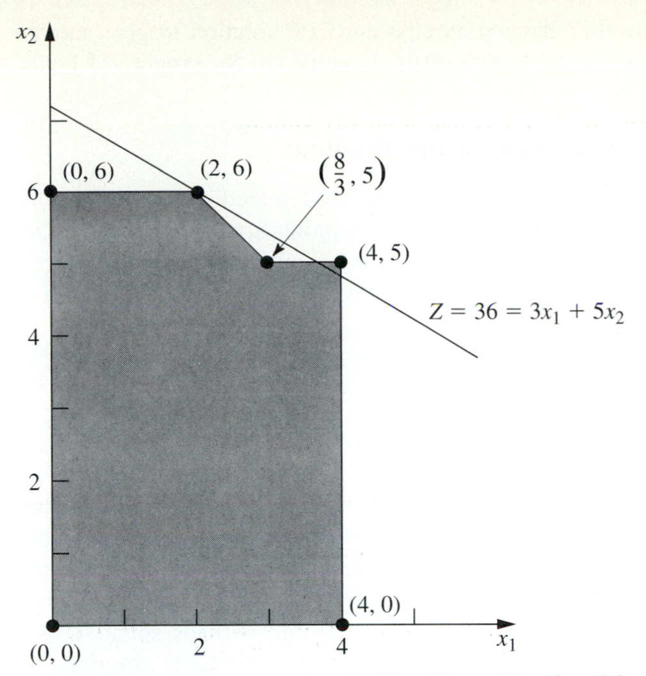

Figure 5.3 Modification of the Wyndor Glass Co. problem that violates both linear programming and Property 3 for CPF solutions in linear programming.

erty 3. The reason is that the boundary of the feasible region goes down from (2, 6) to $(\frac{8}{3}, 5)$ and then "bends outward" to (4, 5), beyond the objective function line passing through (2, 6).

The key point is that the kind of situation illustrated in Fig. 5.3 can never occur in linear programming. The feasible region in Fig. 5.3 implies that the $2x_2 \leq 12$ and $3x_1 + 2x_2 \leq 18$ constraints apply for $0 \leq x_1 \leq \frac{8}{3}$. However, under the condition that $\frac{8}{3} \leq x_1 \leq 4$, the $3x_1 + 2x_2 \leq 18$ constraint is dropped and replaced by $x_2 \leq 5$. Such "conditional constraints" just are not allowed in linear programming.

The basic reason that Property 3 holds for any linear programming problem is that the feasible region always has the property of being a *convex set,* as defined in Appendix 2 and illustrated in several figures there. For two-variable linear programming problems, this convex property means that the *angle* inside the feasible region at *every* CPF solution is less than 180°. This property is illustrated in Fig. 5.1, where the angles at (0, 0), (0, 6), and (4, 0) are 90° and those at (2, 6) and (4, 3) are between 90° and 180°. By contrast, the feasible region in Fig. 5.3 is *not* a convex set, because the angle at $(\frac{8}{3}, 5)$ is more than 180°. This is the kind of "bending outward" at an angle greater than 180° that can never occur in linear programming. In higher dimensions, the same intuitive notion of "never bending outward" continues to apply.

To clarify the significance of a convex feasible region, consider the objective function hyperplane that passes through a CPF solution that has no adjacent CPF solutions that are better. [In the original Wyndor Glass Co. example, this hyperplane is the objective function line passing through (2, 6).] All these adjacent solutions [(0, 6) and (4, 3) in the example] must lie either on the hyperplane or on the unfavorable side (as measured by Z) of the hyperplane. The feasible region being convex means that its boundary cannot "bend outward" beyond an adjacent CPF solution to give another CPF solution that lies on the favorable side of the hyperplane. So Property 3 holds.

Extensions to the Augmented Form of the Problem

For any linear programming problem in our standard form (including functional constraints in $\leq$ form), the appearance of the functional constraints after slack variables are introduced is as follows:

$$
\begin{aligned}
&(1)\ a_{11}x_1 + a_{12}x_2 + \cdots + a_{1n}x_n + x_{n+1} &&= b_1 \\
&(2)\ a_{21}x_1 + a_{22}x_2 + \cdots + a_{2n}x_n \qquad + x_{n+2} &&= b_2 \\
&\cdots\cdots\cdots\cdots\cdots\cdots\cdots\cdots\cdots\cdots\cdots\cdots \\
&(m)\ a_{m1}x_1 + a_{m2}x_2 + \cdots + a_{mn}x_n \qquad\qquad + x_{n+m} &&= b_m,
\end{aligned}
$$

where $x_{n+1}, x_{n+2}, \ldots, x_{n+m}$ are the slack variables. For other linear programming problems, Sec. 4.6 described how essentially this same appearance (proper form from Gaussian elimination) can be obtained by introducing artificial variables, etc. Thus the original solutions $(x_1, x_2, \ldots, x_n)$ now are augmented by the corresponding values of the slack or artificial variables $(x_{n+1}, x_{n+2}, \ldots, x_{n+m})$ and perhaps some surplus variables as well. This augmentation led in Sec. 4.2 to defining **basic solutions** as *augmented corner-point solutions* and **basic feasible solutions (BF solutions)** as *augmented CPF solutions.* Consequently, the preceding three properties of CPF solutions also hold for BF solutions.

Now let us clarify the algebraic relationships between basic solutions and corner-point solutions. Recall that each corner-point solution is the simultaneous solution of a system of n constraint boundary equations, which we called its *defining equations.* The key question is: How do we tell whether a particular constraint boundary equation is one of the defining equations when the problem is in augmented form? The answer, fortunately, is a simple one. Each constraint has an **indicating variable** that completely indicates (by whether its value is zero) whether that constraint's boundary equation is satisfied by the current solution. A summary appears in Table 5.3. For the type of

Table 5.3 **Indicating Variables for Constraint Boundary Equations***

Type of Constraint	Form of Constraint	Constraint in Augmented Form	Constraint Boundary Equation	Indicating Variable
Nonnegativity	$x_j \geq 0$	$x_j \geq 0$	$x_j = 0$	x_j
Functional ($\leq$)	$\sum_{j=1}^{n} a_{ij}x_j \leq b_i$	$\sum_{j=1}^{n} a_{ij}x_j + x_{n+i} = b_i$	$\sum_{j=1}^{n} a_{ij}x_j = b_i$	x_{n+i}
Functional ($=$)	$\sum_{j=1}^{n} a_{ij}x_j = b_i$	$\sum_{j=1}^{n} a_{ij}x_j + \bar{x}_{n+i} = b_i$	$\sum_{j=1}^{n} a_{ij}x_j = b_i$	$\bar{x}_{n+i}$
Functional ($\geq$)	$\sum_{j=1}^{n} a_{ij}x_j \geq b_i$	$\sum_{j=1}^{n} a_{ij}x_j + \bar{x}_{n+i} - x_{s_i} = b_i$	$\sum_{j=1}^{n} a_{ij}x_j = b_i$	$\bar{x}_{n+i} - x_{s_i}$

* Indicating variable $= 0 \Rightarrow$ constraint boundary equation satisfied;
indicating variable $\neq 0 \Rightarrow$ constraint boundary equation violated.

constraint in each row of the table, note that the corresponding constraint boundary equation (fourth column) is satisfied if and only if this constraint's indicating variable (fifth column) equals zero. In the last row (functional constraint in $\geq$ form), the indicating variable $\bar{x}_{n+i} - x_{s_i}$ actually is the difference between the artificial variable $\bar{x}_{n+i}$ and the surplus variable x_{s_i}.

Thus whenever a constraint boundary equation is one of the defining equations for a corner-point solution, its indicating variable has a value of zero in the augmented form of the problem. Each such indicating variable is called a *nonbasic variable* for the corresponding basic solution. The resulting conclusions and terminology (already introduced in Sec. 4.2) are summarized next.

> Each **basic solution** has m **basic variables,** and the rest of the variables are **nonbasic variables** set equal to zero. (The number of nonbasic variables equals n plus the number of surplus variables.) The values of the **basic variables** are given by the simultaneous solution of the system of m equations for the problem in augmented form (after the nonbasic variables are set to zero). This basic solution is the augmented corner-point solution whose n defining equations are those indicated by the nonbasic variables. In particular, whenever an indicating variable in the fifth column of Table 5.3 is a nonbasic variable, the constraint boundary equation in the fourth column is a defining equation for the corner-point solution. (For functional constraints in $\geq$ form, at least one of the two supplementary variables $\bar{x}_{n+i}$ and x_{s_i} always is a nonbasic variable, but the constraint boundary equation becomes a defining equation only if *both* of these variables are nonbasic variables.)

Now consider the basic *feasible* solutions. Note that the only requirements for a solution to be feasible in the augmented form of the problem are that it satisfy the system of equations and that *all* the variables be *nonnegative.*

> A **BF solution** is a basic solution where all m basic variables are nonnegative (≥ 0). A BF solution is said to be **degenerate** if any of these m variables equals zero.

Thus it is possible for a variable to be zero and still not be a nonbasic variable for the current BF solution. (This case corresponds to a CPF solution that satisfies another constraint boundary equation in addition to its n defining equations.) Therefore, it is necessary to keep track of which is the current set of nonbasic variables (or the current set of basic variables) rather than to rely upon their zero values.

We noted earlier that not every system of n constraint boundary equations yields a corner-point solution, because either the system has no solution or it has multiple solutions. For analogous reasons, not every set of n nonbasic variables yields a basic solution. However, these cases are avoided by the simplex method.

To illustrate these definitions, consider the Wyndor Glass Co. example once more. Its constraint boundary equations and indicating variables are shown in Table 5.4.

Augmenting each of the CPF solutions (see Table 5.1) yields the BF solutions listed in Table 5.5. This table places adjacent BF solutions next to each other, except for the pair consisting of the first and last solutions listed. Notice that in each case the nonbasic variables necessarily are the indicating variables for the defining equations. Thus adjacent BF solutions differ by having just one different nonbasic variable. Also notice that each BF solution is the simultaneous solution of the system of equations for the problem in augmented form (see Table 5.4) when the nonbasic variables are set equal to zero.

Table 5.4 **Indicating Variables for the Constraint Boundary Equations of the Wyndor Glass Co. Problem***

Constraint	Constraint in Augmented Form	Constraint Boundary Equation	Indicating Variable
$x_1 \geq 0$	$x_1 \geq 0$	$x_1 = 0$	x_1
$x_2 \geq 0$	$x_2 \geq 0$	$x_2 = 0$	x_2
$x_1 \leq 4$	(1) $x_1 + x_3 = 4$	$x_1 = 4$	x_3
$2x_2 \leq 12$	(2) $2x_2 + x_4 = 12$	$2x_2 = 12$	x_4
$3x_1 + x_2 \leq 18$	(3) $3x_1 + 2x_2 + x_5 = 18$	$3x_1 + 2x_2 = 18$	x_5

* Indicating variable $= 0 \Rightarrow$ constraint boundary equation satisfied;
indicating variable $\neq 0 \Rightarrow$ constraint boundary equation violated.

Table 5.5 **BF Solutions for the Wyndor Glass Co. Problem**

CPF Solution	Defining Equations	BF Solution	Nonbasic Variables
(0, 0)	$x_1 = 0$	(0, 0, 4, 12, 18)	x_1
	$x_2 = 0$		x_2
(0, 6)	$x_1 = 0$	(0, 6, 4, 0, 6)	x_1
	$2x_2 = 12$		x_4
(2, 6)	$2x_2 = 12$	(2, 6, 2, 0, 0)	x_4
	$3x_1 + 2x_2 = 18$		x_5
(4, 3)	$3x_1 + 2x_2 = 18$	(4, 3, 0, 6, 0)	x_5
	$x_1 = 4$		x_3
(4, 0)	$x_1 = 4$	(4, 0, 0, 12, 6)	x_3
	$x_2 = 0$		x_2

Similarly, the other three corner-point solutions (see Table 5.2) yield the remaining basic solutions shown in Table 5.6.

The other two sets of nonbasic variables, (1) x_1 and x_3 and (2) x_2 and x_4, do not yield a basic solution, because setting either pair of variables equal to zero leads to having no solution for the system of Eqs. (1) to (3) given in Table 5.4. This conclusion parallels the observation we made early in this section that the corresponding sets of constraint boundary equations do not yield a solution.

The *simplex method* starts at a BF solution and then iteratively moves to a better adjacent BF solution until an optimal solution is reached. At each iteration, how is the adjacent BF solution reached?

For the original form of the problem, recall that an adjacent CPF solution is reached from the current one by (1) deleting one constraint boundary (defining equation) from the set of n constraint boundaries defining the current solution, (2) moving

Table 5.6 **Basic Infeasible Solutions for the Wyndor Glass Co. Problem**

Corner-Point Infeasible Solution	Defining Equations	Basic Infeasible Solution	Nonbasic Variables
(0, 9)	$x_1 = 0$	(0, 9, 4, −6, 0)	x_1
	$3x_1 + 2x_2 = 18$		x_5
(4, 6)	$2x_2 = 12$	(4, 6, 0, 0, −6)	x_4
	$x_1 = 4$		x_3
(6, 0)	$3x_1 + 2x_2 = 18$	(6, 0, −2, 12, 0)	x_5
	$x_2 = 0$		x_2

Table 5.7 **Sequence of Solutions Obtained by the Simplex Method for the Wyndor Glass Co. Problem**

Iteration	CPF Solution	Defining Equations	BF Solution	Nonbasic Variables	Functional Constraints in Augmented Form
0	(0, 0)	$x_1 = 0$ $x_2 = 0$	(0, 0, 4, 12, 18)	$x_1 = 0$ $x_2 = 0$	$x_1 + \mathbf{x_3} = 4$ $2x_2 + \mathbf{x_4} = 12$ $3x_1 + 2x_2 + \mathbf{x_5} = 18$
1	(0, 6)	$x_1 = 0$ $2x_2 = 12$	(0, 6, 4, 0, 6)	$x_1 = 0$ $x_4 = 0$	$x_1 + \mathbf{x_3} = 4$ $2\mathbf{x_2} + x_4 = 12$ $3x_1 + 2\mathbf{x_2} + \mathbf{x_5} = 18$
2	(2, 6)	$2x_2 = 12$ $3x_1 + 2x_2 = 18$	(2, 6, 2, 0, 0)	$x_4 = 0$ $x_5 = 0$	$\mathbf{x_1} + \mathbf{x_3} = 4$ $2\mathbf{x_2} + x_4 = 12$ $3\mathbf{x_1} + 2\mathbf{x_2} + x_5 = 18$

away from the current solution in the feasible direction along the intersection of the remaining $n - 1$ constraint boundaries (an edge of the feasible region), and (3) stopping when the *first* new constraint boundary (defining equation) is reached.

Equivalently, in our new terminology, the simplex method reaches an adjacent BF solution from the current one by (1) deleting one variable (the entering basic variable) from the set of n nonbasic variables defining the current solution, (2) moving away from the current solution by *increasing* this one variable from zero (and adjusting the other basic variables to still satisfy the system of equations) while keeping the remaining $n - 1$ nonbasic variables at zero, and (3) stopping when the *first* of the basic variables (the leaving basic variable) reaches a value of zero (its constraint boundary). With either interpretation, the choice among the n alternatives in step 1 is made by selecting the one that would give the best rate of improvement in Z (per unit increase in the entering basic variable) during step 2.

Table 5.7 illustrates the close correspondence between these geometric and algebraic interpretations of the simplex method. Using the results already presented in Secs. 4.3 and 4.4, the fourth column summarizes the sequence of BF solutions found for the Wyndor Glass Co. problem, and the second column shows the corresponding CPF solutions. In the third column, note how each iteration results in deleting one constraint boundary (defining equation) and substituting a new one to obtain the new CPF solution. Similarly, note in the fifth column how each iteration results in deleting one nonbasic variable and substituting a new one to obtain the new BF solution. Furthermore, the nonbasic variables being deleted and added are the indicating variables for the defining equations being deleted and added in the third column. The last column displays the initial system of equations [excluding Eq. (0)] for the augmented form of the problem, with the current basic variables shown in bold type. In each case, note how setting the nonbasic variables equal to zero and then solving this system of equations for the basic variables must yield the same solution for (x_1, x_2) as the corresponding pair of defining equations in the third column.

5.2 The Revised Simplex Method

The simplex method as described in Chap. 4 (hereafter called the *original simplex method*) is a straightforward algebraic procedure. However, this way of executing the algorithm (in either algebraic or tabular form) is not the most efficient computational procedure for computers because it computes and stores many numbers that are not needed at the current iteration and that may not even become relevant for decision making at subsequent iterations. The only pieces of information relevant at each iteration are the coefficients of the nonbasic variables in Eq. (0), the coefficients of the entering basic variable in the other equations, and the right-hand sides of the equations. It would be very useful to have a procedure that could obtain this information efficiently without computing and storing all the other coefficients.

As mentioned in Sec. 4.8, these considerations motivated the development of the *revised simplex method.* This method was designed to accomplish exactly the same things as the original simplex method, but in a way that is more efficient for execution on a computer. Thus it is a streamlined version of the original procedure. It computes and stores only the information that is currently needed, and it carries along the essential data in a more compact form.

The revised simplex method explicitly uses *matrix* manipulations, so it is necessary to describe the problem in matrix notation. (See Appendix 4 for a review of matrices.) To help you distinguish between matrices, vectors, and scalars, we consistently use **BOLDFACE CAPITAL** letters to represent matrices, **boldface lowercase** letters to represent vectors, and *italicized* letters in ordinary print to represent scalars. We also use a boldface zero (**0**) to denote a *null vector* (a vector whose elements all are zero) in either column or row form (which one should be clear from the context), whereas a zero in ordinary print (0) continues to represent the number zero.

Using matrices, our standard form for the general linear programming model given in Sec. 3.2 becomes

$$\text{Maximize} \quad Z = \mathbf{c}\mathbf{x},$$

subject to

$$\mathbf{A}\mathbf{x} \le \mathbf{b} \quad \text{and} \quad \mathbf{x} \ge \mathbf{0},$$

where **c** is the row vector

$$\mathbf{c} = [c_1, c_2, \ldots, c_n],$$

x, **b**, and **0** are the column vectors such that

$$\mathbf{x} = \begin{bmatrix} x_1 \\ x_2 \\ \vdots \\ x_n \end{bmatrix}, \quad \mathbf{b} = \begin{bmatrix} b_1 \\ b_2 \\ \vdots \\ b_m \end{bmatrix}, \quad \mathbf{0} = \begin{bmatrix} 0 \\ 0 \\ \vdots \\ 0 \end{bmatrix},$$

and **A** is the matrix

$$\mathbf{A} = \begin{bmatrix} a_{11} & a_{12} & \cdots & a_{1n} \\ a_{21} & a_{22} & \cdots & a_{2n} \\ \cdots & \cdots & \cdots & \cdots \\ a_{m1} & a_{m2} & \cdots & a_{mn} \end{bmatrix}.$$

To obtain the *augmented form* of the problem, introduce the column vector of slack variables

$$\mathbf{x}_s = \begin{bmatrix} x_{n+1} \\ x_{n+2} \\ \vdots \\ x_{n+m} \end{bmatrix},$$

so that the constraints become

$$[\mathbf{A}, \mathbf{I}]\begin{bmatrix} \mathbf{x} \\ \mathbf{x}_s \end{bmatrix} = \mathbf{b} \qquad \text{and} \qquad \begin{bmatrix} \mathbf{x} \\ \mathbf{x}_s \end{bmatrix} \geq \mathbf{0},$$

where $\mathbf{I}$ is the $m \times m$ identity matrix, and the null vector $\mathbf{0}$ now has $n + m$ elements. (We comment at the end of the section about how to deal with problems that are not in our standard form.)

Solving for a Basic Feasible Solution

Recall that the general approach of the simplex method is to obtain a sequence of *improving BF solutions* until an optimal solution is reached. One of the key features of the revised simplex method involves the way in which it solves for each new BF solution after identifying its basic and nonbasic variables. Given these variables, the resulting basic solution is the solution of the m equations

$$[\mathbf{A}, \mathbf{I}]\begin{bmatrix} \mathbf{x} \\ \mathbf{x}_s \end{bmatrix} = \mathbf{b},$$

in which the n *nonbasic variables* from the $n + m$ elements of

$$\begin{bmatrix} \mathbf{x} \\ \mathbf{x}_s \end{bmatrix}$$

are set equal to zero. Eliminating these n variables by equating them to zero leaves a set of m equations in m unknowns (the *basic variables*). This set of equations can be denoted by

$$\mathbf{B}\mathbf{x}_B = \mathbf{b},$$

where the **vector of basic variables**

$$\mathbf{x}_B = \begin{bmatrix} x_{B1} \\ x_{B2} \\ \vdots \\ x_{Bm} \end{bmatrix}$$

is obtained by eliminating the nonbasic variables from

$$\begin{bmatrix} \mathbf{x} \\ \mathbf{x}_s \end{bmatrix},$$

and the **basis matrix**

$$\mathbf{B} = \begin{bmatrix} B_{11} & B_{12} & \cdots & B_{1m} \\ B_{21} & B_{22} & \cdots & B_{2m} \\ \cdots & \cdots & \cdots & \cdots \\ B_{m1} & B_{m2} & \cdots & B_{mm} \end{bmatrix}$$

is obtained by eliminating the columns corresponding to coefficients of nonbasic variables from $[\mathbf{A}, \mathbf{I}]$. (In addition, the elements of $\mathbf{x}_B$ and, therefore, the columns of $\mathbf{B}$ may be placed in a different order when the simplex method is executed.)

The simplex method introduces only basic variables such that $\mathbf{B}$ is *nonsingular,* so that $\mathbf{B}^{-1}$ always will exist. Therefore, to solve $\mathbf{B}\mathbf{x}_B = \mathbf{b}$, both sides are premultiplied by $\mathbf{B}^{-1}$:

$$\mathbf{B}^{-1}\mathbf{B}\mathbf{x}_B = \mathbf{B}^{-1}\mathbf{b}.$$

Since $\mathbf{B}^{-1}\mathbf{B} = \mathbf{I}$, the desired solution for the basic variables is

$$\boxed{\mathbf{x}_B = \mathbf{B}^{-1}\mathbf{b}.}$$

Let $\mathbf{c}_B$ be the vector whose elements are the objective function coefficients (including zeros for slack variables) for the corresponding elements of $\mathbf{x}_B$. The value of the objective function for this basic solution is then

$$\boxed{Z = \mathbf{c}_B\mathbf{x}_B = \mathbf{c}_B\mathbf{B}^{-1}\mathbf{b}.}$$

Example: To illustrate this method of solving for a BF solution, consider again the Wyndor Glass Co. problem presented in Sec. 3.1 and solved by the original simplex method in Table 4.8. In this case,

$$\mathbf{c} = [3, 5], \qquad [\mathbf{A}, \mathbf{I}] = \begin{bmatrix} 1 & 0 & 1 & 0 & 0 \\ 0 & 2 & 0 & 1 & 0 \\ 3 & 2 & 0 & 0 & 1 \end{bmatrix}, \qquad \mathbf{b} = \begin{bmatrix} 4 \\ 12 \\ 18 \end{bmatrix}, \qquad \mathbf{x} = \begin{bmatrix} x_1 \\ x_2 \end{bmatrix}, \qquad \mathbf{x}_s = \begin{bmatrix} x_3 \\ x_4 \\ x_5 \end{bmatrix}.$$

Referring to Table 4.8, we see that the sequence of BF solutions obtained by the simplex method (original or revised) is the following:

Iteration 0

$$\mathbf{x}_B = \begin{bmatrix} x_3 \\ x_4 \\ x_5 \end{bmatrix}, \qquad \mathbf{B} = \begin{bmatrix} 1 & 0 & 0 \\ 0 & 1 & 0 \\ 0 & 0 & 1 \end{bmatrix} = \mathbf{B}^{-1}, \qquad \text{so} \qquad \begin{bmatrix} x_3 \\ x_4 \\ x_5 \end{bmatrix} = \begin{bmatrix} 1 & 0 & 0 \\ 0 & 1 & 0 \\ 0 & 0 & 1 \end{bmatrix}\begin{bmatrix} 4 \\ 12 \\ 18 \end{bmatrix} = \begin{bmatrix} 4 \\ 12 \\ 18 \end{bmatrix},$$

$$\mathbf{c}_B = [0, 0, 0], \qquad \text{so} \qquad Z = [0, 0, 0]\begin{bmatrix} 4 \\ 12 \\ 18 \end{bmatrix} = 0.$$

Iteration 1

$$\mathbf{x}_B = \begin{bmatrix} x_3 \\ x_2 \\ x_5 \end{bmatrix}, \qquad \mathbf{B} = \begin{bmatrix} 1 & 0 & 0 \\ 0 & 2 & 0 \\ 0 & 2 & 1 \end{bmatrix}, \qquad \mathbf{B}^{-1} = \begin{bmatrix} 1 & 0 & 0 \\ 0 & \frac{1}{2} & 0 \\ 0 & -1 & 1 \end{bmatrix},$$

so

$$\begin{bmatrix} x_3 \\ x_2 \\ x_5 \end{bmatrix} = \begin{bmatrix} 1 & 0 & 0 \\ 0 & \frac{1}{2} & 0 \\ 0 & -1 & 1 \end{bmatrix}\begin{bmatrix} 4 \\ 12 \\ 18 \end{bmatrix} = \begin{bmatrix} 4 \\ 6 \\ 6 \end{bmatrix},$$

$$\mathbf{c}_B = [0, 5, 0], \qquad \text{so} \qquad Z = [0, 5, 0]\begin{bmatrix} 4 \\ 6 \\ 6 \end{bmatrix} = 30.$$

Iteration 2

$$\mathbf{x}_B = \begin{bmatrix} x_3 \\ x_2 \\ x_1 \end{bmatrix}, \qquad \mathbf{B} = \begin{bmatrix} 1 & 0 & 1 \\ 0 & 2 & 0 \\ 0 & 2 & 3 \end{bmatrix}, \qquad \mathbf{B}^{-1} = \begin{bmatrix} 1 & \frac{1}{3} & -\frac{1}{3} \\ 0 & \frac{1}{2} & 0 \\ 0 & -\frac{1}{3} & \frac{1}{3} \end{bmatrix},$$

so

$$\begin{bmatrix} x_3 \\ x_2 \\ x_1 \end{bmatrix} = \begin{bmatrix} 1 & \frac{1}{3} & -\frac{1}{3} \\ 0 & \frac{1}{2} & 0 \\ 0 & -\frac{1}{3} & \frac{1}{3} \end{bmatrix} \begin{bmatrix} 4 \\ 12 \\ 18 \end{bmatrix} = \begin{bmatrix} 2 \\ 6 \\ 2 \end{bmatrix},$$

$$\mathbf{c}_B = [0, 5, 3], \qquad \text{so} \qquad Z = [0, 5, 3] \begin{bmatrix} 2 \\ 6 \\ 2 \end{bmatrix} = 36.$$

Matrix Form of the Current Set of Equations

The last preliminary before we summarize the revised simplex method is to show the matrix form of the set of equations appearing in the simplex tableau for any iteration of the original simplex method.

For the *original* set of equations, the matrix form is

$$\begin{bmatrix} 1 & -\mathbf{c} & \mathbf{0} \\ \mathbf{0} & \mathbf{A} & \mathbf{I} \end{bmatrix} \begin{bmatrix} Z \\ \mathbf{x} \\ \mathbf{x}_s \end{bmatrix} = \begin{bmatrix} 0 \\ \mathbf{b} \end{bmatrix}.$$

This set of equations also is exhibited in the first simplex tableau of Table 5.8.

The algebraic operations performed by the simplex method (multiply an equation by a constant and add a multiple of one equation to another equation) are expressed in matrix form by premultiplying both sides of the original set of equations by the appropriate matrix. This matrix would have the same elements as the identity matrix, *except* that each multiple for an algebraic operation would go into the spot needed to have the matrix multiplication perform this operation. Even after a series of algebraic operations over several iterations, we still can deduce what this matrix must be (symbolically) for the entire series by using what we already know about the right-hand sides of the new set of equations. In particular, after any iteration, $\mathbf{x}_B = \mathbf{B}^{-1}\mathbf{b}$ and $Z = \mathbf{c}_B\mathbf{B}^{-1}\mathbf{b}$, so the right-hand sides of the new set of equations have become

Table 5.8 **Initial and Later Simplex Tableaux in Matrix Form**

Iteration	Basic Variable	Eq.	Coefficient of: Z	Original Variables	Slack Variables	Right Side
0	Z	(0)	1	$-\mathbf{c}$	$\mathbf{0}$	0
	$\mathbf{x}_B$	$(1, 2, \ldots, m)$	$\mathbf{0}$	$\mathbf{A}$	$\mathbf{I}$	$\mathbf{b}$
⋮	⋮	⋮	⋮	⋮	⋮	⋮
Any	Z	(0)	1	$\mathbf{c}_B\mathbf{B}^{-1}\mathbf{A} - \mathbf{c}$	$\mathbf{c}_B\mathbf{B}^{-1}$	$\mathbf{c}_B\mathbf{B}^{-1}\mathbf{b}$
	$\mathbf{x}_B$	$(1, 2, \ldots, m)$	$\mathbf{0}$	$\mathbf{B}^{-1}\mathbf{A}$	$\mathbf{B}^{-1}$	$\mathbf{B}^{-1}\mathbf{b}$

$$\begin{bmatrix} Z \\ \mathbf{x}_B \end{bmatrix} = \begin{bmatrix} 1 & \mathbf{c}_B\mathbf{B}^{-1} \\ \mathbf{0} & \mathbf{B}^{-1} \end{bmatrix} \begin{bmatrix} 0 \\ \mathbf{b} \end{bmatrix} = \begin{bmatrix} \mathbf{c}_B\mathbf{B}^{-1}\mathbf{b} \\ \mathbf{B}^{-1}\mathbf{b} \end{bmatrix}.$$

Because we perform the same series of algebraic operations on *both* sides of the original set of operations, we use this same matrix that premultiplies the original right-hand side to premultiply the original left-hand side. Consequently, since

$$\begin{bmatrix} 1 & \mathbf{c}_B\mathbf{B}^{-1} \\ \mathbf{0} & \mathbf{B}^{-1} \end{bmatrix} \begin{bmatrix} 1 & -\mathbf{c} & \mathbf{0} \\ \mathbf{0} & \mathbf{A} & \mathbf{I} \end{bmatrix} = \begin{bmatrix} 1 & \mathbf{c}_B\mathbf{B}^{-1}\mathbf{A} - \mathbf{c} & \mathbf{c}_B\mathbf{B}^{-1} \\ \mathbf{0} & \mathbf{B}^{-1}\mathbf{A} & \mathbf{B}^{-1} \end{bmatrix},$$

the desired matrix form of the *set of equations after any iteration* is

$$\boxed{\begin{bmatrix} 1 & \mathbf{c}_B\mathbf{B}^{-1}\mathbf{A} - \mathbf{c} & \mathbf{c}_B\mathbf{B}^{-1} \\ \mathbf{0} & \mathbf{B}^{-1}\mathbf{A} & \mathbf{B}^{-1} \end{bmatrix} \begin{bmatrix} Z \\ \mathbf{x} \\ \mathbf{x}_s \end{bmatrix} = \begin{bmatrix} \mathbf{c}_B\mathbf{B}^{-1}\mathbf{b} \\ \mathbf{B}^{-1}\mathbf{b} \end{bmatrix}.}$$

The second simplex tableau of Table 5.8 also exhibits this same set of equations.

Example: To illustrate this matrix form for the current set of equations, consider the final set of equations resulting from iteration 2 for the Wyndor Glass Co. problem. Using the $\mathbf{B}^{-1}$ given for iteration 2, we have

$$\mathbf{B}^{-1}\mathbf{A} = \begin{bmatrix} 1 & \frac{1}{3} & -\frac{1}{3} \\ 0 & \frac{1}{2} & 0 \\ 0 & -\frac{1}{3} & \frac{1}{3} \end{bmatrix} \begin{bmatrix} 1 & 0 \\ 0 & 2 \\ 3 & 2 \end{bmatrix} = \begin{bmatrix} 0 & 0 \\ 0 & 1 \\ 1 & 0 \end{bmatrix},$$

$$\mathbf{c}_B\mathbf{B}^{-1} = [0, 5, 3] \begin{bmatrix} 1 & \frac{1}{3} & -\frac{1}{3} \\ 0 & \frac{1}{2} & 0 \\ 0 & -\frac{1}{3} & \frac{1}{3} \end{bmatrix} = [0, \tfrac{3}{2}, 1],$$

$$\mathbf{c}_B\mathbf{B}^{-1}\mathbf{A} - \mathbf{c} = [0, 5, 3] \begin{bmatrix} 0 & 0 \\ 0 & 1 \\ 1 & 0 \end{bmatrix} - [3, 5] = [0, 0].$$

Also, by using the values of $\mathbf{x}_B = \mathbf{B}^{-1}\mathbf{b}$ and $Z = \mathbf{c}_B\mathbf{B}^{-1}\mathbf{b}$ calculated a few pages back, these results give the following set of equations:

$$\left[\begin{array}{c|cc|ccc} 1 & 0 & 0 & 0 & \frac{3}{2} & 1 \\ \hline 0 & 0 & 0 & 1 & \frac{1}{3} & -\frac{1}{3} \\ 0 & 0 & 1 & 0 & \frac{1}{2} & 0 \\ 0 & 1 & 0 & 0 & -\frac{1}{3} & \frac{1}{3} \end{array}\right] \begin{bmatrix} Z \\ x_1 \\ x_2 \\ x_3 \\ x_4 \\ x_5 \end{bmatrix} = \begin{bmatrix} 36 \\ 2 \\ 6 \\ 2 \end{bmatrix},$$

as shown in the final simplex tableau in Table 4.8.

The Overall Procedure

There are two key implications from the matrix form of the current set of equations shown at the bottom of Table 5.8. The first is that *only* $\mathbf{B}^{-1}$ needs to be derived to be able to calculate all the numbers in the simplex tableau from the original parameters ($\mathbf{A}$, $\mathbf{b}$, $\mathbf{c}_B$) of the problem. (This implication is the essence of the **fundamental insight** described in the next section.) The second is that *any one* of these numbers (except $Z = \mathbf{c}_B\mathbf{B}^{-1}\mathbf{b}$) can be obtained by performing *only* a vector multiplication (one row times one column) instead of a complete matrix multiplication. Therefore, the *required numbers* to perform an iteration of the simplex method can be obtained as needed *without* expending the computational effort to obtain *all* the numbers. These two key implications are incorporated into the following summary of the overall procedure.

Summary of the Revised Simplex Method

1. *Initialization:* Same as for the original simplex method.
2. *Iteration:*

   Step 1 Determine the entering basic variable: Same as for the original simplex method.

   Step 2 Determine the leaving basic variable: Same as for the original simplex method, except calculate only the numbers required to do this [the coefficients of the entering basic variable in every equation but Eq. (0), and then, for each strictly positive coefficient, the right-hand side of that equation].[1]

   Step 3 Determine the new BF solution: Derive $\mathbf{B}^{-1}$ and set $\mathbf{x}_B = \mathbf{B}^{-1}\mathbf{b}$. (Calculating $\mathbf{x}_B$ is optional unless the optimality test finds it to be optimal.)
3. *Optimality test:* Same as for the original simplex method, except calculate only the numbers required to do this test, i.e., the coefficients of the *nonbasic variables* in Eq. (0).

In step 3 of an iteration, $\mathbf{B}^{-1}$ could be derived each time by using a standard computer routine for inverting a matrix. However, since $\mathbf{B}$ (and therefore $\mathbf{B}^{-1}$) changes so little from one iteration to the next, it is much more efficient to derive the new $\mathbf{B}^{-1}$ (denote it by $\mathbf{B}_{\text{new}}^{-1}$) from the $\mathbf{B}^{-1}$ at the preceding iteration (denote it by $\mathbf{B}_{\text{old}}^{-1}$). (For the initial BF solution, $\mathbf{B} = \mathbf{I} = \mathbf{B}^{-1}$.) One method for doing this derivation is based directly upon the interpretation of the elements of $\mathbf{B}^{-1}$ [the coefficients of the slack variables in the current Eqs. (1), (2), . . . , (m)] presented in the next section, as well as upon the procedure used by the original simplex method to obtain the new set of equations from the preceding set.

To describe this method formally, let

x_k = entering basic variable,

a'_{ik} = coefficient of x_k in current Eq. (i), for $i = 1, 2, \ldots, m$ (calculated in step 2 of an iteration),

r = number of equation containing the leaving basic variable.

Recall that the new set of equations [excluding Eq. (0)] can be obtained from the preceding set by subtracting a'_{ik}/a'_{rk} times Eq. (r) from Eq. (i), for all $i = 1, 2, \ldots, m$ except $i = r$ and then dividing Eq. (r) by a'_{rk}. Therefore, the element in row i and column j of $\mathbf{B}_{\text{new}}^{-1}$ is

[1] Because the value of $\mathbf{x}_B$ is the entire vector of right-hand sides except for Eq. (0), the relevant right-hand sides need not be calculated here if $\mathbf{x}_B$ was calculated in step 3 of the preceding iteration.

$$(\mathbf{B}_{\text{new}}^{-1})_{ij} = \begin{cases} (\mathbf{B}_{\text{old}}^{-1})_{ij} - \dfrac{a'_{ik}}{a'_{rk}}(\mathbf{B}_{\text{old}}^{-1})_{rj} & \text{if } i \neq r, \\ \dfrac{1}{a'_{rk}}(\mathbf{B}_{\text{old}}^{-1})_{rj} & \text{if } i = r. \end{cases}$$

These formulas are expressed in matrix notation as

$$\mathbf{B}_{\text{new}}^{-1} = \mathbf{E}\mathbf{B}_{\text{old}}^{-1},$$

where matrix $\mathbf{E}$ is an identity matrix except that its rth column is replaced by the vector

$$\boldsymbol{\eta} = \begin{bmatrix} \eta_1 \\ \eta_2 \\ \vdots \\ \eta_m \end{bmatrix}, \qquad \text{where} \qquad \eta_i = \begin{cases} -\dfrac{a'_{ik}}{a'_{rk}} & \text{if } i \neq r, \\ \dfrac{1}{a'_{rk}} & \text{if } i = r. \end{cases}$$

Thus $\mathbf{E} = [\mathbf{U}_1, \mathbf{U}_2, \ldots, \mathbf{U}_{r-1}, \boldsymbol{\eta}, \mathbf{U}_{r+1}, \ldots, \mathbf{U}_m]$, where the m elements of each of the $\mathbf{U}_i$ column vectors are 0 except for a 1 in the ith position.

Example: We shall illustrate the revised simplex method by applying it to the Wyndor Glass Co. problem. The initial basic variables are the slack variables

$$\mathbf{x}_B = \begin{bmatrix} x_3 \\ x_4 \\ x_5 \end{bmatrix}.$$

Iteration 1

Because the initial $\mathbf{B}^{-1} = \mathbf{I}$, no calculations are needed to obtain the numbers required to identify the entering basic variable x_2 $(-c_2 = -5 < -3 = -c_1)$ and the leaving basic variable x_4 $(a_{12} = 0,\ b_2/a_{22} = \frac{12}{2} < \frac{18}{2} = b_3/a_{32}$, so $r = 2)$. Thus the new set of basic variables is

$$\mathbf{x}_B = \begin{bmatrix} x_3 \\ x_2 \\ x_5 \end{bmatrix}.$$

To obtain the new $\mathbf{B}^{-1}$,

$$\boldsymbol{\eta} = \begin{bmatrix} -\dfrac{a_{12}}{a_{22}} \\ \dfrac{1}{a_{22}} \\ -\dfrac{a_{32}}{a_{22}} \end{bmatrix} = \begin{bmatrix} 0 \\ \dfrac{1}{2} \\ -1 \end{bmatrix},$$

so

$$\mathbf{B}^{-1} = \begin{bmatrix} 1 & 0 & 0 \\ 0 & \frac{1}{2} & 0 \\ 0 & -1 & 1 \end{bmatrix} \begin{bmatrix} 1 & 0 & 0 \\ 0 & 1 & 0 \\ 0 & 0 & 1 \end{bmatrix} = \begin{bmatrix} 1 & 0 & 0 \\ 0 & \frac{1}{2} & 0 \\ 0 & -1 & 1 \end{bmatrix},$$

so that

$$\begin{bmatrix} x_3 \\ x_2 \\ x_5 \end{bmatrix} = \begin{bmatrix} 1 & 0 & 0 \\ 0 & \frac{1}{2} & 0 \\ 0 & -1 & 1 \end{bmatrix} \begin{bmatrix} 4 \\ 12 \\ 18 \end{bmatrix} = \begin{bmatrix} 4 \\ 6 \\ 6 \end{bmatrix}.$$

To test whether this solution is optimal, we calculate the coefficients of the nonbasic variables (x_1 and x_4) in Eq. (0). Performing only the relevant parts of the matrix multiplications, we obtain

$$\mathbf{c}_B\mathbf{B}^{-1}\mathbf{A} - \mathbf{c} = [0, 5, 0] \begin{bmatrix} 1 & 0 & 0 \\ 0 & \frac{1}{2} & 0 \\ 0 & -1 & 1 \end{bmatrix} \begin{bmatrix} 1 & — \\ 0 & — \\ 3 & — \end{bmatrix} - [3, —] = [-3, —],$$

$$\mathbf{c}_B\mathbf{B}^{-1} = [0, 5, 0] \begin{bmatrix} — & 0 & — \\ — & \frac{1}{2} & — \\ — & -1 & — \end{bmatrix} = [—, \tfrac{5}{2}, —],$$

so the coefficients of x_1 and x_4 are -3 and $\frac{5}{2}$, respectively. Since x_1 has a negative coefficient, this solution is *not* optimal.

Iteration 2

Using these coefficients of the nonbasic variables, we begin the next iteration by identifying x_1 as the entering basic variable. To determine the leaving basic variable, we must calculate the other coefficients of x_1:

$$\mathbf{B}^{-1}\mathbf{A} = \begin{bmatrix} 1 & 0 & 0 \\ 0 & \frac{1}{2} & 0 \\ 0 & -1 & 1 \end{bmatrix} \begin{bmatrix} 1 & — \\ 0 & — \\ 3 & — \end{bmatrix} = \begin{bmatrix} 1 & — \\ 0 & — \\ 3 & — \end{bmatrix}.$$

By using the *right side* column for the current BF solution (the value of $\mathbf{x}_B$) just given for iteration 1, the ratios $4/1 > 6/3$ indicate that x_5 is the leaving basic variable, so the new set of basic variables is

$$\mathbf{x}_B = \begin{bmatrix} x_3 \\ x_2 \\ x_1 \end{bmatrix} \qquad \text{with} \qquad \boldsymbol{\eta} = \begin{bmatrix} -\dfrac{a'_{11}}{a'_{31}} \\ -\dfrac{a'_{21}}{a'_{31}} \\ \dfrac{1}{a'_{31}} \end{bmatrix} = \begin{bmatrix} -\dfrac{1}{3} \\ 0 \\ \dfrac{1}{3} \end{bmatrix}.$$

Therefore, the new $\mathbf{B}^{-1}$ is

$$\mathbf{B}^{-1} = \begin{bmatrix} 1 & 0 & -\frac{1}{3} \\ 0 & 1 & 0 \\ 0 & 0 & \frac{1}{3} \end{bmatrix} \begin{bmatrix} 1 & 0 & 0 \\ 0 & \frac{1}{2} & 0 \\ 0 & -1 & 1 \end{bmatrix} = \begin{bmatrix} 1 & \frac{1}{3} & -\frac{1}{3} \\ 0 & \frac{1}{2} & 0 \\ 0 & -\frac{1}{3} & \frac{1}{3} \end{bmatrix},$$

so that

$$\begin{bmatrix} x_3 \\ x_2 \\ x_1 \end{bmatrix} = \begin{bmatrix} 1 & \frac{1}{3} & -\frac{1}{3} \\ 0 & \frac{1}{2} & 0 \\ 0 & -\frac{1}{3} & \frac{1}{3} \end{bmatrix} \begin{bmatrix} 4 \\ 12 \\ 18 \end{bmatrix} = \begin{bmatrix} 2 \\ 6 \\ 2 \end{bmatrix}.$$

Applying the optimality test, we find that the coefficients of the nonbasic variables (x_4 and x_5) in Eq. (0) are

$$\mathbf{c}_B\mathbf{B}^{-1} = [0, 5, 3]\begin{bmatrix} — & \frac{1}{3} & -\frac{1}{3} \\ — & \frac{1}{2} & 0 \\ — & -\frac{1}{3} & \frac{1}{3} \end{bmatrix} = [—, \tfrac{3}{2}, 1].$$

Because both coefficients ($\frac{3}{2}$ and 1) are nonnegative, the current solution ($x_1 = 2$, $x_2 = 6$, $x_3 = 2$, $x_4 = 0$, $x_5 = 0$) is optimal and the procedure terminates.

General Observations

Although the preceding pages describe the essence of the revised simplex method, we should point out that minor modifications may be made to improve the efficiency of its execution on computers. For example, $\mathbf{B}^{-1}$ may be obtained as the product of the previous $\mathbf{E}$ matrices. This modification requires storing only the η column of $\mathbf{E}$ and the number of the column, rather than the $\mathbf{B}^{-1}$ matrix, at each iteration.

Also note that the preceding discussion was limited to the case of linear programming problems fitting our standard form given in Sec. 3.2. However, the modifications for other forms are relatively straightforward. The initialization would be conducted just as it would for the original simplex method (see Sec. 4.6). When this step involves introducing artificial variables to obtain an initial BF solution (and thereby to obtain an *identity matrix* as the *initial basis matrix*), these variables are included among the m elements of $\mathbf{x}_s$.

Let us summarize the advantages of the revised simplex method over the original simplex method. One advantage is that the number of arithmetic computations may be reduced. This is especially true when the $\mathbf{A}$ matrix contains a large number of zero elements (which is usually the case for the large problems arising in practice). The amount of information that must be stored at each iteration is less, sometimes considerably so. The revised simplex method also permits the control of the rounding errors inevitably generated by computers. This control can be exercised by periodically obtaining the current $\mathbf{B}^{-1}$ by directly inverting $\mathbf{B}$. Furthermore, some of the post-optimality analysis problems discussed in Sec. 4.7 can be handled more conveniently with the revised simplex method. For all these reasons, the revised simplex method is usually preferable to the original simplex method for computer execution.

5.3 A Fundamental Insight

We shall now focus on a property of the simplex method (in any form) that has been revealed by the revised simplex method in the preceding section.[1] This fundamental insight provides the key to both duality theory and sensitivity analysis (Chap. 6), two very important parts of linear programming.

The insight involves the coefficients of the *slack* variables and the information they give. It is a direct result of the initialization, where the ith slack variable x_{n+i} is given a coefficient of $+1$ in Eq. (i) and a coefficient of 0 in *every other equation*

[1] However, since some instructors do not cover the preceding section, we have written this section in a way that can be understood without first reading Sec. 5.2.

[including Eq. (0)] for $i = 1, 2, \ldots, m$, as shown by the null vector $\mathbf{0}$ and the identity matrix $\mathbf{I}$ in the *slack variables* column for iteration 0 in Table 5.8.[1] The other key factor is that subsequent iterations change the initial equations *only* by

1. Multiplying (or dividing) an *entire* equation by a nonzero constant
2. Adding (or subtracting) a multiple of one *entire* equation to another *entire* equation

As already described in the preceding section, a sequence of these kinds of elementary algebraic operations is equivalent to premultiplying the initial simplex tableau by some matrix. (See Appendix 4 for a review of matrices.) The consequence can be summarized as follows.

Verbal description of fundamental insight: After any iteration, the coefficients of the *slack* variables in each equation immediately reveal how that equation has been obtained from the *initial* equations.

As one example of the importance of this insight, recall from Table 5.8 that the matrix formula for the optimal solution obtained by the simplex method is

$$\mathbf{x}_B = \mathbf{B}^{-1}\mathbf{b},$$

where $\mathbf{x}_B$ is the vector of basic variables, $\mathbf{B}^{-1}$ is the matrix of coefficients of slack variables for rows 1 to m of the final tableau, and $\mathbf{b}$ is the vector of original right-hand sides (resource availabilities). (We soon will denote this particular $\mathbf{B}^{-1}$ by $\mathbf{S}^*$.) Postoptimality analysis normally includes an investigation of possible changes in $\mathbf{b}$. By using this formula, you can see exactly how the optimal BF solution changes (or whether it becomes infeasible because of negative variables) as a function of $\mathbf{b}$. You do *not* have to reapply the simplex method over and over for each new $\mathbf{b}$, because the coefficients of the slack variables tell all! In a similar fashion, this fundamental insight provides a tremendous computational saving for the rest of sensitivity analysis as well.

To spell out the how and the why of this insight, let us look again at the Wyndor Glass Co. example. (The OR Courseware also includes another demonstration example.)

Example: Table 5.9 shows the relevant portion of the simplex tableau for demonstrating this fundamental insight. Darker lines have been drawn around the coefficients of the slack variables in the second and third tableaux because these are the crucial coefficients for applying the insight. To avoid clutter, we then identify the pivot row and pivot column by a single box around the pivot number only.

Iteration 1

To demonstrate the fundamental insight, our focus is on the algebraic operations performed by the simplex method while using Gaussian elimination to obtain the new BF solution. If we do all the algebraic operations with the *old* row 2 (the pivot row) rather than the new one, then the algebraic operations spelled out in Chap. 4 for iteration 1 are

[1] Throughout most of this section, we assume that the problem is in *our standard form,* with $b_i \geq 0$ for all $i = 1, 2, \ldots, m$, so that no additional adjustments are needed in the initialization. We then adapt our conclusions to nonstandard forms late in the section.

Table 5.9 **Simplex Tableaux without Leftmost Columns for the Wyndor Glass Co. Problem**

Iteration	Coefficient of: x_1	x_2	x_3	x_4	x_5	Right Side
0	-3	-5	0	0	0	0
	1	0	1	0	0	4
	0	[2]	0	1	0	12
	3	2	0	0	1	18
1	-3	0	0	$\frac{5}{2}$	0	30
	1	0	1	0	0	4
	0	1	0	$\frac{1}{2}$	0	6
	[3]	0	0	-1	1	6
2	0	0	0	$\frac{3}{2}$	1	36
	0	0	1	$\frac{1}{3}$	$-\frac{1}{3}$	2
	0	1	0	$\frac{1}{2}$	0	6
	1	0	0	$-\frac{1}{3}$	$\frac{1}{3}$	2

$$\begin{aligned}
\text{New row } 0 &= \text{old row } 0 + \left(\tfrac{5}{2}\right)(\text{old row } 2),\\
\text{New row } 1 &= \text{old row } 1 + (0)(\text{old row } 2),\\
\text{New row } 2 &= \left(\tfrac{1}{2}\right)(\text{old row } 2),\\
\text{New row } 3 &= \text{old row } 3 + (-1)(\text{old row } 2).
\end{aligned}$$

Ignoring row 0 for the moment, we see that these algebraic operations amount to premultiplying rows 1 to 3 of the initial tableau by the first matrix shown below.

$$\text{New rows 1–3} = \begin{bmatrix} 1 & 0 & 0 \\ 0 & \frac{1}{2} & 0 \\ 0 & -1 & 1 \end{bmatrix} \left[\begin{array}{cc|ccc|c} 1 & 0 & 1 & 0 & 0 & 4 \\ 0 & 2 & 0 & 1 & 0 & 12 \\ 3 & 2 & 0 & 0 & 1 & 18 \end{array}\right]$$

$$= \left[\begin{array}{cc|ccc|c} 1 & 0 & 1 & 0 & 0 & 4 \\ 0 & 1 & 0 & \frac{1}{2} & 0 & 6 \\ 3 & 0 & 0 & -1 & 1 & 6 \end{array}\right].$$

Note how this first matrix is reproduced exactly as the coefficients of the slack variables in rows 1 to 3 of the new tableau, because the coefficients of the slack variables in rows 1 to 3 of the initial tableau form an *identity matrix.* Thus, just as stated in the verbal description of the fundamental insight, the coefficients of the slack variables in the new tableau do indeed provide a record of the algebraic operations performed.

This insight is not much to get excited about after just one iteration, since you can readily see from the initial tableau what the algebraic operations had to be, but it becomes invaluable after all the iterations are completed.

For row 0, the algebraic operation performed amounts to the following matrix calculations, where now our focus is on the vector $[0, \frac{5}{2}, 0]$ that premultiplies rows 1 to 3 of the initial tableau.

$$\text{New row } 0 = [-3, \quad -5 \,\vdots\, 0, \quad 0, \quad 0 \,\vdots\, 0] + [0, \quad \tfrac{5}{2}, \quad 0]\left[\begin{array}{cc:ccc:c} 1 & 0 & 1 & 0 & 0 & 4 \\ 0 & 2 & 0 & 1 & 0 & 12 \\ 3 & 2 & 0 & 0 & 1 & 18 \end{array}\right]$$

$$= [-3, \quad 0, \quad \boxed{0, \quad \tfrac{5}{2}, \quad 0}, \quad 30].$$

Note how this vector is reproduced exactly as the coefficients of the slack variables in row 0 of the new tableau, just as was claimed in the statement of the fundamental insight. (Once again, the reason is the identity matrix for the coefficients of the slack variables in rows 1 to 3 of the initial tableau, along with the zeros for these coefficients in row 0 of the initial tableau.)

Iteration 2

The algebraic operations performed on the second tableau of Table 5.9 for iteration 2 are

$$\begin{aligned} \text{New row } 0 &= \text{old row } 0 + \quad (1)(\text{old row } 3), \\ \text{New row } 1 &= \text{old row } 1 + (-\tfrac{1}{3})(\text{old row } 3), \\ \text{New row } 2 &= \text{old row } 2 + \quad (0)(\text{old row } 3), \\ \text{New row } 3 &= \qquad\qquad\quad (\tfrac{1}{3})(\text{old row } 3). \end{aligned}$$

Ignoring row 0 for the moment, we see that these operations amount to premultiplying rows 1 to 3 of this tableau by the matrix

$$\begin{bmatrix} 1 & 0 & -\frac{1}{3} \\ 0 & 1 & 0 \\ 0 & 0 & \frac{1}{3} \end{bmatrix}.$$

Writing this second tableau as the matrix product shown for iteration 1 (namely, the corresponding matrix times rows 1 to 3 of the initial tableau) then yields

$$\text{Final rows 1–3} = \begin{bmatrix} 1 & 0 & -\frac{1}{3} \\ 0 & 1 & 0 \\ 0 & 0 & \frac{1}{3} \end{bmatrix}\begin{bmatrix} 1 & 0 & 0 \\ 0 & \frac{1}{2} & 0 \\ 0 & -1 & 1 \end{bmatrix}\left[\begin{array}{cc:ccc:c} 1 & 0 & 1 & 0 & 0 & 4 \\ 0 & 2 & 0 & 1 & 0 & 12 \\ 3 & 2 & 0 & 0 & 1 & 18 \end{array}\right]$$

$$= \begin{bmatrix} 1 & \frac{1}{3} & -\frac{1}{3} \\ 0 & \frac{1}{2} & 0 \\ 0 & -\frac{1}{3} & \frac{1}{3} \end{bmatrix}\left[\begin{array}{cc:ccc:c} 1 & 0 & 1 & 0 & 0 & 4 \\ 0 & 2 & 0 & 1 & 0 & 12 \\ 3 & 2 & 0 & 0 & 1 & 18 \end{array}\right]$$

$$= \left[\begin{array}{ccccc c} 0 & 0 & 1 & \frac{1}{3} & -\frac{1}{3} & 2 \\ 0 & 1 & 0 & \frac{1}{2} & 0 & 6 \\ 1 & 0 & 0 & -\frac{1}{3} & \frac{1}{3} & 2 \end{array}\right].$$

The first two matrices shown on the first line of these calculations summarize the algebraic operations of the second and first iterations, respectively. Their product, shown as the first matrix on the second line, then combines the algebraic operations of the two iterations. Note how this matrix is reproduced exactly as the coefficients of the slack variables in rows 1 to 3 of the new (final) tableau shown on the third line. What this portion of the tableau reveals is how the *entire* final tableau (except row 0) has been obtained from the initial tableau, namely,

$$\begin{aligned}\text{Final row 1} &= (1)(\text{initial row 1}) + (\tfrac{1}{3})(\text{initial row 2}) + (-\tfrac{1}{3})(\text{initial row 3}),\\ \text{Final row 2} &= (0)(\text{initial row 1}) + (\tfrac{1}{2})(\text{initial row 2}) + (0)(\text{initial row 3}),\\ \text{Final row 3} &= (0)(\text{initial row 1}) + (-\tfrac{1}{3})(\text{initial row 2}) + (\tfrac{1}{3})(\text{initial row 3}).\end{aligned}$$

To see why these multipliers of the initial rows are correct, you would have to trace through all the algebraic operations of both iterations. For example, why does final row 1 include ($\frac{1}{3}$)(initial row 2), even though a multiple of row 2 has never been added directly to row 1? The reason is that initial row 2 was subtracted from initial row 3 in iteration 1, and then ($\frac{1}{3}$)(old row 3) was subtracted from old row 1 in iteration 2.

However, there is no need for you to trace through. Even when the simplex method has gone through hundreds or thousands of iterations, the coefficients of the slack variables in the final tableau will reveal how this tableau has been obtained from the initial tableau. Furthermore, the same algebraic operations would give these same coefficients even if the values of some of the parameters in the original model (initial tableau) were changed, so these coefficients also reveal how the *rest* of the final tableau changes with changes in the initial tableau.

To complete this story for row 0, the fundamental insight reveals that the entire final row 0 can be calculated from the initial tableau by using just the coefficients of the slack variables in the final row 0—[0, $\frac{3}{2}$, 1]. This calculation is shown below, where the first vector is row 0 of the initial tableau and the matrix is rows 1 to 3 of the initial tableau.

$$\text{Final row 0} = [-3, \quad -5 \,\vdots\, 0, \quad 0, \quad 0 \,\vdots\, 0] + [0, \quad \tfrac{3}{2}, \quad 1]\begin{bmatrix} 1 & 0 & \vdots & 1 & 0 & 0 & \vdots & 4 \\ 0 & 2 & \vdots & 0 & 1 & 0 & \vdots & 12 \\ 3 & 2 & \vdots & 0 & 0 & 1 & \vdots & 18 \end{bmatrix}$$

$$= [0, \quad 0, \quad \boxed{0, \quad \tfrac{3}{2}, \quad 1}, \quad 36].$$

Note again how the vector premultiplying rows 1 to 3 of the initial tableau is reproduced exactly as the coefficients of the slack variables in the final row 0. These quantities must be identical because of the coefficients of the slack variables in the initial tableau (an identity matrix below a null vector). This conclusion is the row 0 part of the fundamental insight.

Mathematical Summary

Because its primary applications involve the *final* tableau, we shall now give a general mathematical expression for the fundamental insight just in terms of this tableau, using matrix notation. If you have not read Sec. 5.2, you now need to know that the *parameters* of the model are given by the matrix $\mathbf{A} = \|a_{ij}\|$ and the vectors $\mathbf{b} = \|b_i\|$ and $\mathbf{c} = \|c_j\|$, as displayed at the beginning of that section.

The only other notation needed is summarized and illustrated in Table 5.10. Notice how vector $\mathbf{t}$ (representing row 0) and matrix $\mathbf{T}$ (representing the other rows) together correspond to the rows of the initial tableau in Table 5.9, whereas vector $\mathbf{t}^*$ and matrix $\mathbf{T}^*$ together correspond to the rows of the final tableau in Table 5.9. This table also shows these vectors and matrices partitioned into three parts: The coefficients of the original variables, the coefficients of the slack variables (our focus), and the right-hand side. Once again, the notation distinguishes between parts of the initial tableau and the final tableau by using an asterisk only in the latter case.

For the coefficients of the slack variables (the middle part) in the initial tableau of Table 5.10, notice the null vector $\mathbf{0}$ in row 0 and the identity matrix $\mathbf{I}$ below, which

Table 5.10 General Notation for Initial and Final Simplex Tableaux in Matrix Form, Illustrated by the Wyndor Glass Co. Problem

Initial Tableau

Row 0: $\mathbf{t} = [-3, -5 \,\vdots\, 0, 0, 0 \,\vdots\, 0] = [-\mathbf{c} \,\vdots\, \mathbf{0} \,\vdots\, 0]$.

Other rows: $\mathbf{T} = \left[\begin{array}{cc|ccc|c} 1 & 0 & 1 & 0 & 0 & 4 \\ 0 & 2 & 0 & 1 & 0 & 12 \\ 3 & 2 & 0 & 0 & 1 & 18 \end{array}\right] = [\mathbf{A} \,\vdots\, \mathbf{I} \,\vdots\, \mathbf{b}]$.

Combined: $\begin{bmatrix} \mathbf{t} \\ \mathbf{T} \end{bmatrix} = \left[\begin{array}{c|c|c} -\mathbf{c} & \mathbf{0} & 0 \\ \mathbf{A} & \mathbf{I} & \mathbf{b} \end{array}\right]$.

Final Tableau

Row 0: $\mathbf{t}^* = [0, 0 \,\vdots\, 0, \frac{3}{2}, 1 \,\vdots\, 36] = [\mathbf{z}^* - \mathbf{c} \,\vdots\, \mathbf{y}^* \,\vdots\, Z^*]$.

Other rows: $\mathbf{T}^* = \left[\begin{array}{cc|ccc|c} 0 & 0 & 1 & \frac{1}{3} & -\frac{1}{3} & 2 \\ 0 & 1 & 0 & \frac{1}{2} & 0 & 6 \\ 1 & 0 & 0 & -\frac{1}{3} & \frac{1}{3} & 2 \end{array}\right] = [\mathbf{A}^* \,\vdots\, \mathbf{S}^* \,\vdots\, \mathbf{b}^*]$.

Combined: $\begin{bmatrix} \mathbf{t}^* \\ \mathbf{T}^* \end{bmatrix} = \left[\begin{array}{c|c|c} \mathbf{z}^* - \mathbf{c} & \mathbf{y}^* & Z^* \\ \mathbf{A}^* & \mathbf{S}^* & \mathbf{b}^* \end{array}\right]$.

provide the keys for the fundamental insight. The vector and matrix in the same location of the final tableau, $\mathbf{y}^*$ and $\mathbf{S}^*$, then play a prominent role in the equations for the fundamental insight. (This matrix was denoted by $\mathbf{B}^{-1}$ for *any* tableau in Sec. 5.2, but we now are letting $\mathbf{S}^*$ denote this particular matrix for just the final tableau, where *S* stands for slack variable coefficients.) And $\mathbf{A}$ and $\mathbf{b}$ in the initial tableau turn into $\mathbf{A}^*$ and $\mathbf{b}^*$ in the final tableau. For row 0 of the final tableau, the coefficients of the original variables are $\mathbf{z}^* - \mathbf{c}$ (so vector $\mathbf{z}^*$ is what has been added to the vector of initial coefficients, $-\mathbf{c}$), and the right-hand side Z^* denotes the optimal value of Z.

Now suppose that you are given the initial tableau, $\mathbf{t}$ and $\mathbf{T}$, and just $\mathbf{y}^*$ and $\mathbf{S}^*$ from the final tableau. How can this information alone be used to calculate the rest of the final tableau? The answer is provided by Table 5.8. This table includes some information that is not directly relevant to our current discussion, namely, how $\mathbf{y}^*$ and $\mathbf{S}^*$ themselves can be calculated ($\mathbf{y}^* = \mathbf{c}_B\mathbf{B}^{-1}$ and $\mathbf{S}^* = \mathbf{B}^{-1}$) by knowing the current set of basic variables and so the current basis matrix $\mathbf{B}$. However, the lower part of this table (which can represent either an intermediate or a final simplex tableau) also shows how the rest of the tableau can be obtained from the coefficients of the slack variables, which is summarized as follows.

Fundamental Insight

(1) $\mathbf{t}^* = \mathbf{t} + \mathbf{y}^*\mathbf{T} = [\mathbf{y}^*\mathbf{A} - \mathbf{c} \,\vdots\, \mathbf{y}^* \,\vdots\, \mathbf{y}^*\mathbf{b}]$.
(2) $\mathbf{T}^* = \mathbf{S}^*\mathbf{T} = [\mathbf{S}^*\mathbf{A} \,\vdots\, \mathbf{S}^* \,\vdots\, \mathbf{S}^*\mathbf{b}]$.

We already used these two equations when dealing with iteration 2 for the Wyndor Glass Co. problem in the preceding subsection. In particular, the right-hand side of the expression for final row 0 for iteration 2 is just $\mathbf{t} + \mathbf{y}^*\mathbf{T}$, and the second line of the expression for final rows 1 to 3 is just $\mathbf{S}^*\mathbf{T}$.

Now let us summarize the mathematical logic behind the two equations for the fundamental insight. To derive Eq. (2), recall that the entire sequence of algebraic operations performed by the simplex method (excluding those involving row 0) is equivalent to premultiplying **T** by some matrix, call it **M**. Therefore,

$$\mathbf{T}^* = \mathbf{MT},$$

but now we need to identify **M**. By writing out the component parts of **T** and **T***, this equation becomes

$$[\mathbf{A}^* \mid \mathbf{S}^* \mid \mathbf{b}^*] = \mathbf{M}\ [\mathbf{A} \mid \mathbf{I} \mid \mathbf{b}]$$
$$= [\mathbf{MA} \mid \mathbf{M} \mid \mathbf{Mb}].$$

Because the middle (or any other) component of these equal matrices must be the same, it follows that $\mathbf{M} = \mathbf{S}^*$, so Eq. (2) is a valid equation.

Equation (1) is derived in a similar fashion by noting that the entire sequence of algebraic operations involving row 0 amounts to adding some linear combination of the rows in **T** to **t**, which is equivalent to adding to **t** some *vector* times **T**. Denoting this vector by **v**, we thereby have

$$\mathbf{t}^* = \mathbf{t} + \mathbf{vT},$$

but **v** still needs to be identified. Writing out the component parts of **t** and **t*** yields

$$[\mathbf{z}^* - \mathbf{c} \mid \mathbf{y}^* \mid Z^*] = [-\mathbf{c} \mid \mathbf{0} \mid 0] + \mathbf{v}\ [\mathbf{A} \mid \mathbf{I} \mid \mathbf{b}]$$
$$= [-\mathbf{c} + \mathbf{vA} \mid \mathbf{v} \mid \mathbf{vb}].$$

Equating the middle component of these equal vectors gives $\mathbf{v} = \mathbf{y}^*$, which validates Eq. (1).

Thus far, the fundamental insight has been described under the assumption that the original model is in our standard form, described in Sec. 3.2. However, the above mathematical logic now reveals just what adjustments are needed for other forms of the original model. The key is the identity matrix **I** in the initial tableau, which turns into **S*** in the final tableau. If some artificial variables must be introduced into the initial tableau to serve as initial basic variables, then it is the set of columns (appropriately ordered) for *all* the initial basic variables (both slack and artificial) that forms **I** in this tableau. (The columns for any surplus variables are extraneous.) The *same* columns in the final tableau provide **S*** for the $\mathbf{T}^* = \mathbf{S}^*\mathbf{T}$ equation and **y*** for the $\mathbf{t}^* = \mathbf{t} + \mathbf{y}^*\mathbf{T}$ equation. If *M*'s were introduced into the preliminary row 0 as coefficients for artificial variables, then the **t** for the $\mathbf{t}^* = \mathbf{t} + \mathbf{y}^*\mathbf{T}$ equation is the row 0 for the initial tableau after these nonzero coefficients for basic variables are algebraically eliminated. (Alternatively, the preliminary row 0 can be used for **t**, but then these *M*'s must be subtracted from the final row 0 to give **y***.) (See Prob. 5.3-11.)

Applications

The fundamental insight has a variety of important applications in linear programming. One of these applications involves the revised simplex method. As described in the preceding section (see Table 5.8), this method used $\mathbf{S}^* = \mathbf{B}^{-1}$ and the initial tableau to calculate all the relevant numbers in the current tableau for every iteration. It goes even further than the fundamental insight by using $\mathbf{B}^{-1}$ to calculate **y*** itself as $\mathbf{y}^* = c_B\mathbf{B}^{-1}$.

Another application involves the interpretation of the *shadow prices* $(y_1^*, y_2^*, \ldots, y_m^*)$ described in Sec. 4.7. The fundamental insight reveals that Z^* (the value of Z for the optimal solution) is

$$Z^* = \mathbf{y}^*\mathbf{b} = \sum_{i=1}^{m} y_i^* b_i,$$

so, e.g.,

$$Z^* = 0b_1 + \tfrac{3}{2}b_2 + b_3$$

for the Wyndor Glass Co. problem. This equation immediately yields the interpretation for the y_i^* values given in Sec. 4.7.

Another group of extremely important applications involves various *post-optimality tasks* (reoptimization technique, sensitivity analysis, parametric linear programming—described in Sec. 4.7) that investigate the effect of making one or more changes in the original model. In particular, suppose that the simplex method already has been applied to obtain an optimal solution (as well as $\mathbf{y}^*$ and $\mathbf{S}^*$) for the original model, and then these changes are made. If exactly the same sequence of algebraic operations were to be applied to the revised initial tableau, what would be the resulting changes in the final tableau? Because $\mathbf{y}^*$ and $\mathbf{S}^*$ don't change, the fundamental insight reveals the answer immediately.

For example, consider the change from $b_2 = 12$ to $b_2 = 13$ as illustrated in Fig. 4.8 for the Wyndor Glass Co. problem. It is not necessary to *solve* for the new optimal solution $(x_1, x_2) = (\frac{5}{3}, \frac{13}{2})$ because the values of the basic variables in the final tableau ($\mathbf{b}^*$) are immediately revealed by the fundamental insight:

$$\begin{bmatrix} x_3 \\ x_2 \\ x_1 \end{bmatrix} = \mathbf{b}^* = \mathbf{S}^*\mathbf{b} = \begin{bmatrix} 1 & \frac{1}{3} & -\frac{1}{3} \\ 0 & \frac{1}{2} & 0 \\ 0 & -\frac{1}{3} & \frac{1}{3} \end{bmatrix} \begin{bmatrix} 4 \\ 13 \\ 18 \end{bmatrix} = \begin{bmatrix} \frac{7}{3} \\ \frac{13}{2} \\ \frac{5}{3} \end{bmatrix}.$$

There is an even easier way to make this calculation. Since the only change is in the *second* component of $\mathbf{b}$, which gets premultiplied by only the *second* column of $\mathbf{S}^*$, the *change* in $\mathbf{b}^*$ can be calculated as simply

$$\Delta\mathbf{b}^* = \begin{bmatrix} \frac{1}{3} \\ \frac{1}{2} \\ -\frac{1}{3} \end{bmatrix} \Delta b_2 = \begin{bmatrix} \frac{1}{3} \\ \frac{1}{2} \\ -\frac{1}{3} \end{bmatrix},$$

so the original values of the basic variables in the final tableau ($x_3 = 2, x_2 = 6, x_1 = 2$) now become

$$\begin{bmatrix} x_3 \\ x_2 \\ x_1 \end{bmatrix} = \begin{bmatrix} 2 \\ 6 \\ 2 \end{bmatrix} + \begin{bmatrix} \frac{1}{3} \\ \frac{1}{2} \\ -\frac{1}{3} \end{bmatrix} = \begin{bmatrix} \frac{7}{3} \\ \frac{13}{2} \\ \frac{5}{3} \end{bmatrix}.$$

(If any of these new values were *negative,* and thus infeasible, then the reoptimization technique described in Sec. 4.7 would be applied, starting from this revised final tableau.) Applying *incremental analysis* to the preceding equation for Z^* also immediately yields

$$\Delta Z^* = \tfrac{3}{2}\,\Delta b_2 = \tfrac{3}{2}.$$

The fundamental insight can be applied to investigating other kinds of changes in the original model in a very similar fashion; it is the crux of the sensitivity analysis procedure described in the latter part of Chap. 6.

You also will see in the next chapter that the fundamental insight plays a key role in the very useful duality theory for linear programming.

5.4 Conclusions

Although the simplex method is an algebraic procedure, it is based on some fairly simple geometric concepts. These concepts enable one to use the algorithm to examine only a relatively small number of BF solutions before reaching and identifying an optimal solution.

The revised simplex method provides an effective way of adapting the simplex method for computer implementation.

The final simplex tableau includes complete information on how it can be algebraically reconstructed directly from the initial simplex tableau. This fundamental insight has some very important applications, especially for post-optimality analysis.

SELECTED REFERENCES

1. Dantzig, G. B., *Linear Programming and Extensions,* Princeton University Press, Princeton, NJ, 1963.
2. Gass, S. I., *Linear Programming,* 5th ed., McGraw-Hill, New York, 1985.
3. Murty, K. G., *Linear Programming,* 2d ed., Wiley, New York, 1983.
4. Schriver, A., *Theory of Linear and Integer Programming,* Wiley, New York, 1986.

RELEVANT ROUTINES IN THE OR COURSEWARE

A demonstration example: *Fundamental Insight*

Interactive routines: *Enter or Revise a General Linear Programming Model*
Solve Interactively by the Graphical Method

An automatic routine: *Solve Automatically by the Simplex Method*

To access these routines, call the MathProg program and then choose *General Linear Programming* under the Area menu. See Appendix 1 for documentation of the software.

PROBLEMS[1]

To the left of each of the following problems (or their parts), we have inserted a D (for demonstration), I (for interactive routine), or A (for automatic routine) whenever a corresponding routine listed above can be helpful. An asterisk on the I or A indicates that this routine definitely should be used (unless your instructor gives contrary instructions), and the printout from this

[1] Problems 5.1-10, 5.1-11, 5.1-12, 5.3-6, and 5.3-7 have been adapted, with permission, from previous operations research examinations given by the Society of Actuaries.

routine is all that needs to be turned in to show your work in executing the algorithm. An asterisk on the problem number indicates that at least a partial answer is given in the back of the book.

5.1-1.* Consider the following problem.

$$\text{Maximize} \quad Z = 3x_1 + 2x_2,$$

subject to

$$2x_1 + x_2 \leq 6$$
$$x_1 + 2x_2 \leq 6$$

and

$$x_1 \geq 0, \quad x_2 \geq 0.$$

I* (*a*) Solve this problem graphically. Identify the CPF solutions by circling them on the graph.

(*b*) Identify all the sets of two defining equations for this problem. For each set, solve (if a solution exists) for the corresponding corner-point solution, and classify it as a CPF solution or corner-point infeasible solution.

(*c*) Introduce slack variables in order to write the functional constraints in augmented form. Use these slack variables to identify the basic solution that corresponds to each corner-point solution found in part (*b*).

(*d*) Do the following for *each* set of two defining equations from part (*b*): Identify the indicating variable for each defining equation. Display the set of equations from part (*c*) *after* deleting these two indicating (nonbasic) variables. Then use the latter set of equations to solve for the two remaining variables (the basic variables). Compare the resulting basic solution to the corresponding basic solution obtained in part (*c*).

(*e*) Without executing the simplex method, use its geometric interpretation (and the objective function) to identify the path (sequence of CPF solutions) it would follow to reach the optimal solution. For each of these CPF solutions in turn, identify the following decisions being made for the next iteration: (*i*) which defining equation is being deleted and which is being added; (*ii*) which indicating variable is being deleted (the entering basic variable) and which is being added (the leaving basic variable).

5.1-2. Repeat Prob. 5.1-1 for the model in Prob. 3.1-2.

5.1-3. Consider the following problem.

$$\text{Maximize} \quad Z = 2x_1 + 3x_2,$$

subject to

$$-3x_1 + x_2 \leq 1$$
$$4x_1 + 2x_2 \leq 20$$
$$4x_1 - x_2 \leq 10$$
$$-x_1 + 2x_2 \leq 5$$

and

$$x_1 \geq 0, \quad x_2 \geq 0.$$

I* (*a*) Solve this problem graphically. Identify the CPF solutions by circling them on the graph.

(*b*) Develop a table giving each of the CPF solutions and the corresponding defining equations, BF solution, and nonbasic variables. Calculate Z for each of these solutions, and use just this information to identify the optimal solution.

(*c*) Develop the corresponding table for the corner-point infeasible solutions, etc. Also identify the sets of defining equations and nonbasic variables that do not yield a solution.

5.1-4. Consider the following problem.

$$\text{Maximize} \quad Z = 2x_1 - x_2 + x_3,$$

subject to

$$3x_1 + x_2 + x_3 \leq 60$$
$$x_1 - x_2 + 2x_3 \leq 10$$
$$x_1 + x_2 - x_3 \leq 20$$

and

$$x_1 \geq 0, \quad x_2 \geq 0, \quad x_3 \geq 0.$$

After slack variables are introduced and then one complete iteration of the simplex method is performed, the following simplex tableau is obtained.

Iteration	Basic Variable	Eq.	*Coefficient of:*							Right Side
			Z	x_1	x_2	x_3	x_4	x_5	x_6	
1	Z	(0)	1	0	−1	3	0	2	0	20
	x_4	(1)	0	0	4	−5	1	−3	0	30
	x_1	(2)	0	1	−1	2	0	1	0	10
	x_6	(3)	0	0	2	−3	0	−1	1	10

(*a*) Identify the CPF solution obtained at iteration 1.

(*b*) Identify the constraint boundary equations that define this CPF solution.

5.1-5. Consider the three-variable linear programming problem shown in Fig. 5.2.

(*a*) Construct a table like Table 5.1, giving the set of defining equations for each CPF solution.

(*b*) What are the defining equations for the corner-point infeasible solution (6, 0, 5)?

(*c*) Identify one of the systems of three constraint boundary equations that yields neither a CPF solution nor a corner-point infeasible solution. Explain why this occurs for this system.

5.1-6. Consider the linear programming problem given in Table 6.1 as the dual problem for the Wyndor Glass Co. example.

(*a*) Identify the 10 sets of defining equations for this problem. For each one, solve (if a solution exists) for the corresponding corner-point solution, and classify it as a CPF solution or corner-point infeasible solution.

(*b*) For each corner-point solution, give the corresponding basic solution and its set of nonbasic variables. (Compare with Table 6.9.)

5.1-7. Consider the following problem.

$$\text{Minimize} \quad Z = x_1 + 2x_2,$$

subject to

$$-x_1 + x_2 \leq 15$$

$$2x_1 + x_2 \leq 90$$

$$x_2 \geq 30$$

and

$$x_1 \geq 0, \qquad x_2 \geq 0.$$

I* (*a*) Solve this problem graphically.

(*b*) Develop a table giving each of the CPF solutions and the corresponding defining equations, BF solution, and nonbasic variables.

5.1-8. Reconsider the model in Prob. 4.6-3.

(*a*) Identify the 10 sets of defining equations for this problem. For each one, solve (if a solution exists) for the corresponding corner-point solution, and classify it as a CPF solution or a corner-point infeasible solution.

(*b*) For each corner-point solution, give the corresponding basic solution and its set of nonbasic variables.

5.1-9. Reconsider the model in Prob. 3.1-1.

(*a*) Identify the 15 sets of defining equations for this problem. For each one, solve (if a solution exists) for the corresponding corner-point solution, and classify it as a CPF solution or a corner-point infeasible solution.

(*b*) For each corner-point solution, give the corresponding basic solution and its set of nonbasic variables.

5.1-10. Consider the original form (before augmenting) of a linear programming problem with n decision variables (each with a nonnegativity constraint) and m functional constraints. Label each of the following statements as true or false, and then justify your answer with specific references (including page citations) to material in the chapter.

(*a*) If a feasible solution is optimal, it must be a CPF solution.

(*b*) The number of CPF solutions is at least

$$\frac{(m+n)!}{m!n!}.$$

(*c*) If a CPF solution has adjacent CPF solutions that are better (as measured by Z), then one of these adjacent CPF solutions must be an optimal solution.

5.1-11. Label each of the following statements about linear programming problems as true or false, and then justify your answer.

(*a*) If a feasible solution is optimal but not a CPF solution, then infinitely many optimal solutions exist.

(*b*) If the value of the objective function is equal at two different feasible points $\mathbf{x}^*$ and $\mathbf{x}^{**}$, then all points on the line segment connecting $\mathbf{x}^*$ and $\mathbf{x}^{**}$ are feasible and Z has the same value at all those points.

(*c*) If the problem has n variables (before augmenting), then the simultaneous solution of any set of n constraint boundary equations is a CPF solution.

5.1-12. Consider the augmented form of linear programming problems that have feasible solutions and a bounded feasible region. Label each of the following statements as true or false, and then justify your answer by referring to specific statements (with page citations) in the chapter.

(*a*) There must be at least one optimal solution.

(*b*) An optimal solution must be a BF solution.

(*c*) The number of BF solutions is finite.

5.1-13.* Reconsider the model in Prob. 4.6-8. Now you are given the information that the basic variables in the optimal solution are x_2 and x_3. Use this information to identify a system of three constraint boundary equations whose simultaneous solution must be this optimal solution. Then solve this system of equations to obtain this solution.

5.1-14. Reconsider Prob. 4.3-7. Now use the given information and the theory of the simplex method to identify a system of three constraint boundary equations (in x_1, x_2, x_3) whose simultaneous solution must be the optimal solution, without applying the simplex method. Solve this system of equations to find the optimal solution.

5.1-15. Reconsider Prob. 4.3-8. Using the given information and the theory of the simplex method, analyze the constraints of the problem in order to identify a system of three constraint boundary equations whose simultaneous solution must be the optimal solution (not augmented). Then solve this system of equations to obtain this solution.

5.1-16. Consider the following problem.

$$\text{Maximize} \quad Z = 2x_1 + 2x_2 + 3x_3,$$

subject to

$$2x_1 + x_2 + 2x_3 \leq 4$$
$$x_1 + x_2 + x_3 \leq 3$$

and

$$x_1 \geq 0, \quad x_2 \geq 0, \quad x_3 \geq 0.$$

Let x_4 and x_5 be the slack variables for the respective functional constraints. Starting with these two variables as the basic variables for the initial BF solution, you now are given the information that the simplex method proceeds as follows to obtain the optimal solution in two iterations: (1) In iteration 1, the entering basic variable is x_3 and the leaving basic variable is x_4; (2) in iteration 2, the entering basic variable is x_2 and the leaving basic variable is x_5.

(*a*) Develop a three-dimensional drawing of the feasible region for this problem, and show the path followed by the simplex method.
(*b*) Give a geometric interpretation of why the simplex method followed this path.
(*c*) For each of the two edges of the feasible region traversed by the simplex method, give the equation of each of the two constraint boundaries on which it lies, and then give the equation of the additional constraint boundary at each endpoint.
(*d*) Identify the set of defining equations for each of the three CPF solutions (including the initial one) obtained by the simplex method. Use the defining equations to solve for these solutions.
(*e*) For each CPF solution obtained in part (*d*), give the corresponding BF solution and its set of nonbasic variables. Explain how these nonbasic variables identify the defining equations obtained in part (*d*).

5.1-17. Consider the following problem.

$$\text{Maximize} \quad Z = 3x_1 + 4x_2 + 2x_3,$$

subject to

$$x_1 + x_2 + x_3 \leq 20$$
$$x_1 + 2x_2 + x_3 \leq 30$$

and

$$x_1 \geq 0, \quad x_2 \geq 0, \quad x_3 \geq 0.$$

Let x_4 and x_5 be the slack variables for the respective functional constraints. Starting with these two variables as the basic variables for the initial BF solution, you now are given the information that the simplex method proceeds as follows to obtain the optimal solution in two iterations: (1) In iteration 1, the entering basic variable is x_2 and the leaving basic variable is x_5; (2) in iteration 2, the entering basic variable is x_1 and the leaving basic variable is x_4.

Follow the instructions of Prob. 5.1-16 for this situation.

5.1-18. By inspecting Fig. 5.2, explain why Property 1*b* for CPF solutions holds for this problem if it has the following objective function.

(*a*) Maximize $Z = x_3$.
(*b*) Maximize $Z = -x_1 + 2x_3$.

5.1-19. Consider the three-variable linear programming problem shown in Fig. 5.2.

(*a*) Explain in geometric terms why the set of solutions satisfying any individual constraint is a convex set, as defined in Appendix 2.
(*b*) Use the conclusion in part (*a*) to explain why the entire feasible region (the set of solutions that simultaneously satisfies every constraint) is a convex set.

5.1-20. Suppose that the three-variable linear programming problem given in Fig. 5.2 has the objective function

$$\text{Maximize} \qquad Z = 3x_1 + 4x_2 + 3x_3.$$

Without using the algebra of the simplex method, apply just its geometric reasoning (including choosing the edge giving the maximum rate of increase of Z) to determine and explain the path it would follow in Fig. 5.2 from the origin to the optimal solution.

5.1-21. Consider the three-variable linear programming problem shown in Fig. 5.2.

(*a*) Construct a table like Table 5.4, giving the indicating variable for each constraint boundary equation and original constraint.
(*b*) For the CPF solution (2, 4, 3) and its three adjacent CPF solutions (4, 2, 4), (0, 4, 2), and (2, 4, 0), construct a table like Table 5.5, showing the corresponding defining equations, BF solution, and nonbasic variables.
(*c*) Use the sets of defining equations from part (*b*) to demonstrate that (4, 2, 4), (0, 4, 2), and (2, 4, 0) are indeed adjacent to (2, 4, 3), but that none of these three CPF solutions are adjacent to each other. Then use the sets of nonbasic variables from part (*b*) to demonstrate the same thing.

5.1-22. The formula for the line passing through (2, 4, 3) and (4, 2, 4) in Fig. 5.2 can be written as

$$(2, 4, 3) + \alpha[(4, 2, 4) - (2, 4, 3)] = (2, 4, 3) + \alpha(2, -2, 1),$$

where $0 \leq \alpha \leq 1$ for just the line segment between these points. After augmenting with the slack variables x_4, x_5, x_6, x_7 for the respective functional constraints, this formula becomes

$$(2, 4, 3, 2, 0, 0, 0) + \alpha(2, -2, 1, -2, 2, 0, 0).$$

Use this formula directly to answer each of the following questions, and thereby relate the algebra and geometry of the simplex method as it goes through one iteration in moving from (2, 4, 3) to (4, 2, 4). (You are given the information that it is moving along this line segment.)

(*a*) What is the entering basic variable?
(*b*) What is the leaving basic variable?
(*c*) What is the new BF solution?

5.1-23. Consider a two-variable mathematical programming problem that has the feasible region shown on the graph, where the six dots correspond to CPF solutions. The problem has a linear objective function, and the two dashed lines are objective function lines passing through the optimal solution (4, 5) and the second-best CPF solution (2, 5). Note that the nonoptimal solution (2, 5) is better than both of its adjacent CPF solutions, which violates Property 3 in Sec. 5.1 for CPF solutions in linear programming. Demonstrate that this problem *cannot* be a linear

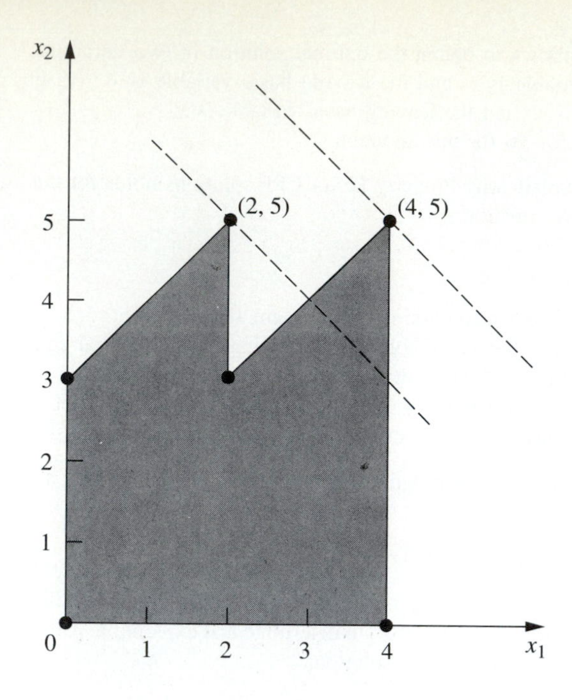

programming problem by constructing the feasible region that would result if the six line segments on the boundary were constraint boundaries for linear programming constraints.

5.2-1. Consider the following problem.

$$\text{Maximize} \quad Z = 8x_1 + 4x_2 + 6x_3 + 3x_4 + 9x_5,$$

subject to

$$x_1 + 2x_2 + 3x_3 + 3x_4 + 3x_5 \leq 180 \qquad \text{(resource 1)}$$
$$4x_1 + 3x_2 + 2x_3 + x_4 + x_5 \leq 270 \qquad \text{(resource 2)}$$
$$x_1 + 3x_2 + 2x_3 + x_4 + 3x_5 \leq 180 \qquad \text{(resource 3)}$$

and

$$x_j \geq 0, \qquad j = 1, \ldots, 5.$$

You are given the facts that the basic variables in the optimal solution are x_3, x_1, and x_5 and that

$$\begin{bmatrix} 3 & 1 & 0 \\ 2 & 4 & 1 \\ 0 & 1 & 3 \end{bmatrix}^{-1} = \frac{1}{27}\begin{bmatrix} 11 & -3 & 1 \\ -6 & 9 & -3 \\ 2 & -3 & 10 \end{bmatrix}.$$

(*a*) Use the given information to identify the optimal solution.
(*b*) Use the given information to identify the shadow prices for the three resources.

A **5.2-2.*** Use the revised simplex method to solve the following problem.

$$\text{Maximize} \quad Z = 5x_1 + 8x_2 + 7x_3 + 4x_4 + 6x_5,$$

subject to

$$2x_1 + 3x_2 + 3x_3 + 2x_4 + 2x_5 \leq 20$$
$$3x_1 + 5x_2 + 4x_3 + 2x_4 + 4x_5 \leq 30$$

and

$$x_j \geq 0, \qquad j = 1, 2, 3, 4, 5.$$

A 5.2-3. Use the revised simplex method to solve the model given in Prob. 4.3-4.

5.2-4. Reconsider Prob. 5.1-1. For the sequence of CPF solutions identified in part (*e*), construct the basis matrix **B** for each of the corresponding BF solutions. For each one, invert **B** manually, use this $\mathbf{B}^{-1}$ to calculate the current solution, and then perform the next iteration (or demonstrate that the current solution is optimal).

A **5.2-5.** Use the revised simplex method to solve the model given in Prob. 4.1-2.

A **5.2-6.** Use the revised simplex method to solve the model given in Prob. 4.7-6.

A **5.2-7.** Use the revised simplex method to solve each of the following models:

(*a*) Model given in Prob. 3.1-2

(*b*) Model given in Prob. 4.7-8

D **5.3-1.*** Consider the following problem.

$$\text{Maximize} \qquad Z = x_1 - x_2 + 2x_3,$$

subject to

$$2x_1 - 2x_2 + 3x_3 \leq 5$$
$$x_1 + x_2 - x_3 \leq 3$$
$$x_1 - x_2 + x_3 \leq 2$$

and

$$x_1 \geq 0, \qquad x_2 \geq 0, \qquad x_3 \geq 0.$$

Let x_4, x_5, and x_6 denote the slack variables for the respective constraints. After you apply the simplex method, a portion of the final simplex tableau is as follows:

		Coefficient of:							
Basic Variable	Eq.	Z	x_1	x_2	x_3	x_4	x_5	x_6	Right Side
Z	(0)	1				1	1	0	
x_2	(1)	0				1	3	0	
x_6	(2)	0				0	1	1	
x_3	(3)	0				1	2	0	

(*a*) Use the fundamental insight presented in Sec. 5.3 to identify the missing numbers in the final simplex tableau. Show your calculations.

(*b*) Identify the defining equations of the CPF solution corresponding to the optimal BF solution in the final simplex tableau.

D **5.3-2.** Consider the following problem.

$$\text{Maximize} \qquad Z = 4x_1 + 3x_2 + x_3 + 2x_4,$$

subject to

$$4x_1 + 2x_2 + x_3 + x_4 \leq 5$$
$$3x_1 + x_2 + 2x_3 + x_4 \leq 4$$

and

$$x_1 \geq 0, \qquad x_2 \geq 0, \qquad x_3 \geq 0, \qquad x_4 \geq 0.$$

Let x_5 and x_6 denote the slack variables for the respective constraints. After you apply the simplex method, a portion of the final simplex tableau is as follows:

Basic Variable	Eq.	Coefficient of:							Right Side
		Z	x_1	x_2	x_3	x_4	x_5	x_6	
Z	(0)	1					1	1	
x_2	(1)	0					1	−1	
x_4	(2)	0					−1	2	

(*a*) Use the fundamental insight presented in Sec. 5.3 to identify the missing numbers in the final simplex tableau. Show your calculations.
(*b*) Identify the defining equations of the CPF solution corresponding to the optimal BF solution in the final simplex tableau.

D **5.3-3.** Consider the following problem.

$$\text{Maximize} \quad Z = 6x_1 + x_2 + 2x_3,$$

subject to

$$2x_1 + 2x_2 + \tfrac{1}{2}x_3 \leq 2$$
$$-4x_1 - 2x_2 - \tfrac{3}{2}x_3 \leq 3$$
$$x_1 + 2x_2 + \tfrac{1}{2}x_3 \leq 1$$

and

$$x_1 \geq 0, \quad x_2 \geq 0, \quad x_3 \geq 0.$$

Let x_4, x_5, and x_6 denote the slack variables for the respective constraints. After you apply the simplex method, a portion of the final simplex tableau is as follows:

Basic Variable	Eq.	Coefficient of:							Right Side
		Z	x_1	x_2	x_3	x_4	x_5	x_6	
Z	(0)	1				2	0	2	
x_5	(1)	0				1	1	2	
x_3	(2)	0				−2	0	4	
x_1	(3)	0				1	0	−1	

Use the fundamental insight presented in Sec. 5.3 to identify the missing numbers in the final simplex tableau. Show your calculations.

D **5.3-4.** Consider the following problem.

$$\text{Maximize} \quad Z = x_1 - x_2 + 2x_3,$$

subject to

$$x_1 + x_2 + 3x_3 \leq 15$$
$$2x_1 - x_2 + x_3 \leq 2$$
$$-x_1 + x_2 + x_3 \leq 4$$

and

$$x_1 \geq 0, \quad x_2 \geq 0, \quad x_3 \geq 0.$$

Let x_4, x_5, and x_6 denote the slack variables for the respective constraints. After the simplex method is applied, a portion of the final simplex tableau is as follows:

Basic Variable	Eq.	Coefficient of:							Right Side
		Z	x_1	x_2	x_3	x_4	x_5	x_6	
Z	(0)	1				0	$\frac{3}{2}$	$\frac{1}{2}$	
x_4	(1)	0				1	-1	-2	
x_3	(2)	0				0	$\frac{1}{2}$	$\frac{1}{2}$	
x_2	(3)	0				0	$-\frac{1}{2}$	$\frac{1}{2}$	

(*a*) Use the fundamental insight presented in Sec. 5.3 to identify the missing numbers in the final simplex tableau. Show your calculations.
(*b*) Identify the defining equations of the CPF solution corresponding to the optimal BF solution in the final simplex tableau.

D **5.3-5.** Consider the following problem.

$$\text{Maximize} \quad Z = 20x_1 + 6x_2 + 8x_3,$$

subject to

$$8x_1 + 2x_2 + 3x_3 \le 200$$
$$4x_1 + 3x_2 + 3x_3 \le 100$$
$$2x_1 + 3x_2 + x_3 \le 50$$
$$x_3 \le 20$$

and

$$x_1 \ge 0, \quad x_2 \ge 0, \quad x_3 \ge 0.$$

Let x_4, x_5, x_6, and x_7 denote the slack variables for the first through fourth constraints, respectively. Suppose that after some number of iterations of the simplex method, a portion of the current simplex tableau is as follows:

Basic Variable	Eq.	Coefficient of:								Right Side
		Z	x_1	x_2	x_3	x_4	x_5	x_6	x_7	
Z	(0)	1				$\frac{9}{4}$	$\frac{1}{2}$	0	0	
x_1	(1)	0				$\frac{3}{16}$	$-\frac{1}{8}$	0	0	
x_2	(2)	0				$-\frac{1}{4}$	$\frac{1}{2}$	0	0	
x_6	(3)	0				$-\frac{3}{8}$	$\frac{1}{4}$	1	0	
x_7	(4)	0				0	0	0	1	

(*a*) Use the fundamental insight presented in Sec. 5.3 to identify the missing numbers in the current simplex tableau. Show your calculations.
(*b*) Indicate which of these missing numbers would be generated by the revised simplex method in order to perform the next iteration.
(*c*) Identify the defining equations of the CPF solution corresponding to the BF solution in the current simplex tableau.

D **5.3-6.** You are using the simplex method to solve the following linear programming problem.

$$\text{Maximize} \quad Z = 6x_1 + 5x_2 - x_3 + 4x_4,$$

subject to

$$3x_1 + 2x_2 - 3x_3 + x_4 \leq 120$$
$$3x_1 + 3x_2 + x_3 + 3x_4 \leq 180$$

and

$$x_1 \geq 0, \quad x_2 \geq 0, \quad x_3 \geq 0, \quad x_4 \geq 0.$$

You have obtained the following final simplex tableau where x_5 and x_6 are the slack variables for the respective constraints.

Basic Variable	Eq.	Coefficient of:							Right Side
		Z	x_1	x_2	x_3	x_4	x_5	x_6	
Z	(0)	1	0	$\frac{1}{4}$	0	$\frac{1}{2}$	$\frac{3}{4}$	$\frac{5}{4}$	Z^*
x_1	(1)	0	1	$\frac{11}{12}$	0	$\frac{5}{6}$	$\frac{1}{12}$	$\frac{1}{4}$	b_1^*
x_3	(2)	0	0	$\frac{1}{4}$	1	$\frac{1}{2}$	$-\frac{1}{4}$	$\frac{1}{4}$	b_2^*

Use the fundamental insight presented in Sec. 5.3 to identify Z^*, b_1^*, and b_2^*. Show your calculations.

D **5.3-7.** Consider the following problem.

$$\text{Maximize} \quad Z = c_1x_1 + c_2x_2 + c_3x_3,$$

subject to

$$x_1 + 2x_2 + x_3 \leq b$$
$$2x_1 + x_2 + 3x_3 \leq 2b$$

and

$$x_1 \geq 0, \quad x_2 \geq 0, \quad x_3 \geq 0.$$

Note that values have not been assigned to the coefficients in the objective function (c_1, c_2, c_3), and that the only specification for the right-hand side of the functional constraints is that the second one ($2b$) be twice as large as the first (b).

Now suppose that your boss has inserted her best estimate of the values of c_1, c_2, c_3, and b without informing you and then has run the simplex method. You are given the resulting final simplex tableau below (where x_4 and x_5 are the slack variables for the respective functional constraints), but you are unable to read the value of Z^*.

Basic Variable	Eq.	Coefficient of:						Right Side
		Z	x_1	x_2	x_3	x_4	x_5	
Z	(0)	1	$\frac{7}{10}$	0	0	$\frac{3}{5}$	$\frac{4}{5}$	Z^*
x_2	(1)	0	$\frac{1}{5}$	1	0	$\frac{3}{5}$	$-\frac{1}{5}$	1
x_3	(2)	0	$\frac{3}{5}$	0	1	$-\frac{1}{5}$	$\frac{2}{5}$	3

(*a*) Use the fundamental insight presented in Sec. 5.3 to identify the value of (c_1, c_2, c_3) that was used.

(*b*) Use the fundamental insight presented in Sec. 5.3 to identify the value of b that was used.

(*c*) Calculate the value of Z^* in two ways, where one way uses your results from part (*a*) and the other way uses your result from part (*b*). Show your two methods for finding Z^*.

5.3-8. For iteration 2 of the example in Sec. 5.3, the following expression was shown:

$$\text{Final row } 0 = [-3, \quad -5 \mid 0, \quad 0, \quad 0 \mid 0] + [0, \quad \tfrac{3}{2}, \quad 1]\begin{bmatrix} 1 & 0 & 1 & 0 & 0 & 4 \\ 0 & 2 & 0 & 1 & 0 & 12 \\ 3 & 2 & 0 & 0 & 1 & 18 \end{bmatrix}.$$

Derive this expression by combining the algebraic operations (in matrix form) for iterations 1 and 2 that affect row 0.

5.3-9. Most of the description of the fundamental insight presented in Sec. 5.3 assumes that the problem is in our standard form. Now consider each of the following other forms, where the additional adjustments in the initialization step are those presented in Sec. 4.6, including the use of artificial variables and the Big M method where appropriate. Describe the resulting adjustments in the fundamental insight.

(*a*) Equality constraints
(*b*) Functional constraints in $\geq$ form
(*c*) Negative right-hand sides
(*d*) Variables allowed to be negative (with no lower bound)

5.3-10. Reconsider the model in Prob. 4.6-6. For this model that is not in our standard form, construct the complete first simplex tableau for the simplex method with the Big M method, and then identify the columns that will contain $\mathbf{S}^*$ for applying the fundamental insight in the final tableau. Explain why these are the appropriate columns.

5.3-11. Consider the following problem.

$$\text{Minimize} \quad Z = 2x_1 + 3x_2 + 2x_3,$$

subject to

$$x_1 + 4x_2 + 2x_3 \geq 8$$
$$3x_1 + 2x_2 + 2x_3 \geq 6$$

and

$$x_1 \geq 0, \quad x_2 \geq 0, \quad x_3 \geq 0.$$

Let x_4 and x_6 be the surplus variables for the first and second constraints, respectively. Let x_5 and x_7 be the corresponding artificial variables. After you make the adjustments described in Sec. 4.6 for this model form when using the Big M method, the initial simplex tableau ready to apply the simplex method is as follows:

		Coefficient of:									
Basic Variable	Eq.	Z	x_1	x_2	x_3	x_4	$\bar{x}_5$	x_6	$\bar{x}_7$	Right Side
Z	(0)	-1	$-4M + 2$	$-6M + 3$	$-2M + 2$	M	0	M	0	$-14M$
$\bar{x}_5$	(1)	0	1	4	2	-1	1	0	0	8
$\bar{x}_7$	(2)	0	3	2	0	0	0	-1	1	6

After you apply the simplex method, a portion of the final simplex tableau is as follows:

Basic Variable	Eq.	Coefficient of:									Right Side
		Z	x_1	x_2	x_3	x_4	$\bar{x}_5$	x_6	$\bar{x}_7$		
Z	(0)	-1					$M - 0.5$		$M - 0.5$		
x_2	(1)	0					0.3		-0.1		
x_1	(2)	0					-0.2		0.4		

(*a*) Based on the above tableaux, use the fundamental insight presented in Sec. 5.3 to identify the missing numbers in the final simplex tableau. Show your calculations.
(*b*) Examine the mathematical logic presented in Sec. 5.3 to validate the fundamental insight (see the $\mathbf{T}^* = \mathbf{MT}$ and $\mathbf{t}^* = \mathbf{t} + \mathbf{vT}$ equations and the subsequent derivations of $\mathbf{M}$ and $\mathbf{v}$). This logic assumes that the original model fits our standard form, whereas the current problem does not fit this form. Show how, with minor adjustments, this same logic applies to the current problem when $\mathbf{t}$ is row 0 and $\mathbf{T}$ is rows 1 and 2 in the initial simplex tableau given above. Derive $\mathbf{M}$ and $\mathbf{v}$ for this problem.
(*c*) When you apply the $\mathbf{t}^* = \mathbf{t} + \mathbf{vT}$ equation, another option is to use $\mathbf{t} = [2, 3, 2, 0, M, 0, M, 0]$, which is the *preliminary* row 0 before the algebraic elimination of the nonzero coefficients of the initial basic variables $\bar{x}_5$ and $\bar{x}_7$. Repeat part (*b*) for this equation with this new $\mathbf{t}$. After you derive the new $\mathbf{v}$, show that this equation yields the same final row 0 for this problem as the equation derived in part (*b*).
(*d*) Identify the defining equations of the CPF solution corresponding to the optimal BF solution in the final simplex tableau.

5.3-12. Consider the following problem.

$$\text{Maximize} \quad Z = 2x_1 + 4x_2 + 3x_3,$$

subject to

$$x_1 + 3x_2 + 2x_3 = 20$$
$$x_1 + 5x_2 \geq 10$$

and

$$x_1 \geq 0, \quad x_2 \geq 0, \quad x_3 \geq 0.$$

Let $\bar{x}_4$ be the artificial variable for the first constraint. Let x_5 and $\bar{x}_6$ be the surplus variable and artificial variable, respectively, for the second constraint.

You are now given the information that a portion of the final simplex tableau is as follows:

Basic Variable	Eq.	Coefficient of:							Right Side
		Z	x_1	x_2	x_3	$\bar{x}_4$	x_5	$\bar{x}_6$	
Z	(0)	1				$M + 2$	0	M	
x_1	(1)	0				1	0	0	
x_5	(2)	0				1	1	-1	

(*a*) Extend the fundamental insight presented in Sec. 5.3 to identify the missing numbers in the final simplex tableau. Show your calculations.
(*b*) Identify the defining equations of the CPF solution corresponding to the optimal solution in the final simplex tableau.

5.3-13. Consider the following problem.

$$\text{Maximize} \quad Z = 3x_1 + 7x_2 + 2x_3,$$

subject to

$$-2x_1 + 2x_2 + x_3 \leq 10$$
$$3x_1 + x_2 - x_3 \leq 20$$

and

$$x_1 \geq 0, \qquad x_2 \geq 0, \qquad x_3 \geq 0.$$

You are given the fact that the basic variables in the optimal solution are x_1 and x_3.

(*a*) Introduce slack variables, and then use the given information to find the optimal solution directly by Gaussian elimination.

(*b*) Extend the work in part (*a*) to find the shadow prices.

(*c*) Use the given information to identify the defining equations of the optimal CPF solution, and then solve these equations to obtain the optimal solution.

(*d*) Construct the basis matrix **B** for the optimal BF solution, invert **B** manually, and then use this $\mathbf{B}^{-1}$ to solve for the optimal solution and the shadow prices $\mathbf{y}^*$. Then apply the optimality test for the revised simplex method to verify that this solution is optimal.

(*e*) Given $\mathbf{B}^{-1}$ and $\mathbf{y}^*$ from part (*d*), use the fundamental insight presented in Sec. 5.3 to construct the complete final simplex tableau.

6

Duality Theory and Sensitivity Analysis

One of the most important discoveries in the early development of linear programming was the concept of duality and its many important ramifications. This discovery revealed that every linear programming problem has associated with it another linear programming problem called the **dual**. The relationships between the dual problem and the original problem (called the **primal**) prove to be extremely useful in a variety of ways. For example, you soon will see that the shadow prices described in Sec. 4.7 actually are provided by the optimal solution for the dual problem. We shall describe many other valuable applications of duality theory in this chapter as well.

One of the key uses of duality theory lies in the interpretation and implementation of *sensitivity analysis.* As we already mentioned in Secs. 2.3, 3.3, and 4.7, sensitivity analysis is a very important part of almost every linear programming study. Because some of or all the parameter values used in the original model are just *estimates* of future conditions, the effect on the optimal solution if other conditions prevail instead needs to be investigated. Furthermore, certain parameter values (such as resource amounts) may represent *managerial decisions,* in which case the choice of the parameter values may be the main issue to be studied, which can be done through sensitivity analysis.

For greater clarity, the first three sections discuss duality theory under the assumption that the *primal* linear programming problem is in *our standard form* (but with no restriction that the b_i values need to be positive). Other forms are then discussed in Sec. 6.4. We begin the chapter by introducing the essence of duality theory and its applications. We then describe the economic interpretation of the dual problem (Sec. 6.2) and delve deeper into the relationships between the primal and dual problems (Sec. 6.3). Section 6.5 focuses on the role of duality theory in sensitivity analysis. The basic procedure for sensitivity analysis (which is based on the fundamental insight of Sec. 5.3) is summarized in Sec. 6.6 and illustrated in Sec. 6.7.

6.1 The Essence of Duality Theory

Given our standard form for the *primal problem* at the left (perhaps after conversion from another form), its *dual problem* has the form shown to the right.

Primal Problem

Maximize $\quad Z = \sum_{j=1}^{n} c_j x_j,$

subject to

$$\sum_{j=1}^{n} a_{ij} x_j \leq b_i \qquad \text{for } i = 1, 2, \dots, m$$

and

$$x_j \geq 0, \qquad \text{for } j = 1, 2, \dots, n.$$

Dual Problem

Minimize $\quad y_0 = \sum_{i=1}^{m} b_i y_i,$

subject to

$$\sum_{i=1}^{m} a_{ij} y_i \geq c_j, \qquad \text{for } j = 1, 2, \dots, n$$

and

$$y_i \geq 0, \qquad \text{for } i = 1, 2, \dots, m.$$

Thus the dual problem uses exactly the same *parameters* as the primal problem, but in different locations. To highlight the comparison, now look at these same two problems in matrix notation (as introduced at the beginning of Sec. 5.2), where $\mathbf{c}$ and $\mathbf{y} = [y_1, y_2, \dots, y_m]$ are row vectors but $\mathbf{b}$ and $\mathbf{x}$ are column vectors.

Primal Problem

Maximize $\quad Z = \mathbf{cx},$

subject to

$$\mathbf{Ax} \leq \mathbf{b}$$

and

$$\mathbf{x} \geq \mathbf{0}.$$

Dual Problem

Minimize $\quad y_0 = \mathbf{yb},$

subject to

$$\mathbf{yA} \geq \mathbf{c}$$

and

$$\mathbf{y} \geq \mathbf{0}.$$

To illustrate, the primal and dual problems for the Wyndor Glass Co. example of Sec. 3.1 are shown in Table 6.1 in both algebraic and matrix form.

The **primal-dual table** for linear programming (Table 6.2) also helps to highlight the correspondence between the two problems. It shows all the linear programming parameters (the a_{ij}, b_i, and c_j) and how they are used to construct the two prob-

Table 6.1 **Primal and Dual Problems for the Wyndor Glass Co. Example**

Primal Problem in Algebraic Form

$$\begin{aligned} \text{Maximize} \quad & Z = 3x_1 + 5x_2, \\ \text{subject to} \quad & \\ & x_1 \leq 4 \\ & 2x_2 \leq 12 \\ & 3x_1 + 2x_2 \leq 18 \\ \text{and} \quad & x_1 \geq 0, \quad x_2 \geq 0. \end{aligned}$$

Dual Problem in Algebraic Form

$$\begin{aligned} \text{Minimize} \quad & y_0 = 4y_1 + 12y_2 + 18y_3, \\ \text{subject to} \quad & \\ & y_1 + 3y_3 \geq 3 \\ & 2y_2 + 2y_3 \geq 5 \\ \text{and} \quad & \\ & y_1 \geq 0, \quad y_2 \geq 0, \quad y_3 \geq 0. \end{aligned}$$

Primal Problem in Matrix Form

$$\text{Maximize} \quad Z = [3, 5]\begin{bmatrix} x_1 \\ x_2 \end{bmatrix},$$

subject to

$$\begin{bmatrix} 1 & 0 \\ 0 & 2 \\ 3 & 2 \end{bmatrix}\begin{bmatrix} x_1 \\ x_2 \end{bmatrix} \leq \begin{bmatrix} 4 \\ 12 \\ 18 \end{bmatrix}$$

and

$$\begin{bmatrix} x_1 \\ x_2 \end{bmatrix} \geq \begin{bmatrix} 0 \\ 0 \end{bmatrix}.$$

Dual Problem in Matrix Form

$$\text{Minimize} \quad y_0 = [y_1, y_2, y_3]\begin{bmatrix} 4 \\ 12 \\ 18 \end{bmatrix},$$

subject to

$$[y_1, y_2, y_3]\begin{bmatrix} 1 & 0 \\ 0 & 2 \\ 3 & 2 \end{bmatrix} \geq [3, 5]$$

and

$$[y_1, y_2, y_3] \geq [0, 0, 0].$$

lems. All the headings for the primal problem are horizontal, whereas the headings for the dual problem are read by turning the book sideways. We suggest that you begin with Table 6.2*a* by looking at each problem *individually* by covering up the headings for the other problem with your hands. Then, after you see what the table is saying for the individual problems, compare them. Next, repeat this process with the example in Table 6.2*b*, while also referring to the corresponding models in Table 6.1.

Particularly notice in Table 6.2 how (1) the parameters for a *constraint* in either problem are the coefficients of a *variable* in the other problem and (2) the coefficients for the *objective function* of either problem are the *right sides* for the other problem. Thus there is a direct correspondence between these entities in the two problems, as summarized in Table 6.3. These correspondences are a key to some of the applications of duality theory, including sensitivity analysis.

Origin of the Dual Problem

Duality theory is based directly on the fundamental insight (particularly with regard to row 0) presented in Sec. 5.3. To see why, we continue to use the notation introduced in Table 5.10 for row 0 of the *final* tableau, except for replacing Z^* by y_0^* and dropping the asterisks from $\mathbf{z}^*$ and $\mathbf{y}^*$ when referring to *any* tableau. Thus, at *any* given iteration of the simplex method for the primal problem, the current numbers in row 0 are denoted as shown in the (partial) tableau given in Table 6.4. Also recall [see Eq. (1) in the "Mathematical Summary" subsection of Sec. 5.3] that the fundamental insight led to

Table 6.2 Primal-Dual Table for Linear Programming, Illustrated by the Wyndor Glass Co. Example

(a) General Case

			Primal Problem					
			Coefficient of:				Right Side	
			x_1	x_2	$\cdots$	x_n		
Dual Problem	Coefficient of:	y_1	a_{11}	a_{12}	$\cdots$	a_{1n}	$\leq b_1$	Coefficients for Objective Function (Minimize)
		y_2	a_{21}	a_{22}	$\cdots$	a_{2n}	$\leq b_2$	
		$\vdots$	$\cdots$	$\cdots$	$\cdots$	$\cdots$	$\vdots$	
		y_m	a_{m1}	a_{m2}	$\cdots$	a_{mn}	$\leq b_m$	
	Right Side		$\geq c_1$	$\geq c_2$	$\cdots$	$\geq c_n$		
			Coefficients for Objective Function (Maximize)					

(b) Wyndor Glass Co. Example

	x_1	x_2	
y_1	1	0	≤ 4
y_2	0	2	≤ 12
y_3	3	2	≤ 18
	≥ 3	≥ 5	

the following relationships between these quantities and the parameters of the original model:

$$y_0 = \mathbf{yb} = \sum_{i=1}^{m} b_i y_i,$$

$$\mathbf{z} = \mathbf{yA}, \qquad \text{so} \qquad z_j = \sum_{i=1}^{m} a_{ij} y_i, \qquad \text{for } j = 1, 2, \ldots, n.$$

The remaining key is to express what the simplex method tries to accomplish (according to the optimality test) in terms of these symbols. Specifically, it seeks a set of basic variables, and the corresponding BF solution, such that *all* coefficients in row 0 are *nonnegative.* It then stops with this optimal solution. This goal is expressed symbolically as follows:

Condition for Optimality:

$$z_j - c_j \geq 0 \qquad \text{for } j = 1, 2, \ldots, n,$$
$$y_i \geq 0 \qquad \text{for } i = 1, 2, \ldots, m.$$

Table 6.3 Correspondence between Entities in Primal and Dual Problems

One Problem		Other Problem
Constraint i	$\longleftrightarrow$	Variable i
Objective function	$\longleftrightarrow$	Right sides

After we substitute the preceding expression for z_j, the condition for optimality says that the simplex method can be interpreted as seeking values for $y_1, y_2, \ldots, y_m$ such that

$$y_0 = \sum_{i=1}^{m} b_i y_i,$$

subject to

$$\sum_{i=1}^{m} a_{ij} y_i \geq c_j, \qquad \text{for } j = 1, 2, \ldots, n$$

and

$$y_i \geq 0, \qquad \text{for } i = 1, 2, \ldots, m.$$

But, except for lacking an objective for y_0, this problem is precisely the *dual problem*! To complete the formulation, let us now explore what the missing objective should be.

Since y_0 is just the current value of Z, and since the objective for the primal problem is to maximize Z, a natural first reaction is that y_0 should be maximized also. However, this is not correct for the following rather subtle reason: The only *feasible* solutions for this new problem are those that satisfy the condition for *optimality* for the primal problem. Therefore, it is *only* the optimal solution for the primal problem that corresponds to a feasible solution for this new problem. As a consequence, the optimal value of Z in the primal problem is the *minimum* feasible value of y_0 in the new problem, so y_0 should be minimized. (The full justification for this conclusion is provided by the relationships we develop in Sec. 6.3.) Adding this objective of minimizing y_0 gives the *complete* dual problem.

Consequently, the dual problem may be viewed as a restatement in linear programming terms of the *goal* of the simplex method, namely, to reach a solution for the primal problem that *satisfies the optimality test. Before* this goal has been reached, the corresponding **y** in row 0 (coefficients of slack variables) of the current tableau must be *infeasible* for the *dual problem.* However, *after* the goal is reached, the corresponding **y** must be an *optimal solution* (labeled **y***) for the *dual problem,* because it is a feasible solution that attains the minimum feasible value of y_0. This optimal solution $(y_1^*, y_2^*, \ldots, y_m^*)$ provides for the primal problem the shadow prices that were described in Sec. 4.7. Furthermore, this optimal y_0 is just the optimal value of Z, so the *optimal objective function values are equal* for the two problems. This fact also implies that $\mathbf{cx} \leq \mathbf{yb}$ for any **x** and **y** that are *feasible* for the primal and dual problems, respectively.

To illustrate, Table 6.5 shows row 0 for the respective iterations when the simplex method is applied to the Wyndor Glass Co. example. In each case, row 0 is partitioned into three parts: the coefficients of the original variables (x_1, x_2), the coeffi-

Table 6.4 **Notation for Entries in Row 0 of a Simplex Tableau**

Iteration	Basic Variable	Eq.	*Coefficient of:*									Right Side
			Z	x_1	x_2	$\cdots$	x_n	x_{n+1}	x_{n+2}	$\cdots$	x_{n+m}	
Any	Z	(0)	1	$z_1 - c_1$	$z_2 - c_2$	$\cdots$	$z_n - c_n$	y_1	y_2	$\cdots$	y_m	y_0

cients of the slack variables (x_3, x_4, x_5), and the right-hand side (value of Z). Each row 0 identifies a solution for the dual problem, as shown to its right in Table 6.5. By using the expression for z_j given earlier, the right side of the table also includes the calculated values of

$$z_1 - c_1 = y_1 + 3y_3 - 3,$$
$$z_2 - c_2 = 2y_2 + 2y_3 - 5,$$

i.e., the values of the *surplus variables* for the functional constraints of the dual problem $y_1 + 3y_3 \geq 3$ and $2y_2 + 2y_3 \geq 5$. Thus a negative value for either surplus variable indicates that the corresponding constraint is violated. Also included on the right side of the table is the calculated value of the dual objective function $y_0 = 4y_1 + 12y_2 + 18y_3$. As displayed in Table 6.4, *all* these quantities already are identified by row 0 without requiring any new calculations. In particular, note in Table 6.5 how *each* number obtained for the dual problem already appears in row 0 in the spot indicated by Table 6.4.

For the initial row 0, Table 6.5 shows that the corresponding dual solution $(y_1, y_2, y_3) = (0, 0, 0)$ is infeasible because both surplus variables are negative. The first iteration succeeds in eliminating one of these negative values, but not the other. After two iterations, the optimality test is satisfied for the primal problem because all the dual variables and surplus variables are nonnegative. This dual solution $(y_1^*, y_2^*, y_3^*) = (0, \frac{3}{2}, 1)$ is optimal (as could be verified by applying the simplex method directly to the dual problem), so the optimal value of Z and y_0 is $Z^* = 36 = y_0^*$.

Summary of Primal-Dual Relationships

Now let us summarize the newly discovered key relationships between the primal and dual problems.

Weak duality property: If **x** is a feasible solution for the primal problem and **y** is a feasible solution for the dual problem, then

$$\mathbf{cx} \leq \mathbf{yb}.$$

For example, for the Wyndor Glass Co. problem, one feasible solution (we use the superscript T to denote the *transpose operation,* described in Appendix 4) is $\mathbf{x} = [3, 3]^T$, which yields $Z = \mathbf{cx} = 24$, and one feasible solution for the dual problem is $\mathbf{y} = [1, 1, 2]$, which yields a larger objective function value $y_0 = \mathbf{yb} = 52$. For *any* such pair of feasible solutions, this inequality must hold because the *maximum* feasible value of $Z = \mathbf{cx}$ (36) *equals* the *minimum* feasible value of the dual objective function $y_0 = \mathbf{yb}$, which is our next property.

Table 6.5 **Row 0 and Corresponding Dual Solution for Each Iteration for the Wyndor Glass Co. Example**

	Primal Problem	Dual Problem					
Iteration	Row 0	y_1	y_2	y_3	$z_1 - c_1$	$z_2 - c_2$	y_0
0	$[-3, \; -5 \,\vdots\, 0, \; 0, \; 0 \,\vdots\, 0]$	0	0	0	−3	−5	0
1	$[-3, \; 0 \,\vdots\, 0, \; \frac{5}{2}, \; 0 \,\vdots\, 30]$	0	$\frac{5}{2}$	0	−3	0	30
2	$[\;0, \; 0 \,\vdots\, 0, \; \frac{3}{2}, \; 1 \,\vdots\, 36]$	0	$\frac{3}{2}$	1	0	0	36

Strong duality property: If $\mathbf{x}^*$ is an optimal solution for the primal problem and $\mathbf{y}^*$ is an optimal solution for the dual problem, then

$$\mathbf{cx}^* = \mathbf{y}^*\mathbf{b}.$$

Thus, these two properties imply that $\mathbf{cx} < \mathbf{yb}$ for feasible solutions if one or both of them are *not optimal* for their respective problems, whereas equality holds when both are optimal.

The *weak duality property* describes the relationship between any pair of solutions for the primal and dual problems where *both* solutions are *feasible* for their respective problems. At each iteration, the simplex method finds a specific pair of solutions for the two problems, where the primal solution is feasible but the dual solution is *not feasible* (except at the final iteration). Our next property describes this situation and the relationship between this pair of solutions.

Complementary-solutions property: At each iteration, the simplex method simultaneously identifies a CPF solution $\mathbf{x}$ for the primal problem and a **complementary solution y** for the dual problem (found in row 0, the coefficients of the slack variables), where

$$\mathbf{cx} = \mathbf{yb}.$$

If $\mathbf{x}$ is *not optimal* for the primal problem, then $\mathbf{y}$ is *not feasible* for the dual problem.

To illustrate, after one iteration for the Wyndor Glass Co. problem, $\mathbf{x} = [0, 6]^T$ and $\mathbf{y} = [0, \frac{5}{2}, 0]$, with $\mathbf{cx} = 30 = \mathbf{yb}$. This $\mathbf{x}$ is feasible for the primal problem, but this $\mathbf{y}$ is not feasible for the dual problem (since it violates the constraint, $y_1 + 3y_3 \geq 3$).

The complementary-solutions property also holds at the final iteration of the simplex method, where an optimal solution is found for the primal problem. However, more can be said about the complementary solution $\mathbf{y}$ in this case, as presented in the next property.

Complementary optimal solutions property: At the final iteration, the simplex method simultaneously identifies an optimal solution $\mathbf{x}^*$ for the primal problem and a **complementary optimal solution** $\mathbf{y}^*$ for the dual problem (found in row 0, the coefficients of the slack variables), where

$$\mathbf{cx}^* = \mathbf{y}^*\mathbf{b}.$$

The y_i^* are the shadow prices for the primal problem.

For the example, the final iteration yields $\mathbf{x}^* = [2, 6]^T$ and $\mathbf{y}^* = [0, \frac{3}{2}, 1]$, with $\mathbf{cx}^* = 36 = \mathbf{y}^*\mathbf{b}$.

We shall take a closer look at some of these properties in Sec. 6.3. There you will see that the complementary solutions property can be extended considerably further. In particular, after slack and surplus variables are introduced to augment the respective problems, every *basic* solution in the primal problem has a complementary *basic* solution in the dual problem. We already have noted that the simplex method identifies the values of the surplus variables for the dual problem as $z_j - c_j$ in Table 6.4. This result then leads to an additional *complementary slackness property* that relates the basic variables in one problem to the nonbasic variables in the other (Tables 6.7 and 6.8), but more about that later.

In Sec. 6.4, after describing how to construct the dual problem when the primal problem is *not* in our standard form, we discuss another very useful property, which is summarized as follows:

Symmetry property: For *any* primal problem and its dual problem, all relationships between them must be *symmetric* because the dual of this dual problem is this primal problem.

Therefore, all the preceding properties hold regardless of which of the two problems is labeled as the primal problem. (The direction of the inequality for the weak duality property does require that the primal problem be expressed or reexpressed in maximization form and the dual problem in minimization form.) Consequently, the simplex method can be applied to either problem, and it simultaneously will identify complementary solutions (ultimately a complementary optimal solution) for the other problem.

So far, we have focused on the relationships between *feasible* or *optimal* solutions in the primal problem and corresponding solutions in the dual problem. However, it is possible that the primal (or dual) problem either has *no feasible solutions* or has feasible solutions but *no optimal solution* (because the objective function is unbounded). Our final property summarizes the primal-dual relationships under all these possibilities.

Duality theorem: The following are the only possible relationships between the primal and dual problems.

1. If one problem has *feasible solutions* and a *bounded* objective function (and so has an optimal solution), then so does the other problem, so both the weak and strong duality properties are applicable.
2. If one problem has *feasible solutions* and an *unbounded* objective function (and so *no optimal solution*), then the other problem has *no feasible solutions.*
3. If one problem has *no feasible solutions,* then the other problem has either *no feasible solutions* or an *unbounded* objective function.

Applications

As we have just implied, one important application of duality theory is that the *dual* problem can be solved directly by the simplex method in order to identify an optimal solution for the primal problem. We discussed in Sec. 4.8 that the number of functional constraints affects the computational effort of the simplex method far more than the number of variables does. If $m > n$, so that the dual problem has fewer functional constraints (n) than the primal problem (m), then applying the simplex method directly to the dual problem instead of the primal problem probably will achieve a substantial reduction in computational effort.

The *weak* and *strong duality properties* describe key relationships between the primal and dual problems. One useful application is for evaluating a proposed solution for the primal problem. For example, suppose that $\mathbf{x}$ is a feasible solution that has been proposed for implementation and that a feasible solution $\mathbf{y}$ has been found by inspection for the dual problem such that $\mathbf{cx} = \mathbf{yb}$. In this case, $\mathbf{x}$ must be *optimal* without the simplex method even being applied! Even if $\mathbf{cx} < \mathbf{yb}$, then $\mathbf{yb}$ still provides an upper bound on the optimal value of Z, so if $\mathbf{yb} - \mathbf{cx}$ is small, intangible factors favoring $\mathbf{x}$ may lead to its selection without further ado.

One of the key applications of the complementary-solutions property is its use in the dual simplex method presented in Sec. 7.1. This algorithm operates on the primal problem exactly as if the simplex method were being applied simultaneously to the dual problem, which can be done because of this property. Because the roles of row 0 and the right side in the simplex tableau have been reversed, the dual simplex method requires that row 0 *begin and remain nonnegative* while the right side *begins* with some *negative* values (subsequent iterations strive to reach a nonnegative right side). Consequently, this algorithm occasionally is used because it is more convenient to set up the initial tableau in this form than in the form required by the simplex method. Furthermore, it frequently is used for reoptimization (discussed in Sec. 4.7), because changes in the original model lead to the revised final tableau fitting this form. This situation is common for certain types of sensitivity analysis, as you will see later in the chapter.

In general terms, duality theory plays a central role in sensitivity analysis. This role is the topic of Sec. 6.5.

Another important application is its use in the economic interpretation of the dual problem and the resulting insights for analyzing the primal problem. You already have seen one example when we discussed shadow prices in Sec. 4.7. The next section describes how this interpretation extends to the entire dual problem and then to the simplex method.

6.2 Economic Interpretation of Duality

The economic interpretation of duality is based directly upon the typical interpretation for the primal problem (linear programming problem in our standard form) presented in Sec. 3.2. To refresh your memory, we have summarized this interpretation of the primal problem in Table 6.6.

Interpretation of the Dual Problem

To see how this interpretation of the primal problem leads to an economic interpretation for the dual problem,[1] note in Table 6.4 that y_0 is the value of Z (total profit) at the current iteration. Because

$$y_0 = b_1 y_1 + b_2 y_2 + \cdots + b_m y_m,$$

each $b_i y_i$ can thereby be interpreted as the current *contribution to profit* by having b_i units of resource i available for the primal problem. Thus

> Variable y_i is interpreted as the contribution to profit per unit of resource i ($i = 1, 2, \ldots, m$), when the current set of basic variables is used to obtain the primal solution.

[1] Actually, several slightly different interpretations have been proposed. The one presented here seems to us to be the most useful because it also directly interprets what the simplex method does in the primal problem.

Table 6.6 Economic Interpretation of the Primal Problem

Quantity	Interpretation
x_j	Level of activity j $\quad (j = 1, 2, \ldots, n)$
c_j	Unit profit from activity j
Z	Total profit from all activities
b_i	Amount of resource i available $\quad (i = 1, 2, \ldots, m)$
a_{ij}	Amount of resource i consumed by each unit of activity j

In other words, the y_i values (or y_i^* values in the optimal solution) are just the shadow prices discussed in Sec. 4.7.

This interpretation of the dual variables leads to our interpretation of the overall dual problem. Specifically, since each unit of activity j in the primal problem consumes a_{ij} units of resource i,

> $\sum_{i=1}^{m} a_{ij} y_i$ is interpreted as the current contribution to profit of the mix of resources that would be consumed if 1 unit of activity j were used ($j = 1, 2, \ldots, n$).

This same mix of resources (and more) probably can be used in other ways as well, but no alternative use should be considered if it is less profitable than 1 unit of activity j. Since c_j is interpreted as the unit profit from activity j, each functional constraint in the dual problem is interpreted as follows:

> $\sum_{i=1}^{m} a_{ij} y_i \geq c_j$ says that the actual contribution to profit of the above mix of resources must be at least as much as if they were used by 1 unit of activity j; otherwise, we would not be making the best possible use of these resources.

Similarly, the interpretation of the nonnegativity constraints is the following:

> $y_i \geq 0$ says that the contribution to profit of resource i ($i = 1, 2, \ldots, m$) must be nonnegative: otherwise, it would be better not to use this resource at all.

The objective

$$\text{Minimize} \quad y_0 = \sum_{i=1}^{m} b_i y_i$$

can be viewed as minimizing the total implicit value of the resources consumed by the activities.

This interpretation can be sharpened somewhat by differentiating between basic and nonbasic variables in the primal problem for any given BF solution $(x_1, x_2, \ldots, x_{n+m})$. Recall that the *basic* variables (the only variables whose values can be nonzero) *always* have a coefficient of *zero* in row 0. Therefore, referring again to Table 6.4 and the accompanying equation for z_j, we see that

$$\sum_{i=1}^{m} a_{ij} y_i = c_j, \qquad \text{if } x_j > 0 \qquad (j = 1, 2, \ldots, n),$$

$$y_i = 0, \qquad \text{if } x_{n+i} > 0 \qquad (i = 1, 2, \ldots, m).$$

(This is one version of the complementary slackness property discussed in the next section.) The economic interpretation of the first statement is that whenever an activity j operates at a strictly positive level ($x_j > 0$), the marginal value of the resources it consumes *must equal* (as opposed to exceeding) the unit profit from this activity. The second statement implies that the marginal value of resource i is *zero* ($y_i = 0$) whenever the supply of this resource is not exhausted by the activities ($x_{n+i} > 0$). In economic terminology, such a resource is a "free good"; the price of goods that are oversupplied must drop to zero by the law of supply and demand. This fact is what justifies interpreting the objective for the dual problem as minimizing the total implicit value of the resources *consumed,* rather than the resources *allocated.*

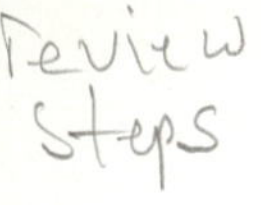

Interpretation of the Simplex Method

The interpretation of the dual problem also provides an economic interpretation of what the simplex method does in the primal problem. The *goal* of the simplex method is to find how to use the available resources in the most profitable feasible way. To attain this goal, we must reach a BF solution that satisfies all the *requirements* on profitable use of the resources (the constraints of the dual problem). These requirements comprise the *condition for optimality* for the algorithm. For any given BF solution, the requirements (dual constraints) associated with the basic variables are automatically satisfied (with equality). However, those associated with nonbasic variables may or may not be satisfied.

In particular, if an original variable x_j is nonbasic so that activity j is not used, then the current contribution to profits of the resources that would be required to undertake each unit of activity j

$$\sum_{i=1}^{m} a_{ij}y_i$$

may be either smaller ($<$) or larger ($\geq$) than the unit profit c_j obtainable from the activity. If it is smaller, so that $z_j - c_j < 0$ in row 0 of the simplex tableau, then these resources can be used more profitably by initiating this activity. If it is larger, then these resources already are being assigned elsewhere in a more profitable way, so they should not be diverted to activity j.

Similarly, if a slack variable x_{n+i} is nonbasic so that the total allocation b_i of resource i is being used, then y_i is the current contribution to profit of this resource on a marginal basis. Hence, if $y_i < 0$, profit can be increased by cutting back on the use of this resource (i.e., increasing x_{n+i}). If $y_i \geq 0$, it is worthwhile to continue fully using this resource.

Therefore, what the simplex method does is to examine all the nonbasic variables in the current BF solution to see which ones can provide a *more profitable use of the resources* by being increased. If *none* can, so that no feasible shifts or reductions in the current proposed use of the resources can increase profit, then the current solution must be optimal. If one or more can, the simplex method selects the variable that, if increased by 1, would *improve the profitability* of the use of the resources the most. It then actually increases this variable (the entering basic variable) as much as it can until the marginal values of the resources change. This increase results in a new BF solution with a new row 0 (dual solution), and the whole process is repeated.

To solidify your understanding of this interpretation of the simplex method, we suggest that you apply it to the Wyndor Glass Co. problem, using both Fig. 3.2 and Table 4.8. (See Prob. 6.2-1.)

The economic interpretation of the dual problem considerably expands our ability to analyze the primal problem. However, you already have seen in Sec. 6.1 that this interpretation is just one ramification of the relationships between the two problems. In the next section, we delve into these relationships more deeply.

6.3 Primal-Dual Relationships

Because the dual problem is a linear programming problem, it also has corner-point solutions. Furthermore, by using the augmented form of the problem, we can express these corner-point solutions as basic solutions. Because the functional constraints have the $\geq$ form, this augmented form is obtained by *subtracting* the surplus (rather than adding the slack) from the left-hand side of each constraint j ($j = 1, 2, \ldots, n$).[1] This surplus is

$$z_j - c_j = \sum_{i=1}^{m} a_{ij} y_i - c_j, \qquad \text{for } j = 1, 2, \ldots, n.$$

Thus $z_j - c_j$ plays the role of the *surplus variable* for constraint j (or its slack variable if the constraint is multiplied through by -1). Therefore, augmenting each corner-point solution $(y_1, y_2, \ldots, y_m)$ yields a basic solution $(y_1, y_2, \ldots, y_m, z_1 - c_1, z_2 - c_2, \ldots, z_n - c_n)$ by using this expression for $z_j - c_j$. Since the augmented form of the dual problem has n functional constraints and $n + m$ variables, each basic solution has n basic variables and m nonbasic variables. (Note how m and n reverse their previous roles here because, as Table 6.3 indicates, dual constraints correspond to primal variables and dual variables correspond to primal constraints.)

Complementary Basic Solutions

One of the important relationships between the primal and dual problems is a direct correspondence between their basic solutions. The key to this correspondence is row 0 of the simplex tableau for the primal basic solution, such as shown in Table 6.4 or 6.5. Such a row 0 can be obtained for *any* primal basic solution, feasible or not, by using the formulas given in the bottom part of Table 5.8.

Note again in Tables 6.4 and 6.5 how a complete solution for the dual problem (including the surplus variables) can be read directly from row 0. Thus, because of its coefficient in row 0, each variable in the primal problem has an associated variable in the dual problem, as summarized in Table 6.7.

A key insight here is that the dual solution read from row 0 must also be a basic solution! The reason is that the m basic variables for the primal problem are required to have a coefficient of zero in row 0, which thereby requires the m associated dual variables to be zero, i.e., nonbasic variables for the dual problem. The values of the remaining n (basic) variables then will be the simultaneous solution to the system of equations given at the beginning of the section. In matrix form, this system of equations is $\mathbf{z} - \mathbf{c} = \mathbf{yA} - \mathbf{c}$, and the fundamental insight of Sec. 5.3 actually identifies its solution for $\mathbf{z} - \mathbf{c}$ and $\mathbf{y}$ as being the corresponding entries in row 0.

Because of the symmetry property quoted in Sec. 6.1 (and the direct association between variables shown in Table 6.7), the correspondence between basic solutions in the primal and dual problems is a symmetric one. Furthermore, a pair of complementary basic solutions has the same objective function value, shown as y_0 in Table 6.4.

[1] You might wonder why we do not also introduce *artificial variables* into these constraints as discussed in Sec. 4.6. The reason is that these variables have no purpose other than to change the feasible region temporarily as a convenience in starting the simplex method. We are not interested now in applying the simplex method to the dual problem, and we do not want to change its feasible region.

Table 6.7 **Association between Variables in Primal and Dual Problems**

Primal Variable	Associated Dual Variable	
(Original variable) x_j	$z_j - c_j$ (surplus variable)	$j = 1, 2, \ldots, n$
(Slack variable) x_{n+i}	y_i (original variable)	$i = 1, 2, \ldots, m$

Let us now summarize our conclusions about the correspondence between primal and dual basic solutions, where the first property extends the complementary solutions property of Sec. 6.1 to the augmented forms of the two problems and then to any basic solution (feasible or not) in the primal problem.

> **Complementary basic solutions property:** Each *basic* solution in the *primal problem* has a **complementary basic solution** in the *dual problem,* where their respective objective function values (Z and y_0) are equal. Given row 0 of the simplex tableau for the primal basic solution, the complementary dual basic solution$(\mathbf{y}, \mathbf{z} - \mathbf{c})$ is found as shown in Table 6.4.

The next property shows how to identify the basic and nonbasic variables in this complementary basic solution.

> **Complementary slackness property:** Given the association between variables in Table 6.7, the variables in the primal basic solution and the complementary dual basic solution satisfy the **complementary slackness** relationship shown in Table 6.8. Furthermore, this relationship is a symmetric one, so that these two basic solutions are complementary to each other.

The reason for using the name *complementary slackness* for this latter property is that it says (in part) that for each pair of associated variables, if one of them has *slack* in its nonnegativity constraint (a basic variable > 0), then the other one must have *no slack* (a nonbasic variable $= 0$). We mentioned in Sec. 6.2 that this property has a useful economic interpretation for linear programming problems.

Example: To illustrate these two properties, again consider the Wyndor Glass Co. problem of Sec. 3.1. All eight of its basic solutions (five feasible and three infeasible) are shown in Tables 5.5 and 5.6 along with the corresponding corner-point solutions.

Thus its dual problem (see Table 6.1) also must have eight basic solutions, each complementary to one of these primal solutions, as shown in Table 6.9.

The three BF solutions obtained by the simplex method for the primal problem are the first, fifth, and sixth primal solutions shown in Table 6.9. You already saw in Table 6.5 how the complementary basic solutions for the dual problem can be read directly from row 0, starting with the coefficients of the slack variables and then the original variables. The other dual basic solutions also could be identified in this way by constructing row 0 for each of the other primal basic solutions, using the formulas given in the bottom part of Table 5.8.

Table 6.8 **Complementary Slackness Relationship for Complementary Basic Solutions**

Primal Variable	Associated Dual Variable	
Basic	Nonbasic	(m variables)
Nonbasic	Basic	(n variables)

Table 6.9 **Complementary Basic Solutions for the Wyndor Glass Co. Example**

	Primal Problem			***Dual Problem***	
No.	Basic Solution	Feasible?	$Z = y_0$	Feasible?	Basic Solution
1	(0, 0, 4, 12, 18)	Yes	0	No	(0, 0, 0, −3, −5)
2	(4, 0, 0, 12, 6)	Yes	12	No	(3, 0, 0, 0, −5)
3	(6, 0, −2, 12, 0)	No	18	No	(0, 0, 1, 0, −3)
4	(4, 3, 0, 6, 0)	Yes	27	No	$(-\frac{9}{2}, 0, \frac{5}{2}, 0, 0)$
5	(0, 6, 4, 0, 6)	Yes	30	No	$(0, \frac{5}{2}, 0, -3, 0)$
6	(2, 6, 2, 0, 0)	Yes	36	Yes	$(0, \frac{3}{2}, 1, 0, 0)$
7	(4, 6, 0, 0, −6)	No	42	Yes	$(3, \frac{5}{2}, 0, 0, 0)$
8	(0, 9, 4, −6, 0)	No	45	Yes	$(0, 0, \frac{5}{2}, \frac{9}{2}, 0)$

Alternatively, for each primal basic solution, the complementary slackness property can be used to identify the basic and nonbasic variables for the complementary dual basic solution, so that the system of equations given at the beginning of the section can be solved directly to obtain this complementary solution. For example, consider the next-to-last primal basic solution in Table 6.9, where x_1, x_2, and x_5 are basic variables. Using Tables 6.7 and 6.8, we see that the complementary slackness property implies that $z_1 - c_1$, $z_2 - c_2$, and y_3 are nonbasic variables for the complementary dual basic solution. Setting these variables equal to zero in the dual problem equations $y_1 + 3y_3 - (z_1 - c_1) = 3$ and $2y_2 + 2y_3 - (z_2 - c_2) = 5$ immediately yields $y_1 = 3$, $y_2 = \frac{5}{2}$.

Finally, notice that Table 6.9 demonstrates that $(0, \frac{3}{2}, 1, 0, 0)$ is the optimal solution for the dual problem, because it is the basic *feasible* solution with minimal y_0 (36).

Relationships between Complementary Basic Solutions

We now turn our attention to the relationships between complementary basic solutions, beginning with their *feasibility* relationships. The middle columns in Table 6.9 provide some valuable clues. For the pairs of complementary solutions, notice how the yes or no answers on feasibility also satisfy a complementary relationship in most cases. In particular, with one exception, whenever one solution is feasible, the other is not. (It also is possible for *neither* solution to be feasible, as happened with the third pair.) The one exception is the sixth pair, where the primal solution is known to be optimal. The explanation is suggested by the $Z = y_0$ column. Because the sixth dual solution also is optimal (by the complementary optimal solutions property), with $y_0 = 36$, the first five dual solutions *cannot be feasible* because $y_0 < 36$ (remember that the dual problem objective is to *minimize* y_0). By the same token, the last two primal solutions cannot be feasible because $Z > 36$.

This explanation is further supported by the strong duality property that optimal primal and dual solutions have $Z = y_0$.

Next, let us state the *extension* of the complementary optimal solutions property of Sec. 6.1 for the augmented forms of the two problems.

Complementary optimal basic solutions property: Each *optimal* basic solution in the *primal problem* has a **complementary optimal basic solution** in the dual problem, where their respective objective function values (Z and y_0) are equal. Given row 0 of the simplex tableau for the optimal primal solution, the complementary optimal dual solution $(\mathbf{y}^*, \mathbf{z}^* - \mathbf{c})$ is found as shown in Table 6.4.

Table 6.10 **Classification of Basic Solutions**

		Satisfies Condition for Optimality?	
		Yes	No
Feasible?	Yes	Optimal	Suboptimal
	No	Superoptimal	Neither feasible nor superoptimal

Table 6.11 **Relationships between Complementary Basic Solutions**

Primal Basic Solution	Complementary Dual Basic Solution
Suboptimal	Superoptimal
Optimal	Optimal
Superoptimal	Suboptimal
Neither feasible nor superoptimal	Neither feasible nor superoptimal

To review the reasoning behind this property, note that the dual solution $(\mathbf{y}^*, \mathbf{z}^* - \mathbf{c})$ must be feasible for the dual problem because the condition for optimality for the primal problem requires that *all* these dual variables (including surplus variables) be *nonnegative.* Since this solution is *feasible,* it must be *optimal* for the dual problem by the weak duality property.

Basic solutions can be classified according to whether they satisfy each of two conditions. One is the *condition for feasibility,* namely, whether *all* the variables (including slack variables) in the augmented solution are *nonnegative.* The other is the *condition for optimality,* namely, whether *all* the coefficients in row 0 (i.e., all the variables in the complementary basic solution) are *nonnegative.* Our names for the different types of basic solutions are summarized in Table 6.10. For example, in Table 6.9, primal basic solutions 1, 2, 4, and 5 are suboptimal, 6 is optimal, 7 and 8 are superoptimal, and 3 is neither feasible nor superoptimal.

Given these definitions, the general relationships between complementary basic solutions are summarized in Table 6.11. The resulting range of possible (common) values for the objective functions ($Z = y_0$) for the first three pairs given in Table 6.11 (the last pair can have any value) is shown in Fig. 6.1. Thus, while the simplex method is dealing directly with suboptimal basic solutions and working toward optimality in the primal problem, it is simultaneously dealing indirectly with complementary superoptimal solutions and working toward feasibility in the dual problem. Conversely, it sometimes is more convenient (or necessary) to work directly with superoptimal basic solutions and to move toward feasibility in the primal problem, which is the purpose of the dual simplex method described in Sec. 7.1.

These relationships prove very useful, particularly in sensitivity analysis, as you will see later in the chapter.

6.4 Adapting to Other Primal Forms

Thus far it has been assumed that the model for the primal problem is in our standard form. However, we indicated at the beginning of the chapter that any linear program-

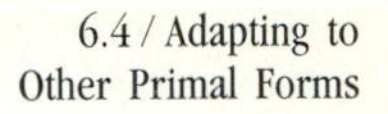

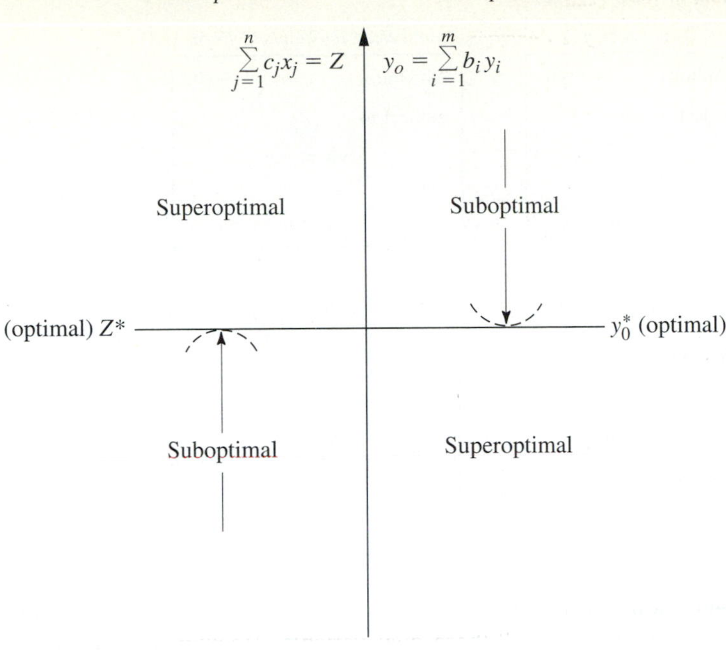

Figure 6.1 Range of possible values of $Z = y_0$ for certain types of complementary basic solutions.

ming problem, whether in our standard form or not, possesses a dual problem. Therefore, this section focuses on how the dual problem changes for other primal forms.

Each nonstandard form was discussed in Sec. 4.6, and we pointed out how it is possible to convert each one to an equivalent standard form if so desired. These conversions are summarized in Table 6.12. Hence you always have the option of converting any model to our standard form and *then* constructing its dual problem in the usual way. To illustrate, we do this for our standard dual problem (it must have a dual also) in Table 6.13. Note that what we end up with is just our standard primal problem! Since any pair of primal and dual problems can be converted to these forms, this fact implies that the dual of the dual problem always is the primal problem. Therefore, for any primal problem and its dual problem, all relationships between them must be symmetric. This is just the symmetry property already stated in Sec. 6.1 (without proof), but now Table 6.13 demonstrates why it holds.

Table 6.12 **Conversions to Standard Form for Linear Programming Models**

Nonstandard Form	Equivalent Standard Form
Minimize Z	Maximize $(-Z)$
$\sum_{j=1}^{n} a_{ij} x_j \ge b_i$	$-\sum_{j=1}^{n} a_{ij} x_j \le -b_i$
$\sum_{j=1}^{n} a_{ij} x_j = b_i$	$\sum_{j=1}^{n} a_{ij} x_j \le b_i$ and $-\sum_{j=1}^{n} a_{ij} x_j \le -b_i$
x_j unconstrained in sign	$x_j^+ - x_j^-$, $x_j^+ \ge 0$, $x_j^- \ge 0$

Table 6.13 Constructing the Dual of the Dual Problem

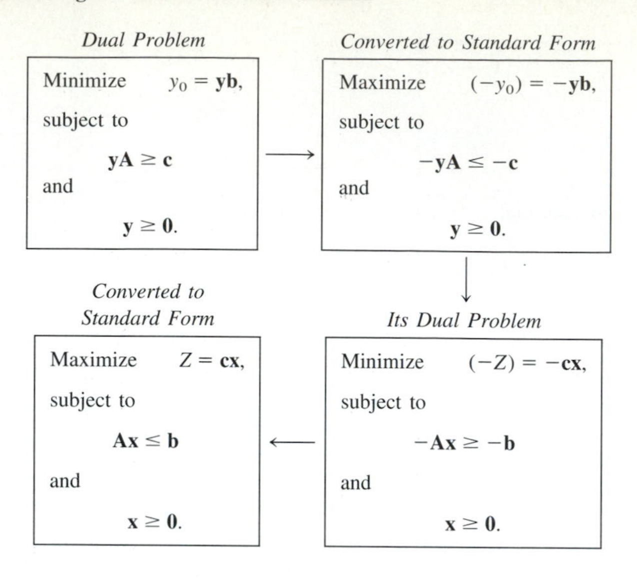

One consequence of the symmetry property is that all the statements made earlier in the chapter about the relationships of the dual problem to the primal problem also hold in reverse.

Another consequence is that it is immaterial which problem is called the primal and which is called the dual. In practice, you might see a linear programming problem fitting our standard form being referred to as the dual problem. The convention is that the model formulated to fit the actual problem is called the primal problem, regardless of its form.

Our illustration of how to construct the dual problem for a nonstandard primal problem did not involve either equality constraints or variables unconstrained in sign. Actually, for these two forms, a shortcut is available. It is possible to show (see Probs. 6.4-7 and 6.4-2*a*) that an *equality constraint* in the primal problem should be treated just like a $\leq$ constraint in constructing the dual problem except that the nonnegativity constraint for the corresponding dual variable should be *deleted* (i.e., this variable is unconstrained in sign). By the symmetry property, deleting a nonnegativity constraint in the primal problem affects the dual problem only by changing the corresponding inequality constraint to an equality constraint.

Another shortcut involves functional constraints in $\geq$ form for a maximum problem. The straightforward (but longer) approach would begin by converting each such constraint to $\leq$ form

$$\sum_{j=1}^{n} a_{ij}x_j \geq b_i \longrightarrow -\sum_{j=1}^{n} a_{ij}x_j \leq -b_i.$$

Constructing the dual problem in the usual way then gives $-a_{ij}$ as the coefficient of y_i in functional constraint j (which has $\geq$ form) and a coefficient of $-b_i$ in the objective function (which is to be minimized), where y_i also has a nonnegativity constraint $y_i \geq 0$. Now suppose we define a new variable $y_i' = -y_i$. The changes caused by expressing the dual problem in terms of y_i' instead of y_i are that (1) the coefficients of the variable become a_{ij} for functional constraint j and b_i for the objective function and

Table 6.14 Corresponding Primal-Dual Forms

Primal Problem (or Dual Problem)		Dual Problem (or Primal Problem)
Maximize Z (or y_0)		Minimize y_0 (or Z)
Constraint i:		Variable y_i (or x_i):
$\leq$form	$\longleftrightarrow$	$y_i \geq 0$
$=$form	$\longleftrightarrow$	Unconstrained
$\geq$form	$\longleftrightarrow$	$y_i' \leq 0$
Variable x_j (or y_j):		Constraint j:
$x_j \geq 0$	$\longleftrightarrow$	$\geq$ form
Unconstrained	$\longleftrightarrow$	$=$ form
$x_j' \leq 0$	$\longleftrightarrow$	$\leq$ form

(2) the constraint on the variable becomes $y_i' \leq 0$ (a *nonpositivity constraint*). The shortcut is to use y_i' instead of y_i as a dual variable so that the parameters in the original constraint (a_{ij} and b_i) immediately become the coefficients of this variable in the dual problem.

Using these shortcuts, Table 6.14 summarizes the form of the dual problem for any primal problem. Choose one column for the primal problem according to whether this problem is in maximization form (left column) or minimization form (right column). Then read off the form of the dual problem from the other column. In particular, the two-sided arrows in the table give the specific correspondence between the form of a functional constraint in one problem and the form of the constraint (if any) on the corresponding variable in the other problem.

To illustrate this procedure, consider the radiation therapy example presented in Sec. 3.4. (Its model is shown on p. 46.) To show the conversion in both directions in Table 6.14, we begin with the maximization form of this model as the primal problem, before using the (original) minimization form.

The primal problem in maximization form is shown on the left side of Table 6.15. By using the left column of Table 6.14 to represent this problem, the arrows in this table indicate the form of the dual problem in the right column. These same arrows are used in Table 6.15 to show the resulting dual problem. (Because of these arrows, we have placed the functional constraints last in the dual problem rather than in their usual top position.)

Table 6.15 One Primal-Dual Form for the Radiation Therapy Example

Primal Problem		*Dual Problem*
Maximize $-Z = -0.4x_1 - 0.5x_2$,		Minimize $y_0 = 2.7y_1 + 6y_2 + 6y_3'$,
subject to		subject to
$0.3x_1 + 0.1x_2 \leq 2.7$	$\longleftrightarrow$	$y_1 \geq 0$
$0.5x_1 + 0.5x_2 = 6$	$\longleftrightarrow$	y_2 unconstrained in sign
$0.6x_1 + 0.4x_2 \geq 6$	$\longleftrightarrow$	$y_3' \leq 0$
and		and
$x_1 \geq 0$	$\longleftrightarrow$	$0.3y_1 + 0.5y_2 + 0.6y_3' \geq -0.4$
$x_2 \geq 0$	$\longleftrightarrow$	$0.1y_1 + 0.5y_2 + 0.4y_3' \geq -0.5$

However, there was no need (other than for illustrative purposes) to convert the primal problem to maximization form. Using the original minimization form, the equivalent primal problem is shown on the left side of Table 6.16. Now we use the right column of Table 6.14 to represent this problem, where the arrows indicate the form of the dual problem in the left column. These same arrows in Table 6.16 show the resulting dual problem on the right side.

Just as the primal problems in Tables 6.15 and 6.16 are equivalent, the two dual problems also are completely equivalent. The key to recognizing this equivalency lies in the fact that the variables in each version of the dual problem are the negative of those in the other version ($y_1' = -y_1, y_2' = -y_2, y_3 = -y_3'$). Therefore, for each version, if the variables in the other version are used instead, and if both the objective function and the constraints are multiplied through by -1, then the other version is obtained. (Problem 6.4-5 asks you to verify this.)

If the simplex method is to be applied to either a primal or a dual problem that has any variables constrained to be *nonpositive* (for example, $y_3' \leq 0$ in the dual problem of Table 6.15), this variable may be replaced by its *nonnegative* counterpart (for example, $y_3 = -y_3'$).

When artificial variables are used to help the simplex method solve a primal problem, the duality interpretation of row 0 of the simplex tableau is the following: Since artificial variables play the role of slack variables, their coefficients in row 0 now provide the values of the corresponding dual variables in the complementary basic solution for the dual problem. Since artificial variables are used to replace the real problem with a more convenient artificial problem, this dual problem actually is the dual of the artificial problem. However, after all the artificial variables become nonbasic, we are back to the real primal and dual problems. With the two-phase method, the artificial variables would need to be retained in phase 2 in order to read off the complete dual solution from row 0. With the Big M method, since M has been added initially to the coefficient of each artificial variable in row 0, the current value of each corresponding dual variable is the current coefficient of this artificial variable *minus M*.

For example, look at row 0 in the final simplex tableau for the radiation therapy example, given at the bottom of Table 4.12 on p. 112. After M is subtracted from the coefficients of the artificial variables $\bar{x}_4$ and x_6, the optimal solution for the corresponding dual problem given in Table 6.15 is read from the coefficients of x_3, $\bar{x}_4$, and $\bar{x}_6$ as $(y_1, y_2, y_3') = (0.5, -1.1, 0)$. As usual, the surplus variables for the two functional constraints are read from the coefficients of x_1 and x_2 as $z_1 - c_1 = 0$ and $z_2 - c_2 = 0$.

Table 6.16 **The Other Primal-Dual Form for the Radiation Therapy Example**

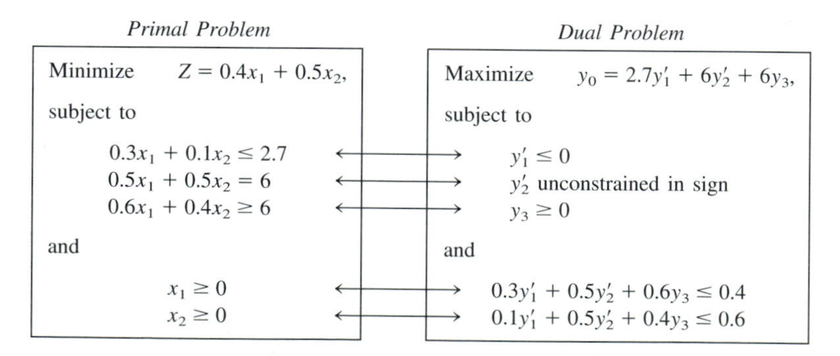

6.5 The Role of Duality Theory in Sensitivity Analysis

As described further in the next two sections, sensitivity analysis basically involves investigating the effect on the optimal solution of making changes in the values of the model parameters a_{ij}, b_i, and c_j. However, changing parameter values in the primal problem also changes the corresponding values in the dual problem. Therefore, you have your choice of which problem to use to investigate each change. Because of the primal-dual relationships presented in Secs. 6.1 and 6.3 (especially the complementary basic solutions property), it is easy to move back and forth between the two problems as desired. In some cases, it is more convenient to analyze the dual problem directly in order to determine the complementary effect on the primal problem. We begin by considering two such cases.

Changes in the Coefficients of a Nonbasic Variable

Suppose that the changes made in the original model occur in the coefficients of a variable that was nonbasic in the original optimal solution. What is the effect of these changes on this solution? Is it still feasible? Is it still optimal?

Because the variable involved is nonbasic (value of zero), changing its coefficients cannot affect the feasibility of the solution. Therefore, the open question in this case is whether it is still optimal. As Tables 6.10 and 6.11 indicate, an equivalent question is whether the complementary basic solution for the dual problem is still feasible after these changes are made. Since these changes affect the dual problem by changing only one constraint, this question can be answered simply by checking whether this complementary basic solution still satisfies this revised constraint.

We shall illustrate this case in the corresponding subsection of Sec. 6.7 after developing a relevant example.

Introduction of a New Variable

As indicated in Table 6.6, the decision variables in the model typically represent the levels of the various activities under consideration. In some situations, these activities were selected from a larger group of *possible* activities, where the remaining activities were not included in the original model because they seemed less attractive. Or perhaps these other activities did not come to light until after the original model was formulated and solved. Either way, the key question is whether any of these previously unconsidered activities are sufficiently worthwhile to warrant initiation. In other words, would adding any of these activities to the model change the original optimal solution?

Adding another activity amounts to introducing a new variable, with the appropriate coefficients in the functional constraints and objective function, into the model. The only resulting change in the dual problem is to add a *new constraint* (see Table 6.3).

After these changes are made, would the original optimal solution, along with the new variable equal to zero (nonbasic), still be optimal for the primal problem? As for the preceding case, an equivalent question is whether the complementary basic solution for the dual problem is still feasible. And, as before, this question can be answered simply by checking whether this complementary basic solution satisfies one constraint, which in this case is the new constraint for the dual problem.

To illustrate, suppose for the Wyndor Glass Co. problem of Sec. 3.1 that a possible third new product now is being considered for inclusion in the product line. Letting x_{new} represent the production rate for this product, we show the resulting revised model as follows:

$$\text{Maximize} \qquad Z = 3x_1 + 5x_2 + 4x_{new},$$

subject to

$$\begin{aligned} x_1 \qquad\quad + 2x_{new} &\leq 4 \\ 2x_2 + 3x_{new} &\leq 12 \\ 3x_1 + 2x_2 + \ \ x_{new} &\leq 18 \end{aligned}$$

and

$$x_1 \geq 0, \qquad x_2 \geq 0, \qquad x_{new} \geq 0.$$

After we introduced slack variables, the original optimal solution for this problem without x_{new} (see Table 4.8) was $(x_1, x_2, x_3, x_4, x_5) = (2, 6, 2, 0, 0)$. Is this solution, along with $x_{new} = 0$, still optimal?

To answer this question, check the complementary basic solution for the dual problem, which Table 6.9 (and Table 4.8) identifies as

$$(y_1, y_2, y_3, z_1 - c_1, z_2 - c_2) = (0, \tfrac{3}{2}, 1, 0, 0).$$

Since this solution was optimal for the original dual problem, it certainly satisfies the original dual constraints shown in Table 6.1. But does it satisfy this new dual constraint?

$$2y_1 + 3y_2 + y_3 \geq 4$$

Plugging in this solution, we see that

$$2(0) + 3(\tfrac{3}{2}) + (1) \geq 4$$

is satisfied, so this dual solution is still feasible (and thus still optimal). Consequently, the original primal solution (2, 6, 2, 0, 0), along with $x_{new} = 0$, is still optimal, so this third possible new product should *not* be added to the product line.

This approach also makes it very easy to conduct sensitivity analysis on the coefficients of the new variable added to the primal problem. By simply checking the new dual constraint, you can immediately see how far any of these parameter values can be changed before they affect the feasibility of the dual solution and so the optimality of the primal solution.

Other Applications

Already we have discussed two other key applications of duality theory to sensitivity analysis, namely, shadow prices and the dual simplex method. As described in Secs. 4.7 and 6.2, the optimal dual solution $(y_1^*, y_2^*, \ldots, y_m^*)$ provides the shadow prices for the respective resources that indicate how Z would change if (small) changes were made in the b_i (the resource amounts). The resulting analysis will be illustrated in some detail in Sec. 6.7.

In more general terms, the economic interpretation of the dual problem and of the simplex method presented in Sec. 6.2 provides some useful insights for sensitivity analysis.

When we investigate the effect of changing the b_i or the a_{ij} values (for basic variables), the original optimal solution may become a *superoptimal* basic solution instead (see Table 6.10). If we then want to *reoptimize* to identify the new optimal solution, the dual simplex method (discussed at the end of Secs. 6.1 and 6.3) should be applied, starting from this basic solution.

We mentioned in Sec. 6.1 that sometimes it is more efficient to solve the dual problem directly by the simplex method in order to identify an optimal solution for the primal problem. When the solution has been found in this way, sensitivity analysis for the primal problem then is conducted by applying the procedure described in the next two sections directly to the dual problem and then inferring the complementary effects on the primal problem (e.g., see Table 6.11). This approach to sensitivity analysis is relatively straightforward because of the close primal-dual relationships described in Secs. 6.1 and 6.3. (See Prob. 6.6-3.)

6.6 The Essence of Sensitivity Analysis

The work of the operations research team usually is not even nearly done when the simplex method has been successfully applied to identify an optimal solution for the model. As we pointed out at the end of Sec. 3.3, one assumption of linear programming is that all the parameters of the model (a_{ij}, b_i, and c_j) are *known constants.* Actually, the parameter values used in the model normally are just *estimates* based on a *prediction of future conditions.* The data obtained to develop these estimates often are rather crude or nonexistent, so that the parameters in the original formulation may represent little more than quick rules of thumb provided by harassed line personnel. The data may even represent deliberate overestimates or underestimates to protect the interests of the estimators.

Thus the successful manager and operations research staff will maintain a healthy skepticism about the original numbers coming out of the computer and will view them in many cases as only a starting point for further analysis of the problem. An "optimal" solution is optimal only with respect to the specific model being used to represent the real problem, and such a solution becomes a reliable guide for action only after it has been verified as performing well for other reasonable representations of the problem. Furthermore, the model parameters (particularly b_i) sometimes are set as a result of managerial policy decisions (e.g., the amount of certain resources to be made available to the activities), and these decisions should be reviewed after their potential consequences are recognized.

For these reasons it is important to perform **sensitivity analysis** to investigate the effect on the optimal solution provided by the simplex method if the parameters take on other possible values. Usually there will be some parameters that can be assigned any reasonable value without the optimality of this solution being affected. However, there may also be parameters with likely alternative values that would yield a new optimal solution. This situation is particularly serious if the original solution would then have a substantially inferior value of the objective function, or perhaps even be infeasible!

Therefore, one main purpose of sensitivity analysis is to identify the **sensitive parameters** (i.e., the parameters whose values cannot be changed without changing the optimal solution). For certain parameters that are not categorized as sensitive, it is also very helpful to determine the *range of values* of the parameter over which the optimal solution will remain unchanged. (We call this range of values the *allowable range to stay optimal.*) In some cases, changing a parameter value can affect the *feasibility* of the optimal BF solution. For such parameters, it is useful to determine the range of values over which the optimal BF solution (with adjusted values for the basic variables) will remain feasible. (We call this range of values the *allowable range to stay feasible.*) In the next section, we will describe the specific procedures for obtaining this kind of information.

Such information is invaluable in two ways. First, it identifies the more important parameters, so that special care can be taken to estimate them closely and to select a solution that performs well for most of their likely values. Second, it identifies the parameters that will need to be monitored particularly closely as the study is implemented. If it is discovered that the true value of a parameter lies outside its allowable range, this immediately signals a need to change the solution.

Sensitivity analysis would require an exorbitant computational effort if it were necessary to reapply the simplex method from the beginning to investigate each new change in a parameter value. Fortunately, the fundamental insight discussed in Sec. 5.3 virtually eliminates computational effort. The basic idea is that the fundamental insight *immediately* reveals just how any changes in the original model would change the numbers in the final simplex tableau (assuming that the *same* sequence of algebraic operations originally performed by the simplex method were to be *duplicated*). Therefore, after making a few simple calculations to revise this tableau, we can check easily whether the original optimal BF solution is now nonoptimal (or infeasible). If so, this solution would be used as the initial basic solution to restart the simplex method (or dual simplex method) to find the new optimal solution, if desired. If the changes in the model are not major, only a very few iterations should be required to reach the new optimal solution from this ''advanced'' initial basic solution.

To describe this procedure more specifically, consider the following situation. The simplex method already has been used to obtain an optimal solution to a linear programming model with specified values for the b_i, c_j, and a_{ij} parameters. To initiate sensitivity analysis, one or more of the parameters is changed. After the changes are made, let $\overline{b}_i$, $\overline{c}_j$, and $\overline{a}_{ij}$ denote the values of the various parameters. Thus, in matrix notation,

$$\mathbf{b} \rightarrow \overline{\mathbf{b}}, \qquad \mathbf{c} \rightarrow \overline{\mathbf{c}}, \qquad \mathbf{A} \rightarrow \overline{\mathbf{A}},$$

for the revised model.

The first step is to revise the final simplex tableau to reflect these changes. Continuing to use the notation presented in Table 5.10, as well as the accompanying formulas for the fundamental insight [(1) $\mathbf{t}^* = \mathbf{t} + \mathbf{y}^*\mathbf{T}$ and (2) $\mathbf{T}^* = \mathbf{S}^*\mathbf{T}$], we see that the revised final tableau is calculated from $\mathbf{y}^*$ and $\mathbf{S}^*$ (which have not changed) and the new initial tableau, as shown in Table 6.17.

To illustrate, suppose that the original model for the Wyndor Glass Co. problem of Sec. 3.1 is revised as shown at the right.

Original Model	*Revised Model*
Maximize $Z = [3, 5]\begin{bmatrix} x_1 \\ x_2 \end{bmatrix}$, subject to $\begin{bmatrix} 1 & 0 \\ 0 & 2 \\ 3 & 2 \end{bmatrix}\begin{bmatrix} x_1 \\ x_2 \end{bmatrix} \le \begin{bmatrix} 4 \\ 12 \\ 18 \end{bmatrix}$ and $\mathbf{x} \ge \mathbf{0}$.	Maximize $Z = [4, 5]\begin{bmatrix} x_1 \\ x_2 \end{bmatrix}$, subject to $\begin{bmatrix} 1 & 0 \\ 0 & 2 \\ 2 & 2 \end{bmatrix}\begin{bmatrix} x_1 \\ x_2 \end{bmatrix} \le \begin{bmatrix} 4 \\ 24 \\ 18 \end{bmatrix}$ and $\mathbf{x} \ge \mathbf{0}$.

Thus the changes from the original model are $c_1 = 3 \rightarrow 4$, $a_{31} = 3 \rightarrow 2$, and $b_2 = 12 \rightarrow 24$. Figure 6.2 shows the graphical effect of these changes. For the original model, the simplex method already has identified the optimal CPF solution as (2, 6), lying at the intersection of the two constraint boundaries, shown as dashed lines $2x_2 = 12$ and $3x_1 + 2x_2 = 18$. Now the revision of the model has shifted both of these constraint boundaries as shown by the dark lines $2x_2 = 24$ and $2x_1 + 2x_2 = 18$. Consequently, the previous CPF solution (2, 6) now shifts to the new intersection (−3, 12), which is a corner-point *infeasible* solution for the revised model. The procedure described in the preceding paragraphs finds this shift *algebraically* (in augmented form). Furthermore, it does so in a manner that is very efficient even for huge problems where graphical analysis is impossible.

To carry out this procedure, we begin by displaying the parameters of the revised model in matrix form:

$$\bar{\mathbf{c}} = [4, 5], \qquad \bar{\mathbf{A}} = \begin{bmatrix} 1 & 0 \\ 0 & 2 \\ 2 & 2 \end{bmatrix}, \qquad \bar{\mathbf{b}} = \begin{bmatrix} 4 \\ 24 \\ 18 \end{bmatrix}.$$

The resulting new initial simplex tableau is shown at the top of Table 6.18. Below this tableau is the original final tableau (as first given in Table 4.8). We have drawn dark boxes around the portions of this final tableau that the changes in the model definitely *do not change,* namely, the coefficients of the slack variables in both row 0 ($\mathbf{y}^*$) and the rest of the rows ($\mathbf{S}^*$). Thus,

$$\mathbf{y}^* = [0, \tfrac{3}{2}, 1], \qquad \mathbf{S}^* = \begin{bmatrix} 1 & \frac{1}{3} & -\frac{1}{3} \\ 0 & \frac{1}{2} & 0 \\ 0 & -\frac{1}{3} & \frac{1}{3} \end{bmatrix}.$$

Table 6.17 **Revised Final Simplex Tableau Resulting from Changes in Original Model**

		Coefficient of:			
	Eq.	Z	Original Variables	Slack Variables	Right Side
New initial tableau	(0)	1	$-\bar{\mathbf{c}}$	$\mathbf{0}$	0
	$(1, 2, \ldots, m)$	$\mathbf{0}$	$\bar{\mathbf{A}}$	$\mathbf{I}$	$\bar{\mathbf{b}}$
Revised final tableau	(0)	1	$\mathbf{z}^* - \bar{\mathbf{c}} = \mathbf{y}^*\bar{\mathbf{A}} - \bar{\mathbf{c}}$	$\mathbf{y}^*$	$Z^* = \mathbf{y}^*\bar{\mathbf{b}}$
	$(1, 2, \ldots, m)$	$\mathbf{0}$	$\mathbf{A}^* = \mathbf{S}^*\bar{\mathbf{A}}$	$\mathbf{S}^*$	$\mathbf{b}^* = \mathbf{S}^*\bar{\mathbf{b}}$

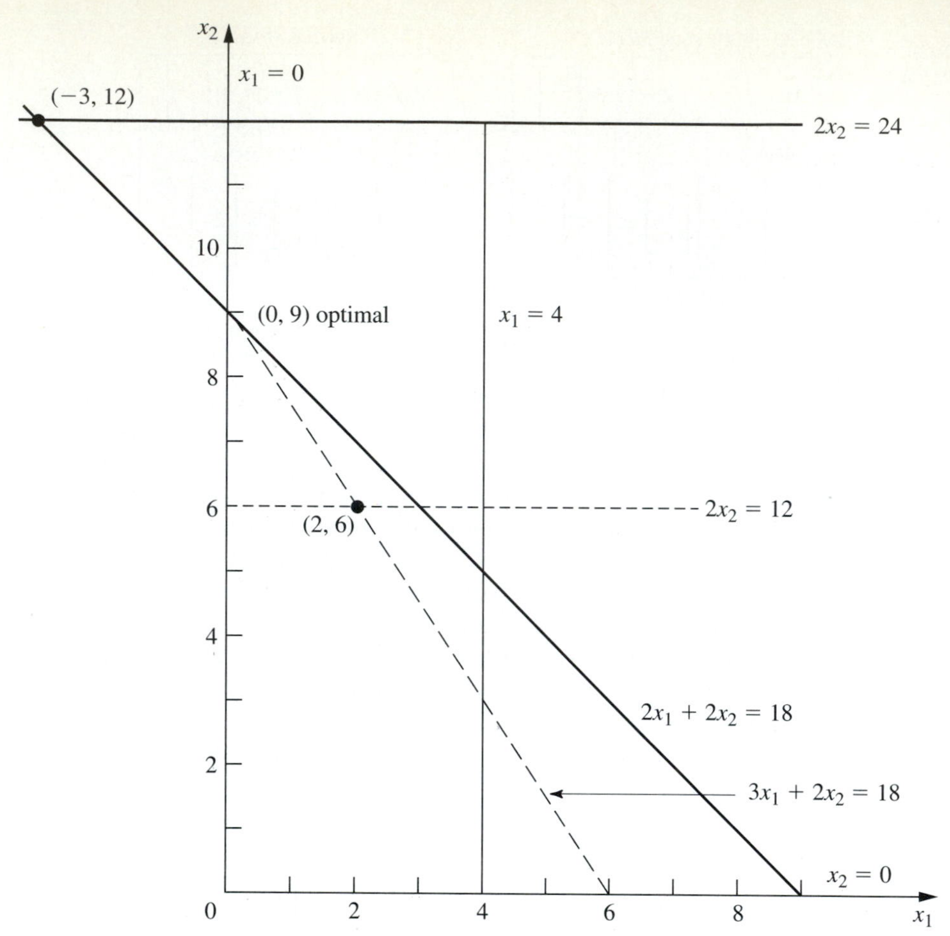

Figure 6.2 Shift of the final corner-point solution from (2, 6) to (−3, 12) for the revision of the Wyndor Glass Co. problem where $c_1 = 3 \rightarrow 4$, $a_{31} = 3 \rightarrow 2$, and $b_2 = 12 \rightarrow 24$.

These coefficients of the slack variables necessarily are unchanged with the same algebraic operations originally performed by the simplex method because the coefficients of these same variables in the initial tableau are unchanged.

However, because other portions of the initial tableau have changed, there will be changes in the rest of the final tableau as well. Using the formulas in Table 6.17, we calculate the revised numbers in the rest of the final tableau as follows:

$$\mathbf{z}^* - \bar{\mathbf{c}} = [0, \tfrac{3}{2}, 1]\begin{bmatrix} 1 & 0 \\ 0 & 2 \\ 2 & 2 \end{bmatrix} - [4, 5] = [-2, 0], \qquad Z^* = [0, \tfrac{3}{2}, 1]\begin{bmatrix} 4 \\ 24 \\ 18 \end{bmatrix} = 54,$$

$$\mathbf{A}^* = \begin{bmatrix} 1 & \frac{1}{3} & -\frac{1}{3} \\ 0 & \frac{1}{2} & 0 \\ 0 & -\frac{1}{3} & \frac{1}{3} \end{bmatrix}\begin{bmatrix} 1 & 0 \\ 0 & 2 \\ 2 & 2 \end{bmatrix} = \begin{bmatrix} \frac{1}{3} & 0 \\ 0 & 1 \\ \frac{2}{3} & 0 \end{bmatrix},$$

$$\mathbf{b}^* = \begin{bmatrix} 1 & \frac{1}{3} & -\frac{1}{3} \\ 0 & \frac{1}{2} & 0 \\ 0 & -\frac{1}{3} & \frac{1}{3} \end{bmatrix}\begin{bmatrix} 4 \\ 24 \\ 18 \end{bmatrix} = \begin{bmatrix} 6 \\ 12 \\ -2 \end{bmatrix}.$$

Table 6.18 **Obtaining the Revised Final Simplex Tableau for the Revised Wyndor Glass Co. Problem**

	Basic Variable	Eq.	Coefficient of: Z	x_1	x_2	x_3	x_4	x_5	Right Side
New initial tableau	Z	(0)	1	−4	−5	0	0	0	0
	x_3	(1)	0	1	0	1	0	0	4
	x_4	(2)	0	0	2	0	1	0	24
	x_5	(3)	0	2	2	0	0	1	18
Final tableau for original model	Z	(0)	1	0	0	0	$\frac{3}{2}$	1	36
	x_3	(1)	0	0	0	1	$\frac{1}{3}$	$-\frac{1}{3}$	2
	x_2	(2)	0	0	1	0	$\frac{1}{2}$	0	6
	x_1	(3)	0	1	0	0	$-\frac{1}{3}$	$\frac{1}{3}$	2
Revised final tableau	Z	(0)	1	−2	0	0	$\frac{3}{2}$	1	54
	x_3	(1)	0	$\frac{1}{3}$	0	1	$\frac{1}{3}$	$-\frac{1}{3}$	6
	x_2	(2)	0	0	1	0	$\frac{1}{2}$	0	12
	x_1	(3)	0	$\frac{2}{3}$	0	0	$-\frac{1}{3}$	$\frac{1}{3}$	−2

The resulting revised final tableau is shown at the bottom of Table 6.18.

Actually, we can substantially streamline these calculations for obtaining the revised final tableau. Because none of the coefficients of x_2 changed in the original model (tableau), none of them can change in the final tableau, so we can delete their calculation. Several other original parameters (a_{11}, a_{21}, b_1, b_3) also were not changed, so another shortcut is to calculate only the *incremental changes* in the final tableau in terms of the incremental changes in the initial tableau, ignoring those terms in the vector or matrix multiplication that involve zero change in the initial tableau. In particular, the only incremental changes in the initial tableau are $\Delta c_1 = 1$, $\Delta a_{31} = -1$, and $\Delta b_2 = 12$, so these are the only terms that need be considered. This streamlined approach is shown below, where a zero or dash appears in each spot where no calculation is needed.

$$\Delta(\mathbf{z}^* - \mathbf{c}) = \mathbf{y}^* \Delta\mathbf{A} - \Delta\mathbf{c} = [0, \tfrac{3}{2}, 1]\begin{bmatrix} 0 & — \\ 0 & — \\ -1 & — \end{bmatrix} - [1, —] = [-2, —].$$

$$\Delta Z^* = \mathbf{y}^* \Delta\mathbf{b} = [0, \tfrac{3}{2}, 1]\begin{bmatrix} 0 \\ 12 \\ 0 \end{bmatrix} = 18.$$

$$\Delta\mathbf{A}^* = \mathbf{S}^* \Delta\mathbf{A} = \begin{bmatrix} 1 & \frac{1}{3} & -\frac{1}{3} \\ 0 & \frac{1}{2} & 0 \\ 0 & -\frac{1}{3} & \frac{1}{3} \end{bmatrix}\begin{bmatrix} 0 & — \\ 0 & — \\ -1 & — \end{bmatrix} = \begin{bmatrix} \frac{1}{3} & — \\ 0 & — \\ -\frac{1}{3} & — \end{bmatrix}.$$

$$\Delta\mathbf{b}^* = \mathbf{S}^* \Delta\mathbf{b} = \begin{bmatrix} 1 & \frac{1}{3} & -\frac{1}{3} \\ 0 & \frac{1}{2} & 0 \\ 0 & -\frac{1}{3} & \frac{1}{3} \end{bmatrix}\begin{bmatrix} 0 \\ 12 \\ 0 \end{bmatrix} = \begin{bmatrix} 4 \\ 6 \\ -4 \end{bmatrix}.$$

Adding these increments to the original quantities in the final tableau (middle of Table 6.18) then yields the revised final tableau (bottom of Table 6.18).

This *incremental analysis* also provides a useful general insight, namely, that changes in the final tableau must be *proportional* to each change in the initial tableau. We illustrate in the next section how this property enables us to use linear interpolation or extrapolation to determine the range of values for a given parameter over which the final basic solution remains both feasible and optimal.

After obtaining the revised final simplex tableau, we next convert the tableau to proper form from Gaussian elimination (as needed). In particular, the basic variable for row i must have a coefficient of 1 in that row and a coefficient of 0 in every other row (including row 0) for the tableau to be in the proper form for identifying and evaluating the current basic solution. Therefore, if the changes have violated this requirement (which can occur only if the original constraint coefficients of a basic variable have been changed), further changes must be made to restore this form. This restoration is done by using Gaussian elimination, i.e., by successively applying step 3 of an iteration for the simplex method (see Chap. 4) as if each violating basic variable were an entering basic variable. Note that these algebraic operations may also cause further changes in the *right side* column, so that the current basic solution can be read from this column only when the proper form from Gaussian elimination has been fully restored.

For the example, the revised final simplex tableau shown in the top half of Table 6.19 is not in proper form from Gaussian elimination because of the column for the basic variable x_1. Specifically, the coefficient of x_1 in *its* row (row 3) is $\frac{2}{3}$ instead of 1, and it has nonzero coefficients (-2 and $\frac{1}{3}$) in rows 0 and 1. To restore proper form, row 3 is multiplied by $\frac{3}{2}$; then 2 times this new row 3 is added to row 0 and $\frac{1}{3}$ times new row 3 is subtracted from row 1. This yields the proper form from Gaussian elimination shown in the bottom half of Table 6.19, which now can be used to identify the new values for the current (previously optimal) basic solution:

$$(x_1, x_2, x_3, x_4, x_5) = (-3, 12, 7, 0, 0).$$

Because x_1 is negative, this basic solution no longer is feasible. However, it is *superoptimal* (see Table 6.10) because *all* the coefficients in row 0 still are *nonnegative.* Therefore, the dual simplex method can be used to reoptimize (if desired), by starting from this basic solution. (The sensitivity analysis routine in the OR Courseware includes this option.) Referring to Fig. 6.2 (and ignoring slack variables), we see that the dual simplex method uses just one iteration to move from the corner-point solution $(-3, 12)$ to the optimal CPF solution $(0, 9)$. (It is often useful in sensitivity analysis to identify the solutions that are optimal for some set of likely values of the model parameters and then to determine which of these solutions most *consistently* performs well for the various likely parameter values.)

If the basic solution $(-3, 12, 7, 0, 0)$ had been *neither* feasible nor superoptimal (i.e., if the tableau had negative entries in *both* the *right side* column and row 0),

Table 6.19 **Converting the Revised Final Simplex Tableau to Proper Form from Gaussian Elimination for the Revised Wyndor Glass Co. Problem**

	Basic Variable	Eq.	*Coefficient of:*						Right Side
			Z	x_1	x_2	x_3	x_4	x_5	
Revised final tableau	Z	(0)	1	-2	0	0	$\frac{3}{2}$	1	54
	x_3	(1)	0	$\frac{1}{3}$	0	1	$\frac{1}{3}$	$-\frac{1}{3}$	6
	x_2	(2)	0	0	1	0	$\frac{1}{2}$	0	12
	x_1	(3)	0	$\frac{2}{3}$	0	0	$-\frac{1}{3}$	$\frac{1}{3}$	-2
Converted to proper form	Z	(0)	1	0	0	0	$\frac{1}{2}$	2	48
	x_3	(1)	0	0	0	1	$\frac{1}{2}$	$-\frac{1}{2}$	7
	x_2	(2)	0	0	1	0	$\frac{1}{2}$	0	12
	x_1	(3)	0	1	0	0	$-\frac{1}{2}$	$\frac{1}{2}$	-3

artificial variables could have been introduced to convert the tableau to the proper form for an initial simplex tableau.[1]

When one is testing to see how *sensitive* the original optimal solution is to the various parameters of the model, the common approach is to check each parameter (or at least c_j and b_i) individually. In addition to finding allowable ranges as described in the next section, this check might include changing the value of the parameter from its initial estimate to other possibilities in the *range of likely values* (including the end-points of this range). Then some combinations of simultaneous changes of parameter values (such as changing an entire functional constraint) may be investigated. *Each* time one (or more) of the parameters is changed, the procedure described and illustrated here would be applied. Let us now summarize this procedure.

Summary of Procedure for Sensitivity Analysis

1. *Revision of model:* Make the desired change or changes in the model to be investigated next.
2. *Revision of final tableau:* Use the fundamental insight to determine the resulting changes in the final simplex tableau.
3. *Conversion to proper form from Gaussian elimination:* Convert this tableau to the proper form for identifying and evaluating the current basic solution by applying (as necessary) Gaussian elimination.
4. *Feasibility test:* Test this solution for feasibility by checking whether all its basic variable values in the right-side column of the tableau still are nonnegative.
5. *Optimality test:* Test this solution for optimality (if feasible) by checking whether all its nonbasic variable coefficients in row 0 of the tableau still are nonnegative.
6. *Reoptimization:* If this solution fails either test, the new optimal solution can be obtained (if desired) by using the current tableau as the initial simplex tableau (and making any necessary conversions) for the simplex method or dual simplex method.

The interactive routine entitled *Sensitivity Analysis* in the OR Courseware will enable you to efficiently practice applying this procedure. In addition, a demonstration (also entitled *Sensitivity Analysis*) provides you with another example.

In the next section, we shall discuss and illustrate the application of this procedure to each of the major categories of revisions in the original model. This discussion will involve, in part, expanding upon the example introduced in this section for investigating changes in the Wyndor Glass Co. model. In fact, we shall begin by *individually* checking each of the preceding changes. At the same time, we shall integrate some of the applications of duality theory to sensitivity analysis discussed in Sec. 6.5.

6.7 Applying Sensitivity Analysis

Sensitivity analysis often begins with the investigation of changes in the values of b_i, the amount of resource i ($i = 1, 2, \ldots, m$) being made available for the activities under consideration. The reason is that there generally is more flexibility in setting and

[1] There also exists a primal-dual algorithm that can be directly applied to such a simplex tableau without any conversion.

adjusting these values than there is for the other parameters of the model. As already discussed in Secs. 4.7 and 6.2, the economic interpretation of the dual variables (the y_i) as shadow prices is extremely useful for deciding which changes should be considered.

Case 1—Changes in b_i

Suppose that the only changes in the current model are that one or more of the b_i parameters ($i = 1, 2, \ldots, m$) has been changed. In this case, the *only* resulting changes in the final simplex tableau are in the *right side* column. Therefore, both the conversion to proper form from Gaussian elimination and the *optimality test* steps of the general procedure can be skipped.

As shown in Table 6.17, when the vector of the b_i values is changed from $\mathbf{b}$ to $\bar{\mathbf{b}}$, the formulas for calculating the new *right side* column in the final tableau are

$$\text{Right side of final row 0:} \qquad Z^* = \mathbf{y}^*\bar{\mathbf{b}},$$

$$\text{Right side of final rows } 1, 2, \ldots, m\text{:} \qquad \mathbf{b}^* = \mathbf{S}^*\bar{\mathbf{b}}.$$

(See the bottom of Table 6.17 for the location of the unchanged vector $\mathbf{y}^*$ and matrix $\mathbf{S}^*$ in the final tableau.)

Example: Sensitivity analysis is begun for the original Wyndor Glass Co. problem of Sec. 3.1 by examining the optimal values of the y_i dual variables ($y_1^* = 0$, $y_2^* = \frac{3}{2}$, $y_3^* = 1$). These shadow prices give the marginal value of each resource i for the activities (two new products) under consideration, where marginal value is expressed in the units of Z (thousands of dollars of profit per week). As discussed in Sec. 4.7 (see Fig. 4.8), the total profit from these activities can be increased \$1,500 per week (y_2^* times \$1,000 per week) for each additional unit of resource 2 (hour of production time per week in Plant 2) that is made available. This increase in profit holds for relatively small changes that do not affect the feasibility of the current basic solution (and so do not affect the y_i^* values).

Consequently, the OR team has investigated the marginal profitability from the other current uses of this resource to determine if any are less than \$1,500 per week. This investigation reveals that one old product is far less profitable. The production rate for this product already has been reduced to the minimum amount that would justify its marketing expenses. However, it can be discontinued altogether, which would provide an additional 12 units of resource 2 for the new products. Thus the next step is to determine the profit that could be obtained from the new products if this shift were made. This shift changes b_2 from 12 to 24 in the linear programming model. Figure 6.3 shows the graphical effect of this change, including the shift in the final corner-point solution from (2, 6) to (−2, 12). (Note that this figure differs from Fig. 6.2 because the constraint $3x_1 + 2x_2 \leq 18$ has not been changed here.)

The change in the vector of the b_i values is

$$\mathbf{b} = \begin{bmatrix} 4 \\ 12 \\ 18 \end{bmatrix} \longrightarrow \bar{\mathbf{b}} = \begin{bmatrix} 4 \\ 24 \\ 18 \end{bmatrix}.$$

Therefore, when the fundamental insight (Table 6.17) is applied, the effect of this change on the original final simplex tableau (middle of Table 6.18) is that the entries in

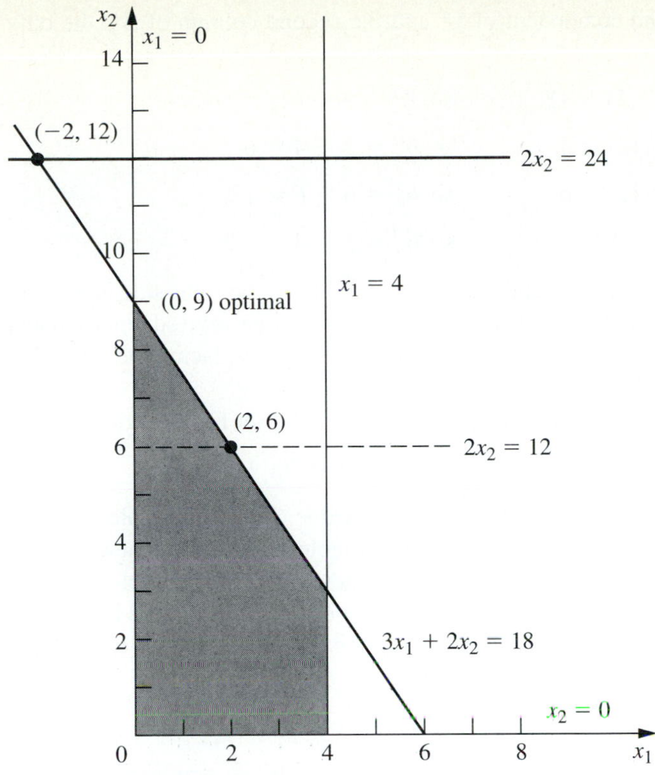

Figure 6.3 Feasible region for the Wyndor Glass Co. problem after b_2 is changed to 24.

the right-side column change to the following values:

$$Z^* = \mathbf{y}^*\bar{\mathbf{b}} = [0, \tfrac{3}{2}, 1]\begin{bmatrix} 4 \\ 24 \\ 18 \end{bmatrix} = 54,$$

$$\mathbf{b}^* = \mathbf{S}^*\bar{\mathbf{b}} = \begin{bmatrix} 1 & \frac{1}{3} & -\frac{1}{3} \\ 0 & \frac{1}{2} & 0 \\ 0 & -\frac{1}{3} & \frac{1}{3} \end{bmatrix}\begin{bmatrix} 4 \\ 24 \\ 18 \end{bmatrix} = \begin{bmatrix} 6 \\ 12 \\ -2 \end{bmatrix}, \qquad \text{so} \begin{bmatrix} x_3 \\ x_2 \\ x_1 \end{bmatrix} = \begin{bmatrix} 6 \\ 12 \\ -2 \end{bmatrix}.$$

Equivalently, because the only change in the original model is $\Delta b_2 = 24 - 12 = 12$, incremental analysis can be used to calculate these same values more quickly. Incremental analysis involves calculating just the *increments* in the tableau values caused by the change (or changes) in the original model, and then adding these increments to the original values. In this case, the increments in Z^* and $\mathbf{b}^*$ are

$$\Delta Z^* = \mathbf{y}^* \,\Delta\bar{\mathbf{b}} = \mathbf{y}^*\begin{bmatrix} \Delta b_1 \\ \Delta b_2 \\ \Delta b_3 \end{bmatrix} = \mathbf{y}^*\begin{bmatrix} 0 \\ 12 \\ 0 \end{bmatrix},$$

$$\Delta \mathbf{b}^* = \mathbf{S}^* \,\Delta\bar{b} = \mathbf{S}^*\begin{bmatrix} \Delta b_1 \\ \Delta b_2 \\ \Delta b_3 \end{bmatrix} = \mathbf{S}^*\begin{bmatrix} 0 \\ 12 \\ 0 \end{bmatrix}.$$

Therefore, using the second component of **y*** and the second column of **S***, the only calculations needed are

$$\Delta Z^* = \tfrac{3}{2}(12) = 18, \qquad \text{so } Z^* = 36 + 18 = 54,$$

$$\Delta b_1^* = \tfrac{1}{3}(12) = 4, \qquad \text{so } b_1^* = 2 + 4 = 6,$$

$$\Delta b_2^* = \tfrac{1}{2}(12) = 6, \qquad \text{so } b_2^* = 6 + 6 = 12,$$

$$\Delta b_3^* = -\tfrac{1}{3}(12) = -4, \qquad \text{so } b_3^* = 2 - 4 = -2,$$

where the original values of these quantities are obtained from the right-side column in the original final tableau (middle of Table 6.18). The resulting revised final tableau corresponds completely to this original final tableau except for replacing the right-side column with these new values.

Therefore, the current (previously optimal) basic solution has become

$$(x_1, x_2, x_3, x_4, x_5) = (-2, 12, 6, 0, 0),$$

which fails the feasibility test because of the negative value. The dual simplex method now can be applied, starting with this revised simplex tableau, to find the new optimal solution. This method leads in just one iteration to the new final simplex tableau shown in Table 6.20. (Alternatively, the simplex method could be applied from the beginning, which also would lead to this final tableau in just one iteration in this case.) This tableau indicates that the new optimal solution is

$$(x_1, x_2, x_3, x_4, x_5) = (0, 9, 4, 6, 0),$$

with $Z = 45$, thereby providing an increase in profit from the new products of 9 units ($9,000 per week) over the previous $Z = 36$. The fact that $x_4 = 6$ indicates that 6 of the 12 additional units of resource 2 are unused by this solution.

Although $\Delta b_2 = 12$ proved to be too large an increase in b_2 to retain feasibility (and so optimality) with the basic solution where x_1, x_2, and x_3 are the basic variables (middle of Table 6.18), the above incremental analysis shows immediately just how large an increase is feasible. In particular, note that

$$b_1^* = 2 + \tfrac{1}{3}\,\Delta b_2,$$

$$b_2^* = 6 + \tfrac{1}{2}\,\Delta b_2,$$

$$b_3^* = 2 - \tfrac{1}{3}\,\Delta b_2,$$

where these three quantities are the values of x_3, x_2, and x_1, respectively, for this basic solution. The solution remains feasible, and so optimal, as long as all three quantities remain nonnegative.

Table 6.20 **Revised Data for Wyndor Glass Co. Problem after Just b_2 Is Changed**

Model Parameters

$c_1 = 3,$	$c_2 = 5$	$(n = 2)$
$a_{11} = 1,$	$a_{12} = 0,$	$b_1 = 4$
$a_{21} = 0,$	$a_{22} = 2,$	$b_2 = 24$
$a_{31} = 3,$	$a_{32} = 2,$	$b_3 = 18$

Final Simplex Tableau after Reoptimization

Basic Variable	Eq.	*Coefficient of:* Z	x_1	x_2	x_3	x_4	x_5	Right Side
Z	(0)	1	$\frac{9}{2}$	0	0	0	$\frac{5}{2}$	45
x_3	(1)	0	1	0	1	0	0	4
x_2	(2)	0	$\frac{3}{2}$	1	0	0	$\frac{1}{2}$	9
x_4	(3)	0	−3	0	0	1	−1	6

$$2 + \tfrac{1}{3}\Delta b_2 \geq 0 \Rightarrow \tfrac{1}{3}\Delta b_2 \geq -2 \Rightarrow \Delta b_2 \geq \ -6,$$
$$6 + \tfrac{1}{2}\Delta b_2 \geq 0 \Rightarrow \tfrac{1}{2}\Delta b_2 \geq -6 \Rightarrow \Delta b_2 \geq -12,$$
$$2 - \tfrac{1}{3}\Delta b_2 \geq 0 \Rightarrow 2 \geq \tfrac{1}{3}\Delta b_2 \quad \Rightarrow \Delta b_2 \leq \quad 6.$$

Therefore, since $b_2 = 12 + \Delta b_2$, the solution remains feasible only if

$$-6 \leq \Delta b_2 \leq 6, \qquad \text{that is,} \qquad 6 \leq b_2 \leq 18.$$

(Verify this graphically in Fig. 6.3.) This range of values for b_2 is referred to as its *allowable range to stay feasible.*

> For any b_i, its **allowable range to stay feasible** is the range of values over which the optimal BF solution[1] (with adjusted values for the basic variables) remains feasible. (It is assumed that the change in this one b_i value is the only change in the model.) The adjusted values for the basic variables are obtained from the formula $\mathbf{b}^* = \mathbf{S}^*\overline{\mathbf{b}}$. The calculation of the allowable range to stay feasible then is based on finding the range of values of b_i such that $\mathbf{b}^* \geq \mathbf{0}$.

Many linear programming software packages, including the automatic routine for the simplex method in the OR Courseware, use this same technique for automatically generating the allowable range to stay feasible for each b_i. (A similar technique, discussed under Cases 2a and 3, also is used to generate an *allowable range to stay optimal* for each c_j.)

Based on the results with $b_2 = 24$, the relatively unprofitable old product will be discontinued and the unused 6 units of resource 2 will be saved for some future use. Since y_3^* still is positive, a similar study is made of the possibility of changing the allocation of resource 3, but the resulting decision is to retain the current allocation. Therefore, the current linear programming model at this point has the parameter values and optimal solution shown in Table 6.20.

Case 2*a*—Changes in the Coefficients of a Nonbasic Variable

Consider a particular variable x_j (fixed j) that is a nonbasic variable in the optimal solution shown by the final simplex tableau. In Case 2a, the only change in the current model is that one or more of the coefficients of this variable—$c_j, a_{1j}, a_{2j}, \ldots, a_{mj}$—have been changed. Thus, letting $\overline{c}_j$ and $\overline{a}_{ij}$ denote the new values of these parameters, with $\overline{\mathbf{A}}_j$ (column j of matrix $\overline{\mathbf{A}}$) as the vector containing the $\overline{a}_{ij}$, we have

$$c_j \longrightarrow \overline{c}_j, \qquad \mathbf{A}_j \longrightarrow \overline{\mathbf{A}}_j$$

for the revised model.

As described at the beginning of Sec. 6.5, duality theory provides a very convenient way of checking these changes. In particular, if the *complementary* basic solution $\mathbf{y}^*$ in the dual problem still satisfies the single dual constraint that has changed, then the original optimal solution in the primal problem *remains optimal* as is. Conversely, if $\mathbf{y}^*$ violates this dual constraint, then this primal solution is *no longer optimal.*

If the optimal solution has changed and you wish to find the new one, you can do so rather easily. Simply apply the fundamental insight to revise the x_j column (the only

[1] When there is more than one optimal BF solution for the current model (before changing b_i), we are referring here to the one obtained by the simplex method.

one that has changed) in the final simplex tableau. Specifically, the formulas in Table 6.17 reduce to the following:

$$\text{Coefficient of } x_j \text{ in final row 0:} \qquad z_j^* - \bar{c}_j = \mathbf{y}^*\overline{\mathbf{A}}_j - \bar{c}_j,$$
$$\text{Coefficient of } x_j \text{ in final rows 1 to } m\text{:} \qquad \mathbf{A}_j^* = \mathbf{S}^*\overline{\mathbf{A}}_j.$$

With the current basic solution no longer optimal, the new value of $z_j^* - c_j$ now will be the one negative coefficient in row 0, so restart the simplex method with x_j as the initial entering basic variable.

Note that this procedure is a streamlined version of the general procedure summarized at the end of Sec. 6.6. Steps 3 and 4 (conversion to proper form from Gaussian elimination and the feasibility test) have been deleted as irrelevant, because the only column being changed in the revision of the final tableau (before reoptimization) is for the nonbasic variable x_j. Step 5 (optimality test) has been replaced by a quicker test of optimality to be performed right after step 1 (revision of model). It is only if this test reveals that the optimal solution has changed, and you wish to find the new one, that steps 2 and 6 (revision of final tableau and reoptimization) are needed.

Example: Since x_1 is nonbasic in the current optimal solution (see Table 6.20) for the Wyndor Glass Co. problem, the next step in its sensitivity analysis is to check whether any reasonable changes in the estimates of the coefficients of x_1 could still make it advisable to introduce product 1. The set of changes that goes as far as realistically possible to make product 1 more attractive would be to reset $c_1 = 4$ and $a_{31} = 2$ (as was done in Sec. 6.6). Rather than exploring each of these changes independently (as is often done in sensitivity analysis), we will consider them together. Thus, the changes under consideration are

$$c_1 = 3 \longrightarrow \bar{c}_1 = 4, \qquad \mathbf{A}_1 = \begin{bmatrix} 1 \\ 0 \\ 3 \end{bmatrix} \longrightarrow \overline{\mathbf{A}}_1 = \begin{bmatrix} 1 \\ 0 \\ 2 \end{bmatrix}.$$

This change in a_{31} revises the feasible region from that shown in Fig. 6.3 to the corresponding region in Fig. 6.2 when $3x_1 + 2x_2 = 18$ is replaced by $2x_1 + 2x_2 = 18$. (Ignore the $2x_2 = 12$ line, because the $2x_2 \leq 12$ constraint already has been replaced by $2x_2 \leq 24$.) The change in c_1 revises the objective function from $Z = 3x_1 + 5x_2$ to $Z = 4x_1 + 5x_2$. By using Fig. 6.2 to draw the objective function line $Z = 45 = 4x_1 + 5x_2$ through the current optimal solution (0, 9), you can verify that (0, 9) remains optimal after these changes in a_{31} and c_1.

To use duality theory to draw this same conclusion, observe that the changes in c_1 and a_{31} lead to a single revised constraint for the dual problem (see Table 6.1). Both this revised constraint and the current $\mathbf{y}^*$ (coefficients of the slack variables in row 0 of Table 6.20) are shown below.

$$y_1^* = 0, \qquad y_2^* = 0, \qquad y_3^* = \tfrac{5}{2},$$
$$y_1 + 3y_3 \geq 3 \longrightarrow y_1 + 2y_3 \geq 4,$$
$$0 + 2(\tfrac{5}{2}) \geq 4.$$

Since $\mathbf{y}^*$ *still* satisfies the revised constraint, the current primal solution (Table 6.20) is still optimal.

Because this solution is still optimal, there is no need to revise the x_j column in the final tableau (step 2). Nevertheless, we do so below for illustrative purposes.

$$z_j^* - \overline{c}_j = \mathbf{y}^*\overline{\mathbf{A}}_j - c_j = [0, 0, \tfrac{5}{2}]\begin{bmatrix}1\\0\\2\end{bmatrix} - 4 = 1.$$

$$\mathbf{A}_j^* = \mathbf{S}^*\overline{\mathbf{A}}_j = \begin{bmatrix}1 & 0 & 0\\0 & 0 & \frac{1}{2}\\0 & 1 & -1\end{bmatrix}\begin{bmatrix}1\\0\\2\end{bmatrix} = \begin{bmatrix}1\\1\\-2\end{bmatrix}.$$

The fact that $z_j^* - \overline{c}_j \geq 0$ again confirms the optimality of the current solution. Since $z_j^* - \overline{c}_j$ is the surplus variable for the revised constraint in the dual problem, this way of testing for optimality is equivalent to the one used above.

This completes the analysis of the effect of the indicated changes in c_1 and a_{31}. Because any larger changes in the original estimates of the coefficients of x_1 would be unrealistic, the OR team concludes that these coefficients are *insensitive* parameters in the current model. Therefore, they will be kept fixed at their best estimates shown in Table 6.20—$c_1 = 3$ and $a_{31} = 3$—for the remainder of the sensitivity analysis.

The Allowable Range to Stay Optimal: We have just described and illustrated how to analyze *simultaneous* changes in the coefficients of a nonbasic variable x_j. It is common practice in sensitivity analysis to also focus on the effect of changing just *one* parameter, c_j. This involves streamlining the above approach to find the *allowable range to stay optimal* for c_j.

> For any c_j, its **allowable range to stay optimal** is the range of values over which the current optimal solution (as obtained by the simplex method for the current model before c_j is changed) remains optimal. (It is assumed that the change in this one c_j is the only change in the current model.) When x_j is a nonbasic variable for this solution, the solution remains optimal as long as $z_j^* - c_j \geq 0$, where $z_j^* = \mathbf{y}^*\mathbf{A}_j$ is a constant unaffected by any change in the value of c_j. Therefore, the allowable range to stay optimal for c_j can be calculated as $c_j \leq \mathbf{y}^*\mathbf{A}_j$.

For example, consider the current model for the Wyndor Glass Co. problem summarized on the left side of Table 6.20, where the current optimal solution (with $c_1 = 3$) is given on the right side. When just c_1 is changed, this solution remains optimal as long as

$$c_1 \leq \mathbf{y}^*\mathbf{A}_1 = [0, \quad 0, \quad \tfrac{5}{2}]\begin{bmatrix}1\\0\\3\end{bmatrix} = 7\tfrac{1}{2},$$

so $c_1 \leq 7\frac{1}{2}$ is the allowable range to stay optimal.

An alternative to performing this vector multiplication is to note in Table 6.20 that $z_1^* - c_1 = \frac{9}{2}$ (the coefficient of x_1 in row 0) when $c_1 = 3$, so $z_1^* = 3 + \frac{9}{2} = 7\frac{1}{2}$. Since $z_1^* = \mathbf{y}^*\mathbf{A}_1$, this immediately yields the same allowable range.

For any nonbasic decision variable x_j, the value of $z_j^* - c_j$ sometimes is referred to as the **reduced cost** for x_j, because it is the minimum amount by which the unit *cost* of activity j would have to be *reduced* to make it worthwhile to undertake activity j (increase x_j from zero). Interpreting c_j as the unit profit of activity j (so reducing the unit cost increases c_j by the same amount), the value of $z_j^* - c_j$ thereby is the maximum allowable increase in c_j to keep the current BF solution optimal.

Case 2*b*—Introduction of a New Variable

After solving for the optimal solution, we may discover that the linear programming formulation did not consider all the attractive alternative activities. Considering a new activity requires introducing a new variable with the appropriate coefficients into the objective function and constraints of the current model—which is Case 2*b*.

The convenient way to deal with this case is to treat it just as if it were Case 2*a*! This is done by pretending that the new variable x_j actually was in the original model with all its coefficients equal to zero (so that they still are zero in the final simplex tableau) and that x_j is a nonbasic variable in the current BF solution. Therefore, if we change these zero coefficients to their actual values for the new variable, the procedure (including any reoptimization) does indeed become identical to that for Case 2*a*.

In particular, all you have to do to check whether the current solution still is optimal is to check whether the complementary basic solution $\mathbf{y}^*$ satisfies the one new dual constraint that corresponds to the new variable in the primal problem. We already have described this approach and then illustrated it for the Wyndor Glass Co. problem in Sec. 6.5.

Case 3—Changes in the Coefficients of a Basic Variable

Now suppose that the variable x_j (fixed j) under consideration is a *basic* variable in the optimal solution shown by the final simplex tableau. Case 3 assumes that the only changes in the current model are made to the coefficients of this variable.

Case 3 differs from Case 2*a* because of the requirement that a simplex tableau be in proper form from Gaussian elimination. This requirement allows the column for a nonbasic variable to be anything, so it does not affect Case 2*a*. However, for Case 3, the basic variable x_j must have a coefficient of 1 in its row of the simplex tableau and a coefficient of 0 in every other row (including row 0). Therefore, after the changes in the x_j column of the final simplex tableau have been calculated,[1] it probably will be necessary to apply Gaussian elimination to restore this form, as illustrated in Table 6.19. In turn, this step probably will change the value of the current basic solution and may make it either infeasible or nonoptimal (so reoptimization may be needed). Consequently, all the steps of the overall procedure summarized at the end of Sec. 6.6 are required for Case 3.

Before Gaussian elimination is applied, the formulas for revising the x_j column are the same as for Case 2*b*, as summarized below.

Coefficient of x_j in final row 0: $\quad z_j^* - \bar{c}_j = \mathbf{y}^*\bar{\mathbf{A}}_j - \bar{c}_j.$

Coefficient of x_j in final rows 1 to m: $\quad \mathbf{A}_j^* = \mathbf{S}^*\bar{\mathbf{A}}_j.$

Example: Because x_2 is a basic variable in Table 6.20 for the Wyndor Glass Co. problem, sensitivity analysis of its coefficients fits Case 3. Given the current optimal solution ($x_1 = 0, x_2 = 9$), product 2 is the *only* new product that should be introduced, and its production rate should be relatively large. Therefore, the key question now is

[1] For the relatively sophisticated reader, we should point out a possible pitfall for Case 3 that would be discovered at this point. Specifically, the changes in the initial tableau can destroy the linear independence of the columns of coefficients of basic variables. This event occurs only if the unit coefficient of the basic variable x_j in the final tableau has been changed to zero at this point, in which case more extensive simplex method calculations must be used for Case 3.

whether the initial estimates that led to the coefficients of x_2 in the current model could have *overestimated* the attractiveness of product 2 so much as to invalidate this conclusion. This question can be tested by checking the *most pessimistic* set of reasonable estimates for these coefficients, which turns out to be $c_2 = 3$, $a_{22} = 3$, and $a_{32} = 4$. Consequently, the changes to be investigated are

$$c_2 = 5 \longrightarrow \bar{c}_2 = 3, \qquad \mathbf{A}_2 = \begin{bmatrix} 0 \\ 2 \\ 2 \end{bmatrix} \longrightarrow \overline{\mathbf{A}}_2 = \begin{bmatrix} 0 \\ 3 \\ 4 \end{bmatrix}.$$

The graphical effect of these changes is that the feasible region changes from the one shown in Fig. 6.3 to the one in Fig. 6.4. The optimal solution in Fig. 6.3 is $(x_1, x_2) = (0, 9)$, which is the corner-point solution lying at the intersection of the $x_1 = 0$ and $3x_1 + 2x_2 = 18$ constraint boundaries. With the revision of the constraints, the corresponding corner-point solution in Fig. 6.4 is $(0, \frac{9}{2})$. However, this solution no longer is optimal, because the revised objective function of $Z = 3x_1 + 3x_2$ now yields a new optimal solution of $(x_1, x_2) = (4, \frac{3}{2})$.

Now let us see how we draw these same conclusions algebraically. Because the only changes in the model are in the coefficients of x_2, the *only* resulting changes in the final simplex tableau (Table 6.20) are in the x_2 column. Therefore, the above formulas are used to recompute just this column.

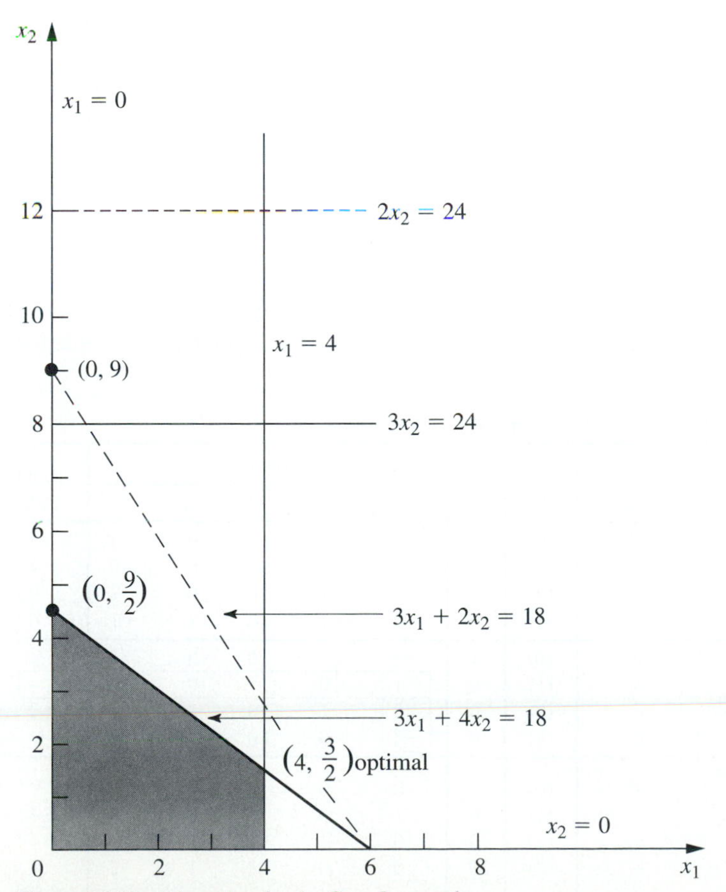

Figure 6.4 Feasible region for the Case 3 example.

$$z_2 - \bar{c}_2 = \mathbf{y}^*\bar{\mathbf{A}}_2 - \bar{c}_2 = \left[0, 0, \tfrac{5}{2}\right] \begin{bmatrix} 0 \\ 3 \\ 4 \end{bmatrix} - 3 = 7.$$

$$\mathbf{A}_2^* = \mathbf{S}^*\bar{\mathbf{A}}_2 = \begin{bmatrix} 1 & 0 & 0 \\ 0 & 0 & \tfrac{1}{2} \\ 0 & 1 & -1 \end{bmatrix} \begin{bmatrix} 0 \\ 3 \\ 4 \end{bmatrix} = \begin{bmatrix} 0 \\ 2 \\ -1 \end{bmatrix}.$$

(Equivalently, incremental analysis with $\Delta c_2 = -2$, $\Delta a_{22} = 1$, and $\Delta a_{32} = 2$ can be used in the same way to obtain this column.)

The resulting revised final tableau is shown at the top of Table 6.21. Note that the new coefficients of the basic variable x_2 do not have the required values, so the conversion to proper form from Gaussian elimination must be applied next. This step involves dividing row 2 by 2, subtracting 7 times the new row 2 from row 0, and adding the new row 2 to row 3.

The resulting second tableau in Table 6.21 gives the new value of the current basic solution, namely, $x_3 = 4$, $x_2 = \frac{9}{2}$, $x_4 = \frac{21}{2}$ $(x_1 = 0, x_5 = 0)$. Since all these variables are nonnegative, the solution is still feasible. However, because of the negative coefficient of x_1 in row 0, we know that it is no longer optimal. Therefore, the simplex method would be applied to this tableau, with this solution as the initial BF solution, to find the new optimal solution. The initial entering basic variable is x_1, with x_3 as the leaving basic variable. Just one iteration is needed in this case to reach the new optimal solution $x_1 = 4$, $x_2 = \frac{3}{2}$, $x_4 = \frac{39}{2}$ $(x_3 = 0, x_5 = 0)$, as shown in the last tableau of Table 6.21.

All this analysis suggests that c_2, a_{22}, and a_{32} are relatively sensitive parameters. However, additional data for estimating them more closely can be obtained only by conducting a pilot run. Therefore, the OR team recommends that production of product 2 be initiated immediately on a small scale ($x_2 = \frac{3}{2}$) and that this experience be used to guide the decision on whether the remaining production capacity should be allocated to product 2 or product 1.

The Allowable Range to Stay Optimal: For Case 2a, we described how to find the allowable range to stay optimal for any c_j such that x_j is a nonbasic variable for

Table 6.21 **Sensitivity Analysis Procedure Applied to Case 3 Example**

	Basic Variable	Eq.	Coefficient of: Z	x_1	x_2	x_3	x_4	x_5	Right Side
Revised final tableau	Z	(0)	1	$\frac{9}{2}$	7	0	0	$\frac{5}{2}$	45
	x_3	(1)	0	1	0	1	0	0	4
	x_2	(2)	0	$\frac{3}{2}$	2	0	0	$\frac{1}{2}$	9
	x_4	(3)	0	-3	-1	0	1	-1	6
Converted to proper form	Z	(0)	1	$-\frac{3}{4}$	0	0	0	$\frac{3}{4}$	$\frac{27}{2}$
	x_3	(1)	0	1	0	1	0	0	4
	x_2	(2)	0	$\frac{3}{4}$	1	0	0	$\frac{1}{4}$	$\frac{9}{2}$
	x_4	(3)	0	$-\frac{9}{4}$	0	0	1	$-\frac{3}{4}$	$\frac{21}{2}$
New final tableau after reoptimization (only one iteration of the simplex method needed in this case)	Z	(0)	1	0	0	$\frac{3}{4}$	0	$\frac{3}{4}$	$\frac{33}{2}$
	x_1	(1)	0	1	0	1	0	0	4
	x_2	(2)	0	0	1	$-\frac{3}{4}$	0	$\frac{1}{4}$	$\frac{3}{2}$
	x_4	(3)	0	0	0	$\frac{9}{4}$	1	$-\frac{3}{4}$	$\frac{39}{2}$

the current optimal solution (before c_j is changed). When x_j is a basic variable instead, the procedure is somewhat more involved because of the need to convert to proper form from Gaussian elimination before testing for optimality.

To illustrate the procedure, consider the new version of the Wyndor Glass Co. model (with $c_2 = 3$, $a_{22} = 3$, $a_{23} = 4$) that is graphed in Fig. 6.4 and solved in Table 6.21. Since x_2 is a basic variable for the optimal solution (with $c_2 = 3$) given at the bottom of this table, the steps needed to find the allowable range to stay optimal for c_2 are the following:

1. Since x_2 is a basic variable, note that its coefficient in the new final row 0 is automatically $z_2^* - c_2 = 0$ before c_2 is changed from its current value of 3.
2. Now increment $c_2 = 3$ by Δc_2 (so $c_2 = 3 + \Delta c_2$). This changes the coefficient noted in step 1 to $z_2^* - c_2 = -\Delta c_2$.
3. With this coefficient now not zero, we must perform elementary row operations to restore proper form from Gaussian elimination. In particular, add to row 0 the product, Δc_2 times row 2, which gives a new row 0 of $\left[0, 0, \frac{3}{4} - \frac{3}{4}\Delta c_2, 0, \frac{3}{4} + \frac{1}{4}\Delta c_2 \,\vdots\, \frac{33}{2} + \frac{3}{2}\Delta c_2\right]$.
4. Using this new row 0, solve for the range of values of Δc_2 that keeps the coefficients of the nonbasic variables (x_3 and x_5) nonnegative.

$$\frac{3}{4} - \frac{3}{4}\Delta c_2 \geq 0 \quad \Rightarrow \quad \frac{3}{4} \geq \frac{3}{4}\Delta c_2 \quad \Rightarrow \quad \Delta c_2 \leq 1.$$
$$\frac{3}{4} + \frac{1}{4}\Delta c_2 \geq 0 \quad \Rightarrow \quad \frac{1}{4}\Delta c_2 \geq -\frac{3}{4} \quad \Rightarrow \quad \Delta c_2 \geq -3.$$

   Thus, the range of values is $-3 \leq \Delta c_2 \leq 1$.
5. Since $c_2 = 3 + \Delta c_2$, add 3 to this range of values, which yields

$$0 \leq c_2 \leq 4$$

   as the allowable range to stay optimal for c_2.

With just two decision variables, this allowable range can be verified graphically by using Fig. 6.4 with an objective function of $Z = 3x_1 + c_2x_2$. With the current value of $c_2 = 3$, the optimal solution is $(4, \frac{3}{2})$. When c_2 is increased, this solution remains optimal only for $c_2 \leq 4$. For $c_2 \geq 4$, $(0, \frac{9}{2})$ becomes optimal (with a tie at $c_2 = 4$), because of the constraint boundary $3x_1 + 4x_2 = 18$. When c_2 is decreased instead, $(4, \frac{3}{2})$ remains optimal only for $c_2 \geq 0$. For $c_2 \leq 0$, $(4, 0)$ becomes optimal because of the constraint boundary $x_1 = 4$.

In a similar manner, the allowable range to stay optimal for c_1 (with c_2 fixed at 3) can be derived either algebraically or graphically to be $c_1 \geq \frac{9}{4}$. (Problem 6.7-9 asks you to verify this both ways.)

Case 4—Introduction of a New Constraint

In the last case, a new constraint must be introduced to the model after it has already been solved. This case may occur because the constraint was overlooked initially or because new considerations have arisen since the model was formulated. Another possibility is that the constraint was deleted purposely to decrease computational effort because it appeared to be less restrictive than other constraints already in the model, but now this impression needs to be checked with the optimal solution actually obtained.

To see if the current optimal solution would be affected by a new constraint, all you have to do is to check directly whether the optimal solution satisfies the constraint. If it does, then it would still be the *best feasible solution* (i.e., the optimal solution), even if the constraint were added to the model. The reason is that a new constraint can only eliminate some previously feasible solutions without adding any new ones.

If the new constraint does eliminate the current optimal solution, and if you want to find the new solution, then introduce this constraint into the final simplex tableau (as an additional row) *just* as if this were the initial tableau, where the usual additional variable (slack variable or artificial variable) is designated to be the basic variable for this new row. Because the new row probably will have *nonzero* coefficients for some of the other basic variables, the conversion to proper form from Gaussian elimination is applied next, and then the reoptimization step is applied in the usual way.

Just as for some of the preceding cases, this procedure for Case 4 is a streamlined version of the general procedure summarized at the end of Sec. 6.6. The only question to be addressed for this case is whether the previously optimal solution still is *feasible,* so step 5 (optimality test) has been deleted. Step 4 (feasibility test) has been replaced by a much quicker test of feasibility (does the previously optimal solution satisfy the new constraint?) to be performed right after step 1 (revision of model). It is only if this test provides a negative answer, and you wish to reoptimize, that steps 2, 3, and 6 are used (revision of final tableau, conversion to proper form from Gaussian elimination, and reoptimization).

Example: To illustrate this case, suppose that the new constraint

$$2x_1 + 3x_2 \leq 24$$

is introduced into the model given in Table 6.20. The graphical effect is shown in Fig. 6.5. The previous optimal solution (0, 9) violates the new constraint, so the optimal solution changes to (0, 8).

To analyze this example algebraically, note that (0, 9) yields $2x_1 + 3x_2 = 27 > 24$, so this previous optimal solution is no longer feasible. To find the new optimal solution, add the new constraint to the current final simplex tableau as just described, with the slack variable x_6 as its initial basic variable. This step yields the first tableau shown in Table 6.22. The conversion to proper form from Gaussian elimination then requires subtracting from the new row the product, 3 times row 2, which identifies the current basic solution $x_3 = 4$, $x_2 = 9$, $x_4 = 6$, $x_6 = -3$ ($x_1 = 0$, $x_5 = 0$), as shown in the second tableau. Applying the dual simplex method (described in Sec. 7.1) to this tableau then leads in just one iteration (more are sometimes needed) to the new optimal solution in the last tableau of Table 6.22.

Systematic Sensitivity Analysis—Parametric Programming

So far we have described how to test specific changes in the model parameters. Another common approach to sensitivity analysis is to vary one or more parameters continuously over some interval(s) to see when the optimal solution changes.

For example, with the Wyndor Glass Co. problem, rather than beginning by testing the specific change from $b_2 = 12$ to $\bar{b}_2 = 24$, we might instead set

$$\bar{b}_2 = 12 + \theta$$

and then vary θ continuously from 0 to 12 (the maximum value of interest). The geometric interpretation in Fig. 6.3 is that the $2x_2 = 12$ constraint line is being shifted

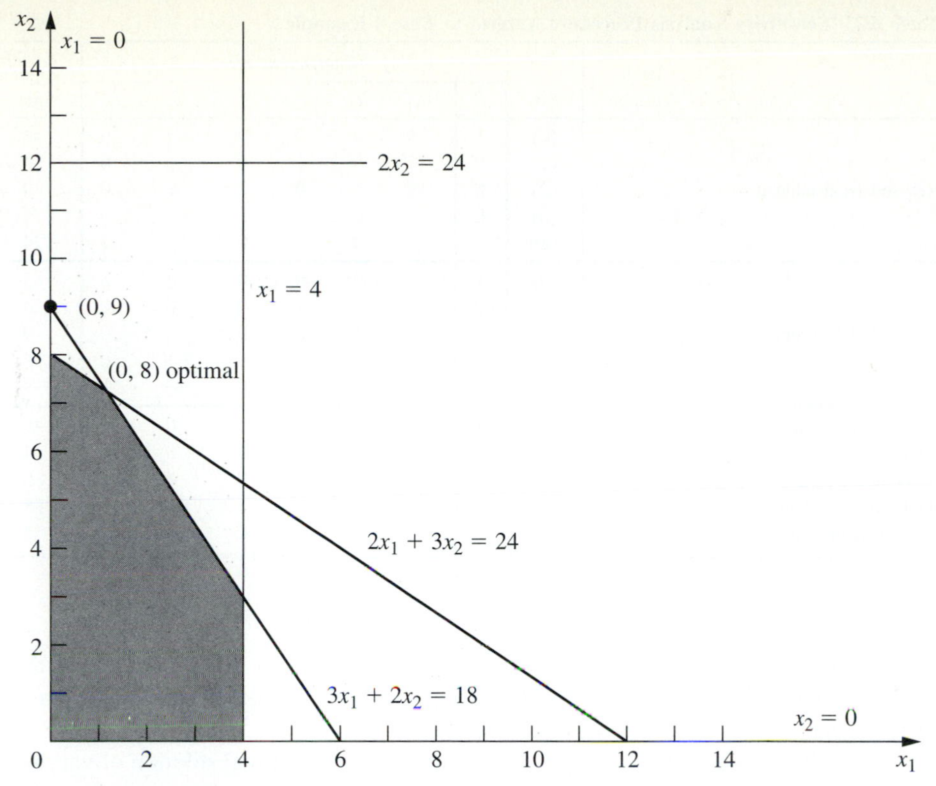

Figure 6.5 Feasible region for the Case 4 example.

upward to $2x_2 = 12 + \theta$, with θ being increased from 0 to 12. The result is that the original optimal CPF solution (2, 6) shifts up the $3x_1 + 2x_2 = 18$ constraint line toward (−2, 12). This corner-point solution remains optimal as long as it is still feasible ($x_1 \geq 0$), after which (0, 9) becomes the optimal solution.

The algebraic calculations of the effect of having $\Delta b_2 = \theta$ are directly analogous to those for the Case 1 example where $\Delta b_2 = 12$. In particular, by using the expressions for Z^* and $\mathbf{b}^*$ given for Case 1, the middle tableau in Table 6.18 indicates that the corresponding optimal solution is

$$Z = 36 + \tfrac{3}{2}\theta$$

$$x_3 = 2 + \tfrac{1}{3}\theta$$

$$x_2 = 6 + \tfrac{1}{2}\theta \qquad (x_4 = 0, x_5 = 0)$$

$$x_1 = 2 - \tfrac{1}{3}\theta$$

for θ small enough that this solution still is feasible, i.e., for $\theta \leq 6$. For $\theta > 6$, the dual simplex method (described in Sec. 7.1) yields the tableau shown in Table 6.20 except for the value of x_4. Thus, $Z = 45$, $x_3 = 4$, $x_2 = 9$ (along with $x_1 = 0$, $x_5 = 0$), and the expression for $\mathbf{b}^*$ yields

$$x_4 = b_3^* = 0(4) + 1(12 + \theta) - 1(18) = -6 + \theta.$$

Table 6.22 **Sensitivity Analysis Procedure Applied to Case 4 Example**

	Basic Variable	Eq.	Coefficient of:							Right Side
			Z	x_1	x_2	x_3	x_4	x_5	x_6	
Revised final tableau	Z	(0)	1	$\frac{9}{2}$	0	0	0	$\frac{5}{2}$	0	45
	x_3	(1)	0	1	0	1	0	0	0	4
	x_2	(2)	0	$\frac{3}{2}$	1	0	0	$\frac{1}{2}$	0	9
	x_4	(3)	0	-3	0	0	1	-1	0	6
	x_6	New	0	2	3	0	0	0	1	24
Converted to proper form	Z	(0)	1	$\frac{9}{2}$	0	0	0	$\frac{5}{2}$	0	45
	x_3	(1)	0	1	0	1	0	0	0	4
	x_2	(2)	0	$\frac{3}{2}$	1	0	0	$\frac{1}{2}$	0	9
	x_4	(3)	0	-3	0	0	1	-1	0	6
	x_6	New	0	$-\frac{5}{2}$	0	0	0	$-\frac{3}{2}$	1	-3
New final tableau after reoptimization (only one iteration of dual simplex method needed in this case)	Z	(0)	1	$\frac{1}{3}$	0	0	0	0	$\frac{5}{3}$	40
	x_3	(1)	0	1	0	1	0	0	0	4
	x_2	(2)	0	$\frac{2}{3}$	1	0	0	0	$\frac{1}{3}$	8
	x_4	(3)	0	$-\frac{4}{3}$	0	0	1	0	$-\frac{2}{3}$	8
	x_5	New	0	$\frac{5}{3}$	0	0	0	1	$-\frac{2}{3}$	2

This information can then be used (along with other data not incorporated into the model on the effect of increasing b_2) to decide whether to retain the original optimal solution and, if not, how much to increase b_2.

In a similar way, we can investigate the effect on the optimal solution of varying several parameters simultaneously. When we vary just the b_i parameters, we express the new value $\bar{b}_i$ in terms of the original value b_i as follows:

$$\bar{b}_i = b_i + \alpha_i \theta, \qquad \text{for } i = 1, 2, \ldots, m,$$

where the α_i values are input constants specifying the desired rate of increase (positive or negative) of the corresponding right-hand side as θ is increased.

For example, suppose that it is possible to shift some of the production of a current Wyndor Glass Co. product from Plant 2 to Plant 3, thereby increasing b_2 by decreasing b_3. Also suppose that b_3 decreases twice as fast as b_2 increases. Then

$$\bar{b}_2 = 12 + \theta$$

$$\bar{b}_3 = 18 - 2\theta,$$

where the (nonnegative) value of θ measures the amount of production shifted. (Thus $\alpha_1 = 0$, $\alpha_2 = 1$, and $\alpha_3 = -2$ in this case.) In Fig. 6.3, the geometric interpretation is that as θ is increased from 0, the $2x_2 = 12$ constraint line is being pushed up to $2x_2 = 12 + \theta$ (ignore the $2x_2 = 24$ line) and simultaneously the $3x_1 + 2x_2 = 18$ constraint line is being pushed down to $3x_1 + 2x_2 = 18 - 2\theta$. The original optimal CPF solution (2, 6) lies at the intersection of the $2x_2 = 12$ and $3x_1 + 2x_2 = 18$ lines, so shifting these lines causes this corner-point solution to shift. However, with the objective function of $Z = 3x_1 + 5x_2$, this corner-point solution will remain optimal as long as it is still feasible ($x_1 \geq 0$).

An algebraic investigation of simultaneously changing b_2 and b_3 in this way again involves using the formulas for Case 1 (treating θ as representing an unknown number) to calculate the resulting changes in the final tableau (middle of Table 6.18), namely,

$$Z^* = \mathbf{y}^*\overline{\mathbf{b}} = \left[0, \tfrac{3}{2}, 1\right] \begin{bmatrix} 4 \\ 12 + \theta \\ 18 - 2\theta \end{bmatrix} = 36 - \tfrac{1}{2}\theta,$$

$$\mathbf{b}^* = \mathbf{S}^*\overline{\mathbf{b}} = \begin{bmatrix} 1 & \frac{1}{3} & -\frac{1}{3} \\ 0 & \frac{1}{2} & 0 \\ 0 & -\frac{1}{3} & \frac{1}{3} \end{bmatrix} \begin{bmatrix} 4 \\ 12 + \theta \\ 18 - 2\theta \end{bmatrix} = \begin{bmatrix} 2 + \theta \\ 6 + \frac{1}{2}\theta \\ 2 - \theta \end{bmatrix}.$$

Therefore, the optimal solution becomes

$$Z = 36 - \tfrac{1}{2}\theta$$

$$x_3 = 2 + \theta$$

$$x_2 = 6 + \tfrac{1}{2}\theta \qquad (x_4 = 0, \qquad x_5 = 0)$$

$$x_1 = 2 - \theta$$

for θ small enough that this solution still is feasible, i.e., for $\theta \leq 2$. (Check this conclusion in Fig. 6.3.) However, the fact that Z decreases as θ increases from 0 indicates that the best choice for θ is $\theta = 0$, so none of the possible shifting of production should be done.

The approach to varying several c_j parameters simultaneously is similar. In this case, we express the new value $\overline{c}_j$ in terms of the original value of c_j as

$$\overline{c}_j = c_j + \alpha_j\theta, \qquad \text{for } j = 1, 2, \ldots, n,$$

where the α_j are input constants specifying the desired rate of increase (positive or negative) of c_j as θ is increased.

To illustrate this case, reconsider the sensitivity analysis of c_1 and c_2 for the Wyndor Glass Co. problem that was performed earlier in this section. Starting with the version of the model presented in Table 6.20 and Fig. 6.3, we separately considered the effect of changing c_1 from 3 to 4 (its most optimistic estimate) and c_2 from 5 to 3 (its most pessimistic estimate). Now we can simultaneously consider both changes, as well as various intermediate cases with smaller changes, by setting

$$\overline{c}_1 = 3 + \theta \qquad \text{and} \qquad \overline{c}_2 = 5 - 2\theta,$$

where the value of θ measures the *fraction* of the maximum possible change that is made. The result is to replace the original objective function $Z = 3x_1 + 5x_2$ by a *function* of θ

$$Z(\theta) = (3 + \theta)x_1 + (5 - 2\theta)x_2,$$

so the optimization now can be performed for any desired (fixed) value of θ between 0 and 1. By checking the effect as θ increases from 0 to 1, we can determine just when and how the optimal solution changes as the error in the original estimates of these parameters increases.

Considering these changes simultaneously is especially appropriate if there are factors that cause the parameters to change together. Are the two products competitive in some sense, so that a larger-than-expected unit profit for one implies a smaller-than-expected unit profit for the other? Are they both affected by some exogenous factor, such as the advertising emphasis of a competitor? Is it possible to simultaneously change both unit profits through appropriate shifting of personnel and equipment?

In the feasible region shown in Fig. 6.3, the geometric interpretation of changing the objective function from $Z = 3x_1 + 5x_2$ to $Z(\theta) = (3 + \theta)x_1 + (5 - 2\theta)x_2$ is that we are changing the *slope* of the original objective function line ($Z = 45 = 3x_1 + 5x_2$)

Table 6.23 **Dealing with $\Delta c_1 = \theta$ and $\Delta c_2 = -2\theta$ for the Model of Table 6.20**

	Basic Variable	Eq.	Coefficient of: Z	x_1	x_2	x_3	x_4	x_5	Right Side
Final tableau	Z	(0)	1	$\frac{9}{2}$	0	0	0	$\frac{5}{2}$	45
	x_2	(2)	0	$\frac{3}{2}$	1	0	0	$\frac{1}{2}$	9
Revised final tableau when $\Delta c_1 = \theta$ and $\Delta c_2 = -2\theta$	$Z(\theta)$	(0)	1	$\frac{9}{2} - \theta$	2θ	0	0	$\frac{5}{2}$	45
	x_2	(2)	0	$\frac{3}{2}$	1	0	0	$\frac{1}{2}$	9
Converted to proper form	$Z(\theta)$	(0)	1	$\frac{9}{2} - 4\theta$	0	0	0	$\frac{5}{2} - \theta$	$45 - 18\theta$
	x_2	(2)	0	$\frac{3}{2}$	1	0	0	$\frac{1}{2}$	9

that can be drawn through the optimal solution (0, 9). If θ is increased enough, this slope will change sufficiently that the optimal solution will switch from (0, 9) to another CPF solution (4, 3). (Check graphically whether this occurs for $\theta \leq 1$.)

The algebraic procedure for dealing simultaneously with these two changes ($\Delta c_1 = \theta$ and $\Delta c_2 = -2\theta$) is shown in Table 6.23. Although the changes now are expressed in terms of θ rather than specific numerical amounts, θ is treated just as an unknown number. The table displays just the relevant rows of the tableaux involved (row 0 and the row for the basic variable x_2). The first tableau shown is just the final tableau for the current version of the model (before c_1 and c_2 are changed) as given in Table 6.20. Refer to the formulas in Table 6.17. The only changes in the *revised* final tableau shown next are that Δc_1 and Δc_2 are subtracted from the row 0 coefficients of x_1 and x_2, respectively. To convert this tableau to proper form from Gaussian elimination, we subtract 2θ times row 2 from row 0, which yields the last tableau shown. The expressions in terms of θ for the coefficients of nonbasic variables x_1 and x_5 in row 0 of this tableau show that the current BF solution remains optimal for $\theta \leq \frac{9}{8}$. Because $\theta = 1$ is the maximum realistic value of θ, this indicates that c_1 and c_2 together are insensitive parameters with respect to the model of Table 6.20. There is no need to try to estimate these parameters more closely unless other parameters change (as occurred for the Case 3 example).

As we discussed in Sec. 4.7, this way of continuously varying several parameters simultaneously is referred to as *parametric linear programming.* Section 7.2 presents the complete parametric linear programming procedure (including identifying new optimal solutions for larger values of θ) when just the c_j parameters are being varied and then when just the b_i parameters are being varied. Some linear programming software packages also include routines for varying just the coefficients of a single variable or just the parameters of a single constraint. In addition to the other applications discussed in Sec. 4.7, these procedures provide a convenient way of conducting sensitivity analysis systematically.

6.8 Conclusions

Every linear programming problem has associated with it a dual linear programming problem. There are a number of very useful relationships between the original (primal) problem and its dual problem that enhance our ability to analyze the primal problem. For example, the economic interpretation of the dual problem gives shadow prices that measure the marginal value of the resources in the primal problem and provides an interpretation of the simplex method. Because the simplex method can be applied directly to either problem in order to solve both of them simultaneously, considerable computational effort sometimes can be saved by dealing directly with the dual problem. Duality theory, including the dual simplex method for working with superoptimal basic solutions, also plays a major role in sensitivity analysis.

The values used for the parameters of a linear programming model generally are just estimates. Therefore, sensitivity analysis needs to be performed to investigate what happens if these estimates are wrong. The fundamental insight of Sec. 5.3 provides the key to performing this investigation efficiently. The general objectives are to identify the sensitive parameters that affect the optimal solution, to try to estimate these sensitive parameters more closely, and then to select a solution that remains good over the range of likely values of the sensitive parameters. This analysis is a very important part of most linear programming studies.

SELECTED REFERENCES

1. Bazaraa, M. S., J. J. Jarvis, and H. D. Sherali, *Linear Programming and Network Flows,* 2d ed., Wiley, New York, 1990.
2. Bradley, S. P., A. C. Hax, and T. L. Magnanti, *Applied Mathematical Programming,* Addison-Wesley, Reading, MA, 1977.
3. Eppen, G. D., F. J. Gould, and C. Schmidt, *Quantitative Concepts for Management,* 3d ed., Prentice-Hall, Englewood Cliffs, NJ, 1988.
4. Murty, K., *Linear Programming,* 2d ed., Wiley, New York, 1983.

RELEVANT ROUTINES IN YOUR OR COURSEWARE

A demonstration example: *Sensitivity Analysis*

Interactive routines: *Enter or Revise a General Linear Programming Model*
Solve Interactively by the Graphical Method
Sensitivity Analysis

An automatic routine: *Solve Automatically by the Simplex Method*

To access these routines, call the MathProg program and then choose *General Linear Programming* under the Area menu. See Appendix 1 for documentation of the software.

PROBLEMS[1]

To the left of each of the following problems (or their parts), we have inserted a D (for demonstration), I (for interactive routine), or A (for automatic routine) whenever a corresponding routine listed above can be helpful. An asterisk on the I or A indicates that this routine definitely should be used (unless your instructor gives contrary instructions), and the printout from this routine is all that needs to be turned in to show your work in executing the algorithm. An asterisk on the problem number indicates that at least a partial answer is given in the back of the book.

[1] Problems 6.1-6 to 6.1-8, 6.1-15, 6.4-4, and 6.4-10 have been adapted, with permission, from previous operations research examinations given by the Society of Actuaries.

6.1-1. Construct the primal-dual table and the dual problem for each of the following linear programming models fitting our standard form.

(*a*) Model in Prob. 4.1-3
(*b*) Model in Prob. 4.7-8

6.1-2.* Construct the dual problem for each of the following linear programming models fitting our standard form.

(*a*) Model in Prob. 3.1-2
(*b*) Model in Prob. 4.7-6

6.1-3. Consider the linear programming model in Prob. 4.5-2.

(*a*) Construct the primal-dual table and the dual problem for this model.
(*b*) What does the fact that Z is unbounded for this model imply about its dual problem?

6.1-4. For each of the following linear programming models, give your recommendation on which is the more efficient way (probably) to obtain an optimal solution: by applying the simplex method directly to this primal problem or by applying the simplex method directly to the dual problem instead. Explain.

(*a*) Maximize

$$Z = 10x_1 - 4x_2 + 7x_3,$$

subject to

$$3x_1 - x_2 + 2x_3 \leq 25$$
$$x_1 - 2x_2 + 3x_3 \leq 25$$
$$5x_1 + x_2 + 2x_3 \leq 40$$
$$x_1 + x_2 + x_3 \leq 90$$
$$2x_1 - x_2 + x_3 \leq 20$$

and

$$x_1 \geq 0, \qquad x_2 \geq 0, \qquad x_3 \geq 0.$$

(*b*) Maximize

$$Z = 2x_1 + 5x_2 + 3x_3 + 4x_4 + x_5,$$

subject to

$$x_1 + 3x_2 + 2x_3 + 3x_4 + x_5 \leq 6$$
$$4x_1 + 6x_2 + 5x_3 + 7x_4 + x_5 \leq 15$$

and

$$x_j \geq 0, \qquad \text{for } j = 1, 2, 3, 4, 5.$$

6.1-5. Consider the following problem.

$$\text{Maximize} \qquad Z = -x_1 - 2x_2 - x_3,$$

subject to

$$x_1 + x_2 + 2x_3 \leq 12$$
$$x_1 + x_2 - x_3 \leq 1$$

and

$$x_1 \geq 0, \qquad x_2 \geq 0, \qquad x_3 \geq 0.$$

(*a*) Construct the dual problem.
(*b*) Use duality theory to show that the optimal solution for the primal problem has $Z \leq 0$.

6.1-6. Consider the following problem.

$$\text{Maximize} \quad Z = 2x_1 + 6x_2 + 9x_3,$$

subject to

$$x_1 + x_3 \leq 3 \quad \text{(resource 1)}$$
$$x_2 + 2x_3 \leq 5 \quad \text{(resource 2)}$$

and

$$x_1 \geq 0, \quad x_2 \geq 0, \quad x_3 \geq 0.$$

(*a*) Construct the dual problem for this primal problem.

I (*b*) Solve the dual problem graphically. Use this solution to identify the shadow prices for the resources in the primal problem.

A* (*c*) Confirm your results from part (*b*) by solving the primal problem automatically by the simplex method and then identifying the shadow prices.

6.1-7. Follow the instructions of Prob. 6.1-6 for the following problem.

$$\text{Maximize} \quad Z = x_1 - 3x_2 + 2x_3,$$

subject to

$$2x_1 + 2x_2 - 2x_3 \leq 6 \quad \text{(resource 1)}$$
$$- x_2 + 2x_3 \leq 4 \quad \text{(resource 2)}$$

and

$$x_1 \geq 0, \quad x_2 \geq 0, \quad x_3 \geq 0.$$

6.1-8. Consider the following problem.

$$\text{Maximize} \quad Z = x_1 + 2x_2,$$

subject to

$$-x_1 + x_2 \leq -2$$
$$4x_1 + x_2 \leq 4$$

and

$$x_1 \geq 0, \quad x_2 \geq 0.$$

I (*a*) Demonstrate graphically that this problem has no feasible solutions.

(*b*) Construct the dual problem.

I (*c*) Demonstrate graphically that the dual problem has an unbounded objective function.

6.1-9. Construct and graph a primal problem with two decision variables and two functional constraints that has feasible solutions and an unbounded objective function. Then construct the dual problem and demonstrate graphically that it has no feasible solutions.

6.1-10. Construct a pair of primal and dual problems, each with two decision variables and two functional constraints, such that both problems have no feasible solutions. Demonstrate this property graphically.

6.1-11. Construct a pair of primal and dual problems, each with two decision variables and two functional constraints, such that the primal problem has no feasible solutions and the dual problem has an unbounded objective function.

6.1-12. Use the weak duality property to prove that if both the primal and the dual problem have feasible solutions, then both must have an optimal solution.

6.1-13. Consider the primal and dual problems in our standard form presented in matrix notation at the beginning of Sec. 6.1. Use only this definition of the dual problem for a primal problem in this form to prove each of the following results.

(*a*) The weak duality property presented in Sec. 6.1.
(*b*) If the primal problem has an unbounded feasible region that permits increasing Z indefinitely, then the dual problem has no feasible solutions.

6.1-14. Consider the primal and dual problems in our standard form presented in matrix notation at the beginning of Sec. 6.1. Let $\mathbf{y}^*$ denote the optimal solution for this dual problem. Suppose that $\mathbf{b}$ is then replaced by $\overline{\mathbf{b}}$. Let $\overline{\mathbf{x}}$ denote the optimal solution for the new primal problem. Prove that

$$\mathbf{c}\overline{\mathbf{x}} \leq \mathbf{y}^*\overline{\mathbf{b}}.$$

6.1-15. For any linear programming problem in our standard form and its dual problem, label each of the following statements as true or false and then justify your answer.
(*a*) The sum of the number of functional constraints and the number of variables (before augmenting) is the same for both the primal and the dual problems.
(*b*) At each iteration, the simplex method simultaneously identifies a CPF solution for the primal problem and a CPF solution for the dual problem such that their objective function values are the same.
(*c*) If the primal problem has an unbounded objective function, then the optimal value of the objective function for the dual problem must be zero.

6.2-1. Consider the simplex tableaux for the Wyndor Glass Co. problem given in Table 4.8. For each tableau, give the economic interpretation of the following items:
(*a*) Each of the coefficients of the slack variables (x_3, x_4, x_5) in row 0
(*b*) Each of the coefficients of the decision variables (x_1, x_2) in row 0
(*c*) The resulting choice for the entering basic variable (or the decision to stop after the final tableau)

6.3-1.* Consider the following problem.

$$\text{Maximize} \quad Z = 6x_1 + 8x_2,$$

subject to

$$5x_1 + 2x_2 \leq 20$$
$$x_1 + 2x_2 \leq 10$$

and

$$x_1 \geq 0, \quad x_2 \geq 0.$$

(*a*) Construct the dual problem for this primal problem.
(*b*) Solve both the primal problem and the dual problem graphically. Identify the CPF solutions and corner-point infeasible solutions for both problems. Calculate the objective function values for all these solutions.
(*c*) Use the information obtained in part (*b*) to construct a table listing the complementary basic solutions for these problems. (Use the same column headings as for Table 6.9.)
I* (*d*) Solve the primal problem by the simplex method. After each iteration (including iteration 0), identify the BF solution for this problem and the complementary basic solution for the dual problem. Also identify the corresponding corner-point solutions.

6.3-2. Consider the model with two functional constraints and two variables given in Prob. 4.1-2. Follow the instructions of Prob. 6.3-1 for this model.

6.3-3. Consider the primal and dual problems for the Wyndor Glass Co. example given in Table 6.1. Using Tables 5.5, 5.6, 6.8, and 6.9, construct a new table showing the eight sets of nonbasic variables for the primal problem in column 1, the corresponding sets of associated variables for the dual problem in column 2, and the set of nonbasic variables for each complementary basic solution of the dual problem in column 3. Explain why this table demonstrates the complementary slackness property for this example.

6.3-4. Suppose that a primal problem has a *degenerate* BF solution (one or more basic variables equal to zero) as its optimal solution. What does this degeneracy imply about the dual problem? Why? Is the converse also true?

6.3-5. Consider the following problem.

$$\text{Maximize} \quad Z = 2x_1 - 4x_2,$$

subject to

$$x_1 - x_2 \leq 1$$

and

$$x_1 \geq 0, \quad x_2 \geq 0.$$

(*a*) Construct the dual problem, and then find its optimal solution by inspection.
(*b*) Use the complementary slackness property and the optimal solution for the dual problem to find the optimal solution for the primal problem.
(*c*) Suppose that c_1, the coefficient of x_1 in the primal objective function, actually can have any value in the model. For what values of c_1 does the dual problem have no feasible solutions? For these values, what does duality theory then imply about the primal problem?

6.3-6. Consider the following problem.

$$\text{Maximize} \quad Z = 2x_1 + 7x_2 + 4x_3,$$

subject to

$$x_1 + 2x_2 + x_3 \leq 10$$
$$3x_1 + 3x_2 + 2x_3 \leq 10$$

and

$$x_1 \geq 0, \quad x_2 \geq 0, \quad x_3 \geq 0.$$

(*a*) Construct the dual problem for this primal problem.
(*b*) Use the dual problem to demonstrate that the optimal value of Z for the primal problem cannot exceed 25.
(*c*) It has been conjectured that x_2 and x_3 should be the basic variables for the optimal solution of the primal problem. Directly derive this basic solution (and Z) by using Gaussian elimination. Simultaneously derive and identify the complementary basic solution for the dual problem by using Eq. (0) for the primal problem. Then draw your conclusions about whether these two basic solutions are optimal for their respective problems.
I (*d*) Solve the dual problem graphically. Use this solution to identify the basic variables and the nonbasic variables for the optimal solution of the primal problem. Directly derive this solution, using Gaussian elimination.

6.3-7.* Reconsider the model of Prob. 6.1-4*b*.
(*a*) Construct its dual problem.
I (*b*) Solve this dual problem graphically.
(*c*) Use the result from part (*b*) to identify the nonbasic variables and basic variables for the optimal BF solution for the primal problem.
(*d*) Use the results from part (*c*) to obtain the optimal solution for the primal problem directly by using Gaussian elimination to solve for its basic variables, starting from the initial system of equations [excluding Eq. (0)] constructed for the simplex method and setting the nonbasic variables to zero.
(*e*) Use the results from part (*c*) to identify the defining equations (see Sec. 5.1) for the optimal CPF solution for the primal problem, and then use these equations to find this solution.

6.3-8. Consider the model given in Prob. 5.3-13.

(*a*) Construct the dual problem.
(*b*) Use the given information about the basic variables in the optimal primal solution to identify the nonbasic variables and basic variables for the optimal dual solution.
(*c*) Use the results from part (*b*) to identify the defining equations (see Sec. 5.1) for the optimal CPF solution for the dual problem, and then use these equations to find this solution.
I (*d*) Solve the dual problem graphically to verify your results from part (*c*).

6.3-9. Consider the model given in Prob. 3.1-1.

(*a*) Construct the dual problem for this model.
(*b*) Use the fact that $(x_1, x_2) = (13, 5)$ is optimal for the primal problem to identify the nonbasic variables and basic variables for the optimal BF solution for the dual problem.
(*c*) Identify this optimal solution for the dual problem by directly deriving Eq. (0) corresponding to the optimal primal solution identified in part (*b*). Derive this equation by using Gaussian elimination.
(*d*) Use the results from part (*b*) to identify the defining equations (see Sec. 5.1) for the optimal CPF solution for the dual problem. Verify your optimal dual solution from part (*c*) by checking to see that it satisfies this system of equations.

6.3-10. Suppose that you also want information about the dual problem when you apply the revised simplex method (see Sec. 5.2) to the primal problem in our standard form.

(*a*) How would you identify the optimal solution for the dual problem?
(*b*) After obtaining the BF solution at each iteration, how would you identify the complementary basic solution in the dual problem?

6.4-1. Consider the following problem.

$$\text{Maximize} \quad Z = x_1 + x_2,$$

subject to

$$x_1 + 2x_2 = 10$$
$$2x_1 + x_2 \geq 2$$

and

$$x_2 \geq 0 \quad (x_1 \text{ unconstrained in sign}).$$

(*a*) Construct the dual problem in the form indicated by Table 6.14.
(*b*) Use Table 6.12 to convert the primal problem to our standard form given at the beginning of Sec. 6.1, and construct the corresponding dual problem. Then show that this dual problem is equivalent to the one obtained in part (*a*).

6.4-2. Consider the primal and dual problems in our standard form presented in matrix notation at the beginning of Sec. 6.1. Use only this definition of the dual problem for a primal problem in this form to prove each of the following results.

(*a*) If the functional constraints for the primal problem $\mathbf{Ax} \leq \mathbf{b}$ are changed to $\mathbf{Ax} = \mathbf{b}$, the only resulting change in the dual problem is to *delete* the nonnegativity constraints, $\mathbf{y} \geq \mathbf{0}$. (*Hint:* The constraints $\mathbf{Ax} = \mathbf{b}$ are equivalent to the set of constraints $\mathbf{Ax} \leq \mathbf{b}$ *and* $\mathbf{Ax} \geq \mathbf{b}$.)
(*b*) If the functional constraints for the primal problem $\mathbf{Ax} \leq \mathbf{b}$ are changed to $\mathbf{Ax} \geq \mathbf{b}$, the only resulting change in the dual problem is that the nonnegativity constraints $\mathbf{y} \geq \mathbf{0}$ are replaced by nonpositivity constraints $\mathbf{y} \leq \mathbf{0}$, where the current dual variables are interpreted as the negative of the original dual variables. (*Hint:* The constraints $\mathbf{Ax} \geq \mathbf{b}$ are equivalent to $-\mathbf{Ax} \leq -\mathbf{b}$.)
(*c*) If the nonnegativity constraints for the primal problem $\mathbf{x} \geq \mathbf{0}$ are deleted, the only resulting change in the dual problem is to replace the functional constraints $\mathbf{yA} \geq \mathbf{c}$ by $\mathbf{yA} = \mathbf{c}$. (*Hint:* A variable unconstrained in sign can be replaced by the difference of two nonnegative variables.)

6.4-3.* Construct the dual problem for the linear programming problem given in Prob. 4.6-4.

6.4-4. Consider the following problem.

$$\text{Minimize} \qquad Z = x_1 + 2x_2,$$

subject to

$$-2x_1 + x_2 \geq 1$$
$$x_1 - 2x_2 \geq 1$$

and

$$x_1 \geq 0, \qquad x_2 \geq 0.$$

(*a*) Construct the dual problem.
(*b*) Use graphical analysis of the dual problem to determine whether the primal problem has feasible solutions and, if so, whether its objective function is bounded.

6.4-5. Consider the two versions of the dual problem for the radiation therapy example that are given in Tables 6.15 and 6.16. Review in Sec. 6.4 the general discussion of why these two versions are completely equivalent. Then fill in the details to verify this equivalency by proceeding step by step to convert the version in Table 6.15 to equivalent forms until the version in Table 6.16 is obtained.

6.4-6. For each of the following linear programming models, convert this primal problem to one of the two forms given in Table 6.14 and then construct its dual problem.
(*a*) Model in Prob. 4.6-3
(*b*) Model in Prob. 4.6-6
(*c*) Model in Prob. 4.6-16

6.4-7. Consider the model with equality constraints given in Prob. 4.6-2.
(*a*) Construct its dual problem.
(*b*) Demonstrate that the answer in part (*a*) is correct (i.e., equality constraints yield dual variables without nonnegativity constraints) by first converting the primal problem to our standard form (see Table 6.12), then constructing its dual problem, and next converting this dual problem to the form obtained in part (*a*).

6.4-8.* Consider the model without nonnegativity constraints given in Prob. 4.6-14.
(*a*) Construct its dual problem.
(*b*) Demonstrate that the answer in part (*a*) is correct (i.e., variables without nonnegativity constraints yield equality constraints in the dual problem) by first converting the primal problem to our standard form (see Table 6.12), then constructing its dual problem, and finally converting this dual problem to the form obtained in part (*a*).

6.4-9. Consider the dual problem for the Wyndor Glass Co. example given in Table 6.1. Demonstrate that *its* dual problem is the primal problem given in Table 6.1 by going through the conversion steps given in Table 6.13.

6.4-10. Consider the following problem.

$$\text{Minimize} \qquad -x_1 - 3x_2,$$

subject to

$$x_1 - 2x_2 \leq 2$$
$$-x_1 + x_2 \leq 4$$

and

$$x_1 \geq 0, \qquad x_2 \geq 0.$$

I (*a*) Demonstrate graphically that this problem has an unbounded objective function.
(*b*) Construct the dual problem.
I (*c*) Demonstrate graphically that the dual problem has no feasible solutions.

6.5-1. Consider the model of Prob. 6.7-1. Use duality theory directly to determine whether the current basic solution remains optimal after each of the following independent changes.

(*a*) The change in part (*e*) of Prob. 6.7-1
(*b*) The change in part (*g*) of Prob. 6.7-1

6.5-2. Consider the model of Prob. 6.7-3. Use duality theory directly to determine whether the current basic solution remains optimal after each of the following independent changes.

(*a*) The change in part (*c*) of Prob. 6.7-3
(*b*) The change in part (*f*) of Prob. 6.7-3

6.5-3. Consider the model of Prob. 6.7-4. Use duality theory directly to determine whether the current basic solution remains optimal after each of the following independent changes.

(*a*) The change in part (*b*) of Prob. 6.7-4
(*b*) The change in part (*d*) of Prob. 6.7-4

6.5-4. Reconsider part (*d*) of Prob. 6.7-6. Use duality theory directly to determine whether the original optimal solution is still optimal.

6.6-1.* Consider the following problem.

$$\text{Maximize} \quad Z = 3x_1 + x_2 + 4x_3,$$

subject to

$$6x_1 + 3x_2 + 5x_3 \le 25$$
$$3x_1 + 4x_2 + 5x_3 \le 20$$

and

$$x_1 \ge 0, \quad x_2 \ge 0, \quad x_3 \ge 0.$$

The corresponding final set of equations yielding the optimal solution is

$$(0) \qquad Z + 2x_2 + \tfrac{1}{5}x_4 + \tfrac{3}{5}x_5 = 17,$$
$$(1) \qquad x_1 - \tfrac{1}{3}x_2 + \tfrac{1}{3}x_4 - \tfrac{1}{3}x_5 = \tfrac{5}{3},$$
$$(2) \qquad x_2 + x_3 - \tfrac{1}{5}x_4 + \tfrac{2}{5}x_5 = 3.$$

(*a*) Identify the optimal solution from this set of equations.
(*b*) Construct the dual problem.
(*c*) Identify the optimal solution for the dual problem from the final set of equations. Verify this solution by solving the dual problem graphically.
(*d*) Suppose that the original problem is changed to

$$\text{Maximize} \quad Z = 3x_1 + 3x_2 + 4x_3,$$

subject to

$$6x_1 + 2x_2 + 5x_3 \le 25$$
$$3x_1 + 3x_2 + 5x_3 \le 20$$

and

$$x_1 \ge 0, \quad x_2 \ge 0, \quad x_3 \ge 0.$$

Use duality theory to determine whether the previous optimal solution is still optimal.
(*e*) Use the fundamental insight presented in Sec. 5.3 to identify the new coefficients of x_2 in the final set of equations after it has been adjusted for the changes in the original problem given in part (*d*).

(f) Now suppose that the only change in the original problem is that a new variable x_{new} has been introduced into the model as follows:

$$\text{Maximize} \quad Z = 3x_1 + x_2 + 4x_3 + 2x_{new},$$

subject to

$$6x_1 + 3x_2 + 5x_3 + 3x_{new} \leq 25$$
$$3x_1 + 4x_2 + 5x_3 + 2x_{new} \leq 20$$

and

$$x_1 \geq 0, \quad x_2 \geq 0, \quad x_3 \geq 0, \quad x_{new} \geq 0.$$

Use duality theory to determine whether the previous optimal solution, along with $x_{new} = 0$, is still optimal.

(g) Use the fundamental insight presented in Sec. 5.3 to identify the coefficients of x_{new} as a nonbasic variable in the final set of equations resulting from the introduction of x_{new} into the original model as shown in part (f).

D, I* **6.6-2.** Reconsider the model of Prob. 6.6-1. You are now to conduct sensitivity analysis by *independently* investigating each of the following six changes in the original model. For each change, use the sensitivity analysis procedure to revise the given final set of equations (in tableau form) and convert it to proper form from Gaussian elimination. Then test this solution for feasibility and for optimality. (Do not reoptimize.)

(a) Change the right-hand side of constraint 1 to $b_1 = 15$.
(b) Change the right-hand side of constraint 2 to $b_2 = 5$.
(c) Change the coefficient of x_2 in the objective function to $c_2 = 4$.
(d) Change the coefficient of x_3 in the objective function to $c_3 = 3$.
(e) Change the coefficient of x_2 in constraint 2 to $a_{22} = 1$.
(f) Change the coefficient of x_1 in constraint 1 to $a_{11} = 10$.

D, I* **6.6-3.** Consider the following problem.

$$\text{Minimize} \quad y_0 = 5y_1 + 4y_2,$$

subject to

$$4y_1 + 3y_2 \geq 4$$
$$2y_1 + y_2 \geq 3$$
$$y_1 + 2y_2 \geq 1$$
$$y_1 + y_2 \geq 2$$

and

$$y_1 \geq 0, \quad y_2 \geq 0.$$

Because this primal problem has more functional constraints than variables, suppose that the simplex method has been applied directly to its dual problem. If we let x_5 and x_6 denote the slack variables for this dual problem, the resulting final simplex tableau is

Basic Variable	Eq.	*Coefficient of:*							Right Side
		Z	x_1	x_2	x_3	x_4	x_5	x_6	
Z	(0)	1	3	0	2	0	1	1	9
x_2	(1)	0	1	1	1	0	1	−1	1
x_4	(2)	0	2	0	3	1	−1	2	3

For each of the following independent changes in the original primal model, you now are to conduct sensitivity analysis by directly investigating the effect on the dual problem and then inferring the complementary effect on the primal problem. For each change, apply the procedure for sensitivity analysis summarized at the end of Sec. 6.6 to the dual problem (do *not* reoptimize), and then give your conclusions as to whether the current basic solution for the primal problem still is feasible and whether it still is optimal. Then check your conclusions by a direct graphical analysis of the primal problem.

(*a*) Change the objective function to $y_0 = 3y_1 + 5y_2$.

(*b*) Change the right-hand sides of the functional constraints to 3, 5, 2, and 3, respectively.

(*c*) Change the first constraint to $2y_1 + 4y_2 \geq 7$.

(*d*) Change the second constraint to $5y_1 + 2y_2 \geq 10$.

D, I* **6.7-1.*** Consider the following problem.

$$\text{Maximize} \quad Z = -5x_1 + 5x_2 + 13x_3,$$

subject to

$$-x_1 + x_2 + 3x_3 \leq 20$$
$$12x_1 + 4x_2 + 10x_3 \leq 90$$

and

$$x_j \geq 0 \quad (j = 1, 2, 3).$$

If we let x_4 and x_5 be the slack variables for the respective constraints, the simplex method yields the following final set of equations:

$$(0) \quad Z \qquad + 2x_3 + 5x_4 \qquad = 100,$$
$$(1) \quad -x_1 + x_2 + 3x_3 + x_4 \qquad = 20,$$
$$(2) \quad 16x_1 \qquad - 2x_3 - 4x_4 + x_5 = 10.$$

Now you are to conduct sensitivity analysis by *independently* investigating each of the following nine changes in the original model. For each change, use the sensitivity analysis procedure to revise this set of equations (in tableau form) and convert it to proper form from Gaussian elimination for identifying and evaluating the current basic solution. Then test this solution for feasibility and for optimality. (Do not reoptimize.)

(*a*) Change the right-hand side of constraint 1 to

$$b_1 = 30.$$

(*b*) Change the right-hand side of constraint 2 to

$$b_2 = 70.$$

(*c*) Change the right-hand sides to

$$\begin{bmatrix} b_1 \\ b_2 \end{bmatrix} = \begin{bmatrix} 10 \\ 100 \end{bmatrix}.$$

(*d*) Change the coefficient of x_3 in the objective function to

$$c_3 = 8.$$

(*e*) Change the coefficients of x_1 to

$$\begin{bmatrix} c_1 \\ a_{11} \\ a_{21} \end{bmatrix} = \begin{bmatrix} -2 \\ 0 \\ 5 \end{bmatrix}.$$

(*f*) Change the coefficients of x_2 to

$$\begin{bmatrix} c_2 \\ a_{12} \\ a_{22} \end{bmatrix} = \begin{bmatrix} 6 \\ 2 \\ 5 \end{bmatrix}.$$

(*g*) Introduce a new variable x_6 with coefficients

$$\begin{bmatrix} c_6 \\ a_{16} \\ a_{26} \end{bmatrix} = \begin{bmatrix} 10 \\ 3 \\ 5 \end{bmatrix}.$$

(*h*) Introduce a new constraint $2x_1 + 3x_2 + 5x_3 \leq 50$. (Denote its slack variable by x_6.)

(*i*) Change constraint 2 to

$$10x_1 + 5x_2 + 10x_3 \leq 100.$$

6.7-2.* Reconsider the model of Prob. 6.7-1. Suppose that we now want to apply parametric linear programming analysis to this problem. Specifically, the right-hand sides of the functional constraints are changed to

$$20 + 2\theta \qquad \text{(for constraint 1)}$$

and

$$90 - \theta \qquad \text{(for constraint 2)}$$

where θ can be assigned any positive or negative values.

Express the basic solution (and Z) corresponding to the original optimal solution as a function of θ. Determine the lower and upper bounds on θ before this solution would become infeasible.

D, I* **6.7-3.** Consider the following problem.

$$\text{Maximize} \qquad Z = 2x_1 - x_2 + x_3,$$

subject to

$$3x_1 + x_2 + x_3 \leq 60$$
$$x_1 - x_2 + 2x_3 \leq 10$$
$$x_1 + x_2 - x_3 \leq 20$$

and

$$x_1 \geq 0, \qquad x_2 \geq 0, \qquad x_3 \geq 0.$$

Let x_4, x_5, and x_6 denote the slack variables for the respective constraints. After we apply the simplex method, the final simplex tableau is

Basic Variable	Eq.	*Coefficient of:*							Right Side
		Z	x_1	x_2	x_3	x_4	x_5	x_6	
Z	(0)	1	0	0	$\frac{3}{2}$	0	$\frac{3}{2}$	$\frac{1}{2}$	25
x_4	(1)	0	0	0	1	1	-1	-2	10
x_1	(2)	0	1	0	$\frac{1}{2}$	0	$\frac{1}{2}$	$\frac{1}{2}$	15
x_2	(3)	0	0	1	$-\frac{3}{2}$	0	$-\frac{1}{2}$	$\frac{1}{2}$	5

Now you are to conduct sensitivity analysis by *independently* investigating each of the following six changes in the original model. For each change, use the sensitivity analysis procedure to revise this final tableau and convert it to proper form from Gaussian elimination for identifying and evaluating the current basic solution. Then test this solution for feasibility and for optimality. If either test fails, reoptimize to find a new optimal solution.

(*a*) Change the right-hand sides

$$\text{from} \quad \begin{bmatrix} b_1 \\ b_2 \\ b_3 \end{bmatrix} = \begin{bmatrix} 60 \\ 10 \\ 20 \end{bmatrix} \quad \text{to} \quad \begin{bmatrix} b_1 \\ b_2 \\ b_3 \end{bmatrix} = \begin{bmatrix} 70 \\ 20 \\ 10 \end{bmatrix}.$$

(*b*) Change the coefficients of x_1

$$\text{from} \quad \begin{bmatrix} c_1 \\ a_{11} \\ a_{21} \\ a_{31} \end{bmatrix} = \begin{bmatrix} 2 \\ 3 \\ 1 \\ 1 \end{bmatrix} \quad \text{to} \quad \begin{bmatrix} c_1 \\ a_{11} \\ a_{21} \\ a_{31} \end{bmatrix} = \begin{bmatrix} 1 \\ 2 \\ 2 \\ 0 \end{bmatrix}.$$

(*c*) Change the coefficients of x_3

$$\text{from} \quad \begin{bmatrix} c_3 \\ a_{13} \\ a_{23} \\ a_{33} \end{bmatrix} = \begin{bmatrix} 1 \\ 1 \\ 2 \\ -1 \end{bmatrix} \quad \text{to} \quad \begin{bmatrix} c_3 \\ a_{13} \\ a_{23} \\ a_{33} \end{bmatrix} = \begin{bmatrix} 2 \\ 3 \\ 1 \\ -2 \end{bmatrix}.$$

(*d*) Change the objective function to $Z = 3x_1 - 2x_2 + 3x_3$.

(*e*) Introduce a new constraint $3x_1 - 2x_2 + x_3 \leq 30$. (Denote its slack variable by x_7.)

(*f*) Introduce a new variable x_8 with coefficients

$$\begin{bmatrix} c_8 \\ a_{18} \\ a_{28} \\ a_{38} \end{bmatrix} = \begin{bmatrix} -1 \\ -2 \\ 1 \\ 2 \end{bmatrix}.$$

D, I* **6.7-4.** Consider the following problem.

$$\text{Maximize} \quad Z = 2x_1 + 7x_2 - 3x_3,$$

subject to

$$x_1 + 3x_2 + 4x_3 \leq 30$$
$$x_1 + 4x_2 - x_3 \leq 10$$

and

$$x_1 \geq 0, \quad x_2 \geq 0, \quad x_3 \geq 0.$$

By letting x_4 and x_5 be the slack variables for the respective constraints, the simplex method yields the following final set of equations:

(0) $\quad Z + x_2 + x_3 + 2x_5 = 20,$

(1) $\quad - x_2 + 5x_3 + x_4 - x_5 = 20,$

(2) $\quad x_1 + 4x_2 - x_3 + x_5 = 10.$

Now you are to conduct sensitivity analysis by *independently* investigating each of the following seven changes in the original model. For each change, use the sensitivity analysis procedure to revise this set of equations (in tableau form) and convert it to proper form from Gaussian elimination for identifying and evaluating the current basic solution. Then test this solution for feasibility and for optimality. If either test fails, reoptimize to find a new optimal solution.

(*a*) Change the right-hand sides to

$$\begin{bmatrix} b_1 \\ b_2 \end{bmatrix} = \begin{bmatrix} 20 \\ 30 \end{bmatrix}.$$

(*b*) Change the coefficients of x_3 to

$$\begin{bmatrix} c_3 \\ a_{13} \\ a_{23} \end{bmatrix} = \begin{bmatrix} -2 \\ 3 \\ -2 \end{bmatrix}.$$

(*c*) Change the coefficients of x_1 to

$$\begin{bmatrix} c_1 \\ a_{11} \\ a_{21} \end{bmatrix} = \begin{bmatrix} 4 \\ 3 \\ 2 \end{bmatrix}.$$

(*d*) Introduce a new variable x_6 with coefficients

$$\begin{bmatrix} c_6 \\ a_{16} \\ a_{26} \end{bmatrix} = \begin{bmatrix} -3 \\ 1 \\ 2 \end{bmatrix}.$$

(*e*) Change the objective function to $Z = x_1 + 5x_2 - 2x_3$.
(*f*) Introduce a new constraint $3x_1 + 2x_2 + 3x_3 \leq 25$.
(*g*) Change constraint 2 to $x_1 + 2x_2 + 2x_3 \leq 35$.

6.7-5. Reconsider the model of Prob. 6.7-4. Suppose that we now want to apply parametric linear programming analysis to this problem. Specifically, the right-hand sides of the functional constraints are changed to

$$30 + 3\theta \qquad \text{(for constraint 1)}$$

and

$$10 - \theta \qquad \text{(for constraint 2)},$$

where θ can be assigned any positive or negative values.

Express the basic solution (and Z) corresponding to the original optimal solution as a function of θ. Determine the lower and upper bounds on θ before this solution would become infeasible.

D, I* **6.7-6.** Consider the following problem.

$$\text{Maximize} \qquad Z = 2x_1 - x_2 + x_3,$$

subject to

$$\begin{aligned} 3x_1 - 2x_2 + 2x_3 &\leq 15 \\ -x_1 + x_2 + x_3 &\leq 3 \\ x_1 - x_2 + x_3 &\leq 4 \end{aligned}$$

and

$$x_1 \geq 0, \qquad x_2 \geq 0, \qquad x_3 \geq 0.$$

If we let x_4, x_5, and x_6 be the slack variables for the respective constraints, the simplex method yields the following final set of equations:

$$\begin{aligned} &(0) \qquad Z + 2x_3 + x_4 + x_5 = 18, \\ &(1) \qquad x_2 + 5x_3 + x_4 + 3x_5 = 24, \\ &(2) \qquad 2x_3 + x_5 + x_6 = 7, \\ &(3) \qquad x_1 + 4x_3 + x_4 + 2x_5 = 21. \end{aligned}$$

Now you are to conduct sensitivity analysis by *independently* investigating each of the following eight changes in the original model. For each change, use the sensitivity analysis procedure to revise this set of equations (in tableau form) and convert it to proper form from Gaussian elimination for identifying and evaluating the current basic solution. Then test this solution for feasibility and for optimality. If either test fails, reoptimize to find a new optimal solution.

(*a*) Change the right-hand sides to

$$\begin{bmatrix} b_1 \\ b_2 \\ b_3 \end{bmatrix} = \begin{bmatrix} 10 \\ 4 \\ 2 \end{bmatrix}.$$

(*b*) Change the coefficient of x_3 in the objective function to $c_3 = 2$.
(*c*) Change the coefficient of x_1 in the objective function to $c_1 = 3$.
(*d*) Change the coefficients of x_3 to

$$\begin{bmatrix} c_3 \\ a_{13} \\ a_{23} \\ a_{33} \end{bmatrix} = \begin{bmatrix} 4 \\ 3 \\ 2 \\ 1 \end{bmatrix}.$$

(*e*) Change the coefficients of x_1 and x_2 to

$$\begin{bmatrix} c_1 \\ a_{11} \\ a_{21} \\ a_{31} \end{bmatrix} = \begin{bmatrix} 1 \\ 1 \\ -2 \\ 3 \end{bmatrix} \quad \text{and} \quad \begin{bmatrix} c_2 \\ a_{12} \\ a_{22} \\ a_{32} \end{bmatrix} = \begin{bmatrix} -2 \\ -2 \\ 3 \\ 2 \end{bmatrix},$$

respectively.
(*f*) Change the objective function to $Z = 5x_1 + x_2 + 3x_3$.
(*g*) Change constraint 1 to $2x_1 - x_2 + 4x_3 \leq 12$.
(*h*) Introduce a new constraint $2x_1 + x_2 + 3x_3 \leq 60$.

6.7-7. Consider the following problem.

$$\text{Maximize} \quad Z = c_1x_1 + c_2x_2,$$

subject to

$$2x_1 - x_2 \leq b_1$$
$$x_1 - x_2 \leq b_2$$

and

$$x_1 \geq 0, \quad x_2 \geq 0.$$

Let x_3 and x_4 denote the slack variables for the respective functional constraints. When $c_1 = 3$, $c_2 = -2$, $b_1 = 30$, and $b_2 = 10$, the simplex method yields the following final simplex tableau.

Basic Variable	Eq.	*Coefficient of:*					Right Side
		Z	x_1	x_2	x_3	x_4	
Z	(0)	1	0	0	1	1	40
x_2	(1)	0	0	1	1	−2	10
x_1	(2)	0	1	0	1	−1	20

(*a*) Use graphical analysis to determine the allowable range to stay optimal for c_1 and c_2.
(*b*) Use algebraic analysis to derive and verify your answers in part (*a*).

(*c*) Use graphical analysis to determine the allowable range to stay feasible for b_1 and b_2.
(*d*) Use algebraic analysis to derive and verify your answers in part (*c*).

6.7-8. Consider the Case 3 example (see Fig. 6.4 and Table 6.21), where the changes in the parameter values given in Table 6.20 are $\bar{c}_2 = 3$, $\bar{a}_{22} = 3$, and $\bar{a}_{32} = 4$. Use the formula $\mathbf{b}^* = \mathbf{S}^*\bar{\mathbf{b}}$ to find the allowable range to stay feasible for each b_i. Then interpret each allowable range graphically.

6.7-9. Consider the Case 3 example (see Fig. 6.4 and Table 6.21), where the changes in the parameter values given in Table 6.20 are $\bar{c}_2 = 3$, $\bar{a}_{22} = 3$, and $\bar{a}_{32} = 4$. Verify both algebraically and graphically that the allowable range to stay optimal for c_1 is $c_1 \geq \frac{9}{4}$.

6.7-10. Consider the following problem.

$$\text{Maximize} \quad Z = 3x_1 + x_2 + 2x_3,$$

subject to

$$x_1 - x_2 + 2x_3 \leq 20$$
$$2x_1 + x_2 - x_3 \leq 10$$

and

$$x_1 \geq 0, \quad x_2 \geq 0, \quad x_3 \geq 0.$$

Let x_4 and x_5 denote the slack variables for the respective functional constraints. After we apply the simplex method, the final simplex tableau is

Basic Variable	Eq.	Coefficient of:						Right Side
		Z	x_1	x_2	x_3	x_4	x_5	
Z	(0)	1	8	0	0	3	4	100
x_3	(1)	0	3	0	1	1	1	30
x_2	(2)	0	5	1	0	1	2	40

(*a*) Perform sensitivity analysis to determine which of the 11 parameters of the model are sensitive parameters in the sense that *any* change in just that parameter's value will change the optimal solution.
(*b*) Find the allowable range to stay optimal for each c_j.
(*c*) Find the allowable range to stay feasible for each b_i.

6.7-11. For the problem given in Table 6.20, find the allowable range to stay optimal for c_2. Show your work algebraically, using the tableau given in Table 6.20. Then justify your answer from a geometric viewpoint, referring to Fig. 6.3.

6.7-12.* For the original Wyndor Glass Co. problem, use the last tableau in Table 4.8 to do the following.
(*a*) Find the allowable range to stay feasible for each b_i.
(*b*) Find the allowable range to stay optimal for c_1 and c_2.

6.7-13. For the Case 4 example presented in Sec. 6.7, use the last tableau in Table 6.22 to do the following.
(*a*) Find the allowable range to stay feasible for each b_i.
(*b*) Find the allowable range to stay optimal for c_1 and c_2.

6.7-14. Consider the following problem.

$$\text{Maximize} \quad Z = 2x_1 + 5x_2,$$

subject to

$$x_1 + 2x_2 \leq 10$$

$$x_1 + 3x_2 \leq 12$$

and

$$x_1 \geq 0, \qquad x_2 \geq 0.$$

Let x_3 and x_4 denote the slack variables for the respective functional constraints. After we apply the simplex method, the final simplex tableau is

Basic Variable	Eq.	*Coefficient of:* Z	x_1	x_2	x_3	x_4	Right Side
Z	(0)	1	0	0	1	1	22
x_1	(1)	0	1	0	3	−2	6
x_2	(2)	0	0	1	−1	1	2

While doing post-optimality analysis, you learn that all four b_i and c_j values used in the original model just given are accurate only to within ±50 percent. In other words, their ranges of *likely values* are $5 \leq b_1 \leq 15$, $6 \leq b_2 \leq 18$, $1 \leq c_1 \leq 3$, and $2.5 \leq c_2 \leq 7.5$. Your job now is to perform sensitivity analysis to determine for each parameter whether this uncertainty might affect either the feasibility or the optimality of the above basic solution (perhaps with new values for the basic variables). Specifically, determine the allowable range to stay feasible for each b_i and the allowable range to stay optimal for each c_j. Then divide up the range of likely values between these allowable values and other values for which the current basic solution will no longer be both feasible and optimal.

(*a*) Perform this sensitivity analysis graphically on the original model.

(*b*) Now perform this sensitivity analysis as described and illustrated in Sec. 6.7 for b_1 and c_1.

(*c*) Repeat part (*b*) for b_2.

(*d*) Repeat part (*b*) for c_2.

6.7-15. Consider the following problem.

$$\text{Maximize} \qquad Z = 3x_1 + 4x_2 + 8x_3,$$

subject to

$$2x_1 + 3x_2 + 5x_3 \leq 9$$

$$x_1 + 2x_2 + 3x_3 \leq 5$$

and

$$x_1 \geq 0, \qquad x_2 \geq 0, \qquad x_3 \geq 0.$$

Let x_4 and x_5 denote the slack variables for the respective functional constraints. After we apply the simplex method, the final simplex tableau is

Basic Variable	Eq.	*Coefficient of:* Z	x_1	x_2	x_3	x_4	x_5	Right Side
Z	(0)	1	0	1	0	1	1	14
x_1	(1)	0	1	−1	0	3	−5	2
x_3	(2)	0	0	1	1	−1	2	1

While doing post-optimality analysis, you learn that some of the parameter values used in the original model just given are just rough estimates, where the range of likely values in each case is within ±50 percent of the value used here. For each of these following parameters, perform sensitivity analysis to determine whether this uncertainty might affect either the feasibility or the optimality of the above basic solution. Specifically, for each parameter, determine the allowable range of values for which the current basic solution (perhaps with new values for the basic variables) will remain both feasible and optimal. Then divide up the range of likely values between these allowable values and other values for which the current basic solution will no longer be both feasible and optimal.

(*a*) Parameter b_2
(*b*) Parameter c_2
(*c*) Parameter a_{22}
(*d*) Parameter c_3
(*e*) Parameter a_{12}
(*f*) Parameter b_1

6.7-16.* Consider the example for Case 3 of sensitivity analysis in Sec. 6.7, where $\bar{c}_2 = 3$, $\bar{a}_{22} = 3$, $\bar{a}_{32} = 4$, and where the other parameters are given in Table 6.20. Starting from the resulting final tableau given at the bottom of Table 6.21, construct a table like Table 6.23 to perform parametric linear programming analysis, where

$$c_1 = 3 + \theta \qquad \text{and} \qquad c_2 = 3 + 2\theta.$$

How far can θ be increased above 0 before the current basic solution is no longer optimal?

6.7-17. Reconsider the model of Prob. 6.7-6. Suppose that you now have the option of making trade-offs in the profitability of the first two activities, whereby the objective function coefficient of x_1 can be increased by any amount by simultaneously decreasing the objective function coefficient of x_2 by the same amount. Thus the alternative choices of the objective function are

$$Z(\theta) = (2 + \theta)x_1 - (1 + \theta)x_2 + x_3,$$

where any nonnegative value of θ can be chosen.

Construct a table like Table 6.23 to perform parametric linear programming analysis on this problem. Determine the upper bound on θ before the original optimal solution would become nonoptimal. Then determine the best choice of θ over this range.

6.7-18. Consider the following parametric linear programming problem.

$$\text{Maximize} \qquad Z(\theta) = (10 - 4\theta)x_1 + (4 - \theta)x_2 + (7 + \theta)x_3,$$

subject to

$$3x_1 + x_2 + 2x_3 \le 7 \qquad \text{(resource 1)},$$

$$2x_1 + x_2 + 3x_3 \le 5 \qquad \text{(resource 2)},$$

and

$$x_1 \ge 0, \qquad x_2 \ge 0, \qquad x_3 \ge 0,$$

where θ can be assigned any positive or negative values. Let x_4 and x_5 be the slack variables for the respective constraints. After we apply the simplex method with $\theta = 0$, the final simplex tableau is

Basic Variable	Eq.	*Coefficient of:*						Right Side
		Z	x_1	x_2	x_3	x_4	x_5	
Z	(0)	1	0	0	3	2	2	24
x_1	(1)	0	1	0	−1	1	−1	2
x_2	(2)	0	0	1	5	−2	3	1

(*a*) Determine the range of values of θ over which the above BF solution will remain optimal. Then find the best choice of θ within this range.
(*b*) Given that θ is within the range of values found in part (*a*), find the allowable range to stay feasible for b_1 (the available amount of resource 1). Then do the same for b_2 (the available amount of resource 2).
(*c*) Given that θ is within the range of values found in part (*a*), identify the shadow prices (as a function of θ) for the two resources. Use this information to determine how the optimal value of the objective function would change (as a function of θ) if the available amount of resource 1 were decreased by 1 and the available amount of resource 2 simultaneously were increased by 1.

I (*d*) Construct the dual of this parametric linear programming problem. Set $\theta = 0$ and solve this dual problem graphically to find the corresponding shadow prices for the two resources of the primal problem. Then find these shadow prices as a function of θ [within the range of values found in part (*a*)] by algebraically solving for this same optimal CPF solution for the dual problem as a function of θ.

6.7-19. Consider the following parametric linear programming problem.

$$\text{Maximize} \quad Z(\theta) = 2x_1 + 4x_2 + 5x_3,$$

subject to

$$x_1 + 3x_2 + 2x_3 \leq 5 + \theta$$
$$x_1 + 2x_2 + 3x_3 \leq 6 + 2\theta$$

and

$$x_1 \geq 0, \quad x_2 \geq 0, \quad x_3 \geq 0,$$

where θ can be assigned any positive or negative values. Let x_4 and x_5 be the slack variables for the respective functional constraints. After we apply the simplex method with $\theta = 0$, the final simplex tableau is

Basic Variable	Eq.	*Coefficient of:*						Right Side
		Z	x_1	x_2	x_3	x_4	x_5	
Z	(0)	0	0	1	0	1	1	11
x_1	(1)	1	1	5	0	3	-2	3
x_3	(2)	2	0	-1	1	-1	1	1

(*a*) Express the BF solution (and Z) given in this tableau as a function of θ. Determine the lower and upper bounds on θ before this optimal solution would become infeasible. Then determine the best choice of θ between these bounds.
(*b*) Given that θ is between the bounds found in part (*a*), determine the allowable range to stay optimal for c_1 (the coefficient of x_1 in the objective function).

6.7-20. Consider the following parametric linear programming problem, where the parameter θ must be nonnegative:

$$\text{Maximize} \quad Z(\theta) = (5 + 2\theta)x_1 + (2 - \theta)x_2 + (3 + \theta)x_3,$$

subject to

$$4x_1 + x_2 \geq 5 + 5\theta$$
$$3x_1 + x_2 + 2x_3 = 10 - 10\theta$$

and

$$x_1 \geq 0, \quad x_2 \geq 0, \quad x_3 \geq 0.$$

Let x_4 be the surplus variable for the first functional constraint, and let $\bar{x}_5$ and $\bar{x}_6$ be the artificial variables for the respective functional constraints. After we apply the simplex method with the Big M method and with $\theta = 0$, the final simplex tableau is

Basic Variable	Eq.	Coefficient of:							Right Side
		Z	x_1	x_2	x_3	x_4	$\bar{x}_5$	$\bar{x}_6$	
Z	(0)	1	1	0	1	0	M	$M + 2$	20
x_2	(1)	0	3	1	2	0	0	1	10
x_4	(2)	0	−1	0	2	1	−1	1	5

(*a*) Use the fundamental insight (Sec. 5.3) to revise this tableau to reflect the inclusion of the parameter θ in the original model. Show the complete tableau needed to apply the feasibility test and the optimality test for any value of θ. Express the corresponding basic solution (and Z) as a function of θ.

(*b*) Determine the range of nonnegative values of θ over which this basic solution is feasible.

(*c*) Determine the range of nonnegative values of θ over which this basic solution is both feasible and optimal. Determine the best choice of θ over this range.

6.7-21. Consider the following problem.

$$\text{Maximize} \quad Z = 10x_1 + 4x_2,$$

subject to

$$3x_1 + x_2 \leq 30$$
$$2x_1 + x_2 \leq 25$$

and

$$x_1 \geq 0, \quad x_2 \geq 0.$$

Let x_3 and x_4 denote the slack variables for the respective functional constraints. After we apply the simplex method, the final simplex tableau is

Basic Variable	Eq.	Coefficient of:					Right Side
		Z	x_1	x_2	x_3	x_4	
Z	(0)	1	0	0	2	2	110
x_2	(1)	0	0	1	−2	3	15
x_1	(2)	0	1	0	1	−1	5

Now suppose that both of the following changes are made simultaneously in the original model:

1. The first constraint is changed to $4x_1 + x_2 \leq 40$.
2. Parametric programming is introduced to change the objective function to the alternative choices of

$$Z(\theta) = (10 - 2\theta)x_1 + (4 + \theta)x_2,$$

where any nonnegative value of θ can be chosen.

(*a*) Construct the resulting revised final tableau (as a function of θ), and then convert this tableau to proper form from Gaussian elimination. Use this tableau to identify the new optimal solution that applies for either $\theta = 0$ or sufficiently small values of θ.
(*b*) What is the upper bound on θ before this optimal solution would become nonoptimal?
(*c*) Over the range of θ from zero to this upper bound, which choice of θ gives the largest value of the objective function?

6.7-22. Consider the following problem.

$$\text{Maximize} \quad Z = 9x_1 + 8x_2 + 5x_3,$$

subject to

$$2x_1 + 3x_2 + x_3 \le 4$$
$$5x_1 + 4x_2 + 3x_3 \le 11$$

and

$$x_1 \ge 0, \quad x_2 \ge 0, \quad x_3 \ge 0.$$

Let x_4 and x_5 denote the slack variables for the respective functional constraints. After we apply the simplex method, the final simplex tableau is

Basic Variable	Eq.	Coefficient of:						Right Side
		Z	x_1	x_2	x_3	x_4	x_5	
Z	(0)	1	0	2	0	2	1	19
x_1	(1)	0	1	5	0	3	−1	1
x_3	(2)	0	0	−7	1	−5	2	2

D, I* (*a*) Suppose that a new technology has become available for conducting the first activity considered in this problem. If the new technology were adopted to replace the existing one, the coefficients of x_1 in the model would change

$$\text{from} \quad \begin{bmatrix} c_1 \\ a_{11} \\ a_{21} \end{bmatrix} = \begin{bmatrix} 9 \\ 2 \\ 5 \end{bmatrix} \quad \text{to} \quad \begin{bmatrix} c_1 \\ a_{11} \\ a_{21} \end{bmatrix} = \begin{bmatrix} 18 \\ 3 \\ 6 \end{bmatrix}.$$

Use the sensitivity analysis procedure to investigate the potential effect and desirability of adopting the new technology. Specifically, assuming it were adopted, construct the resulting revised final tableau, convert this tableau to proper form from Gaussian elimination, and then reoptimize (if necessary) to find the new optimal solution.

(*b*) Now suppose that you have the option of mixing the old and new technologies for conducting the first activity. Let θ denote the fraction of the technology used that is from the new technology, so $0 \le \theta \le 1$. Given θ, the coefficients of x_1 in the model become

$$\begin{bmatrix} c_1 \\ a_{11} \\ a_{21} \end{bmatrix} = \begin{bmatrix} 9 + 9\theta \\ 2 + \theta \\ 5 + \theta \end{bmatrix}.$$

Construct the resulting revised final tableau (as a function of θ), and convert this tableau to proper form from Gaussian elimination. Use this tableau to identify the current basic solution as a function of θ. Over the allowable values of $0 \le \theta \le 1$, give the range of values of θ for which this solution is both feasible and optimal. What is the best choice of θ within this range?

6.7-23. Consider the following problem.

$$\text{Maximize} \quad Z = 3x_1 + 5x_2 + 2x_3,$$

subject to

$$-2x_1 + 2x_2 + x_3 \leq 5$$
$$3x_1 + x_2 - x_3 \leq 10$$

and

$$x_1 \geq 0, \quad x_2 \geq 0, \quad x_3 \geq 0.$$

Let x_4 and x_5 be the slack variables for the respective functional constraints. After we apply the simplex method, the final simplex tableau is

Basic Variable	Eq.	*Coefficient of:*						Right Side
		Z	x_1	x_2	x_3	x_4	x_5	
Z	(0)	1	0	20	0	9	7	115
x_1	(1)	0	1	3	0	1	1	15
x_3	(2)	0	0	8	1	3	2	35

Parametric linear programming analysis now is to be applied simultaneously to the objective function and right-hand sides, where the model in terms of the new parameter is the following:

$$\text{Maximize} \quad Z(\theta) = (3 + 2\theta)x_1 + (5 + \theta)x_2 + (2 - \theta)x_3,$$

subject to

$$-2x_1 + 2x_2 + x_3 \leq 5 + 6\theta$$
$$3x_1 + x_2 - x_3 \leq 10 - 8\theta$$

and

$$x_1 \geq 0, \quad x_2 \geq 0, \quad x_3 \geq 0.$$

Construct the resulting revised final tableau (as a function of θ), and convert this tableau to proper form from Gaussian elimination. Use this tableau to identify the current basic solution as a function of θ. For $\theta \geq 0$, give the range of values of θ for which this solution is both feasible and optimal. What is the best choice of θ within this range?

6.7-24. Consider the Wyndor Glass Co. problem described in Sec. 3.1. Suppose that, in addition to considering the introduction of two new products, management now is considering changing the production rate of a certain old product that is still profitable. Refer to Table 3.1. The number of production hours per week used per unit production rate of this old product is 1, 4, and 3 for Plants 1, 2, and 3, respectively. Therefore, if we let θ denote the *change* (positive or negative) in the production rate of this old product, the right-hand sides of the three functional constraints in Sec. 3.1 become $4 - \theta$, $12 - 4\theta$, and $18 - 3\theta$, respectively. Thus choosing a negative value of θ would free additional capacity for producing more of the two new products, whereas a positive value would have the opposite effect.

(*a*) Use a parametric linear programming formulation to determine the effect of different choices of θ on the optimal solution for the product mix of the two new products given in the final tableau of Table 4.8. In particular, use the fundamental insight of Sec. 5.3 to obtain expressions for Z and the basic variables x_3, x_2, and x_1 in terms of θ, assuming that θ is sufficiently close to zero that this "final" basic solution still is feasible and thus optimal for the given value of θ.

(*b*) Now consider the broader question of the choice of θ along with the product mix for the two new products. What is the breakeven unit profit for the old product (in comparison with the two new products) below which its production rate should be decreased ($\theta < 0$) in favor of the new products and above which its production rate should be increased ($\theta > 0$)?

(*c*) If the unit profit is above this breakeven point, how much can the old product's production rate be increased before the final BF solution would become infeasible?

(*d*) If the unit profit is below this breakeven point, how much can the old product's production rate be decreased (assuming its previous rate was larger than this decrease) before the final BF solution would become infeasible?

6.7-25. Consider the following problem.

$$\text{Maximize} \quad Z = 2x_1 - x_2 + 3x_3,$$

subject to

$$x_1 + x_2 + x_3 = 3$$
$$x_1 - 2x_2 + x_3 \geq 1$$
$$2x_2 + x_3 \leq 2$$

and

$$x_1 \geq 0, \quad x_2 \geq 0, \quad x_3 \geq 0.$$

Suppose that the Big M method (see Sec. 4.6) is used to obtain the initial (artificial) BF solution. Let x_4 be the artificial slack variable for the first constraint, x_5 the surplus variable for the second constraint, $\bar{x}_6$ the artificial variable for the second constraint, and x_7 the slack variable for the third constraint. The corresponding final set of equations yielding the optimal solution is

$$(0) \quad Z + 5x_2 + (M+2)\bar{x}_4 + M\bar{x}_6 + x_7 = 8,$$
$$(1) \quad x_1 - x_2 + \bar{x}_4 - x_7 = 1,$$
$$(2) \quad 2x_2 + x_3 + x_7 = 2,$$
$$(3) \quad 3x_2 + \bar{x}_4 + x_5 - \bar{x}_6 = 2.$$

Suppose that the original objective function is changed to $Z = 2x_1 + 3x_2 + 4x_3$ and that the original third constraint is changed to $2x_2 + x_3 \leq 1$. Use the sensitivity analysis procedure to revise the final set of equations (in tableau form) and convert it to proper form from Gaussian elimination for identifying and evaluating the current basic solution. Then test this solution for feasibility and for optimality. (Do not reoptimize.)

CASE PROBLEM

CP6-1. Refer to Sec. 3.4 (subsection entitled "Controlling Air Pollution") for the Nori & Leets Co. problem. After the OR team obtained an optimal solution, we mentioned that the team then conducted sensitivity analysis. We now continue this story by having you retrace the steps taken by the OR team, after we provide some additional background.

The values of the various parameters in the original formulation of the model are given in Tables 3.12, 3.13, and 3.14. Since the company does not have much prior experience with the pollution abatement methods under consideration, the cost estimates given in Table 3.14 are fairly rough, and each one could easily be off by as much as 10 percent in either direction. There also is some uncertainty about the parameter values given in Table 3.13, but less so than for those in Table 3.14. By contrast, the values in Table 3.12 are policy standards and so are prescribed constants.

However, there still is considerable debate about where to set these policy standards on the required reductions in emission rate of the various pollutants. The numbers in Table 3.12 actually are preliminary values tentatively agreed upon before it is known what the total cost would be to meet these standards. Both city and company officials agree that the final decision on these policy standards should be based on the *trade-off* between costs and benefits. With this in mind, the city has concluded that each 10 percent increase in policy standards over the current values (all the numbers in Table 3.12) would be worth $3.5 million to the city. Therefore, the city has agreed to reduce the company's tax payments to the city by $3.5 million for *each* 10 percent reduction in the policy standards (up to 50 percent) that is accepted by the company.

Finally, there has been some debate about the *relative* values of the policy standards for the three pollutants. As indicated in Table 3.12, the required reduction for particulates now is less than half of that for either sulfur oxides or hydrocarbons. Some have argued for decreasing this disparity. Others contend that an even greater disparity is justified because sulfur oxides and hydrocarbons cause considerably more damage than particulates. Agreement has been reached that this issue will be reexamined after information is obtained about which trade-offs in policy standards (increasing one while decreasing another) are available without increasing the total cost.

A* (*a*) Use the OR Courseware to solve the model for this problem as formulated in Sec. 3.4. In addition to the optimal solution, note the additional output provided for post-optimality analysis. This output provides the basis for the following steps.

(*b*) Ignoring the constraints with no uncertainty about their parameter values (namely, $x_j \leq 1$ for $j = 1, 2, \ldots, 6$), identify the parameters of the model that should be classified as *sensitive parameters*. (*Hint:* See the subsection "Sensitivity Analysis" in Sec. 4.7.) Make a resulting recommendation about which parameters should be estimated more closely, if possible.

(*c*) Analyze the effect of an inaccuracy in estimating each cost parameter given in Table 3.14. If the true value is 10 percent *less* than the estimated value, would this alter the optimal solution? Would it change if the true value were 10 percent *more* than the estimated value? Make a resulting recommendation about where to focus further work in estimating the cost parameters more closely.

(*d*) The automatic routine for the simplex method in your OR Courseware converted the model to maximization form before applying the simplex method. Use Table 6.14 to construct the corresponding dual problem, and use the output from part (*a*) to identify an optimal solution for this dual problem. If the primal problem had been left in minimization form, how would this have affected the form of the dual problem and the sign of the optimal dual variables?

(*e*) For each pollutant, use your results from part (*d*) to specify the rate at which the total cost of an optimal solution would change with any small change in the required reduction of the annual emission rate of the pollutant. Also specify how much this required reduction can be changed (up or down) without affecting the rate of change in the total cost.

(*f*) For each unit change in the policy standard for particulates given in Table 3.12, determine the change in the opposite direction for sulfur oxides that would keep the total cost of an optimal solution unchanged. Repeat this for hydrocarbons instead of sulfur oxides. Then do it for a simultaneous and equal change for both sulfur oxides and hydrocarbons in the opposite direction from particulates.

(*g*) Letting θ denote the percentage increase in all the policy standards given in Table 3.12, formulate the problem of analyzing the effect of simultaneous proportional increases in these standards as a parametric linear programming problem. Then use your results from part (*e*) to determine the rate at which the total cost of an optimal solution would increase with a small increase in θ from zero.

A* (*h*) Use the automatic routine for the simplex method in the OR Courseware to find an optimal solution for the parametric linear programming problem formulated in part (*g*) for each $\theta = 10, 20, 30, 40, 50$. Considering the tax incentive offered by the city, use these results to determine which value of θ (including the option of $\theta = 0$) should be chosen by the company to minimize the total cost of both pollution abatement and taxes.

(*i*) For the value of θ chosen in part (*h*), repeat parts (*e*) and (*f*) so that the decision makers can make a final decision on the relative values of the policy standards for the three pollutants.

7

Other Algorithms for Linear Programming

The key to the widespread use of linear programming is the availability of an exceptionally efficient algorithm—the simplex method—that will routinely solve the large-size problems that typically arise in practice. However, the simplex method is only part of the arsenal of algorithms regularly used by linear programming practitioners. We now turn to these other algorithms.

This chapter focuses first on three particularly important algorithms that are, in fact, *variants* of the simplex method. In particular, the next three sections present the *dual simplex method* (a modification particularly useful for sensitivity analysis), *parametric linear programming* (an extension for systematic sensitivity analysis), and the *upper bound technique* (a streamlined version of the simplex method for dealing with variables having upper bounds).

Section 4.9 introduced an exciting new advancement in linear programming—the development of a powerful new type of algorithm that moves through the interior of the feasible region. We describe this *interior-point approach* further in Sec. 7.4.

We next introduce *linear goal programming* where, rather than having a *single objective* (maximize or minimize Z) as for linear programming, the problem has *sev-*

eral goals toward which we must strive simultaneously. Certain formulation techniques enable us to convert a linear goal programming problem back to a linear programming problem so that solution procedures based on the simplex method can still be used. Section 7.5 describes these techniques and procedures.

7.1 The Dual Simplex Method

The *dual simplex method* can be thought of as the *mirror image* of the simplex method. This interpretation is best explained by referring to Tables 6.10 and 6.11 and Fig. 6.1. The simplex method deals directly with *suboptimal* basic solutions and moves toward an optimal solution by striving to satisfy the *optimality test.* By contrast, the dual simplex method deals directly with *superoptimal* basic solutions and moves toward an optimal solution by striving to achieve *feasibility.* Furthermore, the dual simplex method deals with a problem as if the simplex method were being applied simultaneously to its dual problem. If we make their *initial* basic solutions *complementary,* the two methods move in complete sequence, obtaining *complementary* basic solutions with each iteration.

The dual simplex method is very useful in certain special types of situations. Ordinarily it is easier to find an initial basic feasible (BF) solution than an initial superoptimal basic solution. However, it is occasionally necessary to introduce many *artificial* variables to construct an initial BF solution artificially. In such cases it may be easier to begin with a superoptimal basic solution and use the dual simplex method. Furthermore, fewer iterations may be required when it is not necessary to drive many artificial variables to zero.

As we mentioned several times in Chap. 6 as well as in Sec. 4.7, another important primary application of the dual simplex method is its use in conjunction with sensitivity analysis. Suppose that an optimal solution has been obtained by the simplex method but that it becomes necessary (or of interest for sensitivity analysis) to make minor changes in the model. If the formerly optimal basic solution is *no longer feasible* (but still satisfies the optimality test), you can immediately apply the dual simplex method by starting with this *superoptimal* basic solution. Applying the dual simplex method in this way usually leads to the new optimal solution much more quickly than would solving the new problem from the beginning with the simplex method.

The rules for the dual simplex method are very similar to those for the simplex method. In fact, once the methods are started, the only difference between them is in the criteria used for selecting the entering and leaving basic variables and for stopping the algorithm.

To start the dual simplex method (for a maximization problem), we must have all the coefficients in Eq. (0) *nonnegative* (so that the basic solution is superoptimal). The basic solutions will be infeasible (except for the last one) only because some of the variables are negative. The method continues to decrease the value of the objective function, always retaining *nonnegative coefficients* in Eq. (0), until all the *variables* are nonnegative. Such a basic solution is feasible (it satisfies all the equations) and is, therefore, optimal by the simplex method criterion of nonnegative coefficients in Eq. (0).

The details of the dual simplex method are summarized next.

Summary of the Dual Simplex Method

1. *Initialization:* After converting any functional constraints in $\geq$ form to $\leq$ form (by multiplying through both sides by -1), introduce slack variables as needed to construct a set of equations describing the problem. Find a basic solution such that the coefficients in Eq. (0) are zero for basic variables and nonnegative for nonbasic variables (so the solution is optimal if it is feasible). Go to the feasibility test.
2. *Feasibility test:* Check to see whether all the basic variables are *nonnegative.* If they are, then this solution is feasible, and therefore optimal, so stop. Otherwise, go to an iteration.
3. *Iteration:*

   Step 1 Determine the *leaving basic variable:* Select the *negative* basic variable that has the largest absolute value.

   Step 2 Determine the *entering basic variable:* Select the nonbasic variable whose coefficient in Eq. (0) reaches zero first as an increasing multiple of the equation containing the leaving basic variable is added to Eq. (0). This selection is made by checking the nonbasic variables with *negative coefficients* in that equation (the one containing the leaving basic variable) and selecting the one with the smallest absolute value of the ratio of the Eq. (0) coefficient to the coefficient in that equation.

   Step 3 Determine the *new basic solution:* Starting from the current set of equations, solve for the basic variables in terms of the nonbasic variables by Gaussian elimination. When we set the nonbasic variables equal to zero, each basic variable (and Z) equals the new right-hand side of the one equation in which it appears (with a coefficient of $+1$). Return to the feasibility test.

To fully understand the dual simplex method, you must realize that the method proceeds just as if the *simplex method* were being applied to the complementary basic solutions in the *dual problem.* (In fact, this interpretation was the motivation for constructing the method as it is.) Step 1 of an iteration, determining the leaving basic variable, is equivalent to determining the entering basic variable in the dual problem. The negative variable with the largest absolute value corresponds to the negative coefficient with the largest absolute value in Eq. (0) of the dual problem (see Table 6.3). Step 2, determining the entering basic variable, is equivalent to determining the leaving basic variable in the dual problem. The coefficient in Eq. (0) that reaches zero first corresponds to the variable in the dual problem that reaches zero first. The two criteria for stopping the algorithm are also complementary.

We shall now illustrate the dual simplex method by applying it to the *dual problem* for the Wyndor Glass Co. (see Table 6.1). Normally this method is applied directly to the problem of concern (a primal problem). However, we have chosen this problem because you have already seen the simplex method applied to *its* dual problem (namely, the primal problem[1]) in Table 4.8 so you can compare the two. To facilitate the comparison, we shall continue to denote the decision variables in the problem being solved by y_i rather than x_j.

In *maximization* form, the problem to be solved is

$$\text{Maximize} \quad Z = -4y_1 - 12y_2 - 18y_3,$$

subject to

$$\begin{aligned} y_1 \qquad + 3y_3 &\geq 3 \\ 2y_2 + 2y_3 &\geq 5 \end{aligned}$$

[1] Recall that the symmetry property in Sec. 6.1 points out that the dual of a dual problem is the original primal problem.

Table 7.1 **Dual Simplex Method Applied to the Wyndor Glass Co. Dual Problem**

Iteration	Basic Variable	Eq.	Coefficient of: Z	y_1	y_2	y_3	y_4	y_5	Right Side
0	Z	(0)	1	4	12	18	0	0	0
	y_4	(1)	0	−1	0	−3	1	0	−3
	y_5	(2)	0	0	−2	−2	0	1	−5
1	Z	(0)	1	4	0	6	0	6	−30
	y_4	(1)	0	−1	0	−3	1	0	−3
	y_2	(2)	0	0	1	1	0	$-\frac{1}{2}$	$\frac{5}{2}$
2	Z	(0)	1	2	0	0	2	6	−36
	y_3	(1)	0	$\frac{1}{3}$	0	1	$-\frac{1}{3}$	0	1
	y_2	(2)	0	$-\frac{1}{3}$	1	0	$\frac{1}{3}$	$-\frac{1}{2}$	$\frac{3}{2}$

and

$$y_1 \geq 0, \qquad y_2 \geq 0, \qquad y_3 \geq 0.$$

Since negative right-hand sides are now allowed, we do not need to introduce artifical variables to be the initial basic variables. Instead, we simply convert the functional constraints to $\leq$ form and introduce slack variables to play this role. The resulting initial set of equations is that shown for iteration 0 in Table 7.1. Notice that all the coefficients in Eq. (0) are nonnegative, so the solution is optimal if it is feasible.

The initial basic solution is $y_1 = 0$, $y_2 = 0$, $y_3 = 0$, $y_4 = -3$, $y_5 = -5$, with $Z = 0$, which is not feasible because of the negative values. The leaving basic variable is y_5 ($5 > 3$), and the entering basic variable is y_2 ($12/2 < 18/2$), which leads to the second set of equations, labeled as iteration 1 in Table 7.1. The corresponding basic solution is $y_1 = 0$, $y_2 = \frac{5}{2}$, $y_3 = 0$, $y_4 = -3$, $y_5 = 0$, with $Z = -30$, which is not feasible.

The next leaving basic variable is y_4, and the entering basic variable is y_3 ($6/3 < 4/1$), which leads to the final set of equations in Table 7.1. The corresponding basic solution is $y_1 = 0$, $y_2 = \frac{3}{2}$, $y_3 = 1$, $y_4 = 0$, $y_5 = 0$, with $Z = -36$, which is feasible and therefore optimal.

Notice that the optimal solution for the dual of this problem[1] is $x_1^* = 2$, $x_2^* = 6$, $x_3^* = 2$, $x_4^* = 0$, $x_5^* = 0$, as was obtained in Table 4.8 by the simplex method. We suggest that you now trace through Tables 7.1 and 4.8 simultaneously and compare the complementary steps for the two mirror-image methods.

7.2 Parametric Linear Programming

At the end of Sec. 6.7 we described *parametric linear programming* and its use for conducting sensitivity analysis systematically by gradually changing various model parameters simultaneously. We shall now present the algorithmic procedure, first for the case where the c_j parameters are being changed and then where the b_i parameters are varied.

[1] The *complementary optimal basic solutions property* presented in Sec. 6.3 indicates how to read the optimal solution for the dual problem from row 0 of the final simplex tableau for the primal problem. This same conclusion holds regardless of whether the simplex method or the dual simplex method is used to obtain the final tableau.

Systematic Changes in the c_j Parameters

For the case where the c_j parameters are being changed, the *objective function* of the ordinary linear programming model

$$Z = \sum_{j=1}^{n} c_j x_j$$

is replaced by

$$Z(\theta) = \sum_{j=1}^{n} (c_j + \alpha_j \theta) x_j,$$

where the α_j are given input constants representing the *relative* rates at which the coefficients are to be changed. Therefore, gradually increasing θ from zero changes the coefficients at these relative rates.

The values assigned to the α_j may represent interesting simultaneous changes of the c_j for systematic sensitivity analysis of the effect of increasing the magnitude of these changes. They may also be based on how the coefficients (e.g., unit profits) would change together with respect to some factor measured by θ. This factor might be uncontrollable, e.g., the state of the economy. However, it may also be under the control of the decision maker, e.g., the amount of personnel and equipment to shift from some of the activities to others.

For any given value of θ, the optimal solution of the corresponding linear programming problem can be obtained by the simplex method. This solution may have been obtained already for the original problem where $\theta = 0$. However, the objective is to *find the optimal solution* of the modified linear programming problem [maximize $Z(\theta)$ subject to the original constraints] *as a function of* θ. Therefore, in the solution procedure you need to be able to determine when and how the optimal solution changes (if it does) as θ increases from zero to any specified positive number.

Figure 7.1 illustrates how $Z^*(\theta)$, the objective function value for the optimal solution (given θ), changes as θ increases. In fact, $Z^*(\theta)$ always has this *piecewise linear* and *convex*[1] form (see Prob. 7.2-10). The corresponding optimal solution

[1] See Appendix 2 for a definition and discussion of convex functions.

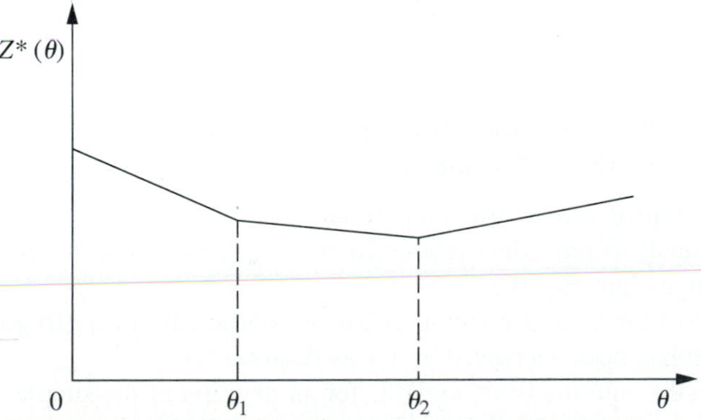

Figure 7.1 The objective function value for an optimal solution as a function of θ for parametric linear programming with systematic changes in the c_j parameters.

changes (as θ increases) *just* at the values of θ where the slope of the $Z^*(\theta)$ function changes. Thus Fig. 7.1 depicts a problem where three different solutions are optimal for different values of θ, the first for $0 \le \theta \le \theta_1$, the second for $\theta_1 \le \theta \le \theta_2$, and the third for $\theta \ge \theta_2$. Because the value of each x_j remains the same within each of these intervals for θ, the value of $Z^*(\theta)$ varies with θ only because the *coefficients* of the x_j are changing as a linear function of θ. The solution procedure is based directly upon the sensitivity analysis procedure for investigating changes in the c_j parameters (Cases 2*a* and 3, Sec. 6.7). As described in the last subsection of Sec. 6.7, the only basic difference with parametric linear programming is that the changes now are expressed in terms of θ rather than as specific numbers.

To illustrate, suppose that $\alpha_1 = 2$ and $\alpha_2 = -1$ for the original Wyndor Glass Co. problem presented in Sec. 3.1, so that

$$Z(\theta) = (3 + 2\theta)x_1 + (5 - \theta)x_2.$$

Beginning with the final simplex tableau for $\theta = 0$ (Table 4.8), we see that its Eq. (0)

$$\text{(0)} \qquad Z + \tfrac{3}{2}x_4 + x_5 = 36$$

would first have these changes from the original ($\theta = 0$) coefficients added into it on the left-hand side:

$$\text{(0)} \qquad Z - 2\theta x_1 + \theta x_2 + \tfrac{3}{2}x_4 + x_5 = 36.$$

Because both x_1 and x_2 are basic variables [appearing in Eqs. (3) and (2), respectively], they both need to be eliminated algebraically from Eq. (0):

$$\begin{array}{ll} & Z - 2\theta x_1 + \theta x_2 + \tfrac{3}{2}x_4 + x_5 = 36 \\ & \quad +2\theta \text{ times Eq. (3)} \\ & \quad -\ \theta \text{ times Eq. (2)} \\ \hline \text{(0)} & Z + (\tfrac{3}{2} - \tfrac{7}{6}\theta)x_4 + (1 + \tfrac{2}{3}\theta)x_5 = 36 - 2\theta. \end{array}$$

The optimality test says that the current BF solution will remain optimal as long as these coefficients of the nonbasic variables remain nonnegative:

$$\tfrac{3}{2} - \tfrac{7}{6}\theta \ge 0, \qquad \text{for } 0 \le \theta \le \tfrac{9}{7},$$

$$1 + \tfrac{2}{3}\theta \ge 0, \qquad \text{for all } \theta \ge 0.$$

Therefore, after θ is increased past $\theta = \frac{9}{7}$, x_4 would need to be the entering basic variable for another iteration of the simplex method to find the new optimal solution. Then θ would be increased further until another coefficient went negative, and so on until θ had been increased as far as desired.

This entire procedure is now summarized, and the example is completed in Table 7.2.

Summary of the Parametric Programming Procedure for Systematic Changes in the c_j Parameters

1. Solve the problem with $\theta = 0$ by the simplex method.
2. Use the sensitivity analysis procedure (Cases 2*a* and 3, Sec. 6.7) to introduce the $\Delta c_j = \alpha_j \theta$ changes into Eq. (0).
3. Increase θ until one of the nonbasic variables has its coefficient in Eq. (0) go negative (or until θ has been increased as far as desired).
4. Use this variable as the entering basic variable for an iteration of the simplex method to find the new optimal solution. Return to step 3.

Table 7.2 **The c_j Parametric Programming Procedure Applied to the Wyndor Glass Co. Example**

Range of θ	Basic Variable	Eq.	Z	x_1	x_2	x_3	x_4	x_5	Right Side	Optimal Solution
	$Z(\theta)$	(0)	1	0	0	0	$\frac{9-7\theta}{6}$	$\frac{3+2\theta}{3}$	$36-2\theta$	$x_4 = 0$
										$x_5 = 0$
$0 \leq \theta \leq \frac{9}{7}$	x_3	(1)	0	0	0	1	$\frac{1}{3}$	$-\frac{1}{3}$	2	$x_3 = 2$
	x_2	(2)	0	0	1	0	$\frac{1}{2}$	0	6	$x_2 = 6$
	x_1	(3)	0	1	0	0	$-\frac{1}{3}$	$\frac{1}{3}$	2	$x_1 = 2$
	$Z(\theta)$	(0)	1	0	0	$\frac{-9+7\theta}{2}$	0	$\frac{5-\theta}{2}$	$27+5\theta$	$x_3 = 0$
										$x_5 = 0$
$\frac{9}{7} \leq \theta \leq 5$	x_4	(1)	0	0	0	3	1	-1	6	$x_4 = 6$
	x_2	(2)	0	0	1	$-\frac{3}{2}$	0	$\frac{1}{2}$	3	$x_2 = 3$
	x_1	(3)	0	1	0	1	0	0	4	$x_1 = 4$
	$Z(\theta)$	(0)	1	0	$-5+\theta$	$3+2\theta$	0	0	$12+8\theta$	$x_2 = 0$
										$x_3 = 0$
$\theta \geq 5$	x_4	(1)	0	0	2	0	1	0	12	$x_4 = 12$
	x_5	(2)	0	0	2	-3	0	1	6	$x_5 = 6$
	x_1	(3)	0	1	0	1	0	0	4	$x_1 = 4$

Systematic Changes in the b_i Parameters

For the case where the b_i parameters change systematically, the one modification made in the original linear programming model is that b_i is replaced by $b_i + \alpha_i\theta$, for $i = 1, 2, \ldots, m$, where the α_i are given input constants. Thus the problem becomes

$$\text{Maximize} \quad Z(\theta) = \sum_{j=1}^{n} c_j x_j,$$

subject to

$$\sum_{j=1}^{n} a_{ij} \leq b_i + \alpha_i\theta \qquad \text{for } i = 1, 2, \ldots, m$$

and

$$x_j \geq 0 \qquad \text{for } j = 1, 2, \ldots, n.$$

The goal is to identify the optimal solution as a function of θ.

With this formulation, the corresponding objective function value $Z^*(\theta)$ always has the *piecewise linear* and *concave*[1] form shown in Fig. 7.2. (See Prob. 7.2-11.) The set of basic variables in the optimal solution still changes (as θ increases) *only* where the slope of $Z^*(\theta)$ changes. However, in contrast to the preceding case, the values of

[1] See Appendix 2 for a definition and discussion of concave functions.

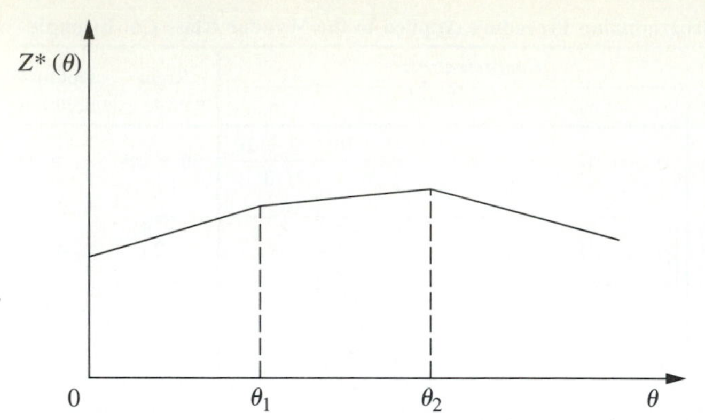

Figure 7.2 The objective function value for an optimal solution as a function of θ for parametric linear programming with systematic changes in the b_i parameters.

these variables now change as a (linear) function of θ between the slope changes. The reason is that increasing θ changes the right-hand sides in the initial set of equations, which then causes changes in the right-hand sides in the final set of equations, i.e., in the values of the final set of basic variables. Figure 7.2 depicts a problem with three sets of basic variables that are optimal for different values of θ, the first for $0 \leq \theta \leq \theta_1$, the second for $\theta_1 \leq \theta \leq \theta_2$, and the third for $\theta \geq \theta_2$. Within each of these intervals of θ, the value of $Z^*(\theta)$ varies with θ despite the fixed coefficients c_j because the x_j values are changing.

The following solution procedure summary is very similar to that just presented for systematic changes in the c_j parameters. The reason is that changing the b_i values is equivalent to changing the coefficients in the objective function of the *dual* model. Therefore, the procedure for the primal problem is exactly *complementary* to applying simultaneously the procedure for systematic changes in the c_j parameters to the *dual* problem. Consequently, the *dual simplex method* (see Sec. 7.1) now would be used to obtain each new optimal solution, and the applicable sensitivity analysis case (see Sec. 6.7) now is Case 1, but these differences are the only major differences.

Summary of the Parametric Programming Procedure for Systematic Changes in the b_i Parameters

1. Solve the problem with $\theta = 0$ by the simplex method.
2. Use the sensitivity analysis procedure (Case 1, Sec. 6.7) to introduce the $\Delta b_i = \alpha_i \theta$ changes to the *right side* column.
3. Increase θ until one of the basic variables has its value in the *right side* column go negative (or until θ has been increased as far as desired).
4. Use this variable as the leaving basic variable for an iteration of the dual simplex method to find the new optimal solution. Return to step 3.

To illustrate this procedure in a way that demonstrates its *duality* relationship with the procedure for systematic changes in the c_j parameters, we now apply it to the dual problem for the Wyndor Glass Co. (see Table 6.1). In particular, suppose that $\alpha_1 = 2$ and $\alpha_2 = -1$ so that the functional constraints become

$$y_1 + 3y_3 \geq 3 + 2\theta \quad \text{or} \quad -y_1 - 3y_3 \leq -3 - 2\theta$$

$$2y_2 + 2y_3 \geq 5 - \theta \quad \text{or} \quad -2y_2 - 2y_3 \leq -5 + \theta.$$

Table 7.3 The b_i Parametric Programming Procedure Applied to the Dual of the Wyndor Glass Co. Example

Range of θ	Basic Variable	Eq.	Z	y_1	y_2	y_3	y_4	y_5	Right Side	Optimal Solution
$0 \le \theta \le \frac{9}{7}$	$Z(\theta)$	(0)	1	2	0	0	2	6	$-36 + 2\theta$	$y_1 = y_4 = y_5 = 0$
	y_3	(1)	0	$\frac{1}{3}$	0	1	$-\frac{1}{3}$	0	$\frac{3 + 2\theta}{3}$	$y_3 = \frac{3 + 2\theta}{3}$
	y_2	(2)	0	$-\frac{1}{3}$	1	0	$\frac{1}{3}$	$-\frac{1}{2}$	$\frac{9 - 7\theta}{6}$	$y_2 = \frac{9 - 7\theta}{6}$
$\frac{9}{7} \le \theta \le 5$	$Z(\theta)$	(0)	1	0	6	0	4	3	$-27 - 5\theta$	$y_2 = y_4 = y_5 = 0$
	y_3	(1)	0	0	1	1	0	$-\frac{1}{2}$	$\frac{5 - \theta}{2}$	$y_3 = \frac{5 - \theta}{2}$
	y_1	(2)	0	1	-3	0	-1	$\frac{3}{2}$	$\frac{-9 + 7\theta}{2}$	$y_1 = \frac{-9 + 7\theta}{2}$
$\theta \ge 5$	$Z(\theta)$	(0)	1	0	12	6	4	0	$-12 - 8\theta$	$y_2 = y_3 = y_4 = 0$
	y_5	(1)	0	0	-2	-2	0	1	$-5 + \theta$	$y_5 = -5 + \theta$
	y_1	(2)	0	1	0	3	-1	0	$3 + 2\theta$	$y_1 = 3 + 2\theta$

Thus the dual of *this* problem is just the example considered in Table 7.2.

This problem with $\theta = 0$ has already been solved in Table 7.1, so we begin with the final simplex tableau given there. Using the sensitivity analysis procedure for Case 1, Sec. 6.7, we find that the entries in the *right side* column of the tableau change to the values given below.

$$y_0^* = \mathbf{y}^*\bar{\mathbf{b}} = [2, 6]\begin{bmatrix} -3 - 2\theta \\ -5 + \theta \end{bmatrix} = -36 + 2\theta,$$

$$\mathbf{b}^* = \mathbf{S}^*\bar{\mathbf{b}} = \begin{bmatrix} -\frac{1}{3} & 0 \\ \frac{1}{3} & -\frac{1}{2} \end{bmatrix}\begin{bmatrix} -3 - 2\theta \\ -5 + \theta \end{bmatrix} = \begin{bmatrix} 1 + \frac{2\theta}{3} \\ \frac{3}{2} - \frac{7\theta}{6} \end{bmatrix}.$$

Therefore, the two basic variables in this tableau

$$y_3 = \frac{3 + 2\theta}{3} \qquad \text{and} \qquad y_2 = \frac{9 - 7\theta}{6}$$

remain nonnegative for $0 \le \theta \le \frac{9}{7}$. Increasing θ past $\theta = \frac{9}{7}$ requires making y_2 a leaving basic variable for another iteration of the dual simplex method, and so on, as summarized in Table 7.3.

We suggest that you now trace through Tables 7.2 and 7.3 simultaneously to note the duality relationship between the two procedures.

7.3 The Upper Bound Technique

It is fairly common in linear programming problems for some of or all the *individual* x_j variables to have *upper bound constraints*

$$x_j \le u_j,$$

where u_j is a positive constant representing the maximum *feasible* value of x_j. We pointed out in Sec. 4.8 that the most important determinant of computation time for the simplex method is the *number of functional constraints,* whereas the number of *nonnegativity* constraints is relatively unimportant. Therefore, having a large number of upper bound constraints among the functional constraints greatly increases the computational effort required.

The *upper bound technique* avoids this increased effort by removing the upper bound constraints from the functional constraints and treating them separately, essentially like nonnegativity constraints. Removing the upper bound constraints in this way causes no problems as long as none of the variables gets increased over its upper bound. The only time the simplex method increases some of the variables is when the entering basic variable is increased to obtain a new BF solution. Therefore, the upper bound technique simply applies the simplex method in the usual way to the *remainder* of the problem (i.e., without the upper bound constraints) but with the one additional restriction that each new BF solution must satisfy the upper bound constraints in addition to the usual lower bound (nonnegativity) constraints.

To implement this idea, note that a decision variable x_j with an upper bound constraint $x_j \leq u_j$ can always be replaced by

$$x_j = u_j - y_j,$$

where y_j would then be the decision variable. In other words, you have a choice between letting the decision variable be the *amount above zero* (x_j) or the *amount below* u_j ($y_j = u_j - x_j$). (We shall refer to x_j and y_j as *complementary* decision variables.) Because

$$0 \leq x_j \leq u_j$$

it also follows that

$$0 \leq y_j \leq u_j.$$

Thus at any point during the simplex method, you can either

1. Use x_j, where $0 \leq x_j \leq u_j$.
2. Replace x_j by $u_j - y_j$, where $0 \leq y_j \leq u_j$.

The upper bound technique uses the following rule to make this choice:

Rule: Begin with choice 1.
Whenever $x_j = 0$, use choice 1, so x_j is *nonbasic.*
Whenever $x_j = u_j$, use choice 2, so $y_j = 0$ is *nonbasic.*
Switch choices only when the other extreme value of x_j is reached.

Therefore, whenever a basic variable reaches its upper bound, you should switch choices and use its complementary decision variable as the new nonbasic variable (the leaving basic variable) for identifying the new BF solution. Thus the one substantive modification being made in the simplex method is in the rule for selecting the leaving basic variable.

Recall that the simplex method selects as the leaving basic variable the one that would be the first to become infeasible by going negative as the entering basic variable is increased. The modification now made is to select instead the variable that would be the first to become infeasible *in any way,* either by going negative or by going over the upper bound, as the entering basic variable is increased. (Notice that one possibility is that the entering basic variable may become infeasible first by going over its upper bound, so that its complementary decision variable becomes the leaving basic variable.) If the leaving basic variable reaches zero, then proceed as usual with the simplex method. However, if it reaches its upper bound instead, then switch choices and make its complementary decision variable the leaving basic variable.

To illustrate, consider this problem:

$$\text{Maximize} \quad Z = 2x_1 + x_2 + 2x_3,$$

subject to

$$\begin{aligned} 4x_1 + x_2 \qquad &= 12 \\ -2x_1 \qquad + x_3 &= 4 \end{aligned}$$

and

$$0 \le x_1 \le 4, \qquad 0 \le x_2 \le 15, \qquad 0 \le x_3 \le 6.$$

Thus all three variables have upper bound constraints ($u_1 = 4, u_2 = 15, u_3 = 6$).

The two equality constraints are already in proper form from Gaussian elimination for identifying the initial BF solution ($x_1 = 0, x_2 = 12, x_3 = 4$), and none of the variables in this solution exceeds its upper bound, so x_2 and x_3 can be used as the initial basic variables without artificial variables being introduced. However, these variables then need to be eliminated algebraically from the objective function to obtain the initial Eq. (0), as follows:

$$\begin{array}{lrl} & Z \quad - 2x_1 - x_2 - 2x_3 &= 0 \\ & + (4x_1 + x_2 \qquad\quad &= 12) \\ & + 2(- 2x_1 \qquad + \; x_3 &= 4) \\ \hline (0) & Z \quad - 2x_1 \qquad\qquad\quad &= 20. \end{array}$$

To start the first iteration, this initial Eq. (0) indicates that the initial *entering* basic variable is x_1. Since the upper bound constraints are not to be included, the entire initial set of equations and the corresponding calculations for selecting the leaving basic variables are those shown in Table 7.4. The second column shows how much the entering basic variable x_1 can be *increased* from zero before some basic variable (including x_1) becomes infeasible. The maximum value given next to Eq. (0) is just the upper bound constraint for x_1. For Eq. (1), since the coefficient of x_1 is *positive, increasing* x_1 to 3 decreases the basic variable in this equation (x_2) from 12 to its *lower* bound of *zero*. For Eq. (2), since the coefficient of x_1 is *negative, increasing* x_1 to 1 *increases* the basic variable in this equation (x_3) from 4 to its *upper* bound of 6.

Table 7.4 **Equations and Calculations for the Initial Leaving Basic Variable in the Example for the Upper Bound Technique**

Initial Set of Equations	Maximum Feasible Value of x_1
(0) $Z - 2x_1 \qquad = 20$	$x_1 \le 4$ (since $u_1 = 4$)
(1) $4x_1 + x_2 \qquad = 12$	$x_1 \le \frac{12}{4} = 3$
(2) $-2x_1 \qquad + x_3 = 4$	$x_1 \le \frac{6-4}{2} = 1 \leftarrow$ minimum (because $u_3 = 6$)

Because this last maximum value of x_1 is the smallest, x_3 provides the *leaving* basic variable. However, because x_3 reached its *upper* bound, replace x_3 by $6 - y_3$, so that $y_3 = 0$ becomes the new nonbasic variable for the next BF solution and x_1 becomes the new basic variable in Eq. (2). This replacement leads to the following changes in this equation:

$$
\begin{array}{lrl}
(2) & -2x_1 + x_3 & = 4 \\
& \rightarrow \quad -2x_1 + 6 - y_3 & = 4 \\
& \rightarrow \quad -2x_1 - y_3 & = -2 \\
& \rightarrow \quad x_1 + \tfrac{1}{2}y_3 & = 1
\end{array}
$$

Therefore, after we eliminate x_1 algebraically from the other equations, the *second* complete set of equations becomes

$$
\begin{array}{lrl}
(0) & Z \qquad + y_3 & = 22 \\
(1) & x_2 - 2y_3 & = 8 \\
(2) & x_1 \qquad + \tfrac{1}{2}y_3 & = 1.
\end{array}
$$

The resulting BF solution is $x_1 = 1, x_2 = 8, y_3 = 0$. By the optimality test, it also is an optimal solution, so $x_1 = 1$, $x_2 = 8$, $x_3 = 6 - y_3 = 6$ is the desired solution for the original problem.

7.4 An Interior-Point Algorithm

In Sec. 4.9 we discussed a dramatic recent development in linear programming, the invention by Narendra Karmarkar of AT&T Bell Laboratories of a powerful algorithm for solving huge linear programming problems. We now introduce the nature of Karmarkar's approach by describing a relatively elementary variant (the *affine* or *affine-scaling* variant) of his algorithm.[1] (Your OR Courseware also includes this variant in the Procedure menu under the title *Solve Automatically by the Interior-Point Algorithm.*)

Throughout this section we shall focus on Karmarkar's main ideas on an intuitive level while avoiding mathematical details. In particular, we shall bypass certain details that are needed for the full implementation of the algorithm (e.g., how to find an initial feasible trial solution) but are not central to a basic conceptual understanding. The ideas to be described can be summarized as follows:

Concept 1: Shoot through the *interior* of the feasible region toward an optimal solution.

Concept 2: Move in a direction that improves the objective function value at the fastest possible rate.

Concept 3: Transform the feasible region to place the current trial solution near its center, thereby enabling a large improvement when concept 2 is implemented.

[1] The basic approach for this variant actually was proposed in 1967 by a Russian mathematician I. I. Dikin and then rediscovered soon after the appearance of Karmarkar's work by a number of researchers, including E. R. Barnes, T. M. Cavalier, and A. L. Soyster. Also see R. J. Vanderbei, M. S. Meketon, and B. A. Freedman, "A Modification of Karmarkar's Linear Programming Algorithm," *Algorithmica,* **1**(4) (Special Issue on New Approaches to Linear Programming): 395–407, 1986.

To illustrate these ideas throughout the section, we shall use the following example:

$$\text{Maximize} \qquad Z = x_1 + 2x_2,$$

subject to

$$x_1 + x_2 \le 8$$

and

$$x_1 \ge 0, \qquad x_2 \ge 0.$$

This problem is depicted graphically in Fig. 7.3, where the optimal solution is seen to be $(x_1, x_2) = (0, 8)$ with $Z = 16$.

The Relevance of the Gradient for Concepts 1 and 2

The algorithm begins with an initial trial solution that (like all subsequent trial solutions) lies in the *interior* of the feasible region, i.e., *inside the boundary* of the feasible region. Thus, for the example, the solution must not lie on any of the three lines ($x_1 = 0, x_2 = 0, x_1 + x_2 = 8$) that form the boundary of this region in Fig. 7.3. (A trial solution that lies on the boundary cannot be used because this would lead to the undefined mathematical operation of division by zero at one point in the algorithm.) We have arbitrarily chosen $(x_1, x_2) = (2, 2)$ to be the initial trial solution.

To begin implementing concepts 1 and 2, note in Fig. 7.3 that the direction of movement from (2, 2) that increases Z at the fastest possible rate is *perpendicular* to (and toward) the objective function line $Z = 16 = x_1 + 2x_2$. We have shown this direction by the arrow from (2, 2) to (3, 4). Using vector addition, we have

$$(3, 4) = (2, 2) + (1, 2),$$

where the vector (1, 2) is the **gradient** of the objective function. (We will discuss

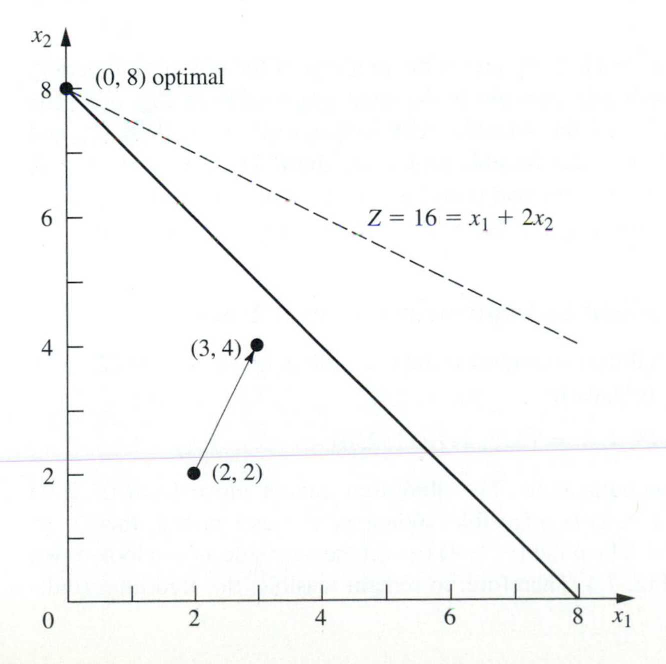

Figure 7.3 Example for the interior-point algorithm.

gradients further in Sec. 13.5 in the broader context of *nonlinear programming,* where algorithms similar to Karmarkar's have long been used.) The components of (1, 2) are just the coefficients in the objective function. Thus, with one subsequent modification, the gradient (1, 2) defines the ideal direction to which to move, where the question of the *distance to move* will be considered later.

The algorithm actually operates on linear programming problems after they have been rewritten in augmented form. Letting x_3 be the slack variable for the functional constraint of the example, we see that this form is

$$\text{Maximize} \qquad Z = x_1 + 2x_2,$$

subject to

$$x_1 + x_2 + x_3 = 8$$

and

$$x_1 \geq 0, \qquad x_2 \geq 0, \qquad x_3 \geq 0.$$

In matrix notation (slightly different from Chap. 5 because the slack variable now is incorporated into the notation), the augmented form can be written in general as

$$\text{Maximize} \qquad Z = \mathbf{c}^T\mathbf{x},$$

subject to

$$\mathbf{A}\mathbf{x} = \mathbf{b}$$

and

$$\mathbf{x} \geq \mathbf{0},$$

where

$$\mathbf{c} = \begin{bmatrix} 1 \\ 2 \\ 0 \end{bmatrix}, \qquad \mathbf{x} = \begin{bmatrix} x_1 \\ x_2 \\ x_3 \end{bmatrix}, \qquad \mathbf{A} = [1, \quad 1, \quad 1], \qquad \mathbf{b} = [8], \qquad \mathbf{0} = \begin{bmatrix} 0 \\ 0 \\ 0 \end{bmatrix}$$

for the example. Note that $\mathbf{c}^T = [1, 2, 0]$ now is the gradient of the objective function.

The augmented form of the example is depicted graphically in Fig. 7.4. The feasible region now consists of the triangle with vertices (8, 0, 0), (0, 8, 0), and (0, 0, 8). Points in the interior of this feasible region are those where $x_1 > 0$, $x_2 > 0$, and $x_3 > 0$. Each of these three $x_j > 0$ conditions has the effect of forcing (x_1, x_2) away from one of the three lines forming the boundary of the feasible region in Fig. 7.3.

Using the Projected Gradient to Implement Concepts 1 and 2

In augmented form, the initial trial solution for the example is $(x_1, x_2, x_3) = (2, 2, 4)$. Adding the gradient (1, 2, 0) leads to

$$(3, 4, 4) = (2, 2, 4) + (1, 2, 0).$$

However, now there is a complication. The algorithm cannot move from (2, 2, 4) toward (3, 4, 4), because (3, 4, 4) is infeasible! When $x_1 = 3$ and $x_2 = 4$, then $x_3 = 8 - x_1 - x_2 = 1$ instead of 4. The point (3, 4, 4) lies on the near side as you look down on the feasible triangle in Fig. 7.4. Therefore, to remain feasible, the algorithm (indi-

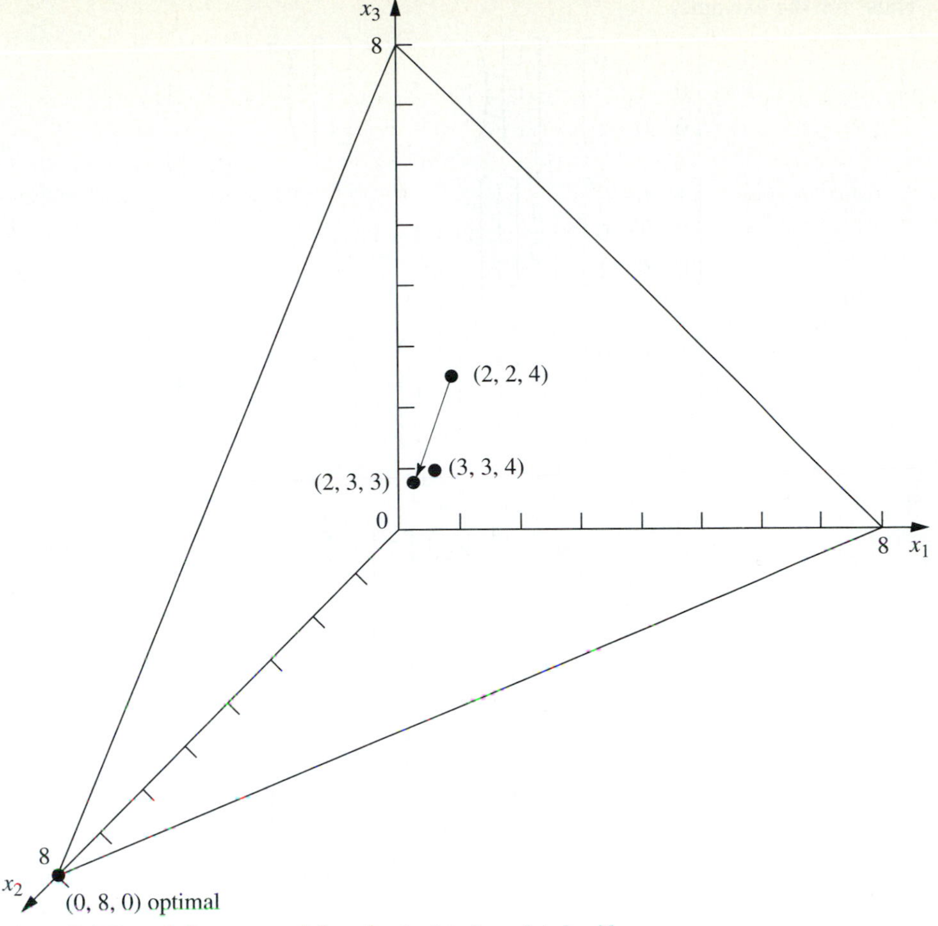

Figure 7.4 Example in augmented form for the interior-point algorithm.

rectly) *projects* the point (3, 4, 4) down onto the feasible triangle by dropping a line that is *perpendicular* to this triangle. A vector from (0, 0, 0) to (1, 1, 1) is perpendicular to this triangle, so the perpendicular line through (3, 4, 4) is given by the equation

$$(x_1, x_2, x_3) = (3, 4, 4) - \theta(1, 1, 1),$$

where θ is a scalar. Since the triangle satisfies the equation $x_1 + x_2 + x_3 = 8$, this perpendicular line intersects the triangle at (2, 3, 3). Because

$$(2, 3, 3) = (2, 2, 4) + (0, 1, -1),$$

the **projected gradient** of the objective function (the gradient projected onto the feasible region) is $(0, 1, -1)$. It is this projected gradient that defines the direction of movement for the algorithm, as shown by the arrow in Fig. 7.4.

A formula is available for computing the projected gradient directly. By defining the *projection matrix* **P** as

$$\mathbf{P} = \mathbf{I} - \mathbf{A}^T(\mathbf{A}\mathbf{A}^T)^{-1}\mathbf{A},$$

the *projected gradient* (in column form) is

$$\mathbf{c}_p = \mathbf{P}\mathbf{c}.$$

Thus, for the example,

$$\mathbf{P} = \begin{bmatrix} 1 & 0 & 0 \\ 0 & 1 & 0 \\ 0 & 0 & 1 \end{bmatrix} - \begin{bmatrix} 1 \\ 1 \\ 1 \end{bmatrix} \left([1 \quad 1 \quad 1] \begin{bmatrix} 1 \\ 1 \\ 1 \end{bmatrix} \right)^{-1} [1 \quad 1 \quad 1]$$

$$= \begin{bmatrix} 1 & 0 & 0 \\ 0 & 1 & 0 \\ 0 & 0 & 1 \end{bmatrix} - \frac{1}{3} \begin{bmatrix} 1 \\ 1 \\ 1 \end{bmatrix} [1 \quad 1 \quad 1]$$

$$= \begin{bmatrix} 1 & 0 & 0 \\ 0 & 1 & 0 \\ 0 & 0 & 1 \end{bmatrix} - \frac{1}{3} \begin{bmatrix} 1 & 1 & 1 \\ 1 & 1 & 1 \\ 1 & 1 & 1 \end{bmatrix} = \begin{bmatrix} \frac{2}{3} & -\frac{1}{3} & -\frac{1}{3} \\ -\frac{1}{3} & \frac{2}{3} & -\frac{1}{3} \\ -\frac{1}{3} & -\frac{1}{3} & \frac{2}{3} \end{bmatrix},$$

so

$$\mathbf{c}_p = \begin{bmatrix} \frac{2}{3} & -\frac{1}{3} & -\frac{1}{3} \\ -\frac{1}{3} & \frac{2}{3} & -\frac{1}{3} \\ -\frac{1}{3} & -\frac{1}{3} & \frac{2}{3} \end{bmatrix} \begin{bmatrix} 1 \\ 2 \\ 0 \end{bmatrix} = \begin{bmatrix} 0 \\ 1 \\ -1 \end{bmatrix}.$$

Moving from (2, 2, 4) in the direction of the projected gradient (0, 1, −1) involves increasing α from zero in the formula

$$\mathbf{x} = \begin{bmatrix} 2 \\ 2 \\ 4 \end{bmatrix} + 4\alpha \mathbf{c}_p = \begin{bmatrix} 2 \\ 2 \\ 4 \end{bmatrix} + 4\alpha \begin{bmatrix} 0 \\ 1 \\ -1 \end{bmatrix},$$

where the coefficient 4 is used simply to give an upper bound of 1 for α to maintain feasibility (all $x_j \geq 0$). Note that increasing α to $\alpha = 1$ would cause x_3 to decrease to $x_3 = 4 + 4(1)(-1) = 0$, where $\alpha > 1$ yields $x_3 < 0$. Thus α measures the fraction used of the distance that could be moved before the feasible region is left.

How large should α be made for moving to the next trial solution? Because the increase in Z is proportional to α, a value close to the upper bound of 1 is good for giving a relatively large step toward optimality on the current iteration. However, the problem with a value too close to 1 is that the next trial solution then is jammed against a constraint boundary, thereby making it difficult to take large improving steps during subsequent iterations. Therefore, it is very helpful for trial solutions to be near the center of the feasible region (or at least near the center of the portion of the feasible region in the vicinity of an optimal solution), and not too close to any constraint boundary. With this in mind, Karmarkar has stated for his algorithm that a value as large as $\alpha = 0.25$ should be "safe." In practice, much larger values (for example, $\alpha = 0.9$) sometimes are used. For the purposes of this example (and the problems at the end of the chapter), we have chosen $\alpha = 0.5$. (Your OR Courseware uses $\alpha = 0.5$ as the default value, but also has $\alpha = 0.9$ available under the Option menu.)

A Centering Scheme for Implementing Concept 3

We now have just one more step to complete the description of the algorithm, namely, a special scheme for transforming the feasible region to place the current trial solution near its center. We have just described the benefit of having the trial solution near the center, but another important benefit of this centering scheme is that it keeps turning the direction of the projected gradient to point more nearly toward an optimal solution as the algorithm converges toward this solution.

The basic idea of the centering scheme is straightforward—simply change the scale (units) for each of the variables so that the trial solution becomes equidistant from the constraint boundaries in the new coordinate system. (Karmarkar's original algorithm uses a more sophisticated centering scheme.)

For the example, there are three constraint boundaries in Fig. 7.3, each one corresponding to a zero value for one of the three variables of the problem in augmented form, namely, $x_1 = 0$, $x_2 = 0$, and $x_3 = 0$. In Fig. 7.4, see how these three constraint boundaries intersect the $\mathbf{Ax} = \mathbf{b}$ $(x_1 + x_2 + x_3 = 8)$ plane to form the boundary of the feasible region. The initial trial solution is $(x_1, x_2, x_3) = (2, 2, 4)$, so this solution is 2 units away from the $x_1 = 0$ and $x_2 = 0$ constraint boundaries and 4 units away from the $x_3 = 0$ constraint boundary, when the units of the respective variables are used. However, whatever these units are in each case, they are quite arbitrary and can be changed as desired without changing the problem. Therefore, let us rescale the variables as follows:

$$\tilde{x}_1 = \frac{x_1}{2}, \qquad \tilde{x}_2 = \frac{x_2}{2}, \qquad \tilde{x}_3 = \frac{x_3}{4}$$

in order to make the current trial solution of $(x_1, x_2, x_3) = (2, 2, 4)$ become

$$(\tilde{x}_1, \tilde{x}_2, \tilde{x}_3) = (1, 1, 1).$$

In these new coordinates (substituting $2\tilde{x}_1$ for x_1, $2\tilde{x}_2$ for x_2, and $4\tilde{x}_3$ for x_3), the problem becomes

$$\text{Maximize} \qquad Z = 2\tilde{x}_1 + 4\tilde{x}_2,$$

subject to

$$2\tilde{x}_1 + 2\tilde{x}_2 + 4\tilde{x}_3 = 8$$

and

$$\tilde{x}_1 \geq 0, \qquad \tilde{x}_2 \geq 0, \qquad \tilde{x}_3 \geq 0,$$

as depicted graphically in Fig. 7.5.

Note that the trial solution (1, 1, 1) in Fig. 7.5 is equidistant from the three constraint boundaries $\tilde{x}_1 = 0, \tilde{x}_2 = 0, \tilde{x}_3 = 0$. For each subsequent iteration as well, the problem is rescaled again to achieve this same property, so that the current trial solution always is (1, 1, 1) in the current coordinates.

Summary and Illustration of the Algorithm

Now let us summarize and illustrate the algorithm by going through the first iteration for the example, then giving a summary of the general procedure, and finally applying this summary to a second iteration.

Iteration 1: Given the initial trial solution $(x_1, x_2, x_3) = (2, 2, 4)$, let $\mathbf{D}$ be the corresponding *diagonal matrix* such that $\mathbf{x} = \mathbf{D}\tilde{\mathbf{x}}$, so that

$$\mathbf{D} = \begin{bmatrix} 2 & 0 & 0 \\ 0 & 2 & 0 \\ 0 & 0 & 4 \end{bmatrix}.$$

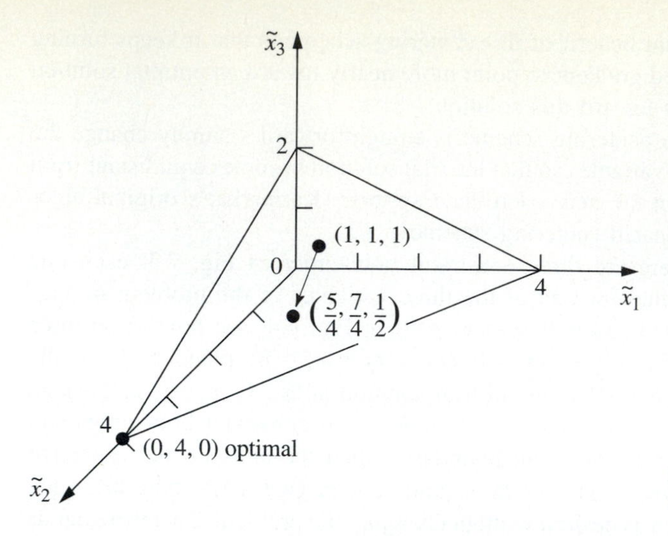

Figure 7.5 Example after rescaling for iteration 1.

The rescaled variables then are the components of

$$\tilde{\mathbf{x}} = \mathbf{D}^{-1}\mathbf{x} = \begin{bmatrix} \frac{1}{2} & 0 & 0 \\ 0 & \frac{1}{2} & 0 \\ 0 & 0 & \frac{1}{4} \end{bmatrix} \begin{bmatrix} x_1 \\ x_2 \\ x_3 \end{bmatrix} = \begin{bmatrix} \frac{x_1}{2} \\ \frac{x_2}{2} \\ \frac{x_3}{4} \end{bmatrix}.$$

In these new coordinates, **A** and **c** have become

$$\tilde{\mathbf{A}} = \mathbf{AD} = [1 \quad 1 \quad 1] \begin{bmatrix} 2 & 0 & 0 \\ 0 & 2 & 0 \\ 0 & 0 & 4 \end{bmatrix} = [2 \quad 2 \quad 4],$$

$$\tilde{\mathbf{c}} = \mathbf{Dc} = \begin{bmatrix} 2 & 0 & 0 \\ 0 & 2 & 0 \\ 0 & 0 & 4 \end{bmatrix} \begin{bmatrix} 1 \\ 2 \\ 0 \end{bmatrix} = \begin{bmatrix} 2 \\ 4 \\ 0 \end{bmatrix}.$$

Therefore, the projection matrix is

$$\mathbf{P} = \mathbf{I} - \tilde{\mathbf{A}}^T(\tilde{\mathbf{A}}\tilde{\mathbf{A}}^T)^{-1}\tilde{\mathbf{A}}$$

$$= \begin{bmatrix} 1 & 0 & 0 \\ 0 & 1 & 0 \\ 0 & 0 & 1 \end{bmatrix} - \begin{bmatrix} 2 \\ 2 \\ 4 \end{bmatrix} \left([2 \quad 2 \quad 4] \begin{bmatrix} 2 \\ 2 \\ 4 \end{bmatrix} \right)^{-1} [2 \quad 2 \quad 4]$$

$$= \begin{bmatrix} 1 & 0 & 0 \\ 0 & 1 & 0 \\ 0 & 0 & 1 \end{bmatrix} - \frac{1}{24} \begin{bmatrix} 4 & 4 & 8 \\ 4 & 4 & 8 \\ 8 & 8 & 16 \end{bmatrix} = \begin{bmatrix} \frac{5}{6} & -\frac{1}{6} & -\frac{1}{3} \\ -\frac{1}{6} & \frac{5}{6} & -\frac{1}{3} \\ -\frac{1}{3} & -\frac{1}{3} & \frac{1}{3} \end{bmatrix},$$

so that the projected gradient is

$$\mathbf{c}_p = \mathbf{P}\tilde{\mathbf{c}} = \begin{bmatrix} \frac{5}{6} & -\frac{1}{6} & -\frac{1}{3} \\ -\frac{1}{6} & \frac{5}{6} & -\frac{1}{3} \\ -\frac{1}{3} & -\frac{1}{3} & \frac{1}{3} \end{bmatrix} \begin{bmatrix} 2 \\ 4 \\ 0 \end{bmatrix} = \begin{bmatrix} 1 \\ 3 \\ -2 \end{bmatrix}.$$

Define v as the *absolute value* of the *negative* component of $\mathbf{c}_p$ having the *largest* absolute value, so that $v = |-2| = 2$ in this case. Consequently, in the current coordinates, the algorithm now moves from the current trial solution $(\tilde{x}_1, \tilde{x}_2, \tilde{x}_3) = (1, 1, 1)$ to the next trial solution

$$\tilde{\mathbf{x}} = \begin{bmatrix} 1 \\ 1 \\ 1 \end{bmatrix} + \frac{\alpha}{v}\mathbf{c}_p = \begin{bmatrix} 1 \\ 1 \\ 1 \end{bmatrix} + \frac{0.5}{2}\begin{bmatrix} 1 \\ 3 \\ -2 \end{bmatrix} = \begin{bmatrix} \frac{5}{4} \\ \frac{7}{4} \\ \frac{1}{2} \end{bmatrix},$$

as shown in Fig. 7.5. (The definition of v has been chosen to make the smallest component of $\tilde{\mathbf{x}}$ equal to zero when $\alpha = 1$ in this equation for the next trial solution.) In the original coordinates, this solution is

$$\begin{bmatrix} x_1 \\ x_2 \\ x_3 \end{bmatrix} = \mathbf{D}\tilde{\mathbf{x}} = \begin{bmatrix} 2 & 0 & 0 \\ 0 & 2 & 0 \\ 0 & 0 & 4 \end{bmatrix} \begin{bmatrix} \frac{5}{4} \\ \frac{7}{4} \\ \frac{1}{2} \end{bmatrix} = \begin{bmatrix} \frac{5}{2} \\ \frac{7}{2} \\ 2 \end{bmatrix}.$$

This completes the iteration, and this new solution will be used to start the next iteration.

These steps can be summarized as follows for any iteration.

Summary of the Interior-Point Algorithm

1. Given the current trial solution $(x_1, x_2, \ldots, x_n)$, set

$$\mathbf{D} = \begin{bmatrix} x_1 & 0 & 0 & \cdots & 0 \\ 0 & x_2 & 0 & \cdots & 0 \\ 0 & 0 & x_3 & \cdots & 0 \\ \cdots & \cdots & \cdots & \cdots & \cdots \\ 0 & 0 & 0 & \cdots & x_n \end{bmatrix}.$$

2. Calculate $\tilde{\mathbf{A}} = \mathbf{A}\mathbf{D}$ and $\tilde{\mathbf{c}} = \mathbf{D}\mathbf{c}$.
3. Calculate $\mathbf{P} = \mathbf{I} - \tilde{\mathbf{A}}^T(\tilde{\mathbf{A}}\tilde{\mathbf{A}}^T)^{-1}\tilde{\mathbf{A}}$ and $\mathbf{c}_p = \mathbf{P}\tilde{\mathbf{c}}$.
4. Identify the negative component of $\mathbf{c}_p$ having the largest absolute value, and set v equal to this absolute value. Then calculate

$$\tilde{\mathbf{x}} = \begin{bmatrix} 1 \\ 1 \\ \vdots \\ 1 \end{bmatrix} + \frac{\alpha}{v}\mathbf{c}_p,$$

   where α is a selected constant between 0 and 1 (for example, $\alpha = 0.5$).
5. Calculate $\mathbf{x} = \mathbf{D}\tilde{\mathbf{x}}$ as the trial solution for the next iteration (step 1). (If this trial solution is virtually unchanged from the preceding one, then the algorithm has virtually converged to an optimal solution, so stop.)

Now let us apply this summary to iteration 2 for the example.

Iteration 2:

Step 1:

Given the current trial solution $(x_1, x_2, x_3) = (\frac{5}{2}, \frac{7}{2}, 2)$, set

$$\mathbf{D} = \begin{bmatrix} \frac{5}{2} & 0 & 0 \\ 0 & \frac{7}{2} & 0 \\ 0 & 0 & 2 \end{bmatrix}.$$

(Note that the rescaled variables are

$$\begin{bmatrix} \tilde{x}_1 \\ \tilde{x}_2 \\ \tilde{x}_3 \end{bmatrix} = \mathbf{D}^{-1}\mathbf{x} = \begin{bmatrix} \frac{2}{5} & 0 & 0 \\ 0 & \frac{2}{7} & 0 \\ 0 & 0 & \frac{1}{2} \end{bmatrix} \begin{bmatrix} x_1 \\ x_2 \\ x_3 \end{bmatrix} = \begin{bmatrix} \frac{2}{5}x_1 \\ \frac{2}{7}x_2 \\ \frac{1}{2}x_3 \end{bmatrix},$$

so that the BF solutions in these new coordinates are

$$\tilde{\mathbf{x}} = \mathbf{D}^{-1}\begin{bmatrix} 8 \\ 0 \\ 0 \end{bmatrix} = \begin{bmatrix} \frac{16}{5} \\ 0 \\ 0 \end{bmatrix}, \qquad \tilde{\mathbf{x}} = \mathbf{D}^{-1}\begin{bmatrix} 0 \\ 8 \\ 0 \end{bmatrix} = \begin{bmatrix} 0 \\ \frac{16}{7} \\ 0 \end{bmatrix},$$

and

$$\tilde{\mathbf{x}} = \mathbf{D}^{-1}\begin{bmatrix} 0 \\ 0 \\ 8 \end{bmatrix} = \begin{bmatrix} 0 \\ 0 \\ 4 \end{bmatrix},$$

as depicted in Fig. 7.6.)

Step 2:

$$\tilde{\mathbf{A}} = \mathbf{AD} = [\tfrac{5}{2}, \tfrac{7}{2}, 2] \qquad \text{and} \qquad \tilde{\mathbf{c}} = \mathbf{Dc} = \begin{bmatrix} \frac{5}{2} \\ 7 \\ 0 \end{bmatrix}.$$

Step 3:

$$\mathbf{P} = \begin{bmatrix} \frac{13}{18} & -\frac{7}{18} & -\frac{2}{9} \\ -\frac{7}{18} & \frac{41}{90} & -\frac{14}{45} \\ -\frac{2}{9} & -\frac{14}{45} & \frac{37}{45} \end{bmatrix} \qquad \text{and} \qquad \mathbf{c}_p = \begin{bmatrix} -\frac{11}{12} \\ \frac{133}{60} \\ -\frac{41}{15} \end{bmatrix}.$$

Step 4:

$|-\frac{41}{15}| > |-\frac{11}{12}|$, so $v = \frac{41}{15}$ and

$$\tilde{\mathbf{x}} = \begin{bmatrix} 1 \\ 1 \\ 1 \end{bmatrix} + \frac{0.5}{\frac{41}{15}} \begin{bmatrix} -\frac{11}{12} \\ \frac{133}{60} \\ -\frac{41}{15} \end{bmatrix} = \begin{bmatrix} \frac{273}{328} \\ \frac{461}{328} \\ \frac{1}{2} \end{bmatrix} \approx \begin{bmatrix} 0.83 \\ 1.40 \\ 0.50 \end{bmatrix}.$$

Step 5:

$$\mathbf{x} = \mathbf{D}\tilde{\mathbf{x}} = \begin{bmatrix} \frac{1365}{656} \\ \frac{3227}{656} \\ 1 \end{bmatrix} \approx \begin{bmatrix} 2.08 \\ 4.92 \\ 1.00 \end{bmatrix}$$

is the trial solution for iteration 3.

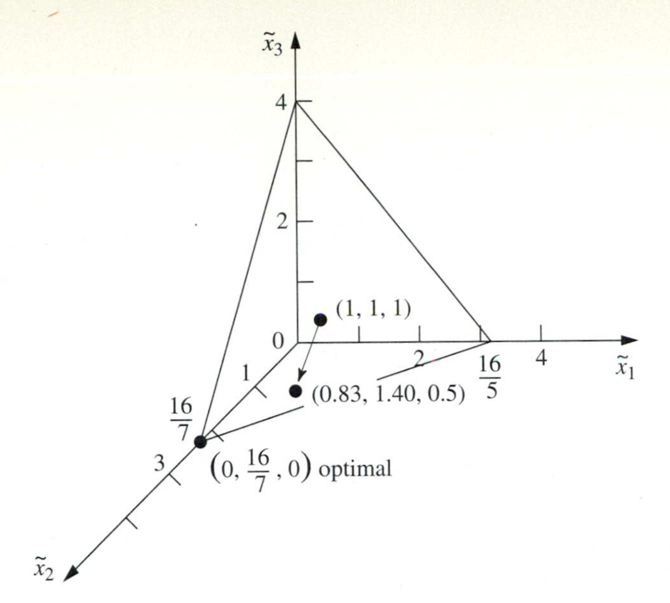

Figure 7.6 Example after rescaling for iteration 2.

Since there is little to be learned by repeating these calculations for additional iterations, we shall stop here. However, we do show in Fig. 7.7 the reconfigured feasible region after rescaling based on the trial solution just obtained for iteration 3. As always, the rescaling has placed the trial solution at $(\tilde{x}_1, \tilde{x}_2, \tilde{x}_3) = (1, 1, 1)$, equidistant from the $\tilde{x}_1 = 0$, $\tilde{x}_2 = 0$, and $\tilde{x}_3 = 0$ constraint boundaries. Note in Figs. 7.5, 7.6, and 7.7 how the sequence of iterations and rescaling have the effect of "sliding" the optimal solution toward (1, 1, 1) while the other BF solutions tend to slide away. Eventually, after enough iterations, the optimal solution will lie very near $(\tilde{x}_1, \tilde{x}_2, \tilde{x}_3) = (0, 1, 0)$ after rescaling, while the other two BF solutions will be *very* far from the origin on the $\tilde{x}_1$ and $\tilde{x}_3$ axes. Step 5 of that iteration then will yield a solution in the original coordinates very near the optimal solution of $(x_1, x_2, x_3) = (0, 8, 0)$.

Figure 7.8 shows the progress of the algorithm in the original $x_1 = x_2$ coordinate system before the problem is augmented. The three points—$(x_1, x_2) = (2, 2)$, (2.5, 3.5), and (2.08, 4.92)—are the trial solutions for initiating iterations 1, 2, and 3, respectively. We then have drawn a smooth curve through and beyond these points to show the trajectory of the algorithm in subsequent iterations as it approaches $(x_1, x_2) = (0, 8)$.

The functional constraint for this particular example happened to be an inequality constraint. However, equality constraints cause no difficulty for the algorithm, since it deals with the constraints only after any necessary augmenting has been done to convert them to equality form ($\mathbf{Ax} = \mathbf{b}$) anyway. To illustrate, suppose that the only change in the example is that the constraint $x_1 + x_2 \leq 8$ is changed to $x_1 + x_2 = 8$. Thus the feasible region in Fig. 7.3 changes to just the line segment between (8, 0) and (0, 8). Given an initial feasible trial solution in the interior ($x_1 > 0$ and $x_2 > 0$) of this line segment—say, $(x_1, x_2) = (4, 4)$—the algorithm can proceed just as presented in the five-step summary with just the two variables and $\mathbf{A} = [1, 1]$. For each iteration, the projected gradient points along this line segment in the direction of (0, 8). With $\alpha = \frac{1}{2}$, iteration 1 leads from (4, 4) to (2, 6), iteration 2 leads from (2, 6) to (1, 7), etc. (Problem 7.4-3 asks you to verify these results.)

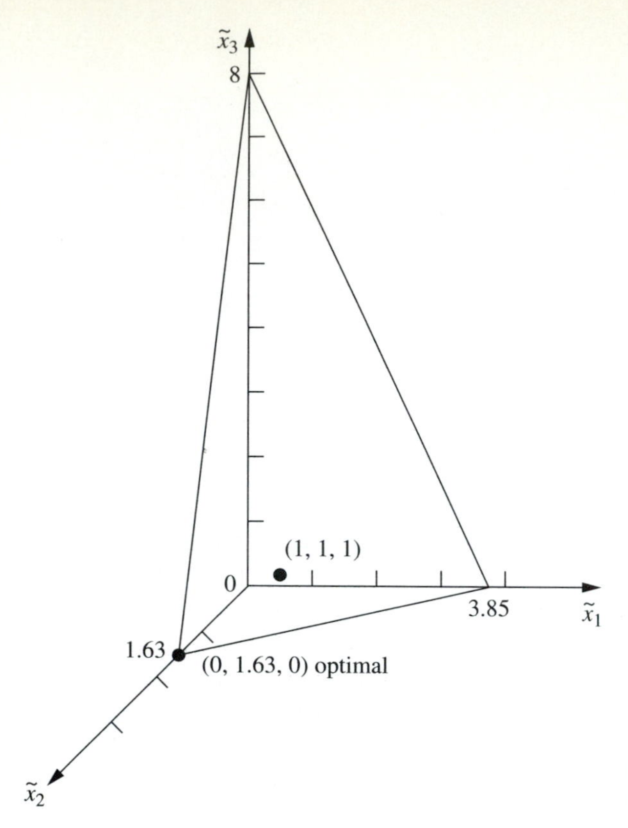

Figure 7.7 Example after rescaling for iteration 3.

Although either version of the example has only one functional constraint, having more than one leads to just one change in the procedure as already illustrated (other than more extensive calculations). Having a single functional constraint in the example meant that **A** had only a single row, so the $(\tilde{\mathbf{A}}\tilde{\mathbf{A}}^T)^{-1}$ term in step 3 only involved taking the reciprocal of the number obtained from the vector product $\tilde{\mathbf{A}}\tilde{\mathbf{A}}^T$. Multiple functional constraints mean that **A** has multiple rows, so then the $(\tilde{\mathbf{A}}\tilde{\mathbf{A}}^T)^{-1}$ term involves finding the *inverse* of the matrix obtained from the matrix product $\tilde{\mathbf{A}}\tilde{\mathbf{A}}^T$.

To conclude, we need to add a comment to place the algorithm into better perspective. For our extremely small example, the algorithm requires relatively extensive calculations and then, after many iterations, obtains only an approximation of the optimal solution. By contrast, the graphical procedure of Sec. 3.1 finds the optimal solution in Fig. 7.3 immediately, and the simplex method requires only one quick iteration. However, do not let this contrast fool you into downgrading the efficiency of the interior-point algorithm. This algorithm is designed for dealing with *big* problems having many hundreds or thousands of functional constraints. The simplex method typically requires thousands of iterations on such problems. By ''shooting'' through the interior of the feasible region, the interior-point algorithm tends to require a substantially smaller number of iterations (although with considerably more work per iteration). Therefore, as discussed in Sec. 4.9, interior-point algorithms similar to the one presented here should play an important role in the future of linear programming.

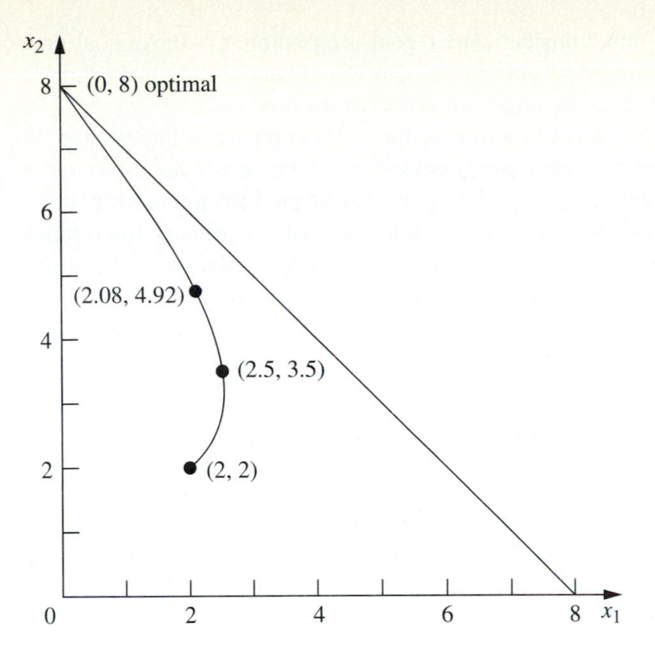

Figure 7.8 Trajectory of the interior-point algorithm for the example in the original (x_1, x_2) coordinate system.

7.5 Linear Goal Programming and Its Solution Procedures

We have assumed throughout the preceding chapters that the objectives of the organization conducting the linear programming study can be encompassed within a single overriding objective, such as maximizing total profit or minimizing total cost. However, this assumption is not always realistic. In fact, as we discussed in Sec. 2.1, studies have found that the management of U.S. corporations frequently focuses on a variety of other objectives, e.g., to maintain stable profits, increase (or maintain) market share, diversify products, maintain stable prices, improve worker morale, maintain family control of the business, and increase company prestige. *Goal programming* provides a way of striving toward several such objectives *simultaneously*.

The basic approach of **goal programming** is to establish a specific numeric goal for each of the objectives, formulate an objective function for each objective, and then seek a solution that minimizes the (weighted) sum of deviations of these objective functions from their respective goals. There are three possible types of goals:

1. A **lower, one-sided goal** sets a *lower limit* that we do not want to fall under (but exceeding the limit is fine).
2. An **upper, one-sided goal** sets an *upper limit* that we do not want to exceed (but falling under the limit is fine).
3. A **two-sided goal** sets a *specific target* that we do not want to miss on either side.

Goal programming problems can be categorized according to the type of mathematical programming model (linear programming, integer programming, nonlinear programming, etc.) that it fits except for having multiple goals instead of a single objective. In this book, we only consider *linear* goal programming—those goal programming problems that fit linear programming otherwise (each objective function is linear, etc.) and so we will drop the adjective *linear* from now on.

Another categorization is according to how the goals compare in importance. In one case, called **nonpreemptive goal programming**, all the goals are of *roughly comparable importance*. In another case, called **preemptive goal programming**, there is a *hierarchy of priority levels* for the goals, so that the goals of primary importance receive first-priority attention, those of secondary importance receive second-priority attention, and so forth (if there are more than two priority levels).

We begin with an example that illustrates the basic features of nonpreemptive goal programming and then discuss the preemptive case.

Prototype Example for Nonpreemptive Goal Programming

The DEWRIGHT COMPANY is considering three new products to replace current models that are being discontinued, so their OR department has been assigned the task of determining which mix of these products should be produced. Management wants primary consideration given to three factors: long-run profit, stability in the workforce, and the level of capital investment that would be required now for new equipment. In particular, management has established the goals of (1) achieving a long-run profit (net present value) of at least \$125 million from these products, (2) maintaining the current employment level of 4,000 employees, and (3) holding the capital investment to less than \$55 million. However, management realizes that it probably will not be possible to attain all these goals simultaneously, so it has discussed priorities with the OR department. This discussion has led to setting *penalty weights* of 5 for missing the profit goal (per \$1 million under), 2 for going over the employment goal (per 100 employees), 4 for going under this same goal, and 3 for exceeding the capital investment goal (per \$1 million over). Each new product's contribution to profit, employment level, and capital investment level is *proportional* to the rate of production. These contributions per unit rate of production are shown in Table 7.5, along with the goals and penalty weights.

Formulation: The Dewright Company problem includes all three possible types of goals: a lower, one-sided goal (long-run profit); a two-sided goal (employment level); and an upper, one-sided goal (capital investment). Letting the decision variables x_1, x_2, x_3 be the production rates of products 1, 2, and 3, respectively, we see that these goals can be stated as

$$
\begin{aligned}
12x_1 + 9x_2 + 15x_3 &\geq 125 \quad && \text{profit goal} \\
5x_1 + 3x_2 + 4x_3 &= 40 && \text{employment goal} \\
5x_1 + 7x_2 + 8x_3 &\leq 55 && \text{investment goal.}
\end{aligned}
$$

Table 7.5 **Data for the Dewright Co. Nonpreemptive Goal Programming Problem**

	Unit Contribution				
	Product:				
Factor	1	2	3	Goal (Units)	Penalty Weight
Long-run profit	12	9	15	≥ 125 (millions of dollars)	5
Employment level	5	3	4	= 40 (hundreds of employees)	2(+), 4(−)
Capital investment	5	7	8	≤ 55 (millions of dollars)	3

More precisely, given the penalty weights in the rightmost column of Table 7.5, let Z be the *number of penalty points* incurred by missing these goals. The overall objective then is to choose the values of x_1, x_2, and x_3 so as to

$$\begin{aligned} \text{Minimize} \quad Z = {} & 5(\text{amount under the long-run profit goal}) \\ & + 2(\text{amount over the employment level goal}) \\ & + 4(\text{amount under the employment level goal}) \\ & + 3(\text{amount over the capital investment goal}), \end{aligned}$$

where no penalty points are incurred for being over the long-run profit goal or for being under the capital investment goal. To express this overall objective mathematically, we introduce some *auxiliary variables* (extra variables that are helpful for formulating the model) y_1, y_2, and y_3, defined as follows:

$$\begin{aligned} y_1 &= 12x_1 + 9x_2 + 15x_3 - 125 && \text{(long-run profit minus the target).} \\ y_2 &= 5x_1 + 3x_2 + 4x_3 - 40 && \text{(employment level minus the target).} \\ y_3 &= 5x_1 + 7x_2 + 8x_3 - 55 && \text{(capital investment minus the target).} \end{aligned}$$

Since each y_i can be either positive or negative, we next use the technique described at the end of Sec. 4.6 for dealing with such variables; namely, we replace each one by the difference of two nonnegative variables:

$$\begin{aligned} y_1 &= y_1^+ - y_1^-, && \text{where } y_1^+ \geq 0,\ y_1^- \geq 0, \\ y_2 &= y_2^+ - y_2^-, && \text{where } y_2^+ \geq 0,\ y_2^- \geq 0, \\ y_3 &= y_3^+ - y_3^-, && \text{where } y_3^+ \geq 0,\ y_3^- \geq 0. \end{aligned}$$

As discussed in Sec. 4.6, for any BF solution, these new auxiliary variables have the interpretation

$$y_j^+ = \begin{cases} y_j & \text{if } y_j \geq 0, \\ 0 & \text{otherwise;} \end{cases}$$

$$y_j^- = \begin{cases} |y_j| & \text{if } y_j \leq 0, \\ 0 & \text{otherwise;} \end{cases}$$

so that y_j^+ represents the positive part of the variable y_j and y_j^- its negative part (as suggested by the superscripts).

Given these new auxiliary variables, the overall objective can be expressed mathematically as

$$\text{Minimize} \quad Z = 5y_1^- + 2y_2^+ + 4y_2^- + 3y_3^+,$$

which now is a legitimate objective function for a linear programming model. (Because there is no penalty for exceeding the profit goal of 125 or being under the investment goal of 55, neither y_1^+ nor y_3^- should appear in this objective function representing the total penalty for deviations from the goals.)

To complete the conversion of this goal programming problem to a linear programming model, we must incorporate the above definitions of the y_j^+ and y_j^- directly into the model. (It is not enough to simply record the definitions, as we just did, because the simplex method considers only the objective function and constraints that constitute the model.) For example, since $y_1^+ - y_1^- = y_1$, the above expression for y_1 gives

$$12x_1 + 9x_2 + 15x_3 - 125 = y_1^+ - y_1^-.$$

After we move the variables $(y_1^+ - y_1^-)$ to the left-hand side and the constant (125) to the right-hand side,

$$12x_1 + 9x_2 + 15x_3 - (y_1^+ - y_1^-) = 125$$

becomes a legitimate equality constraint for a linear programming model. Furthermore, this constraint forces the auxiliary variables $(y_1^+ - y_1^-)$ to satisfy their definition in terms of the decision variables (x_1, x_2, x_3).

Proceeding in the same way for $y_2^+ - y_2^-$ and $y_3^+ - y_3^-$, we obtain the following linear programming formulation of this goal programming problem:

$$\text{Minimize} \quad Z = 5y_1^- + 2y_2^+ + 4y_2^- + 3y_3^+,$$

subject to

$$\begin{aligned} 12x_1 + 9x_2 + 15x_3 - (y_1^+ - y_1^-) &= 125 \\ 5x_1 + 3x_2 + 4x_3 - (y_2^+ - y_2^-) &= 40 \\ 5x_1 + 7x_2 + 8x_3 - (y_3^+ - y_3^-) &= 55 \end{aligned}$$

and

$$x_j \geq 0, \quad y_k^+ \geq 0, \quad y_k^- \geq 0 \quad (j = 1, 2, 3;\ k = 1, 2, 3).$$

(If the original problem had any actual linear programming constraints, such as constraints on fixed amounts of certain resources being available, these would be included in the model.)

Applying the simplex method to this formulation yields an optimal solution $x_1 = \frac{25}{3}$, $x_2 = 0$, $x_3 = \frac{5}{3}$, with $y_1^+ = 0$, $y_1^- = 0$, $y_2^+ = \frac{25}{3}$, $y_2^- = 0$, $y_3^+ = 0$, and $y_3^- = 0$. Therefore, $y_1 = 0$, $y_2 = \frac{25}{3}$, and $y_3 = 0$, so the first and third goals are fully satisfied, but the employment level goal of 40 is exceeded by $8\frac{1}{3}$ (833 employees). The resulting penalty for deviating from the goals is $Z = 16\frac{2}{3}$.

Preemptive Goal Programming

In the preceding example we assume that all the goals are of roughly comparable importance. Now consider the case of *preemptive* goal programming, where there is a hierarchy of priority levels for the goals. Such a case arises when one or more of the goals clearly are far more important than the others. Thus the initial focus should be on achieving as closely as possible these *first-priority* goals. The other goals also might naturally divide further into second-priority goals, third-priority goals, and so on. After we find an optimal solution with respect to the first-priority goals, we can break any ties for the optimal solution by considering the second-priority goals. Any ties that remain after this reoptimization can be broken by considering the third-priority goals, and so on.

When we deal with goals on the *same* priority level, our approach is just like the one described for nonpreemptive goal programming. Any of the same three types of goals (lower one-sided, two-sided, upper one-sided) can arise. Different penalty weights for deviations from different goals still can be included, if desired. The same formulation technique of introducing auxiliary variables again is used to reformulate this portion of the problem to fit the linear programming format.

There are two basic methods based on linear programming for solving preemptive goal programming problems. One is called the *sequential procedure,* and the other is the *streamlined procedure.* We shall illustrate these procedures in turn by solving the following example.

EXAMPLE: Faced with the unpleasant recommendation to increase the company's workforce by more than 20 percent, the management of the Dewright Company has reconsidered the original formulation of the problem that was summarized in Table 7.5. This increase in workforce probably would be a rather temporary one, so the very high cost of training 833 new employees would be largely wasted, and the large (undoubtedly well-publicized) layoffs would make it more difficult for the company to attract high-quality employees in the future. Consequently, management has concluded that a very high priority should be placed on avoiding an increase in the workforce. Furthermore, management has learned that raising *more than* \$55 million for capital investment for the new products would be extremely difficult, so a very high priority also should be placed on avoiding capital investment above this level.

Based on these considerations, management has concluded that a *preemptive goal programming* approach now should be used, where the two goals just discussed should be the first-priority goals, and the other two original goals (exceeding \$125 million in long-run profit and avoiding a decrease in the employment level) should be the second-priority goals. Within the two priority levels, management feels that the relative penalty weights still should be the same as those given in the rightmost column of Table 7.5. This reformulation is summarized in Table. 7.6, where a factor of M (representing a huge positive number) has been included in the penalty weights for the first-priority goals to emphasize that these goals preempt the second-priority goals. (The portions of Table 7.5 that are not included in Table 7.6 are *unchanged.*)

The Sequential Procedure for Preemptive Goal Programming

The *sequential procedure* solves a preemptive goal programming problem by solving a *sequence* of linear programming models.

At the first stage of the sequential procedure, the only goals included in the linear programming model are the first-priority goals, and the simplex method is applied in the usual way. If the resulting optimal solution is *unique,* we adopt it immediately without considering any additional goals.

However, if there are *multiple* optimal solutions with the same optimal value of Z (call it Z^*), we prepare to break the tie among these solutions by moving to the second

Table 7.6 **Revised Formulation for the Dewright Co. Preemptive Goal Programming Problem**

Priority Level	Factor	Goal	Penalty Weight
First priority	Employment level	≤ 40	$2M$
	Capital investment	≤ 55	$3M$
Second priority	Long-run profit	≥ 125	5
	Employment level	≥ 40	4

stage and adding the second-priority goals to the model. If $Z^* = 0$, all the auxiliary variables representing the *deviations from first-priority goals* must equal zero (full achievement of these goals) for the solutions remaining under consideration. Thus, in this case, all these auxiliary variables now can be completely deleted from the model, where the equality constraints that contain these variables are replaced by the mathematical expressions (inequalities or equations) for these first-priority goals, to ensure that they continue to be fully achieved. On the other hand, if $Z^* > 0$, the second-stage model simply adds the second-priority goals to the first-stage model (as if these additional goals actually were first-priority goals), but then it also adds the constraint that the *first-stage objective function* equal Z^* (which enables us again to delete the terms involving first-priority goals from the second-stage objective function). After we apply the simplex method again, if there still are multiple optimal solutions, we repeat the same process for any lower-priority goals.

Example: We now illustrate this procedure by applying it to the example summarized in Table 7.6.

At the first stage, only the two *first-priority* goals are included in the linear programming model. Therefore, we can drop the common factor M for their penalty weights, shown in Table 7.6. By proceeding just as for the nonpreemptive model if these were the only goals, the resulting linear programming model is

$$\text{Minimize} \qquad Z = 2y_2^+ + 3y_3^+,$$

subject to

$$5x_1 + 3x_2 + 4x_3 - (y_2^+ - y_2^-) = 40$$
$$5x_1 + 7x_2 + 8x_3 - (y_3^+ - y_3^-) = 55$$

and

$$x_j \geq 0, \qquad y_k^+ \geq 0, \qquad y_k^- \geq 0 \qquad (j = 1, 2, 3;\ k = 2, 3).$$

(For ease of comparison with the nonpreemptive model with all four goals, we have kept the same subscripts on the auxiliary variables.)

By using the simplex method (or inspection), an optimal solution for this linear programming model has $y_2^+ = 0$ and $y_3^+ = 0$, with $Z = 0$ (so $Z^* = 0$), because there are innumerable soiutions for (x_1, x_2, x_3) that satisfy the relationships

$$5x_1 + 3x_2 + 4x_3 \leq 40$$
$$5x_1 + 7x_2 + 8x_3 \leq 55$$

as well as the nonnegativity constraints. Therefore, these two first-priority goals should be used as *constraints* hereafter. Using them as constraints will force y_2^+ and y_3^+ to remain zero and thereby disappear from the model automatically.

If we drop y_2^+ and y_3^+ but add the second-priority goals, the second-stage linear programming model becomes

$$\text{Minimize} \qquad Z = 5y_1^- + 4y_2^-,$$

subject to

$$\begin{aligned} 12x_1 + 9x_2 + 15x_3 - (y_1^+ - y_1^-) \qquad\qquad &= 125 \\ 5x_1 + 3x_2 + 4x_3 \qquad\qquad + y_2^- \qquad &= 40 \\ 5x_1 + 7x_2 + 8x_3 \qquad\qquad\qquad + y_3^- &= 55 \end{aligned}$$

and

$$x_j \geq 0, \qquad y_1^+ \geq 0, \qquad y_k^- \geq 0 \qquad (j = 1, 2, 3;\ k = 1, 2, 3).$$

Applying the simplex method to this model yields the unique optimal solution $x_1 = 5$, $x_2 = 0$, $x_3 = 3\frac{3}{4}$, $y_1^+ = 0$, $y_1^- = 8\frac{3}{4}$, $y_2^- = 0$, and $y_3^- = 0$, with $Z = 43\frac{3}{4}$.

Because this solution is unique (*or* because there are no more priority levels), the procedure can now stop, with $(x_1, x_2, x_3) = (5, 0, 3\frac{3}{4})$ as the optimal solution for the *overall* problem. This solution fully achieves both first-priority goals as well as one of the second-priority goals (no decrease in employment level), and it falls short of the other second-priority goal (long-run profit ≥ 125) by just $8\frac{3}{4}$.

The Streamlined Procedure for Preemptive Goal Programming

Instead of solving a sequence of linear programming models, like the sequential procedure, the *streamlined procedure* finds an optimal solution for a preemptive goal programming problem by solving just *one* linear programming model. Thus, the streamlined procedure is able to duplicate the work of the sequential procedure with just *one run* of the simplex method. This one run *simultaneously* finds optimal solutions based just on first-priority goals and breaks ties among these solutions by considering lower-priority goals. However, this does require a slight modification of the simplex method.

If there are just *two* priority levels, the modification of the simplex method is one you already have seen, namely, the form of the *Big M method* illustrated throughout Sec. 4.6. In this form, instead of replacing M throughout the model by some huge positive number before running the simplex method, we retain the *symbolic* quantity M in the sequence of simplex tableaux. Each coefficient in row 0 (for each iteration) is some linear function $aM + b$, where a is the current *multiplicative factor* and b is the current *additive term.* The usual decisions based on these coefficients (entering basic variable and optimality test) now are based solely on the *multiplicative* factors, except that any ties would be broken by using the *additive* terms. This is how the OR Courseware operates when solving either automatically or interactively by the simplex method (and choosing the Big M method).

The linear programming formulation for the streamlined procedure with two priority levels would include *all* the goals in the model in the usual manner, but with basic penalty weights of M and 1 assigned to deviations from first-priority and second-priority goals, respectively. If different penalty weights are desired within the same priority level, these basic penalty weights then are multiplied by the individual penalty weights assigned within the level. This approach is illustrated by the following example.

Example: For the Dewright Co. preemptive goal programming problem summarized in Table 7.6, note that (1) different penalty weights are assigned within each of the two priority levels and (2) the individual penalty weights (2 and 3) for the first-priority goals have been multiplied by M. These penalty weights yield the following single linear programming model that incorporates all the goals.

$$\text{Minimize} \qquad Z = 5y_1^- + 2My_2^+ + 4y_2^- + 3My_3^+,$$

subject to

$$\begin{aligned} 12x_1 + 9x_2 + 15x_3 - (y_1^+ - y_1^-) &= 125 \\ 5x_1 + 3x_2 + 4x_3 - (y_2^+ - y_2^-) &= 40 \\ 5x_1 + 7x_2 + 8x_3 - (y_3^+ - y_3^-) &= 55 \end{aligned}$$

and

$$x_j \geq 0, \qquad y_k^+ \geq 0, \qquad y_k^- \geq 0 \qquad (j = 1, 2, 3;\ k = 1, 2, 3).$$

Because this model uses M to symbolize a huge positive number, the simplex method should be applied as described and illustrated throughout Sec. 4.6. Doing this with your OR Courseware naturally yields the same unique optimal solution obtained by the sequential procedure.

More than Two Priority Levels: When there are more than two priority levels (say, p of them), the streamlined procedure generalizes in a straightforward way. The basic penalty weights for the respective levels now are $M_1, M_2, \ldots, M_{p-1}, 1$, where M_1 represents a number that is vastly larger than M_2, M_2 is vastly larger than $M_3, \ldots$, and M_{p-1} is vastly larger than 1. Each coefficient in row 0 of each simplex tableau is now a linear function of all these quantities, where the multiplicative factor of M_1 is used to make the necessary decisions, with tie breakers beginning with the multiplicative factor of M_2 and ending with the additive term.

7.6 Conclusions

The *dual simplex method* and *parametric linear programming* are especially valuable for post-optimality analysis, although they also can be very useful in other contexts.

The *upper bound technique* provides a way of streamlining the simplex method for the common situation in which many of or all the variables have explicit upper bounds. It can greatly reduce the computational effort for large problems.

Mathematical-programming computer packages usually include all three of these procedures, and they are widely used. Because their basic structure is based largely upon the simplex method as presented in Chap. 4, they retain the exceptional computational efficiency to handle very large problems of the sizes described in Sec. 4.8.

Various other special-purpose algorithms also have been developed to exploit the special structure of particular types of linear programming problems (such as those to be discussed in Chaps. 8 and 9). Much research is currently being done in this area.

Karmarkar's interior-point algorithm has been an exciting recent development in linear programming. This algorithm and its variants hold much promise as a powerful new approach for efficiently solving some very large problems.

Linear goal programming and its solution procedures provide an effective way of dealing with problems where management wishes to strive toward several goals simultaneously. The key is a fomulation technique of introducing auxiliary variables that enable one to convert the problem to a linear programming format.

SELECTED REFERENCES

1. Bradley, S. P., A. C. Hax, and T. L. Magnanti, *Applied Mathematical Programming,* Addison-Wesley, Reading, MA, 1977.
2. Hooker, J. N., "Karmarkar's Linear Programming Algorithm," *Interfaces,* **16:** 75–90, July-August 1986.
3. Lustig, I. J., R. E. Marsten, and D. F. Shanno, "Interior-Point Methods for Linear Programming: Computational State of the Art," *ORSA Journal on Computing,* **6:** 1–14, 1994. (Also see pp. 15–86 of this issue for commentaries on this article.)

4. Marsten, R., R. Subramanian, M. Saltzman, I. Lustig, and D. Shanno, ''Interior-Point Methods for Linear Programming: Just Call Newton, Lagrange, and Fiacco and McCormick!'' *Interfaces,* **20:** 105–116, July-August 1990.
5. Steuer, R. E., *Multiple Criteria Optimization: Theory, Computation, and Application,* Wiley, New York, 1985.
6. Vanderbei, R. J., ''Affine-Scaling for Linear Programs with Free Variables,'' *Mathematical Programming,* **43:** 31–44, 1989.

RELEVANT ROUTINES IN YOUR OR COURSEWARE

Interactive routines: *Enter or Revise a General Linear Programming Model*
Solve Interactively by the Graphical Method
Set Up for the Simplex Method—Interactive Only
Solve Interactively by the Simplex Method

Automatic routines: *Solve Automatically by the Simplex Method*
Solve Automatically by the Interior-Point Algorithm

To access these routines, call the MathProg program and then choose *General Linear Programming* under the Area menu. See Appendix 1 for documentation of the software.

PROBLEMS

To the left of each of the following problems (or their parts), we have inserted an I (for interactive routine) or A (for automatic routine) whenever a corresponding routine listed above can be helpful. (For Secs. 7.1 and 7.3, an A signifies that the automatic routine for the simplex method can be used to check your final answer after you have done the assigned work by hand.) An asterisk on the I or A indicates that this routine definitely should be used (unless your instructor gives contrary instructions), and the printout from this routine is all that needs to be turned in to show your work in executing the algorithm. (For Sec. 7.2, an I* signifies that the interactive routine for the simplex method should be used for $\theta = 0$, after which you should apply parametric linear programming by hand for $\theta > 0$.) An asterisk on the problem number indicates that at least a partial answer is given in the back of the book.

7.1-1. Consider the following problem.

$$\text{Maximize} \quad Z = -x_1 - x_2,$$

subject to

$$x_1 + x_2 \leq 8$$
$$x_2 \geq 3$$
$$-x_1 + x_2 \leq 2$$

and

$$x_1 \geq 0, \qquad x_2 \geq 0.$$

I (*a*) Solve this problem graphically.
A (*b*) Use the dual simplex method to solve this problem.
(*c*) Trace graphically the path taken by the dual simplex method.

A **7.1-2.** Use the dual simplex method to solve each of the following linear programming models:

(*a*) Model in Prob. 4.6-3
(*b*) Model in Prob. 4.6-4

7.1-3.* Use the dual simplex method to solve the following problem.

$$\text{Minimize} \quad Z = 5x_1 + 2x_2 + 4x_3,$$

subject to

$$3x_1 + x_2 + 2x_3 \geq 4$$
$$6x_1 + 3x_2 + 5x_3 \geq 10$$

and

$$x_1 \geq 0, \quad x_2 \geq 0, \quad x_3 \geq 0.$$

A **7.1-4.** Use the dual simplex method to solve the following problem.

$$\text{Minimize} \quad Z = 7x_1 + 2x_2 + 5x_3 + 4x_4,$$

subject to

$$2x_1 + 4x_2 + 7x_3 + x_4 \geq 5$$
$$8x_1 + 4x_2 + 6x_3 + 4x_4 \geq 8$$
$$3x_1 + 8x_2 + x_3 + 4x_4 \geq 4$$

and

$$x_j \geq 0, \quad \text{for } j = 1, 2, 3, 4.$$

7.1-5. Consider the following problem.

$$\text{Maximize} \quad Z = 3x_1 + 2x_2,$$

subject to

$$3x_1 + x_2 \leq 12$$
$$x_1 + x_2 \leq 6$$
$$5x_1 + 3x_2 \leq 27$$

and

$$x_1 \geq 0, \quad x_2 \geq 0.$$

I* (*a*) Solve by the original simplex method (in tabular form). Identify the complementary basic solution for the dual problem obtained at each iteration.
(*b*) Solve the dual of this problem by the dual simplex method. Compare the resulting sequence of basic solutions with the complementary basic solutions obtained in part (*a*).

7.1-6. Consider the example for Case 1 of sensitivity analysis given in Sec. 6.7, where the initial simplex tableau of Table 4.8 is modified by changing b_2 from 12 to 24, thereby changing the respective entries in the *right side* column of the final simplex tableau to 54, 6, 12, and -2. Starting from this revised final simplex tableau, use the dual simplex method to obtain the new optimal solution shown in Table 6.20. Show your work.

7.1-7.* Consider parts (*a*) and (*b*) of Prob. 6.7-1. Use the dual simplex method to reoptimize for each of these two cases, starting from the revised final tableau.

7.2-1.* Consider the following problem.

$$\text{Maximize} \quad Z = 8x_1 + 24x_2,$$

subject to

$$x_1 + 2x_2 \le 10$$
$$2x_1 + x_2 \le 10$$

and

$$x_1 \ge 0, \quad x_2 \ge 0.$$

Suppose that Z represents profit and that it is possible to modify the objective function somewhat by an appropriate shifting of key personnel between the two activities. In particular, suppose that the unit profit of activity 1 can be increased above 8 (to a maximum of 18) at the expense of decreasing the unit profit of activity 2 below 24 by twice the amount. Thus Z can actually be represented as

$$Z(\theta) = (8 + \theta)x_1 + (24 - 2\theta)x_2,$$

where θ is also a decision variable such that $0 \le \theta \le 10$.

I (*a*) Solve the original form of this problem graphically. Then extend this graphical procedure to solve the parametric extension of the problem; i.e., find the optimal solution and the optimal value of $Z(\theta)$ as a function of θ, for $0 \le \theta \le 10$.

I* (*b*) Find an optimal solution for the original form of the problem by the simplex method. Then use parametric linear programming to find an optimal solution and the optimal value of $Z(\theta)$ as a function of θ, for $0 \le \theta \le 10$. Plot $Z(\theta)$.

(*c*) Determine the optimal value of θ. Then indicate how this optimal value could have been identified directly by solving only two ordinary linear programming problems. (*Hint:* A convex function achieves its maximum at an endpoint.)

I* **7.2-2.** Use parametric linear programming to find the optimal solution for the following problem as a function of θ, for $0 \le \theta \le 20$.

$$\text{Maximize} \quad Z(\theta) = (20 + 4\theta)x_1 + (30 - 3\theta)x_2 + 5x_3,$$

subject to

$$3x_1 + 3x_2 + x_3 \le 30$$
$$8x_1 + 6x_2 + 4x_3 \le 75$$
$$6x_1 + x_2 + x_3 \le 45$$

and

$$x_1 \ge 0, \quad x_2 \ge 0, \quad x_3 \ge 0.$$

7.2-3. Consider the following problem.

$$\text{Maximize} \quad Z(\theta) = (10 - \theta)x_1 + (12 + \theta)x_2 + (7 + 2\theta)x_3,$$

subject to

$$x_1 + 2x_2 + 2x_3 \le 30$$
$$x_1 + x_2 + x_3 \le 20$$

and

$$x_1 \ge 0, \quad x_2 \ge 0, \quad x_3 \ge 0.$$

I* (*a*) Use parametric linear programming to find an optimal solution for this problem as a function of θ, for $\theta \geq 0$.

(*b*) Construct the dual model for this problem. Then find an optimal solution for this dual problem as a function of θ, for $\theta \geq 0$, by the method described in the latter part of Sec. 7.2. Indicate graphically what this algebraic procedure is doing. Compare the basic solutions obtained with the complementary basic solutions obtained in part (*a*).

7.2-4. Consider Prob. 6.7-17. Use parametric linear programming to find an optimal solution as a function of θ, for $\theta \geq 0$.

7.2-5. Consider Prob. 6.7-22(*b*). Extend the parametric linear programming procedure for making systematic changes in the c_j parameters to consider also systematic changes in the a_{ij} parameters in order to find an optimal solution as a function of θ, for all values of θ such that $0 \leq \theta \leq 1$.

I* **7.2-6.*** Use the parametric linear programming procedure for making systematic changes in the b_i parameters to find an optimal solution for the following problem as a function of θ, for $0 \leq \theta \leq 25$.

$$\text{Maximize} \quad Z(\theta) = 2x_1 + x_2,$$

subject to

$$\begin{aligned} x_1 \qquad &\leq 10 + 2\theta \\ x_1 + x_2 &\leq 25 - \theta \\ x_2 &\leq 10 + 2\theta \end{aligned}$$

and

$$x_1 \geq 0, \qquad x_2 \geq 0.$$

Indicate graphically what this algebraic procedure is doing.

I* **7.2-7.** Use parametric linear programming to find an optimal solution for the following problem as a function of θ, for $0 \leq \theta \leq 30$.

$$\text{Maximize} \quad Z(\theta) = 5x_1 + 6x_2 + 4x_3 + 7x_4,$$

subject to

$$\begin{aligned} 3x_1 - 2x_2 + x_3 + 3x_4 &\leq 135 - 2\theta \\ 2x_1 + 4x_2 - x_3 + 2x_4 &\leq 78 - \theta \\ x_1 + 2x_2 + x_3 + 2x_4 &\leq 30 + \theta \end{aligned}$$

and

$$x_j \geq 0, \qquad \text{for } j = 1, 2, 3, 4.$$

Then identify the value of θ that gives the largest optimal value of $Z(\theta)$.

7.2-8. Consider Prob. 6.7-2. Use parametric linear programming to find the optimal solution as a function of θ over the following ranges of θ.

(*a*) $0 \leq \theta \leq 20$

(*b*) $-20 \leq \theta \leq 0$ (*Hint:* Substitute $-\theta'$ for θ, and then increase θ' from zero.)

7.2-9. Follow the instructions of Prob. 7.2-8 for Prob. 6.7-5.

7.2-10. Consider the $Z^*(\theta)$ function shown in Fig. 7.1 for parametric linear programming with systematic changes in the c_j parameters.

(*a*) Explain why this function is piecewise linear.

(*b*) Show that this function must be convex.

7.2-11. Consider the $Z^*(\theta)$ function shown in Fig. 7.2 for parametric linear programming with systematic changes in the b_i parameters.

(*a*) Explain why this function is piecewise linear.
(*b*) Show that this function must be concave.

7.2-12. Let

$$Z^* = \max\left\{\sum_{j=1}^{n} c_j x_j\right\},$$

subject to

$$\sum_{j=1}^{n} a_{ij} x_j \leq b_i \qquad \text{for } i = 1, 2, \ldots, m$$

and

$$x_j \leq 0 \qquad \text{for } j = 1, 2, \ldots, n,$$

(where a_{ij}, b_i, and c_j are fixed constants), and let $(y_1^*, y_2^*, \ldots, y_m^*)$ be the corresponding optimal dual solution. Then let

$$Z^{**} = \max\left\{\sum_{j=1}^{n} c_j x_j\right\},$$

subject to

$$\sum_{j=1}^{n} a_{ij} x_j \leq b_i + k_i \qquad \text{for } i = 1, 2, \ldots, m$$

and

$$x_j \geq 0 \qquad \text{for } j = 1, 2, \ldots, n,$$

where $k_1, k_2, \ldots, k_m$ are given constants. Show that

$$Z^{**} \leq Z^* + \sum_{i=1}^{m} k_i y_i^*.$$

7.3-1. Use the upper bound technique to solve the Wyndor Glass Co. problem presented in Sec. 3.1.

7.3-2. Consider the following problem.

$$\text{Maximize} \qquad Z = 2x_1 + x_2,$$

subject to

$$\begin{aligned} x_1 - x_2 &\leq 5 \\ x_1 \quad\;\; &\leq 10 \\ x_2 &\leq 10 \end{aligned}$$

and

$$x_1 \geq 0, \qquad x_2 \geq 0.$$

I (*a*) Solve this problem graphically.
(*b*) Use the upper bound technique to solve this problem.
(*c*) Trace graphically the path taken by the upper bound technique.

7.3-3.* Use the upper bound technique to solve the following problem.

$$\text{Maximize} \quad Z = x_1 + 3x_2 - 2x_3,$$

subject to

$$\begin{aligned} x_2 - 2x_3 &\le 1 \\ 2x_1 + x_2 + 2x_3 &\le 8 \\ x_1 \qquad\qquad &\le 1 \\ x_2 \qquad &\le 3 \\ x_3 &\le 2 \end{aligned}$$

and

$$x_1 \ge 0, \qquad x_2 \ge 0, \qquad x_3 \ge 0.$$

A **7.3-4.** Use the upper bound technique to solve the following problem.

$$\text{Maximize} \quad Z = 2x_1 + 3x_2 - 2x_3 + 5x_4,$$

subject to

$$\begin{aligned} 2x_1 + 2x_2 + x_3 + 2x_4 &\le 5 \\ x_1 + 2x_2 - 3x_3 + 4x_4 &\le 5 \end{aligned}$$

and

$$0 \le x_j \le 1, \qquad \text{for } j = 1, 2, 3, 4.$$

A **7.3-5.** Use the upper bound technique to solve the following problem.

$$\text{Maximize} \quad Z = 2x_1 + 5x_2 + 3x_3 + 4x_4 + x_5,$$

subject to

$$\begin{aligned} x_1 + 3x_2 + 2x_3 + 3x_4 + x_5 &\le 6 \\ 4x_1 + 6x_2 + 5x_3 + 7x_4 + x_5 &\le 15 \end{aligned}$$

and

$$0 \le x_j \le 1, \qquad \text{for } j = 1, 2, 3, 4, 5.$$

A **7.3-6.** Use both the upper bound technique and the dual simplex method to solve the following problem.

$$\text{Minimize} \quad Z = 3x_1 + 4x_2 + 2x_3,$$

subject to

$$\begin{aligned} x_1 + x_2 + x_3 &\ge 15 \\ x_2 + x_3 &\ge 10 \end{aligned}$$

and

$$0 \le x_1 \le 25, \qquad 0 \le x_2 \le 5, \qquad 0 \le x_3 \le 15.$$

7.3-7. Use both the upper bound technique and the dual simplex method to solve the Nori & Leets Co. problem given in Sec. 3.4 for controlling air pollution.

A* **7.4-1.** Reconsider the example used to illustrate the interior-point algorithm in Sec. 7.4. Suppose that $(x_1, x_2) = (1, 3)$ were used instead as the initial feasible trial solution. Perform two iterations by hand, starting from this solution. Then use the automatic routine in your OR Courseware to check your work.

7.4-2. Consider the following problem.

$$\text{Maximize} \quad Z = 3x_1 + x_2,$$

subject to

$$x_1 + x_2 \leq 4$$

and

$$x_1 \geq 0, \quad x_2 \geq 0.$$

I (*a*) Solve this problem graphically. Also identify all CPF solutions.

A* (*b*) Starting from the initial trial solution $(x_1, x_2) = (1, 1)$, perform four iterations of the interior-point algorithm presented in Sec. 7.4 by hand. Then use the automatic routine in your OR Courseware to check your work.

(*c*) Draw figures corresponding to Figs. 7.4, 7.5, 7.6, 7.7, and 7.8 for this problem. In each case, identify the basic (or corner-point) feasible solutions in the current coordinate system. (Trial solutions can be used to determine projected gradients.)

7.4-3. Consider the following problem.

$$\text{Maximize} \quad Z = x_1 + 2x_2,$$

subject to

$$x_1 + x_2 = 8$$

and

$$x_1 \geq 0, \quad x_2 \geq 0.$$

A* (*a*) Near the end of Sec. 7.4, there is a discussion of what the interior-point algorithm does on this problem when we start from the initial feasible trial solution $(x_1, x_2) = (4, 4)$. Verify the results presented there by performing two iterations by hand. Then use the automatic routine in your OR Courseware to check your work.

(*b*) Use these results to predict what subsequent trial solutions would be if additional iterations were performed.

(*c*) Suppose that the stopping rule adopted for the algorithm in this application is that the algorithm stops when two successive trial solutions differ by no more than 0.01 in any component. Use your predictions from part (*b*) to predict the final trial solution and the total number of iterations required to get there. How close would this solution be to the optimal solution $(x_1, x_2) = (0, 8)$?

7.4-4. Consider the following problem.

$$\text{Maximize} \quad Z = x_1 + x_2,$$

subject to

$$x_1 + 2x_2 \leq 9$$
$$2x_1 + x_2 \leq 9$$

and

$$x_1 \geq 0, \quad x_2 \geq 0.$$

I (*a*) Solve the problem graphically.

(*b*) Find the gradient of the objective function in the original x_1x_2 coordinate system. If you move from the origin in the direction of the gradient until you reach the boundary of the feasible region, where does it lead relative to the optimal solution?

A* (*c*) Starting from the initial trial solution $(x_1, x_2) = (1, 1)$, use your OR Courseware to perform 10 iterations of the interior-point algorithm presented in Sec. 7.4.

A* (*d*) Repeat part (*c*) with $\alpha = 0.9$.

7.4-5. Consider the following problem.

$$\text{Maximize} \quad Z = 2x_1 + 5x_2 + 7x_3,$$

subject to

$$x_1 + 2x_2 + 3x_3 = 6$$

and

$$x_1 \geq 0, \quad x_2 \geq 0, \quad x_3 \geq 0.$$

I (*a*) Graph the feasible region.

(*b*) Find the gradient of the objective function, and then find the projected gradient onto the feasible region.

(*c*) Starting from the initial trial solution $(x_1, x_2, x_3) = (1, 1, 1)$, perform two iterations of the interior-point algorithm presented in Sec. 7.4 by hand.

A* (*d*) Starting from this same initial trial solution, use your OR Courseware to perform 10 iterations of this algorithm.

A* **7.4-6.** Starting from the initial trial solution $(x_1, x_2) = (2, 2)$, use your OR Courseware to apply 15 iterations of the interior-point algorithm presented in Sec. 7.4 to the Wyndor Glass Co. problem presented in Sec. 3.1. Also draw a figure like Fig. 7.8 to show the trajectory of the algorithm in the original x_1x_2 coordinate system.

7.5-1. The research and development division of a certain company has developed three new products. The problem is to decide which mix of these products should be produced. Management wants primary consideration given to three factors: long-run profit, stability in the workforce, and an increase in the company's earnings next year. In particular, using the units given in the following table, they want to

$$\text{Maximize} \quad Z = 2P - 5C - 3D,$$

where P = total (discounted) profit over life of new products,
C = change (in either direction) in current level of employment,
D = decrease (if any) in next year's earnings from current year's level.

The amount of any increase in earnings does not enter into Z, because management is concerned primarily with just achieving some increase to keep the stockholders happy. (It has mixed feelings about a large increase that then would be difficult to surpass in subsequent years.)

The impact of each of the new products (per unit rate of production) on each of these factors is shown in the following table:

	Unit Contribution				
	Product				
Factor	1	2	3	Goal	(Units)
Long-run profit	20	15	25	Maximize	(millions of dollars)
Employment level	6	4	11	= 50	(hundreds of employees)
Earnings next year	8	7	5	≥ 75	(millions of dollars)

(*a*) Assuming that there are no additional constraints on the production rates not described here, use the goal programming technique to formulate a linear programming model for this problem.

A* (*b*) Use the simplex method to solve this model.

7.5-2.* Consider a preemptive goal programming problem with three priority levels, just one goal for each priority level, and just two activities to contribute toward these goals, as summarized in the following table:

	Unit Contribution		
	Activity		
Priority Level	1	2	Goal
First priority	1	2	$\leq$ 20
Second priority	1	1	$=$ 15
Third priority	2	1	$\geq$ 40

(*a*) Use the goal programming technique to formulate one complete linear programming model for this problem.

(*b*) Construct the initial simplex tableau for applying the streamlined procedure. Identify the initial BF solution and the initial entering basic variable, but do not proceed further.

(*c*) Starting from (*b*), use the streamlined procedure to solve the problem.

I (*d*) Use the logic of preemptive goal programming to solve the problem graphically by focusing on just the two decision variables. Explain the logic used.

I (*e*) Use the sequential procedure to solve this problem. After using the goal programming technique to formulate the linear programming model (including auxiliary variables) at each stage, solve the model graphically by focusing on just the two decision variables. Identify *all* optimal solutions obtained for each stage.

7.5-3. Redo Prob. 7.5-2 with the following revised table:

	Unit Contribution		
	Activity		
Priority Level	1	2	Goal
First priority	1	1	$\leq$ 20
Second priority	1	1	$\geq$ 30
Third priority	1	2	$\geq$ 50

7.5-4. A certain developing country has 15 million acres of publicly controlled agricultural land in active use. Its government currently is planning a way to divide this land among three basic crops (labeled 1, 2, and 3) next year. A certain percentage of each of these crops is exported in order to obtain badly needed foreign capital (dollars), and the rest of these crops is used to feed the populace. Raising these crops also provides employment for a significant proportion of the population. Therefore, the main factors to be considered in allocating the land to these crops are (1) the amount of foreign capital generated, (2) the number of citizens fed, and (3) the number of citizens employed in raising these crops. The following table shows how much each 1,000 acres of crop contributes toward these factors, and the rightmost column gives the goal established by the government for each factor.

Factor	Contribution per 1,000 Acres			Goal
	Crop			
	1	2	3	
Foreign capital, $	3,000	5,000	4,000	≥ 70,000,000
Citizens fed	150	75	100	≥ 1,750,000
Citizens employed	10	15	12	= 200,000

(*a*) In evaluating the relative seriousness of *not* achieving these goals, the government has concluded that the following deviations from the goals should be considered *equally undesirable:* (1) each $100 under the foreign capital goal, (2) each person under the citizens-fed goal, and (3) each deviation of one (in either direction) from the citizens-employed goal. Use the goal programming technique to formulate a linear programming model for this problem.

(*b*) Now suppose that the government concludes that the importance of the various goals differs greatly so that a preemptive goal programming approach should be used. In particular, the first-priority goal is citizens fed ≥ 1,750,000, the second-priority goal is foreign capital ≥ $70,000,000, and the third-priority goal is citizens employed = 200,000. Use the goal programming technique to formulate one complete linear programming model for this problem.

(*c*) Use the streamlined procedure to solve the problem as formulated in part (*b*).

A* (*d*) Use the sequential procedure to solve the problem as presented in part (*b*).

7.5-5. One of the most important problems in the field of statistics is the linear regression problem. Roughly speaking, this problem involves fitting a straight line to statistical data represented by points—(x_1, y_1), $(x_2, y_2, \ldots, (x_n, y_n)$—on a graph. If we denote the line by $y = a + bx$, the objective is to choose the constants a and b to provide the "best" fit according to some criterion. The criterion usually used is the *method of least squares,* but there are other interesting criteria where linear programming can be used to solve for the optimal values of a and b.

Formulate a linear programming model for this problem under the following criterion:

Minimize the sum of the absolute deviations of the data from the line; that is,

$$\text{Minimize} \quad \sum_{i=1}^{n} |y_i - (a + bx_i)|.$$

(*Hint:* Note that this problem can be viewed as a nonpreemptive goal programming problem where each data point represents a "goal" for the regression line.)

8

The Transportation and Assignment Problems

Chapter 3 emphasized the wide applicability of linear programming. We continue to broaden our horizons in this chapter by discussing two particularly important (and related) types of linear programming problems. One type, called the *transportation problem,* received this name because many of its applications involve determining how to optimally transport goods. However, some of its important applications (e.g., production scheduling) actually have nothing to do with transportation.

The second type, called the *assignment problem,* involves such applications as assigning people to tasks. Although its applications appear to be quite different from those for the transportation problem, we shall see that the assignment problem can be viewed as a special type of transportation problem.

The next chapter will introduce additional special types of linear programming problems involving *networks,* including the *minimum cost flow problem* (Sec. 9.6). There we shall see that both the transportation and assignment problems actually are special cases of the minimum cost flow problem. We introduce the network representation of the transportation and assignment problems in this chapter.

Table 8.1 **Table of Constraint Coefficients for Linear Programming**

$$A = \begin{bmatrix} a_{11} & a_{12} & \cdots & a_{1n} \\ a_{21} & a_{22} & \cdots & a_{2n} \\ \cdots & \cdots & \cdots & \cdots \\ a_{m1} & a_{m2} & \cdots & a_{mn} \end{bmatrix}$$

Applications of the transportation and assignment problems tend to require a very large number of constraints and variables, so a straightforward computer application of the simplex method may require an exorbitant computational effort. Fortunately, a key characteristic of these problems is that most of the a_{ij} coefficients in the constraints are zeros, and the relatively few nonzero coefficients appear in a distinctive pattern. As a result, it has been possible to develop special *streamlined* algorithms that achieve dramatic computational savings by exploiting this special structure of the problem. Therefore, it is important to become sufficiently familiar with these special types of problems that you can recognize them when they arise and apply the proper computational procedure.

To describe special structures, we shall introduce the table (matrix) of constraint coefficients shown in Table 8.1, where a_{ij} is the coefficient of the jth variable in the ith functional constraint. Later, portions of the table containing only coefficients equal to zero will be indicated by leaving them blank, whereas blocks containing nonzero coefficients will be shaded.

After presenting a prototype example for the transportation problem, we describe the special structure in its model and give additional examples of its applications. Section 8.2 presents the *transportation simplex method,* a special streamlined version of the simplex method for efficiently solving transportation problems. (You will see in Sec. 9.7 that this algorithm is related to the *network simplex method,* another streamlined version of the simplex method for efficiently solving any minimum cost flow problem, including both transportation and assignment problems.) Section 8.3 then focuses on the assignment problem.

8.1 The Transportation Problem

Prototype Example

One of the main products of the P & T COMPANY is canned peas. The peas are prepared at three canneries (near Bellingham, Washington; Eugene, Oregon; and Albert Lea, Minnesota) and then shipped by truck to four distributing warehouses in the western United States (Sacramento, California; Salt Lake City, Utah; Rapid City, South Dakota; and Albuquerque, New Mexico), as shown in Fig. 8.1. Because the shipping costs are a major expense, management is initiating a study to reduce them as much as possible. For the upcoming season, an estimate has been made of the output from each cannery, and each warehouse has been allocated a certain amount from the total supply of peas. This information (in units of truckloads), along with the shipping cost per

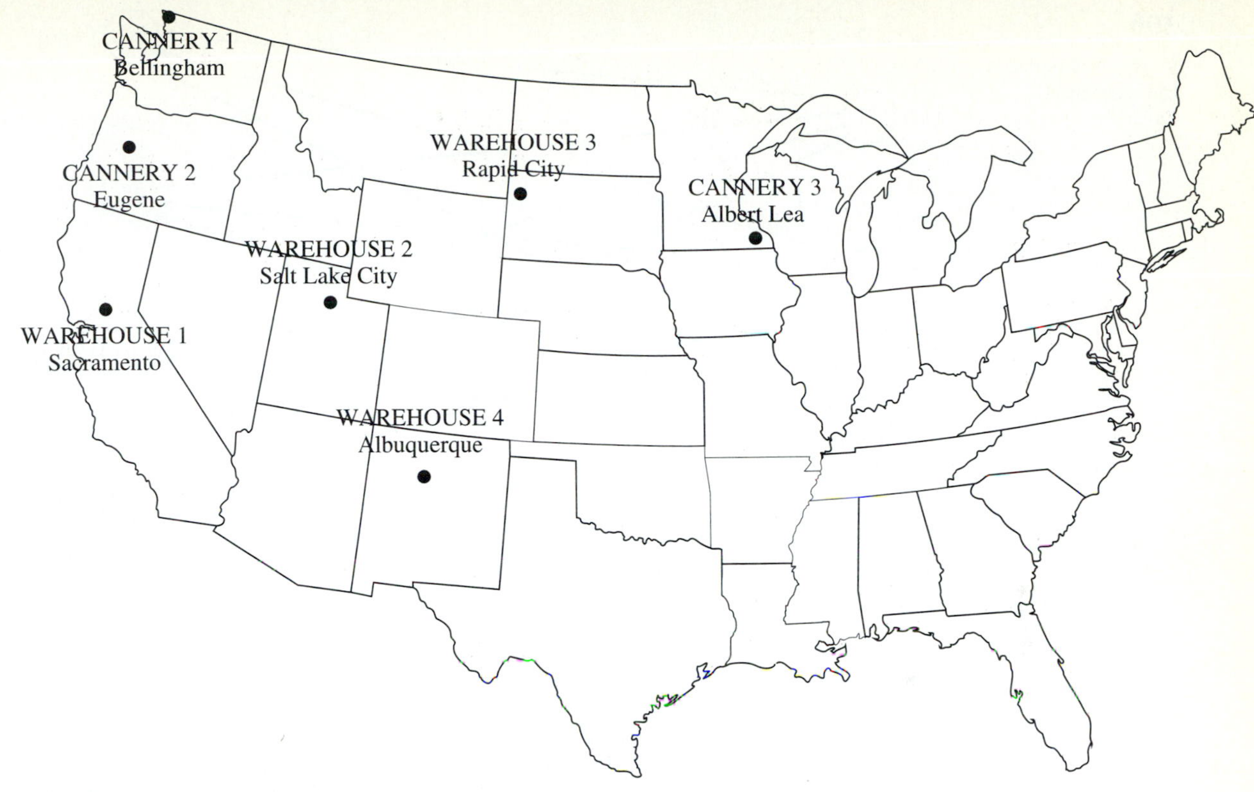

Figure 8.1 Location of canneries and warehouses for the P & T Co. problem.

truckload for each cannery-warehouse combination, is given in Table 8.2. Thus there are a total of 300 truckloads to be shipped. The problem now is to determine which plan for assigning these shipments to the various cannery-warehouse combinations would *minimize the total shipping cost.*

By ignoring the geographical layout of the canneries and warehouses, we can provide a *network representation* of this problem in a simple way by lining up all the canneries in one column on the left and all the warehouses in one column on the right. This representation is shown in Fig. 8.2. The arrows show the possible routes for the truckloads, where the number next to each arrow is the shipping cost per truckload for that route. A square bracket next to each location gives the number of truckloads to be

Table 8.2 **Shipping Data for P & T Co.**

		Shipping Cost ($) per Truckload				
		Warehouse				
		1	2	3	4	Output
	1	464	513	654	867	75
Cannery	2	352	416	690	791	125
	3	995	682	388	685	100
Allocation		80	65	70	85	

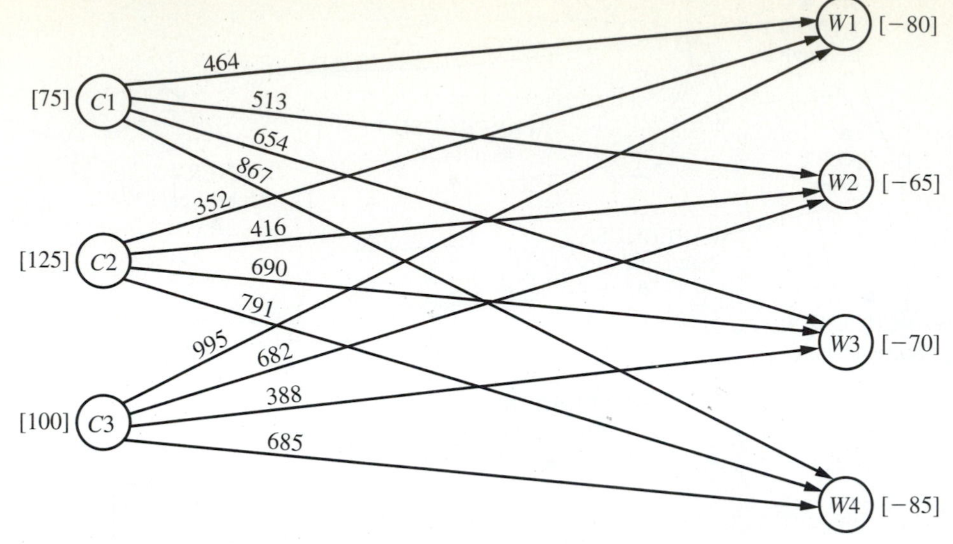

Figure 8.2 Network representation of the P & T Co. problem.

shipped *out* of that location (so that the allocation into each warehouse is given as a negative number).

The problem depicted in Fig. 8.2 is actually a linear programming problem of the *transportation problem* type. To formulate the model, let Z denote total shipping cost, and let x_{ij} $(i = 1, 2, 3; j = 1, 2, 3, 4)$ be the number of truckloads to be shipped from cannery i to warehouse j. Thus the objective is to choose the values of these 12 decision variables (the x_{ij}) so as to

$$\text{Minimize} \quad Z = 464x_{11} + 513x_{12} + 654x_{13} + 867x_{14} + 352x_{21} + 416x_{22} + 690x_{23} + 791x_{24} + 995x_{31} + 682x_{32} + 388x_{33} + 685x_{34},$$

subject to the constraints

$$\begin{aligned}
&x_{11} + x_{12} + x_{13} + x_{14} &&= 75\\
&x_{21} + x_{22} + x_{23} + x_{24} &&= 125\\
&x_{31} + x_{32} + x_{33} + x_{34} &&= 100\\
&x_{11} + x_{21} + x_{31} &&= 80\\
&x_{12} + x_{22} + x_{32} &&= 65\\
&x_{13} + x_{23} + x_{33} &&= 70\\
&x_{14} + x_{24} + x_{34} &&= 85
\end{aligned}$$

and

$$x_{ij} \geq 0 \qquad (i = 1, 2, 3; j = 1, 2, 3, 4).$$

Table 8.3 shows the constraint coefficients. As you will see next, it is the special structure in the pattern of these coefficients that distinguishes this problem as a transportation problem, not its context.

Table 8.3 **Constraint Coefficients for P & T Co.**

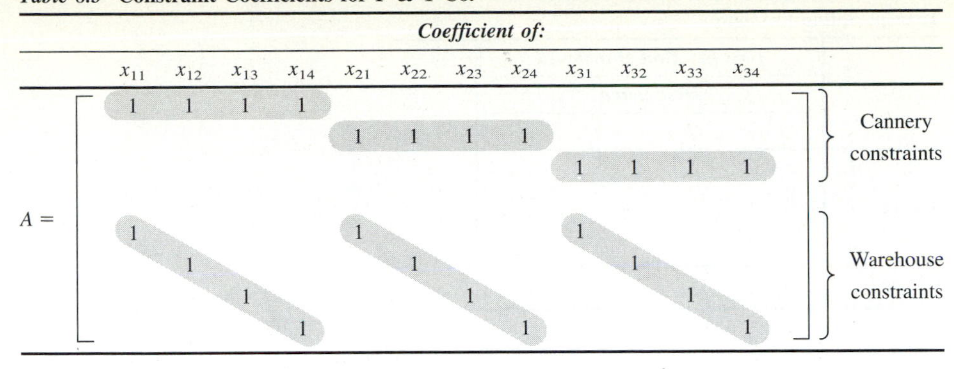

By the way, the optimal solution for this problem is $x_{11} = 0$, $x_{12} = 20$, $x_{13} = 0$, $x_{14} = 55$, $x_{21} = 80$, $x_{22} = 45$, $x_{23} = 0$, $x_{24} = 0$, $x_{31} = 0$, $x_{32} = 0$, $x_{33} = 70$, $x_{34} = 30$. When you learn the optimality test that appears in Sec. 8.2, you will be able to verify this yourself (see Prob. 8.2-5).

The Transportation Problem Model

To describe the general model for the transportation problem, we need to use terms that are considerably less specific than those for the components of the prototype example. In particular, the general transportation problem is concerned (literally or figuratively) with distributing *any* commodity from *any* group of supply centers, called **sources**, to *any* group of receiving centers, called **destinations**, in such a way as to minimize the total distribution cost. The correspondence in terminology between the prototype example and the general problem is summarized in Table 8.4.

Thus, in general, source i ($i = 1, 2, \ldots, m$) has a supply of s_i units to distribute to the destinations, and destination j ($j = 1, 2, \ldots, n$) has a demand for d_j units to be received from the sources. A basic assumption is that the cost of distributing units from source i to destination j is directly proportional to the number distributed, where c_{ij} denotes the cost per unit distributed. As for the prototype example (see Table 8.2), these input data can be summarized conveniently in the **cost and requirements table** shown in Table 8.5. Similarly, the network representation given in Fig. 8.2 for the example now has the similar appearance shown in Fig. 8.3 for the general problem.

By letting Z be total distribution cost and x_{ij} ($i = 1, 2, \ldots, m$; $j = 1, 2, \ldots, n$) be the number of units to be distributed from source i to destination j, the linear

Table 8.4 **Terminology for the Transportation Problem**

Prototype Example	General Problem
Truckloads of canned peas	Units of a commodity
Three canneries	m sources
Four warehouses	n destinations
Output from cannery i	Supply s_i from source i
Allocation to warehouse j	Demand d_j at destination j
Shipping cost per truckload from cannery i to warehouse j	Cost c_{ij} per unit distributed from source i to destination j

Table 8.5 **Cost and Requirements Table for the Transportation Problem**

		Cost per Unit Distributed				
		Destination				
		1	2	$\cdots$	n	Supply
Source	1	c_{11}	c_{12}	$\cdots$	c_{1n}	s_1
	2	c_{21}	c_{22}	$\cdots$	c_{2n}	s_2
	$\vdots$	$\cdots$	$\cdots$	$\cdots$	$\cdots$	$\vdots$
	m	c_{m1}	c_{m2}	$\cdots$	c_{mn}	s_m
Demand		d_1	d_2	$\cdots$	d_n	

programming formulation of this problem becomes

$$\text{Minimize} \quad Z = \sum_{i=1}^{m} \sum_{j=1}^{n} c_{ij} x_{ij},$$

subject to

$$\sum_{j=1}^{n} x_{ij} = s_i \qquad \text{for } i = 1, 2, \ldots, m,$$

$$\sum_{i=1}^{m} x_{ij} = d_j \qquad \text{for } j = 1, 2, \ldots, n,$$

and

$$x_{ij} \geq 0, \qquad \text{for all } i \text{ and } j.$$

Note that the resulting table of constraint coefficients has the special structure shown in Table 8.6. *Any* linear programming problem that fits this special formulation is of the transportation problem type, regardless of its physical context. In fact, there have been numerous applications unrelated to transportation that have been fitted to this special structure, as we shall illustrate in the next example. (The assignment problem described in Sec. 8.3 is an additional example.) This is one of the reasons why the transportation problem is generally considered the most important special type of linear programming problem.

Table 8.6 **Constraint Coefficients for the Transportation Problem**

	Coefficient of:													
	x_{11}	x_{12}	$\cdots$	x_{1n}	x_{21}	x_{22}	$\cdots$	x_{2n}	$\cdots$	x_{m1}	x_{m2}	$\cdots$	x_{mn}	
$A =$	1	1	$\cdots$	1										Supply constraints
					1	1	$\cdots$	1						
									$\ddots$					
										1	1	$\cdots$	1	
	1				1					1				Demand constraints
		1				1			$\cdots$		1			
			$\ddots$				$\ddots$					$\ddots$		
				1				1					1	

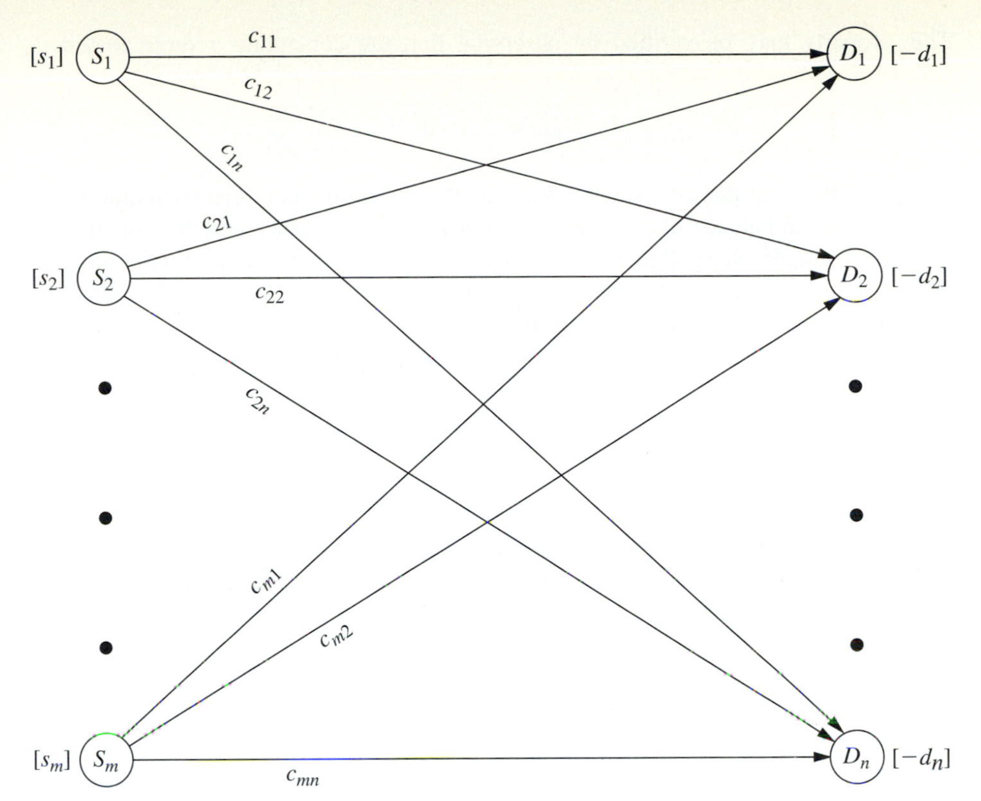

Figure 8.3 Network representation of the transportation problem.

For many applications, the supply and demand quantities in the model (the s_i and d_i) have integer values, and implementation will require that the distribution quantities (the x_{ij}) also have integer values. Fortunately, because of the special structure shown in Table 8.6, all such problems have the following property.

Integer solutions property: For transportation problems where every s_i and d_j have an integer value, all the basic variables (allocations) in *every* basic feasible (BF) solution (including an optimal one) also have *integer* values.

The solution procedure described in Sec. 8.2 deals only with BF solutions, so it automatically will obtain an *integer* optimal solution for this case. (You will be able to see why this solution procedure actually gives a proof of the integer solutions property after you learn the procedure; Prob. 8.2-22 guides you through the reasoning involved.) Therefore, it is unnecessary to add a constraint to the model that the x_{ij} must have integer values.

However, in order to have an optimal solution of any kind, a transportation problem must possess *feasible* solutions. The following property indicates when this will occur.

Feasible solutions property: A necessary and sufficient condition for a transportation problem to have any feasible solutions is that

$$\sum_{i=1}^{m} s_i = \sum_{j=1}^{n} d_j.$$

This property may be verified by observing that the constraints require that both

$$\sum_{i=1}^{m} s_i \quad \text{and} \quad \sum_{j=1}^{n} d_j \quad \text{equal} \quad \sum_{i=1}^{m} \sum_{j=1}^{n} x_{ij}.$$

This condition that the total supply must equal the total demand merely requires that the system be in balance. If the problem has physical significance and this condition is not met, it usually means that either s_i or d_j actually represents a *bound* rather than an exact requirement. If this is the case, a fictitious "source" or "destination" (called the **dummy source** or the **dummy destination**) can be introduced to take up the slack in order to convert the inequalities to equalities and satisfy the feasibility condition. The next two examples illustrate how to do this conversion as well as how to fit some other common variations into the transportation problem formulation.

Example—Production Scheduling: The NORTHERN AIRPLANE COMPANY builds commercial airplanes for various airline companies around the world. The last stage in the production process is to produce the jet engines and then to install them (a very fast operation) in the completed airplane frame. The company has been working under some contracts to deliver a considerable number of airplanes in the near future, and the production of the jet engines for these planes must now be scheduled for the next 4 months.

To meet the contracted dates for delivery, the company must supply engines for installation in the quantities indicated in the second column of Table 8.7. Thus the cumulative number of engines produced by the end of months 1, 2, 3, and 4 must be at least 10, 25, 50, and 70, respectively.

The facilities that will be available for producing the engines vary according to other production, maintenance, and renovation work scheduled during this period. The resulting monthly differences in the maximum number that can be produced and the cost (in millions of dollars) of producing each one are given in the third and fourth columns of Table 8.7.

Because of the variations in production costs, it may well be worthwhile to produce some of the engines a month or more before they are scheduled for installation, and this possibility is being considered. The drawback is that such engines must be stored until the scheduled installation (the airplane frames will not be ready early) at a storage cost of $15,000 per month (including interest on expended capital) for each engine,[1] as shown in the rightmost column of Table 8.7.

[1] For modeling purposes, assume that this storage cost is incurred at the *end of the month* for just those engines that are being held over into the next month. Thus engines that are produced in a given month for installation in the same month are assumed to incur no storage cost.

Table 8.7 **Production Scheduling Data for Northern Airplane Co.**

Month	Scheduled Installations	Maximum Production	Unit Cost* of Production	Unit Cost* of Storage
1	10	25	1.08	0.015
2	15	35	1.11	0.015
3	25	30	1.10	0.015
4	20	10	1.13	

* Cost is expressed in millions of dollars.

The production manager wants a schedule developed for the number of engines to be produced in each of the 4 months so that the total of the production and storage costs will be minimized.

One way to formulate a mathematical model for this problem is to let x_j be the number of jet engines to be produced in month j, for $j = 1, 2, 3, 4$. By using only these four decision variables, the problem can be formulated as a linear programming problem that does *not* fit the transportation problem type. (See Prob. 8.2-20.)

On the other hand, by adopting a different viewpoint, we can instead formulate the problem as a transportation problem that requires *much* less effort to solve. This viewpoint will describe the problem in terms of sources and destinations and then identify the corresponding x_{ij}, c_{ij}, s_i, and d_j. (See if you can do this before reading further.)

Because the units being distributed are jet engines, each of which is to be scheduled for production in a particular month and then installed in a particular (perhaps different) month,

$$\text{Source } i = \text{production of jet engines in month } i \ (i = 1, 2, 3, 4)$$

$$\text{Destination } j = \text{installation of jet engines in month } j \ (j = 1, 2, 3, 4)$$

$$x_{ij} = \text{number of engines produced in month } i \text{ for installation in month } j$$

$$c_{ij} = \text{cost associated with each unit of } x_{ij}$$

$$= \begin{cases} \text{cost per unit for production and any storage} & \text{if } i \le j \\ ? & \text{if } i > j \end{cases}$$

$$s_i = ?$$

$$d_j = \text{number of scheduled installations in month } j.$$

The corresponding (incomplete) cost and requirements table is given in Table 8.8. Thus it remains to identify the missing costs and the supplies.

Since it is impossible to produce engines in one month for installation in an earlier month, x_{ij} must be zero if $i > j$. Therefore, there is no real cost that can be associated with such x_{ij}. Nevertheless, in order to have a well-defined transportation problem to which a standard software package (solution procedure of Sec. 8.2) can be applied, it is necessary to assign some value for the unidentified costs. Fortunately, we can use the *Big M method* introduced in Sec. 4.6 to assign this value. Thus we assign a *very* large number (denoted by M for convenience) to the unidentified cost entries in Table 8.8 to force the corresponding values of x_{ij} to be zero in the final solution.

The numbers that need to be inserted into the supply column of Table 8.8 are not obvious because the "supplies," the amounts produced in the respective months, are

Table 8.8 **Incomplete Cost and Requirements Table for Northern Airplane Co.**

		Cost per Unit Distributed				
		Destination				
		1	2	3	4	Supply
Source	1	1.080	1.095	1.110	1.125	?
	2	?	1.110	1.125	1.140	?
	3	?	?	1.100	1.115	?
	4	?	?	?	1.130	?
Demand		10	15	25	20	

not fixed quantities. In fact, the objective is to solve for the most desirable values of these production quantities. Nevertheless, it is necessary to assign some fixed number to every entry in the table, including those in the supply column, to have a transportation problem. A clue is provided by the fact that although the supply constraints are not present in the usual form, these constraints do exist in the form of upper bounds on the amount that can be supplied, namely,

$$x_{11} + x_{12} + x_{13} + x_{14} \le 25,$$

$$x_{21} + x_{22} + x_{23} + x_{24} \le 35,$$

$$x_{31} + x_{32} + x_{33} + x_{34} \le 30,$$

$$x_{41} + x_{42} + x_{43} + x_{44} \le 10.$$

The only change from the standard model for the transportation problem is that these constraints are in the form of inequalities instead of equalities.

To convert these inequalities to equations in order to fit the transportation problem model, we use the familiar device of *slack variables,* introduced in Sec. 4.2. In this context, the slack variables are allocations to a single *dummy destination* that represent the *unused production capacity* in the respective months. This change permits the supply in the transportation problem formulation to be the total production capacity in the given month. Furthermore, because the demand for the dummy destination is the total unused capacity, this demand is

$$(25 + 35 + 30 + 10) - (10 + 15 + 25 + 20) = 30.$$

With this demand included, the sum of the supplies now equals the sum of the demands, which is the condition given by the *feasible solutions property* for having feasible solutions.

The cost entries associated with the dummy destination should be zero because there is no cost incurred by a fictional allocation. (Cost entries of M would be *inappropriate* for this column because we do not want to force the corresponding values of x_{ij} to be zero. In fact, these values need to sum to 30.)

The resulting final cost and requirements table is given in Table 8.9, with the dummy destination labeled as destination $5(D)$. By using this formulation, it is quite easy to find the optimal production schedule by the solution procedure described in Sec. 8.2. (See Prob. 8.2-11 and its answer in the back of the book.)

Example—Distribution of Water Resources: METRO WATER DISTRICT is an agency that administers water distribution in a large geographic region. The region is fairly arid, so the district must purchase and bring in water from outside the region. The sources of this imported water are the Colombo, Sacron, and Calorie rivers. The district

Table 8.9 **Complete Cost and Requirements Table for Northern Airplane Co.**

		Cost per Unit Distributed					
		Destination					
		1	2	3	4	5(*D*)	Supply
Source	1	1.080	1.095	1.110	1.125	0	25
	2	*M*	1.110	1.125	1.140	0	35
	3	*M*	*M*	1.100	1.115	0	30
	4	*M*	*M*	*M*	1.130	0	10
Demand		10	15	25	20	30	

then resells the water to users in its region. Its main customers are the water departments of the cities of Berdoo, Los Devils, San Go, and Hollyglass.

It is possible to supply any of these cities with water brought in from any of the three rivers, with the exception that no provision has been made to supply Hollyglass with Calorie River water. However, because of the geographic layouts of the viaducts and the cities in the region, the cost to the district of supplying water depends upon both the source of the water and the city being supplied. The variable cost per acre foot of water (in tens of dollars) for each combination of river and city is given in Table 8.10. Despite these variations, the price per acre foot charged by the district is independent of the source of the water and is the same for all cities.

The management of the district is now faced with the problem of how to allocate the available water during the upcoming summer season. In units of 1 million acre feet, the amounts available from the three rivers are given in the rightmost column of Table 8.10. The district is committed to providing a certain minimum amount to meet the essential needs of each city (with the exception of San Go, which has an independent source of water), as shown in the *minimum needed* row of the table. The *requested* row indicates that Los Devils desires no more than the minimum amount, but that Berdoo would like to buy as much as 20 more, San Go would buy up to 30 more, and Hollyglass will take as much as it can get.

Management wishes to allocate *all* the available water from the three rivers to the four cities in such a way as to at least meet the essential needs of each city while minimizing the total cost to the district.

Formulation: Table 8.10 already is close to the proper form for a cost and requirements table, with the rivers being the sources and the cities being the destinations. However, the one basic difficulty is that it is not clear what the demands at the destinations should be. The amount to be received at each destination (except Los Devils) actually is a decision variable, with both a lower bound and an upper bound. This upper bound is the amount requested unless the request exceeds the total supply remaining after the minimum needs of the other cities are met, in which case this *remaining supply* becomes the upper bound. Thus insatiably thirsty Hollyglass has an upper bound of

$$(50 + 60 + 50) - (30 + 70 + 0) = 60.$$

Unfortunately, just like the other numbers in the cost and requirements table of a transportation problem, the demand quantities must be *constants,* not bounded decision variables. To begin resolving this difficulty, temporarily suppose that it is not necessary to satisfy the minimum needs, so that the upper bounds are the only constraints on amounts to be allocated to the cities. In this circumstance, can the requested allocations be viewed as the demand quantities for a transportation problem formulation? After one adjustment, yes! (Do you see already what the needed adjustment is?)

Table 8.10 **Water Resources Data for Metro Water District**

	Cost (Tens of Dollars) per Acre Foot				
	Berdoo	Los Devils	San Go	Hollyglass	Supply
Colombo River	16	13	22	17	50
Sacron River	14	13	19	15	60
Calorie River	19	20	23	—	50
Minimum needed	30	70	0	10	(in units of 1
Requested	50	70	30	∞	million acre feet)

The situation is analogous to Northern Airplane Co.'s production scheduling problem, where there was *excess supply capacity.* Now there is *excess demand capacity.* Consequently, rather than introducing a *dummy destination* to "receive" the unused supply capacity, the adjustment needed here is to introduce a *dummy source* to "send" the *unused demand capacity.* The imaginary supply quantity for this dummy source would be the amount by which the sum of the demands exceeds the sum of the real supplies:

$$(50 + 70 + 30 + 60) - (50 + 60 + 50) = 50.$$

This formulation yields the cost and requirements table shown in Table 8.11, which uses units of million acre feet and tens of millions of dollars. The cost entries in the *dummy* row are zero because there is no cost incurred by the fictional allocations from this dummy source. On the other hand, a huge unit cost of M is assigned to the Calorie River–Hollyglass spot. The reason is that Calorie River water cannot be used to supply Hollyglass, and assigning a cost of M will prevent any such allocation.

Now let us see how we can take each city's minimum needs into account in this kind of formulation. Because San Go has no minimum need, it is all set. Similarly, the formulation for Hollyglass does not require any adjustments because its demand (60) exceeds the dummy source's supply (50) by 10, so the amount supplied to Hollyglass from the *real* sources will be *at least 10* in any feasible solution. Consequently, its minimum need of 10 from the rivers is guaranteed. (If this coincidence had not occurred, Hollyglass would need the same adjustments that we shall have to make for Berdoo.)

Los Devils' minimum need equals its requested allocation, so its *entire* demand of 70 must be filled from the real sources rather than the dummy source. This requirement calls for the Big M method! Assigning a huge unit cost of M to the allocation from the dummy source to Los Devils ensures that this allocation will be zero in an optimal solution.

Finally, consider Berdoo. In contrast to Hollyglass, the dummy source has an adequate (fictional) supply to "provide" at least some of Berdoo's minimum need in addition to its extra requested amount. Therefore, since Berdoo's minimum need is 30, adjustments must be made to prevent the dummy source from contributing more than 20 to Berdoo's total demand of 50. This adjustment is accomplished by splitting Berdoo into two destinations, one having a demand of 30 with a unit cost of M for any allocation from the dummy source and the other having a demand of 20 with a unit cost of zero for the dummy source allocation. This formulation gives the final cost and requirements table shown in Table 8.12.

This problem will be solved in the next section to illustrate the solution procedure presented there.

Table 8.11 **Cost and Requirements Table without Minimum Needs for Metro Water District**

		Cost (Tens of Millions of Dollars) per Unit Distributed				
		Destination				
		Berdoo	Los Devils	San Go	Hollyglass	Supply
Source	Colombo River	16	13	22	17	50
	Sacron River	14	13	19	15	60
	Calorie River	19	20	23	M	50
	Dummy	0	0	0	0	50
Demand		50	70	30	60	

Table 8.12 **Cost and Requirements Table for Metro Water District**

		Cost (Tens of Millions of Dollars) per Unit Distributed					
		Destination					
		Berdoo (min.) 1	Berdoo (extra) 2	Los Devils 3	San Go 4	Hollyglass 5	Supply
Source	Colombo River 1	16	16	13	22	17	50
	Sacron River 2	14	14	13	19	15	60
	Calorie River 3	19	19	20	23	M	50
	Dummy 4(D)	M	0	M	0	0	50
Demand		30	20	70	30	60	

8.2 A Streamlined Simplex Method for the Transportation Problem

Because the transportation problem is just a special type of linear programming problem, it can be solved by applying the simplex method as described in Chap. 4. However, you will see in this section that some tremendous computational shortcuts can be taken in this method by exploiting the special structure shown in Table 8.6. We shall refer to this streamlined procedure as the **transportation simplex method**.

As you read on, note particularly how the special structure is exploited to achieve great computational savings. Then bear in mind that comparable savings sometimes can be achieved by exploiting other types of special structures as well, including those described later in the chapter.

Setting Up the Transportation Simplex Method

To highlight the streamlining achieved by the transportation simplex method, let us first review how in the general (unstreamlined) simplex method the transportation problem would be set up in tabular form. After constructing the table of constraint coefficients (see Table 8.6), converting the objective function to maximization form, and using the Big M method to introduce artificial variables $z_1, z_2, \ldots, z_{m+n}$ into the $m + n$ respective equality constraints (see Sec. 4.6), we see that typical columns of the simplex tableau would have the form shown in Table 8.13, where all entries *not shown* in these columns are *zeros*. [The one remaining adjustment to be made before the first iteration of the simplex method is to algebraically eliminate the nonzero coefficients of the initial (artificial) basic variables in row 0.]

After any subsequent iteration, row 0 then would have the form shown in Table 8.14. Because of the pattern of 0s and 1s for the coefficients in Table 8.13, by the *fundamental insight* presented in Sec. 5.3, u_i and v_j would have the following interpretation:

u_i = multiple of *original* row i that has been subtracted (directly or indirectly) from *original* row 0 by the simplex method during all iterations leading to the current simplex tableau.

v_j = multiple of *original* row $m + j$ that has been subtracted (directly or indirectly) from *original* row 0 by the simplex method during all iterations leading to the current simplex tableau.

Table 8.13 **Original Simplex Tableau before Simplex Method Is Applied to Transportation Problem**

Basic Variable	Eq.	Coefficient of: Z	···	x_{ij}	···	z_i	···	z_{m+j}	···	Right side
Z	(0)	-1		c_{ij}		M		M		0
	(1)									
	⋮									
z_i	(i)	0		1		1				s_i
	⋮									
z_{m+j}	$(m+j)$	0		1				1		d_j
	⋮									
	$(m+n)$									

You might recognize u_i and v_j from Chap. 6 as being the *dual variables.*[1] If x_{ij} is a nonbasic variable, $c_{ij} - u_i - v_j$ is interpreted as the rate at which Z will change as x_{ij} is increased.

To lay the groundwork for simplifying this setup, recall what information is needed by the simplex method. In the initialization, an initial BF solution must be obtained, which is done artificially by introducing artificial variables as the initial basic variables and setting them equal to s_i and d_j. The optimality test and step 1 of an iteration (selecting an entering basic variable) require knowing the current row 0, which is obtained by subtracting a certain multiple of another row from the preceding row 0. Step 2 (determining the leaving basic variable) must identify the basic variable that reaches zero first as the entering basic variable is increased, which is done by comparing the current coefficients of the entering basic variable and the corresponding right side. Step 3 must determine the new BF solution, which is found by subtracting certain multiples of one row from the other rows in the current simplex tableau.

Now, how does the *transportation simplex method* obtain the same information in much simpler ways? This story will unfold fully in the coming pages, but here are some preliminary answers.

First, *no artificial variables* are needed, because a simple and convenient procedure (with several variations) is available for constructing an initial BF solution.

Second, the currrent row 0 can be obtained *without using any other row* simply by calculating the currrent values of u_i and v_j directly. Since each basic variable must

[1] It would be easier to recognize these variables as dual variables by relabeling all these variables as y_i and then changing all the signs in row 0 of Table 8.14 by converting the objective function back to its original minimization form.

Table 8.14 **Row 0 of Simplex Tableau When Simplex Method Is Applied to Transportation Problem**

Basic Variable	Eq.	Coefficient of: Z	···	x_{ij}	···	z_i	···	z_{m+j}	···	Right Side
Z	(0)	-1		$c_{ij} - u_i - v_j$		$M - u_i$		$M - v_j$		$-\sum_{i=1}^{m} s_i u_i - \sum_{j=1}^{n} d_j v_j$

have a coefficient of zero in row 0, the current u_i and v_j are obtained by solving the set of equations

$$c_{ij} - u_i - v_j = 0 \qquad \text{for each } i \text{ and } j \text{ such that } x_{ij} \text{ is a basic variable,}$$

which can be done in a straightforward way. (Note how the special structure in Table 8.13 makes this convenient way of obtaining row 0 possible by yielding $c_{ij} - u_i - v_j$ as the coefficient of x_{ij} in Table 8.14.)

Third, the leaving basic variable can be identified in a simple way without (explicitly) using the coefficients of the entering basic variable. The reason is that the special structure of the problem makes it easy to see how the solution must change as the entering basic variable is increased. As a result, the new BF solution also can be identified immediately *without any algebraic manipulations* on the rows of the simplex tableau.

The grand conclusion is that *almost the entire simplex tableau* (and the work of maintaining it) *can be eliminated*! Besides the input data (the c_{ij}, s_i, and d_j values), the only information needed by the transportation simplex method is the current BF solution,[1] the current values of u_i and v_j, and the resulting values of $c_{ij} - u_i - v_j$ for nonbasic variables x_{ij}. When you solve a problem by hand, it is convenient to record this information for each iteration in a **transportation simplex tableau**, such as shown in Table 8.15. (Note carefully that the values of x_{ij} and $c_{ij} - u_i - v_j$ are distinguished in these tableaux by circling the former but not the latter.)

[1] Since nonbasic variables are automatically zero, the current BF solution is fully identified by recording just the values of the basic variables. We shall use this convention from now on.

Table 8.15 **Format of a Transportation Simplex Tableau**

		Destination					
		1	2	⋯	n	Supply	u_i
Source	1	c_{11}	c_{12}	⋯	c_{1n}	s_1	
	2	c_{21}	c_{22}	⋯	c_{2n}	s_2	
	⋮	⋯	⋯	⋯	⋯	⋮	
	m	c_{m1}	c_{m2}	⋯	c_{mn}	s_m	
Demand		d_1	d_2	⋯	d_n	$Z =$	
	v_j						

Additional information to be added to each cell:

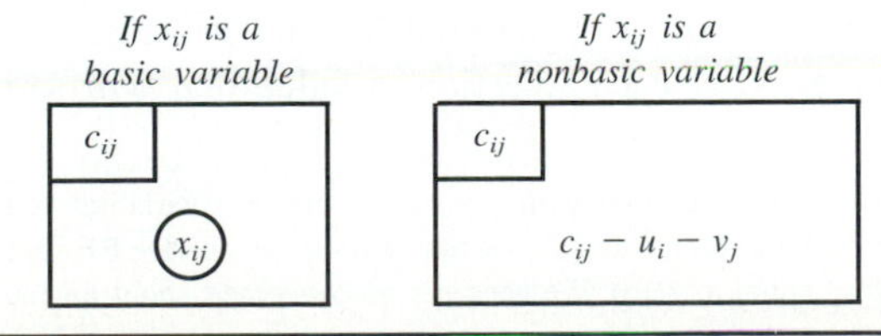

You can gain a fuller appreciation for the great difference in efficiency and convenience between the simplex and the transportation simplex methods by applying both to the same small problem (see Prob. 8.2-19). However, the difference becomes even more pronounced for large problems that must be solved on a computer. This pronounced difference is suggested somewhat by comparing the sizes of the simplex and the transportation simplex tableaux. Thus, for a transportation problem having m sources and n destinations, the simplex tableau would have $m + n + 1$ rows and $(m + 1)(n + 1)$ columns (excluding those to the left of the x_{ij} columns), and the transportation simplex tableau would have m rows and n columns (excluding the two extra informational rows and columns). Now try plugging in various values for m and n (for example, $m = 10$ and $n = 100$ would be a rather typical medium-size transportation problem), and note how the ratio of the number of cells in the simplex tableau to the number in the transportation simplex tableau increases as m and n increase.

Initialization

Recall that the objective of the initialization is to obtain an initial BF solution. Because all the functional constraints in the transportation problem are *equality* constraints, the simplex method would obtain this solution by introducing artificial variables and using them as the initial basic variables, as described in Sec. 4.6. The resulting basic solution actually is feasible only for a revised version of the problem, so a number of iterations are needed to drive these artificial variables to zero in order to reach the real BF solutions. The transportation simplex method bypasses all this by instead using a simpler procedure to directly construct a real BF solution on a transportation simplex tableau.

Before outlining this procedure, we need to point out that the number of basic variables in any basic solution of a transportation problem is one fewer than you might expect. Ordinarily, there is one basic variable for each functional constraint in a linear programming problem. For transportation problems with m sources and n destinations, the number of functional constraints is $m + n$. However,

$$\text{Number of basic variables} = m + n - 1.$$

The reason is that the functional constraints are equality constraints, and this set of $m + n$ equations has one *extra* (or *redundant*) equation that can be deleted without changing the feasible region; i.e., any one of the constraints is automatically satisfied whenever the other $m + n - 1$ constraints are satisfied. (This fact can be verified by showing that any supply constraint exactly equals the sum of the demand constraints minus the sum of the *other* supply constraints, and that any demand equation also can be reproduced by summing the supply equations and subtracting the other demand equations. See Prob. 8.2-21.) Therefore, any *BF solution* appears on a transportation simplex tableau with exactly $m + n - 1$ circled *nonnegative* allocations, where the sum of the allocations for each row or column equals its supply or demand.[1]

The procedure for constructing an initial BF solution selects the $m + n - 1$ basic variables one at a time. After each selection, a value that will satisfy one additional

[1] However, note that any feasible solution with $m + n - 1$ nonzero variables is *not necessarily* a basic solution because it might be the weighted average of two or more degenerate BF solutions (i.e., BF solutions having some basic variables equal to zero). We need not be concerned about mislabeling such solutions as being basic, however, because the transportation simplex method constructs only legitimate BF solutions.

constraint (thereby eliminating that constraint's row or column from further consideration for providing allocations) is assigned to that variable. Thus, after $m + n - 1$ selections, an entire basic solution has been constructed in such a way as to satisfy all the constraints. A number of different criteria have been proposed for selecting the basic variables. We present and illustrate three of these criteria here, after outlining the general procedure.

General Procedure[1] for Constructing an Initial BF Solution

To begin, all source rows and destination columns of the transportation simplex tableau are initially under consideration for providing a basic variable (allocation).

1. From the rows and columns still under consideration, select the next basic variable (allocation) according to some criterion.
2. Make that allocation large enough to exactly use up the remaining supply in its row or the remaining demand in its column (whichever is smaller).
3. Eliminate that row or column (whichever had the smaller remaining supply or demand) from further consideration. (If the row and column have the same remaining supply and demand, then arbitrarily select the *row* as the one to be eliminated. The column will be used later to provide a *degenerate* basic variable, i.e., a circled allocation of zero.)
4. If only one row or only one column remains under consideration, then the procedure is completed by selecting every *remaining* variable (i.e., those variables that were neither previously selected to be basic nor eliminated from consideration by eliminating their row or column) associated with that row or column to be basic with the only feasible allocation. Otherwise, return to step 1.

Alternative Criteria for Step 1

1. *Northwest corner rule:* Begin by selecting x_{11} (that is, start in the northwest corner of the transportation simplex tableau). Thereafter, if x_{ij} was the last basic variable selected, then next select $x_{i,j+1}$ (that is, move one column to the *right*) if source i has any supply remaining. Otherwise, next select $x_{i+1,j}$ (that is, move one row *down*).

Example: To make this description more concrete, we now illustrate the general procedure on the Metro Water District problem (see Table 8.12) with the northwest corner rule being used in step 1. Because $m = 4$ and $n = 5$ in this case, the procedure would find an initial BF solution having $m + n - 1 = 8$ basic variables.

As shown in Table 8.16, the first allocation is $x_{11} = 30$, which exactly uses up the demand in column 1 (and eliminates this column from further consideration). This first iteration leaves a supply of 20 remaining in row 1, so next select $x_{1,1+1} = x_{12}$ to be a basic variable. Because this supply is no larger than the demand of 20 in column 2, all of it is allocated, $x_{12} = 20$, and this row is eliminated from further consideration. (Row 1 is chosen for elimination rather than column 2 because of the parenthetical instruction in step 3.) Therefore, select $x_{1+1,2} = x_{22}$ next. Because the remaining demand of 0 in column 2 is less than the supply of 60 in row 2, allocate $x_{22} = 0$ and eliminate column 2.

[1] In Sec. 4.1 we pointed out that the simplex method is an example of the algorithms (iterative solution procedures) so prevalent in OR work. Note that this procedure also is an algorithm, where each successive execution of the (four) steps constitutes an iteration.

Table 8.16 **Initial BF Solution from the Northwest Corner Rule**

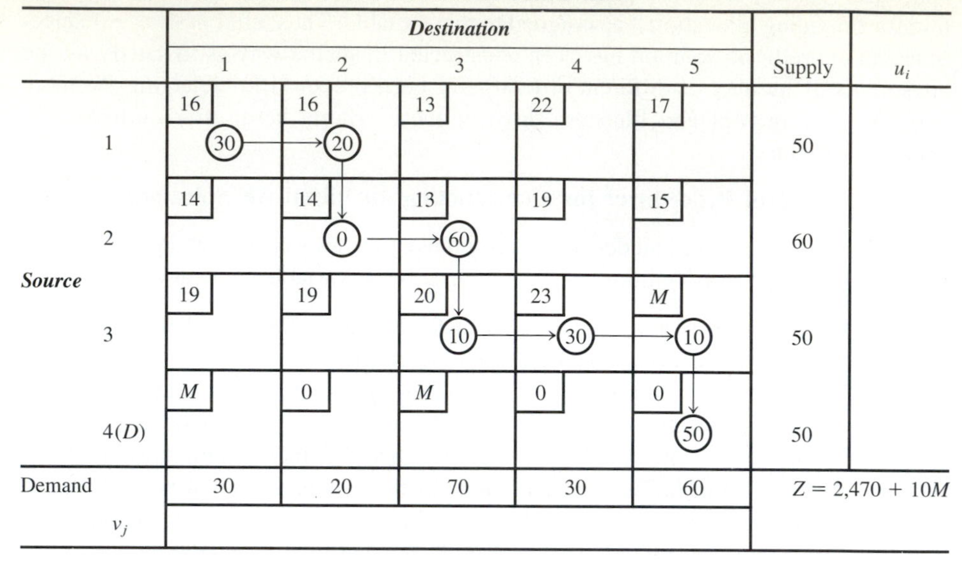

		Destination						
		1	2	3	4	5	Supply	u_i
Source	1	16 (30)	16 (20)	13	22	17	50	
	2	14	14 (0)	13 (60)	19	15	60	
	3	19	19	20 (10)	23 (30)	M (10)	50	
	4(D)	M	0	M	0	0 (50)	50	
Demand		30	20	70	30	60	$Z = 2{,}470 + 10M$	
	v_j							

Continuing in this manner, we eventually obtain the entire *initial BF solution* shown in Table 8.16, where the circled numbers are the values of the basic variables ($x_{11} = 30, \ldots, x_{45} = 50$) and all the other variables (x_{13}, etc.) are nonbasic variables equal to zero. Arrows have been added to show the order in which the basic variables (allocations) were selected. The value of Z for this solution is

$$Z = 16(30) + 16(20) + \cdots + 0(50) = 2{,}470 + 10M.$$

2. *Vogel's approximation method:* For each row and column remaining under consideration, calculate its **difference**, which is defined as *the arithmetic difference between the smallest and next-to-the-smallest unit cost c_{ij} still remaining in that row or column.* (If two unit costs tie for being the smallest remaining in a row or column, then the *difference* is 0.) In that row or column having the *largest difference,* select the variable having the *smallest remaining unit cost.* (Ties for the largest difference, or for the smallest remaining unit cost, may be broken arbitrarily.)

Example: Now let us apply the general procedure to the Metro Water District problem by using the criterion for Vogel's approximation method to select the next basic variable in step 1. With this criterion, it is more convenient to work with cost and requirements tables (rather than with complete transportation simplex tableaux), beginning with the one shown in Table 8.12. At each iteration, after the difference for every row and column remaining under consideration is calculated and displayed, the largest difference is circled and the smallest unit cost in its row or column is enclosed in a box. The resulting selection (and value) of the variable having this unit cost as the next basic variable is indicated in the lower right-hand corner of the current table, along with the row or column thereby being eliminated from further consideration (see steps 2 and 3 of the general procedure). The table for the next iteration is exactly the same except for deleting this row or column and subtracting the last allocation from its supply or demand (whichever remains).

Applying this procedure to the Metro Water District problem yields the sequence of cost and requirements tables shown in Table 8.17, where the resulting initial BF

Table 8.17 Initial BF Solution from Vogel's Approximation Method

		Destination						
		1	2	3	4	5	Supply	Row Difference
Source	1	16	16	13	22	17	50	3
	2	14	14	13	19	15	60	1
	3	19	19	20	23	M	50	0
	4(D)	M	0	M	0	0	50	0
Demand		30	20	70	30	60	Select $x_{44} = 30$	
Column difference		2	14	0	19	15	Eliminate column 4	

		Destination					
		1	2	3	5	Supply	Row Difference
Source	1	16	16	13	17	50	3
	2	14	14	13	15	60	1
	3	19	19	20	M	50	0
	4(D)	M	0	M	0	20	0
Demand		30	20	70	60	Select $x_{45} = 20$	
Column difference		2	14	0	15	Eliminate row 4(D)	

		Destination					
		1	2	3	5	Supply	Row Difference
Source	1	16	16	13	17	50	3
	2	14	14	13	15	60	1
	3	19	19	20	M	50	0
Demand		30	20	70	40	Select $x_{13} = 50$	
Column difference		2	2	0	2	Eliminate row 1	

		Destination					
		1	2	3	5	Supply	Row Difference
Source	2	14	14	13	15	60	1
	3	19	19	20	M	50	0
Demand		30	20	20	40	Select $x_{25} = 40$	
Column difference		5	5	7	$M - 15$	Eliminate column 5	

		Destination				
		1	2	3	Supply	Row Difference
Source	2	14	14	13	20	1
	3	19	19	20	50	0
Demand		30	20	20	Select $x_{23} = 20$	
Column difference		5	5	7	Eliminate row 2	

		Destination				
		1	2	3	Supply	
Source	3	19	19	20	50	
Demand		30	20	0	Select $x_{31} = 30$ $x_{32} = 20$ $x_{33} = 0$	$Z = 2{,}460$

solution consists of the eight basic variables (allocations) given in the lower right-hand corner of the respective cost and requirements tables.

This example illustrates two relatively subtle features of the general procedure that warrant special attention. First, note that the final iteration selects *three* variables (x_{31}, x_{32}, and x_{33}) to become basic instead of the single selection made at the other iterations. The reason is that only *one* row (row 3) remains under consideration at this point. Therefore, step 4 of the general procedure says to select *every* remaining variable associated with row 3 to be basic.

Second, note that the allocation of $x_{23} = 20$ at the next-to-last iteration exhausts *both* the remaining supply in its row *and* the remaining demand in its column. However, rather than eliminate both the row and column from further consideration, step 3 says to eliminate *only the row,* saving the column to provide a *degenerate* basic variable later. Column 3 is, in fact, used for just this purpose at the final iteration when $x_{33} = 0$ is selected as one of the basic variables. For another illustration of this same phenomenon, see Table 8.16 where the allocation of $x_{12} = 20$ results in eliminating only row 1, so that column 2 is saved to provide a degenerate basic variable, $x_{22} = 0$, at the next iteration.

Although a zero allocation might seem irrelevant, it actually plays an important role. You will see soon that the transportation simplex method must know *all* $m + n - 1$ basic variables, including those with value zero, in the current BF solution.

3. *Russell's approximation method:* For each source row i remaining under consideration, determine its $\bar{u}_i$, which is the largest unit cost c_{ij} still remaining in that row. For each destination column j remaining under consideration, determine its $\bar{v}_j$, which is the largest unit cost c_{ij} still remaining in that column. For each variable x_{ij} not previously selected in these rows and columns, calculate $\Delta_{ij} = c_{ij} - \bar{u}_i - \bar{v}_j$. Select the variable having the *largest* (in absolute terms) *negative* value of Δ_{ij}. (Ties may be broken arbitrarily.)

EXAMPLE: Using the criterion for Russell's approximation method in step 1, we again apply the general procedure to the Metro Water District problem (see Table 8.12). The results, including the sequence of basic variables (allocations), are shown in Table 8.18.

At iteration 1, the largest unit cost in row 1 is $\bar{u}_1 = 22$, the largest in column 1 is $\bar{v}_1 = M$, and so forth. Thus

$$\Delta_{11} = c_{11} - \bar{u}_1 - \bar{v}_1 = 16 - 22 - M = -6 - M.$$

Table 8.18 **Initial BF Solution from Russell's Approximation Method**

Iteration	$\bar{u}_1$	$\bar{u}_2$	$\bar{u}_3$	$\bar{u}_4$	$\bar{v}_1$	$\bar{v}_2$	$\bar{v}_3$	$\bar{v}_4$	$\bar{v}_5$	Largest Negative Δ_{ij}	Allocation
1	22	19	M	M	M	19	M	23	M	$\Delta_{45} = -2M$	$x_{45} = 50$
2	22	19	M		19	19	20	23	M	$\Delta_{15} = -5 - M$	$x_{15} = 10$
3	22	19	23		19	19	20	23		$\Delta_{13} = -29$	$x_{13} = 40$
4		19	23		19	19	20	23		$\Delta_{23} = -26$	$x_{23} = 30$
5		19	23		19	19		23		$\Delta_{21} = -24$*	$x_{21} = 30$
6										Irrelevant	$x_{31} = 0$
											$x_{32} = 20$
											$x_{34} = 30$
											$Z = 2,570$

* Tie with $\Delta_{22} = -24$ broken arbitrarily.

Calculating all the Δ_{ij} values for $i = 1, 2, 3, 4$ and $j = 1, 2, 3, 4, 5$ shows that $\Delta_{45} = 0 - 2M$ has the largest negative value, so $x_{45} = 50$ is selected as the first basic variable (allocation). This allocation exactly uses up the supply in row 4, so this row is eliminated from further consideration.

Note that eliminating this row changes $\bar{v}_1$ and $\bar{v}_3$ for the next iteration. Therefore, the second iteration requires recalculating the Δ_{ij} with $j = 1, 3$ as well as eliminating $i = 4$. The largest negative value now is

$$\Delta_{15} = 17 - 22 - M = -5 - M,$$

so $x_{15} = 10$ becomes the second basic variable (allocation), eliminating column 5 from further consideration.

The subsequent iterations proceed similarly, but you may want to test your understanding by verifying the remaining allocations given in Table 8.18. As with the other procedures in this (and other) section(s), you should find your OR Courseware useful for doing the calculations involved and illuminating the approach.

Comparison of Alternative Criteria for Step 1

Now let us compare these three criteria for selecting the next basic variable. The main virtue of the northwest corner rule is that it is quick and easy. However, because it pays no attention to unit costs c_{ij}, usually the solution obtained will be far from optimal. (Note in Table 8.16 that $x_{35} = 10$ even though $c_{35} = M$.) Expending a little more effort to find a good initial BF solution might greatly reduce the number of iterations then required by the transportation simplex method to reach an optimal solution (see Probs. 8.2-10 and 8.2-8). Finding such a solution is the objective of the other two criteria.

Vogel's approximation method has been a popular criterion for many years,[1] partially because it is relatively easy to implement by hand. Because the *difference* represents the minimum extra unit cost incurred by failing to make an allocation to the cell having the smallest unit cost in that row or column, this criterion does take costs into account in an effective way.

Russell's approximation method provides another excellent criterion[2] that is still quick to implement on a computer (but not manually). Although more experimentation is required to determine which is more effective *on average,* this criterion *frequently* does obtain a better solution than Vogel's. (For the example, Vogel's approximation method happened to find the optimal solution with $Z = 2{,}460$, whereas Russell's misses slightly with $Z = 2{,}570$.) For a large problem, it may be worthwhile to apply both criteria and then use the better solution to start the iterations of the transportation simplex method.

One distinct advantage of Russell's approximation method is that it is patterned directly after step 1 for the transportation simplex method (as you will see soon), which somewhat simplifies the overall computer code. In particular, the $\bar{u}_i$ and $\bar{v}_j$ values have been defined in such a way that the relative values of the $c_{ij} - \bar{u}_i - \bar{v}_j$ *estimate* the relative values of $c_{ij} - u_i - v_j$ that will be obtained when the transportation simplex method reaches an optimal solution.

We now shall use the initial BF solution obtained in Table 8.18 by Russell's approximation method to illustrate the remainder of the transportation simplex method. Thus, our *initial transportation simplex tableau* (before we solve for u_i and v_j) is shown in Table 8.19.

[1] N. V. Reinfeld and W. R. Vogel, *Mathematical Programming,* Prentice-Hall, Englewood Cliffs, NJ, 1958.

[2] E. J. Russell, "Extension of Dantzig's Algorithm to Finding an Initial Near-Optimal Basis for the Transportation Problem," *Operations Research,* **17**: 187–191, 1969.

Table 8.19 **Initial Transportation Simplex Tableau (before We Obtain $c_{ij} - u_i - v_j$) from Russell's Approximation Method**

Iteration 0		Destination 1	2	3	4	5	Supply	u_i
Source	1	16	16	13 (40)	22	17 (10)	50	
	2	14 (30)	14	13 (30)	19	15	60	
	3	19 (0)	19 (20)	20	23 (30)	M	50	
	4(*D*)	M	0	M	0	0 (50)	50	
Demand		30	20	70	30	60	$Z = 2{,}570$	
v_j								

The next step is to check whether this initial solution is optimal by applying the *optimality test.*

Optimality Test

Using the notation of Table 8.14, we can reduce the standard optimality test for the simplex method (see Sec. 4.3) to the following for the transportation problem:

> **Optimality test:** A BF solution is optimal if and only if $c_{ij} - u_i - v_j \geq 0$ for every (i, j) such that x_{ij} is nonbasic.[1]

Thus the only work required by the optimality test is the derivation of the values of u_i and v_j for the current BF solution and then the calculation of these $c_{ij} - u_i - v_j$, as described below.

Since $c_{ij} - u_i - v_j$ is required to be zero if x_{ij} is a basic variable, u_i and v_j satisfy the set of equations

$$c_{ij} = u_i + v_j \qquad \text{for each } (i, j) \text{ such that } x_{ij} \text{ is basic.}$$

There are $m + n - 1$ basic variables, and so there are $m + n - 1$ of these equations. Since the number of unknowns (the u_i and v_j) is $m + n$, one of these variables can be assigned a value arbitrarily without violating the equations. The choice of this one variable and its value does not affect the value of any $c_{ij} - u_i - v_j$, even when x_{ij} is nonbasic, so the only (minor) difference it makes is in the ease of solving these equations. A convenient choice for this purpose is to select the u_i that has the *largest number*

[1] The one exception is that two or more equivalent degenerate BF solutions (i.e., identical solutions having different degenerate basic variables equal to zero) can be optimal with only some of these basic solutions satisfying the optimality test. This exception is illustrated later in the example (see the identical solutions in the last two tableaux of Table 8.23, where only the latter solution satisfies the criterion for optimality).

of allocations in its row (break any tie arbitrarily) and to assign to it the value zero. Because of the simple structure of these equations, it is then very simple to solve for the remaining variables algebraically.

To demonstrate, we give each equation that corresponds to a basic variable in our initial BF solution.

$$
\begin{array}{lll}
x_{31}: & 19 = u_3 + v_1. & \text{Set } u_3 = 0, \text{ so } v_1 = 19, \\
x_{32}: & 19 = u_3 + v_2. & v_2 = 19, \\
x_{34}: & 23 = u_3 + v_4. & v_4 = 23. \\
x_{21}: & 14 = u_2 + v_1. & \text{Know } v_1 = 19, \text{ so } u_2 = -5. \\
x_{23}: & 13 = u_2 + v_3. & \text{Know } u_2 = -5, \text{ so } v_3 = 18. \\
x_{13}: & 13 = u_1 + v_3. & \text{Know } v_3 = 18, \text{ so } u_1 = -5. \\
x_{15}: & 17 = u_1 + v_5. & \text{Know } u_1 = -5, \text{ so } v_5 = 22. \\
x_{45}: & 0 = u_4 + v_5. & \text{Know } v_5 = 22, \text{ so } u_4 = -22.
\end{array}
$$

Setting $u_3 = 0$ (since row 3 of Table 8.19 has the largest number of allocations—3) and moving down the equations one at a time immediately give the derivation of values for the unknowns shown to the right of the equations. (Note that this derivation of the u_i and v_j values depends on which x_{ij} variables are *basic variables* in the current BF solution, so this derivation will need to be repeated each time a new BF solution is obtained.)

Once you get the hang of it, you probably will find it even more convenient to solve these equations without writing them down by working directly on the transportation simplex tableau. Thus, in Table 8.19 you begin by writing in the value $u_3 = 0$ and then picking out the circled allocations (x_{31}, x_{32}, x_{34}) in that row. For each one you set $v_j = c_{3j}$ and then look for circled allocations (except in row 3) in these columns (x_{21}). Mentally calculate $u_2 = c_{21} - v_1$, pick out x_{23}, set $v_3 = c_{23} - u_2$, and so on until you have filled in all the values for u_i and v_j. (Try it.) Then calculate and fill in the value of $c_{ij} - u_i - v_j$ for each nonbasic variable x_{ij} (that is, for each cell without a circled allocation), and you will have the completed initial transportation simplex tableau shown in Table 8.20.

We are now in a position to apply the optimality test by checking the value of $c_{ij} - u_i - v_j$ given in Table 8.20. Because two of these values ($c_{25} - u_2 - v_5 = -2$ and $c_{44} - u_4 - v_4 = -1$) are negative, we conclude that the current BF solution is not optimal. Therefore, the transportation simplex method must next go to an iteration to find a better BF solution.

An Iteration

As with the full-fledged simplex method, an iteration for this streamlined version must determine an entering basic variable (step 1), a leaving basic variable (step 2), and then identify the resulting new BF solution (step 3).

Step 1: Since $c_{ij} - u_i - v_j$ represents the rate at which the objective function will change as the nonbasic variable x_{ij} is increased, the entering basic variable must have a *negative* $c_{ij} - u_i - v_j$ value to decrease the total cost Z. Thus the candidates in Table

Table 8.20 **Completed Initial Transportation Simplex Tableau**

Iteration 0		Destination 1	2	3	4	5	Supply	u_i
Source	1	16; +2	16; +2	13; (40)	22; +4	17; (10)	50	−5
	2	14; (30)	14; 0	13; (30)	19; +1	15; −2	60	−5
	3	19; (0)	19; (20)	20; +2	23; (30)	M; $M - 22$	50	0
	4(*D*)	M; $M + 3$	0; +3	M; $M + 4$	0; −1	0; (50)	50	−22
Demand		30	20	70	30	60	$Z = 2{,}570$	
	v_j	19	19	18	23	22		

8.20 are x_{25} and x_{44}. To choose between the candidates, select the one having the larger (in absolute terms) negative value of $c_{ij} - u_i - v_j$ to be the entering basic variable, which is x_{25} in this case.

Step 2: Increasing the entering basic variable from zero sets off a *chain reaction* of compensating changes in other basic variables (allocations), in order to continue satisfying the supply and demand constraints. The first basic variable to be decreased to zero then becomes the leaving basic variable.

With x_{25} as the entering basic variable, the chain reaction in Table 8.20 is the relatively simple one summarized in Table 8.21. (We shall always indicate the entering basic variable by placing a boxed plus sign in the center of its cell while leaving the corresponding value of $c_{ij} - u_i - v_j$ in the lower right-hand corner of this cell.) In-

Table 8.21 **Part of Initial Transportation Simplex Tableau Showing the Chain Reaction Caused by Increasing the Entering Basic Variable x_{25}**

			Destination 3	4	5	Supply
Source	1	...	13; (40) +	22; +4	17; (10) −	50
	2	...	13; (30) −	19; +1	15; [+]; −2	60
		...	...	...	...	
Demand			70	30	60	

creasing x_{25} by some amount requires decreasing x_{15} by the same amount to restore the demand of 60 in column 5. This change then requires increasing x_{13} by this same amount to restore the supply of 50 in row 1. This change then requires decreasing x_{23} by this amount to restore the demand of 70 in column 3. This decrease in x_{23} successfully completes the chain reaction because it also restores the supply of 60 in row 2. (Equivalently, we could have started the chain reaction by restoring this supply in row 2 with the decrease in x_{23}, and then the chain reaction would continue with the increase in x_{13} and decrease in x_{15}.)

The net result is that cells (2, 5) and (1, 3) become **recipient cells**, each receiving its additional allocation from one of the **donor cells**, (1, 5) and (2, 3). (These cells are indicated in Table 8.21 by the plus and minus signs.) Note that cell (1, 5) had to be the donor cell for column 5 rather than cell (4, 5), because cell (4, 5) would have no recipient cell in row 4 to continue the chain reaction. [Similarly, if the chain reaction had been started in row 2 instead, cell (2, 1) could not be the donor cell for this row because the chain reaction could not then be completed successfully after necessarily choosing cell (3, 1) as the next recipient cell and either cell (3, 2) or (3, 4) as its donor cell.] Also note that, except for the entering basic variable, *all* recipient cells and donor cells in the chain reaction must correspond to *basic* variables in the current BF solution.

Each donor cell decreases its allocation by exactly the same amount as the entering basic variable (and other recipient cells) is increased. Therefore, the donor cell that starts with the smallest allocation—cell (1, 5) in this case (since $10 < 30$ in Table 8.21)—must reach a zero allocation first as the entering basic variable x_{25} is increased. Thus x_{15} becomes the leaving basic variable.

In general, there always is just *one* chain reaction (in either direction) that can be completed successfully to maintain feasibility when the entering basic variable is increased from zero. This chain reaction can be identified by selecting from the cells having a basic variable: first the donor cell in the *column* having the entering basic variable, then the recipient cell in the row having this donor cell, then the donor cell in the column having this recipient cell, and so on until the chain reaction yields a donor cell in the *row* having the entering basic variable. When a column or row has more than one additional basic variable cell, it may be necessary to trace them all further to see which one must be selected to be the donor or recipient cell. (All but this one eventually will reach a dead end in a row or column having no additional basic variable cell.) After the chain reaction is identified, the donor cell having the *smallest* allocation automatically provides the leaving basic variable. (In the case of a tie for the donor cell having the smallest allocation, any one can be chosen arbitrarily to provide the leaving basic variable.)

STEP 3: The *new BF solution* is identified simply by adding the value of the leaving basic variable (before any change) to the allocation for each recipient cell and subtracting *this same amount* from the allocation for each donor cell. In Table 8.21 the value of the leaving basic variable x_{15} is 10, so this portion of the transportation simplex tableau changes as shown in Table 8.22 for the new solution. (Since x_{15} is nonbasic in the new solution, its new allocation of zero is no longer shown in this new tableau.)

We can now highlight a useful interpretation of the $c_{ij} - u_i - v_j$ quantities derived during the optimality test. Because of the shift of 10 allocation units from the donor cells to the recipient cells (shown in Tables 8.21 and 8.22), the total cost changes by

$$\Delta Z = 10(15 - 17 + 13 - 13) = 10(-2) = 10(c_{25} - u_2 - v_5).$$

Table 8.22 **Part of Second Transportation Simplex Tableau Showing the Changes in the BF Solution**

		Destination				
			3	4	5	Supply
Source	1	⋯	13 / (50)	22	17	50
	2	⋯	13 / (20)	19	15 / (10)	60
		⋯	⋯	⋯	⋯	
Demand			70	30	60	

Thus the effect of increasing the entering basic variable x_{25} from zero has been a cost change at the rate of -2 per unit increase in x_{25}. This is precisely what the value of $c_{25} - u_2 - v_5 = -2$ in Table 8.20 indicates would happen. In fact, another (but less efficient) way of deriving $c_{ij} - u_i - v_j$ for each nonbasic variable x_{ij} is to identify the chain reaction caused by increasing this variable from 0 to 1 and then to calculate the resulting cost change. This intuitive interpretation sometimes is useful for checking calculations during the optimality test.

Before completing the solution of the Metro Water District problem, we now summarize the rules for the transportation simplex method.

Summary of the Transportation Simplex Method

Initialization: Construct an initial BF solution by the procedure outlined earlier in this section. Go to the optimality test.

Optimality test: Derive u_i *and* v_j by selecting the row having the largest number of allocations, setting its $u_i = 0$, and then solving the set of equations $c_{ij} = u_i + v_j$ for each (i, j) such that x_{ij} is basic. If $c_{ij} - u_i - v_j \geq 0$ for every (i, j) such that x_{ij} is *nonbasic,* then the current solution is optimal, so stop. Otherwise, go to an iteration.

Iteration:

1. Determine the entering basic variable: Select the nonbasic variable x_{ij} having the *largest* (in absolute terms) *negative* value of $c_{ij} - u_i - v_j$.
2. Determine the leaving basic variable: Identify the chain reaction required to retain feasibility when the entering basic variable is increased. From the donor cells, select the basic variable having the *smallest* value.
3. Determine the new BF solution: Add the value of the leaving basic variable to the allocation for each recipient cell. Subtract this value from the allocation for each donor cell.

Continuing to apply this procedure to the Metro Water District problem yields the complete set of transportation simplex tableaux shown in Table 8.23. Since all the $c_{ij} - u_i - v_j$ values are nonnegative in the fourth tableau, the optimality test identifies the set of allocations in this tableau as being optimal, which concludes the algorithm.

It would be good practice for you to derive the values of u_i and v_j given in the second, third, and fourth tableaux. Try doing this by working directly on the tableaux. Also check out the chain reactions in the second and third tableaux, which are somewhat more complicated than the one you have seen in Table 8.21.

Note three special points that are illustrated by this example. First, the initial BF solution is *degenerate* because the basic variable $x_{31} = 0$. However, this degenerate basic variable causes no complication, because cell (3, 1) becomes a *recipient cell* in the second tableau, which increases x_{31} to a value greater than zero.

Second, another degenerate basic variable (x_{34}) arises in the third tableau because the basic variables for *two* donor cells in the second tableau, cells (2, 1) and (3, 4), *tie* for having the smallest value (30). (This tie is broken arbitrarily by selecting x_{21} as the leaving basic variable; if x_{34} had been selected instead, then x_{21} would have become the degenerate basic variable.) This degenerate basic variable does appear to create a complication subsequently, because cell (3, 4) becomes a *donor cell* in the third tableau but has nothing to donate! Fortunately, such an event actually gives no cause for concern. Since zero is the amount to be added to or subtracted from the allocations for the recipient and donor cells, these allocations do not change. However, the degenerate basic variable does become the leaving basic variable, so it is replaced by the entering basic variable as the circled allocation of zero in the fourth tableau. This change in the set of basic variables changes the values of u_i and v_j. Therefore, if any of the $c_{ij} - u_i - v_j$ had been negative in the fourth tableau, the algorithm would have gone on to make *real* changes in the allocations (whenever all donor cells have nondegenerate basic variables).

Third, because none of the $c_{ij} - u_i - v_j$ turned out to be negative in the fourth tableau, the equivalent set of allocations in the third tableau is optimal also. Thus the algorithm executed one more iteration than was necessary. This extra iteration is a flaw that occasionally arises in both the transportation simplex method and the simplex method because of degeneracy, but it is not sufficiently serious to warrant any adjustments to these algorithms.

For another (smaller) example of the application of the transportation simplex method, refer to the demonstration provided for the transportation problem area in your OR Courseware. Also provided under the Procedure menu are both interactive and automatic routines for the transportation simplex method.

Now that you have studied the transportation simplex method, you are in a position to check for yourself how the algorithm actually provides a proof of the *integer solutions property* presented in Sec. 8.1. Problem 8.2–22 helps to guide you through the reasoning.

8.3 The Assignment Problem

The **assignment problem** is a special type of linear programming problem where **assignees** are being assigned to perform **tasks**. For example, the assignees might be employees who need to be given work assignments. Assigning people to jobs is a common application of the assignment problem. However, the assignees need not be people. They also could be machines, or vehicles, or plants, or even time slots to be assigned tasks. The first example below involves machines being assigned to locations, so the tasks in this case simply involve holding a machine. A subsequent example involves plants being assigned products to be produced.

To fit the definition of an assignment problem, these kinds of applications need to be formulated in a way that satisfy the following assumptions.

1. The number of assignees and the number of tasks are the same. (This number is denoted by n.)
2. Each assignee is to be assigned to exactly *one* task.

Table 8.23 **Complete Set of Transportation Simplex Tableaux for the Metro Water District Problem**

Iteration 0		Destination 1	Destination 2	Destination 3	Destination 4	Destination 5	Supply	u_i
Source	1	16; +2	16; +2	13; (40) +	22; +4	17; (10) −	50	−5
	2	14; (30)	14; 0	13; (30) −	19; +1	15; [+]; −2	60	−5
	3	19; (0)	19; (20)	20; +2	23; (30)	M; $M - 22$	50	0
	4(D)	M; $M + 3$	0; +3	M; $M + 4$	0; −1	0; (50)	50	−22
Demand		30	20	70	30	60	$Z = 2{,}570$	
v_j		19	19	18	23	22		

Iteration 1		Destination 1	Destination 2	Destination 3	Destination 4	Destination 5	Supply	u_i
Source	1	16; +2	16; +2	13; (50)	22; +4	17; +2	50	−5
	2	14; (30) −	14; 0	13; (20)	19; +1	15; (10) +	60	−5
	3	19; (0) +	19; (20)	20; +2	23; (30) −	M; $M - 20$	50	0
	4(D)	M; $M + 1$	0; +1	M; $M + 2$	0; [+]; −3	0; (50) −	50	−20
Demand		30	20	70	30	60	$Z = 2{,}550$	
v_j		19	19	18	23	20		

3. Each task is to be performed by exactly *one* assignee.
4. There is a cost c_{ij} associated with assignee i ($i = 1, 2, \ldots, n$) performing task j ($j = 1, 2, \ldots, n$).
5. The objective is to determine how all n assignments should be made in order to minimize the total cost.

Any problem satisfying all these assumptions can be solved extremely efficiently by algorithms designed specifically for assignment problems.

The first three assumptions are fairly restrictive. Many potential applications do not quite satisfy these assumptions. However, it often is possible to reformulate the

Table 8.23 **(*Continued*)**

Iteration 2		Destination 1	2	3	4	5	Supply	u_i
Source	1	16; +5	16; +5	13; (50)	22; +7	17; +2	50	−8
	2	14; +3	14; +3	13; (20)−	19; +4	15; (40)+	60	−8
	3	19; (30)	19; (20)	20; [+] −1	23; (0)−	*M*; *M* −23	50	0
	4(*D*)	*M*; *M* + 4	0; +4	*M*; *M* + 2	0; (30)+	0; (20)−	50	−23
Demand		30	20	70	30	60	$Z = 2,460$	
v_j		19	19	21	23	23		

Iteration 3		Destination 1	2	3	4	5	Supply	u_i
Source	1	16; +4	16; +4	13; (50)	22; +7	17; +2	50	−7
	2	14; +2	14; +2	13; (20)	19; +4	15; (40)	60	−7
	3	19; (30)	19; (20)	20; (0)	23; +1	*M*; *M* − 22	50	0
	4(*D*)	*M*; *M* + 3	0; +3	*M*; *M* + 2	0; (30)	0; (20)	50	−22
Demand		30	20	70	30	60	$Z = 2,460$	
v_j		19	19	20	22	22		

problem to make it fit. For example, *dummy assignees* or *dummy tasks* frequently can be used for this purpose. We illustrate these formulation techniques in the examples.

Prototype Example

The JOB SHOP COMPANY has purchased three new machines of different types. There are four available locations in the shop where a machine could be installed. Some of these locations are more desirable than others for particular machines because of their proximity to work centers that will have a heavy work flow to and from these machines. (There will be no work flow *between* the new machines.) Therefore, the

Table 8.24 **Materials-Handling Cost Data for Job Shop Co.**

		Location			
		1	2	3	4
	1	13	16	12	11
Machine	2	15	—	13	20
	3	5	7	10	6

objective is to assign the new machines to the available locations to minimize the total cost of materials handling. The estimated cost per unit time of materials handling involving each of the machines is given in Table 8.24 for the respective locations. Location 2 is not considered suitable for machine 2, so no cost is given for this case.

To formulate this problem as an assignment problem, we must introduce a *dummy machine* for the extra location. Also, an extremely large cost M should be attached to the assignment of machine 2 to location 2 to prevent this assignment in the optimal solution. The resulting assignment problem *cost table* is shown in Table 8.25. This cost table contains all the necessary data for solving the problem. The optimal solution is to assign machine 1 to location 4, machine 2 to location 3, and machine 3 to location 1, for a total cost of 29. The dummy machine is assigned to location 2, so this location is available for some future real machine.

We shall discuss how this solution is obtained after we formulate the mathematical model for the general assignment problem.

The Assignment Problem Model and Solution Procedures

The mathematical model for the assignment problem uses the following decision variables:

$$x_{ij} = \begin{cases} 1 & \text{if assignee } i \text{ performs task } j, \\ 0 & \text{if not,} \end{cases}$$

for $i = 1, 2, \ldots, n$ and $j = 1, 2, \ldots, n$. Thus each x_{ij} is a *binary variable* (it has value 0 or 1). As discussed at length in the chapter on integer programming (Chap. 12), binary variables are important in OR for representing *yes/no decisions.* In this case, the yes/no decision is: Should assignee i perform task j?

By letting Z denote the total cost, the assignment problem model is

$$\text{Minimize} \quad Z = \sum_{i=1}^{n} \sum_{j=1}^{n} c_{ij} x_{ij},$$

Table 8.25 **Cost Table for the Job Shop Co. Assignment Problem**

		Task (Location)			
		1	2	3	4
	1	13	16	12	11
Assignee	2	15	M	13	20
(Machine)	3	5	7	10	6
	4(*D*)	0	0	0	0

subject to

$$\sum_{j=1}^{n} x_{ij} = 1 \qquad \text{for } i = 1, 2, \ldots, n,$$

$$\sum_{i=1}^{n} x_{ij} = 1 \qquad \text{for } j = 1, 2, \ldots, n,$$

and

$$x_{ij} \geq 0, \qquad \text{for all } i \text{ and } j$$

$$(x_{ij} \text{ binary}, \qquad \text{for all } i \text{ and } j).$$

The first set of functional constraints specifies that each assignee is to perform exactly one task, whereas the second set requires each task to be performed by exactly one assignee. If we delete the parenthetical restriction that the x_{ij} be binary, the model clearly is a special type of linear programming problem and so can be readily solved. Fortunately, for reasons about to unfold, we *can* delete this restriction. (This deletion is the reason that the assignment problem appears in this chapter rather than in the integer programming chapter.)

Now compare this model (without the binary restriction) with the transportation problem model presented in the second subsection of Sec. 8.1 (including Table 8.6). Note how similar their structures are. In fact, the assignment problem is just a special type of transportation problem where the *sources* now are *assignees* and the *destinations* now are *tasks* and where

Number of sources m = number of destinations n,

Every supply $s_i = 1$,

Every demand $d_j = 1$.

Now focus on the *integer solutions property* in the subsection on the transportation problem model. Because s_i and d_j are integers (= 1) now, this property implies that *every BF solution* (including an optimal one) is an *integer* solution for an assignment problem. The functional constraints of the assignment problem model prevent any variable from being greater than 1, and the nonnegativity constraints prevent values less than 0. Therefore, by deleting the binary restriction to enable us to solve an assignment problem as a linear programming problem, the resulting BF solutions obtained (including the final optimal solution) *automatically* will satisfy the binary restriction anyway.

Just as the transportation problem has a network representation (see Fig. 8.3), the assignment problem can be depicted in a very similar way, as shown in Fig. 8.4. The first column now lists the n assignees and the second column the n tasks. Each number in a square bracket indicates the number of assignees being provided at that location in the network, so the values are automatically 1 on the left, whereas the values of -1 on the right indicate that each task is using up one assignee.

For any particular assignment problem, practitioners normally do not bother writing out the full mathematical model. It is simpler to formulate the problem by filling out a cost table (e.g., Table 8.25), including identifying the assignees and tasks, since this table contains all the essential data in a far more compact form.

Because the assignment problem is a special type of transportation problem, one convenient way to solve any particular assignment problem is to apply the transporta-

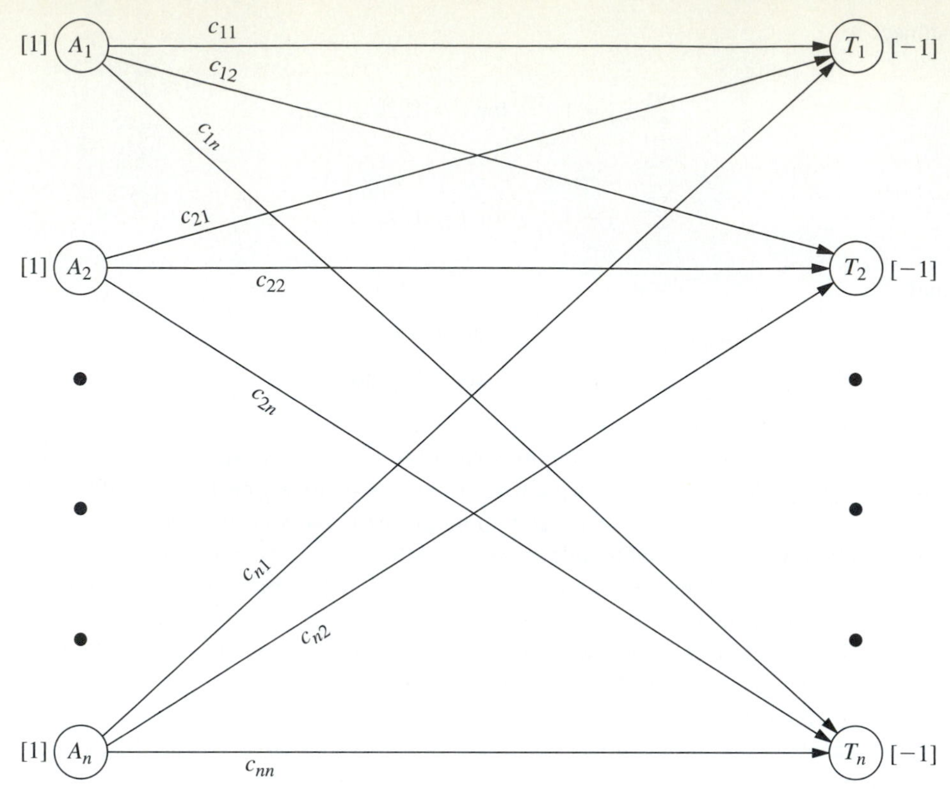

Figure 8.4 Network representation of the assignment problem.

tion simplex method described in Sec. 8.2. This approach requires converting the cost table to a cost and requirements table for the equivalent transportation problem, as shown in Table 8.26*a*.

For example, Table 8.26*b* shows the cost and requirements table for the Job Shop Co. problem that is obtained from the cost table of Table 8.25. When the transportation simplex method is applied to this transportation problem formulation, the resulting optimal solution has basic variables $x_{13} = 0, x_{14} = 1, x_{23} = 1, x_{31} = 1, x_{41} = 0, x_{42} = 1, x_{43} = 0$. (You are asked to verify this solution in Prob. 8.3-6.) The degenerate basic

Table 8.26 **Cost and Requirements Table for the Assignment Problem Formulated as a Transportation Problem, Illustrated by the Job Shop Co. Example**

(a) General Case

		Cost per Unit Distributed				
		Destination				
		1	2	⋯	n	Supply
Source	1	c_{11}	c_{12}	⋯	c_{1n}	1
	2	c_{21}	c_{22}	⋯	c_{2n}	1
	⋮	⋯	⋯	⋯	⋯	⋮
	$m = n$	c_{n1}	c_{n2}	⋯	c_{nn}	1
Demand		1	1	⋯	1	

(b) Job Shop Co. Example

		Cost per Unit Distributed				
		Destination (Location)				
		1	2	3	4	Supply
Source (Machine)	1	13	16	12	11	1
	2	15	M	13	20	1
	3	5	7	10	6	1
	4(D)	0	0	0	0	1
Demand		1	1	1	1	

variables ($x_{ij} = 0$) and the assignment for the dummy machine ($x_{42} = 1$) do not mean anything for the original problem, so the real assignments are machine 1 to location 4, machine 2 to location 3, and machine 3 to location 1.

It is no coincidence that this optimal solution provided by the transportation simplex method has so many degenerate basic variables. For any assignment problem with n assignments to be made, the transportation problem formulation shown in Table 8.26a has $m = n$, that is, both the number of sources (m) and the number of destinations (n) in this formulation equal the number of assignments (n). Transportation problems in general have $m + n - 1$ basic variables (allocations), so every BF solution for this particular kind of transportation problem has $2n - 1$ basic variables, but exactly n of these x_{ij} equal 1 (corresponding to the n assignments being made). Therefore, since all the variables are binary variables, there always are $n - 1$ degenerate basic variables ($x_{ij} = 0$). As discussed at the end of Sec. 8.2, degenerate basic variables do not cause any major complication in the execution of the algorithm. However, they do frequently cause *wasted iterations,* where nothing changes (same allocations) except for the labeling of which allocations of zero correspond to degenerate basic variables rather than nonbasic variables. These wasted iterations are a major drawback to applying the transportation simplex method in this kind of situation, where there *always* are so many degenerate basic variables.

Another drawback of the transportation simplex method here is that it is purely a *general-purpose* algorithm for solving all transportation problems. Therefore, it does nothing to exploit the additional special structure in this special type of transportation problem ($m = n$, every $s_i = 1$, and every $d_j = 1$). Although we will not take the space to describe them,[1] specialized algorithms have been developed to fully streamline the procedure for solving just assignment problems. These algorithms operate directly on the cost table and do not bother with degenerate basic variables. When a computer code is available for one of these algorithms, it generally should be used in preference to the transportation simplex method. Your OR Courseware includes an automatic routine for such an algorithm (under the Procedure menu for *The Transportation Problem*).

Example—Assigning Products to Plants

The BETTER PRODUCTS COMPANY has decided to initiate the production of four new products, using three plants that currently have excess production capacity. The products require a comparable production effort per unit, so the available production capacity of the plants is measured by the number of units of any product that can be produced per day, as given in the rightmost column of Table 8.27. The bottom row gives the required production rate per day to meet projected sales. Each plant can produce any of these products, *except* that Plant 2 *cannot* produce product 3. However, the variable costs per unit of each product differ from plant to plant, as shown in the main body of Table 8.27.

Management now needs to make a decision on how to split up the production of the products among plants. Two kinds of options are available.

Option 1: Permit *product splitting,* where the same product is produced in more than one plant.
Option 2: Prohibit *product splitting.*

[1] For an article comparing various algorithms for the assignment problem, see J. L. Kennington and Z. Wang, ''An Empirical Analysis of the Dense Assignment Problem: Sequential and Parallel Implementations,'' *ORSA Journal on Computing,* **3**: 299–306, 1991.

Table 8.27 **Data for the Better Products Co. Problem**

		Unit Cost for Product				Capacity Available
		1	2	3	4	
	1	41	27	28	24	75
Plant	2	40	29	—	23	75
	3	37	30	27	21	45
Production rate		20	30	30	40	

This second option imposes a constraint that can only increase the cost of an optimal solution based on Table 8.27. On the other hand, the key advantage of Option 2 is that it eliminates some *hidden costs* associated with product splitting that are not reflected in Table 8.27, including extra setup, distribution, and administration. Therefore, management wants both options analyzed before a final decision is made. For Option 2, management further specifies that every plant should be assigned at least one of the products.

We will formulate and solve the model for each option in turn, where Option 1 leads to a transportation problem and Option 2 leads to an assignment problem.

Formulation of Option 1: With product splitting permitted, Table 8.27 can be converted directly to a cost and requirements table for a transportation problem. The plants become the sources, and the products become the destinations (or vice versa), so the supplies are the available production capacities and the demands are the required production rates. Only two changes need to be made in Table 8.27. First, because Plant 2 cannot produce product 3, such an allocation is prevented by assigning to it a huge unit cost of M. Second, the total capacity ($75 + 75 + 45 = 195$) exceeds the total required production ($20 + 30 + 30 + 40 = 120$), so a dummy destination with a demand of 75 is needed to balance these two quantities. The resulting cost and requirements table is shown in Table 8.28.

The optimal solution for this transportation problem has basic variables (allocations) $x_{12} = 30$, $x_{13} = 30$, $x_{15} = 15$, $x_{24} = 15$, $x_{25} = 60$, $x_{31} = 20$, and $x_{34} = 25$, so

Plant 1 produces all of products 2 and 3.
Plant 2 produces 37.5 percent of product 4.
Plant 3 produces 62.5 percent of product 4 and all of product 1.

The total cost is $Z = 3{,}260$.

Table 8.28 **Cost and Requirements Table for the Transportation Problem Formulation of Option 1 for the Better Products Co. Problem**

		Cost per Unit Distributed					
		Destination (Product)					
		1	2	3	4	5(*D*)	Supply
Source (Plant)	1	41	27	28	24	0	75
	2	40	29	M	23	0	75
	3	37	30	27	21	0	45
Demand		20	30	30	40	75	

Formulation of Option 2: Without product splitting, each product must be assigned to just one plant. Therefore, producing the products can be interpreted as the tasks for an assignment problem, where the plants are the assignees.

Management has specified that every plant should be assigned at least one of the products. There are more products (four) than plants (three), so one of the plants will need to be assigned two products. Plant 3 has only enough excess capacity to produce one product (see Table 8.27), so *either* Plant 1 or Plant 2 will take the extra product.

To make this assignment of an extra product possible within an assignment problem formulation, Plants 1 and 2 each are split into two assignees, as shown in Table 8.29.

The number of assignees (now five) must equal the number of tasks (now four), so a dummy task (product) is introduced into Table 8.29 as 5(D). The role of this dummy task is to provide the fictional second product to either Plant 1 or Plant 2, whichever one receives only one real product. There is no cost for producing a fictional product so, as usual, the cost entries for the dummy task are zero. The one exception is the entry of M in the last row of Table 8.29. The reason for M here is that Plant 3 must be assigned a real product (a choice of product 1, 2, 3, or 4), so the Big M method is needed to prevent the assignment of the fictional product to Plant 3 instead.

The remaining cost entries in Table 8.29 are *not* the unit costs shown in Table 8.27 or 8.28. Table 8.28 gives a transportation problem formulation (for Option 1), so unit costs are appropriate here, but now we are formulating an assignment problem (for Option 2). For an assignment problem, the cost c_{ij} is the *total* cost associated with assignee i performing task j. For Table 8.29, the *total* cost (per day) for Plant i to produce product j is the unit cost of production *times* the number of units produced (per day), where these two quantities for the multiplication are given separately in Table 8.27. (As in Table 8.28, again M is used to prevent the infeasible assignment of product 3 to Plant 2.)

The optimal solution for this assignment problem is as follows:

Plant 1 produces products 2 and 3.
Plant 2 produces product 1.
Plant 3 produces product 4.

Here the dummy assignment is given to Plant 2. The total cost is $Z = 3{,}290$.

As usual, one way to obtain this optimal solution is to convert the cost table of Table 8.29 to a cost and requirements table for the equivalent transportation problem (see Table 8.26) and then apply the transportation simplex method. Because of the

Table 8.29 **Cost Table for the Assignment Problem Formulation of Option 2 for the Better Products Co. Problem**

		Task (Product)				
		1	2	3	4	5(D)
Assignee (Plant)	1a	820	810	840	960	0
	1b	820	810	840	960	0
	2a	800	870	M	920	0
	2b	800	870	M	920	0
	3	740	900	810	840	M

identical rows in Table 8.29, this approach can be streamlined by combining the five assignees into three sources with supplies 2, 2, and 1, respectively. (See Prob. 8.3-5.) This streamlining also decreases by two the number of degenerate basic variables in every BF solution. Therefore, even though this streamlined formulation no longer fits the format presented in Table 8.26*a* for an assignment problem, it is a more efficient formulation for applying the transportation simplex method.

Now look back and compare this solution to the one obtained for Option 1, which included the splitting of product 4 between Plants 2 and 3. The allocations are somewhat different for the two solutions, but the total costs are virtually the same ($Z =$ 3,260 for Option 1 versus $Z =$ 3,290 for Option 2). However, there are hidden costs associated with product splitting (including the cost of extra setup, distribution, and administration) that are not included in the objective function for Option 1. As with any application of OR, the mathematical model used can provide only an approximate representation of the total problem, so management needs to consider factors that cannot be incorporated into the model before it makes a final decision. In this case, after evaluating the disadvantages of product splitting, management decided to adopt the Option 2 solution.

8.4 Conclusions

The linear programming model encompasses a wide variety of specific types of problems. The general simplex method is a powerful algorithm that can solve surprisingly large versions of any of these problems. However, some of these problem types have such simple formulations that they can be solved much more efficiently by *streamlined* algorithms that exploit their *special structure.* These streamlined algorithms can cut down tremendously on the computer time required for large problems, and they sometimes make it computationally feasible to solve huge problems. This is particularly true for the two types of linear programming problems studied in this chapter, namely, the transportation problem and the assignment problem. Both types have a number of common applications, so it is important to recognize them when they arise and to use the best available algorithms. These special-purpose algorithms are included in some linear programming software packages.

We shall reexamine the special structure of the transportation and assignment problems in Sec. 9.6. There we shall see that these problems are special cases of an important class of linear programming problems known as the *minimum cost flow problem.* This problem has the interpretation of minimizing the cost for the flow of goods through a network. A streamlined version of the simplex method called the *network simplex method* (described in Sec. 9.7) is widely used for solving this type of problem, including its various special cases.

Much research continues to be devoted to developing streamlined algorithms for special types of linear programming problems, including some not discussed here. At the same time, there is widespread interest in applying linear programming to optimize the operation of complicated large-scale systems, including social systems. The resulting formulations usually have special structures that can be exploited. Being able to recognize and exploit special structures has become a very important factor in the successful application of linear programming.

SELECTED REFERENCES

1. Bazaraa, M. S., and J. J. Jarvis, *Linear Programming and Network Flows,* 2d ed., Wiley, New York, 1990.
2. Gass, S. I., *Linear Programming: Methods and Applications,* 5th ed., McGraw-Hill, New York, 1985.
3. Geoffrion, A. M., "Elements of Large-Scale Mathematical Programming," *Management Science,* **16:** 652–691, 1970.
4. Gribeck, P. R., and K. O. Kortanek, *Extremal Methods of Operations Research,* Marcel Dekker, New York, 1985.

RELEVANT ROUTINES IN YOUR OR COURSEWARE

A demonstration example: *The Transportation Problem*

Interactive routines: *Enter or Revise a Transportation Problem*
Find Initial Basic Feasible Solution—for Interactive Method
Solve Interactively by the Transportation Simplex Method
Enter or Revise an Assignment Problem

Automatic routines: *Solve Automatically by the Transportation Simplex Method*
Solve Automatically by the Assignment Problem Algorithm

To access these routines, call the MathProg program and then choose *The Transportation Problem* under the Area menu. See Appendix 1 for documentation of the software.

PROBLEMS

To the left of each of the following problems (or their parts), we have inserted a D (for demonstration), I (for interactive routine), or A (for automatic routine) whenever a corresponding routine listed above can be helpful. An asterisk on the I or A indicates that this routine definitely should be used (unless your instructor gives contrary instructions) and that the printout from this routine is all that needs to be turned in to show your work in executing the algorithm. An asterisk on the problem number indicates that at least a partial answer is given in the back of the book.

8.1-1. A company has three plants producing a certain product that is to be shipped to four distribution centers. Plants 1, 2, and 3 produce 12, 17, and 11 shipments per month, respectively. Each distribution center needs to receive 10 shipments per month. The distance from each plant to the respective distributing centers is given in miles as follows:

		Distribution Center			
		1	2	3	4
	1	800	1,300	400	700
Plant	2	1,100	1,400	600	1,000
	3	600	1,200	800	900

The freight cost for each shipment is $100 plus $0.50 per mile.

How much should be shipped from each plant to each of the distribution centers to minimize the total shipping cost?

(*a*) Formulate this problem as a transportation problem by constructing the appropriate cost and requirements table.

A* (*b*) Use the automatic routine for the transportation simplex method in your OR Courseware to obtain an optimal solution for this problem.

8.1-2.* Tom would like 3 pints of home brew today and an additional 4 pints of home brew tomorrow. Dick is willing to sell a maximum of 5 pints total at a price of $3.00 per pint today and $2.70 per pint tomorrow. Harry is willing to sell a maximum of 4 pints total at a price of $2.90 per pint today and $2.80 per pint tomorrow.

Tom wishes to know what his purchases should be to minimize his cost while satisfying his thirst requirements.

(*a*) Formulate a linear programming model for this problem, and construct the initial simplex tableau (see Chaps. 3 and 4).

(*b*) Formulate this problem as a transportation problem by constructing the appropriate cost and requirements table.

A* (*c*) Use the automatic routine for the transportation simplex method in your OR Courseware to obtain an optimal solution for this problem.

8.1-3. A corporation has decided to produce three new products. Five branch plants now have excess product capacity. The unit manufacturing cost of the first product would be $31, $29, $32, $28, and $29, in Plants 1, 2, 3, 4 and 5, respectively. The unit manufacturing cost of the second product would be $45, $41, $46, $42, and $43 in Plants 1, 2, 3, 4, and 5, respectively. The unit manufacturing cost of the third product would be $38, $35, and $40 in Plants 1, 2, and 3, respectively, and Plants 4 and 5 do not have the capability for producing this product. Sales forecasts indicate that 600, 1,000, and 800 units of products 1, 2, and 3, respectively, should be produced per day. Plants 1, 2, 3, 4, and 5 have the capacity to produce 400, 600, 400, 600, and 1,000 units daily, respectively, regardless of the product or combinations of products involved. Assume that any plant having the capability and capacity to produce them can produce any combination of the products in any quantity.

Management wishes to know how to allocate the new products to the plants to minimize the total manufacturing cost.

(*a*) Formulate this problem as a transportation problem by constructing the appropriate cost and requirements table.

A* (*b*) Use the automatic routine for the transportation simplex method in your OR Courseware to obtain an optimal solution for this problem.

8.1-4. Suppose that England, France, and Spain produce all the wheat, barley, and oats in the world. The world demand for wheat requires 125 million acres of land devoted to wheat production. Similarly, 60 million acres of land is required for barley and 75 million acres of land for oats. The total amount of land available for these purposes in England, France, and Spain is 70 million, 110 million, and 80 million acres, respectively. The number of hours of labor needed in England, France, and Spain to produce 1 acre of wheat is 18, 13, and 16, respectively. The number of hours of labor needed in England, France, and Spain to produce 1 acre of barley is 15, 12, and 12 respectively. The number of hours of labor needed in England, France, and Spain to produce 1 acre of oats is 12, 10, and 16, respectively. The labor cost per hour in producing wheat is $3.00, $2.40, and $3.30 in England, France, and Spain, respectively. The labor cost per hour in producing barley is $2.70, $3.00, and $2.80 in England, France, and Spain, respectively. The labor cost per hour in producing oats is $2.30, $2.50, and $2.10 in England, France, and Spain, respectively. The problem is to allocate land use in each country so as to meet the world food requirement and to minimize the total labor cost.

(*a*) Formulate this problem as a transportation problem by constructing the appropriate cost and requirements table.

A* (*b*) Use the automatic routine for the transportation simplex method in your OR Courseware to obtain an optimal solution for this problem.

8.1-5. A firm producing a single product has three plants and four customers. The three plants will produce 6, 8, and 4 units, respectively, during the next time period. The firm has made a commitment to sell 4 units to customer 1, 6 units to customer 2, and at least 2 units to customer 3. Both customers 3 and 4 also want to buy as many of the remaining units as possible. The net profit associated with shipping 1 unit from plant i for sale to customer j is given by the following table:

		Customer			
		1	2	3	4
	1	8	7	5	2
Plant	2	5	2	1	3
	3	6	4	3	5

Management wishes to know how many units to sell to customers 3 and 4 and how many units to ship from each of the plants to each of the customers to maximize profit.

(*a*) Formulate this problem as a transportation problem by constructing the appropriate cost and requirements table.

A* (*b*) Use the automatic routine for the transportation simplex method in your OR Courseware to obtain an optimal solution for this problem.

8.1-6. A company has two plants producing a certain product that is to be shipped to three distribution centers. The unit production costs are the same at both plants, and the shipping cost (in hundreds of dollars) per unit of the product is shown for each combination of plant and distribution center as follows:

		Distribution Center		
		1	2	3
Plant	*A*	8	7	4
	B	6	8	5

A total of 60 units is to be produced and shipped per week. Each plant can produce and ship any amount up to a maximum of 50 units per week, so there is considerable flexibility on how to divide the total production between the two plants so as to reduce shipping costs.

Management's objective is to determine both how much should be produced at each plant and what the overall shipping pattern should be in order to minimize total shipping cost.

(*a*) Assume that each distribution center must receive exactly 20 units per week. Formulate this problem as a transportation problem by constructing the appropriate cost and requirements table.

A* (*b*) Use the automatic routine for the transportation simplex method in your OR Courseware to obtain an optimal solution for the problem as formulated in part (*a*).

(*c*) Now assume that any distribution center may receive any quantity between 10 and 30 units per week in order to further reduce the total shipping cost, provided only that the total shipped to all three distribution centers still equals 60 units per week. Formulate this problem as a transportation problem by constructing the appropriate cost and requirements table.

A* (*d*) Repeat part (*b*) for the problem as formulated in part (*c*).

8.1-7. The BUILD-EM-FAST COMPANY has agreed to supply its best customer with three widgets during each of the next 3 weeks, even though producing them will require some overtime work. The relevant production data are as follows:

Week	Maximum Production, Regular Time	Maximum Production, Overtime	Production Cost per Unit, Regular Time ($)
1	2	2	300
2	2	1	500
3	1	2	400

The cost per unit produced with overtime for each week is $100 more than that for regular time. The cost of storage is $50 per unit for each week. There is already an inventory of two widgets on hand currently, but he company does not want to retain any widgets in inventory after 3 weeks.

Management wants to know how many units should be produced in each week in order to maximize profit.

(*a*) Formulate this problem as a transportation problem by constructing the appropriate cost and requirements table.

A* (*b*) Use the automatic routine for the transportation simplex method in your OR Courseware to obtain an optimal solution for this problem.

8.1-8. A manufacturer must produce two products in sufficient quantity to meet contracted sales in each of the next 3 months. The two products share the same production facilities, and each unit of both products requires the same amount of production capacity. The available production and storage facilities are changing month by month, so the production capacities, unit production costs, and unit storage costs vary by month. Therefore, it may be worthwhile to overproduce one or both products in some months and to store them until needed.

For each month, the second column of the following table gives the maximum number of units of the two products combined that can be produced on regular time (RT) and on overtime (OT). For each product, the subsequent columns give (1) the number of units needed for the contracted sales, (2) the cost per unit produced on regular time, (3) the cost per unit produced on overtime, and (4) the cost of storing each extra unit that is held over into the next month. In each case, the numbers for the two products are separated by a slash, with the number for Product 1 on the left and the number for Product 2 on the right.

	Maximum Combined Production		***Product 1/Product 2***			
				Unit Cost of Production		
Month	RT	OT	Sales	RT	OT	Unit Cost of Storage
1	10	3	5/3	15/16	18/20	1/2
2	8	2	3/5	17/15	20/18	2/1
3	10	3	4/4	19/17	22/22	

The production manager wants a schedule developed for the number of units of each product to be produced on regular time and (if regular time production capacity is used up) on overtime in each of the three months. The objective is to minimize the total of the production and storage costs while meeting the contracted sales for each month. There is no initial inventory, and no final inventory is desired after 3 months.

(*a*) Formulate this problem as a transportation problem by constructing the appropriate cost and requirements table.

A* (*b*) Use the automatic routine for the transportation simplex method in your OR Courseware to obtain an optimal solution for this problem.

8.2-1. Consider the transportation problem having the following cost and requirements table:

		Destination			
		1	2	3	Supply
	1	6	3	5	4
Source	2	4	*M*	7	3
	3	3	4	3	2
Demand		4	2	3	

(*a*) Use Vogel's approximation method (by hand) to select the first basic variable for an initial BF solution.
(*b*) Use Russell's approximation method (by hand) to select the first basic variable for an initial BF solution.
(*c*) Use the northwest corner rule (by hand) to construct a complete initial BF solution.

D, I* **8.2-2.*** Consider the transportation problem having the following cost and requirements table:

		Destination					
		1	2	3	4	5	Supply
	1	2	4	6	5	7	4
Source	2	7	6	3	*M*	4	6
	3	8	7	5	2	5	6
	4	0	0	0	0	0	4
Demand		4	4	2	5	5	

Use each of the following criteria to obtain an initial BF solution. Compare the values of the objective function for these solutions.
(*a*) Northwest corner rule
(*b*) Vogel's approximation method
(*c*) Russell's approximation method

D, I* **8.2-3.** Consider the transportation problem having the following cost and requirements table:

		Destination						
		1	2	3	4	5	6	Supply
	1	13	10	22	29	18	0	5
	2	14	13	16	21	*M*	0	6
Source	3	3	0	*M*	11	6	0	7
	4	18	9	19	23	11	0	4
	5	30	24	34	36	28	0	3
Demand		3	5	4	5	6	2	

Use each of the following criteria to obtain an initial BF solution. Compare the values of the objective function for these solutions.

(*a*) Northwest corner rule
(*b*) Vogel's approximation method
(*c*) Russell's approximation method

8.2-4. Consider the transportation problem having the following cost and requirements table:

		Destination				
		1	2	3	4	Supply
	1	7	4	1	4	1
	2	4	6	7	2	1
Source	3	8	5	4	6	1
	4	6	7	6	3	1
Demand		1	1	1	1	

(*a*) Notice that this problem has three special characteristics: (1) number of sources = number of destinations, (2) each supply = 1, and (3) each demand = 1. Transportation problems with these characteristics are a special type called the assignment problem (as described in Sec. 8.3). Use the integer solutions property to explain why this type of transportation problem can be interpreted as assigning sources to destinations on a one-to-one basis.

(*b*) How many basic variables are there in every BF solution? How many of these are degenerate basic variables (= 0)?

D, I* (*c*) Use the northwest corner rule to obtain an initial BF solution.

I* (*d*) Construct an initial BF solution by applying the general procedure for the initialization step of the transportation simplex method. However, rather than using one of the three criteria for step 1 presented in Sec. 8.2, use the minimum cost criterion given below for selecting the next basic variable. (With your OR Courseware, choose the *Northwest Corner Rule* under the Option menu, since this choice actually allows the use of any criterion.)

Minimum cost criterion: From the rows and columns still under consideration, select the variable x_{ij} having the smallest unit cost c_{ij} to be the next basic variable. (Ties may be broken arbitrarily.)

D, I*, A (*e*) Starting with the initial BF solution from part (*c*), interactively apply the transportation simplex method to obtain an optimal solution.

8.2-5. Consider the prototype example for the transportation problem (the P & T Co. problem) presented at the beginning of Sec. 8.1. Verify that the solution given there actually is optimal by applying just the optimality test portion of the transportation simplex method to this solution.

8.2-6. Consider the transportation problem formulation of Option 1 for the Better Products Co. problem presented in Table 8.28. Verify that the optimal solution given in Sec. 8.3 actually is optimal by applying just the optimality test portion of the transportation simplex method to this solution.

8.2-7. Consider the transportation problem having the following cost and requirements table:

		Destination					
		1	2	3	4	5	Supply
Source	1	8	6	3	7	5	20
	2	5	M	8	4	7	30
	3	6	3	9	6	8	30
	4(D)	0	0	0	0	0	20
Demand		25	25	20	10	20	

After several iterations of the transportation simplex method, the following transportation simplex tableau is obtained:

		Destination						
		1	*2*	*3*	*4*	*5*	*Supply*	u_i
Source	1	8	6	3 (20)	7	5	20	
	2	5 (25)	M	8	4 (5)	7	30	
	3	6	3 (25)	9	6 (5)	8	30	
	4(D)	0	0 (0)	0 (0)	0	0 (20)	20	
Demand		25	25	20	10	20		
	v_j							

Continue the transportation simplex method for *two more* iterations by hand. After two iterations, state whether the solution is optimal and, if so, why.

D, I*, A **8.2-8.*** Consider the transportation problem having the following cost and requirements table:

		Destination				
		1	2	3	4	Supply
Source	1	3	7	6	4	5
	2	2	4	3	2	2
	3	4	3	8	5	3
Demand		3	3	2	2	

Use each of the following criteria to obtain an initial BF solution. In each case, interactively apply the transportation simplex method, starting with this initial solution, to obtain an optimal solution. Compare the resulting number of iterations for the transportation simplex method.

(*a*) Northwest corner rule
(*b*) Vogel's approximation method
(*c*) Russell's approximation method

D, I*, A **8.2-9.** Consider the transportation problem having the following cost and requirements table:

		Destination				
		1	2	3	4	Supply
Source	1	5	6	4	2	10
	2	2	*M*	1	3	20
	3	3	4	2	1	20
	4	2	1	3	2	10
Demand		20	10	10	20	

(*a*) Use the northwest corner rule to construct an initial BF solution.
(*b*) Starting with the initial basic solution from part (*a*), interactively apply the transportation simplex method to obtain an optimal solution.

8.2-10. Plans need to be made for the energy systems for a new building. The three possible sources of energy are electricity, natural gas, and a solar heating unit.

Energy needs in the building are for electricity, water heating, and space heating, where the daily requirements (all measured in the same units) are

Electricity	20 units
Water heating	10 units
Space heating	30 units.

The size of the roof limits the solar heater to 30 units, but there is no limit to the electricity and natural gas available. Electricity needs can be met only by purchasing electricity (at a cost of $50 per unit). Both other energy needs can be met by any source or combination of sources. The unit costs are as follows:

	Electricity	Natural Gas	Solar Heater
Water heating	$90	$60	$30
Space heating	$80	$50	$40

The objective is to minimize the total cost of meeting the energy needs.

(*a*) Formulate this problem as a transportation problem by constructing the appropriate cost and requirements table.
D, I* (*b*) Use the northwest corner rule to obtain an initial BF solution for this problem.
D, I*, A (*c*) Starting with the initial BF solution from part (*b*), interactively apply the transportation simplex method to obtain an optimal solution.
D, I* (*d*) Use Vogel's approximation method to obtain an initial BF solution for this problem.
D, I*, A (*e*) Starting with the initial BF solution from part (*d*), interactively apply the transportation simplex method to obtain an optimal solution.

I* (*f*) Use Russell's approximation method to obtain an initial BF solution for this problem.

D, I*, A (*g*) Starting with the initial BF solution obtained from part (*f*), interactively apply the transportation simplex method to obtain an optimal solution. Compare the number of iterations required by the transportation simplex method here and in parts (*c*) and (*e*).

D, I*, A **8.2-11.*** Interactively apply the transportation simplex method to solve the Northern Airplane Co. production scheduling problem as it is formulated in Table 8.9.

D, I*, A **8.2-12.*** Reconsider Prob. 8.1-1.

(*a*) Use the northwest corner rule to obtain an initial BF solution.

(*b*) Starting with the initial BF solution from part (*a*), interactively apply the transportation simplex method to obtain an optimal solution.

D, I*, A **8.2-13.** Reconsider Prob. 8.1-2*b*. Starting with the northwest corner rule, interactively apply the transportation simplex method to obtain an optimal solution for this problem.

D, I*, A **8.2-14.** Reconsider Prob. 8.1-3. Starting with Vogel's approximation method, interactively apply the transportation simplex method to obtain an optimal solution for this problem.

D, I*, A **8.2-15.** Reconsider Prob. 8.1-4. Starting with the northwest corner rule, interactively apply the transportation simplex method to obtain an optimal solution for this problem.

D, I*, A **8.2-16.** Reconsider Prob. 8.1-5. Starting with Russell's approximation method, interactively apply the transportation simplex method to obtain an optimal solution for this problem.

8.2-17. Reconsider the transportation problem formulated in Prob. 8.1-6*a*.

D, I* (*a*) Use each of the three criteria presented in Sec. 8.2 to obtain an initial BF solution, and time how long you spend for each one. Compare both these times and the values of the objective function for these solutions.

A* (*b*) Use the automatic routine for the transportation simplex method in your OR Courseware to obtain an optimal solution for this problem. For each of the three initial BF solutions obtained in part (*a*), calculate the percentage by which its objective function value exceeds the optimal one.

D, I* (*c*) For each of the three initial BF solutions obtained in part (*a*), interactively apply the transportation simplex method to obtain (and verify) an optimal solution. Time how long you spend in each of the three cases. Compare both these times and the number of iterations needed to reach an optimal solution.

8.2-18. Follow the instructions of Prob. 8.2-16 for the transportation problem formulated in Prob. 8.1-6*c*.

8.2-19. Consider the transportation problem having the following cost and requirements table:

		Destination		
		1	2	Supply
Source	1	8	5	4
	2	6	4	2
Demand		3	3	

(*a*) Using your choice of a criterion from Sec. 8.2 for obtaining the initial BF solution, solve this problem manually by the transportation simplex method. (Keep track of your time.)

(*b*) Reformulate this problem as a general linear programming problem, and then solve it manually by the simplex method. [Keep track of how long part (*b*) takes you, and contrast it with the computation time for part (*a*).]

8.2-20. Consider the Northern Airplane Co. production scheduling problem presented in Sec. 8.1 (see Table 8.7). Formulate this problem as a general linear programming problem by letting the decision variables be x_j = number of jet engines to be produced in month j (j = 1, 2, 3, 4). Construct the initial simplex tableau for this formulation, and then contrast the size (number of rows and columns) of this tableau and the corresponding tableaux used to solve the transportation problem formulation of the problem (see Table 8.9).

8.2-21. Consider the general linear programming formulation of the transportation problem (see Table 8.6). Verify the claim in Sec. 8.2 that the set of $m + n$ functional constraint equations (m supply constraints and n demand constraints) has one *redundant* equation; i.e., any one equation can be reproduced from a linear combination of the other $m + n - 1$ equations.

8.2-22. When you deal with a transportation problem where the supply and demand quantities have *integer* values, explain why the steps of the transportation simplex method guarantee that all the basic variables (allocations) in the BF solutions obtained have integer values. Begin with why this occurs with the initialization step when the general procedure for constructing an *initial* BF solution is used (regardless of the criterion for selecting the next basic variable). Then given a *current* BF solution that is integer, next explain why step 3 of an iteration must obtain a new BF solution that also is integer. Finally, explain how the initialization step can be used to construct *any* initial BF solution, so the transportation simplex method actually gives a proof of the integer solutions property presented in Sec. 8.1.

8.2-23. A contractor has to haul gravel to three building sites. She can purchase as much as 18 tons at a gravel pit in the north of the city and 14 tons at one in the south. She needs 10, 5, and 10 tons at sites 1, 2, and 3, respectively. The purchase price per ton at each gravel pit and the hauling cost per ton are given in the table below.

	Hauling Cost per Ton at Site			
Pit	1	2	3	Price per Ton
North	3	6	5	10
South	6	3	4	12

The contractor wishes to determine how much to haul from each pit to each site in order to minimize the total cost for purchasing and hauling gravel.

(*a*) Formulate a linear programming model for this problem. Using the Big M method, construct the initial simplex tableau ready to apply the simplex method (but do not actually solve).

(*b*) Now formulate this problem as a transportation problem by constructing the appropriate cost and requirements table. Compare the size of this table (and the corresponding transportation simplex tableaux) used by the transportation simplex method with the size of the simplex tableaux from part (*a*) that would be needed by the simplex method.

D (*c*) The contractor notices that she can supply sites 1 and 2 completely from the north pit and site 3 completely from the south pit. Use the optimality test (but no iterations) of the transportation simplex method to check whether the corresponding BF solution is optimal.

D, I*, A (*d*) Starting with the northwest corner rule, interactively apply the transportation simplex method to solve the problem as formulated in part (*b*).

(*e*) As usual, let c_{ij} denote the unit cost associated with source i and destination j as given in the cost and requirements table constructed in part (*b*). For the optimal solution obtained in part (d), suppose that the value of c_{ij} for each basic variable x_{ij} is fixed at the value given in the cost and requirements table, but that the value of c_{ij} for each nonbasic variable x_{ij} possibly can be altered through bargaining because the site manager wants to increase business. Use sensitivity analysis to determine the *allowable range to stay optimal* for each of the latter c_{ij} values, and explain how this information is useful to the contractor.

8.2-24. Reconsider the transportation problem given in Prob. 8.2-4. Starting from a certain initial BF solution, the transportation simplex method yields the following *final* transportation simplex tableau:

		Destination					
		1	***2***	***3***	***4***	***Supply***	u_i
Source	1	7 / 1	4 / (0)	1 / (1)	4 / (0)	1	0
	2	4 / (1)	6 / 4	7 / 8	2 / (0)	1	−2
	3	8 / 1	5 / (1)	4 / 2	6 / 1	1	1
	4	6 / 1	7 / 4	6 / 6	3 / (1)	1	−1
Demand		1	1	1	1	$Z = 13$	
	v_j	6	4	1	4		

Adapt the sensitivity analysis procedure for general linear programming presented in Secs. 6.6 and 6.7 to *independently* investigate each of the two changes in the original model indicated below by deducing the resulting change or changes in the above final transportation simplex tableau. Use this approach to determine if the BF solution in this tableau is still optimal.

(a) Change $c_{11} = 7$ to $c_{11} = 5$.
(b) Change $c_{13} = 1$ to $c_{13} = 3$.

8.2-25. Consider the transportation problem formulation and solution of the Metro Water District problem presented in Secs. 8.1 and 8.2 (see Tables 8.12 and 8.23). Adapt the sensitivity analysis procedure presented in Secs. 6.6 and 6.7 to conduct sensitivity analysis on this problem by *independently* investigating each of the following four changes in the original model. For each change, revise the final transportation simplex tableau as needed to identify and evaluate the current basic solution. Then test this solution for feasibility and for optimality. (Do not reoptimize.)

(a) Change c_{34} from 23 to 20.
(b) Change c_{23} from 13 to 16.

(*c*) Decrease the supply from source 2 to 50, and decrease the demand at destination 5 to 50.

(*d*) Increase the supply at source 2 to 80, and increase the demand at destination 2 to 40.

8.3-1. Four cargo ships will be used for shipping goods from one port to four other ports (labeled 1, 2, 3, 4). Any ship can be used for making any one of these four trips. However, because of differences in the ships and cargoes, the total cost of loading, transporting, and unloading the goods for the different ship-port combinations varies considerably, as shown in the following table:

		Port			
		1	2	3	4
Ship	1	5	4	6	7
	2	6	6	7	5
	3	7	5	7	6
	4	5	4	6	6

The objective is to assign the ships to ports on a one-to-one basis in such a way as to minimize the total cost for all four shipments.

(*a*) Describe how this problem fits into the general format for the assignment problem.

A* (*b*) Use the assignment problem algorithm in your OR Courseware to solve this assignment problem.

(*c*) Reformulate this problem as an equivalent transportation problem by constructing the appropriate cost and requirements table.

D, I* (*d*) Use the northwest corner rule to obtain an initial BF solution for the problem as formulated in part (*c*).

D, I*, A (*e*) Starting with the initial BF solution from part (*d*), interactively apply the transportation simplex method to obtain an optimal set of assignments for the original problem.

D, I* (*f*) Are there other optimal solutions in addition to the one obtained in part (*e*)? If so, use the transportation simplex method to identify them.

8.3-2. Reconsider Prob. 8.1-3. Suppose that the sales forecasts have been revised downward to 240, 400, and 320 units per day of products 1, 2, and 3, respectively. Thus each plant now has the capacity to produce all that is required of any one product. Therefore, management has decided that each new product should be assigned to only one plant and that no plant should be assigned more than one product (so that three plants are each to be assigned one product, and two plants are to be assigned none). The objective is to make these assignments so as to minimize the *total* cost of producing these amounts of the three products.

(*a*) Formulate this problem as an assignment problem by constructing the appropriate cost table.

A* (*b*) Use the assignment problem algorithm in your OR Courseware to solve this assignment problem.

(*c*) Reformulate this assignment problem as an equivalent transportation problem by constructing the appropriate cost and requirements table.

D, I*, A (*d*) Starting with Vogel's approximation method, interactively apply the transportation simplex method to solve the problem as formulated in part (*c*).

8.3-3.* The coach of an age-group swim team needs to assign swimmers to a 200-yard medley relay team to send to the Junior Olympics. Since most of his best swimmers are very fast in more than one stroke, it is not clear which swimmer should be assigned to each of the four strokes. The five fastest swimmers and the best times (in seconds) they have achieved in each of the strokes (for 50 yards) are as follows:

Stroke	Carl	Chris	David	Tony	Ken
Backstroke	37.7	32.9	33.8	37.0	35.4
Breaststroke	43.4	33.1	42.2	34.7	41.8
Butterfly	33.3	28.5	38.9	30.4	33.6
Freestyle	29.2	26.4	29.6	28.5	31.1

The coach wishes to determine how to assign four swimmers to the four different strokes to minimize the sum of the corresponding best times.

(*a*) Formulate this problem as an assignment problem.

A* (*b*) Use the assignment problem algorithm in your OR Courseware to solve this assignment problem.

8.3-4. Reconsider Prob. 8.2-23. Now suppose that trucks (and their drivers) need to be hired to do the hauling, where each truck can only be used to haul gravel from a single pit to a single site. Each truck can haul 5 tons, and the cost per truck is 5 times the hauling cost per ton given earlier. Only full trucks are used to supply each site.

(*a*) Formulate this problem as an assignment problem by constructing the appropriate cost table and identifying the assignees and assignments.

A* (*b*) Use the assignment problem algorithm in your OR Courseware to solve this assignment problem.

(*c*) Reformulate this assignment problem as an equivalent transportation problem with two sources and three destinations by constructing the appropriate cost and requirements table.

A* (*d*) Use the automatic routine for the transportation simplex method in your OR Courseware to obtain an optimal solution for the problem as formulated in part (*c*).

8.3-5. Consider the assignment problem formulation of Option 2 for the Better Products Co. problem presented in Table 8.29.

(*a*) Reformulate this problem as an equivalent transportation problem with three sources and five destinations by constructing the appropriate cost and requirements table.

(*b*) Convert the optimal solution given in Sec. 8.3 for this assignment problem to a complete BF solution (including degenerate basic variables) for the transportation problem formulated in part (*a*). Specifically, apply the general procedure for constructing an initial BF solution given in Sec. 8.2. For each iteration of the procedure, rather than use any of the three alternative criteria presented for step 1, select the next basic variable to correspond to the next assignment of a plant to a product given in the optimal solution. When only one row or only one column remains under consideration, use step 4 to select the remaining basic variables.

(*c*) Verify that the optimal solution given in Sec. 8.3 for this assignment problem actually is optimal by applying just the optimality test portion of the transportation simplex method to the complete BF solution obtained in part (*b*).

(*d*) Now reformulate this assignment problem as an equivalent transportation problem with five sources and five destinations by constructing the appropriate cost and requirements table. Compare this transportation problem with the one formulated in part (*a*).

(*e*) Repeat part (*b*) for the problem as formulated in part (*d*). Compare the BF solution obtained with the one from part (*b*).

D, I*, A **8.3-6.** Starting with Vogel's approximation method, interactively apply the transportation simplex method to solve the Job Shop Co. assignment problem as formulated in Table 8.26*b*. (As stated in Sec. 8.3, the resulting optimal solution has basic variables $x_{13} = 0$, $x_{14} = 1$, $x_{23} = 1$, $x_{31} = 1$, $x_{41} = 0$, $x_{42} = 1$, $x_{43} = 0$.)

8.3-7. Reconsider Prob. 8.1-6. Now assume that distribution centers 1, 2, and 3 must receive exactly 10, 20, and 30 units per week, respectively. For administrative convenience, management has decided that each distribution center will be supplied totally by a single plant, so that one plant will supply one distribution center and the other plant will supply the other two distribution centers. The choice of these assignments of plants to distribution centers is to be made solely on the basis of minimizing the total shipping cost.

(*a*) Formulate this problem as an assignment problem by constructing the appropriate cost table and identifying the corresponding assignees and tasks.

A* (*b*) Use the assignment problem algorithm in your OR Courseware to solve this assignment problem.

(*c*) Reformulate this assignment problem as an equivalent transportation problem (with four sources) by constructing the appropriate cost and requirements table.

A* (*d*) Use the automatic routine for the transportation simplex method in your OR Courseware to solve the problem as formulated in part (*c*).

(*e*) Repeat part (*c*) with just two sources.

A* (*f*) Repeat part (*d*) for the problem as formulated in part (*e*).

8.3-8. Consider the assignment problem having the following cost table.

		Job		
		1	2	3
Person	*A*	5	7	4
	B	3	6	5
	C	2	3	4

The optimal solution is *A*-3, *B*-1, *C*-2, with $Z = 10$.

A* (*a*) Use your OR Courseware to verify this optimal solution.

(*b*) Reformulate this problem as an equivalent transportation problem by constructing the appropriate cost and requirements table.

A* (*c*) Use your OR Courseware to obtain an optimal solution (automatically) for the transportation problem formulated in part (*b*).

(*d*) Why does the optimal BF solution obtained in part (*c*) include some (degenerate) basic variables that are not part of the optimal solution for the assignment problem?

(*e*) Now consider the *nonbasic* variables in the optimal BF solution obtained in part (*c*). For each nonbasic variable x_{ij} and the corresponding cost c_{ij}, adapt the sensitivity analysis procedure for general linear programming (see Case 2*a* in Sec. 6.7) to determine the *allowable range to stay optimal* for c_{ij}.

8.3-9. Consider the linear programming model for the general assignment problem given in Sec. 8.3. Construct the table of constraint coefficients for this model. Compare this table with the one for the general transportation problem (Table 8.6). In what ways does the general assignment problem have more special structure than the general transportation problem?

9

Network Analysis, Including PERT-CPM

Networks arise in numerous settings and in a variety of guises. Transportation, electrical, and communication networks pervade our daily lives. Network representations also are widely used for problems in such diverse areas as production, distribution, project planning, facilities location, resource management, and financial planning, to name just a few examples. In fact, a network representation provides such a powerful visual and conceptual aid for portraying the relationships between the components of systems that it is used in virtually every field of scientific, social, and economic endeavor.

One of the most exciting developments in operations research (OR) in recent years has been the unusually rapid advance in both the methodology and the application of network optimization models. A number of algorithmic breakthroughs have had a major impact, as have ideas from computer science concerning data structures and efficient data manipulation. Consequently, algorithms and software now are available *and are being used* to solve huge problems on a routine basis that would have been completely intractable a couple of decades ago.

Many network optimization models actually are special types of *linear programming* problems. For example, both of the special types of linear programming problems discussed in the preceding chapter—the transportation problem and the assignment problem—also are network optimization problems that can be solved with network methodology because of their network representations, presented in Figs. 8.3 and 8.4.

One of the linear programming examples presented in Sec. 3.4 also is a network optimization problem. This is the Distribution Unlimited Co. problem of how to distribute its goods through the distribution network shown in Fig. 3.13. This special type of linear programming problem, called the *minimum cost flow* problem, is presented in Sec. 9.6. We shall return to this specific example in that section and then solve it with network methodology in the following section.

The third linear programming case study presented in Sec. 3.5 also features an application of the minimum cost flow problem. This case study involved planning the supply, distribution, and marketing of goods at Citgo Petroleum Corp. The OR team at Citgo developed an optimization-based decision support system, using a minimum cost flow problem model for each product, and coupled this system with an on-line corporate database. Each product's model has about 3,000 equations (nodes) and 15,000 variables (arcs), which is a very modest size by today's standards for the application of network optimization models. The model takes in all aspects of the business, helping management decide everything from run levels at the various refineries to what prices to pay or charge. A network representation is essential because of the flow of goods through several stages: purchase of crude oil from various suppliers, shipping it to refineries, refining it into various products, and sending the products to distribution centers and product storage terminals for subsequent sale. As discussed in Sec. 3.5, the modeling system enabled the company to reduce its petroleum products inventory by over $116 million with no drop in service levels. This resulted in a savings in annual interest of $14 million as well as improvements in coordination, pricing, and purchasing decisions worth another $2.5 million each year, along with many indirect benefits.

In this one chapter we only scratch the surface of the current state of the art of network methodology. However, we shall introduce you to five important kinds of network problems and some basic ideas of how to solve them (without delving into issues of data structures that are so vital to successful large-scale implementations). Each of the first three problem types—the *shortest-path problem,* the *minimum spanning tree problem,* and the *maximum flow problem*—has a very specific structure that arises frequently in applications.

The fourth type—the *minimum cost flow problem*—provides a unified approach to many other applications because of its far more general structure. In fact, this structure is so general that it includes as special cases both the shortest-path problem and the maximum flow problem as well as the transportation problem and the assignment problem from Chap. 8. Because the minimum cost flow problem is a special type of linear programming problem, it can be solved extremely efficiently by a streamlined version of the simplex method called the *network simplex method.* (We shall not discuss even more general network problems that are more difficult to solve.)

The last problem type considered is *project planning and control* with PERT (program evaluation and review technique) and CPM (critical-path method). Although limited to this one area of application, PERT and CPM have proved to be invaluable tools there. In fact, since their development in the late 1950s, PERT and CPM have been (and probably continue to be) the most widely used kind of network technique in OR.

The first section introduces a prototype example that will be used subsequently to illustrate the approach to the first three of these problems. Section 9.2 presents some basic terminology for networks. The next four sections deal with the first four problems in turn. Section 9.7 then is devoted to the network simplex method, and Sec. 9.8 discusses the last problem type.

9.1 Prototype Example

SEERVADA PARK has recently been set aside for a limited amount of sightseeing and backpack hiking. Cars are not allowed into the park, but there is a narrow, winding road system for trams and for jeeps driven by the park rangers. This road system is shown (without the curves) in Fig. 9.1, where location O is the entrance into the park; other letters designate the locations of ranger stations (and other limited facilities). The numbers give the distances of these winding roads in miles.

The park contains a scenic wonder at station T. A small number of trams are used to transport sightseers from the park entrance to station T and back.

The park management currently faces three problems. One is to determine which route from the park entrance to station T has the *smallest total distance* for the operation of the trams. (This is an example of the shortest-path problem to be discussed in Sec. 9.3.)

A second problem is that telephone lines must be installed under the roads to establish telephone communication among all the stations (including the park entrance). Because the installation is both expensive and disruptive to the natural environment, lines will be installed under just enough roads to provide some connection between every pair of stations. The question is where the lines should be laid to accomplish this with a *minimum* total number of miles of line installed. (This is an example of the minimum spanning tree problem to be discussed in Sec. 9.4.)

The third problem is that more people want to take the tram ride from the park entrance to station T than can be accommodated during the peak season. To avoid unduly disturbing the ecology and wildlife of the region, a strict ration has been placed on the number of tram trips that can be made on each of the roads per day. (These limits differ for the different roads, as we shall describe in detail in Sec. 9.5.) Therefore, during the peak season, various routes might be followed regardless of distance to increase the number of tram trips that can be made each day. The question pertains to how to route the various trips to *maximize* the number of trips that can be made per day without violating the limits on any individual road. (This is an example of the maximum flow problem to be discussed in Sec. 9.5.)

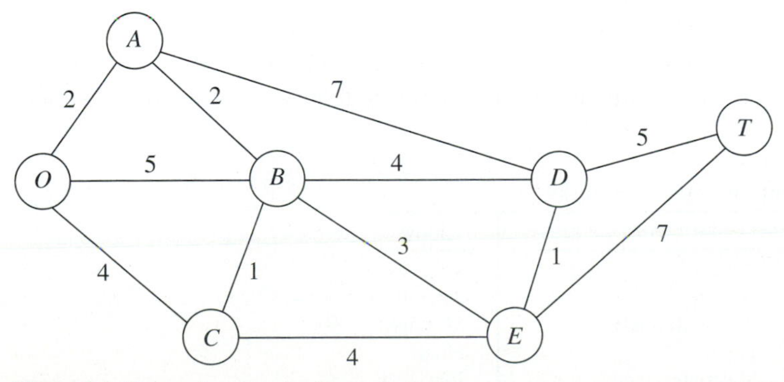

Figure 9.1 The road system for Seervada Park.

9.2 The Terminology of Networks

A relatively extensive terminology has been developed to describe the various kinds of networks and their components. Although we have avoided as much of this special vocabulary as we could, we still need to introduce a considerable number of terms for use throughout the chapter. We suggest that you read through this section once at the outset to understand the definitions and then plan to return to refresh your memory as the terms are used in subsequent sections. To assist you, each term is highlighted in **boldface** at the point where it is defined.

A network consists of a set of *points* and a set of *lines* connecting certain pairs of the points. The points are called **nodes** (or vertices); e.g., the network in Fig. 9.1 has seven nodes designated by the seven circles. The lines are called **arcs** (or links or edges or branches); e.g., the network in Fig. 9.1 has 12 arcs corresponding to the 12 roads in the road system. Arcs are labeled by naming the nodes at either end; for example, *AB* is the arc between nodes *A* and *B* in Fig. 9.1.

The arcs of a network may have a flow of some type through them, e.g., the flow of trams on the roads of Seervada Park in Sec. 9.1. Table 9.1 gives several examples of flow in typical networks. If flow through an arc is allowed in only one direction (e.g., a one-way street), the arc is said to be a **directed arc**. The direction is indicated by adding an arrowhead at the end of the line representing the arc. When a directed arc is labeled by listing two nodes it connects, the *from* node always is given before the *to* node; e.g., an arc that is directed *from* node *A* *to* node *B* must be labeled as *AB* rather than *BA*. Alternatively, this arc may be labeled as $A \rightarrow B$.

If flow through an arc is allowed in either direction (e.g., a pipeline that can be used to pump fluid in either direction), the arc is said to be an **undirected arc**. To help you distinguish between the two kinds of arcs, we shall frequently refer to undirected arcs by the suggestive name of **links**.

Although the flow through an undirected arc is allowed to be in either direction, we do assume that the flow will be one way in the direction of choice rather than having simultaneous flows in opposite directions. (The latter case requires the use of a *pair of directed arcs* in opposite directions.) However, in the process of making the decision on the flow through an undirected arc, it is permissible to make a sequence of assignments of flows in opposite directions, but with the understanding that the actual flow will be the *net flow* (the difference of the assigned flows in the two directions). For example, if a flow of 10 has been assigned in one direction and then a flow of 4 is assigned in the opposite direction, the actual effect is to *cancel* 4 units of the original assignment by reducing the flow in the original direction from 10 to 6. Even for a directed arc, the same technique sometimes is used as a convenient device to reduce a previously assigned flow. In particular, you are allowed to make a fictional assignment

Table 9.1 **Components of Typical Networks**

Nodes	Arcs	Flow
Intersections	Roads	Vehicles
Airports	Air lanes	Aircraft
Switching points	Wires, channels	Messages
Pumping stations	Pipes	Fluids
Work centers	Materials-handling routes	Jobs

of flow in the "wrong" direction through a directed arc to record a reduction of that amount in the flow in the "right" direction.

A network that has only directed arcs is called a **directed network**. Similarly, if all its arcs are undirected, the network is said to be an **undirected network**. A network with a mixture of directed and undirected arcs (or even all undirected arcs) can be converted to a directed network, if desired, by replacing each undirected arc by a pair of directed arcs in opposite directions. (You then have the choice of interpreting the flows through each pair of directed arcs as being simultaneous flows in opposite directions or providing a net flow in one direction, depending on which fits your application.)

When two nodes are not connected by an arc, a natural question is whether they are connected by a series of arcs. A **path** between two nodes is a *sequence of distinct arcs* connecting these nodes. For example, one of the paths connecting nodes O and T in Fig. 9.1 is the sequence of arcs OB–BD–DT ($O \rightarrow B \rightarrow D \rightarrow T$), or vice versa. When some of or all the arcs in the network are directed arcs, we then distinguish between directed paths and undirected paths. A **directed path** from node i to node j is a sequence of connecting arcs whose direction (if any) is *toward* node j, so that flow from node i to node j along this path is feasible. An **undirected path** from node i to node j is a sequence of connecting arcs whose direction (if any) can be *either* toward or away from node j. (Notice that a directed path also satisfies the definition of an undirected path, but not vice versa.) Frequently, an undirected path will have some arcs directed toward node j but others directed away (i.e., toward node i). You will see in Secs. 9.5 and 9.7 that, perhaps surprisingly, *undirected* paths play a major role in the analysis of *directed* networks.

To illustrate these definitions, Fig. 9.2 shows a typical directed network. (Its nodes and arcs are the same as in Fig. 3.13, where nodes A and B represent two factories, nodes D and E represent two warehouses, node C represents a distribution center, and the arcs represent shipping lanes.) The sequence of arcs AB–BC–CE ($A \rightarrow B \rightarrow C \rightarrow E$) is a directed path from node A to E, since flow toward node E along this entire path is feasible. On the other hand, BC–AC–AD ($B \rightarrow C \rightarrow A \rightarrow D$) is *not* a directed path from node B to node D, because the direction of arc AC is away from node D (on this path). However, $B \rightarrow C \rightarrow A \rightarrow D$ is an undirected path from node B

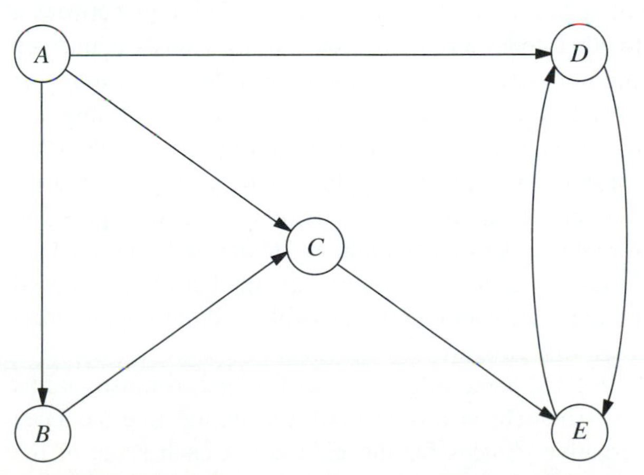

Figure 9.2 The distribution network for Distribution Unlimited Co., first shown in Fig. 3.13, illustrates a directed network.

to node D, because the sequence of arcs BC–AC–AD does *connect* these two nodes (even though the direction of arc AC prevents flow through this path).

As an example of the relevance of undirected paths, suppose that 2 units of flow from node A to node C had previously been assigned to arc AC. Given this previous assignment, it now is feasible to assign a smaller flow, say, 1 unit, to the entire undirected path $B \rightarrow C \rightarrow A \rightarrow D$, even though the direction of arc AC prevents positive flow through $C \rightarrow A$. The reason is that this assignment of flow in the "wrong" direction for arc AC actually just *reduces* the flow in the "right" direction by 1 unit. Sections 9.5 and 9.7 make heavy use of this technique of assigning a flow through an undirected path that includes arcs whose direction is opposite to this flow, where the real effect for these arcs is to reduce previously assigned positive flows in the "right" direction.

A path that begins and ends at the same node is called a **cycle**. In a *directed* network, a cycle is either a directed or an undirected cycle, depending on whether the path involved is a directed or an undirected path. (Since a directed path also is an undirected path, a directed cycle is an undirected cycle, but not vice versa in general.) In Fig. 9.2, for example, DE–ED is a directed cycle. By contrast, AB–BC–AC is *not* a directed cycle, because the direction of arc AC opposes the direction of arcs AB and BC. On the other hand, AB–BC–AC is an undirected cycle, because $A \rightarrow B \rightarrow C \rightarrow A$ is an undirected path. In the undirected network shown in Fig. 9.1, there are many cycles, for example, OA–AB–BC–CO. However, note that the definition of *path* (a sequence of *distinct* arcs) rules out retracing one's steps in forming a cycle. For example, OB–BO in Fig. 9.1 does not qualify as a cycle, because OB and BO are two labels for the *same* arc (link). On the other hand, DE–ED is a (directed) cycle in Fig. 9.2, because DE and ED are distinct arcs.

Two nodes are said to be **connected** if the network contains at least one *undirected* path between them. (Note that the path does not need to be directed even if the network is directed.) A **connected network** is a network where every pair of nodes is connected. Thus the networks in Figs. 9.1 and 9.2 are both connected. However, the latter network would not be connected if arcs AD and CE were removed.

Consider a connected network with n nodes (e.g., the $n = 5$ nodes in Fig. 9.2) where all the arcs have been deleted. A "tree" can then be "grown" by adding one arc (or "branch") at a time from the original network in a certain way. The first arc can go anywhere to connect some pair of nodes. Thereafter, each new arc should be between a node that already is connected to other nodes and a new node not previously connected to any other nodes. Adding an arc in this way avoids creating a cycle and ensures that the number of connected nodes is 1 greater than the number of arcs. Each new arc creates a larger **tree**, which is a *connected network* (for some subset of the n nodes) that contains *no undirected cycles*. Once the $(n - 1)$st arc has been added, the process stops because the resulting tree spans (connects) all n nodes. This tree is called a **spanning tree**, i.e., a *connected network* for all n nodes that contains *no undirected cycles*. Every spanning tree has exactly $n - 1$ arcs, since this is the *minimum* number of arcs needed to have a connected network and the *maximum* number possible without having undirected cycles.

Figure 9.3 uses the five nodes and some of the arcs of Fig. 9.2 to illustrate this process of growing a tree one arc (branch) at a time until a spanning tree has been obtained. There are several alternative choices for the new arc at each stage of the process, so Fig. 9.3 shows only one of many ways to construct a spanning tree in this case. Note, however, how each new added arc satisfies the conditions specified in the preceding paragraph. We shall discuss and illustrate spanning trees further in Sec. 9.4.

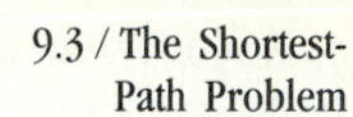

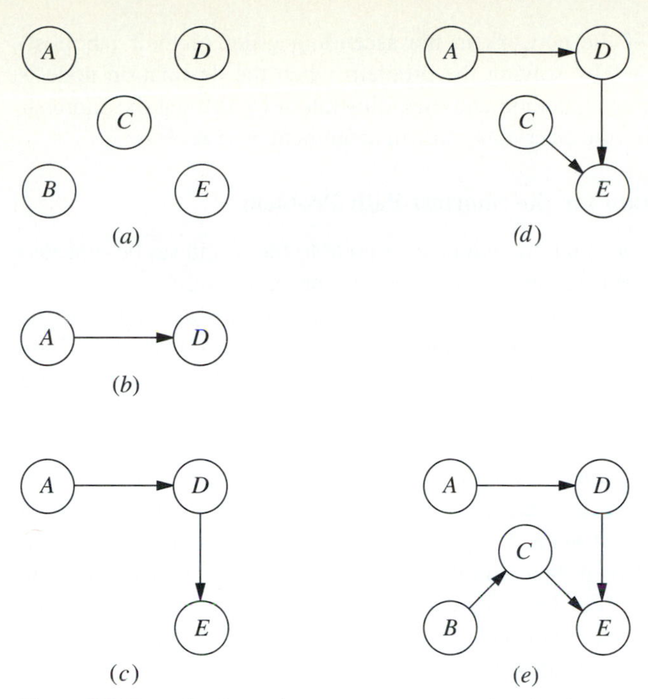

Figure 9.3 Example of growing a tree one arc at a time for the network of Fig. 9.2: (*a*) The nodes without arcs; (*b*) a tree with one arc; (*c*) a tree with two arcs; (*d*) a tree with three arcs; (*e*) a spanning tree.

Spanning trees play a key role in the analysis of many networks. For example, they form the basis for the minimum spanning tree problem discussed in Sec. 9.4. Another prime example is that (feasible) spanning trees correspond to the BF solutions for the network simplex method discussed in Sec. 9.7.

Finally, we shall need a little additional terminology about *flows* in networks. The maximum amount of flow (possibly infinity) that can be carried on a directed arc is referred to as the **arc capacity**. For nodes, a distinction is made among those that are net generators of flow, net absorbers of flow, or neither. A **supply node** (or source node or source) has the property that the flow *out* of the node exceeds the flow *into* the node. The reverse case is a **demand node** (or sink node or sink), where the flow *into* the node exceeds the flow *out* of the node. A **transshipment node** (or intermediate node) satisfies *conservation of flow,* so flow in equals flow out.

9.3 The Shortest-Path Problem

Although several other versions of the shortest-path problem (including some for directed networks) are mentioned at the end of the section, we shall focus on the following simple version. Consider an *undirected* and *connected* network with two special nodes called the *origin* and the *destination.* Associated with each of the *links* (undirected arcs) is a nonnegative *distance.* The objective is to find the shortest path (the path with the minimum total distance) from the origin to the destination.

A relatively straightforward algorithm is available for this problem. The essence of this procedure is that it fans out from the origin, successively identifying the shortest path to each of the nodes of the network in the ascending order of their (shortest) distances from the origin, thereby solving the problem when the destination node is reached. We shall first outline the method and then illustrate it by solving the shortest-path problem encountered by the Seervada Park management in Sec. 9.1.

Algorithm for the Shortest-Path Problem

Objective of n*th iteration:* Find the nth nearest node to the origin (to be repeated for $n = 1, 2, \ldots$ until the nth nearest node is the destination.

Input for n*th iteration:* $n - 1$ nearest nodes to the origin (solved for at the previous iterations), including their shortest path and distance from the origin. (These nodes, plus the origin, will be called *solved nodes;* the others are *unsolved nodes.*)

Candidates for n*th nearest node:* Each solved node that is directly connected by a link to one or more unsolved nodes provides *one* candidate—the unsolved node with the *shortest* connecting link. (Ties provide additional candidates.)

Calculation of n*th nearest node:* For each such solved node and its candidate, add the distance between them and the distance of the shortest path from the origin to this solved node. The candidate with the smallest such total distance is the nth nearest node (ties provide additional solved nodes), and its shortest path is the one generating this distance.

Example: The Seervada Park management needs to find the shortest path from the park entrance (node O) to the scenic wonder (node T) through the road system shown in Fig. 9.1. Applying the preceding algorithm to this problem yields the results shown in Table 9.2 (where the tie for the second nearest node allows one to skip directly to seeking the fourth nearest node next). The first column indicates the iteration count (n). The second column simply lists the *solved nodes* for beginning the current iteration after deleting the irrelevant ones (those not connected directly to any unsolved node). The third column then gives the *candidates* for the nth nearest node (the unsolved nodes with the *shortest* connecting link to a solved node). The fourth column calculates

Table 9.2 Applying the Shortest-Path Algorithm to the Seervada Park Problem

n	Solved Nodes Directly Connected to Unsolved Nodes	Closest Connected Unsolved Node	Total Distance Involved	nth Nearest Node	Minimum Distance	Last Connection
1	O	A	2	A	2	OA
2, 3	O	C	4	C	4	OC
	A	B	$2 + 2 = 4$	B	4	AB
4	A	D	$2 + 7 = 9$			
	B	E	$4 + 3 = 7$	E	7	BE
	C	E	$4 + 4 = 8$			
5	A	D	$2 + 7 = 9$			
	B	D	$4 + 4 = 8$	D	8	BD
	E	D	$7 + 1 = 8$	D	8	ED
6	D	T	$8 + 5 = 13$	T	13	DT
	E	T	$7 + 7 = 14$			

the distance of the shortest path from the origin to each of these candidates (namely, the distance to the solved node plus the link distance to the candidate). The candidate with the smallest such distance is the nth nearest node to the origin, as listed in the fifth column. The last two columns summarize the information for this *newest solved node* that is needed to proceed to subsequent iterations (namely, the distance of the shortest path from the origin to this node and the last link on this shortest path).

The shortest path *from the destination to the origin* can now be traced back through the last column of Table 9.2 as *either* $T \rightarrow D \rightarrow E \rightarrow B \rightarrow A \rightarrow O$ or $T \rightarrow D \rightarrow B \rightarrow A \rightarrow O$. Therefore, the two alternates for the shortest path *from the origin to the destination* have been identified as $O \rightarrow A \rightarrow B \rightarrow E \rightarrow D \rightarrow T$ and $O \rightarrow A \rightarrow B \rightarrow D \rightarrow T$, with a total distance of 13 miles on either path.

Other Applications

Before concluding this discussion of the shortest-path problem, we need to emphasize one point. The problem thus far has been described in terms of minimizing the distance from the origin to the destination. However, in actuality, the network problem being solved involves finding which path connecting two specified nodes minimizes the sum of the *link values* on the path. There is no reason that these link values need to represent distances, even indirectly. For example, the links might correspond to activities of some kind, where the value associated with each link is the cost of that activity. The problem then would be to find which sequence of activities that accomplishes a specified objective minimizes the total cost involved. (See Prob. 9.3-1.) Another alternative is that the value associated with each link is the *time* required for that activity. The problem then would be to find which sequence of activities that accomplishes a specified objective minimizes the total time involved. (See Prob. 9.3-5.) Thus some of the most important applications of the shortest-path problem have nothing to do with distances.

Many of these applications require finding the shortest *directed* path from the origin to the destination through a *directed* network. The algorithm already presented can be easily modified to deal just with directed paths at each iteration. In particular, when candidates for the nth nearest node are identified, only directed arcs *from* a solved node *to* an unsolved node are considered.

Another version of the shortest-path problem is to find the shortest paths from the origin to *all* the other nodes of the network. Notice that the algorithm already solves for the shortest path to each node that is closer to the origin than the destination. Therefore, when all nodes are potential destinations, the only modification needed in the algorithm is that it does not stop until all nodes are solved nodes.

An even more general version of the shortest-path problem is to find the shortest paths from *every* node to every other node. Another option is to drop the restriction that "distances" (arc values) be nonnegative. Constraints also can be imposed on the paths that can be followed. All these variations occasionally arise in applications and so have been studied by researchers.

The algorithms for a wide variety of combinatorial optimization problems, such as certain vehicle routing or network design problems, often call for the solution of a large number of shortest-path problems as subroutines. Although we lack the space to pursue this topic further, this use may now be the most important kind of application of the shortest-path problem.

9.4 The Minimum Spanning Tree Problem

The minimum spanning tree problem bears some similarities to the main version of the shortest-path problem presented in the preceding section. In both cases, an *undirected* and *connected* network is being considered, where the given information includes some measure of the positive *length* (distance, cost, time, etc.) associated with each link. Both problems also involve choosing a set of links that have the *shortest total length* among all sets of links that satisfy a certain property. For the shortest-path problem, this property is that the chosen links must provide a path between the origin and the destination. For the minimum spanning tree problem, the required property is that the chosen links must provide a path between *each* pair of nodes.

A network with n nodes requires only $(n - 1)$ links to provide a path between each pair of nodes. No extra links should be used, since this would needlessly increase the total length of the chosen links. The $(n - 1)$ links need to be chosen in such a way that the resulting network (with just the chosen links) forms a *spanning tree* (as defined in Sec. 9.2). Therefore, the problem is to find the spanning tree with a minimum total length of the links.

Figure 9.4 illustrates this concept of a spanning tree for the Seervada Park problem (see Sec. 9.1). Thus Fig. 9.4*a* is *not* a spanning tree because nodes *O, A, B,* and *C*

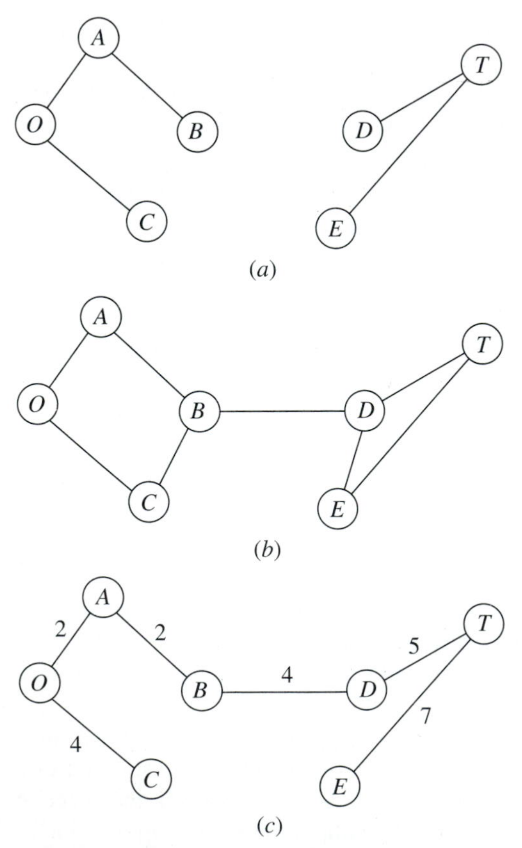

Figure 9.4 Illustrations of the spanning tree concept for the Seervada Park problem: (*a*) Not a spanning tree; (*b*) not a spanning tree; (*c*) a spanning tree.

are not connected with nodes D, E, and T. It needs another link to make this connection. This network actually consists of two trees, one for each of these two sets of nodes. The links in Fig. 9.4*b* do *span* the network (i.e., the network is connected as defined in Sec. 9.2), but it is *not* a tree because there are two *cycles* (O–A–B–C–O and D–T–E–D). It has too many links. Because the Seervada Park problem has $n = 7$ nodes, Sec. 9.2 indicates that the network must have exactly $n - 1 = 6$ links, with *no cycles,* to qualify as a spanning tree. This condition is achieved in Fig. 9.4*c,* so this network is a *feasible* solution (with a value of 24 miles for the total length of the links) for the minimum spanning tree problem. (You soon will see that this solution is not *optimal* because it is possible to construct a spanning tree with only 14 miles of links.)

This problem has a number of important practical applications. For example, it can sometimes be helpful in planning transportation networks that will not be used much, where the primary consideration is to provide *some* path between all pairs of nodes in the most economical way. (See Prob. 9.4-2.) The nodes would be the locations that require access to the other locations, the branches would be transportation lanes (highways, railroad tracks, air lanes, and so forth), and the distances (link values) would be the costs of providing the transportation lanes. In this context, the minimum spanning tree problem seeks to determine which transportation lanes would service all the locations with a minimum total cost. Other examples where a comparable decision arises include the planning of large-scale communication networks and distribution networks. Both represent important application areas.

The minimum spanning tree problem can be solved in a very straightforward way because it happens to be one of the few OR problems where being *greedy* at each stage of the solution procedure still leads to an overall optimal solution at the end! Thus, beginning with any node, the first stage involves choosing the shortest possible link to another node, without worrying about the effect of this choice on subsequent decisions. The second stage involves identifying the unconnected node that is closest to either of these connected nodes and then adding the corresponding link to the network. This process is repeated, per the following summary, until all the nodes have been connected. (Note that this is the same process already illustrated in Fig. 9.3 for constructing a spanning tree, but now with a specific rule for selecting each new link.) The resulting network is guaranteed to be a minimum spanning tree.

Algorithm for the Minimum Spanning Tree Problem

1. Select any node arbitrarily, and then connect it (i.e., add a link) to the nearest distinct node.
2. Identify the unconnected node that is closest to a connected node, and then connect these two nodes (i.e., add a link between them). Repeat this step until all nodes have been connected.
3. Tie breaking: Ties for the nearest distinct node (step 1) or the closest unconnected node (step 2) may be broken arbitrarily, and the algorithm must still yield an optimal solution. However, such ties are a signal that there may be (but need not be) multiple optimal solutions. All such optimal solutions can be identified by pursuing all ways of breaking ties to their conclusion.

The fastest way of executing this algorithm manually is the graphical approach illustrated as follows.

Example: The Seervada Park management (see Sec. 9.1) needs to determine under which roads telephone lines should be installed to connect all stations with a minimum total length of line. Using the data given in Fig. 9.1, we outline the step-by-step solution of this problem.

Nodes and distances for the problem are summarized below, where the thin lines now represent *potential* links.

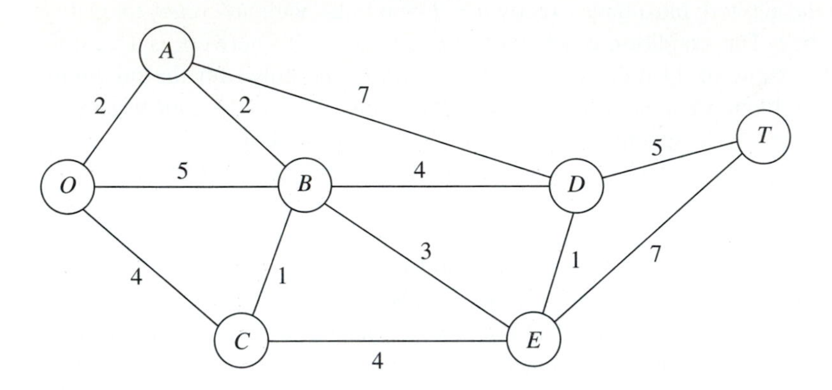

Arbitrarily select node O to start. The unconnected node closest to node O is node A. Connect node A to node O.

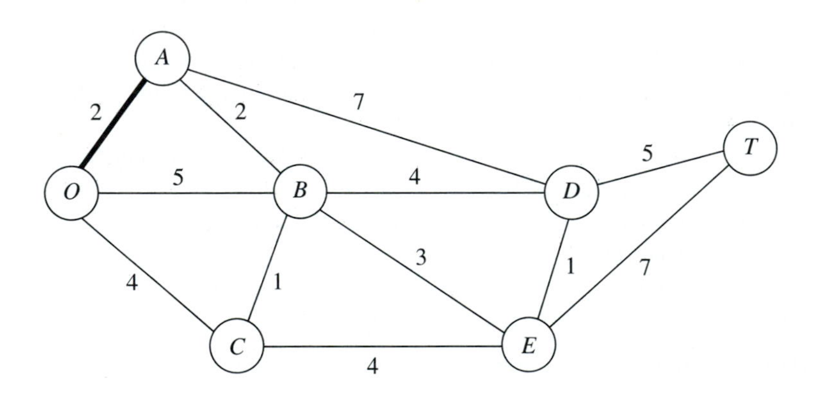

The unconnected node closest to either node O or node A is node B (closest to A). Connect node B to node A.

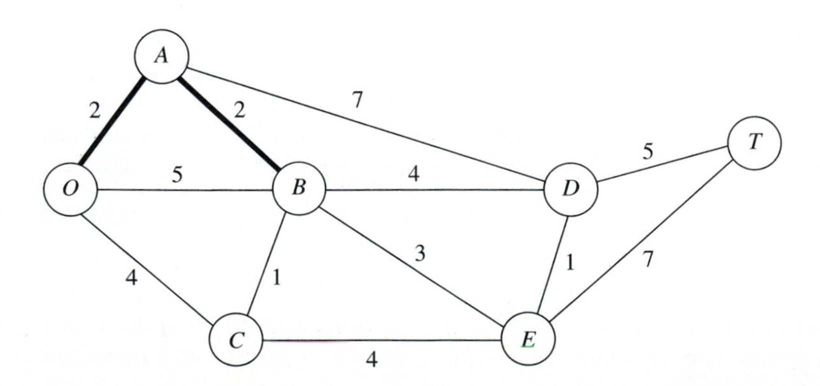

The unconnected node closest to node O, A, or B is node C (closest to B). Connect node C to node B.

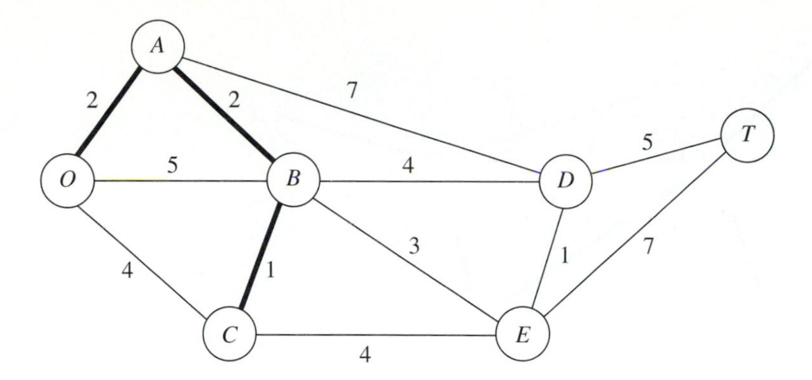

The unconnected node closest to node O, A, B, or C is node E (closest to B). Connect node E to node B.

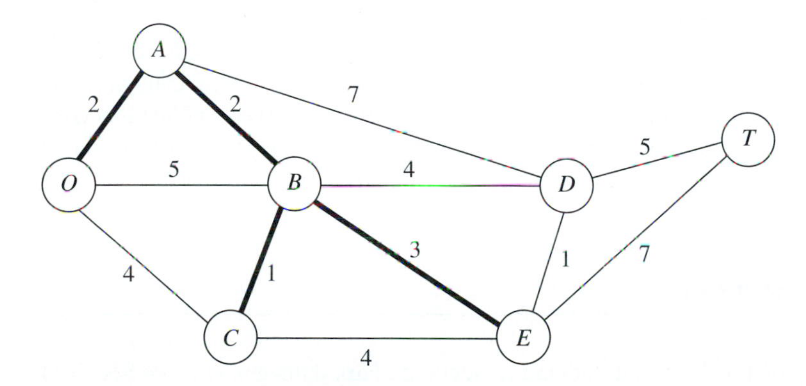

The unconnected node closest to node O, A, B, C, or E is node D (closest to E). Connect node D to node E.

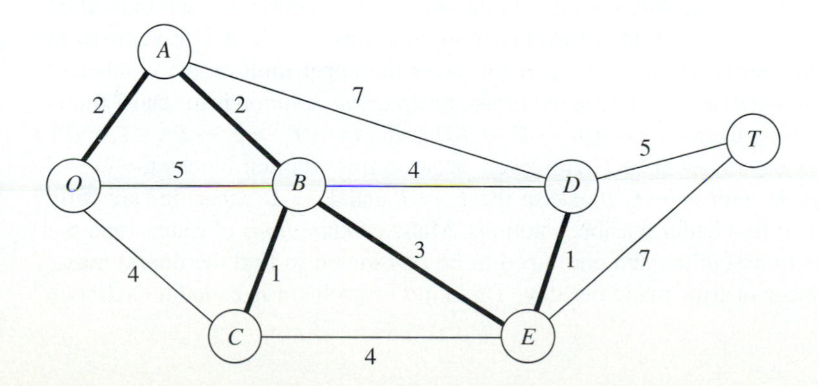

The only remaining unconnected node is node T. It is closest to node D. Connect node T to node D.

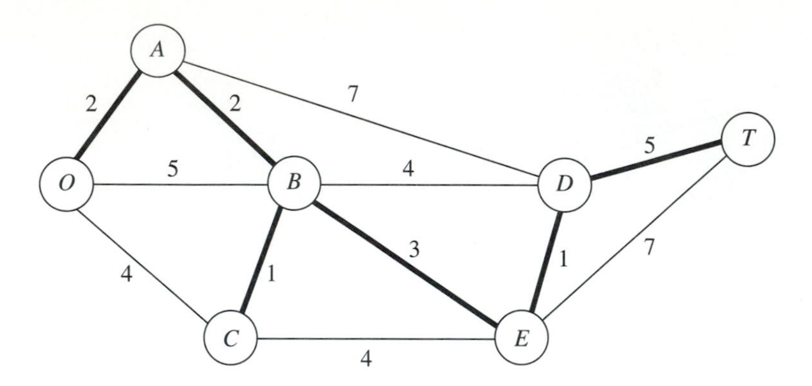

All nodes are now connected, so this solution to the problem is the desired (optimal) one. The total length of the links is 14 miles.

Although it may appear at first glance that the choice of the initial node will affect the resulting final solution (and its total link length) with this procedure, it really does not. We suggest you verify this fact for the example by reapplying the algorithm, starting with nodes other than node O.

The minimum spanning tree problem is the one problem we consider in this chapter that falls into the broad category of *network design.* In this category, the objective is to design the most appropriate network for the given application (frequently involving transportation systems) rather than analyzing an already designed network. Selected Reference 8 provides a survey of this important area.

9.5 The Maximum Flow Problem

Now recall that the third problem facing the Seervada Park management (see Sec. 9.1) during the peak season is to determine how to route the various tram trips from the park entrance (station O in Fig. 9.1) to the scenic wonder (station T) to maximize the number of trips per day. (Each tram will return by the same route it took on the outgoing trip, so the analysis focuses on outgoing trips only.) To avoid unduly disturbing the ecology and wildlife of the region, strict upper limits have been imposed on the number of outgoing trips allowed per day in the outbound direction on each individual road. For each road, the direction of travel for outgoing trips is indicated by an arrow in Fig. 9.5. The number at the base of the arrow gives the upper limit on the number of outgoing trips allowed per day. Given the limits, one *feasible solution* is to send 7 trams per day, with 5 using the route $O \rightarrow B \rightarrow E \rightarrow T$, 1 using $O \rightarrow B \rightarrow C \rightarrow E \rightarrow T$, and 1 using $O \rightarrow B \rightarrow C \rightarrow E \rightarrow D \rightarrow T$. However, because this solution blocks the use of any routes starting with $O \rightarrow C$ (because the $E \rightarrow T$ and $E \rightarrow D$ capacities are fully used), it is easy to find better feasible solutions. Many *combinations* of routes (and the number of trips to assign to each one) need to be considered to find the one(s) maximizing the number of trips made per day. This kind of problem is called a *maximum flow problem.*

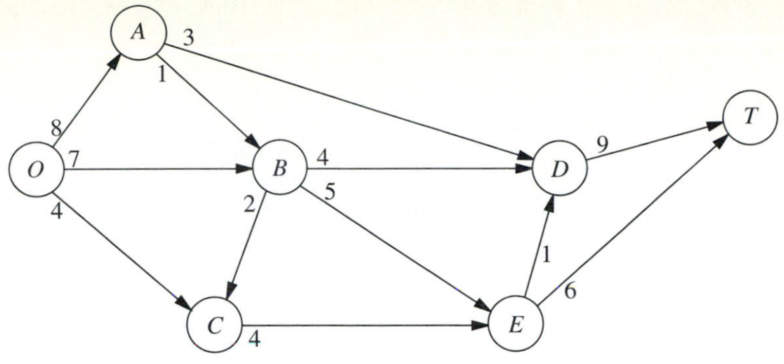

Figure 9.5 The Seervada Park maximum flow problem.

Using the terminology introduced in Sec. 9.2, we can describe the maximum flow problem for Seervada Park formally as follows. Consider a *directed* and *connected* network (Fig. 9.5) where just one node is a *supply node* (node O), one node is a *demand node* (node T), and the rest are transshipment nodes. Given the arc capacities (the numbers in Fig. 9.5), the objective is to determine the feasible pattern of flows through the network that *maximizes the total flow* from the supply node to the demand node.

Because the maximum flow problem can be formulated as a *linear programming problem* (see Prob. 9.5-2), it can be solved by the simplex method. However, an even more efficient *augmenting path algorithm* is available for solving this problem. This algorithm is based on two intuitive concepts, a *residual network* and an *augmenting path.*

After some flows have been assigned to the arcs of the original network, the **residual network** shows the *remaining* arc capacities (called **residual capacities**) for assigning *additional* flows. For example, consider arc $O \rightarrow B$ in Fig. 9.5, which has an arc capacity of 7. Now suppose that the assigned flows include a flow of 5 through this arc, which leaves a residual capacity of $7 - 5 = 2$ for any additional flow assignment through $O \rightarrow B$. This status is depicted as follows in the residual network.

The number on an arc next to a node gives the residual capacity for flow *from* that node *to* the other node. Therefore, in addition to the residual capacity of 2 for flow from O to B, the 5 on the right indicates a residual capacity of 5 for assigning some flow from B to O (that is, for canceling some previously assigned flow from O to B).

Initially, before any flows have been assigned, the residual network has the appearance shown in Fig. 9.6. Every arc in the original network (Fig. 9.5) has been changed from a *directed* arc to an *undirected* arc. However, the arc capacity in the original direction remains the same, and the arc capacity in the opposite direction is zero, so the constraints on flows are unchanged.

Subsequently, whenever some amount of flow is assigned to an arc, that amount is *subtracted* from the residual capacity in the same direction and *added* to the residual capacity in the opposite direction.

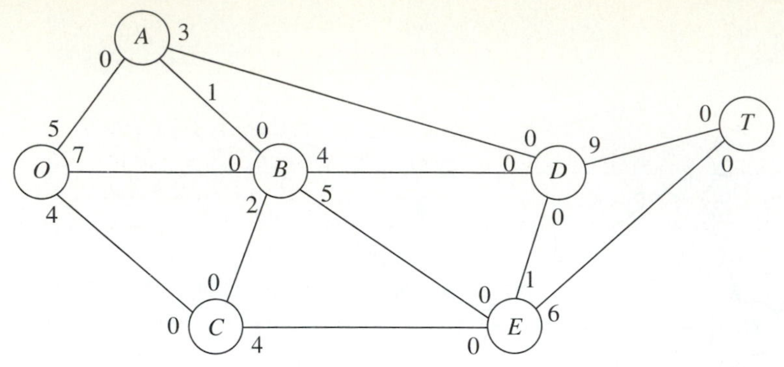

Figure 9.6 The initial residual network for the Seervada Park maximum flow problem.

An **augmenting path** is a directed path from the supply node to the demand node in the residual network such that *every* arc on this path has *strictly positive* residual capacity. The *minimum* of these residual capacities is called the *residual capacity of the augmenting path* because it represents the amount of flow that can feasibly be added to the entire path. Therefore, each augmenting path provides an opportunity to further augment the flow through the original network.

The augmenting path algorithm repeatedly selects some augmenting path and adds a flow equal to its residual capacity to that path in the original network. This process continues until there are no more augmenting paths, so the flow from the supply node to the demand node cannot be increased further. The key to ensuring that the final solution necessarily is optimal is the fact that augmenting paths can cancel some previously assigned flows in the original network, so an indiscriminate selection of paths for assigning flows cannot prevent the use of a better combination of flow assignments.

To summarize, each *iteration* of the algorithm consists of the following three steps.

The Augmenting Path Algorithm for the Maximum Flow Problem[1]

1. Identify an augmenting path by finding some directed path from the supply node to the demand node in the residual network such that every arc on this path has strictly positive residual capacity. (If no augmenting path exists, the net flows already assigned constitute an optimal flow pattern.)
2. Identify the residual capacity c^* of this augmenting path by finding the *minimum* of the residual capacities of the arcs on this path. *Increase* the flow in this path by c^*.
3. *Decrease* by c^* the residual capacity of each arc on this augmenting path. *Increase* by c^* the residual capacity of each arc in the opposite direction on this augmenting path. Return to step 1.

When step 1 is carried out, there often will be a number of alternative augmenting paths from which to choose. Although the algorithmic strategy for making this selection is important for the efficiency of large-scale implementations, we shall not delve into this relatively specialized topic. (Later in the section, we do describe a systematic

[1] It is assumed that the arc capacities are either integers or rational numbers.

procedure for finding some augmenting path.) Therefore, for the following example (and the problems at the end of the chapter), the selection is just made arbitrarily.

Example: Applying this algorithm to the Seervada Park problem (see Fig. 9.5 for the original network) yields the results summarized next. Starting with the initial residual network given in Fig. 9.6, we give the new residual network after each one or two iterations, where the total amount of flow from O to T achieved thus far is shown in **boldface** (next to nodes O and T).

Iteration 1: In Fig. 9.6, one of several augmenting paths is $O \rightarrow B \rightarrow E \rightarrow T$, which has a residual capacity of $\min\{7, 5, 6\} = 5$. By assigning a flow of 5 to this path, the resulting residual network is

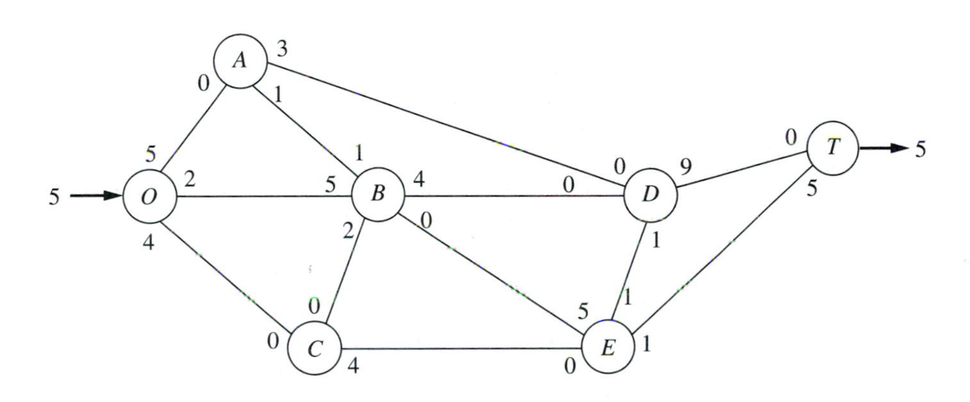

Iteration 2: Assign a flow of 3 to the augmenting path $O \rightarrow A \rightarrow D \rightarrow T$. The resulting residual network is

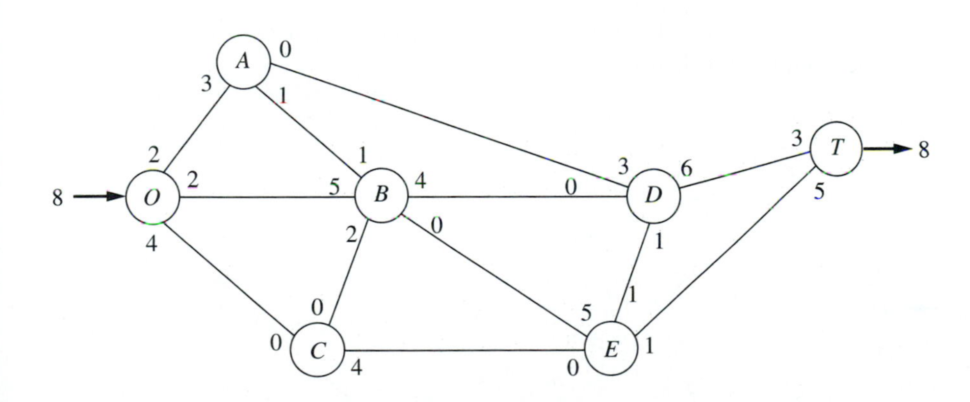

Iteration 3: Assign a flow of 1 to the augmenting path $O \rightarrow A \rightarrow B \rightarrow D \rightarrow T$.

Iteration 4: Assign a flow of 2 to the augmenting path $O \rightarrow B \rightarrow D \rightarrow T$. The resulting residual network is

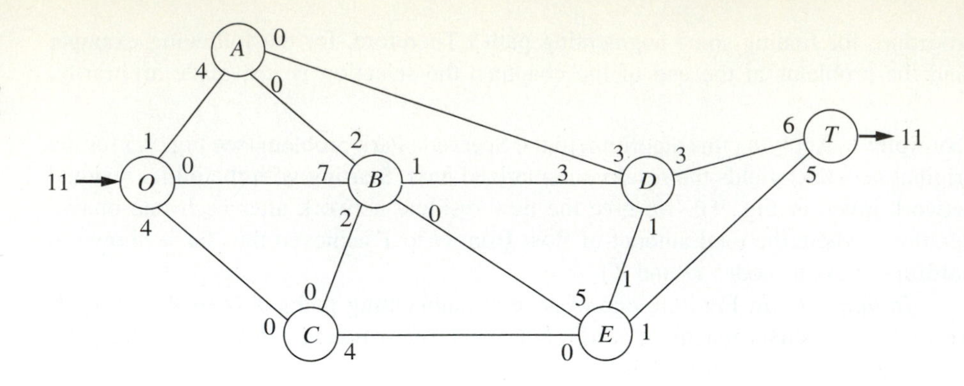

Iteration 5: Assign a flow of 1 to the augmenting path $O \rightarrow C \rightarrow E \rightarrow D \rightarrow T$.

Iteration 6: Assign a flow of 1 to the augmenting path $O \rightarrow C \rightarrow E \rightarrow T$. The resulting residual network is

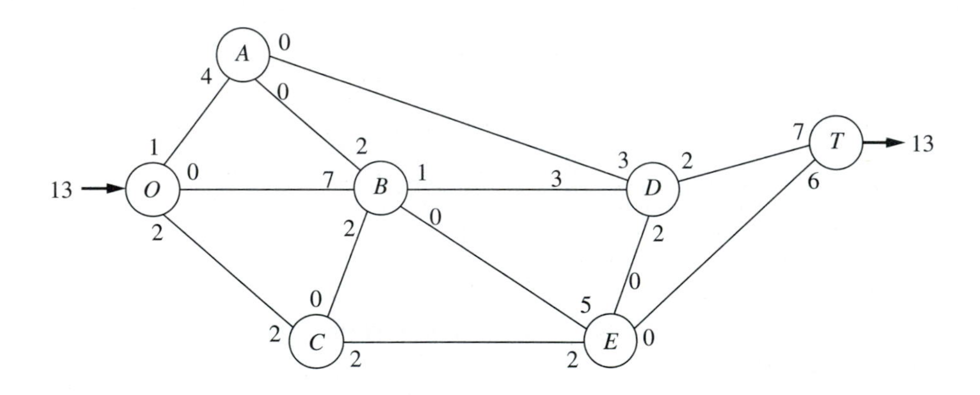

Iteration 7: Assign a flow of 1 to the augmenting path $O \rightarrow C \rightarrow E \rightarrow B \rightarrow D \rightarrow T$. The resulting residual network is

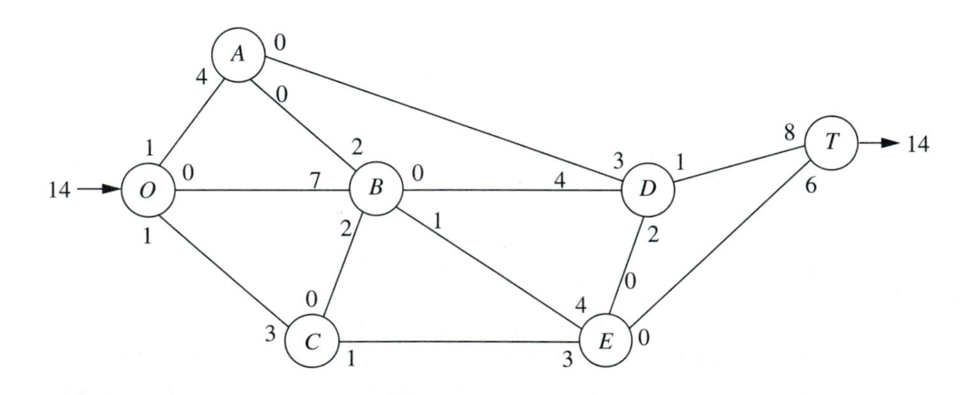

There are no more augmenting paths, so the current flow pattern is optimal.

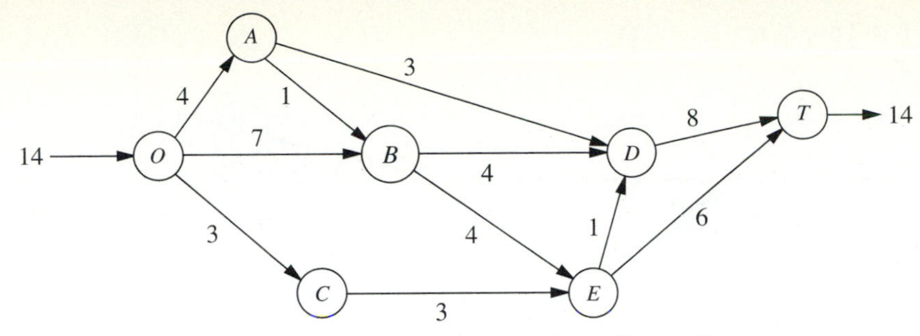

Figure 9.7 Optimal solution for the Seervada Park maximum flow problem.

The current flow pattern may be identified by either cumulating the flow assignments or comparing the final residual capacities with the original arc capacities. If we use the latter method, there is flow along an arc if the final residual capacity is less than the original capacity. The magnitude of this flow equals the difference in these capacities. Applying this method by comparing the residual network obtained from the last iteration with either Fig. 9.5 or 9.6 yields the optimal flow pattern shown in Fig. 9.7.

This example nicely illustrates the reason for replacing each directed arc $i \rightarrow j$ in the original network by an undirected arc in the residual network and then increasing the residual capacity for $j \rightarrow i$ by c^* when a flow of c^* is assigned to $i \rightarrow j$. Without this refinement, the first six iterations would be unchanged. However, at that point it would appear that no augmenting paths remain (because the real unused arc capacity for $E \rightarrow B$ is zero). Therefore, the refinement permits us to add the flow assignment of 1 for $O \rightarrow C \rightarrow E \rightarrow B \rightarrow D \rightarrow T$ in iteration 7. In effect, this additional flow assignment cancels 1 unit of flow assigned at iteration 1 ($O \rightarrow B \rightarrow E \rightarrow T$) and replaces it by assignments of 1 unit of flow to *both* $O \rightarrow B \rightarrow D \rightarrow T$ and $O \rightarrow C \rightarrow E \rightarrow T$.

The most difficult part of this algorithm for *large* networks involves finding an augmenting path. This task may be simplified by the following systematic procedure. Begin by determining all nodes that can be reached from the supply node along a single arc with strictly positive residual capacity. Then, for each of these nodes that were reached, determine all *new* nodes (those not yet reached) that can be reached from this node along an arc with strictly positive residual capacity. Repeat this successively with the new nodes as they are reached. The result will be the identification of a tree of all the nodes that can be reached from the supply node along a path with strictly positive residual flow capacity. Hence this *fanning-out procedure* will always identify an augmenting path if one exists. The procedure is illustrated in Fig. 9.8 for the residual network that results from iteration 6 in the preceding example.

Although the procedure illustrated in Fig. 9.8 is relatively straightforward, it would be helpful to be able to recognize when optimality has been reached without an exhaustive search for a nonexistent path. It is sometimes possible to recognize this event because of an important theorem of network theory known as the *max-flow min-cut theorem.* A **cut** may be defined as any set of directed arcs containing at least one arc from every directed path from the supply node to the demand node. The **cut value** is the sum of the arc capacities of the arcs (in the specified direction) of the cut. The **max-flow min-cut theorem** states that, for any network with a single supply node and demand node, the *maximum feasible flow* from the supply node to the demand node *equals* the *minimum cut value* for all cuts of the network. Thus, if we let F denote the amount of flow from the supply node to the demand node for any feasible flow pattern,

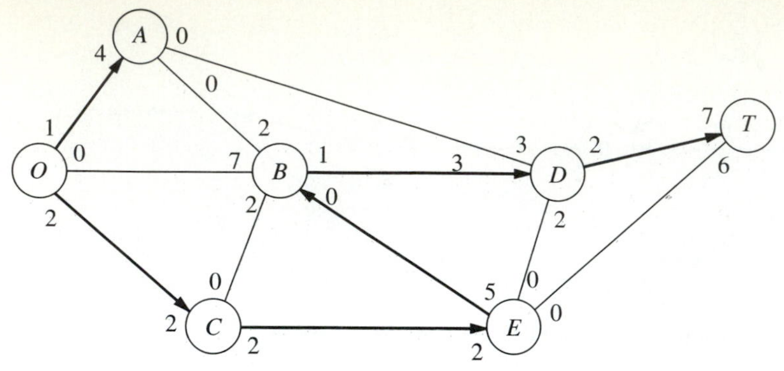

Figure 9.8 Procedure for finding an augmenting path for iteration 7 of the Seervada Park example.

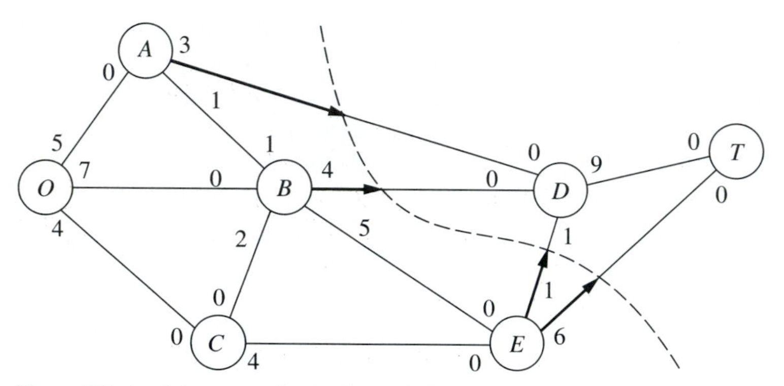

Figure 9.9 A minimum cut for the Seervada Park problem.

the value of any cut provides an upper bound to F, and the smallest of the cut values is equal to the maximum value of F. Therefore, if a cut whose value equals the value of F currently attained by the solution procedure can be found in the original network, the current flow pattern must be *optimal.* Eventually, optimality has been attained whenever there exists a cut in the residual network whose value is zero.

To illustrate, consider the cut in the network of Fig. 9.6 that is indicated in Fig. 9.9. Notice that the value of the cut is $3 + 4 + 1 + 6 = 14$, which was found to be the maximum value of F, so this cut is a minimum cut. Notice also that, in the residual network resulting from iteration 7, where $F = 14$, the corresponding cut has a value of zero. If this had been noticed, it would not have been necessary to search for additional augmenting paths.

9.6 The Minimum Cost Flow Problem

The minimum cost flow problem holds a central position among network optimization models, both because it encompasses such a broad class of applications and because it can be solved extremely efficiently. Like the maximum flow problem, it considers flow through a network with limited arc capacities. Like the shortest-path problem, it considers a cost (or distance) for flow through an arc. Like the transportation problem or assignment problem of Chap. 8, it can consider multiple sources (supply nodes) and multiple destinations (demand nodes) for the flow, again with associated costs. In fact, all four of these previously studied problems are special cases of the minimum cost flow problem, as we will demonstrate shortly.

The reason that the minimum cost flow problem can be solved so efficiently is that it can be formulated as a linear programming problem; so it can be solved by a streamlined version of the simplex method called the *network simplex method.* We describe this algorithm in the next section.

Formulation

Consider a directed and connected network, where the n nodes include at least one supply node and at least one demand node. The decision variables are

$$x_{ij} = \text{flow through arc } i \rightarrow j,$$

and the given information includes

$$c_{ij} = \text{cost per unit flow through arc } i \rightarrow j,$$

$$u_{ij} = \text{arc capacity for arc } i \rightarrow j,$$

$$b_i = \text{net flow generated at node } i.$$

The value of b_i depends on the nature of node i, where

$$b_i > 0 \qquad \text{if node } i \text{ is a supply node,}$$

$$b_i < 0 \qquad \text{if node } i \text{ is a demand node,}$$

$$b_i = 0 \qquad \text{if node } i \text{ is a transshipment node.}$$

The objective is to minimize the total cost of sending the available supply through the network to satisfy the given demand.

By using the convention that summations are taken only over existing arcs, the linear programming formulation of this problem is

$$\text{Minimize} \qquad Z = \sum_{i=1}^{n} \sum_{j=1}^{n} c_{ij} x_{ij},$$

subject to

$$\sum_{j=1}^{n} x_{ij} - \sum_{j=1}^{n} x_{ji} = b_i, \qquad \text{for each node } i,$$

and

$$0 \le x_{ij} \le u_{ij}, \qquad \text{for each arc } i \rightarrow j.$$

The first summation in the node constraints represents the total flow *out* of node i, whereas the second summation represents the total flow *into* node i, so the difference is the net flow generated at this node.

In some applications, it is necessary to have a lower bound $L_{ij} > 0$ for the flow through each arc $i \rightarrow j$. When this occurs, use a translation of variables $x'_{ij} = x_{ij} - L_{ij}$, with $x'_{ij} + L_{ij}$ substituted for x_{ij} throughout the model, to convert the model back to the above format with nonnegativity constraints.

It is not guaranteed that the problem actually will possess *feasible* solutions, depending partially upon which arcs are present in the network and their arc capacities. However, for a reasonably designed network, the main condition needed is the following.

Feasible solutions property: A necessary condition for a minimum cost flow problem to have any feasible solutions is that

$$\sum_{i=1}^{n} b_i = 0.$$

That is, the total flow being generated at the supply nodes equals the total flow being absorbed at the demand nodes.

If the values of b_i provided for some application violate this condition, the usual interpretation is that either the supplies or the demands (whichever are in excess) actually represent upper bounds rather than exact amounts. When this situation arose for the transportation problem in Sec. 8.1, either a dummy destination was added to receive the excess supply or a dummy source was added to send the excess demand. The analogous step now is that either a dummy demand node should be added to absorb the excess supply (with $c_{ij} = 0$ arcs added from every supply node to this node) or a dummy supply node should be added to generate the flow for the excess demand (with $c_{ij} = 0$ arcs added from this node to every demand node).

For many applications, b_i and u_{ij} will have *integer* values, and implementation will require that the flow quantities x_{ij} also be integer. Fortunately, just as for the transportation problem, this outcome is guaranteed without explicitly imposing integer constraints on the variables because of the following property.

Integer solutions property: For minimum cost flow problems where every b_i and u_{ij} have integer values, all the basic variables in *every* basic feasible (BF) solution (including an optimal one) also have integer values.

An example of a minimum cost flow problem is shown in Fig. 9.10. This network actually is the distribution network for the Distribution Unlimited Co. problem pre-

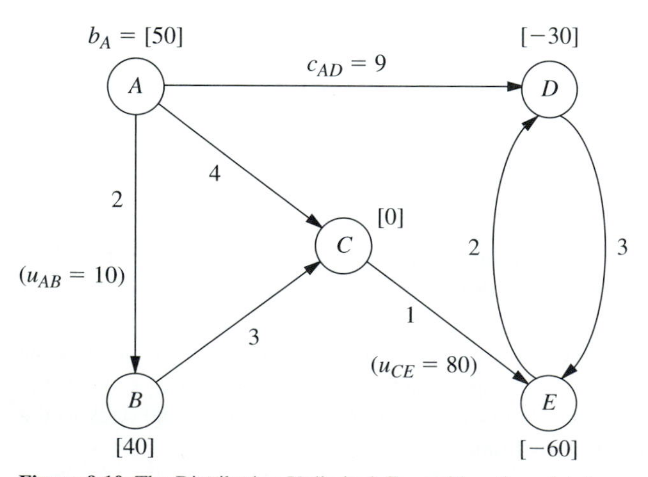

Figure 9.10 The Distribution Unlimited Co. problem formulated as a minimum cost flow problem.

sented in Sec. 3.4 (see Fig. 3.13). The quantities given in Fig. 3.13 provide the values of b_i, c_{ij}, and u_{ij} shown here. The b_i values are shown in square brackets by the nodes, so the supply nodes ($b_i > 0$) are A and B (the company's two factories), the demand nodes ($b_i < 0$) are D and E (two warehouses), and the one transshipment node ($b_i = 0$) is C (a distribution center). The c_{ij} values are shown next to the arcs. In this example, all but two of the arcs have arc capacities exceeding the total flow generated (90), so $u_{ij} = \infty$ for all practical purposes. The two exceptions are arc $A \rightarrow B$, where $u_{AB} = 10$, and arc $C \rightarrow E$, which has $u_{CE} = 80$.

The linear programming model for this example is

$$\text{Minimize} \quad Z = 2x_{AB} + 4x_{AC} + 9x_{AD} + 3x_{BC} + x_{CE} + 3x_{DE} + 2x_{ED},$$

subject to

$$\begin{aligned}
x_{AB} + x_{AC} + x_{AD} \qquad\qquad\qquad\qquad\qquad\qquad &= 50 \\
-x_{AB} \qquad\qquad + x_{BC} \qquad\qquad\qquad\qquad &= 40 \\
- x_{AC} \qquad - x_{BC} + x_{CE} \qquad\qquad\qquad &= 0 \\
- x_{AD} \qquad\qquad + x_{DE} - x_{ED} &= -30 \\
- x_{CE} - x_{DE} + x_{ED} &= -60
\end{aligned}$$

and

$$x_{AB} \leq 10, \qquad x_{CE} \leq 80, \qquad \text{all } x_{ij} \geq 0.$$

Now note the pattern of coefficients for each variable in the set of five link constraints. Each variable has exactly *two* nonzero coefficients, where one is $+1$ and the other is -1. This pattern recurs in *every* minimum cost flow problem, and it is this special structure that leads to the integer solutions property.

Another implication of this special structure is that (any) one of the link constraints is *redundant*. The reason is that summing all these constraint equations yields nothing but zeros on both sides (assuming feasible solutions exist, so the b_i values sum to zero), so the negative of any one of these equations equals the sum of the rest of the equations. With just $n - 1$ nonredundant link constraints, these equations provide just $n - 1$ basic variables for a BF solution. In the next section, you will see that the network simplex method treats the $x_{ij} \leq u_{ij}$ constraints as mirror images of the nonnegativity constraints, so the *total* number of basic variables is $n - 1$. This leads to a direct correspondence between the $n - 1$ arcs of a *spanning tree* and the $n - 1$ basic variables—but more about that story later.

We shall soon solve this example by the network simplex method. However, let us first see how some special cases fit into the network format of the minimum cost flow problem.

Special Cases

The Transportation Problem: To formulate the transportation problem presented in Sec. 8.1 as a minimum cost flow problem, a *supply node* is provided for each *source,* as well as a *demand node* for each *destination,* but no transshipment nodes are included in the network. All the arcs are directed from a supply node to a demand node, where distributing x_{ij} units from source i to destination j corresponds to a flow of x_{ij} through arc $i \rightarrow j$. The cost c_{ij} per unit distributed becomes the cost c_{ij} per unit of flow.

Since the transportation problem does not impose upper bound constraints on individual x_{ij}, all the $u_{ij} = \infty$.

Using this formulation for the P & T Co. transportation problem presented in Table 8.2 yields the network shown in Fig. 8.2. The corresponding network for the general transportation problem is shown in Fig. 8.3.

The Assignment Problem: Since the assignment problem discussed in Sec. 8.3 is a special type of transportation problem, its formulation as a minimum cost flow problem fits into the same format. The additional factors are that (1) the number of supply nodes equals the number of demand nodes, (2) $b_i = 1$ for each supply node, and (3) $b_i = -1$ for each demand node.

Figure 8.4 shows this formulation for the general assignment problem.

The Transshipment Problem: This special case actually includes all the general features of the minimum cost flow problem except for not having (finite) arc capacities. Thus, any minimum cost flow problem where each arc can carry any desired amount of flow is also called a transshipment problem.

For example, the Distribution Unlimited Co. problem shown in Fig. 9.10 would be a transshipment problem if the upper bounds on the flow through arcs $A \rightarrow B$ and $C \rightarrow E$ were removed.

Transshipment problems frequently arise as generalizations of transportation problems where units being distributed from each source to each destination can first pass through intermediate points. These intermediate points may include other sources and destinations, as well as additional transfer points that would be represented by transshipment nodes in the network representation of the problem. For example, the Distribution Unlimited Co. problem can be viewed as a generalization of a transportation problem with two sources (the two factories represented by nodes A and B in Fig. 9.10), two destinations (the two warehouses represented by nodes D and E), and one additional intermediate transfer point (the distribution center represented by node C).

The Shortest-Path Problem: Now consider the main version of the shortest-path problem presented in Sec. 9.3 (finding the shortest path from one origin to one destination through an *undirected* network). To formulate this problem as a minimum cost flow problem, one supply node with a supply of 1 is provided for the origin, one demand node with a demand of 1 is provided for the destination, and the rest of the nodes are transshipment nodes. Because the network of our shortest-path problem is undirected, whereas the minimum cost flow problem is assumed to have a directed network, we replace each link by a pair of directed arcs in opposite directions (depicted by a single line with arrowheads at both ends). The only exceptions are that there is no need to bother with arcs *into* the supply node or *out of* the demand node. The distance between nodes i and j becomes the unit cost c_{ij} or c_{ji} for flow in either direction between these nodes. As with the preceding special cases, no arc capacities are imposed, so all $u_{ij} = \infty$.

Figure 9.11 depicts this formulation for the Seervada Park shortest-path problem shown in Fig. 9.1, where the numbers next to the lines now represent the unit cost of flow in either direction.

The Maximum Flow Problem: The last special case we shall consider is the maximum flow problem described in Sec. 9.5. In this case, a network already is provided with one supply node, one demand node, and various transshipment nodes as

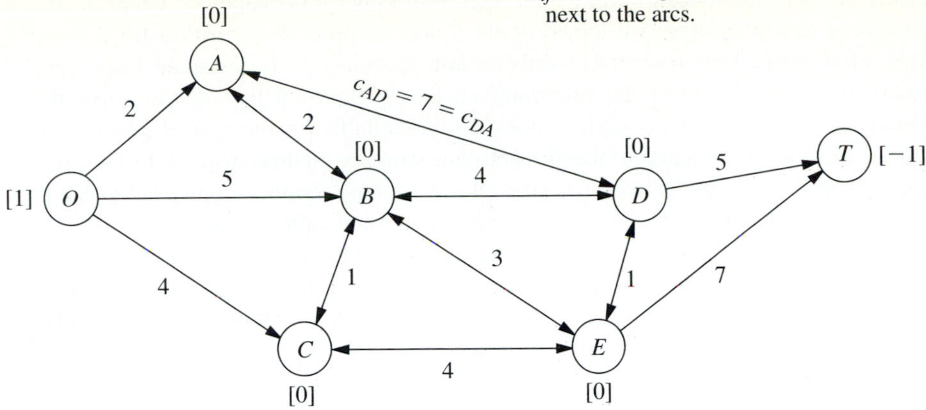

Figure 9.11 Formulation of the Seervada Park shortest-path problem as a minimum cost flow problem.

well as the various arcs and arc capacities. Only three adjustments are needed to fit this problem into the format for the minimum cost flow problem. First, set $c_{ij} = 0$ for all existing arcs to reflect the absence of costs in the maximum flow problem. Second, select a quantity $\bar{F}$, which is a safe upper bound on the maximum feasible flow through the network, and then assign a supply and a demand of $\bar{F}$ to the supply node and the demand node, respectively. (Because all other nodes are transshipment nodes, they automatically have $b_i = 0$.) Third, add an arc going directly from the supply node to the demand node, and assign it an arbitrarily large unit cost of $c_{ij} = M$ as well as an unlimited arc capacity ($u_{ij} = \infty$). Because of this huge cost, the minimum cost flow problem will send the maximum feasible flow through the *other* arcs, which achieves the objective of the maximum flow problem.

Applying this formulation to the Seervada Park maximum flow problem shown in Fig. 9.5 yields the network given in Fig. 9.12, where the numbers given next to the original arcs are the arc capacities.

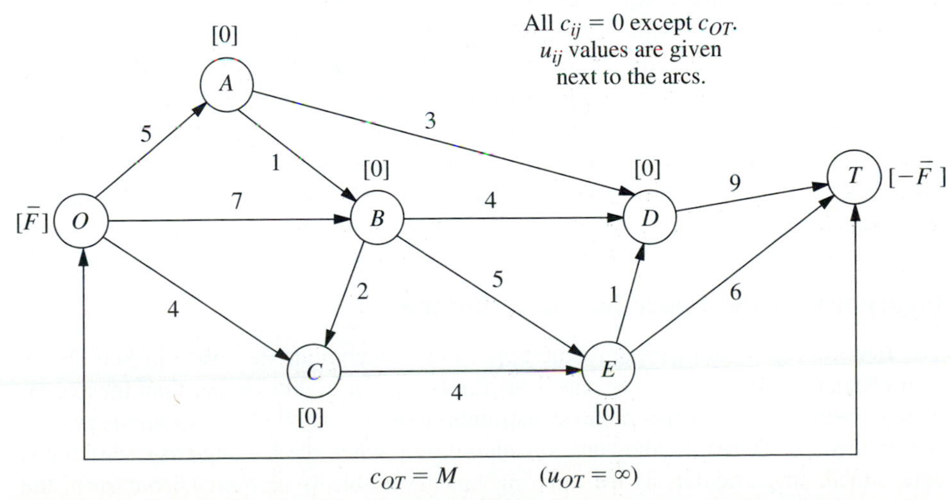

Figure 9.12 Formulation of the Seervada Park maximum flow problem as a minimum cost flow problem.

Final Comments: Except for the transshipment problem, each of these special cases has been the focus of a previous section in either this chapter or Chap. 8. When each was first presented, we talked about a special-purpose algorithm for solving it very efficiently. Therefore, it certainly is not necessary to reformulate these special cases to fit the format of the minimum cost flow problem in order to solve them. However, when a computer code is not readily available for the special-purpose algorithm, it is very reasonable to use the network simplex method instead. In fact, recent implementations of the network simplex method have become so powerful that it now provides an excellent alternative to the special-purpose algorithm.

The fact that these problems are special cases of the minimum cost flow problem is of interest for other reasons as well. One reason is that the underlying theory for the minimum cost flow problem and for the network simplex method provides a unifying theory for all these special cases. Another reason is that some of the many applications of the minimum cost flow problem include features of one or more of the special cases, so it is important to know how to reformulate these features into the broader framework of the general problem.

9.7 The Network Simplex Method

The network simplex method is a highly streamlined version of the simplex method for solving minimum cost flow problems. As such, it goes through the same basic steps at each iteration—finding the entering basic variable, determining the leaving basic variable, and solving for the new BF solution—in order to move from the current BF solution to a better adjacent one. However, it executes these steps in ways that exploit the special network structure of the problem without ever needing a simplex tableau.

You may note some similarities between the network simplex method and the transportation simplex method presented in Sec. 8.2. In fact, both are streamlined versions of the simplex method that provide alternative algorithms for solving transportation problems in similar ways. The network simplex method extends these ideas to solving other types of minimum cost flow problems as well.

In this section, we provide a somewhat abbreviated description of the network simplex method that focuses just on the main concepts. We omit certain details needed for a full computer implementation, including how to construct an initial BF solution and how to perform certain calculations (such as for finding the entering basic variable) in the most efficient manner. These details are provided in various more specialized textbooks, such as Selected References 1, 2, 4, 6, 7, and 9.

Incorporating the Upper Bound Technique

The first concept is to incorporate the upper bound technique described in Sec. 7.3 to deal efficiently with the arc capacity constraints $x_{ij} \leq u_{ij}$. Thus, rather than these constraints being treated as *functional* constraints, they are handled just as *nonnegativity* constraints are. Therefore, they are considered only when the leaving basic variable is determined. In particular, as the entering basic variable is increased from zero, the leaving basic variable is the *first* basic variable that reaches either its lower bound (0) or its upper bound (u_{ij}). A nonbasic variable at its upper bound $x_{ij} = u_{ij}$ is replaced by $x_{ij} = u_{ij} - y_{ij}$, so $y_{ij} = 0$ becomes the nonbasic variable. See Sec. 7.3 for further details.

In our current context, y_{ij} has an interesting network interpretation. Whenever y_{ij} becomes a basic variable with a strictly positive value ($\leq u_{ij}$), this value can be thought of as flow from node j to node i (so in the "wrong" direction through arc $i \rightarrow j$) that, in actuality, is *canceling* that amount of the previously assigned flow ($x_{ij} = u_{ij}$) from node i to node j. Thus, when $x_{ij} = u_{ij}$ is replaced by $x_{ij} = u_{ij} - y_{ij}$, we also replace the *real* arc $i \rightarrow j$ by the **reverse arc** $j \rightarrow i$, where this new arc has arc capacity u_{ij} (the maximum amount of the $x_{ij} = u_{ij}$ flow that can be canceled) and unit cost $-c_{ij}$ (since each unit of flow canceled saves c_{ij}). To reflect the flow of $x_{ij} = u_{ij}$ through the deleted arc, we shift this amount of net flow generated from node i to node j by *decreasing* b_i by u_{ij} and *increasing* b_j by u_{ij}. Later, if y_{ij} becomes the leaving basic variable by reaching its upper bound, then $y_{ij} = u_{ij}$ is replaced by $y_{ij} = u_{ij} - x_{ij}$ with $x_{ij} = 0$ as the new nonbasic variable, so the above process would be reversed (replace arc $j \rightarrow i$ by arc $i \rightarrow j$, etc.) to the original configuration.

To illustrate this process, consider the minimum cost flow problem shown in Fig. 9.10. While the network simplex method is generating a sequence of BF solutions, suppose that x_{AB} has become the leaving basic variable for some iteration by reaching its upper bound of 10. Consequently, $x_{AB} = 10$ is replaced by $x_{AB} = 10 - y_{AB}$, so $y_{AB} = 0$ becomes the new nonbasic variable. At the same time, we replace arc $A \rightarrow B$ by arc $B \rightarrow A$ (with y_{AB} as its flow quantity), and we assign this new arc a capacity of 10 and a unit cost of -2. To take $x_{AB} = 10$ into account, we also decrease b_A from 50 to 40 and increase b_B from 40 to 50. The resulting adjusted network is shown in Fig. 9.13.

We shall soon illustrate the entire network simplex method with this same example, starting with $y_{AB} = 0$ ($x_{AB} = 10$) as a nonbasic variable and so using Fig. 9.13. A later iteration will show x_{CE} reaching its upper bound of 80 and so being replaced by $x_{CE} = 80 - y_{CE}$, and so on, and then the next iteration has y_{AB} reaching its upper bound of 10. You will see that all these operations are performed directly on the network, so we will not need to use the x_{ij} or y_{ij} labels for arc flows or even to keep track of which arcs are *real* arcs and which are *reverse* arcs (except when we record the final solution). Using the upper bound technique leaves the *node constraints* (flow out minus flow in $= b_i$) as the only functional constraints. Minimum cost flow problems tend to have

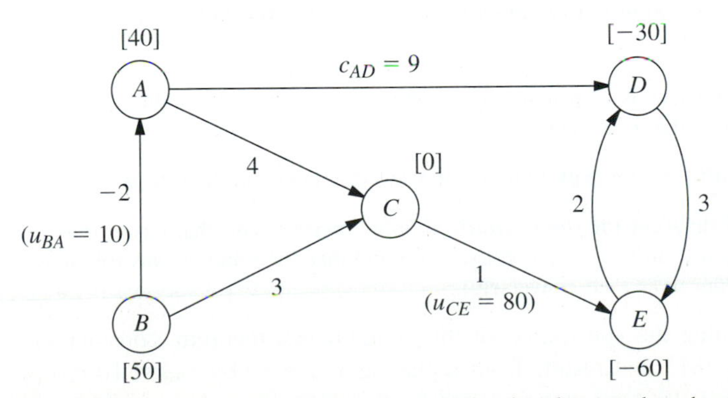

Figure 9.13 The adjusted network for the example when the upper-bound technique leads to replacing $x_{AB} = 10$ by $x_{AB} = 10 - y_{AB}$.

far more arcs than nodes, so the resulting number of functional constraints generally is only a small fraction of what it would have been if the arc capacity constraints had been included. The computation time for the simplex method goes up relatively rapidly with the number of functional constraints, but only slowly with the number of variables (or the number of bounding constraints on these variables). Therefore, incorporating the upper bound technique here tends to provide a tremendous saving in computation time.

However, this technique is not needed for *uncapacitated* minimum cost flow problems (including all but the last special case considered in the preceding section), where there are no arc capacity constraints.

Correspondence between BF Solutions and Feasible Spanning Trees

The most important concept underlying the network simplex method is its network representation of *BF solutions.* Recall from Sec. 9.6 that with n nodes, every BF solution has $(n-1)$ basic variables, where each basic variable x_{ij} represents the flow through arc $i \rightarrow j$. These $(n-1)$ arcs are referred to as **basic arcs**. (Similarly, the arcs corresponding to the *nonbasic* variables $x_{ij} = 0$ or $y_{ij} = 0$ are called **nonbasic arcs**.)

A key property of basic arcs is that they never form undirected *cycles.* (This property prevents the resulting solution from being a weighted average of another pair of feasible solutions, which would violate one of the general properties of BF solutions.) However, *any* set of $n-1$ arcs that contains no undirected cycles forms a *spanning tree.* Therefore, any complete set of $n-1$ basic arcs forms a spanning tree.

Thus BF solutions can be obtained by ''solving'' spanning trees, as summarized below.

A **spanning tree solution** is obtained as follows:

1. For the arcs *not* in the spanning tree (the nonbasic arcs), set the corresponding variables (x_{ij} or y_{ij}) equal to zero.
2. For the arcs that are in the spanning tree (the basic arcs), solve for the corresponding variables (x_{ij} or y_{ij}) in the system of linear equations provided by the node constraints.

(The network simplex method actually solves for the new BF solution from the current one much more efficiently, without solving this system of equations from scratch.) Note that this solution process does not consider either the nonnegativity constraints or the arc capacity constraints for the basic variables, so the resulting spanning tree solution may or may not be feasible with respect to these constraints—which leads to our next definition.

A **feasible spanning tree** is a spanning tree whose solution from the node constraints also satisfies all the other constraints ($0 \leq x_{ij} \leq u_{ij}$ or $0 \leq y_{ij} \leq u_{ij}$).

With these definitions, we now can summarize our key conclusion as follows:

The **fundamental theorem for the network simplex method** says that basic solutions are *spanning tree solutions* (and conversely) and that BF solutions are solutions for *feasible spanning trees* (and conversely).

To begin illustrating the application of this fundamental theorem, consider the network shown in Fig. 9.13 that results from replacing $x_{AB} = 10$ by $x_{AB} = 10 - y_{AB}$ for our example in Fig. 9.10. One spanning tree for this network is the one shown in Fig. 9.3e, where the arcs are $A \rightarrow D$, $D \rightarrow E$, $C \rightarrow E$, and $B \rightarrow C$. With these as the

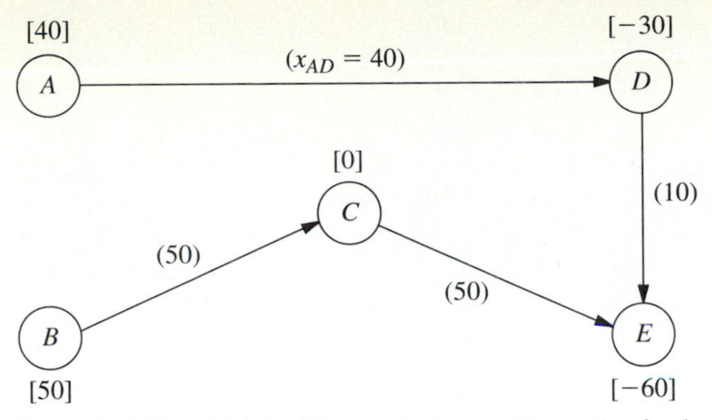

Figure 9.14 The initial feasible spanning tree and its solution for the example.

basic arcs, the process of finding the spanning tree solution is shown below. On the left is the set of node constraints given in Sec. 9.6 after $10 - y_{AB}$ is substituted for x_{AB}, where the *basic* variables are shown in **boldface**. On the right, starting at the top and moving down, is the sequence of steps for setting or calculating the values of the variables.

$$
\begin{array}{lcrcl}
 & & \multicolumn{3}{c}{\underline{y_{AB} = 0,\ x_{AC} = 0,\ x_{ED} = 0}} \\
-y_{AB} + x_{AC} + \mathbf{x_{AD}} & = & 40 & & x_{AD} = 40. \\
y_{AB} + \mathbf{x_{BC}} & = & 50 & & x_{BC} = 50. \\
-x_{AC} - \mathbf{x_{BC}} + \mathbf{x_{CE}} & = & 0 & \text{so} & x_{CE} = 50. \\
-\mathbf{x_{AD}} + \mathbf{x_{DE}} - x_{ED} & = & -30 & \text{so} & x_{DE} = 10. \\
-\mathbf{x_{CE}} - \mathbf{x_{DE}} + x_{ED} & = & -60 & \text{Redundant.} &
\end{array}
$$

Since the values of all these basic variables satisfy the nonnegativity constraints and the one relevant arc capacity constraint ($x_{CE} \leq 80$), the spanning tree is a *feasible spanning tree,* so we have a *BF solution.*

We shall use this solution as the initial BF solution for demonstrating the network simplex method. Figure 9.14 shows its network representation, namely, the feasible spanning tree and its solution. Thus the numbers given next to the arcs now represent *flows* (values of x_{ij}) rather than the unit costs c_{ij} previously given. (To help you distinguish, we shall always put parentheses around flows but not around costs.)

Selecting the Entering Basic Variable

To begin an iteration of the network simplex method, recall that the standard simplex method criterion for selecting the entering basic variable is to choose the nonbasic variable which, when increased from zero, will *improve Z at the fastest rate.* Now let us see how this is done without having a simplex tableau.

To illustrate, consider the nonbasic variable x_{AC} in our initial BF solution, i.e., the nonbasic arc $A \rightarrow C$. Increasing x_{AC} from zero to some value θ means that the arc $A \rightarrow C$ with flow θ must be added to the network shown in Fig. 9.14. Adding a nonbasic arc to a spanning tree *always* creates a unique undirected *cycle,* where the cycle in this case is seen in Fig. 9.15 to be *AC–CE–DE–AD*. Figure 9.15 also shows

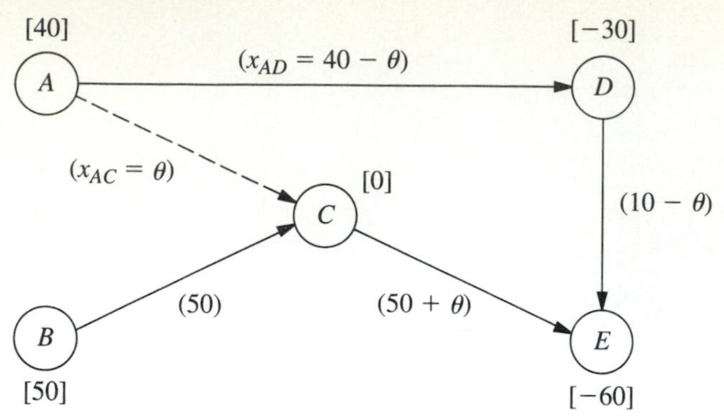

Figure 9.15 The effect on flows of adding arc $A \to C$ with flow θ to the initial feasible spanning tree.

the effect of adding the flow θ to arc $A \to C$ on the other flows in the network. Specifically, the flow is thereby *increased* by θ for other arcs that have the *same* direction as $A \to C$ in the cycle (arc $C \to E$), whereas the *net* flow is *decreased* by θ for other arcs whose direction is *opposite* to $A \to C$ in the cycle (arcs $D \to E$ and $A \to D$). In the latter case, the new flow is, in effect, canceling a flow of θ in the opposite direction. Arcs not in the cycle (arc $B \to C$) are unaffected by the new flow. (Check these conclusions by noting the effect of the change in x_{AC} on the values of the other variables in the solution just derived for the initial feasible spanning tree.)

Now what is the incremental effect on Z (total flow cost) from adding the flow θ to arc $A \to C$? Figure 9.16 shows most of the answer by giving the unit cost times the change in the flow for each arc of Fig. 9.15. Therefore, the overall increment in Z is

$$\begin{aligned}\Delta Z &= c_{AC}\theta + c_{CE}\theta + c_{DE}(-\theta) + c_{AD}(-\theta)\\ &= 4\theta + \theta - 3\theta - 9\theta\\ &= -7\theta.\end{aligned}$$

Setting $\theta = 1$ then gives the *rate* of change of Z as x_{AC} is increased, namely,

$$\Delta Z = -7, \qquad \text{when } \theta = 1.$$

Because the objective is to *minimize* Z, this large rate of decrease in Z by increasing x_{AC} is very desirable, so x_{AC} becomes a prime candidate to be the entering basic variable.

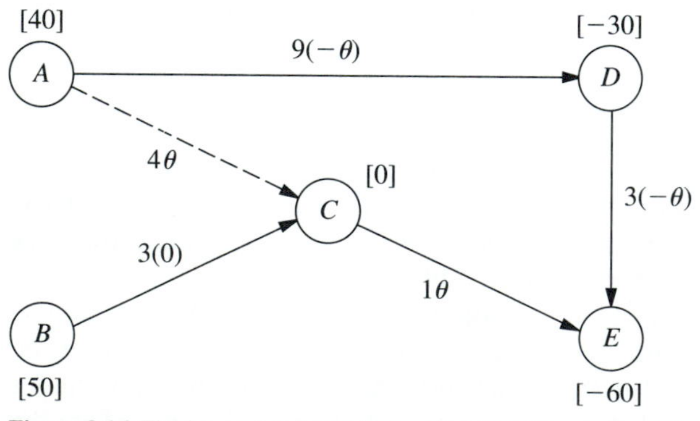

Figure 9.16 The incremental effect on costs of adding arc $A \to C$ with flow θ to the initial feasible spanning tree.

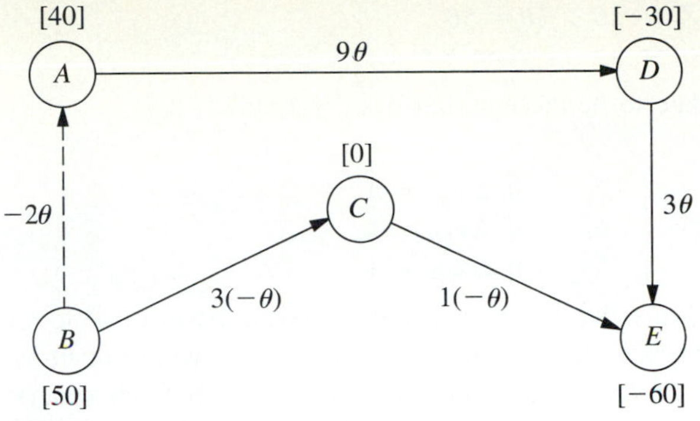

Figure 9.17 The incremental effect on costs of adding arc $B \to A$ with flow θ to the initial feasible spanning tree.

We now need to perform the same analysis for the other nonbasic variables before we make the final selection of the entering basic variable. The only other nonbasic variables are y_{AB} and x_{ED}, corresponding to the two other nonbasic arcs $B \to A$ and $E \to D$ in Fig. 9.13.

Figure 9.17 shows the incremental effect on costs of adding arc $B \to A$ with flow θ to the initial feasible spanning tree given in Fig. 9.14. Adding this arc creates the undirected cycle *BA–AD–DE–CE–BC*, so the flow increases by θ for arcs $A \to D$ and $D \to E$ but decreases by θ for the two arcs in the opposite direction on this cycle, $C \to E$ and $B \to C$. These flow increments, θ and $-\theta$, are the multiplicands for the c_{ij} values in the figure. Therefore,

$$\begin{aligned}\Delta Z &= -2\theta + 9\theta + 3\theta + 1(-\theta) + 3(-\theta) = 6\theta \\ &= 6, \qquad \text{when } \theta = 1.\end{aligned}$$

The fact that Z *increases* rather than decreases when y_{AB} (flow through the reverse arc $B \to A$) is increased from zero rules out this variable as a candidate to be the entering basic variable. (Remember that increasing y_{AB} from zero really means decreasing x_{AB}, flow through the real arc $A \to B$, from its upper bound of 10.)

A similar result is obtained for the last nonbasic arc $E \to D$. Adding this arc with flow θ to the initial feasible spanning tree creates the undirected cycle *ED–DE* shown in Fig. 9.18; so the flow also increases by θ for arc $D \to E$, but no other arcs are affected. Therefore,

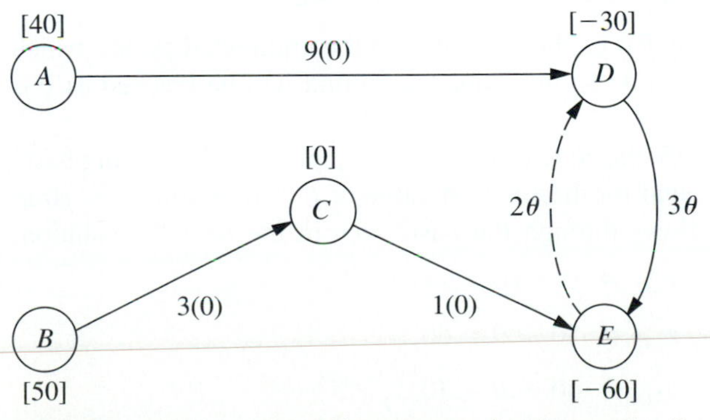

Figure 9.18 The incremental effect on costs of adding arc $E \to D$ with flow θ to the initial feasible spanning tree.

$$\begin{aligned}\Delta Z &= 2\theta + 3\theta = 5\theta \\ &= 5, \qquad \text{when } \theta = 1,\end{aligned}$$

so x_{ED} is ruled out as a candidate to be the entering basic variable.

To summarize,

$$\Delta Z = \begin{cases} -7, & \text{if } \Delta x_{AC} = 1 \\ 6, & \text{if } \Delta y_{AB} = 1 \\ 5, & \text{if } \Delta x_{ED} = 1 \end{cases}$$

so the negative value for x_{AC} implies that x_{AC} becomes the entering basic variable for the first iteration. If there had been more than one nonbasic variable with a *negative* value of ΔZ, then the one having the *largest* absolute value would have been chosen. (If there had been no nonbasic variables with a negative value of ΔZ, the current BF solution would have been optimal.)

Rather than identifying undirected cycles, etc., the network simplex method actually obtains these ΔZ values by an algebraic procedure that is considerably more efficient (especially for large networks). The procedure is analogous to that used by the transportation simplex method (see Sec. 8.2) to solve for u_i and v_j in order to obtain the value of $c_{ij} - u_i - v_j$ for each nonbasic variable x_{ij}. We shall not describe this procedure further, so you should just use the undirected cycles method when you are doing problems at the end of the chapter.

Finding the Leaving Basic Variable and the Next BF Solution

After selection of the entering basic variable, only one more quick step is needed to simultaneously determine the leaving basic variable and solve for the next BF solution. For the first iteration of the example, the key is Fig. 9.15. Since x_{AC} is the entering basic variable, the flow θ through arc $A \rightarrow C$ is to be increased from zero as far as possible until one of the basic variables reaches *either* its lower bound (0) or its upper bound (u_{ij}). For those arcs whose flow *increases* with θ in Fig. 9.15 (arcs $A \rightarrow C$ and $C \rightarrow E$), only the *upper* bounds ($u_{AC} = \infty$ and $u_{CE} = 80$) need to be considered:

$$\begin{aligned} x_{AC} &= \theta \le \infty. \\ x_{CE} &= 50 + \theta \le 80, \qquad \text{so} \qquad \theta \le 30. \end{aligned}$$

For those arcs whose flow *decreases* with θ (arcs $D \rightarrow E$ and $A \rightarrow D$), only the *lower* bound of 0 needs to be considered:

$$\begin{aligned} x_{DE} &= 10 - \theta \ge 0, \qquad \text{so} \qquad \theta \le 10. \\ x_{AD} &= 40 - \theta \ge 0, \qquad \text{so} \qquad \theta \le 40. \end{aligned}$$

Arcs whose flow is unchanged by θ (i.e., those not part of the undirected cycle), which is just arc $B \rightarrow C$ in Fig. 9.15, can be ignored since no bound will be reached as θ is increased.

For the five arcs in Fig. 9.15, the conclusion is that x_{DE} must be the leaving basic variable because it reaches a bound for the smallest value of θ (10). Setting $\theta = 10$ in this figure thereby yields the flows through the basic arcs in the next BF solution:

$$\begin{aligned} x_{AC} &= \theta = 10, \\ x_{CE} &= 50 + \theta = 60, \\ x_{AD} &= 40 - \theta = 30, \\ x_{BC} &= 50. \end{aligned}$$

The corresponding feasible spanning tree is shown in Fig. 9.19.

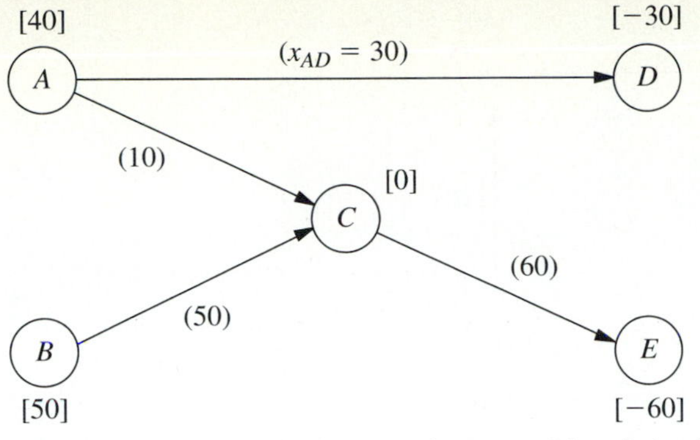

Figure 9.19 The second feasible spanning tree and its solution for the example.

If the leaving basic variable had reached its upper bound, then the adjustments discussed for the upper bound technique would have been needed at this point (as you will see illustrated during the next two iterations). However, because it was the lower bound of 0 that was reached, nothing more needs to be done.

Completing the Example: For the two remaining iterations needed to reach the optimal solution, the primary focus will be on some features of the upper bound technique they illustrate. The pattern for finding the entering basic variable, the leaving basic variable, and the next BF solution will be very similar to that described for the first iteration, so we only summarize these steps briefly.

Iteration 2: Starting with the feasible spanning tree shown in Fig. 9.19 and referring to Fig. 9.13 for the unit costs c_{ij}, we arrive at the calculations for selecting the entering basic variable in Table 9.3. The second column identifies the unique undirected cycle that is created by adding the nonbasic arc in the first column to this spanning tree, and the third column shows the incremental effect on costs because of the changes in flows on this cycle caused by adding a flow of $\theta = 1$ to the nonbasic arc. Arc $E \rightarrow D$ has the largest (in absolute terms) negative value of ΔZ, so x_{ED} is the entering basic variable.

We now make the flow θ through arc $E \rightarrow D$ as large as possible, while satisfying the following flow bounds:

$$
\begin{aligned}
x_{ED} &= \theta \leq u_{ED} = \infty, && \text{so} \quad \theta \leq \infty. \\
x_{AD} &= 30 - \theta \geq 0, && \text{so} \quad \theta \leq 30. \\
x_{AC} &= 10 + \theta \leq u_{AC} = \infty, && \text{so} \quad \theta \leq \infty. \\
x_{CE} &= 60 + \theta \leq u_{CE} = 80, && \text{so} \quad \theta \leq 20. \quad \leftarrow \text{Minimum}
\end{aligned}
$$

Because x_{CE} imposes the smallest upper bound (20) on θ, x_{CE} becomes the leaving basic variable. Setting $\theta = 20$ in the above expressions for x_{ED}, x_{AD}, and x_{AC} then

Table 9.3 **Calculations for Selecting the Entering Basic Variable for Iteration 2**

Nonbasic Arc	Cycle Created	ΔZ When $\theta = 1$	
$B \rightarrow A$	BA–AC–BC	$-2 + 4 - 3 = -1$	
$D \rightarrow E$	DE–CE–AC–AD	$3 - 1 - 4 + 9 = 7$	
$E \rightarrow D$	ED–AD–AC–CE	$2 - 9 + 4 + 1 = -2$	← Minimum

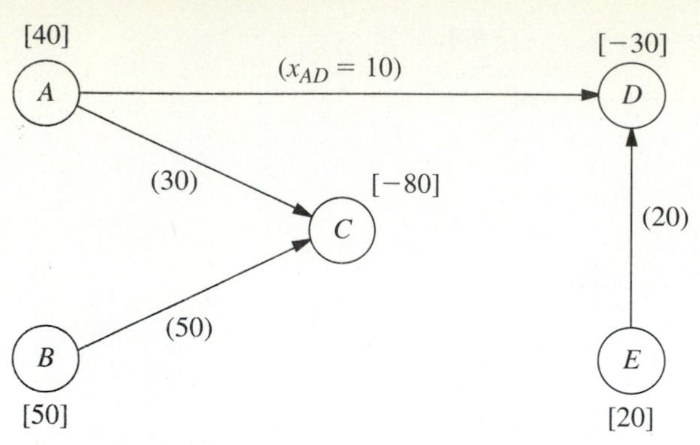

Figure 9.20 The third feasible spanning tree and its solution for the example.

yields the flow through the basic arcs for the next BF solution (with $x_{BC} = 50$ unaffected by θ), as shown in Fig. 9.20.

What is of special interest here is that the leaving basic variable x_{CE} was obtained by the variable reaching its upper bound (80). Therefore, by using the upper bound technique, x_{CE} is replaced by $80 - y_{CE}$, where $y_{CE} = 0$ is the new nonbasic variable. At the same time, the original arc $C \rightarrow E$ with $c_{CE} = 1$ and $u_{CE} = 80$ is replaced by the reverse arc $E \rightarrow C$ with $c_{EC} = -1$ and $u_{EC} = 80$. The values of b_E and b_C also are adjusted by adding 80 to b_E and subtracting 80 from b_C. The resulting adjusted network is shown in Fig. 9.21, where the nonbasic arcs are shown as dashed lines and the numbers by all the arcs are unit costs.

Iteration 3: If Figs. 9.20 and 9.21 are used to initiate the next iteration, Table 9.4 shows the calculations that lead to selecting y_{AB} (reverse arc $B \rightarrow A$) as the entering basic variable. We then add as much flow θ through arc $B \rightarrow A$ as possible while satisfying the flow bounds below:

$y_{AB} = \theta \le u_{BA} = 10,$	so	$\theta \le 10.$	$\leftarrow$ Minimum
$x_{AC} = 30 + \theta \le u_{AC} = \infty,$	so	$\theta \le \infty.$	
$x_{BC} = 50 - \theta \ge 0,$	so	$\theta \le 50.$	

The smallest upper bound (10) on θ is imposed by y_{AB}, so this variable becomes the leaving basic variable. Setting $\theta = 10$ in these expressions for x_{AC} and x_{BC} (along with

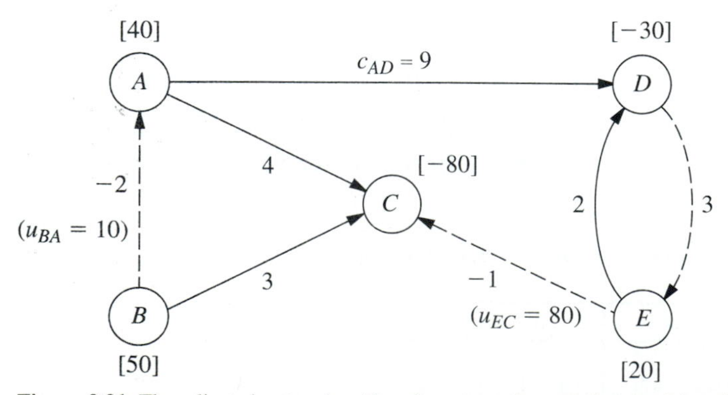

Figure 9.21 The adjusted network with unit costs at the completion of iteration 2.

Table 9.4 **Calculations for Selecting the Entering Basic Variable for Iteration 3**

Nonbasic Arc	Cycle Created	ΔZ When $\theta = 1$	
$B \rightarrow A$	BA–AC–BC	$-2+4-3=-1$	← Minimum
$D \rightarrow E$	DE–ED	$3+2=\ \ 5$	
$E \rightarrow C$	EC–AC–AD–ED	$-1-4+9-2=\ \ 2$	

the unchanged values of $x_{AC} = 10$ and $x_{ED} = 20$) then yields the next BF solution, as shown in Fig. 9.22.

As with iteration 2, the leaving basic variable (y_{AB}) was obtained here by the variable reaching its upper bound. In addition, there are two other points of special interest concerning this particular choice. One is that the *entering* basic variable y_{AB} also became the *leaving* basic variable on the same iteration! This event occurs occasionally with the upper bound technique whenever increasing the entering basic variable from zero causes *its* upper bound to be reached first before any of the other basic variables reach a bound.

The other interesting point is that the arc $B \rightarrow A$ that now needs to be replaced by a *reverse* arc $A \rightarrow B$ (because of the leaving basic variable reaching an upper bound) already is a reverse arc! This is no problem, because the reverse arc for a reverse arc is simply the original *real* arc. Therefore, the arc $B \rightarrow A$ (with $c_{BA} = -2$ and $u_{BA} = 10$) in Fig. 9.21 now is replaced by arc $A \rightarrow B$ (with $c_{AB} = 2$ and $u_{AB} = 10$), which is the arc between nodes A and B in the original network shown in Fig. 9.10, and a generated net flow of 10 is shifted from node B ($b_B = 50 \rightarrow 40$) to node A ($b_A = 40 \rightarrow 50$). Simultaneously, the variable $y_{AB} = 10$ is replaced by $10 - x_{AB}$, with $x_{AB} = 0$ as the new nonbasic variable. The resulting adjusted network is shown in Fig. 9.23.

Passing the Optimality Test: At this point, the algorithm would attempt to use Figs. 9.22 and 9.23 to find the next entering basic variable with the usual calculations shown in Table 9.5. However, *none* of the nonbasic arcs gives a *negative* value of ΔZ, so an improvement in Z *cannot* be achieved by introducing flow through any of them. This means that the current BF solution shown in Fig. 9.22 has *passed* the optimality test, so the algorithm stops.

To identify the flows through real arcs rather than reverse arcs for this optimal solution, the current adjusted network (Fig. 9.23) should be compared with the original network (Fig. 9.10). Note that each of the arcs has the same direction in the two

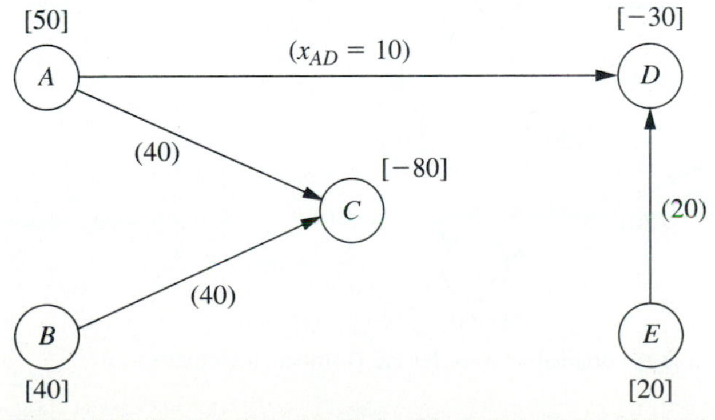

Figure 9.22 The fourth (and final) feasible spanning tree and its solution for the example.

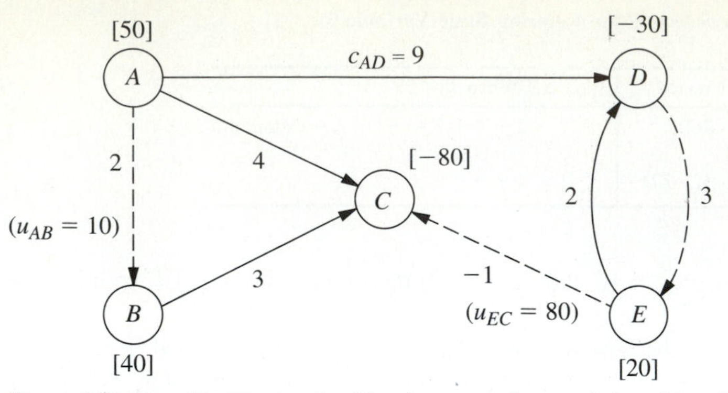

Figure 9.23 The adjusted network with unit costs at the completion of iteration 3.

Table 9.5 **Calculations for the Optimality Test at the End of Iteration 3**

Nonbasic Arc	Cycle Created	ΔZ When $\theta = 1$
$A \to B$	$AB–BC–AC$	$2 + 3 - 4 = 1$
$D \to E$	$DE–EC–AC–AD$	$3 - 1 - 4 + 9 = 7$
$E \to C$	$EC–AC–AD–ED$	$-1 - 4 + 9 - 2 = 2$

networks with the one exception of the arc between nodes C and E. This means that the only reverse arc in Fig. 9.23 is arc $E \to C$, where its flow is given by the variable y_{CE}. Therefore, calculate $x_{CE} = u_{CE} - y_{CE} = 80 - y_{CE}$. Arc $E \to C$ happens to be a nonbasic arc, so $y_{CE} = 0$ and $x_{CE} = 80$ is the flow through the real arc $C \to E$. All the other flows through real arcs are the flows given in Fig. 9.22. Therefore, the optimal solution is the one shown in Fig. 9.24.

Another complete example of applying the network simplex method is provided by the demonstration in the *Network Analysis Area* of your OR Courseware. Also included under the Procedure menu are both an interactive routine and an automatic routine for the network simplex method.

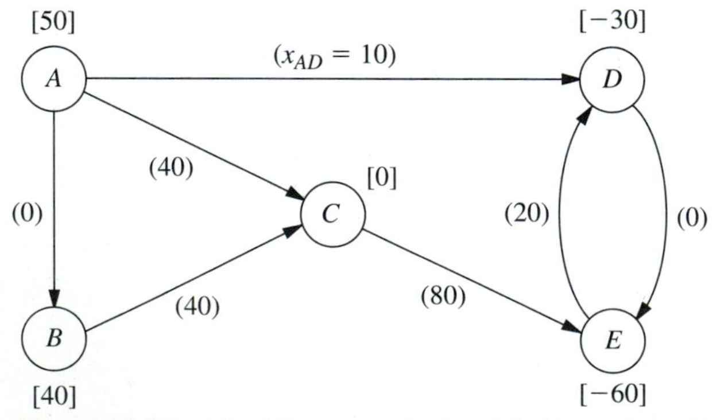

Figure 9.24 The optimal flow pattern in the original network for the Distribution Unlimited Co. example.

9.8 Project Planning and Control with PERT-CPM

The successful management of large-scale projects requires careful *planning, scheduling,* and *coordinating* of numerous interrelated activities. To aid in these tasks, formal procedures based on the use of *networks* and *network techniques* were developed beginning in the late 1950s. The most prominent of these procedures are PERT (program evaluation and review technique) and CPM (critical-path method), although there have been many variants under different names. As you will see later, there are a few important differences between these two procedures. However, in recent years the trend has been to merge the two approaches into what is usually referred to as a *PERT-type system.*

Although the original application of PERT-type systems was for evaluating a schedule for a research and development program, it is also used to measure and control progress on numerous other types of special projects. Examples include construction programs, programming of computers, preparation of bids and proposals, maintenance planning, and installation of computer systems. This kind of approach has even been applied to movie productions, political campaigns, and complex surgery.

A PERT-type system is designed to *aid* in planning and control, so it may not involve much direct *optimization.* Sometimes one of the primary objectives is to determine the probability of meeting specified deadlines. It also identifies the activities that are most likely to be bottlenecks and, therefore, the places where the greatest effort should be made to stay on schedule. A third objective is to evaluate the effect of changes in the program. For example, it will evaluate the effect of a contemplated shift of resources from the less critical activities to the activities identified as probable bottlenecks. Other resource and performance trade-offs may also be evaluated. Another important use is to evaluate the effect of deviations from schedule.

All PERT-type systems use a **project network** to portray graphically the interrelationships among the elements of a project. This network representation of the project plan shows all the *precedence relationships* regarding the order in which tasks must be performed. This feature is illustrated by Fig. 9.25, which shows the initial project network for building a house. This network indicates that the excavation must be done before the foundation is laid, and then the foundation must be completed before the rough wall is put up. Once the rough wall is up, three tasks (rough electrical work, rough exterior plumbing, and putting up the roof) can be done in parallel. Tracing through the network further then spells out the ordering of subsequent tasks.

In the terminology of PERT, each *arc* of the project network represents an **activity** that is one of the tasks required by the project. Each *node* represents an **event** that usually is defined as the time when all activities leading into that node are completed. The *arrowheads* indicate the sequences in which the events must be achieved. Furthermore, an event must precede the initiation of the activities leading out of that node. (In reality, it is often possible to overlap successive phases of a project by starting the next phase before completing the current one, so the network may represent an approximate idealization of the project plan.)

The node toward which all activities lead is the event that corresponds to the completion of the currently planned project. The network may represent either the plan for the project from its inception or, if the project has already begun, the plan for the completion of the project. In the latter case, each node without incoming arcs repre-

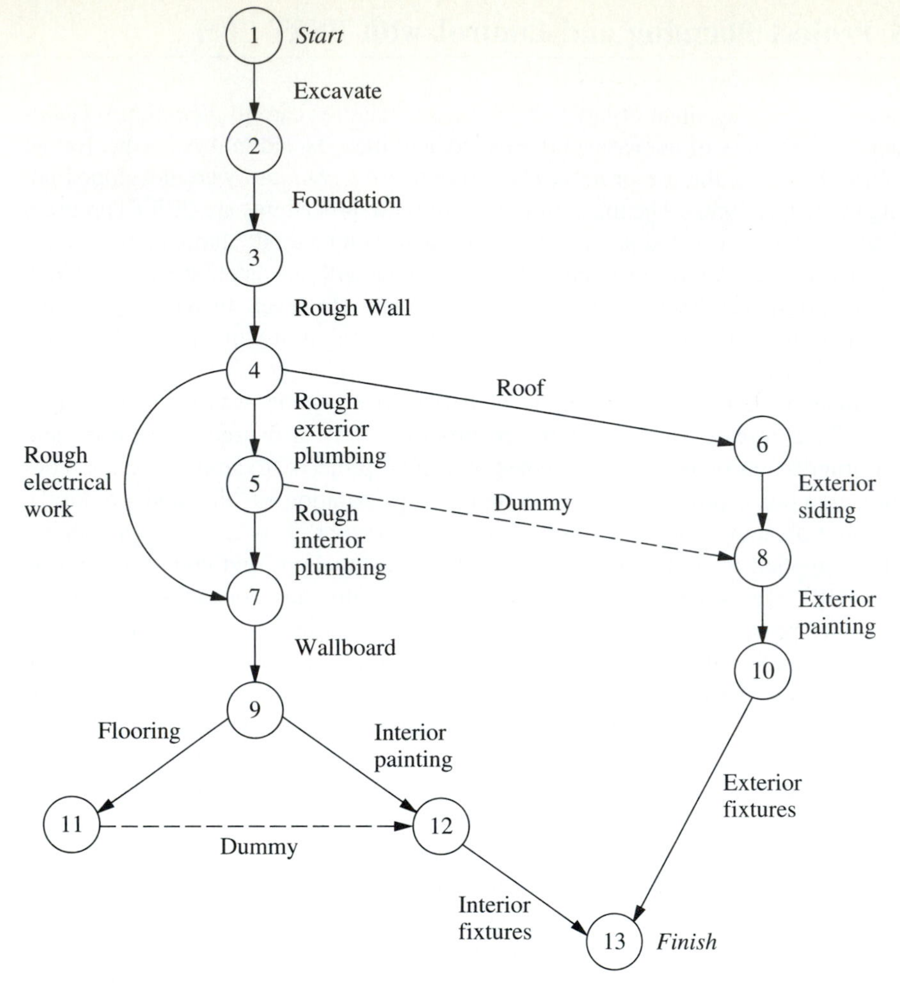

Figure 9.25 Initial project network for constructing a house.

sents either the event of continuing a current activity or the event of initiating a new activity that may begin at any time.

Each arc plays a dual role of representing an activity and helping to show the precedence relationships between the various activities. Occasionally, an arc is needed to further define precedence relationships even when there is no *real* activity to be represented. In this case, a **dummy activity** requiring *zero* time is introduced, where the arc representing this fictional activity is shown as a dashed-line arrow that indicates a precedence relationship. To illustrate, consider arc $5 \rightarrow 8$ representing a dummy activity in Fig. 9.25. The sole purpose of this arc is to indicate that the rough exterior plumbing must be completed before exterior painting can begin.

A common rule for constructing these project networks is that two nodes can be directly connected by *no more than one* arc. Dummy activities can also be used to avoid violating this rule when there are two or more concurrent activities, as illustrated by arc $11 \rightarrow 12$ in Fig. 9.25. The sole purpose of this arc is to indicate that the flooring must be completed before the interior fixtures are installed, without having two arcs from node 9 to node 12.

After the network for a project has been developed, the next step is to estimate the time required for each of the activities. These estimates for the house construction example of Fig. 9.25 are shown by the darker numbers (in units of work days) next to the arcs in Fig. 9.26. These times are used to calculate two basic quantities for each event, namely, its earliest time and its latest time.

> The **earliest time** for an event is the (estimated) time at which the event will occur if the *preceding* activities are started *as early as possible.*

The earliest times are obtained by making a *forward* pass through the network, starting with the initial events and working forward in time toward the final events. For each event, a calculation is made of the time at which that event will occur if each immediately preceding event occurs at its earliest time and if each intervening activity consumes exactly its estimated time. The initiation of the project should be labeled as time 0. This process is shown in Table 9.6 for the example considered in Figs. 9.25 and 9.26. When just one activity leads into an event (as for events 2, 3, 4, 5, 6, 9, 10, and 11), the

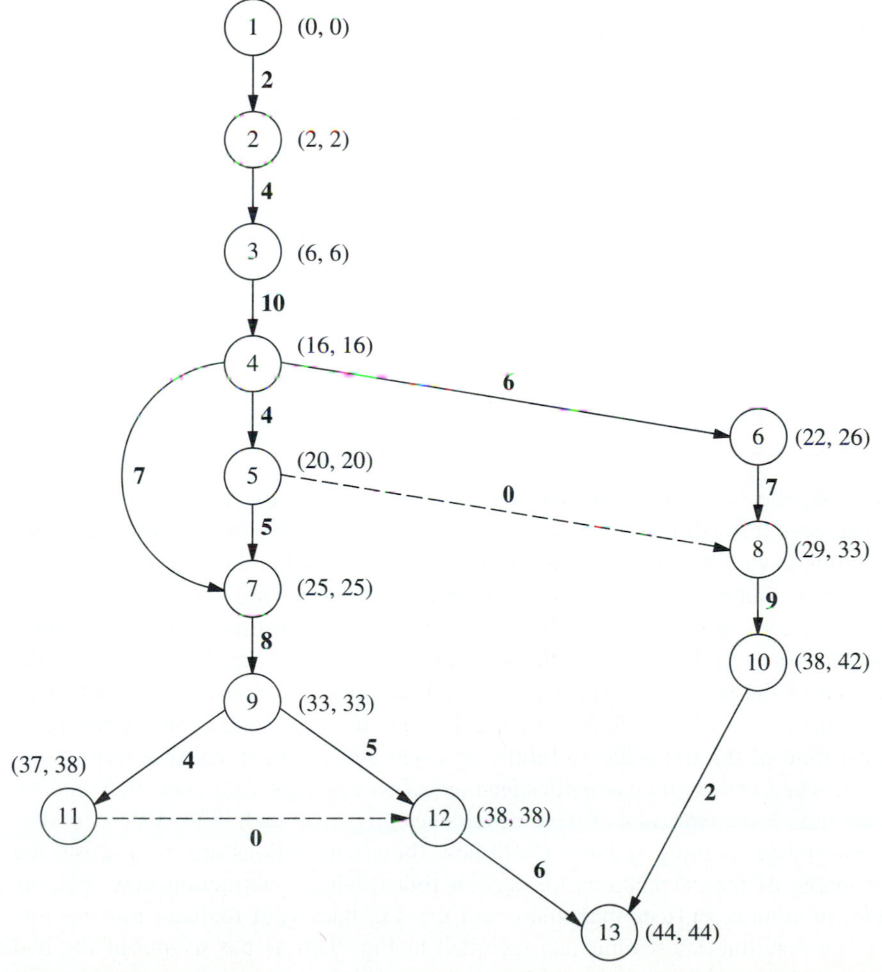

Figure 9.26 Project network for constructing a house after recording the estimated activity times (shown in boldface) as well as the earliest time and latest time for each event (shown in parentheses).

Table 9.6 **Calculation of Earliest Times for the House Construction Example**

Event	Immediately Preceding Event	Earliest Time + Activity Time	Maximum = Earliest Time
1	—	—	0
2	1	0 + 2	2
3	2	2 + 4	6
4	3	6 + 10	16
5	4	16 + 4	20
6	4	16 + 6	22
7	4	16 + 7	25
	5	20 + 5	
8	5	20 + 0	29
	6	22 + 7	
9	7	25 + 8	33
10	8	29 + 9	38
11	9	33 + 4	37
12	9	33 + 5	38
	11	37 + 0	
13	10	38 + 2	44
	12	38 + 6	

earliest time is the *sum* of the earliest time of the immediately preceding event and the intervening activity time. However, when two or more activities lead into an event (as for events 7, 8, 12, and 13), note in Table 9.6 that the earliest time is the *maximum* of these sums involving each immediately preceding event and intervening activity. The resulting earliest times are recorded in Fig. 9.26 as the *first* of the two numbers given in parentheses by each node.

> The **latest time** for an event is the (estimated) last time at which the event can occur *without delaying the completion of the project* beyond its earliest time.

In this case, the latest times are obtained successively for the events by making a *backward* pass through the network, starting with the final events and working backward in time toward the initial events. For each event, a calculation is made of the final time the event can occur in order for each immediately *following* event to occur at its latest time if each intervening activity consumes exactly its estimated time. This process is illustrated in Table 9.7, with 44 as the earliest time *and* latest time for the completion of the house construction project. When just one activity leads out of an event (as for events 12, 11, 10, 8, 7, 6, 3, 2, and 1), the latest time is the *difference* of the latest time of the immediately following event and the intervening activity time. However, when two or more activities lead out of an event (as for events 9, 5, and 4), the latest time is the *minimum* of these differences involving each immediately following event and intervening activity. (Note how these calculations are, in a sense, the *mirror image* of the calculations for earliest times, where subtraction now replaces addition, minimum replaces maximum, and moving backward replaces moving forward.) The resulting latest times are recorded in Fig. 9.26 as the *second* of the two numbers given in parentheses by each node.

Let activity (i, j) denote the activity going from event i to event j in the project network.

Table 9.7 **Calculation of Latest Times for the House Construction Example**

Event	Immediately Following Event	Latest Time − Activity Time	Minimum = Latest Time
13	—	—	44
12	13	44 − 6	38
11	12	38 − 0	38
10	13	44 − 2	42
9	12	38 − 5	33
	11	38 − 4	
8	10	42 − 9	33
7	9	33 − 8	25
6	8	33 − 7	26
5	8	33 − 0	20
	7	25 − 5	
4	7	25 − 7	16
	6	26 − 6	
	5	20 − 4	
3	4	16 − 10	6
2	3	6 − 4	2
1	2	2 − 2	0

The **slack for an event** is the *difference* between its latest and its earliest times. The **slack for an activity** (i, j) is $L_j - (E_i + t_{ij})$, where L_j is the latest time of event j, E_i is the earliest time of event i, and t_{ij} is the estimated time of activity (i, j).

Thus, assuming no other delays, the slack for an event indicates how much delay in reaching the event can be tolerated without delaying the project's completion, and the slack for an activity indicates the same thing regarding a delay in the completion of that activity. The calculation of these slacks is illustrated in Table 9.8 for the house construction project. Figure 9.27 shows the resulting final project network after these slacks are recorded inside a square box to the left of each event and activity.

Table 9.8 **Calculation of Slacks for the House Construction Example**

Event	Slack	Activity	Slack
1	0 − 0 = 0	(1, 2)	2 − (0 + 2) = 0
2	2 − 2 = 0	(2, 3)	6 − (2 + 4) = 0
3	6 − 6 = 0	(3, 4)	16 − (6 + 10) = 0
4	16 − 16 = 0	(4, 5)	20 − (16 + 4) = 0
5	20 − 20 = 0	(4, 6)	26 − (16 + 6) = 4
6	26 − 22 = 4	(4, 7)	25 − (16 + 7) = 2
7	25 − 25 = 0	(5, 7)	25 − (20 + 5) = 0
8	33 − 29 = 4	(6, 8)	33 − (22 + 7) = 4
9	33 − 33 = 0	(7, 9)	33 − (25 + 8) = 0
10	42 − 38 = 4	(8, 10)	42 − (29 + 9) = 4
11	38 − 37 = 1	(9, 11)	38 − (33 + 4) = 1
12	38 − 38 = 0	(9, 12)	38 − (33 + 5) = 0
13	44 − 44 = 0	(10, 13)	44 − (38 + 2) = 4
		(12, 13)	44 − (38 + 6) = 0

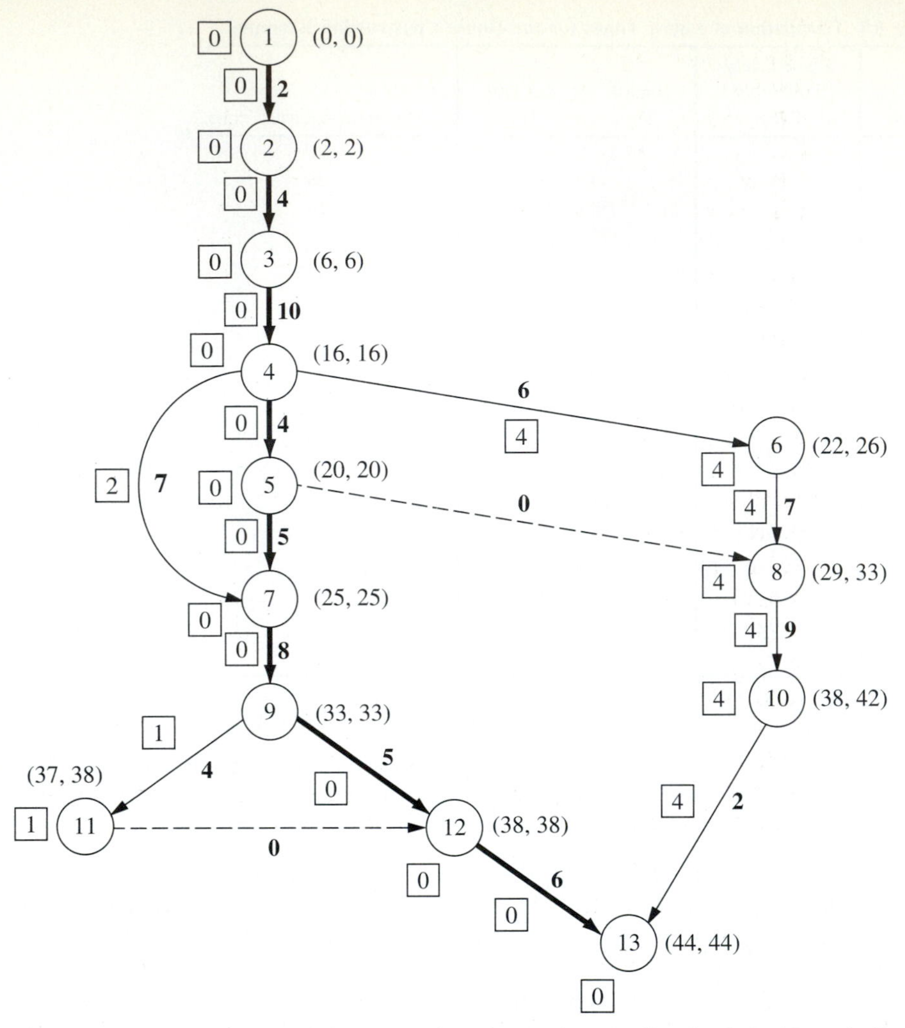

Figure 9.27 Final project network for constructing a house after recording the slacks (shown inside a box) for each event and activity. The dark arrows show the resulting critical path.

Activities having *zero slack* are the particularly critical ones, since any delays in these activities will delay the completion of the project. This fact leads to the key concept of a critical path, as defined below.

> A **critical path** for a project is a path through the network such that all the activities on this path have *zero slack.*

Checking the activities in Fig. 9.27 that have zero slack, we find that the house construction example has the critical path $1 \rightarrow 2 \rightarrow 3 \rightarrow 4 \rightarrow 5 \rightarrow 7 \rightarrow 9 \rightarrow 12 \rightarrow 13$, as shown by the dark arrows. Thus this is the sequence of critical activities that must be kept strictly on schedule in order to avoid slippage in completing the project.

Critical paths have several important properties, as summarized below.

Properties of Critical Paths

1. A project network always has a critical path, and sometimes there is more than one.

2. *All* activities having *zero* slack must lie on a critical path, whereas *no* activities having slack *greater than zero* can lie on a critical path.
3. *All* events having *zero* slack must lie on a critical path, whereas *no* events having slack *greater than zero* can lie on a critical path.
4. A path through the network such that the *events* on this path have *zero slack* need *not* be a critical path because one or more *activities* on the path can have slack *greater than zero.*

Figure 9.27 illustrates all these properties. In this case, $1 \rightarrow 2 \rightarrow 3 \rightarrow 4 \rightarrow 5 \rightarrow 7 \rightarrow 9 \rightarrow 12 \rightarrow 13$ is the only critical path. The activities having zero slack are (1, 2), (2, 3), (3, 4), (4, 5), (5, 7), (7, 9), (9, 12), and (12, 13), and they all lie on this critical path. None of the activities having slack greater than zero lie on the critical path. (If any did, this would violate the definition of critical path.) The events having zero slack are 1, 2, 3, 4, 5, 7, 9, 12, 13, and all lie on the critical path, whereas none of the other events do. To illustrate the fourth property, look at the path $1 \rightarrow 2 \rightarrow 3 \rightarrow 4 \rightarrow 7 \rightarrow 9 \rightarrow 12 \rightarrow 13$, which deviates from the critical path by including activity (4, 7) instead of activities (4, 5) and (5, 7). Every event on this path has zero slack, but the path is not a critical path because activity (4, 7) has slack greater than zero. [The reason this activity has slack greater than zero is that its estimated time is less than the sum of the estimated times for activities (4, 5) and (5, 7).]

To illustrate a project network that has more than one critical path, suppose that the estimated time for activity (4, 6) in Fig. 9.27 is increased from 6 to 10. This decreases the slack from 4 to 0 for activities (6, 8), (8, 10), and (10, 13), as well as for events 6, 8, and 10. Therefore, the path $1 \rightarrow 2 \rightarrow 3 \rightarrow 4 \rightarrow 6 \rightarrow 8 \rightarrow 10 \rightarrow 13$ now is a second critical path.

All the information displayed in the project network (including earliest and latest times, slacks, and the critical paths) is invaluable for the project manager. Among other things, it enables the manager to investigate the effect of possible improvements in the project plan, to determine where special effort should be expended to stay on schedule, and to assess the impact of schedule slippage.

The PERT Three-Estimate Approach

Thus far we have implicitly assumed that reasonably accurate estimates can be made of the time required for each activity of the project. In actuality, there frequently is considerable uncertainty about what the time will be; it really is a *random variable* having some probability distribution. The original version of PERT took this uncertainty into account by using three different types of estimates of the activity time to obtain basic information about its probability distribution. This information for all the activity times is then used to estimate the *probability* of completing the project by the scheduled date.

The three time estimates used by PERT for each activity are a most likely estimate, an optimistic estimate, and a pessimistic estimate. The **most likely estimate**, denoted by m, is intended to be the *most realistic* estimate of the *mode* (the highest point) of the probability distribution for the activity time. The **optimistic estimate**, denoted by a, is intended to be the *unlikely but possible time if everything goes well.* Statistically speaking, it is an estimate of essentially the *lower bound* of the probability distribution. The **pessimistic estimate**, denoted by b, is intended to be the *unlikely but possible time if everything goes badly.* Statistically speaking, it is an estimate of essentially the *upper bound* of the probability distribution. The intended location of these three estimates with respect to the probability distribution is shown in Fig. 9.28.

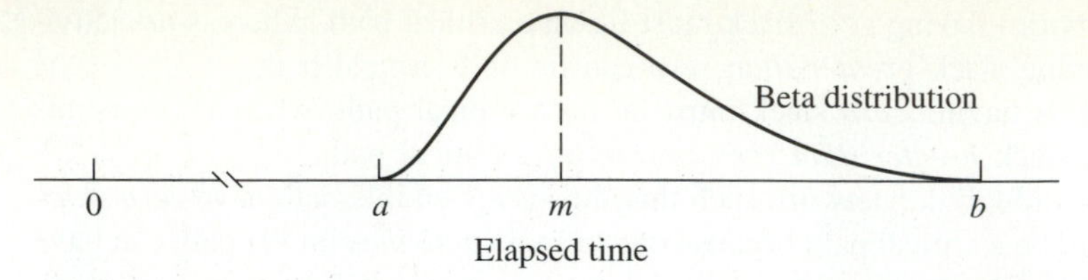

Figure 9.28 Model of the probability distribution of activity times for the PERT three-estimate approach: m = most likely estimate, a = optimistic estimate, and b = pessimistic estimate.

Two assumptions are made to convert m, a, and b to estimates of the *expected value* t_e and *variance* σ^2 of the elapsed time required by the activity.

Assumption 1: The spread between a (the optimistic estimate) and b (the pessimistic estimate) is 6 *standard deviations,* that is, $6\sigma = b - a$. Consequently, the variance of an activity time is

$$\sigma^2 = [\tfrac{1}{6}(b - a)]^2.$$

The rationale for this assumption is that the tails of many probability distributions (such as the normal distribution) are considered to lie about 3 standard deviations from the mean, so that there is a spread of about 6 standard deviations between the tails. For example, the control charts commonly used for statistical quality control are constructed so that the spread between the control limits is estimated to be 6 standard deviations.

To obtain the estimated expected value t_e, we also need an assumption about the *form* of the probability distribution.

Assumption 2: The probability distribution of each activity time is (at least approximately) a **beta distribution**.

Beta distributions have the form shown in Fig. 9.28, with a single mode (m) and two endpoints (a and b), where we assume that $0 \leq a \leq b$. Thus, it fits our definitions of the three time estimates well, and it gives a reasonable shape for a distribution of activity times.

Under these assumptions, the expected value of the activity time is approximately

$$t_e = \tfrac{1}{3}[2m + \tfrac{1}{2}(a + b)].$$

Notice that the midrange $(a + b)/2$ lies midway between a and b, so that t_e is the weighted arithmetic mean of the mode and the midrange, with the mode carrying two-thirds of the entire weight. Although the assumption of a beta distribution is an arbitrary one, it serves its purpose of locating the expected value with respect to m, a, and b in what seems to be a reasonable way.

For example, suppose that the three-estimate approach has been applied to the house construction example shown in Fig. 9.25, where the three estimates for each activity are given in Table 9.9. Plugging these estimates into the above expressions for t_e and σ^2 then gives the expected value and variance of each activity time, as shown in the last two columns.

Table 9.9 **Expected Value and Variance of Each Activity Time for the House Construction Example**

Activity	Optimistic Estimate a	Most Likely Estimate m	Pessimistic Estimate b	Expected Value t_e	Variance σ^2
(1, 2)	1	2	3	2	$\frac{1}{9}$
(2, 3)	2	$3\frac{1}{2}$	8	4	1
(3, 4)	6	9	18	10	4
(4, 5)	1	$4\frac{1}{2}$	5	4	$\frac{4}{9}$
(4, 6)	4	$5\frac{1}{2}$	10	6	1
(4, 7)	3	$7\frac{1}{2}$	9	7	1
(5, 7)	4	4	10	5	1
(6, 8)	5	$6\frac{1}{2}$	11	7	1
(7, 9)	3	9	9	8	1
(8, 10)	5	8	17	9	4
(9, 11)	4	4	4	4	0
(9, 12)	1	$5\frac{1}{2}$	7	5	1
(10, 13)	1	2	3	2	$\frac{1}{9}$
(12, 13)	5	$5\frac{1}{2}$	9	6	$\frac{4}{9}$

Notice that the expected values given in the next-to-last column of Table 9.9 are the same as the times used in Figs. 9.26 and 9.27. Therefore, the critical path (based on expected times) still is the one shown in Fig. 9.27, and the times on this path still add to 44.

However, the actual time required on this critical path now can be different from 44. For example, suppose that every activity time turned out to equal its pessimistic estimate. Then the times on this critical path would add up to 69 days. Furthermore, another path ($1 \rightarrow 2 \rightarrow 3 \rightarrow 4 \rightarrow 6 \rightarrow 8 \rightarrow 10 \rightarrow 13$) would require even more time, namely, 70 days. Thus, the critical path based on expected times can turn out to require less time than another path.

Suppose now that the deadline for completing the house construction project is in 47 workdays. How can the information in Table 9.9 be used to determine the probability of meeting this schedule?

Having used the first two assumptions to obtain the last two columns of Table 9.9, we now need three additional assumptions (or approximations) to enable us to calculate the probability of completing the project on schedule.

Assumption 3: The activity times are *statistically independent* random variables.

Thus, it is being assumed that where a particular activity time turns out to lie in its distribution (Fig. 9.28) does not influence where the times for the other activities will lie in their distributions. In certain cases, the assumption may be violated to some extent because some unexpected event (e.g., the loss of a key employee) can cause the times for several activities to lie on the same side of their expected values.

Assumption 4: As an approximation, assume that the *critical path* (based on expected times) *always* requires a longer total elapsed time than any other path.

This assumption is just an approximation because, in actuality, substantial delays along another path may cause its total elapsed time to be longer than for the critical path. In fact, we just gave an example above where this occurs when every activity time turns out to equal its pessimistic estimate. However, assumption 4 is a reasonable (optimistic) approximation, since the critical path often turns out to be the longest and, when not, usually will be close to the longest.

These two assumptions provide the following convenient *approximation* for estimating the expected value and variance of **project time** (the total time needed to complete the project).

> **Project time:** Under assumption 4, project time equals the *sum* of the activity times on the *critical path* (based on expected times). The expected value of a sum of random variables is the sum of their expected values. Therefore, the expected project time is (approximately) the sum of the expected activity times on the critical path. Furthermore, the variance of a sum of statistically independent random variables is the sum of their variances. Consequently, the variance of the project time is (approximately) the sum of the variances of the activity times on the critical path.

Table 9.10 shows the application of this approach to the example. Summing the second and third columns yields

$$\text{Expected project time} = 44,$$

$$\text{Variance of project time} = 9.$$

Finally, we need an assumption about the *form* of the distribution of project time.

> **Assumption 5:** The probability distribution of project time is (at least approximately) a *normal distribution.*

The rationale for assumption 5 is the *central-limit theorem* from probability theory. Under assumptions 3 and 4, the project time is the sum of a number of independent random variables (the times for the activities on the critical path). Even though these random variables individually have been assumed to have a beta distribution, the *sum* of such random variables does *not* have a beta distribution. In fact, the general version of the central-limit theorem states that the probability distribution of a sum of many independent random variables is approximately *normal* under a wide range of conditions. Thus, assumption 5 holds as a reasonable approximation.

Table 9.10 **Expected Value and Variance of Project Time for the Home Construction Example**

Activities on Critical Path	Expected Value	Variance
(1, 2)	2	$\frac{1}{9}$
(2, 3)	4	1
(3, 4)	10	4
(4, 5)	4	$\frac{4}{9}$
(5, 7)	5	1
(7, 9)	8	1
(9, 12)	5	1
(12, 13)	6	$\frac{4}{9}$
Project time	44	9

Appendix 5 provides a table for finding probabilities from a normal distribution, given its mean and variance. Therefore, after assumptions 3 and 4 are used to obtain the mean and variance, assumption 5 makes it straightforward to find the approximate probability that the project time will be less than the scheduled completion time.[1]

For the example, with a mean of 44 and variance of 9, the standard deviation is $\sqrt{9} = 3$. Thus, the scheduled completion time of 47 is 1.0 standard deviation above the mean. Using $K_\alpha = 1.0$ in Appendix 5 then gives an approximate probability of $1 - 0.1587 \approx 0.84$ that this schedule will be met.

This estimate of the probability of meeting the schedule is a fairly rough one, because of the approximations introduced by the five simplifying assumptions. (For example, assumption 4 always biases the probability in the optimistic direction.) Considerable research has been done on ways to improve the precision of this procedure, and good progress has been made.[2] Unfortunately, these improvements have not yet attracted much attention. The relative simplicity of the above procedure is appealing, and it continues to be widely used.

The CPM Method of Time-Cost Trade-offs

The original versions of CPM and PERT differ in two important ways. First, CPM assumes that activity times are *deterministic* (i.e., they can be reliably predicted without significant uncertainty), so that the three-estimate approach just described is not needed. Second, rather than primarily emphasizing time (explicitly), CPM places equal emphasis on *time and cost.* This dual emphasis is achieved by constructing a **time-cost curve** for *each activity,* such as the one shown in Fig. 9.29. This curve plots the relationship between the budgeted *direct cost*[3] for the activity and its resulting *duration time.* The plot normally is based on two[4] points, the *normal* and the *crash points.* The **normal point** gives the cost and time involved when the activity is performed in the *normal* way *without* any extra costs (overtime labor, special time-saving materials or equipment) being expended to speed up the activity. By contrast, the **crash point** gives the time and cost involved when the activity is performed on a *crash basis,* i.e., it is *fully expedited* with no cost spared to reduce the duration time as much as possible. As an approximation, it is then assumed that *all* intermediate *time-cost trade-offs* also are possible and that they lie on the line segment between these two points (see the solid line segment shown in Fig. 9.29). Thus the only estimates that need to be obtained from the project personnel for this activity are the cost and time for the two points.

The basic objective of CPM is to detemine just which time-cost trade-off should be used for each activity to *meet the scheduled project completion time at a minimum cost.* One way of determining the optimal combination of time-cost trade-offs for all the activities is to use linear programming. To describe this approach, we need to

[1] The same procedure can also be used to find the probaility that an *intermediate* event will be accomplished before a scheduled time.

[2] For example, see D. L. Keefer and W. A. Verdini, "Better Estimation of PERT Activity Time Parameters," *Management Science,* **39:** 1086–1091, 1993.

[3] *Direct* cost includes the cost of the material, equipment, and direct labor required to perform the activity but *excludes* indirect project costs such as supervision and other customary overhead costs, interest charges, and so forth.

[4] More than two points can be used under certain circumstances.

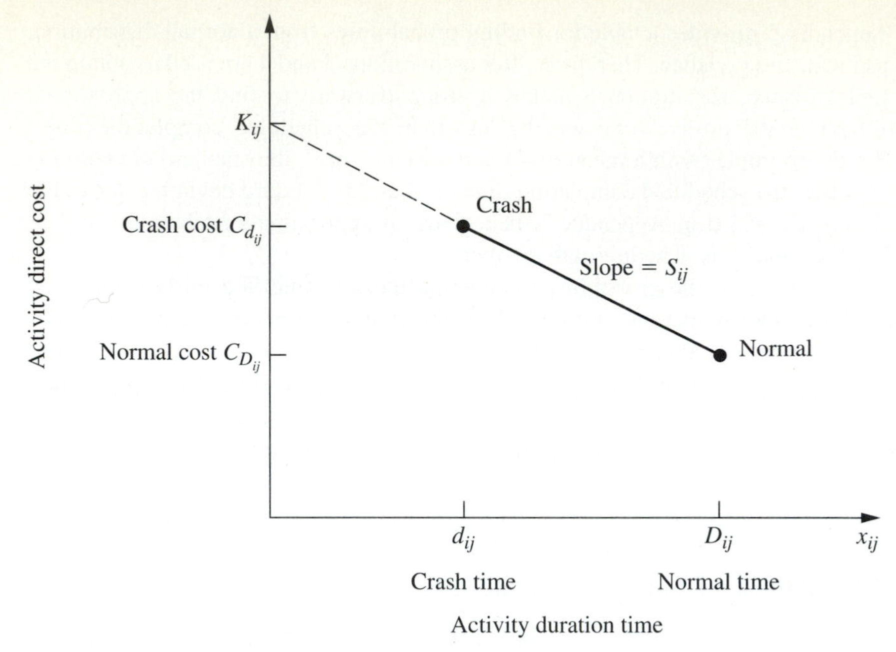

Figure 9.29 Time-cost curve for activity (i, j).

introduce considerable notation, some of which is summarized in Fig. 9.29. Let

$$D_{ij} = \text{normal time for activity } (i, j).$$

$$C_{D_{ij}} = \text{normal (direct) cost for activity } (i, j).$$

$$d_{ij} = \text{crash time for activity } (i, j).$$

$$C_{d_{ij}} = \text{crash (direct) cost for activity } (i, j).$$

The *decision variables* for the problem are the x_{ij}, where

$$x_{ij} = \text{duration time for activity } (i, j).$$

Thus there is one decision variable x_{ij} for each activity, but there is none for those values of i and j that do not have a corresponding activity.

To express the direct cost for activity (i, j) as a (linear) function of x_{ij}, denote the *slope* of the line through the normal and crash points for activity (i, j) by

$$S_{ij} = \frac{C_{D_{ij}} - C_{d_{ij}}}{D_{ij} - d_{ij}}.$$

Also define K_{ij} as the *intercept* with the *direct-cost axis* of this line, as shown in Fig. 9.29. Therefore,

$$\text{Direct cost for activity } (i, j) = K_{ij} + S_{ij}x_{ij}.$$

Consequently,

$$\text{Total direct cost for project} = \sum_{(i, j)} (K_{ij} + S_{ij}x_{ij}),$$

where the summation is over all activities (i, j). We are now ready to state and formulate the problem mathematically.

> **The problem:** For a given (maximum) project completion time T, choose x_{ij} to *minimize the total direct cost* for the project.

Linear Programming Formulation: To take the project completion time into account, we need one more variable for each event in the linear programming formulation of the problem. This additional variable is

y_k = (unknown) earliest time for event k, which is a deterministic function of x_{ij}.

Each y_k is an *auxiliary variable,* i.e., a variable that is introduced into the model as a convenience in the formulation rather than one representing a decision. However, the simplex method treats auxiliary variables just like the regular decision variables (the x_{ij}).

To illustrate how the y_k values are worked into the formulation, consider event 7 in Fig. 9.25. By definition, its earliest time is

$$y_7 = \max\{y_4 + x_{47}, y_5 + x_{57}\}.$$

In other words, y_7 is the *smallest* quantity such that *both* of the following constraints hold:

$$y_4 + x_{47} \leq y_7$$

$$y_5 + x_{57} \leq y_7,$$

so that these two constraints can be incorporated directly into the linear programming formulation (after y_7 is brought to the left-hand side for proper form). Furthermore, we shall soon describe why the optimal solution obtained by the simplex method for the overall model *automatically* will have y_7 at the *smallest* quantity that satisfies these constraints (unless y_7 is not relevant for determining the project completion time), so no further constraints are needed to incorporate the definition of y_7 into the model.

In the process of adding these constraints for all the events, *every* variable x_{ij} will appear in exactly one constraint of this type,

$$y_i + x_{ij} \leq y_j,$$

which then is expressed in proper form as

$$y_i + x_{ij} - y_j \leq 0.$$

To continue the preparations for writing down the complete linear programming model, label

Event 1 = project start
Event n = project completion,

so

$$y_1 = 0$$

y_n = (unknown) project completion time.

Also note that ΣK_{ij} is just a fixed constant that can be dropped from the objective function, so that minimizing total direct cost for the project is *equivalent* (see Sec. 4.6) to *maximizing* $\Sigma\,(-S_{ij})x_{ij}$. Therefore, the linear programming problem is to find the x_{ij} (and the corresponding y_k) that

$$\text{Maximize} \qquad Z = \sum_{(i,j)} (-S_{ij})x_{ij},$$

subject to

$$\left.\begin{aligned} x_{ij} &\geq d_{ij} \\ x_{ij} &\leq D_{ij} \\ y_i + x_{ij} - y_j &\leq 0 \end{aligned}\right\} \quad \text{for all activities } (i, j)$$

$$y_n \leq T.$$

From a computational viewpoint, this formulation can be improved somewhat by replacing each x_{ij} by

$$x_{ij} = d_{ij} + x'_{ij}$$

throughout the model, so that the first set of functional constraints ($x_{ij} \geq d_{ij}$) would be replaced by simple *nonnegativity constraints*

$$x'_{ij} \geq 0.$$

As a convenience we can also introduce nonnegativity constraints for the other variables

$$y_k \geq 0,$$

although these variables already are forced to be nonnegative by setting $y_1 = 0$ because of the $x'_{ij} \geq 0$ and $y_j \geq y_i + d_{ij} + x'_{ij}$ constraints.

Your OR Courseware includes an automatic routine for using this linear programming approach to solving CPM problems after you enter the basic data.

One interesting property of an optimal solution for this model is that (under common circumstances) *every* path through the network may be a critical path requiring a time T. The reason is that such a solution satisfies the $y_n \leq T$ constraint while avoiding the extra cost involved in shortening the time for any path. (The kind of circumstance where this property does not hold occurs where there exists a path shorter than T even when *all* its activities are at their normal point.)

The key to this formulation is the way that the y_k are introduced into the model through the $y_i + x_{ij} - y_j \leq 0$ constraints in order to provide earliest times for the respective events (given the values of x_{ij} in the current BF solution). Since earliest times must be obtained sequentially, all these y_k values are needed for the sole purpose of ultimately obtaining the correct value of y_n (for the current values of the x_{ij}), thereby enabling the $y_n \leq T$ constraint to be enforced. However, obtaining the correct value does require that the value of each y_j (including y_n) be the *smallest* quantity that satisfies all the $y_i + x_{ij} \leq y_j$ constraints (unless event j has *positive* slack so y_j is not relevant for obtaining y_n). Now let us briefly describe why this property holds for an optimal solution.

Consider any solution for the x_{ij} variables such that every path through the network is a critical path requiring a time of T (so every event has *zero* slack). If the values of the y_k variables satisfy the above property, then y_k values are true *earliest times* with $y_n = T$ exactly, and the overall solution for x_{ij} and y_k satisfies all the constraints.

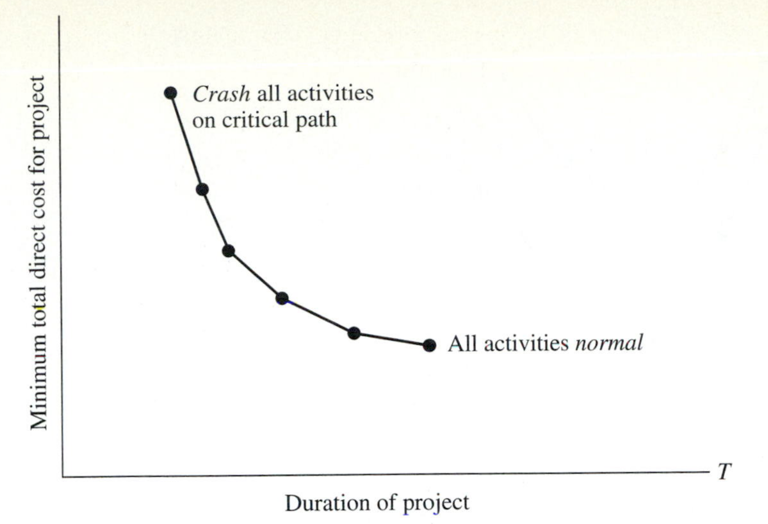

Figure 9.30 Time-cost curve for the overall project.

However, if any particular y_i were made a little larger, this would create a chain reaction whereby some y_j would need to be made a little larger to still satisfy the $y_i + x_{ij} \leq y_j$ constraints, etc., until ultimately y_n must be made a little larger, thereby violating the $y_n \leq T$ constraint. The only way to avoid violating this constraint with the larger y_i is to make the duration times for some activities (subsequent to event i) a little *smaller,* thereby increasing the cost. Therefore, an optimal solution will avoid making any y_k larger than need be to satisfy the $y_i + x_{ij} \leq y_j$ constraints.

The problem as stated here assumes that a specified *deadline* T has been fixed (perhaps by contract) for the completion of the project. In fact, some projects do not have such a deadline, in which case it is not clear what value should be assigned to T in the linear programming formulation. In such situations, the decision on T actually is a question of what is the *best trade-off* between the *total cost* and the *total time* for the project.

The basic information we need to address this question concerns how the *minimum total direct cost* changes as T is changed in the preceding formulation, as illustrated in Fig. 9.30. This information can be obtained by using *parametric linear programming* (see Secs. 4.7, 6.7, and 7.2) to solve for the optimal solution *as a function of* T over its entire range.[1]

Figure 9.30 provides a useful basis for a managerial decision on T (and the corresponding optimal solution for x_{ij}) when the important effects of the project duration (other than direct costs) are largely intangible. However, when these other effects are primarily financial (indirect costs), it is appropriate to combine the minimum total direct cost curve of Fig. 9.30 with a curve of minimum total indirect cost (supervision, facilities, clerical, interest, contractual penalties) versus T, as shown in Fig. 9.31. The sum of these curves thereby gives the minimum total project cost for the various values of T. The optimal value of T is then the one that minimizes this total-cost curve.

[1] The *slope* of the time-cost curve changes at the points shown in Fig. 9.30 because the set of basic variables that give the optimal solution changes at these values of T. This fact is discussed further in a more general context in Sec. 7.2.

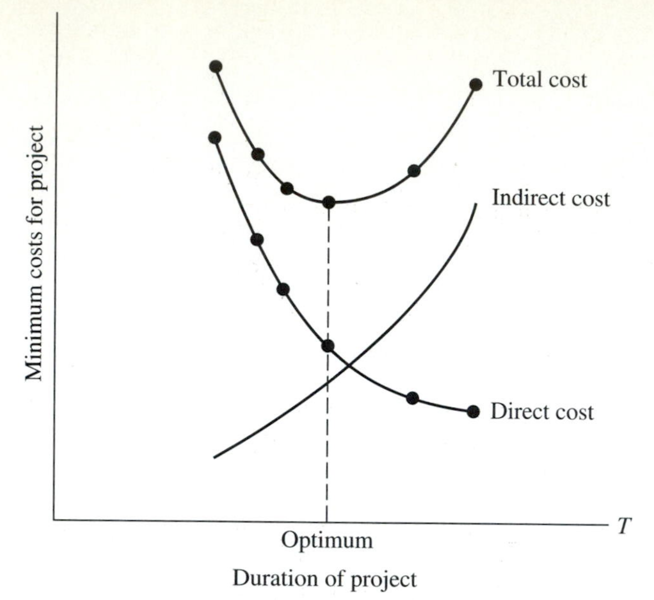

Figure 9.31 Minimum cost curves for the overall project.

Choosing between PERT and CPM

The choice between the PERT *three-esimate approach* and the CPM *method of time-cost trade-offs* depends primarily upon the *type of project* and the *managerial objectives.* PERT is particularly appropriate when there is considerable uncertainty in predicting activity times and when it is important to effectively *control* the project schedule; e.g., most research and development projects fall into this category. On the other hand, CPM is particularly appropriate when activity times can be predicted well (perhaps based on previous experience) but these times can be adjusted readily (e.g., by changing crew sizes), and when it is important to plan an appropriate trade-off between project time and cost. This latter type is typified by most construction and maintenance projects.

Actually, differences between current versions of PERT and CPM are not necessarily as pronounced as we have described them. Most versions of PERT now allow using only a *single* estimate (the most likely estimate) of each activity time and thus omit the probabilistic investigation. A version called *PERT/Cost* also considers *time-cost trade-offs* in a manner similar to CPM.

9.9 Conclusions

Networks of some type arise in a wide variety of contexts. Network representations are very useful for portraying the relationships and connections between the components of systems. Frequently, flow of some type must be sent through a network, so a decision needs to be made about the best way to do this. The kinds of network optimization models and algorithms introduced in this chapter provide a powerful tool for making such decisions.

The minimum cost flow problem plays a central role among these network optimization models, both because it is so broadly applicable and because it can be solved extremely efficiently by the network simplex method. Two of its special cases included in this chapter, the shortest-path problem and the maximum flow problem, also are basic network optimization models, as are additional special cases discussed in Chap. 8 (the transportation problem and the assignment problem).

Whereas all these models are concerned with optimizing the *operation* of an *existing* network, the minimum spanning tree problem is a prominent example of a model for optimizing the *design* of a *new* network.

This chapter has only scratched the surface of the current state of the art of network methodology. Because of their combinatorial nature, network problems often are extremely difficult to solve. However, great progress is being made in developing powerful modeling techniques and solution methodologies that are opening up new vistas for important applications. In fact, recent algorithmic advances are enabling us to solve successfully some complex network problems of enormous size.

The most widely used network technique has been the *PERT-type system* for project planning and control. It has been very valuable for organizing planning effort, testing alternative plans, revealing the overall dimensions and details of the project plan, establishing well-understood management responsibilities, and identifying realistic expectations for the project. It also lays the foundation for *anticipatory* management action against potential trouble spots during the course of the project. Although not a panacea, it has greatly aided project management on numerous occasions.

SELECTED REFERENCES

1. Ahuja, R. K., T. L. Magnanti, and J. B. Orlin, *Network Flows: Theory, Algorithms, and Applications,* Prentice-Hall, Englewood Cliffs, NJ, 1993.
2. Bazaraa, M. S., J. J. Jarvis, and H. D. Sherali, *Linear Programming and Network Flows,* 2d ed., Wiley, New York, 1990.
3. Dreger, J. B., *Project Management,* Van Nostrand Reinhold, New York, 1992.
4. Evans, J., and E. Minieka, *Optimization Algorithms for Networks and Graphs,* 2d ed., Marcel Dekker, New York, 1992.
5. Glover, F., and D. Klingman, ''Network Application in Industry and Government,'' *AIIE Transactions,* **9:** 363–376, 1977.
6. Glover, F., D. Klingman, and N. V. Phillips, *Network Models in Optimization and Their Applications in Practice,* Wiley, New York, 1992.
7. Jensen, P. A., and J. W. Barnes, *Network Flow Programming,* Wiley, New York, 1980.
8. Magnanti, T. L., and R. T. Wong, ''Network Design and Transportation Planning: Models and Algorithms,'' *Transportation Science,* **18:** 1–55, 1984.
9. Murty, K. G., *Network Programming,* Prentice-Hall, Englewood Cliffs, NJ, 1992.
10. Sheffi, Y., *Urban Transportation Networks: Equilibrium Analysis with Mathematical Programming Methods,* Prentice-Hall, Englewood Cliffs, NJ, 1985.

RELEVANT ROUTINES IN YOUR OR COURSEWARE

A demonstration example: *Network Simplex Method*
An interactive routine: *Network Simplex Method—Interactive*
Automatic routines: *Network Simplex Method—Automatic*
CPM Method of Time-Cost Trade-offs—Automatic

To access these routines, call the MathProg program and then choose *Network Analysis* under the Area menu. See Appendix 1 for documentation of the software.

PROBLEMS[1]

To the left of each of the following problems (or their parts), we have inserted a D (for demonstration), I (for interactive routine), or A (for automatic routine) whenever a corresponding routine listed above can be helpful. An asterisk on the I or A indicates that this routine definitely should be used (unless your instructor gives contrary instructions) and that the printout from this routine is all that needs to be turned in to show your work in executing the algorithm. An asterisk on the problem number indicates that at least a partial answer is given in the back of the book.

9.2-1. Consider the following directed network.

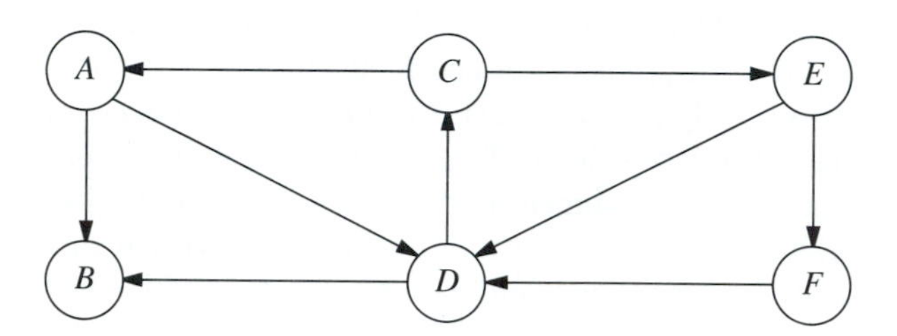

(*a*) Find a directed path from node *A* to node *F*, and then identify three other undirected paths from node *A* to node *F*.
(*b*) Find three directed cycles. Then identify an undirected cycle that includes every node.
(*c*) Identify a set of arcs that forms a spanning tree.
(*d*) Use the process illustrated in Fig. 9.3 to grow a tree one arc at a time until a spanning tree has been formed. Then repeat this process to obtain another spanning tree. [Do not duplicate the spanning tree identified in part (*c*).]

9.3-1. At a small but growing airport, the local airline company is purchasing a new tractor for a tractor-trailer train to bring luggage to and from the airplanes. A new mechanized luggage system will be installed in 3 years, so the tractor will not be needed after that. However, because it will receive heavy use, so that the running and maintenance costs will increase rapidly as the tractor ages, it may still be more economical to replace the tractor after 1 or 2 years. The following table gives the total net discounted cost associated with purchasing a tractor (purchase price minus trade-in allowance, plus running and maintenance costs) at the end of year *i* and trading it in at the end of year *j* (where year 0 is now).

		j		
		1	2	3
	0	8	18	31
i	1		10	21
	2			12

The problem is to determine at what times (if any) the tractor should be replaced to minimize the total cost for the tractors over 3 years.

[1] Problems 9.8-5 to 9.8-12, 9.8-16, 9.8-17, and 9.8-21 have been adapted, with permission, from previous operations research examinations given by the Society of Actuaries.

(*a*) Formulate this problem as a shortest-path problem.
(*b*) Use the algorithm described in Sec. 9.3 to solve this shortest-path problem.

9.3-2.* Use the algorithm described in Sec. 9.3 to find the shortest path through networks *a* and *b*, where the numbers represent actual distances between the corresponding nodes.

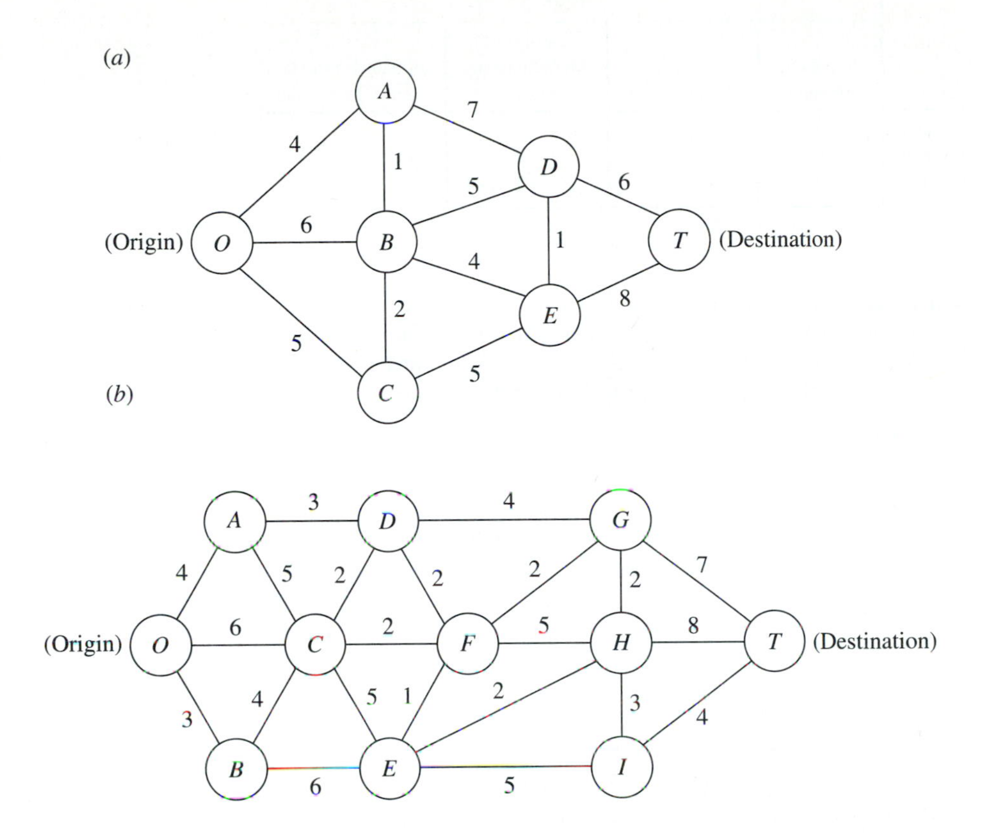

9.3-3. Formulate the shortest-path problem as a linear programming problem.

9.3-4. A company has learned that a competitor is planning to come out with a new kind of product with a great sales potential. This company has been working on a similar product, and research is nearly complete. It now wishes to rush the product out to meet the competition. There are four nonoverlapping phases left to be accomplished, including the remaining research that currently is being conducted at a normal pace. However, each phase can instead be conducted at a priority or crash level to expedite completion. The times required (in months) at these levels are as follows:

	Time			
Level	Remaining Research	Development	Design of Manufacturing System	Initiate Production and Distribution
Normal	5			
Priority	4	3	5	2
Crash	2	2	3	1

For these four phases $30,000,000 is available. The cost (in millions of dollars) at the different levels is as follows:

	Cost			
Level	Remaining Research	Development	Design of Manufacturing System	Initiate Production and Distribution
Normal	3			
Priority	6	6	9	3
Crash	9	9	12	6

Management wishes to determine at which level to conduct each of the four phases to minimize the total time until the product can be marketed subject to the budget restriction.

(*a*) Formulate this problem as a shortest-path problem.

(*b*) Use the algorithm described in Sec. 9.3 to solve this shortest-path problem.

9.4-1.* Reconsider the networks shown in Prob. 9.3-2. Use the algorithm described in Sec. 9.4 to find the minimum spanning tree for each of these networks.

9.4-2. A logging company will soon begin logging eight groves of trees in the same general area. Therefore, it must develop a system of dirt roads that makes each grove accessible from every other grove. The distance (in miles) between every pair of groves is as follows:

		Distance between Pairs of Groves							
		1	2	3	4	5	6	7	8
	1	—	1.3	2.1	0.9	0.7	1.8	2.0	1.5
	2	1.3	—	0.9	1.8	1.2	2.6	2.3	1.1
	3	2.1	0.9	—	2.6	1.7	2.5	1.9	1.0
Grove	4	0.9	1.8	2.6	—	0.7	1.6	1.5	0.9
	5	0.7	1.2	1.7	0.7	—	0.9	1.1	0.8
	6	1.8	2.6	2.5	1.6	0.9	—	0.6	1.0
	7	2.0	2.3	1.9	1.5	1.1	0.6	—	0.5
	8	1.5	1.1	1.0	0.9	0.8	1.0	0.5	—

The problem is to determine between which pairs of groves the roads should be constructed to connect all groves with a minimum total length of road.

(*a*) Describe how this problem fits the network description of the minimum spanning tree problem.

(*b*) Use the algorithm described in Sec. 9.4 to solve the problem.

9.4-3. A bank soon will be hooking up computer terminals at each of its branch offices to the computer at its main office, using special phone lines with telecommunications devices. The phone line from a branch office need not be connected directly to the main office. It can be connected indirectly by being connected to another branch office that is connected (directly or indirectly) to the main office. The only requirement is that every branch office be connected by some route to the main office.

The charge for the special phone lines is directly proportional to the mileage involved, where the distance (in miles) between every pair of offices is as follows:

	Distance between Pairs of Offices					
	Main	B.1	B.2	B.3	B.4	B.5
Main office	—	190	70	115	270	160
Branch 1	190	—	100	110	215	50
Branch 2	70	100	—	140	120	220
Branch 3	115	110	140	—	175	80
Branch 4	270	215	120	175	—	310
Branch 5	160	50	220	80	310	—

Management wishes to determine which pairs of offices should be directly connected by special phone lines in order to connect every branch office (directly or indirectly) to the main office at a minimum total cost.

(*a*) Describe how this problem fits the network description of the minimum spanning tree problem.

(*b*) Use the algorithm described in Sec. 9.4 to solve the problem.

9.5-1.* For networks (*a*) and (*b*), use the augmenting path algorithm described in Sec. 9.5 to find the flow pattern giving the *maximum flow* from the supply node (the leftmost node) to the demand node (the rightmost node), given that the arc capacity from node i to node j is the number nearest node i along the arc between these nodes.

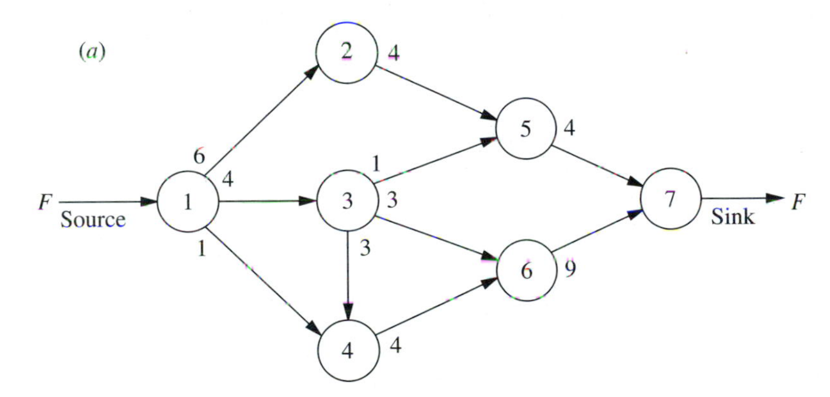

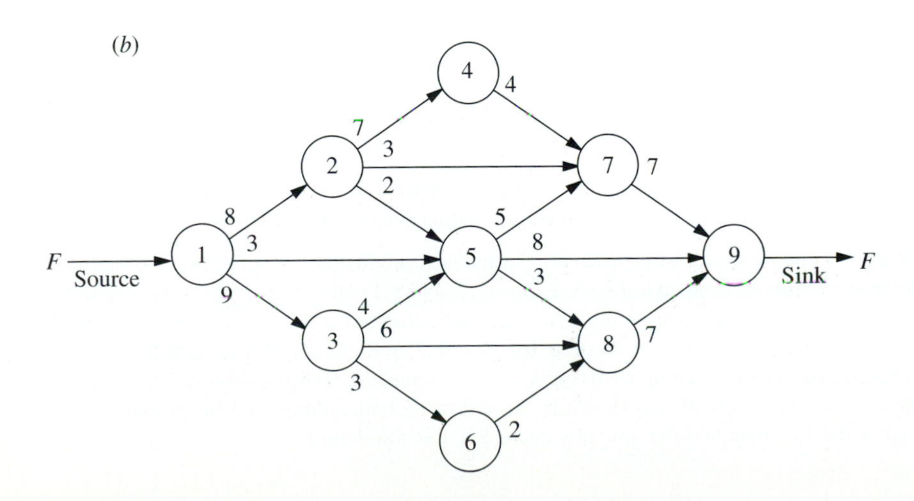

9.5-2. Formulate the general maximum flow problem as a linear programming problem.

9.5-3. One track of the Eura Railroad system runs from the major industrial city of Faireparc to the major port city of Portstown. This track is heavily used by both express passenger and freight trains. The passenger trains are carefully scheduled and have priority over the slow freight trains (this is a European railroad), so that the freight trains must pull over onto a siding whenever a passenger train is scheduled to pass them soon. It is now necessary to increase the freight service, so the problem is to schedule the freight trains so as to maximize the number that can be sent each day without interfering with the fixed schedule for passenger trains.

Consecutive freight trains must maintain a schedule differential of at least 0.1 hour, and this is the time unit used for scheduling them (so that the daily schedule indicates the status of each freight train at times 0.0, 0.1, 0.2, . . . , 23.9). There are S sidings between Faireparc and Portstown, where siding i is long enough to hold n_i freight trains ($i = 1, \ldots, S$). It requires t_i time units (rounded up to an integer) for a freight train to travel from siding i to siding $i + 1$ (where t_0 is the time from the Faireparc station to siding 1 and t_s is the time from siding S to the Portstown station). A freight train is allowed to pass or leave siding i ($i = 0, 1, \ldots, S$) at time j ($j = 0.0, 0.1, \ldots, 23.9$) only if it would not be overtaken by a scheduled passenger train before reaching siding $i + 1$ (let $\delta_{ij} = 1$ if it would not be overtaken, and let $\delta_{ij} = 0$ if it would be). A freight train also is required to stop at a siding if there will not be room for it at all subsequent sidings that it would reach before being overtaken by a passenger train.

Formulate this problem as a maximum flow problem by identifying every node (including the supply node and the demand node) as well as every arc and its arc capacity for the network representation of the problem. (*Hint:* Use a different set of nodes for each of the 240 times.)

9.5-4. Consider the maximum flow problem shown below, where the supply node is node A, the demand node is node F, and the arc capacities are the numbers shown next to these directed arcs. Use the augmenting path algorithm described in Sec. 9.5 to solve this problem.

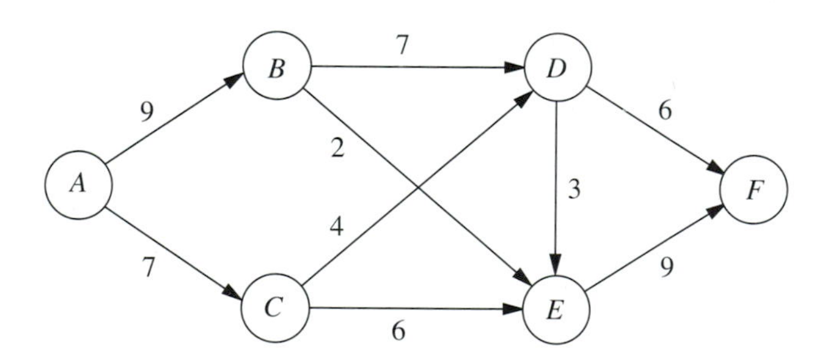

9.6-1. Reconsider the maximum flow problem shown in Prob. 9.5-4. Formulate this problem as a minimum cost flow problem, including adding the arc $A \to F$. Use $\overline{F} = 20$.

9.6-2. A company will be producing the same new product at two different factories, and then the product must be shipped to two warehouses. Factory 1 can send an unlimited amount by rail to warehouse 1 only, whereas factory 2 can send an unlimited amount by rail to warehouse 2 only. However, independent truckers can be used to ship up to 50 units from each factory to a distribution center, from which up to 50 units can be shipped to each warehouse. The shipping cost per unit for each alternative is shown in the following table, along with the amounts to be produced at the factories and the amounts needed at the warehouses.

From \ To	Unit Shipping Cost			Output
	Distribution Center	Warehouse 1	Warehouse 2	
Factory 1	3	7	—	80
Factory 2	4	—	9	70
Distribution center		2	4	
Allocation		60	90	

(*a*) Formulate the network representation of this problem as a minimum cost flow problem.

(*b*) Formulate the linear programming model for this problem.

9.6-3. Reconsider Prob. 9.3-1. Now formulate this problem as a minimum cost flow problem by showing the appropriate network representation.

D **9.7-1.** Without using your OR Courseware, consider the minimum cost flow problem shown below, where the b_i values (net flows generated) are given by the nodes, the c_{ij} values (costs per unit flow) are given by the arcs, and the u_{ij} values (arc capacities) are given to the right of the network.

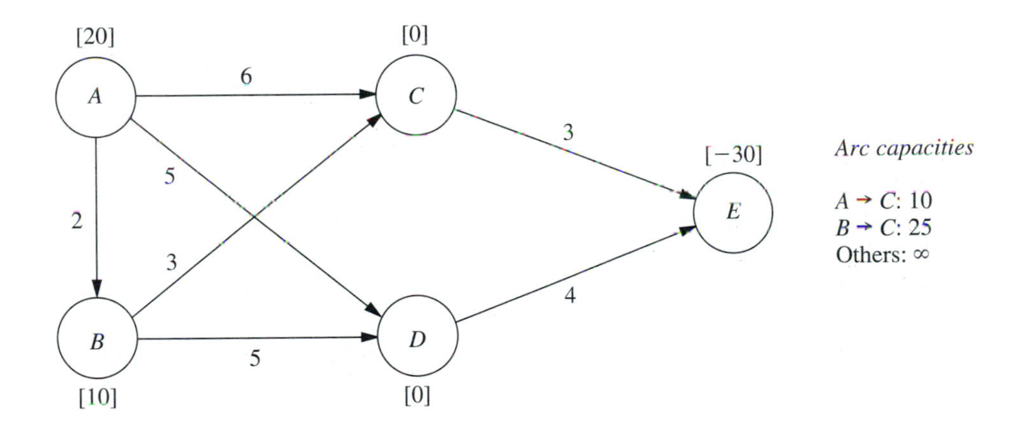

(*a*) Obtain an initial BF solution by solving the feasible spanning tree with basic arcs $A \rightarrow B$, $C \rightarrow E$, $D \rightarrow E$, and $C \rightarrow A$ (a reverse arc), where one of the nonbasic arcs ($C \rightarrow B$) also is a reverse arc. Show the resulting network (including b_i, c_{ij}, and u_{ij}) in the same format as the above one (except use dashed lines to draw the nonbasic arcs), and add the flows in parentheses next to the basic arcs.

(*b*) Use the optimality test to verify that this initial BF solution is optimal and that there are multiple optimal solutions. Apply one iteration of the network simplex method to find the other optimal BF solution, and then use these results to identify the other optimal solutions that are not BF solutions.

(*c*) Now consider the following BF solution.

Basic Arc	Flow	Nonbasic Arc
$A \to D$	20	$A \to B$
$B \to C$	10	$A \to C$
$C \to E$	10	$B \to D$
$D \to E$	20	

Starting from this BF solution, apply *one* iteration of the network simplex method. Identify the entering basic arc, the leaving basic arc, and the next BF solution, but do not proceed further.

9.7-2. Reconsider the minimum cost flow problem formulated in Prob. 9.6-1.

(*a*) Obtain an initial BF solution by solving the feasible spanning tree with basic arcs $A \to B$, $A \to C$, $A \to F$, $B \to D$, and $E \to F$, where two of the nonbasic arcs ($E \to C$ and $F \to D$) are *reverse* arcs.

D, I*, A (*b*) Use the network simplex method to solve this problem.

9.7-3. Reconsider the minimum cost flow problem formulated in Prob. 9.6-2.

(*a*) Obtain an initial BF solution by solving the feasible spanning tree that corresponds to using just the two rail lines plus factory 1 shipping to warehouse 2 via the distribution center.

D, I*, A (*b*) Use the network simplex method to solve this problem.

D, I*, A **9.7-4.** Reconsider the minimum cost flow problem formulated in Prob. 9.6-3. Starting with the initial BF solution that corresponds to replacing the tractor every year, use the network simplex method to solve this problem.

D, I*, A **9.7-5.** For the P & T Co. transportation problem given in Table 8.2, consider its network representation as a minimum cost flow problem presented in Fig. 8.2. Use the northwest corner rule to obtain an initial BF solution from Table 8.2. Then use the network simplex method to solve this problem (and verify the optimal solution given in Sec. 8.1).

9.7-6. Consider the Metro Water District transportation problem presented in Table 8.12.

(*a*) Formulate the network representation of this problem as a minimum cost flow problem. (*Hint:* Arcs where flow is prohibited should be deleted.)

D, I*, A (*b*) Starting with the initial BF solution given in Table 8.19, use the network simplex method to solve this problem. Compare the sequence of BF solutions obtained with the sequence obtained by the transportation simplex method in Table 8.23.

D, I*, A **9.7-7.** Consider the transportation problem having the following cost and requirements table:

		Destination			
		1	2	3	Supply
Source	1	6	7	4	40
	2	5	8	6	60
Demand		30	40	30	

Formulate the network representation of this problem as a minimum cost flow problem. Use the northwest corner rule to obtain an initial BF solution. Then use the network simplex method to solve the problem.

D, I*, A **9.7-8.** Consider the minimum cost flow problem shown below, where the b_i values are given by the nodes, the c_{ij} values are given by the arcs, and the *finite* u_{ij} values are given in parentheses by the arcs. Obtain an initial BF solution by solving the feasible spanning tree with basic arcs $A \to C$, $B \to A$, $C \to D$, and $C \to E$, where one of the nonbasic arcs $(D \to A)$ is a reverse arc. Then use the network simplex method to solve this problem.

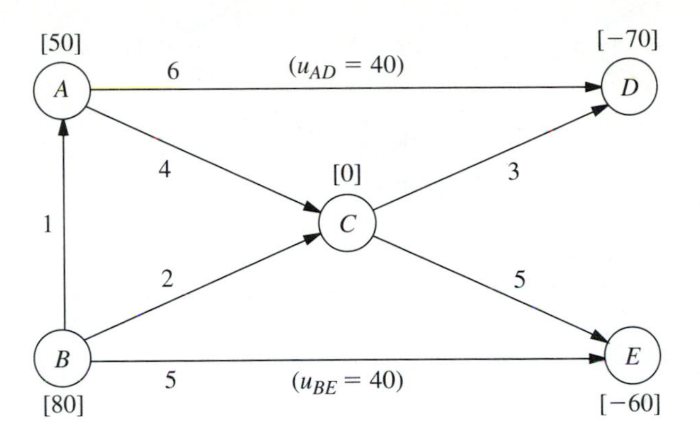

9.8-1.* Consider the following project network. Assume that the time required (in weeks) for each activity is a predictable constant and that it is given by the number along the corresponding arc. Find the earliest time, latest time, and slack for each event as well as the slack for each activity. Also identify the critical path.

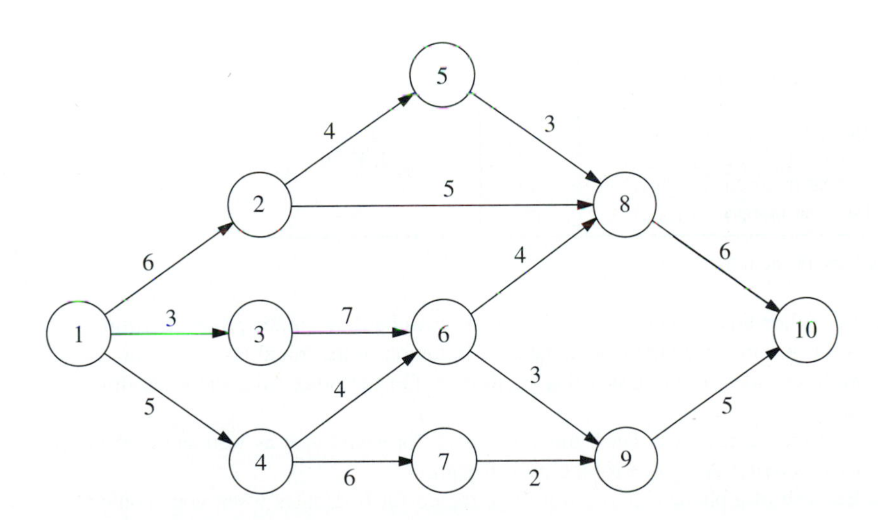

9.8-2. Consider the following project network. Assume that the time required (in days) for each activity is a predictable constant and that it is given by the number along the corresponding arc. Find the earliest time, latest time, and slack for each event as well as the slack for each activity. Also identify the critical path.

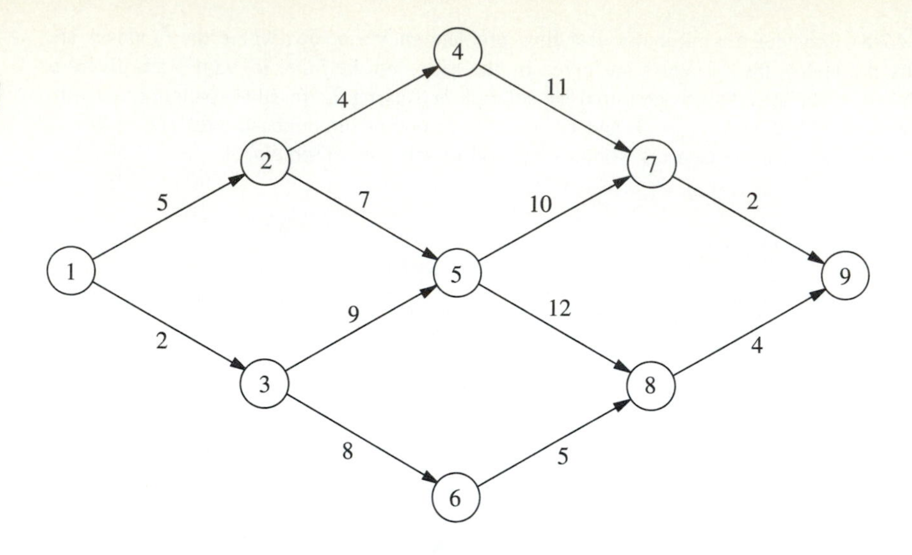

9.8-3. You and several friends are about to prepare a lasagna dinner. The tasks to be performed, their times (in minutes), and the precedence constraints are as follows:

Task	Task Description	Time	Tasks that Must Precede
A	Buy the mozzarella cheese*	30	
B	Slice the mozzarella	5	A
C	Beat 2 eggs	2	
D	Mix eggs and ricotta cheese	3	C
E	Cut up onions and mushrooms	7	
F	Cook the tomato sauce	25	E
G	Boil large quantity of water	15	
H	Boil the lasagna noodles	10	G
I	Drain the lasagna noodles	2	H
J	Assemble all the ingredients	10	I, F, D, B
K	Preheat the oven	15	
L	Bake the lasagna	30	J, K

* There is none in the refrigerator.

(*a*) Formulate this problem as a PERT-type system by drawing the project network. Use one event to represent the simultaneous initiation of the initial tasks. On one side of each arc, identify the task being performed. On the other side, show the times required.

(*b*) Find the earliest time, latest time, and slack for each event as well as the slack for each activity. Also identify the critical path.

(*c*) Because of a phone call you were interrupted for 6 minutes when you should have been cutting the onions and mushrooms. By how much will the dinner be delayed? If you use your food processor, which reduces the cutting time from 7 to 2 minutes, will the dinner still be delayed?

9.8-4. What is the relationship between the critical path for a project network and the longest path (in total time) through the network?

9.8-5. Consider the following project network, where the estimated activity times are given next to the corresponding arcs and (4, 5) is a dummy activity. Find the earliest time, latest time, and slack for each event as well as the slack for each activity. Also identify the critical path.

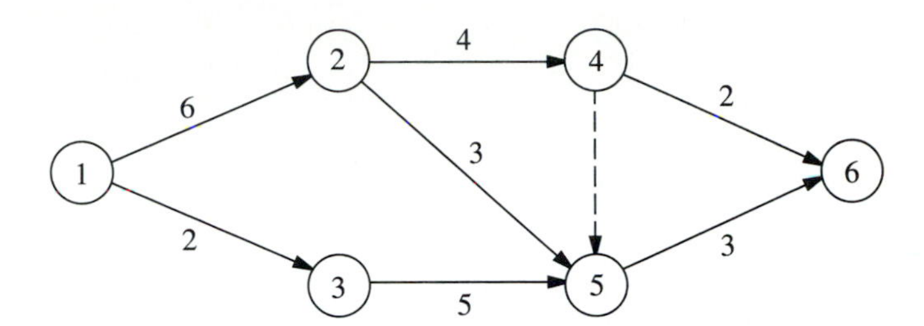

9.8-6. A project consists of 12 activities, represented by the project network below, where the number by each arc represents the duration (in days) of the associated activity.

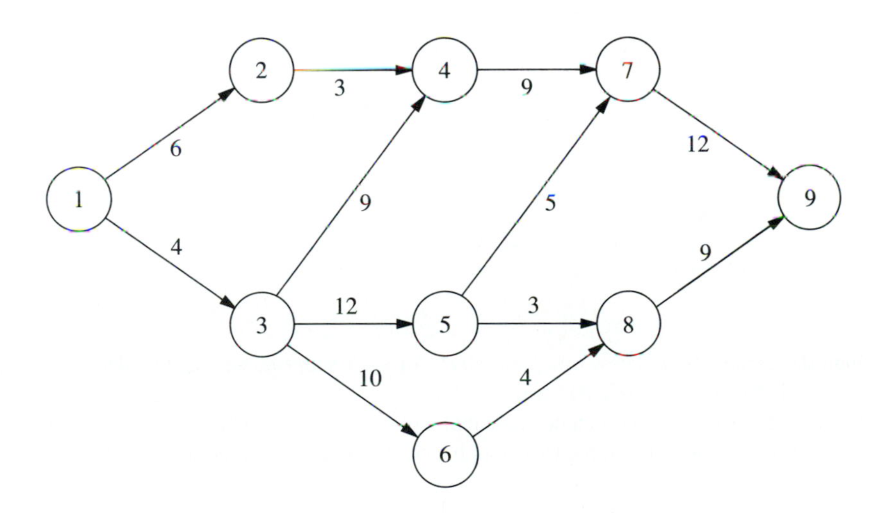

(*a*) Find the earliest time, latest time, and slack for each event as well as the slack for each activity. Also identify the critical path.

(*b*) The project has been scheduled to be completed in the minimum possible time. If all previous activities have begun as early as possible and if activity (3, 5) has just been completed, how many days can the start of activity (5, 8) be delayed without delaying the completion of the project?

9.8-7. Consider a project whose activities and required times are as given:

Activity	Required Time (Minutes)
(1, 2)	5
(2, 3)	5
(3, 4)	5
(4, 5)	5
(5, 9)	5
(2, 6)	6
(6, 7)	2
(7, 8)	5
(8, 9)	6
(4, 7)	0

(*a*) Construct the project network for this project.

(*b*) Find the earliest time, latest time, and slack for each event as well as the slack for each activity. Also identify the critical path.

(*c*) If the precedence relationship implied by dummy activity (4, 7) could be removed, by how much would the project completion time be reduced?

9.8-8. Consider the following project network, where the estimated activity times are given next to the corresponding arcs and (5, 6) is a dummy activity.

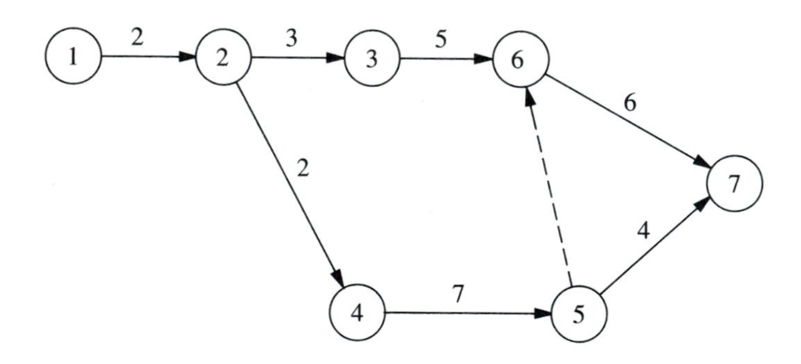

(*a*) Find the earliest time, latest time, and slack for each event as well as the slack for each activity. Also identify the critical path.

(*b*) Suppose that arrangements can be made that would enable one to begin activity (6, 7) before finishing activity (4, 5). How much would this reduce the project completion time?

9.8-9. You are given the following information about a project consisting of six activities:

Activity	Preceding Activities	Estimated Required Time (Minutes)
A	None	5
B	None	1
C	B	2
D	A, C	4
E	A	6
F	D, E	3

(*a*) Construct the project network for this project.
(*b*) Find the earliest time, latest time, and slack for each event as well as the slack for each activity. Also identify the critical path.
(*c*) If all other activities take the estimated amount of time, what is the maximum amount of time that activity D can take without delaying the completion of the project?

9.8-10. Consider a project network with six events (nodes), where event 1 is project start and event 6 is project completion. Based on the estimated times for the various activities (arcs), you are given the following information about the shortest and longest paths (in total time) through this network.

Paths	Shortest	Longest
From event 1 to event 6	5	9
From event 3 to event 6	3	6
From event 1 to event 6 through event 4	8	8

(*a*) Assuming that the estimated activity times do occur, what will be the time required to complete the project? Why?
(*b*) What is the latest time for event 3? Explain your reasoning.
(*c*) What is the slack for event 4? Explain your reasoning.

9.8-11. Consider the following project network, where (3, 5) is a dummy activity. The times for activities (2, 3) and (3, 6) have not yet been estimated, so they are represented here by the unknowns x and y, respectively. Determine the values of x and y for which $1 \rightarrow 2 \rightarrow 3 \rightarrow 6$ is a critical path.

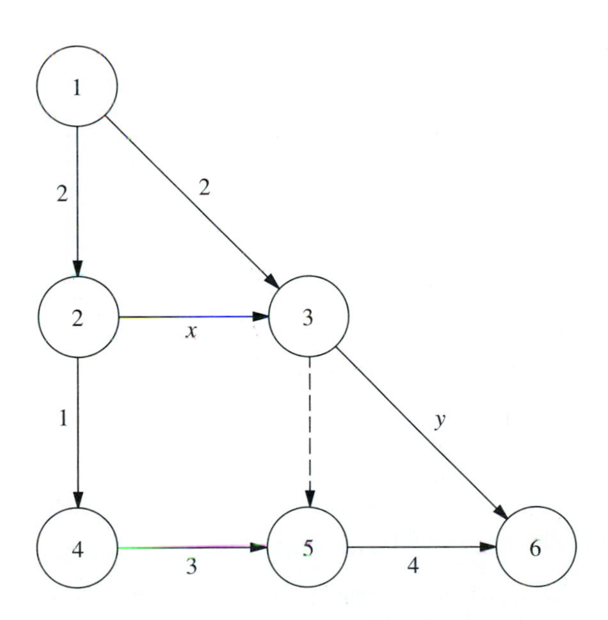

9.8-12. Label each of the following statements about project networks as true or false, and then justify your answer.

(*a*) For each event, an estimate needs to be provided of the *event time*, i.e., the duration of the event.
(*b*) The slack for an event j is at least equal to the maximum of the slacks of all the activities that end at j.
(*c*) In constructing a project network, dummy activities sometimes are used to avoid having two activities with identical beginning and ending nodes.

9.8-13.* By using the PERT three-estimate approach, the three estimates for one of the activities are as follows: optimistic estimate = 30 days, most likely estimate = 36 days, pessimistic estimate = 48 days. What are the resulting estimates of the expected value and variance of the time required by the activity?

9.8-14. Consider the following project network. The PERT three-estimate approach has been used, and it has led to the following estimates of the expected value (in months) and variance of the time required for the respective activities:

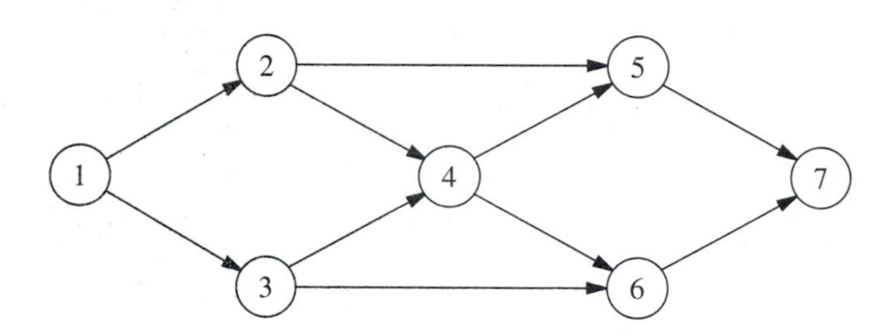

Activity	*Activity Times* Estimated Expected Value	Estimated Variance
1 → 2	4	5
1 → 3	6	10
2 → 4	4	8
2 → 5	8	12
3 → 4	3	6
3 → 6	7	14
4 → 5	5	12
4 → 6	3	5
5 → 7	5	8
6 → 7	5	7

The scheduled project completion time is 22 months after the start of the project.
(*a*) Using expected values, determine the critical path for the project.
(*b*) Using the procedure described in Sec. 9.8, find the approximate probability that the project will be completed by the scheduled time.
(*c*) In addition to the critical path, there are five other paths through the network. For each of these other paths, find the approximate probability that the sum of the activity times along the path is not more than 22 months.

9.8-15. Consider the following project network. By using the PERT three-estimate approach, suppose that the usual three estimates for the time required (in weeks) for each of these activities are as follows:

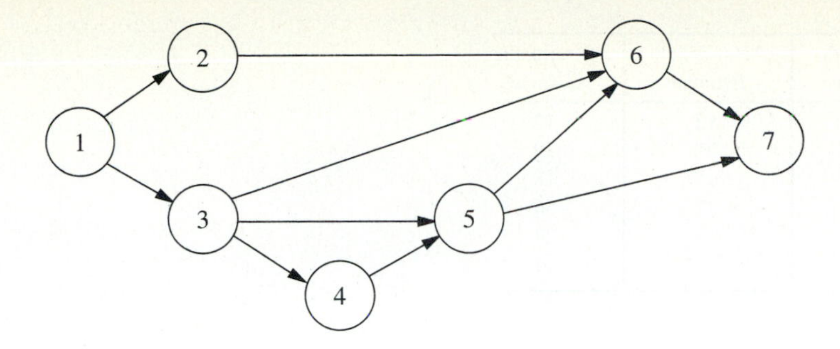

Activity	Optimistic Estimate	Most Likely Estimate	Pessimistic Estimate
$1 \rightarrow 2$	28	32	36
$1 \rightarrow 3$	22	28	32
$2 \rightarrow 6$	26	36	46
$3 \rightarrow 4$	14	16	18
$3 \rightarrow 5$	32	32	32
$3 \rightarrow 6$	40	52	74
$4 \rightarrow 5$	12	16	24
$5 \rightarrow 6$	16	20	26
$5 \rightarrow 7$	26	34	42
$6 \rightarrow 7$	12	16	30

The project is ready to start now, and the deadline for completing the project is 100 weeks hence.

(*a*) On the basis of the estimates just listed, calculate the expected value and standard deviation of the time required for each activity.

(*b*) Using expected times, determine the critical path for the project.

(*c*) Using the procedure described in Sec. 9.8, find the approximate probability that the project will be completed by the deadline.

9.8-16. A project with a deadline for completion 57 days hence has the following project network.

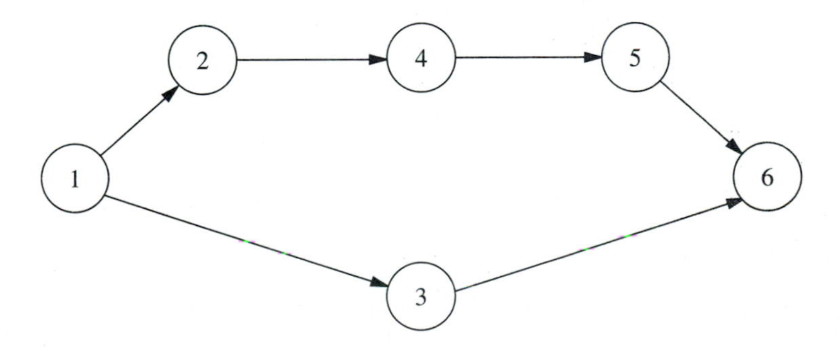

By using the PERT three-estimate approach, the following estimates of activity times (in days) have been obtained.

Activity	Optimistic Estimate	Most Likely Estimate	Pessimistic Estimate
$1 \rightarrow 2$	12	12	12
$1 \rightarrow 3$	15	21	39
$2 \rightarrow 4$	12	15	18
$3 \rightarrow 6$	18	27	36
$4 \rightarrow 5$	12	18	24
$5 \rightarrow 6$	2	5	14

(*a*) Calculate the expected value and standard deviation of each activity time.
(*b*) Using expected times, determine the critical path for the project.
(*c*) Using the procedure described in Sec. 9.8, find the approximate probability that the project will be completed by the deadline.
(*d*) If all the activity times turn out to be equal to their pessimistic estimates, which path will be the resulting critical path?
(*e*) Using the results from part (*a*), what is the approximate probability that the path identified in part (*d*) will be completed by the project deadline?
(*f*) Combine the results from parts (*c*) and (*e*) to develop a better estimate of the probability that the project will be completed by the deadline.

9.8-17. Label each of the following statements about the PERT three-estimate approach as true or false, and then justify your answer by referring to specific statements (with page citations) in the chapter.
(*a*) Activity times are assumed to be no larger than the optimistic estimate and no smaller than the pessimistic estimate.
(*b*) Activity times are assumed to have a normal distribution.
(*c*) The critical path (in terms of expected times) is assumed to always require the minimum elapsed time of any path through the project network.

9.8-18. A project which must be completed in 12 months has the following project network.

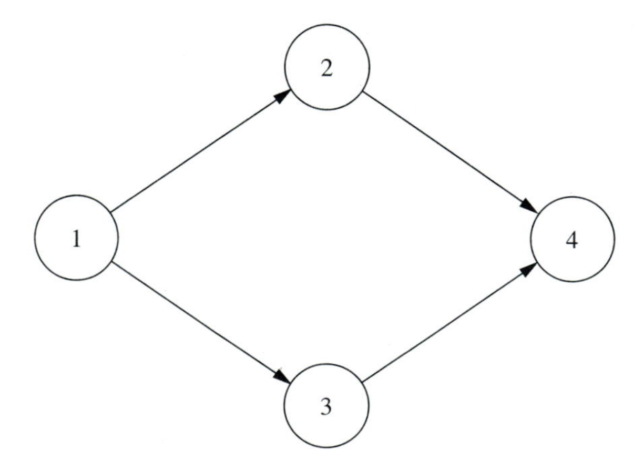

Using the CPM method of time-cost trade-offs yields the following data (in months and dollars).

Activity	Normal Time	Crash Time	Normal Cost	Crash Cost
$1 \rightarrow 2$	8	5	25,000	40,000
$2 \rightarrow 4$	6	4	16,000	24,000
$1 \rightarrow 3$	9	7	20,000	30,000
$3 \rightarrow 4$	7	4	27,000	45,000

(*a*) Consider the path $1 \rightarrow 2 \rightarrow 4$. Formulate a two-variable linear programming model for the problem of how to minimize the cost of performing this sequence of two activities within 12 months. Solve this model graphically.

(*b*) Repeat part (*a*) for the path $1 \rightarrow 3 \rightarrow 4$.

(*c*) Now use the CPM linear programming formulation to formulate a complete model for the problem of how to minimize the cost of completing the project within 12 months.

A* (*d*) Solve this problem automatically with the CPM routine in your OR Courseware.

A* (*e*) Check the effect of changing the deadline by repeating part (*d*) with a deadline of 11 months and then with a deadline of 13 months.

9.8-19. Consider the following project network. The CPM method of time-cost trade-offs is to be used to minimize the cost of completing the project within 15 weeks. The relevant data (in weeks and thousands of dollars) are as follows:

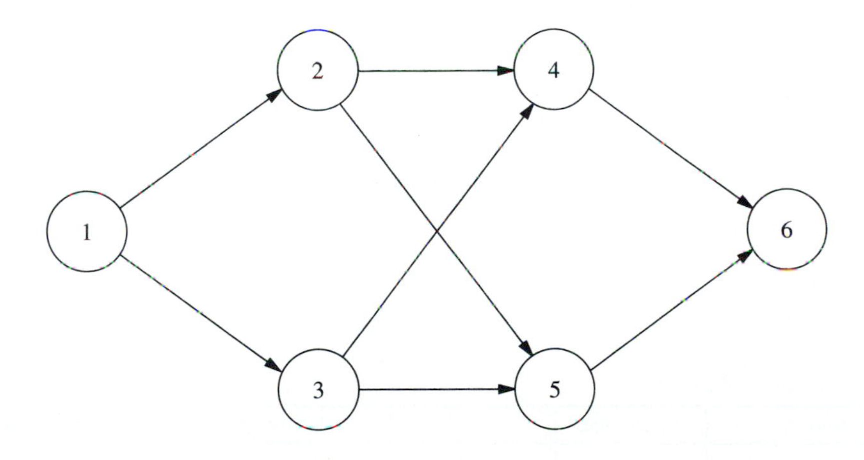

Activity	Normal Time	Crash Time	Normal Cost	Crash Cost
$1 \rightarrow 2$	5	3	20	30
$1 \rightarrow 3$	3	2	10	20
$2 \rightarrow 4$	4	2	16	24
$2 \rightarrow 5$	6	3	25	43
$3 \rightarrow 4$	5	4	22	30
$3 \rightarrow 5$	7	4	30	48
$4 \rightarrow 6$	9	5	25	45
$5 \rightarrow 6$	8	6	30	44

(*a*) Formulate a linear programming model for this problem.

A* (*b*) Solve this problem automatically with the CPM routine in your OR Courseware.

9.8-20. Reconsider the project network shown in Prob. 9.8-15. The CPM method of time-cost trade-offs is to be used to determine how to meet the project deadline (100 weeks hence) in the most economical way. The crash time and normal time for each of the activities correspond to the times shown in the *optimistic estimate* and *pessimistic estimate* columns of the table for Prob. 9.8-15 (except that activity 3 → 5 has a crash time of 28 and a normal time of 36) and that the difference between the crash cost and normal cost is 10 (in units of thousands of dollars) for every activity.

(*a*) Formulate a linear programming model for this problem.

A* (*b*) Solve this problem automatically with the CPM routine in your OR Courseware.

9.8-21. Consider the following project network, where (2, 5) is a dummy activity. The deadline for completion of this project is 13 weeks hence. The CPM method of time-cost trade-offs is to be used to determine how to meet this deadline in the most economical manner. The relevant data (in weeks and dollars) are as follows:

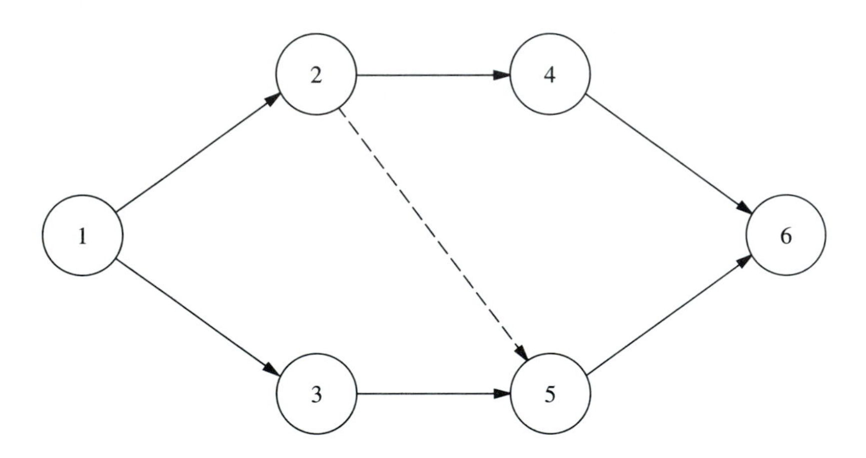

Activity	Normal Time	Crash Time	Normal Cost	Crash Cost
1 → 2	6	4	4,200	6,200
1 → 3	2	1	1,800	3,300
2 → 4	4	2	5,400	8,000
3 → 5	5	3	3,600	5,800
4 → 6	7	4	5,900	9,500
5 → 6	9	6	6,300	8,700

(*a*) Formulate a linear programming model for this problem.

A* (*b*) Use the simplex method to solve this model.

A* (*c*) Reformulate this model to enable one to replace lower-bound constraints on individual variables by nonnegativity constraints. Then solve by the simplex method.

A* (*d*) Solve this problem automatically with the CPM routine in your OR Courseware.

9.8-22. Suppose that the scheduled completion time for the house construction project described in Figs. 9.25, 9.26, and 9.27 has been moved forward to 40. Therefore, the CPM method of time-cost trade-offs is to be used to determine how to accelerate the project to meet this deadline in the most economical way. The relevant data are as follows:

Activity	Normal Time	Crash Time	Normal Cost	Crash Cost
$1 \rightarrow 2$	2	1	1,800	2,300
$2 \rightarrow 3$	4	2	3,200	3,600
$3 \rightarrow 4$	10	7	6,200	7,300
$4 \rightarrow 5$	4	3	4,100	4,900
$4 \rightarrow 6$	6	4	2,600	3,000
$4 \rightarrow 7$	7	5	2,100	2,400
$5 \rightarrow 7$	5	3	1,800	2,200
$6 \rightarrow 8$	7	4	9,000	9,600
$7 \rightarrow 9$	8	6	4,300	4,600
$8 \rightarrow 10$	9	6	2,000	2,500
$9 \rightarrow 11$	4	3	1,600	1,800
$9 \rightarrow 12$	5	3	2,500	3,000
$10 \rightarrow 13$	2	1	1,000	1,500
$12 \rightarrow 13$	6	3	3,300	4,000

(*a*) Formulate a linear programming model for this problem.

A* (*b*) Solve this problem automatically with the CPM routine in your OR Courseware.

10

Dynamic Programming

Dynamic programming is a useful mathematical technique for making a sequence of interrelated decisions. It provides a systematic procedure for determining the optimal combination of decisions.

In contrast to linear programming, there does not exist a standard mathematical formulation of ''the'' dynamic programming problem. Rather, dynamic programming is a general type of approach to problem solving, and the particular equations used must be developed to fit each situation. Therefore, a certain degree of ingenuity and insight into the general structure of dynamic programming problems is required to recognize when and how a problem can be solved by dynamic programming procedures. These abilities can best be developed by an exposure to a wide variety of dynamic programming applications and a study of the characteristics that are common to all these situations. A large number of illustrative examples are presented for this purpose.

10.1 A Prototype Example for Dynamic Programming

Example 1—The Stagecoach Problem

The STAGECOACH PROBLEM is a problem specially constructed[1] to illustrate the features and to introduce the terminology of dynamic programming. It concerns a mythical fortune seeker in Missouri who decided to go west to join the gold rush in California during the mid-19th century. The journey would require traveling by stagecoach through unsettled country where there was serious danger of attack by marauders. Although his starting point and destination were fixed, he had considerable choice as to which states (or territories that subsequently became states) to travel through en route. The possible routes are shown in Fig. 10.1, where each state is represented by a circled letter and the direction of travel is always from left to right in the diagram. Thus four stages (stagecoach runs) were required to travel from his point of embarkation in state A (Missouri) to his destination in state J (California).

This fortune seeker was a prudent man who was quite concerned about his safety. After some thought, he came up with a rather clever way of determining the safest route. Life insurance policies were offered to stagecoach passengers. Because the cost of the **policy** for taking any given stagecoach run was based on a careful evaluation of the safety of that run, the safest route should be the one with the cheapest total life insurance policy.

The cost for the standard policy on the stagecoach run from state i to state j, which will be denoted by c_{ij}, is

	B	C	D
A	2	4	3

	E	F	G
B	7	4	6
C	3	2	4
D	4	1	5

	H	I
E	1	4
F	6	3
G	3	3

	J
H	3
I	4

These costs are also shown in Fig. 10.1.

We shall now focus on the question of which route minimizes the total cost of the policy.

Solving the Problem

First note that the shortsighted approach of selecting the cheapest run offered by each successive stage need not yield an overall optimal decision. Following this strategy would give the route $A \rightarrow B \rightarrow F \rightarrow I \rightarrow J$, at a total cost of 13. However, sacrificing a little on one stage may permit greater savings thereafter. For example, $A \rightarrow D \rightarrow F$ is cheaper overall than $A \rightarrow B \rightarrow F$.

One possible approach to solving this problem is to use trial and error.[2] However, the number of possible routes is large (18), and having to calculate the total cost for each route is not an appealing task.

[1] This problem was developed by Professor Harvey M. Wagner while he was at Stanford University.

[2] This problem also can be formulated as a *shortest-path problem* (see Sec. 9.3), where *costs* here play the role of *distances* in the shortest-path problem. The algorithm presented in Sec. 9.3 actually uses the philosophy of dynamic programming. However, because the present problem has a fixed number of stages, the dynamic programming approach presented here is even better.

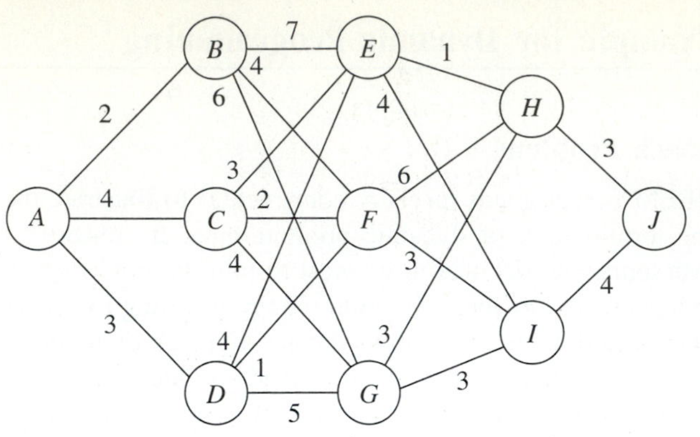

Figure 10.1 The road system and costs for the stagecoach problem.

Fortunately, dynamic programming provides a solution with much less effort than exhaustive enumeration. (The computational savings are enormous for larger versions of this problem.) Dynamic programming starts with a small portion of the original problem and finds the optimal solution for this smaller problem. It then gradually enlarges the problem, finding the current optimal solution from the preceding one, until the original problem is solved in its entirety.

For the stagecoach problem, we start with the smaller problem where the fortune seeker has nearly completed his journey and has only one more stage (stagecoach run) to go. The obvious optimal solution for this smaller problem is to go from his current state (whatever it is) to his ultimate destination (state J). At each subsequent iteration, the problem is enlarged by increasing by 1 the number of stages left to go to complete the journey. For this enlarged problem, the optimal solution for where to go next from each possible state can be found relatively easily from the results obtained at the preceding iteration. The details involved in implementing this approach follow.

FORMULATION: Let the decision variables x_n ($n = 1, 2, 3, 4$) be the immediate destination on stage n (the nth stagecoach run to be taken). Thus the route selected is $A \rightarrow x_1 \rightarrow x_2 \rightarrow x_3 \rightarrow x_4$, where $x_4 = J$.

Let $f_n(s, x_n)$ be the total cost of the best overall *policy* for the *remaining* stages, given that the fortune seeker is in state s, ready to start stage n, and selects x_n as the immediate destination. Given s and n, let x_n^* denote any value of x_n (not necessarily unique) that minimizes $f_n(s, x_n)$, and let $f_n^*(s)$ be the corresponding minimum value of $f_n(s, x_n)$. Thus

$$f_n^*(s) = \min_{x_n} f_n(s, x_n) = f_n(s, x_n^*),$$

where

$$\begin{aligned} f_n(s, x_n) &= \text{immediate cost (stage } n) + \text{minimum future cost (stages } n+1 \text{ onward)} \\ &= c_{sx_n} + f_{n+1}^*(x_n). \end{aligned}$$

The value of c_{sx_n} is given by the preceding tables for c_{ij} by setting $i = s$ (the current state) and $j = x_n$ (the immediate destination). Because the ultimate destination (state J) is reached at the end of stage 4, $f_5^*(J) = 0$.

The objective is to find $f_1^*(A)$ and the corresponding route. Dynamic programming finds it by successively finding $f_4^*(s), f_3^*(s), f_2^*(s)$, for each of the possible states s and then using $f_2^*(s)$ to solve for $f_1^*(A)$.†

Solution Procedure: When the fortune seeker has only one more stage to go ($n = 4$), his route thereafter is determined entirely by his current state s (either H or I) and his final destination $x_4 = J$, so the route for this final stagecoach run is $s \rightarrow J$. Therefore, since $f_4^*(s) = f_4(s, J) = c_{s,J}$, the immediate solution to the $n = 4$ problem is

$n = 4$:

s	$f_4^*(s)$	x_4^*
H	3	J
I	4	J

When the fortune seeker has two more stages to go ($n = 3$), the solution procedure requires a few calculations. For example, suppose that the fortune seeker is in state F. Then, as depicted below, he must next go to either state H or I at an immediate cost of $c_{F,H} = 6$ or $c_{F,I} = 3$, respectively. If he chooses state H, the minimum additional cost after he reaches there is given in the preceding table as $f_4^*(H) = 3$, as shown above the H node in the diagram. Therefore, the total cost for this decision is $6 + 3 = 9$. If he chooses state I instead, the total cost is $3 + 4 = 7$, which is smaller. Therefore, the optimal choice is this latter one, $x_3^* = I$, because it gives the minimum cost $f_3^*(F) = 7$.

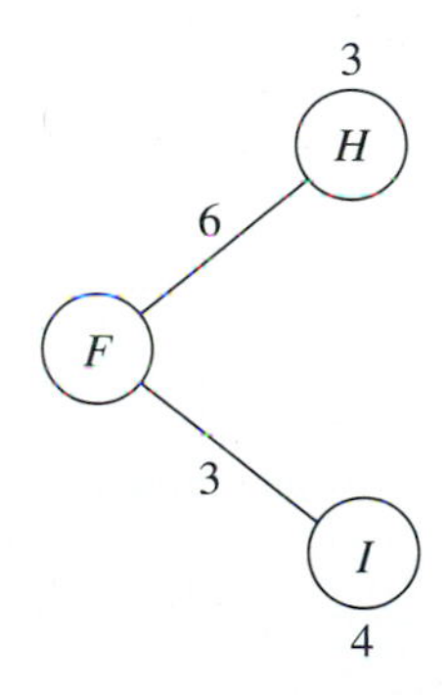

Similar calculations need to be made when you start from the other two possible states $s = E$ and $s = G$ with two stages to go. Try it, proceeding both graphically (Fig. 10.1) and algebraically [combining c_{ij} and $f_4^*(s)$ values], to verify the following complete results for the $n = 3$ problem.

$n = 3$:

x_3 / s	$f_3(s, x_3) = c_{sx_3} + f_4^*(x_3)$: H	I	$f_3^*(s)$	x_3^*
E	4	8	4	H
F	9	7	7	I
G	6	7	6	H

† Because this procedure involves moving *backward* stage by stage, some writers also count n backward to denote the number of *remaining stages* to the destination. We use the more natural *forward counting* for greater simplicity.

The solution for the second-stage problem ($n = 2$), where there are three stages to go, is obtained in a similar fashion. In this case, $f_2(s, x_2) = c_{sx_2} + f_3^*(x_2)$. For example, suppose that the fortune seeker is in state C, as depicted below.

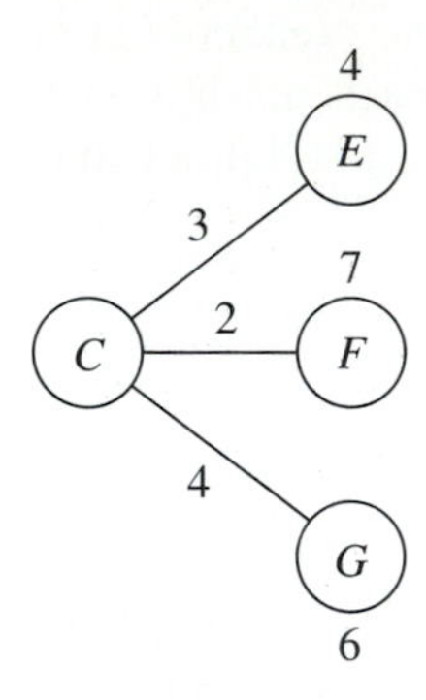

He must next go to state E, F, or G at an immediate cost of $c_{C,E} = 3$, $c_{C,F} = 2$, or $c_{C,G} = 4$, respectively. After getting there, the minimum additional cost for stage 3 to the end is given by the $n = 3$ table as $f_3^*(E) = 4$, $f_3^*(F) = 7$, or $f_3^*(G) = 6$, respectively, as shown above the E and F nodes and below the G node in the above diagram. The resulting calculations for the three alternatives are summarized below.

$$x_2 = E: \quad f_2(C, E) = c_{C,E} + f_3^*(E) = 3 + 4 = 7.$$

$$x_2 = F: \quad f_2(C, F) = c_{C,F} + f_3^*(F) = 2 + 7 = 9.$$

$$x_2 = G: \quad f_2(C, G) = c_{C,G} + f_3^*(G) = 4 + 6 = 10.$$

The minimum of these three numbers is 7, so the minimum total cost from state C to the end is $f_2^*(\text{C}) = 7$, and the immediate destination should be $x_2^* = E$.

Making similar calculations when you start from state B or D (try it) yields the following results for the $n = 2$ problem:

$n = 2$:

s \ x_2	$f_2(s, x_2) = c_{sx_2} + f_3^*(x_2)$			$f_2^*(s)$	x_2^*
	E	F	G		
B	11	11	12	11	E or F
C	7	9	10	7	E
D	8	8	11	8	E or F

In the first and third rows of this table, note that E and F tie as the minimizing value of x_2, so the immediate destination from either state B or D should be $x_2^* = E$ or F.

Moving to the first-stage problem ($n = 1$), with all four stages to go, we see that the calculations are similar to those just shown for the second-stage problem ($n = 2$), except now there is just *one* possible starting state $s = A$, as depicted below.

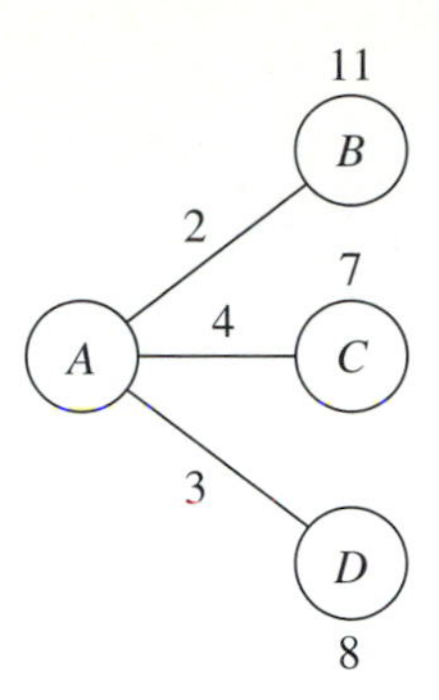

These calculations are summarized next for the three alternatives for the immediate destination:

$$x_1 = B: \quad f_1(A, B) = c_{A,B} + f_2^*(B) = 2 + 11 = 13.$$

$$x_1 = C: \quad f_1(A, C) = c_{A,C} + f_2^*(C) = 4 + 7 = 11.$$

$$x_1 = D: \quad f_1(A, D) = c_{A,D} + f_2^*(D) = 3 + 8 = 11.$$

Since 11 is the minimum, $f_1^*(A) = 11$ and $x_1^* = C$ or D, as shown in the following table

$n = 1$:

s \ x_1	$f_1(s, x_1) = c_{sx_1} + f_2^*(x_1)$				
	B	C	D	$f_1^*(s)$	x_1^*
A	13	11	11	11	C or D

An optimal solution for the entire problem can now be identified from the four tables. Results for the $n = 1$ problem indicate that the fortune seeker should go initially to either state C or state D. Suppose that he chooses $x_1^* = C$. For $n = 2$, the result for $s = C$ is $x_2^* = E$. This result leads to the $n = 3$ problem, which gives $x_3^* = H$ for $s =$ E, and the $n = 4$ problem yields $x_4^* = J$ for $s = H$. Hence one optimal route is $A \rightarrow C \rightarrow E \rightarrow H \rightarrow J$. Choosing $x_1^* = D$ leads to the other two optimal routes $A \rightarrow D \rightarrow E \rightarrow H \rightarrow J$ and $A \rightarrow D \rightarrow F \rightarrow I \rightarrow J$. They all yield a total cost of $f_1^*(A) = 11$.

These results of the dynamic programming analysis also are summarized in Fig. 10.2. Note how the two arrows for stage 1 come from the first and last columns of the $n = 1$ table and the resulting cost comes from the next-to-last column. Each of the other arrows (and the resulting cost) comes from one row in one of the other tables in just the same way.

You will see in the next section that the special terms describing the particular context of this problem—*stage, state,* and *policy*—actually are part of the general terminology of dynamic programming with an analogous interpretation in other contexts.

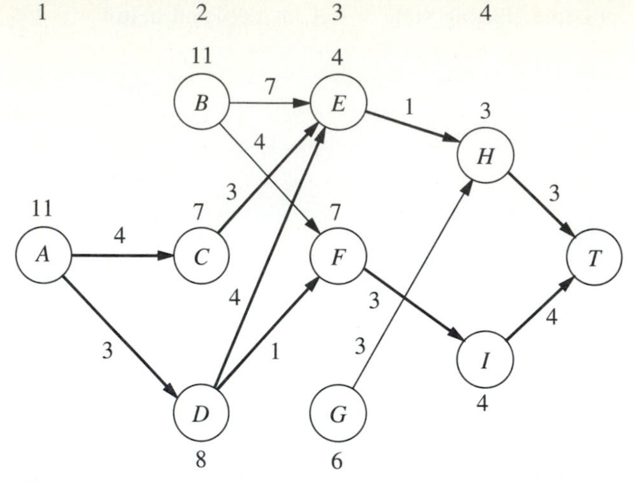

Figure 10.2 Graphical display of the dynamic programming solution of the stagecoach problem. Each arrow shows an optimal policy decision (the best immediate destination) from that state, where the number by the state is the resulting cost from there to the end. Following the boldface arrows from A to T gives the three optimal solutions (the three routes giving the minimum total cost of 11).

10.2 Characteristics of Dynamic Programming Problems

The stagecoach problem is a literal prototype of dynamic programming problems. In fact, this example was purposely designed to provide a literal physical interpretation of the rather abstract structure of such problems. Therefore, one way to recognize a situation that can be formulated as a dynamic programming problem is to notice that its basic structure is analogous to the stagecoach problem.

These basic features that characterize dynamic programming problems are presented and discussed here.

1. The problem can be divided into **stages**, with a **policy decision** required at each stage.

The stagecoach problem was literally divided into its four stages (stagecoaches) that correspond to the four legs of the journey. The policy decision at each stage was which life insurance policy to choose (i.e., which destination to select for the next stagecoach ride). Similarly, other dynamic programming problems require making a *sequence of interrelated decisions,* where each decision corresponds to one stage of the problem.

2. Each stage has a number of **states** associated with the beginning of that stage.

The states associated with each stage in the stagecoach problem were the states (or territories) in which the fortune seeker could be located when embarking on that particular leg of the journey. In general, the states are the various *possible conditions* in which the system might be at that stage of the problem. The number of states may be either finite (as in the stagecoach problem) or infinite (as in some subsequent examples).

3. The effect of the policy decision at each stage is to *transform the current state to a state associated with the beginning of the next stage* (possibly according to a probability distribution).

The fortune seeker's decision as to his next destination led him from his current state to the next state on his journey. This procedure suggests that dynamic programming problems can be interpreted in terms of the *networks* described in Chap. 9. Each *node* would correspond to a *state.* The network would consist of columns of nodes, with each *column* corresponding to a *stage,* so that the flow from a node can go only to a node in the next column to the right. The links from a node to nodes in the next column correspond to the possible policy decisions on which state to go to next. The value assigned to each link usually can be interpreted as the *immediate contribution* to the objective function from making that policy decision. In most cases, the objective corresponds to finding either the *shortest* or the *longest path* through the network.

4. The solution procedure is designed to find an **optimal policy** for the overall problem, i.e., a prescription of the optimal policy decision at each stage for *each* of the possible states.

For the stagecoach problem, the solution procedure constructed a table for each stage (n) that prescribed the optimal decision (x_n^*) for *each* possible state (s). Thus, in addition to identifying three *optimal solutions* (optimal routes) for the overall problem, the results show the fortune seeker how he should proceed if he gets detoured to a state that is not on an optimal route. For any problem, dynamic programming provides this kind of *policy* prescription of what to do under every possible circumstance (which is why the actual decision made upon reaching a particular state at a given stage is referred to as a *policy* decision). Providing this additional information beyond simply specifying an optimal solution (optimal sequence of decisions) can be helpful in a variety of ways, including sensitivity analysis.

5. Given the current state, an *optimal policy for the remaining stages* is *independent* of the policy decisions adopted in *previous stages.* Therefore, the optimal immediate decision depends on only the current state and not on how you got there. This is the **principle of optimality** for dynamic programming.

Given the state in which the fortune seeker is currently located, the optimal life insurance policy (and its associated route) from this point onward is independent of how he got there. For dynamic programming problems in general, knowledge of the current state of the system conveys all the information about its previous behavior necessary for determining the optimal policy henceforth. (This property is the *Markovian property,* discussed in Sec. 14.2.) Any problem lacking this property cannot be formulated as a dynamic programming problem.

6. The solution procedure begins by finding the *optimal policy for the last stage.*

The optimal policy for the last stage prescribes the optimal policy decision for *each* of the possible states at that stage. The solution of this one-stage problem is usually trivial, as it was for the stagecoach problem.

7. A **recursive relationship** that identifies the optimal policy for stage n, given the optimal policy for stage $n + 1$, is available.

For the stagecoach problem, this recursive relationship was

$$f_n^*(s) = \min_{x_n} \{c_{sx_n} + f_{n+1}^*(x_n)\}.$$

Therefore, finding the *optimal policy decision* when you start in state s at stage n requires finding the minimizing value of x_n. For this particular problem, the corresponding minimum cost is achieved by using this value of x_n and then following the optimal policy when you start in state x_n at stage $n + 1$.

The precise form of the recursive relationship differs somewhat among dynamic programming problems. However, notation analogous to that introduced in the preceding section will continue to be used here, as summarized below.

N = number of stages.
n = label for current stage ($n = 1, 2, \ldots, N$).
s_n = current *state* for stage n.
x_n = decision variable for stage n.
x_n^* = optimal value of x_n (given s_n).
$f_n(s_n, x_n)$ = contribution of stages $n, n+1, \ldots, N$ to objective function if system starts in state s_n at stage n, immediate decision is x_n, and optimal decisions are made thereafter.
$f_n^*(s_n) = f_n(s_n, x_n^*)$.

The recursive relationship will always be of the form

$$f_n^*(s_n) = \max_{x_n} \{f_n(s_n, x_n)\} \qquad \text{or} \qquad f_n^*(s_n) = \min_{x_n} \{f_n(s_n, x_n)\},$$

where $f_n(s_n, x_n)$ would be written in terms of s_n, x_n, $f_{n+1}^*(s_{n+1})$, and probably some measure of the immediate contribution of x_n to the objective function. It is the inclusion of $f_{n+1}^*(s_{n+1})$ on the right-hand side, so that $f_n^*(s_n)$ is defined in terms of $f_{n+1}^*(s_{n+1})$, that makes the expression for $f_n^*(s_n)$ a recursive relationship.

The recursive relationship keeps recurring as we move backward stage by stage. When the current stage number n is decreased by 1, the new $f_n^*(s_n)$ function is derived by using the $f_{n+1}^*(s_{n+1})$ function that was just derived during the preceding iteration, and then this process keeps repeating. This property is emphasized in the next (and final) characteristic of dynamic programming.

8. When we use this recursive relationship, the solution procedure starts at the end and moves *backward* stage by stage—each time finding the optimal policy for that stage—until it finds the optimal policy starting at the *initial* stage. This optimal policy immediately yields an optimal solution for the entire problem, namely, x_1^* for the initial state s_1, then x_2^* for the resulting state s_2, then x_3^* for the resulting state s_3, and so forth to x_N^* for the resulting stage s_N.

This backward movement was demonstrated by the stagecoach problem, where the optimal policy was found successively beginning in each state at stages 4, 3, 2, and 1, respectively.[1] For all dynamic programming problems, a table such as the following would be obtained for each stage ($n = N, N-1, \ldots, 1$).

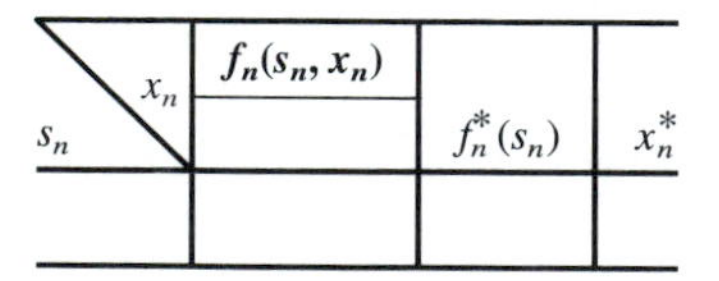

s_n \ x_n	$f_n(s_n, x_n)$	$f_n^*(s_n)$	x_n^*

When this table is finally obtained for the initial stage ($n = 1$), the problem of interest is solved. Because the initial state is known, the initial decision is specified by x_1^* in this

[1] Actually, for this problem the solution procedure can move *either* backward or forward. However, for many problems (especially when the stages correspond to *time periods*), the solution procedure *must* move backward.

table. The optimal value of the other decision variables is then specified by the other tables in turn according to the state of the system that results from the preceding decisions.

To assist you in executing this solution procedure, your OR Courseware includes an interactive routine for *deterministic* dynamic programming problems (the topic of the next section) when s_n and x_n are restricted to integer values. Also included is a demonstration of another dynamic programming example.

10.3 Deterministic Dynamic Programming

This section further elaborates upon the dynamic programming approach to *deterministic* problems, where the *state* at the *next stage* is *completely determined* by the *state* and *policy decision* at the *current stage*. The *probabilistic* case, where there is a probability distribution for what the next state will be, is discussed in the next section.

Deterministic dynamic programming can be described diagrammatically as shown in Fig. 10.3. Thus at stage n the process will be in some state s_n. Making policy decision x_n then moves the process to some state s_{n+1} at stage $n + 1$. The contribution *thereafter* to the objective function under an optimal policy has been previously calculated to be $f^*_{n+1}(s_{n+1})$. The policy decision x_n also makes some contribution to the objective function. Combining these two quantities in an appropriate way provides $f_n(s_n, x_n)$, the contribution of stages n onward to the objective function. Optimizing with respect to x_n then gives $f^*_n(s_n) = f_n(s_n, x^*_n)$. After x^*_n and $f^*_n(s_n)$ are found for each possible value of s_n, the solution procedure is ready to move back one stage.

One way of categorizing deterministic dynamic programming problems is by the *form of the objective function*. For example, the objective might be to minimize the sum of the contributions from the individual stages (as for the stagecoach problem), or to maximize such a sum, or to minimize a product of such terms, and so on. Another categorization is in terms of the nature of the *set of states* for the respective stages. In particular, states s_n might be representable by a *discrete* state variable (as for the stagecoach problem) or by a *continuous* state variable, or perhaps a state *vector* (more than one variable) is required.

Several examples are presented to illustrate these various possibilities. More importantly, they illustrate that these apparently major differences are actually quite inconsequential (except in terms of computational difficulty) because the underlying basic structure shown in Fig. 10.3 always remains the same.

The first new example arises in a much different context from the stagecoach problem, but it has the same *mathematical formulation* except that the objective is to *maximize* rather than minimize a sum.

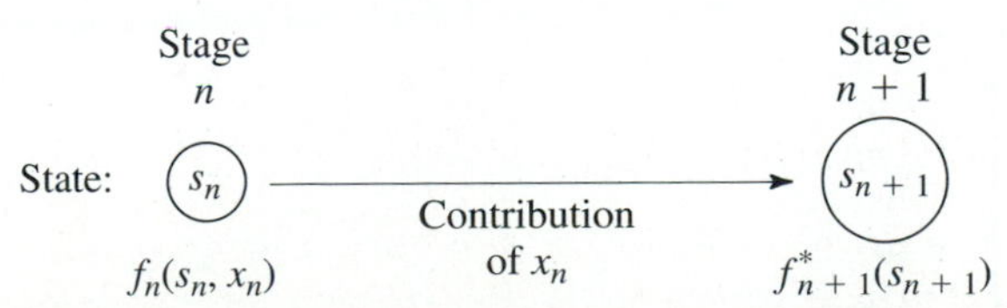

Figure 10.3 The basic structure for deterministic dynamic programming.

Example 2—Distributing Medical Teams to Countries

The WORLD HEALTH COUNCIL is devoted to improving health care in the underdeveloped countries of the world. It now has five medical teams available to allocate among three such countries to improve their medical care, health education, and training programs. Therefore, the council needs to determine how many teams (if any) to allocate to each of these countries to maximize the total effectiveness of the five teams. The teams must be kept intact, so the number allocated to each country must be an integer.

The measure of performance being used is *additional person-years of life.* (For a particular country, this measure equals the *increased life expectancy* in years times the country's population.) Table 10.1 gives the estimated additional person-years of life (in multiples of 1,000) for each country for each possible allocation of medical teams.

Which allocation maximizes the measure of performance?

FORMULATION: This problem requires making three *interrelated decisions,* namely, how many medical teams to allocate to each of the three countries. Therefore, even though there is no fixed sequence, these three countries can be considered as the three stages in a dynamic programming formulation. The decision variables x_n $(n = 1, 2, 3)$ are the number of teams to allocate to stage (country) n.

The identification of the states may not be readily apparent. To determine the states, we ask questions such as the following. What is it that changes from one stage to the next? Given that the decisions have been made at the previous stages, how can the status of the situation at the current stage be described? What information about the current state of affairs is necessary to determine the optimal policy hereafter? On these bases, an appropriate choice for the "state of the system" is

s_n = number of medical teams still available for allocation to remaining countries $(n, \ldots, 3)$.

Thus, at stage 1 (country 1), where all three countries remain under consideration for allocations, $s_1 = 5$. However, at stage 2 or 3 (country 2 or 3), s_n is just 5 minus the number of teams allocated at preceding stages, so that the sequence of states is

$$s_1 = 5, \qquad s_2 = 5 - x_1, \qquad s_3 = s_2 - x_2.$$

With the dynamic programming procedure of solving backward stage by stage, when we are solving at stage 2 or 3, we shall not yet have solved for the allocations at the

Table 10.1 **Data for the World Health Council Problem**

Medical Teams	*Thousands of Additional Person-Years of Life*		
	Country		
	1	2	3
0	0	0	0
1	45	20	50
2	70	45	70
3	90	75	80
4	105	110	100
5	120	150	130

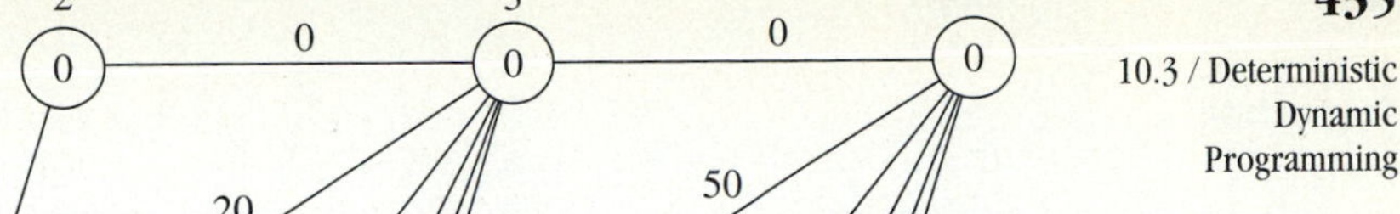
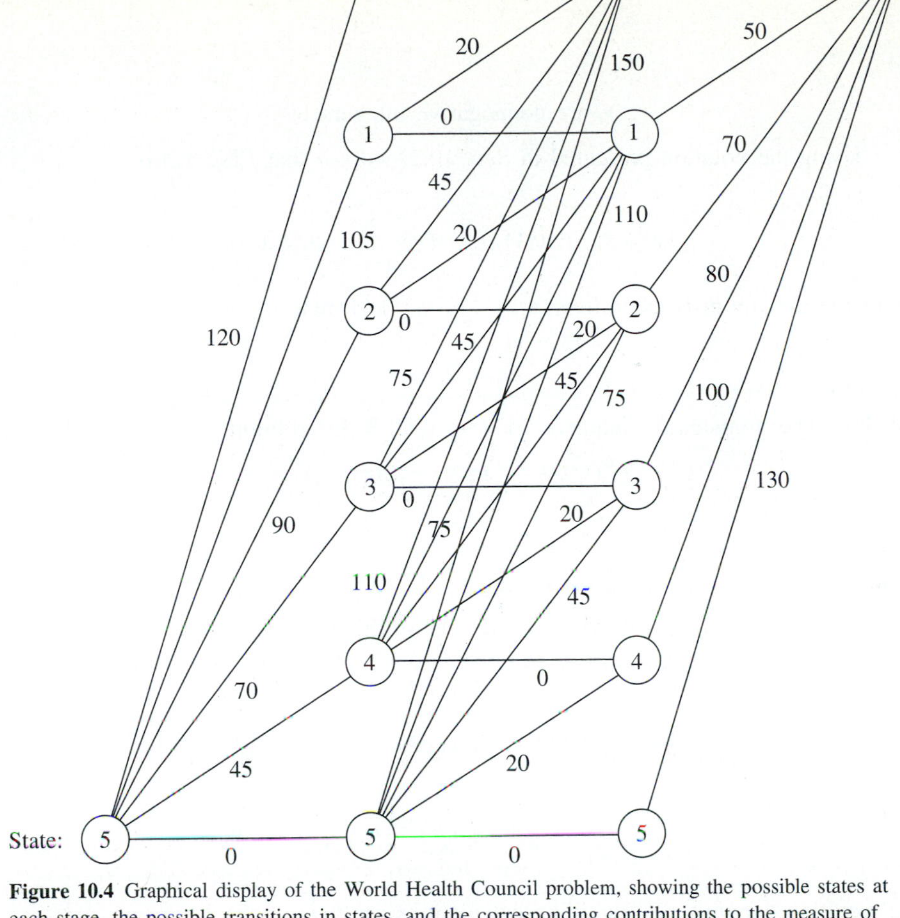

Figure 10.4 Graphical display of the World Health Council problem, showing the possible states at each stage, the possible transitions in states, and the corresponding contributions to the measure of performance.

preceding stages. Therefore, we shall consider every possible state we could be in at stage 2 or 3, namely, $s_n = 0, 1, 2, 3, 4$, or 5.

Figure 10.4 shows the states to be considered at each stage. The links (line segments) show the possible transitions in states from one stage to the next from making a feasible allocation of medical teams to the country involved. The numbers shown next to the links are the corresponding contributions to the measure of performance, where these numbers come from Table 10.1. From the perspective of this figure, the overall problem is to find the path from the initial state 5 (beginning stage 1) to the final state 0 (after stage 3) that maximizes the sum of the numbers along the path.

To state the overall problem mathematically, let $p_i(x_i)$ be the measure of performance from allocating x_i medical teams to country i, as given in Table 10.1. Thus the objective is to choose x_1, x_2, x_3 so as to

$$\text{Maximize} \quad \sum_{i=1}^{3} p_i(x_i),$$

subject to

$$\sum_{i=1}^{3} x_i = 5,$$

and

$$x_i \text{ are nonnegative integers.}$$

Using the notation presented in Sec. 10.2, we see that $f_n(s_n, x_n)$ is

$$f_n(s_n, x_n) = p_n(x_n) + \max \sum_{i=n+1}^{3} p_i(x_i),$$

where the maximum is taken over $x_{n+1}, \ldots, x_3$ such that

$$\sum_{i=n}^{3} x_i = s_n$$

and the x_i are nonnegative integers, for $n = 1, 2, 3$. In addition,

$$f_n^*(s_n) = \max_{x_n=0,1,\ldots,s_n} f_n(s_n, x_n)$$

Therefore,

$$f_n(s_n, x_n) = p_n(x_n) + f_{n+1}^*(s_n - x_n)$$

(with f_4^* defined to be zero). These basic relationships are summarized in Fig. 10.5.

Consequently, the *recursive relationship* relating functions f_1^*, f_2^*, and f_3^* for this problem is

$$f_n^*(s_n) = \max_{x_n=0,1,\ldots,s_n} \{p_n(x_n) + f_{n+1}^*(s_n - x_n)\}, \qquad \text{for } n = 1, 2.$$

For the last stage ($n = 3$),

$$f_3^*(s_3) = \max_{x_3=0,1,\ldots,s_3} p_3(x_3).$$

The resulting dynamic programming calculations are given next.

Solution Procedure: Beginning with the last stage ($n = 3$), we note that the values of $p_3(x_3)$ are given in the last column of Table 10.1 and that these values keep increasing as we move down the column. Therefore, with s_3 medical teams still available for allocation to country 3, the maximum of $p_3(x_3)$ is automatically achieved by allocating all s_3 teams; so $x_3^* = s_3$ and $f_3^*(s_3) = p_3(s_3)$, as shown in the following table.

$n = 3$:

s_3	$f_3^*(s_3)$	x_3^*
0	0	0
1	50	1
2	70	2
3	80	3
4	100	4
5	130	5

We now move backward to start from the next-to-last stage ($n = 2$). Here, finding x_2^* requires calculating and comparing $f_2(s_2, x_2)$ for the alternative values of x_2,

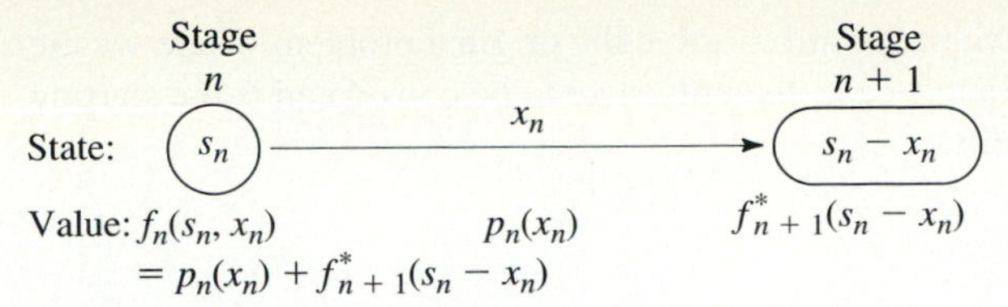

Figure 10.5 The basic structure for the World Health Council problem.

namely, $x_2 = 0, 1, \ldots, s_2$. To illustrate, we depict this situation when $s_2 = 2$ graphically:

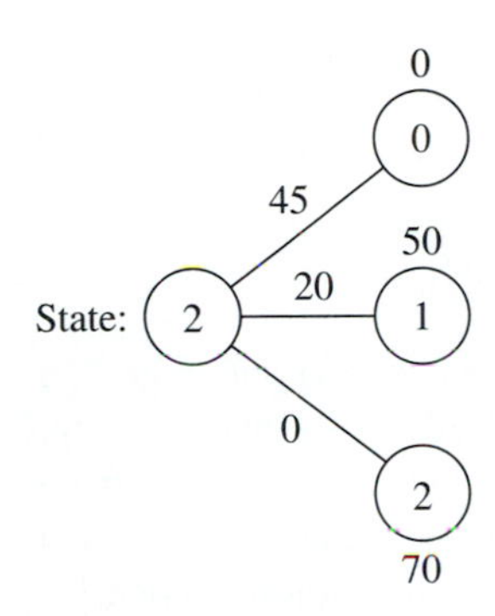

This diagram corresponds to Fig. 10.5 except that all three possible states at stage 3 are shown. Thus, if $x_2 = 0$, the resulting state at stage 3 will be $s_2 - x_2 = 2 - 0 = 2$, whereas $x_2 = 1$ leads to state 1 and $x_2 = 2$ leads to state 0. The corresponding values of $p_2(x_2)$ from the country 2 column of Table 10.1 are shown along the links, and the values of $f^*_3(s_2 - x_2)$ from the $n = 3$ table are given next to the stage 3 nodes. The required calculations for this case of $s_2 = 2$ are summarized below.

Formula: $f_2(2, x_2) = p_2(x_2) + f^*_3(2 - x_2)$.
$p_2(x_2)$ is given in the country 2 column of Table 10.1.
$f^*_3(2 - x_2)$ is given in the $n = 3$ table (bottom of preceding page).

$x_2 = 0$: $f_2(2, 0) = p_2(0) + f^*_3(2) = 0 + 70 = 70$.

$x_2 = 1$: $f_2(2, 1) = p_2(1) + f^*_3(1) = 20 + 50 = 70$.

$x_2 = 2$: $f_2(2, 2) = p_2(2) + f^*_3(0) = 45 + 0 = 45$.

Because the objective is *maximization*, $x^*_2 = 0$ or 1 with $f^*_2(2) = 70$.

Proceeding in a similar way with the other possible values of s_2 (try it) yields the following table.

$n = 2$:

s_2 \ x_2	$f_2(s_2, x_2) = p_2(x_2) + f^*_3(s_2 - x_2)$							
	0	1	2	3	4	5	$f^*_2(s_2)$	x^*_2
0	0						0	0
1	50	20					50	0
2	70	70	45				70	0 or 1
3	80	90	95	75			95	2
4	100	100	115	125	110		125	3
5	130	120	125	145	160	150	160	4

We now are ready to move backward to solve the original problem where we are starting from stage 1 ($n = 1$). In this case, the only state to be considered is the starting state of $s_1 = 5$, as depicted below.

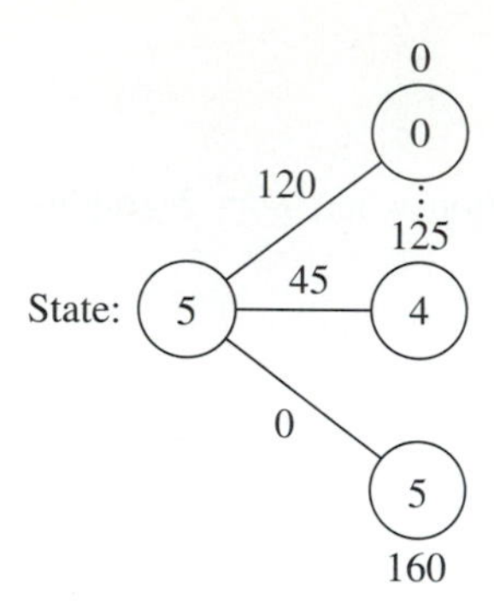

Since allocating x_1 medical teams to country 1 leads to a state of $5 - x_1$ at stage 2, a choice of $x_1 = 0$ leads to the bottom node on the right, $x = 1$ leads to the next node up, and so forth up to the top node with $x_1 = 5$. The corresponding $p_1(x_1)$ values from Table 10.1 are shown next to the links. The numbers next to the nodes are obtained from the $f_2^*(s_2)$ column of the $n = 2$ table. As with $n = 2$, the calculation needed for each alternative value of the decision variable involves adding the corresponding link value and node value, as summarized below.

Formula: $f_1(5, x_1) = p_1(x_1) + f_2^*(5 - x_1)$.
$p_1(x_1)$ is given in the country 1 column of Table 10.1.
$f_2^*(5 - x_1)$ is given in the $n = 2$ table.

$$x_1 = 0: \quad f_1(5, 0) = p_1(0) + f_2^*(5) = 0 + 160 = 160.$$

$$x_1 = 1: \quad f_1(5, 1) = p_1(1) + f_2^*(4) = 45 + 125 = 170.$$

$$\vdots$$

$$x_1 = 5: \quad f_1(5, 5) = p_1(5) + f_2^*(0) = 120 + 0 = 120.$$

The similar calculations for $x_1 = 2, 3, 4$ (try it) verify that $x_1^* = 1$ with $f_1^*(5) = 170$, as shown in the following table.

$n = 1$:

s_1 \ x_2	$f_1(s_1, x_1) = p_1(x_1) + f_2^*(s_1 - x_1)$						$f_1^*(s_1)$	x_1^*
	0	1	2	3	4	5		
5	160	170	165	160	155	120	170	1

Thus the optimal solution has $x_1^* = 1$, which makes $s_2 = 5 - 1 = 4$, so $x_2^* = 3$, which makes $s_3 = 4 - 3 = 1$, so $x_3^* = 1$. Since $f_1^*(5) = 170$, this (1, 3, 1) allocation of medical teams to the three countries will yield an estimated total of 170,000 additional person-years of life, which is at least 5,000 more than for any other allocation.

These results of the dynamic programming analysis also are summarized in Fig. 10.6.

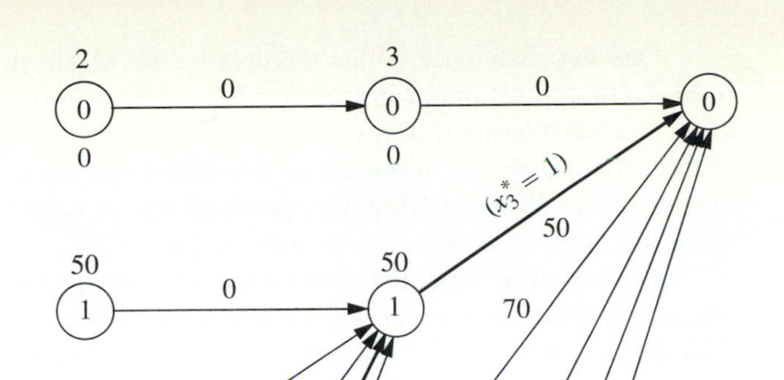
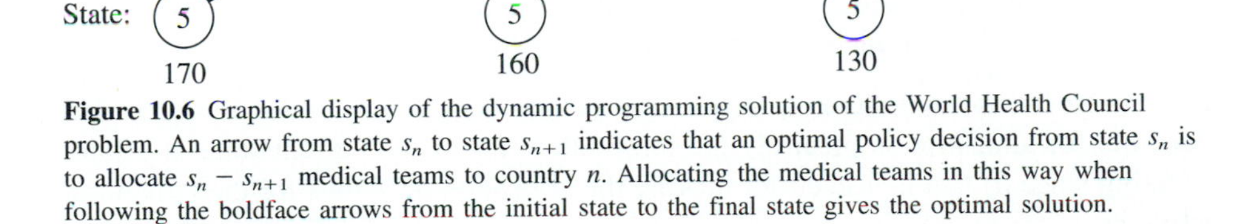

Figure 10.6 Graphical display of the dynamic programming solution of the World Health Council problem. An arrow from state s_n to state s_{n+1} indicates that an optimal policy decision from state s_n is to allocate $s_n - s_{n+1}$ medical teams to country n. Allocating the medical teams in this way when following the boldface arrows from the initial state to the final state gives the optimal solution.

A Prevalent Problem Type—The Distribution of Effort Problem

The preceding example illustrates a particularly common type of dynamic programming problem called the *distribution of effort problem.* For this type of problem, there is just one kind of *resource* that is to be allocated to a number of *activities.* The objective is to determine how to distribute the effort (the resource) among the activities most effectively. For the World Health Council example, the resource involved is the medical teams, and the three activities are the health care work in the three countries.

Assumptions: This interpretation of allocating resources to activities should ring a bell for you, because it is the typical interpretation for linear programming problems given at the beginning of Chap. 3. However, there also are some key differences between the distribution of effort problem and linear programming that help illuminate the general distinctions between dynamic programming and other areas of mathematical programming.

One key difference is that the distribution of effort problem involves only *one resource* (one functional constraint), whereas linear programming can deal with hundreds or even thousands of resources. (In principle, dynamic programming can handle slightly more than one resource, as we shall illustrate in Example 5 by solving the three-resource Wyndor Glass Co. problem, but it quickly becomes very inefficient when the number of resources is increased.)

On the other hand, the distribution of effort problem is far more general than linear programming in other ways. Consider the four assumptions of linear programming presented in Sec. 3.3: proportionality, additivity, divisibility, and certainty. *Proportionality* is routinely violated by nearly all dynamic programming problems, including distribution of effort problems (e.g., Table 10.1 violates proportionality). *Divisibility* also is often violated, as in Example 2, where the decision variables must be integers. In fact, dynamic programming calculations become more complex when divisibility does hold (as in Examples 4 and 5). Although we shall consider the distribution of effort problem only under the assumption of *certainty,* this is not necessary, and many other dynamic programming problems violate this assumption as well (as described in Sec. 10.4).

Of the four assumptions of linear programming, the *only* one needed by the distribution of effort problem (or other dynamic programming problems) is *additivity* (or its analog for functions involving a *product* of terms). This assumption is needed to satisfy the *principle of optimality* for dynamic programming (characteristic 5 in Sec. 10.2).

Formulation: Because they always involve allocating one kind of resource to a number of activities, distribution of effort problems always have the following dynamic programming formulation (where the ordering of the activities is arbitrary):

Stage n = activity n ($n = 1, 2, \ldots, N$).
x_n = amount of resource allocated to activity n.
State s_n = amount of resource still available for allocation to remaining activities ($n, \ldots, N$).

The reason for defining state s_n in this way is that the amount of the resource still available for allocation is precisely the information about the current state of affairs (entering stage n) that is needed for making the allocation decisions for the remaining activities.

When the system starts at stage n in state s_n, the choice of x_n results in the next state at stage $n + 1$ being $s_{n+1} = s_n - x_n$, as depicted below:[1]

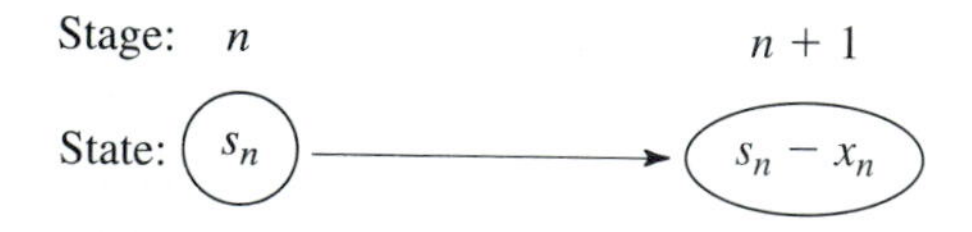

Note how the structure of this diagram corresponds to the one shown in Fig. 10.5 for the World Health Council example of a distribution of effort problem. What will differ

[1] This statement assumes that x_n and s_n are expressed in the same units. If it is more convenient to define x_n as some other quantity such that the amount of the resource allocated to activity n is $a_n x_n$, then $s_{n+1} = s_n - a_n x_n$.

from one such example to the next is the *rest* of what is shown in Fig. 10.5, namely, the relationship between $f_n(s_n, x_n)$ and $f^*_{n+1}(s_n - x_n)$, and then the resulting *recursive relationship* between the f^*_n and f^*_{n+1} functions. These relationships depend on the particular objective function for the overall problem.

The structure of the next example is similar to the one for the World Health Council because it, too, is a distribution of effort problem. However, its recursive relationship differs in that its objective is to minimize a product of terms for the respective stages.

At first glance this example may appear *not* to be a deterministic dynamic programming problem because probabilities are involved. However, it does indeed fit our definition because the state at the next stage is completely determined by the state and policy decision at the current stage.

Example 3—Distributing Scientists to Research Teams

A government space project is conducting research on a certain engineering problem that must be solved before people can fly safely to Mars. Three research teams are currently trying three different approaches for solving this problem. The estimate has been made that, under present circumstances, the probability that the respective teams—call them 1, 2, and 3—will not succeed is 0.40, 0.60, and 0.80, respectively. Thus the current probability that all three teams will fail is $(0.40)(0.60)(0.80) = 0.192$. Because the objective is to minimize the probability of failure, two more top scientists have been assigned to the project.

Table 10.2 gives the estimated probability that the respective teams will fail when 0, 1, or 2 additional scientists are added to that team. Only integer numbers of scientists are considered because each new scientist will need to devote full attention to one team. The problem is to determine how to allocate the two additional scientists to minimize the probability that all three teams will fail.

Formulation: Because both Examples 2 and 3 are distribution of effort problems, their underlying structure is actually very similar. In this case, scientists replace medical teams as the kind of resource involved, and research teams replace countries as the activities. Therefore, instead of medical teams being allocated to countries, scientists are being allocated to research teams. The only basic difference between the two problems is in their objective functions.

With so few scientists and teams involved, this problem could be solved very easily by a process of exhaustive enumeration. However, the dynamic programming solution is presented for illustrative purposes.

In this case, stage n ($n = 1, 2, 3$) corresponds to research team n, and the state s_n is the number of new scientists *still available* for allocation to the remaining teams. The

Table 10.2 **Data for the Government Space Project Problem**

	Probability of Failure		
	Team		
New Scientists	1	2	3
0	0.40	0.60	0.80
1	0.20	0.40	0.50
2	0.15	0.20	0.30

decision variables x_n ($n = 1, 2, 3$) are the number of additional scientists allocated to team n.

Let $p_i(x_i)$ denote the probability of failure for team i if it is assigned x_i additional scientists, as given by Table 10.2. If we let Π denote multiplication, the government's objective is to choose x_1, x_2, x_3 so as to

$$\text{Minimize} \quad \prod_{i=1}^{3} p_i(x_i) = p_1(x_1)p_2(x_2)p_3(x_3),$$

subject to

$$\sum_{i=1}^{3} x_i = 2$$

and

$$x_i \text{ are nonnegative integers.}$$

Consequently, $f_n(s_n, x_n)$ for this problem is

$$f_n(s_n, x_n) = p_n(x_n) \cdot \min \prod_{i=n+1}^{3} p_i(x_i),$$

where the minimum is taken over $x_{n+1}, \ldots, x_3$ such that

$$\sum_{i=n}^{3} x_i = s_n$$

and

$$x_i \text{ are nonnegative integers,}$$

for $n = 1, 2, 3$. Thus

$$f_n^*(s_n) = \min_{x_n=0,1,\ldots,s_n} f_n(s_n, x_n),$$

where

$$f_n(s_n, x_n) = p_n(x_n)f_{n+1}^*(s_n - x_n)$$

(with f_4^* defined to be 1). Figure 10.7 summarizes these basic relationships.

Thus the *recursive relationship* relating the f_1^*, f_2^*, and f_3^* functions in this case is

$$f_n^*(s_n) = \min_{x_n=0,1,\ldots,s_n} \{p_n(x_n) \cdot f_{n+1}^*(s_n - x_n)\}, \qquad \text{for } n = 1, 2,$$

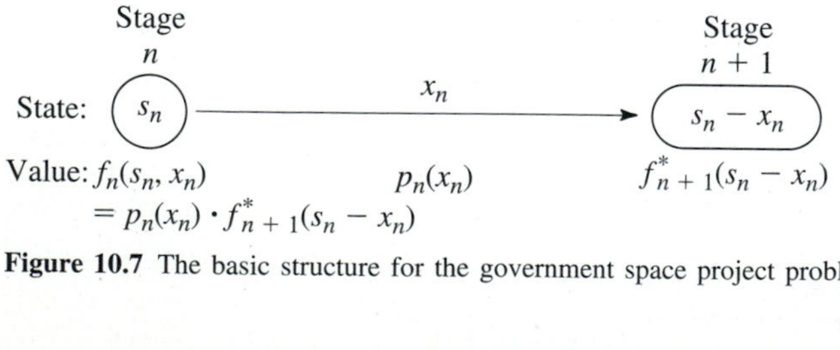

Figure 10.7 The basic structure for the government space project problem.

and, when $n = 3$,

$$f_3^*(s_3) = \min_{x_3=0,1,\ldots,s_3} p_3(x_3).$$

Solution Procedure: The resulting dynamic programming calculations are as follows:

$n = 3$:

s_3	$f_3^*(s_3)$	x_3^*
0	0.80	0
1	0.50	1
2	0.30	2

$n = 2$:

s_2 \ x_2	$f_2(s_2, x_2) = p_2(x_2) \cdot f_3^*(s_2 - x_2)$			$f_2^*(s_2)$	x_2^*
	0	1	2		
0	0.48			0.48	0
1	0.30	0.32		0.30	0
2	0.18	0.20	0.16	0.16	2

$n = 1$:

s_1 \ x_1	$f_1(s_1, x_1) = p_1(x_1) \cdot f_2^*(s_1 - x_1)$			$f_1^*(s_1)$	x_1^*
	0	1	2		
2	0.064	0.060	0.072	0.060	1

Therefore, the optimal solution must have $x_1^* = 1$, which makes $s_2 = 2 - 1 = 1$, so that $x_2^* = 0$, which makes $s_3 = 1 - 0 = 1$, so that $x_3^* = 1$. Thus teams 1 and 3 should each receive one additional scientist. The new probability that all three teams will fail would then be 0.060.

All the examples thus far have had a *discrete* state variable s_n at each stage. Furthermore, they all have been *reversible* in the sense that the solution procedure actually could have moved *either* backward or forward stage by stage. (The latter alternative amounts to renumbering the stages in reverse order and then applying the procedure in the standard way.) This reversibility is a general characteristic of distribution of effort problems such as Examples 2 and 3, since the activities (stages) can be ordered in any desired manner.

The next example is different in both respects. Rather than being restricted to integer values, its state variable s_n at stage n is a *continuous* variable that can take on *any* value over certain intervals. Since s_n now has an infinite number of values, it is no longer possible to consider each of its feasible values individually. Rather, the solution for $f_n^*(s_n)$ and x_n^* must be expressed as *functions* of s_n. Furthermore, this example is *not* reversible because its stages correspond to *time periods,* so the solution procedure *must* proceed backward.

Example 4—Scheduling Employment Levels

The workload for the LOCAL JOB SHOP is subject to considerable seasonal fluctuation. However, machine operators are difficult to hire and costly to train, so the manager is reluctant to lay off workers during the slack seasons. He is likewise reluctant to maintain his peak season payroll when it is not required. Furthermore, he is definitely opposed to overtime work on a regular basis. Since all work is done to custom orders, it is not possible to build up inventories during slack seasons. Therefore, the manger is in a dilemma as to what his policy should be regarding employment levels.

The following estimates are given for the minimum employment requirements during the four seasons of the year for the foreseeable future:

Season	Spring	Summer	Autumn	Winter	Spring
Requirements	255	220	240	200	255

Employment will not be permitted to fall below these levels. Any employment above these levels is wasted at an approximate cost of $2,000 per person per season. It is estimated that the hiring and firing costs are such that the total cost of changing the level of employment from one season to the next is $200 times the square of the difference in employment levels. Fractional levels of employment are possible because of a few part-time employees, and the cost data also apply on a fractional basis.

Formulation: On the basis of the data available, it is not worthwhile to have the employment level go above the peak season requirements of 255. Therefore, spring employment should be at 255, and the problem is reduced to finding the employment level for the other three seasons.

For a dynamic programming formulation, the seasons should be the stages. There are actually an indefinite number of stages because the problem extends into the indefinite future. However, each year begins an identical cycle, and because spring employment is known, it is possible to consider only one cycle of four seasons ending with the spring season, as summarized below.

Stage 1 = summer,
Stage 2 = autumn,
Stage 3 = winter,
Stage 4 = spring.
x_n = employment level for stage n ($n = 1, 2, 3, 4$).
($x_4 = 255$.)

It is necessary that the spring season be the last stage because the optimal value of the decision variable for each state at the last stage must be either known or obtainable without considering other stages. For every other season, the solution for the optimal employment level must consider the effect on costs in the following season.

Let

$$r_n = \text{minimum employment requirement for stage } n,$$

where these requirements were given earlier as $r_1 = 220$, $r_2 = 240$, $r_3 = 200$, and $r_4 = 255$. Thus the only feasible values for x_n are

$$r_n \leq x_n \leq 255.$$

Table 10.3 **Data for the Local Job Shop Problem**

n	r_n	Feasible x_n	Possible $s_n = x_{n-1}$	Cost
1	220	$220 \le x_1 \le 255$	$s_1 = 255$	$200(x_1 - 255)^2 + 2{,}000(x_1 - 220)$
2	240	$240 \le x_2 \le 255$	$220 \le s_2 \le 255$	$200(x_2 - x_1)^2 + 2{,}000(x_2 - 240)$
3	200	$200 \le x_3 \le 255$	$240 \le s_3 \le 255$	$200(x_3 - x_2)^2 + 2{,}000(x_3 - 200)$
4	255	$x_4 = 255$	$200 \le s_4 \le 255$	$200(255 - x_3)^2$

Referring to the cost data given in the problem statement, we have

$$\text{Cost for stage } n = 200(x_n - x_{n-1})^2 + 2{,}000(x_n - r_n).$$

Note that the cost at the current stage depends only upon the current decision x_n and the employment in the preceding season x_{n-1}. Thus the preceding employment level is all the information about the current state of affairs that we need to determine the optimal policy henceforth. Therefore, the state s_n for stage n is

$$\text{State } s_n = x_{n-1}.$$

When $n = 1$, $s_1 = x_0 = x_4 = 255$.

For your ease of reference while working through the problem, a summary of the above data is given in Table 10.3.

The objective for the problem is to choose x_1, x_2, x_3 (with $x_4 = 255$) so as to

$$\text{Minimize} \quad \sum_{i=1}^{4} [200(x_i - x_{i-1})^2 + 2{,}000(x_i - r_i)],$$

subject to

$$r_i \le x_i \le 255, \qquad \text{for } i = 1, 2, 3, 4.$$

Thus for stage n onward ($n = 1, 2, 3, 4$), since $s_n = x_{n-1}$

$$f_n(s_n, x_n) = 200(x_n - s_n)^2 + 2{,}000(x_n - r_n) + \min_{r_i \le x_i \le 255} \sum_{i=n+1}^{4} [200(x_i - x_{i-1})^2 + 2{,}000(x_i - r_i)],$$

where this summation equals zero when $n = 4$ (because it has no terms). Also,

$$f_n^*(s_n) = \min_{r_n \le x_n \le 255} f_n(s_n, x_n).$$

Hence

$$f_n(s_n, x_n) = 200(x_n - s_n)^2 + 2{,}000(x_n - r_n) + f_{n+1}^*(x_n)$$

(with f_5^* defined to be zero because costs after stage 4 are irrelevant to the analysis). A summary of these basic relationships is given in Fig. 10.8.

Consequently, the recursive relationship relating the f_n^* functions is

$$f_n^*(s_n) = \min_{r_n \le x_n \le 255} \{200(x_n - s_n)^2 + 2{,}000(x_n - r_n) + f_{n+1}^*(x_n)\}.$$

The dynamic programming approach uses this relationship to identify successively these functions—$f_4^*(s_4), f_3^*(s_3), f_2^*(s_2), f_1^*(255)$—and the corresponding minimizing x_n.

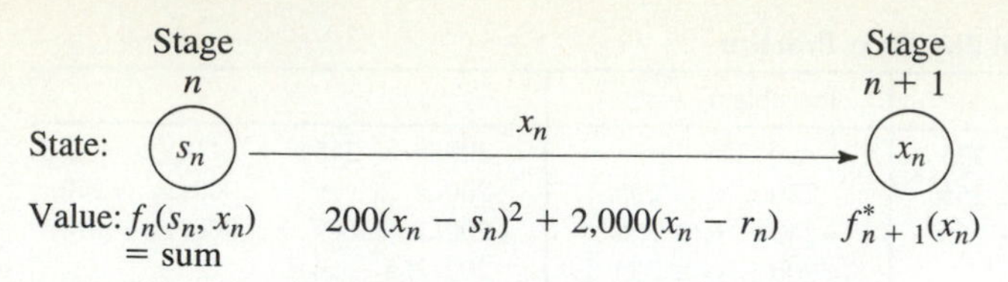

Figure 10.8 The basic structure for the Local Job Shop problem.

Solution Procedure

Stage 4: Beginning at the last stage ($n = 4$), we already know that $x_4^* = 255$, so the necessary results are

$n = 4$:

s_4	$f_4^*(s_4)$	x_4^*
$200 \le s_4 \le 255$	$200(255 - s_4)^2$	255

Stage 3: For the problem consisting of just the last *two* stages ($n = 3$), the recursive relationship reduces to

$$f_3^*(s_3) = \min_{200 \le x_3 \le 255} \{200(x_3 - s_3)^2 + 2{,}000(x_3 - 200) + f_4^*(x_3)\}$$

$$= \min_{200 \le x_3 \le 255} \{200(x_3 - s_3)^2 + 2{,}000(x_3 - 200) + 200(255 - x_3)^2\},$$

where the possible values of s_3 are $240 \le s_3 \le 255$.

One way to solve for the value of x_3 that minimizes $f_3(s_3, x_3)$ for any particular value of s_3 is the graphical approach illustrated in Fig. 10.9.

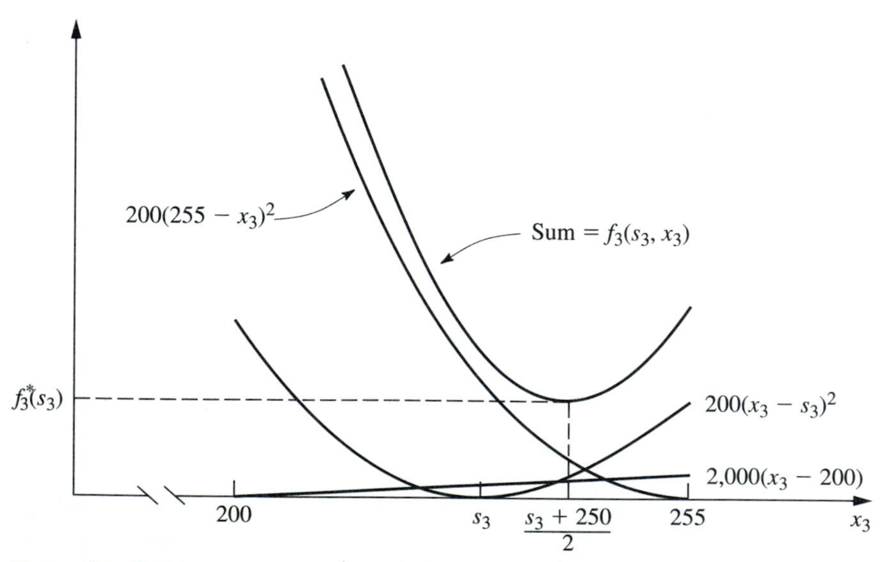

Figure 10.9 Graphical solution for $f_3^*(s_3)$ for the Local Job Shop problem.

However, a faster way is to use *calculus.* We want to solve for the minimizing x_3 in terms of s_3 by considering s_3 to have some fixed (but unknown) value. Therefore, set the first (partial) derivative of $f_3(s_3, x_3)$ with respect to x_3 equal to zero

$$\begin{aligned}\frac{\partial}{\partial x_3} f_3(s_3, x_3) &= 400(x_3 - s_3) + 2{,}000 - 400(255 - x_3)\\ &= 400(2x_3 - s_3 - 250)\\ &= 0,\end{aligned}$$

which yields

$$x_3^* = \frac{s_3 + 250}{2}.$$

Because the second derivative is positive, and because this solution lies in the feasible interval for x_3 ($200 \le x_3 \le 255$) for all possible s_3 ($240 \le s_3 \le 255$), it is indeed the desired minimum.

Note a key difference between the nature of this solution and those obtained for the preceding examples where there were only a few possible states to consider. We now have an *infinite* number of possible states ($240 \le s_3 \le 255$), so it is no longer feasible to solve separately for x_3^* for each possible value of s_3. Therefore, we instead have solved for x_3^* as a *function* of the unknown s_3.

Using

$$\begin{aligned}f_3^*(s_3) = f_3(s_3, x_3^*) &= 200\left(\frac{s_3 + 250}{2} - s_3\right)^2 + 200\left(255 - \frac{s_3 + 250}{2}\right)^2\\ &\quad + 2{,}000\left(\frac{s_3 + 250}{2} - 200\right)\end{aligned}$$

and reducing this expression algebraically complete the required results for the third-stage problem, summarized as follows.

$n = 3$:

s_3	$f_3^*(s_3)$	x_3^*
$240 \le s_3 \le 255$	$50(250 - s_3)^2 + 50(260 - s_3)^2 + 1{,}000(s_3 - 150)$	$\dfrac{s_3 + 250}{2}$

Stage 2: The second-stage ($n = 2$) and first-stage problems ($n = 1$) are solved in a similar fashion. Thus for $n = 2$,

$$\begin{aligned}f_2(s_2, x_2) &= 200(x_2 - s_2)^2 + 2{,}000(x_2 - r_2) + f_3^*(x_2)\\ &= 200(x_2 - s_2)^2 + 2{,}000(x_2 - 240)\\ &\quad + 50(250 - x_2)^2 + 50(260 - x_2)^2 + 1{,}000(x_2 - 150).\end{aligned}$$

The possible values of s_2 are $220 \le s_2 \le 255$, and the feasible region for x_2 is $240 \le x_2 \le 255$. The problem is to find the minimizing value of x_2 in this region, so that

$$f_2^*(s_2) = \min_{240 \le x_2 \le 255} f_2(s_2, x_2).$$

Setting to zero the partial derivative with respect to x_2

$$\frac{\partial}{\partial x_2} f_2(s_2, x_2) = 400(x_2 - s_2) + 2,000 - 100(250 - x_2) - 100(260 - x_2) + 1,000$$
$$= 200(3x_2 - 2s_2 - 240)$$
$$= 0$$

yields

$$x_2 = \frac{2s_2 + 240}{3}.$$

Because

$$\frac{\partial^2}{\partial x_2^2} f_2(s_2, x_2) = 600 > 0,$$

this value of x_2 is the desired minimizing value *if* it is *feasible* ($240 \le x_2 \le 255$). Over the possible s_2 values ($220 \le s_2 \le 255$), this solution actually is feasible only if $240 \le s_2 \le 255$.

Therefore, we still need to solve for the feasible value of x_2 that minimizes $f_2(s_2, x_2)$ when $220 \le s_2 < 240$. The key to analyzing the behavior of $f_2(s_2, x_2)$ over the feasible region for x_2 again is the partial derivative of $f_2(s_2, x_2)$. When $s_2 < 240$,

$$\frac{\partial}{\partial x_2} f_2(s_2, x_2) > 0, \qquad \text{for } 240 \le x_2 \le 255,$$

so that $x_2 = 240$ is the desired minimizing value.

The next step is to plug these values of x_2 into $f_2(s_2, x_2)$ to obtain $f_2^*(s_2)$ for $s_2 \ge 240$ and $s_2 < 240$. This yields

$n = 2$:

s_2	$f_2^*(s_2)$	x_2^*
$220 \le s_2 \le 240$	$200(240 - s_2)^2 + 115,000$	240
$240 \le s_2 \le 255$	$\frac{200}{9}[(240 - s_2)^2 + (255 - s_2)^2 + (270 - s_2)^2] + 2,000(s_2 - 195)$	$\frac{2s_2 + 240}{3}$

Stage 1: For the first-stage problem ($n = 1$),

$$f_1(s_1, x_1) = 200(x_1 - s_1)^2 + 2,000(x_1 - r_1) + f_2^*(x_1).$$

Because $r_1 = 220$, the feasible region for x_1 is $220 \le x_1 \le 255$. The expression for $f_2^*(x_1)$ will differ in the two portions $220 \le x_1 \le 240$ and $240 \le x_1 \le 255$ of this region. Therefore,

$$f_1(s_1, x_1) = \begin{cases} 200(x_1 - s_1)^2 + 2,000(x_1 - 220) + 200(240 - x_1)^2 + 115,000, & \text{if } 220 \le x_1 \le 240 \\ 200(x_1 - s_1)^2 + 2,000(x_1 - 220) + \frac{200}{9}[(240 - x_1)^2 + (255 - x_1)^2 + (270 - x_1)^2] \\ \quad + 2,000(x_1 - 195), & \text{if } 240 \le x_1 \le 255. \end{cases}$$

Considering first the case where $220 \le x_1 \le 240$, we have

$$\frac{\partial}{\partial x_1} f_1(s_1, x_1) = 400(x_1 - s_1) + 2{,}000 - 400(240 - x_1)$$

$$= 400(2x_1 - s_1 - 235).$$

It is known that $s_1 = 255$ (spring employment), so that

$$\frac{\partial}{\partial x_1} f_1(s_1, x_1) = 800(x_1 - 245) < 0$$

for all $x_1 \le 240$. Therefore, $x_1 = 240$ is the minimizing value of $f_1(s_1, x_1)$ over the region $220 \le x_1 \le 240$.

When $240 \le x_1 \le 255$,

$$\frac{\partial}{\partial x_1} f_1(s_1, x_1) = 400(x_1 - s_1) + 2{,}000 - \frac{400}{9}[(240 - x_1) + (255 - x_1) + (270 - x_1)] + 2{,}000$$

$$= \frac{400}{3}(4x_1 - 3s_1 - 225).$$

Because

$$\frac{\partial^2}{\partial x_1^2} f_1(s_1, x_1) > 0 \qquad \text{for all } x_1,$$

set

$$\frac{\partial}{\partial x_1} f_1(s_1, x_1) = 0,$$

which yields

$$x_1 = \frac{3s_1 + 225}{4}.$$

Because $s_1 = 255$, it follows that $x_1 = 247.5$ minimizes $f_1(s_1, x_1)$ over the region $240 \le x_1 \le 255$.

Note that this region ($240 \le x_1 \le 255$) includes $x_1 = 240$, so that $f_1(s_1, 240) > f_1(s_1, 247.5)$. In the next-to-last paragraph, we found that $x_1 = 240$ minimizes $f_1(s_1, x_1)$ over the region $220 \le x_1 \le 240$. Consequently, we now can conclude that $x_1 = 247.5$ also minimizes $f_1(s_1, x_1)$ over the *entire* feasible region $220 \le x_1 \le 255$.

Our final calculation is to find $f_1^*(s_1)$ for $s_1 = 255$ by plugging $x_1 = 247.5$ into the expression for $f_1(255, x_1)$ that holds for $240 \le x_1 \le 255$. Hence

$$f_1^*(255) = 200(247.5 - 255)^2 + 2{,}000(247.5 - 220) + \frac{200}{9}[2(250 - 247.5)^2 + (265 - 247.5)^2 + 30(742.5 - 575)]$$

$$= 185{,}000.$$

These results are summarized as follows:

$n = 1$:

s_1	$f_1^*(s_1)$	x_1^*
255	185,000	247.5

Therefore, by tracing back through the tables for $n = 2$, $n = 3$, and $n = 4$, respectively, and setting $s_n = x_{n-1}^*$ each time, the resulting optimal solution is $x_1^* = 247.5$, $x_2^* = 245$, $x_3^* = 247.5$, $x_4^* = 255$, with a total estimated cost per cycle of \$185,000.

To conclude our illustrations of deterministic dynamic programming, we give one example that requires *more than one* variable to describe the state at each stage.

Example 5—Wyndor Glass Company Problem

Consider the following linear programming problem:

$$\text{Maximize} \quad Z = 3x_1 + 5x_2,$$

subject to

$$\begin{aligned} x_1 \qquad &\leq 4 \\ 2x_2 &\leq 12 \\ 3x_1 + 2x_2 &\leq 18 \end{aligned}$$

and

$$x_1 \geq 0, \qquad x_2 \geq 0.$$

(You might recognize this model as the prototype example for linear programming in Chap. 3.) One way of solving small linear (or nonlinear) programming problems like this one is by dynamic programming, which is illustrated below.

FORMULATION: This problem requires making two interrelated decisions, namely, the level of activity 1, denoted by x_1, and the level of activity 2, denoted by x_2. Therefore, these two activities can be interpreted as the two stages in a dynamic programming formulation. Although they can be taken in either order, let stage n = activity n ($n = 1, 2$). Thus x_n is the decision variable at stage n.

What are the states? In other words, given that the decision had been made at prior stages (if any), what information is needed about the current state of affairs before the decision can be made at stage n? Reflection might suggest that the required information is the *amount of slack* left in the functional constraints. Interpret the right-hand side of these constraints (4, 12, and 18) as the total available amount of resources 1, 2, and 3, respectively (as described in Sec. 3.1). Then state s_n can be defined as

$$\text{State } s_n = \text{amount of respective resources still available for allocation to remaining activities.}$$

(Note that the definition of the state is analogous to that for distribution of effort problems, including Examples 2 and 3, except that there are now three resources to be allocated instead of just one.) Thus

$$s_n = (R_1, R_2, R_3),$$

where R_i is the amount of resource i remaining to be allocated ($i = 1, 2, 3$). Therefore,

$$s_1 = (4, 12, 18),$$

$$s_2 = (4 - x_1, 12, 18 - 3x_1).$$

However, when we begin by solving for stage 2, we do not yet know the value of x_1, and so we use $s_2 = (R_1, R_2, R_3)$ at that point.

Therefore, in contrast to the preceding examples, this problem has *three* state variables (i.e., a *state vector* with three components) at each stage rather than one. From a theoretical standpoint, this difference is not particularly serious. It only means that, instead of considering all possible values of the one state variable, we must consider all possible *combinations* of values of the several state variables. However, from the standpoint of computational efficiency, this difference tends to be a very serious complication. Because the number of combinations, in general, can be as large as the *product* of the number of possible values of the respective variables, the number of required calculations tends to "blow up" rapidly when additional state variables are introduced. This phenomenon has been given the apt name of the **curse of dimensionality**.

Each of the three state variables is *continuous*. Therefore, rather than consider each possible combination of values separately, we must use the approach introduced in Example 4 of solving for the required information as a *function* of the state of the system.

Despite these complications, this problem is small enough that it can still be solved without great difficulty. To solve it, we need to introduce the usual dynamic programming notation. Thus,

$$f_2(R_1, R_2, R_3, x_2) = \text{contribution of activity 2 to } Z \text{ if system starts in state } (R_1, R_2, R_3) \text{ at stage 2 and decision is } x_2$$

$$= 5x_2,$$

$$f_1(4, 12, 18, x_1) = \text{contribution of activities 1 and 2 to } Z \text{ if system starts in state (4, 12, 18) at stage 1, immediate decision is } x_1 \text{, and then optimal decision is made at stage 2,}$$

$$= 3x_1 + \max_{\substack{x_2 \le 12 \\ 2x_2 \le 18 - 3x_1 \\ x_2 \ge 0}} \{5x_2\}.$$

Similarly, for $n = 1, 2$,

$$f_n^*(R_1, R_2, R_3) = \max_{x_n} \quad f_n(R_1, R_2, R_3, x_n),$$

where this maximum is taken over the feasible values of x_n. Consequently, using the relevant portions of the constraints of the problem gives

$$(1) \qquad f_2^*(R_1, R_2, R_3) = \max_{\substack{2x_2 \le R_2 \\ 2x_2 \le R_3 \\ x_2 \ge 0}} \{5x_2\},$$

$$(2) \qquad f_1(4, 12, 18, x_1) = 3x_1 + f_2^*(4 - x_1, 12, 18 - 3x_1),$$

$$(3) \qquad f_1^*(4, 12, 18) = \max_{\substack{x_1 \le 4 \\ 3x_1 \le 18 \\ x_1 \ge 0}} \{3x_1 + f_2^*(4 - x_1, 12, 18 - 3x_1)\}.$$

Equation (1) will be used to solve the stage 2 problem. Equation (2) shows the basic dynamic programming structure for this problem, also depicted in Fig. 10.10. Equation (3) gives the *recursive relationship* between f_1^* and f_2^* that will be used to solve the stage 1 problem.

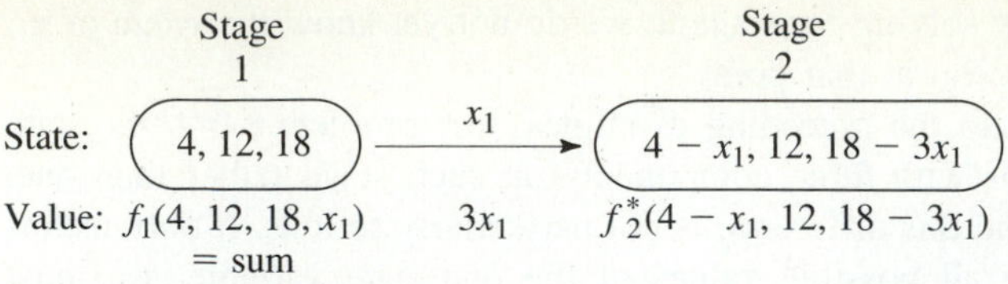

	Stage 1		Stage 2
State:	4, 12, 18	$\xrightarrow{x_1}$	$4 - x_1, 12, 18 - 3x_1$
Value:	$f_1(4, 12, 18, x_1)$ = sum	$3x_1$	$f_2^*(4 - x_1, 12, 18 - 3x_1)$

Figure 10.10 The basic structure for the Wyndor Glass Co. linear programming problem.

Solution Procedure

Stage 2: To solve at the last stage ($n = 2$), Eq. (1) indicates that x_2^* must be the largest value of x_2 that *simultaneously* satisfies $2x_2 \le R_2$, $2x_2 \le R_3$, and $x_2 \ge 0$. Assuming that $R_2 \ge 0$ and $R_3 \ge 0$, so that feasible solutions exist, this largest value is the smaller of $R_2/2$ and $R_3/2$. Thus the solution is

$n = 2$:

(R_1, R_2, R_3)	$f_2^*(R_1, R_2, R_3)$	x_2^*
$R_2 \ge 0,\ R_3 \ge 0$	$5 \min\left\{\frac{R_2}{2}, \frac{R_3}{2}\right\}$	$\min\left\{\frac{R_2}{2}, \frac{R_3}{2}\right\}$

Stage 1: To solve the two-stage problem ($n = 1$), we plug the solution just obtained for $f_2^*(R_1, R_2, R_3)$ into Eq. (3). For stage 2,

$$(R_1, R_2, R_3) = (4 - x_1, 12, 18 - 3x_1),$$

so that

$$f_2^*(4 - x_1, 12, 18 - 3x_1) = 5 \min\left\{\frac{R_2}{2}, \frac{R_3}{2}\right\} = 5 \min\left\{\frac{12}{2}, \frac{18 - 3x_1}{2}\right\}$$

is the specific solution plugged into Eq. (3). After we combine its constraints on x_1, Eq. (3) then becomes

$$f_1^*(4, 12, 18) = \max_{0 \le x_1 \le 4} \left\{3x_1 + 5 \min\left\{\frac{12}{2}, \frac{18 - 3x_1}{2}\right\}\right\}.$$

Over the feasible interval $0 \le x_1 \le 4$, notice that

$$\min\left\{\frac{12}{2}, \frac{18 - 3x_1}{2}\right\} = \begin{cases} 6 & \text{if } 0 \le x_1 \le 2, \\ 9 - \frac{3}{2}x_1 & \text{if } 2 \le x_1 \le 4, \end{cases}$$

so that

$$3x_1 + 5 \min\left\{\frac{12}{2}, \frac{18 - 3x_1}{2}\right\} = \begin{cases} 3x_1 + 30 & \text{if } 0 \le x_1 \le 2, \\ 45 - \frac{9}{2}x_1 & \text{if } 2 \le x_1 \le 4. \end{cases}$$

Because both

$$\max_{0 \le x_1 \le 2} \{3x_1 + 30\} \quad \text{and} \quad \max_{2 \le x_1 \le 4} \left\{45 - \frac{9}{2}x_1\right\}$$

achieve their maximum at $x_1 = 2$, it follows that $x_1^* = 2$ and that this maximum is 36, as given in the following table.

$n = 1$:

(R_1, R_2, R_3)	$f_1^*(R_1, R_2, R_3)$	x_1^*
(4, 12, 18)	36	2

Because $x_1^* = 2$ leads to

$$R_1 = 4 - 2 = 2, \qquad R_2 = 12, \qquad R_3 = 18 - 3(2) = 12$$

for stage 2, the $n = 2$ table yields $x_2^* = 6$. Consequently, $x_1^* = 2$, $x_2^* = 6$ is the optimal solution for this problem (as originally found in Sec. 3.1), and the $n = 1$ table shows that the resulting value of Z is 36.

10.4 Probabilistic Dynamic Programming

Probabilistic dynamic programming differs from deterministic dynamic programming in that the state at the next stage is *not* completely determined by the state and policy decision at the current stage. Rather, there is a *probability distribution* for what the next state will be. However, this probability distribution still is completely determined by the state and policy decision at the current stage. The resulting basic structure for probabilistic dynamic programming is described diagrammatically in Fig. 10.11.

For the purposes of this diagram, we let S denote the number of possible states at stage $n + 1$ and label these states on the right side as $1, 2, \ldots, S$. The system goes to state i with probability p_i $(i = 1, 2, \ldots, S)$ given state s_n and decision x_n at stage n. If the system goes to state i, C_i is the contribution of stage n to the objective function.

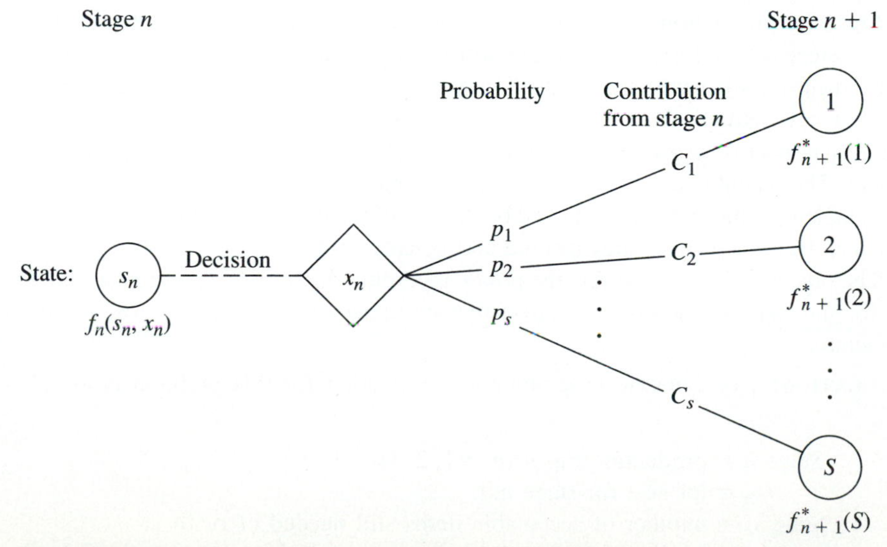

Figure 10.11 The basic structure for probabilistic dynamic programming.

When Fig. 10.11 is expanded to include all the possible states and decisions at all the stages, it is sometimes referred to as a **decision tree**. If the decision tree is not too large, it provides a useful way of summarizing the various possibilities.

Because of the probabilistic structure, the relationship between $f_n(s_n, x_n)$ and the $f^*_{n+1}(s_{n+1})$ necessarily is somewhat more complicated than that for deterministic dynamic programming. The precise form of this relationship will depend upon the form of the overall objective function.

To illustrate, suppose that the objective is to *minimize* the *expected sum* of the contributions from the individual stages. In this case, $f_n(s_n, x_n)$ represents the minimum expected sum from stage n onward, *given* that the state and policy decision at stage n are s_n and x_n, respectively. Consequently,

$$f_n(s_n, x_n) = \sum_{i=1}^{S} p_i[C_i + f^*_{n+1}(i)],$$

with

$$f^*_{n+1}(i) = \min_{x_{n+1}} f_{n+1}(i, x_{n+1}),$$

where this minimization is taken over the *feasible* values of x_{n+1}.

Example 6 has this same form. Example 7 will illustrate another form.

Example 6—Determining Reject Allowances

The HIT-AND-MISS MANUFACTURING COMPANY has received an order to supply one item of a particular type. However, the customer has specified such stringent quality requirements that the manufacturer may have to produce more than one item to obtain an item that is acceptable. The number of *extra* items produced in a production run is called the *reject allowance*. Including a reject allowance is common practice when one is producing for a custom order, and it seems advisable in this case.

The manufacturer estimates that each item of this type that is produced will be *acceptable* with probability $\frac{1}{2}$ and *defective* (without possibility for rework) with probability $\frac{1}{2}$. Thus the number of acceptable items produced in a lot of size L will have a *binomial distribution;* i.e., the probability of producing no acceptable items in such a lot is $(\frac{1}{2})^L$.

Marginal production costs for this product are estimated to be \$100 per item (even if defective), and excess items are worthless. In addition, a setup cost of \$300 must be incurred whenever the production process is set up for this product, and a completely new setup at this same cost is required for each subsequent production run if a lengthy inspection procedure reveals that a completed lot has not yielded an acceptable item. The manufacturer has time to make no more than three production runs. If an acceptable item has not been obtained by the end of the third production run, the cost to the manufacturer in lost sales income and penalty costs will be \$1,600.

The objective is to determine the policy regarding the lot size (1 + reject allowance) for the required production run(s) that minimizes total expected cost for the manufacturer.

FORMULATION: A dynamic programming formulation for this problem is as follows:

Stage n = production run n ($n = 1, 2, 3$),
x_n = lot size for stage n,
State s_n = number of acceptable items still needed (1 or 0) at beginning of stage n.

Thus, at stage 1, state $s_1 = 1$. If at least one acceptable item is obtained subsequently, the state changes to $s_n = 0$, after which no additional costs need to be incurred.

Because of the stated objective for the problem,

$f_n(s_n, x_n)$ = total expected cost for stages $n, \ldots, 3$ if system starts in state s_n at stage n, immediate decision is x_n, and optimal decisions are made thereafter,

$$f_n^*(s_n) = \min_{x_n=0, 1, \ldots} f_n(s_n, x_n),$$

where $f_n^*(0) = 0$. With \$100 as the unit of money, the contribution to cost from stage n is $[K(x_n) + x_n]$ regardless of the next state, where $K(x_n)$ is a function of x_n such that

$$K(x_n) = \begin{cases} 0, & \text{if } x_n = 0 \\ 3, & \text{if } x_n > 0. \end{cases}$$

Therefore, for $s_n = 1$,

$$f_n(1, x_n) = K(x_n) + x_n + \left(\frac{1}{2}\right)^{x_n} f_{n+1}^*(1) + \left[1 - \left(\frac{1}{2}\right)^{x_n}\right] f_{n+1}^*(0)$$

$$= K(x_n) + x_n + \left(\frac{1}{2}\right)^{x_n} f_{n+1}^*(1)$$

[where $f_4^*(1)$ is defined to be 16, the terminal cost if no acceptable items have been obtained]. A summary of these basic relationships is given in Fig. 10.12.

Consequently, the recursive relationship for the dynamic programming calculations is

$$f_n^*(1) = \min_{x_n=0, 1, \ldots} \left\{K(x_n) + x_n + \left(\frac{1}{2}\right)^{x_n} f_{n+1}^*(1)\right\}$$

for $n = 1, 2, 3$.

Solution Procedure: The calculations using this recursive relationship are summarized as follows.

$n = 3$:

s_3 \ x_3	$f_3(1, x_3) = K(x_3) + x_3 + 16(\frac{1}{2})^{x_3}$						$f_3^*(s_3)$	x_3^*
	0	1	2	3	4	5		
0	0						0	0
1	16	12	9	8	8	$8\frac{1}{2}$	8	3 or 4

$n = 2$:

s_2 \ x_2	$f_2(1, x_2) = K(x_2) + x_2 + (\frac{1}{2})^{x_2} f_3^*(1)$					$f_2^*(s_2)$	x_2^*
	0	1	2	3	4		
0	0					0	0
1	8	8	7	7	$7\frac{1}{2}$	7	2 or 3

$n = 1$:

s_1 \ x_1	$f_1(1, x_1) = K(x_1) + x_1 + (\frac{1}{2})^{x_1} f_2^*(1)$					$f_1^*(s_1)$	x_1^*
	0	1	2	3	4		
1	7	$7\frac{1}{2}$	$6\frac{3}{4}$	$6\frac{7}{8}$	$7\frac{7}{16}$	$6\frac{3}{4}$	2

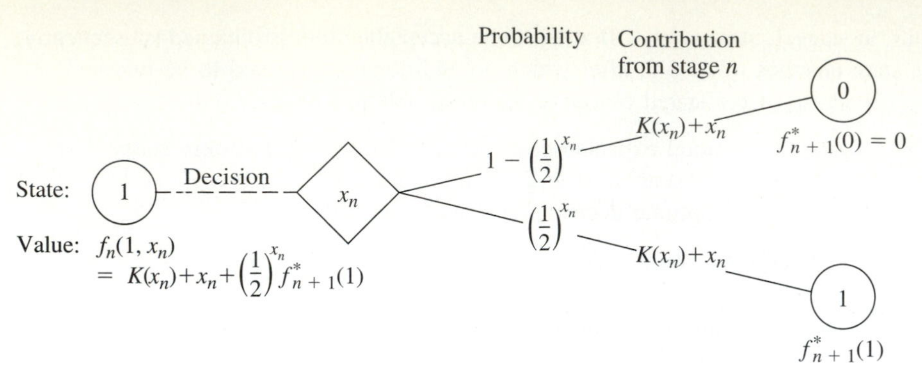

Figure 10.12 The basic structure for the Hit-and-Miss Manufacturing Co. problem.

Thus the optimal policy is to produce two items on the first production run; if none is acceptable, then produce either two or three items on the second production run; if none is acceptable, then produce either three or four items on the third production run. The total expected cost for this policy is \$675.

Example 7—Winning in Las Vegas

An enterprising young statistician believes that she has developed a system for winning a popular Las Vegas game. Her colleagues do not believe that her system works, so they have made a large bet with her that if she starts with three chips, she will not have at least five chips after three plays of the game. Each play of the game involves betting any desired number of available chips and then either winning or losing this number of chips. The statistician believes that her system will give her a probability of $\frac{2}{3}$ of winning a given play of the game.

Assuming the statistician is correct, we now use dynamic programming to determine her optimal policy regarding how many chips to bet (if any) at each of the three plays of the game. The decision at each play should take into account the results of earlier plays. The objective is to maximize the probability of winning her bet with her colleagues.

FORMULATION: The dynamic programming formulation for this problem is

$$\text{Stage } n = n\text{th play of game } (n = 1, 2, 3),$$
$$x_n = \text{number of chips to bet at stage } n,$$
$$\text{State } s_n = \text{number of chips in hand to begin stage } n.$$

This definition of the state is chosen because it provides the needed information about the current situation for making an optimal decision on how many chips to bet next.

Because the objective is to maximize the probability that the statistician will win her bet, the objective function to be maximized at each stage must be the probability of finishing the three plays with at least five chips. (Note that the value of ending with more than five chips is just the same as ending with exactly five, since the bet is won either way.) Therefore,

$f_n(s_n, x_n)$ = probability of finishing three plays with at least five chips, given that statistician starts stage n in state s_n, makes immediate decision x_n, and makes optimal decisions thereafter,

$$f^*_n(s_n) = \max_{x_n=0, 1, \ldots, s_n} f_n(s_n, x_n).$$

The expression for $f_n(s_n, x_n)$ must reflect the fact that it may still be possible to accumulate five chips eventually even if the statistician should lose the next play. If she loses, the state at the next stage will be $s_n - x_n$, and the probability of finishing with at least five chips will then be $f^*_{n+1}(s_n - x_n)$. If she wins the next play instead, the state will become $s_n + x_n$, and the corresponding probability will be $f^*_{n+1}(s_n + x_n)$. Because the assumed probability of winning a given play is $\frac{2}{3}$, it now follows that

$$f_n(s_n, x_n) = \tfrac{1}{3} f^*_{n+1}(s_n - x_n) + \tfrac{2}{3} f^*_{n+1}(s_n + x_n)$$

[where $f^*_4(s_4)$ is defined to be 0 for $s_4 < 5$ and 1 for $s_4 \geq 5$]. Thus there is no direct contribution to the objective function from stage n other than the effect of then being in the next state. These basic relationships are summarized in Fig. 10.13.

Therefore, the recursive relationship for this problem is

$$f^*_n(s_n) = \max_{x_n = 0, 1, \ldots, s_n} \{\tfrac{1}{3} f^*_{n+1}(s_n - x_n) + \tfrac{2}{3} f^*_{n+1}(s_n + x_n)\},$$

for $n = 1, 2, 3$, with $f^*_4(s_4)$ as just defined.

Solution Procedure: This recursive relationship leads to the following computational results.

$n = 3$:

s_3	$f^*_3(s_3)$	x^*_3
0	0	—
1	0	—
2	0	—
3	$\frac{2}{3}$	2 (or more)
4	$\frac{2}{3}$	1 (or more)
≥5	1	0 (or $\leq s_3 - 5$)

$n = 2$:

s_2 \ x_2	$f_2(s_2, x_2) = \frac{1}{3} f^*_3(s_2 - x_2) + \frac{2}{3} f^*_3(s_2 + x_2)$					$f^*_2(s_2)$	x^*_2
	0	1	2	3	4		
0	0					0	—
1	0	0				0	—
2	0	$\frac{4}{9}$	$\frac{4}{9}$			$\frac{4}{9}$	1 or 2
3	$\frac{2}{3}$	$\frac{4}{9}$	$\frac{2}{3}$	$\frac{2}{3}$		$\frac{2}{3}$	0, 2, or 3
4	$\frac{2}{3}$	$\frac{8}{9}$	$\frac{2}{3}$	$\frac{2}{3}$	$\frac{2}{3}$	$\frac{8}{9}$	1
≥5	1					1	0 (or $\leq s_2 - 5$)

$n = 1$:

s_1 \ x_1	$f_1(s_1, x_1) = \frac{1}{3} f^*_2(s_1 - x_1) + \frac{2}{3} f^*_2(s_1 + x_1)$				$f^*_1(s_1)$	x^*_1
	0	1	2	3		
3	$\frac{2}{3}$	$\frac{20}{27}$	$\frac{2}{3}$	$\frac{2}{3}$	$\frac{20}{27}$	1

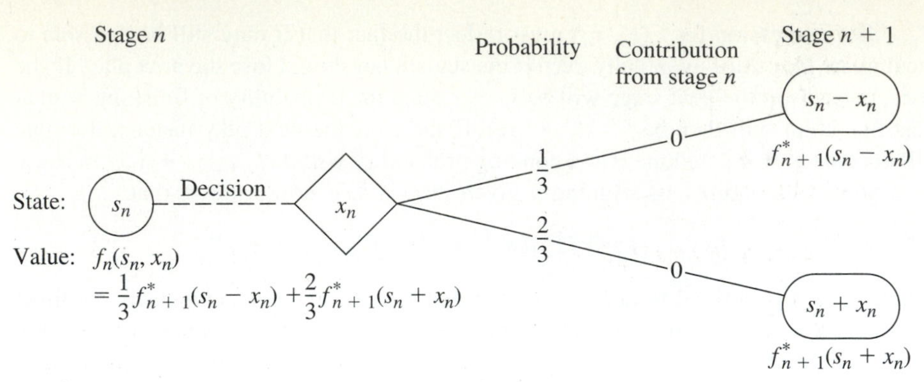

Figure 10.13 The basic structure for the Las Vegas problem.

Therefore, the optimal policy is

$$x_1^* = 1 \begin{cases} \text{if win,} & x_2^* = 1 \begin{cases} \text{if win,} & x_3^* = 0 \\ \text{if lose,} & x_3^* = 2 \text{ or } 3. \end{cases} \\ \text{if lose,} & x_2^* = 1 \text{ or } 2 \begin{cases} \text{if win,} & x_3^* = \begin{cases} 2 \text{ or } 3 & (\text{for } x_2^* = 1) \\ 1, 2, 3, \text{ or } 4 & (\text{for } x_2^* = 2) \end{cases} \\ \text{if lose,} & \text{bet is lost.} \end{cases} \end{cases}$$

This policy gives the statistician a probability of $\frac{20}{27}$ of winning her bet with her colleagues.

10.5 Conclusions

Dynamic programming is a very useful technique for making a *sequence of interrelated decisions.* It requires formulating an appropriate *recursive relationship* for each individual problem. However, it provides a great computational savings over using exhaustive enumeration to find the best combination of decisions, especially for large problems. For example, if a problem has 10 stages with 10 states and 10 possible decisions at each stage, then exhaustive enumeration must consider up to 10 billion combinations, whereas dynamic programming need make no more than a thousand calculations (10 for each state at each stage).

This chapter has considered only dynamic programming with a *finite* number of stages. Chapter 19 is devoted to a general kind of model for probabilistic dynamic programming where the stages continue to recur indefinitely, namely, Markov decision processes.

SELECTED REFERENCES

1. Bertsekas, D. P., *Dynamic Programming: Deterministic and Stochastic Models,* Prentice-Hall, Englewood Cliffs, NJ, 1987.
2. Cooper, L. L., and M. W. Cooper, *Introduction to Dynamic Programming,* Pergamon Press, Elmsford, NY, 1981.
3. Denardo, E. V., *Dynamic Programming Theory and Applications,* Prentice-Hall, Englewood Cliffs, NJ, 1982.

4. Dreyfus, S. E., and A. M. Law, *The Art and Theory of Dynamic Programming,* Academic Press, New York, 1977.
5. Howard, R. A., "Dynamic Programming," *Management Science,* **12:** 317–345, 1966.
6. Smith, D. K., *Dynamic Programming: A Practical Introduction,* Ellis Horwood, London, 1991.
7. Sniedovich, M., *Dynamic Programming,* Marcel Dekker, New York, 1991.

RELEVANT ROUTINES IN YOUR OR COURSEWARE

A Demonstration Example: *Deterministic Dynamic Programming Algorithm*

An Interactive Routine: *Interactive Deterministic Dynamic Programming Algorithm*

To access this routine, call the MathProg program and then choose *Dynamic Programming* under the Area menu. See Appendix 1 for documentation of the software.

PROBLEMS[1]

To the left of each of the following problems (or their parts), we have inserted a D (for demonstration) or I (for interactive routine) whenever a corresponding routine listed above can be helpful. An asterisk on the I indicates that this routine definitely should be used (unless your instructor gives contrary instructions) and that the printout from this routine is all that needs to be turned in to show your work in executing the algorithm. An asterisk on the problem number indicates that at least a partial answer is given in the back of the book.

10.2-1. Consider the following network, where each number along a link represents the actual distance between the pair of nodes connected by that link. The objective is to find the shortest path from the origin to the destination.

(*a*) What are the stages and states for the dynamic programming formulation of this problem?

(*b*) Use dynamic programming to solve this problem. However, instead of using the usual tables, show your work graphically (similar to Fig. 10.2). In particular, start with the above network, where the answers already are given for $f_n^*(s_n)$ for four of the nodes;

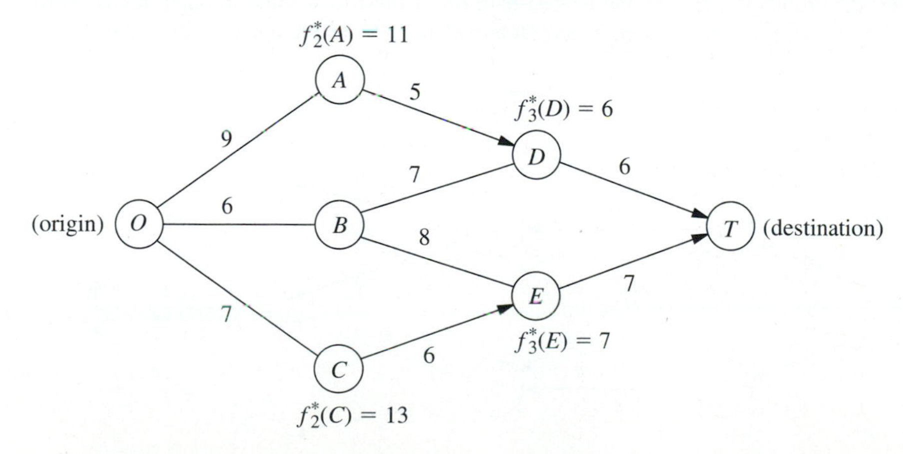

[1] Problems 10.2-4 and 10.3-12 have been adapted, with permission, from previous operations research examinations given by the Society of Actuaries.

then solve for and fill in $f_2^*(B)$ and $f_1^*(O)$. Draw an arrowhead that shows the optimal link to traverse out of each of the latter two nodes. Finally, identify the optimal route by following the arrows from node O onward to node T.

(*c*) Use dynamic programming to solve this problem by manually constructing the usual tables for $n = 3$, $n = 2$, and $n = 1$.

(*d*) Use the shortest-path algorithm presented in Sec. 9.3 to solve this problem. Compare and contrast this approach with the one in parts (*b*) and (*c*).

10.2-2. The sales manager for a publisher of college textbooks has six traveling salespeople to assign to three different regions of the country. She has decided that each region should be assigned at least one salesperson and that each individual salesperson should be restricted to one of the regions; but now she wants to determine how many salespeople should be assigned to the respective regions in order to maximize sales.

The following table gives the estimated increase in sales (in appropriate units) in each region if it were allocated various numbers of salespeople:

	Region		
Salespersons	1	2	3
1	35	21	28
2	48	42	41
3	70	56	63
4	89	70	75

(*a*) Use dynamic programming to solve this problem. Instead of using the usual tables, show your work graphically by constructing and filling in a network such as the one shown for Prob. 10.2-1. Proceed as in Prob. 10.2-1*b* by solving for $f_n^*(s_n)$ for each node (except the terminal node) and writing its value by the node. Draw an arrowhead to show the optimal link (or links in case of a tie) to traverse out of each node. Finally, identify the resulting optimal route (or routes) through the network and the corresponding optimal solution (or solutions).

D, I* (*b*) Use dynamic programming to solve this problem by constructing the usual tables for $n = 3$, $n = 2$, and $n = 1$.

10.2-3. Consider the following project network for a PERT-type system as described in Sec. 9.8, where the number next to each arc is the time required for the corresponding activity. Consider the problem of finding the longest path (the largest total time) through this network from event 1 (project start) to event 9 (project completion), since the longest path is the critical path.

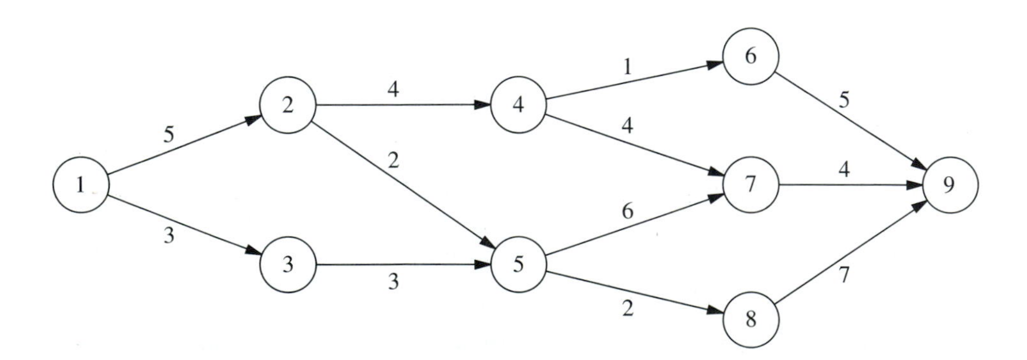

(*a*) What are the stages and states for the dynamic programming formulation of this problem?

(*b*) Use dynamic programming to solve this problem. However, instead of using the usual tables, show your work graphically. In particular, fill in the values of the various $f_n^*(s_n)$ next to the corresponding nodes (except node 9), and show the resulting optimal arc to traverse out of each node by drawing an arrowhead near the beginning of the arc. Then identify the optimal path (the longest path) by following these arrowheads from node 1 to node 9. If there is more than one optimal path, identify them all.

D, I* (*c*) Use dynamic programming to solve this problem by constructing the usual tables for $n = 4$, $n = 3$, $n = 2$, and $n = 1$.

10.2-4. Consider the following statements about solving dynamic programming problems. Label each statement as true or false, and then justify your answer by referring to specific statements (with page citations) in the chapter.

(*a*) The solution procedure uses a recursive relationship that enables solving for the optimal policy for stage $(n + 1)$ given the optimal policy for stage n.

(*b*) After completing the solution procedure , if a nonoptimal decision is made by mistake at some stage, the solution procedure will need to be reapplied to determine the new optimal decisions (given this nonoptimal decision) at the subsequent stages.

(*c*) Once an optimal policy has been found for the overall problem, the information needed to specify the optimal decision at a particular stage is the state at that stage and the decisions made at preceding stages.

D, I* **10.3-1.*** The owner of a chain of three grocery stores has purchased five crates of fresh strawberries. The estimated probability distribution of potential sales of the strawberries before spoilage differs among the three stores. Therefore, the owner wants to know how to allocate five crates to the three stores to maximize expected profit.

For administrative reasons, the owner does not wish to split crates between stores. However, he is willing to distribute no crates to any of his stores.

The following table gives the estimated expected profit at each store when it is allocated various numbers of crates:

	Store		
Crates	1	2	3
0	0	0	0
1	5	6	4
2	9	11	9
3	14	15	13
4	17	19	18
5	21	22	20

Use dynamic programming to determine how many of the five crates should be assigned to each of the three stores to maximize the total expected profit.

D, I* **10.3-2.** A college student has 7 days remaining before final examinations begin in her four courses, and she wants to allocate this study time as effectively as possible. She needs at least 1 day on each course, and she likes to concentrate on just one course each day, so she wants to allocate 1, 2, 3, or 4 days to each course. Having recently taken an operations research course, she decides to use dynamic programming to make these allocations to maximize the total grade points to be obtained from the four courses. She estimates that the alternative allocations for each course would yield the number of grade points shown in the following table:

	Estimated Grade Points			
	Course			
Study Days	1	2	3	4
1	3	5	2	6
2	5	5	4	7
3	6	6	7	9
4	7	9	8	9

Solve this problem by dynamic programming.

D, I* **10.3-3.** A company is planning its advertising strategy for next year for its three major products. Since the three products are quite different, each advertising effort will focus on a single product. In units of millions of dollars, a total of 6 is available for advertising next year, where the advertising expenditure for each product must be an integer greater than or equal to 1. The vice-president for marketing has established the objective: Determine how much to spend on each product in order to maximize total sales. The following table gives the estimated increase in sales (in appropriate units) for the different advertising expenditures:

Advertising	*Product*		
Expenditure	1	2	3
1	7	4	6
2	10	8	9
3	14	11	13
4	17	14	15

Use dynamic programming to solve this problem.

D, I* **10.3-4.** A political campaign is entering its final stage, and polls indicate a very close election. One of the candidates has enough funds left to purchase TV time for a total of five prime-time commercials on TV stations located in four different areas. Based on polling information, an estimate has been made of the number of additional votes that can be won in the different broadcasting areas depending upon the number of commercials run. These estimates are given in the following table in thousands of votes:

	Area			
Commercials	1	2	3	4
0	0	0	0	0
1	4	6	5	3
2	7	8	9	7
3	9	10	11	12
4	12	11	10	14
5	15	12	9	16

Use dynamic programming to determine how the five commercials should be distributed among the four areas in order to maximize the estimated number of votes won.

D, I* **10.3-5.** A county chairwoman of a certain political party is making plans for an upcoming presidential election. She has received the services of six volunteer workers for precinct work, and she wants to assign them to four precincts in such a way as to maximize their effectiveness. She feels that it would be inefficient to assign a worker to more than one precinct, but she is willing to assign no workers to any one of the precincts if they can accomplish more in other precincts.

The following table gives the estimated increase in the number of votes for the party's candidate in each precinct if it were allocated various numbers of workers:

	Precinct			
Workers	1	2	3	4
0	0	0	0	0
1	4	7	5	6
2	9	11	10	11
3	15	16	15	14
4	18	18	18	16
5	22	20	21	17
6	24	21	22	18

This problem has several optimal solutions for how many of the six workers should be assigned to each of the four precincts to maximize the total estimated increase in the plurality of the party's candidate. Use dynamic programming to find all of them so the chairwoman can make the final selection based on other factors.

10.3-6. Use dynamic programming to solve the Northern Airplane Co. production scheduling problem presented in Sec. 8.1 (see Table 8.7). Assume that production quantities must be integer multiples of 5.

D, I* **10.3-7.** Reconsider the Build-Em-Fast Co. problem described in Prob. 8.1-7. Use dynamic programming to solve this problem.

10.3-8.* A company will soon be introducing a new product into a very competitive market and is currently planning its marketing strategy. The decision has been made to introduce the product in three phases. Phase 1 will feature making a special introductory offer of the product to the public at a greatly reduced price to attract first-time buyers. Phase 2 will involve an intensive advertising campaign to persuade these first-time buyers to continue purchasing the product at a regular price. It is known that another company will be introducing a new competitive product at about the time that phase 2 will end. Therefore, phase 3 will involve a follow-up advertising and promotion campaign to try to keep the regular purchasers from switching to the competitive product.

A total of $4 million has been budgeted for this marketing campaign. The problem now is to determine how to allocate this money most effectively to the three phases. Let m denote the initial share of the market (expressed as a percentage) attained in phase 1, f_2 the fraction of this market share that is retained in phase 2, and f_3 the fraction of the remaining market share that is retained in phase 3. Given the following data, use dynamic programming to determine how to allocate the $4 million to maximize the final share of the market for the new product, i.e., to maximize mf_2f_3.

D, I* (*a*) Assume that the money must be spent in integer multiples of $1 million in each phase, where the minimum permissible multiple is 1 for phase 1 and 0 for phases 2 and 3.

The following table gives the estimated effect of expenditures in each phase:

Millions of Dollars Expended	Effect on Market Share		
	m	f_2	f_3
0	—	0.2	0.3
1	20	0.4	0.5
2	30	0.5	0.6
3	40	0.6	0.7
4	50	—	—

(*b*) Now assume that *any* amount within the total budget can be spent in each phase, where the estimated effect of spending an amount x_i (in units of *millions* of dollars) in phase i ($i = 1, 2, 3$) is

$$m = 10x_1 - x_1^2$$

$$f_2 = 0.40 + 0.10x_2$$

$$f_3 = 0.60 + 0.07x_3.$$

[*Hint:* After solving for the $f_2^*(s)$ and $f_3^*(s)$ functions analytically, solve for x_1^* graphically.]

10.3-9. The management of a company is considering three possible new products for next year's product line. A decision now needs to be made regarding which products to market and at what production levels.

Initiating the production of two of these products would require a substantial start-up cost, as shown in the first row of the table below. Once production is under way, the marginal net revenue from each unit produced is shown in the second row. The third row gives the percentage of the available production capacity that would be used for each unit produced.

	Product		
	1	2	3
Start-up cost	3	2	0
Marginal net revenue	2	3	1
Capacity used per unit, %	20	40	20

Only 3 units of product 1 could be sold, whereas all units that could be produced of the other two products could be sold. The objective is to determine the number of units of each product to produce in order to maximize the total profit (total net revenue minus start-up costs).

D, I* (*a*) Assuming that production quantities must be integers, use dynamic programming to solve this problem.

(*b*) Now consider the case where the divisibility assumption holds, so that the variables representing production quantities are treated as *continuous* variables. Assuming that proportionality holds for both net revenues and capacities used, use dynamic programming to solve this problem.

10.3-10. Consider an electronic system consisting of four components, each of which must work for the system to function. The reliability of the system can be improved by installing several parallel units in one or more of the components. The following table gives the probability that the respective components will function if they consist of one, two, or three parallel units:

	Probability of Functioning			
Parallel Units	Component 1	Component 2	Component 3	Component 4
1	0.5	0.6	0.7	0.5
2	0.6	0.7	0.8	0.7
3	0.8	0.8	0.9	0.9

The probability that the system will function is the product of the probabilities that the respective components will function.

The cost (in hundreds of dollars) of installing one, two, or three parallel units in the respective components is given by the following table:

	Cost			
Parallel Units	Component 1	Component 2	Component 3	Component 4
1	1	2	1	2
2	2	4	3	3
3	3	5	4	4

Because of budget limitations, a maximum of $1,000 can be expended.

Use dynamic programming to determine how many parallel units should be installed in each of the four components to maximize the probability that the system will function.

D, I* **10.3-11.** Consider the following integer nonlinear programming problem.

$$\text{Maximize} \qquad Z = 3x_1^2 - x_1^3 + 5x_2^2 - x_2^3,$$

subject to

$$x_1 + 2x_2 \leq 4$$

and

$$x_1 \geq 0, \qquad x_2 \geq 0$$

$$x_1, x_2 \text{ are integers.}$$

Use dynamic programming to solve this problem.

10.3-12. Consider the following integer nonlinear programming problem.

$$\text{Maximize} \qquad Z = 18x_1 - x_1^2 + 20x_2 + 10x_3,$$

subject to

$$2x_1 + 4x_2 + 3x_3 \leq 11$$

and

$$x_1, x_2, x_3 \text{ are nonnegative integers.}$$

Use dynamic programming to solve this problem.

10.3-13. Consider the following integer nonlinear programming problem.

$$\text{Maximize} \quad Z = x_1 x_2^2 x_3^3,$$

subject to

$$x_1 + 2x_2 + 3x_3 \leq 10$$

$$x_1 \geq 1, \quad x_2 \geq 1, \quad x_3 \geq 1,$$

and

$$x_1, x_2, x_3 \text{ are integers.}$$

Use dynamic programming to solve this problem.

10.3-14.* Consider the following nonlinear programming problem.

$$\text{Maximize} \quad Z = 36x_1 + 9x_1^2 - 6x_1^3 + 36x_2 - 3x_2^3,$$

subject to

$$x_1 + x_2 \leq 3$$

and

$$x_1 \geq 0, \quad x_2 \geq 0.$$

Use dynamic programming to solve this problem.

10.3-15. Resolve the Local Job Shop employment scheduling problem (Example 4) when the total cost of changing the level of employment from one season to the next is changed to $100 times the square of the difference in employment levels.

10.3-16. Consider the following nonlinear programming problem.

$$\text{Maximize} \quad Z = 2x_1^2 + 2x_2 + 4x_3 - x_3^2$$

subject to

$$2x_1 + x_2 + x_3 \leq 4$$

and

$$x_1 \geq 0, \quad x_2 \geq 0, \quad x_3 \geq 0.$$

Use dynamic programming to solve this problem.

10.3-17. Consider the following nonlinear programming problem.

$$\text{Maximize} \quad Z = 2x_1 + x_2^2,$$

subject to

$$x_1^2 + x_2^2 \leq 4$$

and

$$x_1 \geq 0, \quad x_2 \geq 0.$$

Use dynamic programming to solve this problem.

10.3-18. Consider the following nonlinear programming problem.

$$\text{Minimize} \quad Z = x_1^4 + 2x_2^2$$

subject to

$$x_1^2 + x_2^2 \geq 2$$

(there are no nonnegativity constraints). Use dynamic programming to solve this problem.

10.3-19. Consider the following nonlinear programming problem.

$$\text{Maximize} \quad Z = x_1^2 x_2,$$

subject to

$$x_1^2 + x_2 \leq 2.$$

(Note that there are no nonnegativity constraints.) Use dynamic programming to solve this problem.

10.3-20. Consider the following nonlinear programming problem.

$$\text{Maximize} \quad Z = x_1^3 + 4x_2^2 + 16x_3,$$

subject to

$$x_1 x_2 x_3 = 4$$

and

$$x_1 \geq 1, \quad x_2 \geq 1, \quad x_3 \geq 1.$$

(*a*) Solve by dynamic programming when, in addition to the given constraints, all three variables also are required to be integer.
(*b*) Use dynamic programming to solve the problem as given (continuous variables).

10.3-21. Consider the following nonlinear programming problem.

$$\text{Maximize} \quad Z = x_1(1 - x_2)x_3,$$

subject to

$$x_1 - x_2 + x_3 \leq 1$$

and

$$x_1 \geq 0, \quad x_2 \geq 0, \quad x_3 \geq 0.$$

Use dynamic programming to solve this problem.

10.3-22. Consider the following linear programming problem.

$$\text{Maximize} \quad Z = 15x_1 + 10x_2,$$

subject to

$$x_1 + 2x_2 \leq 6$$
$$3x_1 + x_2 \leq 8$$

and

$$x_1 \geq 0, \quad x_2 \geq 0.$$

Use dynamic programming to solve this problem.

10.3-23. Consider the following nonlinear programming problem.

$$\text{Maximize} \quad Z = 5x_1 + x_2,$$

subject to

$$2x_1^2 + x_2 \leq 13$$
$$x_1^2 + x_2 \leq 9$$

and

$$x_1 \geq 0, \quad x_2 \geq 0.$$

Use dynamic programming to solve this problem.

10.3-24. Consider the following "fixed-charge" problem.

$$\text{Maximize} \quad Z = 3x_1 + 7x_2 + 6f(x_3),$$

subject to

$$x_1 + 3x_2 + 2x_3 \leq 6$$
$$x_1 + x_2 \leq 5$$

and

$$x_1 \geq 0, \quad x_2 \geq 0, \quad x_3 \geq 0,$$

where

$$f(x_3) = \begin{cases} 0 & \text{if } x_3 = 0, \\ -1 + x_3 & \text{if } x_3 > 0. \end{cases}$$

Use dynamic programming to solve this problem.

10.4-1. A backgammon player will be playing three consecutive matches with friends tonight. For each match, he will have the opportunity to place an even bet that he will win; the amount bet can be *any* quantity of his choice between zero and the amount of money he still has left after the bets on the preceding matches. For each match, the probability is $\frac{1}{2}$ that he will win the match and thus win the amount bet, whereas the probability is $\frac{1}{2}$ that he will lose the match and thus lose the amount bet. He will begin with \$75, and his goal is to have \$100 at the end. (Because these are friendly matches, he does not want to end up with more than \$100.) Therefore, he wants to find the optimal betting policy (including all ties) that maximizes the probability that he will have exactly \$100 after the three matches.

Use dynamic programming to solve this problem.

10.4-2. Imagine that you have \$5,000 to invest and that you will have an opportunity to invest that amount in either of two investments (*A* or *B*) at the beginning of each of the next 3 years. Both investments have uncertain returns. For investment *A* you will either lose your money entirely or (with higher probability) get back \$10,000 (a profit of \$5,000) at the end of the year. For investment *B* you will get back either just your \$5,000 or (with low probability) \$10,000 at the end of the year. The probabilities for these events are as follows:

Investment	Amount Returned (\$)	Probability
A	0	0.3
	10,000	0.7
B	5,000	0.9
	10,000	0.1

You are allowed to make only (at most) *one* investment each year, and you can invest only \$5,000 each time. (Any additional money accumulated is left idle.)

(*a*) Use dynamic programming to find the investment policy that maximizes the expected amount of money you will have after 3 years.

(*b*) Use dynamic programming to find the investment policy that maximizes the probability that you will have at least \$10,000 after 3 years.

10.4-3.* Suppose that the situation for the Hit-and-Miss Manufacturing Co. problem (Example 6) has changed somewhat. After a more careful analysis, you now estimate that each item produced will be acceptable with probability $\frac{2}{3}$, rather than $\frac{1}{2}$, so that the probability of producing *zero* acceptable items in a lot of size L is $(\frac{1}{3})^L$. Furthermore, there now is only enough time available to make two production runs. Use dynamic programming to determine the new optimal policy for this problem.

10.4-4. Reconsider Example 7. Suppose that the bet is changed as follows: ''Starting with two chips, she will not have at least five chips after five plays of the game.'' By referring to the previous computational results, make additional calculations to determine the new optimal policy for the enterprising young statistician.

10.4-5. The Profit & Gambit Co. has a major product that has been losing money recently because of declining sales. In fact, during the current quarter of the year, sales will be 4 million units below the break-even point. Because the marginal revenue for each unit sold exceeds the marginal cost by $5, this amounts to a loss of $20 million for the quarter. Therefore, management must take action quickly to rectify this situation. Two alternative courses of action are being considered. One is to abandon the product immediately, incurring a cost of $20 million for shutting down. The other alternative is to undertake an intensive advertising campaign to increase sales and then abandon the product (at the cost of $20 million) only if the campaign is not sufficiently successful. Tentative plans for this advertising campaign have been developed and analyzed. It would extend over the next three quarters (subject to early cancellation), and the cost would be $30 million in each of the three quarters. It is estimated that the increase in sales would be approximately 3 million units in the first quarter, another 2 million units in the second quarter, and another 1 million units in the third quarter. However, because of a number of unpredictable market variables, there is considerable uncertainty as to what impact the advertising actually would have; and careful analysis indicates that the estimates for each quarter could turn out to be off by as much as 2 million units in either direction. (To quantify this uncertainty, assume that the additional increases in sales in the three quarters are independent random variables having a uniform distribution with a range from 1 to 5 million, from 0 to 4 million, and from -1 to 3 million, respectively.) If the actual increases are too small, the advertising campaign can be discontinued and the product abandoned at the end of either of the next two quarters.

If the intensive advertising campaign were initiated and continued to its completion, it is estimated that the sales for some time thereafter would continue to be at about the same level as in the third (last) quarter of the campaign. Therefore, if the sales in that quarter still were below the break-even point, the product would be abandoned. Otherwise, it is estimated that the expected discounted profit thereafter would be $40 for each unit sold over the break-even point in the third quarter.

Use dynamic programming to determine the optimal policy maximizing the expected profit.

11

Game Theory

Life is full of conflict and competition. Numerous examples involving adversaries in conflict include parlor games, military battles, political campaigns, athletic competitions, advertising and marketing campaigns by competing business firms, and so forth. A basic feature in many of these situations is that the final outcome depends primarily upon the combination of strategies selected by the adversaries. Game theory is a mathematical theory that deals with the general features of competitive situations like these in a formal, abstract way. It places particular emphasis on the decision-making processes of the adversaries.

As briefly surveyed in Sec. 11.6, research on game theory continues to delve into rather complicated types of competitive situations. However, the focus in this chapter is on the simplest case, called **two-person, zero-sum games**. As the name implies, these games involve only two adversaries, or *players* (who may be armies, teams, firms, and so on). They are called *zero-sum* games because one player wins whatever the other player loses, so that the sum of their net winnings is zero.

After we introduce the basic model for two-person, zero-sum games in Sec. 11.1, in the next four sections we describe and illustrate different approaches to solving such games. We then conclude the chapter by mentioning some other kinds of competitive situations that are dealt with by other branches of game theory.

11.1 The Formulation of Two-Person, Zero-Sum Games

To illustrate the basic characteristics of two-person, zero-sum games, consider the game called *odds and evens.* This game consists simply of each player simultaneously showing either one finger or two fingers. If the number of fingers matches, so that the total number for both players is even, then the player taking evens (say, player 1) wins the bet (say, $1) from the player taking odds (player 2). If the number does not match, player 1 pays $1 to player 2. Thus each player has two *strategies:* to show either one finger or two fingers. The resulting payoff to player 1 in dollars is shown in the *payoff table* given in Table 11.1.

In general, a two-person game is characterized by

1. The strategies of player 1
2. The strategies of player 2
3. The payoff table

Before the game begins, each player knows the strategies she or he has available, the ones the opponent has available, and the payoff table. The actual play of the game consists of each player simultaneously choosing a strategy without knowing the opponent's choice.

A strategy may involve only a simple action, such as showing a certain number of fingers in the odds and evens game. On the other hand, in more complicated games involving a series of moves, a **strategy** is a predetermined rule that specifies completely how one intends to respond to each possible circumstance at each stage of the game. For example, a strategy for one side in chess would indicate how to make the next move for *every* possible position on the board, so the total number of possible strategies would be astronomical. Applications of game theory normally involve far less complicated competitive situations than chess does, but the strategies involved can be fairly complex.

The **payoff table** shows the gain (positive or negative) for player 1 that would result from each combination of strategies for the two players. It is given only for player 1 because the table for player 2 is just the negative of this one, due to the zero-sum nature of the game.

Table 11.1 **Payoff Table for the Odds and Evens Game**

		Player 2	
	Strategy	1	2
Player 1	1	1	−1
	2	−1	1

The entries in the payoff table may be in any units desired, such as dollars, provided that they accurately represent the *utility* to player 1 of the corresponding outcome. However, utility is not necessarily proportional to the amount of money (or any other commodity) when large quantities are involved. For example, $2 million (after taxes) is probably worth much less than twice as much as $1 million to a poor person. In other words, given the choice between (1) a 50 percent chance of receiving $2 million rather than nothing and (2) being sure of getting $1 million, a poor person probably would much prefer the latter. On the other hand, the outcome corresponding to an entry of 2 in a payoff table should be ''worth twice as much'' to player 1 as the outcome corresponding to an entry of 1. Thus, given the choice, he or she should be indifferent between a 50 percent chance of receiving the former outcome (rather than nothing) and definitely receiving the latter outcome instead.[1]

A primary objective of game theory is the development of *rational criteria* for selecting a strategy. Two key assumptions are made:

1. *Both* players are *rational.*
2. *Both* players choose their strategies solely to *promote their own welfare* (no compassion for the opponent).

Game theory contrasts with *decision analysis* (see Chap. 20), where the assumption is that the decision maker is playing a game with a passive opponent—nature—which chooses its strategies in some random fashion.

We shall develop the standard game theory criteria for choosing strategies by means of illustrative examples. In particular, the next section presents a prototype example that illustrates the formulation of a two-person, zero-sum game and its solution in some simple situations. A more complicated variation of this game is then carried into Sec. 11.3 to develop a more general criterion. Sections 11.4 and 11.5 describe a graphical procedure and a linear programming formulation for solving such games.

11.2 Solving Simple Games—A Prototype Example

Two politicians are running against each other for the U.S. Senate. Campaign plans must now be made for the final 2 days, which are expected to be crucial because of the closeness of the race. Therefore, both politicians want to spend these days campaigning in two key cities, Bigtown and Megalopolis. To avoid wasting campaign time, they plan to travel at night and spend either 1 full day in each city or 2 full days in just one of the cities. However, since the necessary arrangements must be made in advance, neither politician will learn his (or her)[2] opponent's campaign schedule until after he has finalized his own. Therefore, each politician has asked his campaign manager in each of these cities to assess what the impact would be (in terms of votes won or lost) from the various possible combinations of days spent there by himself and by his opponent. He then wishes to use this information to choose his best strategy on how to use these 2 days.

[1] See Sec. 20.5 for a further discussion of the concept of utility.

[2] We use only *his* or only *her* in some examples and problems for ease of reading: we do not mean to imply that only men or only women are engaged in the various activities.

Formulation: To formulate this problem as a two-person, zero-sum game, we must identify the two *players* (obviously the two politicians), the *strategies* for each player, and the *payoff table.*

As the problem has been stated, each player has the following three strategies:

Strategy 1 = spend 1 day in each city.
Strategy 2 = spend both days in Bigtown.
Strategy 3 = spend both days in Megalopolis.

By contrast, the strategies would be more complicated in a different situation where each politician learns where his opponent will spend the first day before he finalizes his own plans for his second day. In that case, a typical strategy would be: Spend the first day in Bigtown; if the opponent also spends the first day in Bigtown, then spend the second day in Bigtown; however, if the opponent spends the first day in Megalopolis, then spend the second day in Megalopolis. There would be eight such strategies, one for each combination of the two first-day choices, the opponent's two first-day choices, and the two second-day choices.

Each entry in the payoff table for player 1 represents the *utility* to player 1 (or the negative utility to player 2) of the outcome resulting from the corresponding strategies used by the two players. From the politician's viewpoint, the objective is to *win votes,* and each additional vote (before he learns the outcome of the election) is of equal value to him. Therefore, the appropriate entries for the payoff table are the *total net votes won* from the opponent (i.e., the sum of the net vote changes in the two cities) resulting from these 2 days of campaigning. This formulation is summarized in Table 11.2.

However, we should also point out that this payoff table would *not* be appropriate if additional information were available to the politicians. In particular, assume that they know exactly how the populace is planning to vote 2 days before the election, so that each politician knows exactly how many net votes (positive or negative) he needs to switch in his favor during the last 2 days of campaigning in order to win the election. Consequently, the only significance of the data prescribed by Table 11.2 would be to indicate which politician would win the election with each combination of strategies. Because the ultimate goal is to win the election and because the size of the plurality is relatively inconsequential, the utility entries in the table then should be some positive constant (say, +1) when politician 1 wins and −1 when he loses. Even if only a *probability* of winning can be determined for each combination of strategies, the appropriate entries would be the probability of winning minus the probability of losing because they then would represent *expected* utilities. However, sufficiently accurate data to make such determinations usually are not available.

Table 11.2 **Form of the Payoff Table for the Political Campaign Problem**

		Total Net Votes Won by Politician 1 (in Units of 1,000 Votes)		
		Politician 2		
	Strategy	1	2	3
	1			
Politician 1	2			
	3			

Table 11.3 **Payoff Table for Variation 1 of the Political Campaign Problem**

	Strategy	Player 2: 1	2	3
	1	1	2	4
Player 1	2	1	0	5
	3	0	1	−1

Using the form given in Table 11.2, we give three alternative sets of data for the payoff table to illustrate how to solve three different kinds of games.

Variation 1: Given that Table 11.3 is the payoff table for the two politicians (players), which strategy should each select?

This situation is a rather special one, where the answer can be obtained just by applying the concept of **dominated strategies** to rule out a succession of inferior strategies until only one choice remains. Specifically, a strategy can be eliminated from further consideration if it is *dominated* by another strategy, i.e., if there is another strategy that is *always at least as good* regardless of what the opponent does.

At the outset, Table 11.3 includes no dominated strategies for player 2. However, for player 1, strategy 3 is dominated by strategy 1 because the latter has larger payoffs ($1 \geq 0$, $2 \geq 1$, $4 \geq -1$) regardless of what player 2 does. Eliminating strategy 3 from further consideration yields the following reduced payoff table:

	1	2	3
1	1	2	4
2	1	0	5

Because both players are assumed to be rational, player 2 also can deduce that player 1 has only these two strategies remaining under consideration. Therefore, player 2 now *does* have a dominated strategy—strategy 3, which is dominated by both strategies 1 and 2 because they always have smaller losses (payoffs to player 1) in this reduced payoff table (for strategy 1: $1 \leq 4$, $1 \leq 5$; for strategy 2: $2 \leq 4$, $0 \leq 5$). Eliminating this strategy yields

	1	2
1	1	2
2	1	0

At this point, strategy 2 for player 1 becomes dominated by strategy 1 because the latter is better in column 2 ($2 \geq 0$) and equally good in column 1 ($1 \geq 1$). Eliminating the dominated strategy leads to

	1	2
1	1	2

where strategy 2 for player 2 is dominated by strategy 1 ($1 \leq 2$). Consequently, both players should select their strategy 1. Player 1 then will receive a payoff of 1 from player 2 (that is, politician 1 will gain 1,000 votes from politician 2).

In general, the payoff to player 1 when both players play optimally is referred to as the **value of the game**. A game that has a value of 0 is said to be a **fair game**. Since this particular game has a value of 1, it is *not* a fair game.

The concept of a dominated strategy is a very useful one for reducing the size of the payoff table that needs to be considered and, in unusual cases like this one, actually identifying the optimal solution for the game. However, most games require another approach to at least finish solving, as illustrated by the next two variations of the example.

Variation 2: Now suppose that the current data give Table 11.4 as the payoff table for the politicians (players). This game does not have dominated strategies, so it is not obvious what the players should do. What line of reasoning does game theory say they should use?

Consider player 1. By selecting strategy 1, he could win 6 or could lose as much as 3. However, because player 2 is rational and thus will seek a strategy that will protect himself from large payoffs to player 1, it seems likely that player 1 would incur a loss by playing strategy 1. Similarly, by selecting strategy 3, player 1 could win 5, but more probably his rational opponent would avoid this loss and instead administer him a loss, which could be as large as 4. On the other hand, if player 1 selects strategy 2, he is guaranteed not to lose anything and he could even win something. Therefore, because it provides the *best guarantee* (a payoff of 0), strategy 2 seems to be a ''rational'' choice for player 1 against his rational opponent. (This line of reasoning assumes that both players are adverse to risking larger losses than necessary, in contrast to those individuals who enjoy gambling for a large payoff against long odds.)

Now consider player 2. He could lose as much as 5 or 6 by using strategy 1 or 3, but is guaranteed at least breaking even with strategy 2. Therefore, by the same reasoning of seeking the best guarantee against a rational opponent, his apparent choice is strategy 2.

If both players choose their strategy 2, the result is that both break even. Thus, in this case, neither player improves upon his best guarantee, but both also are forcing the opponent into the same position. Even when the opponent deduces a player's strategy, the opponent cannot exploit this information to improve his position. Stalemate.

The end product of this line of reasoning is that each player should play in such a way as to *minimize his maximum losses* whenever the resulting choice of strategy cannot be exploited by the opponent to then improve his position. This so-called **minimax criterion** is a standard criterion proposed by game theory for selecting a strategy. In effect, this criterion says to select a strategy that would be best even if the selection

Table 11.4 **Payoff Table for Variation 2 of the Political Campaign Problem**

		Player 2			
	Strategy	1	2	3	Minimum
	1	−3	−2	6	−3
Player 1	2	2	0	2	0 ← Maximin value
	3	5	−2	−4	−4
	Maximum:	5	0	6	
			↑ Minimax value		

were being announced to the opponent before the opponent chooses a strategy. In terms of the payoff table, it implies that *player 1* should select the strategy whose *minimum payoff* is *largest,* whereas *player 2* should choose the one whose *maximum payoff to player 1* is the *smallest.* This criterion is illustrated in Table 11.4 where strategy 2 is identified as the maximin strategy for player 1 and strategy 2 is the minimax strategy for player 2. The resulting payoff of 0 is the value of the game, so this is a fair game.

Notice the interesting fact that the same entry in this payoff table yields both the maximin and minimax values. The reason is that this entry is both the minimum in its row and the maximum of its column. The position of any such entry is called a **saddle point**.

The fact that this game possesses a saddle point was actually crucial in determining how it should be played. Because of the saddle point, neither player can take advantage of the opponent's strategy to improve his own position. In particular, when player 2 predicts or learns that player 1 is using strategy 2, player 2 would only increase his losses if he were to change from his original plan of using his strategy 2. Similarly, player 1 would only worsen his position if he were to change his plan. Thus neither player has any motive to consider changing strategies, either to take advantage of his opponent or to prevent the opponent from taking advantage of him. Therefore, since this is a **stable solution** (also called an *equilibrium solution*), players 1 and 2 should exclusively use their maximin and minimax strategies, respectively.

As the next variation illustrates, some games do not possess a saddle point, in which case a more complicated analysis is required.

Variation 3: Late developments in the campaign result in the final payoff table for the two politicians (players) given by Table 11.5. How should this game be played?

Suppose that both players attempt to apply the minimax criterion in the same way as in variation 2. Player 1 can guarantee that he will lose no more than 2 by playing strategy 1. Similarly, player 2 can guarantee that he will lose no more than 2 by playing strategy 3.

However, notice that the maximin value (-2) and the minimax value (2) do not coincide in this case. The result is that there is no saddle point.

What are the resulting consequences if both players plan to use the strategies just derived? It can be seen that player 1 would win 2 from player 2, which would make player 2 unhappy. Because player 2 is rational and can therefore foresee this outcome, he would then conclude that he can do much better, actually winning 2 rather than losing 2, by playing strategy 2 instead. Because player 1 is also rational, he would anticipate this switch and conclude that he can improve considerably, from -2 to 4, by changing to strategy 2. Realizing this, player 2 would then consider switching back to

Table 11.5 **Payoff Table for Variation 3 of the Political Campaign Problem**

		Player 2			
	Strategy	1	2	3	Minimum
	1	0	-2	2	$-2 \leftarrow$ Maximin value
Player 1	2	5	4	-3	-3
	3	2	3	-4	-4
	Maximum:	5	4	2	
				$\uparrow$ Minimax value	

strategy 3 to convert a loss of 4 to a gain of 3. This possibility of a switch would cause player 1 to consider again using strategy 1, after which the whole cycle would start over again. Therefore, even though this game is being played only once, *any* tentative choice of a strategy leaves that player with a motive to consider changing strategies, either to take advantage of his opponent or to prevent the opponent from taking advantage of him.

In short, the originally suggested solution (player 1 to play strategy 1 and player 2 to play strategy 3) is an **unstable solution**, so it is necessary to develop a more satisfactory solution. But what kind of solution should it be?

The key fact seems to be that whenever one player's strategy is predictable, the opponent can take great advantage of this information to improve his position. Therefore, an essential feature of a rational plan for playing a game such as this one is that neither player should be able to deduce which strategy the other will use. Hence, in this case, rather than applying some known criterion for determining a single strategy that will definitely be used, it is necessary to choose among alternative acceptable strategies on some kind of random basis. By doing this, neither player knows in advance which of his own strategies will be used, let alone what his opponent will do.

This suggests, in very general terms, the kind of approach that is required for games lacking a saddle point. In the next section we discuss the approach more fully. Given this foundation, we turn our attention to procedures for finding an optimal way of playing such games. This particular variation of the political campaign problem will continue to be used to illustrate these ideas as they are developed.

11.3 Games with Mixed Strategies

Whenever a game does not possess a saddle point, game theory advises each player to assign a probability distribution over her set of strategies. To express this mathematically, let

$$x_i = \text{probability that player 1 will use strategy } i \ (i = 1, 2, \ldots, m),$$
$$y_j = \text{probability that player 2 will use strategy } j \ (j = 1, 2, \ldots, n),$$

where m and n are the respective numbers of available strategies. Thus player 1 would specify her plan for playing the game by assigning values to $x_1, x_2, \ldots, x_m$. Because these values are probabilities, they would need to be nonnegative and add to 1. Similarly, the plan for player 2 would be described by the values she assigns to her decision variables $y_1, y_2, \ldots, y_n$. These plans $(x_1, x_2, \ldots, x_m)$ and $(y_1, y_2, \ldots, y_n)$ are usually referred to as **mixed strategies**, and the original strategies are then called **pure strategies**.

When the game is actually played, it is necessary for each player to use one of her pure strategies. However, this pure strategy would be chosen by using some random device to obtain a random observation from the probability distribution specified by the mixed strategy, where this observation would indicate which particular pure strategy to use.

To illustrate, suppose that players 1 and 2 in variation 3 of the political campaign problem (see Table 11.5) select the mixed strategies $(x_1, x_2, x_3) = (\frac{1}{2}, \frac{1}{2}, 0)$ and $(y_1, y_2, y_3) = (0, \frac{1}{2}, \frac{1}{2})$, respectively. This selection would say that player 1 is giving an equal chance (probability of $\frac{1}{2}$) to choosing either (pure) strategy 1 or 2, but he is discarding strategy 3 entirely. Similarly, player 2 is randomly choosing between his last two pure strategies. To play the game, each player could then flip a coin to determine which of his two acceptable pure strategies he will actually use.

Although no completely satisfactory measure of performance is available for evaluating mixed strategies, a very useful one is the *expected payoff.* By applying the probability theory definition of expected value, this quantity is

$$\text{Expected payoff for player 1} = \sum_{i=1}^{m} \sum_{j=1}^{n} p_{ij} x_i y_j,$$

where p_{ij} is the payoff if player 1 uses pure strategy i and player 2 uses pure strategy j. In the example of mixed strategies just given, there are four possible payoffs $(-2, 2, 4, -3)$, each occurring with a probability of $\frac{1}{4}$, so the expected payoff is $\frac{1}{4}(-2 + 2 + 4 - 3) = \frac{1}{4}$. Thus this measure of performance does not disclose anything about the risks involved in playing the game, but it does indicate what the average payoff will tend to be if the game is played many times.

By using this measure, game theory extends the concept of the minimax criterion to games that lack a saddle point and thus need mixed strategies. In this context, the **minimax criterion** says that a given player should select the mixed strategy that *minimizes* the *maximum expected loss* to himself. Equivalently, when we focus on payoffs (player 1) rather than losses (player 2), this criterion says to *maximin* instead, i.e., *maximize* the *minimum expected payoff* to the player. By the *minimum expected payoff* we mean the smallest possible expected payoff that can result from any mixed strategy with which the opponent can counter. Thus the mixed strategy for player 1 that is *optimal* according to this criterion is the one that provides the *guarantee* (minimum expected payoff) that is *best* (maximal). (The value of this best guarantee is the *maximin value,* denoted by $\underline{v}$.) Similarly, the *optimal* strategy for player 2 is the one that provides the *best guarantee,* where *best* now means *minimal* and *guarantee* refers to the *maximum expected loss* that can be administered by any of the opponent's mixed strategies. (This best guarantee is the *minimax value,* denoted by $\overline{v}$.)

Recall that when only pure strategies were used, games not having a saddle point turned out to be *unstable* (no stable solutions). The reason was essentially that $\underline{v} < \overline{v}$, so that the players would want to change their strategies to improve their positions. Similarly, for games with mixed strategies, it is necessary that $\underline{v} = \overline{v}$ for the optimal solution to be *stable.* Fortunately, according to the minimax theorem of game theory, this condition always holds for such games.

> **Minimax theorem:** If mixed strategies are allowed, the pair of mixed strategies that is optimal according to the minimax criterion provides a *stable solution* with $\underline{v} = \overline{v} = v$ (the value of the game), so that neither player can do better by unilaterally changing her or his strategy.

One proof of this theorem is included in Sec. 11.5.

Although the concept of mixed strategies becomes quite intuitive if the game is played *repeatedly,* it requires some interpretation when the game is to be played just *once.* In this case, using a mixed strategy still involves selecting and using *one* pure strategy (randomly selected from the specified probability distribution), so it might seem more sensible to ignore this randomization process and just choose the one "best" pure strategy to be used. However, we have already illustrated for variation 3 in the preceding section that a player must *not* allow the opponent to deduce what his strategy will be (i.e., the solution procedure under the rules of game theory must not *definitely* identify which pure strategy will be used when the game is unstable). Furthermore, even if the opponent is able to use only his knowledge of the tendencies of the first player to deduce probabilities (for the pure strategy chosen) that are different from those for the optimal mixed strategy, then he still can take advantage of this knowledge to reduce the expected payoff to the first player. Therefore, the only way to guarantee attaining the optimal expected payoff v is to randomly select the pure strategy to be used from the probability distribution for the optimal mixed strategy. (Valid statistical procedures for making such a random selection are discussed in Sec. 21.2.)

Now we must show how to find the optimal mixed strategy for each player. There are several methods of doing this. One is a graphical procedure that may be used whenever one of the players has only two (undominated) pure strategies; this approach is described in the next section. When larger games are involved, the usual method is to transform the problem to a linear programming problem that then can be solved by the simplex method on a computer; Sec. 11.5 discusses this approach.

11.4 Graphical Solution Procedure

Consider any game with mixed strategies such that, after dominated strategies are eliminated, one of the players has only two pure strategies. To be specific, let this player be player 1. Because her mixed strategies are (x_1, x_2) and $x_2 = 1 - x_1$, it is necessary for her to solve only for the optimal value of x_1. However, it is straightforward to plot the expected payoff as a function of x_1 for each of her opponent's pure strategies. This graph can then be used to identify the point that maximizes the minimum expected payoff. The opponent's minimax mixed strategy can also be identified from the graph.

To illustrate this procedure, consider variation 3 of the political campaign problem (see Table 11.5). Notice that the third pure strategy for player 1 is dominated by her second, so the payoff table can be reduced to the form given in Table 11.6. Therefore, for each of the pure strategies available to player 2, the expected payoff for player 1 will be

(y_1, y_2, y_3)	Expected Payoff
(1, 0, 0)	$0x_1 + 5(1 - x_1) = 5 - 5x_1$
(0, 1, 0)	$-2x_1 + 4(1 - x_1) = 4 - 6x_1$
(0, 0, 1)	$2x_1 - 3(1 - x_1) = -3 + 5x_1$

Now plot these expected-payoff lines on a graph, as shown in Fig. 11.1. For any given values of x_1 and (y_1, y_2, y_3), the expected payoff will be the appropriate weighted average of the corresponding points on these three lines. In particular,

$$\text{Expected payoff for player 1} = y_1(5 - 5x_1) + y_2(4 - 6x_1) + y_3(-3 + 5x_1).$$

Table 11.6 **Reduced Payoff Table for Variation 3 of the Political Campaign Problem**

			Player 2		
		Probability	y_1	y_2	y_3
	Probability	Pure Strategy	1	2	3
Player 1	x_1	1	0	−2	2
	$1 - x_1$	2	5	4	−3

Remember that player 2 wants to minimize this expected payoff for player 1. Given x_1, player 2 can minimize this expected payoff by choosing the pure strategy that corresponds to the "bottom" line for that x_1 in Fig. 11.1 (either $-3 + 5x_1$ or $4 - 6x_1$, but never $5 - 5x_1$). According to the minimax (or maximin) criterion, player 1 wants to maximize this minimum expected payoff. Consequently, player 1 should select the value of x_1 where the bottom line peaks, i.e., where the $(-3 + 5x_1)$ and $(4 - 6x_1)$ lines intersect, which yields an expected payoff of

$$\underline{v} = v = \max_{0 \le x_1 \le 1} \{\min\{-3 + 5x_1, 4 - 6x_1\}\}.$$

Therefore, the optimal value of x_1 is the one at the intersection of the two lines $-3 + 5x_1$ and $4 - 6x_1$. Solving algebraically gives

$$-3 + 5x_1 = 4 - 6x_1,$$

so that $x_1 = \frac{7}{11}$; thus $(x_1, x_2) = (\frac{7}{11}, \frac{4}{11})$ is the *optimal mixed strategy* for player 1, and

$$\underline{v} = v = -3 + 5(\tfrac{7}{11}) = \tfrac{2}{11}$$

is the value of the game.

To find the corresponding optimal mixed strategy for player 2, we now reason as follows. According to the definition of the minimax value $\overline{v}$ and the minimax theorem, the expected payoff resulting from the optimal strategy $(y_1, y_2, y_3) = (y_1^*, y_2^*, y_3^*)$ will satisfy the condition

$$y_1^*(5 - 5x_1) + y_2^*(4 - 6x_1) + y_3^*(-3 + 5x_1) \le \overline{v} = v = \tfrac{2}{11}$$

for all values of x_1 $(0 \le x_1 \le 1)$. Furthermore, when player 1 is playing optimally (that is, $x_1 = \frac{7}{11}$), this inequality will be an equality (by the minimax theorem), so that

$$\tfrac{20}{11}y_1^* + \tfrac{2}{11}y_2^* + \tfrac{2}{11}y_3^* = v = \tfrac{2}{11}.$$

Because (y_1, y_2, y_3) is a probability distribution, it is also known that

$$y_1^* + y_2^* + y_3^* = 1.$$

Therefore, $y_1^* = 0$ because $y_1^* > 0$ would violate the next-to-last equation; i.e., the expected payoff on the graph at $x_1 = \frac{7}{11}$ would be above the maximin point. (In general, any line that does not pass through the maximin point must be given a zero weight to avoid increasing the expected payoff above this point.)

Hence

$$y_2^*(4 - 6x_1) + y_3^*(-3 + 5x_1) \begin{cases} \le \frac{2}{11} & \text{for } 0 \le x_1 \le 1, \\ = \frac{2}{11} & \text{for } x_1 = \frac{7}{11}. \end{cases}$$

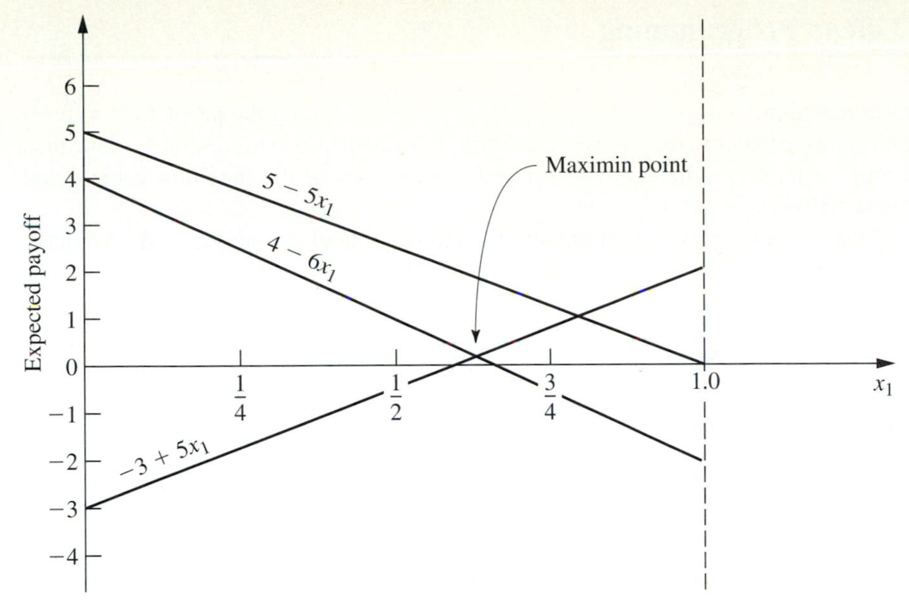

Figure 11.1 Graphical procedure for solving games.

But y_2^* and y_3^* are numbers, so the left-hand side is the equation of a straight line, which is a fixed weighted average of the two "bottom" lines on the graph. Because the ordinate of this line must equal $\frac{2}{11}$ at $x_1 = \frac{7}{11}$, and because it must never exceed $\frac{2}{11}$, the line necessarily is horizontal. (This conclusion is always true unless the optimal value of x_1 is either 0 or 1, in which case player 2 also should use a single pure strategy.) Therefore,

$$y_2^*(4 - 6x_1) + y_3^*(-3 + 5x_1) = \tfrac{2}{11}, \qquad \text{for } 0 \leq x_1 \leq 1.$$

Hence, to solve for y_2^* and y_3^*, select two values of x_1 (say, 0 and 1), and solve the resulting two simultaneous equations. Thus

$$4y_2^* - 3y_3^* = \tfrac{2}{11},$$

$$-2y_2^* + 2y_3^* = \tfrac{2}{11},$$

so that $y_3^* = \frac{6}{11}$ and $y_2^* = \frac{5}{11}$. Therefore, the *optimal mixed strategy* for player 2 is $(y_1, y_2, y_3) = (0, \frac{5}{11}, \frac{6}{11})$.

If, in another problem, there should happen to be more than two lines passing through the maximin point, so that more than two of the y_j^* values can be greater than zero, this condition would imply that there are many ties for the optimal mixed strategy for player 2. One such strategy can then be identified by setting all but two of these y_j^* values equal to zero and solving for the remaining two in the manner just described. For the remaining two, the associated lines must have positive slope in one case and negative slope in the other.

Although this graphical procedure has been illustrated for only one particular problem, essentially the same reasoning can be used to solve any game with mixed strategies that has only two undominated pure strategies for one of the players.

11.5 Solving by Linear Programming

Any game with mixed strategies can be solved by transforming the problem to a linear programming problem. As you will see, this transformation requires little more than applying the minimax theorem and using the definitions of the maximin value $\underline{v}$ and minimax value $\overline{v}$.

First, consider how to find the optimal mixed strategy for player 1. As indicated in Sec. 11.3,

$$\text{Expected payoff for player 1} = \sum_{i=1}^{m} \sum_{j=1}^{n} p_{ij} x_i y_j$$

and the strategy $(x_1, x_2, \ldots, x_m)$ is optimal if

$$\sum_{i=1}^{m} \sum_{j=1}^{n} p_{ij} x_i y_j \geq \underline{v} = v$$

for all opposing strategies $(y_1, y_2, \ldots, y_n)$. Thus this inequality will need to hold, e.g., for each of the pure strategies of player 2, that is, for each of the strategies $(y_1, y_2, \ldots, y_n)$ where one $y_j = 1$ and the rest equal 0. Substituting these values into the inequality yields

$$\sum_{i=1}^{m} p_{ij} x_i \geq v \qquad \text{for } j = 1, 2, \ldots, n,$$

so that the inequality *implies* this set of n inequalities. Furthermore, this set of n inequalities *implies* the original inequality (rewritten)

$$\sum_{j=1}^{n} y_j \left(\sum_{i=1}^{m} p_{ij} x_i \right) \geq \sum_{j=1}^{n} y_j v = v,$$

since

$$\sum_{j=1}^{n} y_j = 1.$$

Because the implication goes in both directions, it follows that imposing this set of n linear inequalities is *equivalent* to requiring the original inequality to hold for all strategies $(y_1, y_2, \ldots, y_n)$. But these n inequalities are legitimate linear programming constraints, as are the additional constraints

$$x_1 + x_2 + \cdots + x_m = 1$$
$$x_i \geq 0, \qquad \text{for } i = 1, 2, \ldots, m$$

that are required to ensure that the x_i are probabilities. Therefore, any solution $(x_1, x_2, \ldots, x_m)$ that satisfies this entire set of linear programming constraints is the desired optimal mixed strategy.

Consequently, the problem of finding an optimal mixed strategy has been reduced to finding a feasible solution for a linear programming problem, which can be done as described in Chap. 4. The two remaining difficulties are that (1) v is unknown and (2) the linear programming problem has no objective function. Fortunately, both these difficulties can be resolved at one stroke by replacing the unknown constant v by the variable x_{m+1} and then *maximizing* x_{m+1}, so that x_{m+1} automatically will equal v (by definition) at the *optimal* solution for the linear programming problem!

To summarize, player 1 would find his optimal mixed strategy by using the simplex method to solve the linear programming problem:

$$\text{Maximize} \quad x_{m+1},$$

subject to

$$\begin{aligned} p_{11}x_1 + p_{21}x_2 + \cdots + p_{m1}x_m - x_{m+1} &\geq 0 \\ p_{12}x_1 + p_{22}x_2 + \cdots + p_{m2}x_m - x_{m+1} &\geq 0 \\ \cdots\cdots\cdots\cdots\cdots\cdots\cdots\cdots\cdots\cdots\cdots \\ p_{1n}x_1 + p_{2n}x_2 + \cdots + p_{mn}x_m - x_{m+1} &\geq 0 \\ x_1 + x_2 + \cdots + x_m &= 1 \end{aligned}$$

and

$$x_i \geq 0, \qquad \text{for } i = 1, 2, \ldots, m.$$

Note that x_{m+1} is not restricted to be nonnegative, whereas the simplex method can be applied only after *all* the variables have nonnegativity constraints. However, this matter can be easily rectified, as will be discussed shortly.

Now consider player 2. He could find his optimal mixed strategy by rewriting the payoff table as the payoff to himself rather than to player 1 and then by proceeding exactly as just described. However, it is enlightening to summarize his formulation in terms of the original payoff table. By proceeding in a way that is completely analogous to that just described, player 2 would conclude that his optimal mixed strategy is given by an optimal solution to the linear programming problem:

$$\text{Minimize} \quad y_{n+1},$$

subject to

$$\begin{aligned} p_{11}y_1 + p_{12}y_2 + \cdots + p_{1n}y_n - y_{n+1} &\leq 0 \\ p_{21}y_1 + p_{22}y_2 + \cdots + p_{2n}y_n - y_{n+1} &\leq 0 \\ \cdots\cdots\cdots\cdots\cdots\cdots\cdots\cdots\cdots\cdots\cdots \\ p_{m1}y_1 + p_{m2}y_2 + \cdots + p_{mn}y_n - y_{n+1} &\leq 0 \\ y_1 + y_2 + \cdots + y_n &= 1 \end{aligned}$$

and

$$y_j \geq 0, \qquad \text{for } j = 1, 2, \ldots, n.$$

It is easy to show (see Prob. 11.5-5 and its hint) that this linear programming problem and the one given for player 1 are *dual* to each other in the sense described in Secs. 6.1 and 6.4. This fact has several important implications. One implication is that the optimal mixed strategies for both players can be found by solving only one of the linear programming problems because the optimal dual solution is an automatic by-product of the simplex method calculations to find the optimal primal solution. A second implication is that this brings all *duality theory* (described in Chap. 6) to bear upon the interpretation and analysis of games.

A related implication is that this provides a simple proof of the minimax theorem. Let x^*_{m+1} and y^*_{n+1} denote the value of x_{m+1} and y_{n+1} in the optimal solution of the respective linear programming problems. It is known from the *strong duality property* given in Sec. 6.1 that $-x^*_{m+1} = -y^*_{n+1}$, so that $x^*_{m+1} = y^*_{n+1}$. However, it is evident from the definition of $\underline{v}$ and $\overline{v}$ that $\underline{v} = x^*_{m+1}$ and $\overline{v} = y^*_{n+1}$, so it follows that $\underline{v} = \overline{v}$, as claimed by the minimax theorem.

One remaining loose end needs to be tied up, namely, what to do about x_{m+1} and y_{n+1} being unrestricted in sign in the linear programming formulations. If it is clear that $v \geq 0$ so that the optimal values of x_{m+1} and y_{n+1} are nonnegative, then it is safe to introduce nonnegativity constraints for these variables for the purpose of applying the simplex method. However, if $v < 0$, then an adjustment needs to be made. One possibility is to use the approach described in Sec. 4.6 for replacing a variable without a nonnegativity constraint by the difference of two nonnegative variables. Another is to reverse players 1 and 2 so that the payoff table would be rewritten as the payoff to the original player 2, which would make the corresponding value of v positive. A third, and the most commonly used, procedure is to add a sufficiently large fixed constant to all the entries in the payoff table that the new value of the game will be positive. (For example, setting this constant equal to the absolute value of the largest negative entry will suffice.) Because this same constant is added to every entry, this adjustment cannot alter the optimal mixed strategies in any way, so they can now be obtained in the usual manner. The indicated value of the game would be increased by the amount of the constant, but this value can be readjusted after the solution has been obtained.

To illustrate this linear programming approach, consider again variation 3 of the political campaign problem after dominated strategy 3 for player 1 is eliminated (see Table 11.6). Because there are some negative entries in the reduced payoff table, it is unclear at the outset whether the *value* of the game v is *nonnegative* (it turns out to be). For the moment, let us assume that $v \geq 0$ and proceed without making any of the adjustments discussed in the preceding paragraph.

To write out the linear programming model for player 1 for this example, note that p_{ij} in the general model is the entry in row i and column j of Table 11.6, for $i = 1, 2$ and $j = 1, 2, 3$. The resulting model is

$$\text{Maximize} \quad x_3,$$

subject to

$$\begin{aligned} 5x_2 - x_3 &\geq 0 \\ -2x_1 + 4x_2 - x_3 &\geq 0 \\ 2x_1 - 3x_2 - x_3 &\geq 0 \\ x_1 + x_2 &= 1 \end{aligned}$$

and

$$x_1 \geq 0, \qquad x_2 \geq 0.$$

Applying the simplex method to this linear programming problem (after adding the constraint $x_3 \geq 0$) yields $x_1^* = \frac{7}{11}$, $x_2^* = \frac{4}{11}$, $x_3^* = \frac{2}{11}$ as the optimal solution. (See Probs. 11.5-7 and 11.5-8.) Consequently, the optimal mixed strategy for player 1 according to the minimax criterion is $(x_1, x_2) = (\frac{7}{11}, \frac{4}{11})$, and the value of the game is $v = x_3^* = \frac{2}{11}$. The simplex method also yields the optimal solution to the dual (given next) of this problem, namely, $y_1^* = 0$, $y_2^* = \frac{5}{11}$, $y_3^* = \frac{6}{11}$, $y_4^* = \frac{2}{11}$, so the optimal mixed strategy for player 2 is $(y_1, y_2, y_3) = (0, \frac{5}{11}, \frac{6}{11})$. (The automatic routine for the simplex method in your OR Courseware gives the optimal value of each dual variable as the *shadow price* for the corresponding constraint in the primal problem, where the sign of this value should be adjusted as needed to reflect any constraint on the sign of the dual variable.)

The dual of the preceding problem is just the linear programming model for player 2 (the one with variables $y_1, y_2, \ldots, y_n, y_{n+1}$) shown earlier in this section. (See Prob. 11.5-6.) By plugging in the values of p_{ij} from Table 11.6, this model is

$$\text{Minimize} \quad y_4,$$

subject to

$$\begin{aligned} -2y_2 + 2y_3 - y_4 &\leq 0 \\ 5y_1 + 4y_2 - 3y_3 - y_4 &\leq 0 \\ y_1 + y_2 + y_3 \quad &= 1 \end{aligned}$$

and

$$y_1 \geq 0, \qquad y_2 \geq 0, \qquad y_3 \geq 0.$$

Applying the simplex method directly to this model (after adding the constraint $y_4 \geq 0$) yields the optimal solution: $y_1^* = 0$, $y_2^* = \frac{5}{11}$, $y_3^* = \frac{6}{11}$, $y_4^* = \frac{2}{11}$ (as well as the optimal dual solution $x_1^* = \frac{7}{11}$, $x_2^* = \frac{4}{11}$, $x_3^* = \frac{2}{11}$). Thus the optimal mixed strategy for player 2 is $(y_1, y_2, y_3) = (0, \frac{5}{11}, \frac{6}{11})$, and the value of the game is again seen to be $v = y_4^* = \frac{2}{11}$.

Because we already had found the optimal mixed strategy for player 2 while dealing with the first model, we did not have to solve the second one. In general, you always can find optimal mixed strategies for *both* players by choosing just one of the models (either one) and then using the simplex method to solve for both an optimal solution and an optimal dual solution.

When the simplex method was applied to both of these linear programming models, a nonnegativity constraint was added that assumed that $v \geq 0$. If this assumption were violated, both models would have no feasible solutions, so the simplex method would stop quickly with this message. To avoid this risk, we could have added a positive constant, say, 3 (the absolute value of the largest negative entry), to all the entries in Table 11.6. This then would increase by 3 all the coefficients of x_1, x_2, y_1, y_2, and y_3 in the inequality constraints of the two models. (See Prob. 11.5-1.)

11.6 Extensions

Although this chapter has considered only two-person, zero-sum games with a finite number of pure strategies, game theory extends far beyond this kind of game. In fact, extensive research has been done on a number of more complicated types of games, including the ones summarized in this section.

The simplest generalization is to the *two-person, constant-sum game.* In this case, the sum of the payoffs to the two players is a fixed constant (positive or negative) regardless of which combination of strategies is selected. The only difference from a two-person, zero-sum game is that, in the latter case, the constant must be zero. A nonzero constant may arise instead because, in addition to one player winning whatever the other one loses, the two players may share some reward (if the constant is positive) or some cost (if the constant is negative) for participating in the game. Adding this fixed constant does nothing to affect which strategies should be chosen. Therefore, the analysis for determining optimal strategies is exactly the same as described in this chapter for two-person, zero-sum games.

A more complicated extension is to the *n-person game,* where more than two players may participate in the game. This generalization is particularly important because, in many kinds of competitive situations, frequently more than two competitors are involved. This may occur, e.g., in competition among business firms, in international diplomacy, and so forth. Unfortunately, the existing theory for such games is less satisfactory than it is for two-person games.

Another generalization is the *nonzero-sum game,* where the sum of the payoffs to the players need not be 0 (or any other fixed constant). This case reflects the fact that many competitive situations include noncompetitive aspects that contribute to the mutual advantage or mutual disadvantage of the players. For example, the advertising strategies of competing companies can affect not only how they will split the market but also the total size of the market for their competing products.

Because mutual gain is possible, nonzero-sum games are further classified in terms of the degree to which the players are permitted to cooperate. At one extreme is the *noncooperative game,* where there is no preplay communication between the players. At the other extreme is the *cooperative game,* where preplay discussions and binding agreements are permitted. For example, competitive situations involving trade regulations between countries, or collective bargaining between labor and management, might be formulated as cooperative games. When there are more than two players, cooperative games also allow some of or all the players to form coalitions.

Still another extension is to the class of *infinite games,* where the players have an infinite number of pure strategies available to them. These games are designed for the kind of situation where the strategy to be selected can be represented by a *continuous* decision variable. For example, this decision variable might be the time at which to take a certain action, or the proportion of one's resources to allocate to a certain activity, in a competitive situation. Much research has been concentrated on such games in recent years.

However, the analysis required in these extensions beyond the two-person, zero-sum, finite game is relatively complex and will not be pursued further here.

11.7 Conclusions

The general problem of how to make decisions in a competitive environment is a very common and important one. The fundamental contribution of game theory is that it provides a basic conceptual framework for formulating and analyzing such problems in simple situations. However, there is a considerable gap between what the theory can handle and the complexity of most competitive situations arising in practice. Therefore, the conceptual tools of game theory usually play just a supplementary role in dealing with these situations.

Because of the importance of the general problem, research is continuing with some success to extend the theory to more complex situations.

SELECTED REFERENCES

1. Davis, M., *Game Theory: An Introduction,* Basic Books, New York, 1983.
2. Meyerson, R. B., *Game Theory: Analysis of Conflict,* Harvard University Press, Cambridge, MA, 1991.

3. Owen, G., *Game Theory,* 2d ed., Academic Press, New York, 1982.
4. Shubik, M., *Game Theory in the Social Sciences,* vols. 1 (1982) and 2 (1987), MIT Press, Cambridge, MA.
5. Szep, J., and F. Forgo, *Introduction to the Theory of Games,* D. Reidel, Boston, 1985.

RELEVANT ROUTINES IN YOUR OR COURSEWARE

Automatic routines: *Enter or Revise a General Linear Programming Model*
Solve Automatically by the Simplex Method

To access these routines, call the MathProg program and then choose *General Linear Programming* under the Area menu. See Appendix 1 for documentation of the software.

PROBLEMS

To the left of each of the following problems (or their parts), we have inserted an A (for automatic routine) whenever a corresponding routine listed above can be helpful. An asterisk on the A indicates that this routine definitely should be used (unless your instructor gives contrary instructions) and that the printout from this routine is all that needs to be turned in to show your work in executing the algorithm. An asterisk on the problem number indicates that at least a partial answer is given in the back of the book.

11.1-1. The labor union and management of a particular company have been negotiating a new labor contract. However, negotiations have now come to an impasse, with management making a ''final'' offer of a wage increase of $1.10 per hour and the union making a ''final'' demand of a $1.60 per hour increase. Therefore, both sides have agreed to let an impartial arbitrator set the wage increase somewhere between $1.10 and $1.60 per hour (inclusively).

The arbitrator has asked each side to submit to her a confidential proposal for a fair and economically reasonable wage increase (rounded to the nearest dime). From past experience, both sides know that this arbitrator normally accepts the proposal of the side that gives the most from its final figure. If neither side changes its final figure, or if they both give in the same amount, then the arbitrator normally compromises halfway between ($1.35 in this case). Each side now needs to determine what wage increase to propose for its own maximum advantage.

Formulate this problem as a two-person, zero-sum game.

11.1-2. Two manufacturers currently are competing for sales in two different but equally profitable product lines. In both cases the sales volume for manufacturer 2 is three times as large as that for manufacturer 1. Because of a recent technological breakthrough, both manufacturers will be making a major improvement in both products. However, they are uncertain as to what development and marketing strategy to follow.

If both product improvements are developed simultaneously, either manufacturer can have them ready for sale in 12 months. Another alternative is to have a ''crash program'' to develop only one product first to try to get it marketed ahead of the competition. By doing this, manufacturer 2 could have one product ready for sale in 9 months, whereas manufacturer 1 would require 10 months (because of previous commitments for its production facilities). For either manufacturer, the second product could then be ready for sale in an additional 9 months.

For either product line, if both manufacturers market their improved models simultaneously, it is estimated that manufacturer 1 would increase its share of the total future sales of this product by 8 percent of the total (from 25 to 33 percent). Similarly, manufacturer 1 would increase its share by 20, 30, and 40 percent of the total if it marketed the product sooner than manufacturer 2 by 2, 6, and 8 months, respectively. On the other hand, manufacturer 1 would lose 4, 10, 12, and 14 percent of the total if manufacturer 2 marketed it sooner by 1, 3, 7, and 10 months, respectively.

Formulate this problem as a two-person, zero-sum game, and then determine which strategy the respective manufacturers should use according to the minimax criterion.

11.1-3. Consider the following parlor game to be played between two players. Each player begins with three chips: one red, one white, and one blue. Each chip can be used only once.

To begin, each player selects one of her chips and places it on the table, concealed. Both players then uncover the chips and determine the payoff to the winning player. In particular, if both players play the same kind of chip, it is a draw; otherwise, the following table indicates the winner and how much she receives from the other player. Next, each player selects one of her two remaining chips and repeats the procedure, resulting in another payoff according to the following table. Finally, each player plays her one remaining chip, resulting in the third and final payoff.

Winning Chip	Payoff ($)
Red beats white	50
White beats blue	40
Blue beats red	30
Matching colors	0

Formulate this problem as a two-person, zero-sum game by identifying the form of the strategies and payoffs.

11.2-1. Reconsider Prob. 11.1-1.

(*a*) Use the concept of dominated strategies to determine the best strategy for each side.
(*b*) Without eliminating dominated strategies, use the minimax criterion to determine the best strategy for each side.

11.2-2.* For each of the following payoff tables, determine the optimal strategy for each player by successively eliminating dominated strategies. (Indicate the order in which you eliminated strategies.)

(*a*)

		Player 2		
	Strategy	1	2	3
	1	−3	1	2
Player 1	2	1	2	1
	3	1	0	−2

(*b*)

		Player 2		
	Strategy	1	2	3
	1	1	2	0
Player 1	2	2	−3	−2
	3	0	3	−1

11.2-3. Consider the game having the following payoff table.

		Player 2			
	Strategy	1	2	3	4
	1	2	−3	−1	1
Player 1	2	−1	1	−2	2
	3	−1	2	−1	3

Determine the optimal strategy for each player by successively eliminating dominated strategies. Give a list of the dominated strategies (and the corresponding dominating strategies) in the order in which you were able to eliminate them.

11.2-4. Find the saddle point for the game having the following payoff table.

	Strategy	*Player 2* 1	2	3
	1	1	−1	1
Player 1	2	−2	0	3
	3	3	1	2

11.2-5. Find the saddle point for the game having the following payoff table.

	Strategy	*Player 2* 1	2	3	4
	1	3	−3	−2	−4
Player 1	2	−4	−2	−1	1
	3	1	−1	2	0

11.2-6. Two companies share the bulk of the market for a particular kind of product. Each is now planning its new marketing plans for the next year in an attempt to wrest some sales away from the other company. (The total sales for the product are relatively fixed, so one company can increase its sales only by winning them away from the other.) Each company is considering three possibilities: (1) better packaging of the product, (2) increased advertising, and (3) a slight reduction in price. The costs of the three alternatives are quite comparable and sufficiently large that each company will select just one. The estimated effect of each combination of alternatives on the *increased percentage of the sales* for company 1 is as follows:

	Strategy	*Player 2* 1	2	3
	1	2	3	1
Player 1	2	1	4	0
	3	3	−2	−1

Each company must make its selection before learning the decision of the other company.

(*a*) Without eliminating dominated strategies, use the minimax (or maximin) criterion to determine the best strategy for each side.

(*b*) Now identify and eliminate dominated strategies as far as possible. Make a list of the dominated strategies, showing the order in which you were able to eliminate them. Then show the resulting reduced payoff table with no remaining dominated strategies.

11.2-7.* Two politicians soon will be starting their campaigns against each other for a certain political office. Each must now select the main issue she will emphasize as the theme of her campaign. Each has three advantageous issues from which to choose, but the relative effectiveness of each one would depend upon the issue chosen by the opponent. In particular, the estimated increase in the vote for politician 1 (expressed as a percentage of the total vote) resulting from each combination of issues is as follows:

		Issue for Politician 2		
		1	2	3
Issue for Politician 1	1	7	−1	3
	2	1	0	2
	3	−5	−3	−1

However, because considerable staff work is required to research and formulate the issue chosen, each politician must make her own choice before learning the opponent's choice. Which issue should she choose?

For each of the situations described here, formulate this problem as a two-person, zero-sum game, and then determine which issue should be chosen by each politician according to the specified criterion.

(*a*) The current preferences of the voters are very uncertain, so each additional percent of votes won by one of the politicians has the same value to her. Use the minimax criterion.

(*b*) A reliable poll has found that the percentage of the voters currently preferring politician 1 (before the issues have been raised) lies between 45 and 50 percent. (Assume a uniform distribution over this range.) Use the concept of dominated strategies, beginning with the strategies for politician 1.

(*c*) Suppose that the percentage described in part (*b*) actually were 45 percent. Should politician 1 use the minimax criterion? Explain. Which issue would you recommend? Why?

11.2-8. Briefly describe what you feel are the advantages and disadvantages of the minimax criterion.

11.3-1. Consider the following parlor game between two players. It begins when a referee flips a coin, notes whether it comes up heads or tails, and then shows this result to player 1 only. Player 1 may then (1) pass and thereby pay $5 to player 2 or (2) bet. If player 1 passes, the game is terminated. However, if he bets, the game continues, in which case player 2 may then either (1) pass and thereby pay $5 to player 1 or (2) call. If player 2 calls, the referee then shows him the coin; if it came up heads, player 2 pays $10 to player 1; if it came up tails, player 2 receives $10 from player 1.

(*a*) Give the pure strategies for each player. (*Hint:* Player 1 will have four pure strategies, each one specifying how he would respond to each of the two results the referee can show him; player 2 will have two pure strategies, each one specifying how he will respond if player 1 bets.)

(*b*) Develop the payoff table for this game, using expected values for the entries when necessary. Then identify and eliminate any dominated strategies.

(*c*) Show that none of the entries in the resulting payoff table are a saddle point. Then explain why any fixed choice of a pure strategy for each of the two players must be an unstable solution, so mixed strategies should be used instead.

(*d*) Write an expression for the expected payoff in terms of the probabilities of the two players using their respective pure strategies. Then show what this expression reduces to for the following three cases: (i) Player 2 definitely uses his first strategy, (ii) players 2 definitely uses his second strategy, (iii) player 2 assigns equal probabilities to using his two strategies.

11.4-1. Reconsider Prob. 11.3-1. Use the graphical procedure described in Sec. 11.4 to determine the optimal mixed strategy for each player according to the minimax criterion. Also give the corresponding value of the game.

11.4-2. Consider the game having the following payoff table.

	Strategy	*Player 2* 1	2
Player 1	1	3	−2
	2	−1	2

Use the graphical procedure described in Sec. 11.4 to determine the value of the game and the optimal mixed strategy for each player according to the minimax criterion. Check your answer for player 2 by constructing *his* payoff table and applying the graphical procedure directly to this table.

11.4-3.* For each of the following payoff tables, use the graphical procedure described in Sec. 11.4 to determine the value of the game and the optimal mixed strategy for each player according to the minimax criterion.

(*a*)

	Strategy	*Player 2* 1	2	3
Player 1	1	4	3	1
	2	0	1	2

(*b*)

	Strategy	*Player 2* 1	2	3
Player 1	1	1	−1	3
	2	0	4	1
	3	3	−2	5
	4	−3	6	−2

11.4-4. The A. J. Swim Team soon will have an important swim meet with the G. N. Swim Team. Each team has a star swimmer (John and Mark, respectively) who can swim very well in the 100-yard butterfly, backstroke, and breaststroke events. However, the rules prevent them from being used in more than two of these events. Therefore, their coaches now need to decide how to use them to maximum advantage.

Each team will enter three swimmers per event (the maximum allowed). For each event, the following table gives the best time previously achieved by John and Mark as well as the best time for each of the other swimmers who will definitely enter that event. (Whichever event John or Mark does not swim, his team's third entry for that event will be slower than the two shown in the table.)

	A. J. Swim Team			*G. N. Swim Team*		
	Entry			*Entry*		
	1	2	John	Mark	1	2
Butterfly stroke	1:01.6	59.1	57.5	58.4	1:03.2	59.8
Backstroke	1:06.8	1:05.6	1:03.3	1:02.6	1:04.9	1:04.1
Breaststroke	1:13.9	1:12.5	1:04.7	1:06.1	1:15.3	1:11.8

The points awarded are 5 points for first place, 3 points for second place, 1 point for third place, and none for lower places. Both coaches believe that all swimmers will essentially equal their best times in this meet. Thus John and Mark each will definitely be entered in two of these three events.

(*a*) The coaches must submit all their entries before the meet without knowing the entries for the other team, and no changes are permitted later. The outcome of the meet is very uncertain, so each additional point has equal value for the coaches. Formulate this problem as a two-person, zero-sum game. Eliminate dominated strategies, and then use the graphical procedure described in Sec. 11.4 to find the optimal mixed strategy for each team according to the minimax criterion.

(*b*) The situation and assignment are the same as in part (*a*), except that both coaches now believe that the A. J. team will win the swim meet if it can win 13 or more points in these three events, but will lose with less than 13 points. [Compare the resulting optimal mixed strategies with those obtained in part (*a*).]

(*c*) Now suppose that the coaches submit their entries during the meet one event at a time. When submitting his entries for an event, the coach does not know who will be swimming that event for the other team, but he does know who has swum in *preceding* events. The three key events just discussed are swum in the order listed in the table. Once again, the A. J. team needs 13 points in these events to win the swim meet. Formulate this problem as a two-person, zero-sum game. Then use the concept of dominated strategies to determine the best strategy for the G. N. team that actually "guarantees" it will win under the assumptions being made.

(*d*) The situation is the same as in part (*c*). However, now assume that the coach for the G. N. team does not know about game theory and so may, in fact, choose any of his available strategies that have Mark swimming two events. Use the concept of dominated strategies to determine the best strategies from which the coach for the A. J. team should choose. If this coach knows that the other coach has a tendency to enter Mark in the butterfly and the backstroke more often than in the breaststroke, which strategy should she choose?

11.5-1. Refer to the last paragraph of Sec. 11.5. Suppose that 3 were added to all the entries of Table 11.6 in order to ensure that the corresponding linear programming models for both players have feasible solutions with $x_3 \geq 0$ and $y_4 \geq 0$. Write out these two models. Based on the information given in Sec. 11.5, what are the optimal solutions for these two models? What is the relationship between x_3^* and y_4^*? What is the relationship between the *value* of the original game v and the values of x_3^* and y_4^*?

11.5-2.* Consider the game having the following payoff table.

		Player 2			
	Strategy	1	2	3	4
	1	5	0	3	1
Player 1	2	2	4	3	2
	3	3	2	0	4

(*a*) Use the approach described in Sec. 11.5 to formulate the problem of finding optimal mixed strategies according to the minimax criterion as a linear programming problem.

A* (*b*) Use the simplex method to find these optimal mixed strategies.

11.5-3. Follow the instructions of Prob. 11.5-2 for the game having the following payoff table.

Strategy		Player 2 1	2	3
	1	4	2	−3
Player 1	2	−1	0	3
	3	2	3	−2

11.5-4. Follow the instructions of Prob. 11.5-2 for the game having the following payoff table.

Strategy		Player 2 1	2	3	4	5
	1	1	−3	2	−2	1
Player 1	2	2	3	0	3	−2
	3	0	4	−1	−3	2
	4	−4	0	−2	2	−1

11.5-5. Section 11.5 presents a general linear programming formulation for finding an optimal mixed strategy for player 1 and for player 2. Using Table 6.14, show that the linear programming problem given for player 2 is the dual of the problem given for player 1. (*Hint:* Remember that a dual variable with a nonpositivity constraint $y_i' \leq 0$ can be replaced by $y_i = -y_i'$ with a nonnegativity constraint $y_i \geq 0$.)

11.5-6. Consider the linear programming models for players 1 and 2 given near the end of Sec. 11.5 for variation 3 of the political campaign problem (see Table 11.6). Follow the instructions of Prob. 11.5-5 for these two models.

11.5-7. Consider variation 3 of the political campaign problem (see Table 11.6). Refer to the resulting linear programming model for player 1 given near the end of Sec. 11.5. Ignoring the objective function variable x_3, plot the *feasible region* for x_1 and x_2 graphically (as described in Sec. 3.1). (*Hint:* This feasible region consists of a single line segment.) Next, write an algebraic expression for the maximizing value of x_3 for any point in this feasible region. Finally, use this expression to demonstrate that the optimal solution must, in fact, be the one given in Sec. 11.5.

A* **11.5-8.** Consider the linear programming model for player 1 given near the end of Sec. 11.5 for variation 3 of the political campaign problem (see Table 11.6). Verify the optimal mixed strategies for both players given in Sec. 11.5 by applying the automatic routine for the simplex method in your OR Courseware to this model to find both its optimal solution and its optimal dual solution.

11.5-9. Consider the general $m \times n$, two-person, zero-sum game. Let p_{ij} denote the payoff to player 1 if he plays his strategy i ($i = 1, \ldots, m$) and player 2 plays her strategy j ($j = 1, \ldots, n$). Strategy 1 (say) for player 1 is said to be *weakly dominated* by strategy 2 (say) if $p_{1j} \leq p_{2j}$ for $j = 1, \ldots, n$ and $p_{1j} = p_{2j}$ for one or more values of j.

(*a*) Assume that the payoff table possesses one or more saddle points, so that the players have corresponding optimal pure strategies under the minimax criterion. Prove that eliminating *weakly dominated* strategies from the payoff table cannot eliminate all these saddle points and cannot produce any new ones.

(*b*) Assume that the payoff table does not possess any saddle points, so that the optimal strategies under the minimax criterion are mixed strategies. Prove that eliminating weakly dominated pure strategies from the payoff table cannot eliminate all optimal mixed strategies and cannot produce any new ones.

12 Integer Programming

In Chap. 3 you saw several examples of the numerous and diverse applications of linear programming. However, one key limitation that prevents many more applications is the assumption of divisibility (see Sec. 3.3), which requires that noninteger values be permissible for decision variables. In many practical problems, the decision variables actually make sense only if they have integer values. For example, it is often necessary to assign people, machines, and vehicles to activities in integer quantities. If requiring integer values is the only way in which a problem deviates from a linear programming formulation, then it is an *integer programming* (IP) problem. (The more complete name is *integer linear programming,* but the adjective *linear* normally is dropped except when this problem is contrasted with the more esoteric integer nonlinear programming problem, which is beyond the scope of this book.)

The mathematical model for integer programming is the linear programming model (see Sec. 3.2) with the one additional restriction that the variables must have integer values. If only *some* of the variables are required to have integer values (so the divisibility assumption holds for the rest), this model is referred to as **mixed integer**

programming (MIP). When distinguishing the all-integer problem from this mixed case, we call the former *pure* integer programming.

For example, the Wyndor Glass Co. problem presented in Sec. 3.1 actually would have been an IP problem if the two decision variables x_1 and x_2 had represented the total number of units to be produced of products 1 and 2, respectively, instead of the production rates. Because both products (glass doors and wood-framed windows) necessarily come in whole units, x_1 and x_2 would have to be restricted to integer values.

There have been numerous such applications of integer programming that involve a direct extension of linear programming where the divisibility assumption must be dropped. However, another area of application may be of even greater importance, namely, problems involving a number of interrelated "yes-or-no decisions." In such decisions, the only two possible choices are *yes* and *no*. For example, should we undertake a particular fixed project? Should we make a particular fixed investment? Should we locate a facility in a particular site?

yes / no
1 / 0

With just two choices, we can represent such decisions by decision variables that are restricted to just two values, say 0 and 1. Thus the jth yes-or-no decision would be represented by, say, x_j such that

$$x_j = \begin{cases} 1 & \text{if decision } j \text{ is yes,} \\ 0 & \text{if decision } j \text{ is no.} \end{cases}$$

Such variables are called **binary variables** (or 0–1 variables). Consequently, IP problems that contain only binary variables sometimes are called **binary integer programming (BIP)** problems (or 0–1 integer programming problems).

Section 12.1 presents a miniature version of a typical BIP problem. Additional formulation possibilities with binary variables are discussed in Sec. 12.2, and Sec. 12.3 presents a series of formulation examples. The remaining sections then deal with ways to solve IP problems, including both BIP and MIP problems.

12.1 Prototype Example

The CALIFORNIA MANUFACTURING COMPANY is considering expansion by building a new factory in either Los Angeles or San Francisco, or perhaps even in both cities. It also is considering building at most one new warehouse, but the choice of location is restricted to a city where a new factory is being built. The *net present value* (total profitability considering the time value of money) of each of these alternatives is shown in the fourth column of Table 12.1. The rightmost column gives the capital required (already included in the net present value) for the respective investments, where the total capital available is $10 million. The objective is to find the feasible combination of alternatives that maximizes the total net present value.

1 2 3 4 5

Table 12.1 **Data for the California Manufacturing Co. Example**

Decision Number	Yes-or-No Question	Decision Variable	Net Present Value	Capital Required
1	Build factory in Los Angeles?	x_1	$9 million	$6 million
2	Build factory in San Francisco?	x_2	$5 million	$3 million
3	Build warehouse in Los Angeles?	x_3	$6 million	$5 million
4	Build warehouse in San Francisco?	x_4	$4 million	$2 million
			Capital available:	$10 million

Although this problem is small enough that it can be solved very quickly by inspection (build factories in both cities but no warehouse), let us formulate the IP model for illustrative purposes. All the decision variables have the *binary* form

$$x_j = \begin{cases} 1 & \text{if decision } j \text{ is yes,} \\ 0 & \text{if decision } j \text{ is no,} \end{cases} \qquad (j = 1, 2, 3, 4).$$

Because the last two decisions represent *mutually exclusive alternatives* (the company wants *at most* one new warehouse), we need the constraint

$$x_3 + x_4 \leq 1.$$

Furthermore, decisions 3 and 4 are *contingent decisions,* because they are contingent on decisions 1 and 2, respectively (the company would consider building a warehouse in a city only if a new factory also were going there). Thus, in the case of decision 3, we require that $x_3 = 0$ if $x_1 = 0$. This restriction on x_3 (when $x_1 = 0$) is imposed by adding the constraint

$$x_3 \leq x_1.$$

Similarly, the requirement that $x_4 = 0$ if $x_2 = 0$ is imposed by adding the constraint

$$x_4 \leq x_2.$$

Therefore, after we rewrite these two constraints to bring all variables to the left-hand side, the complete BIP model is

$$\text{Maximize} \qquad Z = 9x_1 + 5x_2 + 6x_3 + 4x_4,$$

subject to

$$\begin{array}{rrrrl} 6x_1 & + 3x_2 & + 5x_3 & + 2x_4 & \leq 10 \\ & & x_3 & + \ x_4 & \leq \ 1 \\ -x_1 & & + \ x_3 & & \leq \ 0 \\ & - \ x_2 & & + \ x_4 & \leq \ 0 \\ & & & x_j & \leq \ 1 \\ & & & x_j & \geq \ 0 \end{array}$$

and

$$x_j \text{ is an integer,} \qquad \text{for } j = 1, 2, 3, 4.$$

Equivalently, the last three lines of this model can be replaced by the single restriction

$$x_j \text{ is binary,} \qquad \text{for } j = 1, 2, 3, 4.$$

Except for its small size, this example is typical of many real applications of integer programming where the basic decisions to be made are of the yes-or-no type. Like the second pair of decisions for this example, groups of yes-or-no decisions often constitute groups of **mutually exclusive alternatives** such that *only one* decision in the group can be yes. Each group requires a constraint that the sum of the corresponding binary variables must be equal to 1 (if *exactly one* decision in the group must be yes) or less than or equal to 1 (if *at most one* decision in the group can be yes). Occasionally, decisions of the yes-or-no type are **contingent decision**, i.e., decisions that depend upon previous decisions. For example, one decision is said to be *contingent* on another decision if it is allowed to be yes *only if* the other is yes. This situation occurs when the contingent decision involves a follow-up action that would become irrelevant, or even impossible, if the other decision were no. The form that the resulting constraint takes always is that illustrated by the third and fourth constraints in the example.

12.2 Some Other Formulation Possibilities with Binary Variables

You have just seen a prototype example where the *basic decisions* of the problem are of the *yes-or-no type,* so that *binary variables* are introduced to represent these decisions. We now will look at some other ways in which binary variables can be very useful. In particular, we will see that these variables sometimes enable us to take a problem whose natural formulation is intractable and *reformulate* it as a pure or mixed IP problem.

This kind of situation arises when the original formulation of the problem fits either an IP or a linear programming format *except* for minor disparities involving combinatorial relationships in the model. By expressing these combinatorial relationships in terms of questions that must be answered yes or no, *auxiliary* binary variables can be introduced to the model to represent these yes-or-no decisions. Introducing these variables reduces the problem to an MIP problem (or a *pure* IP problem if all the original variables also are required to have integer values).

Some cases that can be handled by this approach are discussed next, where the x_j denote the *original* variables of the problem (they may be either continuous or integer variables) and the y_i denote the *auxiliary* binary variables that are introduced for the reformulation.

Either-Or Constraints

Consider the important case where a choice can be made between two constraints, so that *only one* (either one) must hold (whereas the other one can hold but is not required to do so). For example, there may be a choice as to which of two resources to use for a certain purpose, so that it is necessary for only one of the two resource availability constraints to hold mathematically. To illustrate the approach to such situations, suppose that one of the requirements in the overall problem is that

$$\text{Either} \quad 3x_1 + 2x_2 \le 18$$

$$\text{or} \quad x_1 + 4x_2 \le 16,$$

i.e., at least one of these two inequalities must hold but not necessarily both. This requirement must be reformulated to fit it into the linear programming format where *all* specified constraints must hold. Let M be a very large positive number. Then this requirement can be rewritten as

$$\text{Either} \quad \begin{aligned} 3x_1 + 2x_2 &\le 18 \\ x_1 + 4x_2 &\le 16 + M \end{aligned}$$

$$\text{or} \quad \begin{aligned} 3x_1 + 2x_2 &\le 18 + M \\ x_1 + 4x_2 &\le 16. \end{aligned}$$

The key is that adding M to the right-hand side of such constraints has the effect of eliminating them, because they would be satisfied automatically by any solutions that satisfy the other constraints of the problem. (This formulation assumes that the set of feasible solutions for the overall problem is a bounded set and that M is large enough that it will not eliminate any feasible solutions.) This formulation is equivalent to the set of constraints

$$3x_1 + 2x_2 \leq 18 + My$$
$$x_1 + 4x_2 \leq 16 + M(1 - y).$$

Because the *auxiliary variable* y must be either 0 or 1, this formulation guarantees that one of the original constraints must hold while the other is, in effect, eliminated. This new set of constraints would then be appended to the other constraints in the overall model to give a pure or mixed IP problem (depending upon whether the x_j are integer or continuous variables).

This approach is related directly to our earlier discussion about expressing combinatorial relationships in terms of questions that must be answered yes or no. The combinatorial relationship involved concerns the combination of the *other* constraints of the model with the *first* of the two *alternative* constraints and then with the *second*. Which of these two combinations of constraints is *better* (in terms of the value of the objective function that then can be achieved)? To rephrase this question in yes-or-no terms, we ask two complementary questions:

1. Should $x_1 + 4x_2 \leq 16$ be selected as the constraint that must hold?
2. Should $3x_1 + 2x_2 \leq 18$ be selected as the constraint that must hold?

Because exactly one of these questions is to be answered affirmatively, we let the binary terms y and $1 - y$, respectively, represent these yes-or-no decisions. Thus, $y = 1$ if the answer is yes to the first question (and no to the second), whereas $1 - y = 1$ (that is, $y = 0$) if the answer is yes to the second question (and no to the first). Since $y + 1 - y = 1$ (one yes) automatically, there is no need to add another constraint to force these two decisions to be mutually exclusive. (If separate binary variables y_1 and y_2 had been used instead to represent these yes-or-no decisions, then an additional constraint $y_1 + y_2 = 1$ would have been needed to make them mutually exclusive.)

A formal presentation of this approach is given next for a more general case.

K out of *N* Constraints Must Hold

Consider the case where the overall model includes a set of N possible constraints such that only some K of these constraints *must* hold. (Assume that $K < N$.) Part of the optimization process is to choose the *combination* of K constraints that permits the objective function to reach its best possible value. The $N - K$ constraints *not* chosen are, in effect, eliminated from the problem, although feasible solutions might coincidentally still satisfy some of them.

This case is a direct generalization of the preceding case, which had $K = 1$ and $N = 2$. Denote the N possible constraints by

$$f_1(x_1, x_2, \ldots, x_n) \leq d_1$$
$$f_2(x_1, x_2, \ldots, x_n) \leq d_2$$
$$\vdots$$
$$f_N(x_1, x_2, \ldots, x_n) \leq d_N.$$

Then, applying the same logic as for the preceding case, we find that an equivalent formulation of the requirement that some K of these constraints *must* hold is

$$f_1(x_1, x_2, \ldots, x_n) \leq d_1 + My_1$$
$$f_2(x_1, x_2, \ldots, x_n) \leq d_2 + My_2$$
$$\vdots$$
$$f_N(x_1, x_2, \ldots, x_n) \leq d_N + My_N$$

$$\sum_{i=1}^{N} y_i = N - K,$$

and

$$y_i \text{ is binary,} \qquad \text{for } i = 1, 2, \ldots, N,$$

where M is an extremely large positive number. For each binary variable y_i ($i = 1, 2, \ldots, N$), note that $y_i = 0$ makes $My_i = 0$, which reduces the new constraint i to the original constraint i. On the other hand, $y_i = 1$ makes $(d_i + My_i)$ so large that (again assuming a bounded feasible region) the new constraint i is automatically satisfied by any solution that satisfies the other new constraints, which has the effect of eliminating the original constraint i. Therefore, because the constraints on the y_i guarantee that K of these variables will equal 0 and those remaining will equal 1, K of the original constraints will be unchanged and the other $(N - K)$ original constraints will, in effect, be eliminated. The choice of *which* K constraints should be retained is made by applying the appropriate algorithm to the overall problem so it finds an optimal solution for *all* the variables simultaneously.

Functions with *N* Possible Values

Consider the situation where a given function is required to take on any one of N given values. Denote this requirement by

$$f(x_1, x_2, \ldots, x_n) = d_1 \qquad \text{or} \qquad d_2, \ldots, \qquad \text{or} \qquad d_N.$$

One special case is where this function is

$$f(x_1, x_2, \ldots, x_n) = \sum_{j=1}^{n} a_j x_j,$$

as on the left-hand side of a linear programming constraint. Another special case is where $f(x_1, x_2, \ldots, x_n) = x_j$ for a given value of j, so the requirement becomes that x_j must take on any one of N given values.

The equivalent IP formulation of this requirement is the following:

$$f(x_1, x_2, \ldots, x_n) = \sum_{i=1}^{N} d_i y_i$$

$$\sum_{i=1}^{N} y_i = 1$$

and

$$y_i \text{ is binary,} \qquad \text{for } i = 1, 2, \ldots, N,$$

so this new set of constraints would replace this requirement in the statement of the overall problem. This set of constraints provides an *equivalent* formulation because exactly one y_i must equal 1 and the others must equal 0, so exactly one d_i is being chosen as the value of the function. In this case, there are N yes-or-no questions being asked, namely, should d_i be the value chosen ($i = 1, 2, \ldots, N$)? Because the y_i respectively represent these *yes-or-no decisions,* the second constraint makes them *mutually exclusive alternatives.*

To illustrate how this case can arise, reconsider the Wyndor Glass Co. problem presented in Sec. 3.1. Eighteen hours of production time per week in Plant 3 currently is unused and available for the two new products *or* for certain future products that will be ready for production soon. In order to leave any remaining capacity in usable blocks for these future products, management now wants to impose the restriction that the production time used by the two current new products be 6 *or* 12 *or* 18 hours per week. Thus the third constraint of the original model ($3x_1 + 2x_2 \leq 18$) now becomes

$$3x_1 + 2x_2 = 6 \quad \text{or} \quad 12 \quad \text{or} \quad 18.$$

In the preceding notation, $N = 3$ with $d_1 = 6$, $d_2 = 12$, and $d_3 = 18$. Consequently, management's new requirement should be formulated as follows:

$$3x_1 + 2x_2 = 6y_1 + 12y_2 + 18y_3$$

$$y_1 + y_2 + y_3 = 1$$

and

$$y_1, y_2, y_3 \text{ are binary.}$$

The overall model for this new version of the problem then consists of the original model (see Sec. 3.1) plus this new set of constraints that replaces the original third constraint. This replacement yields a very tractable MIP formulation.

The Fixed-Charge Problem

It is quite common to incur a fixed charge or setup cost when one is undertaking an activity. For example, such a charge occurs when a production run to produce a batch of a particular product is undertaken and the required production facilities must be set up to initiate the run. In such cases, the total cost of the activity is the sum of a variable cost related to the level of the activity and the setup cost required to initiate the activity. Frequently the variable cost will be at least roughly proportional to the level of the activity. If this is the case, the *total cost* of the activity (say, activity j) can be represented by a function of the form

$$f_j(x_j) = \begin{cases} k_j + c_j x_j & \text{if } x_j > 0, \\ 0 & \text{if } x_j = 0, \end{cases}$$

where x_j denotes the level of activity j ($x_j \geq 0$), k_j denotes the setup cost, and c_j denotes the cost for each incremental unit. Were it not for the setup cost k_j, this cost structure would suggest the possibility of a *linear programming* formulation to determine the optimal levels of the competing activities. Fortunately, even with the k_j, MIP can still be used.

To formulate the overall model, suppose that there are n activities, each with the preceding cost structure (with $k_j \geq 0$ in every case and $k_j > 0$ for some $j = 1, 2, \ldots, n$), and that the problem is to

$$\text{Minimize} \quad Z = f_1(x_1) + f_2(x_2) + \cdots + f_n(x_n),$$

subject to

$$\text{given linear programming constraints.}$$

To convert this problem to an MIP format, we begin by posing n questions that must be answered yes or no; namely, for each $j = 1, 2, \ldots, n$, should activity j be undertaken ($x_j > 0$)? Each of these *yes-or-no decisions* is then represented by an auxiliary *binary variable* y_j, so that

$$Z = \sum_{j=1}^{n} (c_j x_j + k_j y_j),$$

where

$$y_j = \begin{cases} 1 & \text{if } x_j > 0, \\ 0 & \text{if } x_j = 0. \end{cases}$$

Therefore, the y_j can be viewed as *contingent decisions* similar to (but not identical to) the type considered in Sec. 12.1. Let M be an extremely large positive number that exceeds the maximum feasible value of any x_j ($j = 1, 2, \ldots, n$). Then the constraints

$$x_j \leq M y_j \qquad \text{for } j = 1, 2, \ldots, n$$

will ensure that $y_j = 1$ rather than 0 whenever $x_j > 0$. The one difficulty remaining is that these constraints leave y_j free to be either 0 or 1 when $x_j = 0$. Fortunately, this difficulty is automatically resolved because of the nature of the objective function. The case where $k_j = 0$ can be ignored because y_j can then be deleted from the formulation. So we consider the only other case, namely, where $k_j > 0$. When $x_j = 0$, so that the constraints permit a choice between $y_j = 0$ and $y_j = 1$, $y_j = 0$ must yield a smaller value of Z than $y_j = 1$. Therefore, because the objective is to minimize Z, an algorithm yielding an optimal solution would always choose $y_j = 0$ when $x_j = 0$.

To summarize, the MIP formulation of the fixed-charge problem is

$$\text{Minimize} \qquad Z = \sum_{j=1}^{n} (c_j x_j + k_j y_j),$$

subject to

the original constraints, plus

$$x_j - M y_j \leq 0$$

and

$$y_j \text{ is binary}, \qquad \text{for } j = 1, 2, \ldots, n.$$

If the x_j also had been restricted to be integer, then this would be a *pure* IP problem.

To illustrate this approach, look again at the Nori & Leets Co. air pollution problem described in Sec. 3.4. The first of the abatement methods considered—increasing the height of the smokestacks—actually would involve a substantial *fixed charge* to get ready for *any* increase in addition to a variable cost that would be roughly proportional to the amount of increase. After conversion to the equivalent annual costs used in the formulation, this fixed charge would be \$2 million each for the blast furnaces and the open-hearth furnaces, whereas the variable costs are those identified in Table 3.14. Thus, in the preceding notation, $k_1 = 2$, $k_2 = 2$, $c_1 = 8$, and $c_2 = 10$, where the objective function is expressed in units of *millions* of dollars. Because the other abatement methods do not involve any fixed charges, $k_j = 0$ for $j = 3, 4, 5, 6$. Consequently, the new MIP formulation of this problem is

$$\text{Minimize} \quad Z = 8x_1 + 10x_2 + 7x_3 + 6x_4 + 11x_5 + 9x_6 + 2y_1 + 2y_2,$$

subject to

the constraints given in Sec. 3.4, plus

$$x_1 - My_1 \le 0,$$

$$x_2 - My_2 \le 0,$$

and

$$y_1, y_2 \text{ are binary.}$$

Binary Representation of General Integer Variables

Suppose that you have a pure IP problem where most of the variables are *binary* variables, but the presence of a few *general* integer variables prevents you from solving the problem by one of the very efficient BIP algorithms now available. A nice way to circumvent this difficulty is to use the *binary representation* for each of these general integer variables. Specifically, if the bounds on an integer variable x are

$$0 \le x \le u$$

and if N is defined as the integer such that

$$2^N \le u < 2^{N+1},$$

then the **binary representation** of x is

$$x = \sum_{i=0}^{N} 2^i y_i,$$

where the y_i variables are (auxiliary) binary variables. Substituting this binary representation for each of the general integer variables (with a different set of auxiliary binary variables for each) thereby reduces the entire problem to a BIP model.

For example, suppose that an IP problem has just two general integer variables x_1 and x_2 along with many binary variables. Also suppose that the problem has nonnegativity constraints for both x_1 and x_2 and that the functional constraints include

$$\begin{aligned} x_1 \qquad\quad &\le 5 \\ 2x_1 + 3x_2 &\le 30. \end{aligned}$$

These constraints imply that $u = 5$ for x_1 and $u = 10$ for x_2, so the above definition of N gives $N = 2$ for x_1 (since $2^2 \le 5 < 2^3$) and $N = 3$ for x_2 (since $2^3 \le 10 < 2^4$). Therefore, the binary representations of these variables are

$$x_1 = y_0 + 2y_1 + 4y_2$$

$$x_2 = y_3 + 2y_4 + 4y_5 + 8y_6.$$

After we substitute these expressions for the respective variables throughout all the functional constraints and the objective function, the two functional constraints noted above become

$$\begin{aligned} y_0 + 2y_1 + 4y_2 \qquad\qquad\qquad\qquad\qquad\qquad &\le 5 \\ 2y_0 + 4y_1 + 8y_2 + 3y_3 + 6y_4 + 12y_5 + 24y_6 &\le 30. \end{aligned}$$

Observe that each feasible value of x_1 corresponds to one of the feasible values of the vector (y_0, y_1, y_2), and similarly for x_2 and (y_3, y_4, y_5, y_6). For example, $x_1 = 3$ corresponds to $(y_0, y_1, y_2) = (1, 1, 0)$, and $x_2 = 5$ corresponds to $(y_3, y_4, y_5, y_6) = (1, 0, 1, 0)$.

For an IP problem where *all* the variables are (bounded) general integer variables, it is possible to use this same technique to reduce the problem to a BIP model. However, this is not advisable for most cases because of the explosion in the number of variables involved. Applying a good IP algorithm to the original IP model generally should be more efficient than applying a good BIP algorithm to the much larger BIP model.

In general terms, for *all* the formulation possibilities with auxiliary binary variables discussed in this section, we need to strike the same note of caution. This approach sometimes requires adding a relatively large number of such variables, which can make the model *computationally infeasible*. In fact, as Sec. 12.4 explains, having just 100 binary variables can be computationally challenging (although considerably larger problems now can sometimes be solved).

12.3 Some Formulation Examples

We now present a series of examples that illustrate a variety of formulation techniques with binary variables, including those discussed in the preceding two sections. For the sake of clarity, these examples have been kept very small. In actual applications, these formulations typically would be just a small part of a vastly larger model.

Example 1: Making Choices When the Decision Variables Are Continuous

The research and development division of a manufacturing company has developed three possible new products. However, to avoid undue diversification of the company's product line, management has imposed the following restriction:

> **Restriction 1:** From the three possible new products, *at most two* should be chosen to be produced.

Two plants are available that could produce the chosen products. For administrative reasons, management has imposed a second restriction in this regard:

> **Restriction 2:** Just one of the two plants should be chosen to produce the new products.

The production cost per unit of each product would be essentially the same in the two plants. However, because of differences in their production facilities, the number of hours of production time needed per unit of each product might differ between the two plants. These data are given in Table 12.2, along with the total number of production hours available per week for these products in each plant, unit profit for each product, and marketing estimates of the number of units of each product that could be sold per week if it is produced. The objective is to choose the products, the plant, and the production rates of the chosen products so as to maximize the total profit.

In some ways, this problem resembles a standard *product-mix problem* such as the Wyndor Glass Co. example described in Sec. 3.1. In fact, if we changed the problem by dropping the two restrictions *and* by requiring each unit of a product to use the

Table 12.2 **Data for Example 1**

		Product			
		1	2	3	Available Hours per Week
Plant	1	3	4	2	30
	2	4	6	2	40
Unit profit		5	7	3	(thousands of dollars)
Sales potential		7	5	9	(units per week)

production hours given in Table 12.2 in *both plants,* it would become just such a problem. In particular, if we let x_1, x_2, and x_3 be the production rates of the respective products, the model then becomes

$$\text{Maximize} \quad Z = 5x_1 + 7x_2 + 3x_3,$$

subject to

$$\begin{aligned} 3x_1 + 4x_2 + 2x_3 &\leq 30 \\ 4x_1 + 6x_2 + 2x_3 &\leq 40 \\ x_1 \qquad\qquad &\leq 7 \\ x_2 \qquad &\leq 5 \\ x_3 &\leq 9 \end{aligned}$$

and

$$x_1 \geq 0, \qquad x_2 \geq 0, \qquad x_3 \geq 0.$$

For the real problem, however, restriction 1 necessitates adding to the model the constraint

The number of strictly positive decision variables (x_1, x_2, x_3) must be ≤ 2.

This constraint does not fit into a linear or an integer programming format, so the key question is how to convert it to such a format so that a corresponding algorithm can be used to solve the overall model. If the decision variables were binary variables, then the constraint would be expressed in this format as $x_1 + x_2 + x_3 \leq 2$. However, with *continuous* decision variables, a more complicated approach involving the introduction of auxiliary binary variables is needed.

Requirement 2 necessitates replacing the first two functional constraints ($3x_1 + 4x_2 + 2x_3 \leq 30$ and $4x_1 + 6x_2 + 2x_3 \leq 40$) by the restriction

$$\begin{aligned} &\text{Either} \quad 3x_1 + 4x_2 + 2x_3 \leq 30 \\ &\text{Or} \quad\quad\ 4x_1 + 6x_2 + 2x_3 \leq 40 \end{aligned}$$

must hold, where the choice of which constraint must hold corresponds to the choice of which plant will be used to produce the new products. We discussed in the preceding section how such an either-or constraint can be converted to a linear or an integer programming format, again with the help of an auxiliary binary variable.

Formulation with Auxiliary Binary Variables: To deal with requirement 1, we introduce three auxiliary binary variables (y_1, y_2, y_3) with the interpretation

$$y_j = \begin{cases} 1 & \text{if } x_j > 0 \text{ can hold (can produce product } j), \\ 0 & \text{if } x_j = 0 \text{ must hold (cannot produce product } j), \end{cases}$$

for $j = 1, 2, 3$. To enforce this interpretation in the model with the help of M (an extremely large positive number), we add the constraints

$$x_1 \le My_1$$
$$x_2 \le My_2$$
$$x_3 \le My_3$$
$$y_1 + y_2 + y_3 \le 2$$
$$y_j \text{ is binary}, \qquad \text{for } j = 1, 2, 3.$$

The either-or constraint and nonnegativity constraints give a *bounded* feasible region for the decision variables (so each $x_j \le M$ throughout this region). Therefore, in the $x_j \le My_j$ constraint, $y_j = 1$ allows any value of x_j in the feasible region, whereas $y_j = 0$ forces $x_j = 0$. (Conversely, $x_j > 0$ forces $y_j = 1$, whereas $x_j = 0$ allows either value of y_j but the fourth constraint may force $y_j = 0$ in an optimal solution.) Consequently, when the fourth constraint forces choosing at most two of the y_j to equal 1, this amounts to choosing at most two of the new products as the ones that can be produced.

To deal with requirement 2, we introduce another auxiliary binary variable y_4 with the interpretation

$$y_4 = \begin{cases} 1 & \text{if } 4x_1 + 6x_2 + 2x_3 \le 40 \text{ must hold (choose Plant 2)} \\ 0 & \text{if } 3x_1 + 4x_2 + 2x_3 \le 30 \text{ must hold (choose Plant 1).} \end{cases}$$

As discussed in Sec. 12.2, this interpretation is enforced by adding the constraints

$$3x_1 + 4x_2 + 2x_3 \le 30 + My_4$$
$$4x_1 + 6x_2 + 2x_3 \le 40 + M(1 - y_4)$$
$$y_4 \text{ is binary.}$$

Consequently, after we move all variables to the left-hand side of the constraints, the complete model is

$$\text{Maximize} \qquad Z = 5x_1 + 7x_2 + 3x_3,$$

subject to

$$x_1 \le 7$$
$$x_2 \le 5$$
$$x_3 \le 9$$
$$x_1 - My_1 \le 0$$
$$x_2 - My_2 \le 0$$
$$x_3 - My_3 \le 0$$
$$y_1 + y_2 + y_3 \le 2$$
$$3x_1 + 4x_2 + 2x_3 - My_4 \le 30$$
$$4x_1 + 6x_2 + 2x_3 + My_4 \le 40 + M$$

and

$$x_1 \ge 0, \qquad x_2 \ge 0, \qquad x_3 \ge 0$$
$$y_j \text{ is binary}, \qquad \text{for } j = 1, 2, 3, 4.$$

This now is an MIP model, with three variables (the x_j) not required to be integer and four binary variables, so an MIP algorithm can be used to solve the model. When this is done (after substituting a large numerical value for M)[1], the optimal solution is $y_1 = 1$, $y_2 = 0$, $y_3 = 1$, $y_4 = 1$, $x_1 = 5\frac{1}{2}$, $x_2 = 0$, and $x_3 = 9$; that is, choose products 1 and 3 to produce, choose Plant 2 for the production, and choose the production rates of $5\frac{1}{2}$ units per week for product 1 and 9 units per week for product 3. The resulting total profit is \$54,500 per week.

Example 2: Violating Proportionality

A corporation is developing its marketing plans for next year's new products. For three of these products, it is considering purchasing a total of five TV spots for commercials on national television networks. The problem we will focus on is how to allocate the five spots to these three products, with a maximum of three spots (and a minimum of zero) for each product.

Table 12.3 shows the estimated impact of allocating zero, one, two, or three spots to each product. This impact is measured in terms of the *profit* (in units of millions of dollars) from the *additional sales* that would result from the spots, considering also the cost of producing the commercial and purchasing the spots. The objective is to allocate five spots to the products so as to maximize the total profit.

This small problem can be solved easily by dynamic programming (Chap. 10) or even by inspection. (The optimal solution is to allocate two spots to product 1, no spots to product 2, and three spots to product 3.) However, we will show two different BIP formulations for illustrative purposes. Such a formulation would become necessary if this small problem needed to be incorporated into a larger IP model involving the allocation of resources to marketing activities for all the corporation's new products.

One Formulation with Auxiliary Binary Variables: A natural formulation would be to let x_1, x_2, x_3 be the number of TV spots allocated to the respective products. The contribution of each x_j to the objective function then would be given by the corresponding column in Table 12.3. However, each of these columns violates the assumption of proportionality described in Sec. 3.3. Therefore, we cannot write a *linear* objective function in terms of these integer decision variables.

Now see what happens when we introduce an *auxiliary binary variable* y_{ij} for each positive integer value of $x_i = j$ $(j = 1, 2, 3)$, where y_{ij} has the interpretation

[1] In practice, some care is taken to choose a value for M that definitely is large enough to avoid eliminating any feasible solutions, but as small as possible otherwise in order to avoid unduely enlarging the feasible region for the LP-relaxation (and to avoid numerical instability). For this example, a careful examination of the constraints reveals that the minimum feasible value of M is $M = 9$.

Table 12.3 **Data for Example 2**

Number of TV spots	*Profit* — *Product* 1	2	3
0	0	0	0
1	1	0	−1
2	3	2	2
3	3	3	4

$$y_{ij} = \begin{cases} 1 & \text{if } x_i = j, \\ 0 & \text{otherwise.} \end{cases}$$

(For example, $y_{21} = 0$, $y_{22} = 0$, and $y_{23} = 1$ mean that $x_2 = 3$.) The resulting *linear* BIP model is

$$\text{Maximize} \quad Z = y_{11} + 3y_{12} + 3y_{13} + 2y_{22} + 3y_{23} - y_{31} + 2y_{32} + 4y_{33},$$

subject to

$$y_{11} + y_{12} + y_{13} \leq 1$$

$$y_{21} + y_{22} + y_{23} \leq 1$$

$$y_{31} + y_{32} + y_{33} \leq 1$$

$$y_{11} + 2y_{12} + 3y_{13} + y_{21} + 2y_{22} + 3y_{23} + y_{31} + 2y_{32} + 3y_{33} = 5$$

and

$$\text{each } y_{ij} \text{ is binary.}$$

Note that the first three functional constraints ensure that each x_i will be assigned just one of its possible values. (Here $y_{i1} + y_{i2} + y_{i3} = 0$ corresponds to $x_i = 0$, which contributes nothing to the objective function.) The last functional constraint ensures that $x_1 + x_2 + x_3 = 5$. The *linear* objective function then gives the total profit according to Table 12.3.

Solving this BIP model gives an optimal solution of

$$y_{11} = 0, \quad y_{12} = 1, \quad y_{13} = 0, \quad \text{so} \quad x_1 = 2$$

$$y_{21} = 0, \quad y_{22} = 0, \quad y_{23} = 0, \quad \text{so} \quad x_2 = 0$$

$$y_{31} = 0, \quad y_{32} = 0, \quad y_{33} = 1, \quad \text{so} \quad x_3 = 3.$$

Another Formulation with Auxiliary Binary Variables: We now redefine the above auxiliary binary variables y_{ij} as follows:

$$y_{ij} = \begin{cases} 1 & \text{if } x_i \geq j, \\ 0 & \text{otherwise.} \end{cases}$$

Thus, the difference is that $y_{ij} = 1$ now if $x_i \geq j$ instead of $x_i = j$. Therefore,

$$x_i = 0 \quad \Rightarrow \quad y_{i1} = 0, \quad y_{i2} = 0, \quad y_{i3} = 0,$$

$$x_i = 1 \quad \Rightarrow \quad y_{i1} = 1, \quad y_{i2} = 0, \quad y_{i3} = 0,$$

$$x_i = 2 \quad \Rightarrow \quad y_{i1} = 1, \quad y_{i2} = 1, \quad y_{i3} = 0,$$

$$x_i = 3 \quad \Rightarrow \quad y_{i1} = 1, \quad y_{i2} = 1, \quad y_{i3} = 1,$$

$$\text{so } x_i = y_{i1} + y_{i2} + y_{i3}$$

for $i = 1, 2, 3$. Because allowing $y_{i2} = 1$ is contingent upon $y_{i1} = 1$ and allowing $y_{i3} = 1$ is contingent upon $y_{i2} = 1$, these definitions are enforced by adding the constraints

$$y_{i2} \leq y_{i1} \quad \text{and} \quad y_{i3} \leq y_{i2}, \qquad \text{for } i = 1, 2, 3.$$

The new definition of the y_{ij} also changes the objective function, as illustrated in Fig. 12.1 for the product 1 portion of the objective function. Since y_{11}, y_{12}, y_{13} provide the successive increments (if any) in the value of x_1 (starting from a value of 0), the coefficients of y_{11}, y_{12}, y_{13} are given by the respective *increments* in the product 1

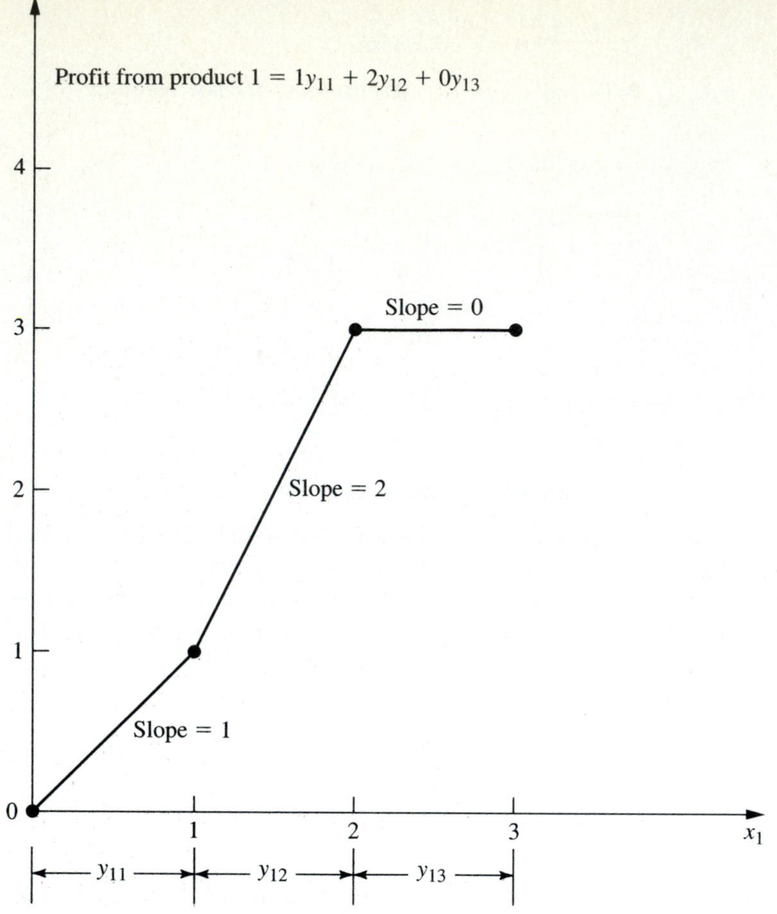

Figure 12.1 The profit from the additional sales of product 1 that would result from x_1 TV spots, where the slopes given the corresponding coefficients in the objective function for the second BIP formulation.

column of Table 12.3 ($1 - 0 = 1$, $3 - 1 = 2$, $3 - 3 = 0$). These *increments* are the *slopes* in Fig. 12.1, yielding $1y_{11} + 2y_{12} + 0y_{13}$ for the product 1 portion of the objective function. Note that applying this approach to all three products still must lead to a *linear* objective function.

After we bring all variables to the left-hand side of the constraints, the resulting complete BIP model is

$$\text{Maximize} \quad Z = y_{11} + 2y_{12} + 2y_{22} + y_{23} - y_{31} + 3y_{32} + 2y_{33},$$

subject to

$$\begin{aligned} y_{12} - y_{11} &\le 0 \\ y_{13} - y_{12} &\le 0 \\ y_{22} - y_{21} &\le 0 \\ y_{23} - y_{22} &\le 0 \\ y_{32} - y_{31} &\le 0 \\ y_{33} - y_{32} &\le 0 \end{aligned}$$

$$y_{11} + y_{12} + y_{13} + y_{21} + y_{22} + y_{23} + y_{31} + y_{32} + y_{33} = 5$$

and

$$\text{each } y_{ij} \text{ is binary.}$$

Solving this BIP model gives an optimal solution of

$$y_{11} = 1, \quad y_{12} = 1, \quad y_{13} = 0, \quad \text{so} \quad x_1 = 2$$
$$y_{21} = 0, \quad y_{22} = 0, \quad y_{23} = 0, \quad \text{so} \quad x_2 = 0$$
$$y_{31} = 1, \quad y_{32} = 1, \quad y_{33} = 1, \quad \text{so} \quad x_3 = 3.$$

There is little to choose between this BIP model and the preceding one other than personal taste. They have the same number of binary variables (the prime consideration in determining computational effort for BIP problems). They also both have some *special structure* (constraints for *mutually exclusive alternatives* in the first model and constraints for *contingent decisions* in the second) that can lead to speedup. The second model does have more functional constraints than the first.

Example 3: Covering All Characteristics

An airline needs to assign its crews to cover all its upcoming flights. We will focus on the problem of assigning three crews based in San Francisco (SF) to the flights listed in the first column of Table 12.4. The other 12 columns show the 12 feasible sequences of flights for a crew. (The numbers in each column indicate the order of the flights.) Exactly three of the sequences need to be chosen (one per crew) in such a way that every flight is covered. (It is permissible to have more than one crew on a flight, where the extra crews would fly as passengers, but union contracts require that the extra crews still be paid for their time as if they were working.) The cost of assigning a crew to a particular sequence of flights is given (in thousands of dollars) in the bottom row of the table. The objective is to minimize the total cost of the three crew assignments that cover all the flights.

Formulation with Binary Variables: With 12 feasible sequences of flights, we have 12 yes-or-no decisions:

Should sequence j be assigned to a crew? $\quad (j = 1, 2, \ldots, 12)$

Therefore, we use 12 binary variables to represent these respective decisions:

$$x_j = \begin{cases} 1 & \text{if sequence } j \text{ is assigned to a crew,} \\ 0 & \text{otherwise.} \end{cases}$$

Table 12.4 **Data for Example 3**

	Feasible Sequence of Flights											
Flight	1	2	3	4	5	6	7	8	9	10	11	12
San Francisco to Los Angeles	1			1			1			1		
San Francisco to Denver		1			1			1			1	
San Francisco to Seattle			1			1			1			1
Los Angeles to Chicago				2			2		3	2		3
Los Angeles to San Francisco	2					3				5	5	
Chicago to Denver				3	3				4			
Chicago to Seattle							3	3		3	3	4
Denver to San Francisco		2		4	4				5			
Denver to Chicago					2			2			2	
Seattle to San Francisco			2				4	4				5
Seattle to Los Angeles						2			2	4	4	2
Cost, $1,000	2	3	4	6	7	5	7	8	9	9	8	9

The most interesting part of this formulation is the nature of each constraint that ensures that a corresponding flight is covered. For example, consider the last flight in Table 12.4 [Seattle to Los Angeles (LA)]. Five sequences (namely, sequences 6, 9, 10, 11, and 12) include this flight. Therefore, at least one of these five sequences must be chosen. The resulting constraint is

$$x_6 + x_9 + x_{10} + x_{11} + x_{12} \geq 1.$$

The complete BIP model is

Minimize $\quad Z = 2x_1 + 3x_2 + 4x_3 + 6x_4 + 7x_5 + 5x_6 + 7x_7 + 8x_8 + 9x_9 + 9x_{10} + 8x_{11} + 9x_{12}.$

subject to

$$x_1 + x_4 + x_7 + x_{10} \geq 1 \qquad \text{(SF to LA)}$$

$$x_2 + x_5 + x_8 + x_{11} \geq 1 \qquad \text{(SF to Denver)}$$

$$x_3 + x_6 + x_9 + x_{12} \geq 1 \qquad \text{(SF to Seattle)}$$

$$x_4 + x_7 + x_9 + x_{10} + x_{12} \geq 1 \qquad \text{(LA to Chicago)}$$

$$x_1 + x_6 + x_{10} + x_{11} \geq 1 \qquad \text{(LA to SF)}$$

$$x_4 + x_5 + x_9 \geq 1 \qquad \text{(Chicago to Denver)}$$

$$x_7 + x_8 + x_{10} + x_{11} + x_{12} \geq 1 \qquad \text{(Chicago to Seattle)}$$

$$x_2 + x_4 + x_5 + x_9 \geq 1 \qquad \text{(Denver to SF)}$$

$$x_5 + x_8 + x_{11} \geq 1 \qquad \text{(Denver to Chicago)}$$

$$x_3 + x_7 + x_8 + x_{12} \geq 1 \qquad \text{(Seattle to SF)}$$

$$x_6 + x_9 + x_{10} + x_{11} + x_{12} \geq 1 \qquad \text{(Seattle to LA)}$$

$$\sum_{j=1}^{12} x_j = 3 \qquad \text{(assign three crews)}$$

and

$$x_j \text{ is binary,} \qquad \text{for } j = 1, 2, \ldots, 12.$$

Your OR Courseware provides the following optimal solution:

$$x_3 = 1 \qquad \text{(assign sequence 3 to a crew)}$$

$$x_4 = 1 \qquad \text{(assign sequence 4 to a crew)}$$

$$x_{11} = 1 \qquad \text{(assign sequence 11 to a crew)}$$

and all other $x_j = 0$, for a total cost of $18,000.

This example illustrates a broader class of problems called **set-covering problems**. Any set-covering problem can be described in general terms as involving a number of potential *activities* (such as flight sequences) and *characteristics* (such as flights). Each activity possesses some of but not all the characteristics. The objective is to determine the least costly combination of activities that collectively possess (cover) each characteristic at least once. Thus, let S_i be the set of all activities that possess characteristic i. At least one member of set S_i must be included among the chosen activities, so the constraint

$$\sum_{j \in S_i} x_j \geq 1$$

is included[1] for each characteristic i.

A related class of problems, called **set-partitioning problems**, change each such constraint to

$$\sum_{j \in S_i} x_j = 1,$$

so now *exactly* one member of each set S_i must be included among the chosen activities. For the crew-scheduling example, this means that each flight must be included *exactly* once among the chosen flight sequences, which rules out having extra crews (as passengers) on any flight.

Airline crew scheduling has become one important application of IP in recent years. Problems involving thousands of possible flight sequences now are being solved. For example, American Airlines is using this approach. Several years ago, it was estimated that the resulting savings at American Airlines is $18 million per year.[2]

12.4 Some Perspectives on Solving Integer Programming Problems

It may seem that IP problems should be relatively easy to solve. After all, *linear programming* problems can be solved extremely efficiently, and the only difference is that IP problems have far fewer solutions to be considered. In fact, *pure* IP problems with a bounded feasible region are guaranteed to have just a *finite* number of feasible solutions.

Unfortunately, there are two fallacies in this line of reasoning. One is that having a finite number of feasible solutions ensures that the problem is readily solvable. Finite numbers can be astronomically large. For example, consider the simple case of BIP problems. With n variables, there are 2^n solutions to be considered (where some of these solutions can subsequently be discarded because they violate the functional constraints). Thus, each time n is increased by 1, the number of solutions is *doubled*. This pattern is referred to as the **exponential growth** of the difficulty of the problem. With $n = 10$, there are more than 1,000 solutions (1,024); with $n = 20$, there are more than 1,000,000; with $n = 30$, there are more than 1 billion; and so forth. Therefore, even the fastest computers are incapable of performing exhaustive enumeration (checking each solution for feasibility and, if it is feasible, calculating the value of the objective value) for BIP problems with more than a few dozen variables, let alone for *general* IP problems with the same number of integer variables. Sophisticated algorithms, such as

[1] Strictly speaking, a set-covering problem does not include any *other* functional constraints such as the last functional constraint in the above crew-scheduling example. It also is sometimes assumed that every coefficient in the objective function being minimized equals 1, and then the name *weighted set-covering problem* is used when this assumption does not hold.

[2] I. Gershkoff, "Optimizing Flight Crew Schedules," *Interfaces*, **19**(4): 29–43, July–August 1989.

those described in subsequent sections, can do somewhat better. In fact, Sec. 12.7 discusses how recently developed algorithms have successfully solved certain *vastly* larger BIP problems (into the thousands of variables). Nevertheless, because of exponential growth, even the best algorithms cannot be guaranteed to solve every relatively small problem (less than 100 binary or integer variables).

The second fallacy is that removing some feasible solutions (the noninteger ones) from a linear programming problem will make it easier to solve. To the contrary, it is only because all these feasible solutions are there that the guarantee can be given (see Sec. 5.1) that there will be a corner-point feasible (CPF) solution [and so a corresponding basic feasible (BF) solution] that is optimal for the overall problem. *This* guarantee is the key to the remarkable efficiency of the simplex method. As a result, linear programming problems generally are *much* easier to solve than IP problems.

Consequently, most successful algorithms for integer programming incorporate the simplex method (or dual simplex method) as much as they can by relating portions of the IP problem under consideration to the corresponding linear programming problem (i.e., the same problem except that the integer restriction is deleted). For any given IP problem, this corresponding linear programming problem commonly is referred to as its **LP relaxation**. The algorithms presented in the next two sections illustrate how a sequence of LP relaxations for portions of an IP problem can be used to solve the overall IP problem effectively.

There is one special situation where solving an IP problem is no more difficult than solving its LP relaxation once by the simplex method, namely, when the optimal solution to the latter problem turns out to satisfy the integer restriction of the IP problem. When this situation occurs, this solution *must* be optimal for the IP problem as well, because it is the best solution among all the feasible solutions for the LP relaxation, which includes all the feasible solutions for the IP problem. Therefore, it is common for an IP algorithm to begin by applying the simplex method to the LP relaxation to check whether this fortuitous outcome has occurred.

Although it generally is quite fortuitous indeed for the optimal solution to the LP relaxation to be integer as well, there actually exist several *special types* of IP problems for which this outcome is *guaranteed.* You already have seen the most prominent of these special types in Chaps. 8 and 9, namely, the *minimum cost flow problem* (with integer parameters) and its special cases (including the *transportation problem,* the *assignment problem,* the *shortest-path problem,* and the *maximum flow problem*). This guarantee can be given for these types of problems because they possess a certain *special structure* (e.g., see Table 8.6) that ensures that every BF solution is integer, as stated in the integer solutions property given in Secs. 8.1 and 9.6. Consequently, these special types of IP problems can be treated as linear programming problems, because they can be solved completely by a streamlined version of the simplex method.

Although this much simplification is somewhat unusual, in practice IP problems frequently have *some* special structure that can be exploited to simplify the problem. (Examples 2 and 3 in the preceding section fit into this category, because of their *mutually exclusive alternatives* constraints or *contingent decisions* constraints or *set-covering* constraints.) Sometimes, very large versions of these problems can be solved successfully. Special-purpose algorithms designed specifically to exploit certain kinds of special structures are becoming increasingly important in integer programming.

Thus the two primary determinants of *computational difficulty* for an IP problem are (1) the *number of integer variables* and (2) any *special structure* in the problem. This situation is in contrast to linear programming, where the number of (functional) constraints is much more important than the number of variables. In integer programming, the number of constraints is of *some* importance (especially if LP relaxations are being solved), but it is strictly secondary to the other two factors. In fact, there occasionally are cases where *increasing* the number of constraints *decreases* the computation time because the number of feasible solutions has been reduced. For MIP problems, it is the number of *integer* variables rather than the *total* number of variables that is important, because the continuous variables have almost no effect on the computational effort.

Because IP problems are, in general, much more difficult to solve than linear programming problems, sometimes it is tempting to use the approximate procedure of simply applying the simplex method to the LP relaxation and then *rounding* the noninteger values to integers in the resulting solution. This approach may be adequate for some applications, especially if the values of the variables are quite large so that rounding creates relatively little error. However, you should beware of two pitfalls involved in this approach.

One pitfall is that an optimal linear programming solution is *not necessarily feasible* after it is rounded. Often it is difficult to see in which way the rounding should be done to retain feasibility. It may even be necessary to change the value of some variables by one or more units after rounding. To illustrate, consider the following problem:

$$\text{Maximize} \quad Z = x_2,$$

subject to

$$-x_1 + x_2 \le \tfrac{1}{2}$$
$$x_1 + x_2 \le 3\tfrac{1}{2}$$

and

$$x_1 \ge 0, \qquad x_2 \ge 0$$
$$x_1, x_2 \text{ are integers.}$$

As Fig. 12.2 shows, the optimal solution for the LP relaxation is $x_1 = 1\frac{1}{2}$, $x_2 = 2$, but it is impossible to round the noninteger variable x_1 to 1 or 2 (or any other integer) and retain feasibility. Feasibility can be retained only by also changing the integer value of x_2. It is easy to imagine how such difficulties can be compounded when there are tens or hundreds of constraints and variables.

Even if an optimal solution for the LP relaxation is rounded successfully, there remains another pitfall. There is no guarantee that this rounded solution will be the optimal integer solution. In fact, it may even be far from optimal in terms of the value of the objective function. This fact is illustrated by the following problem:

$$\text{Maximize} \quad Z = x_1 + 5x_2,$$

subject to

$$x_1 + 10x_2 \le 20$$
$$x_1 \qquad\quad \le 2$$

and

$$x_1 \ge 0, \qquad x_2 \ge 0$$
$$x_1, x_2 \text{ are integers.}$$

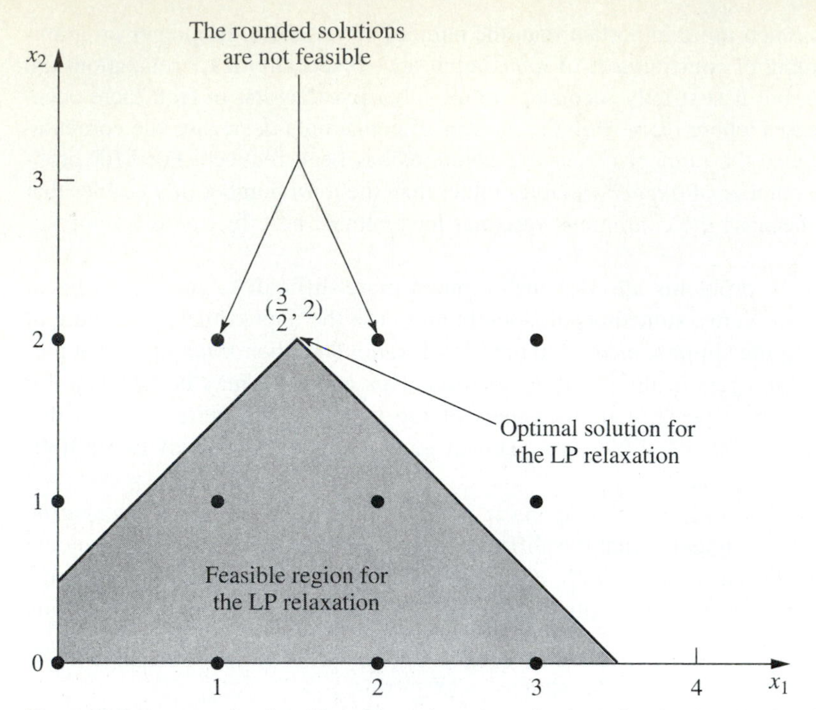

Figure 12.2 An example of an IP problem where the optimal solution for the LP relaxation cannot be rounded in any way that retains feasibility.

Because there are only two decision variables, this problem can be depicted graphically as shown in Fig. 12.3. Either the graph or the simplex method may be used to find that the optimal solution for the LP relaxation is $x_1 = 2, x_2 = \frac{9}{5}$, with $Z = 11$. If a graphical solution were not available (which would be the case with more decision variables), then the variable with the noninteger value $x_2 = \frac{9}{5}$ would normally be rounded in the feasible direction to $x_2 = 1$. The resulting integer solution is $x_1 = 2, x_2 = 1$, which yields $Z = 7$. Notice that this solution is far from the optimal solution $(x_1, x_2) = (0, 2)$, where $Z = 10$.

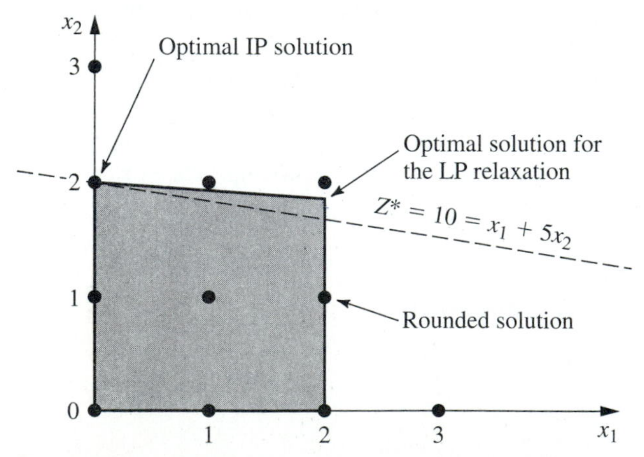

Figure 12.3 An example where rounding the optimal solution for the LP relaxation is far from optimal for the IP problem.

Because of these two pitfalls, a better approach for dealing with IP problems that are too large to be solved exactly is to use one of the available *heuristic algorithms*. These algorithms are extremely efficient for large problems, but they are not guaranteed to find an optimal solution. However, they do tend to be considerably more effective than the rounding approach just discussed in finding very good feasible solutions.

For IP problems that are small enough to be solved to optimality, a considerable number of algorithms now are available. However, no IP algorithm possesses computational efficiency that is even remotely comparable to the *simplex method* (except on special types of problems). Therefore, the development of IP algorithms has continued to be an active area of research. Fortunately, some exciting algorithmic advances have been made during the last decade, and additional progress can be anticipated during the next decade. These recent advances are discussed further in Sec. 12.7.

The most popular mode for IP algorithms is to use the *branch-and-bound technique* and related ideas to *implicitly enumerate* the feasible integer solutions, and we shall focus on this approach. The next section presents the branch-and-bound technique in a general context and illustrates it with a basic branch-and-bound algorithm for BIP problems. Section 12.6 presents another algorithm of the same type for general MIP problems.

12.5 The Branch-and-Bound Technique and Its Application to Binary Integer Programming

Because any bounded *pure* IP problem has only a finite number of feasible solutions, it is natural to consider using some kind of *enumeration procedure* for finding an optimal solution. Unfortunately, as we discussed in the preceding section, this finite number can be, and usually is, very large. Therefore, it is imperative that any enumeration procedure be cleverly structured so that only a tiny fraction of the feasible solutions actually need to be examined. For example, dynamic programming (see Chap. 10) provides one such kind of procedure for many problems having a finite number of feasible solutions (although it is not particularly efficient for most IP problems). Another such approach is provided by the *branch-and-bound technique.* This technique and variations of it have been applied with some success to a variety of OR problems, but it is especially well known for its application to IP problems.

The basic concept underlying the branch-and-bound technique is to *divide and conquer.* Since the original ''large'' problem is too difficult to be solved directly, it is divided into smaller and smaller subproblems until these subproblems can be conquered. The dividing (*branching*) is done by partitioning the entire set of feasible solutions into smaller and smaller subsets. The conquering (*fathoming*) is done partially by *bounding* how good the best solution in the subset can be and then discarding the subset if its bound indicates that it cannot possibly contain an optimal solution for the original problem.

We shall now describe in turn these three basic steps—branching, bounding, and fathoming—and illustrate them by applying a branch-and-bound algorithm to the prototype example (the California Manufacturing Co. problem) presented in Sec. 12.1 and repeated here (with the constraints numbered for later reference).

$$\text{Maximize} \quad Z = 9x_1 + 5x_2 + 6x_3 + 4x_4,$$

subject to

$$
\begin{array}{ll}
(1) & 6x_1 + 3x_2 + 5x_3 + 2x_4 \le 10 \\
(2) & x_3 + x_4 \le 1 \\
(3) & -x_1 + x_3 \le 0 \\
(4) & -x_2 + x_4 \le 0
\end{array}
$$

and

$$(5) \quad x_j \text{ is binary}, \quad \text{for } j = 1, 2, 3, 4.$$

Branching

When you are dealing with binary variables, the most straightforward way to partition the set of feasible solutions into subsets is to fix the value of one of the variables (say, x_1) at $x_1 = 0$ for one subset and at $x_1 = 1$ for the other subset. Doing this for the prototype example divides the whole problem into the two smaller subproblems, shown below.

Subproblem 1:

Fix $x_1 = 0$ so the resulting subproblem is

$$\text{Maximize} \quad Z = 5x_2 + 6x_3 + 4x_4,$$

subject to

$$
\begin{array}{ll}
(1) & 3x_2 + 5x_3 + 2x_4 \le 10 \\
(2) & x_3 + x_4 \le 1 \\
(3) & x_3 \le 0 \\
(4) & -x_2 + x_4 \le 0 \\
(5) & x_j \text{ is binary}, \quad \text{for } j = 2, 3, 4.
\end{array}
$$

Subproblem 2:

Fix $x_1 = 1$ so the resulting subproblem is

$$\text{Maximize} \quad Z = 9 + 5x_2 + 6x_3 + 4x_4,$$

subject to

$$
\begin{array}{ll}
(1) & 3x_2 + 5x_3 + 2x_4 \le 4 \\
(2) & x_3 + x_4 \le 1 \\
(3) & x_3 \le 1 \\
(4) & -x_2 + x_4 \le 0 \\
(5) & x_j \text{ is binary}, \quad \text{for } j = 2, 3, 4.
\end{array}
$$

Figure 12.4 portrays this dividing (branching) into subproblems by a *tree* (defined in Sec. 9.2) with *branches* (arcs) from the *All* node (corresponding to the whole problem having *all* feasible solutions) to the two nodes corresponding to the two subproblems.

This tree, which will continue "growing branches" iteration by iteration, is referred to as the **solution tree** (or **enumeration tree**) for the algorithm. The variable used to do this branching at any iteration by assigning values to the variable (as with x_1 above) is called the **branching variable**. (Sophisticated methods for selecting branching variables are an important part of some branch-and-bound algorithms but, for simplicity, we always select them in their natural order—$x_1, x_2, \ldots, x_n$—throughout this section.)

Later in the section you will see that one of these subproblems can be conquered (fathomed) immediately, whereas the other subproblem will need to be divided further into smaller subproblems by setting $x_2 = 0$ or $x_2 = 1$.

For other IP problems where the integer variables have more than two possible values, the branching can still be done by setting the branching variable at its respective individual values, thereby creating more than two new subproblems. However, a good alternate approach is to specify a *range* of values (for example, $x_j \leq 2$ or $x_j \geq 3$) for the branching variable for each new subproblem. This is the approach used for the algorithm presented in Sec. 12.6.

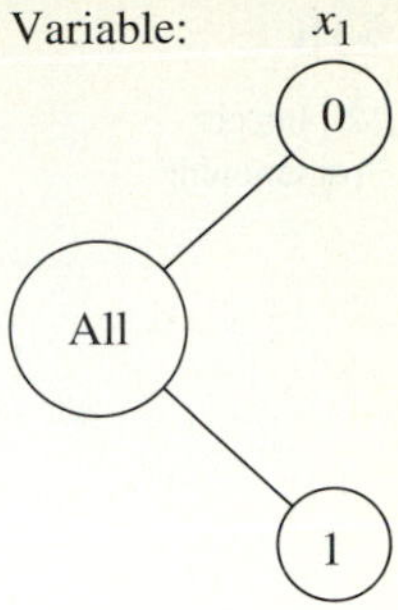

Variable: x_1

0

All

1

Figure 12.4 The solution tree created by the branching for the first iteration of the BIP branch-and-bound algorithm for the example in Sec. 12.1.

Bounding

For each of these subproblems, we now need to obtain a *bound* on how good its best feasible solution can be. The standard way of doing this is to quickly solve a simpler *relaxation* of the subproblem. In most cases, a **relaxation** of a problem is obtained simply by *deleting* ("relaxing") one set of constraints that had made the problem difficult to solve. For IP problems, the most troublesome constraints are those requiring the respective variables to be integer. Therefore, the most widely used relaxation is the *LP relaxation* that deletes this set of constraints.

To illustrate for the example, consider first the whole problem given in Sec. 12.1. Its LP relaxation is obtained by deleting the last line of the model (x_j is an integer, for $j = 1, 2, 3, 4$), but retaining the $x_j \leq 1$ and $x_j \geq 0$ constraints. Using the simplex method to quickly solve this LP relaxation yields its optimal solution

$$(x_1, x_2, x_3, x_4) = (\tfrac{5}{6}, 1, 0, 1), \qquad \text{with } Z = 16\tfrac{1}{2}.$$

Therefore, $Z \leq 16\frac{1}{2}$ for all feasible solutions for the original BIP problem (since these solutions are a subset of the feasible solutions for the LP relaxation). In fact, as summarized below, this *bound* of $16\frac{1}{2}$ can be rounded down to 16, because all coefficients in the objective function are integer, so all integer solutions must have an integer value for Z.

$$\text{Bound for whole problem:} \qquad Z \leq 16.$$

Now let us obtain the bounds for the two subproblems in the same way. Their LP relaxations are obtained from the models in the preceding subsection by replacing the constraint that x_j is binary for $j = 2, 3, 4$ by the constraint $0 \leq x_j \leq 1$ for $j = 2, 3, 4$. Applying the simplex method then yields their optimal solutions (plus the fixed value of x_1) shown below.

LP relaxation of subproblem 1: $(x_1, x_2, x_3, x_4) = (0, 1, 0, 1)$ with $Z = 9$.
LP relaxation of subproblem 2: $(x_1, x_2, x_3, x_4) = (1, \frac{4}{5}, 0, \frac{4}{5})$ with $Z = 16\frac{1}{5}$.

The resulting bounds for the subproblems then are

Bound for subproblem 1: $Z \leq 9$,
Bound for subproblem 2: $Z \leq 16$.

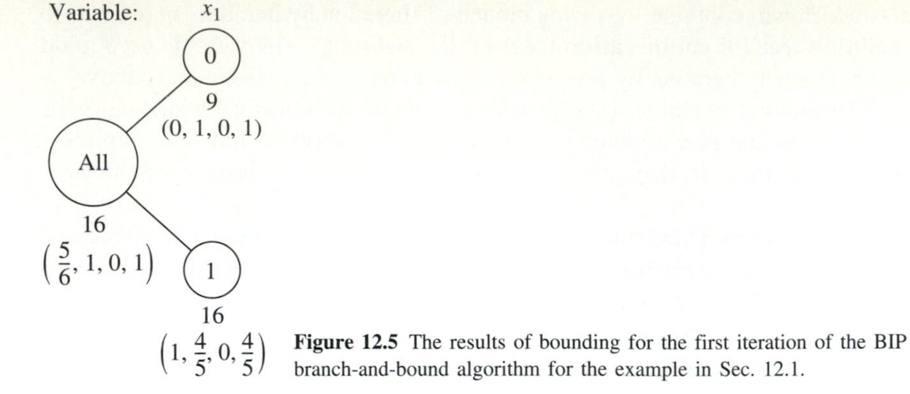

Figure 12.5 The results of bounding for the first iteration of the BIP branch-and-bound algorithm for the example in Sec. 12.1.

Figure 12.5 summarizes these results, where the numbers given just below the nodes are the bounds and below each bound is the optimal solution obtained for the LP relaxation.

Fathoming

A subproblem can be conquered (fathomed), and thereby dismissed from further consideration, in the three ways described below.

One way is illustrated by the results for subproblem 1 given by the $x_1 = 0$ node in Fig. 12.5. Note that the (unique) optimal solution for its LP relaxation, $(x_1, x_2, x_3, x_4) = (0, 1, 0, 1)$, is an *integer* solution. Therefore, this solution must also be the optimal solution for subproblem 1 itself. This solution should be stored as the first **incumbent** (the best feasible solution found so far) for the whole problem, along with its value of Z. This value is denoted by

$$Z^* = \text{value of } Z \text{ for current incumbent,}$$

so $Z^* = 9$ at this point. Since this solution has been stored, there is no reason to consider subproblem 1 any further by branching from the $x_1 = 0$ node, etc. Doing so could only lead to other feasible solutions that are inferior to the incumbent, and we have no interest in such solutions. Because it has been solved, we **fathom** (dismiss) subproblem 1 now.

The above results suggest a second key fathoming test. Since $Z^* = 9$, there is no reason to consider further any subproblem whose *bound* ≤ 9, since such a subproblem cannot have a feasible solution better than the *incumbent.* Stated more generally, a subproblem is fathomed whenever its

$$\text{Bound} \leq Z^*.$$

This outcome does not occur in the current iteration of the example because subproblem 2 has a bound of 16 that is larger than 9. However, it might occur later for **descendants** of this subproblem (new smaller subproblems created by branching on this subproblem, and then perhaps branching further through subsequent ''generations''). Furthermore, as new incumbents with larger values of Z^* are found, it will become easier to *fathom* in this way.

The third way of fathoming is quite straightforward. If the simplex method finds that a subproblem's LP relaxation has *no feasible solutions,* then the subproblem itself must have *no feasible solutions,* so it can be dismissed (fathomed).

In all three cases, we are conducting our search for an optimal solution by retaining for further investigation only those subproblems that could possibly have a feasible solution better than the current incumbent.

Summary of Fathoming Tests

A subproblem is *fathomed* (dismissed from further consideration) if

Test 1: Its bound $\leq Z^*$,

or

Test 2: Its LP relaxation has no feasible solutions,

or

Test 3: The optimal solution for its LP relaxation is *integer.* (If this solution is better than the incumbent, it becomes the new incumbent, and test 1 is reapplied to all unfathomed subproblems with the new larger Z^*.)

Figure 12.6 summarizes the results of applying these three tests to subproblems 1 and 2 by showing the current *solution tree.* Only subproblem 1 has been fathomed, by test 3, as indicated by $F(3)$ next to the $x_1 = 0$ node. The resulting incumbent also is identified below this node.

The subsequent iterations will illustrate successful applications of all three tests. However, before continuing the example, we summarize the algorithm being applied to this BIP problem. (This algorithm assumes that all coefficients in the objective function are integer and that the ordering of the variables for branching is $x_1, x_2, \ldots, x_n$.)

Summary of the BIP Branch-and-Bound Algorithm

Initialization: Set $Z^* = -\infty$. Apply the bounding step, fathoming step, and optimality test described below to the whole problem. If not fathomed, classify this problem as the one remaining "subproblem" for performing the first full iteration below.

Steps for each iteration:

1. *Branching:* Among the *remaining* (unfathomed) subproblems, select the one that was created *most recently.* (Break ties according to which has the *larger bound.*) Branch from the node for this subproblem to create two new subproblems by fixing the next variable (the branching variable) at either 0 or 1.

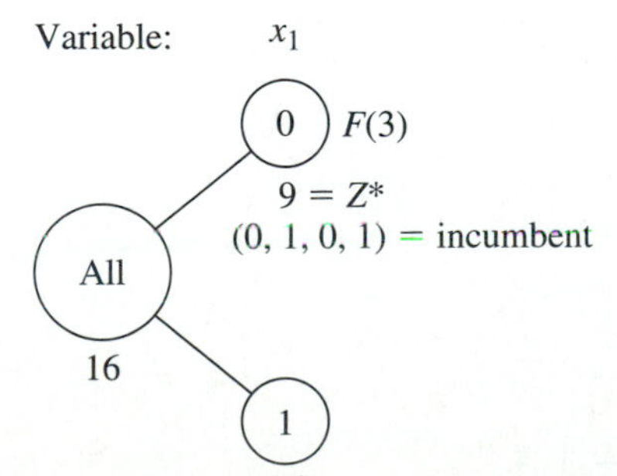

Figure 12.6 The solution tree after the first iteration of the BIP branch-and-bound algorithm for the example in Sec. 12.1.

2. *Bounding:* For each new subproblem, obtain its *bound* by applying the simplex method to its LP relaxation and rounding down the value of Z for the resulting optimal solution.
3. *Fathoming:* For each new subproblem, apply the three fathoming tests summarized above, and discard those subproblems that are fathomed by any of the tests.

Optimality test: Stop when there are *no remaining* subproblems; the current *incumbent* is optimal.[1] Otherwise, return to perform another iteration.

The branching step for this algorithm warrants a comment as to why the subproblem to branch from is selected in this way. One option not used would have been always to select the remaining subproblem with the *best bound,* because this subproblem would be the most promising one to contain an optimal solution for the whole problem. The reason for instead selecting the *most recently created* subproblem is that *LP relaxations* are being solved in the bounding step. Rather than start the simplex method from scratch each time, each LP relaxation generally is solved by *reoptimization* in large-scale implementations of this algorithm. This reoptimization involves revising the final simplex tableau from the preceding LP relaxation as needed because of the few differences in the model (just as for sensitivity analysis) and then applying a few iterations of perhaps the dual simplex method. This reoptimization tends to be *much* faster than starting from scratch, *provided* the preceding and current models are closely related. The models will tend to be closely related under the branching rule used, but *not* when you are skipping around in the solution tree by selecting the subproblem with the best bound.

Completing the Example

The pattern for the remaining iterations will be quite similar to that for the first iteration described above except for the ways in which fathoming occurs. Therefore, we shall summarize the branching and bounding steps fairly briefly and then focus on the fathoming step.

ITERATION 2: The only remaining subproblem corresponds to the $x_1 = 1$ node in Fig. 12.6, so we shall branch from this node to create the two new subproblems given below.

Subproblem 3:

Fix $x_1 = 1$, $x_2 = 0$ so the resulting subproblem is

$$\text{Maximize} \quad Z = 9 + 6x_3 + 4x_4,$$

subject to

(1) $5x_3 + 2x_4 \leq 4$

(2) $x_3 + x_4 \leq 1$

(3) $x_3 \leq 1$

(4) $x_4 \leq 0$

(5) x_j is binary, for $j = 3, 4$.

[1] If there is no incumbent, the conclusion is that the problem has no feasible solutions.

Fix $x_1 = 1$, $x_2 = 1$ so the resulting subproblem is

$$\text{Maximize} \quad Z = 14 + 6x_3 + 4x_4,$$

subject to

(1) $5x_3 + 2x_4 \le 1$

(2) $x_3 + x_4 \le 1$

(3) $x_3 \le 1$

(4) $x_4 \le 1$

(5) x_j is binary, for $j = 3, 4$.

The LP relaxations of these subproblems are obtained by replacing the constraint x_j is binary for $j = 3, 4$ by the constraint $0 \le x_j \le 1$ for $j = 3, 4$. Their optimal solutions (plus the fixed values of x_1 and x_2) are

LP relaxation of subproblem 3: $(x_1, x_2, x_3, x_4) = (1, 0, \frac{4}{5}, 0)$ with $Z = 13\frac{4}{5}$,
LP relaxation of subproblem 4: $(x_1, x_2, x_3, x_4) = (1, 1, 0, \frac{1}{2})$ with $Z = 16$.

The resulting bounds for the subproblems are

Bound for subproblem 3: $Z \le 13$,
Bound for subproblem 4: $Z \le 16$.

Note that both these bounds are larger than $Z^* = 9$, so fathoming test 1 fails in both cases. Test 2 also fails, since both LP relaxations have feasible solutions (as indicated by the existence of an optimal solution). Alas, test 3 fails as well, because both optimal solutions include variables with noninteger values.

Figure 12.7 shows the resulting solution tree at this point. The lack of an F to the right of either new node indicates that both remain unfathomed.

ITERATION 3: So far, the algorithm has created four subproblems. Subproblem 1 has been fathomed, and subproblem 2 has been replaced by (separated into) subproblems 3 and 4, but these last two remain under consideration. Because they were created simultaneously, but subproblem 4 ($x_1 = 1$, $x_2 = 1$) has the larger *bound* ($16 > 13$), the next branching is done from the $(x_1, x_2) = (1, 1)$ node in the solution tree, which creates the

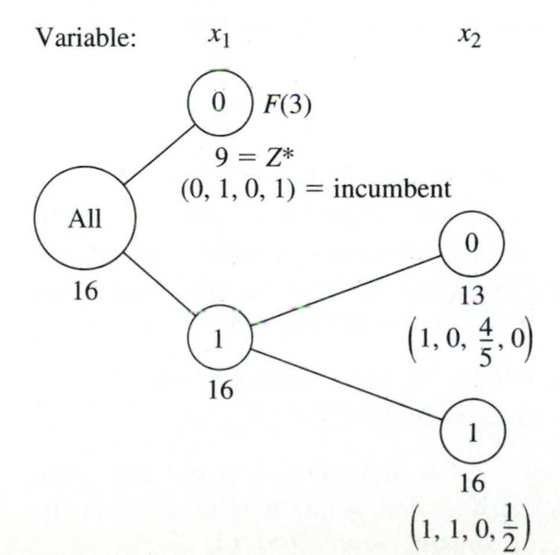

Figure 12.7 The solution tree after iteration 2 of the BIP branch-and-bound algorithm for the example in Sec. 12.1.

following new subproblems (where constraint 3 disappears because it does not contain x_4).

Subproblem 5:

Fix $x_1 = 1$, $x_2 = 1$, $x_3 = 0$ so the resulting subproblem is

$$\text{Maximize} \quad Z = 14 + 4x_4,$$

subject to

(1) $\quad 2x_4 \leq 1$

(2), (4) $\quad x_4 \leq 1 \quad$ (twice)

(5) $\quad x_4$ is binary.

Subproblem 6:

Fix $x_1 = 1$, $x_2 = 1$, $x_3 = 1$ so the resulting subproblem is

$$\text{Maximize} \quad Z = 20 + 4x_4,$$

subject to

(1) $\quad 2x_4 \leq -4$

(2) $\quad x_4 \leq 0$

(4) $\quad x_4 \leq 1$

(5) $\quad x_4$ is binary.

If we form their LP relaxations by replacing constraint 5 by

(5) $\quad 0 \leq x_4 \leq 1,$

the following results are obtained:

LP relaxation of subproblem 5: $(x_1, x_2, x_3, x_4) = (1, 1, 0, \frac{1}{2})$, with $Z = 16$.
LP relaxation of subproblem 6: No feasible solutions.
Bound for subproblem 5: $Z \leq 16$.

Note how the combination of constraints 1 and 5 in the LP relaxation of subproblem 6 prevents any feasible solutions. Therefore, this subproblem is fathomed by test 2. However, subproblem 5 fails this test, as well as test 1 ($16 > 9$) and test 3 ($x_4 = \frac{1}{2}$ is not integer), so it remains under consideration.

We now have the solution tree shown in Fig. 12.8.

Iteration 4: The subproblems corresponding to nodes (1, 0) and (1, 1, 0) in Fig. 12.8 remain under consideration, but the latter node was created more recently, so it is selected for branching from next. Since the resulting branching variable x_4 is the *last* variable, fixing its value at either 0 or 1 actually creates a *single solution* rather than subproblems requiring fuller investigation. These single solutions are

$x_4 = 0$: $\quad (x_1, x_2, x_3, x_4) = (1, 1, 0, 0)$ is feasible, with $Z = 14$,
$x_4 = 1$: $\quad (x_1, x_2, x_3, x_4) = (1, 1, 0, 1)$ is infeasible.

Formally applying the fathoming tests, we see that the first solution passes test 3 and the second passes test 2. Furthermore, this feasible first solution is better than the incumbent ($14 > 9$), so it becomes the new incumbent, with $Z^* = 14$.

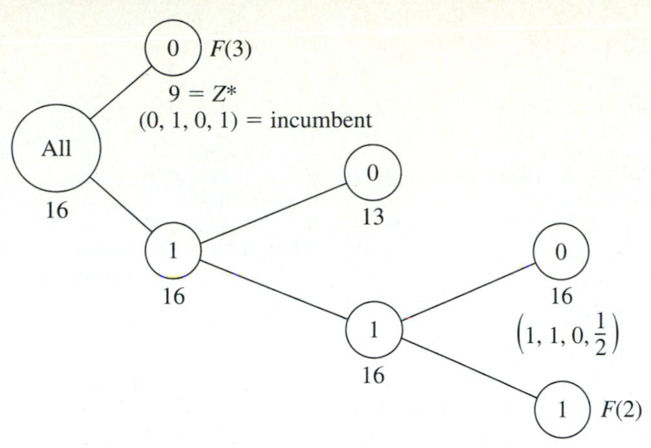

Figure 12.8 The solution tree after iteration 3 of the BIP branch-and-bound algorithm for the example in Sec. 12.1.

Because a new incumbent has been found, we now reapply fathoming test 1 with the new larger value of Z^* to the only remaining subproblem, the one at node (1, 0).

Subproblem 3:

$$\text{Bound} = 13 \leq Z^* = 14.$$

Therefore, this subproblem now is fathomed.

We now have the solution tree shown in Fig. 12.9. Note that there are *no remaining* (unfathomed) subproblems. Consequently, the optimality test indicates that the current incumbent

$$(x_1, x_2, x_3, x_4) = (1, 1, 0, 0)$$

is optimal, so we are done.

Your OR Courseware includes another example of applying this algorithm under the Demo menu. Also included under the IP Procedure menu are both an interactive routine and an automatic routine for executing this algorithm.

Other Options with the Branch-and-Bound Technique

This section has illustrated the branch-and-bound technique by describing a basic branch-and-bound algorithm for solving BIP problems. However, the general framework of the branch-and-bound technique provides a great deal of flexibility in how to design a specific algorithm for any given type of problem such as BIP. There are many options available, and constructing an efficient algorithm requires tailoring the specific design to fit the specific structure of the problem type.

Every branch-and-bound algorithm has the same three basic steps of *branching, bounding,* and *fathoming.* The flexibility lies in how these steps are performed.

Branching always involves *selecting* one remaining subproblem and *dividing* it into smaller subproblems. The flexibility here is found in the rules for selecting and dividing. Our BIP algorithm selected the *most recently created* subproblem, because this is very efficient for *reoptimizing* each LP relaxation from the preceding one.

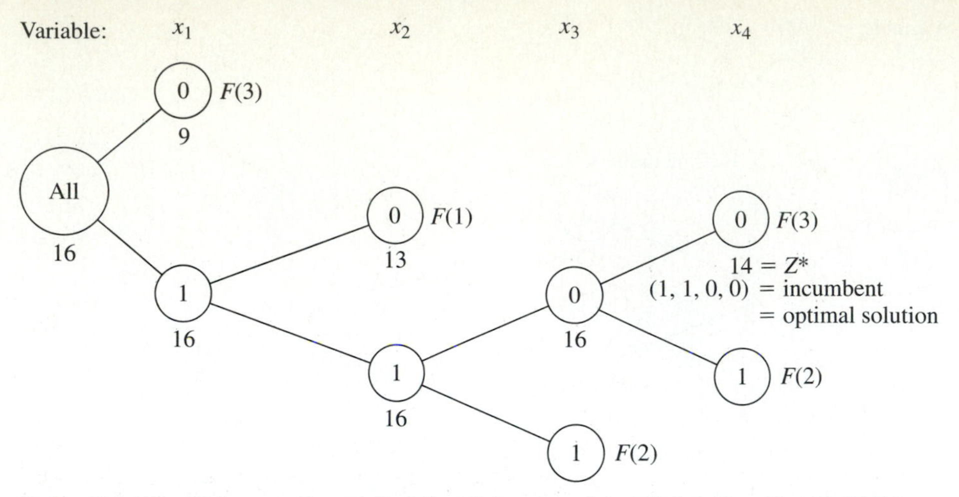

Figure 12.9 The solution tree after the final (fourth) iteration of the BIP branch-and-bound algorithm for the example in Sec. 12.1.

Selecting the subproblem with the *best bound* is the other most popular rule, because it tends to lead more quickly to better incumbents and so more fathoming. Combinations of the two rules also can be used. The *dividing* typically (but not always) is done by choosing a *branching variable* and assigning it either individual values (e.g., our BIP algorithm) or ranges of values (e.g., the algorithm in the next section). More sophisticated algorithms generally use a rule for strategically choosing a branching variable that should tend to lead to early fathoming.

Bounding usually is done by solving a *relaxation*. However, there are a variety of ways to form relaxations. For example, consider the **Lagrangian relaxation**, where the entire set of functional constraints $\mathbf{Ax} \le \mathbf{b}$ (in matrix notation) is *deleted* (except possibly for any ''convenient'' constraints) and then the objective function

$$\text{Maximize} \qquad Z = \mathbf{cx},$$

is replaced by

$$\text{Maximize} \qquad Z_R = \mathbf{cx} - \boldsymbol{\lambda}(\mathbf{Ax} - \mathbf{b}),$$

where the fixed vector $\boldsymbol{\lambda} \ge \mathbf{0}$. If $\mathbf{x}^*$ is an optimal solution for the original problem, its $Z \le Z_R$, so solving the Lagrangian relaxation for the optimal value of Z_R provides a valid *bound*. If $\boldsymbol{\lambda}$ is chosen well, this bound tends to be a reasonably tight one (at least comparable to the bound from the LP relaxation). Without any functional constraints, this relaxation also can be solved extremely quickly. The drawbacks are that fathoming tests 2 and 3 (revised) are not as powerful as for the LP relaxation.

In general terms, two features are sought in choosing a relaxation: it can be solved relatively quickly, and provides a relatively tight bound. Neither alone is adequate. The LP relaxation is popular because it provides an excellent trade-off between these two factors.

One option occasionally employed is to use a quickly solved relaxation and then if fathoming is not achieved, to tighten the relaxation in some way to obtain a somewhat tighter bound.

Fathoming generally is done pretty much as described for the BIP algorithm. The three fathoming criteria can be stated in more general terms as follows.

Summary of Fathoming Criteria

A subproblem is *fathomed* if an analysis of its *relaxation* reveals that

Criterion 1: Feasible solutions of the subproblem must have $Z \leq Z^*$, or
Criterion 2: The subproblem has no feasible solutions, or
Criterion 3: An optimal solution of the subproblem has been found.

Just as for the BIP algorithm, the first two criteria usually are applied by solving the relaxation to obtain a bound for the subproblem and then checking whether this bound is $\leq Z^*$ (test 1) or whether the relaxation has no feasible solutions (test 2). If the relaxation differs from the subproblem *only* by the deletion (or loosening) of some constraints, then the third criterion usually is applied by checking whether the optimal solution for the relaxation is *feasible* for the subproblem, in which case it must be *optimal* for the subproblem. For other relaxations (such as the Lagrangian relaxation), additional analysis is required to determine whether the optimal solution for the relaxation is also optimal for the subproblem.

If the original problem involves *minimization* rather than maximization, two options are available. One is to convert to maximization in the usual way (see Sec. 4.6). The other is to convert the branch-and-bound algorithm directly to minimization form, which requires changing the direction of the inequality for fathoming test 1 from

$$\text{Is the subproblem's bound} \leq Z^*?$$

to

$$\text{Is the subproblem's bound} \geq Z^*?$$

So far, we have described how to use the branch-and-bound technique to find only *one* optimal solution. However, in the case of ties for the optimal solution, it is sometimes desirable to identify *all* these optimal solutions so that the final choice among them can be made on the basis of intangible factors not incorporated into the mathematical model. To find them all, you need to make only a few slight alterations in the procedure. First, change the weak inequality for fathoming test 1 (Is the subproblem's bound $\leq Z^*$?) to a strict inequality (Is the subproblem's bound $< Z^*$?), so that fathoming will not occur if the subproblem can have a feasible solution *equal* to the incumbent. Second, if fathoming test 3 passes and the optimal solution for the subproblem has $Z = Z^*$, then store this solution as *another* (tied) incumbent. Third, if test 3 provides a new incumbent (tied or otherwise), then check whether the optimal solution obtained for the *relaxation* is *unique.* If it is not, then identify the other optimal solutions for the relaxation and check whether they are optimal for the subproblem as well, in which case they also become incumbents. Finally, when the *optimality test* finds that there are *no remaining* (unfathomed) subsets, *all* the current *incumbents* will be the *optimal* solutions.

Finally, note that rather than find an optimal solution, the branch-and-bound technique can be used to find a *nearly optimal* solution, generally with much less computational effort. For some applications, a solution is ''good enough'' if its Z is ''close enough'' to the value of Z for an optimal solution (call it Z^{**}). *Close enough* can be defined in either of two ways as either

$$Z^{**} - K \leq Z \qquad \text{or} \qquad (1 - \alpha)Z^{**} \leq Z$$

for a specified (positive) constant K or α. For example, if the second definition is chosen and $\alpha = 0.05$, then the solution is required to be within 5 percent of optimal. Consequently, if it were known that the value of Z for the current incumbent (Z^*) satisfies either

$$Z^{**} - K \leq Z^* \qquad \text{or} \qquad (1 - \alpha)Z^{**} \leq Z^*$$

then the procedure could be terminated immediately by choosing the incumbent as the desired nearly optimal solution. Although the procedure does not actually identify an optimal solution and the corresponding Z^{**}, if this (unknown) solution is feasible (and so optimal) for the subproblem currently under investigation, then fathoming test 1 finds an upper bound such that

$$Z^{**} \leq \text{bound}$$

so that either

$$\text{Bound} - K \leq Z^* \qquad \text{or} \qquad (1 - \alpha)\text{bound} \leq Z^*$$

would imply that the corresponding inequality in the preceding sentence is satisfied. Even if this solution is not feasible for the current subproblem, a valid upper bound is still obtained for the value of Z for the subproblem's optimal solution. Thus, satisfying either of these last two inequalities is sufficient to fathom this subproblem because the incumbent must be "close enough" to the subproblem's optimal solution.

Therefore, to find a solution that is close enough to being optimal, only one change is needed in the usual branch-and-bound procedure. This change is to replace the usual fathoming test 1 for a subproblem

$$\text{Bound} \leq Z^*?$$

by either

$$\text{Bound} - K \leq Z^*?$$

or

$$(1 - \alpha)(\text{bound}) \leq Z^*?$$

and then perform this test *after* test 3 (so that a feasible solution found with $Z > Z^*$ is still kept as the new incumbent). The reason this weaker test 1 suffices is that regardless of how close Z for the subproblem's (unknown) optimal solution is to the subproblem's bound, the incumbent is still close enough to this solution (if the new inequality holds) that the subproblem does not need to be considered further. When there are no remaining subproblems, the current incumbent will be the desired *nearly optimal* solution. However, it is much easier to fathom with this new fathoming test (in either form), so the algorithm should run much faster. For a large problem, this acceleration may make the difference between finishing with a solution guaranteed to be close to optimal and never terminating.

12.6 A Branch-and-Bound Algorithm for Mixed Integer Programming

We shall now consider the general MIP problem, where *some* of the variables (say, I of them) are restricted to integer values (but not necessarily just 0 and 1) but the rest are ordinary continuous variables. For notational convenience, we shall order the variables so that the first I variables are the *integer-restricted* variables. Therefore, the general form of the problem being considered is

$$\text{Maximize} \qquad Z = \sum_{j=1}^{n} c_j x_j,$$

subject to

$$\sum_{j=1}^{n} a_{ij} x_j \le b_i, \qquad \text{for } i = 1, 2, \ldots, m,$$

and

$$x_j \ge 0, \qquad \text{for } j = 1, 2, \ldots, n,$$

$$x_j \text{ is integer}, \qquad \text{for } j = 1, 2, \ldots, I;\ I \le n.$$

(When $I = n$, this problem becomes the pure IP problem.)

We shall describe a basic branch-and-bound algorithm for solving this problem that, with a variety of refinements, has provided the standard approach to MIP. The structure of this algorithm was first developed by R. J. Dakin,[1] based on a pioneering branch-and-bound algorithm by A. H. Land and A. G. Doig.[2]

This algorithm is quite similar in structure to the BIP algorithm presented in the preceding section. Solving *LP relaxations* again provides the basis for both the *bounding* and *fathoming* steps. In fact, only four changes are needed in the BIP algorithm to deal with the generalizations from *binary* to *general* integer variables and from *pure* IP to *mixed* IP.

One change involves the choice of the *branching variable.* Before, the *next* variable in the natural ordering—$x_1, x_2, \ldots, x_n$—was chosen automatically. Now, the only variables considered are the *integer-restricted* variables that have a *noninteger* value in the optimal solution for the LP relaxation of the current subproblem. Our rule for choosing among these variables is to select the *first* one in the natural ordering. (Production codes generally use a more sophisticated rule.)

The second change involves the values assigned to the branching variable for creating the new smaller subproblems. Before, the *binary* variable was fixed at 0 and 1, respectively, for the two new subproblems. Now, the *general* integer-restricted variable could have a very large number of possible integer values, and it would be inefficient to create *and* analyze *many* subproblems by fixing the variable at its individual integer

[1] R. J. Dakin, "A Tree Search Algorithm for Mixed Integer Programming Problems," *Computer Journal,* **8**(3): 250–255, 1965.

[2] A. H. Land and A. G. Doig, "An Automatic Method of Solving Discrete Programming Problems," *Econometrica,* **28:** 497–520, 1960.

values. Therefore, what is done instead is to create just *two* new subproblems (as before) by specifying two *ranges* of values for the variable.

To spell out how this is done, let x_j be the current branching variable, and let x_j^* be its (noninteger) value in the optimal solution for the LP relaxation of the current subproblem. Using square brackets to denote

$$[x_j^*] = \text{greatest integer} \le x_j^*,$$

we have for the range of values for the two new subproblems

$$x_j \le [x_j^*] \qquad \text{and} \qquad x_j \ge [x_j^*] + 1,$$

respectively. Each inequality becomes an *additional constraint* for that new subproblem. For example, if $x_j^* = 3\frac{1}{2}$, then

$$x_j \le 3 \qquad \text{and} \qquad x_j \ge 4$$

are the respective additional constraints for the new subproblem.

When the two changes to the BIP algorithm described above are combined, an interesting phenomenon of a *recurring branching variable* can occur. To illustrate, as shown in Fig. 12.10, let $j = 1$ in the above example where $x_j^* = 3\frac{1}{2}$, and consider the new subproblem where $x_1 \le 3$. When the LP relaxation of a descendant of this subproblem is solved, suppose that $x_1^* = 1\frac{1}{4}$. Then x_1 *recurs* as the branching variable, and the two new subproblems created have the additional constraint $x_1 \le 1$ and $x_1 \ge 2$, respectively (as well as the previous additional constraint $x_1 \le 3$). Later, when the LP relaxation for a descendant of, say, the $x_1 \le 1$ subproblem is solved, suppose that $x_1^* = \frac{3}{4}$. Then x_1 *recurs* again as the branching variable, and the two new subproblems created have $x_1 = 0$ (because of the new $x_1 \le 0$ constraint and the nonnegativity constraint on x_1) and $x_1 = 1$ (because of the new $x_1 \ge 1$ constraint and the previous $x_1 \le 1$ constraint).

The third change involves the *bounding step.* Before, with a *pure* IP problem and integer coefficients in the objective function, the value of Z for the optimal solution for the subproblem's LP relaxation was *rounded down* to obtain the bound, because any feasible solution for the subproblem must have an *integer* Z. Now, with some of the variables *not* integer-restricted, the bound is the value of Z *without* rounding down.

The fourth (and final) change to the BIP algorithm to obtain our MIP algorithm involves fathoming test 3. Before, with a *pure* IP problem, the test was that the optimal solution for the subproblem's LP relaxation is *integer,* since this ensures that the solution is feasible, and therefore optimal, for the subproblem. Now, with a *mixed* IP problem, the test requires only that the *integer-restricted* variables be *integer* in the optimal solution for the subproblem's LP relaxation, because this suffices to ensure that the solution is feasible, and therefore optimal, for the subproblem.

Incorporating these four changes into the summary presented in the preceding section for the BIP algorithm yields the following summary for the new algorithm for MIP.

Summary of the MIP Branch-and-Bound Algorithm

Initialization: Set $Z^* = -\infty$. Apply the bounding step, fathoming step, and optimality test described below to the whole problem. If not fathomed, classify

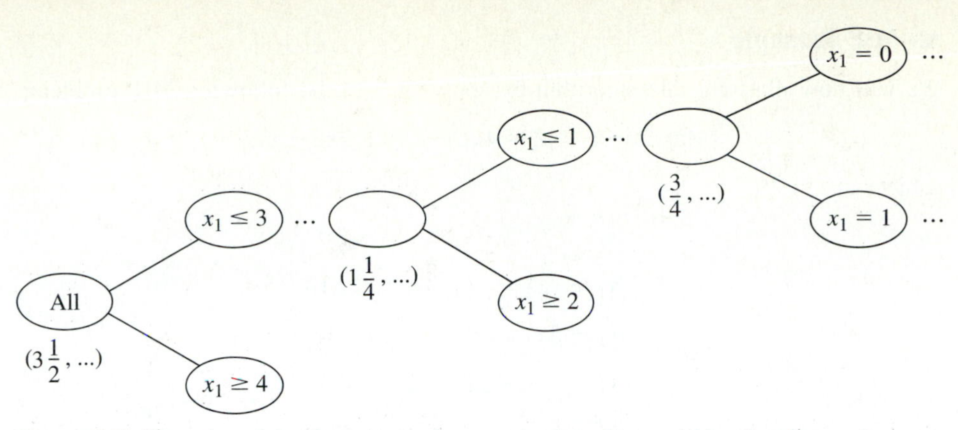

Figure 12.10 Illustration of the phenomenon of a *recurring branching variable,* where here x_1 becomes a branching variable three times because it has a noninteger value in the optimal solution for the LP relaxation at three nodes.

this problem as the one remaining subproblem for performing the first full iteration below.

Steps for each iteration:

1. *Branching:* Among the *remaining* (unfathomed) subproblems, select the one that was created *most recently.* (Break ties according to which has the *larger bound.*) Among the *integer-restricted* variables that have a *noninteger* value in the optimal solution for the LP relaxation of the subproblem, choose the *first one* in the natural ordering of the variables to be the *branching variable.* Let x_j be this variable and x_j^* its value in this solution. Branch from the node for the subproblem to create two new subproblems by adding the respective constraints $x_j \leq [x_j^*]$ and $x_j \geq [x_j^*] + 1$.
2. *Bounding:* For each new subproblem, obtain its bound by applying the simplex method (or the dual simplex method when reoptimizing) to its LP relaxation and using the value of Z for the resulting optimal solution.
3. *Fathoming:* For each new subproblem, apply the three fathoming tests given below, and discard those subproblems that are fathomed by any of the tests.

   Test 1: Its bound $\leq Z^*$, where Z^* is the value of Z for the current *incumbent.*

   Test 2: Its LP relaxation has no feasible solutions.

   Test 3: The optimal solution for its LP relaxation has *integer* values for the *integer-restricted* variables. (If this solution is better than the incumbent, it becomes the new incumbent and test 1 is reapplied to all unfathomed subproblems with the new larger Z^*.)

Optimality test: Stop when there are no remaining subproblems; the current *incumbent* is optimal.[1] Otherwise, perform another iteration.

[1] If there is no incumbent, the conclusion is that the problem has no feasible solutions.

An MIP Example

We will now illustrate this algorithm by applying it to the following MIP problem:

$$\text{Maximize} \quad Z = 4x_1 - 2x_2 + 7x_3 - x_4,$$

subject to

$$\begin{aligned} x_1 \qquad + 5x_3 \qquad &\leq 10 \\ x_1 + x_2 - x_3 \qquad &\leq 1 \\ 6x_1 - 5x_2 \qquad\qquad &\leq 0 \\ -x_1 \qquad + 2x_3 - 2x_4 &\leq 3 \end{aligned}$$

and

$$x_j \geq 0, \qquad \text{for } j = 1, 2, 3, 4$$

$$x_j \text{ is an integer}, \qquad \text{for } j = 1, 2, 3.$$

Note that the number of integer-restricted variables is $I = 3$, so x_4 is the only continuous variable.

Initialization: After setting $Z^* = -\infty$, we form the LP relaxation of this problem by *deleting* the set of constraints that x_j is an integer for $j = 1, 2, 3$. Applying the simplex method to this LP relaxation yields its optimal solution below.

LP relaxation of whole problem: $(x_1, x_2, x_3, x_4) = (\frac{5}{4}, \frac{3}{2}, \frac{7}{4}, 0)$, with $Z = 14\frac{1}{4}$.

Because it has *feasible* solutions and this optimal solution has *noninteger* values for its integer-restricted variables, the whole problem is not fathomed, so the algorithm continues with the first full iteration below.

Iteration 1: In this optimal solution for the LP relaxation, the *first* integer-restricted variable that has a noninteger value is $x_1 = \frac{5}{4}$, so x_1 becomes the branching variable. Branching from the *All* node (*all* feasible solutions) with this branching variable then creates the following two subproblems:

Subproblem 1:

Original problem plus additional constraint

$$x_1 \leq 1.$$

Subproblem 2:

Original problem plus additional constraint

$$x_1 \geq 2.$$

Deleting the set of integer constraints again and solving the resulting LP relaxations of these two subproblems yield the following results.

LP relaxation of subproblem 1: $(x_1, x_2, x_3, x_4) = (1, \frac{6}{5}, \frac{9}{5}, 0)$, with $Z = 14\frac{1}{5}$.
Bound for subproblem 1: $Z \leq 14\frac{1}{5}$.
LP relaxation of subproblem 2: No feasible solutions.

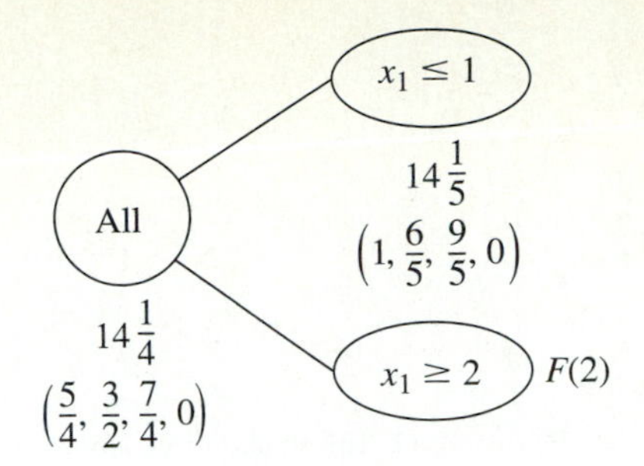

Figure 12.11 The solution tree after the first iteration of the MIP branch-and-bound algorithm for the MIP example.

This outcome for subproblem 2 means that it is fathomed by test 2. However, just as for the whole problem, subproblem 1 fails all fathoming tests.

These results are summarized in the solution tree shown in Fig. 12.11.

Iteration 2: With only one remaining subproblem, corresponding to the $x_1 \le 1$ node in Fig. 12.11, the next branching is from this node. Examining its LP relaxation's optimal solution given below, we see that this node reveals that the *branching variable* is x_2, because $x_2 = \frac{6}{5}$ is the first integer-restricted variable that has a noninteger value. Adding one of the constraints $x_2 \le 1$ or $x_2 \ge 2$ then creates the following two new subproblems.

Subproblem 3:

Original problem plus additional constraints

$$x_1 \le 1, \qquad x_2 \le 1.$$

Subproblem 4:

Original problem plus additional constraints

$$x_1 \le 1, \qquad x_2 \ge 2.$$

Solving their LP relaxations gives the following results.

LP relaxation of subproblem 3:	$(x_1, x_2, x_3, x_4) = (\frac{5}{6}, 1, \frac{11}{6}, 0)$,	with $Z = 14\frac{1}{6}$.
Bound for subproblem 3:	$Z \le 14\frac{1}{6}$.	
LP relaxation of subproblem 4:	$(x_1, x_2, x_3, x_4) = (\frac{5}{6}, 2, \frac{11}{6}, 0)$,	with $Z = 12\frac{1}{6}$.
Bound for subproblem 4:	$Z \le 12\frac{1}{6}$.	

Because both solutions exist (feasible solutions) and have noninteger values for integer-restricted variables, neither subproblem is fathomed. (Test 1 still is not operational, since $Z^* = -\infty$ until the first incumbent is found.)

The solution tree at this point is given in Fig. 12.12.

Iteration 3: With two remaining subproblems (3 and 4) that were created simultaneously, the one with the larger bound (subproblem 3, with $14\frac{1}{6} > 12\frac{1}{6}$) is selected for the next branching. Because $x_1 = \frac{5}{6}$ has a noninteger value in the optimal solution for this subproblem's LP relaxation, x_1 becomes the branching variable. (Note that x_1 now is a *recurring* branching variable, since it also was chosen at iteration 1.) This leads to the following new subproblems.

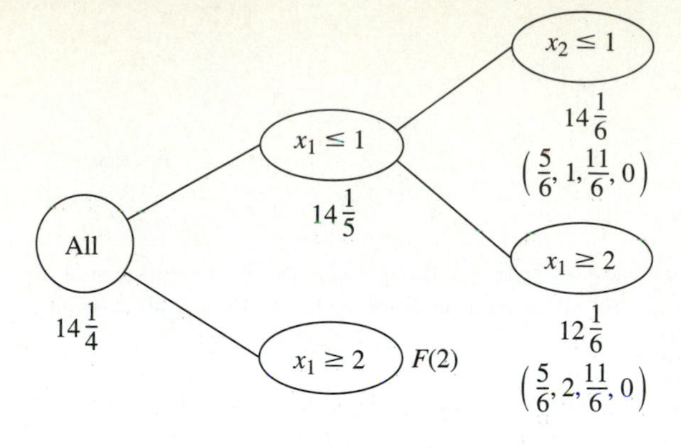

Figure 12.12 The solution tree after the second iteration of the MIP branch-and-bound algorithm for the MIP example.

Subproblem 5:

Original problem plus additional constraints

$$x_1 \le 1$$

$$x_2 \le 1$$

$$x_1 \le 0 \qquad (\text{so } x_1 = 0).$$

Subproblem 6:

Original problem plus additional constraints

$$x_1 \le 1$$

$$x_2 \le 1$$

$$x_1 \ge 1 \qquad (\text{so } x_1 = 1).$$

The results from solving their LP relaxations are given below.

LP relaxation of subproblem 5: $(x_1, x_2, x_3, x_4) = (0, 0, 2, \frac{1}{2})$, with $Z = 13\frac{1}{2}$.
Bound for subproblem 5: $Z \le 13\frac{1}{2}$.
LP relaxation of subproblem 6: No feasible solutions.

Subproblem 6 is immediately fathomed by test 2. However, note that subproblem 5 also can be fathomed. Test 3 passes because the optimal solution for its LP relaxation has integer values ($x_1 = 0$, $x_2 = 0$, $x_3 = 2$) for all three integer-restricted variables. (It does not matter that $x_4 = \frac{1}{2}$, since x_4 is not integer-restricted.) This *feasible* solution for the original problem becomes our first incumbent:

$$\text{Incumbent} = (0, 0, 2, \tfrac{1}{2}) \qquad \text{with } Z^* = 13\tfrac{1}{2}.$$

Using this Z^* to reapply fathoming test 1 to the only other subproblem (subproblem 4) is successful, because its bound $12\frac{1}{6} \le Z^*$.

This iteration has succeeded in fathoming subproblems in all three possible ways. Furthermore, there now are no remaining subproblems, so the current incumbent is optimal.

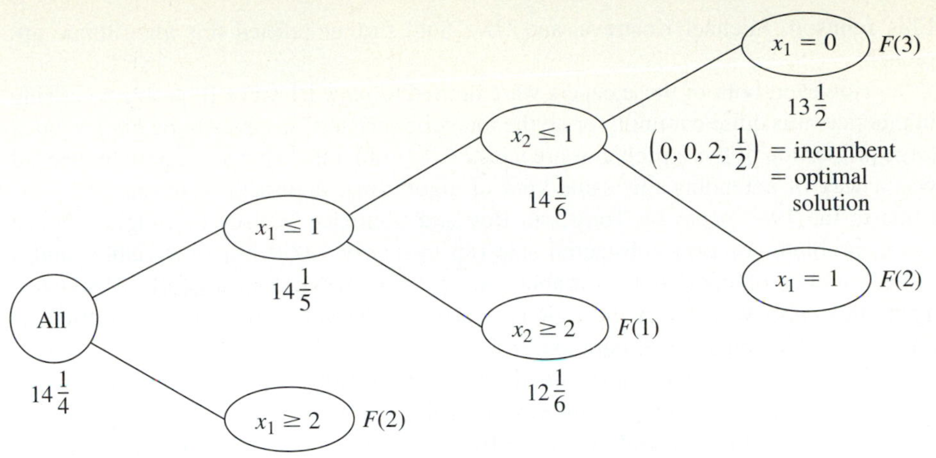

Figure 12.13 The solution tree after the final (third) iteration of the MIP branch-and-bound algorithm for the MIP example.

$$\text{Optimal solution} = (0, 0, 2, \tfrac{1}{2}) \qquad \text{with } Z = 13\tfrac{1}{2}.$$

These results are summarized by the final solution tree given in Fig. 12.13.

Another example of applying the MIP algorithm is presented in your OR Courseware under the Demo menu. This software also includes both an interactive routine and an automatic routine for executing this algorithm.

12.7 Recent Developments in Solving BIP Problems

Integer programming has been an especially exciting area of OR in recent years because of the dramatic progress being made in its solution methodology.

Background

To place this progress into perspective, consider the historical background. One big breakthrough had come in the 1960s and early 1970s with the development and refinement of the branch-and-bound approach. But then the state of the art seemed to hit a plateau. Relatively small problems (well under 100 variables) could be solved very efficiently, but even a modest increase in problem size might cause an explosion in computation time beyond feasible limits. Little progress was being made in overcoming this exponential growth in computation time as the problem size was increased. Many important problems arising in practice could not be solved.

Then came the next breakthrough in the mid-1980s, as reported largely in four papers published in 1983, 1985, 1987, and 1991. (See Selected References 2, 5, 11, and 4.) In the 1983 paper, Harlan Crowder, Ellis Johnson, and Manfred Padberg presented a new algorithmic approach to solving *pure* BIP problems that had successfully solved problems with no apparent special structure having up to 2,756 variables! This paper won the Lanchester Prize, awarded by the Operations Research Society of America for the most notable publication in operations research during 1983. In the 1985 paper,

Ellis Johnson, Michael Kostreva, and Uwe Suhl further refined this algorithmic approach.

However, both of these papers were limited to *pure* BIP. For IP problems arising in practice, it is quite common for all the integer-restricted variables to be *binary,* but a large proportion of these problems are *mixed* BIP problems. What was critically needed was a way of extending this same kind of algorithmic approach to *mixed* BIP. This came in the 1987 paper by Tony Van Roy and Laurence Wolsey of Belgium. Once again, problems of *very* substantial size (up to nearly 1,000 binary variables and a larger number of continuous variables) were being solved successfully. And once again, this paper won a very prestigious award, the Orchard-Hays Prize given triannually by the Mathematical Programming Society.

In the 1991 paper, Karla Hoffman and Manfred Padberg followed up on the 1983 and 1985 papers by developing improved techniques for solving pure BIP problems. Using the name **branch-and-cut algorithm** for this algorithmic approach, they reported successfully solving problems with as many as 6,000 variables!

We do need to add one note of caution. It is not yet clear just how *consistently* this algorithmic approach can successfully solve a wide variety of problems of this kind of very substantial size. The very large pure BIP problems solved had *sparse* **A** matrices; i.e., the percentage of coefficients in the functional constraints that were *nonzeros* was quite small (perhaps less than 5 percent). In fact, the approach depends heavily upon this sparsity. (Fortunately, this kind of sparsity is typical in large practical problems.) Furthermore, there are other important factors besides sparsity and size that affect just how difficult a given IP problem will be to solve. It appears that IP formulations of fairly substantial size should still be approached with considerable caution.

On the other hand, each new algorithmic breakthrough in OR always generates a flurry of new research and development activity to try to refine the new approach. We already have seen substantial effort to develop sophisticated software packages for *IP* that build on recent improvements in linear programming software. For example, some of the new IP techniques already have been incorporated into the IP module of IBM's Optimization Subroutine Library (OSL). The developers of the powerful mathematical programming package CPLEX have an ongoing project to further develop a fully state-of-the-art IP module. (See Selected Reference 9 for a survey of IP software packages.) Theoretical research also continues. We will undoubtedly see some further fruits of intensified research and development in integer programming over the next decade. Perhaps through these efforts the gap in efficiency between integer programming and linear programming algorithms can be further closed.

Although it would be beyond the scope and level of this book to describe the new algorithmic approach fully, we now give a brief overview. (You are encouraged to read Selected References 2, 4, 5, and 11 for further information.) This overview is limited to *pure* BIP, so *all* variables introduced later in this section are *binary* variables.

The approach mainly uses a combination of three kinds[1] of techniques: *automatic problem preprocessing,* the *generation of cutting planes,* and clever *branch-and-bound* techniques. You already are familiar with branch-and-bound techniques, and we will not elaborate further on the more advanced versions incorporated here. An introduction to the other two kinds of techniques is given below.

[1] Still another technique that has played a significant secondary role in the recent progress has been the use of *heuristics* for quickly finding good feasible solutions.

Automatic Problem Preprocessing for Pure BIP

Automatic problem preprocessing involves a "computer inspection" of the user-supplied formulation of the IP problem in order to spot reformulations that make the problem quicker to solve without eliminating any feasible solutions. These reformulations fall into three categories:

1. *Fixing variables:* Identify variables that can be fixed at one of their possible values (either 0 or 1) because the other value cannot possibly be part of a solution that is both feasible and optimal.
2. *Eliminating redundant constraints:* Identify and eliminate *redundant constraints* (constraints that automatically are satisfied by solutions that satisfy all the other constraints).
3. *Tightening constraints:* Tighten some constraints in a way that reduces the feasible region for the LP relaxation without eliminating any feasible solutions for the BIP problem.

These categories are described in turn.

Fixing Variables: One general principle for fixing variables is the following.

> If one value of a variable cannot satisfy a certain constraint, even when the other variables equal their best values for trying to satisfy the constraint, then that variable should be fixed at its other value.

For example, *each* of the following $\le$ constraints would enable us to fix x_1 at $x_1 = 0$, since $x_1 = 1$ with the best values of the other variables (0 with a nonnegative coefficient and 1 with a negative coefficient) would violate the constraint.

$$3x_1 \le 2 \quad \Rightarrow \quad x_1 = 0, \quad \text{since} \quad 3(1) > 2.$$

$$3x_1 + x_2 \le 2 \quad \Rightarrow \quad x_1 = 0, \quad \text{since} \quad 3(1) + 1(0) > 2.$$

$$5x_1 + x_2 - 2x_3 \le 2 \quad \Rightarrow \quad x_1 = 0, \quad \text{since} \quad 5(1) + 1(0) - 2(1) > 2.$$

The general procedure for checking any $\le$ constraint is to identify the variable with the *largest positive coefficient,* and if the *sum* of *that coefficient* and any *negative coefficients* exceeds the right-hand side, then that variable should be fixed at 0. (Once the variable has been fixed, the procedure can be repeated for the variable with the next largest positive coefficient, etc.)

An analogous procedure with $\ge$ constraints can enable us to fix a variable at 1 instead, as illustrated below three times.

$$3x_1 \ge 2 \quad \Rightarrow \quad x_1 = 1, \quad \text{since} \quad 3(0) < 2.$$

$$3x_1 + x_2 \ge 2 \quad \Rightarrow \quad x_1 = 1, \quad \text{since} \quad 3(0) + 1(1) < 2.$$

$$3x_1 + x_2 - 2x_3 \ge 2 \quad \Rightarrow \quad x_1 = 1, \quad \text{since} \quad 3(0) + 1(1) - 2(0) < 2.$$

A $\ge$ constraint also can enable us to fix a variable at 0, as illustrated next.

$$x_1 + x_2 - 2x_3 \ge 1 \quad \Rightarrow \quad x_3 = 0, \quad \text{since} \quad 1(1) + 1(1) - 2(1) < 1.$$

The next example shows a $\geq$ constraint fixing one variable at 1 and another at 0.

$$3x_1 + x_2 - 3x_3 \geq 2 \quad \Rightarrow \quad x_1 = 1, \quad \text{since} \quad 3(0) + 1(1) - 3(0) < 2$$

$$\textit{and} \quad \Rightarrow \quad x_3 = 0, \quad \text{since} \quad 3(1) + 1(1) - 3(1) < 2.$$

Similarly, a $\leq$ constraint with a *negative* right-hand side can result in either 0 or 1 becoming the fixed value of a variable. For example, both happen with the following constraint.

$$3x_1 - 2x_2 \leq -1 \quad \Rightarrow \quad x_1 = 0, \quad \text{since} \quad 3(1) - 2(1) > -1$$

$$\textit{and} \quad \Rightarrow \quad x_2 = 1, \quad \text{since} \quad 3(0) - 2(0) > -1.$$

Fixing a variable from one constraint can sometimes generate a chain reaction of then being able to fix other variables from other constraints. For example, look at what happens with the following three constraints.

$$3x_1 + x_2 - 2x_3 \geq 2 \quad \Rightarrow \quad x_1 = 1 \quad \text{(as above)}.$$

Then

$$x_1 + x_4 + x_5 \leq 1 \quad \Rightarrow \quad x_4 = 0, \quad x_5 = 0.$$

Then

$$-x_5 + x_6 \leq 0 \quad \Rightarrow \quad x_6 = 0.$$

In some cases, it is possible to combine one or more *mutually exclusive alternatives* constraints with another constraint to fix a variable, as illustrated below,

$$\left.\begin{array}{r} 8x_1 - 4x_2 - 5x_3 + 3x_4 \leq 2 \\ x_2 + \quad x_3 \qquad\quad \leq 1 \end{array}\right\} \quad \Rightarrow \quad x_1 = 0,$$

$$\text{since} \quad 8(1) - \max\{4, 5\}(1) + 3(0) > 2.$$

There are additional techniques for fixing variables, including some involving optimality considerations, but we will not delve further into this topic.

Fixing variables can have a dramatic impact on reducing the size of a problem. One example is the problem with 2,756 variables reported in Selected Reference 2. A major factor in being able to solve this problem is that the algorithm succeeded in fixing 1,341 variables, thereby eliminating essentially half of the problem's variables from further consideration.

Eliminating Redundant Constraints: Here is one easy way to detect a redundant constraint.

> If a functional constraint satisfies even the most challenging binary solution, then it has been made redundant by the binary constraints and can be eliminated from further consideration. For a $\leq$ constraint, the most challenging binary solution has variables equal to 1 when they have nonnegative coefficients and other variables equal to 0. (Reverse these values for a $\geq$ constraint.)

Some examples are given below.

$$3x_1 + 2x_2 \le 6 \quad \text{is redundant, since } 3(1) + 2(1) \le 6.$$

$$3x_1 - 2x_2 \le 3 \quad \text{is redundant, since } 3(1) - 2(0) \le 3.$$

$$3x_1 - 2x_2 \ge -3 \quad \text{is redundant, since } 3(0) - 2(1) \ge -3.$$

In most cases where a constraint has been identified as redundant, it was not redundant in the original model but became so after fixing some variables. Of the 11 examples of fixing variables given above, *all* but the last one left a constraint that then was redundant.

Tightening Constraints:[1] Consider the following problem.

$$\text{Maximize} \quad Z = 3x_1 + 2x_2,$$

subject to

$$2x_1 + 3x_2 \le 4$$

and

$$x_1, x_2 \text{ binary}.$$

This BIP problem has just three feasible solutions—(0, 0), (1, 0), and (0, 1)—where the optimal solution is (1, 0) with $Z = 3$. The feasible region for the LP relaxation of this problem is shown in Fig. 12.14. The optimal solution for this LP relaxation is $(1, \frac{2}{3})$ with $Z = 4\frac{1}{3}$, which is not very close to the optimal solution for the BIP problem. A branch-and-bound algorithm would have some work to do to identify the optimal BIP solution.

Now look what happens when the functional constraint $2x_1 + 3x_2 \le 4$ is replaced by

$$x_1 + x_2 \le 1.$$

The feasible solutions for the BIP problem remain exactly the same—(0, 0), (1, 0), and (0, 1)—so the optimal solution still is (1, 0). However, the feasible region for the LP relaxation has been greatly reduced, as shown in Fig. 12.15. In fact, this feasible region has been reduced so much that the optimal solution for the LP relaxation now is (1, 0), so the optimal solution for the BIP problem has been found without needing any additional work.

This is an example of tightening a constraint in a way that reduces the feasible region for the LP relaxation without eliminating any feasible solutions for the BIP problem. It was easy to do for this tiny two-variable problem that could be displayed graphically. However, with application of the same principles for tightening a constraint without eliminating any feasible BIP solutions, the following algebraic procedure can be used to do this for any $\le$ constraint with any number of variables.

[1] Also commonly called *coefficient reduction*.

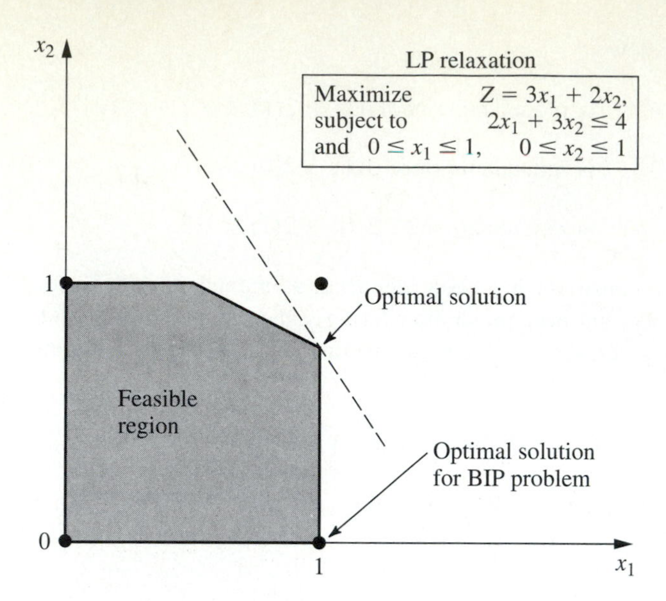

Figure 12.14 The LP relaxation (including its feasible region and optimal solution) for the BIP example used to illustrate tightening a constraint.

Procedure for Tightening a $\leq$ *Constraint*

Denote the constraint by $a_1x_1 + a_2x_2 + \cdots + a_nx_n \leq b$

1. Calculate S = sum of the *positive* a_j.
2. Identify any $a_j \neq 0$ such that $S < b + |a_j|$.
   (*a*) If none, stop; the constraint cannot be tightened further.
   (*b*) If $a_j > 0$, go to step 3.
   (*c*) If $a_j < 0$, go to step 4.
3. ($a_j > 0$) Calculate $\bar{a}_j = S - b$ and $\bar{b} = S - a_j$. Reset $a_j = \bar{a}_j$ *and* $b = \bar{b}$. Return to step 1.

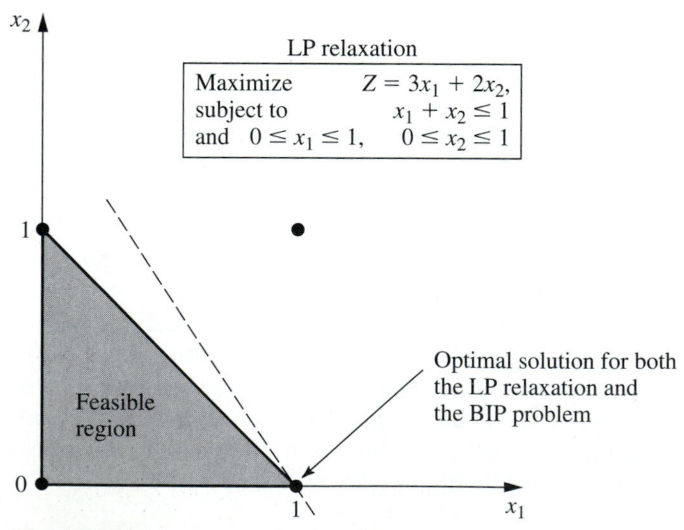

Figure 12.15 The LP relaxation after tightening the constraint, $2x_1 + 3x_2 \leq 4$, to $x_1 + x_2 \leq 1$ for the example of Fig. 12.14.

4. ($a_j < 0$) Increase a_j to $a_j = b - S$. Return to step 1.

Applying this procedure to the functional constraint in the above example flows as follows:

The constraint is $2x_1 + 3x_2 \leq 4$ ($a_1 = 2, a_2 = 3, b = 4$).

1. $S = 2 + 3 = 5$.
2. a_1 satisfies $S < b + |a_1|$, since $5 < 4 + 2$. Also a_2 satisfies $S < b + |a_2|$, since $5 < 4 + 3$. Choose a_1 arbitrarily.
3. $\bar{a} = 5 - 4 = 1$ and $\bar{b} = 5 - 2 = 3$, so reset $a_1 = 1$ and $b = 3$. The new tighter constraint is

$$x_1 + 3x_2 \leq 3 \qquad (a_1 = 1, a_2 = 3, b = 3).$$

1. $S = 1 + 3 = 4$.
2. a_2 satisfies $S < b + |a_2|$, since $4 < 3 + 3$.
3. $\bar{a}_2 = 4 - 3 = 1$ and $\bar{b} = 4 - 3 = 1$, so reset $a_2 = 1$ and $b = 1$. The new tighter constraint is

$$x_1 + x_2 \leq 1 \qquad (a_1 = 1, a_2 = 1, b = 1).$$

1. $S = 1 + 1 = 2$.
2. No $a_j \neq 0$ satisfies $S < b + |a_j|$, so stop; $x_1 + x_2 \leq 1$ is the desired tightened constraint.

If the first execution of step 2 in the above example had chosen a_2 instead, then the first tighter constraint would have been $2x_1 + x_2 \leq 2$. The next series of steps again would have led to $x_1 + x_2 \leq 1$.

Here is another example, where the procedure tightens the constraint on the left to the one on its right and then tightens further to the second one on the right.

$$4x_1 - 3x_2 + x_3 + 2x_4 \leq 5 \qquad \Rightarrow \qquad 2x_1 - 3x_2 + x_3 + 2x_4 \leq 3$$

$$\Rightarrow \qquad 2x_1 - 2x_2 + x_3 + 2x_4 \leq 3.$$

(Practice applying the procedure to confirm these results.)

A constraint in $\geq$ form can be converted to $\leq$ form (by multiplying through both sides by -1) to apply this procedure directly.

Generating Cutting Planes for Pure BIP

A **cutting plane** (or **cut**) for any IP problem is a new functional constraint that reduces the feasible region for the LP relaxation without eliminating any feasible solutions for the IP problem. In fact, you have just seen one way of generating cutting planes for pure BIP problems, namely, apply the above procedure for tightening constraints. Thus, $x_1 + x_2 \leq 1$ is a cutting plane for the BIP problem considered in Fig. 12.14, which leads to the reduced feasible region for the LP relaxation shown in Fig. 12.15.

In addition to this procedure, a number of other techniques have been developed for generating cutting planes that will tend to accelerate how quickly a branch-and-bound algorithm can find an optimal solution for a pure BIP problem. We will focus on just one of these techniques.

To illustrate this technique, consider the California Manufacturing Co. pure BIP problem presented in Sec. 12.1 and used to illustrate the BIP branch-and-bound algorithm in Sec. 12.5. The optimal solution for its LP relaxation is given in Fig. 12.5 as $(x_1, x_2, x_3, x_4) = (\frac{5}{6}, 1, 0, 1)$. One of the functional constraints is

$$6x_1 + 3x_2 + 5x_3 + 2x_4 \leq 10.$$

Now note that the binary constraints and this constraint together imply that

$$x_1 + x_2 + x_4 \leq 2.$$

This new constraint is a *cutting plane.* It eliminates part of the feasible region for the LP relaxation, including what had been the optimal solution, $(\frac{5}{6}, 1, 0, 1)$, but it does not eliminate any feasible *integer* solutions. Adding just this one cutting plane to the original model would improve the performance of the BIP branch-and-bound algorithm in Sec. 12.5 (see Fig. 12.9) in two ways. First, the optimal solution for the new (tighter) LP relaxation would be $(1, 1, \frac{1}{5}, 0)$, with $Z = 15\frac{1}{5}$, so the bounds for the *All* node, $x_1 = 1$ node, and $(x_1, x_2) = (1, 1)$ node now would be 15 instead of 16. Second, one less iteration would be needed because the optimal solution for the LP relaxation at the $(x_1, x_2, x_3) = (1, 1, 0)$ node now would be $(1, 1, 0, 0)$, which provides a new *incumbent* with $Z^* = 14$. Therefore, on the *third* iteration (see Fig. 12.8), this node would be fathomed by test 3, and the $(x_1, x_2) = (1, 0)$ node would be fathomed by test 1, thereby revealing that this incumbent is the optimal solution for the original BIP problem.

Here is the general procedure used to generate this cutting plane.

A Procedure for Generating Cutting Planes

1. Consider any functional constraint in $\leq$ form with only nonnegative coefficients.
2. Find a group of variables (called a **minimum cover** of the constraint) such that
   (*a*) The constraint is violated if every variable in the group equals 1 and all other variables equal 0.
   (*b*) But the constraint becomes satisfied if the value of *any one* of these variables is changed from 1 to 0.
3. By letting N denote the number of variables in the group, the resulting cutting plane has the form

$$\text{Sum of variables in group} \leq N - 1.$$

Applying this procedure to the constraint $6x_1 + 3x_2 + 5x_3 + 2x_4 \leq 10$, we see that the group of variables $\{x_1, x_2, x_4\}$ is a *minimal cover* because

(*a*) $(1, 1, 0, 1)$ violates the constraint.
(*b*) But the constraint becomes satisfied if the value of *any one* of these three variables is changed from 1 to 0.

Since $N = 3$ in this case, the resulting cutting plane is $x_1 + x_2 + x_4 \leq 2$.

This same constraint also has a second minimal cover $\{x_1, x_3\}$, since $(1, 0, 1, 0)$ violates the constraint but both $(0, 0, 1, 0)$ and $(1, 0, 0, 0)$ satisfy the constraint. Therefore, $x_1 + x_3 \leq 1$ is another valid cutting plane.

The new algorithmic approach presented in Selected References 2, 5, 11, and 4 involves generating *many* cutting planes in a similar manner before then applying clever branch-and-bound techniques. The results of including the cutting planes can be quite dramatic in tightening the LP relaxations. For example, for the test problem with 2,756 binary variables considered in the 1983 paper, 326 cutting planes were generated.

The result was that the *gap* between Z for the optimal solution for the LP relaxation of the whole BIP problem and Z for this problem's optimal solution was reduced by 98 percent. Similar results were obtained on about half the problems.

Ironically, the very first algorithms developed for integer programming, including Ralph Gomory's celebrated algorithm announced in 1958, were based on cutting planes (generated in a different way), but this approach proved to be unsatisfactory in practice (except for special classes of problems). However, these algorithms relied solely on cutting planes. We now know that judiciously *combining* cutting planes and branch-and-bound techniques (along with automatic problem preprocessing) provides a powerful algorithmic approach for solving large-scale BIP problems. This is one reason that the name *branch-and-cut algorithm* has been given to this new approach.

12.8 Conclusions

IP problems arise frequently because some of or all the decision variables must be restricted to integer values. There also are many applications involving yes-or-no decisions (including combinatorial relationships expressible in terms of such decisions) that can be represented by binary (0–1) variables. These problems are more difficult than they would be without the integer restriction, so the algorithms available for integer programming are generally much less efficient than the simplex method. The most important determinants of computation time are the *number of integer variables* and whether the problem has some *special structure* that can be exploited. For a fixed number of integer variables, BIP problems generally are much easier to solve than problems with general integer variables, but adding continuous variables (MIP) may not increase the computation time substantially. For special types of BIP problems containing a special structure that can be exploited by a *special-purpose algorithm,* it may be possible to solve very large problems (well over 1,000 binary variables) routinely. Other much smaller problems without such special structure may not be solvable.

Computer codes for IP algorithms now are commonly available in mathematical programming software packages. These algorithms usually are based on the *branch-and-bound* technique and variations thereof.

It appears that a new era in IP solution methodology has now been ushered in by a series of landmark papers since the mid-1980s. The new algorithmic approach involves combining automatic problem preprocessing, the generation of cutting planes, and clever branch-and-bound techniques. Research in this area is continuing, along with the development of sophisticated new software packages that incorporate these new techniques.

IP problems arising in practice sometimes are so large that they cannot be solved by even the latest algorithms. In these cases, it is common to simply apply the simplex method to the LP relaxation and then round the optimal solution to a feasible integer solution. However, this approach is sometimes quite unsatisfactory because it may be difficult (or impossible) to find a feasible integer solution in this way; and the solution found may be far from optimal. This is especially true when you are dealing with binary variables or even general integer variables with small values.

To circumvent these difficulties with rounding, considerable progress has been made in developing *efficient heuristic algorithms.* Even with very large IP problems, these algorithms generally will quickly find very good feasible solutions that are not necessarily optimal but usually are better than those that can be found by simple rounding.

In recent years, there has been considerable investigation into the development of algorithms for integer *nonlinear* programming, and this area continues to be a very active target of research.

SELECTED REFERENCES

1. Barnhart, C., E. L. Johnson, G. L. Nemhauser, M. W. P. Savelsbergh, and P. H. Vance, "Branch-and-Price: Column Generation for Solving Huge Integer Programs," pp. 186–207 in J. R. Birge and K. G. Murty (eds.), *Mathematical Programming: State of the Art 1994,* 15th International Symposium on Mathematical Programming, University of Michigan, Ann Arbor, Mich., 1994.
2. Crowder, H., E. L. Johnson, and M. Padberg, "Solving Large-Scale Zero-One Linear Programming Problems," *Operations Research,* **31:** 803–834, 1983.
3. Geoffrion, A. M., and R. E. Marsten, "Integer Programming Algorithms: A Framework and State-of-the-Art Survey," *Management Science,* **18:** 465–491, 1972.
4. Hoffman, K. L., and M. Padberg, "Improving LP-Representations of Zero-One Linear Programs for Branch-and-Cut," *ORSA Journal on Computing,* **3:** 121–134, 1991.
5. Johnson, E. L., M. M. Kostreva, and U. H. Suhl, "Solving 0–1 Integer Programming Problems Arising from Large Scale Planning Models," *Operations Research,* **33:** 803–819, 1985.
6. Nemhauser, G. L., and L. A. Wolsey, *Integer and Combinatorial Optimization,* Wiley, New York, 1988.
7. Parker, R. G., and R. L. Rardin, *Discrete Optimization,* Academic Press, San Diego, CA, 1988.
8. Salkin, H., and K. Mathur, *Foundations of Integer Programming,* North-Holland, New York, 1989.
9. Saltzman, M. J., "Broad Selection of Software Packages Available," *OR/MS Today,* April 1994, pp. 42–51.
10. Schriver, A., *Theory of Linear and Integer Programming,* Wiley, New York, 1986.
11. Van Roy, T. J., and L. A. Wolsey, "Solving Mixed 0–1 Programs by Automatic Reformulation," *Operations Research,* **35:** 45–57, 1987.
12. Williams, H. P., *Model Building in Mathematical Programming,* 3d ed., Wiley, New York, 1990.

RELEVANT ROUTINES IN YOUR OR COURSEWARE

Demonstration Examples:
Binary Integer Programming Branch-and-Bound Algorithm
Mixed Integer Programming Branch-and-Bound Algorithm

Interactive Routines:
Enter or Revise a General Linear Programming Model
Solve Interactively by the Graphical Method
Enter or Revise an Integer Programming Model
Solve Binary Integer Program Interactively
Solve Mixed Integer Program Interactively

Automatic Routines:
Solve Binary Integer Program Automatically
Solve Mixed Integer Program Automatically

To access these routines, call the MathProg program and then choose *Integer Programming* under the Area menu. (For the first two interactive routines above, choose *General Linear Programming* under the Area menu.) See App. 1 for documentation of the software.

PROBLEMS[1]

To the left of each of the following problems (or their parts), we have inserted a D (for demonstration), I (for interactive routine), or A (for automatic routine) whenever a corresponding routine listed above can be helpful. An asterisk on the I or A indicates that this routine definitely should be used (unless your instructor gives contrary instructions) and that the printout from this routine is all that needs to be turned in to show your work in executing the algorithm. An asterisk on the problem number indicates that at least a partial answer is given in the back of the book.

12.1-1.* A young couple, Eve and Steven, want to divide their main household chores (marketing, cooking, dishwashing, and laundry) between them so that each has two tasks but the total time they spend on household duties is kept to a minimum. Their efficiencies on these tasks differ, and the time each needs to perform the task is given by the following table:

	Hours per Week Needed			
	Marketing	Cooking	Dishwashing	Laundry
Eve	4.5	7.8	3.6	2.9
Steven	4.9	7.2	4.3	3.1

(*a*) Formulate a BIP model for this problem.

A* (*b*) Use the automatic BIP routine in your OR Courseware to solve this problem.

12.1-2. The board of directors of General Wheels Co. is considering seven large capital investments. Each investment can be made only once. These investments differ in the estimated long-run profit (net present value) that they will generate as well as in the amount of capital required, as shown by the following table (in units of millions of dollars):

	Investment Opportunity						
	1	2	3	4	5	6	7
Estimated profit	17	10	15	19	7	13	9
Capital required	43	28	34	48	17	32	23

The total amount of capital available for these investments is $100 million. Investment opportunities 1 and 2 are mutually exclusive, and so are 3 and 4. Furthermore, neither 3 nor 4 can be undertaken unless one of the first two opportunities is undertaken. There are no such restrictions on investment opportunities 5, 6, and 7. The objective is to select the combination of capital investments that will maximize the total estimated long-run profit (net present value).

(*a*) Formulate a BIP model for this problem.

A* (*b*) Use the automatic BIP routine in your OR Courseware to solve this problem.

[1] Problems 12.4-3, 12.5-6, and 12.6-7 have been adapted, with permission, from previous operations research examinations given by the Society of Actuaries.

12.1-3. Reconsider Prob. 8.2-23. Now suppose that trucks (and their drivers) need to be hired to do the hauling, and each truck can only be used to haul gravel from a single pit to a single site. In addition to the hauling and gravel costs specified previously, there now is a fixed cost of 5 associated with hiring each truck. A truck can haul 5 tons, but it is not required to go full. For each combination of pit and site, there now are two decisions to be made: the number of trucks to be used and the amount of gravel to be hauled.

(*a*) Formulate an MIP model for this problem.

A* (*b*) Use the automatic MIP routine in your OR Courseware to solve the problem.

12.1-4. Reconsider Prob. 8.3-3. Formulate a BIP model for this problem.

12.2-1. Reconsider the product-mix problem with start-up costs presented in Prob. 10.3-9.

(*a*) For the case where the production quantities must be *integer,* formulate a pure IP model for this problem.

A* (*b*) Use the automatic MIP routine in your OR Courseware to solve the model formulated in part (*a*).

(*c*) For the case where the production quantities are *continuous* decision variables, formulate an MIP model for this problem.

A* (*d*) Use the automatic routine in your OR Courseware to solve the model formulated in part (*c*).

12.2-2. The research and development division of a company has been developing four possible new product lines. Management must now make a decision as to which of these four products actually will be produced and at what levels. Therefore, they have asked the OR department to formulate a mathematical programming model to find the most profitable product mix.

A substantial cost is associated with beginning the production of any product, as given in the first row of the following table. The marginal net revenue from each unit produced is given in the second row of the table.

	Product			
	1	2	3	4
Start-up cost, $	50,000	40,000	70,000	60,000
Marginal revenue, $	70	60	90	80

Let the continuous decision variables x_1, x_2, x_3, and x_4 be the production levels of products 1, 2, 3, and 4, respectively. Management has imposed the following policy constraints on these variables:

1. No more than two of the products can be produced.
2. Either product 3 or 4 can be produced only if either product 1 or 2 is produced.
3. Either $5x_1 + 3x_2 + 6x_3 + 4x_4 \le 6000$
   or $4x_1 + 6x_2 + 3x_3 + 5x_4 \le 6000$.

Introduce auxiliary binary variables to formulate an MIP model for this problem.

12.2-3. Consider the following mathematical model.

$$\text{Minimize} \quad Z = f_1(x_1) + f_2(x_2),$$

subject to the restrictions

1. Either $x_1 \ge 3$ or $x_2 \ge 3$.
2. At least one of the following inequalities holds:

$$2x_1 + x_2 \ge 7$$
$$x_1 + x_2 \ge 5$$
$$x_1 + 2x_2 \ge 7.$$

3. $|x_1 - x_2| = 0$, or 3, or 6.
4. $x_1 \geq 0$, $x_2 \geq 0$,

where
$$f_1(x_1) = \begin{cases} 7 + 5x_1 & \text{if } x_1 > 0, \\ 0 & \text{if } x_1 = 0, \end{cases}$$

$$f_2(x_2) = \begin{cases} 5 + 6x_2 & \text{if } x_2 > 0, \\ 0 & \text{if } x_2 = 0. \end{cases}$$

Formulate this problem as an MIP problem.

12.2-4. Consider the following mathematical model.

$$\text{Maximize} \quad Z = 3x_1 + 2f(x_2) + 2x_3 + 3g(x_4),$$

subject to the restrictions

1. $2x_1 - x_2 + x_3 + 3x_4 \leq 15$.
2. At least one of the following two inequalities holds:

$$x_1 + x_2 + x_3 + x_4 \leq 4$$

$$3x_1 - x_2 - x_3 + x_4 \leq 3.$$

3. At least two of the following four inequalities holds:

$$5x_1 + 3x_2 + 3x_3 - x_4 \leq 10$$

$$2x_1 + 5x_2 - x_3 + 3x_4 \leq 10$$

$$-x_1 + 3x_2 + 5x_3 + 3x_4 \leq 10$$

$$3x_1 - x_2 + 3x_3 + 5x_4 \leq 10.$$

4. $x_3 = 1$, or 2, or 3.
5. $x_j \geq 0$ $(j = 1, 2, 3, 4)$,

where
$$f(x_2) = \begin{cases} -5 + 3x_2 & \text{if } x_2 > 0, \\ 0 & \text{if } x_2 = 0, \end{cases}$$

and
$$g(x_4) = \begin{cases} -3 + 5x_4, & \text{if } x_4 > 0 \\ 0 & \text{if } x_4 = 0. \end{cases}$$

Formulate this problem as an MIP problem.

12.2-5. A toy manufacturer has developed two new toys for possible inclusion in its product line for the upcoming Christmas season. Setting up the production facilities to begin production would cost \$50,000 for toy 1 and \$80,000 for toy 2. Once these costs are covered, the toys would generate a unit profit of \$10 for toy 1 and \$15 for toy 2.

The company has two factories that are capable of producing these toys. However, to avoid doubling the start-up costs, just one factory would be used and the choice would be based on maximizing profit. For administrative reasons, the same factory would be used for both new toys if both are produced.

Toy 1 can be produced at the rate of 50 per hour in factory 1 and 40 per hour in factory 2. Toy 2 can be produced at the rate of 40 per hour in factory 1 and 25 per hour in factory 2. Factories 1 and 2, respectively, have 500 and 700 hours of production time available before Christmas that could be used to produce these toys.

It is not known whether these two toys would be continued after Christmas. Therefore, the problem is to determine how many units (if any) of each new toy should be produced before Christmas in order to maximize the total profit.

(*a*) Formulate an IP model for this problem.
A* (*b*) Use the automatic MIP routine in your OR Courseware to solve this problem.

12.2-6.* An airline company is considering the purchase of new long-, medium-, and short-range jet passenger airplanes. The purchase price would be \$33.5 million for each long-range plane, \$25 million for each medium-range plane, and \$17.5 million for each short-range plane. The board of directors has authorized a maximum commitment of $750 million for these purchases. Regardless of which airplanes are purchased, air travel of all distances is expected to be sufficiently large that these planes would be utilized at essentially maximum capacity. It is estimated that the net annual profit (after capital recovery costs are subtracted) would be $2.1 million per long-range plane, $1.5 million per medium-range plane, and $1.15 million per short-range plane.

It is predicted that enough trained pilots will be available to the company to crew 30 new airplanes. If only short-range planes were purchased, the maintenance facilities would be able to handle 40 new planes. However, each medium-range plane is equivalent to $1\frac{1}{3}$ short-range planes, and each long-range plane is equivalent to $1\frac{2}{3}$ short-range planes in terms of their use of the maintenance facilities.

The information given here was obtained by a preliminary analysis of the problem. A more detailed analysis will be conducted subsequently. However, using the preceding data as a first approximation, management wishes to know how many planes of each type should be purchased to maximize profit.

(*a*) Formulate an IP model for this problem.

A* (*b*) Use the automatic MIP routine in your OR Courseware to solve this problem.

(*c*) Use a binary representation of the variables to reformulate the IP model in part (*a*) as a BIP problem.

A* (*d*) Use the automatic BIP routine in your OR Courseware to solve the BIP model formulated in part (*c*). Then use this optimal solution to identify an optimal solution for the IP model formulated in part (*a*).

12.2-7. Consider the two-variable IP example discussed in Sec. 12.4 and illustrated in Fig. 12.3.

(*a*) Use a binary representation of the variables to reformulate this model as a BIP problem.

A* (*b*) Use the automatic BIP routine in your OR Courseware to solve this BIP problem. Then use this optimal solution to identify an optimal solution for the original IP model.

12.3-1. Reconsider the Wyndor Glass Co. problem presented in Sec. 3.1. Management now has decided that only one of the two new products should be produced, and the choice is to be made on the basis of maximizing profit. Introduce *auxiliary binary variables* to formulate an MIP model for this new version of the problem.

12.3-2. Reconsider Prob. 3.1-3. (See the Answers to Selected Problems in the back of the book for the linear programming model for this problem.) Now add the managerial restriction that no more than two of the three prospective new products should be produced. Introduce auxiliary binary variables to formulate an MIP model for this new version of the problem.

12.3-3. Reconsider Probs. 12.2-1*a* and 10.3-9*a*. Now suppose that a more detailed analysis of the various cost and revenue factors has revealed that the profit from each product is *not* a linear function of the production quantity. The profits are instead given by the following table.

	Profit		
	Product		
Units Produced	1	2	3
0	0	0	0
1	−1	1	1
2	2	5	3
3	4		5
4			6
5			7

(*a*) Formulate a BIP model for this problem that includes constraints for mutually exclusive alternatives.

A* (*b*) Use the automatic BIP routine in your OR Courseware to solve the model formulated in part (*a*). Then use this optimal solution to identify the optimal number of units of the respective products to produce.

(*c*) Formulate another BIP model for this model that includes constraints for contingent decisions.

A* (*d*) Repeat part (*b*) for the model formulated in part (*c*).

12.3-4. Consider the following integer nonlinear programming problem.

$$\text{Maximize} \quad Z = 4x_1^2 - x_1^3 + 10x_2^2 - x_2^4,$$

subject to

$$x_1 + x_2 \le 3$$

and

$$x_1 \ge 0, \qquad x_2 \ge 0$$

$$x_1 \text{ and } x_2 \text{ are integers.}$$

This problem can be reformulated in two different ways as an equivalent pure BIP problem (with a linear objective function) with six binary variables (y_{1j} and y_{2j} for $j = 1, 2, 3$), depending on the interpretation given to the binary variables.

(*a*) Formulate a BIP model for this problem where the binary variables have the interpretation

$$y_{ij} = \begin{cases} 1 & \text{if } x_i = j, \\ 0 & \text{otherwise.} \end{cases}$$

A* (*b*) Use the automatic BIP routine in your OR Courseware to solve the model formulated in part (*a*), and thereby identify an optimal solution for (x_1, x_2) for the original problem.

(*c*) Formulate a BIP model for this problem where the binary variables have the interpretation

$$y_{ij} = \begin{cases} 1 & \text{if } x_i \ge j, \\ 0 & \text{otherwise.} \end{cases}$$

A* (*d*) Use the automatic BIP routine in your OR Courseware to solve the model formulated in part (*c*), and thereby identify an optimal solution for (x_1, x_2) for the original problem.

12.3-5. Consider the following discrete nonlinear programming problem.

$$\text{Maximize} \quad Z = 2x_1 - x_1^2 + 3x_2 - 3x_2^2,$$

subject to

$$x_1 + x_2 \le 0.75$$

and

each variable is restricted to the values $\frac{1}{2}, \frac{1}{3}, \frac{1}{4}, \frac{1}{5}$.

(*a*) Reformulate this problem as a pure binary integer linear programming problem.

A* (*b*) Use the automatic BIP routine in your OR Courseware to solve the model formulated in part (*a*), and thereby identify an optimal solution for (x_1, x_2) for the original problem.

12.3-6.* Consider the following special type of shortest-path problem (see Sec. 9.3) where the nodes are in columns and the only paths considered always move forward one column at a time.

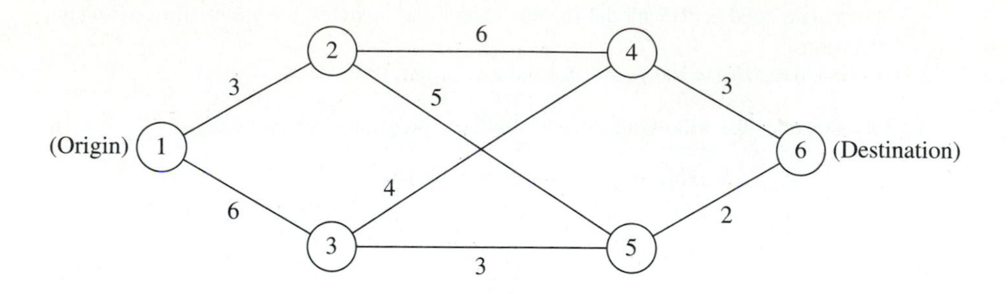

The numbers along the links represent distances, and the objective is to find the shortest path from the origin to the destination.

This problem also can be formulated as a BIP model involving both mutually exclusive alternatives and contingent decisions.

(*a*) Formulate this model.

A* (*b*) Use the automatic BIP routine in your OR Courseware to solve this problem.

12.3-7. Consider the project network for a PERT-type system shown in Prob. 10.2-3. Formulate a BIP model for the problem of finding a *critical path* (i.e., a *longest path*) for this project network.

12.3-8. A new planned community is being developed, and one of the decisions to be made is where to locate the two fire stations that have been allocated to the community. For planning purposes, the community has been divided into five tracts, with no more than one fire station to be located in any given tract. Each station is to respond to *all* the fires that occur in the tract in which it is located as well as in the other tracts that are assigned to this station. Thus the decisions to be made consist of (1) the tracts to receive a fire station and (2) the assignment of each of the other tracts to one of the fire stations. The objective is to minimize the overall average of the *response times* to fires.

The following table gives the average response time (in minutes) to a fire in each tract (the columns) if that tract is served by a station in a given tract (the rows). The bottom row gives the forecasted average number of fires that will occur in each of the tracts each day.

	Response Times				
	Fire in Tract				
Assigned Station Located in Tract	1	2	3	4	5
1	5	12	30	20	15
2	20	4	15	10	25
3	15	20	6	15	12
4	25	15	25	4	10
5	10	25	15	12	5
Frequency of emergencies	2	1	3	1	3

Formulate a complete BIP model for this problem. Identify any constraints that correspond to mutually exclusive alternatives or contingent decisions.

12.3-9. Reconsider Prob. 12.3-8. Suppose now that the cost (in thousands of dollars) of locating a fire station in a tract is 200 for tract 1; 250 for tract 2; 400 for tract 3; 300 for tract 4; and 500 for tract 5. Also suppose that the objective has been changed to the following:

> Determine which tracts should receive a station in order to minimize the total cost of stations while ensuring that each tract has at least one station close enough to respond to a fire in no more than 15 minutes (on the average).

In contrast to the original problem, note that the total number of fire stations is no longer fixed. Furthermore, if a tract without a station has more than one station within 15 minutes, it is no longer necessary to assign this tract to just one of these stations.

(*a*) Formulate a complete pure BIP model with five binary variables for this problem.

(*b*) Is this a *set-covering problem?* Explain, and identify the relevant sets.

A* (*c*) Use the automatic BIP routine in your OR Courseware to solve the model formulated in part (*a*).

12.3-10. Suppose that a state sends R persons to the U.S. House of Representatives. There are D counties in the state ($D > R$), and the state legislature wants to group these counties into R distinct electoral districts, each of which sends a delegate to Congress. The total population of the state is P, and the legislature wants to form districts whose population approximates $p = P/R$. Suppose that the appropriate legislative committee studying the electoral districting problem generates a long list of N *candidates* to be districts ($N > R$). Each of these candidates contains contiguous counties and a total population p_j ($j = 1, 2, \ldots, N$) that is acceptably close to p. Define $c_j = |p_j - p|$. Each county i ($i = 1, 2, \ldots, D$) is included in at least one candidate and typically will be included in a considerable number of candidates (in order to provide many feasible ways of selecting a set of R candidates that includes each county exactly once). Define

$$a_{ij} = \begin{cases} 1 & \text{if county } i \text{ is included in candidate } j, \\ 0 & \text{if not.} \end{cases}$$

Given the values of the c_j and the a_{ij}, the objective is to select R of these N possible districts such that each county is contained in a single district and such that the largest of the associated c_j is as small as possible.

Formulate a BIP model for this problem.

12.3-11 A U.S. professor will be spending a short sabbatical leave at the University of Iceland. She wishes to bring all needed items with her on the airplane. After collecting the professional items that she must have, she finds that airline regulations on space and weight for checked luggage will severely limit the clothes she can take. (She plans to carry on a warm coat and then purchase a warm Icelandic sweater upon arriving in Iceland.) Clothes under consideration for checked luggage include 3 skirts, 3 slacks, 4 tops, and 3 dresses. The professor wants to maximize the number of outfits she will have in Iceland (including the special dress she will wear on the airplane). Each dress constitutes an outfit. Other outfits consist of a combination of a top and either a skirt or slacks. However, certain combinations are not fashionable and so will not qualify as an outfit.

In the following table, the combinations that will make an outfit are marked with an x.

		Top				
		1	2	3	4	Icelandic Sweater
	1	x	x			x
Skirt	2	x			x	
	3		x	x	x	x
	1	x		x		
Slacks	2	x	x		x	x
	3			x	x	x

The weight (in grams) and volume (in cubic centimeters) of each item are shown in the following table:

		Weight	Volume
	1	600	5,000
Skirt	2	450	3,500
	3	700	3,000
	1	600	3,500
Slacks	2	550	6,000
	3	500	4,000
	1	350	4,000
Top	2	300	3,500
	3	300	3,000
	4	450	5,000
	1	600	6,000
Dress	2	700	5,000
	3	800	4,000
Total allowed		4,000	32,000

Formulate a BIP model to choose which items of clothing to take. (*Hint:* After using binary decision variables to represent the individual items, you should introduce *auxiliary* binary variables to represent outfits involving combinations of items. Then use constraints and the objective function to ensure that these auxiliary variables have the correct values, given the values of the decision variables.)

12.4-1.* Consider the following IP problem.

$$\text{Maximize } Z = 5x_1 + x_2,$$

subject to

$$-x_1 + 2x_2 \le 4$$
$$x_1 - x_2 \le 1$$
$$4x_1 + x_2 \le 12$$

and

$$x_1 \ge 0, \qquad x_2 \ge 0$$
$$x_1, x_2 \text{ are integers.}$$

(*a*) Solve this problem graphically.

I (*b*) Solve the LP relaxation graphically. Round this solution to the *nearest* integer solution and check whether it is feasible. Then enumerate *all* the rounded solutions by rounding this solution for the LP relaxation in *all* possible ways (i.e., by rounding each noninteger value both up and down). For each rounded solution, check for feasibility and, if feasible, calculate Z. Are any of these feasible rounded solutions optimal for the IP problem?

12.4-2. Follow the instructions of Prob. 12.4-1 for the following IP problem.

$$\text{Maximize } Z = 220x_1 + 80x_2,$$

subject to

$$5x_1 + 2x_2 \leq 16$$
$$2x_1 - x_2 \leq 4$$
$$-x_1 + 2x_2 \leq 4$$

and

$$x_1 \geq 0, \quad x_2 \geq 0$$
$$x_1, x_2 \text{ are integers.}$$

12.4-3. Label each of the following statements as true or false, and then justify your answer by referring to specific statements (with page citations) in the chapter.

(*a*) Linear programming problems are generally much easier to solve than IP problems.

(*b*) For IP problems, the number of integer variables is generally more important in determining the computational difficulty than is the number of functional constraints.

(*c*) To solve an IP problem with an approximate procedure, one may apply the simplex method to the LP relaxation problem and then round each noninteger value to the nearest integer. The result will be a feasible but not necessarily optimal solution for the IP problem.

D, I*, A **12.5-1.*** Use the BIP branch-and-bound algorithm presented in Sec. 12.5 to solve the following problem interactively.

$$\text{Maximize } Z = 2x_1 - x_2 + 5x_3 - 3x_4 + 4x_5,$$

subject to

$$3x_1 - 2x_2 + 7x_3 - 5x_4 + 4x_5 \leq 6$$
$$x_1 - x_2 + 2x_3 - 4x_4 + 2x_5 \leq 0$$

and

$$x_j \text{ is binary,} \quad \text{for } j = 1, 2, \ldots, 5.$$

D, I*, A **12.5-2.** Use the BIP branch-and-bound algorithm presented in Sec. 12.5 to solve the following problem interactively.

$$\text{Minimize} \quad Z = 5x_1 + 6x_2 + 7x_3 + 8x_4 + 9x_5,$$

subject to

$$3x_1 - x_2 + x_3 + x_4 - 2x_5 \geq 2$$
$$x_1 + 3x_2 - x_3 - 2x_4 + x_5 \geq 0$$
$$-x_1 - x_2 + 3x_3 + x_4 + x_5 \geq 1$$

and

$$x_j \text{ is binary,} \quad \text{for } j = 1, 2, \ldots, 5.$$

D, I*, A **12.5-3.** Use the BIP branch-and-bound algorithm presented in Sec. 12.5 to solve the following problem interactively.

$$\text{Maximize} \quad Z = 5x_1 + 5x_2 + 8x_3 - 2x_4 - 4x_5,$$

subject to

$$-3x_1 + 6x_2 - 7x_3 + 9x_4 + 9x_5 \geq 10$$
$$x_1 + 2x_2 - x_4 - 3x_5 \leq 0$$

and

$$x_j \text{ is binary,} \quad \text{for } j = 1, 2, \ldots, 5.$$

D, I*, A **12.5-4.** Reconsider Prob. 12.2-7*a*. Use the BIP branch-and-bound algorithm presented in Sec. 12.5 to solve this BIP model interactively.

D, I*, A **12.5-5.** Reconsider Prob. 12.3-9*a*. Use the BIP algorithm presented in Sec. 12.5 to solve this problem interactively.

12.5-6. Consider the following statements about any pure IP problem (in maximization form) and its LP relaxation. Label each of the statements as true or false, and then justify your answer.

(*a*) The feasible region for the LP relaxation is a subset of the feasible region for the IP problem.

(*b*) If an optimal solution for the LP relaxation is an integer solution, then the optimal value of the objective function is the same for both problems.

(*c*) If a noninteger solution is feasible for the LP relaxation, then the nearest-integer solution (rounding each variable to the nearest integer) is a feasible solution for the IP problem.

12.5-7.* Consider the assignment problem with the following cost table:

		Task				
		1	2	3	4	5
	1	39	65	69	66	57
	2	64	84	24	92	22
Assignee	3	49	50	61	31	45
	4	48	45	55	23	50
	5	59	34	30	34	18

(*a*) Design a branch-and-bound algorithm for solving such assignment problems by specifying how the branching, bounding, and fathoming steps would be performed. (*Hint:* For the assignees not yet assigned for the current subproblem, form the relaxation by deleting the constraints that each of these assignees must perform exactly one task.)

(*b*) Use this algorithm to solve this problem.

12.5-8. Five jobs need to be done on a certain machine. However, the setup time for each job depends upon which job immediately preceded it, as shown by the following table:

		Setup Time				
		Job				
		1	2	3	4	5
	None	4	5	8	9	4
	1	—	7	12	10	9
Immediately	2	6	—	10	14	11
Preceding Job	3	10	11	—	12	10
	4	7	8	15	—	7
	5	12	9	8	16	—

The objective is to schedule the *sequence* of jobs that minimizes the sum of the resulting setup times.

(*a*) Design a branch-and-bound algorithm for sequencing problems of this type by specifying how the branch, bound, and fathoming steps would be performed.

(*b*) Use this algorithm to solve this problem.

12.5-9.* Consider the following *nonlinear* BIP problem.

$$\text{Maximize} \quad Z = 80x_1 + 60x_2 + 40x_3 + 20x_4 - (7x_1 + 5x_2 + 3x_3 + 2x_4)^2,$$

subject to

$$x_j \text{ is binary}, \quad \text{for } j = 1, 2, 3, 4.$$

Given the value of the first k variables $x_1, \ldots, x_k$, where $k = 0, 1, 2$, or 3, an upper bound on the value of Z that can be achieved by the corresponding feasible solutions is

$$\sum_{j=1}^{k} c_j x_j - \left(\sum_{j=1}^{k} d_j x_j\right)^2 + \sum_{j=k+1}^{4} \max\left\{0, c_j - \left[\left(\sum_{i=1}^{k} d_i x_i + d_j\right)^2 - \left(\sum_{i=1}^{k} d_i x_i\right)^2\right]\right\},$$

where $c_1 = 80$, $c_2 = 60$, $c_3 = 40$, $c_4 = 20$, $d_1 = 7$, $d_2 = 5$, $d_3 = 3$, $d_4 = 2$. Use this bound to solve the problem by the branch-and-bound technique.

12.5-10. Consider the Lagrangian relaxation described near the end of Sec. 12.5.

(*a*) If $\mathbf{x}$ is a feasible solution for an MIP problem, show that $\mathbf{x}$ also must be a feasible solution for the corresponding Lagrangian relaxation.

(*b*) If $\mathbf{x}^*$ is an optimal solution for an MIP problem, with an objective function value of Z, show that $Z \leq Z_R^*$, where Z_R^* is the optimal objective function value for the corresponding Lagrangian relaxation.

12.6-1.* Consider the following IP problem.

$$\text{Maximize} \quad Z = -3x_1 + 5x_2,$$

subject to

$$5x_1 - 7x_2 \geq 3$$

and

$$x_j \leq 3$$
$$x_j \geq 0$$
$$x_j \text{ is integer}, \quad \text{for } j = 1, 2.$$

(*a*) Solve this problem graphically.

I (*b*) Use the MIP branch-and-bound algorithm presented in Sec. 12.6 to solve this problem by hand. For each subproblem, solve its LP relaxation *graphically* (either by hand or with your OR Courseware).

(*c*) Use the binary representation for integer variables to reformulate this problem as a BIP problem.

D, I*, A (*d*) Use the BIP branch-and-bound algorithm presented in Sec. 12.5 to solve the problem as formulated in part (*c*) interactively.

12.6-2. Follow the instructions of Prob. 12.6-1 for the following IP model.

$$\text{Minimize} \quad Z = 2x_1 + 3x_2,$$

subject to

$$x_1 + x_2 \geq 3$$
$$x_1 + 3x_2 \geq 6$$

and

$$x_1 \geq 0, \quad x_2 \geq 0$$
$$x_1, x_2 \text{ are integers}.$$

12.6-3. Reconsider the IP model of Prob. 12.4-1.

I (*a*) Use the MIP branch-and-bound algorithm presented in Sec. 12.6 to solve this problem by hand. For each subproblem, solve its LP relaxation *graphically* (either by hand or with your OR Courseware).

D, I* (*b*) Now use the interactive routine for this algorithm in your OR Courseware to solve this problem.

A* (*c*) Check your answer by using the automatic routine for this algorithm in your OR Courseware.

12.6-4. Follow the instructions of Prob. 12.6-3 for the IP model of Prob. 12.4-2.

D, I*, A **12.6-5.** Consider the IP example discussed in Sec. 12.4 and illustrated in Fig. 12.3. Use the MIP branch-and-bound algorithm presented in Sec. 12.6 to solve this problem interactively.

D, I*, A **12.6-6.** Reconsider Prob. 12.2-6*a*. Use the MIP branch-and-bound algorithm presented in Sec. 12.6 to solve this IP problem interactively.

12.6-7. A machine shop makes two products. Each unit of the first product requires 3 hours on machine 1 and 2 hours on machine 2. Each unit of the second product requires 2 hours on machine 1 and 3 hours on machine 2. Machine 1 is available only 8 hours per day and machine 2 only 7 hours per day. The profit per unit sold is 16 for the first product and 10 for the second. The amount of each product produced per day must be an integral multiple of 0.25. The objective is to determine the mix of production quantities that will maximize profit.

(*a*) Formulate an IP model for this problem.

(*b*) Solve this model graphically.

(*c*) Use graphical analysis to apply the MIP branch-and-bound algorithm presented in Sec. 12.6 to solve this model.

D, I* (*d*) Apply this algorithm interactively with your OR Courseware.

A* (*e*) Apply this same algorithm automatically with your OR Courseware.

D, I*, A **12.6-8.** Use the MIP branch-and-bound algorithm presented in Sec. 12.6 to solve the following MIP problem interactively.

$$\text{Maximize } Z = 5x_1 + 4x_2 + 4x_3 + 2x_4.$$

subject to

$$x_1 + 3x_2 + 2x_3 + x_4 \leq 10$$
$$5x_1 + x_2 + 3x_3 + 2x_4 \leq 15$$
$$x_1 + x_2 + x_3 + x_4 \leq 6$$

and

$$x_j \geq 0, \qquad \text{for } j = 1, 2, 3, 4$$
$$x_j \text{ is integer}, \qquad \text{for } j = 1, 2, 3.$$

D, I*, A **12.6-9.** Use the MIP branch-and-bound algorithm presented in Sec. 12.6 to solve the following MIP problem interactively.

$$\text{Maximize} \quad Z = 3x_1 + 4x_2 + 2x_3 + x_4 + 2x_5,$$

subject to

$$2x_1 - x_2 + x_3 + x_4 + x_5 \leq 3$$
$$-x_1 + 3x_2 + x_3 - x_4 - 2x_5 \leq 2$$
$$2x_1 + x_2 - x_3 + x_4 + 3x_5 \leq 1$$

and

$$x_j \geq 0, \qquad \text{for } j = 1, 2, 3, 4, 5$$
$$x_j \text{ is binary}, \qquad \text{for } j = 1, 2, 3.$$

D, I*, A **12.6-10.** Use the MIP branch-and-bound algorithm presented in Sec. 12.6 to solve the following MIP problem interactively.

$$\text{Minimize} \quad Z = 5x_1 + x_2 + x_3 + 2x_4 + 3x_5,$$

subject to

$$x_2 - 5x_3 + x_4 + 2x_5 \geq -2$$
$$5x_1 - x_2 + x_5 \geq 7$$
$$x_1 + x_2 + 6x_3 + x_4 \geq 4$$

and

$$x_j \geq 0, \quad \text{for } j = 1, 2, 3, 4, 5$$
$$x_j \text{ is integer}, \quad \text{for } j = 1, 2, 3.$$

12.6-11. Reconsider the discrete nonlinear programming problem given in Prob. 12.3-5.

(*a*) Use the following outline in designing the main features of a branch-and-bound algorithm for solving this problem (and similar problems) directly without reformulation.
  (i) Specify the tightest possible nonlinear programming relaxation that has only continuous variables and so can be solved efficiently by nonlinear programming techniques. (The next chapter will describe how such nonlinear programming problems can be solved efficiently.)
  (ii) Specify the fathoming tests.
  (iii) Specify a branching procedure that involves specifying two ranges of values for a single variable.

(*b*) Use the algorithm designed in part (*a*) to solve this problem by using the routine *Solve Quadratic Programming Model Automatically* in the Nonlinear Programming area of your OR Courseware to solve the relaxation at each iteration.

12.7-1. For each of the following constraints of pure BIP problems, use the constraint to fix as many variables as possible.

(*a*) $4x_1 + x_2 + 3x_3 + 2x_4 \leq 2$
(*b*) $4x_1 - x_2 + 3x_3 + 2x_4 \leq 2$
(*c*) $4x_1 - x_2 + 3x_3 + 2x_4 \geq 7$

12.7-2. For each of the following constraints of pure BIP problems, use the constraint to fix as many variables as possible.

(*a*) $20x_1 - 7x_2 + 5x_3 \leq 10$
(*b*) $10x_1 - 7x_2 + 5x_3 \geq 10$
(*c*) $10x_1 - 7x_2 + 5x_3 \leq -1$

12.7-3. Use the following set of constraints for the *same* pure BIP problem to fix as many variables as possible. Also identify the constraints which become redundant because of the fixed variables.

$$3x_3 - x_5 + x_7 \leq 1$$
$$x_2 + x_4 + x_6 \leq 1$$
$$x_1 - 2x_5 + 2x_6 \geq 2$$
$$x_1 + x_2 - x_4 \leq 0$$

12.7-4. For each of the following constraints of pure BIP problems, identify which ones are made redundant by the binary constraints. Explain why each one is, or is not, redundant.

(*a*) $2x_1 + x_2 + 2x_3 \leq 5$
(*b*) $3x_1 - 4x_2 + 5x_3 \leq 5$
(*c*) $x_1 + x_2 + x_3 \geq 2$
(*d*) $3x_1 - x_2 - 2x_3 \geq -4$

12.7-5. Apply the procedure for tightening constraints to the following constraint for a pure BIP problem.

$$3x_1 - 2x_2 + x_3 \leq 3.$$

12.7-6. Apply the procedure for tightening constraints to the following constraint for a pure BIP problem.

$$x_1 - x_2 + 3x_3 + 4x_4 \geq 1.$$

12.7-7. Apply the procedure for tightening constraints to each of the following constraints for a pure BIP problem.

(*a*) $x_1 + 3x_2 - 4x_3 \leq 2$
(*b*) $3x_1 - x_2 + 4x_3 \geq 1$

12.7-8. In Sec. 12.7, a pure BIP example with the constraint $2x_1 + 3x_2 \leq 4$ was used to illustrate the procedure for tightening constraints. Show that applying the procedure for generating cutting planes to this constraint yields the same new constraint $x_1 + x_2 \leq 1$.

12.7-9. One of the constraints of a certain pure BIP problem is

$$x_1 + 3x_2 + 2x_3 + 4x_4 \leq 5.$$

Identify all the minimal covers for this constraint, and then give the corresponding cutting planes.

12.7-10. One of the constraints of a certain pure BIP problem is

$$3x_1 + 4x_2 + 2x_3 + 5x_4 \leq 7.$$

Identify all the minimal covers for this constraint, and then give the corresponding cutting planes.

12.7-11. Generate as many cutting planes as possible from the following constraint for a pure BIP problem.

$$3x_1 + 5x_2 + 4x_3 + 8x_4 \leq 10.$$

12.7-12. Generate as many cutting planes as possible from the following constraint for a pure BIP problem.

$$5x_1 + 3x_2 + 7x_3 + 4x_4 + 6x_5 \leq 9.$$

12.7-13. Consider the following BIP problem.

$$\text{Maximize} \quad Z = 2x_1 + 3x_2 + x_3 + 4x_4 + 3x_5 + 2x_6 + 2x_7 + x_8 + 3x_9,$$

subject to

$$3x_2 + x_4 + x_5 \geq 3$$
$$x_1 + x_2 \leq 1$$
$$x_2 + x_4 - x_5 - x_6 \leq -1$$
$$x_2 + 2x_6 + 3x_7 + x_8 + 2x_9 \geq 4$$
$$-x_3 + 2x_5 + x_6 + 2x_7 - 2x_8 + x_9 \leq 5$$

and

$$\text{all } x_j \text{ binary.}$$

Develop the tightest possible formulation of this problem by using the techniques of automatic problem preprocessing (fixing variables, deleting redundant constraints, and tightening constraints). Then use this tightened formulation to determine an optimal solution by inspection.

CP12-1. Reconsider Prob. CP3-1 at the end of Chap. 3.

The school board now has made the decision to prohibit the splitting of residential areas among multiple schools. Thus, each of the six areas now must be assigned to a single school. To provide more flexibility for doing this, the capacity of each school is being increased by 50 students by providing portable classrooms as needed. (*Hint:* Constraints that each grade cannot constitute more than 35 percent of each school's population also ensure that no grade can fall below 30 percent, so separate constraints are not needed for the latter restriction.)

(*a*) Formulate a BIP model for this problem.

(*b*) Referring to part (*a*) of Prob. CP3-1, explain why that linear programming model and the BIP model just formulated are so different given that they deal with nearly the same problem.

A* (*c*) Use a computer to solve the model formulated in part (*a*) by a BIP algorithm.

(*d*) Referring to part (*c*) of Prob. CP3-1, determine how much the total busing cost increases because of the decision to prohibit the splitting of residential areas among multiple schools.

A* (*e*), (*f*), (*g*), (*h*) Repeat parts (*e*), (*f*), (*g*), and (*h*) of Prob. CP3-1 under the new school board decision.

13

Nonlinear Programming

The fundamental role of linear programming in operations research (OR) is accurately reflected by the fact that it is the focus of *one-third* of this book. A key assumption of linear programming is that *all its functions* (objective function and constraint functions) are linear. Although this assumption essentially holds for numerous practical problems, frequently it does not hold. In fact, many economists have found that some degree of nonlinearity is the rule and not the exception in economic planning problems.[1] Therefore it often is necessary to deal directly with nonlinear programming problems, so we turn our attention to this important area.

In one general form,[2] the *nonlinear programming problem* is to find $\mathbf{x} = (x_1, x_2, \ldots, x_n)$ so as to

$$\text{Maximize} \qquad f(\mathbf{x}),$$

[1] For example, see W. J. Baumol and R. C. Bushnell, "Error Produced by Linearization in Mathematical Programming," *Econometrica,* **35:** 447–471, 1967.

[2] The other *legitimate forms* correspond to those for *linear programming* listed in Sec. 3.2. Section 4.6 describes how to convert these other forms to the form given here.

subject to

$$g_i(\mathbf{x}) \le b_i, \qquad \text{for } i = 1, 2, \ldots, m,$$

and

$$\mathbf{x} \ge \mathbf{0},$$

where $f(\mathbf{x})$ and the $g_i(\mathbf{x})$ are given functions of the n decision variables.[1]

No algorithm that will solve *every* specific problem fitting this format is available. However, substantial progress has been made for some important special cases of this problem by making various assumptions about these functions, and research is continuing very actively. This area is a large one, and we do not have the space to survey it completely. However, we do present a few sample applications and then introduce some of the basic ideas for solving certain important types of nonlinear programming problems.

Both Appendixes 2 and 3 provide useful background for this chapter, and we recommend that you review these appendixes as you study the next few sections.

13.1 Sample Applications

The following examples illustrate a few of the many important types of problems to which nonlinear programming has been applied.

The Product-Mix Problem with Price Elasticity

In *product-mix* problems, such as the Wyndor Glass Co. problem of Sec. 3.1, the goal is to determine the optimal mix of production levels for a firm's products, given limitations on the resources needed to produce those products, in order to maximize the firm's total profit. In some cases, there is a fixed unit profit associated with each product, so the resulting objective function will be linear. However, in many product-mix problems, certain factors introduce *nonlinearities* into the objective function. For example, a large manufacturer may encounter *price elasticity,* whereby the amount of a product that can be sold has an inverse relationship to the price charged. Thus the *price-demand curve* for a typical product might look like the one shown in Fig. 13.1, where $p(x)$ is the price required in order to be able to sell x units. The firm's profit from producing and selling x units of the product then is the sales revenue $xp(x)$ minus the production and distribution costs. Therefore, if the unit cost for producing and distributing the product is fixed at c (see the dashed line in Fig. 13.1), the firm's profit from producing and selling x units is given by the nonlinear function

$$P(x) = xp(x) - cx,$$

as plotted in Fig. 13.2. If *each* of the firm's n products has a similar profit function, say, $P_j(x_j)$ for producing and selling x_j units of product j ($j = 1, 2, \ldots, n$), then the overall objective function is

$$f(\mathbf{x}) = \sum_{j=1}^{n} P_j(x_j),$$

a sum of nonlinear functions.

[1] For simplicity, we assume throughout the chapter that *all* these functions either are *differentiable* everywhere or are *piecewise linear functions* (discussed in Secs. 13.1 and 13.8).

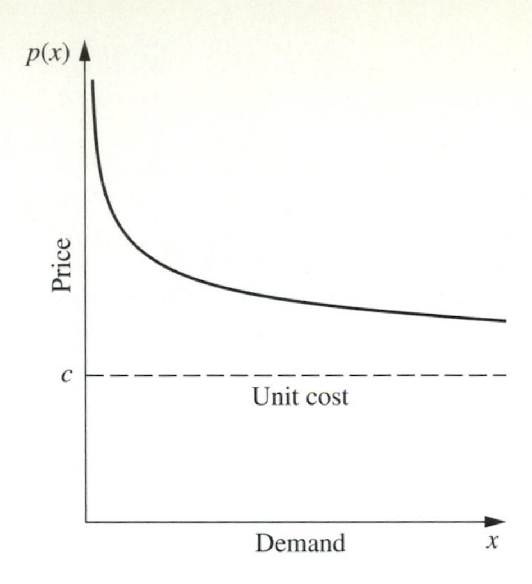

Figure 13.1 Price-demand curve.

Another reason that nonlinearities can arise in the objective function is due to the fact that the *marginal cost* of producing another unit of a given product varies with the production level. For example, the marginal cost may decrease when the production level is increased because of a *learning-curve effect* (more efficient production with more experience). On the other hand, it may increase instead, because special measures such as overtime or more expensive production facilities may be needed to increase production further.

Nonlinearities also may arise in the $g_i(\mathbf{x})$ constraint functions in a similar fashion. For example, if there is a budget constraint on total production cost, the cost function will be nonlinear if the marginal cost of production varies as just described. For constraints on the other kinds of resources, $g_i(\mathbf{x})$ will be nonlinear whenever the use of the corresponding resource is not strictly proportional to the production levels of the respective products.

The Transportation Problem with Volume Discounts on Shipping Costs

As illustrated by the P & T Company example in Sec. 8.1, a typical application of the transportation problem is to determine an optimal plan for shipping goods from various

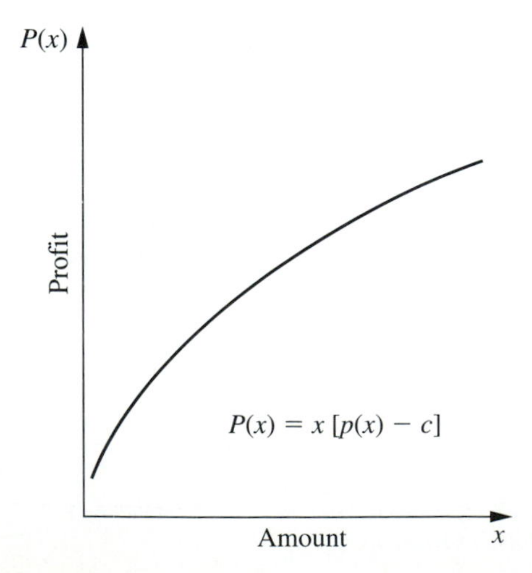

Figure 13.2 Profit function.

sources to various destinations, given supply and demand constraints, in order to minimize total shipping cost. It was assumed in Chap. 8 that the *cost per unit shipped* from a given source to a given destination is *fixed,* regardless of the amount shipped. In actuality, this cost may not be fixed. *Volume discounts* sometimes are available for large shipments, so that the *marginal cost* of shipping one more unit might follow a pattern like the one shown in Fig. 13.3. The resulting cost of shipping x units then is given by a *nonlinear* function $C(x)$, which is a *piecewise linear function* with slope equal to the marginal cost, like the one shown in Fig. 13.4. [The function in Fig. 13.4 consists of a line segment with slope 6.5 from (0, 0) to (0.6, 3.9), a second line segment with slope 5 from (0.6, 3.9) to (1.5, 8.4), a third line segment with slope 4 from (1.5, 8.4) to (2.7, 13.2), and a fourth line segment with slope 3 from (2.7, 13.2) to (4.5, 18.6).] Consequently, if *each* combination of source and destination has a similar shipping cost function, so that the cost of shipping x_{ij} units from source i ($i = 1, 2, \ldots, m$) to destination j ($j = 1, 2, \ldots, n$) is given by a nonlinear function $C_{ij}(x_{ij})$, then the overall objective function to be *minimized* is

$$f(\mathbf{x}) = \sum_{i=1}^{m} \sum_{j=1}^{n} C_{ij}(x_{ij}).$$

Even with this nonlinear objective function, the constraints normally are still the special linear constraints that fit the transportation problem model in Sec. 8.1.

Portfolio Selection with Risky Securities

It now is common practice for professional managers of large stock portfolios to use computer models based partially on nonlinear programming to guide them. Because investors are concerned about both the *expected return* (gain) and the *risk* associated with their investments, nonlinear programming is used to determine a portfolio that, under certain assumptions, provides an optimal trade-off between these two factors.

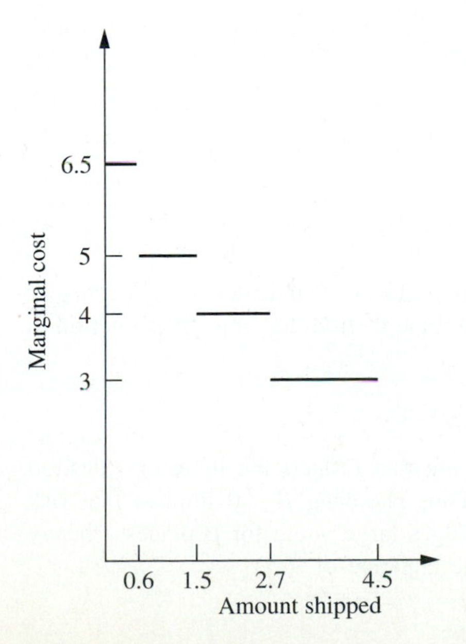

Figure 13.3 Marginal shipping cost.

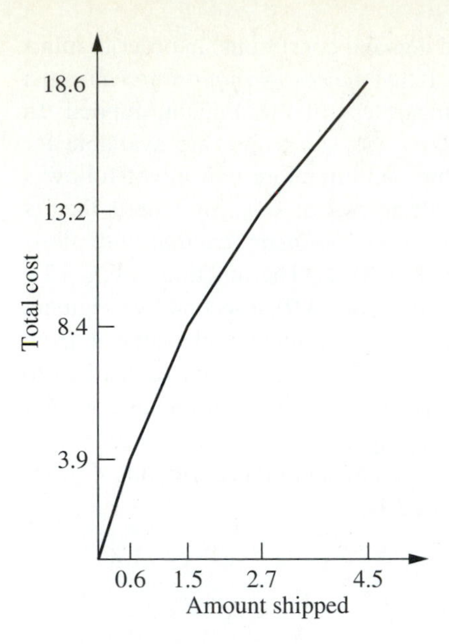

Figure 13.4 Shipping cost function.

This approach is based largely on path-breaking research done by Harry Markowitz and William Sharpe that helped them win the 1991 Nobel Prize in Economics.

A nonlinear programming model can be formulated for this problem as follows. Suppose that n stocks (securities) are being considered for inclusion in the portfolio, and let the decision variables x_j ($j = 1, 2, \ldots, n$) be the number of shares of stock j to be included. Let μ_j and σ_{jj} be the (estimated) *mean* and *variance,* respectively, of the return on each share of stock j, where σ_{jj} measures the risk of this stock. For $i = 1, 2, \ldots, n$ ($i \neq j$), let σ_{ij} be the *covariance* of the return on one share each of stock i and stock j. (Because it would be difficult to estimate all the σ_{ij} values, the usual approach is to make certain assumptions about market behavior that enable us to calculate σ_{ij} directly from σ_{ii} and σ_{jj}.) Then the expected value $R(\mathbf{x})$ and the variance $V(\mathbf{x})$ of the total return from the entire portfolio are

$$R(\mathbf{x}) = \sum_{j=1}^{n} \mu_j x_j$$

and

$$V(\mathbf{x}) = \sum_{i=1}^{n} \sum_{j=1}^{n} \sigma_{ij} x_i x_j,$$

where $V(\mathbf{x})$ measures the risk associated with the portfolio. The device used to consider the trade-off between these two factors is to combine them in the objective function to be maximized

$$f(\mathbf{x}) = R(\mathbf{x}) - \beta V(\mathbf{x}),$$

where the parameter β is a nonnegative constant that reflects the investor's desired trade-off between expected return and risk. Thus choosing $\beta = 0$ implies that risk should be ignored completely, whereas choosing a large value for β places a heavy weight on minimizing risk [by maximizing the negative of $V(\mathbf{x})$].

The complete nonlinear programming model might be

$$\text{Maximize} \qquad f(\mathbf{x}) = \sum_{j=1}^{n} \mu_j x_j - \beta \sum_{i=1}^{n} \sum_{j=1}^{n} \sigma_{ij} x_i x_j,$$

subject to

$$\sum_{j=1}^{n} P_j x_j \le B$$

and

$$x_j \ge 0, \qquad \text{for } j = 1, 2, \ldots, n,$$

where P_j is the price for each share of stock j and B is the amount of money budgeted for the portfolio. Under certain assumptions about the investor's *utility function* (measuring the relative value to the investor of different total returns), it can be shown that an optimal solution for this nonlinear programming problem maximizes the investor's *expected utility.*[1]

One drawback of the preceding formulation is that, because $R(\mathbf{x})$ and $V(\mathbf{x})$ are somewhat incommensurable, it is relatively difficult to choose an appropriate value for β. Therefore, rather than stopping with one choice of β, it is common to use a *parametric* (nonlinear) programming approach to generate the optimal solution as a function of β over a wide range of values of β. The next step is to examine the values of $R(\mathbf{x})$ and $V(\mathbf{x})$ for these solutions that are optimal for some value of β and then to choose the solution that seems to give the best trade-off between these two quantities. This procedure often is referred to as generating the solutions on the *efficient frontier* of the two-dimensional graph of $(R(\mathbf{x}), V(\mathbf{x}))$ points for feasible $\mathbf{x}$. The reason is that the $(R(\mathbf{x}), V(\mathbf{x}))$ point for an optimal $\mathbf{x}$ (for some β) lies on the *frontier* (boundary) of the feasible points. Furthermore, each optimal $\mathbf{x}$ is *efficient* in the sense that no other feasible solution is at least equally good with one measure (R or V) and strictly better with the other measure (smaller V or larger R).

13.2 Graphical Illustration of Nonlinear Programming Problems

When a nonlinear programming problem has just one or two variables, it can be represented graphically much like the Wyndor Glass Co. example for linear programming in Sec. 3.1. Because such a graphical representation gives considerable insight into the properties of optimal solutions for linear and nonlinear programming, let us look at a few examples. To highlight the difference between linear and nonlinear programming, we shall use some *nonlinear* variations of the Wyndor Glass Co. problem.

Figure 13.5 shows what happens to this problem if the only changes in the model shown in Sec. 3.1 are that both the second and the third functional constraints are replaced by the single nonlinear constraint $9x_1^2 + 5x_2^2 \le 216$. Compare Fig. 13.5 with Fig. 3.3. The optimal solution still happens to be $(x_1, x_2) = (2, 6)$. Furthermore, it still lies on the boundary of the feasible region. However, it is *not* a corner-point feasible

[1] See Selected Reference 1 for further details.

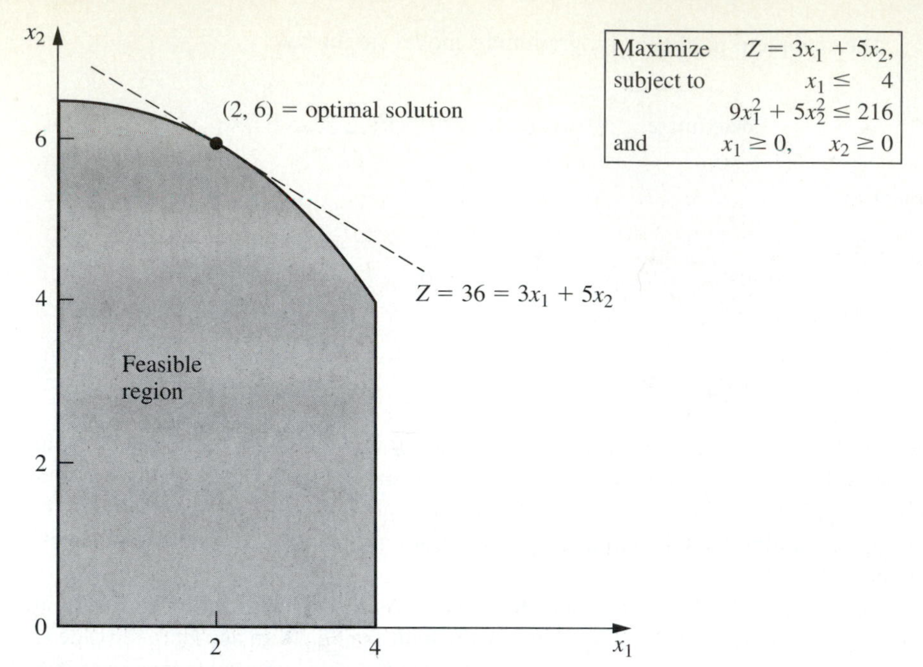

Figure 13.5 The Wyndor Glass Co. example with the nonlinear constraint $9x_1^2 + 5x_2^2 \leq 216$ replacing the original second and third functional constraints.

(CPF) solution. The optimal solution could have been a CPF solution with a different objective function (check $Z = 3x_1 + x_2$), but the fact that it need not be one means that we no longer have the tremendous simplification used in linear programming of limiting the search for an optimal solution to just the CPF solutions.

Now suppose that the linear constraints of Sec. 3.1 are kept unchanged, but the objective function is made nonlinear. For example, if

$$Z = 126x_1 - 9x_1^2 + 182x_2 - 13x_2^2,$$

then the graphical representation in Fig. 13.6 indicates that the optimal solution is $x_1 = \frac{8}{3}$, $x_2 = 5$, which again lies on the boundary of the feasible region. (The value of Z for this optimal solution is $Z = 857$, so Fig. 13.6 depicts the fact that the locus of all points with $Z = 857$ intersects the feasible region at just this one point, whereas the locus of points with any larger Z does not intersect the feasible region at all.) On the other hand, if

$$Z = 54x_1 - 9x_1^2 + 78x_2 - 13x_2^2,$$

then Fig. 13.7 illustrates that the optimal solution turns out to be $(x_1, x_2) = (3, 3)$, which lies *inside* the boundary of the feasible region. (You can check that this solution is optimal by using calculus to derive it as the unconstrained global maximum; because it also satisfies the constraints, it must be optimal for the constrained problem.) Therefore, a general algorithm for solving similar problems needs to consider *all* solutions in the feasible region, not just those on the boundary.

Another complication that arises in nonlinear programming is that a *local* maximum need not be a *global* maximum (the overall optimal solution). For example, consider the function of a single variable plotted in Fig. 13.8. Over the interval $0 \leq x \leq$

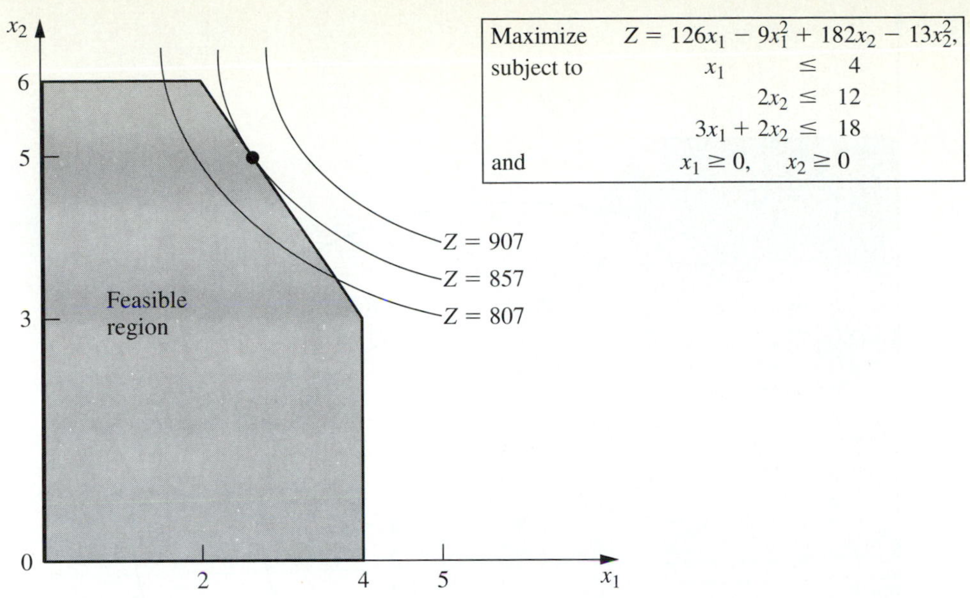

Figure 13.6 The Wyndor Glass Co. example with the original feasible region but with the nonlinear objective function $Z = 126x_1 - 9x_1^2 + 182x_2 - 13x_2^2$ replacing the original objective function.

5, this function has three local maxima—$x = 0$, $x = 2$, and $x = 4$—but only one of these—$x = 4$—is a *global maximum.* (Similarly, there are local minima at $x = 1$, 3, and 5, but only $x = 5$ is a *global minimum.*)

Nonlinear programming algorithms generally are unable to distinguish between a local maximum and a global maximum (except by finding another *better* local maximum). Therefore, it becomes crucial to know the conditions under which any local maximum is *guaranteed* to be a global maximum over the feasible region. You may recall from calculus that when we maximize an ordinary (doubly differentiable) function of a single variable $f(x)$ without any constraints, this guarantee can be given when

$$\frac{d^2f}{dx^2} \leq 0 \qquad \text{for all } x.$$

Such a function that is always ''curving downward'' (or not curving at all) is called a **concave** function.[1] Similarly, if $\leq$ is replaced by $\geq$, so that the function is always ''curving upward'' (or not curving at all), it is called a **convex** function.[2] (Thus a *linear* function is both concave and convex.) See Fig. 13.9 for examples. Then note that Fig. 13.8 illustrates a function that is neither concave nor convex because it alternates between curving upward and curving downward.

Functions of multiple variables also can be characterized as concave or convex if they always curve downward or curve upward. These intuitive definitions are restated in precise terms, along with further elaboration on these concepts, in Appendix 2. Appendix 2 also provides a convenient test for checking whether a function of two variables is concave, convex, or neither.

[1] Concave functions sometimes are referred to as *concave downward.*

[2] Convex functions sometimes are referred to as *concave upward.*

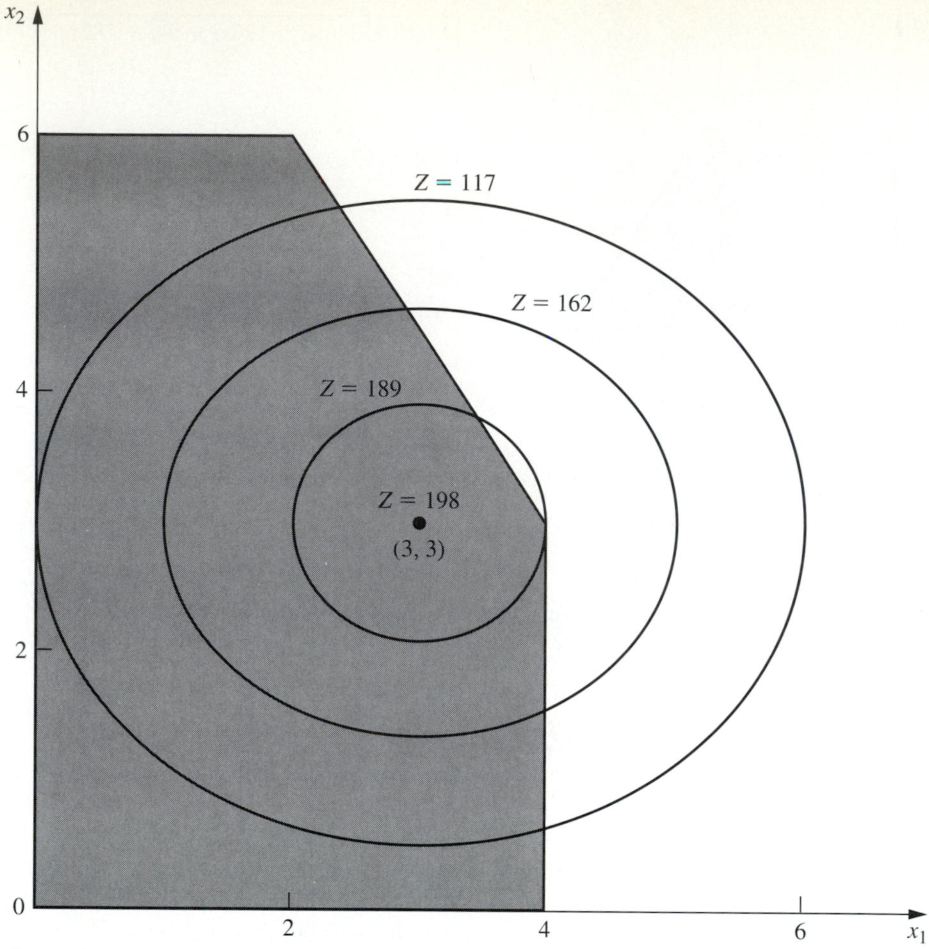

Figure 13.7 The Wyndor Glass Co. example with the original feasible region but with another nonlinear objective function, $Z = 54x_1 - 9x_1^2 + 78x_2 - 13x_2^2$, replacing the original objective function.

Here is a convenient way of checking this for a function of more than two variables when the function consists of a sum of smaller functions of just one or two variables each. If each smaller function is concave, then the overall function is concave. Similarly, the overall function is convex if each smaller function is convex.

To illustrate, consider the function

$$\begin{aligned} f(x_1, x_2, x_3) &= 4x_1 - x_1^2 - (x_2 - x_3)^2 \\ &= [4x_1 - x_1^2] + [-(x_2 - x_3)^2], \end{aligned}$$

which is the sum of the two smaller functions given in square brackets. The first smaller function $4x_1 - x_1^2$ is a function of the single variable x_1, so it can be found to be concave by noting that its second derivative is negative. The second smaller function $-(x_2 - x_3)^2$ is a function of just x_2 and x_3, so the test for functions of two variables given in Appendix 2 is applicable. In fact, Appendix 2 uses this particular function to illustrate the test and finds that the function is concave. Because both smaller functions are concave, the overall function $f(x_1, x_2, x_3)$ must be concave.

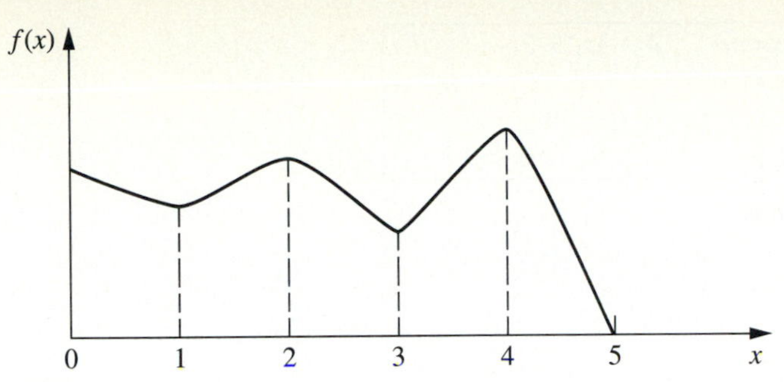

Figure 13.8 A function with several local maxima.

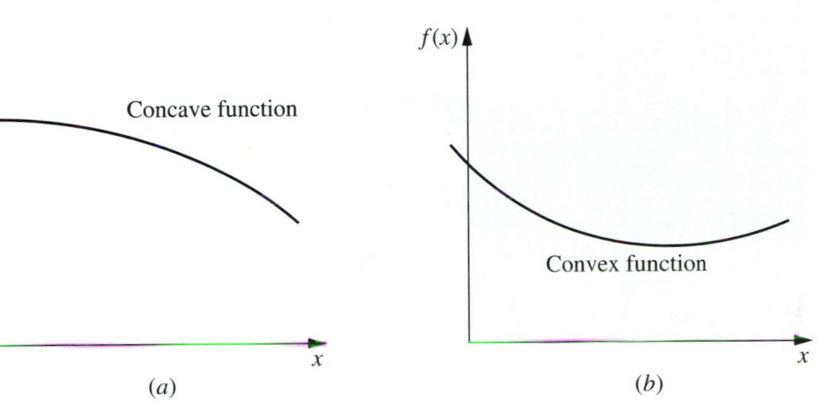

Figure 13.9 Examples of (*a*) a concave function and (*b*) a convex function.

If a nonlinear programming problem has no constraints, the objective function being *concave* guarantees that a local maximum is a *global maximum.* (Similarly, the objective function being *convex* ensures that a local minimum is a *global minimum.*) If there are constraints, then one more condition will provide this guarantee, namely, that the *feasible region* is a **convex set**. As discussed in Appendix 2, a convex set is simply a set of points such that, for each pair of points in the collection, the entire line segment joining these two points is also in the collection. Thus the feasible region for the original Wyndor Glass Co. problem (see Fig. 13.6 or 13.7) is a convex set. In fact, the feasible region for *any* linear programming problem is a convex set. Similarly, the feasible region in Fig. 13.5 is a convex set.

In general, the feasible region for a nonlinear programming problem is a convex set whenever all the $g_i(\mathbf{x})$ [for the constraints $g_i(\mathbf{x}) \leq b_i$] are convex. For the example of Fig. 13.5, both of its $g_i(\mathbf{x})$ are convex, since $g_1(\mathbf{x}) = x_1$ (a linear function is automatically both concave and convex) and $g_2(\mathbf{x}) = 9x_1^2 + 5x_2^2$ (both $9x_1^2$ and $5x_2^2$ are convex functions, so their sum is a convex function). These two convex $g_i(\mathbf{x})$ lead to the feasible region of Fig. 13.5 being a convex set.

Now let's see what happens when just one of these $g_i(\mathbf{x})$ is a concave function instead. In particular, suppose that the only change made in the example of Fig. 13.5 is that its nonlinear constraint is replaced by $8x_1 - x_1^2 + 14x_2 - x_2^2 \leq 49$. Therefore, the new $g_2(\mathbf{x}) = 8x_1 - x_1^2 + 14x_2 - x_2^2$, which is a concave function since both $8x_1 - x_1^2$ and $14x_2 - x_2^2$ are concave functions. The new feasible region shown in Fig. 13.10 is *not* a convex set. Why? Because this feasible region contains pairs of points, for example, (0, 7) and (4, 3), such that part of the line segment joining these two points is

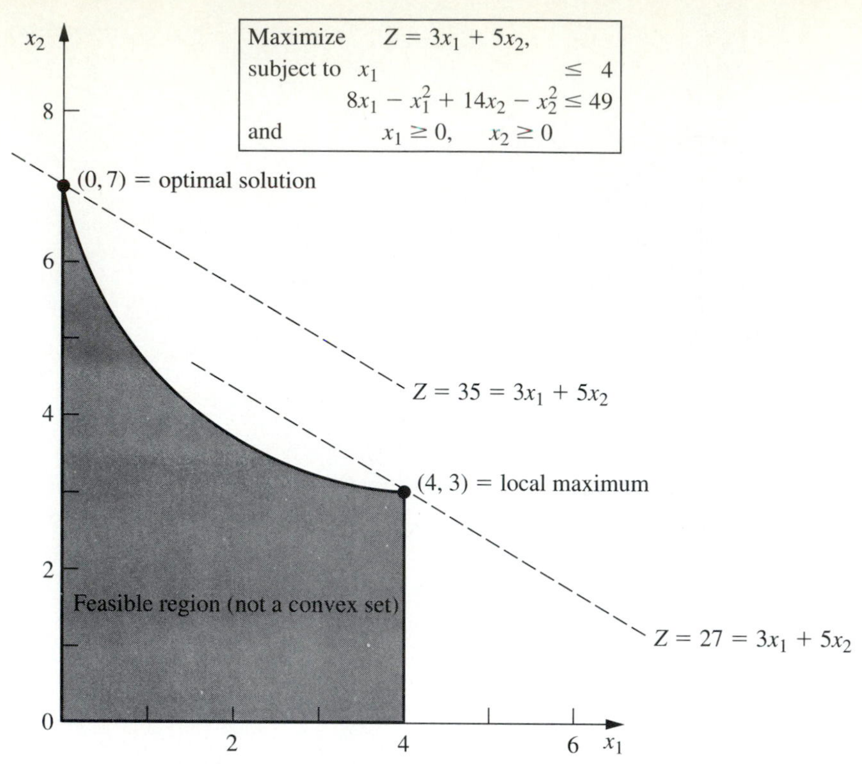

Figure 13.10 The Wyndor Glass Co. example with another nonlinear constraint, $8x_1 - x_1^2 + 14x_2 - x_2^2 \leq 49$, replacing the original second and third functional constraints.

not in the feasible region. Consequently, we cannot guarantee that a local maximum is a global maximum. In fact, this example has two local maxima, (0, 7) and (4, 3), but only (0, 7) is a global maximum.

Therefore, to guarantee that a local maximum is a global maximum for a nonlinear programming problem with constraints $g_i(\mathbf{x}) \leq b_i$ $(i = 1, 2, \ldots, m)$ and $\mathbf{x} \geq \mathbf{0}$, the objective function $f(\mathbf{x})$ must be *concave* and each $g_i(\mathbf{x})$ must be *convex*. Such a problem is called a *convex programming problem*, which is one of the key types of nonlinear programming problems discussed in the next section.

13.3 Types of Nonlinear Programming Problems

Nonlinear programming problems come in many different shapes and forms. Unlike the simplex method for linear programming, no single algorithm can solve all these different types of problems. Instead, algorithms have been developed for various individual *classes* (special types) of nonlinear programming problems. The most important classes are introduced briefly in this section. The subsequent sections then describe how some problems of these types can be solved.

Unconstrained Optimization

Unconstrained optimization problems have *no* constraints, so the objective is simply to

$$\text{Maximize} \quad f(\mathbf{x})$$

over *all* values of $\mathbf{x} = (x_1, x_2, \ldots, x_n)$. As reviewed in Appendix 3, the *necessary* condition that a particular solution $\mathbf{x} = \mathbf{x}^*$ be optimal when $f(\mathbf{x})$ is a differentiable function is

$$\frac{\partial f}{\partial x_j} = 0 \qquad \text{at } \mathbf{x} = \mathbf{x}^*, \text{ for } j = 1, 2, \ldots, n.$$

When $f(\mathbf{x})$ is *concave,* this condition also is *sufficient,* so then solving for $\mathbf{x}^*$ reduces to solving the system of n equations obtained by setting the n partial derivatives equal to zero. Unfortunately, for *nonlinear* functions $f(\mathbf{x})$, these equations often are going to be nonlinear as well, in which case you are unlikely to be able to solve analytically for their simultaneous solution. What then? Sections 13.4 and 13.5 describe *algorithmic search procedures* for finding $\mathbf{x}^*$, first for $n = 1$ and then for $n > 1$. These procedures also play an important role in solving many of the problem types described next, where there are constraints. The reason is that many algorithms for constrained problems are designed so that they can focus on an unconstrained version of the problem during a portion of each iteration.

When a variable x_j does have a nonnegativity constraint $x_j \geq 0$, the preceding necessary and (perhaps) sufficient condition changes slightly to

$$\frac{\partial f}{\partial x_j} \begin{cases} \leq 0 & \text{at } \mathbf{x} = \mathbf{x}^*, & \text{if } x_j^* = 0 \\ = 0 & \text{at } \mathbf{x} = \mathbf{x}^*, & \text{if } x_j^* > 0 \end{cases}$$

for each such j. This condition is illustrated in Fig. 13.11, where the optimal solution for a problem with a single variable is at $x = 0$ even though the derivative there is negative rather than zero. Because this example has a concave function to be maximized subject to a nonnegativity constraint, having the derivative less than or equal to 0 at $x = 0$ is both a necessary and sufficient condition for $x = 0$ to be optimal.

A problem that has some nonnegativity constraints but no functional constraints is one special case ($m = 0$) of the next class of problems.

Linearly Constrained Optimization

Linearly constrained optimization problems are characterized by constraints that completely fit linear programming, so that *all* the $g_i(\mathbf{x})$ constraint functions are linear, but the objective function $f(\mathbf{x})$ is nonlinear. The problem is considerably simplified by having just one nonlinear function to take into account, along with a linear programming feasible region. A number of special algorithms based upon *extending* the simplex method to consider the nonlinear objective function have been developed.

One important special case, which we consider next, is quadratic programming.

Quadratic Programming

Quadratic programming problems again have linear constraints, but now the objective function $f(\mathbf{x})$ must be *quadratic*. Thus, the only difference between them and a linear programming problem is that some of the terms in the objective function involve the *square* of a variable or the *product* of two variables.

Many algorithms have been developed for this case under the additional assumption that $f(\mathbf{x})$ is concave. Section 13.7 presents an algorithm that involves a direct extension of the simplex method.

Quadratic programming is very important, partially because such formulations arise naturally in many applications. For example, the problem of portfolio selection

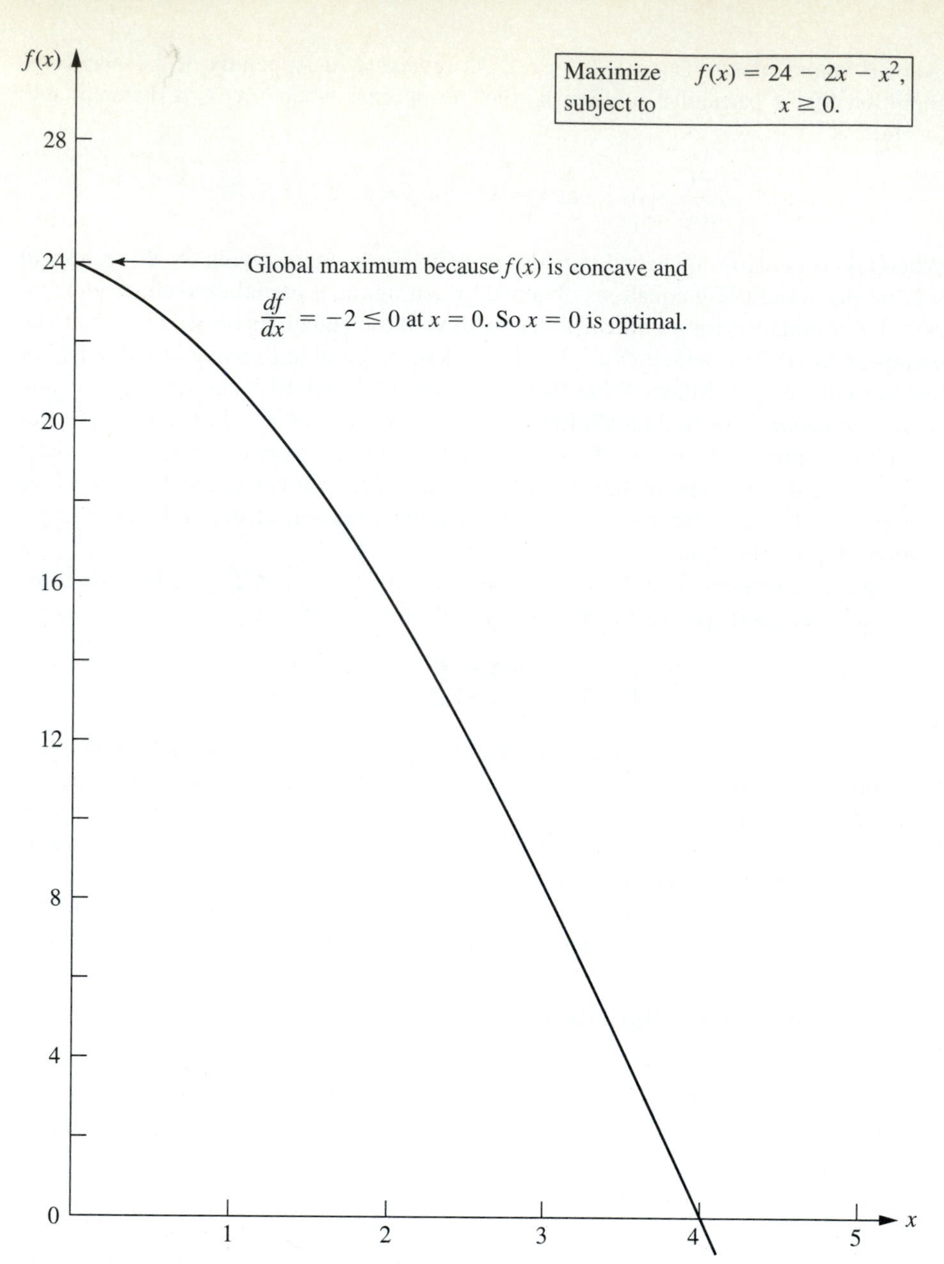

Figure 13.11 An example that illustrates how an optimal solution can lie at a point where a derivative is negative instead of zero, because that point lies at the boundary of a nonnegativity constraint.

with risky securities described in Sec. 13.1 fits into this format. However, another major reason for its importance is that a common approach to solving general linearly constrained optimization problems is to solve a sequence of quadratic programming approximations.

Convex Programming

Convex programming covers a broad class of problems that actually encompasses as special cases all the preceding types when $f(\mathbf{x})$ is concave. The assumptions are that

1. $f(\mathbf{x})$ is concave.
2. Each $g_i(\mathbf{x})$ is convex.

As discussed at the end of Sec. 13.2, these assumptions are enough to ensure that a local maximum is a global maximum. You will see in Sec. 13.6 that the necessary and sufficient conditions for such an optimal solution are a natural generalization of the conditions just given for *unconstrained optimization* and its extension to include *non-negativity constraints.* Section 13.9 then describes algorithmic approaches to solving convex programming problems.

Separable Programming

Separable programming is a special case of convex programming, where the one additional assumption is that

3. All the $f(\mathbf{x})$ and $g_i(\mathbf{x})$ functions are separable functions.

A **separable function** is a function where *each term* involves just a *single variable,* so that the function is separable into a sum of functions of individual variables. For example, if $f(\mathbf{x})$ is a separable function, it can be expressed as

$$f(\mathbf{x}) = \sum_{j=1}^{n} f_j(x_j),$$

where each $f_j(x_j)$ function includes only the terms involving just x_j. In the terminology of linear programming (see Sec. 3.3), separable programming problems satisfy the assumption of additivity but not the assumption of proportionality (for nonlinear functions).

It is important to distinguish these problems from other convex programming problems, because any separable programming problem can be closely approximated by a linear programming problem so that the extremely efficient simplex method can be used. This approach is described in Sec. 13.8. (For simplicity, we focus there on the *linearly constrained* case where the special approach is needed only on the objective function.)

Nonconvex Programming

Nonconvex programming encompasses all nonlinear programming problems that do not satisfy the assumptions of convex programming. Now, even if you are successful in finding a *local maximum,* there is no assurance that it also will be a *global maximum.* Therefore, there is no algorithm that will guarantee finding an optimal solution for all such problems. However, there do exist some algorithms that are relatively well suited for finding local maxima, especially when the forms of the nonlinear functions do not deviate too strongly from those assumed for convex programming. One such algorithm is presented in Sec. 13.10.

However, certain specific types of nonconvex programming problems can be solved without great difficulty by special methods. Two especially important such types are discussed briefly next.

Geometric Programming

When we apply nonlinear programming to engineering design problems, the objective function and the constraint functions frequently take the form

$$g(\mathbf{x}) = \sum_{i=1}^{N} c_i P_i(\mathbf{x}),$$

where

$$P_i(\mathbf{x}) = x_1^{a_{i1}} x_2^{a_{i2}} \cdots x_n^{a_{in}}, \qquad \text{for } i = 1, 2, \ldots, N.$$

In such cases, the c_i and a_{ij} typically represent physical constants, and the x_j are design variables. These functions generally are neither convex nor concave, so the techniques of convex programming cannot be applied directly to these *geometric programming* problems. However, there is one important case where the problem can be transformed to an equivalent convex programming problem. This case is where *all* the c_i coefficients in each function are strictly positive, so that the functions are *generalized positive polynomials*—(now called **posynomials**)—and the objective function is to be minimized. The equivalent convex programming problem with decision variables $y_1, y_2, \ldots, y_n$ is then obtained by setting

$$x_j = e^{y_j}, \qquad \text{for } j = 1, 2, \ldots, n$$

throughout the original model, so now a convex programming algorithm can be applied. Alternative solution procedures also have been developed for solving these *posynomial programming* problems, as well as for geometric programming problems of other types.[1]

Fractional Programming

Suppose that the objective function is in the form of a *fraction,* i.e., the ratio of two functions,

$$\text{Maximize} \qquad f(\mathbf{x}) = \frac{f_1(\mathbf{x})}{f_2(\mathbf{x})}.$$

Such *fractional programming* problems arise, e.g., when one is maximizing the ratio of output to person-hours expended (productivity), or profit to capital expended (rate of return), or expected value to standard deviation of some measure of performance for an investment portfolio (return/risk). Some special solution procedures[2] have been developed for certain forms of $f_1(\mathbf{x})$ and $f_2(\mathbf{x})$.

When it can be done, the most straightforward approach to solving a fractional programming problem is to transform it to an equivalent problem of a standard type for which effective solution procedures already are available. To illustrate, suppose that $f(\mathbf{x})$ is of the *linear fractional programming* form

$$f(\mathbf{x}) = \frac{\mathbf{cx} + c_0}{\mathbf{dx} + d_0},$$

where $\mathbf{c}$ and $\mathbf{d}$ are row vectors, $\mathbf{x}$ is a column vector, and c_0 and d_0 are scalars. Also assume that the constraint functions $g_i(\mathbf{x})$ are linear, so that the constraints in matrix form are $\mathbf{Ax} \leq \mathbf{b}$ and $\mathbf{x} \geq \mathbf{0}$.

[1] R. J. Duffin, E. L. Peterson, and C. M. Zehner, *Geometric Programming,* Wiley, New York, 1967; C. Beightler and D. T. Phillips, *Applied Geometric Programming,* Wiley, New York, 1976.

[2] The pioneering work on fractional programming was done by A. Charnes and W. W. Cooper, "Programming with Linear Fractional Functionals," *Naval Research Logistics Quarterly,* **9**: 181–186, 1962. Also see S. Schaible, "A Survey of Fractional Programming," in S. Schaible and W. T. Ziemba (eds.), *Generalized Concavity in Optimization and Economics,* Academic Press, New York, 1981, pp. 417–440.

Under mild additional assumptions, we can transform the problem to an equivalent *linear programming* problem by letting

$$\mathbf{y} = \frac{\mathbf{x}}{\mathbf{dx} + d_0} \qquad \text{and} \qquad t = \frac{1}{\mathbf{dx} + d_0},$$

so that $\mathbf{x} = \mathbf{y}/t$. This result yields

$$\text{Maximize} \qquad Z = \mathbf{cy} + c_0 t,$$

subject to

$$\mathbf{Ay} - \mathbf{b}t \leq \mathbf{0},$$
$$\mathbf{dy} + d_0 t = 1,$$

and

$$\mathbf{y} \geq \mathbf{0}, \qquad t \geq 0,$$

which can be solved by the simplex method. More generally, the same kind of transformation can be used to convert a fractional programming problem with concave $f_1(\mathbf{x})$, convex $f_2(\mathbf{x})$, and convex $g_i(\mathbf{x})$ to an equivalent convex programming problem.

The Complementarity Problem

When we deal with quadratic programming in Sec. 13.7, you will see one example of how solving certain nonlinear programming problems can be reduced to solving the complementarity problem. Given variables $w_1, w_2, \ldots, w_p$ and $z_1, z_2, \ldots, z_p$, the **complementarity problem** is to find a *feasible* solution for the set of constraints

$$\mathbf{w} = F(\mathbf{z}), \qquad \mathbf{w} \geq \mathbf{0}, \qquad \mathbf{z} \geq \mathbf{0}$$

that also satisfies the **complementarity constraint**

$$\mathbf{w}^T\mathbf{z} = 0.$$

Here, **w** and **z** are column vectors, F is a given vector-valued function, and the superscript T denotes the transpose (see Appendix 4). The problem has no objective function, so technically it is not a full-fledged nonlinear programming problem. It is called the complementarity problem because of the complementary relationships that either

$$w_i = 0 \qquad \text{or} \qquad z_i = 0 \qquad \text{(or both)} \qquad \text{for each } i = 1, 2, \ldots, p.$$

An important special case is the **linear complementarity problem,** where

$$F(\mathbf{z}) = \mathbf{q} + \mathbf{Mz},$$

where **q** is a given column vector and **M** is a given $p \times p$ matrix. Efficient algorithms have been developed for solving this problem under suitable assumptions[1] about the properties of the matrix **M**. One type involves pivoting from one basic feasible (BF) solution to the next, much like the simplex method for linear programming.

In addition to having applications in nonlinear programming, complementarity problems have applications in game theory, economic equilibrium problems, and engineering equilibrium problems.

[1] See R. W. Cottle and G. B. Dantzig, "Complementary Pivot Theory of Mathematical Programming," *Linear Algebra and Its Applications,* **1**: 103–125, 1966; and R. W. Cottle, J.-S. Pang, and R. E. Stone, *The Linear Complementarity Problem,* Academic Press, Boston, 1992.

13.4 One-Variable Unconstrained Optimization

We now begin discussing how to solve some of the types of problems just described by considering the simplest case—*unconstrained optimization* with just a single variable x ($n = 1$), where the differentiable function $f(x)$ to be maximized is *concave.*[1] Thus the *necessary and sufficient condition* for a particular solution $x = x^*$ to be optimal (a global maximum) is

$$\frac{df}{dx} = 0 \qquad \text{at } x = x^*,$$

as depicted in Fig. 13.12. If this equation can be solved directly for x^*, you are done. However, if $f(x)$ is not a particularly simple function, so the derivative is not just a linear or quadratic function, you may not be able to solve the equation *analytically.* If not, the *one-dimensional search procedure* provides a straightforward way of solving the problem *numerically.*

The One-Dimensional Search Procedure

Like other search procedures in nonlinear programming, the *one-dimensional* search procedure finds a sequence of *trial solutions* that leads toward an optimal solution. At each iteration, you begin at the current trial solution to conduct a systematic search that culminates by identifying a new *improved* trial solution.

The idea behind the one-dimensional search procedure is a very intuitive one, namely, that whether the slope (derivative) is positive or negative at a trial solution definitely indicates whether improvement lies immediately to the right or left, respectively. Thus, if the derivative evaluated at a particular value of x is *positive,* then x^* must be larger than this x (see Fig. 13.12), so this x becomes a *lower bound* on the trial solutions that need to be considered thereafter. Conversely, if the derivative is *negative,*

[1] See the beginning of Appendix 3 for a review of the corresponding case when $f(x)$ is not concave.

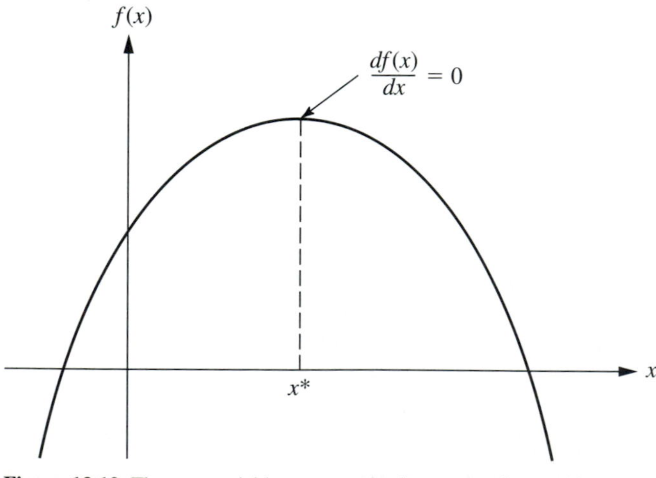

Figure 13.12 The one-variable unconstrained programming problem when the function is concave.

then x^* must be *smaller* than this x, so x would become an *upper bound.* Therefore, after both types of bounds have been identified, each new trial solution selected between the current bounds provides a new tighter bound of one type, thereby narrowing the search further. As long as a reasonable rule is used to select each trial solution in this way, the resulting *sequence* of trial solutions must *converge* to x^*. In practice, this means continuing the sequence until the distance between the bounds is sufficiently small that the next trial solution must be within a prespecified *error tolerance* of x^*.

This entire process is summarized next, given the notation

$$x' = \text{current trial solution,}$$

$$\underline{x} = \text{current lower bound on } x^*,$$

$$\overline{x} = \text{current upper bound on } x^*,$$

$$\epsilon = \text{error tolerance for } x^*.$$

Although there are several reasonable rules for selecting each new trial solution, the one used in the following procedure is the **midpoint rule** (traditionally called the *Bolzano search plan*), which says simply to select the midpoint between the two current bounds.

Summary of the One-Dimensional Search Procedure

Initialization: Select ϵ. Find an initial $\underline{x}$ and $\overline{x}$ by inspection (or by respectively finding any value of x at which the derivative is positive and then negative). Select an initial trial solution

$$x' = \frac{\underline{x} + \overline{x}}{2}.$$

Iteration: **1.** Evaluate $\dfrac{df(x)}{dx}$ at $x = x'$.

2. If $\dfrac{df(x)}{dx} \geq 0$, reset $\underline{x} = x'$.

3. If $\dfrac{df(x)}{dx} \leq 0$, reset $\overline{x} = x'$.

4. Select a new $x' = \dfrac{\underline{x} + \overline{x}}{2}$.

Stopping rule: If $\overline{x} - \underline{x} \leq 2\epsilon$, so that the new x' must be within ϵ of x^*, stop. Otherwise, perform another iteration.

We shall now illustrate this procedure by applying it to the following example.

Example: Suppose that the function to be maximized is

$$f(x) = 12x - 3x^4 - 2x^6,$$

as plotted in Fig. 13.13. Its first two derivatives are

$$\frac{df(x)}{dx} = 12(1 - x^3 - x^5),$$

$$\frac{d^2f(x)}{dx^2} = -12(3x^2 + 5x^4).$$

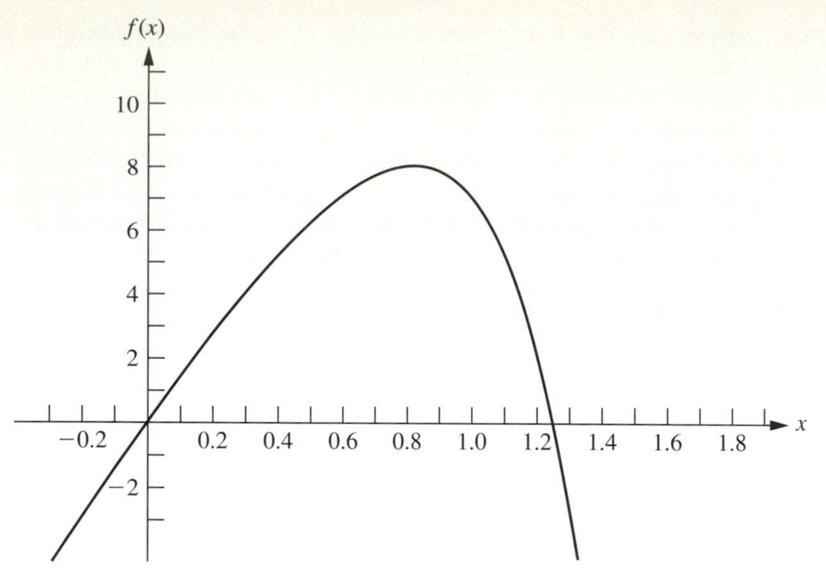

Figure 13.13 Example for the one-dimensional search procedure.

Because the second derivative is nonpositive everywhere, $f(x)$ is a concave function, so the one-dimensional search procedure can be applied safely to find its global maximum (assuming a global maximum exists). A quick inspection of this function (without even constructing its graph as shown in Fig. 13.13) indicates that $f(x)$ is positive for small positive values of x, but it is negative for $x < 0$ or $x > 2$. Therefore, $\underline{x} = 0$ and $\overline{x} = 2$ can be used as the initial bounds, with their midpoint $x' = 1$ as the initial trial solution. (The existence of these bounds ensures that a global maximum exists, since this rules out the only contrary possibilities that the first derivative is either positive everywhere or negative everywhere.) Let 0.01 be the error tolerance for x^* in the stopping rule, so the final $\overline{x} - \underline{x} \leq 0.02$ with the final x' at the midpoint. Applying the one-dimensional search procedure then yields the sequence of results shown in Table 13.1. [This table includes both the function and derivative values for your information, where the derivative is evaluated at the trial solution generated at the *preceding* iteration. However, note that the algorithm actually does not need to calculate $f(x')$ at all and that it only needs to

Table 13.1 **Application of the One-Dimensional Search Procedure to the Example**

Iteration	$\frac{df(x)}{dx}$	$\underline{x}$	$\overline{x}$	New x'	$f(x')$
0		0	2	1	7.0000
1	−12	0	1	0.5	5.7812
2	+10.12	0.5	1	0.75	7.6948
3	+4.09	0.75	1	0.875	7.8439
4	−2.19	0.75	0.875	0.8125	7.8672
5	+1.31	0.8125	0.875	0.84375	7.8829
6	−0.34	0.8125	0.84375	0.828125	7.8815
7	+0.51	0.828125	0.84375	0.8359375	7.8839
Stop					

calculate the derivative far enough to determine its sign.] The conclusion is that

$$x^* \approx 0.836,$$

$$0.828125 < x^* < 0.84375.$$

Your OR Courseware includes both an interactive routine and an automatic routine for executing the one-dimensional search procedure.

13.5 Multivariable Unconstrained Optimization

Now consider the problem of maximizing a *concave* function $f(\mathbf{x})$ of *multiple* variables $\mathbf{x} = (x_1, x_2, \ldots, x_n)$ when there are no constraints on the feasible values. Suppose again that the necessary and sufficient condition for optimality, given by the system of equations obtained by setting the respective partial derivatives equal to zero (see Sec. 13.3), cannot be solved analytically, so that a numerical search procedure must be used. How can the preceding one-dimensional search procedure be extended to this multidimensional problem?

In Sec. 13.4, the value of the ordinary derivative was used to select one of just two possible directions (increase x or decrease x) in which to move from the current trial solution to the next one. The goal was to reach a point eventually where this derivative is (essentially) 0. Now, there are *innumerable* possible directions in which to move; they correspond to the possible *proportional rates* at which the respective variables can be changed. The goal is to reach a point eventually where all the partial derivatives are (essentially) 0. Therefore, extending the one-dimensional search procedure requires using the values of the *partial* derivatives to select the specific direction in which to move. This selection involves using the gradient of the objective function, as described next.

Because the objective function $f(\mathbf{x})$ is assumed to be differentiable, it possesses a gradient, denoted by $\nabla f(\mathbf{x})$, at each point $\mathbf{x}$. In particular, the **gradient** at a specific point $\mathbf{x} = \mathbf{x}'$ is the *vector* whose elements are the respective *partial derivatives* evaluated at $\mathbf{x} = \mathbf{x}'$, so that

$$\nabla f(\mathbf{x}') = \left(\frac{\partial f}{\partial x_1}, \frac{\partial f}{\partial x_2}, \ldots, \frac{\partial f}{\partial x_n}\right) \qquad \text{at } \mathbf{x} = \mathbf{x}'.$$

The significance of the gradient is that the (infinitesimal) change in $\mathbf{x}$ that *maximizes* the rate at which $f(\mathbf{x})$ increases is the change that is *proportional* to $\nabla f(\mathbf{x})$. To express this idea geometrically, the "direction" of the gradient $\nabla f(\mathbf{x}')$ is interpreted as the *direction* of the directed line segment (arrow) from the origin $(0, 0, \ldots, 0)$ to the point $(\partial f/\partial x_1, \partial f/\partial x_2, \ldots, \partial f/\partial x_n)$, where $\partial f/\partial x_j$ is evaluated at $x_j = x_j'$. Therefore, it may be said that the rate at which $f(\mathbf{x})$ increases is maximized if (infinitesimal) changes in $\mathbf{x}$ are in the *direction* of the gradient $\nabla f(\mathbf{x})$. Because the objective is to find the feasible solution maximizing $f(\mathbf{x})$, it would seem expedient to attempt to move in the direction of the gradient as much as possible.

The Gradient Search Procedure

Because the current problem has no constraints, this interpretation of the gradient suggests that an efficient search procedure should keep moving in the direction of the gradient until it (essentially) reaches an optimal solution $\mathbf{x}^*$, where $\nabla f(\mathbf{x}^*) = \mathbf{0}$. However, normally it would not be practical to change $\mathbf{x}$ *continuously* in the direction of $\nabla f(\mathbf{x})$, because this series of changes would require continuously *reevaluating* the $\partial f/\partial x_j$ and changing the direction of the path. Therefore, a better approach is to keep moving in a *fixed* direction from the current trial solution, not stopping until $f(\mathbf{x})$ stops increasing. This stopping point would be the next trial solution, so the gradient then would be recalculated to determine the new direction in which to move. With this approach, each iteration involves changing the current trial solution $\mathbf{x}'$ as follows:

$$\text{Reset} \qquad \mathbf{x}' = \mathbf{x}' + t^* \, \nabla f(\mathbf{x}'),$$

where t^* is the positive value of t that *maximizes* $f(\mathbf{x}' + t\, \nabla f(\mathbf{x}'))$; that is,

$$f(\mathbf{x}' + t^* \, \nabla f(\mathbf{x}')) = \max_{t \geq 0} f(\mathbf{x}' + t\, \nabla f(\mathbf{x}')).$$

[Note that $f(\mathbf{x}' + t\, \nabla f(\mathbf{x}'))$ is simply $f(\mathbf{x})$ where

$$x_j = x'_j + t\left(\frac{\partial f}{\partial x_j}\right)_{\mathbf{x}=\mathbf{x}'}, \qquad \text{for } j = 1, 2, \ldots, n,$$

and that these expressions for the x_j involve only constants and t, so $f(\mathbf{x})$ becomes a function of just the single variable t.] The iterations of this gradient search procedure continue until $\nabla f(\mathbf{x}) = 0$ within a small tolerance ϵ, that is, until

$$\left|\frac{\partial f}{\partial x_j}\right| \leq \epsilon \qquad \text{for } j = 1, 2, \ldots, n.^{\dagger}$$

An analogy may help to clarify this procedure. Suppose that you need to climb to the top of a hill. You are nearsighted, so you cannot see the top of the hill in order to walk directly in that direction. However, when you stand still, you can see the ground around your feet well enough to determine the direction in which the hill is sloping upward most sharply. You are able to walk in a straight line. While walking, you also are able to tell when you stop climbing (zero slope in your direction). Assuming that the hill is *concave,* you now can use the *gradient search procedure* for climbing to the top efficiently. This problem is a *two-variable* problem, where (x_1, x_2) represents the coordinates (ignoring height) of your current location. The function $f(x_1, x_2)$ gives the height of the hill at (x_1, x_2). You start each iteration at your current location (current trial solution) by determining the direction [in the (x_1, x_2) coordinate system] in which the hill is sloping upward most sharply (the direction of the gradient) at this point. You then begin walking in this fixed direction and continue as long as you still are climbing. You eventually stop at a new trial location (solution) when the hill becomes level in your direction, at which point you prepare to do another iteration in another direction. You continue these iterations, following a zigzag path up the hill, until you reach a trial location where the slope is essentially zero in all directions. Under the assumption that the hill $[f(x_1, x_2)]$ is concave, you must then be essentially at the top of the hill.

The most difficult part of the gradient search procedure usually is to find t^*, the value of t that maximizes f in the direction of the gradient, at each iteration. Because $\mathbf{x}$ and $\nabla f(\mathbf{x})$ have fixed values for the maximization, and because $f(\mathbf{x})$ is concave, this

† This stopping rule generally will provide a solution $\mathbf{x}$ that is close to an optimal solution $\mathbf{x}^*$, with a value of $f(\mathbf{x})$ that is very close to $f(\mathbf{x}^*)$. However, this cannot be guaranteed, since it is possible that the function maintains a very small positive slope ($\leq \epsilon$) over a great distance from $\mathbf{x}$ to $\mathbf{x}^*$.

problem should be viewed as maximizing a *concave* function of a *single variable t*. Therefore, it can be solved by the one-dimensional search procedure of Sec. 13.4 (where the initial lower bound on t must be nonnegative because of the $t \geq 0$ constraint). Alternatively, if f is a simple function, it may be possible to obtain an analytical solution by setting the derivative with respect to t equal to zero and solving.

Summary of the Gradient Search Procedure

Initialization: Select ϵ and any initial trial solution x'. Go first to the stopping rule.

Iteration: **1.** Express $f(\mathbf{x}' + t\,\nabla f(\mathbf{x}'))$ as a function of t by setting

$$x_j = x_j' + t\left(\frac{\partial f}{\partial x_j}\right)_{\mathbf{x}=\mathbf{x}'}, \qquad \text{for } j = 1, 2, \ldots, n,$$

and then substituting these expressions into $f(\mathbf{x})$.

2. Use the one-dimensional search procedure (or calculus) to find $t = t^*$ that maximizes $f(\mathbf{x}' + t\,\nabla f(\mathbf{x}'))$ over $t \geq 0$.

3. Reset $\mathbf{x}' = \mathbf{x}' + t^*\,\nabla f(\mathbf{x}')$. Then go to the stopping rule.

Stopping rule: Evaluate $\nabla f(\mathbf{x}')$ at $\mathbf{x} = \mathbf{x}'$. Check if

$$\left|\frac{\partial f}{\partial x_j}\right| \leq \epsilon \qquad \text{for all } j = 1, 2, \ldots, n.$$

If so, stop with the current $\mathbf{x}'$ as the desired approximation of an optimal solution $\mathbf{x}^*$. Otherwise, perform another iteration.

Now let us illustrate this procedure.

Example: Consider the following two-variable problem:

$$\text{Maximize} \qquad f(\mathbf{x}) = 2x_1x_2 + 2x_2 - x_1^2 - 2x_2^2.$$

Thus

$$\frac{\partial f}{\partial x_1} = 2x_2 - 2x_1,$$

$$\frac{\partial f}{\partial x_2} = 2x_1 + 2 - 4x_2.$$

We also can verify (see Appendix 2) that $f(\mathbf{x})$ is concave. To begin the gradient search procedure, suppose that $\mathbf{x} = (0, 0)$ is selected as the initial trial solution. Because the respective partial derivatives are 0 and 2 at this point, the gradient is

$$\nabla f(0, 0) = (0, 2).$$

Therefore, to begin the first iteration, set

$$x_1 = 0 + t(0) = 0,$$

$$x_2 = 0 + t(2) = 2t,$$

and then substitute these expressions into $f(\mathbf{x})$ to obtain

$$\begin{aligned} f(\mathbf{x}' + t\,\nabla f(\mathbf{x}')) &= f(0, 2t) \\ &= 2(0)(2t) + 2(2t) - 0^2 - 2(2t)^2 \\ &= 4t - 8t^2. \end{aligned}$$

Because

$$f(0, 2t^*) = \max_{t \geq 0} f(0, 2t) = \max_{t \geq 0} \{4t - 8t^2\}$$

and

$$\frac{d}{dt}(4t - 8t^2) = 4 - 16t = 0,$$

it follows that

$$t^* = \tfrac{1}{4},$$

so

$$\text{Reset} \quad \mathbf{x}' = (0, 0) + \tfrac{1}{4}(0, 2) = (0, \tfrac{1}{2}).$$

For this new trial solution, the gradient is

$$\nabla f(0, \tfrac{1}{2}) = (1, 0).$$

Thus for the second iteration, set

$$\mathbf{x} = (0, \tfrac{1}{2}) + t(1, 0) = (t, \tfrac{1}{2}),$$

so

$$\begin{aligned} f(\mathbf{x}' + t\,\nabla f(\mathbf{x}')) &= f(0 + t, \tfrac{1}{2} + 0t) = f(t, \tfrac{1}{2}) \\ &= (2t)(\tfrac{1}{2}) + 2(\tfrac{1}{2}) - t^2 - 2(\tfrac{1}{2})^2 \\ &= t - t^2 + \tfrac{1}{2}. \end{aligned}$$

Because

$$f(t^*, \tfrac{1}{2}) = \max_{t \geq 0} f(t, \tfrac{1}{2}) = \max_{t \geq 0} \{t - t^2 + \tfrac{1}{2}\}$$

and

$$\frac{d}{dt}(t - t^2 + \tfrac{1}{2}) = 1 - 2t = 0,$$

then

$$t^* = \tfrac{1}{2},$$

so

$$\text{Reset} \quad \mathbf{x}' = (0, \tfrac{1}{2}) + \tfrac{1}{2}(1, 0) = (\tfrac{1}{2}, \tfrac{1}{2}).$$

A nice way of organizing this work is to write out a table such as Table 13.2 which summarizes the preceding two iterations. At each iteration, the second column shows the current trial solution, and the rightmost column shows the eventual new trial solution, which then is carried down into the second column for the next iteration. The fourth column gives the expressions for the x_j in terms of t that need to be substituted into $f(\mathbf{x})$ to give the fifth column.

By continuing in this fashion, the subsequent trial solutions would be $(\tfrac{1}{2}, \tfrac{3}{4})$, $(\tfrac{3}{4}, \tfrac{3}{4})$, $(\tfrac{3}{4}, \tfrac{7}{8})$, $(\tfrac{7}{8}, \tfrac{7}{8})$, . . . , as shown in Fig. 13.14. Because these points are converging to $\mathbf{x}^* = (1, 1)$, this solution is the optimal solution, as verified by the fact that

$$\nabla f(1, 1) = (0, 0).$$

Table 13.2 **Application of the Gradient Search Procedure to the Example**

Iteration	$\mathbf{x}'$	$\nabla f(\mathbf{x}')$	$\mathbf{x}' + t\,\nabla f(\mathbf{x}')$	$f(\mathbf{x}' + t\,\nabla f(\mathbf{x}'))$	t^*	$\mathbf{x}' + t^*\,\nabla f(\mathbf{x}')$
1	$(0, 0)$	$(0, 2)$	$(0, 2t)$	$4t - 8t^2$	$\frac{1}{4}$	$(0, \frac{1}{2})$
2	$(0, \frac{1}{2})$	$(1, 0)$	$(t, \frac{1}{2})$	$t - t^2 + \frac{1}{2}$	$\frac{1}{2}$	$(\frac{1}{2}, \frac{1}{2})$

However, because this converging sequence of trial solutions never reaches its limit, the procedure actually will stop somewhere (depending on ϵ) slightly below (1, 1) as its final approximation of $\mathbf{x}^*$.

As Fig. 13.14 suggests, the gradient search procedure zigzags to the optimal solution rather than moving in a straight line. Some modifications of the procedure have been developed that *accelerate* movement toward the optimum by taking this zigzag behavior into account.

If $f(\mathbf{x})$ were *not* a concave function, the gradient search procedure still would converge to a *local* maximum. The only change in the description of the procedure for this case is that t^* now would correspond to the *first local maximum* of $f(\mathbf{x}' + t\,\nabla f(\mathbf{x}'))$ as t is increased from 0.

If the objective were to *minimize* $f(\mathbf{x})$ instead, one change in the procedure would be to move in the *opposite* direction of the gradient at each iteration. In other words, the rule for obtaining the next point would be

$$\text{Reset} \qquad \mathbf{x}' = \mathbf{x}' - t^*\,\nabla f(\mathbf{x}').$$

The only other change is that t^* now would be the nonnegative value of t that *minimizes* $f(\mathbf{x}' - t\,\nabla f(\mathbf{x}'))$; that is,

$$f(\mathbf{x}' - t^*\,\nabla f(\mathbf{x}')) = \min_{t \geq 0} f(\mathbf{x}' - t\,\nabla f(\mathbf{x}')).$$

Another example of an application of the gradient search procedure is included under the Demo menu in your OR Courseware. The Procedure menu includes both an interactive routine and an automatic routine for applying this algorithm.

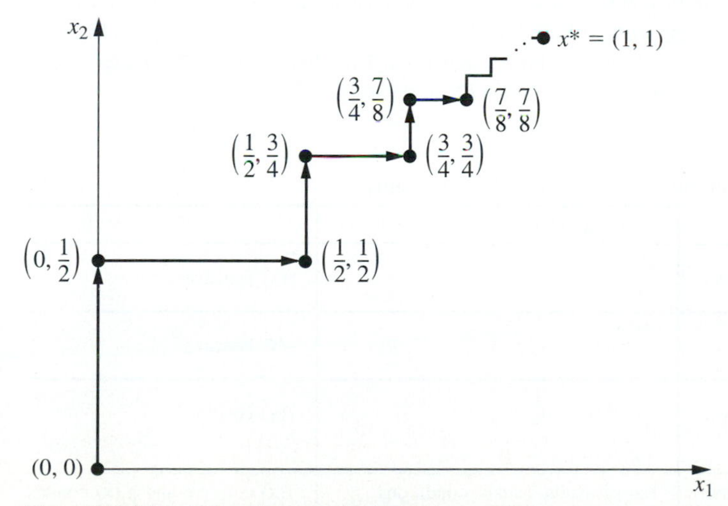

Figure 13.14 Illustration of the gradient search procedure.

13.6 The Karush-Kuhn-Tucker (KKT) Conditions for Constrained Optimization

We now focus on the question of how to recognize an *optimal solution* for a nonlinear programming problem (with differentiable functions). What are the necessary and (perhaps) sufficient conditions that such a solution must satisfy?

In the preceding sections we already noted these conditions for *unconstrained optimization,* as summarized in the first two rows of Table 13.3. Early in Sec. 13.3 we also gave these conditions for the slight *extension* of unconstrained optimization where the *only* constraints are nonnegativity constraints. These conditions are shown in the third row of Table 13.3. As indicated in the last row of the table, the conditions for the general case are called the **Karush-Kuhn-Tucker conditions** (or **KKT conditions**), because they were derived independently by Karush[1] and by Kuhn and Tucker.[2] Their basic result is embodied in the following theorem.

THEOREM: Assume that $f(\mathbf{x})$, $g_1(\mathbf{x})$, $g_2(\mathbf{x})$, . . . , $g_m(\mathbf{x})$ are *differentiable* functions satisfying certain regularity conditions.[3] Then

$$\mathbf{x}^* = (x_1^*, x_2^*, \ldots, x_n^*)$$

can be an *optimal solution* for the nonlinear programming problem only if there exist m numbers $u_1, u_2, \ldots, u_m$ such that *all* the following *KKT conditions* are satisfied:

$$\left.\begin{aligned} &\mathbf{1.}\ \frac{\partial f}{\partial x_j} - \sum_{i=1}^{m} u_i \frac{\partial g_i}{\partial x_j} \le 0 \\ &\mathbf{2.}\ x_j^* \left(\frac{\partial f}{\partial x_j} - \sum_{i=1}^{m} u_i \frac{\partial g_i}{\partial x_j} \right) = 0 \end{aligned}\right\} \quad \text{at } \mathbf{x} = \mathbf{x}^*, \text{ for } j = 1, 2, \ldots, n.$$

$$\left.\begin{aligned} &\mathbf{3.}\ g_i(\mathbf{x}^*) - b_i \le 0 \\ &\mathbf{4.}\ u_i[g_i(\mathbf{x}^*) - b_i] = 0 \end{aligned}\right\} \quad \text{for } i = 1, 2, \ldots, m.$$

$$\mathbf{5.}\ x_j^* \ge 0, \qquad \text{for } j = 1, 2, \ldots, n.$$

$$\mathbf{6.}\ u_i \ge 0, \qquad \text{for } i = 1, 2, \ldots, m.$$

[1] W. Karush, "Minima of Functions of Several Variables with Inequalities as Side Conditions," M.S. thesis, Department of Mathematics, University of Chicago, 1939.

[2] H. W. Kuhn and A. W. Tucker, "Nonlinear Programming," in Jerzy Neyman (ed.), *Proceedings of the Second Berkeley Symposium,* University of California Press, Berkeley, 1951, pp. 481–492.

[3] Ibid., p. 483.

Table 13.3 **Necessary and Sufficient Conditions for Optimality**

Problem	Necessary Conditions for Optimality	Also Sufficient if:
One-variable unconstrained	$\frac{df}{dx} = 0$	$f(x)$ concave
Multivariable unconstrained	$\frac{\partial f}{\partial x_j} = 0 \quad (j = 1, 2, \ldots, n)$	$f(\mathbf{x})$ concave
Constrained, nonnegativity constraints only	$\frac{\partial f}{\partial x_j} = 0 \quad (j = 1, 2, \ldots, n)$ (or ≤ 0 if $x_j = 0$)	$f(\mathbf{x})$ concave
General constrained problem	Karush-Kuhn-Tucker conditions	$f(\mathbf{x})$ concave and $g_i(\mathbf{x})$ convex $(i = 1, 2, \ldots, m)$

Note that both conditions 2 and 4 require that the product of two quantities be zero. Therefore, each of these conditions really is saying that at least one of the two quantities must be zero. Consequently, condition 4 can be combined with condition 3 to express them in another equivalent form as

$$(3, 4) \qquad g_i(\mathbf{x}^*) - b_i = 0$$
$$(\text{or} \le 0 \quad \text{if } u_i = 0), \qquad \text{for } i = 1, 2, \ldots, m.$$

Similarly, condition 2 can be combined with condition 1 as

$$(1, 2) \qquad \frac{\partial f}{\partial x_j} - \sum_{i=1}^{m} u_i \frac{\partial g_i}{\partial x_j} = 0$$
$$(\text{or} \le 0 \quad \text{if } x_j^* = 0), \qquad \text{for } j = 1, 2, \ldots, n.$$

When $m = 0$ (no functional constraints), this summation drops out and the combined condition (1, 2) reduces to the condition given in the third row of Table 13.3. Thus, for $m > 0$, each term in the summation modifies the $m = 0$ condition to incorporate the effect of the corresponding functional constraint.

In conditions 1, 2, 4, and 6, the u_i correspond to the *dual variables* of linear programming (we expand on this correspondence at the end of the section), and they have a comparable economic interpretation. However, the u_i actually arose in the mathematical derivation as *Lagrange multipliers* (discussed in Appendix 3). Conditions 3 and 5 do nothing more than ensure the feasibility of the solution. The other conditions eliminate most of the feasible solutions as possible candidates for an optimal solution.

However, note that satisfying these conditions does not guarantee that the solution is optimal. As summarized in the rightmost column of Table 13.3, certain additional *convexity* assumptions are needed to obtain this guarantee. These assumptions are spelled out in the following extension of the theorem.

Corollary: Assume that $f(\mathbf{x})$ is a *concave* function and that $g_1(\mathbf{x}), g_2(\mathbf{x}), \ldots, g_m(\mathbf{x})$ are *convex* functions (i.e., this problem is a convex programming problem), where all these functions satisfy the regularity conditions. Then $\mathbf{x}^* = (x_1^*, x_2^*, \ldots, x_n^*)$ is an *optimal solution* if and only if all the conditions of the theorem are satisfied.

Example: To illustrate the formulation and application of the *KKT conditions,* we consider the following two-variable nonlinear programming problem:

$$\text{Maximize} \qquad f(\mathbf{x}) = \ln(x_1 + 1) + x_2,$$

subject to

$$2x_1 + x_2 \le 3$$

and

$$x_1 \ge 0, \qquad x_2 \ge 0,$$

where ln denotes the natural logarithm. Thus $m = 1$ (one functional constraint) and $g_1(\mathbf{x}) = 2x_1 + x_2$, so $g_1(\mathbf{x})$ is convex. Furthermore, it can be easily verified (see Appendix 2) that $f(\mathbf{x})$ is concave. Hence the corollary applies, so any solution that satisfies the KKT conditions will definitely be an optimal solution. Applying the formulas given in the theorem yields the following KKT conditions for this example:

1(j = 1). $\frac{1}{x_1 + 1} - 2u_1 \leq 0.$

2(j = 1). $x_1\left(\frac{1}{x_1 + 1} - 2u_1\right) = 0.$

1(j = 2). $1 - u_1 \leq 0.$

2(j = 2). $x_2(1 - u_1) = 0.$

3. $2x_1 + x_2 - 3 \leq 0.$

4. $u_1(2x_1 + x_2 - 3) = 0.$

5. $x_1 \geq 0,\ x_2 \geq 0.$

6. $u_1 \geq 0.$

The steps in solving the KKT conditions for this particular example are outlined below.

1. $u_1 \geq 1$, from condition $1(j = 2)$.
   $x_1 \geq 0$, from condition 5.
2. Therefore, $\frac{1}{x_1 + 1} - 2u_1 < 0.$
3. Therefore, $x_1 = 0$, from condition $2(j = 1)$.
4. $u_1 \neq 0$ implies that $2x_1 + x_2 - 3 = 0$, from condition 4.
5. Steps 3 and 4 imply that $x_2 = 3$.
6. $x_2 \neq 0$ implies that $u_1 = 1$, from condition $2(j = 2)$.
7. No conditions are violated by $x_1 = 0$, $x_2 = 3$, $u_1 = 1$.

Therefore, there exists a number $u_1 = 1$ such that $x_1 = 0$, $x_2 = 3$, and $u_1 = 1$ satisfy all the conditions. Consequently, $\mathbf{x}^* = (0, 3)$ is an optimal solution for this problem.

The particular progression of steps needed to solve the KKT conditions will differ from one problem to the next. When the logic is not apparent, it is sometimes helpful to consider separately the different cases where each x_j and u_i are specified to be either equal to or greater than 0 and then trying each case until one leads to a solution. In the example, there are eight such cases corresponding to the eight combinations of $x_1 = 0$ versus $x_1 > 0$, $x_2 = 0$ versus $x_2 > 0$, and $u_1 = 0$ versus $u_1 > 0$. Each case leads to a simpler statement and analysis of the conditions. To illustrate, consider first the case shown next, where $x_1 = 0$, $x_2 = 0$, and $u_1 = 0$.

KKT Conditions for the Case $x_1 = 0, x_2 = 0, u_1 = 0$

1(j = 1). $\frac{1}{0 + 1} \leq 0.$ Contradiction.

1(j = 2). $1 - 0 \leq 0.$ Contradiction.

3. $0 + 0 \leq 3.$

(All the other conditions are redundant.)

As listed below, the other three cases where $u_1 = 0$ also give immediate contradictions in a similar way, so no solution is available.

Case $x_1 = 0$, $x_2 > 0$, $u_1 = 0$ contradicts conditions $1(j = 1)$, $1(j = 2)$, and $2(j = 2)$.

Case $x_1 > 0$, $x_2 = 0$, $u_1 = 0$ contradicts conditions $1(j = 1)$, $2(j = 1)$, and $1(j = 2)$.

Case $x_1 > 0$, $x_2 > 0$, $u_1 = 0$ contradicts conditions $1(j = 1)$, $2(j = 1)$, $1(j = 2)$, and $2(j = 2)$.

The case $x_1 > 0$, $x_2 > 0$, $u_1 > 0$ enables one to delete these nonzero multipliers from conditions $2(j = 1)$, $2(j = 2)$, and 4, which then enables deletion of conditions $1(j = 1)$, $1(j = 2)$, and 3 as redundant, as summarized next.

KKT Conditions for the Case $x_1 > 0$, $x_2 > 0$, $u_1 > 0$

1(j = 1). $\dfrac{1}{x_1 + 1} - 2u_1 = 0.$

2(j = 2). $1 - u_1 = 0.$

4. $0 + x_2 - 3 = 0.$

(All the other conditions are redundant.)

Therefore, $u_1 = 1$, so $x_1 = -\frac{1}{2}$, which contradicts $x_1 > 0$.

Now suppose that the case $x_1 = 0$, $x_2 > 0$, $u_1 > 0$ is tried next.

KKT Conditions for the Case $x_1 = 0$, $x_2 > 0$, $u_1 > 0$

1(j = 1). $\dfrac{1}{0 + 1} - 2u_1 = 0.$

2(j = 2). $1 - u_1 = 0.$

4. $0 + x_2 - 3 = 0.$

(All the other conditions are redundant.)

Therefore, $x_1 = 0$, $x_2 = 3$, $u_1 = 1$. Having found a solution, we know that no additional cases need be considered.

For problems more complicated than this example, it may be difficult, if not essentially impossible, to derive an optimal solution *directly* from the KKT conditions. Nevertheless, these conditions still provide valuable clues as to the identity of an optimal solution, and they also permit us to check whether a proposed solution may be optimal.

There also are many valuable *indirect* applications of the KKT conditions. One of these applications arises in the *duality theory* that has been developed for nonlinear programming to parallel the duality theory for linear programming presented in Chap. 6. In particular, for any given constrained maximization problem (call it the *primal problem*), the KKT conditions can be used to define a closely associated dual problem that is a constrained minimization problem. The variables in the dual problem[1] consist of both the Lagrange multipliers u_i ($i = 1, 2, \ldots, m$) and the primal variables x_j ($j = 1, 2, \ldots, n$). In the special case where the primal problem is a linear programming problem, the x_j variables drop out of the dual problem and it becomes the familiar dual problem of linear programming (where the u_i variables here correspond to the y_i variables in Chap. 6). When the primal problem is a convex programming problem, it is possible to establish relationships between the primal problem and the dual problem that are similar to those for linear programming. For example, the *strong duality property* of Sec. 6.1, which states that the optimal objective function values of the two problems are equal, also holds here. Furthermore, the values of the u_i variables in an optimal solution for the dual problem can again be interpreted as *shadow prices* (see Secs. 4.7 and 6.2); i.e., they give the rate at which the optimal objective function value for the primal problem could be increased by (slightly) increasing the right-hand side of the corresponding constraint. Because duality theory for nonlinear programming is a relatively advanced topic, the interested reader is referred elsewhere for further information.[2]

You will see another indirect application of the KKT conditions in the next section.

[1] For details on this formulation, see O. T. Mangasarian, *Nonlinear Programming*, McGraw-Hill, New York, 1969, chap 8. For a unified survey of various approaches to duality in nonlinear programming, see A. M. Geoffrion, "Duality in Nonlinear Programming: A Simplified Applications-Oriented Development," *SIAM Review*, **13**: 1–37, 1971.

[2] Ibid.

13.7 Quadratic Programming

As indicated in Sec. 13.3, the quadratic programming problem differs from the linear programming problem only in that the objective function also includes x_j^2 and $x_i x_j$ $(i \neq j)$ terms. Thus, if we use matrix notation like that introduced at the beginning of Sec. 5.2, the problem is to find $\mathbf{x}$ so as to

$$\text{Maximize} \qquad f(\mathbf{x}) = \mathbf{cx} - \tfrac{1}{2}\mathbf{x}^T\mathbf{Qx},$$

subject to

$$\mathbf{Ax} \leq \mathbf{b} \qquad \text{and} \qquad \mathbf{x} \geq \mathbf{0},$$

where $\mathbf{c}$ is a row vector, $\mathbf{x}$ and $\mathbf{b}$ are column vectors, $\mathbf{Q}$ and $\mathbf{A}$ are matrices, and the superscript T denotes the transpose (see Appendix 4). The q_{ij} (elements of Q) are given constants such that $q_{ij} = q_{ji}$ (which is the reason for the factor of $\frac{1}{2}$ in the objective function). By performing the indicated vector and matrix multiplications, the objective function then is expressed in terms of these q_{ij}, the c_j (elements of $\mathbf{c}$), and the variables as follows:

$$f(\mathbf{x}) = \mathbf{cx} - \tfrac{1}{2}\mathbf{x}^T\mathbf{Qx} = \sum_{j=1}^{n} c_j x_j - \tfrac{1}{2}\sum_{i=1}^{n}\sum_{j=1}^{n} q_{ij} x_i x_j.$$

If $i = j$ in this double summation, then $x_i x_j = x_j^2$, so $-\frac{1}{2}q_{jj}$ is the coefficient of x_j^2. If $i \neq j$, then $-\frac{1}{2}(q_{ij}x_i x_j + q_{ji}x_j x_i) = -q_{ij}x_i x_j$, so $-q_{ij}$ is the total coefficient for the product of x_i and x_j.

To illustrate this notation, consider the following example of a quadratic programming problem.

$$\text{Maximize} \qquad f(x_1, x_2) = 15x_1 + 30x_2 + 4x_1x_2 - 2x_1^2 - 4x_2^2,$$

subject to

$$x_1 + 2x_2 \leq 30$$

and

$$x_1 \geq 0, \qquad x_2 \geq 0.$$

In this case,

$$\mathbf{c} = [15 \quad 30], \qquad \mathbf{x} = \begin{bmatrix} x_1 \\ x_2 \end{bmatrix}, \qquad \mathbf{Q} = \begin{bmatrix} 4 & -4 \\ -4 & 8 \end{bmatrix},$$

$$\mathbf{A} = [1 \quad 2], \qquad \mathbf{b} = [30].$$

Note that

$$\begin{aligned}
\mathbf{x}^T\mathbf{Qx} &= [x_1 \quad x_2]\begin{bmatrix} 4 & -4 \\ -4 & 8 \end{bmatrix}\begin{bmatrix} x_1 \\ x_2 \end{bmatrix} \\
&= [(4x_1 - 4x_2) \quad (-4x_1 + 8x_2)]\begin{bmatrix} x_1 \\ x_2 \end{bmatrix} \\
&= 4x_1^2 - 4x_2x_1 - 4x_1x_2 + 8x_2^2 \\
&= q_{11}x_1^2 + q_{21}x_2x_1 + q_{12}x_1x_2 + q_{22}x_2^2.
\end{aligned}$$

Multiplying through by $-\frac{1}{2}$ gives

$$-\tfrac{1}{2}\mathbf{x}^T\mathbf{Q}\mathbf{x} = -2x_1^2 + 4x_1x_2 - 4x_2^2,$$

which is the nonlinear portion of the objective function for this example.

Several algorithms have been developed for the special case of the quadratic programming problem where the objective function is a *concave* function. (A way to verify that the objective function is concave is to verify the equivalent condition that

$$\mathbf{x}^T\mathbf{Q}\mathbf{x} \geq \mathbf{0}$$

for all $\mathbf{x}$, that is, $\mathbf{Q}$ is a *positive semidefinite* matrix.) We shall describe one[1] of these algorithms, the *modified simplex method,* that has been quite popular because it requires using only the simplex method with a slight modification. The key to this approach is to construct the KKT conditions from the preceding section and then to reexpress these conditions in a convenient form that closely resembles linear programming. Therefore, before describing the algorithm, we shall develop this convenient form.

The KKT Conditions for Quadratic Programming

For concreteness, let us first consider the above example. Starting with the form given in the preceding section, its KKT conditions are the following.

1(j = 1). $15 + 4x_2 - 4x_1 - u_1 \leq 0.$
2(j = 1). $x_1(15 + 4x_2 - 4x_1 - u_1) = 0.$
1(j = 2). $30 + 4x_1 - 8x_2 - 2u_1 \leq 0.$
2(j = 2). $x_2(30 + 4x_1 - 8x_2 - 2u_1) = 0.$
3. $x_1 + 2x_2 - 30 \leq 0.$
4. $u_1(x_1 + 2x_2 - 30) = 0.$
5. $x_1 \geq 0, \quad x_2 \geq 0.$
6. $u_1 \geq 0.$

To begin reexpressing these conditions in a more convenient form, we move the constants in conditions 1(j = 1), 1(j = 2), and 3 to the right-hand side and then introduce nonnegative *slack variables* (denoted by y_1, y_2, and v_1, respectively) to convert these inequalities to equations.

1(j = 1). $-4x_1 + 4x_2 - u_1 + y_1 = -15$
1(j = 2). $4x_1 - 8x_2 - 2u_1 + y_2 = -30$
3. $x_1 + 2x_2 + v_1 = 30$

Note that condition 2(j = 1) can now be reexpressed as simply requiring that either $x_1 = 0$ or $y_1 = 0$; that is,

2(j = 1). $x_1y_1 = 0.$

In just the same way, conditions 2(j = 2) and 4 can be replaced by

2(j = 2). $x_2y_2 = 0,$
4. $u_1v_1 = 0.$

[1] P. Wolfe, "The Simplex Method for Quadratic Programming," *Econometrics,* **27**: 382–398, 1959. This paper develops both a short form and a long form of the algorithm. We present a version of the *short form,* which assumes further that *either* $\mathbf{c} = \mathbf{0}$ *or* the objective function is *strictly* concave.

For each of these three pairs—(x_1, y_1), (x_2, y_2), (u_1, v_1)—the two variables are called **complementary variables**, because only one of the two variables can be nonzero. Since all six variables are required to be nonnegative, these new forms of conditions $2(j = 1)$, $2(j = 2)$, and 4 can be combined into one constraint

$$x_1y_1 + x_2y_2 + u_1v_1 = 0,$$

called the **complementarity constraint**.

After multiplying through the equations for conditions $1(j = 1)$ and $1(j = 2)$ by -1 to obtain nonnegative right-hand sides, we now have the desired convenient form for the entire set of conditions shown here:

$$\begin{aligned} 4x_1 - 4x_2 + u_1 - y_1 \qquad &= 15 \\ -4x_1 + 8x_2 + 2u_1 \qquad - y_2 \qquad &= 30 \\ x_1 + 2x_2 \qquad\qquad + v_1 &= 30 \end{aligned}$$

$$x_1 \geq 0, \quad x_2 \geq 0, \quad u_1 \geq 0, \quad y_1 \geq 0, \quad y_2 \geq 0, \quad v_1 \geq 0$$

$$x_1y_1 + x_2y_2 + u_1v_1 = 0$$

This form is particularly convenient because, except for the complementarity constraint, these conditions are *linear programming constraints.*

For *any* quadratic programming problem, its KKT conditions can be reduced to this same convenient form containing just linear programming constraints plus one complementarity constraint. In matrix notation again, this general form is

$$\mathbf{Qx} + \mathbf{A}^T\mathbf{u} - \mathbf{y} = \mathbf{c}^T,$$

$$\mathbf{Ax} + \mathbf{v} = \mathbf{b},$$

$$\mathbf{x} \geq \mathbf{0}, \quad \mathbf{u} \geq \mathbf{0}, \quad \mathbf{y} \geq \mathbf{0}, \quad \mathbf{v} \geq \mathbf{0},$$

$$\mathbf{x}^T\mathbf{y} + \mathbf{u}^T\mathbf{v} = 0,$$

where the elements of the column vector **u** are the u_i of the preceding section and the elements of the column vectors **y** and **v** are slack variables.

Because the objective function of the original problem is assumed to be concave and because the constraint functions are linear and therefore convex, the corollary to the theorem of Sec. 13.6 applies. Thus **x** is *optimal* if and only if there exist values of **y**, **u**, and **v** such that all four vectors together satisfy all these conditions. The original problem is thereby reduced to the equivalent problem of finding a *feasible solution* to these *constraints.*

It is of interest to note that this equivalent problem is one example of the *linear complementarity problem* introduced in Sec. 13.3 (see Prob. 13.3-6), and that a key constraint for the linear complementarity problem is its *complementarity constraint.*

The Modified Simplex Method

The *modified simplex method* exploits the key fact that, with the exception of the complementarity constraint, the KKT conditions in the convenient form obtained above are nothing more than linear programming constraints. Furthermore, the complementarity constraint simply implies that it is not permissible for *both* complementary variables of any pair to be (nondegenerate) basic variables (the only variables > 0) when (nondegenerate) BF solutions are considered. Therefore, the problem reduces to finding an initial BF solution to any linear programming problem that has these constraints, subject to this additional restriction on the identity of the basic variables. (This initial BF solution may be the only feasible solution in this case.)

As we discussed in Sec. 4.6, finding such an initial BF solution is relatively straightforward. In the simple case where $\mathbf{c}^T \leq \mathbf{0}$ (unlikely) and $\mathbf{b} \geq \mathbf{0}$, the initial basic variables are the elements of $\mathbf{y}$ and $\mathbf{v}$ (multiply through the first set of equations by -1), so that the desired solution is $\mathbf{x} = \mathbf{0}$, $\mathbf{u} = \mathbf{0}$, $\mathbf{y} = -\mathbf{c}^T$, $\mathbf{v} = \mathbf{b}$. Otherwise, you need to revise the problem by introducing an *artificial variable* into each of the equations where $c_j > 0$ (add the variable on the left) or $b_i < 0$ (subtract the variable on the left and then multiply through by -1), in order to use these artificial variables (call them z_1, z_2, and so on) as initial basic variables for the revised problem. (Note that this choice of initial basic variables satisfies the complementarity constraint, because as nonbasic variables $\mathbf{x} = \mathbf{0}$ and $\mathbf{u} = \mathbf{0}$ automatically.)

Next, use phase 1 of the *two-phase method* (see Sec. 4.6) to find a BF solution for the real problem; i.e., apply the simplex method (with one modification) to the following linear programming problem

$$\text{Minimize} \qquad Z = \sum_j z_j,$$

subject to the linear programming constraints obtained from the KKT conditions, but with these artificial variables included.

The one modification in the simplex method is the following change in the procedure for selecting an entering basic variable.

Restricted-Entry Rule: When you are choosing an entering basic variable, exclude from consideration any nonbasic variable whose *complementary variable* already is a basic variable; the choice should be made from the other nonbasic variables according to the usual criterion for the simplex method.

This rule keeps the complementarity constraint satisfied throughout the course of the algorithm. When an optimal solution

$$\mathbf{x}^*, \mathbf{u}^*, \mathbf{y}^*, \mathbf{v}^*, z_1 = 0, \ldots, z_n = 0$$

is obtained for the phase 1 problem, $\mathbf{x}^*$ is the desired optimal solution for the original quadratic programming problem. Phase 2 of the two-phase method is not needed.

Example: We shall now illustrate this approach on the example given at the beginning of the section. As can be verified from the results in Appendix 2 (see Prob. 13.7-1*a*), $f(x_1, x_2)$ is *strictly concave;* i.e.,

$$\mathbf{Q} = \begin{bmatrix} 4 & -4 \\ -4 & 8 \end{bmatrix}$$

is positive definite, so the algorithm can be applied.

The starting point for solving this example is its KKT conditions in the convenient form obtained earlier in the section. After the needed artificial variables are introduced, the linear programming problem to be addressed explicitly by the modified simplex method then is

$$\text{Minimize} \qquad Z = z_1 + z_2,$$

subject to

$$
\begin{aligned}
4x_1 - 4x_2 + u_1 - y_1 \qquad + z_1 \quad &= 15 \\
-4x_1 + 8x_2 + 2u_1 \qquad - y_2 \qquad + z_2 &= 30 \\
x_1 + 2x_2 \qquad + v_1 \qquad &= 30
\end{aligned}
$$

and

$$
x_1 \geq 0, \quad x_2 \geq 0, \quad u_1 \geq 0, \quad y_1 \geq 0, \quad y_2 \geq 0, \quad v_1 \geq 0, \quad z_1 \geq 0, \quad z_2 \geq 0.
$$

The additional complementarity constraint

$$
x_1 y_1 + x_2 y_2 + u_1 v_1 = 0,
$$

is not included explicitly, because the algorithm automatically enforces this constraint because of the *restricted-entry rule.* In particular, for each of the three pairs of complementary variables—(x_1, y_1), (x_2, y_2), (u_1, v_1)—whenever one of the two variables already is a basic variable, the other variable is *excluded* as a candidate for the entering basic variable. Remember that the only *nonzero* variables are basic variables. Because the initial set of basic variables for the linear programming problem—z_1, z_2, v_1—gives an initial BF solution that satisfies the complementarity constraint, there is no way that this constraint can be violated by any subsequent BF solution.

Table 13.4 shows the results of applying the modified simplex method to this problem. The first simplex tableau exhibits the initial system of equations *after* converting from minimizing Z to maximizing $-Z$ *and* algebraically eliminating the initial basic variables from Eq. (0), just as was done for the radiation therapy example in Sec. 4.6. The three iterations proceed just as for the regular simplex method, *except* for

Table 13.4 **Application of the Modified Simplex Method to the Quadratic Programming Example**

Iteration	Basic Variable	Eq.	Z	x_1	x_2	u_1	y_1	y_2	v_1	z_1	z_2	Right Side
0	Z	(0)	−1	0	−4	−3	1	1	0	0	0	−45
	z_1	(1)	0	4	−4	1	−1	0	0	1	0	15
	z_2	(2)	0	−4	8	2	0	−1	0	0	1	30
	v_1	(3)	0	1	2	0	0	0	1	0	0	30
1	Z	(0)	−1	−2	0	−2	1	$\frac{1}{2}$	0	0	$\frac{1}{2}$	−30
	z_1	(1)	0	2	0	2	−1	$-\frac{1}{2}$	0	1	$\frac{1}{2}$	30
	x_2	(2)	0	$-\frac{1}{2}$	1	$\frac{1}{4}$	0	$-\frac{1}{8}$	0	0	$\frac{1}{8}$	$3\frac{3}{4}$
	v_1	(3)	0	2	0	$-\frac{1}{2}$	0	$\frac{1}{4}$	1	0	$-\frac{1}{4}$	$22\frac{1}{2}$
2	Z	(0)	−1	0	0	$-\frac{5}{2}$	1	$\frac{3}{4}$	1	0	$\frac{1}{4}$	$-7\frac{1}{2}$
	z_1	(1)	0	0	0	$\frac{5}{2}$	−1	$-\frac{3}{4}$	−1	1	$\frac{3}{4}$	$7\frac{1}{2}$
	x_2	(2)	0	0	1	$\frac{1}{8}$	0	$-\frac{1}{16}$	$\frac{1}{4}$	0	$\frac{1}{16}$	$9\frac{3}{8}$
	x_1	(3)	0	1	0	$-\frac{1}{4}$	0	$\frac{1}{8}$	$\frac{1}{2}$	0	$-\frac{1}{8}$	$11\frac{1}{4}$
3	Z	(0)	−1	0	0	0	0	0	0	1	1	0
	u_1	(1)	0	0	0	1	$-\frac{2}{5}$	$-\frac{3}{10}$	$-\frac{2}{5}$	$\frac{2}{5}$	$\frac{3}{10}$	3
	x_2	(2)	0	0	1	0	$\frac{1}{20}$	$-\frac{1}{40}$	$\frac{3}{10}$	$-\frac{1}{20}$	$\frac{1}{40}$	9
	x_1	(3)	0	1	0	0	$-\frac{1}{10}$	$\frac{1}{20}$	$\frac{2}{5}$	$\frac{1}{10}$	$-\frac{1}{20}$	12

eliminating certain candidates for the entering basic variable because of the restricted-entry rule. In the first tableau, u_1 is eliminated as a candidate because its complementary variable (v_1) already is a basic variable (but x_2 would have been chosen anyway because $-4 < -3$). In the second tableau, both u_1 and y_2 are eliminated as candidates (because v_1 and x_2 are basic variables), so x_1 automatically is chosen as the only candidate with a negative coefficient in row 0 (whereas the *regular* simplex method would have permitted choosing *either* x_1 or u_1 because they are tied for having the largest negative coefficient). In the third tableau, both y_1 and y_2 are eliminated (because x_1 and x_2 are basic variables). However, u_1 is *not* eliminated because v_1 no longer is a basic variable, so u_1 is chosen as the entering basic variable in the usual way.

The resulting optimal solution for this phase 1 problem is $x_1 = 12, x_2 = 9, u_1 = 3$, with the rest of the variables zero. (Problem 13.7-1*c* asks you to verify that this solution is optimal by showing that $x_1 = 12, x_2 = 9, u_1 = 3$ satisfy the KKT conditions for the original problem when they are written in the form given in Sec. 13.6.) Therefore, the optimal solution for the quadratic programming problem (which includes only the x_1 and x_2 variables) is $(x_1, x_2) = (12, 9)$.

13.8 Separable Programming

The preceding section showed how one class of nonlinear programming problems can be solved by an extension of the simplex method. We now consider another class, called *separable programming,* that actually can be solved by the simplex method itself, because any such problem can be approximated as closely as desired by a linear programming problem with a larger number of variables.

As indicated in Sec. 13.3, in separable programming it is assumed that the objective function $f(\mathbf{x})$ is concave, that each of the constraint functions $g_i(\mathbf{x})$ is convex, and that all these functions are separable functions (functions where each term involves just a single variable). However, to simplify the discussion, we focus here on the special case where the convex and separable $g_i(\mathbf{x})$ are, in fact, *linear functions,* just as for linear programming. Thus only the objective function requires special treatment.

Under the preceding assumptions, the objective function can be expressed as a sum of concave functions of individual variables

$$f(\mathbf{x}) = \sum_{j=1}^{n} f_j(x_j),$$

so that each $f_j(x_j)$ has a shape[1] such as the one shown in Fig. 13.15 (either case) over the feasible range of values of x_j. Because $f(\mathbf{x})$ represents the measure of performance (say, profit) for all the activities together, $f_j(x_j)$ represents the *contribution to profit* from activity j when it is conducted at level x_j. The condition of $f(\mathbf{x})$ being separable simply implies additivity (see Sec. 3.3); i.e., there are no interactions between the activities (no cross-product terms) that affect total profit beyond their independent contributions. The assumption that each $f_j(x_j)$ is concave says that the *marginal profitability* (slope of the profit curve) either stays the same or decreases (*never* increases) as x_j is increased.

[1] $f(\mathbf{x})$ is concave if and only if *every* $f_j(x_j)$ is concave.

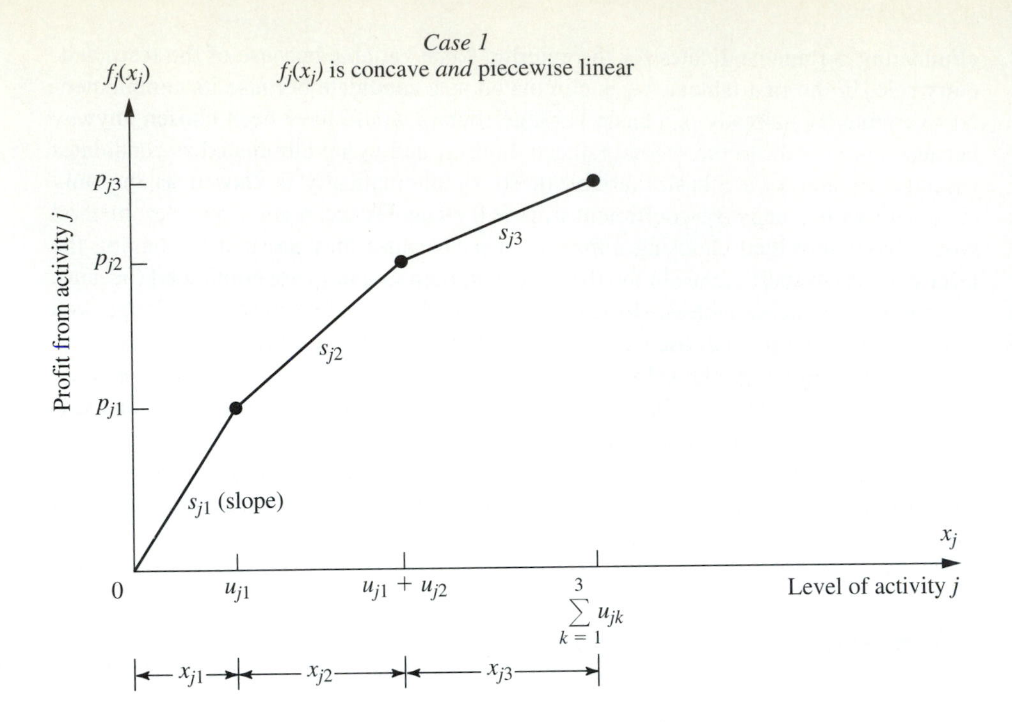

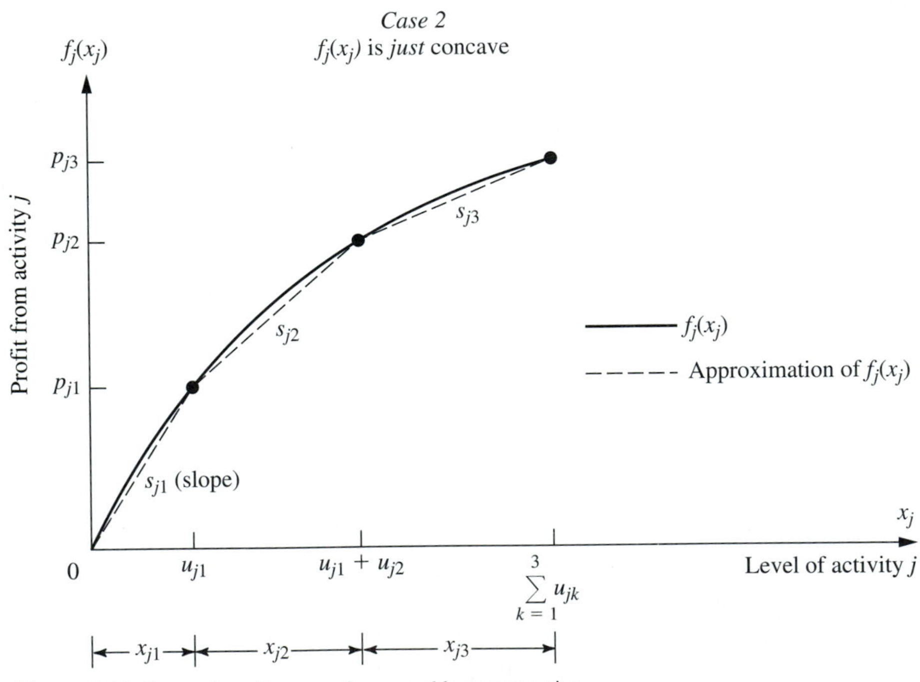

Figure 13.15 Shape of profit curves for separable programming.

Concave profit curves occur quite frequently. For example, it may be possible to sell a limited amount of some product at a certain price, then a further amount at a lower price, and perhaps finally a further amount at a still lower price. Similarly, it may be necessary to purchase raw materials from increasingly expensive sources. In another common situation, a more expensive production process must be used (e.g., overtime rather than regular-time work) to increase the production rate beyond a certain point.

These kinds of situations can lead to either type of profit curve shown in Fig. 13.15. In case 1, the slope decreases only at certain *breakpoints,* so that $f_j(x_j)$ is a *piecewise linear function* (a sequence of connected line segments). For case 2, the slope may decrease continuously as x_j increases, so that $f_j(x_j)$ is a general concave function. Any such function can be approximated as closely as desired by a piecewise linear function, and this kind of approximation is used as needed for separable programming problems. (Figure 13.15 shows an approximating function that consists of just three line segments, but the approximation can be made even better just by introducing additional breakpoints.) This approximation is very convenient because a piecewise linear function of a single variable can be rewritten as a *linear function* of several variables, with one special restriction on the values of these variables, as described next.

Reformulation as a Linear Programming Problem

The key to rewriting a piecewise linear function as a linear function is to use a separate variable for each line segment. To illustrate, consider the piecewise linear function $f_j(x_j)$ shown in Fig. 13.15, case 1 (or the approximating piecewise linear function for case 2), which has three line segments over the feasible range of values of x_j. Introduce the three new variables x_{j1}, x_{j2}, and x_{j3} and set

$$x_j = x_{j1} + x_{j2} + x_{j3},$$

where

$$0 \le x_{j1} \le u_{j1}, \qquad 0 \le x_{j2} \le u_{j2}, \qquad 0 \le x_{j3} \le u_{j3}.$$

Then use the slopes s_{j1}, s_{j2}, and s_{j3} to rewrite $f_j(x_j)$ as

$$f_j(x_j) = s_{j1}x_{j1} + s_{j2}x_{j2} + s_{j3}x_{j3},$$

with the *special restriction* that

$$x_{j2} = 0 \qquad \text{whenever} \qquad x_{j1} < u_{j1},$$

$$x_{j3} = 0 \qquad \text{whenever} \qquad x_{j2} < u_{j2}.$$

To see why this special restriction is required, suppose that $x_j = 1$, where $u_{jk} > 1$ $(k = 1, 2, 3)$, so that $f_j(1) = s_{j1}$. Note that

$$x_{j1} + x_{j2} + x_{j3} = 1$$

permits

$$x_{j1} = 1, \quad x_{j2} = 0, \quad x_{j3} = 0 \quad \Rightarrow \quad f_j(1) = s_{j1},$$

$$x_{j1} = 0, \quad x_{j2} = 1, \quad x_{j3} = 0 \quad \Rightarrow \quad f_j(1) = s_{j2},$$

$$x_{j1} = 0, \quad x_{j2} = 0, \quad x_{j3} = 1 \quad \Rightarrow \quad f_j(1) = s_{j3},$$

and so on, where

$$s_{j1} > s_{j2} > s_{j3}.$$

However, the special restriction permits only the first possibility, which is the only one giving the correct value for $f_j(1)$.

Unfortunately, the special restriction does not fit into the required format for linear programming constraints, so *some* piecewise linear functions cannot be rewritten in a linear programming format. However, *our* $f_j(x_j)$ are assumed to be concave, so $s_{j1} > s_{j2} > \cdots$, so that an algorithm for maximizing $f(\mathbf{x})$ *automatically* gives the highest priority to using x_{j1} when (in effect) increasing x_j from zero, the next highest priority to using x_{j2}, and so on, without even including the special restriction explicitly in the model. This observation leads to the following key property.

Key Property of Separable Programming: When $f(\mathbf{x})$ and the $g_i(\mathbf{x})$ satisfy the assumptions of separable programming, and when the resulting piecewise linear functions are rewritten as linear functions, deleting the *special restriction* gives a *linear programming model* whose optimal solution automatically satisfies the special restriction.

We shall elaborate further on the logic behind this key property later in this section in the context of a specific example. (Also see Prob. 13.8-7*a*.)

To write down the complete linear programming model in the above notation, let n_j be the number of line segments in $f_j(x_j)$ (or the piecewise linear function approximating it), so that

$$x_j = \sum_{k=1}^{n_j} x_{jk}$$

would be substituted throughout the original model and

$$f_j(x_j) = \sum_{k=1}^{n_j} s_{jk} x_{jk}$$

would be substituted[1] into the objective function for $j = 1, 2, \ldots, n$. The resulting model is

$$\text{Maximize} \qquad Z = \sum_{j=1}^{n} \left(\sum_{k=1}^{n_j} s_{jk} x_{jk} \right),$$

subject to

$$\sum_{j=1}^{n} a_{ij} \left(\sum_{k=1}^{n_j} x_{jk} \right) \le b_i, \qquad \text{for } i = 1, 2, \ldots, m$$

$$x_{jk} \le u_{jk}, \qquad \text{for } k = 1, 2, \ldots, n_j;\ j = 1, 2, \ldots, n$$

and

$$x_{jk} \ge 0, \qquad \text{for} \qquad k = 1, 2, \ldots, n_j;\ j = 1, 2, \ldots, n.$$

[1] If one or more of the $f_j(x_j)$ already are *linear* functions $f_j(x_j) = c_j x_j$, then $n_j = 1$ so neither of these substitutions will be made for j.

(The $\Sigma_{k=1}^{n_j} x_{jk} \geq 0$ constraints are deleted because they are ensured by the $x_{jk} \geq 0$ constraints.) If some original variable x_j has no upper bound, then $u_{jn_j} = \infty$, so the constraint involving this quantity will be deleted.

An efficient way of solving this model[1] is to use the streamlined version of the simplex method for dealing with upper bound constraints (described in Sec. 7.3). After obtaining an optimal solution for this model, you then would calculate

$$x_j = \sum_{k=1}^{n_j} x_{jk},$$

for $j = 1, 2, \ldots, n$ in order to identify an optimal solution for the original separable programming program (or its piecewise linear approximation).

Example

The Wyndor Glass Co. (see Sec. 3.1) has received a special order for handcrafted goods to be made in Plants 1 and 2 throughout the next 4 months. Filling this order will require borrowing certain employees from the work crews for the regular products, so the remaining workers will need to work overtime to utilize the full production capacity of the plant's machinery and equipment for these regular products. In particular, for the two new regular products discussed in Sec. 3.1, overtime will be required to utilize the last 25 percent of the production capacity available in Plant 1 for product 1 and for the last 50 percent of the capacity available in Plant 2 for product 2. The additional cost of using overtime work will reduce the profit for each unit involved from \$3 to \$2 for product 1 and from \$5 to \$1 for product 2, giving the *profit curves* of Fig. 13.16, both of which fit the form for case 1 of Fig. 13.15.

[1] For a specialized algorithm for solving this model very efficiently, see R. Fourer: ''A Specialized Algorithm for Piecewise-Linear Programming III: Computational Analysis and Applications,'' *Mathematical Programming*, **53**: 213–235, 1992. Also see A. M. Geoffrion: ''Objective Function Approximations in Mathematical Programming,'' *Mathematical Programming*, **13**: 23–37, 1977.

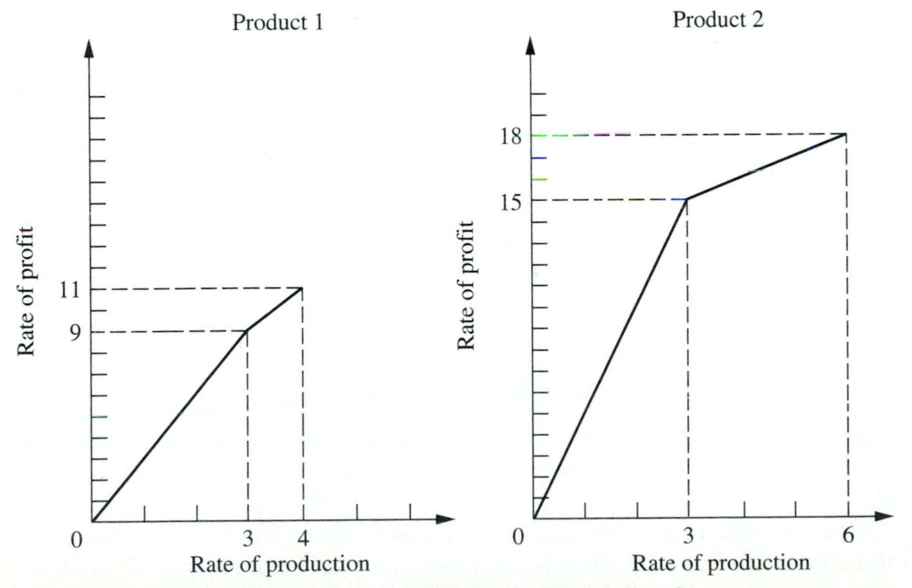

Figure 13.16 Profit data during the next 4 months for the Wyndor Glass Co.

Management has decided to go ahead and use overtime work rather than hire additional workers during this temporary situation. However, it does insist that the work crew for each product be fully utilized on regular time before any overtime is used. Furthermore, it feels that the current production rates ($x_1 = 2$ for product 1 and $x_2 = 6$ for product 2) should be changed temporarily if this would improve overall profitability. Therefore, it has instructed the OR team to review products 1 and 2 again to determine the most profitable product mix during the next 4 months.

Formulation: To refresh your memory, the linear programming model for the original Wyndor Glass Co. problem in Sec. 3.1 is

$$\text{Maximize} \quad Z = 3x_1 + 5x_2,$$

subject to

$$\begin{aligned} x_1 \qquad &\leq 4 \\ 2x_2 &\leq 12 \\ 3x_1 + 2x_2 &\leq 18 \end{aligned}$$

and

$$x_1 \geq 0, \qquad x_2 \geq 0.$$

We now need to modify this model to fit the new situation described above. For this purpose, let the production rate for product 1 be $x_1 = x_{1R} + x_{1O}$, where x_{1R} is the production rate achieved on regular time and x_{1O} is the incremental production rate from using overtime. Define $x_2 = x_{2R} + x_{2O}$ in the same way for product 2. Thus, in the notation of the general linear programming model for separable programming given just before this example, $n = 2$, $n_1 = 2$, and $n_2 = 2$. Plugging the data given in Fig. 13.16 (including maximum rates of production on regular time and on overtime) into this general model gives the specific model for this application. In particular, the new linear programming problem is to determine the values of x_{1R}, x_{1O}, x_{2R}, and x_{2O} so as to

$$\text{Maximize} \quad Z = 3x_{1R} + 2x_{1O} + 5x_{2R} + x_{2O},$$

subject to

$$\begin{aligned} x_{1R} + x_{1O} \qquad\qquad\qquad &\leq 4 \\ 2(x_{2R} + x_{2O}) &\leq 12 \\ 3(x_{1R} + x_{1O}) + 2(x_{2R} + x_{2O}) &\leq 18 \end{aligned}$$

$$x_{1R} \leq 3, \qquad x_{1O} \leq 1, \qquad x_{2R} \leq 3, \qquad x_{2O} \leq 3$$

and

$$x_{1R} \geq 0, \qquad x_{1O} \geq 0, \qquad x_{2R} \geq 0, \qquad x_{2O} \geq 0.$$

(Note that the upper bound constraints in the next-to-last row of the model make the first two functional constraints *redundant,* so these two functional constraints can be deleted.)

However, there is one important factor that is not taken into account explicitly in this formulation. Specifically, there is nothing in the model that requires all available regular time for a product to be fully utilized before any overtime is used for that product. In other words, it may be feasible to have $x_{1O} > 0$ even when $x_{1R} < 3$ and to have $x_{2O} > 0$ even when $x_{2R} < 3$. Such solutions would not, however, be acceptable to management. (Prohibiting such solutions is the *special restriction* discussed earlier in this section.)

Now we come to the *key property of separable programming.* Even though the model does not take this factor into account explicitly, the model does take it into account implicitly! Despite the model's having excess "feasible" solutions that actually are unacceptable, any *optimal* solution for the model is *guaranteed* to be a legitimate one that does not replace any available regular-time work with overtime work. (The reasoning here is analogous to that for the Big M method discussed in Sec. 4.6, where excess feasible but *nonoptimal* solutions also were allowed in the model as a matter of convenience.) Therefore, the simplex method can be safely applied to this model to find the most profitable acceptable product mix. The reasons are twofold. First, the two decision variables for each product *always* appear together as a *sum,* $x_{1R} + x_{1O}$ or $x_{2R} + x_{2O}$, in *each* functional constraint (one in this case) other than the upper bound constraints on individual variables. Therefore, it *always* is possible to convert an unacceptable feasible solution to an acceptable one having the same total production rates $x_1 = x_{1R} + x_{1O}$ and $x_2 = x_{2R} + x_{2O}$, merely by replacing overtime production by regular-time production as much as possible. Second, overtime production is less profitable than regular-time production (i.e., the slope of each profit curve in Fig. 13.16 is a monotonically *decreasing* function of the rate of production), so converting an unacceptable feasible solution to an acceptable one in this way *must* increase the total rate of profit Z. Consequently, any feasible solution that uses overtime production for a product when regular-time production is still available *cannot* be optimal with respect to the model.

For example, consider the unacceptable feasible solution $x_{1R} = 1$, $x_{1O} = 1$, $x_{2R} = 1$, $x_{2O} = 3$, which yields a total rate of profit $Z = 13$. The acceptable way of achieving the same total production rates $x_1 = 2$ and $x_2 = 4$ is $x_{1R} = 2, x_{1O} = 0, x_{2R} = 3, x_{2O} = 1$. This latter solution is still feasible, but it also increases Z by $(3 - 2)(1) + (5 - 1)(2) = 9$.

Similarly, the optimal solution for this model turns out to be $x_{1R} = 3$, $x_{1O} = 1$, $x_{2R} = 3$, $x_{2O} = 0$, which is an acceptable feasible solution.

Notice that most of the functional constraints in the model are upper-bound constraints, i.e., constraints that simply specify the maximum value allowed for an individual variable. When a computer code is available for the special streamlined version of the simplex method for dealing with such constraints (see Sec. 7.3), it provides a very efficient way of solving even extremely large problems of this type.

Extensions

Thus far we have focused on the special case of separable programming where the only nonlinear function is the objective function $f(\mathbf{x})$. Now consider briefly the general case where the constraint functions $g_i(\mathbf{x})$ need not be linear but are convex and separable, so that each $g_i(\mathbf{x})$ can be expressed as a sum of functions of individual variables

$$g_i(\mathbf{x}) = \sum_{j=1}^{n} g_{ij}(x_j),$$

where each $g_{ij}(x_j)$ is a *convex* function. Once again, each of these new functions may be approximated as closely as desired by a *piecewise linear* function (if it is not already in that form). The one new restriction is that for each variable x_j ($j = 1, 2, \ldots, n$), all the piecewise linear approximations of the functions of this variable $[f_j(x_j), g_{1j}(x_j), \ldots, g_{mj}(x_j)]$ must have the *same* breakpoints so that the same new variables $(x_{j1}, x_{j2}, \ldots, x_{jn_j})$ can be used for all these piecewise linear functions. This formulation leads to a linear programming model just like the one given for the special case except that for each i and j, the x_{jk} variables now have different coefficients in constraint i [where these coefficients are the corresponding slopes of the piecewise linear function approximating $g_{ij}(x_j)$]. Because the $g_{ij}(x_j)$ are required to be convex, essentially the same logic as before implies that the key property of separable programming still must hold. (See Prob. 13.8-7*b*.)

One drawback of approximating functions by piecewise linear functions as described in this section is that achieving a close approximation requires a large number of line segments (variables), whereas such a fine grid for the breakpoints is needed only in the immediate neighborhood of an optimal solution. Therefore, more sophisticated approaches that use a succession of *two-segment* piecewise linear functions have been developed[1] to obtain *successively closer approximations* within this immediate neighborhood. This kind of approach tends to be both faster and more accurate in closely approximating an optimal solution.

13.9 Convex Programming

We already have discussed some special cases of convex programming in Secs. 13.4 and 13.5 (unconstrained problems), 13.7 (quadratic objective function with linear constraints), and 13.8 (separable functions). You also have seen some theory for the general case (necessary and sufficient conditions for optimality) in Sec. 13.6. In this section, we briefly discuss some types of approaches used to solve the general convex programming problem [where the objective function $f(\mathbf{x})$ to be maximized is concave and the $g_i(\mathbf{x})$ constraint functions are convex], and then we present one example of an algorithm for convex programming.

There is no single standard algorithm that always is used to solve convex programming problems. Many different algorithms have been developed, each with its own advantages and disadvantages, and research continues to be active in this area. Roughly speaking, most of these algorithms fall into one of the following three categories.

The first category is **gradient algorithms,** where the gradient search procedure of Sec. 13.5 is modified in some way to keep the search path from penetrating any constraint boundary. For example, one popular gradient method is the *generalized reduced gradient* (GRG) method.[2]

The second category—**sequential unconstrained algorithms**—includes *penalty function* and *barrier function* methods. These algorithms convert the original constrained optimization problem to a sequence of *unconstrained optimization* problems whose optimal solutions converge to the optimal solution for the original problem.

[1] R. R. Meyer, "Two-Segment Separable Programming," *Management Science,* **25**: 385–395, 1979.

[2] See Selected Reference 5 for software that implements this method.

Each of these unconstrained optimization problems can be solved by the gradient search procedure of Sec. 13.5. This conversion is accomplished by incorporating the constraints into a penalty function (or barrier function) that is subtracted from the objective function in order to impose large penalties for violating constraints (or even being near constraint boundaries). You will see one example of this category of algorithms in the next section.

The third category—**sequential-approximation algorithms**—includes *linear approximation* and *quadratic approximation* methods. These algorithms replace the nonlinear objective function by a succession of linear or quadratic approximations. For linearly constrained optimization problems, these approximations allow repeated application of linear or quadratic programming algorithms. This work is accompanied by other analysis that yields a sequence of solutions that converges to an optimal solution for the original problem. Although these algorithms are particularly suitable for linearly constrained optimization problems, some also can be extended to problems with nonlinear constraint functions by the use of appropriate linear approximations.

As one example of a *sequential-approximation* algorithm, we present here the **Frank-Wolfe algorithm**[1] for the case of *linearly constrained* convex programming (so the constraints are $\mathbf{Ax} \le \mathbf{b}$ and $\mathbf{x} \ge \mathbf{0}$ in matrix form). This procedure is particularly straightforward; it combines *linear* approximations of the objective function (enabling us to use the simplex method) with the one-dimensional search procedure of Sec. 13.4.

A Sequential Linear Approximation Algorithm (Frank-Wolfe)

Given a feasible trial solution $\mathbf{x}'$, the linear approximation used for the objective function $f(\mathbf{x})$ is the first-order Taylor series expansion of $f(\mathbf{x})$ around $\mathbf{x} = \mathbf{x}'$, namely,

$$f(\mathbf{x}') \approx f(\mathbf{x}') + \sum_{j=1}^{n} \frac{\partial f(\mathbf{x}')}{\partial x_j}(x_j - x_j') = f(\mathbf{x}') + \nabla f(\mathbf{x}')(\mathbf{x} - \mathbf{x}'),$$

where these partial derivatives are evaluated at $\mathbf{x} = \mathbf{x}'$. Because $f(\mathbf{x}')$ and $\nabla f(\mathbf{x}')\mathbf{x}'$ have fixed values, they can be dropped to give an equivalent linear objective function

$$g(\mathbf{x}) = \nabla f(\mathbf{x}')\mathbf{x} = \sum_{j=1}^{n} c_j x_j, \qquad \text{where } c_j = \frac{\partial f(\mathbf{x})}{\partial x_j} \qquad \text{at } \mathbf{x} = \mathbf{x}'.$$

The simplex method (or the graphical procedure if $n = 2$) then is applied to the resulting linear programming (LP) problem [maximize $g(\mathbf{x})$ subject to the original constraints] to find *its* optimal solution $\mathbf{x}_{\text{LP}}$. Note that the linear objective function necessarily increases steadily as one moves along the line segment from $\mathbf{x}'$ to $\mathbf{x}_{\text{LP}}$ (which is on the boundary of the feasible region). However, the linear approximation may not be a particularly close one for $\mathbf{x}$ far from $\mathbf{x}'$, so the *nonlinear* objective function may not continue to increase all the way from $\mathbf{x}'$ to $\mathbf{x}_{\text{LP}}$. Therefore, rather than just accepting $\mathbf{x}_{\text{LP}}$ as the next trial solution, we choose the point that maximizes the nonlinear objective function along this line segment. This point may be found by conducting the one-dimensional search procedure of Sec. 13.4, where the one variable for purposes of this search is the fraction t of the total distance from $\mathbf{x}'$ to $\mathbf{x}_{\text{LP}}$. This point then becomes the

[1] M. Frank and P. Wolfe, "An Algorithm for Quadratic Programming," *Naval Research Logistics Quarterly*, **3**:95–110, 1956. Although originally designed for quadratic programming, this algorithm is easily adapted to the case of a general concave objective function considered here.

new trial solution for initiating the next iteration of the algorithm, as just described. The sequence of trial solutions generated by repeated iterations converges to an optimal solution for the original problem, so the algorithm stops as soon as the successive trial solutions are close enough together to have essentially reached this optimal solution.

Summary of the Frank-Wolfe Algorithm

Initialization: Find a feasible initial trial solution $\mathbf{x}^{(0)}$, for example, by applying linear programming procedures to find an initial BF solution. Set $k = 1$.

Iteration

1. For $j = 1, 2, \ldots, n$, evaluate

$$\frac{\partial f(\mathbf{x})}{\partial x_j} \qquad \text{at } \mathbf{x} = \mathbf{x}^{(k-1)}$$

and set c_j equal to this value.

2. Find an optimal solution $\mathbf{x}_{\text{LP}}^{(k)}$ for the following linear programming problem.

$$\text{Maximize} \qquad g(\mathbf{x}) = \sum_{j=1}^{n} c_j x_j,$$

subject to

$$\mathbf{Ax} \leq \mathbf{b} \qquad \text{and} \qquad \mathbf{x} \geq \mathbf{0}.$$

3. For the variable t $(0 \leq t \leq 1)$, set

$$h(t) = f(\mathbf{x}) \qquad \text{for } \mathbf{x} = \mathbf{x}^{(k-1)} + t(\mathbf{x}_{\text{LP}}^{(k)} - \mathbf{x}^{(k-1)}),$$

so that $h(t)$ gives the value of $f(\mathbf{x})$ on the line segment between $\mathbf{x}^{(k-1)}$ (where $t = 0$) and $\mathbf{x}_{\text{LP}}^{(k)}$ (where $t = 1$). Use some procedure such as the one-dimensional search procedure (see Sec. 13.4) to maximize $h(t)$ over $0 \leq t \leq 1$, and set $\mathbf{x}^{(k)}$ equal to the corresponding $\mathbf{x}$. Go to the stopping rule.

Stopping rule: If $\mathbf{x}^{(k-1)}$ and $\mathbf{x}^{(k)}$ are sufficiently close, stop and use $\mathbf{x}^{(k)}$ (or some extrapolation of $\mathbf{x}^{(0)}, \mathbf{x}^{(1)}, \ldots, \mathbf{x}^{(k-1)}, \mathbf{x}^{(k)}$) as your estimate of an optimal solution. Otherwise, reset $k = k + 1$ and perform another iteration.

Now let us illustrate this procedure.

Example: Consider the following linearly constrained convex programming problem:

$$\text{Maximize} \qquad f(\mathbf{x}) = 5x_1 - x_1^2 + 8x_2 - 2x_2^2,$$

subject to

$$3x_1 + 2x_2 \leq 6$$

and

$$x_1 \geq 0, \qquad x_2 \geq 0.$$

Note that

$$\frac{\partial f}{\partial x_1} = 5 - 2x_1, \qquad \frac{\partial f}{\partial x_2} = 8 - 4x_2,$$

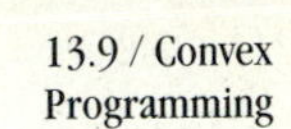

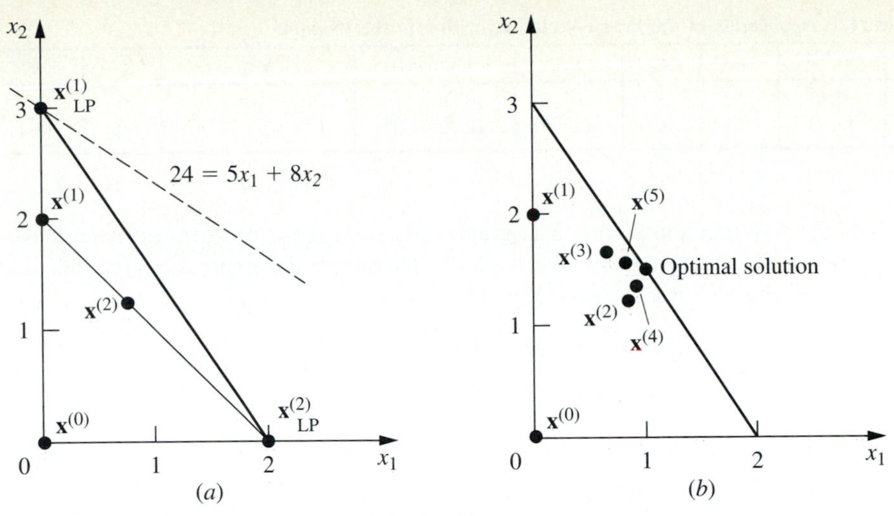

Figure 13.17 Illustration of the Frank-Wolfe algorithm.

so that the *unconstrained* maximum $\mathbf{x} = (\frac{5}{2}, 2)$ violates the functional constraint. Thus more work is needed to find the *constrained* maximum.

Because $\mathbf{x} = (0, 0)$ is clearly feasible (and corresponds to the initial BF solution for the linear programming constraints), let us choose it as the initial trial solution $\mathbf{x}^{(0)}$ for the Frank-Wolfe algorithm. Plugging $x_1 = 0$ and $x_2 = 0$ into the expressions for the partial derivatives gives $c_1 = 5$ and $c_2 = 8$, so that $g(\mathbf{x}) = 5x_1 + 8x_2$ is the initial linear approximation of the objective function. Graphically, solving this linear programming problem (see Fig. 13.17*a*) yields $\mathbf{x}_{\text{LP}}^{(1)} = (0, 3)$. For step 3 of the first iteration, the points on the line segment between (0, 0) and (0, 3) shown in Fig. 13.17*a* are expressed by

$$(x_1, x_2) = (0, 0) + t[(0, 3) - (0, 0)] \qquad \text{for } 0 \leq t \leq 1$$
$$= (0, 3t)$$

as shown in the sixth column of Table 13.5. This expression then gives

$$h(t) = f(0, 3t) = 8(3t) - 2(3t)^2$$
$$= 24t - 18t^2,$$

so that the value $t = t^*$ that maximizes $h(t)$ over $0 \leq t \leq 1$ may be obtained in this case by setting

$$\frac{dh(t)}{dt} = 24 - 36t = 0,$$

so that $t^* = \frac{2}{3}$. This result yields the next trial solution

$$\mathbf{x}^{(1)} = (0, 0) + \tfrac{2}{3}[(0, 3) - (0, 0)]$$
$$= (0, 2),$$

which completes the first iteration.

To sketch the calculations that lead to the results in the second row of Table 13.5, note that $\mathbf{x}^{(1)} = (0, 2)$ gives

$$c_1 = 5 - 2(0) = 5,$$
$$c_2 = 8 - 4(2) = 0.$$

Table 13.5 **Application of the Frank-Wolfe Algorithm to the Example**

k	$\mathbf{x}^{(k-1)}$	c_1	c_2	$\mathbf{x}_{LP}^{(k)}$	$\mathbf{x}$ for $h(t)$	$h(t)$	t^*	$\mathbf{x}^{(k)}$
1	(0, 0)	5	8	(0, 3)	$(0, 3t)$	$24t - 18t^2$	$\frac{2}{3}$	(0, 2)
2	(0, 2)	5	0	(2, 0)	$(2t, 2 - 2t)$	$8 + 10t - 12t^2$	$\frac{5}{12}$	$(\frac{5}{6}, \frac{7}{6})$

For the objective function $g(\mathbf{x}) = 5x_1$, graphically solving the problem over the feasible region in Fig. 13.17*a* gives $\mathbf{x}_{LP}^{(2)} = (2, 0)$. Therefore, the expression for the line segment between $\mathbf{x}^{(1)}$ and $\mathbf{x}_{LP}^{(2)}$ (see Fig. 13.17*a*) is

$$\begin{aligned}\mathbf{x} &= (0, 2) + t[(2, 0) - (0, 2)]\\ &= (2t, 2 - 2t),\end{aligned}$$

so that

$$\begin{aligned}h(t) &= f(2t, 2 - 2t)\\ &= 5(2t) - (2t)^2 + 8(2 - 2t) - 2(2 - 2t)^2\\ &= 8 + 10t - 12t^2.\end{aligned}$$

Setting

$$\frac{dh(t)}{dt} = 10 - 24t = 0$$

yields $t^* = \frac{5}{12}$. Hence

$$\begin{aligned}\mathbf{x}^{(2)} &= (0, 2) + \tfrac{5}{12}[(2, 0) - (0, 2)]\\ &= (\tfrac{5}{6}, \tfrac{7}{6}).\end{aligned}$$

You can see in Fig. 13.17*b* how the trial solutions keep alternating between two trajectories that appear to intersect at approximately the point $\mathbf{x} = (1, \frac{3}{2})$. This point is, in fact, the optimal solution, as can be verified by applying the KKT conditions from Sec. 13.6.

This example illustrates a common feature of the Frank-Wolfe algorithm, namely, that the trial solutions alternate between two (or more) trajectories. When they alternate in this way, we can extrapolate the trajectories to their approximate point of intersection to estimate an optimal solution. This estimate tends to be better than using the last trial solution generated. The reason is that the trial solutions tend to converge rather slowly toward an optimal solution, so the last trial solution may still be quite far from optimal.

Another example illustrating the application of the Frank-Wolfe algorithm is provided under the Demo menu of your OR Courseware. The Procedure menu also includes an interactive routine for this algorithm.

In conclusion, we emphasize that the Frank-Wolfe algorithm is just one example of sequential-approximation algorithms. Many of these algorithms use *quadratic* instead of *linear* approximations at each iteration because quadratic approximations provide a considerably closer fit to the original problem and thus enable the sequence of solutions to converge considerably more rapidly toward an optimal solution than was the case in Fig. 13.17*b*. For this reason, even though sequential linear approximation methods such as the Frank-Wolfe algorithm are relatively straightforward to use, *se-*

quential quadratic approximation methods[1] now are generally preferred in actual applications. Popular among these are the *quasi-Newton* (or *variable metric*) methods, which compute a quadratic approximation to the curvature of a nonlinear function without explicitly calculating second (partial) derivatives. (For linearly constrained optimization problems, this nonlinear function is just the objective function; whereas with nonlinear constraints, it is the Lagrangian function described in Appendix 3.) Some quasi-Newton algorithms do not even explicitly form and solve an approximating quadratic programming problem at each iteration, but instead incorporate some of the basic ingredients of *gradient algorithms.*

For further information about the state of the art in convex programming algorithms, see Selected References 3, 4, and 10.

13.10 Nonconvex Programming

The assumptions of convex programming are very convenient ones, because they ensure that any *local maximum* also is a *global maximum.* Unfortunately, the nonlinear programming problems that arise in practice frequently only come fairly close to satisfying these assumptions, but they have some relatively minor disparities. What kind of approach can be used to deal with such *nonconvex programming* problems?

A common approach is to apply an algorithmic *search procedure* that will stop when it finds a *local maximum* and then to restart it a number of times from a variety of initial trial solutions in order to find as many distinct local maxima as possible. The best of these local maxima is then chosen for implementation. Normally, the search procedure is one that has been designed to find a global maximum when all the assumptions of convex programming hold, but it also can operate to find a local maximum when they do not.

One such search procedure that has been widely used since its development in the 1960s is the *sequential unconstrained minimization technique* (or *SUMT* for short).[2] There actually are two main versions of SUMT, one of which is an *exterior-point* algorithm that deals with *infeasible* solutions while using a *penalty function* to force convergence to the feasible region. We shall describe the other version, which is an *interior-point* algorithm that deals directly with *feasible* solutions while using a *barrier function* to force staying inside the feasible region. Although SUMT was originally presented as a minimization technique, we shall convert it to a maximization technique in order to be consistent with the rest of the chapter. Therefore, we continue to assume that the problem is in the form given at the beginning of the chapter and that all the functions are differentiable.

Sequential Unconstrained Minimization Technique (SUMT)

As the name implies, SUMT replaces the original problem by a *sequence* of *unconstrained* optimization problems whose solutions *converge* to a solution (local maximum) of the original problem. This approach is very attractive because unconstrained

[1] For a survey of these methods, see M. J. D. Powell, "Variable Metric Methods for Constrained Optimization," in A. Bachem, M. Grotschel, and B. Korte (eds.), *Mathematical Programming: The State of the Art,* Springer-Verlag, Berlin, 1983, pp. 288–311.

[2] See Selected Reference 2.

optimization problems are much easier to solve (see the gradient search procedure in Sec. 13.5) than those with constraints. Each of the unconstrained problems in this sequence involves choosing a (successively smaller) strictly positive value of a scalar r and then solving for $\mathbf{x}$ so as to

$$\text{Maximize} \qquad P(\mathbf{x}; r) = f(\mathbf{x}) - rB(\mathbf{x}).$$

Here $B(\mathbf{x})$ is a **barrier function** that has the following properties (for $\mathbf{x}$ that are feasible for the original problem):

1. $B(\mathbf{x})$ is *small* when $\mathbf{x}$ is *far* from the boundary of the feasible region.
2. $B(\mathbf{x})$ is *large* when $\mathbf{x}$ is *close* to the boundary of the feasible region.
3. $B(\mathbf{x}) \to \infty$ as the distance from the (nearest) boundary of the feasible region $\to 0$.

Thus, by starting the search procedure with a *feasible* initial trial solution and then attempting to increase $P(\mathbf{x}; r)$, $B(\mathbf{x})$ provides a *barrier* that prevents the search from ever crossing (or even reaching) the boundary of the feasible region for the original problem.

The most common choice of $B(\mathbf{x})$ is

$$B(\mathbf{x}) = \sum_{i=1}^{m} \frac{1}{b_i - g_i(\mathbf{x})} + \sum_{j=1}^{n} \frac{1}{x_j}.$$

For feasible values of $\mathbf{x}$, note that the denominator of each term is proportional to the distance of $\mathbf{x}$ from the constraint boundary for the corresponding functional or nonnegativity constraint. Consequently, *each* term is a *boundary repulsion term* that has all the preceding three properties with respect to this particular constraint boundary. Another attractive feature of this $B(\mathbf{x})$ is that when all the assumptions of *convex programming* are satisfied, $P(\mathbf{x}; r)$ is a *concave* function.

Because $B(\mathbf{x})$ keeps the search away from the boundary of the feasible region, you probably are asking the very legitimate question: What happens if the desired solution lies there? This concern is the reason that SUMT involves solving a *sequence* of these unconstrained optimization problems for successively smaller values of r approaching zero (where the final trial solution from each one becomes the initial trial solution for the next). For example, each new r might be obtained from the preceding one by multiplying by a constant θ $(0 < \theta < 1)$, where a typical value is $\theta = 0.01$. As r approaches 0, $P(\mathbf{x}; r)$ approaches $f(\mathbf{x})$, so the corresponding local maximum of $P(\mathbf{x}; r)$ converges to a local maximum of the original problem. Therefore, it is necessary to solve only enough unconstrained optimization problems to permit extrapolating their solutions to this limiting solution.

How many are enough to permit this extrapolation? When the original problem satisfies the assumptions of convex programming, useful information is available to guide us in this decision. In particular, if $\bar{x}$ is a global maximizer of $P(\mathbf{x}; r)$, then

$$f(\bar{\mathbf{x}}) \le f(\mathbf{x}^*) \le f(\bar{\mathbf{x}}) + rB(\bar{\mathbf{x}}),$$

where $\mathbf{x}^*$ is the (unknown) *optimal* solution for the original problem. Thus, $rB(\bar{\mathbf{x}})$ is the *maximum error* (in the value of the objective function) that can result by using $\bar{\mathbf{x}}$ to approximate $\mathbf{x}^*$, and extrapolating beyond $\bar{\mathbf{x}}$ to increase $f(\mathbf{x})$ further decreases this error. If an *error tolerance* is established in advance, then you can stop as soon as $rB(\bar{\mathbf{x}})$ is less than this quantity.

Unfortunately, no such guarantee for the maximum error can be given for nonconvex programming problems. However, $rB(\bar{\mathbf{x}})$ still is *likely* to exceed the actual error when $\bar{\mathbf{x}}$ and $\mathbf{x}^*$ now are corresponding *local maxima of* $P(\mathbf{x}; r)$ and the original problem, respectively.

Summary of SUMT

Initialization: Identify a *feasible* initial trial solution $\mathbf{x}^{(0)}$ that is not on the boundary of the feasible region. Set $k = 1$ and choose appropriate strictly positive values for the initial r and for $\theta < 1$ (say, $r = 1$ and $\theta = 0.01$).[1]

Iteration: Starting from $\mathbf{x}^{(k-1)}$, apply the gradient search procedure described in Sec. 13.5 (or some similar method) to find a local maximum $\mathbf{x}^{(k)}$ of

$$P(\mathbf{x}; r) = f(\mathbf{x}) - r\left[\sum_{i=1}^{m} \frac{1}{b_i - g_i(\mathbf{x})} + \sum_{j=1}^{n} \frac{1}{x_j}\right].$$

Stopping rule: If the change from $\mathbf{x}^{(k-1)}$ to $\mathbf{x}^{(k)}$ is negligible, stop and use $\mathbf{x}^{(k)}$ (or an extrapolation of $\mathbf{x}^{(0)}, \mathbf{x}^{(1)}, \ldots, \mathbf{x}^{(k-1)}, \mathbf{x}^{(k)}$) as your estimate of a *local maximum* of the original problem. Otherwise, reset $k = k + 1$ and $r = \theta r$ and perform another iteration.

When the assumptions of convex programming are not satisfied, this algorithm should be repeated a number of times by starting from a variety of feasible initial trial solutions. The best of the *local maxima* thereby obtained for the original problem should be used as the best available approximation of a *global maximum.*

Finally, note that SUMT also can be extended to accommodate *equality* constraints $g_i(\mathbf{x}) = b_i$. One standard way is as follows. For each equality constraint,

$$\frac{-[b_i - g_i(\mathbf{x})]^2}{\sqrt{r}} \qquad \text{replaces} \qquad \frac{-r}{b_i - g_i(\mathbf{x})}$$

in the expression for $P(\mathbf{x}; r)$ given under "Summary of SUMT," and then the same procedure is used. The numerator $-[b_i - g_i(\mathbf{x})]^2$ imposes a large penalty for deviating substantially from satisfying the equality constraint, and then the denominator tremendously increases this penalty as r is decreased to a tiny amount, thereby forcing the sequence of trial solutions to converge toward a point that satisfies the constraint.

SUMT has been widely used because of its simplicity and versatility. However, numerical analysts have found that it is relatively prone to *numerical instability,* so considerable caution is advised. For further information on this issue as well as similar analyses for alternative algorithms, see Selected Reference 4.

Example: To illustrate SUMT, consider the following two-variable problem:

$$\text{Maximize} \qquad f(\mathbf{x}) = x_1 x_2,$$

subject to

$$x_1^2 + x_2 \le 3$$

and

$$x_1 \ge 0, \qquad x_2 \ge 0.$$

[1] A reasonable criterion for choosing the initial r is one that makes $rB(\mathbf{x})$ about the same order of magnitude as $f(\mathbf{x})$ for feasible solutions $\mathbf{x}$ that are not particularly close to the boundary.

Table 13.6 **Illustration of SUMT**

k	r	$x_1^{(k)}$	$x_2^{(k)}$
0		1	1
1	1	0.90	1.36
2	10^{-2}	0.983	1.933
3	10^{-4}	0.998	1.994
		↓	↓
		1	2

Even though $g_1(\mathbf{x}) = x_1^2 + x_2$ is convex (because each term is convex), this problem is a *nonconvex* programming problem because $f(\mathbf{x}) = x_1x_2$ is *not* concave (see Appendix 2.)

For the initialization, $(x_1, x_2) = (1, 1)$ is one obvious feasible solution that is not on the boundary of the feasible region, so we can set $\mathbf{x}^{(0)} = (1, 1)$. Reasonable choices for r and θ are $r = 1$ and $\theta = 0.01$.

For each iteration,

$$P(\mathbf{x}; r) = x_1x_2 - r\left(\frac{1}{3 - x_1^2 - x_2} + \frac{1}{x_1} + \frac{1}{x_2}\right).$$

With $r = 1$, applying the gradient search procedure starting from (1, 1) to maximize this expression eventually leads to $\mathbf{x}^{(1)} = (0.90, 1.36)$. Resetting $r = 0.01$ and restarting the gradient search procedure from (0.90, 1.36) then lead to $\mathbf{x}^{(2)} = (0.983, 1.933)$. One more iteration with $r = 0.01(0.01) = 0.0001$ leads from $\mathbf{x}^{(2)}$ to $\mathbf{x}^{(3)} =$ (0.998, 1.994). This sequence of points, summarized in Table 13.6, quite clearly is converging to (1, 2). Applying the KKT conditions to this solution verifies that it does indeed satisfy the necessary condition for optimality. Graphical analysis demonstrates that $(x_1, x_2) = (1, 2)$ is, in fact, a global maximum. (See Prob. 13.10-4*b*)

For this problem, there are no local maxima other than $(x_1, x_2) = (1, 2)$, so reapplying SUMT from various feasible initial trial solutions always leads to this same solution.[1]

The Demo menu in your OR Courseware includes another example illustrating the application of SUMT. The Procedure menu includes an automatic routine for executing SUMT.

13.11 Conclusions

Practical optimization problems frequently involve *nonlinear* behavior that must be taken into account. It is sometimes possible to *reformulate* these nonlinearities to fit into a linear programming format, as can be done for *separable programming* problems. However, it is frequently necessary to use a *nonlinear programming* formulation.

In contrast to the case of the simplex method for linear programming, there is no efficient all-purpose algorithm that can be used to solve all nonlinear programming

[1] The technical reason is that $f(\mathbf{x})$ is a (strictly) *quasiconcave* function that shares the property of concave functions that a local maximum always is a global maximum. For further information, see M. Avriel, W. E. Diewert, S. Schaible, and I. Zang, *Generalized Concavity,* Plenum, New York, 1985.

problems. In fact, some of these problems cannot be solved in a very satisfactory manner by any method. However, considerable progress has been made for some important classes of problems, including *quadratic programming, convex programming,* and certain special types of *nonconvex programming.* A variety of algorithms that frequently perform well are available for these cases. Some of these algorithms incorporate highly efficient procedures for *unconstrained optimization* for a portion of each iteration, and some use a succession of linear or quadratic approximations to the original problem.

There has been a strong emphasis in recent years on developing high-quality, reliable software packages for general use in applying the best of these algorithms on mainframe computers. (See Selected Reference 9 for a comprehensive survey of the available software packages for the various areas of nonlinear programming as well as other areas of mathematical programming.) For example, several powerful software packages such as MINOS have been developed in the Systems Optimization Laboratory at Stanford University. These packages are widely used elsewhere for solving many of the types of problems discussed in this chapter (as well as linear programming problems). The steady improvements being made in both algorithmic techniques and software now are bringing some rather large problems into the range of computational feasibility.

With the current rapid growth in the use and power of personal computers, good progress is being made in nonlinear programming software development for microcomputers. For example, the GAMS/MINOS package (a combination of the GAMS modeling language introduced in Sec. 4.8 and MINOS) is available for use on IBM personal computers. Another prominent package called GINO (see Selected Reference 5) was developed specifically for microcomputers.

Research in nonlinear programming remains very active.

SELECTED REFERENCES

1. Bazaraa, M. S., H. D. Sherali, and C. M. Shetty, *Nonlinear Programming: Theory and Algorithms,* 2d ed., Wiley, New York, 1993.
2. Fiacco, A. V., and G. P. McCormick, *Nonlinear Programming: Sequential Unconstrained Minimization Techniques,* Classics in Applied Mathematics 4, Society for Industrial and Applied Mathematics, Philadelphia, 1990. (Reprint of a classic book published in 1968.)
3. Fletcher, R., *Practical Methods of Optimization,* 2d ed., Wiley, New York, 1988.
4. Gill, P. E., W. Murray, and M. H. Wright, *Practical Optimization,* Academic Press, London, 1981.
5. Liebman, J., L. Lasdon, A. Waren, and L. Schrage, *Modeling and Optimization with GINO,* Boyd & Fraser, Danvers, MA, 1986.
6. Luenberger, D. G., *Introduction to Linear and Nonlinear Programming,* 2d ed., Addison-Wesley, Reading, MA, 1984.
7. McCormick, G. P., *Nonlinear Programming: Theory, Algorithms and Applications,* Wiley, New York, 1983.
8. Minoux, M., *Mathematical Programming: Theory and Algorithms,* Wiley-Interscience, New York, 1986.
9. Moré, J. J., and S. J. Wright, *Optimization Software Guide,* Frontiers in Applied Mathematics 14, Society for Industrial and Applied Mathematics, Philadelphia, 1993.
10. Murray, W., ''Algorithms for Large Nonlinear Programming Problems,'' pp. 172–185 in J. R. Birge and K. G. Murty (eds.), *Mathematical Programming: State of the Art 1994,* 15th

International Symposium on Mathematical Programming, University of Michigan, Ann Arbor, Mich., 1994.

11. Nesterov, Y., and A. Nemirovskii, *Interior-Point Polynomial Algorithms in Convex Programming,* Studies in Applied Mathematics 13, Society for Industrial and Applied Mathematics, Philadelphia, 1993.
12. Nocedal, J., "Recent Advances in Large-Scale Nonlinear Optimization," pp. 208–219 in J. R. Birge and K. G. Murty (eds.), *Mathematical Programming: State of the Art 1994,* 15th International Symposium on Mathematical Programming, University of Michigan, Ann Arbor, Mich., 1994.
13. Reklaitis, G. V., A. Ravindran, and K. M. Ragsdell, *Engineering Optimization: Methods and Applications,* Wiley, New York, 1983.

RELEVANT ROUTINES IN YOUR OR COURSEWARE

Demonstration Examples:
Gradient Search Procedure
Frank-Wolfe Algorithm
Sequential Unconstrained Minimization Technique—SUMT

Interactive Routines:
Interactive One-Dimensional Search Procedure
Interactive Gradient Search Procedure
Interactive Modified Simplex Method
Interactive Frank-Wolfe Algorithm

Automatic Routines:
Enter or Revise a General Linear Programming Model
Solve Automatically by the Simplex Method
Automatic One-Dimensional Search Procedure
Automatic Gradient Search Procedure
Solve Quadratic Programming Model Automatically
Sequential Unconstrained Minimization Technique—SUMT

To access these routines, call the MathProg program and then choose *Nonlinear Programming* under the Area menu. (For the first two automatic routines listed, choose *General Linear Programming* under the Area menu.) See Appendix 1 for documentation of the software.

PROBLEMS

To the left of each of the following problems (or their parts), we have inserted a D (for demonstration), I (for interactive routine), or A (for automatic routine) whenever a corresponding routine listed above can be helpful. An asterisk on the I or A indicates that this routine definitely should be used (unless your instructor gives contrary instructions) and that the printout from this routine is all that needs to be turned in to show your work in executing the algorithm. An asterisk on the problem number indicates that at least a partial answer is given in the back of the book.

13.1-1. Consider the product-mix problem describe in Prob. 3.1-3. Suppose that this manufacturing firm actually encounters *price elasticity* in selling the three products, so that the profits will be different from those stated in Chap. 3. In particular, suppose that the unit costs for producing products 1, 2, and 3 are \$25, \$10, and \$15, respectively, and that the prices required (in dollars) in order to be able to see x_1, x_2, and x_3 units are $35 + 100x_1^{-1/3}$, $15 + 40x_2^{-1/4}$, and $20 + 50x_3^{-1/2}$, respectively.

Formulate a nonlinear programming model for the problem of determining how many units of each product the firm should produce to maximize profit.

13.1-2. For the P & T Co. problem described in Sec. 8.1, suppose that there is a 10 percent discount in the shipping cost for all truckloads *beyond* the first 40 for each combination of cannery and warehouse. Draw figures like Figs. 13.3 and 13.4, showing the marginal cost and total cost for shipments of truckloads of peas from cannery 1 to warehouse 1. Then describe the overall nonlinear programming model for this problem.

13.1-3. Consider the following example of the portfolio selection problem with risky securities described in Sec. 13.1. Just two stocks are being considered for inclusion in the portfolio. The estimated mean and variance of the return on each share of stock 1 are 5 and 4, respectively, whereas the corresponding quantities for stock 2 are 10 and 100, respectively. The covariance of the return on one share each of the two stocks is 5. The price per share is 20 for stock 1 and 30 for stock 2, where the total amount budgeted for the portfolio is 50.

(*a*) Without assigning a specific numerical value to β, formulate a nonlinear programming model for this problem.

A* (*b*) Use the routine *Solve Quadratic Programming Model Automatically* in your OR Courseware to solve the model formulated in part (*a*) when $\beta = 0.1$ (moderate aversion to risk), $\beta = 0.02$ (low aversion to risk), and $\beta = 0.5$ (high aversion to risk). Also solve by inspection the extreme cases of $\beta = 0$ (no aversion to risk) and $\beta = \infty$ (complete aversion to risk), where the latter case means to minimize $V(\mathbf{x})$. Characterize the solutions for the five values of β in terms of the relative concentration on the conservative investment (stock 1) or the risky investment (stock 2).

13.2-1. Reconsider Prob. 13.1-1. Verify that this problem is a convex programming problem.

13.2-2. Reconsider Prob. 13.1-3. Show that the model formulated in part (*a*) is a convex programming problem by using the test in Appendix 2 to show that the objective function is concave.

13.2-3. Consider the variation of the Wyndor Glass Co. example represented in Fig. 13.5, where the second and third functional constraints of the original problem (see Sec. 3.1) have been replaced by $9x_1^2 + 5x_2^2 \leq 216$. Demonstrate that $(x_1, x_2) = (2, 6)$ with $Z = 36$ is indeed optimal by showing that the objective function line $36 = 3x_1 + 5x_2$ is *tangent* to this constraint boundary at (2, 6). (*Hint:* Express x_2 in terms of x_1 on this boundary, and then differentiate this expression with respect to x_1 to find the slope of the boundary.)

13.2-4. Consider the variation of the Wyndor Glass Co. problem represented in Fig. 13.6, where the original objective function (see Sec. 3.1) has been replaced by $Z = 126x_1 - 9x_1^2 + 182x_2 - 13x_2^2$. Demonstrate that $(x_1, x_2) = (\frac{8}{3}, 5)$ with $Z = 857$ is indeed optimal by showing that the ellipse $857 = 126x_1 - 9x_1^2 + 182x_2 - 13x_2^2$ is *tangent* to the constraint boundary $3x_1 + 2x_2 = 18$ at $(\frac{8}{3}, 5)$. (*Hint:* Solve for x_2 in terms of x_1 for the ellipse, and then differentiate this expression with respect to x_1 to find the slope of the ellipse.)

13.2-5. Consider the following function:

$$f(x) = 48x - 60x^2 + x^3.$$

(*a*) Use the first and second derivatives to find the local maxima and local minima of $f(x)$.

(*b*) Use the first and second derivatives to show that $f(x)$ has neither a global maximum nor a global minimum because it is unbounded in both directions.

13.2-6. For each of the following functions, show whether it is convex, concave, or neither.

(*a*) $f(x) = 10x - x^2$
(*b*) $f(x) = x^4 + 6x^2 + 12x$
(*c*) $f(x) = 2x^3 - 3x^2$
(*d*) $f(x) = x^4 + x^2$
(*e*) $f(x) = x^3 + x^4$

13.2-7.* For each of the following functions, use the test given in Appendix 2 to determine whether it is convex, concave, or neither.

(*a*) $f(\mathbf{x}) = x_1x_2 - x_1^2 - x_2^2$
(*b*) $f(\mathbf{x}) = 3x_1 + 2x_1^2 + 4x_2 + x_2^2 - 2x_1x_2$
(*c*) $f(\mathbf{x}) = x_1^2 + 3x_1x_2 + 2x_2^2$
(*d*) $f(\mathbf{x}) = 20x_1 + 10x_2$
(*e*) $f(\mathbf{x}) = x_1x_2$

13.2-8. Consider the following function:

$$f(\mathbf{x}) = 5x_1 + 2x_2^2 + x_3^2 - 3x_3x_4 + 4x_4^2 + 2x_5^4 + x_5^2 + 3x_5x_6 + 6x_6^2 + 3x_6x_7 + x_7^2.$$

Show that $f(\mathbf{x})$ is convex by expressing it as a sum of functions of one or two variables and then showing (see Appendix 2) that all these functions are convex.

13.2-9. Consider the following nonlinear programming problem:

$$\text{Maximize} \quad f(\mathbf{x}) = x_1 + x_2,$$

subject to

$$x_1^2 + x_2^2 \le 1$$

and

$$x_1 \ge 0, \qquad x_2 \ge 0.$$

(*a*) Verify that this is a convex programming problem.
(*b*) Solve this problem graphically.

13.2-10. Consider the following nonlinear programming problem:

$$\text{Minimize} \quad Z = x_1^4 + 2x_2^2,$$

subject to

$$x_1^2 + x_2^2 \ge 2.$$

(No nonnegativity constraints.)

(*a*) Use geometric analysis to determine whether the feasible region is a convex set.
(*b*) Now use algebra and calculus to determine whether the feasible region is a convex set.

13.3-1. Reconsider Prob. 13.1-2. Show that this problem is a nonconvex programming problem.

13.3-2. Consider the following constrained optimization problem:

$$\text{Maximize} \quad f(x) = -6x + 3x^2 - 2x^3,$$

subject to

$$x \ge 0.$$

Use just the first and second derivatives of $f(x)$ to derive an optimal solution.

13.3-3. Consider the following nonlinear programming problem:

$$\text{Minimize} \quad Z = x_1^4 + 2x_1^2 + 2x_1x_2 + 4x_2^2,$$

subject to

$$2x_1 + x_2 \geq 10$$
$$x_1 + 2x_2 \geq 10$$

and

$$x_1 \geq 0, \qquad x_2 \geq 0.$$

(*a*) Of the special types of nonlinear programming problems described in Sec. 13.3, to which type or types can this particular problem be fitted? Justify your answer.
(*b*) Now suppose that the problem is changed slightly by replacing the nonnegativity constraints by $x_1 \geq 1$ and $x_2 \geq 1$. Convert this new problem to an equivalent problem that has just two functional constraints, two variables, and two nonnegativity constraints.

13.3-4. Consider the following geometric programming problem:

$$\text{Minimize} \quad f(\mathbf{x}) = 2x_1^{-2}x_2^{-1} + x_2^{-2},$$

subject to

$$4x_1x_2 + x_1^2x_2^2 \leq 12$$

and

$$x_1 \geq 0, \qquad x_2 \geq 0.$$

(*a*) Transform this problem to an equivalent convex programming problem.
(*b*) Use the test given in Appendix 2 to verify that the model formulated in part (*a*) is indeed a convex programming problem.

13.3-5. Consider the following linear fractional programming problem:

$$\text{Maximize} \quad f(\mathbf{x}) = \frac{10x_1 + 20x_2 + 10}{3x_1 + 4x_2 + 20},$$

subject to

$$x_1 + 3x_2 \leq 50$$
$$3x_1 + 2x_2 \leq 80$$

and

$$x_1 \geq 0, \qquad x_2 \geq 0.$$

(*a*) Transform this problem to an equivalent linear programming problem.
A* (*b*) Use the simplex method (the *automatic* routine in your OR Courseware) to solve the model formulated in part (*a*). What is the resulting optimal solution for the original problem?

13.3-6. Consider the expressions in matrix notation given in Sec. 13.7 for the general form of the KKT conditions for the quadratic programming problem. Show that the problem of finding a feasible solution for these conditions is a linear complementarity problem, as introduced in Sec. 13.3, by identifying **w**, **z**, **q**, and **M** in terms of the vectors and matrices in Sec. 13.7.

I*, A **13.4-1.*** Use the one-dimensional search procedure to interactively solve (approximately) the following problem:

$$\text{Maximize} \quad f(x) = x^3 + 2x - 2x^2 - 0.25x^4.$$

Use an error tolerance $\epsilon = 0.04$ and initial bounds $\underline{x} = 0$, $\overline{x} = 2.4$.

I*, A **13.4-2.** Use the one-dimensional search procedure with an error tolerance $\epsilon = 0.04$ and with the following initial bounds to interactively solve (approximately) each of the following problems.

(*a*) Maximize $f(x) = 6x - x^2$, with $\underline{x} = 0$, $\overline{x} = 4.8$.
(*b*) Minimize $f(x) = 6x + 7x^2 + 4x^3 + x^4$, with $\underline{x} = -4$, $\overline{x} = 1$.

I*, A **13.4-3.** Use the one-dimensional search procedure to interactively solve (approximately) the following problem:

$$\text{Maximize} \quad f(x) = 48x^5 + 42x^3 + 3.5x - 16x^6 - 61x^4 - 16.5x^2.$$

Use an error tolerance $\epsilon = 0.08$ and initial bounds $\underline{x} = -1$, $\overline{x} = 4$.

I*, A **13.4-4.** Use the one-dimensional search procedure to interactively solve (approximately) the following problem:

$$\text{Maximize} \quad f(x) = x^3 + 30x - x^6 - 2x^4 - 3x^2.$$

Use an error tolerance $\epsilon = 0.07$ and find appropriate initial bounds by inspection.

13.4-5. Consider the following convex programming problem:

$$\text{Minimize} \quad Z = x^4 + x^2 - 4x,$$

subject to

$$x \leq 2 \quad \text{and} \quad x \geq 0.$$

(*a*) Use one simple calculation *just* to check whether the optimal solution lies in the interval $0 \leq x \leq 1$ or the interval $1 \leq x \leq 2$. (Do *not* actually solve for the optimal solution in order to determine the interval in which it must lie.) Explain your logic.

I*, A (*b*) Use the one-dimensional search procedure with initial bounds $\underline{x} = 0$, $\overline{x} = 2$ and with an error tolerance $\epsilon = 0.02$ to interactively solve (approximately) this problem.

13.4-6. Consider the problem of maximizing a differential function $f(x)$ of a single unconstrained variable x. Let $\underline{x}_0$ and $\overline{x}_0$, respectively, be a valid lower bound and upper bound on the same global maximum (if one exists). Prove the following general properties of the one-dimensional search procedure (as presented in Sec. 13.4) for attempting to solve such a problem.

(*a*) Given $\underline{x}_0$, $\overline{x}_0$, and $\epsilon = 0$, the sequence of trial solutions selected by the *midpoint rule* must *converge* to a limiting solution. [*Hint:* First show that $\lim_{n\to\infty}(\overline{x}_n - \underline{x}_n) = 0$, where $\overline{x}_n$ and $\underline{x}_n$ are the upper and lower bounds identified at iteration n.]
(*b*) If $f(x)$ is concave [so that $df(x)/dx$ is a monotone decreasing function of x], then the limiting solution in part (*a*) must be a global maximum.
(*c*) If $f(x)$ is not concave everywhere, but would be concave if its domain were restricted to the interval between $\underline{x}_0$ and $\overline{x}_0$, then the limiting solution in part (*a*) must be a global maximum.
(*d*) If $f(x)$ is not concave even over the interval between $\underline{x}_0$ and $\overline{x}_0$, then the limiting solution in part (*a*) need not be a global maximum. (Prove this by graphically constructing a counterexample.)
(*e*) If $df(x)/dx < 0$ for all x, then no $\underline{x}_0$ exists. If $df(x)/dx > 0$ for all x, then no $\overline{x}_0$ exists. In either case, $f(x)$ does not possess a global maximum.
(*f*) If $f(x)$ is concave and $\lim_{x\to-\infty} df(x)/dx < 0$, then no $\underline{x}_0$ exists. If $f(x)$ is concave and $\lim_{x\to\infty} df(x)/dx > 0$, then no $\overline{x}_0$ exists. In either case, $f(x)$ does not possess a global maximum.

A* **13.4-7.** Consider the following linearly constrained convex programming problem:

$$\text{Maximize} \quad f(\mathbf{x}) = 32x_1 + 50x_2 - 10x_2^2 + x_2^3 - x_1^4 - x_2^4,$$

subject to

$$3x_1 + x_2 \leq 11$$

$$2x_1 + 5x_2 \leq 16$$

and

$$x_1 \geq 0, \qquad x_2 \geq 0.$$

Ignore the constraints and solve the resulting two one-variable unconstrained optimization problems. Use calculus to solve the problem involving x_1, and use the one-dimensional search procedure with $\epsilon = 0.001$ and initial bounds 0 and 4 to solve the problem involving x_2. Show that the resulting solution for (x_1, x_2) satisfies all the constraints, so it is actually optimal for the original problem.

13.5-1. Consider the following unconstrained optimization problem:

$$\text{Maximize} \quad f(\mathbf{x}) = 2x_1x_2 + x_2 - x_1^2 - 2x_2^2.$$

D, I*, A (*a*) Starting from the initial trial solution $(x_1, x_2) = (1, 1)$, interactively apply the gradient search procedure with $\epsilon = 0.25$ to obtain an approximate solution.

(*b*) Solve the system of linear equations obtained by setting $\nabla f(\mathbf{x}) = \mathbf{0}$ to obtain the exact solution.

(*c*) Referring to Fig. 13.14 as a sample for a similar problem, draw the path of trial solutions you obtained in part (*a*). Then show the apparent *continuation* of this path with your best guess for the next three trial solutions [based on the pattern in part (*a*) and Fig. 13.14]. Also show the exact solution from part (*b*) toward which this sequence of trial solutions is converging.

A* (*d*) Apply the automatic routine for the gradient search procedure (with $\epsilon = 0.01$) in your OR Courseware to this problem.

13.5-2. Repeat the four parts of Prob. 13.5-1 (except with $\epsilon = 0.5$) for the following unconstrained optimization problem:

$$\text{Maximize} \quad f(\mathbf{x}) = 2x_1x_2 - 2x_1^2 - x_2^2.$$

D, I*, A* **13.5-3.** Starting from the initial trial solution $(x_1, x_2) = (1, 1)$, interactively apply two iterations of the gradient search procedure to begin solving the following problem, and then apply the automatic routine for this procedure (with $\epsilon = 0.01$).

$$\text{Maximize} \quad f(\mathbf{x}) = 4x_1x_2 - 2x_1^2 - 3x_2^2.$$

Then solve $\nabla f(\mathbf{x}) = \mathbf{0}$ directly to obtain the exact solution.

D, I*, A* **13.5-4.*** Starting from the initial trial solution $(x_1, x_2) = (0, 0)$, interactively apply the gradient search procedure with $\epsilon = 0.3$ to obtain an approximate solution for the following problem, and then apply the automatic routine for this procedure (with $\epsilon = 0.01$).

$$\text{Maximize} \quad f(\mathbf{x}) = 8x_1 - x_1^2 - 12x_2 - 2x_2^2 + 2x_1x_2.$$

Then solve $\nabla f(\mathbf{x}) = \mathbf{0}$ directly to obtain the exact solution.

D, I*, A* **13.5-5.** Starting from the initial trial solution $(x_1, x_2) = (0, 0)$, interactively apply two iterations of the gradient search procedure to begin solving the following problem, and then apply the automatic routine for this procedure (with $\epsilon = 0.01$).

$$\text{Maximize} \quad f(\mathbf{x}) = 6x_1 + 2x_1x_2 - 2x_2 - 2x_1^2 - x_2^2.$$

Then solve $\nabla f(\mathbf{x}) = \mathbf{0}$ directly to obtain the exact solution.

13.5-6. Starting from the initial trial solution $(x_1, x_2) = (0, 0)$, apply *one* iteration of the gradient search procedure to the following problem by hand:

$$\text{Maximize} \qquad f(\mathbf{x}) = 4x_1 + 2x_2 + x_1^2 - x_1^4 - 2x_1x_2 - x_2^2.$$

To complete this iteration, approximately solve for t^* by manually applying *two* iterations of the one-dimensional search procedure with initial bounds $\underline{t} = 0$, $\bar{t} = 1$.

13.5-7. Consider the following unconstrained optimization problem:

$$\text{Maximize} \qquad f(\mathbf{x}) = 3x_1x_2 + 3x_2x_3 - x_1^2 - 6x_2^2 - x_3^2.$$

(*a*) Describe how solving this problem can be reduced to solving a *two-variable* unconstrained optimization problem.

D, I*, A (*b*) Starting from the initial trial solution $(x_1, x_2, x_3) = (1, 1, 1)$, interactively apply the gradient search procedure with $\epsilon = 0.05$ to solve (approximately) the two-variable problem identified in part (*a*).

A* (*c*) Repeat part (*b*) with the automatic routine for this procedure (with $\epsilon = 0.005$).

D, I*, A* **13.5-8.*** Starting from the initial trial solution $(x_1, x_2) = (0, 0)$, interactively apply the gradient search procedure with $\epsilon = 1$ to solve (approximately) each of the following problems, and then apply the automatic routine for this procedure (with $\epsilon = 0.01$).

(*a*) Maximize $f(\mathbf{x}) = x_1x_2 + 3x_2 - x_1^2 - x_2^2$.

(*b*) Minimize $f(\mathbf{x}) = x_1^2x_2^2 + 2x_1^2 + 2x_2^2 - 4x_1 + 4x_2$.

13.6-1. Reconsider the one-variable convex programming model given in Prob. 13.4-5. Use the KKT conditions to derive an optimal solution for this model.

13.6-2. Reconsider Prob. 13.2-9. Use the KKT conditions to check whether $(x_1, x_2) = (1/\sqrt{2}, 1/\sqrt{2})$ is optimal.

13.6-3.* Reconsider the model given in Prob. 13.3-3. What are the KKT conditions for this model? Use these conditions to determine whether $(x_1, x_2) = (0, 10)$ can be optimal.

13.6-4. Consider the following convex programming problem:

$$\text{Maximize} \qquad f(\mathbf{x}) = 24x_1 - x_1^2 + 10x_2 - x_2^2,$$

subject to

$$x_1 \le 8,$$

$$x_2 \le 7,$$

and

$$x_1 \ge 0, \qquad x_2 \ge 0.$$

(*a*) Use the KKT conditions for this problem to derive an optimal solution.

(*b*) Decompose this problem into two separate constrained optimization problems involving just x_1 and just x_2, respectively. For each of these two problems, plot the objective function over the feasible region in order to *demonstrate* that the value of x_1 or x_2 derived in part (*a*) is indeed optimal. Then *prove* that this value is optimal by using just the first and second derivatives of the objective function and the constraints for the respective problems.

13.6-5. Consider the following linearly constrained optimization problem:

$$\text{Maximize } f(\mathbf{x}) = \ln(1 + x_1 + x_2),$$

subject to

$$x_1 + 2x_2 \le 5$$

and

$$x_1 \geq 0, \qquad x_2 \geq 0,$$

where ln denotes the natural logarithm.

(*a*) Verify that this problem is a convex programming problem.
(*b*) Use the KKT conditions to derive an optimal solution.
(*c*) Use intuitive reasoning to demonstrate that the solution obtained in part (*b*) is indeed optimal. [*Hint:* Note that $\ln(1 + x_1 + x_2)$ is a monotonic strictly increasing function of $1 + x_1 + x_2$.]

13.6-6. Consider the following linearly constrained optimization problem:

$$\text{Maximize} \qquad f(\mathbf{x}) = \ln(x_1 + 1) - x_2^2,$$

subject to

$$x_1 + 2x_2 \leq 3$$

and

$$x_1 \geq 0, \qquad x_2 \geq 0,$$

where ln denotes the natural logarithm,

(*a*) Verify that this problem is a convex programming problem.
(*b*) Use the KKT conditions to derive an optimal solution.
(*c*) Use intuitive reasoning to demonstrate that the solution obtained in part (*b*) is indeed optimal.

13.6-7. Consider the following convex programming problem:

$$\text{Maximize} \qquad f(\mathbf{x}) = 10x_1 - 2x_1^2 - x_1^3 + 8x_2 - x_2^2,$$

subject to

$$x_1 + x_2 \leq 2$$

and

$$x_1 \geq 0, \qquad x_2 \geq 0.$$

(*a*) Use the KKT conditions to demonstrate that $(x_1, x_2) = (1, 1)$ is *not* an optimal solution.
(*b*) Use the KKT conditions to derive an optimal solution.

13.6-8.* Consider the nonlinear programming problem given in Prob. 10.3-14. Determine whether $(x_1, x_2) = (1, 2)$ can be optimal by applying the KKT conditions.

13.6-9. Consider the following nonlinear programming problem:

$$\text{Maximize} \qquad f(\mathbf{x}) = \frac{x_1}{x_2 + 1},$$

subject to

$$x_1 - x_2 \leq 2$$

and

$$x_1 \geq 0, \qquad x_2 \geq 0.$$

(*a*) Use the KKT conditions to demonstrate that $(x_1, x_2) = (4, 2)$ is *not* optimal.
(*b*) Derive a solution that does satisfy the KKT conditions.
(*c*) Show that this problem is *not* a convex programming problem.

(*d*) Despite the conclusion in part (*c*), use *intuitive* reasoning to show that the solution obtained in part (*b*) is, in fact, optimal. [The theoretical reason is that $f(\mathbf{x})$ is *pseudo-concave.*]

(*e*) Use the fact that this problem is a linear fractional programming problem to transform it into an equivalent linear programming problem. Solve the latter problem and thereby identify the optimal solution for the original problem. (*Hint:* Use the equality constraint in the linear programming problem to substitute one of the variables out of the model, and then solve the model graphically.)

13.6-10.* Use the KKT conditions to derive an optimal solution for each of the following problems.

(*a*)

$$\text{Maximize} \quad f(\mathbf{x}) = x_1 + 2x_2 - x_2^3,$$

subject to

$$x_1 + x_2 \leq 1$$

and

$$x_1 \geq 0, \quad x_2 \geq 0.$$

(*b*)

$$\text{Maximize} \quad f(\mathbf{x}) = 20x_1 + 10x_2,$$

subject to

$$x_1^2 + x_2^2 \leq 1$$

$$x_1 + 2x_2 \leq 2$$

and

$$x_1 \geq 0, \quad x_2 \geq 0.$$

13.6-11. Reconsider the nonlinear programming model given in Prob. 10.3-16.

(*a*) Use the KKT conditions to determine whether $(x_1, x_2, x_3) = (1, 1, 1)$ can be optimal.

(*b*) If a specific solution satisfies the KKT conditions for this problem, can you draw the definite conclusions that this solution is optimal? Why?

13.6-12. What are the KKT conditions for nonlinear programming problems of the following form?

$$\text{Minimize} \quad f(\mathbf{x}),$$

subject to

$$g_i(\mathbf{x}) \geq b_i, \quad \text{for } i = 1, 2, \ldots, m$$

and

$$\mathbf{x} \geq \mathbf{0}.$$

(*Hint:* Convert this form to our standard form assumed in this chapter by using the techniques presented in Sec. 4.6 and then applying the KKT conditions as given in Sec. 13.6.)

13.6-13. Consider the following nonlinear programming problem:

$$\text{Minimize} \quad Z = 2x_1^2 + x_2^2,$$

subject to

$$x_1 + x_2 = 10$$

and

$$x_1 \geq 0, \quad x_2 \geq 0.$$

(*a*) Of the special types of nonlinear programming problems described in Sec. 13.3, to which type or types can this particular problem be fitted? Justify your answer. (*Hint:* First convert this problem to an equivalent nonlinear programming problem that fits the form given in the second paragraph of the chapter, with $m = 2$ and $n = 2$.)
(*b*) Obtain the KKT conditions for this problem.
(*c*) Use the KKT conditions to derive an optimal solution.

13.6-14. Consider the following linearly constrained programming problem:

$$\text{Minimize} \qquad f(\mathbf{x}) = x_1^3 + 4x_2^2 + 16x_3,$$

subject to

$$x_1 + x_2 + x_3 = 5$$

and

$$x_1 \geq 1, \qquad x_2 \geq 1, \qquad x_3 \geq 1.$$

(*a*) Convert this problem to an equivalent nonlinear programming problem that fits the form given at the beginning of the chapter (second paragraph), with $m = 2$ and $n = 3$.
(*b*) Use the form obtained in part (*a*) to construct the KKT conditions for this problem.
(*c*) Use the KKT conditions to check whether $(x_1, x_2, x_3) = (2, 1, 2)$ is optimal.

13.6-15. Consider the following linearly constrained convex programming problem:

$$\text{Minimize} \qquad Z = x_1^2 - 6x_1 + x_2^3 - 3x_2,$$

subject to

$$x_1 + x_2 \leq 1$$

and

$$x_1 \geq 0, \qquad x_2 \geq 0.$$

(*a*) Obtain the KKT conditions for this problem.
(*b*) Use the KKT conditions to check whether $(x_1, x_2) = (\frac{1}{2}, \frac{1}{2})$ is an optimal solution.
(*c*) Use the KKT conditions to derive an optimal solution.

13.6-16. Consider the following linearly constrained convex programming problem:

$$\text{Maximize} \qquad f(\mathbf{x}) = 8x_1 - x_1^2 + 2x_2 + x_3,$$

subject to

$$x_1 + 3x_2 + 2x_3 \leq 12$$

and

$$x_1 \geq 0, \qquad x_2 \geq 0, \qquad x_3 \geq 0.$$

(*a*) Use the KKT conditions to demonstrate that $(x_1, x_2, x_3) = (2, 2, 2)$ is *not* an optimal solution.
(*b*) Use the KKT conditions to derive an optimal solution. (*Hint:* Do some preliminary intuitive analysis to determine the most promising case regarding which variables are nonzero and which are zero.)

13.6-17. Use the KKT conditions to determine whether $(x_1, x_2, x_3) = (1, 1, 1)$ can be optimal for the following problem:

$$\text{Minimize} \qquad Z = 2x_1 + x_2^3 + x_3^2,$$

subject to

$$x_1^2 + 2x_2^2 + x_3^2 \geq 4$$

and

$$x_1 \geq 0, \quad x_2 \geq 0, \quad x_3 \geq 0.$$

13.6-18. Reconsider the model given in Prob. 13.2-10. What are the KKT conditions for this problem? Use these conditions to determine whether $(x_1, x_2) = (1, 1)$ can be optimal.

13.6-19. Reconsider the linearly constrained convex programming model given in Prob. 13.4-7. Use the KKT conditions to determine whether $(x_1, x_2) = (2, 2)$ can be optimal.

13.7-1. Consider the quadratic programming example presented in Sec. 13.7.

(*a*) Use the test given in Appendix 2 to show that the objective function is *strictly concave.*

(*b*) Verify that the objective function is strictly concave by demonstrating that $\mathbf{Q}$ is a *positive definite* matrix; that is, $\mathbf{x}^T\mathbf{Q}\mathbf{x} > \mathbf{0}$ for all $\mathbf{x} \neq \mathbf{0}$. (*Hint:* Reduce $\mathbf{x}^T\mathbf{Q}\mathbf{x}$ to a sum of squares.)

(*c*) Show that $x_1 = 12$, $x_2 = 9$, and $u_1 = 3$ satisfy the KKT conditions when they are written in the form given in Sec. 13.6.

13.7-2.* Consider the following quadratic programming problem:

$$\text{Maximize} \quad f(\mathbf{x}) = 8x_1 - x_1^2 + 4x_2 - x_2^2,$$

subject to

$$x_1 + x_2 \leq 2$$

and

$$x_1 \geq 0, \quad x_2 \geq 0.$$

(*a*) Use the KKT conditions to derive an optimal solution.

(*b*) Now suppose that this problem is to be solved by the modified simplex method. Formulate the linear programming problem that is to be addressed *explicitly,* and then identify the additional complementarity constraint that is enforced automatically by the algorithm.

I* (*c*) Apply the modified simplex method to the problem as formulated in part (*b*).

13.7-3. Consider the following quadratic programming problem:

$$\text{Maximize} \quad f(\mathbf{x}) = 20x_1 - 20x_1^2 + 50x_2 - 5x_2^2 + 18x_1x_2,$$

subject to

$$x_1 + x_2 \leq 6$$
$$x_1 + 4x_2 \leq 18$$

and

$$x_1 \geq 0, \quad x_2 \geq 0.$$

Suppose that this problem is to be solved by the modified simplex method.

(*a*) Formulate the linear programming problem that is to be addressed explicitly, and then identify the additional complementarity constraint that is enforced automatically by the algorithm.

I* (*b*) Apply the modified simplex method to the problem as formulated in part (*a*).

13.7-4. Consider the following quadratic programming problem.

$$\text{Maximize} \quad f(\mathbf{x}) = 2x_1 + 3x_2 - x_1^2 - x_2^2,$$

subject to

$$x_1 + x_2 \leq 2$$

and

$$x_1 \geq 0, \qquad x_2 \geq 0.$$

(*a*) Use the KKT conditions to derive an optimal solution directly.

(*b*) Now suppose that this problem is to be solved by the modified simplex method. Formulate the linear programming problem that is to be addressed explicitly, and then identify the additional complementarity constraint that is enforced automatically by the algorithm.

(*c*) Without applying the modified simplex method, show that the solution derived in part (*a*) is indeed optimal ($Z = 0$) for the equivalent problem formulated in part (*b*).

I* (*d*) Apply the modified simplex method to the problem as formulated in part (*b*).

13.7-5. Reconsider the first quadratic programming variation of the Wyndor Glass Co. problem presented in Sec. 13.2 (see Fig. 13.6). Analyze this problem by following the instructions of parts (*a*), (*b*), and (*c*) of Prob. 13.7-4.

13.7-6. Consider the following quadratic programming problem:

$$\text{Minimize} \qquad f(x_1, x_2) = (x_1 - 1)^2 + (x_2 - 2)^2 - 3(x_1 + x_2),$$

subject to

$$4x_1 + x_2 \leq 20$$
$$x_1 + 4x_2 \leq 20$$

and

$$x_1 \geq 0, \qquad x_2 \geq 0.$$

(*a*) Obtain the KKT conditions for this problem in the form given in Sec. 13.6. (*Hint:* These conditions assume that the objective function is to be maximized.)

(*b*) You are given the information that the optimal solution does *not* lie on the boundary of the feasible region. Use this information to derive the optimal solution from the KKT conditions.

(*c*) Now suppose that this problem is to be solved by the modified simplex method. Formulate the linear programming problem that is to be addressed explicitly, and then identify the additional complementarity constraint that is enforced automatically by the algorithm.

(*d*) Apply the modified simplex method to the problem as formulated in part (*c*).

13.8-1. Reconsider the quadratic programming model given in Prob. 13.7-6.

(*a*) Use the separable programming formulation presented in Sec. 13.8 to formulate an approximate linear programming model for this problem. Use $x_1, x_2 = 0, 2.5, 5$ as the breakpoints of the piecewise linear functions.

A* (*b*) Use the simplex method to solve the model formulated in part (*a*). Then reexpress this solution in terms of the *original* variables of the problem.

A* (*c*) To improve the approximation, now use $x_1, x_2 = 0, 1, 2, 3, 4, 5$ as the breakpoints of the piecewise linear functions and repeat parts (*a*) and (*b*).

13.8-2. A certain corporation is planning to produce and market three different products. Let x_1, x_2, and x_3 denote the number of units of the three respective products to be produced. The preliminary estimates of their potential profitability are as follows.

For the first 15 units produced of product 1, the unit profit would be approximately \$36. The unit profit would be only \$3 for any additional units of product 1. For the first 20 units produced of product 2, the unit profit is estimated at \$24. The unit profit would be \$12 for each of the next 20 units and \$9 for any additional units. For the first 10 units of product 3, the unit profit would be \$45. The unit profit would be \$30 for each of the next 5 units and \$18 for any additional units.

Certain limitations on the use of needed resources impose the following constraints on the production of the three products:

$$x_1 + x_2 + x_3 \le 60$$
$$3x_1 + 2x_2 \le 200$$
$$x_1 + 2x_3 \le 70.$$

Management wants to know what values of x_1, x_2 and x_3 should be chosen to maximize the total profit.

(*a*) Use the separable programming technique presented in Sec. 13.8 to formulate a linear programming model for this problem.

A* (*b*) Use the simplex method to solve the model formulated in part (*a*). Then reexpress this solution in terms of the *original* variables of the problem.

(*c*) Now suppose that there is an additional constraint that the profit from products 1 and 2 must total at least $900. Use the technique presented in the "Extensions" subsection of Sec. 13.8 to add this constraint to the model formulated in part (*a*).

A* (*d*) Repeat part (*b*) for the model formulated in part (*c*).

13.8-3.* Consider the following convex programming problem:

$$\text{Maximize} \quad f(\mathbf{x}) = 4x_1 + 6x_2 - x_1^3 - 2x_2^2,$$

subject to

$$x_1 + 3x_2 \le 8$$
$$5x_1 + 2x_2 \le 14$$

and

$$x_1 \ge 0, \qquad x_2 \ge 0.$$

(*a*) Verify that $(x_1, x_2) = (2/\sqrt{3}, \frac{3}{2})$ is an optimal solution by applying the KKT conditions.

(*b*) Use the separable programming technique presented in Sec. 13.8 to formulate an *approximate* linear programming model for this problem. Use $x_1, x_2 = 0, 1, 2, 3$ as the breakpoints of the piecewise linear functions.

(*c*) Use the simplex method to solve the approximate model formulated in part (*b*). Verify that the optimal solution satisfies the special restriction for the model. Compare this solution with the exact optimal solution for the original problem [see part (*a*)].

13.8-4. Reconsider the production scheduling problem of the Build-Em-Fast Company described in Prob. 8.1-7. The special restriction for such a situation is that overtime should not be used in any particular period unless regular time in that period is completely used up. Explain why the logic of separable programming implies that this restriction will be satisfied automatically by any optimal solution for the transportation problem formulation of the problem.

13.8-5. Reconsider the linearly constrained convex programming model given in Prob. 13.4-7.

(*a*) Use the separable programming technique presented in Sec. 13.8 to formulate an approximate linear programming model for this problem. Use $x_1 = 0, 1, 2, 3$ and $x_2 = 0, 1, 2, 3$ as the breakpoints of the piecewise linear functions.

A* (*b*) Use the simplex method to solve the model formulated in part (*a*). Then reexpress this solution in terms of the *original* variables of the problem.

13.8-6. Suppose that the separable programming technique has been applied to a certain problem (the "original problem") to convert it to the following equivalent linear programming problem:

$$\text{Maximize} \quad Z = 5x_{11} + 4x_{12} + 2x_{13} + 4x_{21} + x_{22},$$

subject to

$$3x_{11} + 3x_{12} + 3x_{13} + 2x_{21} + 2x_{22} \leq 25$$

$$2x_{11} + 2x_{12} + 2x_{13} - x_{21} - x_{22} \leq 10$$

and

$$0 \leq x_{11} \leq 2 \qquad 0 \leq x_{21} \leq 3$$

$$0 \leq x_{12} \leq 3 \qquad 0 \leq x_{22} \leq 1.$$

$$0 \leq x_{13}$$

What was the mathematical model for the original problem? (You may define the objective function either algebraically or graphically, but express the constraints algebraically.)

13.8-7. For each of the following cases, *prove* that the key property of separable programming given in Sec. 13.8 must hold. (*Hint:* Assume that there exists an optimal solution that violates this property, and then contradict this assumption by showing that there exists a better feasible solution.)

(*a*) The special case of separable programming where all the $g_i(\mathbf{x})$ are linear functions.

(*b*) The general case of separable programming where all the functions are nonlinear functions of the designated form. [*Hint:* Think of the functional constraints as constraints on resources, where $g_{ij}(x_j)$ represents the amount of resource i used by running activity j at level x_j, and then use what the convexity assumption implies about the slopes of the approximating piecewise linear function.]

13.8-8. The MFG Company produces a certain subassembly in each of two separate plants. These subassemblies are then brought to a third nearby plant where they are used in the production of a certain product. The peak season of demand for this product is approaching, so to maintain the production rate within a desired range, it is necessary to use temporarily some overtime in making the subassemblies. The cost per subassembly on regular time (RT) and on overtime (OT) is shown in the following table for both plants, along with the maximum number of subassemblies that can be produced on RT and on OT each day.

	Unit Cost		*Capacity*	
	RT	OT	RT	OT
Plant 1	\$15	\$25	2,000	1,000
Plant 2	\$16	\$24	1,000	500

Let x_1 and x_2 denote the total number of subassemblies produced per day at plants 1 and 2, respectively. Suppose that the objective is to maximize $Z = x_1 + x_2$, subject to the constraint that the total daily cost not exceed \$60,000. Note that the mathematical programming formulation of this problem (with x_1 and x_2 as decision variables) has the same form as the main case of the separable programming model described in Sec. 13.8, except that the separable functions appear in a constraint function rather than the objective function. However, if it is allowable to use OT even when the RT capacity at that plant is not fully used, the same approach can be used to reformulate the problem as a linear programming problem.

(*a*) Formulate this linear programming problem.

(*b*) Explain why the logic of separable programming also applies here to guarantee that an optimal solution for the model formulated in part (*a*) never uses OT unless the RT capacity at that plant has been fully used.

13.8-9. Consider the following nonlinear programming problem (first considered in Prob. 10.3-23):

$$\text{Maximize} \quad Z = 5x_1 + x_2,$$

subject to

$$2x_1^2 + x_2 \le 13$$
$$x_1^2 + x_2 \le 9$$

and

$$x_1 \ge 0, \qquad x_2 \ge 0.$$

(*a*) Show that this problem is a convex programming problem.
(*b*) Use the separable programming technique discussed at the end of Sec. 13.8 to formulate an approximate linear programming model for this problem. Use the integers as the breakpoints of the piecewise linear function.
A* (*c*) Use the simplex method to solve the model formulated in part (*b*). Then reexpress this solution in terms of the original variables of the problem.

13.8-10. Consider the following convex programming problem:

$$\text{Maximize} \quad Z = 32x_1 - x_1^4 + 4x_2 - x_2^2,$$

subject to

$$x_1^2 + x_2^2 \le 9$$

and

$$x_1 \ge 0, \qquad x_2 \ge 0.$$

(*a*) Apply the separable programming technique discussed at the end of Sec. 13.8, with $x_1 = 0, 1, 2, 3$ and $x_2 = 0, 1, 2, 3$ as the breakpoints of the piecewise linear functions, to formulate an approximate linear programming model for this problem.
A* (*b*) Use the simplex method to solve the model formulated in part (*a*). Then reexpress this solution in terms of the original variables of the problem.
(*c*) Use the KKT conditions to determine whether the solution for the original variables obtained in part (*b*) actually is optimal for the original problem (not the approximate model).

13.8-11. Reconsider the integer nonlinear programming model given in Prob. 10.3-11.
(*a*) Show that the objective function is not concave.
(*b*) Formulate an equivalent *pure binary* integer *linear* programming model for this problem as follows. Apply the separable programming technique with the feasible integers as the breakpoints of the piecewise linear functions, so that the auxiliary variables are binary variables. Then add some linear programming constraints on these binary variables to enforce the *special restriction* of separable programming. (Note that the *key property* of separable programming does not hold for this problem because the objective function is not concave.)
A* (*c*) Use the automatic routine for solving binary integer programming problems in the integer programming area of your OR Courseware to solve this problem as formulated in part (*b*). Then reexpress this solution in terms of the original variables of the problem.

D, I* **13.9-1.*** Reconsider the linearly constrained convex programming model given in Prob. 13.6-5. Starting from the initial trial solution $(x_1, x_2) = (1, 1)$, use one iteration of the Frank-Wolfe algorithm to obtain exactly the same solution you found in part (*b*) of Prob. 13.6-5, and then use a second iteration to verify that it is an optimal solution (because it is replicated exactly).

Explain why exactly the same results would be obtained on these two iterations with any other initial trial solution except (0, 0). What complication arises with (0, 0)?

D, I* **13.9-2.** Reconsider the linearly constrained convex programming model given in Prob. 13.6-2. Starting from the initial trial solution $(x_1, x_2) = (0, 0)$, use one iteration of the Frank-Wolfe algorithm to obtain exactly the same solution you found in part (*b*) of Prob. 13.6-2, and then use a second iteration to verify that it is an optimal solution (because it is replicated exactly).

D, I* **13.9-3.** Reconsider the linearly constrained convex programming model given in Prob. 13.6-15. Starting from the initial trial solution $(x_1, x_2) = (0, 0)$, use one iteration of the Frank-Wolfe algorithm to obtain exactly the same solution you found in part (*c*) of Prob. 13.6-15, and then use a second iteration to verify that it is an optimal solution (because it is replicated exactly). Explain why exactly the same results would be obtained on these two iterations with any other trial solution.

D, I* **13.9-4.** Reconsider the linearly constrained convex programming model given in Prob. 13.6-16. Starting from the initial trial solution $(x_1, x_2, x_3) = (0, 0, 0)$, apply two iterations of the Frank-Wolfe algorithm.

D, I* **13.9-5.** Consider the quadratic programming example presented in Sec. 13.7. Starting from the initial trial solution $(x_1, x_2) = (5, 5)$, apply seven iterations of the Frank-Wolfe algorithm.

13.9-6. Reconsider the quadratic programming model given in Prob. 13.7-4.

D, I* (*a*) Starting from the initial trial solution $(x_1, x_2) = (0, 0)$, use the Frank-Wolfe algorithm (six iterations) to solve the problem (approximately).

(*b*) Show graphically how the sequence of trial solutions obtained in part (*a*) can be extrapolated to obtain a closer approximation of an optimal solution. What is your resulting estimate of this solution?

D, I* **13.9-7.** Reconsider the first quadratic programming variation of the Wyndor Glass Co. problem presented in Sec. 13.2 (see Fig. 13.6). Starting from the initial trial solution $(x_1, x_2) = (0, 0)$, use three iterations of the Frank-Wolfe algorithm to obtain and verify the optimal solution.

D, I* **13.9-8.** Reconsider the linearly constrained convex programming model given in Prob. 13.4-7. Starting from the initial trial solution $(x_1, x_2) = (0, 0)$, use the Frank-Wolfe algorithm (four iterations) to solve this model (approximately).

D, I* **13.9-9.** Consider the following linearly constrained convex programming problem:

$$\text{Maximize} \quad f(\mathbf{x}) = 3x_1x_2 + 40x_1 + 30x_2 - 4x_1^2 - x_1^4 - 3x_2^2 - x_2^4,$$

subject to

$$4x_1 + 3x_2 \leq 12$$
$$x_1 + 2x_2 \leq 4$$

and

$$x_1 \geq 0, \qquad x_2 \geq 0.$$

Starting from the initial trial solution $(x_1, x_2) = (0, 0)$, apply two iterations of the Frank-Wolfe algorithm.

D, I* **13.9-10.*** Consider the following linearly constrained convex programming problem:

$$\text{Maximize} \quad f(\mathbf{x}) = 3x_1 + 4x_2 - x_1^3 - x_2^2,$$

subject to

$$x_1 + x_2 \leq 1$$

and

$$x_1 \geq 0, \qquad x_2 \geq 0.$$

(*a*) Starting from the initial trial solution $(x_1, x_2) = (\frac{1}{4}, \frac{1}{4})$, apply three iterations of the Frank-Wolfe algorithm.
(*b*) Use the KKT conditions to check whether the solution obtained in part (*a*) is, in fact, optimal.

13.9-11. Consider the following linearly constrained convex programming problem:

$$\text{Maximize} \qquad f(\mathbf{x}) = 4x_1 - x_1^4 + 2x_2 - x_2^2,$$

subject to

$$4x_1 + 2x_2 \leq 5$$

and

$$x_1 \geq 0, \qquad x_2 \geq 0.$$

(*a*) Starting from the initial trial solution $(x_1, x_2) = (\frac{1}{2}, \frac{1}{2})$, apply four iterations of the Frank-Wolfe algorithm.
(*b*) Show graphically how the sequence of trial solutions obtained in part (*a*) can be extrapolated to obtain a closer approximation of an optimal solution. What is your resulting estimate of this solution?
(*c*) Use the KKT conditions to check whether the solution you obtained in part (*b*) is, in fact, optimal. If not, use these conditions to derive the exact optimal solution.

13.10-1. Reconsider the linearly constrained convex programming model given in Prob. 13.9-10.
(*a*) If SUMT were applied to this problem, what would be the unconstrained function $P(\mathbf{x}; r)$ to be maximized at each iteration?
(*b*) Setting $r = 1$ and using $(\frac{1}{4}, \frac{1}{4})$ as the initial trial solution, manually apply one iteration of the gradient search procedure to begin maximizing the function $P(\mathbf{x}; r)$ you obtained in part (*a*).
D, A* (*c*) Beginning with the same initial trial solution as in part (*b*), use a computer to apply SUMT to this problem with $r = 1, 10^{-2}, 10^{-4}$,
(*d*) Compare the final solution obtained in part (*c*) to the true optimal solution for Prob. 13.9-10 given in the back of the book. What is the percentage error in x_1, in x_2, and in $f(\mathbf{x})$?

13.10-2. Reconsider the linearly constrained convex programming model given in Prob. 13.9-11. Follow the instructions of parts (*a*), (*b*), and (*c*) of Prob. 13.10-1 for this model, except use $(x_1, x_2) = (\frac{1}{2}, \frac{1}{2})$ as the initial trial solution and use $r = 1, 10^{-2}, 10^{-4}, 10^{-6}$.

13.10-3. Reconsider the model given in Prob. 13.3-3.
(*a*) If SUMT were applied directly to this problem, what would be the unconstrained function $P(\mathbf{x}; r)$ to be *minimized* at each iteration?
(*b*) Setting $r = 100$ and using $(x_1, x_2) = (5, 5)$ as the initial trial solution, manually apply one iteration of the gradient search procedure to begin minimizing the function $P(\mathbf{x}; r)$ you obtained in part (*a*).
D, A* (*c*) Beginning with the same initial trial solution as in part (*b*), use a computer to apply SUMT to this problem with $r = 100, 1, 10^{-2}, 10^{-4}$. (*Hint:* The computer routine assumes that the problem has been converted to *maximization* form with the functional constraints in $\leq$ form.)

13.10-4. Consider the example for applying SUMT given in Sec. 13.10.
(*a*) Show that $(x_1, x_2) = (1, 2)$ satisfies the KKT conditions.

(*b*) Display the feasible region graphically, and then plot the locus of points $x_1 x_2 = 2$ to demonstrate that $(x_1, x_2) = (1, 2)$ with $f(1, 2) = 2$ is, in fact, a *global maximum.*

13.10-5.* Consider the following convex programming problem:

$$\text{Maximize} \quad f(\mathbf{x}) = -2x_1 - (x_2 - 3)^2,$$

subject to

$$x_1 \geq 3 \quad \text{and} \quad x_2 \geq 3.$$

(*a*) If SUMT were applied to this problem, what would be the unconstrained function $P(\mathbf{x}; r)$ to be maximized at each iteration?

(*b*) Derive the maximizing solution of $P(\mathbf{x}; r)$ analytically, and then give this solution for $r = 1, 10^{-2}, 10^{-4}, 10^{-6}$.

D, A* (*c*) Beginning with the initial trial solution $(x_1, x_2) = (4, 4)$, use a computer to apply SUMT to this problem with $r = 1, 10^{-2}, 10^{-4}, 10^{-6}$.

13.10-6. Use SUMT to solve the following convex programming problem:

$$\text{Minimize} \quad f(\mathbf{x}) = \frac{(x_1 + 1)^3}{3} + x_2,$$

subject to

$$x_1 \geq 1 \quad \text{and} \quad x_2 \geq 0.$$

(*a*) If SUMT were applied directly to this problem, what would be the unconstrained function $P(\mathbf{x}; r)$ to be minimized at each iteration?

(*b*) Derive the minimizing solution of $P(\mathbf{x}; r)$ analytically, and then give this solution for $r = 1, 10^{-2}, 10^{-4}, 10^{-6}$.

D, A* (*c*) Beginning with the initial trial solution $(x_1, x_2) = (2, 1)$, use a computer to apply SUMT to this problem (in maximization form) with $r = 1, 10^{-2}, 10^{-4}, 10^{-6}$.

D, A* **13.10-7.** Consider the following convex programming problem:

$$\text{Maximize} \quad f(\mathbf{x}) = x_1 x_2 - x_1 - x_1^2 - x_2 - x_2^2,$$

subject to

$$x_2 \geq 0.$$

Beginning with the initial trial solution $(x_1, x_2) = (1, 1)$, use the computer to apply SUMT to this problem with $r = 1, 10^{-2}, 10^{-4}$.

D, A* **13.10-8.** Reconsider the quadratic programming model given in Prob. 13.7-4. Beginning with the initial trial solution $(x_1, x_2) = (\frac{1}{2}, \frac{1}{2})$, use a computer to apply SUMT to this model with $r = 1, 10^{-2}, 10^{-4}, 10^{-6}$.

D, A* **13.10-9.** Reconsider the first quadratic programming variation of the Wyndor Glass Co. problem presented in Sec. 13.2 (see Fig. 13.6). Beginning with the initial trial solution $(x_1, x_2) = (2, 3)$, use a computer to apply SUMT to this problem with $r = 10^2, 1, 10^{-2}, 10^{-4}$.

13.10-10. Consider the following nonconvex programming problem:

$$\text{Maximize} \quad f(x) = 1{,}000x - 400x^2 + 40x^3 - x^4,$$

subject to

$$x^2 + x \leq 500$$

and

$$x \geq 0.$$

(*a*) Identify the feasible values for x. Obtain general expressions for the first three derivatives of $f(x)$. Use this information to help you draw a rough sketch of $f(x)$ over the feasible region for x. Without calculating their values, mark the points on your graph that correspond to *local* maxima and minima.

A* (*b*) Use the one-dimensional search procedure with $\epsilon = 0.05$ to find each of the local maxima. Use your sketch from part (*a*) to identify appropriate initial bounds for each of these searches. Which of the local maxima is a global maximum?

D, A* (*c*) Use a computer to apply SUMT to this problem with $r = 10^3, 10^2, 10, 1$ to find each of the local maxima. Use $x = 3$ and $x = 15$ as the initial trial solutions for these searches. Which of the local maxima is a global maximum?

13.10-11. Consider the following nonconvex programming problem:

$$\text{Maximize} \quad f(\mathbf{x}) = 3x_1x_2 - 2x_1^2 - x_2^2,$$

subject to

$$x_1^2 + 2x_2^2 \le 4$$
$$2x_1 - x_2 \le 3$$
$$x_1x_2^2 + x_1^2x_2 = 2$$

and

$$x_1 \ge 0, \qquad x_2 \ge 0.$$

(*a*) If SUMT were applied to this problem, what would be the unconstrained function $P(\mathbf{x}; r)$ to be minimized at each iteration?

D, A* (*b*) Starting from the initial trial solution $(x_1, x_2) = (1, 1)$, use a computer to apply SUMT to this problem with $r = 1, 10^{-2}, 10^{-4}$.

13.10-12. Reconsider the convex programming model with an equality constraint given in Prob. 13.6-14.

(*a*) If SUMT were applied to this model, what would be the unconstrained function $P(\mathbf{x}; r)$ to be minimized at each iteration?

D, A* (*b*) Starting from the initial trial solution $(x_1, x_2, x_3) = (\frac{3}{2}, \frac{3}{2}, 2)$, use a computer to apply SUMT to this model with $r = 10^{-2}, 10^{-4}, 10^{-6}, 10^{-8}$.

13.10-13. Consider the following nonconvex programming problem.

$$\text{Minimize} \quad f(\mathbf{x}) = \sin 3x_1 + \cos 3x_2 + \sin(x_1 + x_2),$$

subject to

$$x_1^2 - 10x_2 \ge -1$$
$$10x_1 + x_2^2 \le 100$$

and

$$x_1 \ge 0, \qquad x_2 \ge 0.$$

(*a*) If SUMT were applied to this problem, what would be the unconstrained function $P(\mathbf{x}; r)$ to be minimized at each iteration?

(*b*) Describe how SUMT should be applied to attempt to obtain a global minimum. (Do not actually solve.)

13.11-1. Consider the following problem:

$$\text{Maximize} \quad Z = 4x_1 - x_1^2 + 10x_2 - x_2^2,$$

subject to

$$x_1^2 + 4x_2^2 \leq 16$$

and

$$x_1 \geq 0, \qquad x_2 \geq 0.$$

(*a*) Is this a convex programming problem? Answer yes or no, and then justify your answer.

(*b*) Can the modified simplex method be used to solve this problem? Answer yes or no, and then justify your answer (but do not actually solve.)

(*c*) Can the Frank-Wolfe algorithm be used to solve this problem? Answer yes or no, and then justify your answer (but do not actually solve).

(*d*) What are the KKT conditions for this problem? Use these conditions to determine whether $(x_1, x_2) = (1, 1)$ can be optimal.

(*e*) Use the separable programming technique to formulate an *approximate* linear programming model for this problem. Use the feasible integers as the breakpoints for each piecewise linear function.

A* (*f*) Use the simplex method to solve the problem as formulated in part (*e*).

(*g*) Give the function $P(\mathbf{x}; r)$ to be maximized at each iteration when SUMT is applied to this problem. (Do not actually solve.)

D, A* (*h*) Use SUMT to solve the problem as formulated in part (*g*). Begin with the initial trial solution $(x_1, x_2) = (2, 1)$, and use $r = 1, 10^{-2}, 10^{-4}, 10^{-6}$.

14

Markov Chains

In decision-making problems, we often are faced with making decisions based upon phenomena that have uncertainty associated with them. This uncertainty is caused by inherent variation due to sources of variation that elude control or due to the inconsistency of natural phenomena. Rather than treat this variability qualitatively, we can incorporate it into the mathematical model and thus handle it quantitatively. This treatment generally can be accomplished if the natural phenomena exhibit some degree of regularity, so that their variation can be described by a probability model.

This chapter presents probability models for processes that *evolve over time* in a probabilistic manner. Such processes are called *stochastic processes.* After we briefly introduce general stochastic processes in the first section, the remainder of the chapter focuses on a special kind called a *Markov chain.* Markov chains have the special property that probabilities involving how the process will evolve in the future depend only on the present state of the process and so are independent of events in the past. Many processes fit this description, so Markov chains provide an especially important kind of probability model.

Both this chapter and those that follow assume that the reader has a basic knowledge of probability theory.

14.1 Stochastic Processes

A **stochastic process** is defined to be an indexed collection of random variables $\{X_t\}$, where the index t runs through a given set T. Often T is taken to be the set of nonnegative integers, and X_t represents a measurable characteristic of interest at time t. For example, the stochastic process $X_1, X_2, X_3, \ldots$ can represent the collection of weekly (or monthly) inventory levels of a given product, or it can represent the collection of weekly (or monthly) demands for this product.

There are many stochastic processes that are of interest. A consideration of the behavior of a system operating for some period of time often leads to the analysis of a stochastic process with the following structure. At particular points of time t labeled 0, 1, . . . , the system is found in exactly one of a finite number of mutually exclusive and exhaustive categories or *states* labeled 0, 1, . . . , M. The points in time may be spaced equally, or their spacing may depend upon the overall behavior of the physical system in which the stochastic process is *embedded,* e.g., the time between occurrences of some phenomenon of interest. Although the states may constitute a qualitative as well as a quantitative characterization of the system, no loss of generality is entailed by the numerical labels 0, 1, . . . , M, which are used henceforth to denote the possible states of the system. Thus the mathematical representation of the physical system is that of a stochastic process $\{X_t\}$, where the random variables are observed at $t = 0, 1, 2, \ldots$ and where each random variable may take on any one of the $M + 1$ integers 0, 1, . . . , M. These integers are a characterization of the $M + 1$ states of the process. It must be emphasized that each state that the stochastic process reaches is given a label that denotes the physical state of the system. It is only for notational convenience that this set is labeled 0, 1, . . . , M.

As an example, consider the following inventory problem. A camera store stocks a particular model camera that can be ordered weekly. Let $D_1, D_2, \ldots$ represent the demand for this camera during the first week, second week, . . . , respectively. It is assumed that the D_i are independent and identically distributed random variables having a known probability distribution. Let X_0 represent the number of cameras on hand at the outset, X_1 the number of cameras on hand at the end of week one, X_2 the number of cameras on hand at the end of week two, and so on. Assume that $X_0 = 3$. On Saturday night the store places an order that is delivered in time for the opening of the store on Monday. The store uses the following (s, S) ordering policy:[1] If the number of cameras on hand at the end of the week is less than $s = 1$ (no cameras in stock), the store orders (up to) $S = 3$. Otherwise, the store does not order (if there are any cameras in stock, no order is placed). It is assumed that sales are lost when demand exceeds the inventory on hand. Thus $\{X_t\}$ for $t = 0, 1, \ldots$ is a stochastic process of the form just described. The possible states of the process are the integers 0, 1, 2, 3, representing the

[1] In general, an (s, S) policy is a periodic review policy that calls for ordering up to S units whenever the inventory level dips below s $(S \geq s)$. If the inventory level is s or greater, then no order is placed. These policies are discussed in detail in Chap. 17.

possible number of cameras on hand at the end of the week. In fact, the random variables X_t are clearly dependent and may be evaluated iteratively by the expression

$$X_{t+1} = \begin{cases} \max\{3 - D_{t+1}, 0\} & \text{if } X_t < 1, \\ \max\{X_t - D_{t+1}, 0\} & \text{if } X_t \geq 1, \end{cases}$$

for $t = 0, 1, 2, \ldots$ This example is used for illustrative purposes throughout many of the following sections. Section 14.2 further defines the type of stochastic process considered in this chapter.

14.2 Markov Chains

Assumptions regarding the joint distribution of $X_0, X_1, \ldots$ are necessary to obtain analytical results. One assumption that leads to analytical tractability is that the stochastic process is a Markov chain (defined later), which has the following key property: A stochastic process $\{X_t\}$ is said to have the **Markovian property** if $P\{X_{t+1} = j | X_0 = k_0, X_1 = k_1, \ldots, X_{t-1} = k_{t-1}, X_t = i\} = P\{X_{t+1} = j | X_t = i\}$, for $t = 0, 1, \ldots$ and every sequence $i, j, k_0, k_1, \ldots, k_{t-1}$.

This Markovian property can be shown to be equivalent to stating that the conditional probability of any future "event," given any past "event" and the present state $X_t = i$, is *independent* of the past event and depends only upon the present state. The conditional probabilities $P\{X_{t+1} = j | X_t = i\}$ are called **transition probabilities**. If, for each i and j,

$$P\{X_{t+1} = j | X_t = i\} = P\{X_1 = j | X_0 = i\}, \qquad \text{for all } t = 0, 1, \ldots,$$

then the (one-step) transition probabilities are said to be *stationary* and are usually denoted by p_{ij}. Thus having **stationary transition probabilities** implies that the transition probabilities do not change in time. The existence of stationary (one-step) transition probabilities also implies that, for each i, j, and n ($n = 0, 1, 2, \ldots$),

$$P\{X_{t+n} = j | X_t = i\} = P\{X_n = j | X_0 = i\},$$

for all $t = 0, 1, \ldots$ These conditional probabilities are usually denoted by $p_{ij}^{(n)}$ and are called n-step transition probabilities.[1] Thus $p_{ij}^{(n)}$ is just the conditional probability that the random variable X, starting in state i, will be in state j after exactly n steps (time units).

Because the $p_{ij}^{(n)}$ are conditional probabilities, they must be nonnegative, and since the process must make a transition into some state, they must satisfy the properties

$$p_{ij}^{(n)} \geq 0, \qquad \text{for all } i \text{ and } j;\ n = 0, 1, 2, \ldots,$$

and

$$\sum_{j=0}^{M} p_{ij}^{(n)} = 1 \qquad \text{for all } i;\ n = 0, 1, 2, \ldots$$

[1] For $n = 0$, $p_{ij}^{(0)}$ is just $P\{X_0 = j | X_0 = i\}$ and hence is 1 when $i = j$ and is 0 when $i \neq j$. For $n = 1$, $p_{ij}^{(1)}$ is just the (one-step) transition probability and is denoted by p_{ij}.

A convenient notation for representing the n-step transition probabilities is the matrix form

$$\mathbf{P}^{(n)} = \begin{array}{c|cccc} \text{State} & 0 & 1 & \cdots & M \\ \hline 0 & p_{00}^{(n)} & & \cdots & p_{0M}^{(n)} \\ 1 & & & & \\ \vdots & \cdots & \cdots & \cdots & \cdots \\ M & p_{M0}^{(n)} & & \cdots & p_{MM}^{(n)} \end{array}, \qquad \text{for } n = 0, 1, 2, \ldots$$

or, equivalently,

$$\mathbf{P}^{(n)} = \begin{bmatrix} p_{00}^{(n)} & \cdots & p_{0M}^{(n)} \\ \cdots & \cdots & \cdots \\ p_{M0}^{(n)} & \cdots & p_{MM}^{(n)} \end{bmatrix}$$

We drop the superscript n when $n = 1$. It is now possible to define the Markov chains to be considered in this chapter (except Sec. 14.8).

A stochastic process $\{X_t\}$ ($t = 0, 1, \ldots$) is a **Markov chain** if it has the *Markovian property.*

We will consider only Markov chains with the following properties:

1. A finite number of states
2. Stationary transition probabilities.

We also will assume that we know a set of initial probabilities $P\{X_0 = i\}$ for all i.

Returning to the inventory example developed in the preceding section, note that $\{X_t\}$, where X_t is the number of cameras in stock at the end of week t (before an order is received), is a Markov chain. Now consider how to obtain the (one-step) transition probabilities, i.e., the elements of the (one-step) *transition matrix*

$$\mathbf{P} = \begin{bmatrix} p_{00} & p_{01} & p_{02} & p_{03} \\ p_{10} & p_{11} & p_{12} & p_{13} \\ p_{20} & p_{21} & p_{22} & p_{23} \\ p_{30} & p_{31} & p_{32} & p_{33} \end{bmatrix}$$

assuming that each D_t has a Poisson distribution with parameter $\lambda = 1$.

To obtain p_{00}, it is necessary to evaluate $P\{X_t = 0 | X_{t-1} = 0\}$. If $X_{t-1} = 0$, then $X_t = \max\{3 - D_t, 0\}$. Therefore, if $X_t = 0$, then the demand during the week has to be 3 or more. Hence $p_{00} = P\{D_t \geq 3\} = 1 - P\{D_t \leq 2\} = 0.080$, the probability that a Poisson random variable with parameter $\lambda = 1$ takes on a value of 3 or more. And $p_{10} = P\{X_t = 0 | X_{t-1} = 1\}$ can be obtained in a similar way. If $X_{t-1} = 1$, then $X_t = \max\{1 - D_t, 0\}$. To have $X_t = 0$, the demand during the week has to be 1 or more. Hence $p_{10} = P\{D_t \geq 1\} = 1 - P\{D_t = 0\} = 0.632$. To find $p_{21} = P\{X_t = 1 | X_{t-1} = 2\}$, note that $X_t = \max\{2 - D_t, 0\}$ if $X_{t-1} = 2$. Therefore, if $X_t = 1$, then the demand during the week has to be exactly 1. Hence $p_{21} = P\{D_t = 1\} = 0.368$. The remaining entries are obtained in a similar manner and yield the following (one-step) transition matrix:

$$\mathbf{P} = \begin{bmatrix} 0.080 & 0.184 & 0.368 & 0.368 \\ 0.632 & 0.368 & 0 & 0 \\ 0.264 & 0.368 & 0.368 & 0 \\ 0.080 & 0.184 & 0.368 & 0.368 \end{bmatrix}$$

Additional Examples of Markov Chains

Consider the following model for the value of a stock. At the end of a given day, the price is recorded. If the stock has gone up, the probability that it will go up tomorrow is 0.7. If the stock has gone down, the probability that it will go up tomorrow is only 0.5. This is a Markov chain, where state 0 represents the stock's going up and state 1 represents the stock's going down. The transition matrix is given by

$$\mathbf{P} = \begin{bmatrix} 0.7 & 0.3 \\ 0.5 & 0.5 \end{bmatrix}$$

Suppose now that the stock market model is changed so that the stock's going up tomorrow depends upon whether it increased today *and* yesterday. In particular, if the stock has increased for the past two days, it will increase tomorrow with probability 0.9. If the stock increased today but decreased yesterday, then it will increase tomorrow with probability 0.6. If the stock decreased today but increased yesterday, then it will increase tomorrow with probability 0.5. Finally, if the stock decreased for the past two days, then it will increase tomorrow with probability 0.3. If we define the state as representing whether the stock goes up or down today, the system is no longer a Markov chain. However, we can transform the system to a Markov chain by defining the states as follows:[1]

State 0: The stock increased both today and yesterday.
State 1: The stock increased today and decreased yesterday.
State 2: The stock decreased today and increased yesterday.
State 3: The stock decreased both today and yesterday.

This leads to a four-state Markov chain with the following transition matrix:

$$\mathbf{P} = \begin{bmatrix} 0.9 & 0 & 0.1 & 0 \\ 0.6 & 0 & 0.4 & 0 \\ 0 & 0.5 & 0 & 0.5 \\ 0 & 0.3 & 0 & 0.7 \end{bmatrix}$$

One row in the matrix will be verified, say, the second. This corresponds to state 1, which represents the stock increasing today and decreasing yesterday. The first element in the row represents the probability of the stock's increasing tomorrow, having increased today and given that the stock increased today but decreased yesterday. This is just the probability of the stock's increasing tomorrow given that it increased today but decreased yesterday, that is, 0.6. Similarly, the third element in the row represents the probability of the stock's decreasing tomorrow, having increased today and given that the stock increased today and decreased yesterday. This is just the probability of the stock's decreasing tomorrow given that it increased today but decreased yesterday, that is, $1 - 0.6 = 0.4$. The other two elements are 0 because they pertain to events that are

[1] This example demonstrates that Markov chains are able to incorporate arbitrary amounts of history, but at the cost of significantly increasing the number of states.

contradictory; i.e., they correspond to instances where the stock decreased today even though the current state includes that the stock increased today.

Another example involves gambling. Suppose that a player has \$1 and with each play of the game wins \$1 with probability $p > 0$ or loses \$1 with probability $1 - p$. The game ends when the player either accumulates \$3 or goes broke. This model is a Markov chain with the states representing the player's current holding of money, that is, 0, \$1, \$2, or \$3, and with the transition matrix given by

$$\mathbf{P} = \begin{bmatrix} 1 & 0 & 0 & 0 \\ 1-p & 0 & p & 0 \\ 0 & 1-p & 0 & p \\ 0 & 0 & 0 & 1 \end{bmatrix}$$

Note that in both the inventory and gambling examples, the numeric labeling of the states that the process reaches coincides with the physical expression of the system—i.e., actual inventory levels and the player's holding of money, respectively—whereas the numeric labeling of the states in the stock examples has no physical significance.

14.3 Chapman-Kolmogorov Equations

Section 14.2 introduced the n-step transition probability $p_{ij}^{(n)}$. This transition probability can be useful when the process is in state i and the probability that the process will be in state j after n periods is desired. The *Chapman-Kolmogorov equations* provide a method for computing these n-step transition probabilities:

$$p_{ij}^{(n)} = \sum_{k=0}^{M} p_{ik}^{(m)} p_{kj}^{(n-m)}, \qquad \text{for all } i, j, n \text{ and } 0 \leq m \leq n.$$

These equations merely point out that in going from state i to state j in n steps, the process will be in some state k after exactly m (less than n) steps. Thus $p_{ik}^{(m)} p_{kj}^{(n-m)}$ is just the conditional probability that, starting from state i, the process goes to state k after m steps and then to state j in $n - m$ steps. Therefore, summing these conditional probabilities over all possible k must yield $p_{ij}^{(n)}$. The special cases of $m = 1$ and $m = n - 1$ lead to the expressions

$$p_{ij}^{(n)} = \sum_{k=0}^{M} p_{ik} p_{kj}^{(n-1)}$$

and

$$p_{ij}^{(n)} = \sum_{k=0}^{M} p_{ik}^{(n-1)} p_{kj},$$

for all i, j, n. It then becomes evident that the n-step transition probabilities can be obtained from the one-step transition probabilities recursively. This recursive relationship is best explained in matrix notation (see Appendix 4). For $n = 2$, these expressions become

$$p_{ij}^{(2)} = \sum_{k=0}^{M} p_{ik} p_{kj}, \qquad \text{for all } i, j.$$

Note that the $p_{ij}^{(2)}$ are the elements of matrix $\mathbf{P}^{(2)}$. However, also note that these elements

$$\sum_{k=0}^{M} p_{ik} p_{kj}$$

are obtained by multiplying the matrix of one-step transition probabilities by itself; i.e,

$$\mathbf{P}^{(2)} = \mathbf{P} \cdot \mathbf{P} = \mathbf{P}^2.$$

More generally, it follows that the matrix of n-step transition probabilities can be obtained from the expression

$$\begin{aligned} \mathbf{P}^{(n)} &= \mathbf{P} \cdot \mathbf{P} \cdots \mathbf{P} = \mathbf{P}^n \\ &= \mathbf{P}\mathbf{P}^{n-1} = \mathbf{P}^{n-1}\mathbf{P}. \end{aligned}$$

Thus the n-step transition probability matrix can be obtained by computing the nth power of the one-step transition matrix. For values of n that are not too large, the n-step transition matrix can be calculated in the manner just described. However, when n is large, such computations are often tedious, and furthermore, round-off errors may cause inaccuracies.

Returning to the inventory example, we see that the two-step transition matrix is given by[1]

$$\mathbf{P}^{(2)} = \mathbf{P}^2 = \begin{bmatrix} 0.080 & 0.184 & 0.368 & 0.368 \\ 0.632 & 0.368 & 0 & 0 \\ 0.264 & 0.368 & 0.368 & 0 \\ 0.080 & 0.184 & 0.368 & 0.368 \end{bmatrix} \begin{bmatrix} 0.080 & 0.184 & 0.368 & 0.368 \\ 0.632 & 0.368 & 0 & 0 \\ 0.264 & 0.368 & 0.368 & 0 \\ 0.080 & 0.184 & 0.368 & 0.368 \end{bmatrix}$$

$$= \begin{bmatrix} 0.249 & 0.286 & 0.300 & 0.165 \\ 0.283 & 0.252 & 0.233 & 0.233 \\ 0.351 & 0.319 & 0.233 & 0.097 \\ 0.249 & 0.286 & 0.300 & 0.165 \end{bmatrix}$$

Thus, given that there is one camera left in stock at the end of a week, the probability is 0.283 that there will be no cameras in stock 2 weeks later, that is, $p_{10}^{(2)} = 0.283$. Similarly, given that there are two cameras left in stock at the end of a week, the probability is 0.097 that there will be three cameras in stock 2 weeks later, that is, $p_{23}^{(2)} = 0.097$.

The four-step transition matrix can also be obtained as follows:

$$\mathbf{P}^{(4)} = \mathbf{P}^4 = \mathbf{P}^{(2)} \cdot \mathbf{P}^{(2)}$$

$$= \begin{bmatrix} 0.249 & 0.286 & 0.300 & 0.165 \\ 0.283 & 0.252 & 0.233 & 0.233 \\ 0.351 & 0.319 & 0.233 & 0.097 \\ 0.249 & 0.286 & 0.300 & 0.165 \end{bmatrix} \begin{bmatrix} 0.249 & 0.286 & 0.300 & 0.165 \\ 0.283 & 0.252 & 0.233 & 0.233 \\ 0.351 & 0.319 & 0.233 & 0.097 \\ 0.249 & 0.286 & 0.300 & 0.165 \end{bmatrix}$$

$$= \begin{bmatrix} 0.289 & 0.286 & 0.261 & 0.164 \\ 0.282 & 0.285 & 0.268 & 0.166 \\ 0.284 & 0.283 & 0.263 & 0.171 \\ 0.289 & 0.286 & 0.261 & 0.164 \end{bmatrix}$$

[1] Note that round-off errors already appear in the row corresponding to state 1.

Thus, given that there is one camera left in stock at the end of a week, the probability is 0.282 that there will be no cameras in stock 4 weeks later, that is, $p_{10}^{(4)} = 0.282$. Similarly, given that there are two cameras left in stock at the end of a week, the probability is 0.171 that there will be three cameras in stock 4 weeks later, that is, $p_{23}^{(4)} = 0.171$.

Your OR Courseware includes a routine for calculating $\mathbf{P}^{(n)} = \mathbf{P}^n$ for any positive integer $n \leq 99$.

It was pointed out that the one- or n-step transition probabilities are conditional probabilities; for example, $P\{X_n = j | X_0 = i\} = p_{ij}^{(n)}$. If the unconditional probability $P\{X_n = j\}$ is desired, it is necessary to have specified the probability distribution of the initial state. Denote this probability distribution by $Q_{X_0}(i)$, where

$$Q_{X_0}(i) = P\{X_0 = i\}, \qquad \text{for } i = 0, 1, \ldots, M.$$

It then follows that

$$P\{X_n = j\} = Q_{X_0}(0)p_{0j}^{(n)} + Q_{X_0}(1)p_{1j}^{(n)} + \cdots + Q_{X_0}(M)p_{Mj}^{(n)}.$$

In the inventory example it was assumed that initially there were 3 units in stock, that is, $X_0 = 3$. Thus

$$Q_{X_0}(0) = Q_{X_0}(1) = Q_{X_0}(2) = 0,$$

and

$$Q_{X_0}(3) = 1.$$

Hence the (unconditional) probability that there will be three cameras in stock 2 weeks after the inventory system began is 0.165; that is, $P\{X_2 = 3\} = (1)p_{33}^{(2)}$. If instead, it were given that $Q_{X_0}(i) = \frac{1}{4}$, for $i = 0, 1, 2, 3$, then

$$P\{X_2 = 3\} = \tfrac{1}{4}(0.165) + \tfrac{1}{4}(0.233) + \tfrac{1}{4}(0.097) + \tfrac{1}{4}(0.165) = 0.165.$$

The fact that the same answer is obtained by using these two initial probability distributions is purely coincidental.

14.4 Classification of States of a Markov Chain

It is evident that the transition probabilities associated with the states play an important role in the study of Markov chains. To further describe the properties of Markov chains, it is necessary to present some concepts and definitions concerning these states.

State j is said to be **accessible** from state i if $p_{ij}^{(n)} > 0$ for some $n > 0$. (Recall that $p_{ij}^{(n)}$ is just the conditional probability of being in state j after n steps, starting in state i.) Thus, state j being accessible from state i means that it is possible for the system to enter state j eventually when it starts from state i. In the inventory example, $p_{ij}^{(2)} > 0$ for all i, j, so every state is accessible from every other state. In general, a sufficient condition for *all* states to be accessible is that there exist a value of n for which $p_{ij}^{(n)} > 0$ for all i and j.

In the gambling example, state 2 is not accessible from state 3. This can be deduced from the context of the game (once the player reaches state 3, the player never leaves this state) or by noting that the n-step transition matrix for the game $\mathbf{P}^{(n)}$ is of the form

$$\mathbf{P}^{(n)} = \begin{bmatrix} 1 & 0 & 0 & 0 \\ * & * & * & * \\ * & * & * & * \\ 0 & 0 & 0 & 1 \end{bmatrix}$$

for all n, where each asterisk represents a nonnegative number. However, even though state 2 is *not* accessible from state 3, state 3 *is* accessible from state 2 since, for $n = 1$, the transition matrix given at the end of Sec. 13.2 indicates that $p_{23} = p > 0$.

If state j is accessible from state i and state i is accessible from state j, then states i and j are said to **communicate**. In the inventory example, all states communicate. In the gambling example, states 2 and 3 do not. In general, (1) any state communicates with itself (because $p_{ii}^{(0)} = P\{X_0 = i | X_0 = i\} = 1$); (2) if state i communicates with state j, then state j communicates with state i; and (3) if state i communicates with state j and state j communicates with state k, then state i communicates with state k. Properties 1 and 2 follow from the definition of states communicating, whereas property 3 follows from the Chapman-Kolmogorov equations.

As a result of these three properties of communication, the state space may be partitioned into disjoint classes, with two communicating states said to belong to the same class. Thus the states of a Markov chain may consist of one or more disjoint classes (a class may consist of a single state). If there is only one class, i.e., all the states communicate, the Markov chain is said to be **irreducible**. In the inventory example, the Markov chain is irreducible. In the first stock example, the Markov chain is irreducible. The gambling example contains three classes. State 0 forms a class, state 3 forms a class, and states 1 and 2 form a class.

It is often useful to talk about whether a process, starting in state i, will ever return to this state. Let f_{ii} denote the probability that the process will return to state i given that it starts in state i. State i is called a **recurrent** state if $f_{ii} = 1$ and **transient** if $f_{ii} < 1$. A special case of a recurrent state is an *absorbing* state. A state i is said to be an **absorbing** state if the (one-step) transition probability p_{ii} equals 1.

Determining whether a state is recurrent or transient by evaluating f_{ii} is not generally simple. Hence, it is not always evident whether a state should be classified as recurrent or transient. For example, although all states in the inventory example are recurrent (as shown later in this section), it is not simple to prove that f_{ii} equals 1 for all i.

In the gambling example, state 0 and state 3 are absorbing states (each belonging to a separate class) and hence are recurrent, so f_{00} and f_{33} must equal 1. However, states 1 and 2 are transient, and it is not simple to show that f_{11} and f_{22} are less than 1.

It is possible to determine some properties of f_{ii} that may be useful in determining its value. If a Markov process is in state i and i is recurrent, the probability is 1 that the process will return to that state. Since the process is a Markov chain, this is equivalent to the process beginning once more from state i, and with probability 1, it will once again return to that state. Repeating this argument leads to the conclusion that state i will be entered infinitely often. Hence, a recurrent state has the property that the expected number of time periods that the process is in state i is infinite.

If a Markov process is in state i and i is transient, then the probability of reentering state i is f_{ii} (less than 1) and the probability of not reentering state i is $1 - f_{ii}$. It then can be shown that the expected number of time periods that the process is in state i is finite and given by

$$\frac{1}{1 - f_{ii}}.$$

Therefore, it follows that state i is recurrent if and only if the expected number of time periods that the process is in state i is infinite, given that the process started in state i.

To calculate the expected number of time periods that the process is in state i given that $X_0 = i$, define

$$B_n = \begin{cases} 1 & \text{if } X_n = i, \\ 0 & \text{if } X_n \neq i. \end{cases}$$

The quantity

$$\sum_{n=1}^{\infty} B_n | X_0 = i$$

represents the number of time periods that the process is in state i given that $X_0 = i$. Therefore, its expectation is given by

$$E\left(\sum_{n=1}^{\infty} B_n | X_0 = i\right) = \sum_{n=1}^{\infty} E(B_n | X_0 = i)$$

$$= \sum_{n=1}^{\infty} P\{X_n = i | X_0 = i\}$$

$$= \sum_{n=1}^{\infty} p_{ii}^{(n)}.$$

Thus, it has been shown that a state i is recurrent if and only if

$$\sum_{n=1}^{\infty} p_{ii}^{(n)} = \infty.$$

This result can be used to show that recurrence is a class property. That is, all states in a class are either recurrent or transient. Furthermore, in a finite-state Markov chain, not all states can be transient. Therefore, all states in an irreducible finite-state Markov chain are recurrent. Indeed, one can identify an irreducible finite-state Markov chain (and therefore conclude that all states are recurrent) by showing that all states of the process communicate. It has already been pointed out that a sufficient condition for *all* states to be accessible (and therefore communicate with each other) is that there exist a value of n, not dependent upon i and j, for which $p_{ij}^{(n)} > 0$ for all i and j. Thus, all states in the inventory example are recurrent, since $p_{ij}^{(2)}$ is positive for all i and j. Similarly, the first stock example contains only recurrent states, since p_{ij} is positive for all i and j. By calculating $p_{ij}^{(2)}$ for all i and j in the second stock example, it follows that all states are recurrent since $p_{ij}^{(2)} > 0$.

As another example, suppose that a Markov process has the following transition matrix:

$$\mathbf{P} = \text{State} \begin{array}{c} \\ 0 \\ 1 \\ 2 \\ 3 \\ 4 \end{array} \overset{\text{State}}{\begin{array}{c} \begin{array}{ccccc} 0 & 1 & 2 & 3 & 4 \end{array} \\ \begin{bmatrix} \frac{1}{4} & \frac{3}{4} & 0 & 0 & 0 \\ \frac{1}{2} & \frac{1}{2} & 0 & 0 & 0 \\ 0 & 0 & 1 & 0 & 0 \\ 0 & 0 & \frac{1}{3} & \frac{2}{3} & 0 \\ 1 & 0 & 0 & 0 & 0 \end{bmatrix} \end{array}}$$

It is evident that state 2 is an absorbing state (and hence a recurrent state) because once the process enters state 2 (row 3 of the matrix), it will never leave. States 3 and 4 are transient states because once the process is in state 3, there is a positive probability that it will never return. The probability is $\frac{1}{3}$ that the process will go from state 3 to state 2 on the first step. Once the process is in state 2, it remains in state 2. Once a process leaves state 4, it can never return. States 0 and 1 are recurrent states. As indicated earlier, to show that states 0 and 1 are recurrent, it is sufficient to show that $f_{00} = 1$ and $f_{11} = 1$, which is not a simple task. Hence, an alternative method is desirable. Observe that the n-step transition matrix of this example always has the appearance

$$\mathbf{P}^{(n)} = \begin{bmatrix} * & * & 0 & 0 & 0 \\ * & * & 0 & 0 & 0 \\ 0 & 0 & 1 & 0 & 0 \\ 0 & 0 & * & * & 0 \\ * & * & 0 & 0 & 0 \end{bmatrix},$$

where each asterisk here represents a strictly positive number. Hence it is intuitively evident that once the process is in state 0, it will return to state 0 (possibly passing through state 1) after some number of steps. A similar argument holds for state 1.

Another useful property of Markov chains is the property of *periodicities.* The **period** of state i is defined to be the integer $t(t > 1)$ such that $p_{ii}^{(n)} = 0$ for all values of n other than $t, 2t, 3t, \ldots$ and t is the largest integer with this property. In the gambling example, starting in state 1, it is possible for the process to enter state 1 only at times 2, 4, . . . , in which case state 1 is said to have period 2. This is evident by noting that the player can break even (be neither winning nor losing) only at times 2, 4, . . . , or by calculating $p_{11}^{(n)}$ for all n and noting that $p_{11}^{(n)} = 0$ for n odd.

If there are two consecutive numbers s and $s + 1$ such that the process can be in state i at times s and $s + 1$, the state is said to have period 1 and is called an **aperiodic** state.

Just as recurrence is a class property, it can be shown that periodicity is a class property. That is, if state i in a class has period t, then all states in that class have period t. In the gambling example, state 2 also has period 2.

A final property of Markov chains pertains to further classifying recurrent states. A recurrent state i is said to be **positive recurrent** if, starting in state i, the expected time for the process to reenter state i is finite. Similarly, a recurrent state i is said to be **null recurrent** if, starting in state i, the expected time for the process to reenter state i is infinite. It can be shown that for a finite-state Markov chain, all recurrent states are positive recurrent states. Positive recurrent states that are aperiodic are called **ergodic** states. A Markov chain is said to be ergodic if all its states are ergodic states.

14.5 First Passage Times

Section 14.3 dealt with finding n-step transition probabilities [i.e., given that the process is in state i, determining the (conditional) probability that the process will be in state j after n periods]. It is often desirable to make probability statements about the number of transitions made by the process in going from state i to state j *for the first time.* This length of time is called the **first passage time** in going from state i to state j. When $j = i$, this first passage time is just the number of transitions until the process returns to the initial state i. In this case, the first passage time is called the **recurrence time** for state i.

To illustrate these definitions, reconsider the inventory example introduced in Sec. 14.1, where X_t is the number of cameras on hand at the end of week t and we start with $X_0 = 3$. Suppose that it turns out that

$$X_0 = 3, \quad X_1 = 2, \quad X_2 = 1, \quad X_3 = 0, \quad X_4 = 3, \quad X_5 = 1.$$

In this case, the first passage time in going from state 3 to state 1 is 2 weeks, the first passage time in going from state 3 to state 0 is 3 weeks, and the recurrence time for state 3 is 4 weeks.

In general, the first passage times are random variables and hence have probability distributions associated with them. These probability distributions depend upon the transition probabilities of the process. In particular, let $f_{ij}^{(n)}$ denote the probability that the first passage time from state i to j is equal to n. It can be shown that these probabilities satisfy the following recursive relationships:

$$\begin{aligned}
f_{ij}^{(1)} &= p_{ij}^{(1)} = p_{ij},\\
f_{ij}^{(2)} &= p_{ij}^{(2)} - f_{ij}^{(1)}p_{jj},\\
&\vdots\\
f_{ij}^{(n)} &= p_{ij}^{(n)} - f_{ij}^{(1)}p_{jj}^{(n-1)} - f_{ij}^{(2)}p_{jj}^{(n-2)} \cdots - f_{ij}^{(n-1)}p_{jj}.
\end{aligned}$$

Thus the probability of a first passage time from state i to state j in n steps can be computed recursively from the one-step transition probabilities. In the inventory example, the probability distribution of the first passage time in going from state 3 to state 0 is obtained as follows:

$$\begin{aligned}
f_{30}^{(1)} &= 0.080,\\
f_{30}^{(2)} &= (0.249) - (0.080)(0.080) = 0.243.\\
&\vdots
\end{aligned}$$

For fixed i and j, the $f_{ij}^{(n)}$ are nonnegative numbers such that

$$\sum_{n=1}^{\infty} f_{ij}^{(n)} \le 1.$$

Unfortunately, this sum may be strictly less than 1, which implies that a process initially in state i may never reach state j. When the sum does equal 1, $f_{ij}^{(n)}$ (for $n = 1, 2, \ldots$) can be considered as a probability distribution for the random variable, the first passage time.

Whereas calculating $f_{ij}^{(n)}$ for all n may be difficult, it is relatively simple to obtain the expected first passage time from state i to state j. Denote this expectation by μ_{ij}, which is defined by

$$\mu_{ij} = \begin{cases} \infty & \text{if } \sum_{n=1}^{\infty} f_{ij}^{(n)} < 1,\\ \sum_{n=1}^{\infty} n f_{ij}^{(n)} & \text{if } \sum_{n=1}^{\infty} f_{ij}^{(n)} = 1. \end{cases}$$

Whenever

$$\sum_{n=1}^{\infty} f_{ij}^{(n)} = 1,$$

then μ_{ij} satisfies uniquely the equation

$$\mu_{ij} = 1 + \sum_{k \neq j} p_{ik} \mu_{kj}.$$

When $i = j$, μ_{ii} is called the **expected recurrence time**.

For the inventory example, these equations can be used to compute the expected time until the cameras are out of stock, assuming the process is started when three cameras are available; i.e., the expected first passage time, μ_{30}, can be obtained. Since all the states are recurrent, the system of equations leads to the expressions

$$\mu_{30} = 1 + p_{31}\mu_{10} + p_{32}\mu_{20} + p_{33}\mu_{30},$$

$$\mu_{20} = 1 + p_{21}\mu_{10} + p_{22}\mu_{20} + p_{23}\mu_{30},$$

$$\mu_{10} = 1 + p_{11}\mu_{10} + p_{12}\mu_{20} + p_{13}\mu_{30},$$

or

$$\mu_{30} = 1 + 0.184\mu_{10} + 0.368\mu_{20} + 0.368\mu_{30},$$

$$\mu_{20} = 1 + 0.368\mu_{10} + 0.368\mu_{20},$$

$$\mu_{10} = 1 + 0.368\mu_{10}.$$

The simultaneous solution to this system of equations is

$$\mu_{10} = 1.58 \text{ weeks},$$

$$\mu_{20} = 2.51 \text{ weeks},$$

$$\mu_{30} = 3.50 \text{ weeks},$$

so that the expected time until the cameras are out of stock is 3.50 weeks. In making these calculations, we also obtain μ_{20} and μ_{10}.

14.6 Long-Run Properties of Markov Chains

Steady-State Probabilities

In Sec. 14.3 the four-step transition matrix for the inventory example was obtained. It will now be instructive to examine the eight-step transition probabilities given by the matrix

$$\mathbf{P}^{(8)} = \mathbf{P}^8 = \mathbf{P}^4 \cdot \mathbf{P}^4 = \begin{bmatrix} 0.286 & 0.285 & 0.264 & 0.166 \\ 0.286 & 0.285 & 0.264 & 0.166 \\ 0.286 & 0.285 & 0.264 & 0.166 \\ 0.286 & 0.285 & 0.264 & 0.166 \end{bmatrix}$$

Notice the rather remarkable fact that each of the four rows has identical entries. This implies that the probability of being in state j after 8 weeks appears to be independent of the initial level of inventory. In other words, it appears that there is a limiting probability that the system will be in state j after a large number of transitions, and this probability is independent of the initial state. These properties of the long-run behavior of finite-state Markov processes do, in fact, hold under relatively general conditions, as summarized below.

For any irreducible ergodic Markov chain, $\lim_{n\to\infty} p_{ij}^{(n)}$ exists and is independent of i.

Furthermore,

$$\lim_{n\to\infty} p_{ij}^{(n)} = \pi_j > 0,$$

where the π_j uniquely satisfy the following **steady-state equations**

$$\pi_j = \sum_{i=0}^{M} \pi_i p_{ij}, \qquad \text{for } j = 0, 1, \ldots, M,$$

$$\sum_{j=0}^{M} \pi_j = 1.$$

The π_j are called the **steady-state probabilities** of the Markov chain and are equal to the reciprocal of the expected recurrence time, i.e.,

$$\pi_j = \frac{1}{\mu_{jj}}, \qquad \text{for } j = 0, 1, \ldots, M.$$

The term *steady-state* probability means that the probability of finding the process in a certain state, say j, after a large number of transitions tends to the value π_j, independent of the initial probability distribution defined over the states. It is important to note that the steady-state probability does *not* imply that the process settles down into one state. On the contrary, the process continues to make transitions from state to state, and at any step n the transition probability from state i to state j is still p_{ij}.

The π_j can also be interpreted as *stationary probabilities* (not to be confused with stationary transition probabilities) in the following sense. If the *initial* probability of being in state j is given by π_j (that is, $P\{X_0 = j\} = \pi_j$) for all j, then the probability of finding the process in state j at time $n = 1, 2, \ldots$ is also given by π_j (that is, $P\{X_n = j\} = \pi_j$).

Note that the steady-state equations consist of $M + 2$ equations in $M + 1$ unknowns. Because it has a unique solution, at least one equation must be redundant and can, therefore, be deleted. It cannot be the equation

$$\sum_{j=0}^{M} \pi_j = 1,$$

because $\pi_j = 0$ for all j will satisfy the other $M + 1$ equations. Furthermore, the solutions to the other $M + 1$ steady-state equations have a unique solution up to a multiplicative constant, and it is the final equation that forces the solution to be a probability distribution.

Returning to the inventory example, we see that the steady-state equations can be expressed as

$$\begin{aligned}
\pi_0 &= \pi_0 p_{00} + \pi_1 p_{10} + \pi_2 p_{20} + \pi_3 p_{30},\\
\pi_1 &= \pi_0 p_{01} + \pi_1 p_{11} + \pi_2 p_{21} + \pi_3 p_{31},\\
\pi_2 &= \pi_0 p_{02} + \pi_1 p_{12} + \pi_2 p_{22} + \pi_3 p_{32},\\
\pi_3 &= \pi_0 p_{03} + \pi_1 p_{13} + \pi_2 p_{23} + \pi_3 p_{33},\\
1 &= \pi_0 \quad + \pi_1 \quad + \pi_2 \quad + \pi_3.
\end{aligned}$$

Substituting values for p_{ij} into these equations leads to the equations

$$\begin{aligned}
\pi_0 &= 0.080\pi_0 + 0.632\pi_1 + 0.264\pi_2 + 0.080\pi_3,\\
\pi_1 &= 0.184\pi_0 + 0.368\pi_1 + 0.368\pi_2 + 0.184\pi_3,\\
\pi_2 &= 0.368\pi_0 \qquad\qquad + 0.368\pi_2 + 0.368\pi_3,\\
\pi_3 &= 0.368\pi_0 \qquad\qquad\qquad\qquad + 0.368\pi_3,\\
1 &= \pi_0 + \pi_1 + \pi_2 + \pi_3.
\end{aligned}$$

Solving the last four equations simultaneously provides the solution

$$\pi_0 = 0.285, \qquad \pi_2 = 0.264,$$
$$\pi_1 = 0.285, \qquad \pi_3 = 0.166,$$

which is essentially the result that appears in matrix $\mathbf{P}^{(8)}$. Thus after many weeks the probability of finding zero, one, two, and three cameras in stock tends to 0.285, 0.285, 0.264, and 0.166, respectively. The corresponding expected recurrence times are

$$\mu_{00} = \frac{1}{\pi_0} = 3.51 \text{ weeks}, \qquad \mu_{22} = \frac{1}{\pi_2} = 3.79 \text{ weeks},$$

$$\mu_{11} = \frac{1}{\pi_1} = 3.51 \text{ weeks}, \qquad \mu_{33} = \frac{1}{\pi_3} = 6.02 \text{ weeks}.$$

Your OR Courseware includes a routine for solving the steady-state equations to obtain the steady-state probabilities.

There are other important results concerning steady-state probabilities. In particular, if i and j are recurrent states belonging to different classes, then

$$p_{ij}^{(n)} = 0, \qquad \text{for all } n.$$

This result follows from the definition of a class.

Similarly, if j is a transient state, then

$$\lim_{n\to\infty} p_{ij}^{(n)} = 0, \qquad \text{for all } i.$$

This result implies that the probability of finding the process in a transient state after a large number of transitions tends to zero.

Expected Average Cost per Unit Time

The preceding subsection dealt with Markov chains whose states were ergodic (positive recurrent and aperiodic). If the requirement that the states be aperiodic is relaxed, then the limit

$$\lim_{n\to\infty} p_{ij}^{(n)}$$

may not exist. To illustrate this point, consider the two-state transition matrix

$$\mathbf{P} = \begin{bmatrix} 0 & 1 \\ 1 & 0 \end{bmatrix}.$$

If the process starts in state 0 at time 0, it will be in state 0 at times 2, 4, 6, . . . and in state 1 at times 1, 3, 5, ... Thus $p_{00}^{(n)} = 1$ if n is even and $p_{00}^{(n)} = 0$ if n is odd, so that

$$\lim_{n\to\infty} p_{00}^{(n)}$$

does not exist. However, the following limit always exists for an irreducible Markov chain with positive recurrent states:

$$\lim_{n\to\infty} \left(\frac{1}{n} \sum_{k=1}^{n} p_{ij}^{(k)} \right) = \pi_j,$$

where the π_j satisfy the steady-state equations given in the preceding subsection.

This result is important in computing the *long-run average cost per unit time* associated with a Markov chain. Suppose that a cost (or other penalty function) $C(X_t)$ is incurred when the process is in state X_t at time t, for $t = 0, 1, 2, \ldots$. Note that $C(X_t)$ is a random variable that takes on any one of the values $C(0), C(1), \ldots, C(M)$ and that the function $C(\cdot)$ is independent of t. The expected average cost incurred over the first n periods is given by

$$E\left[\frac{1}{n} \sum_{t=1}^{n} C(X_t)\right].$$

By using the result that

$$\lim_{n\to\infty} \left(\frac{1}{n} \sum_{k=1}^{n} p_{ij}^{(k)} \right) = \pi_j,$$

it can be shown that the (long-run) *expected average cost per unit time* is given by

$$\lim_{n\to\infty} E\left[\frac{1}{n} \sum_{t=1}^{n} C(X_t)\right] = \sum_{j=0}^{M} \pi_j C(j).$$

To illustrate, consider the inventory example introduced in Sec. 14.1. Suppose the camera store finds that a storage charge is being allocated for each camera remaining on the shelf at the end of the week. The cost is charged as follows:

$$C(x_t) = \begin{cases} 0 & \text{if} \quad x_t = 0 \\ 2 & \text{if} \quad x_t = 1 \\ 8 & \text{if} \quad x_t = 2 \\ 18 & \text{if} \quad x_t = 3 \end{cases}$$

The long-run expected average storage cost per week can then be obtained from the preceding equation, i.e.,

$$\lim_{n\to\infty} E\left[\frac{1}{n} \sum_{t=1}^{n} C(X_t)\right] = 0(0.285) + 2(0.285) + 8(0.264) + 18(0.166) = 5.67.$$

Note that an alternative measure to the (long-run) expected average cost per unit time is the (long-run) *actual average cost per unit time*. It can be shown that this latter measure is given by

$$\lim_{n\to\infty} \left[\frac{1}{n}\sum_{t=1}^{n} C(X_t)\right] = \sum_{j=0}^{M} \pi_j C(j)$$

for essentially all paths of the process. Thus either measure leads to the same result. These results can also be used to interpret the meaning of the π_j. To interpret them, let

$$C(X_t) = \begin{cases} 1 & \text{if } X_t = j, \\ 0 & \text{if } X_t \neq j. \end{cases}$$

The (long-run) expected fraction of times the system is in state j is then given by

$$\lim_{n\to\infty} E\left[\frac{1}{n}\sum_{t=1}^{n} C(X_t)\right] = \lim_{n\to\infty} E(\text{fraction of times system is in state } j) = \pi_j.$$

Similarly, π_j can also be interpreted as the (long-run) actual fraction of times that the system is in state j.

Expected Average Cost per Unit Time for Complex Cost Functions

In the preceding subsection, the cost function was based solely on the state that the process is in at time t. In many important problems encountered in practice, the cost may depend upon another random variable as well as upon the state that the process is in.

For example, in the inventory example of Sec. 14.1, suppose that the costs to be considered are the ordering cost and the penalty cost for unsatisfied demand (storage costs are small and so will be ignored for simplicity). It is reasonable to assume that the number of cameras ordered depends only upon the state of the process (the number of cameras in stock) when the order is placed. The cost for unsatisfied demand may be assumed to depend upon the demand during the week as well as upon the state of the process at the beginning of the week. The charges for period t will be made at the end of the week and will include the cost of the order delivered at the beginning of that week and the cost of unsatisfied demand during the week. Thus the cost incurred for period t can be described as a function of X_{t-1} and D_t, that is, $C(X_{t-1}, D_t)$.

For this example, the demands $D_t, D_{t+1}, \ldots$ during successive weeks are assumed to be independent and identically distributed random variables. Furthermore, recall that the (s, S) policy $(1, 3)$ is being used. And X_{t-1}, the stock level at the end of period $t-1$ (before ordering), is defined iteratively by the expression given in Sec. 14.1. Thus it follows that $X_0, X_1, X_2, \ldots, X_{t-1}$ and D_t are independent random variables because $X_0, X_1, X_2, \ldots, X_{t-1}$ are functions only of $X_0, D_1, \ldots, D_{t-1}$, which are independent of D_t. Under these conditions, it can be shown that the (long-run) *expected average cost per unit time* is given by

$$\lim_{n\to\infty} E\left[\frac{1}{n}\sum_{t=1}^{n} C(X_{t-1}, D_t)\right] = \sum_{j=0}^{M} k(j)\pi_j,$$

where

$$k(j) = E[C(j, D_t)],$$

and this latter (conditional) expectation is taken with respect to the probability distribution of the random variable D_t (given the state). Similarly, the (long-run) actual average cost per unit time is given by

$$\lim_{n\to\infty}\left[\frac{1}{n}\sum_{t=1}^{n} C(X_{t-1}, D_t)\right] = \sum_{j=0}^{M} k(j)\pi_j.$$

Suppose that the following costs are associated with the (s, S) inventory policy being used (storage charges are now neglected). If $z > 0$ cameras are ordered, the cost incurred is $(10 + 25z)$ dollars. If no cameras are ordered, no ordering cost is incurred. For each unit of unsatisfied demand (lost sales), there is a penalty of \$50. If the $(s = 1, S = 3)$ ordering policy is followed, then the cost in week t is given by $C(X_{t-1}, D_t)$, where

$$C(X_{t-1}, D_t) = \begin{cases} 10 + (25)(3) + 50 \max\{D_t - 3, 0\} & \text{if } X_{t-1} < 1, \\ 50 \max\{D_t - X_{t-1}, 0\} & \text{if } X_{t-1} \geq 1, \end{cases}$$

for $t = 1, 2, \ldots$. Hence

$$C(0, D_t) = 85 + 50 \max\{D_t - 3, 0\},$$

so that

$$\begin{aligned} k(0) = E[C(0, D_t)] &= 85 + 50E(\max\{D_t - 3, 0\}) \\ &= 85 + 50[P_D(4) + 2P_D(5) + 3P_D(6) + \cdots], \end{aligned}$$

where $P_D(i)$ is the probability that the demand equals i, as given by a Poisson distribution with parameter $\lambda = 1$, so that $P_D(i)$ becomes negligible for i larger than about 6. Hence $k(0) = 86.2$. Similar calculations lead to the results

$$\begin{aligned} k(1) = E[C(1, D_t)] &= 50E(\max\{D_t - 1, 0\}) \\ &= 50[P_D(2) + 2P_D(3) + 3P_D(4) + \cdots] \\ &= 18.4, \end{aligned}$$

$$\begin{aligned} k(2) = E[C(2, D_t)] &= 50E(\max\{D_t - 2, 0\}) \\ &= 50[P_D(3) + 2P_D(4) + 3P_D(5) + \cdots] \\ &= 5.2, \end{aligned}$$

and

$$\begin{aligned} k(3) = E[C(3, D_t)] &= 50E(\max\{D_t - 3, 0\}) \\ &= 50[P_D(4) + 2P_D(5) + \cdots] \\ &= 1.2. \end{aligned}$$

Thus the (long-run) expected average cost per week is given by

$$\sum_{j=0}^{3} k(j)\pi_j = 86.2(0.285) + 18.4(0.285) + 5.2(0.264) + 1.2(0.166) = 31.4.$$

This cost is the cost associated with the (s, S) policy $(s, S) = (1, 3)$. The cost of other (s, S) policies can be evaluated in a similar way to identify the policy that minimizes the expected average cost per week.

The results of this subsection were presented only in terms of the inventory example. However, the (nonnumerical) results still hold for other problems as long as the following conditions are satisfied:

1. $\{X_t\}$ is an irreducible Markov chain whose states are positive recurrent.
2. Associated with this Markov chain is a sequence of random variables $\{D_t\}$ which are independent and identically distributed.

3. For a fixed $m = 0, \pm 1, \pm 2, \ldots$, a cost $C(X_t, D_{t+m})$ is incurred at time t, for $t = 0, 1, 2, \ldots$.
4. The sequence $X_0, X_1, X_2, \ldots, X_t$ must be independent of D_{t+m}.

In particular, if these conditions are satisfied, then

$$\lim_{n\to\infty} E\left[\frac{1}{n}\sum_{t=1}^{n} C(X_t, D_{t+m})\right] = \sum_{j=0}^{M} k(j)\pi_j,$$

where

$$k(j) = E[C(j, D_{t+m})],$$

and this latter conditional expectation is taken with respect to the probability distribution of the random variable D_t (given the state). Furthermore,

$$\lim_{n\to\infty} \left[\frac{1}{n}\sum_{t=1}^{n} C(X_t, D_{t+m})\right] = \sum_{j=0}^{M} k(j)\pi_j$$

for essentially all paths of the process.

14.7 Absorption States

It was pointed out in Sec. 14.4 that a state k is called an *absorbing state* if $p_{kk} = 1$, so that once the chain visits k, it remains there forever. If k is an absorbing state and the process starts in state i, the probability of *ever* going to state k is called the **probability of absorption** into k, given that the system started at i. This probability is denoted by f_{ik}. When there are two or more absorbing states in a Markov chain and when it is evident that the process will be absorbed into one of these states, it is desirable to find these probabilities of absorption. These probabilities can be obtained by solving a system of linear equations. In particular, if state k is an absorbing state, then the set of absorption probabilities f_{ik} satisfies the system of equations

$$f_{ik} = \sum_{j=0}^{M} p_{ij} f_{jk}, \qquad \text{for } i = 0, 1, \ldots, M,$$

subject to the conditions

$$f_{kk} = 1,$$

$$f_{ik} = 0, \qquad \text{if state } i \text{ is recurrent and } i \neq k.$$

Absorption probabilities are important in random walks. A **random walk** is a Markov chain with the property that if the system is in a state i, then in a single transition the system either remains at i or moves to one of the states immediately adjacent to i. For example, a random walk often is used as a model for situations involving gambling.

To illustrate, consider a gambling example similar to that presented in Sec. 14.2. However, suppose now that two players, each having \$2, agree to keep playing the game and betting \$1 at a time until one player is broke. The number of dollars that player A has before each bet (0, 1, 2, 3, or 4) provides the states of a Markov chain with transition matrix

$$\mathbf{P} = \begin{bmatrix} 1 & 0 & 0 & 0 & 0 \\ 1-p & 0 & p & 0 & 0 \\ 0 & 1-p & 0 & p & 0 \\ 0 & 0 & 1-p & 0 & p \\ 0 & 0 & 0 & 0 & 1 \end{bmatrix}$$

where p represents the probability of A winning a single bet. The probability of absorption into state 0 (A losing all her money) can be obtained from the preceding system of equations. It can be shown that these equations then result in the alternate expressions (for general M rather than $M = 4$ as in this example)

$$1 - f_{i0} = \frac{\sum_{m=0}^{i-1} \rho^m}{\sum_{m=0}^{M-1} \rho^m} \qquad \text{for } i = 1, 2, \ldots, M,$$

$$= \frac{1 - \rho^i}{1 - \rho^M} \qquad \text{for } p \neq \frac{1}{2},$$

$$= \frac{i}{M} \qquad \text{for } p = \frac{1}{2},$$

where $\rho = (1 - p)/p$.

For $M = 4$, $i = 2$, and $p = \frac{2}{3}$, the probability of A going broke is given by

$$f_{20} = 1 - \left[\frac{1 - \rho^2}{1 - \rho^4}\right] = \frac{1}{5},$$

and the probability of A winning \$4 (B going broke) is given by

$$f_{24} = 1 - f_{20} = \tfrac{4}{5}.$$

There are many other situations where absorbing states play an important role. Consider a department store that classifies the balance of a customer's bill as fully paid (state 0), 1 to 30 days in arrears (state 1), 31 to 60 days in arrears (state 2), or bad debt (state 3). The accounts are checked monthly, and the state of each customer is determined. In general, credit is not extended, and customers are expected to pay their bills within 30 days. Occasionally, customers pay only portions of their bill. If this occurs when the balance is within 30 days in arrears (state 1), the store views the customer as remaining in state 1. If this occurs when the balance is between 31 and 60 days in arrears, the store views the customer as moving to state 1 (1 to 30 days in arrears). Customers that are more than 60 days in arrears are put into the bad-debt category (state 3), and then bills are sent to a collection agency. After examining data over the past several years, the store has developed the following transition matrix:[1]

State \ State	0: Fully Paid	1: 1 to 30 Days in Arrears	2: 31 to 60 Days in Arrears	3: Bad Debt
0: fully paid	1	0	0	0
1: 1 to 30 days in arrears	0.7	0.2	0.1	0
2: 31 to 60 days in arrears	0.5	0.1	0.2	0.2
3: bad debt	0	0	0	1

[1] Customers that are fully paid (in state 0) may be viewed as "new" customers if they return for repeat purchases.

Although each customer ends up in state 0 or 3, the store is interested in determining the probability that a customer will end up as a bad debt given that the account belongs to the 1 to 30 days in arrears state, and similarly, given that the account belongs to the 31 to 60 days in arrears state.

To obtain this information, the set of equations presented at the beginning of this section must be solved. In particular, f_{13} and f_{23} are to be obtained. By substituting, the following two equations are obtained:

$$f_{13} = p_{10}f_{03} + p_{11}f_{13} + p_{12}f_{23} + p_{13}f_{33},$$

$$f_{23} = p_{20}f_{03} + p_{21}f_{13} + p_{22}f_{23} + p_{23}f_{33}.$$

Noting that $f_{03} = 0$ and $f_{33} = 1$, we now have two equations in two unknowns, namely,

$$(1 - p_{11})f_{13} = p_{13} + p_{12}f_{23},$$

$$(1 - p_{22})f_{23} = p_{23} + p_{21}f_{13}.$$

Substituting the values from the transition matrix leads to

$$0.8f_{13} = 0.1f_{23},$$

$$0.8f_{23} = 0.2 + 0.1f_{13},$$

and the solution is

$$f_{13} = 0.032,$$

$$f_{23} = 0.254.$$

Thus, approximately 3 percent of the customers whose accounts are 1 to 30 days in arrears end up as bad debts, whereas 25 percent of the customers whose accounts are 31 to 60 days in arrears end up as bad debts.

14.8 Continuous Time Markov Chains

In all the previous sections, we assumed that the time parameter t was discrete (that is, $t = 0, 1, 2, \ldots$). Such an assumption is suitable for many problems, but there are certain cases (such as for some queueing models considered in the next chapter) where a continuous time parameter (call it t') is required, because the evolution of the process is being observed *continuously* over time. The definition of a Markov chain given in Sec. 14.1 also extends to such continuous processes. This section focuses on describing these continuous time Markov chains and their properties.

Formulation

As before, we label the possible **states** of the system $0, 1, \ldots, M$. Starting at time 0 and letting the time parameter t' run continuously for $t' \geq 0$, we let the random variable $X(t')$ be the state of the system at time t'. Thus, $X(t')$ will take on one of its possible $M + 1$ values over some interval $0 \leq t' < t_1$, then will jump to another value over the next interval $t_1 \leq t' < t_2$, and so on, where these transit points $t_1, t_2, \ldots$ are random points in time (*not* necessarily integer).

Now consider the three points in time (1) $t' = r$ (where $r \geq 0$), (2) $t' = s$ (where $s > r$), and (3) $t' = s + t$ (where $t > 0$), interpreted as follows:

$t' = r$ is a past time,
$t' = s$ is the current time,
$t' = s + t$ is t time units into the future.

Therefore, the state of the system now has been observed at times $t' = s$ and $t' = r$. Label these states as

$$X(s) = i \qquad \text{and} \qquad X(r) = x(r).$$

Given this information, it now would be natural to seek the probability distribution of the state of the system at time $t' = s + t$. In other words, what is

$$P\{X(s + t) = j \mid X(s) = i \text{ and } X(r) = x(r)\}, \qquad \text{for } j = 0, 1, \ldots, M?$$

Deriving this conditional probability often is very difficult. However, this task is considerably simplified if the stochastic process involved possesses the following key property.

A continuous time stochastic process $\{X(t'); t' \geq 0\}$ has the **Markovian property** if

$$P\{X(t + s) = j \mid X(s) = i \text{ and } X(r) = x(r)\} = P\{X(t + s) = j \mid X(s) = i\},$$

for all $i, j = 0, 1, \ldots, M$ and for all $r \geq 0$, $s > r$, and $t > 0$.

Note that $P\{X(t + s) = j \mid X(s) = i\}$ is a **transition probability**, just like the transition probabilities for discrete time Markov chains considered in the preceding sections, where the only difference is that t now need not be an integer.

If the transition probabilities are independent of s, so that

$$P\{X(t + s) = j \mid X(s) = i\} = P\{X(t) = j \mid X(0) = i\}$$

for all $s > 0$, they are called **stationary transition probabilities**.

To simplify notation, we shall denote these stationary transition probabilities by

$$p_{ij}(t) = P\{X(t) = j \mid X(0) = i\},$$

where $p_{ij}(t)$ is referred to as the **continuous time transition probability function**. We assume that

$$\lim_{t \to 0} p_{ij}(t) = \begin{cases} 1 & \text{if } i = j, \\ 0 & \text{if } i \neq j. \end{cases}$$

Now we are ready to define the continuous time Markov chains to be considered in this section.

A continuous time stochastic process $\{X(t'); t' \geq 0\}$ is a **continuous time Markov chain** if it has the *Markovian property.*

We shall restrict our consideration to continuous time Markov chains with the following properties:

1. A finite number of states
2. Stationary transition probabilities.

Some Key Random Variables

In the analysis of continuous time Markov chains, one key set of random variables is the following.

Each time the process enters state i, the amount of time it spends in that state before moving to a different state is a random variable T_i, where $i = 0, 1, \ldots, M$.

Suppose that the process enters state i at time $t' = s$. Then, for any fixed amount of time $t > 0$, note that $T_i > t$ if and only if $X(t') = i$ for all t' over the interval $s \leq t' \leq s + t$. Therefore, the Markovian property (with stationary transition probabilities) implies that

$$P\{T_i > t + s \mid T_i > s\} = P\{T_i > t\}.$$

This is a rather unusual property for a probability distribution to possess. It says that the probability distribution of the *remaining* time until the process transits out of a given state always is the same, regardless of how much time the process has already spent in that state. In effect, the random variable is memoryless; the process forgets its history. There is only one (continuous) probability distribution that possesses this property—the *exponential distribution.* The exponential distribution has a single parameter, call it q, where the mean is $1/q$ and the cumulative distribution function is

$$P\{T_i \leq t\} = 1 - e^{-qt}, \qquad \text{for } t \geq 0.$$

(We shall describe the properties of the exponential distribution in detail in Sec. 15.4.)

This result leads to an equivalent way of describing a continuous time Markov chain:

1. The random variable T_i has an exponential distribution with a mean of $1/q_i$.
2. When leaving state i, the process moves to a state j with probability p_{ij}, where the p_{ij} satisfy the conditions

$$p_{ii} = 0 \qquad \text{for all } i,$$

and

$$\sum_{j=0}^{M} p_{ij} = 1 \qquad \text{for all } i.$$

3. The next state visited after state i is independent of the time spent in state i.

Just as the one-step transition probabilities played a major role in describing discrete time Markov chains, the analogous role for a continuous time Markov chain is played by the transition intensities.

The **transition intensities** are

$$q_i = -\frac{d}{dt}p_{ii}(0) = \lim_{t \to 0} \frac{1 - p_{ii}(t)}{t}, \qquad \text{for } i = 0, 1, 2, \ldots, M,$$

and

$$q_{ij} = \frac{d}{dt}p_{ij}(0) = \lim_{t \to 0} \frac{p_{ij}(t)}{t} = q_i p_{ij}, \qquad \text{for all } j \neq i,$$

where $p_{ij}(t)$ is the *continuous time transition probability function* introduced at the beginning of the section and p_{ij} is the probability described in property 2 of the preceding paragraph. Furthermore, q_i as defined here turns out to still be the parameter of the exponential distribution for T_i as well (see property 1 of the preceding paragraph).

The intuitive interpretation of the q_i and q_{ij} is that they are *transition rates.* In particular, q_i is the *transition rate out of state i* in the sense that q_i is the expected number of times that the process leaves state i per unit of time spent in state i. (Thus, q_i is the reciprocal of the expected time that the process spends in state i per visit to state i; that is, $q_i = 1/E[T_i]$.) Similarly, q_{ij} is the *transition rate from state i to state j* in the sense that q_{ij} is the expected number of times that the process transits from state i to state j per unit of time spent in state i. Thus,

$$q_i = \sum_{j \neq i} q_{ij}.$$

Just as q_i is the parameter of the exponential distribution for T_i, each q_{ij} is the parameter of an exponential distribution for a related random variable described below.

> Each time the process enters state i, the amount of time it will spend in state i before a transition to state j occurs (if a transition to some other state does not occur first) is a random variable T_{ij}, where $i, j = 0, 1, \ldots, M$ and $j \neq i$. The T_{ij} are independent random variables, where each T_{ij} has an *exponential distribution* with parameter q_{ij}, so $E[T_{ij}] = 1/q_{ij}$. The time spent in state i until a transition occurs (T_i) is the *minimum* (over $j \neq i$) of the T_{ij}. When the transition occurs, the probability that it is to state j is $p_{ij} = q_{ij}/q_i$.

Steady-State Probabilities

Just as the transition probabilities for a discrete time Markov chain satisfy the Chapman-Kolmogorov equations, the continuous time transition probability function also satisfies these equations. Therefore, for any states i and j and nonnegative numbers t and s ($0 \leq s \leq t$),

$$p_{ij}(t) = \sum_{k=1}^{M} p_{ik}(s)p_{kj}(t-s).$$

A pair of states i and j are said to *communicate* if there are times t_1 and t_2 such that $p_{ij}(t_1) > 0$ and $p_{ji}(t_2) > 0$. All states that communicate are said to form a *class*. If all states form a single class, i.e., if the Markov chain is *irreducible* (hereafter assumed), then

$$p_{ij}(t) > 0, \qquad \text{for all } t > 0 \text{ and all states } i \text{ and } j.$$

Furthermore,

$$\lim_{t \to \infty} p_{ij}(t) = \pi_j$$

always exists and is independent of the initial state of the Markov chain, for $j = 0, 1, \ldots, M$. These limiting probabilities are commonly referred to as the **steady-state probabilities** (or *stationary probabilities*) of the Markov chain.

The π_j satisfy the equations

$$\pi_j = \sum_{i=0}^{M} \pi_i p_{ij}(t), \qquad \text{for } j = 0, 1, \ldots, M \text{ and every } t \geq 0.$$

However, the following **steady-state equations** provide a more useful system of equations for solving for the steady-state probabilities:

$$\pi_j q_j = \sum_{i \neq j} \pi_i q_{ij}, \qquad \text{for } j = 0, 1, \ldots, M,$$

and

$$\sum_{j=0}^{M} \pi_j = 0.$$

The steady-state equation for state j has an intuitive interpretation. The left-hand side ($\pi_j q_j$) is the *rate* at which the process *leaves* state j, since π_j is the (steady-state) probability that the process is in state j and q_j is the transition rate out of state j given that the process is in state j. Similarly, each term on the right-hand side ($\pi_i q_{ij}$) is the *rate* at which the process *enters* state j from state i, since q_{ij} is the transition rate from state i to state j given that the state is in state i. By summing over all $i \neq j$, the entire right-hand side then gives the rate at which the process enters state j from any other state. The overall equation thereby states that the rate at which the process leaves state j must equal the rate at which the process enters state j. Thus, this equation is analogous to the conservation of flow equations encountered in many engineering and science courses.

Because each of the first $M + 1$ *steady-state equations* requires that two rates be *in balance* (equal), these equations sometimes are called the **balance equations**.

Example

A certain shop has two identical machines that are operated continuously except when they are broken down. Because they break down fairly frequently, the top-priority assignment for a full-time maintenance person is to repair them whenever needed.

The time required to repair a machine has an exponential distribution with a mean of $\frac{1}{2}$ day. Once the repair is completed, the time until the next breakdown has an exponential distribution with a mean of 1 day. These distributions are independent.

Define the random variable $X(t')$ as

$$X(t') = \text{number of machines broken down at time } t',$$

so the possible values of $X(t')$ are 0, 1, 2. Therefore, by letting the time parameter t' run continuously from time 0, the continuous time stochastic process $\{X(t'); t' \geq 0\}$ gives the evolution of the number of machines broken down.

Because both the repair time and the time until a breakdown have exponential distributions, $\{X(t'); t' \geq 0\}$ is a *continuous time Markov chain*[1] with states 0, 1, 2. Consequently, we can use the steady-state equations given in the preceding subsection to find the steady-state probability distribution of the number of machines broken down. To do this, we need to determine all the *transition rates,* i.e., the q_i and q_{ij} for $i, j = 0, 1, 2$.

The state (number of machines broken down) increases by 1 when a breakdown occurs and decreases by 1 when a repair occurs. Since both breakdowns and repairs occur one at a time, $q_{02} = 0$ and $q_{20} = 0$. The expected repair time is $\frac{1}{2}$ day, so the rate at which repairs are completed (when any machines are broken down) is 2 per day, which implies that $q_{21} = 2$ and $q_{10} = 2$. Similarly, the expected time until a particular operational machine breaks down is 1 day, so the rate at which it breaks down (when operational) is 1 per day, which implies that $q_{12} = 1$. During times when both machines are operational, breakdowns occur at the rate of $1 + 1 = 2$ per day, so $q_{01} = 2$.

[1] Proving this fact requires the use of two properties of the exponential distribution discussed in Sec. 15.4 (*lack of memory* and *the minimum of exponentials is exponential*), since these properties imply that the T_{ij} random variables introduced earlier do indeed have exponential distributions.

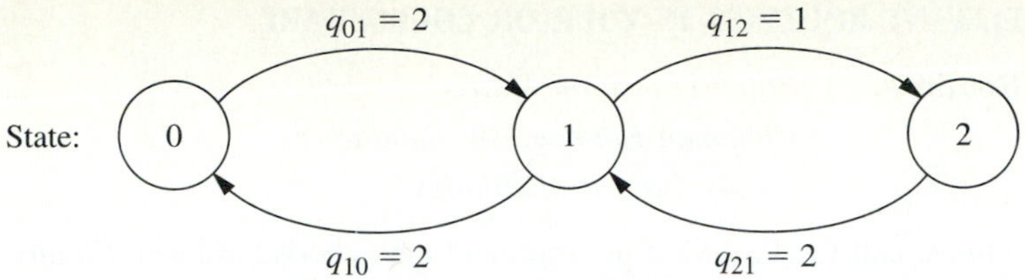

State: 0, 1, 2
$q_{01} = 2$, $q_{12} = 1$, $q_{10} = 2$, $q_{21} = 2$

Figure 14.1 Rate diagram for the example of a continuous time Markov chain.

These transition rates are summarized in the rate diagram shown in Fig. 14.1. These rates now can be used to calculate the *total transition rate* out of each state.

$$q_0 = q_{01} = 2.$$

$$q_1 = q_{10} + q_{12} = 3.$$

$$q_2 = q_{21} = 2.$$

Plugging all the rates into the steady-state equations given in the preceding subsection then yields

Balance equation for state 0:	$2\pi_0 = 2\pi_1$
Balance equation for state 1:	$3\pi_1 = 2\pi_0 + 2\pi_2$
Balance equation for state 2:	$2\pi_2 = \pi_1$
Probabilities sum to 1:	$\pi_0 + \pi_1 + \pi_2 = 1$

Any one of the balance equations (say, the second) can be deleted as redundant, and the simultaneous solution of the remaining equations gives the steady-state distribution as

$$(\pi_0, \pi_1, \pi_2) = (\tfrac{2}{5}, \tfrac{2}{5}, \tfrac{1}{5}).$$

Thus, in the long run, both machines will be broken down simultaneously 20 percent of the time, and one machine will be broken down another 40 percent of the time.

The next chapter (on queueing theory) features many more examples of continuous time Markov chains. In fact, most of the basic models of queueing theory fall into this category. The current example actually fits one of these models (the finite calling population variation of the *M/M/s* model included in Sec. 15.6).

SELECTED REFERENCES

1. Bhattacharva, R. N., and E. C. Waymire, *Stochastic Processes with Applications,* Wiley, New York, 1990.
2. Grassmann, W. K., M. I. Taksar, and D. P. Heyman, ''Regenerative Analysis and Steady State Distributions for Markov Chains,'' *Operations Research,* **33**: 1107–1116, 1985.
3. Heyman, D. P., ''Approximating the Stationary Distribution of an Infinite Stochastic Matrix,'' *Journal of Applied Probability,* **28**: 96–103, 1991.
4. Heyman, D., and M. Sobel, *Stochastic Models in Operations Research,* vol. 1, McGraw-Hill, New York, 1982.
5. Ross, S., *Introduction to Probability Models,* 5th ed., Academic Press, New York, 1993.
6. Ross, S., *Stochastic Processes,* 2d ed., Wiley, New York, 1995.
7. Stewart, W. J. (ed.), *Numerical Solution of Markov Chains,* Marcel Dekker, New York, 1991.
8. Taylor, H., and S. Karlin, *An Introduction to Stochastic Modeling,* Academic Press, New York, 1984.

Automatic Routines: *Enter Transition Matrix*
Chapman-Kolmogorov Equations
Steady-State Probabilities

To access these routines, call the ProbMod program and then choose *Markov Chains* under the Area menu. See Appendix 1 for documentation of the software.

PROBLEMS

To the left of each of the following problems (or their parts), we have inserted an A (for automatic routine) whenever a corresponding routine listed above can be helpful. An asterisk on the A indicates that this routine definitely should be used (unless your instructor gives contrary instructions) and that the printout from this routine is all that needs to be turned in to show your work in executing the procedure. An asterisk on the problem number indicates that at least a partial answer is given in the back of the book.

14.2-1. Assume that the probability of rain tomorrow is 0.5 if it is raining today, and assume that the probability of its being clear (no rain) tomorrow is 0.9 if it is clear today. Also assume that these probabilities do not change if information is also provided about the weather before today.

(*a*) Explain why the stated assumptions imply that the Markovian property holds for the evolution of the weather.

(*b*) Formulate the evolution of the weather as a Markov chain by defining its states and giving its (one-step) transition matrix.

14.2-2. Consider the second version of the stock market model presented as an example in Sec. 14.2. Whether the stock goes up tomorrow depends upon whether it increased today *and* yesterday. If the stock increased today and yesterday, it will increase tomorrow with probability α_1. If the stock increased today and decreased yesterday, it will increase tomorrow with probability α_2. If the stock decreased today and increased yesterday, it will increase tomorrow with probability α_3. Finally, if the stock decreased today and yesterday, it will increase tomorrow with probability α_4.

Determine the (one-step) transition matrix of the Markov chain.

14.2-3. Reconsider Prob. 14.2-2. Suppose now that the stock's going up tomorrow depends upon whether it increased today, yesterday, *and* the day before yesterday. Can this problem be formulated as a Markov chain? If so, what are the possible states? Explain why these states give the process the Markovian property whereas the states in Prob. 14.2-2 do not.

14.3-1. Reconsider Prob. 14.2-1.

A* (*a*) Use the routine *Chapman-Kolmogorov Equations* in your OR Courseware to find the n-step transition matrix $\mathbf{P}^{(n)}$ for $n = 2, 5, 10, 20$.

(*b*) The probability that it will rain today is 0.5. Use the results from part (*a*) to determine the probability that it will rain n days from now, for $n = 2, 5, 10, 20$.

A* (*c*) Use the routine *Steady-State Probabilities* in your OR Courseware to determine the steady-state probabilities of the state of the weather. Describe how the probabilities in the n-step transition matrices obtained in part (*a*) compare to these steady-state probabilities as n grows large.

14.3-2. Suppose that a communications network transmits binary digits, 0 or 1, where each digit is transmitted 10 times in succession. During each transmission, the probability is 0.99 that the digit will be transmitted accurately. In other words, the probability is 0.01 that the digit being transmitted will be recorded with the opposite value at the end of the transmission. If X_0 denotes the binary digit entering the system, X_1 the binary digit recorded after the first transmission, X_2 the binary digit recorded after the second transmission, . . . , then $\{X_n\}$ is a Markov chain.

(*a*) Determine the (one-step) transition matrix.

A* (*b*) Use your OR Courseware to find the 10-step transition matrix $\mathbf{P}^{(10)}$. Use this result to identify the probability that a digit entering the network will be recorded accurately after the last transmission.

(*c*) Suppose that the network is redesigned to improve the probability that a single transmission will be accurate from 0.99 to 0.999. Repeat part (*b*) to find the new probability that a digit entering the network will be recorded accurately after the last transmission.

14.3-3.* A particle moves on a circle through points that have been marked 0, 1, 2, 3, 4 (in clockwise order). The particle starts at point 0. At each step it has probability 0.5 of moving one point clockwise (0 follows 4) and probability 0.5 of moving one point counterclockwise. Let X_n $(n \geq 0)$ denote its location on the circle after step n, and $\{X_n\}$ is a Markov chain.

(*a*) Find the (one-step) transition matrix.

A* (*b*) Use your OR Courseware to determine the n-step transition matrix $\mathbf{P}^{(n)}$ for $n = 5$, 10, 20, 40, 80.

A* (*c*) Use your OR Courseware to determine the steady-state probabilities of the state of the Markov chain. Describe how the probabilities in the n-step transition matrices obtained in part (*b*) compare to these steady-state probabilities as n grows large.

14.4-1.* Given each of the following (one-step) transition matrices of a Markov chain, determine the classes of the Markov chain and whether they are recurrent.

$$(a)\ \mathbf{P} = \begin{bmatrix} 0 & 0 & \frac{1}{3} & \frac{2}{3} \\ 1 & 0 & 0 & 0 \\ 0 & 1 & 0 & 0 \\ 0 & 1 & 0 & 0 \end{bmatrix} \qquad (b)\ \mathbf{P} = \begin{bmatrix} 1 & 0 & 0 & 0 \\ 0 & \frac{1}{2} & \frac{1}{2} & 0 \\ 0 & \frac{1}{2} & \frac{1}{2} & 0 \\ \frac{1}{2} & 0 & 0 & \frac{1}{2} \end{bmatrix}$$

14.4-2. Given each of the following (one-step) transition matrices of a Markov chain, determine the classes of the Markov chain and whether they are recurrent.

$$(a)\ \mathbf{P} = \begin{bmatrix} 0 & \frac{1}{3} & \frac{1}{3} & \frac{1}{3} \\ \frac{1}{3} & 0 & \frac{1}{3} & \frac{1}{3} \\ \frac{1}{3} & \frac{1}{3} & 0 & \frac{1}{3} \\ \frac{1}{3} & \frac{1}{3} & \frac{1}{3} & 0 \end{bmatrix} \qquad (b)\ \mathbf{P} = \begin{bmatrix} 0 & 0 & 1 \\ \frac{1}{2} & \frac{1}{2} & 0 \\ 0 & 1 & 0 \end{bmatrix}$$

14.4-3. Given the following (one-step) transition matrix of a Markov chain, determine the classes of the Markov chain and whether they are recurrent.

$$\begin{bmatrix} \frac{1}{4} & \frac{3}{4} & 0 & 0 & 0 \\ \frac{3}{4} & \frac{1}{4} & 0 & 0 & 0 \\ \frac{1}{3} & \frac{1}{3} & \frac{1}{3} & 0 & 0 \\ 0 & 0 & 0 & \frac{3}{4} & \frac{1}{4} \\ 0 & 0 & 0 & \frac{1}{4} & \frac{3}{4} \end{bmatrix}$$

14.5-1. A computer is inspected at the end of every hour. It is found to be either working (up) or failed (down). If the computer is found to be up, the probability of its remaining up for the next hour is 0.90. If it is down, the computer is repaired, which may require more than 1 hour. Whenever the computer is down (regardless of how long it has been down), the probability of its still being down 1 hour later is 0.35.

(*a*) Construct the (one-step) transition matrix for this Markov chain.

(*b*) Use the approach described in Sec. 14.5 to find the μ_{ij} (the expected first passage time from state i to state j) for all i and j.

14.5-2. A manufacturer has a machine that, when operational at the beginning of a day, has a probability of 0.1 of breaking down sometime during the day. When this happens, the repair is made the next day and is completed at the end of that day.

(*a*) Formulate the evolution of the status of the machine as a Markov chain, by identifying three possible states at the end of each day and then constructing the (one-step) transition matrix.

(*b*) Use the approach described in Sec. 14.5 to find the μ_{ij} (the expected first passage time from state i to state j) for all i and j. Use these results to identify the expected number of full days that the machine will remain operational before the next breakdown after a repair is completed.

(*c*) Now suppose that the machine already has gone 20 full days without a breakdown since the last repair was completed. How does the expected number of full days *hereafter* that the machine will remain operational before the next breakdown compare with the corresponding result from part (*b*) when the repair had just been completed? Explain.

14.5-3. Reconsider Prob. 14.5-2. Now suppose that the manufacturer keeps a spare machine that is used only when the primary machine is being repaired. During a repair day, the spare machine has a probability of 0.1 of breaking down, in which case it is repaired the next day. Denote the state of the system by (x, y), where x and y take on the values 1 or 0 depending upon whether the primary machine (x) and the spare machine (y) are operational at the end of the day. [*Hint:* Note that (0, 0) is not a possible state.]

(*a*) Construct the (one-step) transition matrix for this Markov chain.

(*b*) Find the *expected recurrence time* for state (1, 0).

14.6-1. Reconsider Prob. 14.2-1. Suppose now that the given probabilities, 0.5 and 0.9, are replaced by arbitrary values α and β, respectively. Solve for the steady-state probabilities of the state of the weather in terms of α and β.

14.6-2. A transition matrix **P** is said to be doubly stochastic if the sum over each column equals 1, that is,

$$\sum_{i=0}^{M} p_{ij} = 1, \qquad \text{for all } j.$$

If such a chain is irreducible and aperiodic and consists of $M + 1$ states, show that

$$\pi_j = \frac{1}{M+1}, \qquad \text{for } j = 0, 1, \ldots, M.$$

14.6-3. Reconsider Prob. 14.3-3. Use the results given in Prob. 14.6-2 to find the steady-state probabilities for this Markov chain. Then find out what happens to these steady-state probabilities if, at each step, the probability of moving one point clockwise changes to 0.9 and the probability of moving one point counterclockwise changes to 0.1.

A* **14.6-4.** The leading brewery on the west coast (labeled A) has hired an OR analyst to analyze its market position. It is particularly concerned about its major competitor (labeled B). The analyst believed that brand switching can be modeled as a Markov chain by using three states, with states A and B representing customers drinking beer produced from the aforementioned breweries and state C representing all other brands. Data are taken monthly, and the analyst has constructed the following transition matrix from past data.

	A	B	C
A	0.7	0.2	0.1
B	0.2	0.75	0.05
C	0.1	0.1	0.8

What are the steady-state market shares for the two major breweries?

14.6-5. Consider the following blood inventory problem facing a hospital. There is need for a rare blood type, namely, type AB, Rh-negative blood. The demand D (in pints) over any 3-day period is given by

$$P\{D = 0\} = 0.4, \quad P\{D = 1\} = 0.3, \quad P\{D = 2\} = 0.2, \quad \text{and} \quad P\{D = 3\} = 0.1.$$

Note that the expected demand is 1 pint, since $E(D) = 0.3(1) + 0.2(2) + 0.1(3) = 1$. Suppose that there are 3 days between deliveries. The hospital proposes a policy of receiving 1 pint at each delivery and using the oldest blood first. If more blood is required than is on hand, an expensive emergency delivery is made. Blood is discarded if it is still on the shelf after 21 days. Denote the state of the system as the number of pints on hand just after a delivery. Thus, because of the discarding policy, the largest possible state is 7.

(*a*) Find the (one-step) transition matrix for this Markov chain.
(*b*) Find the steady-state probabilities of the state of the Markov chain.
(*c*) Use the results from part (*b*) to find the steady-state probability that a pint of blood will need to be discarded during a 3-day period. (*Hint:* Because the oldest blood is used first, a pint reaches 21 days only if the state was 7 and then $D = 0$.)
(*d*) Use the results from part (*b*) to find the steady-state probability that an emergency delivery will be needed during the 3-day period between regular deliveries.

14.6-6. A soap company specializes in a luxury type of bath soap. The sales of this soap fluctuate between two levels—low and high—depending upon two factors: (1) whether they advertise and (2) the advertising and marketing of new products being done by competitors. The second factor is out of the company's control, but it is trying to determine what its own advertising policy should be. For example, the marketing manager's proposal is to advertise when sales are low but not to advertise when sales are high. Advertising in any quarter of a year has its primary impact on sales in the *following* quarter. Therefore, at the beginning of each quarter, the needed information is available to forecast accurately whether sales will be low or high that quarter and to decide whether to advertise that quarter.

The cost of advertising is \$1 million for each quarter of a year in which it is done. When advertising is done during a quarter, the probability of having high sales the next quarter is $\frac{1}{2}$ or $\frac{3}{4}$, depending upon whether the current quarter's sales are low or high. These probabilities go down to $\frac{1}{4}$ or $\frac{1}{2}$ when advertising is not done during the current quarter. The company's quarterly profits (excluding advertising costs) are \$4 million when sales are high but only \$2 million when sales are low. (Hereafter, use units of millions of dollars.)

(*a*) Construct the (one-step) transition matrix for each of the following advertising strategies: (i) never advertise, (ii) always advertise, (iii) follow the marketing manager's proposal.
A* (*b*) Determine the steady-state probabilities for each of the three cases in part (*a*).
(*c*) Find the long-run expected average profit (including a deduction for advertising costs) per quarter for each of the three advertising strategies in part (*a*). Which of these strategies is best according to this measure of performance?

14.6-7. Consider the camera inventory problem presented in Sec. 14.1 except that demand now has the following probability distribution:

$$P\{D = 0\} = \tfrac{1}{4}, \qquad P\{D = 2\} = \tfrac{1}{4},$$
$$P\{D = 1\} = \tfrac{1}{2}, \qquad P\{D \geq 3) = 0.$$

The inventory policy is still an (s, S) policy, but now $s = 1$ and $S = 2$. Assume that there is one camera in stock at the time (the end of a week) that the policy is instituted.

(*a*) Find the (one-step) transition matrix.
A* (*b*) Find the probability distribution of the state of this Markov chain n weeks after the new inventory policy is instituted, for $n = 2, 5, 10$.
(*c*) Find the μ_{ij} (the expected first passage time from state i to state j) for all i and j.
A* (*d*) Find the steady-state probabilities of the state of this Markov chain.

(*e*) Assuming that the store pays a storage cost for each camera remaining on the shelf at the end of the week according to the function $C(0) = 0$, $C(1) = \$2$, and $C(2) = \$8$, find the long-run expected average storage cost per week.

14.6-8. A production process contains a machine that deteriorates rapidly in both quality and output under heavy use, so that it is inspected at the end of each day. Immediately after inspection, the condition of the machine is noted and classified into one of four possible states:

State	Condition
0	Good as new
1	Operable—minimum deterioration
2	Operable—major deterioration
3	Inoperable and replaced by a good-as-new machine

The process can be modeled as a Markov chain with its (one-step) transition matrix **P** given by

State	0	1	2	3
0	0	$\frac{7}{8}$	$\frac{1}{16}$	$\frac{1}{16}$
1	0	$\frac{3}{4}$	$\frac{1}{8}$	$\frac{1}{8}$
2	0	0	$\frac{1}{2}$	$\frac{1}{2}$
3	1	0	0	0

A* (*a*) Find the steady-state probabilities.

(*b*) If the costs of being in states 0, 1, 2, 3, are 0, \$1,000, \$3,000, and \$6,000, respectively, what is the long-run expected average cost per day?

(*c*) Find the *expected recurrence time* for state 0 (i.e., the expected length of time a machine can be used before it must be replaced).

A* **14.6-9.** In the last subsection of Sec. 14.6, the (long-run) expected average cost per week (based on just ordering costs and unsatisfied demand costs) is calculated for the camera inventory example of Sec. 14.1. This example uses an (s, S) inventory policy where $s = 1$ and $S = 3$. Suppose now that this inventory policy is changed by increasing s to $s = 2$ (with $S = 3$ still). Therefore, whenever the number of cameras on hand at the end of the week is less than $s = 2$, an order is placed that will bring this number up to $S = 3$.

Recalculate the (long-run) expected average cost per week under this new inventory policy.

14.6-10.* Consider the inventory example introduced in Sec. 14.1. Instead of following an (s, S) policy, suppose that the inventory policy is changed to the following (q, Q) policy. If the stock level at the end of each week is less than $q = 2$ cameras, $Q = 2$ additional cameras will be ordered. Otherwise, no ordering will take place. Assume again that unfilled demand results in lost sales. Let X_n denote the number of cameras on hand at the end of week n (where $X_0 = 0$); and $\{X_n\}$ is a Markov chain. Assume that the demand distribution each week is the same as given in Sec. 14.2 (a Poisson distribution with parameter $\lambda = 1$) and that the storage costs are the same as given in the second subsection of Sec. 14.6.

A* (*a*) Find the steady-state probabilities of the state of this Markov chain.

(*b*) Find the long-run expected average storage cost per week.

14.6-11. Consider the following (k, Q) inventory policy. Let $D_1, D_2, \ldots$ be the demand for a product in periods $1, 2, \ldots$, respectively. If the demand during a period exceeds the number of items available, this unsatisfied demand is backlogged; i.e., it is filled when the next order is received. Let Z_n ($n = 0, 1, \ldots$) denote the amount of inventory on hand minus the number of units backlogged before ordering at the end of period n ($Z_0 = 0$). If Z_n is zero or positive, no orders are backlogged. If Z_n is negative, then $-Z_n$ represents the number of backlogged units and no inventory is on hand. At the end of period n, if $Z_n < k = 1$, an order is placed for $2m$ (Qm in general) units, where m is the smallest integer such that $Z_n + 2m \geq 1$. Orders are filled immediately. Let D_n be independent and identically distributed random variables taking on the values 0, 1, 2, 3, 4, each with probability $\frac{1}{5}$. Let X_n denote the amount of stock on hand *after* ordering at the end of period n (where $X_0 = 2$), so that

$$X_n = \begin{cases} X_{n-1} - D_n + 2m & \text{if } X_{n-1} - D_n < 1 \\ X_{n-1} - D_n & \text{if } X_{n-1} - D_n \geq 1 \end{cases} \qquad (n = 1, 2, \ldots),$$

where $\{X_n\}$ ($n = 0, 1, \ldots$) is a Markov chain. It has only two states, 1 and 2, because ordering will be done only when $Z_n = 0, -1, -2,$ or -3, in which case 2, 2, 4, and 4 units are ordered, respectively, leaving $X_n = 2, 1, 2, 1$, respectively. [In general, for any (k, Q) policy, the possible states are $k, k+1, k+2, \ldots, k+Q-1$.]

(*a*) Find the (one-step) transition matrix.

(*b*) Use the steady-state equations to solve by hand for the steady-state probabilities.

(*c*) Now use the result given in Prob. 14.6-2 to find the steady-state probabilities.

(*d*) Suppose that the ordering cost is given by $2 + 2m$ if an order is placed and 0 otherwise. The holding cost per period is Z_n if $Z_n \geq 0$ and is 0 otherwise. The shortage cost per period is $-4Z_n$ if $Z_n < 0$ and is 0 otherwise. Find the (long-run) expected average cost per unit time.

14.6-12. An important unit consists of two components placed in parallel. The unit performs satisfactorily if one of the two components is operating. Therefore, only one component is operated at a time, but both components are kept operational (capable of being operated) as often as possible by repairing them as needed. An operating component breaks down in a given period with probability 0.2. Assume that the component breaks down only at the end of a period. When this occurs, the parallel component takes over, if it is operational, at the beginning of the next period. The repair of a component requires two periods, and only one component can be repaired at a time. Let X_t be a vector consisting of two elements U and V, where U represents the number of components that are operational at the end of period t and V takes on the value 1 if one additional period is needed to complete a repair and the value 0 otherwise. Thus the state space consists of the four states (2, 0), (1, 0), (0, 1), and (1, 1). For example, state (1, 1) implies that one component is operational and the other component needs one additional period of repair before becoming operational. Denote these four states by 0, 1, 2, 3, respectively. Now $\{X_t\}$ ($t = 0, 1, \ldots$) is a Markov chain (assume that $X_0 = 0$) with transition matrix

$$\mathbf{P} = \begin{bmatrix} 0.8 & 0.2 & 0 & 0 \\ 0 & 0 & 0.2 & 0.8 \\ 0 & 1 & 0 & 0 \\ 0.8 & 0.2 & 0 & 0 \end{bmatrix}$$

A* (*a*) What is the probability that the unit will be inoperable (because both components are down) after n periods, for $n = 2, 5, 10, 20$?

A* (*b*) What are the steady-state probabilities of the state of this Markov chain?

(*c*) If it costs $30,000 per period when the unit is inoperable (both components down) and zero otherwise, what is the (long-run) expected average cost per period?

14.7-1. Consider the following gambler's ruin problem. A gambler bets 1 unit on each play of a game. He has a probability p of winning and probability $q = 1 - p$ of losing. He will continue to play until he goes broke or nets a fortune of T units. Let X_n denote the gambler's fortune (holding of money) after the nth play of the game. Then

$$X_{n+1} = \begin{cases} X_n + 1 & \text{with probability } p \\ X_n - 1 & \text{with probability } q = 1 - p \end{cases} \qquad \text{for } 0 < X_n < T,$$

$$X_{n+1} = X_n, \qquad \text{for } X_n = 0 \text{ or } T.$$

And $\{X_n\}$ is a Markov chain. Assume that successive plays of the game are independent and that the gambler has an initial fortune of X_0 units, where $0 < X_0 < T$.

(*a*) Determine the (one-step) transition matrix of the Markov chain.
(*b*) Find the classes of the Markov chain.
(*c*) Let $T = 3$ and $p = 0.3$. Using the notation of Sec. 14.7, find $f_{10}, f_{1T}, f_{20}, f_{2T}$.
(*d*) Let $T = 3$ and $p = 0.7$. Find $f_{10}, f_{1T}, f_{20}, f_{2T}$.

14.7-2. A video cassette recorder manufacturer is so certain of its quality control that it is offering a complete replacement warranty if a recorder fails within 2 years. Based upon compiled data, the company has noted that only 1 percent of its recorders fail during the first year, whereas 5 percent of the recorders that survive the first year will fail during the second year. The warranty does not cover replacement recorders.

(*a*) Formulate the evolution of the status of a recorder as a Markov chain whose states include two absorption states that involve needing to honor the warranty or having the recorder survive the warranty period. Then construct the (one-step) transition matrix.
(*b*) Use the approach described in Sec. 14.7 to find the probability that the manufacturer will have to honor the warranty.

14.8-1. Reconsider the example presented at the end of Sec. 14.8. Suppose now that a third machine, identical to the first two, has been added to the shop. The one maintenance person still must maintain all the machines.

(*a*) Develop the rate diagram for this Markov chain.
(*b*) Construct the steady-state equations.
(*c*) Solve these equations for the steady-state probabilities.

14.8-2. The state of a particular continuous time Markov chain is defined as the number of jobs currently at a certain work center, where a maximum of three jobs are allowed. Jobs arrive individually. Whenever less than three jobs are present, the time until the next arrival has an exponential distribution with a mean of $\frac{1}{2}$ day. Jobs are processed at the work center one at a time and then leave immediately. Processing times have an exponential distribution with a mean of $\frac{1}{4}$ day.

(*a*) Construct the rate diagram for this Markov chain.
(*b*) Write the steady-state equations.
(*c*) Solve these equations for the steady-state probabilities.

15

Queueing Theory

Queueing theory involves the mathematical study of queues, or waiting lines. The formation of waiting lines is, of course, a common phenomenon that occurs whenever the current demand for a service exceeds the current capacity to provide that service. Decisions regarding the amount of capacity to provide must be made frequently in industry and elsewhere. However, because it is often impossible to predict accurately when units will arrive to seek service and/or how much time will be required to provide that service, these decisions often are difficult ones. Providing too much service involves excessive costs. And not providing enough service capacity causes the waiting line to become excessively long at times. Excessive waiting also is costly in some sense, whether it be a social cost, the cost of lost customers, the cost of idle employees, or some other important cost. Therefore, the ultimate goal is to achieve an economic balance between the cost of service and the cost associated with waiting for that service. Queueing theory itself does not solve this problem directly; however, it does contribute vital information required for such decisions by predicting various characteristics of the waiting line such as the average waiting time.

Queueing theory provides a large number of alternative mathematical models for describing a waiting-line situation. Mathematical results that predict some of the characteristics of the waiting line often are available for these models. After some general discussion, this chapter presents most of the more elementary models and their basic results. Chapter 16 discusses how the information provided by queueing theory might be used for making decisions.

15.1 Prototype Example

The emergency room of COUNTY HOSPITAL provides quick medical care for emergency cases brought to the hospital by ambulance or private automobile. At any hour there is always one doctor on duty in the emergency room. However, because of a growing tendency for emergency cases to use these facilities rather than go to a private physician, the hospital has been experiencing a continuing increase in the number of emergency room visits each year. As a result, it has become quite common for patients arriving during peak usage hours (the early evening) to have to wait until it is their turn to be treated by the doctor. Therefore, a proposal has been made that a second doctor should be assigned to the emergency room during these hours, so that two emergency cases can be treated simultaneously. The hospital's management engineer has been assigned to study this question.[1]

The management engineer began by gathering the relevant historical data and then projecting these data into the next year. Recognizing that the emergency room is a queueing system, she applied several alternative queueing theory models to predict the waiting characteristics of the system with one doctor and with two doctors, as you will see in the latter sections of this chapter (see Tables 15.2, 15.3, and 15.4).

15.2 Basic Structure of Queueing Models

The Basic Queueing Process

The basic process assumed by most queueing models is the following. *Customers* requiring service are generated over time by an *input source*. These customers enter the *queueing system* and join a *queue*. At certain times, a member of the queue is selected for service by some rule known as the *queue discipline*. The required service is then performed for the customer by the *service mechanism*, after which the customer leaves the queueing system. This process is depicted in Fig. 15.1.

Many alternative assumptions can be made about the various elements of the queueing process; they are discussed next.

Input Source (Calling Population)

One characteristic of the input source is its size. The *size* is the total number of customers that might require service from time to time, i.e., the total number of distinct potential customers. This population from which arrivals come is referred to as the **calling population**. The size may be assumed to be either *infinite* or *finite* (so that the

[1] For one actual case study of this kind, see W. Blaker Bolling, ''Queueing Model of a Hospital Emergency Room,'' *Industrial Engineering,* September 1972, pp. 26–31.

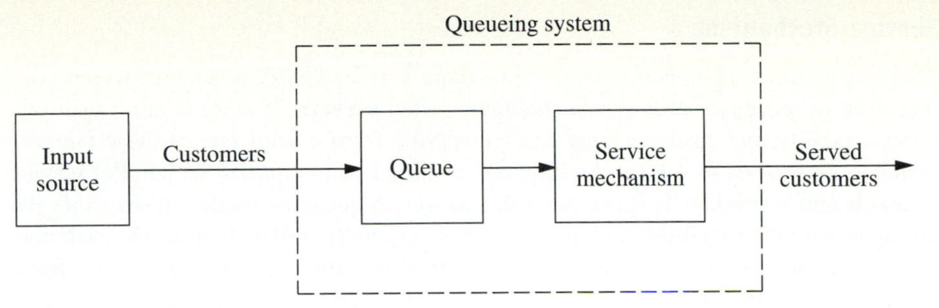

Figure 15.1 The basic queueing process.

input source also is said to be either *unlimited* or *limited*). Because the calculations are far easier for the infinite case, this assumption often is made even when the actual size is some relatively large finite number; and it should be taken to be the implicit assumption for any queueing model that does not state otherwise. The finite case is more difficult analytically because the number of customers in the queueing system affects the number of potential customers outside the system at any time. However, the finite assumption must be made if the rate at which the input source generates new customers is significantly affected by the number of customers in the queueing system.

The statistical pattern by which customers are generated over time must also be specified. The common assumption is that they are generated according to a *Poisson process;* i.e., the number of customers generated until any specific time has a Poisson distribution. As we discuss in Sec. 15.4, this case is the one where arrivals to the queueing system occur randomly but at a certain fixed mean rate, regardless of how many customers already are there (so the *size* of the input source is *infinite*). An equivalent assumption is that the probability distribution of the time between consecutive arrivals is an *exponential* distribution. (The properties of this distribution are described in Sec. 15.4.) The time between consecutive arrivals is referred to as the **interarrival time**.

Any unusual assumptions about the behavior of the customers must also be specified. One example is *balking,* where the customer refuses to enter the system and is lost if the queue is too long.

Queue

A queue is characterized by the maximum permissible number of customers that it can contain. Queues are called *infinite* or *finite,* according to whether this number is infinite or finite. The assumption of an *infinite queue* is the standard one for most queueing models, even for situations where there actually is a (relatively large) finite upper bound on the permissible number of customers, because dealing with such an upper bound would be a complicating factor in the analysis. However, for queueing systems where this upper bound is small enough that it actually would be reached with some frequency, it becomes necessary to assume a *finite queue*.

Queue Discipline

The queue discipline refers to the order in which members of the queue are selected for service. For example, it may be first-come-first-served, random, according to some priority procedure, or some other order. First-come-first-served usually is assumed by queueing models, unless it is stated otherwise.

Service Mechanism

The service mechanism consists of one or more *service facilities,* each of which contains one or more *parallel service channels,* called **servers**. If there is more than one service facility, the customer may receive service from a sequence of these (*service channels in series*). At a given facility, the customer enters one of the parallel service channels and is completely serviced by that server. A queueing model must specify the arrangement of the facilities and the number of servers (parallel channels) at each one. Most elementary models assume one service facility with either one server or a finite number of servers.

The time elapsed from the commencement of service to its completion for a customer at a service facility is referred to as the **service time** (or *holding time*). A model of a particular queueing system must specify the probability distribution of service times for each server (and possibly for different types of customers), although it is common to assume the *same* distribution for all servers (all models in this chapter make this assumption). The service-time distribution that is most frequently assumed in practice (largely because it is far more tractable than any other) is the *exponential* distribution discussed in Sec. 15.4, and most of our models will be of this type. Other important service-time distributions are the *degenerate* distribution (constant service time) and the *Erlang* (gamma) distribution, as illustrated by models in Sec. 15.7.

An Elementary Queueing Process

As we have already suggested, queueing theory has been applied to many different types of waiting-line situations. However, the most prevalent type of situation is the following: A single waiting line (which may be empty at times) forms in the front of a single service facility, within which are stationed one or more servers. Each customer generated by an input source is serviced by one of the servers, perhaps after some waiting in the queue (waiting line). The queueing system involved is depicted in Fig. 15.2.

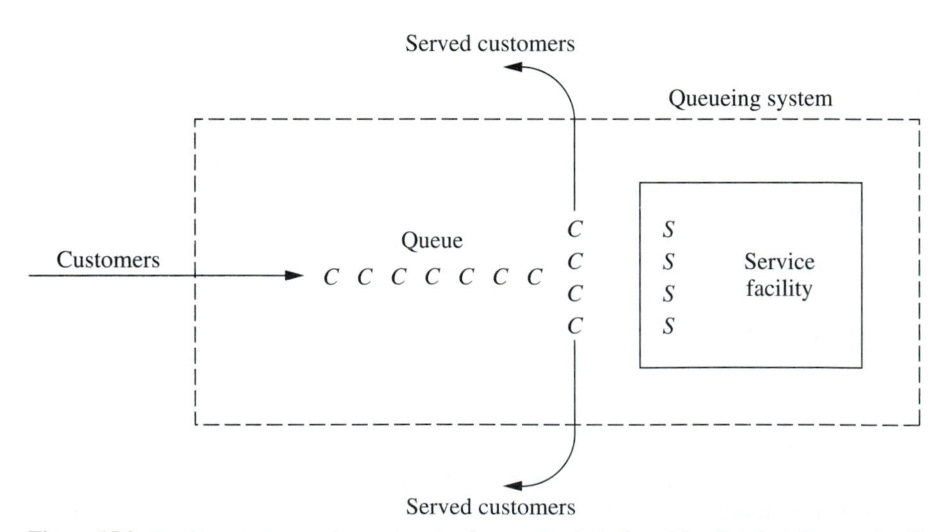

Figure 15.2 An elementary queueing system (each customer is indicated by C and each server by S).

Notice that the queueing process in the illustrative example of Sec. 15.1 is of this type. The input source generates customers in the form of emergency cases requiring medical care. The emergency room is the service facility, and the doctors are the servers.

A server need not be a single individual; it may be a group of persons, e.g., a repair crew that combines forces to perform simultaneously the required service for a customer. Furthermore, servers need not even be people. In many cases, a server may be a machine or a piece of equipment, e.g., a forklift that performs a given service on call (although probably with human guidance). By the same token, the customers in the waiting line need not be people. For example, they may be items waiting for a certain operation by a given type of machine, or they may be cars waiting in front of a tollbooth.

It is not necessary for there actually to be a physical waiting line forming in front of a physical structure that constitutes the service facility; i.e., the members of the queue may be scattered throughout an area, waiting for a server to come to them, e.g., machines waiting to be repaired. The server or group of servers assigned to a given area constitutes the service facility for that area. Queueing theory still gives the average number waiting, the average waiting time, and so on, because it is irrelevant whether the customers wait together in a group. The only essential requirement for queueing theory to be applicable is that changes in the number of customers waiting for a given service occur just as though the physical situation described in Fig. 15.2 (or a legitimate counterpart) prevailed.

Except for Sec. 15.9, all the queueing models discussed in this chapter are of the elementary type depicted in Fig. 15.2. Many of these models further assume that all *interarrival times* are independent and identically distributed and that all *service times* are independent and identically distributed. Such models conventionally are labeled as follows:

Distribution of service times

$- / - / -$ Number of servers

Distribution of interarrival times,

where M = exponential distribution (Markovian), as described in Sec. 15.4,
D = degenerate distribution (constant times), as discussed in Sec. 15.7,
E_k = Erlang distribution (shape parameter = k), as described in Sec. 15.7,
G = general distribution (any arbitrary distribution allowed),[1] as discussed in Sec. 15.7.

For example, the $M/M/s$ model discussed in Sec. 15.6 assumes that both interarrival times and service times have an exponential distribution and that the number of servers is s (any positive integer). The $M/G/1$ model discussed again in Sec. 15.7 assumes that interarrival times have an exponential distribution, but it places no restriction on what the distribution of service times must be, whereas the number of servers is restricted to be exactly 1. Various other models that fit this labeling scheme also are introduced in Sec. 15.7.

[1] When we refer to interarrival times, it is conventional to replace the symbol G by GI = general *i*ndependent distribution.

Terminology and Notation

Unless otherwise noted, the following standard terminology and notation will be used:

State of system = number of customers in queueing system.

Queue length = number of customers waiting for service
= state of system minus number of customers being served.

$N(t)$ = number of customers in queueing system at time t ($t \geq 0$).

$P_n(t)$ = probability of exactly n customers in queueing system at time t, given number at time 0.

s = number of servers (parallel service channels) in queueing system.

λ_n = mean arrival rate (expected number of arrivals per unit time) of new customers when n customers are in system.

μ_n = mean service rate for overall system (expected number of customers completing service per unit time) when n customers are in system. *Note:* μ_n represents *combined* rate at which all *busy* servers (those serving customers) achieve service completions.

λ, μ, ρ = see following paragraph.

When λ_n is a constant for all n, this constant is denoted by λ. When the mean service rate *per busy server* is a constant for all $n \geq 1$, this constant is denoted by μ. (In this case, $\mu_n = s\mu$ when $n \geq s$, that is, when all s servers are busy.) Under these circumstances, $1/\lambda$ and $1/\mu$ are the *expected interarrival time* and the *expected service time,* respectively. Also, $\rho = \lambda/(s\mu)$ is the **utilization factor** for the service facility, i.e., the expected fraction of time the individual servers are busy, because $\lambda/(s\mu)$ represents the fraction of the system's service capacity ($s\mu$) that is being *utilized* on the average by arriving customers (λ).

Certain notation also is required to describe *steady-state* results. When a queueing system has recently begun operation, the state of the system (number of customers in the system) will be greatly affected by the initial state and by the time that has since elapsed. The system is said to be in a **transient condition**. However, after sufficient time has elapsed, the state of the system becomes essentially independent of the initial state and the elapsed time (except under unusual circumstances).[1] The system has now essentially reached a **steady-state condition**, where the probability distribution of the state of the system remains the same (the *steady-state* or *stationary* distribution) over time. Queueing theory has tended to focus largely on the steady-state condition, partially because the transient case is more difficult analytically. (Some transient results exist, but they are generally beyond the technical scope of this book.) The following notation assumes that the system is in a *steady-state condition:*

[1] When λ and μ are defined, these unusual circumstances are that $\rho \geq 1$, in which case the state of the system tends to grow continually larger as time goes on.

P_n = probability of exactly n customers in queueing system.

L = expected number of customers in queueing system.

L_q = expected queue length (excludes customers being served).

$\mathcal{W}$ = waiting time in system (includes service time) for each individual customer.

$W = E(\mathcal{W})$.

$\mathcal{W}_q$ = waiting time in queue (excludes service time) for each individual customer.

$W_q = E(\mathcal{W}_q)$.

Relationships between L, W, L_q, and W_q

Assume that λ_n is a constant λ for all n. It has been proved that in a steady-state queueing process,

$$L = \lambda W.$$

(Because John D. C. Little[1] provided the first rigorous proof, this equation sometimes is referred to as **Little's formula**.) Furthermore, the same proof also shows that

$$L_q = \lambda W_q.$$

If the λ_n are not equal, then λ can be replaced in these equations by $\bar{\lambda}$, the *average* arrival rate over the long run. (We shall show later how $\bar{\lambda}$ can be determined for some basic cases.)

Now assume that the mean service time is a constant, $1/\mu$ for all $n \geq 1$. It then follows that

$$W = W_q + \frac{1}{\mu}.$$

These relationships are extremely important because they enable all four of the fundamental quantities—L, W, L_q, and W_q—to be immediately determined as soon as one is found analytically. This situation is fortunate because some of these quantities often are much easier to find than others when a queueing model is solved from basic principles.

15.3 Examples of Real Queueing Systems

Our description of queueing systems in the preceding section may appear relatively abstract and applicable to only rather special practical situations. On the contrary, queueing systems are surprisingly prevalent in a wide variety of contexts. To broaden your horizons on the applicability of queueing theory, we shall briefly mention various examples of real queueing systems.

One important class of queueing systems that we all encounter in our daily lives is **commercial service systems**, where outside customers receive service from commercial organizations. Many of these involve person-to-person service at a fixed loca-

[1] J. D. C. Little, "A Proof for the Queueing Formula: $L = \lambda W$," *Operations Research*, **9**(3): 383–387, 1961; also see S. Stidham, Jr., "A Last Word on $L = \lambda W$," *Operations Research*, **22**(2): 417–421, 1974.

tion, such as a barber shop (the barbers are the servers), bank teller service, checkout stands at a grocery store, and a cafeteria line (service channels in series). However, many others do not, such as home appliance repairs (the server travels to the customers), a vending machine (the server is a machine), and a gas station (the cars are the customers).

Another important class is **transportation service systems**. For some of these systems the vehicles are the customers, such as cars waiting at a tollbooth or traffic light (the server), a truck or ship waiting to be loaded or unloaded by a crew (the server), and airplanes waiting to land or take off from a runway (the server). (An unusual example of this kind is a parking lot, where the cars are the customers and the parking spaces are the servers, but there is no queue because arriving customers go elsewhere to park if the lot is full.) In other cases, the vehicles, such as taxicabs, fire trucks, and elevators, are the servers.

In recent years, queueing theory probably has been applied most to **business-industrial internal service systems**, where the customers receiving service are *internal* to the organization. Examples include materials-handling systems, where materials-handling units (the servers) move loads (the customers); maintenance systems, where maintenance crews (the servers) repair machines (the customers); and inspection stations, where quality control inspectors (the servers) inspect items (the customers). Employee facilities and departments servicing employees also fit into this category. In addition, machines can be viewed as servers whose customers are the jobs being processed. A related example of great importance is a computer facility, where the computer is viewed as the server.

There is now growing recognition that queueing theory also is applicable to **social service systems**. For example, a judicial system is a queueing network, where the courts are service facilities, the judges (or panels of judges) are the servers, and the cases waiting to be tried are the customers. A legislative system is a similar queueing network, where the customers are the bills waiting to be processed. Various health-care systems also are queueing systems. You already have seen one example in Sec. 15.1 (a hospital emergency room), but you can also view ambulances, x-ray machines, and hospital beds as servers in their own queueing systems. Similarly, families waiting for low- and moderate-income housing, or other social services, can be viewed as customers in a queueing system.

Although these are four broad classes of queueing systems, they still do not exhaust the list. In fact, queueing theory first began early in this century with applications to telephone engineering (the founder of queueing theory, A. K. Erlang, was an employee of the Danish Telephone Company in Copenhagen), and telephone engineering still is an important application. Furthermore, we all have our own personal queues—homework assignments, books to be read, and so forth. However, these examples are sufficient to suggest that queueing systems do indeed pervade many areas of society.

15.4 The Role of the Exponential Distribution

The operating characteristics of queueing systems are determined largely by two statistical properties, namely, the probability distribution of *interarrival times* (see ''Input Source'' in Sec. 15.2) and the probability distribution of *service times* (see ''Service Mechanism'' in Sec. 15.2). For real queueing systems, these distributions can take on almost any form. (The only restriction is that negative values cannot occur.) However, to formulate a queueing theory *model* as a representation of the real system, it is necessary to specify the assumed form of each of these distributions. To be useful, the assumed form should be *sufficiently realistic* that the model provides *reasonable predictions* while, at the same time, being *sufficiently simple* that the model is *mathematically tractable.* Based on these considerations, the most important probability distribution in queueing theory is the *exponential distribution.*

Suppose that a random variable T represents either interarrival or service times. (We shall refer to the occurrences marking the end of these times—arrivals or service completions—as *events.*) This random variable is said to have an *exponential distribution with parameter* α if its probability density function is

$$f_T(t) = \begin{cases} \alpha e^{-\alpha t} & \text{for } t \geq 0, \\ 0 & \text{for } t < 0, \end{cases}$$

as shown in Fig. 15.3. In this case, the cumulative probabilities are

$$\begin{aligned} P\{T \leq t\} &= 1 - e^{-\alpha t} \\ P\{T > t\} &= e^{-\alpha t} \end{aligned} \qquad (t \geq 0),$$

and the expected value and variance of T are, respectively,

$$E(T) = \frac{1}{\alpha},$$

$$\text{var}(T) = \frac{1}{\alpha^2}.$$

What are the implications of assuming that T has an exponential distribution for a queueing model? To explore this question, let us examine six key properties of the exponential distribution.

Property 1: $f_T(t)$ is a strictly *decreasing* function of t ($t \geq 0$).

One consequence of Property 1 is that

$$P\{0 \leq T \leq \Delta t\} > P\{t \leq T \leq t + \Delta t\}$$

for any strictly positive values of Δt and t. [This consequence follows from the fact that these probabilities are the area under the $f_T(t)$ curve over the indicated interval of length Δt, and the average height of the curve is less for the second probability than for

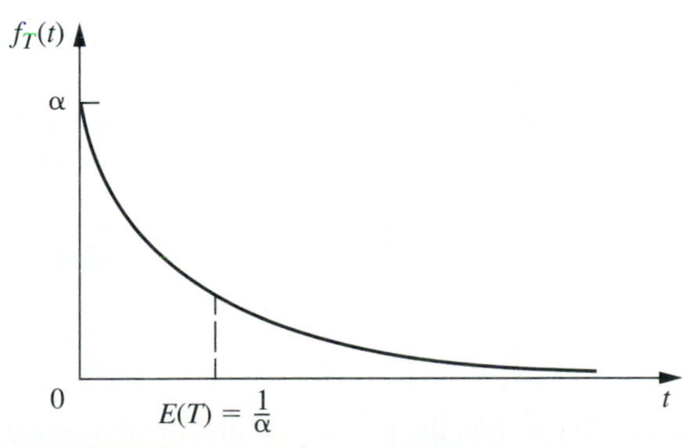

Figure 15.3 Probability density function for the exponential distribution.

the first.] Therefore, it is not only possible but also relatively likely that T will take on a small value near zero. In fact,

$$P\left\{0 \leq T \leq \frac{1}{2}\frac{1}{\alpha}\right\} = 0.393$$

whereas

$$P\left\{\frac{1}{2}\frac{1}{\alpha} \leq T \leq \frac{3}{2}\frac{1}{\alpha}\right\} = 0.383,$$

so that the value T takes on is more likely to be "small" [i.e., less than half of $E(T)$] than "near" its expected value [i.e., no further away than half of $E(T)$], even though the second interval is twice as wide as the first.

Is this really a reasonable property for T in a queueing model? If T represents *service times,* the answer depends upon the general nature of the service involved, as discussed next.

If the service required is essentially identical for each customer, with the server always performing the same sequence of service operations, then the actual service times tend to be near the expected service time. Small deviations from the mean may occur, but usually because of only minor variations in the efficiency of the server. A small service time far below the mean is essentially impossible, because a certain minimum time is needed to perform the required service operations even when the server is working at top speed. The exponential distribution clearly does not provide a close approximation to the service-time distribution for this type of situation.

On the other hand, consider the type of situation where the specific tasks required of the server differ among customers. The broad nature of the service may be the same, but the specific type and amount of service differ. For example, this is the case in the County Hospital emergency room problem discussed in Sec. 15.1. The doctors encounter a wide variety of medical problems. In most cases, they can provide the required treatment rather quickly, but an occasional patient requires extensive care. Similarly, bank tellers and grocery store checkout clerks are other servers of this general type, where the required service is often brief but must occasionally be extensive. An exponential service-time distribution would seem quite plausible for this type of service situation.

If T represents *interarrival times,* Property 1 rules out situations where potential customers approaching the queueing system tend to postpone their entry if they see another customer entering ahead of them. On the other hand, it is entirely consistent with the common phenomenon of arrivals occurring "randomly," described by subsequent properties. Thus, when arrival times are plotted on a time line, they sometimes have the appearance of being clustered with occasional large gaps separating clusters, because of the substantial probability of small interarrival times and the small probability of large interarrival times, but such an irregular pattern is all part of true randomness.

Property 2: Lack of memory.

This property can be stated mathematically as

$$P\{T > t + \Delta t \mid T > \Delta t\} = P\{T > t\}$$

for any positive quantities t and Δt. In other words, the probability distribution of the *remaining* time until the event (arrival or service completion) occurs always is the same, regardless of how much time (Δt) already has passed. In effect, the process "forgets" its history. This surprising phenomenon occurs with the exponential distribution because

$$P\{T > t + \Delta t \mid T > \Delta t\} = \frac{P\{T > \Delta t, T > t + \Delta t\}}{P\{T > \Delta t\}}$$

$$= \frac{P\{T > t + \Delta t\}}{P\{T > \Delta t\}}$$

$$= \frac{e^{-\alpha(t+\Delta t)}}{e^{-\alpha \Delta t}}$$

$$= e^{-\alpha t}.$$

For *interarrival times,* this property describes the common situation where the time until the next arrival is completely uninfluenced by when the last arrival occurred. For *service times,* the property is more difficult to interpret. We should not expect it to hold in a situation where the server must perform the same fixed sequence of operations for each customer, because then a long elapsed service should imply that probably little remains to be done. However, in the type of situation where the required service operations differ among customers, the mathematical statement of the property may be quite realistic. For this case, if considerable service has already elapsed for a customer, the only implication may be that this particular customer requires more extensive service than most.

Property 3: The *minimum* of several independent exponential random variables has an exponential distribution.

To state this property mathematically, let $T_1, T_2, \ldots, T_n$ be *independent* exponential random variables with parameters $\alpha_1, \alpha_2, \ldots, \alpha_n$, respectively. Also let U be the random variable that takes on the value equal to the *minimum* of the values actually taken on by $T_1, T_2, \ldots, T_n$; that is,

$$U = \min\{T_1, T_2, \ldots, T_n\}.$$

Thus, if T_i represents the time until a particular kind of event occurs, then U represents the time until the *first* of the n different events occurs. Now note that for any $t \geq 0$,

$$P\{U > t\} = P\{T_1 > t, T_2 > t, \ldots, T_n > t\}$$

$$= P\{T_1 > t\}P\{T_2 > t\} \cdots P\{T_n > t\}$$

$$= e^{-\alpha_1 t} e^{-\alpha_2 t} \cdots e^{\alpha_n t}$$

$$= \exp\left(-\sum_{i=1}^{n} \alpha_i t\right),$$

so that U indeed has an exponential distribution with parameter

$$\alpha = \sum_{i=1}^{n} \alpha_i.$$

This property has some implications for interarrival times in queueing models. In particular, suppose that there are several (n) *different* types of customers, but the interarrival times for *each* type (type i) have an exponential distribution with parameter α_i ($i = 1, 2, \ldots, n$). By Property 2, the *remaining* time from any specified instant until the next arrival of a customer of type i has this same distribution. Therefore, let T_i be this remaining time, measured from the instant a customer of *any* type arrives. Property 3 then tells us that U, the interarrival times for the queueing system as a whole, has an exponential distribution with parameter α defined by the last equation. As a result, you can choose to ignore the distinction between customers and still have exponential interarrival times for the queueing model.

However, the implications are even more important for *service times* in queueing models having more than one server than for interarrival times. For example, consider the situation where all the servers have the same exponential service-time distribution with parameter μ. For this case, let n be the number of servers *currently* providing service, and let T_i be the *remaining* service time for server i ($i = 1, 2, \ldots, n$), which also has an exponential distribution with parameter $\alpha_i = \mu$. It then follows that U, the time until the *next* service completion from any of these servers, has an exponential distribution with parameter $\alpha = n\mu$. In effect, the queueing system *currently* is performing just like a *single*-server system, where service times have an exponential distribution with parameter $n\mu$. We shall make frequent use of this implication for analyzing multiple-server models later in the chapter.

Property 4: Relationship to the Poisson distribution.

Suppose that the *time* between consecutive occurrences of some particular kind of event (e.g., arrivals or service completions by a continuously busy server) has an exponential distribution with parameter α. Property 4 then has to do with the resulting implication about the probability distribution of the *number* of times this kind of event occurs over a specified time. In particular, let $X(t)$ be the number of occurrence by time t ($t \geq 0$), where time 0 designates the instant at which the count begins. The implication is that

$$P\{X(t) = n\} = \frac{(\alpha t)^n e^{-\alpha t}}{n!}, \qquad \text{for } n = 0, 1, 2, \ldots;$$

that is, $X(t)$ has a Poisson distribution with parameter αt. For example, with $n = 0$,

$$P\{X(t) = 0\} = e^{-\alpha t},$$

which is just the probability from the exponential distribution that the *first* event occurs after time t. The mean of this Poisson distribution is

$$E\{X(t)\} = \alpha t,$$

so that the expected number of events *per unit time* is α. Thus α is said to be the *mean rate* at which the events occur. When the events are counted on a continuing basis, the counting process $\{X(t); t \geq 0\}$ is said to be a **Poisson process** with parameter α (the mean rate).

This property provides useful information about *service completions* when service times have an exponential distribution with parameter μ. We obtain this information by defining $X(t)$ as the number of service completions achieved by a *continuously busy* server in elapsed time t, where $\alpha = \mu$. For *multiple*-server queueing models, $X(t)$ can also be defined as the number of service completions achieved by n continuously busy servers in elapsed time t, where $\alpha = n\mu$.

The property is particularly useful for describing the probabilistic behavior of *arrivals* when interarrival times have an exponential distribution with parameter λ. In this case, $X(t)$ is the *number* of arrivals in elapsed time t, where $\alpha = \lambda$ is the *mean arrival rate.* Therefore, arrivals occur according to a **Poisson input process** with parameter λ. Such queueing models also are described as assuming a *Poisson input.*

Arrivals sometimes are said to occur *randomly,* meaning that they occur in accordance with a Poisson input process. One intuitive interpretation of this phenomenon is that every time period of fixed length has the *same* chance of having an arrival regardless of when the preceding arrival occurred, as suggested by the following property.

Property 5: For all positive values of t, $P\{T \le t + \Delta t \mid T > t\} \approx \alpha\,\Delta t$, for small Δt.

Continuing to interpret T as the time from the last event of a certain type (arrival or service completion) until the next such event, we suppose that a time t already has elapsed without the event's occurring. We know from Property 2 that the probability that the event will occur within the next time interval of fixed length Δt is a *constant* (identified in the next paragraph), regardless of how large or small t is. Property 5 goes further to say that when the value of Δt is small, this constant probability can be approximated very closely by $\alpha\,\Delta t$. Furthermore, when considering different small values of Δt, this probability is essentially *proportional* to Δt, with proportionality factor α. In fact, α is the *mean rate* at which the events occur (see Property 4), so that the *expected number* of events in the interval of length Δt is *exactly* $\alpha\,\Delta t$. The only reason that the probability of an event's occurring differs slightly from this value is the possibility that *more than one* event will occur, which has negligible probability when Δt is small.

To see why Property 5 holds mathematically, note that the constant value of our probability (for a fixed value of $\Delta t > 0$) is just

$$P\{T \le t + \Delta t \mid T > t\} = P\{T \le \Delta t\}$$
$$= 1 - e^{-\alpha\,\Delta t},$$

for any $t \ge 0$. Therefore, because the series expansion of e^x for any exponent x is

$$e^x = 1 + x + \sum_{n=2}^{\infty} \frac{x^n}{n!},$$

it follows that

$$P\{T \le t + \Delta t \mid T > t\} = 1 - 1 + \alpha\,\Delta t - \sum_{n=2}^{\infty} \frac{(-\alpha\,\Delta t)^n}{n!}$$
$$\approx \alpha\,\Delta t, \qquad \text{for small } \Delta t,\dagger$$

because the summation terms become relatively negligible for sufficiently small values of $\alpha\,\Delta t$.

† More precisely,

$$\lim_{\Delta t \to 0} \frac{P\{T \le t + \Delta t \mid T > t\}}{\Delta t} = \alpha.$$

Because T can represent either interarrival or service times in queueing models, this property provides a convenient approximation of the probability that the event of interest occurs in the next small interval (Δt) of time. An analysis based on this approximation also can be made exact by taking appropriate limits as $\Delta t \to 0$.

Property 6: Unaffected by aggregation or disaggregation.

This property is relevant primarily for verifying that the *input process* is *Poisson.* Therefore, we shall describe it in these terms, although it also applies directly to the exponential distribution (exponential interarrival times) because of Property 4.

Suppose that there are several (n) *different* types of customers, where the customers of each type (type i) arrive according to a *Poisson input process* with parameter λ_i ($i = 1, 2, \ldots, n$). Assuming that these are *independent* Poisson processes, the property says that the *aggregate* input process (arrival of all customers without regard to type) also must be Poisson, with parameter (arrival rate) $\lambda = \lambda_1 + \lambda_2 + \cdots + \lambda_n$. In other words, having a Poisson process is *unaffected by aggregation.*

This part of the property follows directly from Properties 3 and 4. The latter property implies that the interarrival times for customers of type i have an exponential distribution with parameter λ_i. For this identical situation, we already discussed for Property 3 that it implies that the interarrival times for all customers also must have an exponential distribution, with parameter $\lambda = \lambda_1 + \lambda_2 + \cdots + \lambda_n$. Using Property 4 again then implies that the aggregate input process is Poisson.

The second part of Property 6 (''unaffected by disaggregation'') refers to the reverse case, where the *aggregate* input process is known to be Poisson with parameter λ, but the question now concerns the nature of the *disaggregated* input processes for the individual customer types. Assuming that each arriving customer has a *fixed* probability p_i of being of type i ($i = 1, 2, \ldots, n$), with

$$\lambda_i = p_i \lambda \qquad \text{and} \qquad \sum_{i=1}^{n} p_i = 1,$$

the property says that the input process for customers of type i also must be Poisson with parameter λ_i. In other words, having a Poisson process is *unaffected by disaggregation.*

As one example of the usefulness of this second part of the property, consider the following situation. Indistinguishable customers arrive according to a Poisson process with parameter λ. Each arriving customer has a fixed probability p of *balking* (leaving without entering the queueing system), so the probability of entering the system is $1 - p$. Thus there are two types of customers—those who balk and those who enter the system. The property says that each type arrives according to a Poisson process, with parameters $p\lambda$ and $(1 - p)\lambda$, respectively. Therefore, by using the latter Poisson process, queueing models that assume a Poisson input process can still be used to analyze the performance of the queueing system for those customers who enter the system.

15.5 The Birth-and-Death Process

Most elementary queueing models assume that the inputs (arriving customers) and outputs (leaving customers) of the queueing system occur according to the *birth-and-death process.* This important process in probability theory has applications in various areas. However, in the context of queueing theory, the term **birth** refers to the *arrival* of a new customer into the queueing system, and **death** refers to the *departure* of a served customer. The *state* of the system at time t ($t \geq 0$), denoted by $N(t)$, is the number of customers in the queueing system at time t. The birth-and-death process describes *probabilistically* how $N(t)$ changes as t increases. Broadly speaking, it says that *individual* births and deaths occur *randomly,* where their mean occurrence rates depend only upon the current state of the system. More precisely, the assumptions of the birth-and-death process are the following:

ASSUMPTION 1: Given $N(t) = n$, the current probability distribution of the *remaining* time until the next *birth* (arrival) is *exponential* with parameter λ_n ($n = 0, 1, 2, \ldots$).

ASSUMPTION 2: Given $N(t) = n$, the current probability distribution of the *remaining* time until the next *death* (service completion) is *exponential* with parameter μ_n ($n = 1, 2, \ldots$).

ASSUMPTION 3: The random variable of assumption 1 (the remaining time until the next *birth*) and the random variable of assumption 2 (the remaining time until the next *death*) are mutually independent. The next transition in the state of the process is either

$$n \rightarrow n + 1 \qquad \text{(a single birth)}$$

or

$$n \rightarrow n - 1 \qquad \text{(a single death),}$$

depending on whether the former or latter random variable is smaller.

Because of these assumptions, the birth-and-death process is a special type of *continuous time Markov chain.* (See Sec. 14.8 for a description of continuous time Markov chains and their properties, including an introduction to the general procedure for finding steady-state probabilities that will be applied in the remainder of this section.) Queueing models that can be represented by a continuous time Markov chain are far more tractable analytically than any other.

Because Property 4 for the exponential distribution (see Sec. 15.4) implies that the λ_n and μ_n are mean rates, we can summarize these assumptions by the rate diagram shown in Fig. 15.4. The arrows in this diagram show the only possible *transitions* in the state of the system (as specified by assumption 3), and the entry for each arrow gives the mean rate for that transition (as specified by assumptions 1 and 2) when the system is in the state at the base of the arrow.

Except for a few special cases, analysis of the birth-and-death process is very difficult when the system is in a *transient* condition. Some results about the probability

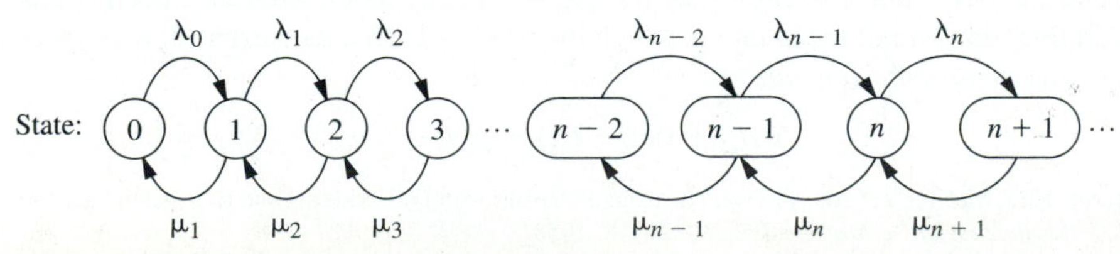

Figure 15.4 Rate diagram for the birth-and-death process.

distribution of $N(t)$ have been obtained,[1] but they are too complicated to be of much practical use. On the other hand, it is relatively straightforward to derive this distribution *after* the system has reached a *steady-state* condition (assuming that this condition can be reached). This derivation can be done directly from the rate diagram, as outlined next.

Consider any particular state of the system n ($n = 0, 1, 2, \ldots$). Starting at time 0, suppose that a count is made of the number of times that the process enters this state and the number of times it leaves this state, as denoted below:

$$E_n(t) = \text{number of times that process enters state } n \text{ by time } t.$$

$$L_n(t) = \text{number of times that process leaves state } n \text{ by time } t.$$

Because the two types of events (entering and leaving) must alternate, these two numbers must always either be equal or differ by just 1; that is,

$$|E_n(t) - L_n(t)| \leq 1.$$

Dividing through both sides by t and then letting $t \to \infty$ give

$$\left|\frac{E_n(t)}{t} - \frac{L_n(t)}{t}\right| \leq \frac{1}{t}, \qquad \text{so} \qquad \lim_{t\to\infty} \left|\frac{E_n(t)}{t} - \frac{L_n(t)}{t}\right| = 0.$$

Dividing $E_n(t)$ and $L_n(t)$ by t gives the actual rate (number of events per unit time) at which these two kinds of events have occurred, and letting $t \to \infty$ then gives the mean rate (expected number of events per unit time):

$$\lim_{t\to\infty} \frac{E_n(t)}{t} = \text{mean rate at which process enters state } n.$$

$$\lim_{t\to\infty} \frac{L_n(t)}{t} = \text{mean rate at which process leaves state } n.$$

These results yield the following key principle:

RATE IN = RATE OUT PRINCIPLE: For any state of the system n ($n = 0, 1, 2, \ldots$), mean entering rate = mean leaving rate.

The equation expressing this principle is called the **balance equation** for state n. After constructing the balance equations for all the states in terms of the *unknown* P_n probabilities, we can solve this system of equations (plus an equation stating that the probabilities must sum to 1) to find these probabilities.

To illustrate a balance equation, consider state 0. The process enter this state *only* from state 1. Thus the steady-state probability of being in state 1 (P_1) represents the proportion of time that it would be *possible* for the process to enter state 0. Given that the process is in state 1, the mean rate of entering state 0 is μ_1. (In other words, for each cumulative unit of time that the process spends in state 1, the expected number of times that it would leave state 1 to enter state 0 is μ_1.) From any *other* state, this mean rate is 0. Therefore, the overall mean rate at which the process leaves its current state to enter state 0 (the *mean entering rate*) is

$$\mu_1 P_1 + 0(1 - P_1) = \mu_1 P_1.$$

[1] S. Karlin and J. McGregor, "Many Server Queueing Processes with Poisson Input and Exponential Service Times," *Pacific Journal of Mathematics,* **8**: 87–118, 1958.

Table 15.1 **Balance Equations for the Birth-and-Death Process**

State	Rate In = Rate Out
0	$\mu_1 P_1 = \lambda_0 P_0$
1	$\lambda_0 P_0 + \mu_2 P_2 = (\lambda_1 + \mu_1)P_1$
2	$\lambda_1 P_1 + \mu_3 P_3 = (\lambda_2 + \mu_2)P_2$
$\vdots$	$\vdots$
$n-1$	$\lambda_{n-2} P_{n-2} + \mu_n P_n = (\lambda_{n-1} + \mu_{n-1})P_{n-1}$
n	$\lambda_{n-1} P_{n-1} + \mu_{n+1} P_{n+1} = (\lambda_n + \mu_n)P_n$
$\vdots$	$\vdots$

By the same reasoning, the *mean leaving rate* must be $\lambda_0 P_0$, so the balance equation for state 0 is

$$\mu_1 P_1 = \lambda_0 P_0.$$

For every other state there are two possible transitions both into and out of the state. Therefore, each side of the balance equations for these states represents the *sum* of the mean rates for the two transitions involved. Otherwise, the reasoning is just the same as for state 0. These balance equations are summarized in Table 15.1.

Notice that the first balance equation contains two variables for which to solve (P_0 and P_1), the first two equations contain three variables (P_0, P_1, and P_2), and so on, so that there always is one "extra" variable. Therefore, the procedure in solving these equations is to solve in terms of one of the variables, the most convenient one being P_0. Thus the first equation is used to solve for P_1 in terms of P_0; this result and the second equation are then used to solve for P_2 in terms of P_0; and so forth. At the end, the requirement that the sum of all the probabilities equal 1 can be used to evaluate P_0.

Applying this procedure yields the following results:

State:

0: $$P_1 = \frac{\lambda_0}{\mu_1} P_0$$

1: $$P_2 = \frac{\lambda_1}{\mu_2} P_1 + \frac{1}{\mu_2}(\mu_1 P_1 - \lambda_0 P_0) = \frac{\lambda_1}{\mu_2} P_1 = \frac{\lambda_1 \lambda_0}{\mu_2 \mu_1} P_0$$

2: $$P_3 = \frac{\lambda_2}{\mu_3} P_2 + \frac{1}{\mu_3}(\mu_2 P_2 - \lambda_1 P_1) = \frac{\lambda_2}{\mu_3} P_2 = \frac{\lambda_2 \lambda_1 \lambda_0}{\mu_3 \mu_2 \mu_1} P_0$$

$\vdots$ $\qquad \vdots$

$n-1$: $$P_n = \frac{\lambda_{n-1}}{\mu_n} P_{n-1} + \frac{1}{\mu_n}(\mu_{n-1} P_{n-1} - \lambda_{n-2} P_{n-2}) = \frac{\lambda_{n-1}}{\mu_n} P_{n-1} = \frac{\lambda_{n-1}\lambda_{n-2}\cdots\lambda_0}{\mu_n \mu_{n-1}\cdots\mu_1} P_0$$

n: $$P_{n+1} = \frac{\lambda_n}{\mu_{n+1}} P_n + \frac{1}{\mu_{n+1}}(\mu_n P_n - \lambda_{n-1} P_{n-1}) = \frac{\lambda_n}{\mu_{n+1}} P_n = \frac{\lambda_n \lambda_{n-1}\cdots\lambda_0}{\mu_{n+1}\mu_n\cdots\mu_1} P_0$$

$\vdots$ $\qquad \vdots$

To simplify notation, let

$$C_n = \frac{\lambda_{n-1}\lambda_{n-2}\cdots\lambda_0}{\mu_n \mu_{n-1}\cdots\mu_1}, \qquad \text{for } n = 1, 2, \ldots,$$

and then define $C_n = 1$ for $n = 0$. Thus the steady-state probabilities are

$$P_n = C_n P_0, \qquad \text{for } n = 0, 1, 2, \ldots.$$

The requirement that

$$\sum_{n=0}^{\infty} P_n = 1$$

implies that

$$\left(\sum_{n=0}^{\infty} C_n \right) P_0 = 1,$$

so that

$$P_0 = \left(\sum_{n=0}^{\infty} C_n \right)^{-1}.$$

Given this information,

$$L = \sum_{n=0}^{\infty} n P_n.$$

Also, because the number of servers s represents the number of customers that can be served (and thus are not in the queue) simultaneously,

$$L_q = \sum_{n=s}^{\infty} (n - s) P_n.$$

Furthermore, the relationships given in Sec. 15.2 yield

$$W = \frac{L}{\bar{\lambda}}, \qquad W_q = \frac{L_q}{\bar{\lambda}},$$

where $\bar{\lambda}$ is the *average* arrival rate over the long run. Because λ_n is the mean arrival rate while the system is in state n ($n = 0, 1, 2, \ldots$) and P_n is the proportion of time that the system is in this state,

$$\bar{\lambda} = \sum_{n=0}^{\infty} \lambda_n P_n.$$

Several of the expressions just given involve summations with an infinite number of terms. Fortunately, these summations have analytic solutions for a number of interesting special cases,[1] as seen in the next section. Otherwise, they can be approximated by summing a finite number of terms on a computer.

These steady-state results have been derived under the assumption that the λ_n and μ_n parameters have values such that the process actually can *reach* a steady-state condition. This assumption *always* holds if $\lambda_n = 0$ for some value of n greater than the initial state, so that only a finite number of states (those less than this n) are possible. It also *always* holds when λ and μ are defined (see "Terminology and Notation" in Sec. 15.2) and $\rho = \lambda/(s\mu) < 1$. It does *not* hold if $\sum_{n=1}^{\infty} C_n = \infty$.

The following section describes several queueing models that are special cases of the birth-and-death process. Therefore, the general steady-state results just given in boxes will be used over and over again to obtain the specific steady-state results for these models.

15.6 Queueing Models Based on the Birth-and-Death Process

Because each of the mean rates $\lambda_0, \lambda_1, \ldots$ and $\mu_1, \mu_2, \ldots$ for the birth-and-death process can be assigned any nonnegative value, we have great flexibility in modeling a queueing system. Probably the most widely used models in queueing theory are based directly upon this process. Because of assumptions 1 and 2 (and Property 4 for the exponential distribution), these models are said to have a **Poisson input** and **exponential service times**. The models differ only in their assumptions about how the λ_n and μ_n change with n. We present four of these models in this section for four important types of queueing systems.

The *M/M/s* Model

As described in Sec. 15.2, the *M/M/s* model assumes that all *interarrival times* are independently and identically distributed according to an exponential distribution (i.e., the input process is Poisson), that all *service times* are independent and identically distributed according to another exponential distribution, and that the number of servers is s (any positive integer). Consequently, this model is just the special case of the birth-and-death process where the queueing system's *mean arrival rate* and *mean service rate per busy server* are constant (λ and μ, respectively) regardless of the state of the system. When the system has just a *single server* ($s = 1$), the implication is that the parameters for the birth-and-death process are $\lambda_n = \lambda$ $(n = 0, 1, 2, \ldots)$ and $\mu_n = \mu$ $(n = 1, 2, \ldots)$. The resulting rate diagram is shown in Fig. 15.5*a*.

However, when the system has *multiple servers* ($s > 1$), the μ_n cannot be expressed this simply. Keep in mind that μ_n represents the mean service rate for the *overall* queueing system (i.e., the mean rate at which service completions occur, so that customers leave the system) when there are n customers currently in the system. As

[1] These solutions are based on the following known results for the sum of any geometric series:

$$\sum_{n=0}^{N} x^n = \frac{1 - x^{N+1}}{1 - x}, \qquad \text{for any } x \neq 1,$$

$$\sum_{n=0}^{\infty} x^n = \frac{1}{1 - x}, \qquad \text{if } |x| < 1.$$

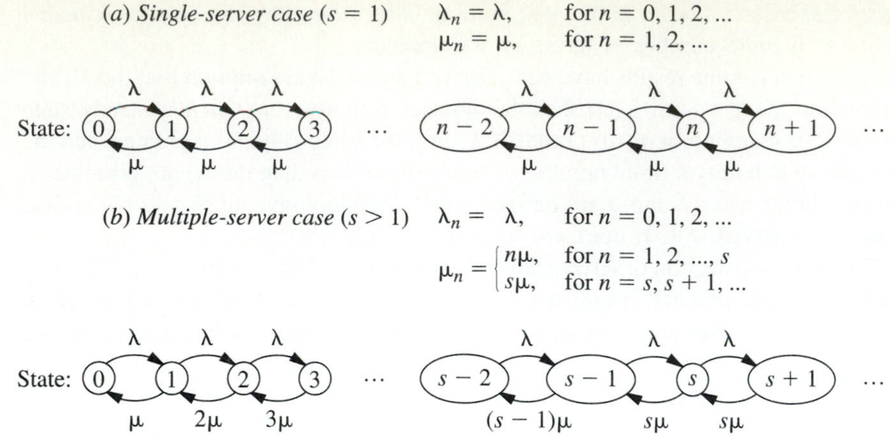

Figure 15.5 Rate diagrams for the *M/M/s* model.

mentioned for Property 4 of the exponential distribution (see Sec. 15.4), when the mean service rate per busy server is μ, the overall mean service rate for n busy servers must be $n\mu$. Therefore, $\mu_n = n\mu$ when $n \le s$, whereas $\mu_n = s\mu$ when $n \ge s$ so that all s servers are busy. The rate diagram for this case is shown in Fig. 15.5*b*.

When the maximum mean service rate $s\mu$ exceeds the mean arrival rate λ, that is, when

$$\rho = \frac{\lambda}{s\mu} < 1,$$

a queueing system fitting this model will eventually reach a steady-state condition. In this situation, the steady-state results derived in Sec. 15.5 for the general birth-and-death process are directly applicable. However, these results simplify considerably for this model and yield closed-form expressions for P_n, L, L_q, and so forth, as shown next.

Results for the Single-Server Case (*M/M/*1): For $s = 1$, the C_n factors for the birth-and-death process reduce to

$$C_n = \left(\frac{\lambda}{\mu}\right)^n = \rho^n, \qquad \text{for } n = 0, 1, 2, \ldots$$

Therefore,

$$P_n = \rho^n P_0, \qquad \text{for } n = 0, 1, 2, \ldots,$$

where

$$P_0 = \left(\sum_{n=0}^{\infty} \rho^n\right)^{-1} = \left(\frac{1}{1-\rho}\right)^{-1} = 1 - \rho.$$

Thus

$$P_n = (1 - \rho)\rho^n, \qquad \text{for } n = 0, 1, 2, \ldots.$$

Consequently,

$$L = \sum_{n=0}^{\infty} n(1-\rho)\rho^n$$

$$= (1-\rho)\rho \sum_{n=0}^{\infty} \frac{d}{d\rho}(\rho^n)$$

$$= (1-\rho)\rho \frac{d}{d\rho}\left(\sum_{n=0}^{\infty} \rho^n\right)$$

$$= (1-\rho)\rho \frac{d}{d\rho}\left(\frac{1}{1-\rho}\right)$$

$$= \frac{\rho}{1-\rho} = \frac{\lambda}{\mu-\lambda}.$$

$$L_q = \sum_{n=1}^{\infty} (n-1)P_n$$

$$= L - 1(1 - P_0)$$

$$= \frac{\lambda^2}{\mu(\mu-\lambda)}.$$

When $\lambda \geq \mu$, so that the mean arrival rate exceeds the mean service rate, the preceding solution "blows up" (because the summation for computing P_0 diverges). For this case, the queue would "explode" and grow without bound. If the queueing system begins operation with no customers present, the server might succeed in keeping up with arriving customers over a short period of time, but this is impossible in the long run. (Even when $\lambda = \mu$, the *expected* number of customers in the queueing system slowly grows without bound over time because, even though a temporary return to no customers present always is possible, the probabilities of huge numbers of customers present become increasingly significant over time.)

Assuming again that $\lambda < \mu$, we now can derive the probability distribution of the waiting time in the system (*including* service) $\mathcal{W}$ for a random arrival when the queue discipline is first-come-first-served. If this arrival finds n customers already in the system, then the arrival will have to wait through $n + 1$ exponential service times, including his or her own. (For the customer currently being served, recall the lack-of-memory property for the exponential distribution discussed in Sec. 15.4.) Therefore, let $T_1, T_2, \ldots$ be independent service-time random variables having an exponential distribution with parameter μ, and let

$$S_{n+1} = T_1 + T_2 + \cdots + T_{n+1}, \qquad \text{for } n = 0, 1, 2, \ldots,$$

so that S_{n+1} represents the *conditional* waiting time given n customers already in the system. As discussed in Sec. 15.7, S_{n+1} is known to have an *Erlang distribution.*[1] Because the probability that the random arrival will find n customers in the system is P_n, it follows that

$$P\{\mathcal{W} > t\} = \sum_{n=0}^{\infty} P_n P\{S_{n+1} > t\},$$

which reduces after considerable manipulation (see Prob. 15.6-15) to

$$P\{\mathcal{W} > t\} = e^{-\mu(1-\rho)t}, \qquad \text{for } t \geq 0.$$

[1] Outside queueing theory, this distribution is known as the *gamma distribution.*

The surprising conclusion is that $\mathcal{W}$ has an *exponential* distribution with parameter $\mu(1-\rho)$. Therefore,

$$W = E(\mathcal{W}) = \frac{1}{\mu(1-\rho)}$$

$$= \frac{1}{\mu - \lambda}.$$

These results *include* service time in the waiting time. In some contexts (e.g., the County Hospital emergency room problem), the more relevant waiting time is just until service begins. Thus consider the waiting time in the queue (so *excluding* service time) $\mathcal{W}_q$ for a random arrival when the queue discipline is first-come-first-served. If this arrival finds no customers already in the system, then the arrival is served immediately, so that

$$P\{\mathcal{W}_q = 0\} = P_0 = 1 - \rho.$$

If this arrival finds $n > 0$ customers already there instead, then the arrival has to wait through n exponential service times until his or her own service begins, so that

$$P\{\mathcal{W}_q > t\} = \sum_{n=1}^{\infty} P_n P\{S_n > t\}$$

$$= \sum_{n=1}^{\infty} (1-\rho)\rho^n P\{S_n > t\}$$

$$= \rho \sum_{n=0}^{\infty} P_n P\{S_{n+1} > t\}$$

$$= \rho P\{\mathcal{W} > t\}$$

$$= \rho e^{-\mu(1-\rho)t}, \qquad \text{for } t \geq 0.$$

Note that Wq does not quite have an exponential distribution, because $P\{\mathcal{W}_q = 0\} > 0$. However, the *conditional* distribution of $\mathcal{W}_q$, given that $\mathcal{W}_q > 0$, does have an exponential distribution with parameter $\mu(1-\rho)$, just as $\mathcal{W}$ does, because

$$P\{\mathcal{W}_q > t \mid \mathcal{W}_q > 0\} = \frac{P\{\mathcal{W}_q > t\}}{P\{\mathcal{W}_q > 0\}} = e^{-\mu(1-\rho)t}, \qquad \text{for } t \geq 0.$$

By deriving the mean of the (unconditional) distribution of $\mathcal{W}_q$ (or applying either $L_q = \lambda W_q$ or $W_q = W - 1/\mu$),

$$W_q = E(\mathcal{W}_q) = \frac{\lambda}{\mu(\mu - \lambda)}.$$

Results for the Multiple-Server Case ($s > 1$): When $s > 1$, the C_n factors become

$$C_n = \begin{cases} \dfrac{(\lambda/\mu)^n}{n!} & \text{for } n = 1, 2, \ldots, s, \\[2ex] \dfrac{(\lambda/\mu)^s}{s!}\left(\dfrac{\lambda}{s\mu}\right)^{n-s} = \dfrac{(\lambda/\mu)^n}{s!s^{n-s}} & \text{for } n = s, s+1, \ldots. \end{cases}$$

Consequently, if $\lambda < s\mu$ [so that $\rho = \lambda/(s\mu) < 1$], then

$$P_0 = 1\Big/\left[1 + \sum_{n=1}^{s-1} \frac{(\lambda/\mu)^n}{n!} + \frac{(\lambda/\mu)^s}{s!} \sum_{n=s}^{\infty} \left(\frac{\lambda}{s\mu}\right)^{n-s}\right]$$

$$= 1\Big/\left[\sum_{n=0}^{s-1} \frac{(\lambda/\mu)^n}{n!} + \frac{(\lambda/\mu)^s}{s!} \frac{1}{1 - \lambda/(s\mu)}\right],$$

where the $n = 0$ term in the last summation yields the correct value of 1 because of the convention that $n! = 1$ when $n = 0$. These C_n factors also give

$$P_n = \begin{cases} \dfrac{(\lambda/\mu)^n}{n!} P_0 & \text{if } 0 \le n \le s, \\ \dfrac{(\lambda/\mu)^n}{s!s^{n-s}} P_0 & \text{if } n \ge s. \end{cases}$$

Furthermore,

$$L_q = \sum_{n=s}^{\infty} (n - s)P_n$$

$$= \sum_{j=0}^{\infty} jP_{s+j}$$

$$= \sum_{j=0}^{\infty} j\frac{(\lambda/\mu)^s}{s!} \rho^j P_0$$

$$= P_0 \frac{(\lambda/\mu)^s}{s!} \rho \sum_{j=0}^{\infty} \frac{d}{d\rho}(\rho^j)$$

$$= P_0 \frac{(\lambda/\mu)^s}{s!} \rho \frac{d}{d\rho}\left(\sum_{j=0}^{\infty} \rho^j\right)$$

$$= P_0 \frac{(\lambda/\mu)^s}{s!} \rho \frac{d}{d\rho}\left(\frac{1}{1 - \rho}\right)$$

$$= \frac{P_0(\lambda/\mu)^s \rho}{s!(1 - \rho)^2};$$

$$W_q = \frac{L_q}{\lambda};$$

$$W = W_q + \frac{1}{\mu};$$

$$L = \lambda\left(W_q + \frac{1}{\mu}\right) = L_q + \frac{\lambda}{\mu}.$$

Figures 15.6 and 15.7 show how P_0 and L change with ρ for various values of s.

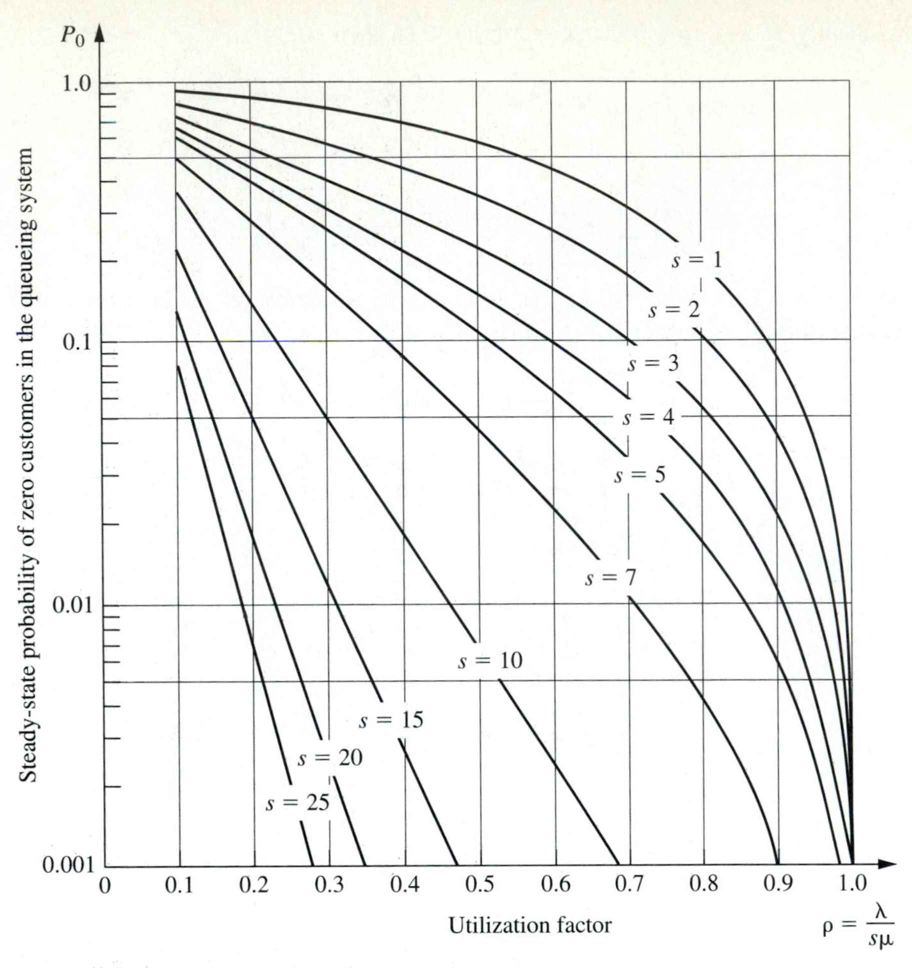

Figure 15.6 Values of P_0 for the $M/M/s$ model (Sec. 15.6).

The single-server method for finding the probability distribution of waiting times also can be extended to the multiple-server case. This yields[1] (for $t \geq 0$)

$$P\{\mathcal{W} > t\} = e^{-\mu t}\left[1 + \frac{P_0(\lambda/\mu)^s}{s!(1-\rho)}\left(\frac{1 - e^{-\mu t(s-1-\lambda/\mu)}}{s - 1 - \lambda/\mu}\right)\right]$$

and
$$P\{\mathcal{W}_q > t\} = (1 - P\{\mathcal{W}_q = 0\})e^{-s\mu(1-\rho)t},$$

where
$$P\{\mathcal{W}_q = 0\} = \sum_{n=0}^{s-1} P_n.$$

The above formulas for the various measures of performance (including the P_n) are relatively imposing for hand calculations. However, your OR Courseware includes a convenient routine for performing all these calculations simultaneously for any values of t, s, λ, and μ you want, provided that $\lambda < s\mu$.

[1] When $s - 1 - \lambda/\mu = 0$, $(1 - e^{-\mu t(s-1-\lambda/\mu)})/(s - 1 - \lambda/\mu)$ should be replaced by μt.

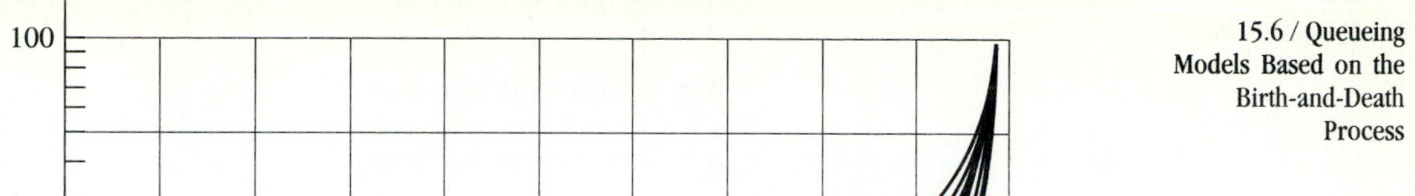

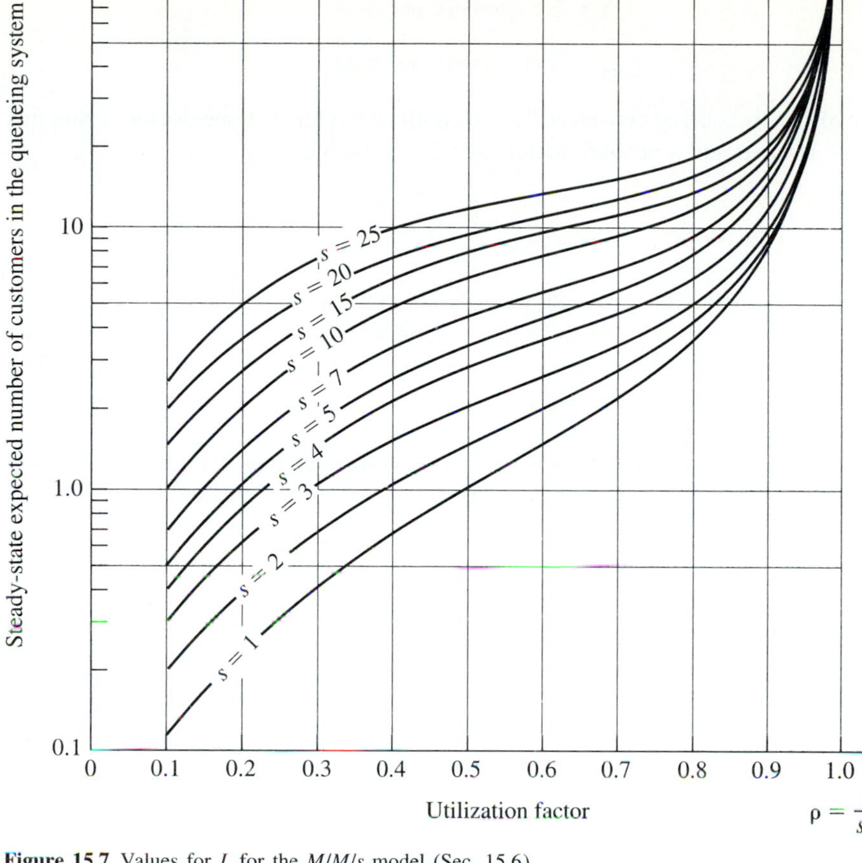

Figure 15.7 Values for L for the $M/M/s$ model (Sec. 15.6).

If $\lambda \geq s\mu$, so that the mean arrival rate exceeds the maximum mean service rate, then the queue grows without bound, so the preceding steady-state solutions are not applicable.

The County Hospital Example with the *M/M/s* Model: For the County Hospital emergency room problem (see Sec. 15.1), the management engineer has concluded that the emergency cases arrive pretty much at random (a *Poisson input process*), so that interarrival times have an exponential distribution. She also has concluded that the time spent by a doctor treating the cases approximately follows an *exponential distribution.* Therefore, she has chosen the $M/M/s$ model for a preliminary study of this queueing system.

By projecting the available data for the early evening shift into next year, she estimates that patients will arrive at an *average* rate of 1 ever $\frac{1}{2}$ hour. A doctor requires an average of 20 minutes to treat each patient. Thus, with one hour as the unit of time,

$$\frac{1}{\lambda} = \frac{1}{2} \text{ hour per customer}$$

and $\frac{1}{\mu} = \frac{1}{3}$ hour per customer,

so that $\lambda = 2$ customers per hour

and $\mu = 3$ customers per hour.

The two alternatives being considered are to continue having just one doctor during this shift ($s = 1$) or to add a second doctor ($s = 2$). In both cases,

$$\rho = \frac{\lambda}{s\mu} < 1,$$

so that the system should approach a steady-state condition. (Actually, because λ is somewhat different during other shifts, the system will never truly reach a steady-state condition, but the management engineer feels that steady-state results will provide a good approximation.) Therefore, the preceding equations are used to obtain the results shown in Table 15.2.

On the basis of these results, she tentatively concluded that a single doctor would be inadequate next year for providing the relatively prompt treatment needed in a hospital emergency room. You will see later how the management engineer checked this conclusion by applying two other queueing models that provide better representations of the real queueing system in some ways.

The Finite Queue Variation of the *M/M/s* Model (Called the *M/M/s/K* Model)

We mentioned in the discussion of queues in Sec. 15.2 that queueing systems sometimes have a *finite queue;* i.e., the number of customers in the system is not permitted to exceed some specified number (denoted by K). Any customer that arrives while the queue is "full" is refused entry into the system and so leaves forever. From the viewpoint of the birth-and-death process, the mean input rate into the system becomes

Table 15.2 **Steady-State Results from the *M/M/s* Model for the County Hospital Problem**

	$s = 1$	$s = 2$
ρ	$\frac{2}{3}$	$\frac{1}{3}$
P_0	$\frac{1}{3}$	$\frac{1}{2}$
P_1	$\frac{2}{9}$	$\frac{1}{3}$
P_n for $n \geq 2$	$\frac{1}{3}(\frac{2}{3})^n$	$(\frac{1}{3})^n$
L_q	$\frac{4}{3}$	$\frac{1}{12}$
L	2	$\frac{3}{4}$
W_q	$\frac{2}{3}$	$\frac{1}{24}$ (in hours)
W	1	$\frac{3}{8}$ (in hours)
$P\{\mathcal{W}_q > 0\}$	0.667	0.167
$P\{\mathcal{W}_q > \frac{1}{2}\}$	0.404	0.022
$P\{\mathcal{W}_q > 1\}$	0.245	0.003
$P\{\mathcal{W}_q > t\}$	$\frac{2}{3}e^{-t}$	$\frac{1}{6}e^{-4t}$
$P\{\mathcal{W} > t\}$	e^{-t}	$\frac{1}{2}e^{-3t}(3 - e^{-t})$

zero at these times. Therefore, the one modification needed in the $M/M/s$ model to introduce a finite queue is to change the λ_n parameters to

$$\lambda_n = \begin{cases} \lambda & \text{for } n = 0, 1, 2, \ldots, K-1, \\ 0 & \text{for } \quad n \geq K. \end{cases}$$

Because $\lambda_n = 0$ for some values of n, a queueing system that fits this model will eventually reach a steady-state condition.

This model commonly is labeled $M/M/s/K$, where the presence of the fourth symbol distinguishes it from the $M/M/s$ model. The single difference in the formulation of these two models is that K is finite for the $M/M/s/K$ model and $K = \infty$ for the $M/M/s$ model.

The usual physical interpretation for the $M/M/s/K$ model is that there is only *limited waiting room* that will accommodate a maximum of K customers in the system. For example, for the County Hospital emergency room problem, this system actually would have a finite queue if there were only K cots for the patients and if the policy were to send arriving patients to another hospital whenever there were no empty cots.

Another possible interpretation is that arriving customers will leave and "take their business elsewhere" whenever they find too many customers (K) ahead of them in the system because they are not willing to incur a long wait. This balking phenomenon is quite common in commercial service systems. However, there are other models available (e.g., see Prob. 15.5-5) that fit this interpretation even better.

The rate diagram for this model is identical to that shown in Fig. 15.5 for the $M/M/s$ model, *except* that it stops with state K.

Results for the Single-Server Case ($M/M/1/K$): For this case,

$$C_n = \begin{cases} \left(\dfrac{\lambda}{\mu}\right)^n = \rho^n & \text{for } n = 0, 1, 2, \ldots, K, \\ 0 & \text{for } \quad n > K. \end{cases}$$

Therefore, for $\rho \neq 1$,†

$$P_0 = \frac{1}{\sum_{n=0}^{K} (\lambda/\mu)^n}$$

$$= 1 \bigg/ \left[\frac{1 - (\lambda/\mu)^{K+1}}{1 - \lambda/\mu}\right]$$

$$= \frac{1 - \rho}{1 - \rho^{K+1}},$$

so that

$$P_n = \frac{1 - \rho}{1 - \rho^{K+1}} \rho^n, \qquad \text{for } n = 0, 1, 2, \ldots, K.$$

† If $\rho = 1$, then $P_n = 1/(K + 1)$ for $n = 0, 1, 2, \ldots, K$, so that $L = K/2$.

Hence,

$$L = \sum_{n=0}^{K} nP_n$$

$$= \frac{1-\rho}{1-\rho^{K+1}} \rho \sum_{n=0}^{K} \frac{d}{d\rho}(\rho^n)$$

$$= \frac{1-\rho}{1-\rho^{K+1}} \rho \frac{d}{d\rho}\left(\sum_{n=0}^{K} \rho^n\right)$$

$$= \frac{1-\rho}{1-\rho^{K+1}} \rho \frac{d}{d\rho}\left(\frac{1-\rho^{K+1}}{1-\rho}\right)$$

$$= \rho \frac{-(K+1)\rho^K + K\rho^{K+1} + 1}{(1-\rho^{K+1})(1-\rho)}$$

$$= \frac{\rho}{1-\rho} - \frac{(K+1)\rho^{K+1}}{1-\rho^{K+1}}.$$

As usual (when $s = 1$),

$$L_q = L - (1 - P_0).$$

Notice that the preceding results do not require that $\lambda < \mu$ (i.e., that $\rho < 1$).

When $\rho < 1$, it can be verified that the second term in the final expression for L converges to 0 as $K \to \infty$, so that *all* the preceding results do indeed converge to the corresponding results given earlier for the $M/M/1$ model.

The waiting-time distributions can be derived by using the same reasoning as for the $M/M/1$ model (see Prob. 15.6-29). However, no simple expressions are obtained in this case, so computer calculations are required. Fortunately, even though $L \neq \lambda W$ and $L_q \neq \lambda W_q$ for the current model because the λ_n are not equal for all n (see the end of Sec. 15.2), the *expected* waiting times for customers entering the system still can be obtained directly from the expressions given at the end of Sec. 15.5

$$W = \frac{L}{\bar{\lambda}}, \qquad W_q = \frac{L_q}{\bar{\lambda}},$$

where

$$\bar{\lambda} = \sum_{n=0}^{\infty} \lambda_n P_n$$

$$= \sum_{n=0}^{K-1} \lambda P_n$$

$$= \lambda(1 - P_K).$$

Results for the Multiple-Server Case ($s > 1$): Because this model does not allow more than K customers in the system, K is the maximum number of servers that could ever be used. Therefore, assume that $s \leq K$. In this case, C_n becomes

$$C_n = \begin{cases} \dfrac{(\lambda/\mu)^n}{n!} & \text{for } n = 0, 1, 2, \ldots, s, \\ \dfrac{(\lambda/\mu)^s}{s!}\left(\dfrac{\lambda}{s\mu}\right)^{n-s} = \dfrac{(\lambda/\mu)^n}{s!s^{n-s}} & \text{for } n = s, s+1, \ldots, K, \\ 0 & \text{for } \quad n > K. \end{cases}$$

Hence,

$$P_n = \begin{cases} \dfrac{(\lambda/\mu)^n}{n!} P_0 & \text{for } n = 1, 2, \ldots, s, \\ \dfrac{(\lambda/\mu)^n}{s!s^{n-s}} P_0 & \text{for } n = s, s+1, \ldots, K, \\ 0 & \text{for } n > K, \end{cases}$$

where $$P_0 = 1 \bigg/ \left[\sum_{n=0}^{s} \frac{(\lambda/\mu)^n}{n!} + \frac{(\lambda/\mu)^s}{s!} \sum_{n=s+1}^{K} \left(\frac{\lambda}{s\mu}\right)^{n-s} \right].$$

Adapting the derivation of L_q for the $M/M/s$ model to this case (see Prob. 15.6-28) yields

$$L_q = \frac{P_0(\lambda/\mu)^s \rho}{s!(1-\rho)^2} [1 - \rho^{K-s} - (K-s)\rho^{K-s}(1-\rho)],$$

where $\rho = \lambda/(s\mu)$.† It can then be shown (see Prob. 15.2-5) that

$$L = \sum_{n=0}^{s-1} nP_n + L_q + s\left(1 - \sum_{n=0}^{s-1} P_n\right).$$

And W and W_q are obtained from these quantities just as shown for the single-server case.

Your OR Courseware has a routine for calculating the above measures of performance (including the P_n) for this model.

The Finite Calling Population Variation of the *M/M/s* Model

Now assume that the only deviation from the $M/M/s$ model is that (as defined in Sec. 15.2) the *input source* is *limited;* i.e., the size of the *calling population* is *finite.* For this case, let N denote the size of the calling population. Thus, when the number of customers in the queueing system is n ($n = 0, 1, 2, \ldots, N$), there are only $N - n$ *potential* customers remaining in the input source.

The most important application of this model has been to the machine repair problem, where one or more maintenance persons are assigned the responsibility of maintaining in operational order a certain group of N machines by repairing each one that breaks down. (The example given at the end of Sec. 15.8 illustrates this application

† If $\rho = 1$, it is necessary to apply L'Hôpital's rule twice to this expression for L_q. Otherwise, all these multiple-server results hold for all $\rho > 0$. The reason that this queueing system can reach steady-state condition even when $\rho \geq 1$ is that $\lambda_n = 0$ for $n \geq K$, so that the number of customers in the system cannot continue to grow indefinitely.

(*a*) *Single-server case* ($s = 1$)

$$\lambda_n = \begin{cases} (N-n)\lambda, & \text{for } n = 0, 1, 2, \ldots, N \\ 0, & \text{for } n \geq N \end{cases}$$

$$\mu_n = \mu, \quad \text{for } n = 1, 2, \ldots$$

$N\lambda$ $(N-1)\lambda$ ... $(N-n+2)\lambda$ $(N-n+1)\lambda$ λ

State: 0 1 2 ... $n-2$ $n-1$ n ... $N-1$ N

μ μ μ μ μ

(*b*) *Multiple-server case* ($s > 1$)

$$\lambda_n = \begin{cases} (N-n)\lambda, & \text{for } n = 0, 1, 2, \ldots, N \\ 0, & \text{for } n \geq N \end{cases}$$

$$\mu_n = \begin{cases} n\mu, & \text{for } n = 1, 2, \ldots, s \\ s\mu, & \text{for } n = s, s+1, \ldots \end{cases}$$

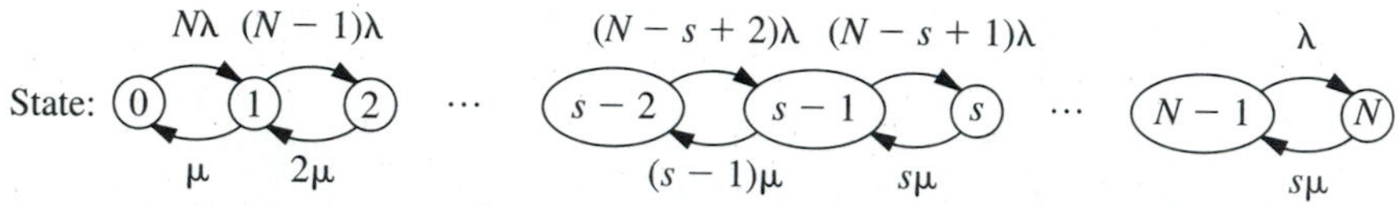

Figure 15.8 Rate diagrams for the finite calling population variation of the *M/M/s* model.

when the general procedures for solving any *continuous time Markov chain* are used rather than the specific formulas available for the birth-and-death process.) The maintenance people are considered to be individual servers in the queueing system if they work individually on different machines, whereas the entire crew is considered to be a single server if crew members work together on each machine. The machines constitute the calling population. Each one is considered to be a customer in the queueing system when it is down waiting to be repaired, whereas it is outside the queueing system while it is operational.

Note that each member of the calling population alternates between being *inside* and *outside* the queueing system. Therefore, the analog of the *M/M/s* model that fits this situation assumes that *each* member's *outside time* (i.e., the elapsed time from leaving the system until returning for the next time) has an *exponential distribution* with parameter λ. When n of the members are *inside,* and so $N - n$ members are *outside,* the current probability distribution of the *remaining* time until the next arrival to the queueing system is the distribution of the *minimum* of the *remaining outside times* for the latter $N - n$ members. Properties 2 and 3 for the exponential distribution imply that this distribution must be exponential with parameter $\lambda_n = (N - n)\lambda$. Hence this model is just the special case of the birth-and-death process that has the rate diagram shown in Fig. 15.8.

Because $\lambda_n = 0$ for $n = N$, any queueing system that fits this model will eventually reach a steady-state condition. The available steady-state results are summarized as follows:

Results for the Single-Server Case ($s = 1$): When $s = 1$, the C_n factors in Sec. 15.5 reduce to

$$C_n = \begin{cases} N(N-1)\cdots(N-n+1)\left(\dfrac{\lambda}{\mu}\right)^n = \dfrac{N!}{(N-n)!}\left(\dfrac{\lambda}{\mu}\right)^n & \text{for } n \leq N, \\ 0 & \text{for } n > N, \end{cases}$$

for this model. Therefore,

$$P_0 = 1 \Big/ \sum_{n=0}^{N} \left[\frac{N!}{(N-n)!} \left(\frac{\lambda}{\mu} \right)^n \right];$$

$$P_n = \frac{N!}{(N-n)!} \left(\frac{\lambda}{\mu} \right)^n P_0, \qquad \text{if } n = 1, 2, \ldots, N;$$

$$L_q = \sum_{n=1}^{N} (n-1) P_n,$$

which can be reduced to

$$L_q = N - \frac{\lambda + \mu}{\lambda}(1 - P_0);$$

$$L = \sum_{n=0}^{N} n P_n = L_q + 1 - P_0$$

$$= N - \frac{\mu}{\lambda}(1 - P_0).$$

Finally, $$W = \frac{L}{\bar{\lambda}} \quad \text{and} \quad W_q = \frac{L_q}{\bar{\lambda}},$$

where $$\bar{\lambda} = \sum_{n=0}^{\infty} \lambda_n P_n = \sum_{n=0}^{N} (N-n)\lambda P_n = \lambda(N - L).$$

Results for the Multiple-Server Case ($s > 1$): For $N \geq s > 1$,

$$C_n = \begin{cases} \dfrac{N!}{(N-n)!n!} \left(\dfrac{\lambda}{\mu} \right)^n & \text{for } n = 0, 1, 2, \ldots, s, \\ \dfrac{N!}{(N-n)!s!s^{n-s}} \left(\dfrac{\lambda}{\mu} \right)^n & \text{for } n = s, s+1, \ldots, N, \\ 0 & \text{for } \quad n > N. \end{cases}$$

Hence,

$$P_n = \begin{cases} \dfrac{N!}{(N-n)!n!} \left(\dfrac{\lambda}{\mu} \right)^n P_0 & \text{if } 0 \leq n \leq s, \\ \dfrac{N!}{(N-n)!s!s^{n-s}} \left(\dfrac{\lambda}{\mu} \right)^n P_0 & \text{if } s \leq n \leq N, \\ 0 & \text{if } \quad n > N, \end{cases}$$

where

$$P_0 = 1 \Big/ \left[\sum_{n=0}^{s-1} \frac{N!}{(N-n)!n!} \left(\frac{\lambda}{\mu} \right)^n + \sum_{n=s}^{N} \frac{N!}{(N-n)!s!s^{n-s}} \left(\frac{\lambda}{\mu} \right)^n \right].$$

Finally,

$$L_q = \sum_{n=s}^{N} (n - s)P_n$$

and

$$L = \sum_{n=0}^{s-1} nP_n + L_q + s\left(1 - \sum_{n=0}^{s-1} P_n\right),$$

which then yield W and W_q by the same equations as in the single-server case.

A routine for performing all the above calculations is available in your OR Courseware.

Extensive tables of computational results also are available[1] for this model for both the single-server and multiple-server cases.

For both cases, it has been shown[2] that the preceding formulas for P_n and P_0 (and so for L_q, L, W, and W_q) *also* hold for a generalization of this model. In particular, we can *drop* the assumption that the times spent *outside* the queueing system by the members of the calling population have an *exponential distribution,* even though this takes the model outside the realm of the birth-and-death process. As long as these times are identically distributed with mean $1/\lambda$ (and the assumption of exponential service times still holds), these outside times can have *any* probability distribution!

A Model with State-Dependent Service Rate and/or Arrival Rate

All the models thus far have assumed that the mean service rate is always a constant, regardless of how many customers are in the system. Unfortunately, this rate often is not a constant in real queueing systems, particularly when the servers are people. When there is a large backlog of work (i.e., a long queue), it is quite likely that such servers will tend to work faster than they do when the backlog is small or nonexistent. This increase in the service rate may result merely because the servers increase their efforts when they are under the pressure of a long queue. However, it may also result partly because the quality of the service is compromised or because assistance is obtained on certain service phases.

Given that the mean service rate does increase as the queue size increases, it is desirable to develop a theoretical model that seems to describe the pattern by which it increases. This model not only should be a reasonable approximation of the actual pattern but also should be simple enough to be practical for implementation. One such model is formulated next. (You have the flexibility to formulate many similar models within the framework of the birth-and-death process.) We then show how the same results apply when the arrival rate is affected by the queue size in an analogous way.

[1] L. G. Peck and R. N. Hazelwood, *Finite Queueing Tables,* Wiley, New York, 1958.

[2] B. D. Bunday and R. E. Scraton, "The G/M/r Machine Interference Model," *European Journal of Operational Research,* **4:** 399–402, 1980.

Formulation for the Single-Server Case ($s = 1$): Let

$$\mu_n = n^c \mu_1, \qquad \text{for } n = 1, 2, \ldots,$$

where n = number of customers in system,
μ_n = mean service rate when n customers are in system,
$1/\mu_1$ = expected "normal" service time—expected time to service customer when that customer is only one in system,
c = pressure coefficient—positive constant that indicates degree to which service rate of system is affected by system state.

Thus, by selecting $c = 1$, for example, we hypothesize that the mean service rate is directly proportional to n; $c = \frac{1}{2}$ implies that the mean service rate is proportional to the square root of n; and so on. The preceding queueing models in this section have implicitly assumed that $c = 0$.

Now assume additionally that the queueing system has a Poisson input with $\lambda_n = \lambda$ (for $n = 0, 1, 2, \ldots$) and exponential service times with μ_n as just given. This case is now a special case of the birth-and-death process, where

$$C_n = \frac{(\lambda/\mu_1)^n}{(n!)^c}, \qquad \text{for } n = 0, 1, 2, \ldots.$$

Thus all the steady-state results given in Sec. 15.5 are applicable to this model. (A steady-state condition always can be reached when $c > 0$.) Unfortunately, analytical expressions are not available for the summations involved. However, nearly exact values of P_0 and L have been tabulated[1] for various values of c and λ/μ_1 by summing a finite number of terms on a computer. A small portion of these results also is shown in Figs. 15.9 and 15.10.

A queueing system may react to a long queue by decreasing the arrival rate instead of increasing the service rate. (The arrival rate may be decreased, e.g., by diverting some of the customers requiring service to another service facility.) The corresponding model for describing mean arrival rates for this case lets

$$\lambda_n = (n + 1)^{-b}\lambda_0, \qquad \text{for } n = 0, 1, 2, \ldots,$$

where b is a constant whose interpretation is analogous to that for c. The C_n values for the birth-and-death process with these λ_n (and with $\mu_n = \mu$ for $n = 1, 2, \ldots$) are *identical* to those just shown (replacing λ by λ_0) for the state-dependent service rate model when $c = b$ and $\lambda/\mu_1 = \lambda_0/\mu$, so the steady-state results also are the same.

A more general model that combines these two patterns can also be used when both the mean arrival rates and the mean service rates are state-dependent. Thus let

$$\mu_n = n^a \mu_1 \qquad \text{for } n = 1, 2, \ldots$$

and

$$\lambda_n = (n + 1)^{-b}\lambda_0 \qquad \text{for } n = 0, 1, 2, \ldots.$$

Once again, the C_n values for the birth-and-death process with these parameters are identical to those shown for the state-dependent service rate model when $c = a + b$ and $\lambda/\mu_1 = \lambda_0/\mu_1$, so the tabulated steady-state results actually are applicable to this general model.

[1] R. W. Conway and W. L. Maxwell, "A Queueing Model with State Dependent Service Rate," *Journal of Industrial Engineering,* **12:** 132–136, 1961.

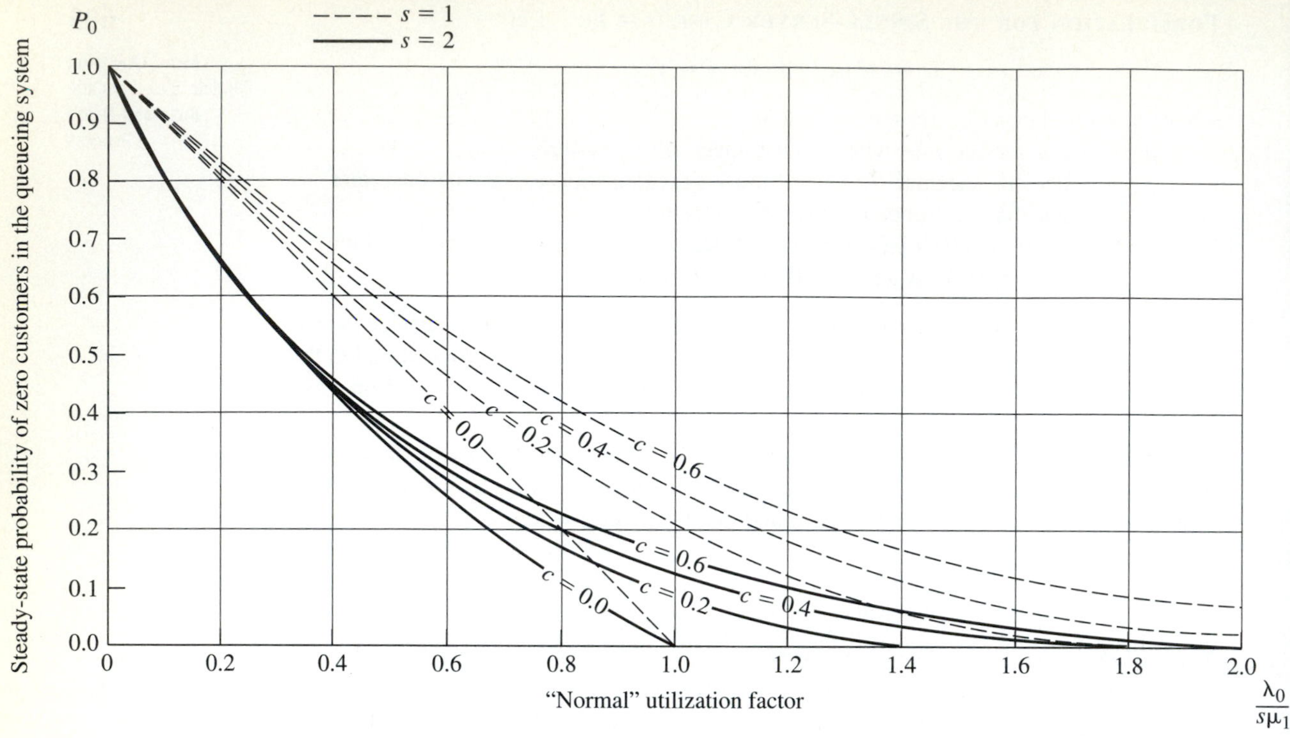

Figure 15.9 Values of P_0 for the state-dependent model (Sec. 15.6).

Formulation for the Multiple-Server Case ($s > 1$): To generalize this combined model further to the multiple-server case, it seems natural to have the μ_n and λ_n vary with the number of customers *per server* (n/s) in essentially the same way that they vary with n for the single-server case. Thus let

$$\mu_n = \begin{cases} n\mu_1 & \text{if } n \leq s \\ \left(\dfrac{n}{s}\right)^a s\mu_1 & \text{if } n \geq s \end{cases}$$

and

$$\lambda_n = \begin{cases} \lambda_0 & \text{if } n \leq s - 1 \\ \left(\dfrac{s}{n+1}\right)^b \lambda_0 & \text{if } n \geq s - 1. \end{cases}$$

Therefore, the birth-and-death process with these parameters has

$$C_n = \begin{cases} \dfrac{(\lambda_0/\mu_1)^n}{n!} & \text{for } n = 0, 1, 2, \ldots, s \\ \dfrac{(\lambda_0/\mu_1)^n}{s!(n!/s!)^c s^{(1-c)(n-s)}} & \text{for } n = s, s+1, \ldots, \end{cases}$$

where $c = a + b$.

Computational results for P_0, L_q, and L have been tabulated[1] for various values of c, λ_0/μ_1, and s. Some of these results also are given in Figs. 15.9 and 15.10.

[1] F. S. Hillier, R. W. Conway, and W. L. Maxwell, "A Multiple Server Queueing Model with State Dependent Service Rate," *Journal of Industrial Engineering,* **15:** 153–157, 1964.

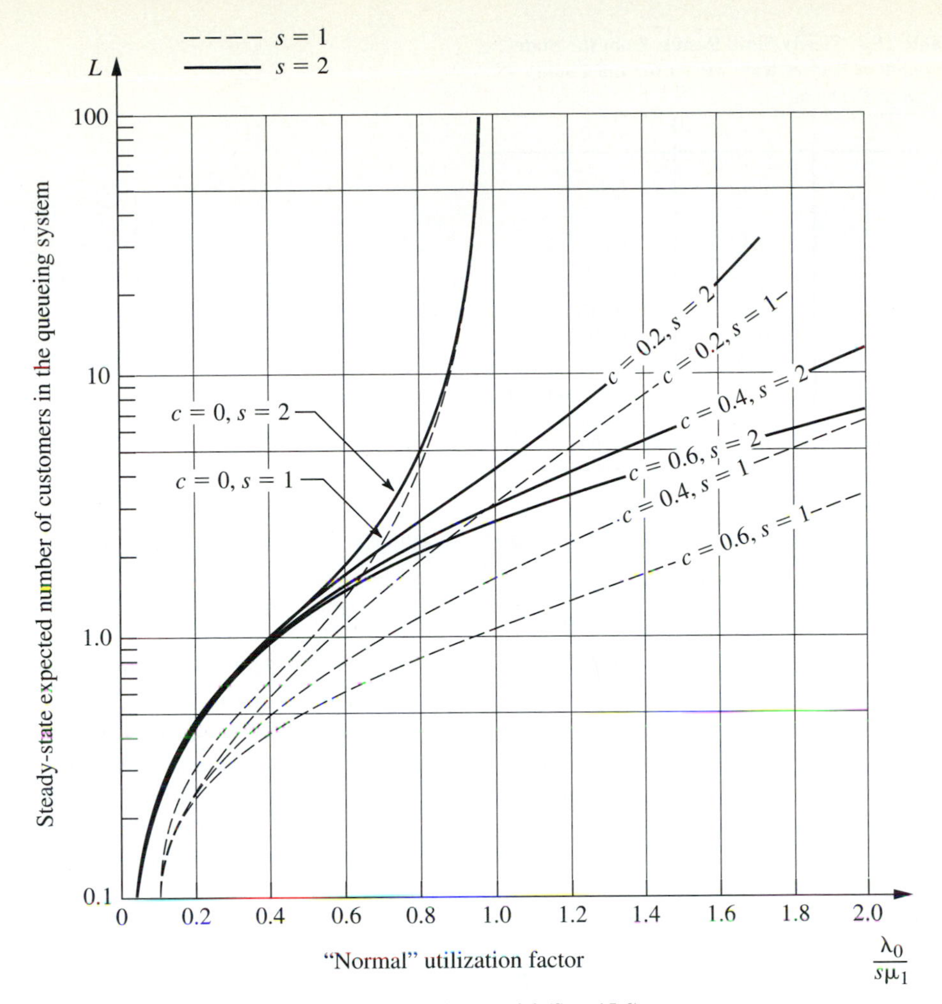

Figure 15.10 Values of L for the state-dependent model (Sec. 15.6).

The County Hospital Example with State-Dependent Service Rates: After gathering additional data for the County Hospital emergency room, the management engineer found that the time a doctor spends with a patient tends to decrease as the number of patients waiting increases. Part of the explanation is simply that the doctor works faster, but the main reason is that more of the treatment is turned over to a nurse for completion. The pattern of the μ_n (the mean rate at which a doctor treats patients while there are a total of n patients to be treated in the emergency room) seems to fit reasonably the state-dependent service rate model presented here. Therefore, the management engineer has decided to apply this model.

The new data indicate that the average time a doctor spends treating a patient is 24 minutes if no other patients are waiting, whereas this average becomes 12 minutes when each doctor has six patients (so five are waiting their turn). Thus, with a single doctor on duty,

$$\mu_1 = 2\tfrac{1}{2} \text{ customers per hour,}$$

$$\mu_6 = 5 \text{ customers per hour.}$$

Table 15.3 Steady-State Results from the State-Dependent Service Rate Model for the County Hospital Problem

	$s = 1$	$s = 2$
$\frac{\lambda}{s\mu_1}$	0.8	0.4
$\frac{\lambda}{s\mu_{6s}}$	0.4	0.2
P_0	0.367	0.440
P_1	0.294	0.352
L_q	0.618	0.095
L	1.251	0.864
W_q	0.309	0.048 (in hours)
W	0.626	0.432 (in hours)
$P\{\mathcal{W}_q > 0\}$	0.633	0.208

Therefore, the pressure coefficient c (or a in the general model) must satisfy the relationship

$$\mu_6 = 6^c \mu_1, \qquad \text{so} \qquad 6^c = 2.$$

Using logarithms to solve for c yields $c = 0.4$. Because $\lambda = 2$ from before, this solution for c completes the specification of parameter values for this model.

To compare the two alternatives of having one doctor ($s = 1$) or two doctors ($s = 2$) on duty, the management engineer developed the various measures of performance shown in Table 15.3. The values of P_0, L, and (for $s = 2$) L_q were obtained directly from the tabulated results for this model. (Except for this L_q, you can approximate the same values from Figs. 15.9 and 15.10.) These values were then used to calculate

$$P_1 = C_1 P_0,$$

$$L_q = L - (1 - P_0), \qquad \text{if } s = 1,$$

$$L_q = L - P_1 - 2(1 - P_0 - P_1), \qquad \text{if } s = 2,$$

$$W_q = \frac{L_q}{\lambda}, \qquad W = \frac{L}{\lambda},$$

$$P\{\mathcal{W}_q > 0\} = 1 - \sum_{n=0}^{s-1} P_n.$$

The fact that some of the results in Table 15.3 do not deviate substantially from those in Table 15.2 reinforces the tentative conclusion that a single doctor will be inadequate next year.

15.7 Queueing Models Involving Nonexponential Distributions

Because all the queueing theory models in the preceding section (except for one generalization) are based on the birth-and-death process, both their interarrival and service times are required to have *exponential* distributions. As discussed in Sec. 15.4, this type of probability distribution has many convenient properties for queueing theory, but it provides a reasonable fit for only certain kinds of queueing systems. In particular, the assumption of exponential interarrival times implies that arrivals occur randomly (a Poisson input process), which is a reasonable approximation in many situations but *not* when the arrivals are carefully scheduled or regulated. Furthermore, the actual service-time distribution frequently deviates greatly from the exponential form, particularly when the service requirements of the customers are quite similar. Therefore, it is important to have available other queueing models that use alternative distributions.

Unfortunately, the mathematical analysis of queueing models with nonexponential distributions is much more difficult. However, it has been possible to obtain some useful results for a few such models. This analysis is beyond the level of this book, but in this section we shall summarize the models and describe their results.

The *M/G/*1 Model

As introduced in Sec. 15.2, the *M/G/*1 model assumes that the queueing system has a *single server* and a *Poisson input process* (exponential interarrival times) with a *fixed* mean arrival rate λ. As usual, it is assumed that the customers have *independent* service times with the *same* probability distribution. However, no restrictions are imposed on what this service-time distribution can be. In fact, it is only necessary to know (or estimate) the mean $1/\mu$ and variance σ^2 of this distribution.

Any such queueing system can eventually reach a steady-state condition if $\rho = \lambda/\mu < 1$. The readily available steady-state results[1] for this general model are the following:

$$P_0 = 1 - \rho,$$

$$L_q = \frac{\lambda^2\sigma^2 + \rho^2}{2(1-\rho)},$$

$$L = \rho + L_q,$$

$$W_q = \frac{L_q}{\lambda},$$

$$W = W_q + \frac{1}{\mu}.$$

Considering the complexity involved in analyzing a model that permits *any* service-time distribution, it is remarkable that such a simple formula can be obtained for L_q. This formula is one of the most important results in queueing theory because of its ease of use and the prevalence of *M/G/*1 queueing systems in practice. This equation for L_q (or its counterpart for W_q) commonly is referred to as the **Pollaczek-Khintchine formula**, named after the Frenchman and the Russian who derived it more than 50 years ago.

For any fixed expected service time $1/\mu$, notice that L_q, L, W_q, and W all increase as σ^2 is increased. This result is important because it indicates that the consistency of the server has a major bearing on the performance of the service facility—not just the server's average speed. This key point is illustrated in the next subsection.

[1] A recursion formula also is available for calculating the probability distribution of the number of customers in the system; see A. Hordijk and H. C. Tijms: "A Simple Proof of the Equivalence of the Limiting Distribution of the Continuous-Time and the Embedded Process of the Queue Size in the *M/G/*1 Queue," *Statistica Neerlandica,* **36:** 97–100, 1976.

When the service-time distribution is exponential, $\sigma^2 = 1/\mu^2$, and the preceding results will reduce to the corresponding results for the $M/M/1$ model given at the beginning of Sec. 15.6.

The complete flexibility in the service-time distribution provided by this model is extremely useful, so it is unfortunate that efforts to derive similar results for the multiple-server case have been unsuccessful. However, some multiple-server results have been obtained for the important special cases described by the following two models.

The *M/D/s* Model

When the service consists of essentially the same routine task to be performed for all customers, there tends to be little variation in the service time required. The $M/D/s$ model often provides a reasonable representation for this kind of situation, because it assumes that all service times actually equal some fixed *constant* (the *degenerate* service-time distribution) and that we have a *Poisson* input process with a fixed mean arrival rate λ.

When there is just a single server, the $M/D/1$ model is just the special case of the $M/G/1$ model where $\sigma^2 = 0$, so that the *Pollaczek-Khintchine formula* reduces to

$$L_q = \frac{\rho^2}{2(1-\rho)},$$

where L, W_q, and W are obtained from L_q as just shown. Notice that these L_q and W_q are exactly *half* as large as those for the exponential service-time case of Sec. 15.6 (the $M/M/1$ model), where $\sigma^2 = 1/\mu^2$, so decreasing σ^2 can *greatly* improve the measures of performance of a queueing system.

For the multiple-server version of this model ($M/D/s$), a complicated method is available[1] for deriving the steady-state probability distribution of the number of customers in the system and its mean [assuming $\rho = \lambda/(s\mu) < 1$]. However, these results have been tabulated for numerous cases,[2] and the means (L) also are given graphically in Fig. 15.11.

The *M/E$_k$/s* Model

The $M/D/s$ model assumes *zero* variation in the service times ($\sigma = 0$), whereas the *exponential* service-time distribution assumes a very large variation ($\sigma = 1/\mu$). Between these two rather extreme cases lies a long middle ground ($0 < \sigma < 1/\mu$), where most *actual* service-time distributions fall. Another kind of theoretical service-time distribution that fills this middle ground is the **Erlang distribution** (named after the founder of queueing theory).

The probability density function for the Erlang distribution is

$$f(t) = \frac{(\mu k)^k}{(k-1)!} t^{k-1} e^{-k\mu t}, \qquad \text{for } t \geq 0,$$

[1] See N. U. Prabhu: *Queues and Inventories,* Wiley, New York, 1965, pp. 32–34; also see pp. 344–346 in Selected Reference 3.

[2] F. S. Hillier and O. S. Yu, with D. Avis, L. Fossett, F. Lo, and M. Reiman, *Queueing Tables and Graphs,* Elsevier North-Holland, New York, 1981.

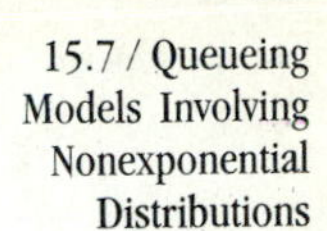

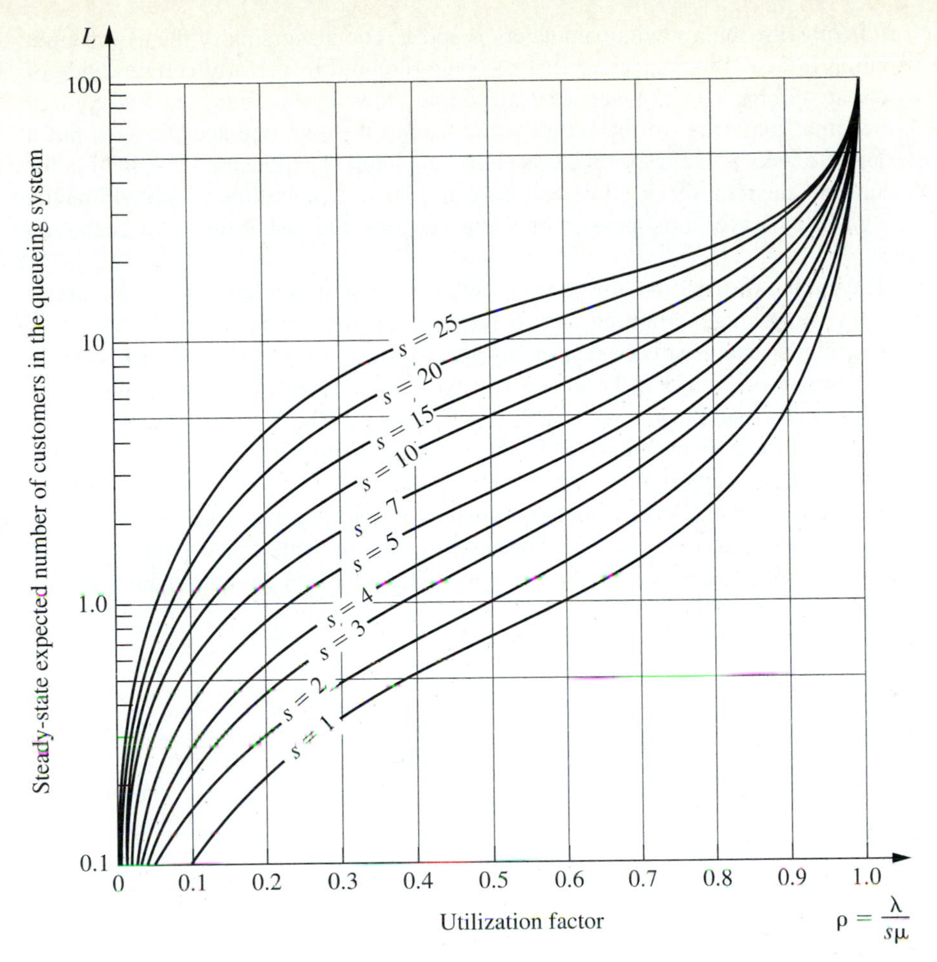

Figure 15.11 Values of L for the $M/D/s$ model (Sec. 15.7).

where μ and k are strictly positive parameters of the distribution and k is further restricted to be integer. (Except for this integer restriction and the definition of the parameters, this distribution is *identical* to the *gamma distribution*.) Its mean and standard deviation are

$$\text{Mean} = \frac{1}{\mu}$$

and

$$\text{Standard deviation} = \frac{1}{\sqrt{k}}\frac{1}{\mu}.$$

Thus k is the parameter that specifies the degree of variability of the service times relative to the mean. It usually is referred to as the *shape parameter*.

The Erlang distribution is a very important distribution in queueing theory for two reasons. To describe the first one, suppose that $T_1, T_2, \ldots, T_k$ are k independent random variables with an identical exponential distribution whose mean is $1/(k\mu)$. Then their sum

$$T = T_1 + T_2 + \cdots + T_k$$

has an *Erlang* distribution with parameters μ and k. The discussion of the exponential distribution in Sec. 15.4 suggested that the time required to perform certain kinds of tasks might well have an exponential distribution. However, the total service required by a customer may involve the server's performing not just one specific task but a sequence of k tasks. If the respective tasks have an identical exponential distribution for their duration, the total service time will have an Erlang distribution, which will be the case, e.g., if the server must perform the *same* exponential task k times for each customer.

The Erlang distribution also is very useful because it is a large (two-parameter) family of distributions permitting only nonnegative values. Hence empirical service-time distributions can usually be reasonably approximated by an Erlang distribution. In fact, both the *exponential* and the *degenerate* (constant) distributions are special cases of the Erlang distribution, with $k = 1$ and $k = \infty$, respectively. Intermediate values of k provide intermediate distributions with mean $= 1/\mu$, mode $= (k - 1)/(k\mu)$, and variance $= 1/(k\mu^2)$, as suggested by Fig. 15.12.

Now consider the $M/E_k/1$ model, which is just the special case of the $M/G/1$ model where service times have an Erlang distribution with shape parameter $= k$. Applying the Pollaczek-Khintchine formula with $\sigma^2 = 1/(k\mu^2)$ (and the accompanying results given for $M/G/1$) yields

$$L_q = \frac{\lambda^2/(k\mu^2) + \rho^2}{2(1 - \rho)} = \frac{1 + k}{2k} \frac{\lambda^2}{\mu(\mu - \lambda)},$$

$$W_q = \frac{1 + k}{2k} \frac{\lambda}{\mu(\mu - \lambda)},$$

$$W = W_q + \frac{1}{\mu},$$

$$L = \lambda W.$$

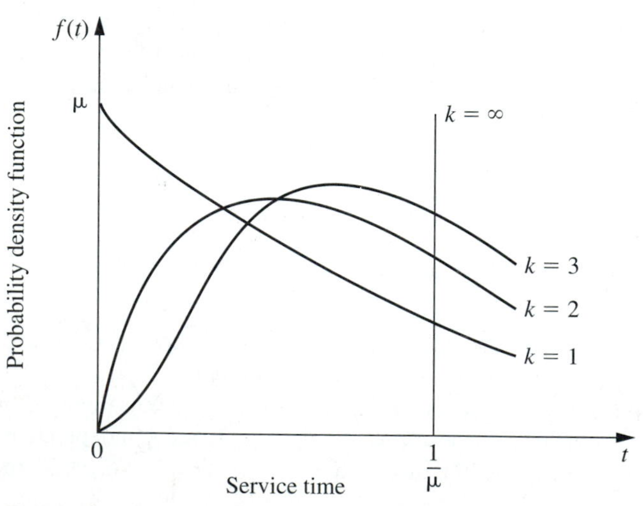

Figure 15.12 A family of Erlang distributions with constant mean $1/\mu$.

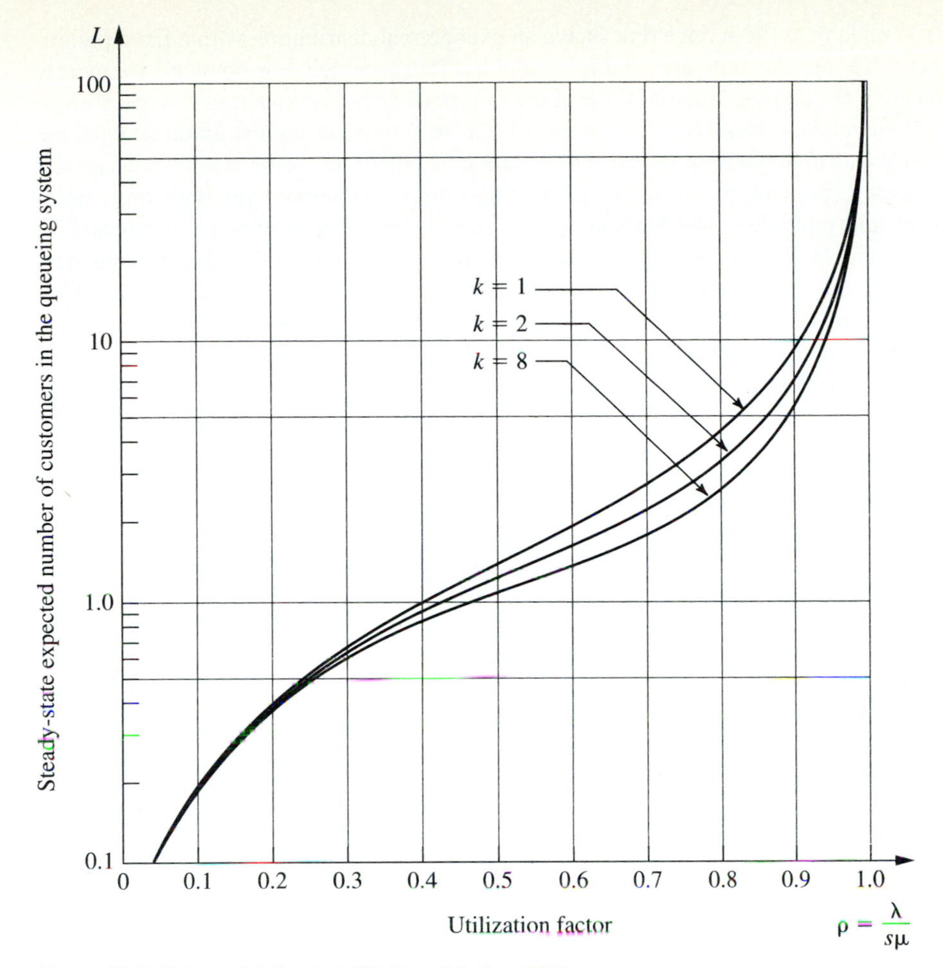

Figure 15.13 Values of L for the $M/E_k/2$ model (Sec. 15.7).

With multiple servers ($M/E_k/s$), the relationship of the Erlang distribution to the exponential distribution just described can be exploited to formulate a *modified* birth-and-death process (continuous time Markov chain) in terms of individual exponential service phases (k per customer) rather than complete customers. However, it has not been possible to derive a general steady-state solution [when $\rho = \lambda/(s\mu) < 1$] for the probability distribution of the number of customers in the system as we did in Sec. 15.5. Instead, advanced theory is required to solve individual cases numerically. Once again, these results have been obtained and tabulated for numerous cases.[1] The means (L) also are given graphically in Fig. 15.13 for some cases where $s = 2$.

Models without a Poisson Input

All the queueing models presented thus far have assumed a Poisson input process (exponential interarrival times). However, this assumption is violated if the arrivals are scheduled or regulated in some way that prevents them from occurring randomly, in which case another model is needed.

[1] Ibid.

As long as the service times have an exponential distribution with a fixed parameter, three such models are readily available. These models are obtained by merely *reversing* the assumed distributions of the *interarrival* and *service times* in the preceding three models. Thus the first new model ($GI/M/s$) imposes no restriction on what the *interarrival time* distribution can be. In this case, there are some steady-state results available[1] (particularly in regard to waiting-time distributions) for both the single-server and multiple-server versions of the model, but these results are not nearly as convenient as the simple expressions given for the $M/G/1$ model. The second new model ($D/M/s$) assumes that all interarrival times equal some fixed *constant,* which would represent a queueing system where arrivals are *scheduled* at regular intervals. The third new model ($E_k/M/s$) assumes an *Erlang* interarrival time distribution, which provides a middle ground between *regularly scheduled* (constant) and *completely random* (exponential) arrivals. Extensive computational results have been tabulated[2] for these latter two models, including the values of L given graphically in Figs. 15.14 and 15.15.

[1] For example, see pp. 304–320 of Selected Reference 3.

[2] Hillier and Yu, op. cit.

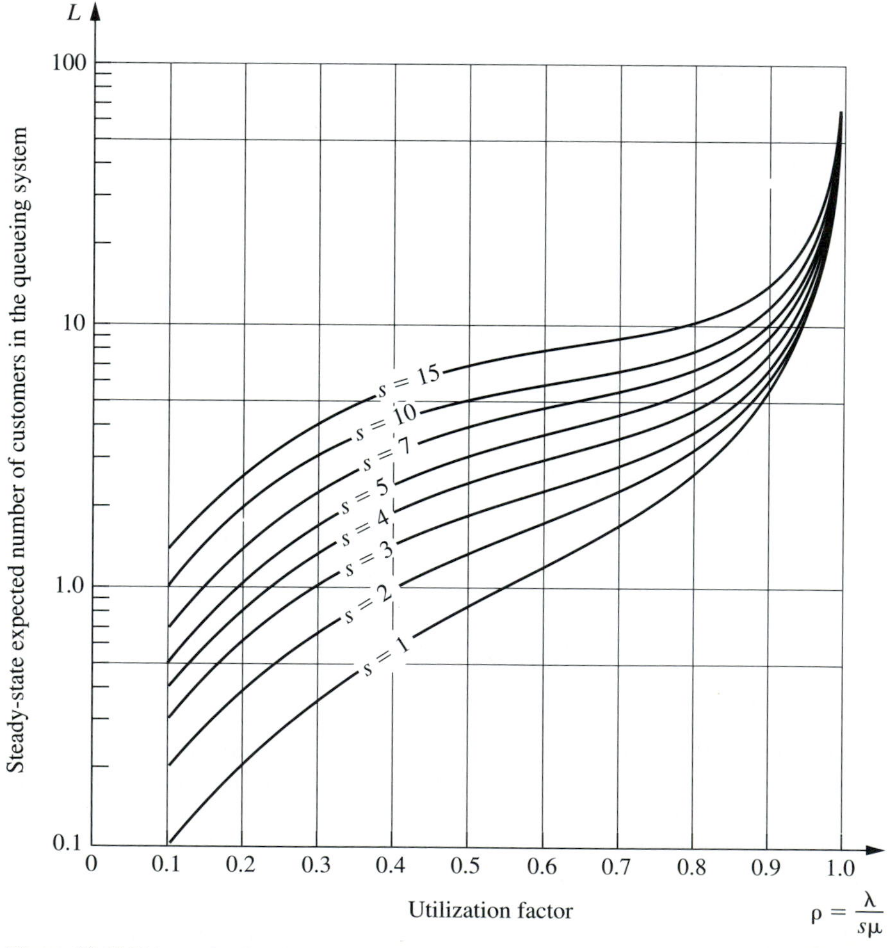

Figure 15.14 Values of L for the $D/M/s$ model (Sec. 15.7).

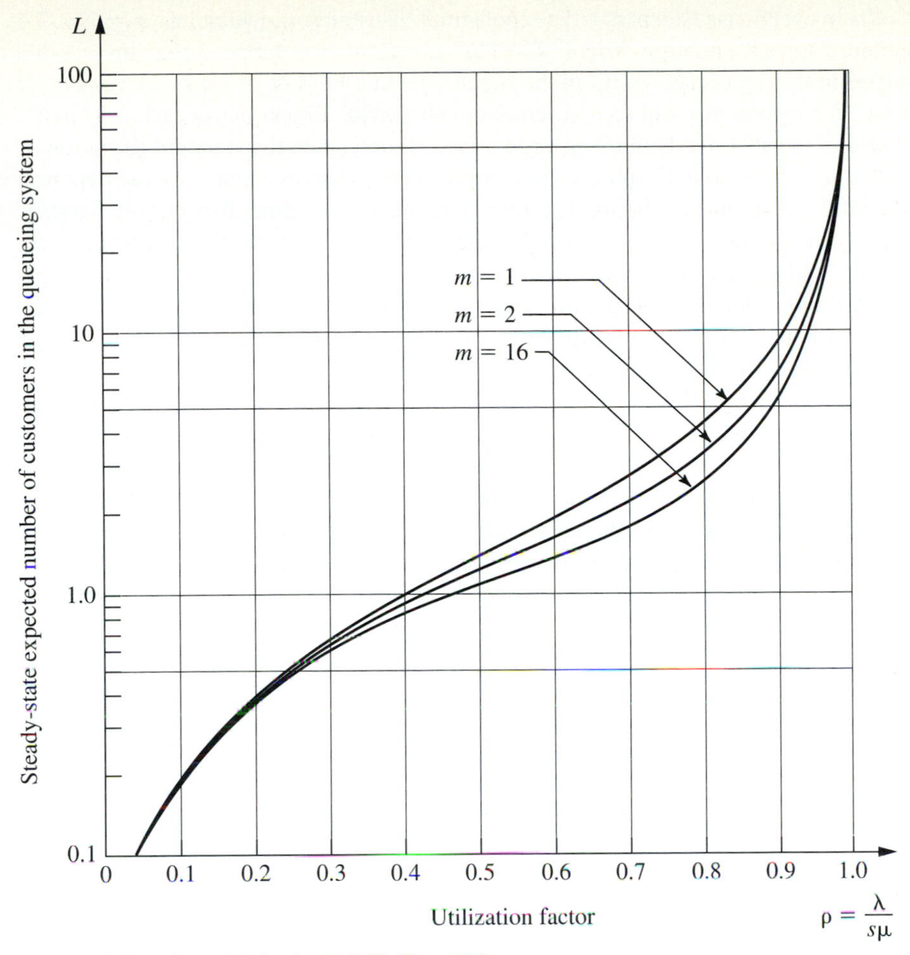

Figure 15.15 Values of L for the $E_k/M/2$ (Sec. 15.7).

If neither the interarrival times nor the service times for a queueing system have an exponential distribution, then there are three additional queueing models for which computational results also are available.[1] One of these models ($E_m/E_k/s$) assumes an Erlang distribution for both these times. The other two models ($E_k/D/s$ and $D/E_k/s$) assume that one of these times has an Erlang distribution and the other time equals some fixed constant.

Other Models

Although you have seen in this section a large number of queueing models that involve nonexponential distributions, we have far from exhausted the list. For example, another distribution that occasionally is used for either interarrival times or service times is the **hyperexponential distribution**. The key characteristic of this distribution is that even though only nonnegative values are allowed, its standard deviation σ actually is larger than its mean $1/\mu$. This characteristic is in contrast to the Erlang distribution, where

[1] Ibid.

$\sigma < 1/\mu$ in every case except $k = 1$ (exponential distribution), which has $\sigma = 1/\mu$. To illustrate a typical situation where $\sigma > 1/\mu$ can occur, we suppose that the service involved in the queueing system is the repair of some kind of machine or vehicle. If many of the repairs turn out to be routine (small service times) but occasional repairs require an extensive overhaul (very large service times), then the standard deviation of service times will tend to be quite large relative to the mean, in which case the hyperexponential distribution may be used to represent the service-time distribution. Specifically, this distribution would assume that there are fixed probabilities, p and $(1 - p)$, for which kind of repair will occur, that the time required for each kind has an exponential distribution, but that the parameters for these two exponential distributions are different. (In general, the hyperexponential distribution is such a composite of two or more exponential distributions.)

Another family of distributions coming into general use consists of **phase-type distributions** (some of which also are called *generalized Erlangian distributions*). These distributions are obtained by breaking down the total time into a number of phases, each having an exponential distribution, where the parameters of these exponential distributions may be different and the phases may be either in series or in parallel (or both). A group of phases being *in parallel* means that the process randomly selects *one* of the phases to go through each time according to specified probabilities. This approach is, in fact, how the hyperexponential distribution is derived, so this distribution is a special case of the phase-type distributions. Another special case is the Erlang distribution, which has the restrictions that all its k phases are in series and that these phases have the *same* parameter for their exponential distributions. Removing these restrictions means that phase-type distributions in general can provide considerably more flexibility than the Erlang distribution in fitting the actual distribution of interarrival times or service times observed in a real queueing system. This flexibility is especially valuable when using the actual distribution directly in the model is not analytically tractable, whereas the ratio of the *mean* to the *standard deviation* for the actual distribution does not closely match the available ratios ($\sqrt{k}$ for $k = 1, 2, \ldots$) for the Erlang distribution.

Because they built up from combinations of exponential distributions, queueing models using phase-type distributions still can be represented by a *continuous time Markov chain.* This Markov chain generally will have an infinite number of states, so solving for the steady-state distribution of the state of the system requires solving an infinite system of linear equations that has a relatively complicated structure. Solving such a system is far from a routine thing, but recent theoretical advances have enabled us to solve these queueing models numerically in some cases. An extensive tabulation of these results for models with various phase-type distributions (including the hyperexponential distribution) is available.[1]

15.8 Priority-Discipline Queueing Models

In priority-discipline queueing models, the queue discipline is based on a *priority system.* Thus the order in which members of the queue are selected for service is based on their assigned priorities.

[1] L. P. Seelen, H. C. Tijms, and M. H. Van Hoorn, *Tables for Multi-Server Queues,* North-Holland, Amsterdam, 1985.

Many real queueing systems fit these priority-discipline models much more closely than other available models. Rush jobs are taken ahead of other jobs, and important customers may be given precedence over others. Therefore, the use of priority-discipline models often provides a very welcome refinement over the more usual queueing models.

We present two basic priority-discipline models here. Since both models make the same assumptions, except for the nature of the priorities, we first describe the models together and then summarize their results separately.

The Models

Both models assume that there are *N priority classes* (class 1 has the highest priority and class *N* has the lowest) and that whenever a server becomes free to begin serving a new customer from the queue, the one customer selected is that member of the *highest* priority class represented in the queue who has waited longest. In other words, customers are selected to begin service in the order of their priority classes, but on a first-come-first-served basis within each priority class. A *Poisson* input process and *exponential* service times are assumed for each priority class. The models also make the somewhat restrictive assumption that the expected service time is the *same* for all priority classes. However, the models do permit the mean arrival rate to differ among priority classes.

The distinction between the two models is whether the priorities are *nonpreemptive* or *preemptive.* With **nonpreemptive priorities**, a customer being served cannot be ejected back into the queue (preempted) if a higher-priority customer enters the queueing system. Therefore, once a server has begun serving a customer, the service must be completed without interruption. The first model assumes nonpreemptive priorities.

With **preemptive priorities**, the lowest-priority customer being served is *preempted* (ejected back into the queue) whenever a higher-priority customer enters the queueing system. A server is thereby freed to begin serving the new arrival immediately. (When a server does succeed in *finishing* a service, the next customer to begin receiving service is selected just as described at the beginning of this subsection, so a preempted customer normally will get back into service again and, after enough tries, will eventually finish.) Because of the lack-of-memory property of the exponential distribution (see Sec. 15.4), we do not need to worry about defining the point at which service begins when a preempted customer returns to service; the distribution of the *remaining* service time *always* is the same. (For any other service-time distribution, it becomes important to distinguish between *preemptive-resume* systems, where service for a preempted customer resumes at the point of interruption, and *preemptive-repeat* systems, where service must start at the beginning again.) The second model assumes preemptive priorities.

For both models, if the distinction between customers in different priority classes were ignored, Property 6 for the exponential distribution (see Sec. 15.4) implies that *all* customers would arrive according to a Poisson input process. Furthermore, all customers have the *same* exponential distribution for service times. Consequently, the two models actually are identical to the $M/M/s$ model studied in Sec. 15.6 *except* for the order in which customers are served. Therefore, when we count just the *total* number of customers in the system, the steady-state distribution for the $M/M/s$ model also applies to both models. Consequently, the formulas for L and L_q also carry over, as do the expected waiting-time results (by Little's formula) W and W_q, for a randomly selected customer. What changes is the *distribution* of waiting times, which was derived in Sec.

15.6 under the assumption of a first-come-first-served queue discipline. With a priority discipline, this distribution has a much larger *variance,* because the waiting times of customers in the highest priority classes tend to be much smaller than those under a first-come-first-served discipline, whereas the waiting times in the lowest priority classes tend to be much larger. By the same token, the breakdown of the total number of customers in the system tends to be disproportionately weighted toward the lower-priority classes. But this condition is just the reason for imposing priorities on the queueing system in the first place. We want to *improve* the *measures of performance* for each of the higher-priority classes at the expense of performance for the lower-priority classes. To determine how much improvement is being made, we need to obtain such measures as *expected waiting time in the system* and *expected number of customers in the system* for the individual priority classes. Expressions for these measures are given next for the two models in turn.

Results for the Nonpreemptive Priorities Model

Let W_k be the steady-state expected waiting time in the system (including service time) for a member of priority class k. Then

$$W_k = \frac{1}{AB_{k-1}B_k} + \frac{1}{\mu}, \qquad \text{for } k = 1, 2, \ldots, N,$$

where
$$A = s!\,\frac{s\mu - \lambda}{r^s} \sum_{j=0}^{s-1} \frac{r^j}{j!} + s\mu,$$
$$B_0 = 1,$$
$$B_k = 1 - \frac{\sum_{i=1}^{k} \lambda_i}{s\mu}, \qquad \text{for } k = 1, 2, \ldots, N,$$
s = number of servers,
μ = mean service rate per busy server,
λ_i = mean arrival rate for priority class i, for $i = 1, 2, \ldots, N$,
$$\lambda = \sum_{i=1}^{N} \lambda_i,$$
$$r = \frac{\lambda}{\mu}.$$

(These results assume that

$$\sum_{i=1}^{k} \lambda_i < s\mu,$$

so that priority class k can reach a steady-state condition.) *Little's formula* still applies to individual priority classes, so L_k, the steady-state expected number of members of priority class k in the queueing system (including those being served), is

$$L_k = \lambda_k W_k, \qquad \text{for } k = 1, 2, \ldots, N.$$

To determine the expected waiting time in the queue (excluding service time) for priority class k, merely subtract $1/\mu$ from W_k; the corresponding expected queue length is again obtained by multiplying by λ_k. For the special case where $s = 1$, the expression for A reduces to $A = \mu^2/\lambda$.

A routine is available in your OR Courseware for performing the above calculations.

Results for the Preemptive Priorities Model

Given the same notation as for the preceding model, having preemption changes the *total* expected waiting time in the system (including the total service time) to

$$W_k = \frac{1/\mu}{B_{k-1}B_k}, \qquad \text{for } k = 1, 2, \dots, N,$$

for the *single-server* case ($s = 1$). When $s > 1$, W_k can be calculated by an iterative procedure that will be illustrated soon by the County Hospital example. The L_k just defined continue to satisfy the relationship

$$L_k = \lambda_k W_k, \qquad \text{for } k = 1, 2, \dots, N.$$

The corresponding results for the queue (excluding customers in service) also can be obtained from W_k and L_k as just described for the case of nonpreemptive priorities. Because of the lack-of-memory property of the exponential distribution (see Sec. 15.4), preemptions do not affect the service process (occurrence of service completions) in any way. The expected total service time for any customer still is $1/\mu$.

Your OR Courseware includes a routine for this model also for calculating the above measures of performance for the single-server case.

The County Hospital Example with Priorities

For the County Hospital emergency room problem, the management engineer has noticed that the patients are not treated on a first-come-first-served basis. Rather, the admitting nurse seems to divide the patients into roughly three categories: (1) *critical* cases, where prompt treatment is vital for survival; (2) *serious* cases, where early treatment is important to prevent further deterioration; and (3) *stable* cases, where treatment can be delayed without adverse medical consequences. Patients are then treated in this order of priority, where those in the same category are normally taken on a first-come-first-served basis. A doctor will interrupt treatment of a patient if a new case in a higher-priority category arrives. Approximately 10 percent of the patients fall into the first category, 30 percent into the second, and 60 percent into the third. Because the more serious cases will be sent to the hospital for further care after receiving emergency treatment, the average treatment time by a doctor in the emergency room actually does not differ greatly among these categories.

The management engineer has decided to use a priority-discipline queueing model as a reasonable representation of this queueing system, where the three categories of patients constitute the three priority classes in the model. Because treatment is interrupted by the arrival of a higher-priority case, the preemptive priorities model is the appropriate one. Given the previously available data ($\mu = 3$ and $\lambda = 2$), the preceding percentages yield $\lambda_1 = 0.2$, $\lambda_2 = 0.6$, and $\lambda_3 = 1.2$. Table 15.4 gives the resulting expected waiting times in the queue (so *excluding* treatment time) in *hours* for the respective priority classes[1] when there is one ($s = 1$) or two ($s = 2$) doctors on duty.

[1] Note that these expected times can no longer be interpreted as the expected time before treatment begins when $k > 1$, because treatment may be interrupted at least once, causing additional waiting time before service is completed.

Table 15.4 **Steady-State Results from the Priority-Discipline Models for the County Hospital Problem**

	Preemptive Priorities		*Nonpreemptive Priorities*	
	$s = 1$	$s = 2$	$s = 1$	$s = 2$
A	—	—	4.5	36
B_1	0.933	—	0.933	0.967
B_2	0.733	—	0.733	0.867
B_3	0.333	—	0.333	0.667
$W_1 - \frac{1}{\mu}$	0.024	0.00037	0.238	0.029
$W_2 - \frac{1}{\mu}$	0.154	0.00793	0.325	0.033
$W_3 - \frac{1}{\mu}$	1.033	0.06542	0.889	0.048

(The corresponding results for the nonpreemptive priorities model also are given in Table 15.4 to show the effect of preempting.)

These preemptive priority results for $s = 2$ were obtained as follows. Because the waiting times for priority class 1 customers are completely unaffected by the presence of customers in lower-priority classes, W_1 will be the same for any other values of λ_2 and λ_3, including $\lambda_2 = 0$ and $\lambda_3 = 0$. Therefore, W_1 must equal W for the corresponding *one-class* model (the $M/M/s$ model in Sec. 15.6) with $s = 2$, $\mu = 3$, and $\lambda = \lambda_1 = 0.2$, which yields

$$W_1 = W = 0.33370, \qquad \text{for } \lambda = 0.2$$

so

$$W_1 - \frac{1}{\mu} = 0.33370 - 0.33333 = 0.00037.$$

Now consider the first two priority classes. Again note that customers in these classes are completely unaffected by lower-priority classes (just priority class 3 in this case), which can therefore be ignored in the analysis. Let $\overline{W}_{1-2}$ be the expected waiting time in the system (so including service time) of a *random arrival* in *either* of these two classes, so the probability is $\lambda_1/(\lambda_1 + \lambda_2) = \frac{1}{4}$ that this arrival is in class 1 and $\lambda_2/(\lambda_1 + \lambda_2) = \frac{3}{4}$ that it is in class 2. Therefore,

$$\overline{W}_{1-2} = \tfrac{1}{4}W_1 + \tfrac{3}{4}W_2.$$

Furthermore, because the *expected* waiting time is the same for *any* queue discipline, $\overline{W}_{1-2}$ must also equal W for the $M/M/s$ model in Sec. 15.6, with $s = 2$, $\mu = 3$, and $\lambda = \lambda_1 + \lambda_2 = 0.8$, which yields

$$\overline{W}_{1-2} = W = 0.33937, \qquad \text{for } \lambda = 0.8.$$

Combining these facts gives

$$W_2 = \frac{4}{3}\left[0.33937 - \frac{1}{4}(0.33370)\right] = 0.34126.$$

$$\left(W_2 - \frac{1}{\mu} = 0.00793.\right)$$

Finally, let $\overline{W}_{1-3}$ be the expected waiting time in the system (so including service time) for a *random arrival* in *any* of the three priority classes, so the probabilities are 0.1, 0.3, and 0.6 that it is in classes 1, 2, and 3, respectively. Therefore,

$$\overline{W}_{1-3} = 0.1W_1 + 0.3W_2 + 0.6W_3.$$

Furthermore, $\overline{W}_{1-3}$ must also equal W for the $M/M/s$ model in Sec. 15.6, with $s = 2$, $\mu = 3$, and $\lambda = \lambda_1 + \lambda_2 + \lambda_3 = 2$, so that (from Table 15.2)

$$\overline{W}_{1-3} = W = 0.375, \qquad \text{for } \lambda = 2.$$

Consequently,

$$W_3 = \frac{1}{0.6}[0.375 - 0.1(0.33370) - 0.3(0.34126)]$$

$$= 0.39875.$$

$$\left(W_3 - \frac{1}{\mu} = 0.06542.\right)$$

The corresponding W_q results for the $M/M/s$ model in Sec. 15.6 also could have been used in exactly the same way to derive the $W_k - 1/\mu$ quantities directly.

When $s = 1$, the $W_k - 1/\mu$ values in Table 15.4 for the preemptive priorities case indicate that providing just a single doctor would cause critical cases to wait about $1\frac{1}{2}$ minutes (0.024 hour) on the average, serious cases to wait more than 9 minutes, and stable cases to wait more than 1 hour. (Contrast these results with the average wait of $W_q = \frac{2}{3}$ hour for all patients that was obtained in Table 15.2 under the first-come-first-served queue discipline.) However, these values represent *statistical expectations,* so some patients have to wait considerably longer than the average for their priority class. This wait would not be tolerable for the critical and serious cases, where a few minutes can be vital. By contrast, the $s = 2$ results in Table 15.4 (preemptive priorities case) indicate that adding a second doctor would virtually eliminate waiting for all but the stable cases. Therefore, the management engineer recommended that there be two doctors on duty in the emergency room during the early evening hours next year. The board of directors for County Hospital adopted this recommendation and simultaneously raised the charge for using the emergency room!

15.9 Queueing Networks

Thus far we have considered only queueing systems that have a *single* service facility with one or more servers. However, queueing systems encountered in OR studies are sometimes actually *queueing networks,* i.e., networks of service facilities where customers must receive service at some of or all these facilities. For example, orders being processed through a job shop must be routed through a sequence of machine groups (service facilities). It is therefore necessary to study the entire network to obtain such information as the expected total waiting time, expected number of customers in the entire system, and so forth.

Because of the importance of queueing networks, research into this area has been very active. However, this is a difficult area, so we limit ourselves to a brief introduction.

One result is of such fundamental importance for queueing networks that this finding and its implications warrant special attention here. This fundamental result is the following *equivalence property* for the *input process* of arriving customers and the *output process* of departing customers for certain queueing systems.

Equivalence property: Assume that a service facility with s servers and an infinite queue has a Poisson input with parameter λ and the same exponential service-time distribution with parameter μ for each server (the $M/M/s$ model), where $s\mu > \lambda$. Then the steady-state *output* of this service facility is also a Poisson process[1] with parameter λ.

Notice that this property makes no assumption about the type of queue discipline used. Whether it is first-come-first-served, random, or even a priority discipline as in Sec. 15.8, the served customers will leave the service facility according to a Poisson process. The crucial implication of this fact for queueing networks is that if these customers must then go to another service facility for further service, this second facility *also* will have a Poisson input. With an exponential service-time distribution, the equivalence property will hold for this facility as well, which can then provide a Poisson input for a third facility, etc. We discuss the consequences for two basic kinds of networks next.

Infinite Queues in Series

Suppose that customers must all receive service at a series of m service facilities in a fixed sequence. Assume that each facility has an infinite queue (no limitation on the number of customers allowed in the queue), so that the series of facilities form a system of *infinite queues in series.* Assume further that the customers arrive at the first facility according to a Poisson process with parameter λ and that each facility i ($i = 1, 2, \ldots, m$) has an exponential service-time distribution with parameter μ_i for its s_i servers, where $s_i\mu_i > \lambda$. It then follows from the equivalence property that (under steady-state conditions) each service facility has a Poisson input with parameter λ. Therefore, the elementary $M/M/s$ model of Sec. 15.6 (or its priority-discipline counterparts in Sec. 15.8) can be used to analyze each service facility independently of the others!

Being able to use the $M/M/s$ model to obtain all measures of performance for each facility independently, rather than analyzing interactions between facilities, is a tremendous simplification. For example, the probability of having n customers at a given facility is given by the formula for P_n in Sec. 15.6 for the $M/M/s$ model. The *joint probability* of n_1 customers at facility 1, n_2 customers at facility 2, . . . , then, is the *product* of the individual probabilities obtained in this simple way. In particular, this joint probability can be expressed as

$$P\{(N_1, N_2, \ldots, N_m) = (n_1, n_2, \ldots, n_m)\} = P_{n_1}P_{n_2}\cdots P_{n_m}.$$

(This simple form for the solution is called the **product form solution.**) Similarly, the expected total waiting time and the expected number of customers in the entire system can be obtained by merely summing the corresponding quantities obtained at the respective facilities.

Unfortunately, the equivalence property and its implications do not hold for the case of *finite* queues discussed in Sec. 15.6. This case is actually quite important in practice, because there is often a definite limitation on the queue length in front of service facilities in networks. For example, only a small amount of buffer storage space is typically provided in front of each facility (station) in a production-line system. For

[1] For a proof, see P. J. Burke: "The Output of a Queueing System," *Operations Research,* **4**(6): 699–704, 1956.

such systems of finite queues in series, no simple product form solution is available. The facilities must be analyzed jointly instead, and only limited results have been obtained.

Jackson Networks

Systems of infinite queues in series are not the only queueing networks where the *M/M/s* model can be used to analyze each service facility independently of the others. Another prominent kind of network with this property (a product form solution) is the *Jackson network,* named after the individual who first characterized the network and showed that this property holds.[1]

The characteristics of a Jackson network are the same as assumed above for the system of infinite queues in series, except now the customers visit the facilities in different orders (and may not visit them all). For each facility, its arriving customers come from *both* outside the system (according to a Poisson process) and the other facilities. These characteristics are summarized below.

A **Jackson network** is a system of m service facilities where facility i ($i = 1, 2, \ldots, m$) has

1. An infinite queue
2. Customers arriving from outside the system according to a Poisson input process with parameter a_i
3. s_i servers with an exponential service-time distribution with parameter μ_i.

A customer leaving facility i is routed next to facility j ($j = 1, 2, \ldots, m$) with probability p_{ij} or departs the system with probability

$$q_i = 1 - \sum_{j=1}^{m} p_{ij}.$$

Any such network has the following key property.

Under steady-state conditions, each facility j ($j = 1, 2, \ldots, m$) in a Jackson network behaves as if it were an *independent M/M/s* queueing system with arrival rate

$$\lambda_j = a_j + \sum_{i=1}^{m} \lambda_i p_{ij},$$

where $s_j\mu_j > \lambda_j$.

This key property cannot be *proved* directly from the equivalence property this time (the reasoning would become circular), but its *intuitive underpinning* is still provided by the latter property. The intuitive viewpoint (not quite technically correct) is that, for each facility i, its input processes from the various sources (outside and other facilities) are *independent Poisson processes,* so the *aggregate* input process is Poisson with parameter λ_i (Property 6 in Sec. 15.4). The equivalence property than says that the *aggregate output* process for facility i must be Poisson with parameter λ_i. By disaggregating this output process (Property 6 again), the process for customers going from facility i to facility j must be Poisson with parameter $\lambda_i p_{ij}$. This process becomes one of the Poisson *input* processes for facility j, thereby helping to maintain the series of Poisson processes in the overall system.

[1] See J. R. Jackson, "Jobshop-Like Queueing Systems," *Management Science,* **10**(1): 131–142, 1963.

The above equation for obtaining λ_j is based on the fact that λ_i is the *departure rate* as well as the arrival rate for all customers using facility i. Because p_{ij} is the proportion of customers departing from facility i who go next to facility j, the rate at which customers from facility i arrive at facility j is $\lambda_i p_{ij}$. Summing this product over all i, and then adding this sum to a_j, gives the *total arrival rate* to facility j from all sources.

To calculate λ_j from this equation requires knowing the λ_i for $i \neq j$, but these λ_i also are unknowns given by the corresponding equations. Therefore, the procedure is to solve *simultaneously* for $\lambda_1, \lambda_2, \ldots, \lambda_m$ by obtaining the simultaneous solution of the entire system of linear equations for λ_j for $j = 1, 2, \ldots, m$. Your OR Courseware includes a routine for solving for the λ_j in this way.

To illustrate these calculations, consider a Jackson network with three service facilities that have the parameters shown in Table 15.5. Plugging into the formula for λ_j for $j = 1, 2, 3$, we obtain

$$\begin{aligned} \lambda_1 &= 1 + 0.1\lambda_2 + 0.4\lambda_3 \\ \lambda_2 &= 4 + 0.6\lambda_1 + 0.4\lambda_3 \\ \lambda_3 &= 3 + 0.3\lambda_1 + 0.3\lambda_2. \end{aligned}$$

(Reason through each equation to see why it gives the total arrival rate.) The simultaneous solution for this system is

$$\lambda_1 = 5, \qquad \lambda_2 = 10, \qquad \lambda_3 = 7\tfrac{1}{2}.$$

Given this simultaneous solution, each of the three service facilities now can be analyzed *independently* by using the formulas for the *M/M/s* model given in Sec. 15.6. For example, to obtain the distribution of the number of customers $N_i = n_i$ at facility i, note that

$$\rho_i = \frac{\lambda_i}{s_i \mu_i} = \begin{cases} \frac{1}{2} & \text{for } i = 1, \\ \frac{1}{2} & \text{for } i = 2, \\ \frac{3}{4} & \text{for } i = 3. \end{cases}$$

Plugging these values (and the parameters in Table 15.5) into the formula for P_n gives

$$P_{n_1} = \tfrac{1}{2}\left(\tfrac{1}{2}\right)^{n_1} \qquad \text{for facility 1,}$$

$$P_{n_2} = \begin{cases} \frac{1}{3} & \text{for } n_2 = 0, \\ \frac{1}{3} & \text{for } n_2 = 1 \\ \frac{1}{3}\left(\frac{1}{2}\right)^{n_2 - 1} & \text{for } n_2 \geq 2, \end{cases} \qquad \text{for facility 2,}$$

$$P_{n_3} = \tfrac{1}{4}\left(\tfrac{3}{4}\right)^{n_3} \qquad \text{for facility 3.}$$

Table 15.5 **Data for the Example of a Jackson Network**

				p_{ij}		
Facility j	s_j	μ_j	a_j	$i = 1$	$i = 2$	$i = 3$
$j = 1$	1	10	1	0	0.1	0.4
$j = 2$	2	10	4	0.6	0	0.4
$j = 3$	1	10	3	0.3	0.3	0

The *joint probability* of (n_1, n_2, n_3) then is given simply by the product form solution

$$P\{(N_1, N_2, N_3) = (n_1, n_2, n_3)\} = P_{n_1} P_{n_2} P_{n_3}.$$

In a similar manner, the expected number of customers L_i at facility i can be calculated from Sec. 15.6 as

$$L_1 = 1, \qquad L_2 = \tfrac{4}{3}, \qquad L_3 = 3.$$

The expected *total* number of customers in the entire system then is

$$L = L_1 + L_2 + L_3 = 5\tfrac{1}{3}.$$

Obtaining W, the expected *total* waiting time in the system (including service times) for a customer, is a little trickier. You cannot simply add the expected waiting times at the respective facilities, because a customer does not necessarily visit each facility exactly once. However, Little's formula can still be used, where the system arrival rate λ is the sum of the arrival rates *from outside* to the facilities, $\lambda = a_1 + a_2 + a_3 = 8$. Thus

$$W = \frac{L}{a_1 + a_2 + a_3} = \frac{2}{3}.$$

In conclusion, we should point out that there do exist other (more complicated) kinds of queueing networks where the individual service facilities can be analyzed independently from the others. In fact, finding queueing networks with a product form solution has been the Holy Grail for research on queueing networks. One source of additional information is Selected Reference 7.

15.10 Conclusions

Queueing systems are prevalent throughout society. The adequacy of these systems can have an important effect on the quality of life and productivity.

Queueing theory studies queueing systems by formulating mathematical models of their operation and then using these models to derive measures of performance. This analysis provides vital information for effectively designing queueing systems that achieve an appropriate balance between the cost of providing a service and the cost associated with waiting for that service.

This chapter presented the most basic models of queueing theory for which particularly useful results are available. However, many other interesting models could be considered if space permitted. In fact, several thousand research papers formulating and/or analyzing queueing models have already appeared in the technical literature, and many more are being published each year!

The *exponential distribution* plays a fundamental role in queueing theory for representing the distribution of interarrival and service times, because this assumption enables us to represent the queueing system as a *continuous time Markov chain.* For the same reason, *phase-type distributions* such as the *Erlang distribution,* where the total time is broken down into individual phases having an exponential distribution, are very useful. Useful analytical results have been obtained for only a relatively few queueing models making other assumptions.

Priority-discipline queueing models are useful for the common situation where some categories of customers are given priority over others for receiving service.

In another common situation, customers must receive service at several different service facilities. Models for queueing networks are gaining widespread use for such situations. This is an area of especially active ongoing research.

When no tractable model that provides a reasonable representation of the queueing system under study is available, a common approach is to obtain relevant performance data by developing a computer program for simulating the operation of the system. This technique is discussed in Chap. 21.

Chapter 16 describes how queueing theory can be used to help design effective queueing systems.

SELECTED REFERENCES

1. Cooper, R. B., *Introduction to Queueing Theory,* 2d ed., Elsevier North-Holland, New York, 1981. (Also distributed by the George Washington University Continuing Engineering Education Program, Washington, D.C.)
2. Cooper, R. B., "Queueing Theory," chap. 10 in D. P. Heyman and M. J. Sobel (eds.), *Stochastic Models,* North Holland, Amsterdam, 1990. (This survey paper also is distributed by the George Washington University Continuing Engineering Education Program, Washington, D.C.)
3. Gross, D., and C. M. Harris, *Fundamentals of Queueing Theory,* 2d ed., Wiley, New York, 1985.
4. Kashyap, B. R. K., and M. L. Chandhry, *An Introduction to Queueing Theory,* A & A Publications, Kingston, Ontario, Canada, 1988.
5. Kleinrock, L., *Queueing Systems,* vol. 1: *Theory,* Wiley, New York, 1975.
6. Medhi, J., *Stochastic Models in Queueing Theory,* Academic Press, San Diego, 1991.
7. Walrand, J., *An Introduction to Queueing Networks,* Prentice-Hall, Englewood Cliffs, NJ, 1988.
8. Wolff, R. W., *Stochastic Modeling and the Theory of Queues,* Prentice-Hall, Englewood Cliffs, NJ, 1989.

RELEVANT ROUTINES IN YOUR OR COURSEWARE

Automatic routines: *The M/M/s Model*
The Finite Queue Variation of the M/M/s Model
The Finite Calling Population Variation of the M/M/s Model
Nonpreemptive Priorities Model
Preemptive Priorities Model
Jackson Network

To access these routines, call the ProbMod program and then choose *Queueing Theory* under the Area menu. See Appendix 1 for documentation of the software.

PROBLEMS[1]

To the left of each of the following problems (or their parts), we have inserted an A (for automatic routine) whenever a corresponding routine listed above can be helpful. An asterisk on the A indicates that this routine definitely should be used (unless your instructor gives contrary instructions) and that the printout from this routine is all that needs to be turned in to show your

[1] See also the end of Chap. 16 for problems involving the application of queueing theory. Problems 15.2-2, 15.2-3, 15.4-3 to 15.4-5, 15.4-7 to 15.4-9, 15.5-2, 15.5-3, 15.5-6, 15.5-10, 15.5-12, 15.6-1 to 15.6-3, 15.6-6 to 15.6-8, 15.6-10, 15.6-12 to 15.6-14, 15.6-17 to 15.6-21, and 15.6-24 have been adapted, with permission, from previous operations research examinations given by the Society of Actuaries.

work in executing the procedure. An asterisk on the problem number indicates that at least a partial answer is given in the back of the book.

15.2-1.* Consider a typical barber shop. Demonstrate that it is a queueing system by describing its components.

15.2-2. You are given a two-server queueing system in a steady-state condition where the number of customers in the system varies between 0 and 4. For $n = 0, 1, \ldots, 4$, the probability P_n that exactly n customers are in the system is $P_0 = \frac{1}{16}$, $P_1 = \frac{4}{16}$, $P_2 = \frac{6}{16}$, $P_3 = \frac{4}{16}$, $P_4 = \frac{1}{16}$.

(*a*) Determine L, the expected number of customers in the system.
(*b*) Determine L_q, the expected number of customers in the queue.
(*c*) Determine the expected number of customers being served.
(*d*) Given that the mean arrival rate is 2 customers per hour, determine the expected waiting time in the system, W, and the expected waiting time in the queue, W_q.
(*e*) Given that both servers have the same expected service time, use the results from part (*d*) to determine this expected service time.

15.2-3. You are given two queueing systems Q_1 and Q_2. The mean customer arrival rate, the mean service rate per busy server, and the steady-state expected number of customers for Q_2 are twice the corresponding values for Q_1. Let W_i = the steady-state expected waiting time in the system for Q_i, for $i = 1, 2$. Determine W_2/W_1.

15.2-4. Consider a single-server queueing system with *any* service-time distribution and *any* distribution of interarrival times (the *GI/G/*1 model). Use only basic definitions and the relationships given in Sec. 15.2 to verify the following general relationships.

(*a*) $L = L_q + 1 - P_0$
(*b*) $L = L_q + \rho$
(*c*) $P_0 = 1 - \rho$

15.2-5. Show that

$$L = \sum_{n=0}^{s-1} nP_n + L_q + s\left(1 - \sum_{n=0}^{s-1} P_n\right)$$

by using the statistical definitions of L and L_q in terms of the P_n.

15.3-1. Identify the customers and the servers in the queueing system in each of the following situations.

(*a*) The checkout stand in a grocery store
(*b*) A fire station
(*c*) The tollbooth for a bridge
(*d*) A bicycle repair shop
(*e*) A shipping dock
(*f*) A group of semiautomatic machines assigned to one operator
(*g*) The materials-handling equipment in a factory area
(*h*) A plumbing shop
(*i*) A job shop producing custom orders
(*j*) A secretarial typing pool

15.4-1. Suppose that a queueing system has two servers, an exponential interarrival time distribution with a mean of 2 hours, and an exponential service-time distribution with a mean of 2 hours for each server. Furthermore, a customer has just arrived at 12:00 noon.

(*a*) What is the probability that the next arrival will come (*i*) before 1:00 p.m., (*ii*) between 1:00 and 2:00 p.m., and (*iii*) after 2:00 p.m.?
(*b*) Suppose that no additional customers arrive before 1:00 p.m. Now what is the probability that the next arrival will come between 1:00 and 2:00 p.m.?
(*c*) What is the probability that the number of arrivals between 1:00 and 2:00 p.m. will be (*i*) 0, (*ii*) 1, and (*iii*) 2 or more?

(*d*) Suppose that both servers are serving customers at 1:00 p.m. What is the probability that *neither* customer will have service completed (*i*) before 2:00 p.m., (*ii*) before 1:10 p.m., and (*iii*) before 1:01 p.m.?

15.4-2.* The jobs to be performed on a particular machine arrive according to a Poisson input process with a mean rate of 2 per hour. Suppose that the machine breaks down and will require 1 hour to be repaired. What is the probability that the number of new jobs that will arrive during this time is (*a*) 0, (*b*) 2, and (*c*) 5 or more?

15.4-3. The time required by a mechanic to repair a machine has an exponential distribution with a mean of 4 hours. However, a special tool would reduce this mean to 2 hours. If the mechanic repairs a machine in less than 2 hours, he is paid $100; otherwise, he is paid $80. Determine the mechanic's expected increase in pay per machine repaired if he uses the special tool.

15.4-4. A three-server queueing system has a controlled arrival process that provides customers in time to keep the servers continuously busy. Service times have an exponential distribution with mean 0.5.

You observe the queueing system starting up with all three servers beginning service at time $t = 0$. You then note that the first completion occurs at time $t = 1$. Given this information, determine the expected amount of time after $t = 1$ until the next service completion occurs.

15.4-5. A queueing system has three servers with expected service times of 20, 15, and 10 minutes. The service times have an exponential distribution. Each server has been busy with a current customer for 5 minutes. Determine the expected remaining time until the next service completion.

15.4-6. Consider a queueing system with two types of customers. Type 1 customers arrive according to a Poisson process with a mean rate of 5 per hour. Type 2 customers also arrive according to a Poisson process with a mean rate of 5 per hour. The system has two servers, both of whom serve both types of customers. For both types, service times have an exponential distibution with a mean of 10 minutes. Service is provided on a first-come-first-served basis.

(*a*) What is the probability distribution (including its mean) of the time between consecutive arrivals of customers of any type?

(*b*) When a particular type 2 customer arrives, she finds two type 1 customers there in the process of being served but no other customers in the system. What is the probability distribution (including its mean) of this type 2 customer's waiting time in the queue?

15.4-7. Consider a two-server queueing system where all service times are independent and identically distributed according to an exponential distribution with a mean of 10 minutes. When a particular customer arrives, he finds that both servers are busy and no one is waiting in the queue.

(*a*) What is the probability distribution (including its mean and standard deviation) of this customer's waiting time in the queue?

(*b*) Determine the expected value and standard deviation of this customer's waiting time in the system.

(*c*) Suppose that this customer still is waiting in the queue 5 minutes after his arrival. Given this information, how does this change the expected value and the standard deviation of this customer's total waiting time in the system from the answers obtained in part (*b*)?

15.4-8. A queueing system has two servers whose service times are independent random variables with an exponential distribution with a mean of 15 minutes. Customer X arrives when both servers are idle. Five minutes later, customer Y arrives while customer X still is being served. Another 10 minutes later, customer Z arrives and both customers X and Y are still being served. No other customers arrived during this 15-minute interval.

(*a*) What is the probability that customer X will complete service before customer Y?
(*b*) What is the probability that customer Z will complete service before customer X?
(*c*) What is the probability that customer Z will complete service before customer Y?
(*d*) Determine the cumulative distribution function of the waiting time in the system for customer X. Also find the mean and standard deviation.
(*e*) Repeat part (*d*) for customer Y.
(*f*) Determine the expected value and standard deviation of the waiting time in the system for customer Z.
(*g*) Determine the probability of exactly 2 more customers arriving during the next 15-minute interval.

15.4-9. For each of the following statements regarding service times modeled by the exponential distribution, label the statement as true or false and then justify your answer by referring to specific statements (with page citations) in the chapter.
(*a*) The expected value and variance of the service times are always equal.
(*b*) The exponential distribution always provides a good approximation of the actual service-time distribution when each customer requires the same service operations.
(*c*) At an s-server facility, $s > 1$, with exactly s customers already in the system, a new arrival would have an expected waiting time before entering service of $1/\mu$ time units, where μ is the mean service rate for each busy server.

15.5-1. Consider the birth-and-death process with all $\mu_n = 2$ ($n = 1, 2, \ldots$), $\lambda_0 = 3$, $\lambda_1 = 2$, $\lambda_2 = 1$, and $\lambda_n = 0$ for $n = 3, 4, \ldots$.
(*a*) Display the rate diagram.
(*b*) Calculate P_0, P_1, P_2, P_3, and P_n for $n = 4, 5, \ldots$.
(*c*) Calculate L, L_q, W, and W_q.

15.5-2. Consider a birth-and-death process with just three attainable states (0, 1, and 2), for which the steady-state probabilities are P_0, P_1, and P_2, respectively. The birth and death rates are summarized in the following table:

State	Birth Rate	Death Rate
0	1	—
1	1	2
2	0	2

(*a*) Construct the rate diagram for this birth-and-death process.
(*b*) Develop the balance equations.
(*c*) Solve these equations to find P_0, P_1, and P_2.
(*d*) Use the general formulas for the birth-and-death process to calculate P_0, P_1, and P_2. Also calculate L, L_q, W, and W_q.

15.5-3. Consider the birth-and-death process with the following mean rates. The birth rates are $\lambda_0 = 2$, $\lambda_1 = 3$, $\lambda_2 = 2$, $\lambda_3 = 1$, and $\lambda_n = 0$ for $n > 3$. The death rates are $\mu_1 = 3$, $\mu_2 = 4$, $\mu_3 = 1$, and $\mu_n = 2$ for $n > 3$.
(*a*) Construct the rate diagram for this birth-and-death process.
(*b*) Develop the balance equations.
(*c*) Solve these equations to find the steady-state probability distribution $P_0, P_1, \ldots$.
(*d*) Use the general formulas for the birth-and-death process to calculate $P_0, P_1, \ldots$. Also calculate L, L_q, W, and W_q.

15.5-4. Consider the birth-and-death process with all $\lambda_n = 2$ ($n = 0, 1, \ldots$), $\mu_1 = 2$, and $\mu_n = 4$ for $n = 2, 3, \ldots$.

(*a*) Display the rate diagram.
(*b*) Calculate P_0 and P_1. Then give a general expression for P_n in terms of P_0 for $n = 2, 3, \ldots$.
(*c*) If a two-server queueing system fits this process, what must be the mean arrival rate and the mean service rate for each server when it is busy serving customers?

15.5-5.* A service station has one gasoline pump. Cars wanting gasoline arrive according to a Poisson process at a mean rate of 15 per hour. However, if the pump already is being used, these potential customers may balk (drive to another service station). In particular, if there are n cars already at the service station, the probability that an arriving potential customer will balk is $n/3$ for $n = 1, 2, 3$. The time required to service a car has an exponential distribution with a mean of 4 minutes.
(*a*) Construct the rate diagram for this queueing system.
(*b*) Develop the balance equations.
(*c*) Solve these equations to find the steady-state probability distribution of the number of cars at the station. Verify that this solution is the same as that given by the general solution for the birth-and death process.
(*d*) Find the expected waiting time (including service) for those cars that stay.

15.5-6. A maintenance person has the job of keeping two machines in working order. The amount of time that a machine works before breaking down has an exponential distribution with a mean of 10 hours. The time then spent by the maintenance person to repair the machine has an exponential distribution with a mean of 8 hours.
(*a*) Show that this process fits the birth-and-death process by defining the states, specifying the values of the λ_n and μ_n, and then constructing the rate diagram.
(*b*) Calculate the P_n.
(*c*) Calculate L, L_q, W, and W_q.
(*d*) Determine the proportion of time that the maintenance person is busy.
(*e*) Determine the proportion of time that any given machine is working.
(*f*) Refer to the nearly identical example of a continuous time Markov chain given at the end of Sec. 15.8. Describe the relationship between continuous time Markov chains and the birth-and-death process that enables both to be applied to this same problem.

15.5-7. Consider a single-server queueing system where interarrival times have an exponential distribution with parameter λ and service times have an exponential distribution with parameter μ. In addition, customers *renege* (leave the queueing system without being served) if their waiting time in the queue grows too large. In particular, assume that the time each customer is willing to wait in the queue before reneging has an exponential distribution with a mean of $1/\theta$.
(*a*) Construct the rate diagram for this queueing system.
(*b*) Develop the balance equations.

15.5-8. Consider a single-server queueing system where some potential customers balk (refuse to enter the system) and some customers who enter the system later get inpatient and renege (leave without being served). Potential customers arrive according to a Poisson process with a mean rate of 4 per hour. An arriving potential customer who finds n customers already there will balk with the following probabilities:

$$P\{\text{balk} \mid n \text{ already there}\} = \begin{cases} 0 & \text{if} \quad n = 0, \\ \frac{1}{2} & \text{if} \quad n = 1, \\ \frac{3}{4} & \text{if} \quad n = 2, \\ 1 & \text{if} \quad n = 3. \end{cases}$$

Service times have an exponential distribution with a mean of 1 hour.

A customer already in service never reneges, but the customers in the queue may renege. In particular, the remaining time that the customer at the front of the queue is willing to wait in the queue before reneging has an exponential distribution with a mean of 1 hour. For a customer in the second position in the queue, the time that she or he is willing to wait in this position before reneging has an exponential distribution with a mean of $\frac{1}{2}$ hour.

(*a*) Construct the rate diagram for this queueing system.
(*b*) Obtain the steady-state distribution of the number of customers in the system.
(*c*) Find the expected fraction of arriving potential customers who are lost due to balking.
(*d*) Find L_q and L.

15.5-9.* A certain small grocery store has a single checkout stand with a full-time cashier. Customers arrive at the stand ''randomly'' (i.e., a Poisson input process) at a mean rate of 30 per hour. When there is only one customer at the stand, he is processed by the cashier alone, with an expected service time 1.5 minutes. However, the stock boy has been given standard instructions that whenever there is more than one customer at the stand, he is to help the cashier by bagging the groceries. This help reduces the expected time required to process a customer to 1 minute. In both cases, the service-time distribution is exponential.

(*a*) Construct the rate diagram for this queueing system.
(*b*) What is the steady-state probability distribution of the number of customers at the checkout stand?
(*c*) Derive L for this system. (*Hint*: Refer to the derivation of L for the *M/M/*1 model at the beginning of Sec. 15.6.) Use this information to determine L_q, W, and W_q.

15.5-10. A department has one word-processing operator. Documents produced in the department are delivered for word processing according to a Poisson process with an expected interarrival time of 20 minutes. When the operator has just one document to process, the expected service time is 15 minutes. When she has more than one document, then editing assistance that is available reduces the expected service time for each document to 10 minutes. In both cases, the service times have an exponential distribution.

(*a*) Construct the rate diagram for this queueing system.
(*b*) Find the steady-state distribution of the number of documents that the operator has received but not yet completed.
(*c*) Derive L for this system. (*Hint*: Refer to the derivation of L for the *M/M/*1 model at the beginning of Sec. 15.6.) Use this information to determine L_q, W, and W_q.

15.5-11. Consider a self-service model in which the customer is also the server. Note that this corresponds to having an infinite number of servers available. Customers arrive according to a Poisson process with parameter λ, and service times have an exponential distribution with parameter μ.

(*a*) Find L_q, and W_q.
(*b*) Construct the rate diagram for this queueing system.
(*c*) Use the balance equations to find the expression for P_n in terms of P_0.
(*d*) Find P_0.
(*e*) Find L and W.

15.5-12. Customers arrive at a queueing system according to a Poisson process with a mean arrival rate of 2 customers per minute. The service time has an exponential distribution with a mean of 1 minute. An unlimited number of servers are available as needed so customers never wait for service to begin. Calculate the steady-state probability that exactly 1 customer is in the system.

15.5-13. Suppose that a single-server queueing system fits all the assumptions of the birth-and-death process *except* that customers always arrive in *pairs*. The mean arrival rate is 2 pairs per hour (4 customers per hour), and the mean service rate (when the server is busy) is 5 customers per hour.

(*a*) Construct the rate diagram for this queueing system.
(*b*) Develop the balance equations.

(*c*) For comparison purposes, display the rate diagram for the corresponding queueing system that completely fits the birth-and-death process, i.e., where customers arrive *individually* at a mean rate of 4 per hour.

15.5-14. The Copy Shop is open 5 days per week for copying materials that are brought to the shop. It has three identical copying machines; but only two operators are kept on duty to run the machines, so the third machine is a spare that is used only when one of the other machines breaks down. When a machine is being used, the time until it breaks down has an exponential distribution with a mean of 2 weeks. If one machine breaks down while the other two are operational, a service representative is called in to repair it, in which case the total time from the breakdown until the repair is completed has an exponential distribution with a mean of 0.2 week. However, if a second machine breaks down before the first one has been repaired, the third machine is shut off while the two operators work together to repair this second machine quickly, in which case its repair time has an exponential distribution with a mean of only $\frac{1}{15}$ week. If the service representative finishes repairing the first machine before the two operators complete the repair of the second, the operators go back to running the two operational machines while the representative finishes the second repair, in which case the remaining repair time has an exponential distribution with a mean of 0.2 week.

(*a*) Letting the state of the system be the number of machines not working, construct the rate diagram for this queueing system.
(*b*) Find the steady-state distribution of the number of machines not working.
(*c*) What is the expected number of operators available for copying?

15.5-15. Consider a single-server queueing system with a finite queue that can hold a maximum of 2 customers *excluding* any being served. The server can provide *batch service* to 2 customers simultaneously, where the service time has an exponential distribution with a mean of 1 unit of time regardless of the number being served. Whenever the queue is not full, customers arrive individually according to a Poisson process at a mean rate of 1 per unit of time.

(*a*) Assume that the server *must* serve 2 customers simultaneously. Thus, if the server is idle when only 1 customer is in the system, the server must wait for another arrival before beginning service. Formulate the queueing model as a continuous time Markov chain by defining the states and then constructing the rate diagram. Give the balance equations, but do not solve further.
(*b*) Now assume that the batch size for a service is 2 only if 2 customers are in the queue when the server finishes the preceding service. Thus, if the server is idle when only 1 customer is in the system, the server must serve this single customer, and any subsequent arrivals must wait in the queue until service is completed for this customer. Formulate the resulting queueing model as a continuous time Markov chain by defining the states and then constructing the rate diagram. Give the balance equations, but do not solve further.

15.5-16. Consider a queueing system that has two classes of customers, two clerks providing service, and no *queue*. Potential customers from each class arrive according to a Poisson process, with a mean arrival rate of 10 customers per hour for class 1 and 5 customers per hour for class 2, but these arrivals are lost to the system if they cannot immediately enter service.

Each customer of class 1 that enters the system will receive service from either one of the clerks who is free, where the service times have an exponential distribution with a mean of 5 minutes.

Each customer of class 2 that enters the system requires the *simultaneous use of both clerks* (the two clerks work together as a single server), where the service times have an exponential distribution with a mean of 5 minutes. Thus, an arriving customer of this kind would be lost to the system unless both clerks were free to begin service immediately.

(*a*) Formulate the queueing model as a continuous time Markov chain by defining the states and constructing the rate diagram.
(*b*) Now describe how the formulation in part (*a*) can be fitted into the format of the birth-and-death process.

(*c*) Use the results for the birth-and-death process to calculate the steady-state joint distribution of the number of customers of each class in the system.

(*d*) For each of the two classes of customers, what is the expected fraction of arrivals who are unable to enter the system?

15.6-1. Customers arrive at a single-server queueing system according to a Poisson process at a mean rate of 10 per hour. If the server works continuously, the number of customers that can be served in an hour has a Poisson distribution with a mean of 15. Determine the proportion of time during which no one is waiting to be served.

15.6-2. Consider the *M/M*/1 model, with $\lambda < \mu$.

(*a*) Determine the steady-state probability that a customer's actual waiting time in the system is longer than the expected waiting time in the system, that is, $P\{\mathcal{W} > W\}$.

(*b*) Determine the steady-state probability that a customer's actual waiting time in the queue is longer than the expected waiting in the queue, that is, $P\{\mathcal{W}_q > W_q\}$.

15.6-3. Verify the following relationships for an *M/M*/1 queueing system:

$$\lambda = \frac{(1 - P_0)^2}{W_q P_0}, \qquad \mu = \frac{1 - P_0}{W_q P_0}.$$

15.6-4.* Jobs arrive at a particular work center according to a Poisson input process at a mean rate of 2 per day, and the operation time has an exponential distribution with a mean of $\frac{1}{4}$ day. Enough in-process storage space is provided at the work center to accommodate 3 jobs in addition to the one being processed, whereas excess jobs are stored temporarily in a less convenient location. For what proportion of the time will this storage space at the work center be adequate to accommodate all waiting jobs?

15.6-5. It is necessary to determine how much in-process storage space to allocate to a particular work center in a new factory. Jobs arrive at this work center according to a Poisson process with a mean rate of 3 per hour, and the time required to perform the necessary work has an exponential distribution with a mean of 0.25 hour. Whenever the waiting jobs require more in-process storage space than has been allocated, the excess jobs are stored temporarily in a less convenient location. If each job requires 1 square foot of floor space while it is in in-process storage at the work center, how much space must be provided to accommodate all waiting jobs (*a*) 50 percent of the time, (*b*) 90 percent of the time, and (*c*) 99 percent of the time? Derive an analytical expression to answer these three questions, and then use your OR Courseware to verify your answers. *Hint*: The sum of a geometric series is

$$\sum_{n=0}^{N} x^n = \frac{1 - x^{N+1}}{1 - x}.$$

15.6-6. Consider the following statements about an *M/M*/1 queueing system and its utilization factor ρ. Lable each of the statements as true or false, and then justify your answer.

(*a*) The probability that a customer has to wait before service begins is proportional to ρ.

(*b*) The expected number of customers in the system is proportional to ρ.

(*c*) If ρ has been increased from $\rho = 0.9$ to $\rho = 0.99$, the effect of any further increase in ρ on L, L_q, W, and W_q will be relatively small as long as $\rho < 1$.

15.6-7. Customers arrive at a single-server queueing system in accordance with a Poisson process with an expected interarrival time of 25 minutes. Service times have an exponential distribution with a mean of 30 minutes.

Label each of the following statements about this system as true or false, and then justify your answer.

(*a*) The server definitely will be busy forever after the first customer arrives.

(*b*) The queue will grow without bound.

(*c*) If a second server with the same service-time distribution is added, the system can reach a steady-state condition.

15.6-8. For each of the following statements about an $M/M/1$ queueing system, label the statement as true or false and then justify your answer by referring to specific statements (with page citations) in the chapter.

(*a*) The waiting time in the system has an exponential distribution.
(*b*) The waiting time in the queue has an exponential distribution.
(*c*) The conditional waiting time in the system, given the number of customers already in the system, has an Erlang (gamma) distribution.

15.6-9. Consider the $M/M/s$ model.

A (*a*) Suppose there is one server and the expected service time is exactly 1 minute. Compare L for the cases where the mean arrival rate is 0.5, 0.9, and 0.99 customer per minute, respectively. Do the same for L_q, W, W_q, and $P\{\mathcal{W} > 5\}$. What conclusions do you draw about the impact of increasing the utilization factor ρ from small values (e.g., $\rho = 0.5$) to fairly large values (e.g., $\rho = 0.9$) and then to even larger values very close to 1 (e.g., $\rho = 0.99$)?

A* (*b*) Now suppose there are two servers and the expected service time is exactly 2 minutes. Follow the instructions for part (*a*).

A* **15.6-10.** Customers arrive at a queueing system according to a Poisson process at a mean rate of 10 per hour. Service times have an exponential distribution with a mean of 5 minutes.

Print out the measures of performance provided by your OR Courseware for this system (with $t = 10$ and $t = 0$, respectively, for the last two lines) when the number of servers is 1, 2, 3, 4, and 5. Then, for each of the following possible criteria for a satisfactory level of service (where the unit of time is 1 minute), use the printed results to determine how many servers are needed to satisfy this criterion.

(*a*) $L_q \leq 0.25$
(*b*) $L \leq 0.9$
(*c*) $W_q \leq 0.1$
(*d*) $W \leq 6$
(*e*) $P\{\mathcal{W}_q > 0\} \leq 0.01$
(*f*) $P\{\mathcal{W} > 10\} \leq 0.2$
(*g*) $\sum_{n=0}^{s} P_n \geq 0.95$

15.6-11. A bank employs 4 tellers to serve its customers. Customers arrive according to a Poisson process at a mean rate of 3 per minute. If a customer finds all tellers busy, he joins a queue that is serviced by all tellers; i.e., there are no lines in front of each teller, but rather one line waiting for the first available teller. The transaction time between the teller and customer has an exponential distribution with a mean of 1 minute.

(*a*) Construct the rate diagram for this queueing system.

A* (*b*) Find the steady-state probability distribution of the number of customers in the bank.

A* (*c*) Find L_q, W_q, W, and L.

15.6-12. Airplanes arrive for takeoff at the runway of an airport according to a Poisson process at a mean rate of 20 per hour. The time required for an airplane to take off has an exponential distribution with a mean of 2 minutes, and this process must be completed before the next airplane can begin takeoff.

Because a brief thunderstorm has just begun, all airplanes which have not commenced takeoff have just been grounded temporarily. However, airplanes continue to arrive at the runway during the thunderstorm to await its end.

Assuming steady-state operation before the thunderstorm, determine the expected number of airplanes that will be waiting to take off at the end of the thunderstorm if it lasts 30 minutes.

15.6-13. A gas station with only one gas pump employs the following policy: If a customer has to wait, the price is \$1 per gallon; if she does not have to wait, the price is \$1.20 per gallon. Customers arrive according to a Poisson process with a mean rate of 15 per hour. Service times at the pump have an exponential distribution with a mean of 3 minutes. Arriving customers always wait until they can eventually buy gasoline. Determine the expected price of gasoline per gallon.

15.6-14. You are given an *M/M/*1 queueing system with mean arrival rate λ and mean service rate μ. An arriving customer receives n dollars if customers are already in the system. Determine the expected cost in dollars per customer.

15.6-15. Section 15.6 gives the following equations for the *M/M/*1 model:

(1) $$P\{\mathcal{W} > t\} = \sum_{n=0}^{\infty} P_n P\{S_{n+1} > t\}.$$

(2) $$P\{\mathcal{W} > t\} = e^{-\mu(1-\rho)t}.$$

Show that Eq. (1) reduces algebraically to Eq. (2). (*Hint:* Use differentiation, algebra, and integration.)

15.6-16. Derive W_q directly for the following cases by developing and reducing an expression analogous to Eq. (1) in Prob. 15.6-15. (*Hint:* Use the *conditional* expected waiting time in the queue, given that a random arrival finds n customers already in the system.)

(*a*) The *M/M/*1 model
(*b*) The *M/M/s* model

A* **15.6-17.** Consider an *M/M/*2 queueing system with $\lambda = 4$ and $\mu = 3$. Determine the mean rate at which service completions occur during the periods when no customers are waiting in the queue.

A* **15.6-18.** You are given an *M/M/*2 queueing system with $\lambda = 4$ per hour and $\mu = 6$ per hour. Determine the probability that an arriving customer will wait more than 30 minutes in the queue, given that at least 2 customers are already in the system.

15.6-19. Consider an *M/M/*2 queueing system where $\lambda = 10$ and $\mu = 7.5$. It is estimated that the mean arrival rate will decline to $\lambda = 5$ next year, at which time the number of servers will be reduced to one.

A* (*a*) Assuming that μ will continue to be 7.5 for next year's *M/M/*1 queueing system, determine L, L_q, W, and W_q for both the current system and next year's system. For each of these four measures of performance, which system yields the smaller value?
(*b*) Now assume that μ will be adjustable when the number of servers is reduced to one. Solve algebraically for the value of μ that would yield the same value of W as for the current system.
(*c*) Repeat part (*b*) with W_q instead of W.

15.6-20. In the Blue Chip Life Insurance Company, the deposit and withdrawal functions associated with a certain investment product are separated between two clerks. Deposit slips arrive randomly at Clara's desk at a mean rate of 16 per hour. Withdrawal slips arrive randomly at Clarence's desk at a mean rate of 14 per hour. The time required to process either transaction has an exponential distribution with a mean of 3 minutes. To reduce the expected waiting time in the system for both deposit slips and withdrawal slips, the actuarial department has made the following recommendations: (1) Train each clerk to handle both deposits and withdrawals, and (2) put both deposit and withdrawal slips into a single queue that is accessed by both clerks.

A (*a*) Determine the expected waiting time in the system under current procedures for each type of slip. Then combine these results to calculate the expected waiting time in the system for a random arrival of either type of slip.
A* (*b*) If the recommendations are adopted, determine the expected waiting time in the system for arriving slips.
A* (*c*) Now suppose that adopting the recommendations would result in a slight increase in the expected processing time. Use your OR Courseware to determine by trial and

error the expected processing time (within 0.001 hour) that would cause the expected waiting time in the system for a random arrival to be essentially the same under current procedures and under the recommendations.

15.6-21. You are given an $M/M/1$ queueing system in which the expected waiting time and expected number in the system are 120 minutes and 8 customers, respectively. Determine the probability that a customer's service time exceeds 20 minutes.

15.6-22. Consider a generalization of the $M/M/1$ model where the server needs to "warm up" at the beginning of a busy period and so serves the first customer of a busy period at a slower rate than other customers. In particular, if an arriving customer finds the server idle, that customer experiences a service time that has an exponential distribution with parameter μ_1. However, if an arriving customer finds the server busy, that customer joins the queue and subsequently experiences a service time that has an exponential distribution with parameter μ_2, where $\mu_1 < \mu_2$. Customers arrive according to a Poisson process with rate λ.

(*a*) Formulate this model as a continuous time Markov chain by defining the states and constructing the rate diagram accordingly.
(*b*) Develop the balance equations.
(*c*) Suppose that numerical values are specified for μ_1, μ_2, and λ and that $\lambda < \mu_2$ (so that a steady-state distribution exists). Since this model has an infinite number of states, the steady-state distribution is the simultaneous solution of an infinite number of balance equations (plus the equation specifying that the sum of the probabilities equals 1). Suppose that you are unable to obtain this solution analytically, so you wish to use a computer to solve the model numerically. Considering that it is impossible to solve an infinite number of equations numerically, briefly describe what still can be done with these equations to obtain an approximation of the steady-state distribution. Under what circumstances will this approximation be essentially exact?
(*d*) Given that the steady-state distribution has been obtained, give explicit expressions for calculating L, L_q, W, and W_q.
(*e*) Given this steady-state distribution, develop an expression for $P\{\mathcal{W} > t\}$ that is analogous to Eq. (1) in Prob. 15.6-15.

15.6-23. For each of the following models, write the balance equations and show that they are satisfied by the solution given in Sec. 15.6 for the steady-state distribution of the number of customers in the system.

(*a*) The $M/M/1$ model
(*b*) The finite queue variation of the $M/M/1$ model, with $K = 2$
(*c*) The finite calling population variation of the $M/M/1$ model, with $N = 2$

A* **15.6-24.** Consider a telephone system with three lines. Calls arrive according to a Poisson process at a mean rate of 6 per hour. The duration of each call has an exponential distribution with a mean of 15 minutes. If all lines are busy, calls will be put on hold until a line becomes available.

(*a*) Print out the measures of performance provided by your OR Courseware for this queueing system (with $t = 1$ hour and $t = 0$, respectively, for the last two lines).
(*b*) Use the printed result giving $P\{\mathcal{W}_q > 0\}$ to identify the steady-state probability that a call will be answered immediately (not put on hold). Then verify this probability by using the printed results for the P_n.
(*c*) Use the printed results to identify the steady-state probability distribution of the number of calls on hold.
(*d*) Print out the new measures of performance if arriving calls are lost whenever all lines are busy. Use these results to identify the steady-state probability that an arriving call is lost.

15.6-25. Reconsider the specific birth-and-death process described in Prob. 15.5-1.

(*a*) Identify a queueing model (and its parameter values) in Sec. 15.6 that fits this process.

A* (*b*) Use the corresponding routine in your OR Courseware to obtain the answers for parts (*b*) and (*c*) of Prob. 15.5-1.

15.6-26. An airline ticket office has two ticket agents answering incoming phone calls for flight reservations. In addition, one caller can be put on hold until one of the agents is available to take the call. If all three phone lines (both agent lines and the hold line) are busy, a potential customer gets a busy signal, and it is assumed that the call goes to another ticket office and that the business is lost. The calls and attempted calls occur randomly (i.e., according to a Poisson process) at a mean rate of 15 per hour. The length of a telephone conversation has an exponential distribution with a mean of 4 minutes.

(*a*) Construct the rate diagram for this queueing system.

A* (*b*) Find the steady-state probability that
- (i) a caller will get to talk to an agent immediately,
- (ii) the caller will be put on hold, and
- (iii) the caller will get a busy signal.

A **15.6-27.*** Plans are being made to open a small car-wash operation, and the owner must decide how much space to provide for waiting cars. It is estimated that customers will arrive randomly (i.e., a Poisson input process) with a mean rate of 1 every 4 minutes, unless the waiting area is full, in which case the arriving customers will take their cars elsewhere. The time that can be attributed to washing one car has an exponential distribution with a mean of 3 minutes. Compare the expected fraction of potential customers that will be *lost* because of inadequate waiting space if (*a*) 0 spaces (not including the car being washed), (*b*) 2 spaces, and (*c*) 4 spaces were provided.

15.6-28. Consider the finite queue variation of the $M/M/s$ model. Derive the expression for L_q given in Sec. 15.6 for this model.

15.6-29. For the finite queue variation of the $M/M/1$ model, develop an expression analogous to Eq. (1) in Prob. 15.6-15 for the following probabilities.

(*a*) $P\{\mathcal{W} > t\}$

(*b*) $\{\mathcal{W}_q > t\}$

[*Hint:* Arrivals can occur only when the system is not full, so the probability that a random arrival finds n customers already there is $P_n/(1 - P_K)$.]

A* **15.6-30.** Suppose that one repair technician has been assigned the responsibility of maintaining three machines. For each machine, the probability distribution of the running time before a breakdown is exponential, with a mean of 9 hours. The repair time also has an exponential distribution, with a mean of 2 hours.

(*a*) Obtain the steady-state probability distribution and the expected number of machines that are not running.

(*b*) As a crude approximation, assume that the calling population is infinite, so that the input process is Poisson with a mean arrival rate of 3 every 9 hours. Compare the result from part (*a*) with that obtained by making this approximation while using (i) the corresponding infinite queue model and (ii) the corresponding finite queue model.

(*c*) Now suppose that a second maintenance person is available whenever more than one of these three machines requires repair. Obtain the information specified in part (*a*).

A* **15.6-31.*** Plans are currently being developed for a new factory. One department has been allocated a large number of automatic machines of a certain type, and we need to determine how many machines should be assigned to each operator for servicing (loading, unloading, adjusting, setup, and so on). For the purpose of this analysis, the following information has been provided.

The running time (time between completing service and the machine's requiring service again) of each machine has an exponential distribution, with a mean of 150 minutes. The service time has an exponential distribution, with a mean of 15 minutes. Each operator attends to her own machine; she does not give help to or receive help from other operators. For the department to achieve the required production rate, the machines must be running at least 89 percent of the time on average.

(*a*) What is the maximum number of machines that can be assigned to an operator while still achieving the required production rate?

(*b*) Given that the maximum number found in part (*a*) is assigned to each operator, what is the expected fraction of time that the operators will be busy servicing machines?

15.6-32. A shop contains three identical machines that are subject to failure of a certain kind. Therefore, a maintenance system is provided to perform the maintenance operation (recharging) required by a failed machine. The time required by each operation has an exponential distribution with a mean of 30 minutes. However, with probability $\frac{1}{3}$, the operation must be performed a second time (with the same distribution of time) in order to bring the failed machine back to a satisfactory operational state. The maintenance system works on only one failed machine at a time, performing all the operations (one or two) required by that machine, on a first-come-first-served basis. After a machine is repaired, the time until its next failure has an exponential distribution with a mean of 3 hours.

(*a*) How should the states of the system be defined in order to formulate this queueing system as a continuous time Markov chain? (*Hint:* Given that a first operation is being performed on a failed machine, completing this operation *successfully* and completing it *unsuccessfully* are two separate events of interest. Then use Property 6 regarding disaggregation for the exponential distribution.)

(*b*) Construct the corresponding rate diagram.

(*c*) Develop the balance equations.

15.6-33. Consider a single-server queueing system. It has been observed that (1) this server seems to speed up as the number of customers in the system increases and (2) the pattern of acceleration seems to fit the state-dependent model presented at the end of Sec. 15.6. Furthermore, it is estimated that the expected service time is 8 minutes when there is only 1 customer in the system. Determine the pressure coefficient c for this model for the following cases.

(*a*) The expected service time is estimated to be 4 minutes when there are 4 customers in the system.

(*b*) The expected service time is estimated to be 5 minutes when there are 4 customers in the system.

A* **15.6-34.** For the state-dependent model presented at the end of Sec. 15.6, show the effect of the *pressure coefficient c* by using Fig. 15.10 to construct a table giving the ratio (expressed as a decimal number) of L for this model to L for the corresponding $M/M/s$ model (i.e., with $c = 0$). Tabulate these ratios for $\lambda_0/(s\mu_1) = 0.5, 0.9, 0.99$ when $c = 0.2, 0.4, 0.6$ and $s = 1, 2$.

15.7-1.* Consider the $M/G/1$ model.

(*a*) Compare the expected waiting time in the queue if the service-time distribution is (i) exponential, (ii) constant, and (iii) Erlang with the amount of variation (i.e, the standard deviation) halfway between the constant and exponential cases.

(*b*) What is the effect on the expected waiting time in the queue and on the expected queue length if both λ and μ are doubled and the scale of the service-time distribution is changed accordingly?

15.7-2. Consider the following statements about an $M/G/1$ queueing system, where σ^2 is the variance of service times. Label each statement as true or false, and then justify your answer.

(*a*) Increasing σ^2 (with fixed λ and μ) will increase L_q and L, but will not change W_q and W.

(*b*) When the choice is between a tortoise (small μ and σ^2) and a hare (large μ and σ^2) to be the server, the tortoise always wins by providing a smaller L_q.

(*c*) With λ and μ fixed, the value of L_q with an exponential service-time distribution is twice as large as with constant service times.

(*d*) Among all possible service-time distributions (with λ and μ fixed), the exponential distribution yields the largest value of L_q.

15.7-3. Consider a queueing system with a Poisson input, where the server must perform two distinguishable tasks in sequence for each customer, so the total service time is the sum of the two task times (which are statistically independent).

(*a*) Suppose that the first task time has an exponential distribution with a mean of 3 minutes and that the second task time has an Erlang distribution with a mean of 9 minutes and with the shape parameter $k = 3$. Which queueing theory model should be used to represent this system?

(*b*) Suppose that part (*a*) is modified so that the first task time also has an Erlang distribution with the shape parameter $k = 3$ (but with the mean still equal to 3 minutes). Which queueing theory model should be used to represent this system?

15.7-4.* An airline maintenance base has facilities for overhauling only one airplane engine at a time. Therefore, to return the airplanes to use as soon as possible, the policy has been to stagger the overhauling of the four engines of each airplane. In other words, only one engine is overhauled each time an airplane comes into the shop. Under this policy, airplanes have arrived according to a Poisson process at a mean rate of 1 per day. The time required for an engine overhaul (once work has begun) has an exponential distribution with a mean of $\frac{1}{2}$ day.

A proposal has been made to change the policy so that all four engines are overhauled consecutively each time an airplane comes into the shop. Although this would quadruple the expected service time, each plane would need to come into the shop only one-fourth as often.

Use queueing theory to compare the two alternatives on a meaningful basis.

15.7-5. Consider a shoe repair shop with a single employee. Pairs of shoes are brought in to be repaired (on a first-come-first-served basis) according to a Poisson process at a mean rate of 1 pair per hour. The time required to repair each individual shoe has an exponential distribution with a mean of 15 minutes.

(*a*) Consider the formulation of this queueing system where the individual shoes (not pairs of shoes) are considered to be the customers. For this formulation, construct the rate diagram and develop the balance equations, but do not solve further.

(*b*) Now consider the formulation of this queueing system where the pairs of shoes are considered to be the customers. Identify the specific queueing model that fits this formulation, and then use the available results for this model to calculate the expected waiting time W. (Interpret W to be the expected waiting time until both shoes in a pair have been repaired.)

15.7-6. Airplanes arrive at a certain maintenance base for overhaul according to a Poisson process with parameter$\lambda = 3$ (per week). Service times have an Erlang distribution with parameters $\mu = 4$ (per week) and $k = 2$. Only one airplane can be overhauled at a time.

(*a*) How should the states of the system be defined in order to formulate the queueing model as a continuous time Markov chain?

(*b*) Construct the corresponding rate diagram.

(*c*) What are L_q, L, W_q, and W for this queueing system? [*Hint:* Do not use parts (*a*) and (*b*).]

15.7-7. A company currently has *two* tool cribs, each with a *single* clerk, in its manufacturing area. One tool crib handles only the tools for the heavy machinery; the second one handles all other tools. However, for each crib the mechanics arrive to obtain tools at a mean rate of 24 per hour, and the expected service time is 2 minutes.

Because of complaints that the mechanics coming to the tool cribs have to wait too long, it has been proposed that the two tool cribs be combined so that either clerk can handle either kind of tool as the demand arises. It is believed that the mean arrival rate to the combined two-clerk tool crib would double to 48 per hour and that the expected service time would continue to be 2 minutes. However, information is not available on the *form* of the probability distributions for interarrival and service times, so it is not clear which queueing model would be most appropriate.

Compare the status quo and the proposal with respect to the total expected number of mechanics at the tool crib(s) and the expected waiting time (including service) for each mechanic. Do this by tabulating these data for the four queueing models considered in Figs. 15.7, 15.11, 15.13, and 15.14 (use $k = 2$ when an Erlang distribution is appropriate).

15.7-8.* Consider a single-server queueing system with a Poisson input, Erlang service times, and a finite queue. In particular, suppose that $k = 2$, the mean arrival rate is 2 customers per hour, the expected service time is 0.25 hour, and the maximum permissible number of customers in the system is 2. This system can be formulated as a continuous time Markov chain by dividing each service time into two consecutive phases, each having an exponential distribution with a mean of 0.125 hour, and then defining the state of the system as (n, p), where n is the number of customers in the system ($n = 0, 1, 2$), and p indicates the phase of the customer being served ($p = 0, 1, 2$, where $p = 0$ means that no customer is being served).

(*a*) Construct the corresponding rate diagram. Write the balance equations, and then use these equations to solve for the steady-state distribution of the state of this Markov chain.

(*b*) Use the steady-state distribution obtained in part (*a*) to identify the steady-state distribution of the number of customers in the system (P_0, P_1, P_2) and the steady-state expected number of customers in the system (L).

A (*c*) Compare the results from part (*b*) with the corresponding results when the service-time distribution is exponential.

15.7-9. Consider the $E_2/M/1$ model with $\lambda = 4$ and $\mu = 5$. This model can be formulated as a continuous time Markov chain by dividing each interarrival time into two consecutive phases, each having an exponential distribution with a mean of $1/(2\lambda) = 0.125$, and then defining the state of the system as (n, p), where n is the number of customers in the system ($n = 0, 1, 2, \ldots$) and p indicates the phase of the *next* arrival (not yet in the system) ($p = 1, 2$). Construct the corresponding rate diagram (but do not solve further).

15.7-10. A company has one repair technician to keep a large group of machines in running order. Treating this group as an infinite calling population, individual breakdowns occur according to a Poisson process at a mean rate of 1 per hour. For each breakdown, the probability is 0.9 that only a minor repair is needed, in which case the repair time has an exponential distribution with a mean of $\frac{1}{2}$ hour. Otherwise, a major repair is needed, in which case the repair time has an exponential distribution with a mean of 5 hours. Because both of these *conditional* distributions are exponential, the *unconditional* (combined) distribution of repair times is *hyperexponential.*

(*a*) Compute the mean and standard deviation of this hyperexponential distribution. [*Hint:* Use the general relationships from probability theory that, for any random variable X and any pair of mutually exclusive events E_1 and E_2, $E(X) = E(X \mid E_1)P(E_1) + E(X \mid E_2)P(E_2)$ and $\text{var}(X) = E(X^2) - E(X)^2$.] Compare this standard deviation with that for an exponential distribution having this mean.

(*b*) What are P_0, L_q, L, W_q, and W for this queueing system?

(*c*) What is the conditional value of W, given that the machine involved requires major repair? A minor repair? What is the division of L between machines requiring the two types of repairs? (*Hint:* Little's formula still applies for the individual categories of machines.)

(*d*) How should the states of the system be defined in order to formulate this queueing system as a continuous time Markov chain? (*Hint:* Consider what additional information must be given, besides the number of machines down, for the conditional distribution of the time remaining until the next event of each kind to be exponential.)

(*e*) Construct the corresponding rate diagram.

15.7-11. Consider the finite queue variation of the $M/G/1$ model, where K is the maximum number of customers allowed in the system. For $n = 1, 2, \ldots$, let the random variable X_n be the number of customers in the system at the moment t_n when the nth customer has just finished being served. (Do not count the departing customer.) The times $\{t_1, t_2, \ldots\}$ are called *regeneration points*. Furthermore, $\{X_n\}$ $(n = 1, 2, \ldots)$ is a discrete time Markov chain and is known as an *embedded Markov chain*. Embedded Markov chains are useful for studying the properties of continuous time stochastic processes such as for an $M/G/1$ model.

Now consider the particular special case where $K = 4$, the service time of successive customers is a fixed constant, say, 10 minutes, and the mean arrival rate is 1 every 50 minutes. Therefore, $\{X_n\}$ is an embedded Markov chain with states 0, 1, 2, 3. (Because there are never more than 4 customers in the system, there can never be more than 3 in the system at a regeneration point.) Because the system is observed at successive departures, X_n can never decrease by more than 1. Furthermore, the probabilities of transitions that result in increases in X_n are obtained directly from the Poisson distribution.

(*a*) Find the one-step transition matrix for the embedded Markov chain. (*Hint:* In obtaining the transition probability from state 3 to state 3, use the probability of 1 or more arrivals rather than just 1 arrival, and similarly for other transitions to state 3.)

A* (*b*) Use the corresponding routine in the Markov chains area of your OR Courseware to find the steady-state probabilities for the number of customers in the system at regeneration points.

(*c*) Compute the expected number of customers in the system at regeneration points, and compare it to the value of L for the $M/D/1$ model (with $K = \infty$) in Sec. 15.7.

A* **15.8-1.** Consider the model with nonpreemptive priorities presented in Sec. 15.8. Suppose there are just two priority classes, with $\lambda_1 = 4$ and $\lambda_2 = 4$. In designing this queueing system, you are offered the choice between the following alternatives: (1) one fast server ($\mu = 10$) and (2) two slow servers ($\mu = 5$).

Compare these alternatives with the usual four mean measures of performance (W, L, W_q, L_q) for the individual priority classes (W_1, W_2, L_1, L_2, and so forth). Which alternative is preferred if your primary concern is the expected waiting time in the *system* for priority class 1 (W_1)? Which is preferred if your primary concern is the expected waiting time in the *queue* for priority class 1?

A* **15.8-2.*** A particular work center in a job shop can be represented as a single-server queueing system, where jobs arrive according to a Poisson process, with a mean rate of 8 per day. Although the arriving jobs are three distinct types, the time required to perform any of these jobs has the same exponential distribution, with a mean of 0.1 working day. The practice has been to work on arriving jobs on a first-come-first-served basis. However, it is important that jobs of type 1 not wait very long, whereas the wait is only moderately important for jobs of type 2 and is relatively unimportant for jobs of type 3. These three types arrive with a mean rate of 2, 4, and 2 per day, respectively. Because all three types have experienced rather long delays on average, it has been proposed that the jobs be selected according to an appropriate priority discipline instead.

Compare the expected waiting time (including service) for each of the three types of jobs if the queue discipline is (*a*) first-come-first-served, (*b*) nonpreemptive priority, and (*c*) preemptive priority.

15.8-3. Reconsider the County Hospital emergency room problem as analyzed in Sec. 15.8. Suppose that the definitions of the three categories of patients are tightened somewhat in order to move marginal cases into a lower category. Consequently, only 5 percent of the patients will quality as critical cases, 20 percent as serious cases, and 75 percent as stable cases. Develop a table showing the data presented in Table 15.4 for this revised problem.

15.8-4. Reconsider the queueing system described in Prob. 15.4-6. Suppose now that type 1 customers are more important than type 2 customers. If the queue discipline were changed from first-come-first-served to a priority system with type 1 customers being given nonpreemptive priority over type 2 customers, would this increase, decrease, or keep unchanged the expected total number of customers in the system?

(*a*) Determine the answer without any calculations, and then present the reasoning that led to your conclusion.

A* (*b*) Verify your conclusion in part (*a*) by finding the expected total number of customers in the system under each of these two queue disciplines.

15.8-5. Consider the queueing model with a preemptive priority queue discipline presented in Sec. 15.8. Suppose that $s = 1$, $N = 2$, and $\lambda_1 + \lambda_2 < \mu$; and let P_{ij} be the steady-state probability that there are i members of the higher-priority class and j members of the lower-priority class in the queueing system ($i = 0, 1, 2, \ldots; j = 0, 1, 2, \ldots$). Use a method analogous to that presented in Sec. 15.5 to derive a system of linear equations whose simultaneous solution is the P_{ij}. Do not actually obtain this solution.

15.9-1. Consider a queueing system with two servers, where the customers arrive from two different sources. From source 1, the customers always arrive 2 at a time, where the time between consecutive arrivals of pairs of customers has an exponential distribution with a mean of 20 minutes. Source 2 is itself a two-server queueing system, which has a Poisson input process with a mean rate of 7 customers per hour, and the service time from each of these two servers has an exponential distribution with a mean of 15 minutes. When a customer completes service at source 2, he or she immediately enters the queueing system under consideration for another type of service. In the latter queueing system, the queue discipline is preemptive priority where customers from source 1 always have preemptive priority over customers from source 2. However, service times are independent and identically distributed for both types of customers according to an exponential distribution with a mean of 6 minutes.

(*a*) First focus on the problem of deriving the steady-state distribution of *only* the number of source 1 customers in the queueing system under consideration. Using a continuous time Markov chain formulation, define the states and construct the rate diagram for most efficiently deriving this distribution (but do not actually derive it).

(*b*) Now focus on the problem of deriving the steady-state distribution of the *total* number of customers of both types in the queueing system under consideration. Using a continuous time Markov chain formulation, define the states and construct the rate diagram for most efficiently deriving this distribution (but do not actually derive it).

(*c*) Now focus on the problem of deriving the steady-state *joint* distribution of the number of customers of each type in the queueing system under consideration. Using a continuous time Markov chain formulation, define the states and construct the rate diagram for deriving this distribution (but do not actually derive it).

15.9-2. Consider a system of two infinite queues in series, where each of the two service facilities has a single server. All service times are independent and have an exponential distribution, with a mean of 3 minutes at facility 1 and 4 minutes at facility 2. Facility 1 has a Poisson input process with a mean rate of 10 per hour.

(*a*) Find the steady-state distribution of the number of customers at facility 1 and then at facility 2. Then show the product form solution for the *joint* distribution of the number at the respective facilities.

(*b*) What is the probability that both servers are idle?

(*c*) Find the expected *total* number of customers in the system and the expected *total* waiting time (including service times) for a customer.

15.9-3. Under the assumptions specified in Sec. 15.9 for a system of infinite queues in series, this kind of queueing network actually is a special case of a Jackson network. Demonstrate that this is true by describing this system as a Jackson network, including specifying the values of the a_j and the P_{ij}, given λ for this system.

15.9-4. Consider a Jackson network with three service facilities having the parameter values shown below.

Facility j	s_j	μ_j	a_j	p_{ij}		
				$i = 1$	$i = 2$	$i = 3$
$j = 1$	1	40	10	0	0.3	0.4
$j = 2$	1	50	15	0.5	0	0.5
$j = 3$	1	30	3	0.3	0.2	0

A* (*a*) Find the total arrival rate at each of the facilities.

(*b*) Find the steady-state distribution of the number of customers at facility 1, facility 2, and facility 3. Then show the product form solution for the joint distribution of the number at the respective facilities.

(*c*) What is the probability that all the facilities have empty queues (no customers waiting to begin service)?

(*d*) Find the expected total number of customers in the system.

(*e*) Find the expected total waiting time (including service times) for a customer.

CASE PROBLEM

CP15-1. One inspector has been assigned the full-time task of inspecting the output from a group of 10 identical machines. Jobs to be done by any one of the machines arrive according to a Poisson process at a mean rate of 70 per hour. The time required by a machine to perform each job has an exponential distribution with a mean of 6 minutes. Thus, whenever all 10 machines are busy, the jobs are being completed ready for inspection at a mean rate of 100 per hour. Unfortunately, the inspector is able to inspect them at a mean rate of only 80 per hour. (In particular, her inspection time has an Erlang distribution with a mean of 0.75 minute and a shape parameter $k = 25$.) This inspection rate has resulted in a substantial average amount of in-process inventory at the inspection station (i.e., the expected number of jobs waiting to complete inspection is fairly large), in addition to that already found at the group of machines.

Management feels that there is too much in-process inventory. The cost of this in-process inventory is estimated to be $5 per hour for each job in process, whether at the machines or at the inspection station. Therefore, management has instructed the production manager to cut down on such inventory. In response, the production manager has made two alternative proposals to reduce the average level of in-process inventory.

Proposal 1 is to use slightly less power for the machines (which would increase their expected time to perform a job to 7 minutes), so that the inspector can keep up with their output better. This also would reduce the cost for each machine (operating cost plus capital recovery cost) from $7.00 to $6.50 per hour. (By contrast, increasing to maximum power would increase this cost to $7.50 per hour while decreasing the expected time to perform a job to 5 minutes.)

Proposal 2 is to substitute a certain younger inspector for this task. He is somewhat faster (albeit more variable in his inspection times because of less experience), so he should keep up better. (His inspection time would have an Erlang distribution with a mean of 0.72 minute and a shape parameter $k = 2$.) This inspector is in a job classification that calls for a total compensation (including benefits) of $16 per hour, whereas the current inspector is in a lower job classification where the compensation is $14 per hour. (The probability distribution of inspection time for each of these inspectors is typical of those in the same job classification.)

The production manager has asked you to "use the latest OR techniques to see how much each proposal would cut down on in-process inventory and then make your recommendations."

A* (*a*) To provide a basis of comparison, begin by evaluating the status quo. Determine the expected amount of in-process inventory (number of jobs) at the machines and at the inspection station. Then calculate the expected total cost per hour of the in-process inventory, the machines, and the inspector.

(*b*) What would be the effect of proposal 1? Why? Make specific comparisons to the results from part (*a*). How would you explain this outcome to the production manager?

(*c*) Determine the effect of proposal 2. Make specific comparisons to the results from part (*a*). How would you explain this outcome to the production manager?

(*d*) Make your recommendations for reducing the average level of in-process inventory at the inspection station and at the group of machines. Be specific in your recommendations, and support them with quantitative analysis like that done in part (*a*). (*Hint:* Data given in the figures of this chapter can be helpful.) Make specific comparisons to the results from part (*a*), and cite the improvements that your recommendations would yield.

16

The Application of Queueing Theory

Queueing theory has enjoyed a prominent place among the modern analytical techniques of OR. However, the emphasis thus far has been on developing a descriptive mathematical theory. Thus queueing theory is not directly concerned with achieving the goal of OR: optimal decision making. Rather, it develops information on the behavior of queueing systems. This theory provides part of the information needed to conduct an OR study attempting to find the best design for a queueing system.

This chapter discusses the *application* of queueing theory in the broader context of an overall OR study. It begins by introducing three examples that will be used for illustration throughout the chapter. Section 16.2 discusses the basic considerations for decision making in this context. The following two sections then develop decision models for the *optimal* design of queueing systems.

16.1 Examples

Example 1—How Many Repairers?

SIMULATION, INC., a small company that makes widgets for analog computers, has 10 widget-making machines. However, because these machines break down and require repair frequently, the company has only enough operators to operate eight machines at a time, so two machines are available on a standby basis for use while other machines are down. Thus eight machines are always operating whenever no more than two machines are waiting to be repaired, but the number of operating machines is reduced by 1 for each additional machine waiting to be repaired.

The time until any given operating machine breaks down has an exponential distribution, with a mean of 20 days. The time required to repair a machine also has an exponential distribution, with a mean of 2 days. Until now the company has had just one repairer to fix these machines, which has frequently resulted in reduced productivity because fewer than eight machines are operating. Therefore, the company is considering hiring a second repairer, so that two machines can be repaired simultaneously.

Thus the queueing system to be studied has the repairers as its servers and the machines requiring repair as its customers, where the problem is to choose between having one or two servers. (Notice the analogy between this problem and the County Hospital emergency room problem described in Sec. 15.1.) With one slight exception, this system fits the *finite calling population variation* of the *M/M/s* model presented in Sec. 15.6, where $N = 10$ machines, $\lambda = \frac{1}{20}$ customer per day (for each operating machine), and $\mu = \frac{1}{2}$ customer per day. The exception is that the λ_0 and λ_1 parameters of the birth-and-death process are changed from $\lambda_0 = 10\lambda$ and $\lambda_1 = 9\lambda$ to $\lambda_0 = 8\lambda$ and $\lambda_1 = 8\lambda$. (All the other parameters are the same as those given in Sec. 15.6.) Therefore, the C_n factors for calculating the P_n probabilities change accordingly (see Sec. 15.5).

Each repairer costs the company approximately $280 per day. However, the estimated *lost profit* from having fewer than eight machines operating to produce widgets is $400 per day for each machine down. (The company can sell the full output from eight operating machines, but not much more.)

The analysis of this problem will be pursued in Secs. 16.3 and 16.4.

Example 2—Which Computer?

EMERALD UNIVERSITY is making plans to lease a supercomputer to be used for scientific research by the faculty and students. Two models are being considered: one from the MBI Corporation and the other from the CRAB Company. The MBI computer costs more but is somewhat faster than the CRAB computer. In particular, if a sequence of typical jobs were run continuously for one 24-hour day, the number completed would have a Poisson distribution with a mean of 30 and 25 for the MBI and the CRAB computers, respectively. It is estimated that an average of 20 jobs will be submitted per day and that the time from one submission to the next will have an exponential distribution with a mean of 0.05 day. The leasing cost per day would be $5,000 for the MBI computer and $3,750 for the CRAB computer.

Thus the queueing system of concern has the computer as its (single) server and the jobs to be run as its customers. Furthermore, this system fits the *M/M/*1 model presented at the beginning of Sec. 15.6. With 1 day as the unit of time, $\lambda = 20$ customers per day, and $\mu = 30$ and 25 customers per day with the MBI and the CRAB computers, respectively. You will see in Secs. 16.3 and 16.4 how the choice was made between the two computers.

Example 3—How Many Tool Cribs?

The MECHANICAL COMPANY is designing a new plant. This plant will need to include one or more tool cribs in the factory area to store tools required by the shop mechanics. The tools will be handed out by clerks as the mechanics arrive and request them and will be returned to the clerks when they are no longer needed. In existing plants, there have been frequent complaints from supervisors that their mechanics have had to waste too much time traveling to tool cribs and waiting to be served, so it appears that there should be *more* tool cribs and *more* clerks in the new plant. On the other hand, management is exerting pressure to reduce overhead in the new plant, and this reduction would lead to *fewer* tool cribs and *fewer* clerks. To resolve these conflicting pressures, an OR study is to be conducted to determine just how many tool cribs and clerks the new plant should have.

Each tool crib constitutes a queueing system, with the clerks as its servers and the mechanics as its customers. Based on previous experience, it is estimated that the time required by a tool crib clerk to service a mechanic has an exponential distribution, with a mean of $\frac{1}{2}$ minute. Judging from the anticipated number of mechanics in the entire factory area, it is also predicted that they would require this service randomly but at a mean rate of 2 mechanics per minute. Therefore, it was decided to use the $M/M/s$ model of Sec. 15.6 to represent each queueing system. With 1 hour as the unit of time, $\mu = 120$. If only one tool crib were to be provided, λ also would be 120. With more than one tool crib, this mean arrival rate would be divided among the different queueing systems.

The total cost to the company of each tool crib clerk is about \$20 per hour. The capital recovery costs, upkeep costs, and so forth associated with each tool crib provided are estimated to be \$16 per working hour. While a mechanic is busy, the value to the company of his or her output averages about \$48 per hour.

Sections 16.3 and 16.4 include discussions of how this (and additional) information was used to make the required decisions.

16.2 Decision Making

Queueing-type situations that require decision making arise in a wide variety of contexts. For this reason, it is not possible to present a meaningful decision-making procedure that is applicable to all these situations. Instead, this section attempts to give a broad conceptual picture of a typical approach.

Designing a queueing system typically involves making one or a combination of the following decisions:

1. Number of servers at a service facility
2. Efficiency of the servers
3. Number of service facilities.

When such problems are formulated in terms of a queueing model, the corresponding decision variables usually are s (number of servers at each facility), μ (mean service rate per busy server), and λ (mean arrival rate at each facility). The *number of service facilities* is directly related to λ because, assuming a uniform workload among the facilities, λ equals the total mean arrival rate to all facilities divided by the number of facilities.

Refer to Sec. 16.1 and note how the three examples there respectively illustrate situations involving these three decisions. In particular, the decision facing Simulation, Inc., is *how many repairers* (servers) to provide. The problem for Emerald University is *how fast a computer* (server) is needed. The problem facing Mechanical Company is *how many tool cribs* (service facilities) to install as well as *how many clerks* (servers) to provide at each facility.

The first kind of decision is particularly common in practice. However, the other two also arise frequently, particularly for the business-industrial internal service systems described in Sec. 15.3. One example illustrating a decision on the efficiency of the servers is the selection of the type of materials-handling equipment (the servers) to purchase to transport certain kinds of loads (the customers). Another such example is the determination of the size of a maintenance crew (where the entire crew is one server). Other decisions concern the number of service facilities, such as restrooms, first-aid centers, drinking fountains, storage areas, and so on, to distribute throughout an area.

All the specific decisions discussed here involve the general question of the *appropriate level of service* to provide in a queueing system. As mentioned at the beginning of Chap. 15, decisions regarding the amount of service capacity to provide usually are based primarily on two considerations: (1) the cost incurred by providing the service, as shown in Fig. 16.1, and (2) the amount of waiting for that service, as suggested in Fig. 16.2. Figure 16.2 can be obtained by using the appropriate waiting-time equation from queueing theory.

These two considerations create conflicting pressures on the decision maker. The objective of reducing service costs recommends a minimal level of service. On the other hand, long waiting times are undesirable, which recommends a high level of service. Therefore, it is necessary to strive for some type of compromise. To assist in finding this compromise, Figs. 16.1 and 16.2 may be combined, as shown in Fig. 16.3. The problem is thereby reduced to selecting the point on the curve of Fig. 16.3 that gives the best balance between the average delay in being serviced and the cost of providing that service. Reference to Figs. 16.1 and 16.2 indicates the corresponding level of service.

Obtaining the proper balance between delays and service costs requires answers to such questions as, How much expenditure on service is equivalent (in its detrimental

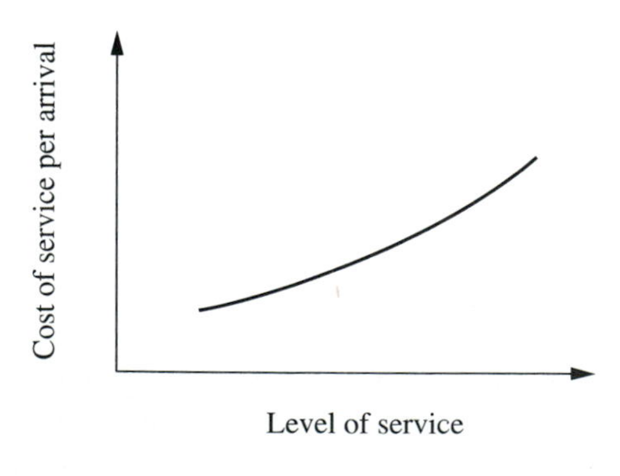

Figure 16.1 Service cost as a function of service level.

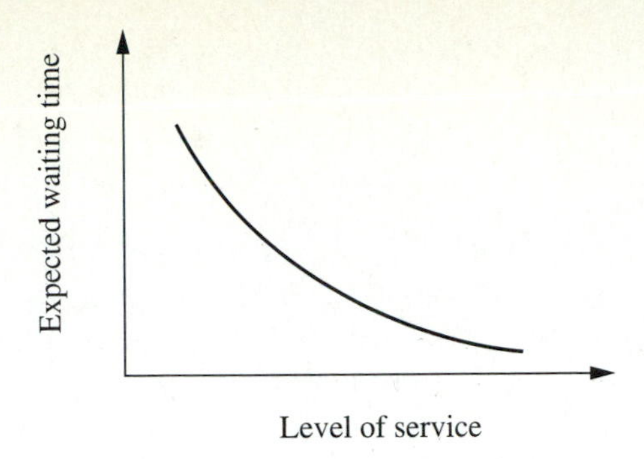

Figure 16.2 Expected waiting time as a function of service level.

impact) to a customer's being delayed 1 unit of time? Thus, to compare service costs and waiting times, it is necessary to adopt (explicitly or implicitly) a common measure of their impact. The natural choice for this common measure is cost, which then requires estimation of the cost of waiting.

Because of the diversity of waiting-line situations, no single process for estimating the cost of waiting is generally applicable. However, we shall discuss the basic considerations involved for several types of situations.

One broad category is where the customers are *external* to the organization providing the service; i.e., they are *outsiders* bringing their business to the organization. Consider first the case of *profit-making* organizations (typified by the commercial service systems described in Sec. 15.3). From the viewpoint of the decision maker, the cost of waiting probably consists primarily of the *lost profit* from *lost business.* This loss of business may occur immediately (because the customer grows impatient and leaves) or in the future (because the customer is sufficiently irritated that he or she does not come again). This kind of cost is quite difficult to estimate, and it may be necessary to revert to other criteria, such as a tolerable probability distribution of waiting times. When the customer is not a human being, but a job being performed on order, there may be more readily identifiable costs incurred, such as those caused by idle in-process inventories or increased expediting and administrative effort.

Now consider the type of situation where service is provided on a *nonprofit* basis to customers *external* to the organization (typical of social service systems and some transportation service systems described in Sec. 15.3). In this case, the cost of waiting usually is a *social cost* of some kind. Thus it is necessary to evaluate the consequences of the waiting for the individuals involved and/or for society as a whole and to try to

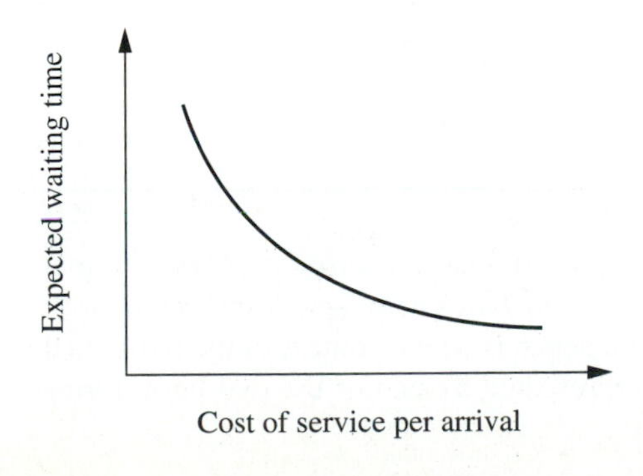

Figure 16.3 Relationship between average delay and service cost.

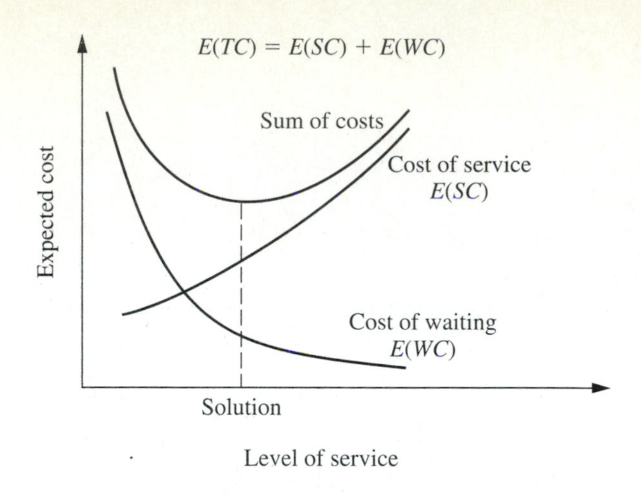

Figure 16.4 Conceptual solution procedure for many waiting-line problems.

impute a monetary value to avoiding these consequences. Once again, this kind of cost is quite difficult to estimate, and it may be necessary to revert to other criteria.

A situation may be more amenable to estimating waiting costs if the customers are *internal* to the organization providing the service (as for business-industrial internal service systems). For example, the customers may be machines (as in Example 1) or employees (as in Example 3) of a firm. Therefore, it may be possible to identify directly some of or all the costs associated with the idleness of these customers. Typically, what is being wasted by this idleness is *productive output,* in which case the waiting cost becomes the *lost profit* from *all lost productivity.*

Given that the cost of waiting has been evaluated explicitly, the remainder of the analysis is conceptually straightforward. The objective is to determine the level of service that minimizes the total of the expected cost of service and the expected cost of waiting for that service. This concept is depicted in Fig. 16.4, where WC denotes *waiting cost,* SC denotes *service cost,* and TC denotes *total cost.* Thus the mathematical statement of the objective is to

$$\text{Minimize} \qquad E(\text{TC}) = E(\text{SC}) + E(\text{WC}).$$

The next two sections are concerned with the application of this concept to various types of problems. Thus Sec. 16.3 describes how $E(\text{WC})$ can be expressed mathematically. Section 16.4 then focuses on $E(\text{SC})$ to formulate the overall objective function $E(\text{TC})$ for several basic design problems (including some with multiple decision variables, so that the level-of-service axis in Fig. 16.4 then requires more than one dimension).

16.3 Formulation of Waiting-Cost Functions

To express $E(\text{WC})$ mathematically, we must first formulate a *waiting-cost function* that describes how the actual waiting cost being incurred varies with the current behavior of the queueing system. The form of this function depends on the context of the individual problem. However, most situations can be represented by one of the two basic forms described next.

The $g(N)$ Form

Consider first the situation discussed in the preceding section where the queueing system *customers* are *internal* to the organization providing the service, and so the primary cost of waiting may be the *lost profit from lost productivity*. The *rate* at which productive output is lost sometimes is essentially *proportional* to the number of customers in the queueing system. However, in many cases there is not enough productive work available to keep all the members of the calling population continuously busy. Therefore, little productive output may be lost by having just a few members idle, waiting for service in the queueing system, whereas the loss may increase greatly if a few more members are made idle because they require service. Consequently, the primary property of the queueing system that determines the *current rate* at which waiting costs are being incurred is N, the number of customers in the system. Thus the form of the waiting-cost function for this kind of situation is that illustrated in Fig. 16.5, namely, a function of N. We shall denote this form by $g(N)$.

The $g(N)$ function is constructed for a particular situation by estimating $g(n)$, the waiting-cost rate incurred when $N = n$, for $n = 1, 2, \ldots,$ where $g(0) = 0$. After computing the P_n probabilities for a given design of the queueing system, we can calculate

$$E(\text{WC}) = E(g(N)).$$

Because N is a random variable, this calculation is made by using the expression for the expected value of a *function* of a *discrete* random variable

$$E(\text{WC}) = \sum_{n=0}^{\infty} g(n)P_n.$$

When $g(N)$ is a linear function (i.e., when the waiting-cost rate is proportional to N), then

$$g(N) = C_w N,$$

where C_w is the cost of waiting per unit time for each customer. In this case, $E(\text{WC})$ reduces to

$$E(\text{WC}) = C_w \sum_{n=0}^{\infty} nP_n = C_w L.$$

Example 1—How Many Repairers? For Example 1 of Sec. 16.1, Simulation, Inc., has two standby widget-making machines, so there is no lost productivity as long as the number of customers (machines requiring repair) in the system does not exceed 2. However, for each *additional* customer (up to the maximum of 10 total), the estimated lost profit is \$400 per day. Therefore,

$$g(n) = \begin{cases} 0 & \text{for } n = 0, 1, 2, \\ 400(n-2) & \text{for } n = 3, 4, \ldots, 10, \end{cases}$$

as shown in Table 16.1. Consequently, $E(\text{WC})$ is calculated by summing the rightmost column of Table 16.1 for each of the two cases of interest, namely, having one repairer ($s = 1$) or two repairers ($s = 2$).

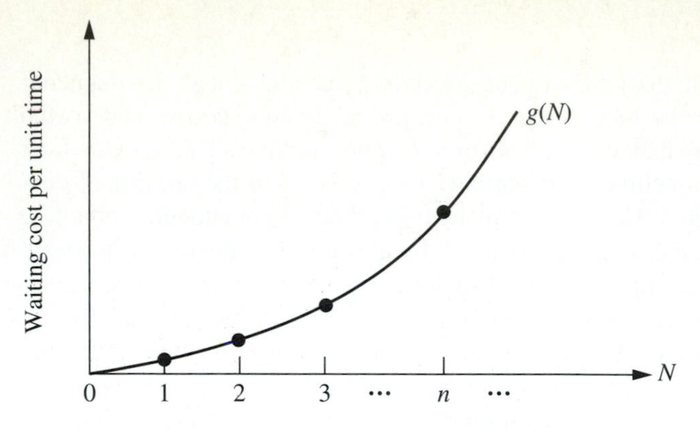

Figure 16.5 The waiting-cost function as a function of N.

The $h(\mathcal{W})$ Form

Now consider the cases discussed in Sec. 16.2 where the queueing system *customers* are *external* to the organization providing the service. Three major types of queueing systems described in Sec. 15.3—commercial service systems, transportation service systems, and social service systems—typically fall into this category. In the case of commercial service systems, the primary cost of waiting may be the lost profit from lost future business. For transportation service systems and social systems, the primary cost of waiting may be in the form of a social cost. However, for either type of cost, its magnitude tends to be affected greatly by the size of the waiting times experienced by the customers. Thus the primary property of the queueing system that determines the waiting cost currently being incurred is $\mathcal{W}$, the waiting time in the system for the *individual* customers. Consequently, the form of the waiting-cost function for this kind of situation is that illustrated in Fig. 16.6, namely, a function of $\mathcal{W}$. We shall denote this form by $h(\mathcal{W})$.

One way of constructing the $h(\mathcal{W})$ function is to estimate $h(w)$ (the waiting cost incurred when a customer's waiting time $\mathcal{W} = w$) for several values of w and then to

Table 16.1 **Calculation of E(WC) for Example 1**

		$s = 1$		$s = 2$	
$N = n$	$g(n)$	P_n	$g(n)P_n$	P_n	$g(n)P_n$
0	0	0.271	0	0.433	0
1	0	0.217	0	0.346	0
2	0	0.173	0	0.139	0
3	400	0.139	56	0.055	24
4	800	0.097	78	0.019	16
5	1,200	0.058	70	0.006	8
6	1,600	0.029	46	0.001	0
7	2,000	0.012	24	3×10^{-4}	0
8	2,400	0.003	7	4×10^{-5}	0
9	2,800	7×10^{-4}	0	4×10^{-6}	0
10	3,200	7×10^{-5}	0	2×10^{-7}	0
E(WC)			\$281 per day		\$48 per day

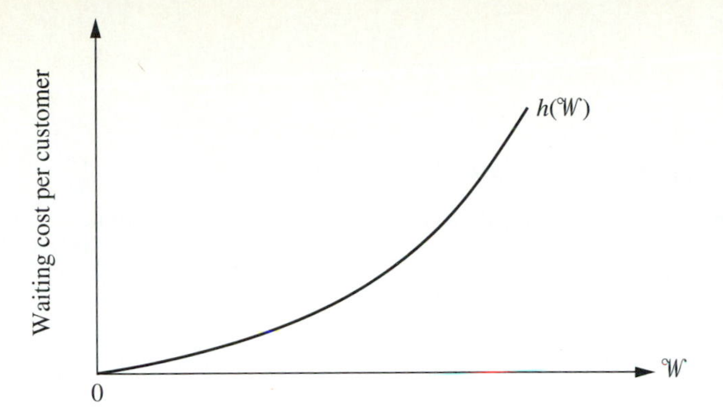

Figure 16.6 The waiting-cost function as a function of $\mathcal{W}$.

fit a polynomial to these points. The expectation of this function of a continuous random variable is then defined as

$$E(h(\mathcal{W})) = \int_0^\infty h(w) f_{\mathcal{W}}(w)\, dw,$$

where $f_{\mathcal{W}}(w)$ is the probability density function of $\mathcal{W}$. However, because $E(h(\mathcal{W}))$ is the expected waiting cost *per customer* and $E(\text{WC})$ is the expected waiting cost *per unit time,* these two quantities are not equal in this case. To relate them, it is necessary to multiply $E(h(\mathcal{W}))$ by the expected *number of customers per unit time* entering the queueing system. In particular, if the mean arrival rate is a constant λ, then

$$E(\text{WC}) = \lambda E(h(\mathcal{W})) = \lambda \int_0^\infty h(w) f_{\mathcal{W}}(w)\, dw.$$

Example 2—Which Computer? Because the faculty and students of Emerald University would experience different turnaround times with the two computers under consideration (see Sec. 16.1), the choice between the computers required an evaluation of the consequences of making them wait for their jobs to be run. Therefore, several leading scientists on the faculty were asked to evaluate these consequences.

The scientists agreed that one major consequence is a *delay in getting research done.* Little effective progress can be made while one is awaiting the results from the run. The scientists estimated that it would be worth \$500 to reduce this delay by 1 day. Therefore, this component of waiting cost was estimated to be \$500 per day, that is, $500\mathcal{W}$, where $\mathcal{W}$ is expressed in days.

The scientists also pointed out that a second major consequence of waiting is a *break in the continuity of the research.* Although a short delay (a fraction of a day) causes little problem in this regard, a longer delay causes significant wasted time in having to gear up to resume the research. The scientists estimated that this wasted time would be roughly proportional to the *square* of the delay time. Dollar figures of \$100 and \$400 were then imputed to the value of being able to avoid this consequence entirely rather than having a wait of $\frac{1}{2}$ day and 1 day, respectively. Therefore, this component of the waiting cost was estimated to be $400\mathcal{W}^2$.

This analysis yields

$$h(\mathcal{W}) = 500\mathcal{W} + 400\mathcal{W}^2$$

Because

$$f_{\mathcal{W}}(w) = \mu(1 - \rho)e^{-\mu(1-\rho)w}$$

for the $M/M/1$ model (see Sec. 15.6) fitting this single-server queueing system,

$$E(h(\mathcal{W})) = \int_0^\infty (500w + 400w^2)\mu(1 - \rho)e^{-\mu(1-\rho)w}\, dw.$$

By using the fact that $\mu(1 - \rho) = \mu - \lambda$ for a single-server system, the values of μ and λ presented in Sec. 16.1 give

$$\mu(1 - \rho) = \begin{cases} 10 & \text{for MBI computer,} \\ 5 & \text{for CRAB computer.} \end{cases}$$

Evaluating the integral for these two cases yields

$$E(h(\mathcal{W})) = \begin{cases} 58 & \text{for MBI computer,} \\ 132 & \text{for CRAB computer.} \end{cases}$$

The result represents the expected waiting cost (in dollars) for each scientist arriving with a job to be run. Because $\lambda = 20$, the total expected waiting cost per day becomes

$$E(\text{WC}) = \begin{cases} \$1{,}160 \text{ per day} & \text{for MBI computer,} \\ \$2{,}640 \text{ per day} & \text{for CRAB computer.} \end{cases}$$

The Linear Case: When $h(\mathcal{W})$ is a linear function

$$h(\mathcal{W}) = C_w\mathcal{W},$$

then $E(\text{WC})$ reduces to

$$E(\text{WC}) = \lambda E(C_w\mathcal{W}) = C_w(\lambda W) = C_w L.$$

Note that this result is identical to the result when $g(N)$ is a linear function. Consequently, when the total waiting cost incurred by the queueing system is simply *proportional* to the total waiting time, it does not matter whether the $g(N)$ or the $h(\mathcal{W})$ form is used for the waiting-cost function.

Example 3—How Many Tool Cribs? As indicated in Sec. 16.1, the value to the Mechanical Company of a busy mechanic's output averages about $48 per hour. Thus $C_w = 48$. Consequently, for each tool crib the expected waiting cost per hour is

$$E(\text{WC}) = 48L,$$

where L represents the expected number of mechanics waiting (or being served) at the tool crib.

16.4 Decision Models

We mentioned in Sec. 16.2 that three common decision variables in designing queueing systems are s (number of servers), μ (mean service rate for each server), and λ (mean arrival rate at each service facility). We shall now formulate models for making some of these decisions.

Model 1—Unknown s

Model 1 is designed for the case where both μ and λ are fixed at a particular service facility, but where a decision must be made on the number of servers to have on duty at the facility.

FORMULATION OF MODEL 1

Definition: C_s = marginal cost of a server per unit time.
Given: μ, λ, C_s.
To find: s.
Objective: Minimize $E(\text{TC}) = sC_s + E(\text{WC})$.

Because only a few alternative values of s normally need to be considered, the usual way of solving this model is to calculate E(TC) for these values of s and select the minimizing one.

EXAMPLE 1—HOW MANY REPAIRERS? For Example 1 of Sec. 16.1, each repairer (server) costs Simulation, Inc., approximately \$280 per day. Thus, with 1 day as the unit of time, $C_s = 280$. Using the values of E(WC) calculated in Table 16.1 then yields the results shown in Table 16.2, which indicate that the company should continue having just one repairer.

Model 2—Unknown μ and s

Model 2 is designed for the case where both the efficiency of service, measured by μ, and the number of servers s at a service facility need to be selected.

Alternative values of μ may be available because there is a choice on the *quality* of the servers. For example, when the servers will be materials-handling units, the quality of the units to be purchased affects their service rate for moving loads.

Another possibility is that the *speed* of the servers can be adjusted mechanically. For example, the speed of machines frequently can be adjusted by changing the amount of power consumed, which also changes the cost of operation.

Still another type of example is the selection of the number of crews (the servers) and the size of each crew (which determines μ) for jointly performing a certain task. The task might be maintenance work, or loading and unloading operations, or inspection work, or setup of machines, and so forth.

In many cases, only a few alternative values of μ are available, e.g., the efficiency of the alternative types of materials-handling equipment or the efficiency of the alternative crew sizes.

Table 16.2 **Calculation of E(TC) for Example 1**

s	sC_s	E(WC)	E(TC)
1	280	281	\$561 per day ← minimum
2	560	48	\$608 per day
≥3	≥840	≥0	≥\$840 per day

Definitions: $f(\mu)$ = marginal cost of server per unit time when mean service rate is μ.

A = set of feasible values of μ.

Given: $\lambda, f(\mu), A$.

To find: μ, s.

Objective: Minimize $E(\text{TC}) = s\,f(\mu) + E(\text{WC})$, subject to $\mu \in A$.

EXAMPLE 2—WHICH COMPUTER? As indicated in Sec. 16.1, $\mu = 30$ for the MBI computer and $\mu = 25$ for the CRAB computer, where 1 day is the unit of time. These computers are the only two being considered by Emerald University, so

$$A = \{25, 30\}.$$

Because the leasing cost per day is \$3,750 for the CRAB computer ($\mu = 25$) and \$5,000 for the MBI computer ($\mu = 30$),

$$f(\mu) = \begin{cases} 3{,}750 & \text{for } \mu = 25, \\ 5{,}000 & \text{for } \mu = 30. \end{cases}$$

The supercomputer chosen will be the only one available to the faculty and students, so the number of servers (supercomputers) for this queueing system is restricted to $s = 1$. Hence

$$E(\text{TC}) = f(\mu) + E(\text{WC}),$$

where $E(\text{WC})$ is given in Sec. 16.3 for the two alternatives. Thus

$$E(\text{TC}) = \begin{cases} 3{,}750 + 2{,}640 = \$6{,}390 \text{ per day} & \text{for CRAB computer,} \\ 5{,}000 + 1{,}160 = \$6{,}160 \text{ per day} & \text{for MBI computer.} \end{cases}$$

Consequently, the decision was made to lease the MBI supercomputer.

This example illustrates a case where the number of feasible values of μ is *finite* but the value of s is fixed. If s were not fixed, a two-stage approach could be used to solve such a problem. First, for each individual value of μ, set $C_s = f(\mu)$, and solve for the value of s that minimizes $E(\text{TC})$ for model 1. Second, compare these minimum $E(\text{TC})$ for the alternative values of μ, and select the one giving the overall minimum.

When the number of feasible values of μ is *infinite* (such as when the speed of a machine or piece of equipment is set mechanically within some feasible interval), another two-stage approach sometimes can be used to solve the problem. First, for each individual value of s, *analytically* solve for the value of μ that minimizes $E(\text{TC})$. [This approach requires setting to zero the derivative of $E(\text{TC})$ with respect to μ and then solving this equation for μ, which can be done only when analytical expressions are available for both $f(\mu)$ and $E(\text{WC})$.] Second, compare these minimum $E(\text{TC})$ for the alternative values of s, and select the one giving the overall minimum.

This analytical approach frequently is relatively straightforward for the case of $s = 1$ (see Prob. 16.4-15). However, because far fewer and less convenient analytical results are available for multiple-server versions of queueing models, this approach is either difficult (requiring computer calculations with numerical methods to solve the equation for μ) or completely impossible when $s > 1$. Therefore, a more practical approach is to consider only a relatively small number of representative values of μ and to use available tabulated results for the appropriate queueing model to obtain (or approximate) $E(\text{TC})$ for these μ values.

Fortunately, under certain fairly common circumstances described next, $s = 1$ (and its minimizing value of μ) *must* yield the overall minimum $E(\text{TC})$ for model 2, so $s > 1$ cases need not be considered at all.

Optimality of a single server: Under certain conditions, $s = 1$ necessarily is *optimal* for model 2.

The primary conditions[1] are that

1. The value of μ minimizing $E(\text{TC})$ for $s = 1$ is feasible.
2. Function $f(\mu)$ is either *linear* or *concave* (as defined in Appendix 2).

In effect, this optimality result indicates that it is better to concentrate service capacity into one fast server rather than dispersing it among several slow servers. Condition 2 says that this concentrating of a given amount of service capacity can be done without increasing the cost of service. Condition 1 says that it must be possible to make μ sufficiently large that a single server can be used to full advantage.

To understand why this result holds, consider any other solution to model 2, $(s, \mu) = (s^*, \mu^*)$, where $s^* > 1$. The service capacity of this system (as measured by the mean rate of service completions when all servers are working) is $s^*\mu^*$. We shall now compare this solution with the corresponding single-server solution $(s, \mu) = (1, s^*\mu^*)$ having the *same* service capacity. In particular, Table 16.3 compares the mean rate at which service completions occur for each given number of customers in the system $N = n$. This table shows that the service efficiency of the (s^*, μ^*) solution sometimes is worse but never is better than for the $(1, s^*\mu^*)$ solution because it can use the full service capacity only when there are at least s^* customers in the system, whereas the single-server solution uses the full capacity whenever there are *any* customers in the system. Because this lower service efficiency can only increase waiting in the system, $E(\text{WC})$ must be larger for (s^*, μ^*) than for $(1, s^*\mu^*)$. Furthermore, the expected service cost must be at least as large because condition 2 [and $f(0) = 0$] implies that

$$s^*f(\mu^*) \geq f(s^*\mu^*).$$

Therefore, $E(\text{TC})$ is larger for (s^*, μ^*) than $(1, s^*\mu^*)$. Finally, note that condition 1 implies that there is a feasible solution with $s = 1$ that is at least as good as $(1, s^*\mu^*)$. The conclusion is that *any* $s > 1$ solution *cannot* be optimal for model 2, so $s = 1$ must be optimal.[2]

[1] There also are minor restrictions on the queueing model and the waiting-cost function. However, any of the constant service-rate queueing models presented in Chap. 15 for $s \geq 1$ are allowed. If the $g(N)$ form is used for the waiting-cost function, it can be any *increasing* function. If the $h(\mathcal{W})$ form is used, it can be any linear function or any convex function (as defined in Appendix 2), which fits most cases of interest.

[2] For a rigorous proof of this result, see S. Stidham, Jr., "On the Optimality of Single-Server Queueing Systems," *Operations Research,* **18:** 708–732, 1970.

Table 16.3 **Comparison of Service Efficiency for Model 2 Solutions**

$N = n$	*Mean Rate of Service Completions* $(s, \mu) = (s^*, \mu^*)$ versus $(s, \mu) = (1, s^*\mu^*)$		
$n = 0$	0	$=$	0
$n = 1, 2, \ldots, s^* - 1$	$n\mu^*$	$<$	$s^*\mu^*$
$n \geq s^*$	$s^*\mu^*$	$=$	$s^*\mu^*$

This result is still of some use even when one or both conditions fail to hold. If μ cannot be made sufficiently large to permit a single server, it still suggests that a *few* fast servers should be preferred to many slow ones. If condition 2 does not hold, we still know that $E(\text{WC})$ is minimized by concentrating any given amount of service capacity into a single server, so the best $s = 1$ solution must be at least nearly optimal unless it causes a *substantial* increase in service cost.

Model 3—Unknown λ and s

Model 3 is designed especially for the case where it is necessary to select both the *number of service facilities* and the *number of servers s* at each facility. In the typical situation, a population (such as the employees in an industrial building) must be provided with a certain service, so a decision must be made as to what proportion of the population (and therefore what value of λ) should be assigned to each service facility. Examples of such facilities include employee facilities (drinking fountains, vending machines, and restrooms), storage facilities, and reproduction equipment facilities. It may sometimes be clear that only a single server should be provided at each facility (e.g., one drinking fountain or one copy machine), but s often is also a decision variable.

To simplify our presentation, we shall require in model 3 that λ and s be the same for all service facilities. However, it should be recognized that a slight improvement in the indicated solution might be achieved by permitting minor deviations in these parameters at individual facilities. This should be investigated as part of the detailed analysis that generally follows the application of the mathematical model.

Formulation of Model 3

Definitions: C_s = marginal cost of server per unit time.

C_f = fixed cost of service per service facility per unit time.

λ_p = mean arrival rate for entire calling population.

n = number of service facilities = λ_p/λ.

Given: μ, C_s, C_f, λ_p.

To find: λ, s.

Objective: Minimize $E(\text{TC})$, subject to $\lambda = \lambda_p/n$, where $n = 1, 2, \ldots$.

It might appear at first glance that the appropriate expression for the expected total cost per unit time of all the facilities should be

$$E(\text{TC}) \stackrel{?}{=} n[C_f + sC_s) + E(\text{WC})],$$

where $E(\text{WC})$ here represents the expected waiting cost per unit time for *each* facility. However, if this expression actually were valid, it would imply that $n = 1$ *necessarily* is optimal for model 3. The reasoning is completely analogous to that for the optimality of a single-server result for model 2; namely, any solution $(n, s) = (n^*, s^*)$ with $n^* > 1$ has higher service costs than the $(n, s) = (1, n^*s^*)$ solution, and it *also* has a higher expected waiting cost because it sometimes makes less effective use of the available service capacity. In particular, it sometimes has idle servers at one facility while customers are waiting at another facility, so the mean rate of service completions would be less than if the customers had access to *all* the servers at one common facility.

Because there are many situations where it obviously would *not* be optimal to have just one service facility (e.g., the number of restrooms in a 50-story building), something must be wrong with this expression. Its deficiency is that it considers only the cost of service and the cost of waiting *at the service facilities* while totally ignoring the cost of the time wasted in *traveling* to and from the facilities. Because travel time would be prohibitive with only one service facility for a large population, enough separate facilities must be distributed throughout the calling population to hold travel time down to a reasonable level.

Thus, letting the random variable T be the round-trip travel time for a customer coming to and going back from one of the service facilities, we see that the total time lost by the customer actually is $\mathcal{W} + T$. (Recall from Chap. 15 that $\mathcal{W}$ is the waiting time in the queueing system *after* the customer arrives.) Therefore, a customer's *total* cost for time lost should be based on $\mathcal{W} + T$ rather than just $\mathcal{W}$. To simplify the analysis, let us separate this total cost into the sum of the waiting-time cost based on $\mathcal{W}$ (or N) and the travel-time cost based on T. We shall also assume that the travel-time cost is proportional to T, where C_t is the cost of each unit of travel time for each customer. For ease of presentation, suppose that the probability distribution of T is the same for each service facility, so that $C_t E(T)$ is the *expected travel cost* for each arrival at any of the service facilities. The resulting expression for $E(\text{TC})$ is

$$E(\text{TC}) = n[(C_f + sC_s) + E(\text{WC}) + \lambda C_t E(T)]$$

because λ is the expected number of arrivals *per unit time* at each facility. Consequently, by evaluating (or estimating) $E(T)$ for each case of interest, model 3 can be solved by calculating $E(\text{TC})$ for various values of s for each n and then selecting the solution giving the overall minimum.

Example 3—How Many Tool Cribs? For the new plant being designed for the Mechanical Company (see Sec. 16.1), the layout of the portion of the factory area where the mechanics will work is shown in Fig. 16.7. The three possible locations for tool cribs are identified as locations 1, 2, and 3, where access to these locations will be provided by a system of orthogonal aisles parallel to the sides of the indicated area. The coördinates are given in units of feet.

The three basic alternatives being considered are these:

Alternative 1: Have one tool crib—use location 2.

Alternative 2: Have two tool cribs—use locations 1 and 3.

Alternative 3: Have three tool cribs—use locations 1, 2, and 3.

The mechanics will be distributed quite uniformly throughout the area shown, and each mechanic will be assigned to the *nearest* tool crib. It is estimated that the mechanics will walk to and from a tool crib at an average speed of slightly less than 3 miles per hour. Based on this information, $E(T)$ is estimated to be 0.04, 0.0278, and 0.02 hour for alternatives 1, 2, and 3, respectively.

The stage now is set for using model 3 to choose from these alternatives. Most of the data required for this model are given in Sec. 16.1, namely,

$$\mu = 120 \text{ per hour}, \qquad C_f = \$16 \text{ per hour},$$
$$C_s = \$20 \text{ per hour},$$
$$\lambda_p = 120 \text{ per hour}, \qquad C_t = \$48 \text{ per hour},$$

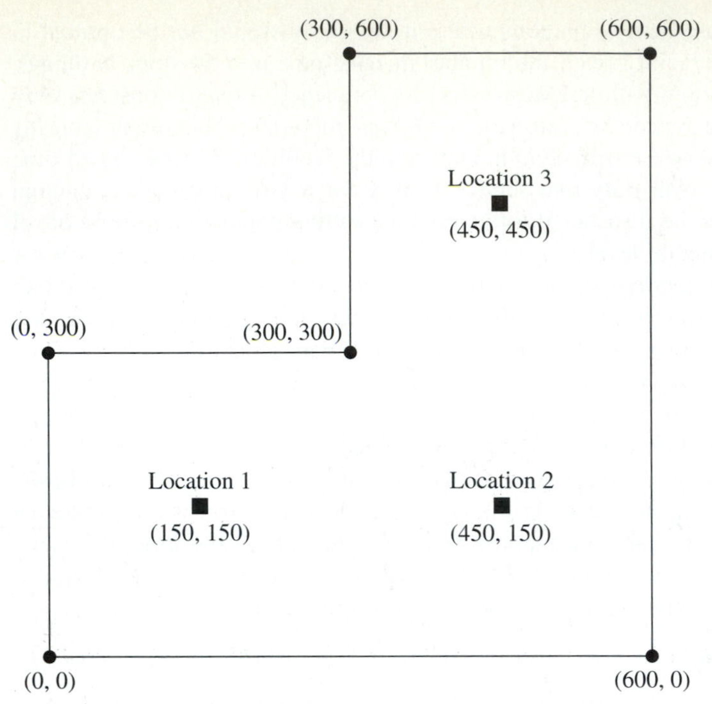

Figure 16.7 Layout for Example 3.

Table 16.4 **Calculation of $E(T)$, Dollars per Hour, for Example 3**

n	λ	s	L	$E(T)$	$C_f + sC_s$	E(WC)	$\lambda C_t E(T)$	E(TC)
1	120	1	∞	0.04	36	∞	230.40	∞
1	120	2	1.333	0.04	56	64.00	230.40	350.40
1	120	3	1.044	0.04	76	50.11	230.40	356.51
2	60	1	1.000	0.0278	36	48.00	80.00	328.00
2	60	2	0.534	0.0278	56	25.63	80.00	323.26
3	40	1	0.500	0.02	36	24.00	38.40	295.20
3	40	2	0.344	0.02	56	16.51	38.40	332.73

where the $M/M/s$ model given in Sec. 15.6 is used to calculate L and so on. In addition, the end of Sec. 16.3 gives $E(\text{WC}) = 48L$ in dollars per hour. Therefore,

$$E(\text{TC}) = n\left[(16 + 20s) + 48L + \frac{120}{n}48E(T)\right].$$

The resulting calculation of E(TC) for various s values for each n is given in Table 16.4, which indicates that the *overall minimum* E(TC) of \$295.20 per hour is obtained by having three tool cribs (so $\lambda = 40$ for each), with one clerk at each tool crib.

16.5 Conclusions

This chapter has discussed the application of queueing theory for *designing* queueing systems. Every individual problem has its own special characteristics, so no standard procedure can be prescribed to fit every situation. Therefore, the emphasis has been on introducing fundamental considerations and approaches that can be adapted to most cases. We have focused on three particularly common decision variables (s, μ, and λ) as a vehicle for introducing and illustrating these concepts. However, there are many other possible decision variables (e.g., the size of a waiting room for a queueing system) and many more complicated situations (e.g., designing a priority queueing system) that can also be analyzed in a similar way.

Another useful area for the application of queueing theory is the development of policies for *controlling* queueing systems, e.g., for *dynamically* adjusting the number of servers or the service rate to compensate for changes in the number of customers in the system. Considerable research is being conducted in this area.

Queueing theory has proved to be a very useful tool, and we anticipate that its use will continue to grow as recognition of the many guises of queueing systems grows.

SELECTED REFERENCES

1. Allen, A. O., *Probability, Statistics, and Queueing Theory with Computer Science Applications,* 2d ed., Academic Press, New York, 1990, chaps. 5, 6.
2. Carmichael, D. G., *Engineering Queues in Construction and Mining,* Wiley, Chichester, England, 1987.
3. Hall, R. W., *Queueing Methods: For Services and Manufacturing,* Prentice-Hall, Englewood Cliffs, NJ, 1991.
4. Kleinrock, L., *Queueing Systems,* vol. 2: *Computer Applications,* Wiley, New York, 1976.
5. Lazowska, E. D., J. Zahorjan, G. S. Graham, and K. C. Sevcik, *Quantitative System Performance: Computer System Analysis Using Queueing Network Models,* Prentice-Hall, Englewood Cliffs, NJ, 1984.
6. Lee, A. M., *Applied Queueing Theory,* St. Martin's Press, New York, 1966.
7. Newell, G. F., *Applications of Queueing Theory,* 2d ed., Chapman and Hall, London, 1982.

RELEVANT ROUTINES IN YOUR OR COURSEWARE

Automatic routines: *The M/M/s Model*
The Finite Queue Variation of the M/M/s Model
The Finite Calling Population Variation of the M/M/s Model
Nonpreemptive Priorities Model
Preemptive Priorities Model

To access these routines, call the ProbMod program and then choose *Queueing Theory* under the Area menu. See Appendix 1 for documentation of the software.

PROBLEMS[1]

To the left of each of the following problems (or their parts), we have inserted an A (for automatic routine) whenever a corresponding routine listed above can be helpful. An asterisk on the A indicates that this routine definitely should be used (unless your instructor gives contrary instructions) and that the printout from this routine is all that needs to be turned in to show your work in executing the procedure. An asterisk on the problem number indicates that at least a partial answer is given in the back of the book.

[1] Problems 16.4-3, 16.4-11, and 16.4-12 have been adapted, with permission, from previous operations research examinations given by the Society of Actuaries.

16.2-1. For each kind of queueing system listed in Prob. 15.3-1, briefly describe the nature of the cost of service and the cost of waiting that would need to be considered in designing the system.

16.3-1.* Suppose that a queueing system fits the $M/M/1$ model described in Sec. 15.6, with $\lambda = 2$ and $\mu = 4$. Evaluate the expected waiting cost per unit time $E(\text{WC})$ for this system when its waiting-cost function has the form

(*a*) $g(N) = 10N + 2N^2$.
(*b*) $h(\mathcal{W}) = 25\mathcal{W} + \mathcal{W}^3$.

16.3-2. Follow the instructions of Prob. 16.3-1 for the following waiting-cost functions.

(*a*) $g(N) = \begin{cases} 10N & \text{for } N = 0, 1, 2, \\ 6N^2 & \text{for } N = 3, 4, 5, \\ N^3 & \text{for } N > 5. \end{cases}$

(*b*) $h(\mathcal{W}) = \begin{cases} \mathcal{W} & \text{for } 0 \le \mathcal{W} \le 1, \\ \mathcal{W}^2 & \text{for } \mathcal{W} \ge 1. \end{cases}$

16.4-1. A certain queueing system has a Poisson input, with a mean arrival rate of 4 customers per hour. The service-time distribution is exponential, with a mean of 0.2 hour. The marginal cost of providing each server is $20 per hour, where it is estimated that the cost incurred by having each customer *idle* (i.e., in the queueing system) is $120 per hour for the first customer and $180 per hour for each additional customer. Determine the number of servers that should be assigned to the system to minimize the expected total cost per hour. [*Hint:* Express $E(\text{WC})$ in terms of L, P_0, and ρ, and then use Figs. 15.6 and 15.7.]

16.4-2.* A small grocery store has a single checkout stand with a full-time cashier. Customers arrive at the stand according to a Poisson process at a mean rate of 30 per hour. The service-time distribution is exponential, with a mean of 1.5 minutes. This situation has resulted in occasional long lines and complaints from customers. Therefore, because there is no room for a second checkout stand, the proposal has been made that another person be hired to help the cashier by bagging the groceries. This help would reduce the expected time required to process a customer to 1 minute, but the distribution still would be exponential.

The total compensation for the new employee would be $8 per hour, which is just half that for the cashier. It is estimated that the grocery store incurs lost profit due to lost future business of $0.08 for each minute that each customer has to wait (including service time). The manager wants to determine on an expected total-cost basis whether it would be worthwhile to hire the new person.

(*a*) Which decision model presented in Sec. 16.4 applies to this problem? Why?
(*b*) Use this model to determine whether to continue the status quo or to adopt the proposal.

A **16.4-3.** Customers arrive at a fast-food restaurant with one server according to a Poisson process at a mean rate of 30 per hour. The server has just resigned, and the two candidates for the replacement are X (fast but expensive) and Y (slow but inexpensive). Both candidates would have an exponential distribution for service times, with X having a mean of 1.2 minutes and Y having a mean of 1.5 minutes. Restaurant revenue per month is given by $6,000/$W$, where W is the expected waiting time (in minutes) of a customer in the system.

Determine the upper bound on the difference in their monthly compensations that would justify hiring X rather than Y.

16.4-4. The problem is to choose between two types of materials-handling equipment, A and B, for transporting certain types of goods between certain producing centers in a job shop. Calls for the materials-handling unit to move a load would come essentially at random (i.e., according to a Poisson input process) at a mean rate of 4 per hour. The total time required to move a load has an exponential distribution, where the expected time is 12 minutes for A and 9 minutes for B. The total equivalent uniform hourly cost (capital recovery cost plus operating cost) would be $50 for A and $150 for B. The estimated cost of idle goods (waiting to be moved or in transit) because of increased in-process inventory is $20 per load per hour. Furthermore, the scheduling of the work at the producing centers allows for just 1 hour from the completion of a load at one center to the arrival of that load at the next center. Therefore, an additional $100 per load per hour of delay (including transit time) *after the first hour* is to be charged for lost production because of idle personnel and equipment, extra costs of expediting and supervision, and so forth.

Assuming that only one materials-handling unit is to be purchased, which type of unit should be selected?

16.4-5. A railroad company paints its own railroad cars as needed. Alternative 1 is to provide two paint shops, where painting is done be hand (one car at a time in each shop), for a total annual cost of $300,000. The painting time for a car is 6 hours. Alternative 2 is to provide one spray shop involving an annual cost of $400,000. In this case, the painting time for a car (again done one at a time) is 3 hours. For both alternatives, the cars arrive according to a Poisson input, with a mean arrival rate of 1 every 5 hours. The cost of idle time per car is $50 per hour. Which alternative should the railroad choose? Assume that the paint shops are always open; i.e., they work (24)(365) = 8,760 hours per year.

16.4-6. An airline maintenance base wants to make a change in its overhaul operation. The present situation is that only one airplane can be repaired at a time, and the expected repair time is 36 hours, whereas the expected time between arrivals is 45 hours. This situation has led to frequent and prolonged delays in repairing incoming planes, even though the base operates continuously. The average cost of an idle plane to the airline is $3,000 per hour. It is estimated that each plane goes into the maintenance shop 5 times per year. It is believed that the input process for the base is essentially Poisson and that the probability distribution of repair times is Erlang, with shape parameter $k = 2$. Alternative A is to provide a duplicate maintenance shop, so that two planes can be repaired simultaneously. The cost, amortized over 5 years, is $400,000 per year for each of the airline's airplanes.

Alternative B is to replace the present maintenance equipment by the most efficient (and expensive) equipment available, thereby reducing the expected repair time to 18 hours. The cost, amortized over 5 years, is $550,000 per year for each airplane.

Which alternative should the airline choose?

16.4-7.* A particular in-process inspection station is used to inspect subassemblies of a certain kind. At present there are two inspectors at the station, and they work together to inspect each subassembly. The inspection time has an exponential distribution, with a mean of 15 minutes. The cost of providing this inspection system is $20 per hour.

A proposal has been made to streamline the inspection procedure so that it can be handled by only one inspector. This inspector would begin by visually inspecting the exterior of the subassembly, and she would then use new efficient equipment to complete the inspection. The times required for these two phases of the inspection have independent Erlang distributions, with shape parameter $k = 2$ and means of 6 and 12 minutes, respectively. The capitalized cost of providing this inspection system would be $15 per hour.

The subassemblies arrive at the inspection station according to a Poisson process at a mean rate of 3 per hour. The cost of having the subassemblies wait at the inspection station (thereby increasing in-process inventory and disrupting subsequent production) is estimated to be $10 per hour for each subassembly.

Determine whether to continue the status quo or adopt the proposal in order to minimize the expected total cost per hour.

A* **16.4-8.** A car rental agency has been subcontracting for the maintenance and repair of its cars. However, due to long delays in getting its cars back, the agency has decided to open its own maintenance shop to do this work more quickly. This shop will operate 42 hours per week.

Alternative 1 is to hire two mechanics (at a cost of $1,500 per week each), so that two cars can be worked on at a time. The time required by a mechanic to service a car has an exponential distribution with a mean of 5 hours.

Alternative 2 is to hire just one mechanic but to provide some additional special equipment (at a capitalized cost of $750 per week) to speed up the work. In this case, the maintenance work on each car is done in stages, where the time required for each stage has an Erlang distribution with the shape parameter $k = 4$, where the mean is 2 hours for the first stage and 1 hour for the second stage.

For both alternatives, the cars arrive according to a Poisson process, with a mean arrival rate of 0.3 car per hour (during work hours). The agency estimates that its net lost revenue due to having its cars unavailable for rental is $150 per week per car.

Which alternative should the agency choose to minimize the expected total cost per week?

A **16.4-9.** A certain small car-wash business is currently being analyzed to see if costs can be reduced. Customers arrive according to a Poisson process at a mean rate of 15 per hour, and only one car can be washed at a time. At present the time required to wash a car has an exponential distribution, with a mean of 4 minutes. It also has been noticed that if there are already 4 cars waiting (including the one being washed), then any additional arriving customers leave and take their business elsewhere. The lost incremental profit from each such lost customer is $6.

Two proposals have been made. Proposal 1 is to add certain equipment, at a capitalized cost of $6 per hour, which would reduce the expected washing time to 3 minutes. In addition, each arriving customer would be given a guarantee that if she had to wait longer than $\frac{1}{2}$ hour (according to a time slip she received upon arrival) before her car was ready, then she would receive a free car wash (at a marginal cost of $4 for the company). This guarantee would be well posted and advertised, so it is believed that no arriving customers would be lost.

Proposal 2 is to obtain the most advanced equipment available, at an increased cost of $20 per hour, and each car would be sent through two cycles of the process in succession. The time required for a cycle has an exponential distribution, with a mean of 1 minute, so the total expected washing time would be 2 minutes. Because of the increased speed and effectiveness, it is believed that essentially no arriving customers would be lost.

The owner also feels that because of the loss of customer goodwill (and consequent lost future business) when customers have to wait, a cost of $0.20 for each minute that a customer has to wait before her car wash begins should be included in the analysis of all alternatives.

Evaluate the expected total cost per hour $E(\text{TC})$ of the status quo, proposal 1, and proposal 2 to determine which should be chosen.

A **16.4-10.*** A single crew is provided for unloading and/or loading each truck that arrives at the loading dock of a warehouse. These trucks arrive according to a Poisson input process at a mean rate of 1 per hour. The time required by a crew to unload and/or load a truck has an exponential distribution (regardless of the crew size). The expected time required by a one-person crew is 1 hour.

The cost of providing each additional member of the crew is $20 per hour. The cost that is attributable to having a truck not in use (i.e., a truck standing at the loading dock) is estimated to be $30 per hour.

(*a*) Assume that the mean service rate of the crew is proportional to its size. What should the size be to minimize the expected total cost per hour?

(*b*) Assume that the mean service rate of the crew is proportional to the square root of its size. What should the size be to minimize expected total cost per hour?

A **16.4-11.** Trucks arrive at a warehouse according to a Poisson process with a mean rate of 4 per hour. Only one truck can be loaded at a time. The time required to load a truck has an exponential distribution with a mean of $10/n$ minutes, where n is the number of loaders ($n = 1, 2, 3, \ldots$). The costs are (i) $18 per hour for each loader and (ii) $20 per hour for each truck being loaded or waiting in line to be loaded. Determine the number of loaders that minimizes the expected hourly cost.

16.4-12. A company's machines break down according to a Poisson process at a mean rate of 3 per hour. Nonproductive time on any machine costs the company $60 per hour.

The company employs a maintenance person who repairs machines at a mean rate of μ machines per hour (when continuously busy) if the company pays that person a wage of 5μ per hour. The repair time has an exponential distribution.

Determine the hourly wage that minimizes the company's total expected cost.

16.4-13. Reconsider Prob. 15.7-6. Suppose that the waiting cost for airplanes at the maintenance base depends linearly on the number in the system with a cost of $90,000 per week per airplane. Suppose that the service rate is continuously adjustable with the cost per week of providing service rate μ given by the expression

$$40{,}000 + 10{,}000\mu.$$

Find the optimal value of μ.

A **16.4-14.*** A machine shop contains a grinder for sharpening the machine cutting tools. A decision must now be made on the speed at which to set the grinder.

The grinding time required by a machine operator to sharpen the cutting tool has an exponential distribution, where the mean $1/\mu$ can be set at anything from $\frac{1}{2}$ minute to 2 minutes, depending upon the speed of the grinder. The running and maintenance costs go up rapidly with the speed of the grinder, so the estimated cost per minute for providing a mean of $1/\mu$ is $\$0.40\mu^2$.

The machine operators arrive to sharpen their tools according to a Poisson process at a mean rate of 1 every 2 minutes. The estimated cost of an operator being away from the machine to the grinder is $0.80 per minute.

Plot the expected total cost per minute $E(\text{TC})$ versus μ over the feasible range for μ to solve graphically for the minimizing value of μ.

16.4-15. Consider the special case of model 2 where (1) any $\mu > \lambda/s$ is feasible and (2) both $f(\mu)$ and the waiting-cost function are linear functions, so that

$$E(\text{TC}) = sC_r\mu + C_w L,$$

where C_r is the marginal cost per unit time for each unit of a server's mean service rate and C_w is the cost of waiting per unit time for each customer. The optimal solution is $s = 1$ (by the optimality of a single-server result), and

$$\mu = \lambda + \sqrt{\frac{\lambda C_w}{C_r}}$$

for any queueing system fitting the $M/M/1$ model presented in Sec. 15.6.

Show that this μ is indeed optimal for the $M/M/1$ model.

16.4-16. Consider a harbor with a single dock for unloading ships. The ships arrive according to a Poisson process at a mean rate of λ ships per week, and the service-time distribution is exponential with a mean rate of μ unloadings per week. Assume that harbor facilities are owned by the shipping company, so that the objective is to balance the cost associated with idle ships with the cost of running the dock. The shipping company has no control over the arrival rate λ (that is, λ is fixed); however, by changing the size of the unloading crew, and so on, the shipping company can adjust the value of μ as desired.

Suppose that the expected cost per unit time of running the unloading dock is $D\mu$. The waiting cost for each idle ship is some constant (C) times the *square* of the total waiting time (including loading time). The shipping company wishes to adjust μ so that the expected total cost (including the waiting cost for idle ships) per unit time is minimized. Derive this optimal value of μ in terms of D and C.

16.4-17. Consider a queueing system with two types of customers. Type 1 customers arrive according to a Poisson process with a mean rate of 5 per hour. Type 2 customers also arrive according to a Poisson process with a mean rate of 5 per hour. The system has two servers, and both serve both types of customers. For types 1 and 2, service times have an exponential distribution with a mean of 10 minutes. Service is provided on a first-come-first-served basis.

Management now wants you to compare this system's design of having both servers serve both types of customers with the alternative design of having one server serve just type 1 customers and the other server serve just type 2 customers. Assume that this alternative design would not change the probability distribution of service times.

(*a*) Without doing any calculations, indicate which design would give a smaller expected total number of customers in the system. What result are you using to draw this conclusion?

A* (*b*) Verify your conclusion in part (*a*) by finding the expected total number of customers in the system under the original design and then under the alternative design.

16.4-18. Reconsider Prob 15.6-31.

(*a*) Formulate part (*a*) to fit as closely as possible a special case of one of the decision models presented in Sec. 16.4. (Do not solve.)

(*b*) Because the answer for part (*b*) reveals a very low utilization of the machine operators (who are relatively expensive employees), the department manager has asked you to analyze each of the following alternative ways of organizing the work of the operators: (i) Pool the operators so that *any* idle operator can take the next machine needing servicing, (ii) combine the operators into small *crews* to work together on any machine needing servicing within their assigned group of machines, and (iii) keep each operator assigned to her own group of machines but allow idle operators to assist busy operators on their machines. Describe each of these alternatives in queueing theory terms, including their relationship (if any) to the decision models presented in Sec. 16.4. Briefly indicate why each of these alternatives might decrease the total number of operators (thereby increasing their utilization) needed to achieve the required production rate. Also point out any dangers that might prevent this decrease.

16.4-19. Consider a factory whose floor area is a square with 600 feet on each side. Suppose that one service facility of a certain kind is provided in the center of the factory. The employees are distributed uniformly throughout the factory, and they walk to and from the facility at an average speed of 3 miles per hour along a system of orthogonal aisles.

Find the expected travel time $E(T)$ per arrival.

16.4-20. A certain large shop doing light fabrication work uses a single central storage facility (dispatch station) for material in in-process storage. The typical procedure is that each employee personally delivers his finished work (by hand, tote box, or hand cart) and receives new work and materials at the facility. Although this procedure worked well in earlier years when the shop was smaller, it appears that it may now be advisable to divide the shop into two semi-independent parts, with a separate storage facility for each one. You have been assigned the job of comparing the use of two facilities and of one facility from a cost standpoint.

The factory has the shape of a rectangle 150 by 100 yards. Thus, by letting 1 yard be the unit of distance, the (x, y) coordinates of the corners are (0, 0), (150, 0), (150, 100), and (0, 100). With this coordinate system, the existing facility is located at (50, 50), and the location available for the second facility is (100, 50).

Each facility would be operated by a single clerk. The time required by a clerk to service a caller has an exponential distribution, with a mean of 2 minutes. Employees arrive at the present facility according to a Poisson input process at a mean rate of 24 per hour. The employees are rather uniformly distributed throughout the shop, and if the second facility were installed, each employee would normally use the nearer of the two facilities. Employees walk at an average speed of about 5,000 yards per hour. All aisles are parallel to the outer walls of the shop. The net cost of providing each facility is estimated to be about $20 per hour, plus $15 per hour for the clerk. The estimated total cost of an employee being idled by traveling or waiting at the facility is $25 per hour.

Given the preceding cost factors, which alternative minimizes the expected total cost?

16.4-21.* Consider the formulation of the County Hospital emergency room problem as a preemptive priority queueing system, as presented in Sec. 15.8. Suppose that the following imputed costs are assigned to making patients wait (*excluding* treatment time): $10 per hour for stable cases, $1,000 per hour for serious cases, and $100,000 per hour for critical cases. The cost associated with having an additional doctor on duty would be $40 per hour. Referring to Table 15.4, determine on an expected-total-cost basis whether there should be one or two doctors on duty.

A* **16.4-22.** A certain job shop has been experiencing long delays in jobs going through the turret lathe department because of inadequate capacity. The supervisor contends that five machines are required, as opposed to the three machines now in place. However, because of pressure from management to hold down capital expenditures, only one additional machine will be authorized unless there is solid evidence that a second one is necessary.

This shop does three kinds of jobs, namely, government jobs, commercial jobs, and standard products. Whenever a turret lathe operator finishes a job, he starts a government job if one is waiting; if not, he starts a commercial job if any are waiting; if not, he starts on a standard product if any are waiting. Jobs of the same type are taken on a first-come-first-served basis.

Although much overtime work is required currently, management wants the turret lathe department to operate on an 8-hour, 5-day-per-week basis. The probability distribution of the time required by a turret lathe operator for a job appears to be approximately exponential, with a mean of 10 hours. Jobs come into the shop according to a Poisson input process, but at a mean rate of 6 per week for government jobs, 4 per week for commercial jobs, and 2 per week for standard products. (These figures are expected to remain the same for the indefinite future.)

It is worth about $750, $450, and $150 to avoid a delay of 1 additional (working) day in a government, commercial, and standard job, respectively. The incremental capitalized cost of providing each turret lathe (including the operator and so on) is estimated to be $250 per working day.

Determine the number of additional turret lathes that should be obtained to minimize the expected total cost.

17

Inventory Theory

Keeping an inventory (stock of goods) for future sale or use is very common in business. Retail firms, wholesalers, manufacturing companies—and even blood banks—generally have a stock of goods on hand. How does such a facility decide upon its "inventory policy," i.e., its policy for when to replenish inventory and by how much? In a small firm, the manager may keep track of inventory and make these decisions. However, since this may not be feasible even in small firms, many companies have saved large sums of money by using *scientific inventory management.* In particular, they

1. Formulate a mathematical model describing the behavior of the inventory system.
2. Derive an optimal inventory policy with respect to this model.
3. Frequently use a computer to maintain a record of the inventory levels and to signal when and how much to replenish.

One example of a company (IBM) that has used this approach very successfully is described at the end of Secs. 2.4, 2.5, and 2.6. By using inventory theory to integrate its national network of spare-parts inventories, IBM improved its service to customers *and* reduced the capital tied up in inventory by over $250 million *and* saved an additional $20 million per year through improved operational efficiency. Section 2.5 describes the massive computer system used to control this inventory system.

There are several basic considerations involved in determining an inventory policy that must be reflected in the mathematical inventory model. These are illustrated in the examples presented in the first section and then are described in general terms in Sec. 17.2. Section 17.3 develops and analyzes *deterministic* inventory models, i.e., models where the future demand for withdrawing the product from inventory is assumed to be known. Section 17.4 deals with *stochastic* models where this demand is a random variable.

17.1 Examples

We present two examples in rather different contexts (a manufacturer and a wholesaler) where an inventory policy needs to be developed.

Example 1: Manufacturing Speakers for TV Sets

A television manufacturing company produces its own speakers, which are used in the production of its television sets. The television sets are assembled on a continuous production line at a rate of 8,000 per month, with one speaker needed per set. The speakers are produced in batches because they do not warrant setting up a continuous production line, and relatively large quantities can be produced in a short time. Therefore, the speakers are placed into inventory until they are needed for assembly into television sets on the production line. The company is interested in determining when to produce a batch of speakers and how many speakers to produce in each batch. Several costs must be considered:

1. Each time a batch is produced, a **setup cost** of $12,000 is incurred. This cost includes the cost of "tooling up," administrative costs, record keeping, and so forth. Note that the existence of this cost argues for producing speakers in large batches.
2. The **unit production cost** of a single speaker (excluding the setup cost) is $10, independent of the batch size produced. (In general, however, the unit production cost need not be constant and may decrease with batch size.)
3. The production of speakers in large batches leads to a large inventory. The estimated **holding cost** of keeping a speaker in stock is $0.30 per month. This cost includes the cost of capital tied up, storage space, insurance, taxes, protection, and so on. The existence of a holding cost argues for producing small batches.
4. Company policy prohibits deliberately planning for shortages of any of its components. However, a shortage of speakers occasionally crops up, and it has been estimated that each speaker that is not available when required costs $1.10 per month. This **shortage cost** includes the cost of installing speakers after the television set is fully assembled, storage space, delayed revenue, record keeping, and so forth.

We will develop the inventory policy for this example with the help of the first inventory model presented in Sec. 17.3.

Example 2: Wholesale Distribution of Bicycles

A wholesale distributor of bicycles is having trouble with shortages of a popular model (a small, one-speed girl's bicycle) and is currently reviewing the inventory policy for this model. The distributor purchases this model bicycle from the manufacturer monthly and then supplies it to various bicycle shops in the western United States in response to purchase orders. What the total demand from bicycle shops will be in any given month is quite uncertain. Therefore, the question is, How many bicycles should be ordered from the manufacturer for any given month, given the stock level leading into that month?

The distributor has analyzed her costs and has determined that the following are important:

1. The **ordering cost**, i.e., the cost of placing an order plus the cost of the bicycles being purchased, has two components: The cost of paperwork involved in placing an order is estimated as $200, and the actual cost of each bicycle is $35 for this wholesaler.
2. The *holding cost,* i.e., the cost of maintaining an inventory, is $1 per bicycle remaining at the end of the month. This cost represents the costs of capital tied up, warehouse space, insurance, taxes, and so on.
3. The *shortage cost* is the cost of not having a bicycle on hand when needed. Most models are easily reordered from the manufacturer, and stores usually accept a delay in delivery. Still, although shortages are permissible, the distributor feels that she incurs a loss, which she estimates to be $15 per bicycle. This cost represents an evaluation of the cost of the loss of goodwill, additional clerical costs incurred, and the cost of the delay in revenue received. On a very few competitive (in price) models, stores do not accept a delay, which results in lost sales. In this case, the cost of lost revenue must be included in the shortage cost.

We will return to this example again in Sec. 17.4.

These examples illustrate that there are two possibilities for how a firm *replenishes inventory,* depending on the situation. One possibility is that the firm *produces* the needed units itself (like the television manufacturer producing speakers). The other is that the firm *orders* the units from a supplier (like the bicycle distributor ordering bicycles from the manufacturer). Inventory models do not need to distinguish between these two ways of replenishing inventory, so we will use such terms as *producing* and *ordering* interchangeably.

Both examples deal with one specific product (speakers for a certain kind of television set or a certain bicycle model). In most inventory models, just one product is being considered at a time. Except for one subsection at the end of the chapter, all the inventory models presented here assume a single product.

Both examples indicate that there exists a trade-off between the costs involved. The next section discusses the basic cost components of inventory models for determining the optimal trade-off between these costs.

17.2 Components of Inventory Models

Because inventory policies affect profitability, the choice among policies depends upon their relative profitability. As already seen in Examples 1 and 2, some of the costs that determine this profitability are (1) the costs of ordering or manufacturing, (2) holding costs, and (3) shortage costs. Other relevant factors include (4) revenues, (5) salvage costs, and (6) discount rates. These six factors are described in turn below.

The **cost of ordering or manufacturing** an amount z can be represented by a function $c(z)$. The simplest form of this function is one that is directly proportional to the amount ordered or produced, that is, $c \cdot z$, where c represents the unit price paid. Another common assumption is that $c(z)$ is composed of two parts: a term that is directly proportional to the amount ordered or produced and a term that is a constant K for z positive and is 0 for $z = 0$. For this case,

$$c(z) = \text{cost of ordering or manufacturing } z \text{ units}$$
$$= \begin{cases} 0 & \text{if } z = 0, \\ K + cz & \text{if } z > 0, \end{cases}$$

where K = setup cost and c = unit cost.

The constant K includes the administrative cost of ordering or, when manufacturing, the preliminary labor and other expenses of starting a production run.

There are other assumptions that can be made about the cost of ordering or manufacturing, but this chapter is restricted to the cases just described.

In Example 1, the speakers are manufactured, and the setup cost for a production run is \$12,000. Furthermore, each speaker costs \$10, so that the *production* cost is given by

$$c(z) = 12{,}000 + 10z, \qquad \text{for } z > 0.$$

In Example 2, the distributor orders bicycles from the manufacturer, and the *ordering* cost is given by

$$c(z) = 200 + 35z, \qquad \text{for } z > 0.$$

The *holding cost* (sometimes called the *storage cost*) represents all the costs associated with the storage of the inventory until it is sold or used. Included are the cost of capital tied up, space, insurance, protection, and taxes attributed to storage. These costs may be a function of the maximum quantity held during a period, the average amount held, or the cumulated excess of supply over the amount required (demand). The last viewpoint is usually taken in this chapter.

In the bicycle example, the holding cost is \$1 per bicycle remaining at the end of the month. This cost can be interpreted as the interest lost in keeping capital tied up in an ''unnecessary'' bicycle for a month, the cost of extra storage space, insurance, and so forth.

The *shortage cost*—sometimes called the *unsatisfied demand cost*—is incurred when the amount of the commodity required (demand) exceeds the available stock. This cost depends upon which of the following two cases applies.

In one case, called **backlogging**, the excess demand is not lost, but instead is held until it can be satisfied when the next normal delivery replenishes the inventory. For a firm incurring a temporary shortage in supplying its customers (as for the bicycle example in Sec. 17.1), the shortage cost then can be interpreted as the loss of customers' goodwill due to the delay, their subsequent reluctance to do business with the firm, the cost of delayed revenue, and extra record keeping. For a manufacturer incurring a temporary shortage in materials needed for production (such as a shortage of speakers for assembly into television sets), the shortage cost becomes the cost associated with delaying the completion of the production process.

In the second case, called **no backlogging**, if any excess of demand over available stock occurs, the distributor cannot wait for the next normal delivery to replenish inventory. Either (1) the excess demand is met by a priority shipment, or (2) it is not met at all. For situation 1, the shortage cost can be viewed as the cost of the priority shipment. For situation 2, the shortage cost can be viewed as the loss in current revenue from not meeting the demand, plus the cost of losing future business because of lost goodwill.

Revenue may or may not be included in the model. If both the price and the demand for the product are established by the market and so are outside the control of the company, the revenue from sales (assuming demand is met) is independent of the firm's inventory policy and may be neglected. However, if revenue is neglected in the model, the *loss in revenue* must then be included in the shortage cost whenever the firm cannot meet the demand and the sale is lost. Furthermore, even in the case where demand is backlogged, the cost of the delay in revenue must also be included in the shortage cost. With these interpretations, revenue will not be considered explicitly in the remainder of this chapter.

The **salvage value** of an item is the value of a leftover item when no further inventory is desired. The salvage value represents the disposal value of the item to the firm, perhaps through a discounted sale. The negative of the salvage value is called the **salvage cost**. If there is a cost associated with the disposal of an item, the salvage cost may be positive. We assume hereafter that any salvage cost is incorporated into the *holding cost.*

Finally, the **discount rate** takes into account the time value of money. When a firm ties up capital in inventory, the firm is prevented from using this money for alternative purposes. For example, it could invest this money in secure investments, say, government bonds, and have a return on investment 1 year hence of, say, 7 percent. Thus \$1 invested today would be worth \$1.07 in year 1, or alternatively, a \$1 profit 1 year hence is equivalent to $\alpha = \$1/1.07$ today. The quantity α is known as the **discount factor**. Thus, in considering the profitability of an inventory policy, the profit or costs 1 year hence should be multiplied by α; in 2 years hence, by α^2; and so on. (Units of time other than 1 year also can be used.)

In problems having short time horizons, α may be assumed to be 1 (and thereby neglected) because the current value of \$1 delivered during this short time horizon does not change very much. However, in problems having long time horizons, the discount factor must be included.

In using quantitative techniques to seek optimal inventory policies, we use the criterion of minimizing the total (expected) discounted cost. Under the assumptions that the price and demand for the product are not under the control of the company and that the lost or delayed revenue is included in the shortage penalty cost, minimizing cost is equivalent to maximizing net income. Another useful criterion is to keep the inventory policy simple, i.e., keep the rule for indicating *when to order* and *how much to order* both understandable and easy to implement. Most of the policies considered in this chapter possess this property.

Inventory models are usually classified according to whether the demand for a period is known (deterministic demand) or is a random variable having a known probability distribution (nondeterministic or random demand). The production of batches of speakers in Example 1 of Sec. 17.1 illustrates deterministic demand because the speakers are used in television assemblies at a fixed rate of 8,000 per month. The bicycle shops' purchases of bicycles from the wholesale distributor in Example 2 of Sec. 17.1 illustrates random demand. This classification is frequently coupled with whether or not there exist time lags in the delivery of the items ordered or produced. In both examples in Sec. 17.1, there was an implication that the items appeared immediately after an order was placed. In fact, the production of speakers may require some time, and similarly, the delivery of bicycles to the wholesaler may not be instantaneous, so that time lags may have to be incorporated into the inventory model. However, for simplicity, all the models considered in this chapter assume instantaneous deliver.[1]

Another possible classification relates to the way the inventory is reviewed, either continuously or periodically. In **continuous review**, an order is placed as soon as the stock level falls below the prescribed reorder point, whereas in **periodic review** the inventory level is checked at discrete intervals, e.g., at the end of each week, and ordering decisions are made only at these times even if the inventory level dips below the reorder point between the preceding and current review times. (In practice, a periodic review policy can be used to approximate a continuous review policy by making the time interval sufficiently small.)

17.3 Deterministic Models

This section is concerned with inventory problems where the actual demand in the future is assumed to be known. Several models are considered, including the well-known economic lot-size formulation, considered first.

Continuous Review—Uniform Demand

The most common inventory problem faced by manufacturers, retailers, and wholesalers is that stock levels are depleted over time and then are replenished by the arrival of new units. A simple model representing this situation is the following **economic lot-size model**. (It sometimes is also referred to as the **economic order quantity model** or, for short, the **EOQ model**.)

Units of the product under consideration are assumed to be withdrawn from inventory continuously at a *known constant rate*, denoted by a; that is, the demand is a units per unit time. It is further assumed that inventory is replenished when needed by producing or ordering a batch of fixed size (Q units), where all Q units arrive simultaneously at the desired time. The only costs to be considered are

$$K = \text{setup cost for producing or ordering one batch,}$$

$$c = \text{unit cost for producing or purchasing each unit,}$$

$$h = \text{holding cost per unit per unit of time held in inventory.}$$

[1] Results for the delivery-lag case often can be obtained from the corresponding instantaneous delivery model by a simple modification in the calculation of some of the inventory costs.

The objective is to determine when and by how much to replenish inventory so as to minimize the sum of these costs per unit time.

We assume *continuous review,* so that inventory can be replenished whenever the inventory level drops sufficiently low. We shall first assume that shortages are not allowed (but later we will relax this assumption). With the fixed demand rate, shortages can be avoided by replenishing inventory each time the inventory level drops to zero, and this also will minimize the holding cost. Figure 17.1 depicts the resulting pattern of inventory levels over time when we start at time 0 by producing or ordering a batch of Q units in order to increase the initial inventory level from 0 to Q.

Example 1 in Sec. 17.1 (manufacturing speakers for TV sets) fits this model and will be used to illustrate the following discussion.

Shortages Not Permitted: For the speaker example, a *cycle* can be viewed as the time between production runs. Thus, if 24,000 speakers are produced in each production run and are used at the rate of 8,000 per month, then the cycle length is 24,000/8,000 = 3 months. In general, the cycle length is Q/a, as illustrated in Fig. 17.1.

The total cost per unit time T is obtained from the following components.

$$\text{Production or ordering cost per cycle} = K + cQ.$$

The average inventory level during a cycle is $(Q + 0)/2 = Q/2$ units per unit time, and the corresponding cost is $hQ/2$ per unit time. Because the cycle length is Q/a,

$$\text{Holding cost per cycle} = \frac{hQ^2}{2a}.$$

Therefore,

$$\text{Total cost per cycle} = K + cQ + \frac{hQ^2}{2a},$$

so the total cost per unit time is

$$T = \frac{K + cQ + hQ^2/(2a)}{Q/a} = \frac{aK}{Q} + ac + \frac{hQ}{2}.$$

The value of Q, say Q^*, that minimizes T is found by setting the first derivative to zero (and noting that the second derivative is positive).

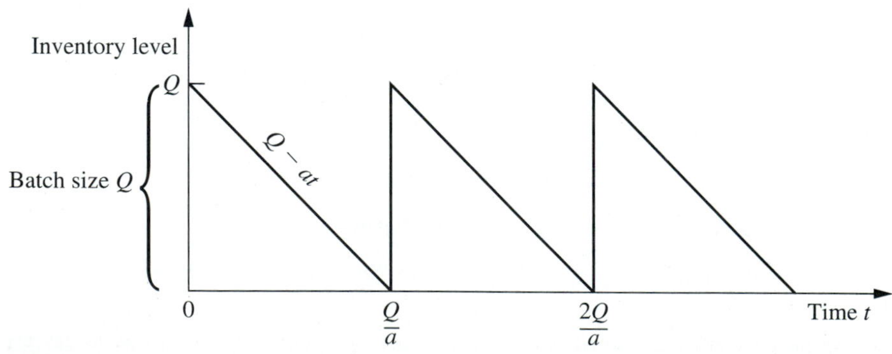

Figure 17.1 Diagram of inventory level as a function of time when no shortages are permitted.

$$\frac{dT}{dQ} = -\frac{aK}{Q^2} + \frac{h}{2} = 0,$$

so that

$$Q^* = \sqrt{\frac{2ak}{h}},$$

which is the well-known *economic lot-size formula*.[1] Similarly, the time it takes to withdraw this optimal value of Q^*, say t^*, is given by

$$t^* = \frac{Q^*}{a} = \sqrt{\frac{2K}{ah}}.$$

It is interesting to observe that Q^* and t^* change in intuitively plausible ways when a change is made in K, h, or a. As the setup cost K increases, both Q^* and t^* increase (fewer setups). When the unit holding cost h increases, both Q^* and t^* decrease (smaller inventory levels). As the demand rate a increases, Q^* increases (larger batches) but t^* decreases (more frequent setups).

These formulas for Q^* and t^* will now be applied to the speaker example. The appropriate parameter values from Sec. 17.1 are

$$K = 12{,}000, \qquad h = 0.30, \qquad a = 8{,}000,$$

so that

$$Q^* = \sqrt{\frac{(2)(8{,}000)(12{,}000)}{0.30}} = 25{,}298$$

and

$$t^* = \frac{25{,}298}{8{,}000} = 3.2 \text{ months.}$$

Hence, the optimal solution is to set up the production facilities to produce speakers once every 3.2 months and to produce 25,298 speakers each time. (The total cost curve is rather flat near this optimal value, so any similar production run that might be more convenient, say 24,000 speakers every 3 months, would be nearly optimal.)

SHORTAGES PERMITTED: It may be worthwhile to permit small shortages to occur because the cycle length can then be increased with a resulting saving in setup costs. However, this benefit may be offset by the *shortage cost,* so a detailed analysis is required. Let

p = shortage cost per unit short per unit of time short,

S = inventory level just after a batch of Q units is added to inventory,

$Q - S$ = shortage in inventory just before a batch of Q units is added.

The resulting pattern of inventory levels over time is shown in Fig. 17.2 when one starts at time 0 with an inventory level of S.

[1] An interesting historical account of this model and formula, including a reprint of a 1913 paper that started it all, is given by D. Erlenkotter, ''Ford Whitman Harris and the Economic Order Quantity Model,'' *Operations Research,* **38**: 937–950, 1990.

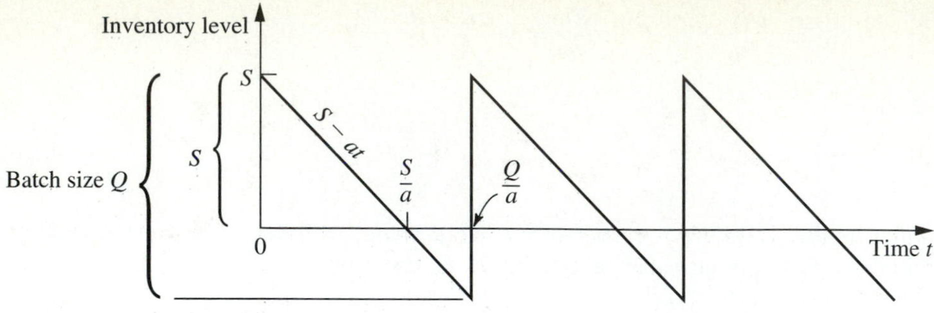

Figure 17.2 Diagram of inventory level as a function of time when shortages are permitted.

The total cost per unit time now is obtained from the following components.

$$\text{Production or ordering cost per cycle} = K + cQ.$$

During each cycle, the inventory level is positive for a time S/a. The average inventory level *during this time* is $(S + 0)/2 = S/2$ units per unit time, and the corresponding cost is $hS/2$ per unit time. Hence,

$$\text{Holding cost per cycle} = \frac{hS}{2}\frac{S}{a} = \frac{hS^2}{2a}.$$

Similarly, shortages occur for a time $(Q - S)/a$. The average amount of shortages *during this time* is $(0 + Q - S)/2 = (Q - S)/2$ units per unit time, and the corresponding cost is $p(Q - S)/2$ per unit time. Hence,

$$\text{Shortage cost per cycle} = \frac{p(Q - S)}{2}\frac{Q - S}{a} = \frac{p(Q - S)^2}{2a}.$$

Therefore,

$$\text{Total cost per cycle} = K + cQ + \frac{hS^2}{2a} + \frac{p(Q - S)^2}{2a},$$

and the *total cost per unit time* is

$$T = \frac{K + cQ + hS^2/(2a) + p(Q - S)^2/(2a)}{Q/a}$$

$$= \frac{aK}{Q} + ac + \frac{hS^2}{2Q} + \frac{p(Q - S)^2}{2Q}.$$

In this model, there are two decision variables (S and Q), so the optimal values (S^* and Q^*) are found by setting the partial derivatives $\partial T/\partial S$ and $\partial T/\partial Q$ equal to zero. Thus,

$$\frac{\partial T}{\partial S} = \frac{hS}{Q} - \frac{p(Q - S)}{Q} = 0.$$

$$\frac{\partial T}{\partial Q} = -\frac{aK}{Q^2} - \frac{hS^2}{2Q^2} + \frac{p(Q - S)}{Q} - \frac{p(Q - S)^2}{2Q^2} = 0.$$

Solving these equations simultaneously leads to

$$S^* = \sqrt{\frac{2aK}{h}}\sqrt{\frac{p}{p+h}}, \qquad Q^* = \sqrt{\frac{2aK}{h}}\sqrt{\frac{p+h}{p}}.$$

The optimal cycle length t^* is given by

$$t^* = \frac{Q^*}{a} = \sqrt{\frac{2K}{ah}}\sqrt{\frac{p+h}{p}}.$$

The maximum shortage is

$$Q^* - S^* = \sqrt{\frac{2aK}{p}}\sqrt{\frac{h}{p+h}}.$$

Further, from Fig. 17.2, the fraction of time that no shortage exists is given by

$$\frac{S^*/a}{Q^*/a} = \frac{p}{p+h},$$

which is independent of K.

When either p or h is made much larger than the other, the above quantities behave in intuitive ways. In particular, when $p \to \infty$ with h constant (so shortage costs dominate holding costs), $Q^* - S^* \to 0$ whereas both Q^* and t^* converge to their values for the preceding model (shortages not permitted). Even though the current model permits shortages, $p \to \infty$ implies that having them is not worthwhile.

On the other hand, when $h \to \infty$ with p constant (so holding costs dominate storage costs), $S^* \to 0$. Thus, having $h \to \infty$ makes it uneconomical to have positive inventory levels, so each new batch of Q^* units goes no further than removing the current shortage in inventory.

If shortages are permitted in the speaker example, the *shortage cost* is estimated in Sec. 17.1 as

$$p = 1.10.$$

As before,

$$K = 12{,}000, \qquad h = 0.30, \qquad a = 8{,}000,$$

so now

$$S^* = \sqrt{\frac{(2)(8{,}000)(12{,}000)}{0.30}}\sqrt{\frac{1.1}{1.1+0.3}} = 22{,}424,$$

$$Q^* = \sqrt{\frac{(2)(8{,}000)(12{,}000)}{0.30}}\sqrt{\frac{1.1+0.3}{1.1}} = 28{,}540,$$

and

$$t^* = \frac{28{,}540}{8{,}000} = 3.6 \text{ months.}$$

Hence, the production facilities are to be set up every 3.6 months to produce 28,540 speakers. The maximum shortage is 6,116 speakers. Note that Q^* and t^* are not very different from the no-shortage case.

Quantity Discounts, Shortages Not Permitted: The above models have assumed that the unit cost of an item is the same regardless of the quantity in the batch. In fact, this assumption resulted in the optimal solutions being independent of this unit cost. Suppose, however, that there exist cost breaks: i.e., the unit cost varies with the batch size. For example, suppose the unit cost for *every* speaker is $c_1 = \$11$ if less than 10,000 speakers are produced, $c_2 = \$10$ if production falls between 10,000 and 80,000 speakers, and $c_3 = \$9.50$ if production exceeds 80,000 speakers. What is the optimal policy? The solution to this specific problem will reveal the general method.

From the results for the first economic lot-size model (shortages not permitted), the total cost per unit time T_j if the unit cost is c_j is given by

$$T_j = \frac{aK}{Q} + ac_j + \frac{hQ}{2}, \qquad \text{for } j = 1, 2, 3.$$

A plot of T_j versus Q is shown in Fig. 17.3.

The feasible values of Q are shown by the solid lines, and it is only these regions that must be investigated. For each curve, the value of Q that minimizes T_j is found just as for the first economic lot-size model. For $K = 12{,}000$, $h = 0.30$, and $a = 8{,}000$, this value is

$$\sqrt{\frac{(2)(8{,}000)(12{,}000)}{0.30}} = 25{,}298.$$

This number is a feasible value for the cost function T_2. For any fixed Q, $T_j < T_{j-1}$ for all j, so T_1 can be eliminated from further consideration. However, T_3 cannot be immediately discarded. Its minimum feasible value (which occurs at $Q = 80{,}000$) must be compared to T_2 evaluated at 25,298 (which is \$87,589). Because T_3 evaluated at 80,000 equals \$89,200, it is better to produce in quantities of 25,298, so this quantity is the optimal value for this set of quantity discounts.

If the quantity discount led to a unit cost of \$9 (instead of \$9.50) when production exceeded 80,000, then T_3 evaluated at 80,000 would equal 85,200, and the optimal production quantity would become 80,000.

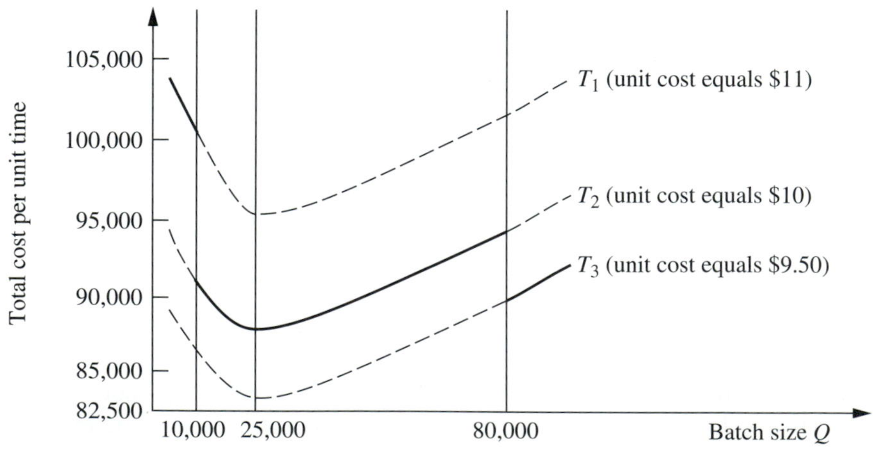

Figure 17.3 Total cost per unit time for the speaker example with quantity discounts.

Although this analysis concerned a specific problem, the same approach is applicable to any similar problem. Furthermore, a similar analysis can be used for other types of quantity discounts, such as incremental quantity discounts where a cost c_0 is incurred for the first q_0 units, c_1 for the next q_1 units, and so on.

Observations about Economic Lot-Size Models

1. If it is assumed that the unit cost of an item is constant throughout time, the unit cost does not appear in the optimal solution for the batch size. This result occurs because no matter what policy is used, the same number of units is required per unit time, so this cost per unit time is fixed.
2. It was assumed that the batch size Q *is* constant from cycle to cycle. The resulting *optimal* batch size Q^* actually minimizes the total cost per unit time for any cycle, so the analysis shows that this constant batch size should be used from cycle to cycle even if a constant batch size is not required.
3. The optimal inventory level at which inventory should be replenished can never be greater than zero under these models. Waiting until the inventory level drops to zero (or less than zero when shortages are permitted) reduces both holding costs and the frequency of incurring the setup cost K. However, if the assumptions of *a known constant demand rate* and *instantaneous delivery* are not completely satisfied, it may become prudent to plan to have some "safety stock" left when the inventory is scheduled to be replenished.
4. The models assume *instantaneous delivery*. However, a *delivery lag* can be accommodated if the lag time is *known*. Simply order the delivery far enough in advance to have it arrive when needed.

Periodic Review—A General Model for Production Planning

The preceding analysis explored the economic lot-size model. The results were dependent upon the assumption of a constant demand rate. When this assumption is relaxed, i.e., when the amounts required from period to period are allowed to vary, the *economic lot-size formula* no longer ensures a minimum cost solution.

Consider the following model. Planning is to be done for the next n periods regarding how much (if any) to produce or order to replenish inventory in each of the periods. The demands for the respective periods are *known* (but *not* the same in every period) and are denoted by

$$r_i = \text{demand in period } i, \qquad \text{for } i = 1, 2, \ldots, n.$$

These demands must be met on time. There is no stock on hand initially, but there is still time for a delivery at the beginning of period 1.

The costs included in this model are similar to those for the first economic lot-size model:

K = setup cost for producing or ordering any units to replenish inventory at beginning of period,

c = unit cost for producing or ordering each unit,

h = holding cost for each unit left in inventory at end of period.

The objective is to minimize the total cost over n periods.

We will illustrate this model by the following variation of the speaker example introduced in Sec. 17.1.

EXAMPLE: A market survey conducted by the television manufacturer has indicated that the demand for television sets is seasonal rather than uniform. In particular, sales of 30,000 sets is forecast for the Christmas season (October to December), 20,000 for the winter slack season (January to March), 30,000 for the "new model" season (April to June), and 20,000 for the summer season (July to September). To meet these sales in the respective seasons, the speakers must be available for assembly into the television set at the beginning of the season. Since all the sales forecasts are *integer multiples* of 10,000, it has been decided to produce the speakers in integer multiples of 10,000 as well. In these units, the demands for speakers at the beginning of the four upcoming periods (seasons) are

$$r_1 = 3, \qquad r_2 = 2, \qquad r_3 = 3, \qquad r_4 = 2.$$

The revised cost estimates for the speakers now are

$$K = 20{,}000, \qquad c = 1, \qquad h = 0.20.$$

The problem is to determine how many speakers to produce (if any) for the beginning of each of the four periods in order to minimize the total cost.

The high setup cost K gives a strong incentive not to produce speakers every period and preferably just once. However, the significant holding cost h makes it undesirable to carry a large inventory by producing the entire demand for all four periods (100,000 speakers) at the beginning. Perhaps the best approach would be an intermediate strategy where speakers are produced more than once but less than four times. For example, one such feasible solution (but not an optimal one) is depicted in Fig. 17.4, which shows the evolution of the inventory level (in units of 10,000 speakers) over the next year that results from producing 30,000 speakers at the beginning of the first period (Christmas season), 60,000 speakers at the beginning of the second period, and 10,000 speakers at the beginning of the fourth period. The dots give the inventory levels after any production at the beginning of the four periods.

How can the optimal production schedule be found? For this model in general, production (or ordering) is automatic in period 1, but a decision on whether to produce must be made for each of the other $n - 1$ periods. Therefore, one approach to solving this model is to enumerate, for each of the 2^{n-1} combinations of production decisions, the possible quantities that can be produced in each period where production is to

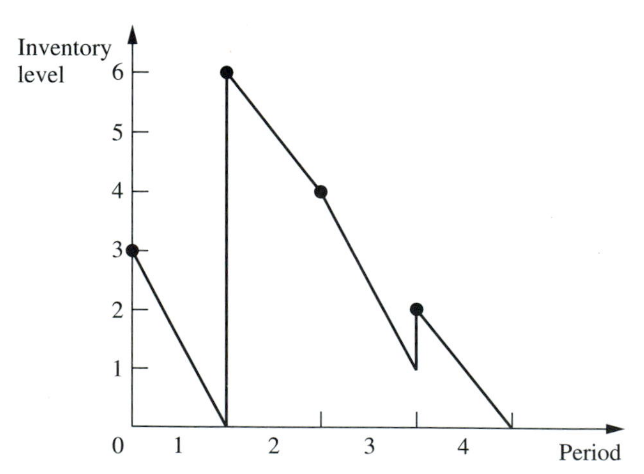

Figure 17.4 The inventory levels that result from one sample production schedule for the revised speaker example.

occur. This approach is rather cumbersome, even for moderate-sized n, so a more efficient method is desirable. Such a method is described next in general terms, and then we will return to finding the optimal production schedule for the example.

Periodic Review—Production Planning: An Algorithm

The key to developing an efficient algorithm for finding an *optimal inventory policy* (or equivalently, an *optimal production schedule*) for the above model is the following insight into the nature of an optimal policy.

> An optimal policy (production schedule) produces *only* when the inventory level is *zero*.

To illustrate why this result is true, consider the policy shown in Fig. 17.4 for the example. (Call it policy A.) Policy A violates the above characterization of an optimal policy because production occurs at the beginning of period 4 when the inventory level is *greater than zero* (namely, one unit of 10,000 speakers). However, this policy can easily be adjusted to satisfy the above characterization by simply producing one less unit in period 2 and one more unit in period 4. This adjusted policy (call it B) is shown by the dashed line in Fig. 17.5 wherever B differs from A (the solid line). Now note that policy B *must* have less total cost than policy A. The setup and the production costs for both policies are the same. However, the holding cost is smaller for B than for A because B has less inventory than A in periods 2 and 3 (and the same inventory in the other periods). Therefore, B is better than A, so A cannot be optimal.

This characterization of optimal policies can be used to identify policies that are not optimal. In addition, because it implies that the only choices for the amount produced at the beginning of the ith period are $0, r_i, r_i + r_{i+1}, \ldots,$ or $r_i + r_{i+1} + \cdots + r_n$, it can be exploited to obtain an efficient algorithm that is related to the *deterministic dynamic programming* approach described in Sec. 10.3.

In particular, define

C_i = total cost of an optimal policy for periods $i, i + 1, \ldots, n$ when period i starts with zero inventory (before producing), for $i = 1, 2, \ldots, n$.

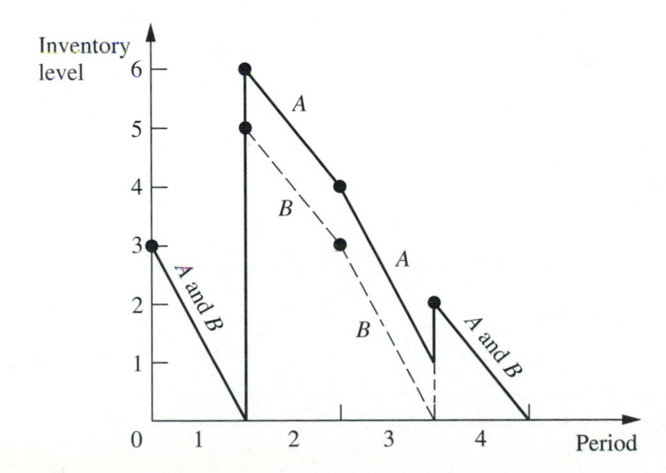

Figure 17.5 Comparison of two inventory policies (production schedules) for the revised speaker example.

By using the dynamic programming approach of solving *backward* period by period, these C_i values can be found by first finding C_n, then finding C_{n-1}, and so on. Thus, after $C_n, C_{n-1}, \ldots, C_{i+1}$ are found, then C_i can be found from the *recursive relationship*

$$C_i = \underset{j=i,\, i+1, \ldots, n}{\text{minimum}} \{C_{j+1} + K + c(r_i + r_{i+1} + \cdots + r_j) + h[r_{i+1} + 2r_{i+2} + 3r_{i+3} + \cdots + (j-i)r_j]\},$$

where j can be viewed as an index that denotes the (end of the) period when the inventory reaches a zero level for the first time after production at the beginning of period i. In the time interval from period i through period j, the term with coefficient c represents the total *production cost* over this interval, and the term with coefficient h represents the total *holding cost* over the same interval. When $j = n$, the term $C_{n+1} = 0$. The *minimizing value* of j indicates that if the inventory level does indeed drop to zero upon entering period i, then the production in period i should cover all demand form period i through this period j.

The algorithm for solving the model consists basically of solving for $C_n, C_{n-1}, \ldots, C_1$ in turn. For $i = 1$, the minimizing value of j then indicates that the production in period 1 should cover the demand through period j, so the second production will be in period $j + 1$. For $i = j + 1$, the new minimizing value of j identifies the time interval covered by the second production, and so forth to the end. We will illustrate this approach with the example.

The application of this algorithm is much quicker than the full dynamic programming approach.[1] As in dynamic programming, $C_n, C_{n-1}, \ldots, C_2$ must be found before C_1 is obtained. However, the number of calculations is much smaller, and the number of possible production quantities is greatly reduced.

Example: Returning to the speaker example, first we consider the case of finding C_4, the cost of the optimal policy from the beginning of period 4 to the end of the planning horizon:

$$C_4 = C_5 + 2 + 1(2) = 0 + 2 + 2 = 4.0.$$

To find C_3, we must consider two cases, namely, the first time after period 3 when the inventory reaches a zero level, which occurs at (1) the end of the third period or (2) the end of the fourth period. In the recursive relationship for C_3, these two cases correspond to (1) $j = 3$ and (2) $j = 4$. Denote the corresponding costs (the right-hand side of the recursive relationship with this j) by $C_3^{(3)}$ and $C_3^{(4)}$, respectively. The policy associated with $C_3^{(3)}$ calls for producing only for period 3 and then following the optimal policy for period 4, whereas the policy associated with $C_3^{(4)}$ calls for producing for periods 3 and 4. The cost C_3 is then the minimum of $C_3^{(3)}$ and $C_3^{(4)}$. These cases are reflected by the policies given in Fig. 17.6.

$$C_3^{(3)} = C_4 + 2 + 1(3) = 4 + 2 + 3 = 9.$$

$$C_3^{(4)} = C_5 + 2 + 1(3 + 2) + \tfrac{1}{5}(2) = 0 + 2 + 5 + 0.4 = 7.4.$$

$$C_3 = \min\{7.4, 9.0\} = 7.4.$$

[1] The full dynamic programming approach is useful, however, for solving *generalizations* of the production planning model (e.g., *nonlinear* production cost and holding cost functions) where the above algorithm is no longer applicable. (See Prob. 17.3-13 for an example.)

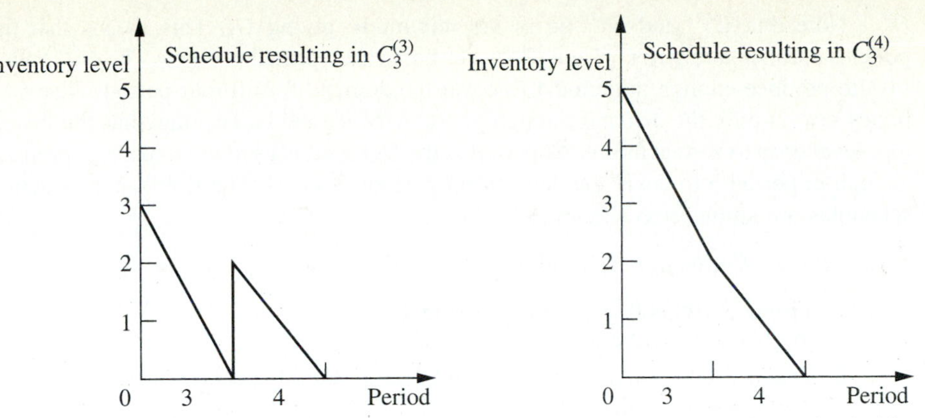

Figure 17.6 Alternative production schedules when production is required at the beginning of period 3.

Therefore, if the inventory level drops to zero upon entering period 3 (so production should occur then), the production in period 3 should cover the demand for both periods 3 and 4.

To find C_2, we must consider three cases, namely, the first time after period 2 when the inventory reaches a zero level, which occurs at (1) the end of the second period, (2) the end of the third period, or (3) the end of the fourth period. In the recursive relationship for C_2, these cases correspond to (1) $j = 2$, (2) $j = 3$, and (3) $j = 4$, where the corresponding costs are $C_2^{(2)}$, $C_2^{(3)}$, and $C_2^{(4)}$, respectively. The cost C_2 is then the minimum of $C_2^{(2)}$, $C_2^{(3)}$, and $C_2^{(4)}$.

$$C_2^{(2)} = C_3 + 2 + 1(2) = 7.4 + 2 + 2 = 11.4,$$

$$C_2^{(3)} = C_4 + 2 + 1(2 + 3) + \tfrac{1}{5}(3) = 4 + 2 + 5 + 0.6 = 11.6,$$

$$C_2^{(4)} = C_5 + 2 + 1(2 + 3 + 2) + \tfrac{1}{5}[3 + 2(2)] = 0 + 2 + 7 + 1.4 = 10.4.$$

$$C_2 = \min\{11.4, 11.6, 10.4\} = 10.4.$$

Consequently, if production occurs in period 2 (because the inventory level drops to zero), this production should cover the demand for all the remaining periods.

Finally, to find C_1, we must consider four cases, namely, the first time after period 1 when the inventory reaches zero, which occurs at the end of (1) the first period, (2) the second period, (3) the third period, or (4) the fourth period. These cases correspond to $j = 1, 2, 3, 4$ and to the costs $C_1^{(1)}$, $C_1^{(2)}$, $C_1^{(3)}$, $C_1^{(4)}$, respectively. The cost C_1 is then the minimum of $C_1^{(1)}$, $C_1^{(2)}$, $C_1^{(3)}$, and $C_1^{(4)}$.

$$C_1^{(1)} = C_2 + 2 + 1(3) = 10.4 + 2 + 3 = 15.4,$$

$$C_1^{(2)} = C_3 + 2 + 1(3 + 2) + \tfrac{1}{5}(2) = 7.4 + 2 + 5 + 0.4 = 14.8.$$

$$C_1^{(3)} = C_4 + 2 + 1(3 + 2 + 3) + \tfrac{1}{5}[2 + 2(3)] = 4 + 2 + 8 + 1.6 = 15.6,$$

$$C_1^{(4)} = C_5 + 2 + 1(3 + 2 + 3 + 2) + \tfrac{1}{5}[2 + 2(3) + 3(2)]$$

$$= 0 + 2 + 10 + 2.8 = 14.8.$$

$$C_1 = \min\{15.4, 14.8, 15.6, 14.8\} = 14.8,$$

Note that $C_1^{(2)}$ and $C_1^{(4)}$ tie as the minimum, giving C_1. This means that the policies corresponding to $C_1^{(2)}$ and $C_1^{(4)}$ tie as being the optimal policies. The $C_1^{(4)}$ policy says to produce enough in period 1 to cover the demand for all four periods. The $C_1^{(2)}$ policy covers only the demand through period 2. Since the latter policy has the inventory level drop to zero at the end of period 2, the C_3 result is used next, namely, produce enough in period 3 to cover the demand for periods 3 and 4. The resulting production schedules are summarized below.

Optimal Production Schedules

1. Produce 100,000 speakers in period 1.
   Total cost = $148,000.
2. Produce 50,000 speakers in period 1 and 50,000 speakers in period 3.
   Total cost = $148,000.

Finally, note that the unit cost c is irrelevant to the problem because, over all the time periods, all policies produce the same number of units at the same cost. Hence, this cost could have been neglected, and the same optimal policies would have been obtained.

17.4 Stochastic Models

This section is concerned with inventory problems where the demand for a period is a random variable having a known probability distribution. Both single-period and multiperiod models are analyzed.

A Single-Period Model with No Setup Cost

The second example discussed in Sec. 17.1 is concerned with a wholesale distributor of bicycles. Suppose that this distributor is offered very favorable terms on the purchase of the model under consideration (a small, one-speed girl's bicycle), whose production is to be discontinued. This opportunity appears to be ideal for the forthcoming Christmas season. Because production has been discontinued, the stores have been informed that one order is welcome but no reorders will be possible. The *unit cost* of each bicycle is $20. There is *no setup cost* for the distributor to submit her order to the manufacturer. The cost of maintaining an inventory is −$9 per bicycle. This cost includes $1, which represents the cost of capital tied up, warehouse space, and so on; and −$10, which is what the distributor can get for each bicycle remaining in the inventory after the Christmas season (the salvage value). Note that this results in a *negative holding cost.* Each bicycle is sold by the wholesaler for $45, for a profit of $25.

Two remaining cost components still require discussion, the shortage cost and the revenue. If the demand exceeds the supply, those customers who fail to purchase a bicycle may bear some ill will, thereby resulting in a ''cost'' to the distributor. This cost is the per-item quantification of the loss of goodwill times the unsatisfied demand whenever a shortage occurs. The distributor considers this cost to be negligible.

If we adopt the criterion of maximizing net income, we must include revenue in the model. Indeed, net income is equal to total revenue minus the costs incurred (the ordering, holding, and shortage costs). Assuming no initial inventory, this net income for the distributor (in dollars) is

$$\text{Net income} = \$45 \times \text{number sold by distributor}$$
$$- \$20 \times \text{number purchased by distributor}$$
$$+ \$9 \times \text{number unsold and so disposed of for salvage value.}$$

Let

$$y = \text{number purchased by distributor}$$

and

$$D = \text{demand by bicycle shops (a random variable),}$$

so that

$$\min \{D, y\} = \text{number sold,}$$
$$\max \{0, y - D\} = \text{number unsold.}$$

Then

$$\text{Net income} = 45 \min \{D, y\} - 20y + 9 \max \{0, y - D\}.$$

The first term also can be written as

$$45 \min \{D, y\} = 45D - 45 \max \{0, D - y\}.$$

The term $45 \max \{0, D - y\}$ represents the *lost revenue from unsatisfied demand.* This lost revenue, plus any cost of the loss of customer goodwill due to unsatisfied demand (assumed negligible in this example), will be interpreted as the *shortage cost* throughout this section.

Now note that $45D$ is independent of the inventory policy (the value of y chosen) and so can be deleted from the objective function, which leaves

$$\text{Relevant net income} = -45 \max \{0, D - y\} - 20y + 9 \max \{0, y - D\}$$

to be maximized. All the terms on the right are the *negative* of *costs,* where these costs are the *shortage cost,* the *ordering cost,* and the *holding cost* (which has a negative value here), respectively. Rather than *maximizing* the *negative* of *total cost,* we instead will do the equivalent of *minimizing*

$$\text{Total cost} = 45 \max \{0, D - y\} + 20y - 9 \max \{0, y - D\}.$$

More precisely, since total cost is a random variable (because D is a random variable), the objective adopted for the model is to *minimize the expected total cost.*

In the discussion about the interpretation of the shortage cost, we assumed that the unsatisfied demand was lost (no backlogging). For the case where this unsatisfied demand is met by a priority shipment, similar reasoning applies. The revenue component of net income becomes the sales price of a bicycle ($45) times the demand *minus* the unit cost of the priority shipment times the unsatisfied demand whenever a shortage occurs. If our wholesale distributor is forced to meet the unsatisfied demand by purchasing bicycles from the midwest distributor for $35 each plus an air freight charge of, say, $2 each, then the appropriate shortage cost is $37 per bicycle. (If there were any costs associated with loss of goodwill, these also would be added to this amount.)

The distributor does not know what the demand for these bicycles will be; i.e., demand D is a random variable. However, an optimal inventory policy can be obtained if information about the probability distribution of D is available. Let

$$P_D(d) = P\{D = d\}.$$

It will be assumed that $P_D(d)$ is known for all values of d.

We now are in a position to summarize the model in general terms.

Summary of the One-Period Model with No Setup Cost

1. Planning is being done for just a single period.
2. The demand D in this period is a random variable with a known probability distribution.
3. There is no initial inventory on hand.
4. The decision to be made is the value of y, the number of units to purchase or produce for inventory at the beginning of the period.
5. The objective is to minimize the expected total cost, where the cost components are

c = unit cost for purchasing or producing each unit,

h = holding cost per unit remaining at end of period (includes storage cost minus salvage value),

p = shortage cost per unit of unsatisfied demand (includes lost revenue and cost of loss of customer goodwill).

This single-period model may represent the inventory of an item that (1) becomes obsolete quickly, such as the bicycle in the example or a daily newspaper; (2) spoils quickly, such as vegetables; (3) is stocked only once, such as spare parts for a single production run of a new model airplane; or (4) has a future that is uncertain beyond a single period.

Analysis of the Model: The decision on the value of y, the amount of inventory to acquire, depends heavily on the probability distribution of demand D. More than the expected demand may be desirable, but probably less than the maximum possible demand. A trade-off is needed between (1) the risk of being short and thereby incurring shortage costs and (2) the risk of having an excess and thereby incurring wasted costs of ordering and holding excess units. This is accomplished by minimizing the expected value (in the statistical sense) of the sum of these costs.

The amount sold is given by

$$\min\{D, y\} = \begin{cases} D & \text{if } D < y, \\ y & \text{if } D \geq y. \end{cases}$$

Hence the cost incurred if the demand is D and y is stocked is given by

$$C(D, y) = cy + p \max\{0, D - y\} + h \max\{0, y - D\}.$$

Because the demand is a random variable [with probability distribution $P_D(d)$], this cost is also a random variable. The expected cost is then given by $C(y)$, where

$$C(y) = E[C(D, y)] = \sum_{d=0}^{\infty} (cy + p \max\{0, d - y\} + h \max\{0, y - d\})P_D(d)$$

$$= cy + \sum_{d=y}^{\infty} p(d - y)P_D(d) + \sum_{d=0}^{y-1} h(y - d)P_D(d).$$

The function $C(y)$ depends upon the probability distribution of D. Frequently, a representation of this probability distribution is difficult to find, particularly when the demand ranges over a large number of possible values. Hence, this *discrete random variable* is often approximated by a *continuous random variable.* Furthermore, when demand ranges over a large number of possible values, this approximation will generally yield a nearly exact value of the optimal amount of inventory to stock. In addition, when discrete demand is used, the resulting expressions may become slightly more difficult to solve analytically. Hence, unless otherwise stated, *continuous demand* is assumed throughout the remainder of this chapter.

For this continuous random variable D, let

$$\varphi_D(\xi) = \text{probability density function of } D$$

and

$$\Phi(a) = \text{cumulative distribution function of } D,$$

so

$$\Phi(a) = \int_0^a \varphi_D(\xi)\, d\xi.$$

The expected cost $C(y)$ is then expressed as

$$\begin{aligned} C(y) = E[C(D, y)] &= \int_0^\infty C(\xi, y)\varphi_D(\xi)\, d\xi \\ &= \int_0^\infty (cy + p \max\{0, \xi - y\} + h \max\{0, y - \xi\})\, \varphi_D(\xi)\, d\xi \\ &= cy + \int_y^\infty p(\xi - y)\varphi_D(\xi)\, d\xi + \int_0^y h(y - \xi)\varphi_D(\xi)\, d\xi \\ &= cy + L(y), \end{aligned}$$

where $L(y)$ is often called the *expected shortage plus holding cost.* It then becomes necessary to find the value of y, say y^0, which minimizes $C(y)$. First we give the answer, and then we will show the derivation a little later.

The optimal quantity to order y^0 is that value which satisfies

$$\Phi(y^0) = \frac{p - c}{p + h}.$$

When the demand has either a uniform or an exponential distribution, a routine is available in your OR Courseware for calculating y^0.

If D is assumed to be a discrete random variable having the cumulative distribution function

$$F_D(b) = \sum_{d=0}^{b} P_D(d),$$

a similar result for the optimal order quantity is obtained. In particular, the optimal quantity to order y^0 is the smallest integer such that

$$F_D(y^0) \geq \frac{p - c}{p + h}.$$

EXAMPLE: Returning to the bicycle example described in this section, we assume that the demand has an exponential distribution with a mean of 10,000, so that its probability density function is

$$\varphi_D(\xi) = \begin{cases} \dfrac{1}{10{,}000} e^{-\xi/10{,}000} & \text{if } \xi \geq 0, \\ 0 & \text{otherwise.} \end{cases}$$

From the data given,

$$c = 20, \qquad p = 45, \qquad h = -9.$$

Therefore,

$$\Phi(a) = \int_0^a \frac{1}{10{,}000} e^{-\xi/10{,}000}\, d\xi = 1 - e^{-a/10{,}000}.$$

The optimal quantity to order y^0 is that value which satisfies

$$1 - e^{-y^0/10{,}000} = \frac{45 - 20}{45 - 9} = 0.69444.$$

By using the natural logarithm (denoted by ln), this equation can be solved as follows:

$$e^{-y^0/10{,}000} = 0.30556,$$

$$\ln e^{-y^0/10{,}000} = \ln 0.30556,$$

$$\frac{-y_0}{10{,}000} = -1.1856,$$

$$y^0 = 11{,}856.$$

Therefore, the distributor should stock 11,856 bicycles in the Christmas season. Note that this number is slightly more than the expected demand of 10,000.

Whenever the demand is exponential with expectation λ, then y^0 can be obtained from the relation

$$y^0 = -\lambda \ln \frac{c + h}{p + h}.$$

MODEL WITH INITIAL STOCK LEVEL: In the above model we assume that there is no initial inventory. As a slight variation, suppose now that the distributor begins with 500 bicycles on hand. How does this stock influence the optimal inventory policy?

In general terms, suppose that the initial stock level is given by x, and the decision to be made is the value of y, the inventory level *after replenishment* by ordering (or producing) additional units. Thus, $y - x$ is to be ordered, so that

$$\text{Amount available } (y) = \text{initial stock } (x) + \text{amount ordered } (y - x).$$

The cost equation presented earlier remains identical except for the term that was previously cy. This term now becomes $c(y - x)$, so that minimizing the expected cost is given by

$$\min_{y \geq x} \left[c(y - x) + \int_y^\infty p(\xi - y)\varphi_D(\xi)\, d\xi + \int_0^y h(y - \xi)\varphi_D(\xi)\, d\xi \right].$$

The constraint $y \geq x$ must be added because the inventory level y after replenishing cannot be less than the initial inventory level x.

The optimal inventory policy is the following:

$$\text{If} \quad x \begin{cases} < y^0 & \text{order } y^0 - x \text{ to bring inventory level up to } y^0, \\ \geq y^0 & \text{do not order,} \end{cases}$$

where y^0 satisfies

$$\Phi(y^0) = \frac{p - c}{p + h}.$$

Thus, in the bicycle example, if there are 500 bicycles on hand, the optimal policy is to bring the inventory level up to 11,856 bicycles (which implies ordering 11,356 additional bicycles). On the other hand, if there were 12,000 bicycles already on hand, the optimal policy would be not to order.

Derivation of the Optimal Policy:[1] We start by assuming that the initial stock level is zero.

For any positive constants c_1 and c_2, define $g(\xi, y)$ as

$$g(\xi, y) = \begin{cases} c_1(y - \xi) & \text{if } y > \xi, \\ c_2(\xi - y) & \text{if } y \leq \xi, \end{cases}$$

and let

$$G(y) = \int_0^\infty g(\xi, y)\varphi_D(\xi)\, d\xi + cy,$$

where $c > 0$. Then $G(y)$ is minimized at $y = y^0$, where y^0 is the solution to

$$\Phi(y^0) = \frac{c_2 - c}{c_2 + c_1}.$$

To see why this value of y^0 minimizes $G(y)$, note that, by definition,

$$G(y) = c_1 \int_0^y (y - \xi)\varphi_D(\xi)\, d\xi + c_2 \int_y^\infty (\xi - y)\varphi_D(\xi)\, d\xi + cy.$$

Taking the derivative (see the end of App. 3) and setting it equal to zero lead to

$$\frac{dG(y)}{dy} = c_1 \int_0^y \varphi_D(\xi)\, d\xi - c_2 \int_y^\infty \varphi_D(\xi)\, d\xi + c = 0.$$

This expression implies that

$$c_1\Phi(y^0) - c_2[1 - \Phi(y^0)] + c = 0,$$

because

$$\int_0^\infty \varphi_D(\xi)\, d\xi = 1.$$

Solving this expression results in

$$\Phi(y^0) = \frac{c_2 - c}{c_2 + c_1}.$$

[1] This subsection may be omitted by the less mathematically inclined reader.

The solution of this equation minimizes $G(y)$ because

$$\frac{d^2G(y)}{dy^2} = (c_1 + c_2)\varphi_D(y) \geq 0$$

for all y.

To apply this result, it is sufficient to show that

$$C(y) = cy + \int_y^{\infty} p(\xi - y)\varphi_D(\xi)\, d\xi + \int_0^y h(y - \xi)\varphi_D(\xi)\, d\xi$$

has the form of $G(y)$. Clearly, $c_1 = h$, $c_2 = p$, and $c = c$, so that the optimal quantity to order y^0 is that value which satisfies

$$\Phi(y^0) = \frac{p - c}{p + h}.$$

To derive the results for the case where the initial stock level is $x > 0$, recall that it is necessary to solve the relationship

$$\min_{y \geq x}\left\{-cx + \left[\left[\int_y^{\infty} p(\xi - y)\varphi_D(\xi)\, d\xi + \int_0^y h(y - \xi)\varphi_D(\xi)\, d\xi + cy\right]\right\}.$$

Note that the expression in brackets has the form of $G(y)$, with $c_1 = h$, $c_2 = p$, and $c = c$. Hence the cost function to be minimized can be written as

$$\min_{y \geq x}\{-cx + G(y)\}.$$

It is clear that $-cx$ is a constant, so that it is sufficient to find the y that satisfies the expression

$$\min_{y \geq x} G(y).$$

Hence the value of y^0 that minimizes $G(y)$ satisfies

$$\Phi(y^0) = \frac{p - c}{p + h}.$$

Furthermore, $G(y)$ must be a convex function, because

$$\frac{d^2G(y)}{dy^2} \geq 0.$$

Also,

$$\lim_{y \to 0} \frac{dG(y)}{dy} = c - p,$$

which is negative,[1] and

$$\lim_{y \to \infty} \frac{dG(y)}{dy} = h + c,$$

[1] If $c - p$ is nonnegative, $G(y)$ will be a monotonic increasing function. This implies that the item should not be stocked, that is, $y^0 = 0$.

which is positive. Hence $G(y)$ must be as shown in Fig. 17.7. Thus the optimal policy must be given by the following:

> If $x < y^0$, order $y^0 - x$ to bring the inventory level up to y^0, because y^0 can be achieved together with the minimum value $G(y^0)$. If $x \geq y^0$, do not order because any $g(y)$, with $y > x$, must exceed $G(x)$.

A similar argument can be constructed for obtaining optimal policies for the following model with nonlinear costs.

Model with Nonlinear Costs: Similar results for these models can be obtained for other than linear holding and shortage costs. Denote the holding cost by

$$\begin{array}{ll} h[y - D] & \text{if } y \geq D, \\ 0 & \text{if } y < D, \end{array}$$

where $h[\cdot]$ is a mathematical function, not necessarily linear.

Similarly, the shortage cost can be denoted by

$$\begin{array}{ll} p[D - y] & \text{if } D \geq y, \\ 0 & \text{if } D < y, \end{array}$$

where $p[\cdot]$ is also a function, not necessarily linear.

Thus the total expected cost is given by

$$c(y - x) + \int_y^\infty p[\xi - y]\varphi_D(\xi)\, d\xi + \int_0^y h[y - \xi]\varphi_D(\xi)\, d\xi,$$

where x is the amount on hand.

If $L(y)$ is defined as the *expected shortage plus holding cost*, i.e.,

$$L(y) = \int_y^\infty p[\xi - y]\varphi_D(\xi)\, d\xi + \int_0^y h[y - \xi]\varphi_D(\xi)\, d\xi,$$

then the total expected cost can be written as

$$c(y - x) + L(y).$$

The optimal policy is obtained by minimizing this expression, subject to the constraint that $y \geq x$, that is,

$$\min_{y \geq x}\{c(y - x) + L(y)\}.$$

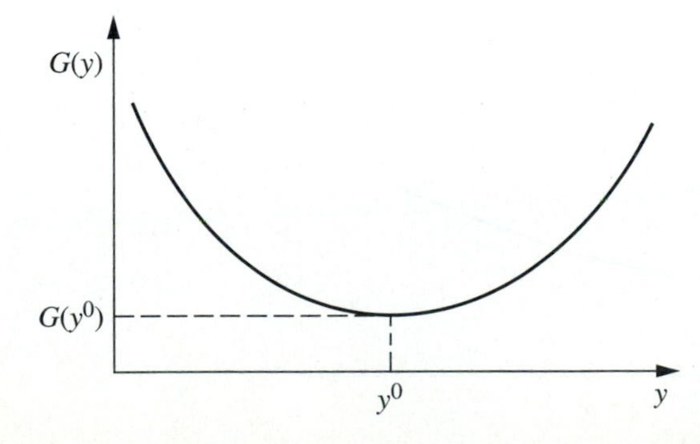

Figure 17.7 Graph of $G(y)$.

If $L(y)$ is strictly convex[1] [a sufficient condition being that the shortage and holding costs each are convex and $\varphi_D(\xi) > 0$], then the optimal policy is the following:

$$\text{If} \quad x \begin{cases} < y^0 & \text{order } y^0 - x \text{ to bring inventory level up to } y^0, \\ \geq y^0 & \text{do not order,} \end{cases}$$

where y^0 is the value of y that satisfies the expression

$$\frac{dL(y)}{dy} + c = 0.$$

A Single-Period Model with a Setup Cost

In discussing the bicycle example previously in this section, we assumed that there was no setup cost incurred in ordering the bicycles for the Christmas season. Suppose now that the cost of placing this special order is $800, so this cost should be included in the analysis of the model. In fact, inclusion of the setup cost generally causes major changes in the results.

In general, the setup cost will be denoted by K. To begin, the shortage and holding costs will each be assumed to be linear. Their resulting effect is then given by $L(y)$, where

$$L(y) = p\int_y^\infty (\xi - y)\varphi_D(\xi)\,d\xi + h\int_0^y (y - \xi)\varphi_D(\xi)\,d\xi.$$

Thus the total expected cost incurred by bringing the inventory level up to y is given by

$$\begin{array}{ll} K + c(y - x) + L(y) & \text{if } y > x, \\ L(x) & \text{if } y = x. \end{array}$$

Note that $cy + L(y)$ is the same expected cost considered earlier when the setup cost was omitted. If $cy + L(y)$ is drawn as a function of y, it will appear as shown in Fig. 17.8.[2] Define S as the value of y that minimizes $cy + L(y)$, and define s as the smallest value of y for which $cs + L(s) = K + cS + L(S)$. From Fig. 17.8, it can be seen that

$$\text{If} \quad x > S, \quad \text{then} \quad K + cy + L(y) > cx + L(x), \quad \text{for all } y > x,$$

[1] See App. 2 for the definition of a strictly convex function.

[2] In the derivation of the optimal policy for the single-period model with no setup cost, $cy + L(y)$ was denoted by $G(y)$ and was rigorously shown to be a convex function of the form plotted in Fig. 17.8.

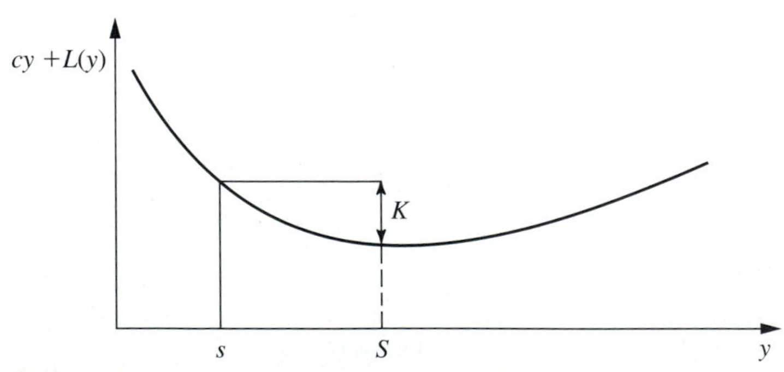

Figure 17.8 Graph of $cy + L(y)$.

so that

$$K + c(y - x) + L(y) > L(x).$$

The left-hand side of the last inequality represents the expected total cost of ordering $y - x$ to bring the inventory level up to y, and the right-hand side of this inequality represents the expected total cost if no ordering occurs. Hence, the optimal policy indicates that if $x > S$, do not order.

If $s \le x \le S$, it can again be seen from Fig. 17.8 that

$$K + cy + L(y) \ge cx + L(x), \qquad \text{for all } y > x,$$

so that

$$K + c(y - x) + L(y) \ge L(x).$$

Again, no ordering is less expensive than ordering.

Finally, if $x < s$, it follows from Fig. 17.8 that

$$\min_{y \ge x} \{K + cy + L(y)\} = K + cS + L(S) < cx + L(x),$$

or

$$\min_{y \ge x} \{K + c(y - x) + L(y)\} = K + c(S - x) + L(S) < L(x),$$

so that it pays to order. The minimum cost is incurred by bringing the inventory level up to S.

The optimal inventory policy is the following:

$$\text{If} \quad x \begin{cases} < s & \text{order } S - x \text{ to bring inventory level up to } S, \\ \ge s & \text{do not order.} \end{cases}$$

The value of S is obtained from

$$\Phi(S) = \frac{p - c}{p + h},$$

and s is the smallest value that satisfies the expression

$$cs + L(s) = K + cS + L(S).$$

When the demand has either a uniform or an exponential distribution, a routine is available in your OR Courseware for calculating s and S.

This kind of policy is referred to as an **(s, S) policy**. It has had extensive use in industry.

Example: Referring to the bicycle example, we found earlier that

$$y^0 = S = 11{,}856.$$

If $K = 800$, $c = 20$, $p = 45$, and $h = -9$, then s is obtained from

$$20s + 45\int_s^{\infty} (\xi - s)\frac{1}{10{,}000} e^{-\xi/10{,}000}\, d\xi - 9\int_0^s (s - \xi)\frac{1}{10{,}000} e^{-\xi/10{,}000}\, d\xi$$

$$= 800 + 20(11{,}856) + 45\int_{11{,}856}^{\infty} (\xi - 11{,}856)\frac{1}{10{,}000} e^{-\xi/10{,}000}\, d\xi$$

$$-9\int_0^{11{,}856} (11{,}856 - \xi)\frac{1}{10{,}000} e^{-\xi/10{,}000}\, d\xi,$$

so that

$$s = 10{,}674.$$

Hence, the optimal policy calls for bringing the inventory level up to $S = 11{,}856$ bicycles if the amount on hand is less than $s = 10{,}674$. Otherwise, no order is placed.

Solution When the Demand Distribution Is Exponential: Now consider the special case where the distribution of demand D is exponential, i.e.,

$$\varphi_D(\xi) = \alpha e^{-\alpha\xi}, \qquad \text{for } \xi \geq 0.$$

From the no-setup-cost results,

$$1 - e^{-\alpha S} = \frac{p - c}{p + h},$$

so

$$S = \frac{1}{\alpha} \ln \frac{h + p}{h + c}.$$

For any y,

$$cy + L(y) = cy + h \int_0^y (y - \xi)\alpha e^{-\alpha\xi}\, d\xi + p \int_y^\infty (\xi - y)\alpha e^{-\alpha\xi}\, d\xi$$

$$= (c + h)y + \frac{1}{\alpha}(h + p)e^{-\alpha y} - \frac{h}{\alpha}.$$

Evaluating $cy + L(y)$ at the points $y = s$ and $y = S$ leads to

$$(c + h)s + \frac{1}{\alpha}(h + p)e^{-\alpha s} - \frac{h}{\alpha} = K + (c + h)S + \frac{1}{\alpha}(h + p)e^{-\alpha S} - \frac{h}{\alpha},$$

or

$$(c + h)s + \frac{1}{\alpha}(h + p)e^{-\alpha s} = K + (c + h)S + \frac{1}{\alpha}(c + h).$$

Although this last equation does not have a closed-form solution, it can be solved numerically quite easily. An approximate analytical solution can be obtained as follows. By letting

$$\Delta = S - s,$$

the last equation yields

$$e^{\alpha\Delta} = \frac{\alpha K}{c + h} + \alpha\Delta + 1.$$

If $\alpha\Delta$ is close to zero, $e^{\alpha\Delta}$ can be expanded into a Taylor series around zero. If the terms beyond the quadratic term are neglected, the result becomes

$$1 + \alpha\Delta + \frac{a^2\Delta^2}{2} \cong \frac{\alpha K}{c + h} + \alpha\Delta + 1,$$

so that

$$\Delta = \sqrt{\frac{2K}{\alpha(c + h)}}.$$

Using this approximation in the bicycle example results in

$$\Delta = \sqrt{\frac{(2)(10{,}000)(800)}{20 - 9}} = 1{,}206,$$

which is quite close to the exact value of $\Delta = 1{,}182$.

Model with Nonlinear Costs: These results can be extended to the case where the expected shortage plus holding cost $L(y)$ is a *strictly convex* function. This extension results in a strictly convex $cy + L(y)$, similar to Fig. 17.8.

For this model, the optimal inventory policy has the following form:

$$\text{If} \quad x \begin{cases} < s & \text{order } S - x \text{ to bring inventory level up to } S, \\ \geq s & \text{do not order,} \end{cases}$$

where S is the value of y that satisfies

$$c + \frac{dL(y)}{dy} = 0$$

and s is the smallest value that satisfies the expression

$$cs + L(s) = K + cS + L(S).$$

A Two-Period Inventory Model with No Setup Cost

The single-period model was illustrated with the bicycle example when the distributor had only one opportunity to place an order. In many situations an opportunity to place an order occurs periodically, e.g., monthly, and the inventory manager must make a decision for each period about whether to replenish inventory then and, if so, by how much. If the probability distribution of demand in each period can be forecasted multiple periods into the future, better decisions can be made by coordinating the plans for all these periods than by planning ahead just one period at a time.

Even for a planning horizon of two periods, using the optimal one-period solution twice is not generally the optimal policy for the two-period problem. Smaller costs can usually be achieved by viewing the problem from a two-period viewpoint and then using the methods of probabilistic dynamic programming introduced in Sec. 10.4 to obtain the best inventory policy.

Assumptions: Except for having two periods, the assumptions for this model are basically the same as for the corresponding one-period model, as summarized below.

Summary of the Two-Period Model with No Setup Cost

1. Planning is being done for two periods, where unsatisfied demand in period 1 is backlogged to be met in period 2, but there is no backlogging of unsatisfied demand in period 2.
2. The demands D_1 and D_2 for periods 1 and 2 are *independent and identically distributed* random variables. Their common probability distribution has probability density function $\phi_D(\xi)$ and cumulative distribution function $\Phi(\xi)$.
3. The initial inventory level (before replenishing) at the beginning of period 1 is x_1 ($x_1 \geq 0$).

4. The objective is to *minimize the expected total cost for both periods,* where the cost components for each period are

c = unit cost for purchasing or producing each unit,

h = holding cost per unit remaining at end of each period,

p = shortage cost per unit of unsatisfied demand at end of each period.

For simplicity, we are assuming that the demand distributions for the two periods are the same and that the values of the above cost components also are the same for the two periods. In many applications, there will be differences between the periods that should be incorporated into the analysis. For example, because of assumption 1, the value of p may well be different for the two periods. Such extensions of the model can be incorporated into the dynamic programming analysis presented below, but we will not delve into these extensions.

Analysis: To begin the analysis, let

y_i^0 = optimal value of y_i, for $i = 1, 2$,

$C_1(x_1)$ = expected total cost for both periods when following an optimal policy given that x_1 is initial inventory level (before replenishing) at beginning of period 1,

$C_2(x_2)$ = expected total cost for just period 2 when following an optimal policy given that x_2 is inventory level (before replenishing) at beginning of period 2.

To use the dynamic programming approach, we begin by solving for $C_2(x_2)$ and y_2^0, where there is just one period to go. Then we will use these results to find $C_1(x_1)$ and y_1^0.

From the results for the single-period model, y_2^0 is found by solving the equation

$$\Phi(y_2^0) = \frac{p - c}{p + h}.$$

Given x_2, the resulting optimal policy then is the following:

$$\text{If} \quad x_2 \begin{cases} < y_2^0 & \text{order } y_2^0 - x_2 \text{ to bring inventory level up to } y_2^0, \\ \geq y_2^0 & \text{do not order.} \end{cases}$$

The cost of this optimal policy can be expressed as

$$C_2(x_2) = \begin{cases} L(x_2) & \text{if } x_2 \geq y_2^0, \\ c(y_2^0 - x_2) + L(y_2^0) & \text{if } x_2 < y_2^0, \end{cases}$$

where $L(z)$ is the expected shortage plus holding cost for a single period when the inventory level (after replenishing) is z. Now $L(z)$ can be expressed as

$$L(z) = \int_z^{\infty} p(\xi - z)\varphi_D(\xi)\, d\xi + \int_0^z h(z - \xi)\varphi_D(\xi)\, d\xi.$$

When both periods 1 and 2 are considered, the costs incurred consist of the purchase cost $c(y_1 - x_1)$, the expected shortage plus holding cost $L(y_1)$, and the costs associated with following an optimal policy during the second period. Thus the expected cost of following the optimal policy for two periods is given by

$$C_1(x_1) = \min_{y_1 \geq x_1} \{c(y_1 - x_1) + L(y_1) + E[C_2(x_2)]\},$$

where $E[C_2(x_2)]$ is obtained as follows. Note that

$$x_2 = y_1 - D_1,$$

so x_2 is a random variable when beginning period 1. Thus,

$$C_2(x_2) = C_2(y_1 - D_1) = \begin{cases} L(y_1 - D_1) & \text{if } y_1 - D_1 \geq y_2^0, \\ c(y_2^0 - y_1 + D_1) + L(y_2^0) & \text{if } y_1 - D_1 < y_2^0. \end{cases}$$

Hence, $C_2(x_2)$ is a random variable, and its expected value is given by

$$\begin{aligned} E[C_2(x_2)] &= \int_0^\infty C_2(y_1 - \xi)\varphi_D(\xi)\, d\xi \\ &= \int_0^{y_1 - y_2^0} L(y_1 - \xi)\varphi_D(\xi)\, d\xi \\ &\quad + \int_{y_1 - y_2^0}^\infty [c(y_2^0 - y_1 + \xi) + L(y_2^0)]\varphi_D(\xi)\, d\xi. \end{aligned}$$

Therefore,

$$\begin{aligned} C_1(x_1) = \min_{y_1 \geq x_1} \bigg\{ & c(y_1 - x_1) + L(y_1) + \int_0^{y_1 - y_2^0} L(y_1 - \xi)\varphi_D(\xi)\, d\xi \\ & + \int_{y_1 - y_2^0}^\infty [(y_2^0 - y_1 + \xi) + L(y_2^0)]\varphi_D(\xi)\, d\xi \bigg\}. \end{aligned}$$

It can be shown that $C_1(x_1)$ has a unique minimum and that the optimal value of y_1, denoted by y_1^0, satisfies the equation

$$\begin{aligned} -p + (p + h)\Phi(y_1^0) + (c - p)\Phi(y_1^0 - y_2^0) & \\ + (p + h) \int_0^{y_1^0 - y_2^0} \Phi(y_1^0 - \xi)\varphi_D(\xi)\, d\xi &= 0. \end{aligned}$$

The resulting optimal policy for period 1 then is the following:

$$\text{If} \quad x_1 \begin{cases} < y_1^0 & \text{order } y_1^0 - x_1 \text{ to bring inventory level up to } y_1^0, \\ \geq y_1^0 & \text{do not order.} \end{cases}$$

The procedure for finding y_1^0 reduces to a simpler result for certain demand distributions. We summarize two such cases next.

Suppose that the demand in each period has a *uniform distribution* over the range 0 to t, that is,

$$\varphi_D(\xi) = \begin{cases} \dfrac{1}{t} & \text{if } 0 \leq \xi \leq t, \\ 0 & \text{otherwise.} \end{cases}$$

Then y_1^0 can be obtained from the expression

$$y_1^0 = \sqrt{(y_2^0)^2 + \frac{2t(c-p)}{p+h}y_2^0 + \frac{t^2[2p(p+h)+(h+c)^2]}{(p+h)^2}} - \frac{t(h+c)}{p+h}.$$

Now suppose that the demand in each period has an exponential distribution, i.e.,

$$\phi(\xi) = \alpha e^{-\alpha\xi}, \qquad \text{for } \xi \geq 0.$$

Then y_1^0 satisfies the relationship

$$(h+c)e^{-\alpha(y_1^0-y_2^0)} + (p+h)e^{-ay_1^0} + \alpha(p+h)(y_1^0 - y_2^0)e^{-ay_1^0} = 2h + c.$$

An alternative way of finding y_1^0 is to let z^0 denote $\alpha(y_1^0 - y_2^0)$. Then z^0 satisfies the relationship

$$e^{-z^0}\left[(h+c) + (p+h)e^{-\alpha y_2^0} + z^0(p+h)e^{-\alpha y_2^0}\right] = 2h + c,$$

and

$$y_1^0 = \frac{1}{\alpha}z^0 + y_2^0.$$

When the demand has either a uniform or an exponential distribution, a routine is available in your OR Courseware for calculating y_1^0 and y_2^0.

Example: Consider a two-period problem where

$$c = 10, \qquad h = 10, \qquad p = 15,$$

and where the probability density function of the demand in each period is given by

$$\varphi_D(\xi) = \begin{cases} \frac{1}{10} & \text{if } 0 \leq \xi \leq 10, \\ 0 & \text{otherwise,} \end{cases}$$

so that the cumulative distribution function of demand is

$$\Phi(\xi) = \begin{cases} 0 & \text{if } \xi < 0, \\ \frac{\xi}{10} & \text{if } 0 \leq \xi \leq 10, \\ 1 & \text{if } \xi > 10. \end{cases}$$

We find y_2^0 from the equation

$$\Phi(y_2^0) = \frac{p-c}{p+h} = \frac{15-10}{15+10} = \frac{1}{5},$$

so that

$$y_2^0 = 2.$$

To find y_1^0, we plug into the expression given for y_1^0 for the case of a *uniform* demand distribution, and we obtain

$$y_1^0 = \sqrt{2^2 + \frac{2(10)(10-15)}{15+10}(2) + 10^2\frac{2(15)(15+10)+(10+10)^2}{(15+10)^2}} - \frac{10(10+10)}{15+10}$$

$$= \sqrt{4 - 8 + 184} - 8 = 13.42 - 8 = 5.42.$$

Substituting $y_1^0 = 5$ and $y_1^0 = 6$ into $C_1(x_1)$ leads to a smaller value with $y_1^0 = 5$. Thus the optimal policy can be described as follows:

If $x_1 < 5$, order $5 - x_1$ to bring inventory level up to 5.

If $x_1 \geq 5$, do not order in period 1.

If $x_2 < 2$, order $2 - x_2$ to bring inventory level up to 2.

If $x_2 \geq 2$, do not order in period 2.

Since unsatisfied demand in period 1 is backlogged to be met in period 2, $x_2 = 5 - D$ can turn out to be either positive or negative.

Multiperiod Models—An Overview

The two-period model can be extended to several periods or to an infinite number of periods. This section presents a summary of multiperiod results that have practical importance.

Multiperiod Model with No Setup Cost: Consider the direct extension of the above two-period model to n periods ($n > 2$) with the identical assumptions. The only difference is that a *discount factor* α (described in Sec. 17.2), with $0 < \alpha < 1$, now will be used in calculating the expected total cost for n periods. The problem still is to find the critical numbers $y_1^0, y_2^0, \ldots, y_n^0$ that describe the optimal inventory policy. As in the two-period model, these values are difficult to obtain numerically, but it can be shown[1] that the optimal policy has the following form.

For each period i ($i = 1, 2, \ldots, n$), with x_i as the inventory level entering that period (before replenishing), do the following

If $x_i \begin{cases} < y_i^0 & \text{order } y_i^0 - x_i \text{ to bring inventory level up to } y_i^0, \\ \geq y_i^0 & \text{do not order in period } i. \end{cases}$

Furthermore,

$$y_n^0 \leq y_{n-1}^0 \leq \cdots \leq y_2^0 \leq y_1^0.$$

For the *infinite-period* case (where $n = \infty$), all these critical numbers $y_1^0, y_2^0, \ldots$ are *equal.* Let y^0 denote this constant value. It can be shown that y^0 satisfies the equation

$$\Phi(y^0) = \frac{p - c(1 - \alpha)}{p + h}.$$

When the demand has either a uniform or an exponential distribution, a routine is available in your OR Courseware for calculating y^0.

A Variation of the Multiperiod Inventory Model with No Setup Cost: These results for the infinite-period case (all the critical numbers equal the same value y^0 and y^0 satisfies the above equation) also apply when n is finite if two new assumptions are made about what happens at the end of the last period. One new

[1] See Theorem 4 in R. Bellman, I. Glicksberg, and O. Gross, "On the Optimal Inventory Equation," *Management Science,* **2**: 83–104, 1955. Also see p. 163 in K. J. Arrow, S. Karlin, and H. Scarf (eds.), *Studies in the Mathematical Theory of Inventory and Production,* Stanford University Press, Stanford, CA, 1958.

assumption is that each unit left over at the end of the final period can be salvaged with a return of the initial purchase cost c. Similarly, if there is a shortage at this time, assume that the shortage is met by an emergency shipment with the same unit purchase cost c.

Example: Consider again the bicycle example introduced in Example 2 of Sec. 17.1. The cost estimates given there imply that

$$c = 35, \qquad h = 1, \qquad p = 15.$$

Suppose now that the distributor places an order with the manufacturer for various model bicycles on the first working day of each month. Because of this routine, she is willing to assume that the marginal setup cost is zero for including an order for the model bicycle under consideration. The appropriate discount factor is $\alpha = 0.995$. From past history, the distribution of demand can be approximated by a uniform distribution with the probability density function

$$\varphi_D(\xi) = \begin{cases} \frac{1}{800} & \text{if } 0 \leq \xi \leq 800, \\ 0 & \text{otherwise,} \end{cases}$$

so the cumulative distribution function over this interval is

$$\Phi(\xi) = \tfrac{1}{800}\xi, \qquad \text{if } 0 \leq \xi \leq 800.$$

The distributor expects to stock this model indefinitely, so that the infinite-period model is appropriate.

For this model, the critical number y^0 for every period satisfies the equation

$$\Phi(y^0) = \frac{p - c(1 - \alpha)}{p + h},$$

so

$$\frac{y^0}{800} = \frac{15 - 35(1 - 0.995)}{15 + 1} = 0.9266,$$

and $y^0 = 741$. Thus, if the number of bicycles on hand x at the first of each month is fewer than 741, the optimal policy calls for bringing the inventory level up to 741 (ordering $741 - x$ bicycles). Otherwise, no order is placed.

Multiperiod Model with Setup Cost: The introduction of a fixed setup cost K that is incurred when ordering (or producing) often adds more realism to the model. For the single-period model with a setup cost, we found that an (s, S) policy is optimal, so that the two critical numbers s and S indicate when to order (if the inventory level is less than s) and how much to order (bring the inventory level up to S). Now with multiple periods, an (s, S) policy again is optimal, but the value of each critical number may be different in different periods. Let s_i and S_i denote these critical numbers for period i, and again let x_i be the inventory level (before replenishing) at the beginning of period i.

> The optimal policy is to do the following at the beginning of each period i $(i = 1, 2, \ldots, n)$:
>
> $$\text{If} \quad x_i \begin{cases} < s_i & \text{order } S_i - x_i \text{ to bring inventory level up to } S_i, \\ \geq s_i & \text{do not order.} \end{cases}$$

Unfortunately, computing exact values of the s_i and S_i is extremely difficult.

A Multiperiod Model with Batch Orders and No Setup Cost: In the preceding models, *any quantity* could be ordered (or produced) at the beginning of each period. However, in some applications, the product may come in a standard batch size, e.g., a case or a truckload. Let Q be the number of units in each batch. In our next model we assume that the number of units ordered must be a *nonnegative integer multiple* of Q.

This model makes the same assumptions about what happens at the end of the last period as the variation of the multiperiod model with no setup cost presented earlier. Thus, we assume that each unit left over at the end of the final period can be salvaged with a return of the initial purchase cost c. Similarly, if there is a shortage at this time, we assume that the shortage is met by an emergency shipment with the same unit purchase cost c.

Otherwise, the assumptions are the same as for our standard multiperiod model with no setup cost.

The optimal policy for this model is known as a **(k, Q) policy** because it uses a critical number k and the quantity Q as described below.

> If at the beginning of a period the inventory level (before replenishing) is less than k, an order should be placed for the smallest integer multiple of Q that will bring the inventory level up to at least k (and probably higher). Otherwise, an order should not be placed. The same critical number k is used in each period.

The critical number k is chosen as follows. Plot the function

$$G(y) = (1 - \alpha)cy + h \int_0^y (y - \xi)\varphi_D(\xi)\, d\xi + p \int_y^\infty (\xi - y)\varphi_D(\xi)\, d\xi,$$

as shown in Fig. 17.9. This function necessarily has the convex shape shown in the figure. As before, the minimizing value y^0 satisfies the equation

$$\Phi(y^0) = \frac{p - c(1 - \alpha)}{p + h}.$$

As shown in Fig. 17.9, if a "ruler" of length Q is placed horizontally into the "valley," k is found to be that value of the abscissa to the left of y^0 where the ruler intersects the valley. If the inventory level lies in R_1, then Q is ordered; if it lies in R_2, then $2Q$ is ordered; and so on.

These results hold regardless of whether the number of periods n is finite or infinite.

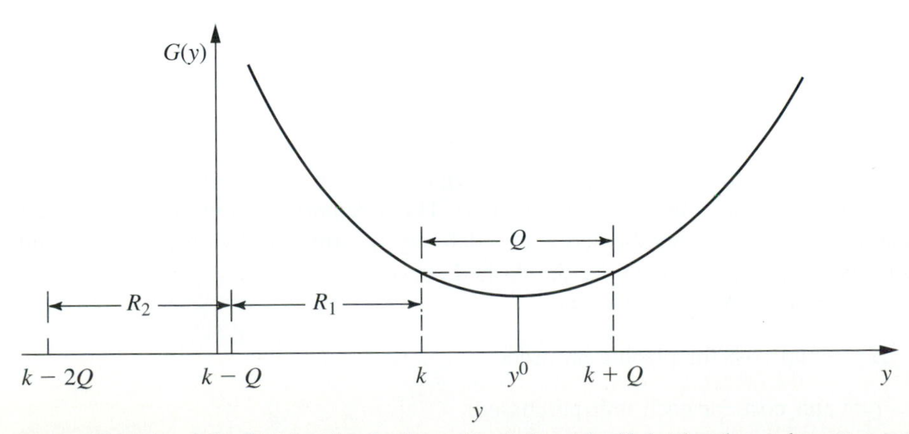

Figure 17.9 Plot of the $G(y)$ function for the multiperiod model with batch orders and no setup cost.

Continuous Review Model with Fixed Delivery Lag and Backlogging: In Sec. 17.3, a deterministic continuous review model, i.e., the economic lot-size model, was considered. The demand was assumed to be at a known constant rate. This model was classified as continuous review in that the inventory was continuously monitored and orders were placed at any time; i.e., orders were placed whenever the inventory level dropped to the *order point* (sometimes called the *trigger point*). This ordering procedure is in contrast to the models considered so far in this section, which assume stochastic demand and periodic review; i.e., the inventory is monitored only at the beginning of each period, and orders are placed only at these times. The model considered now is analogous to the economic lot-size model (continuous review), but where the demand for the item is *stochastic* and there is a *fixed delivery lead time* before an order is received. Only (s, S)-type policies are considered; i.e., when the inventory level falls to a level s, an order is placed to bring the inventory level up to S (a quantity $Q = S - s$ is ordered). This model is often called a *lot-size reorder-point model;* a quantity Q is ordered whenever the inventory level reaches the reorder level s. Unsatisfied demand is assumed to be filled immediately upon replenishment of the inventory; i.e., unsatisfied demand is backlogged.

To describe the model further, we first need to summarize three different ways of measuring the amount of inventory.

1. The **inventory on hand** is the number of units physically located in inventory. Thus, this quantity must be *nonnegative.*
2. The **inventory level** is the inventory on hand *minus* the amount of (backlogged) unsatisfied demand. Unsatisfied demand can occur (temporarily) only after the inventory on hand has dropped to zero, so unsatisfied demand causes the inventory level to be *negative.*
3. The **inventory position** is the inventory level *plus* the amount ordered but not yet received. The inventory position normally will be kept *nonnegative.*

The model can be described in detail as follows. Inventory is stockpiled and used as demand dictates. When the inventory position reaches s, an order is placed for Q units to bring the inventory position up to level S. There is a fixed delivery lead time (often called *lag time*) of length λ before the order is received. The demand for units from inventory during time λ is assumed to be a continuous random variable D having a probability density function denoted by $\varphi_D(\xi)$ and mean

$$E(D) = a\lambda,$$

where a is the expected number of items demanded per unit time.

Figure 17.10 illustrates how both the inventory level (the solid curve) and the inventory position vary over time. Note that this diagram can be viewed as a series of cycles, with a *cycle* defined as the time between receipt of consecutive orders. The figure includes a cycle where the demand during period λ is relatively large, which eventually causes the inventory level to go negative (where this unsatisfied demand is backlogged to be met when the order arrives). The inventory position differs from the inventory level only during the period of a delivery lag time, so the inventory position during these periods is shown by the dashed curve.

The costs to be considered are

$K =$ setup cost for placing an order,

$c =$ unit cost for each unit purchased,

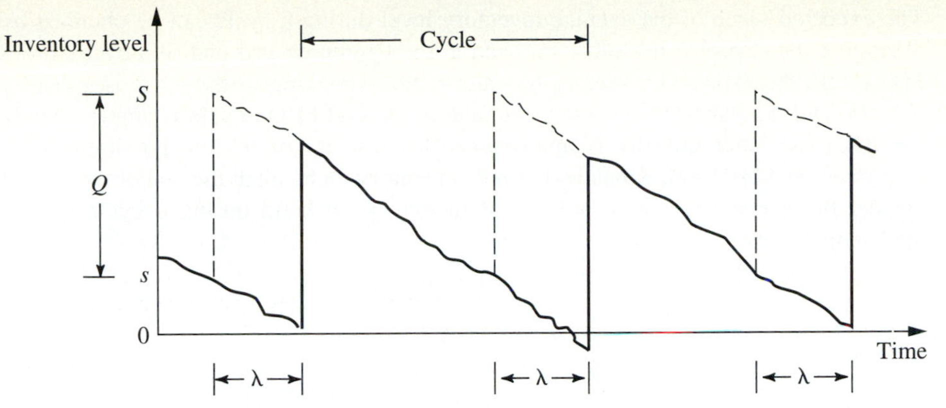

Figure 17.10 Diagram of the inventory level (the solid curve) as a function of time for the stochastic continuous review model. When the inventory position differs from the inventory level, it is shown by a dashed curve.

h = holding cost per unit of inventory on hand per unit time.

p = shortage cost per unit short (until next order arrives), independent of duration of shortage.

The inventory policy is to track the inventory position so that when the inventory position reaches s, an order of size Q is placed; this order will be delivered after a period of length λ. The problem is to determine when to place an order (find the order point s) and what size it should be (find the order quantity Q), in order to minimize the expected total cost per unit time, $C(Q, s)$

$$C(Q, s) = E(\text{OC}) + E(\text{HC}) + E(\text{SC}),$$

where $E(\text{OC})$ = expected ordering cost per unit time,
$E(\text{HC})$ = expected holding cost per unit time,
$E(\text{SC})$ = expected shortage cost per unit time.

To evaluate these terms, we assume initially that λ is sufficiently small that there is never more than a single order outstanding and that the reorder point s (based on the inventory position) is always nonnegative. The first assumption guarantees that the inventory on hand when an order is received will always fall above the reorder point, because otherwise more than one order would be outstanding. If p/h is sufficiently large, as is usually the case in practice, these assumptions are generally satisfied.

The expected ordering cost per unit time $E(\text{OC})$ is simply the ordering cost incurred per cycle *times* the expected number of cycles per unit of time:

$$\text{Ordering cost per cycle} = K + cQ.$$

Because a is the expected demand per unit time,

$$\text{Expected number of cycles per unit time} = \frac{a}{Q}.$$

Therefore,

$$E(\text{OC}) = \frac{a}{Q}(K + cQ).$$

The expected holding cost per unit time is

$$E(\text{HC}) = h\,E(\text{average inventory on hand during a cycle}).$$

The expected value of the average inventory level during a cycle can be obtained by averaging the expected inventory on hand at the beginning and end of a cycle. From Fig. 17.10, the expected inventory on hand at the beginning of the cycle is given by $S - a\lambda$, and the expected inventory on hand at the end of the cycle is approximately $s - a\lambda$. (The latter quantity is approximate because it ignores the possibility of a *negative* inventory level, which leaves *zero* inventory on hand, at the end of the cycle.) Hence, the expected average amount of inventory on hand during a cycle can be approximated by

$$\frac{(S - a\lambda) + (s - a\lambda)}{2} = \frac{Q + s - a\lambda + s - a\lambda}{2} = \frac{Q}{2} + s - a\lambda,$$

so that

$$E(\text{HC}) = h\left(\frac{Q}{2} + s - a\lambda\right).$$

The expected shortage cost per unit time $E(\text{SC})$ is the expected shortage cost incurred per cycle *times* the expected number of cycles per unit time (already obtained as a/Q). Since a shortage can occur only when the demand during the delivery lag time exceeds s, the expected shortage cost per cycle is

$$p\int_s^\infty (\xi - s)\varphi_D(\xi)\, d\xi,$$

so that

$$E(\text{SC}) = \frac{a}{Q}p\int_s^\infty (\xi - s)\varphi_D(\xi)\, d\xi.$$

Adding the expressions for $E(\text{OC})$, $E(\text{HC})$, and $E(\text{SC})$ leads to

$$C(Q, s) = \frac{aK}{Q} + ac + h\left(\frac{Q}{2} + s - a\lambda\right) + \frac{pa}{Q}\int_s^\infty (\xi - s)\varphi_D(\xi)\, d\xi.$$

Because there are two decision variables (Q and s), the optimal values (Q^* and s^*) are found by setting the corresponding partial derivatives equal to zero, i.e.,

$$\frac{\partial C(Q, s)}{\partial Q} = \frac{-aK}{Q^2} + \frac{h}{2} - \frac{pa\int_s^\infty (\xi - s)\varphi_D(\xi)\, d\xi}{Q^2} = 0$$

$$\frac{\partial C(Q, s)}{\partial s} = h - \frac{pa\int_s^\infty \varphi_D(\xi)\, d\xi}{Q} = 0.$$

Solving these equations simultaneously[1] leads to

$$\text{(1)} \qquad Q^* = \sqrt{\frac{2a\left[K + p\int_{s^*}^\infty (\xi - s^*)\varphi_D(\xi)\, d\xi\right]}{h}},$$

$$\text{(2)} \qquad \int_{s^*}^\infty \varphi_D(\xi)\, d\xi = \frac{hQ^*}{pa}.$$

[1] Note that the optimal values of Q and s are independent of c, the unit cost of the units ordered. The total number of units ordered is independent of the values of Q and s, so c can be neglected in determining the optimal values of these parameters.

Unfortunately, solving these equations simultaneously and obtaining a general closed-form expression for Q^* and s^* are not possible, but the following iterative procedure will lead to close approximations of these quantities.

1. As an initial step, assume that p equals zero and obtain a value of Q from Eq. (1). (Note that this equation with $p = 0$ is just the expression for Q in the deterministic economic lot-size model when shortages are not permitted.)
2. Solve for s in Eq. (2), using the value of Q found in step 1.
3. Using the value of s found in step 2, solve for a new Q, using Eq. (1).
4. Repeat steps 2 and 3 until successive values of Q and of s are sufficiently close.

In practice, this procedure will generally converge in just a few iterations.

Remarks: Several remarks can be made about this continuous review model.

1. Note that in Eq. (2), the integral $\int_{s^*}^{\infty} \varphi_D(\xi)\, d\xi$ is just the probability that the random variable demand D during the lead time exceeds s^*, that is, $P\{D > s^*\}$. Hence, $hQ^*/(pa)$ must fall between 0 and 1. If the algorithm ever leads to a value of $hQ/(pa) > 1$, this is an indication that the shortage cost (relative to the holding cost) is too small, with the result that the shortage by the end of a cycle will tend to be large. This would contradict the approximation of neglecting stockouts that was made in the calculation of the expected holding cost per unit time, so that the derived formulas would become inappropriate.

2. If the lead time is close to the average cycle length Q/a, more than one order may be outstanding. Using the inventory position as the measure of inventory still leads to an operational rule, i.e., order when the inventory position reaches s. However, the number of stockouts may become too large to be neglected in the expected holding cost calculations.

3. The quantity $(s - a\lambda)$ is known as the *safety stock,* and it represents "protection" against a stockout during the delivery lag time. The probability of a stockout is $P\{D > s\} = \int_{s}^{\infty} \varphi_D(\xi)\, d\xi$. By Eq. (2), this probability also equals $hQ^*/(pa)$ when s^* and Q^* are used.

4. Because there is no closed-form solution of Eqs. (1) and (2), it is worth considering some special cases for the probability distribution of demand. Consider the *uniform distribution* over the range from 0 to t, so that the probability density function is

$$\varphi_D(\xi) = \begin{cases} \dfrac{1}{t}, & \text{if } 0 \le \xi \le t, \\ 0 & \text{otherwise.} \end{cases}$$

Then Eq. (2) leads to

$$\int_{s}^{\infty} \varphi_D(\xi)\, d\xi = 1 - \frac{s}{t},$$

so that

$$s^* = \frac{t(pa - hQ^*)}{pa} = t\left(1 - \frac{hQ^*}{pa}\right).$$

Furthermore, from Eq. (1),

$$\int_s^\infty (\xi - s)\varphi_D(\xi)\, d\xi = \frac{t}{2} + \frac{s^2}{2t} - s$$

and

$$Q^* = \sqrt{\frac{2aK + apt + aps^{*2}/t - 2aps^*}{h}}.$$

By substituting the expression for s^* into the right-hand side of this last equation and then squaring both sides, the equation reduces algebraically to

$$Q^* = \sqrt{\frac{ap}{ap - ht}}\sqrt{\frac{2aK}{h}}.$$

Notice the close similarity to the economic lot-size formula ($Q^* = \sqrt{2aK/h}$) found at the beginning of Sec. 17.3 for the *deterministic* continuous review model with no shortages permitted.

5. Now consider the case where the demand during the delivery lag time λ has an exponential distribution with mean $a\lambda$, so that

$$\varphi_D(\xi) = \left(\frac{1}{a\lambda}\right)e^{-\xi/(a\lambda)}, \qquad \text{for } \xi \geq 0.$$

Since

$$\int_s^\infty \varphi_D(\xi)\, d\xi = e^{-s/(a\lambda)},$$

Eq. (2) yields

$$s^* = -a\lambda \ln \frac{hQ^*}{pa}.$$

Furthermore, from Eq. (1),

$$\int_s^\infty (\xi - s)\varphi_D(\xi)\, d\xi = a\lambda e^{-s/(a\lambda)},$$

and

$$Q^* = \sqrt{\frac{2a}{h}\left(K + a\lambda p e^{-s^*/(a\lambda)}\right)}.$$

Proceeding just as for the preceding case of a uniform distribution of demand, we now substitute the expression for s^* into this expression for Q^*, square both sides of this latter expression, and reduce it algebraically to obtain

$$Q^* = a\lambda + \sqrt{a^2\lambda^2 + \frac{2aK}{h}}.$$

Once again, note the similarity to the economic lot-size formula.

When the demand has either a uniform or an exponential distribution, a routine is available in your OR Courseware for obtaining s^* and Q^*.

Example: Consider the speaker example presented in Sec. 17.1. It was assumed that $K = 12{,}000$ and $h = 0.30$ per speaker per month, with a fixed demand rate of $a = 8{,}000$ per month and with instantaneous production of the speakers each time a production run is scheduled.

Now suppose that there actually is a *lag time* of $\lambda = 1$ month between ordering a production run to produce speakers and having the speakers ready for assembly into television sets. Also suppose that interruptions in the production of television sets cause the demand for speakers during this lag time to be a random variable, but still with a mean of $a = 8{,}000$ per month. The shortage cost is $p = \$5$ per speaker not ready as soon as it is needed.

If demand has a uniform distribution over the range from 0 to $t = 16{,}000$, the corresponding formulas give

$$Q^* = \sqrt{\frac{8{,}000(5)}{8{,}000(5) - 0.3(16{,}000)}}\sqrt{\frac{2(8{,}000)(12{,}000)}{0.3}} = 26{,}968,$$

$$s^* = 16{,}000\left[1 - \frac{0.3(26{,}968)}{5(8{,}000)}\right] = 12{,}764.$$

The probability of a stockout during a cycle then is

$$P\{D > s^*\} = \frac{hQ^*}{pa} = 0.20.$$

If demand instead has an exponential distribution with a mean of 8,000, the corresponding formulas give

$$Q^* = 8{,}000 + \sqrt{8{,}000^2 + \frac{2(8{,}000)(12{,}000)}{0.3}} = 34{,}533,$$

$$s^* = -8{,}000 \ln \frac{0.3(34{,}533)}{5(8{,}000)} = 10{,}807.$$

The probability of a stockout during a cycle then is given by

$$P\{D > s^*\} = \frac{hQ^*}{pa} = 0.26.$$

Continuous Review Model with Fixed Delivery Lag and No Backlogging: This model is identical to the preceding model except that unsatisfied demand now is assumed to be lost, i.e., unsatisfied demand will not be backlogged. Therefore, lost revenue now is included in the shortage cost.

The derivation of the costs contains the same approximations that were made in the backlogging case, so the subsequent expressions will lead to approximate results. The expressions for the expected ordering cost per unit of time $E(\text{OC})$ and the expected shortage cost per unit of time $E(\text{SC})$ are the same for both models. The only cost that differs is the expected holding cost per unit of time $E(\text{HC})$.

Recall that the backlogging model approximated the average inventory on hand during a cycle as the average of the inventory level at the beginning of the cycle and at the end of the cycle. Without backlogging, the inventory level cannot go negative, so the inventory level at the beginning of the cycle now will be larger by the number of units short (if any) at the end of the preceding cycle. Similarly, the inventory level at the end of a cycle now will be larger than for the backlogging model by the number of units short (if any) at that time. Consequently, the current model needs to adjust the expression for *expected average* inventory level for the backlogging model by adding

$$\text{Expected number of units short at the end of a cycle} = \int_s^\infty (\xi - s)\varphi_D(\xi)\, d\xi.$$

This adjustment yields

$$E(\text{HC}) = h\left[\frac{Q}{2} + s - a\lambda + \int_s^\infty (\xi - s)\varphi_D(\xi)\, d\xi\right].$$

Therefore, the expected total cost per unit time $C(Q, s)$ is

$$C(Q, s) = \frac{aK}{Q} + ac + h\left[\frac{Q}{2} + s - a\lambda + \int_s^\infty (\xi - s)\varphi_D(\xi)\, d\xi\right] + \frac{pa}{Q}\int_s^\infty (\xi - s)\varphi_D(\xi)\, d\xi.$$

Because there are two decision variables (Q and s), the optimal values (Q^* and s^*) are found by setting the corresponding partial derivatives equal to zero, i.e.,

$$\frac{\partial C(Q, s)}{\partial Q} = \frac{-aK}{Q^2} + \frac{h}{2} - \frac{pa\displaystyle\int_s^\infty (\xi - s)\varphi_D(\xi)\, d\xi}{Q^2} = 0$$

$$\frac{\partial C(Q, s)}{\partial s} = h\int_0^s \varphi_D(\xi)\, d\xi - \frac{pa\displaystyle\int_s^\infty \varphi_D(\xi)\, d\xi}{Q} = 0.$$

Solving these equations simultaneously[1] leads to

$$Q^* = \sqrt{\frac{2a\left[K + p\displaystyle\int_{s^*}^\infty (\xi - s^*)\varphi_D(\xi)\, d\xi\right]}{h}} \tag{3}$$

$$\int_{s^*}^\infty \varphi_D(\xi)\, d\xi = \frac{hQ^*}{hQ^* + pa}. \tag{4}$$

Unfortunately, solving these equations simultaneously to obtain a general closed-form expression for Q^* and s^* is not possible. However, the same iterative procedure as given for the backlogging model [but now using Eqs. (3) and (4) in place of Eqs. (1) and (2)] can be used to closely approximate Q^* and s^*. In practice, this procedure will generally converge in just a few iterations.

When the demand has either a uniform or an exponential distribution, a routine is available in your OR Courseware for obtaining s^* and Q^*.

Multiproduct Inventory Systems: In all the models presented in this chapter, we assume that just one product is being considered at a time. However, many real inventory systems deal with multiple products (sometimes even hundreds or thousands

[1] Note that the optimal values of Q and s are independent of c, the unit cost of the items ordered. The reason is that the expected number of units ordered per unit time is independent of the values of Q and s.

of products) simultaneously. There may be various interactions between these products. For example, the products may be sharing the same storage space. There may exist joint budget limitations involving all the products. Certain similar products may be *substitutable* in the sense that one product can be substituted for another that is in short supply.

Single-product models are helpful for providing insight into how to solve multiproduct problems. Furthermore, it is often possible to ''factor'' an N-product problem into N one-product problems without loss of optimality. This factoring can be done if the demand and cost for each product can be treated independently of the other products.

There has been some work on *multiproduct models* in which such factorization is not possible. Considerable research also has been done on *multiechelon models* dealing with one or more products stored at different locations or echelons of a supply system. Managing multiple echelons of a supply system (often called *supply chain management*) has become an increasingly important part of inventory management. This certainly is true for IBM's inventory system integrating its national network of spare-parts inventories that was mentioned at the beginning of the chapter (and referenced at the end of Sec. 2.4).

However, multiproduct and multiechelon models are beyond the scope of this book. (Further information about such models and their application is provided by the IBM study and its references.)

17.5 Conclusions

The inventory models presented here are rather simplified, but they serve the purpose of introducing the general nature of inventory models. Furthermore, they are sufficiently accurate representations of many actual inventory situations that they frequently are useful in practice. For example, the economic lot-size formulas have been particularly widely used, although they are sometimes modified to include some type of stochastic demand. The multiperiod models with stochastic demand have been important in characterizing the types of policies to follow, for example, (s, S) policies, even though the optimal values of s and S may be difficult to obtain.

Nevertheless, many inventory situations possess complications that are not taken into account by the models in this chapter, e.g., interactions between products or multiple echelons of a supply system. More complex models have been formulated in an attempt to fit such situations, but the models still sometimes leave a wide gap between practice and theory.

Continued growth is occurring in the computerization of inventory data processing, along with an accompanying growth in scientific inventory management.

SELECTED REFERENCES

1. Hax, A., and D. Candea, *Production and Inventory Management,* Prentice-Hall, Englewood Cliffs, NJ, 1984.
2. McClain, J. O., L. J. Thomas, and J. B. Mazzola, *Operations Management: Production of Goods and Services,* Prentice-Hall, Englewood Cliffs, NJ, 1992.
3. Nahmias, S., *Production and Operations Analysis,* Irwin, Homewood, IL, 1989.
4. Peterson, R., and E. A. Silver, *Decision Systems for Inventory Management and Production Planning,* 2d ed., Wiley, New York, 1985.

5. Schonberger, R., *Japanese Manufacturing Techniques: Nine Hidden Lessons in Simplicity,* Free Press, New York, 1982.
6. Storm Software, Inc., *STORM: Personal Version 3.0,* Prentice-Hall, Englewood Cliffs, NJ, 1992.
7. Tersine, R. J., *Principles of Inventory and Materials Management,* 3d rev. ed., North-Holland, New York, 1988.
8. Vollmann, T., W. Berry, and C. Whybark, *Manufacturing Planning and Control Systems,* 3d ed., Irwin, Homewood, IL, 1992.
9. Waters, C. D. J., *Inventory Control and Management,* Wiley, Chichester, England, 1992.

RELEVANT ROUTINES IN YOUR OR COURSEWARE

Automatic Routines:
Single-Period Model, No Setup
Single-Period Model, with Setup
Two-Period Model, No Setup
Infinite-Period Model, No Setup
Continuous Review Model, with Fixed Delivery Lag—Backlogging
Continuous Review Model, with Fixed Delivery Lag—No Backlogging

To access these routines, call the ProbMod program and then choose *Inventory Theory* under the Area menu. See App. 1 for documentation of the software.

PROBLEMS

To the left of each of the following problems (or their parts), we have inserted an A (for automatic routine) whenever a routine listed above can be helpful. An asterisk on the A indicates that this routine definitely should be used (unless your instructor gives contrary instructions) and that the printout from this routine is all that needs to be turned in to show your work in executing the procedure. An asterisk on the problem number indicates that at least a partial answer is given in the back of the book.

17.3-1.* Suppose that the demand for a product is 30 units per month and the items are withdrawn at a constant rate. The setup cost each time a production run is undertaken to replenish inventory is \$15. The production cost is \$1 per item, and the inventory holding cost is \$0.30 per item per month.

(*a*) Assuming shortages are not allowed, determine how often to make a production run and what size it should be.

(*b*) If shortages are allowed but cost \$3 per item per month, determine how often to make a production run and what size it should be.

17.3-2. The demand for a product is 600 units per week, and the items are withdrawn at a constant rate. The setup cost for placing an order to replenish inventory is \$25. The unit cost of each item is \$3, and the inventory holding cost is \$0.05 per item per week.

(*a*) Assuming shortages are not allowed, determine how often to order and what size the order should be.

(*b*) If shortages are allowed but cost \$2 per item per week, determine how often to order and what size the order should be.

17.3-3. Solve Prob. 17.3-2*b* when there is a delivery lag of 1 week.

17.3-4.* Consider the speaker example introduced in Sec. 17.1 and used in Sec. 17.3 to illustrate the economic lot-size model. Use this model, with shortages permitted, to solve this example when the shortage cost is changed to \$5 per speaker short per month short.

17.3-5. A taxi company uses gasoline at a constant rate of 8,500 gallons per month. The company purchases and stores huge quantities of gasoline at a discount every few months. The gasoline costs \$1.05 per gallon, with a setup cost of \$1,000 for each order. The inventory holding cost is \$0.01 per gallon per month.

(*a*) Assuming shortages are not allowed, determine how often and how much to order.
(*b*) If shortages cost \$0.50 per gallon per month, determine how often and how much to order.

17.3-6. Solve Prob. 17.3-5*a* when the cost of all the gasoline drops to \$1.00 per gallon if at least 50,000 gallons is purchased.

17.3-7. Solve Prob. 17.3-5*a* when the cost of gasoline is \$1.20 per gallon for the first 20,000 gallons purchased, \$1.10 per gallon for the next 20,000 gallons, and \$1.00 per gallon thereafter.

17.3-8. Consider the economic lot-size model with shortages permitted, as presented in Sec. 17.3. Suppose, however, that a constraint is added $S/Q = 0.8$. Derive the expression for the optimal value of Q.

17.3-9. In the economic lot-size model, suppose the stock is replenished uniformly (rather than instantaneously) at the rate of b items per unit time until the lot size Q is fulfilled. Withdrawals from the inventory are made at the rate of a items per unit time, where $a < b$. Replenishments and withdrawals of the inventory are made simultaneously. For example, if Q is 60, b is 3 per day, and a is 2 per day, then 3 units of stock arrive each day for days 1 to 20, 31 to 50, and so on, whereas units are withdrawn at the rate of 2 per day every day. The diagram of inventory level versus time is given below for this example.

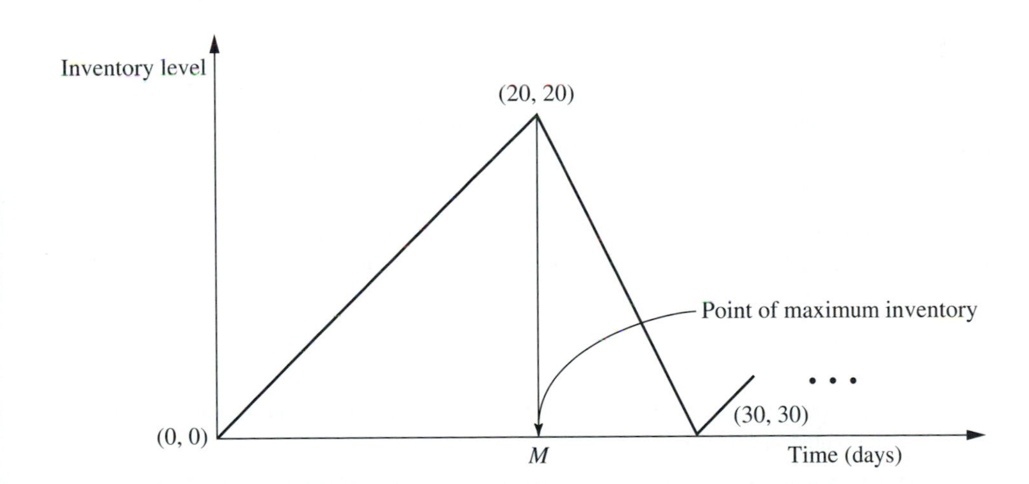

(*a*) Find the total cost per unit time in terms of the setup cost K, production quantity Q, unit cost c, holding cost h, withdrawal rate a, and replenishment rate b.
(*b*) Determine the economic lot size Q^*.

17.3-10. Suppose that production planning is to be done for the next 5 months, where the respective demands are $r_1 = 2$, $r_2 = 4$, $r_3 = 2$, $r_4 = 2$, and $r_5 = 3$. The setup cost is \$4,000, the unit production cost is $1,000, and the unit holding cost is $300. Determine the optimal production schedule that satisfies the monthly requirements.

17.3-11.* Reconsider the example used to illustrate the production planning model. Solve this problem when the demands are increased by 1 unit (10,000 speakers) in each period.

17.3-12. Reconsider the example used to illustrate the production planning model. Solve this problem when the unit cost per speaker during the first and third periods is increased to $1.40.

17.3-13.* Consider a situation where a particular product is produced and placed in in-process inventory until it is needed in a subsequent production process. The number of units required in each of the next 3 months, the setup cost, and the regular-time unit production cost (in units of thousands of dollars) that would be incurred in each month are as follows:

Month	Requirement	Setup Cost	Regular-Time Unit Cost
1	1	5	8
2	3	10	10
3	2	5	9

There currently is 1 unit in inventory, and we want to have 2 units in inventory at the end of 3 months. A maximum of 3 units can be produced on regular-time production in each month, although 1 additional unit can be produced on overtime at a cost that is 2 larger than the regular-time unit production cost. The holding cost is 2 per unit for each extra month that it is stored.

Use dynamic programming to determine how many units should be produced in each month to minimize the total cost.

17.3-14. Consider a situation where a particular product is produced and placed in in-process inventory until it is needed in a subsequent production process. The number of units required in each of the next 2 months, the setup cost, the holding cost for each unit left in inventory at the end of a month, and the unit production cost (all in units of thousands of dollars) are as follows:

Month	Requirement	Setup Cost	Holding Cost	Unit Cost
1	3	5	0.30	9
2	4	5	0.30	9

Determine the optimal production schedule that satisfies the monthly requirements by using the algorithm presented in Sec. 17.3.

A **17.4-1.** A newspaper stand purchases newspapers for $0.36 and sells them for $0.50. The shortage cost is $0.50 per newspaper (because the dealer buys papers at retail price to satisfy shortages). The holding cost $0.002 per newspaper left at the end of the day. The demand distribution is a uniform distribution between 200 and 300. Find the optimal number of papers to buy.

A **17.4-2.** Suppose that the demand D for a spare airplane part has an exponential distribution with mean 50, that is,

$$\varphi_D(\xi) = \begin{cases} \frac{1}{50}e^{-\xi/50} & \text{for } \xi \geq 0, \\ 0 & \text{otherwise.} \end{cases}$$

This airplane will be obsolete in 1 year, so all production of the spare part is to take place at present. The production costs now are $1,000 per item—that is, $c = 1{,}000$—but they become $10,000 per item if they must be supplied at later dates—that is, $p = 10{,}000$. The holding costs, charged on the excess after the end of the period, are $300 per item.

(*a*) Determine the optimal number of spare parts to produce.

(*b*) Suppose that the manufacturer has 23 parts already in inventory (from a similar, but now obsolete airplane). Determine the optimal inventory policy.

(*c*) Suppose that p cannot be determined now, but the manufacturer wishes to order a quantity so that the probability of a shortage equals 0.1. How many units should be ordered?

(*d*) If the manufacturer were following an optimal policy that resulted in ordering the quantity found in part (*c*), what is the implied value of p?

A **17.4-3.*** A baking company distributes bread to grocery stores daily. The company's cost for the bread is $0.80 per loaf. The company sells the bread to the stores for $1.20 per loaf sold, provided that it is disposed of as fresh bread (sold on the day it is baked). Bread not sold is returned to the company. The company has a store outlet that sells bread that is 1 day or more old for $0.60 per loaf. No significant storage cost is incurred for this bread. The cost of the loss of customer goodwill due to a shortage is estimated to be $0.80 per loaf. The daily demand has a uniform distribution between 1,000 and 2,000 loaves. Find the optimal daily number of loaves that the manufacturer should produce.

A **17.4-4.** A student majoring in operations research enjoys optimizing his personal decisions. He is analyzing one such decision currently, namely, how much money (if any) to take out of his savings account to buy traveler's checks before leaving on a summer vacation trip to Europe.

He already has used the money he had in his checking account to buy traveler's checks worth $1,200, but this may not be enough. In fact, he has estimated the probability distribution of what he will need shown in the following table:

Amount needed, $	1,000	1,100	1,200	1,300	1,400	1,500	1,600	1,700
Probability	0.05	0.10	0.15	0.25	0.20	0.10	0.10	0.05

If he turns out to have less than he needs, then he will have to leave Europe 1 week early for every $100 short. Because he places a value of $150 on each week in Europe, each week lost would thereby represent a net imputed loss of $50 to him. However, every $100 traveler's check costs an extra $1. Furthermore, *each* such check left over at the end of the trip (which would be redeposited in the savings account) represents a loss of $2 in interest that could have been earned in the savings account during the trip, so he does not want to purchase too many.

Using these data, determine the optimal decision on how many additional $100 traveler's checks (if any) the student should purchase from his savings account money.

17.4-5. The campus bookstore must decide how many textbooks to order for a course that will be offered only once. The number of students who will take the course is a random variable D, whose distribution can be approximated by a (continuous) uniform distribution on the interval [40, 60]. After the quarter starts, the value of D becomes known. If D exceeds the number of books available, the known shortfall is made up by placing a rush order at a cost of $14 plus $2 per book over the normal ordering cost. If D is less than the stock on hand, the extra books are returned for their original ordering cost less $1 each. What is the order quantity that minimizes the expected cost?

17.4-6. Consider the following inventory model, which is a single-period model with known density of demand $\varphi_D(\xi) = e^{-\xi}$ for $\xi \geq 0$ and $\varphi_D(\xi) = 0$ elsewhere. There are two costs connected with the model. The first is the purchase cost, given by $c(y - x)$. The second is a cost p that is incurred once if there is *any* unsatisfied demand (independent of the amount of unsatisfied demand).

(*a*) If x units are available and goods are ordered to bring the inventory level up to y (if $x < y$), write the expression for the expected loss and describe completely the optimal policy.
(*b*) If a fixed cost K is also incurred whenever an order is placed, describe the optimal policy.

A* **17.4-7.** Find the optimal ordering policy for a one-period model where the demand has the probability density function

$$\varphi_D(\xi) = \begin{cases} \frac{1}{20} & \text{for } 0 \le \xi \le 20, \\ 0 & \text{otherwise,} \end{cases}$$

and the costs are

$$\text{Holding cost} = \$1 \text{ per item,}$$

$$\text{Shortage cost} = \$3 \text{ per item,}$$

$$\text{Setup cost} = \$1.50,$$

$$\text{Production cost} = \$2 \text{ per item.}$$

Show your work, and then check your answer by using the corresponding routine in your OR Courseware.

A* **17.4-8.** Using the approximation for finding the optimal policy for a single-period model when demand has an exponential distribution, find this policy when

$$\varphi_D(\xi) = \begin{cases} \frac{1}{25}e^{-\xi/25} & \text{for } \xi \ge 0, \\ 0 & \text{otherwise,} \end{cases}$$

and the costs are

$$\text{Holding cost} = 40 \text{ cents per item,}$$

$$\text{Shortage cost} = \$1.50 \text{ per item,}$$

$$\text{Purchase price} = \$1 \text{ per item,}$$

$$\text{Setup cost} = \$10.$$

Show your work, and then check your answer by using the corresponding routine in your OR Courseware.

17.4-9.* Consider a one-period model where the only two costs are the holding cost, given by

$$h(y - D) = \tfrac{3}{10}(y - D), \qquad \text{for } y \ge D,$$

and the shortage cost, given by

$$p(D - y) = 2.5(D - y), \qquad \text{for } D \ge y.$$

The probability density function for demand is given by

$$\varphi_D(\xi) = \begin{cases} \dfrac{e^{-\xi/25}}{25} & \text{for } \xi \ge 0, \\ 0 & \text{otherwise.} \end{cases}$$

If you order, you must order an *integer* number of *batches* of 100 units each, and this quantity is delivered immediately. Let $G(y)$ denote the total expected cost when there are y units available for the period (after ordering).
(*a*) Write the expression for $G(y)$.
(*b*) What is the optimal ordering policy?

17.4-10. Consider the following inventory situation. Demands in different periods are independent but with a common probability density function given by

$$\varphi_D(\xi) = \begin{cases} \dfrac{e^{-\xi/25}}{25} & \text{for } \xi \geq 0, \\ 0 & \text{otherwise.} \end{cases}$$

Orders may be placed at the start of each period without setup cost at a unit cost of $c = 10$. There are a holding cost of 6 per unit remaining in stock at the end of each period and a shortage cost of 15 per unit of unsatisfied demand at the end of each period (with backlogging except for the final period).

A (*a*) Find the optimal one-period policy.

A* (*b*) Find the optimal two-period policy.

17.4-11. Consider the following inventory situation. Demands in different periods are independent but with a common probability density function $\phi_D(\xi) = \frac{1}{50}$ for $0 \leq \xi \leq 50$. Orders may be placed at the start of each period without setup cost at a unit cost of $c = 10$. There are a holding cost of 8 per unit remaining in stock at the end of each period and a penalty cost of 15 per unit of unsatisfied demand at the end of each period (with backlogging except for the final period).

A (*a*) Find the optimal one-period policy.

A* (*b*) Find the optimal two-period policy.

A **17.4-12.*** Find the optimal inventory policy for the following two-period model by using a discount factor of $\alpha = 0.9$. The demand D has the probability density function

$$\varphi_D(\xi) = \begin{cases} \frac{1}{25}e^{-\xi/25} & \text{for } \xi \geq 0, \\ 0 & \text{otherwise,} \end{cases}$$

and the costs are

$$\text{Holding cost} = \$0.25 \text{ per item,}$$
$$\text{Shortage cost} = \$2 \text{ per item,}$$
$$\text{Purchase price} = \$1 \text{ per item.}$$

Stock left over at the end of the final period is salvaged for $1 per item, and shortages remaining at this time are met by purchasing the needed items at $1 per item.

A* **17.4-13.** Solve Prob. 17.4-12 for a two-period model, assuming no salvage value, no backlogging at the end of the second period, and no discounting.

A **17.4-14.** Solve Prob. 17.4-12 for an infinite-period model.

A **17.4-15.*** Determine the optimal inventory policy when the goods are to be ordered at the end of every month from now on. The cost of bringing the inventory level up to y when x already is available is given by $2(y - x)$. Similarly, the cost of having the monthly demand D exceed y is given by $5(D - y)$. The probability density function for D is given by $\phi_D(\xi) = e^{-\xi}$. The holding cost when y exceeds D is given by $y - D$. A monthly discount factor of 0.95 is used.

A **17.4-16.** Solve the inventory problem given in Prob. 17.4-15, but assume that the policy is to be used for only 1 year (a 12-period model). Shortages are backlogged each month, except that any shortages remaining at the end of the year are made up by purchasing similar items at a unit cost of $2. Any remaining inventory at the end of the year can be sold at a unit price of $2.

A **17.4-17.** A supplier of high-fidelity receiver kits is interested in using an optimal inventory policy. The distribution of demand per month is uniform between 2,000 and 3,000 kits. The supplier's cost for each kit is $150. The holding cost is estimated to be $2 per kit remaining at the end of a month, and the shortage cost is $30 per kit of unsatisfied demand at the end of a month. Using a monthly discount factor of $\alpha = 0.99$, find the optimal inventory policy for this infinite-period problem.

17.4-18. The weekly demand for a certain type of electronic calculator is estimated to be

$$\varphi_D(\xi) = \begin{cases} \dfrac{1}{1{,}000} e^{-\xi/1{,}000} & \text{for } \xi \geq 0, \\ 0 & \text{otherwise.} \end{cases}$$

The unit cost of these calculators is \$80. The holding cost is \$0.70 per calculator remaining at the end of a week. The shortage cost is \$2 per calculator of unsatisfied demand at the end of a week. Using a weekly discount factor of $\alpha = 0.998$, find the optimal inventory policy for this infinite-period problem.

17.4-19. Find the optimal (k, Q) policy for Prob. 17.4-9 for an infinite-period model with a discount factor of $\alpha = 0.90$.

17.4-20. For the infinite-period model with no setup cost, show that the value of y^0 that satisfies

$$\Phi(y^0) = \frac{p - c(1 - \alpha)}{p + h}$$

is equivalent to the value of y that satisfies

$$\frac{dL(y)}{dy} + c(1 - \alpha) = 0,$$

where $L(y)$, the expected shortage plus holding cost, is given by

$$L(y) = \int_y^\infty p(\xi - y)\varphi_D(\xi)\, d\xi + \int_0^y h(y - \xi)\varphi_D(\xi) d\xi.$$

A **17.4-21.*** Solve Prob. 17.3-1*b*, assuming now that the demand in each half-month period is random with a uniform distribution over the interval (0, 30), unsatisfied demand is backlogged, and delivery lead time is $\frac{1}{2}$ month.

A **17.4-22.** Solve Prob. 17.3-2*b*, assuming now that the weekly demand is random with an exponential distribution with mean 600, unsatisfied demand is backlogged, and delivery lead time is 1 week.

A **17.4-23.** Solve Prob. 17.3-5*b*, assuming now that the weekly demand for gasoline is random with a uniform distribution over the interval (0, 4,250), unsatisfied demand is backlogged, and delivery lead time is $\frac{1}{4}$ month.

A **17.4-24.** Solve Prob. 17.3-5*b*, assuming now that the weekly demand for gasoline is random with an exponential distribution with mean 2,125, unsatisfied demand is backlogged and delivery lead time is $\frac{1}{4}$ month.

17.4-25. Consider the continuous review model with fixed delivery lag and backlogging. Given that the order quantity Q is fixed in advance, derive the expression for the optimal reorder point s^* when the distribution of demand is exponential with mean $a\lambda$.

17.4-26. Suppose that the criterion of minimizing the expected total cost per unit time subject to having a satisfactory service level is chosen for the continuous review model with fixed delivery lag and backlogging. That is, the decision maker wishes to choose an ''optimal'' policy subject to being assured that the probability of a shortage occurring during a cycle will not exceed V, or

$$P\{D > s\} \leq V.$$

(*a*) Assuming that the distribution of demand is uniform over the range from 0 to t and that the inventory policy is to place an order of size Q whenever the inventory position reaches s, find the relation between s and V.

(*b*) Using the smallest value of s in part (*a*), find the expression for $C(Q, s)$, the expected total cost per unit time, in terms of the only remaining variable Q. Call this cost $C(Q)$.
(*c*) Show that the optimal order size Q is given by

$$Q = \sqrt{\frac{2a}{h}\left(K + \frac{ptV^2}{2}\right)}.$$

17.4-27. Follow the instructions of Prob. 17.4-26 when the distribution of demand is exponential with mean $a\lambda$, so that

$$\varphi_D(\xi) = \frac{1}{\alpha\lambda} e^{-\xi/(a\lambda)}, \qquad \text{for } \xi \geq 0.$$

For part (*c*), show that

$$Q = \sqrt{\frac{2a}{h}(K + \alpha\lambda pV)}.$$

17.4-28. Consider the continuous review model with fixed delivery lag and no backlogging. Suppose the distribution of demand is uniform over the range from 0 to t. Show that Eq. (4) becomes

$$s^* = t\frac{pa}{hQ^* + pa}$$

and Eq. (3) becomes

$$Q^* = \sqrt{\frac{2aK + apt + aps^{*2}/t - 2aps^*}{h}}.$$

17.4-29. Consider the continuous review model with fixed delivery lag and no backlogging. Suppose that the distribution of demand is exponential with mean $a\lambda$. Show that Eq. (4) becomes

$$s^* = -a\lambda \ln \frac{hQ^*}{hQ^* + pa}$$

and Eq. (3) becomes

$$Q^* = \sqrt{\frac{2a}{h}(K + a\lambda pe^{-s^*/(a\lambda)})}.$$

A* **17.4-30.** Use your OR Courseware to solve Prob. 17.4-21 when unsatisfied demand is not backlogged.

A* **17.4-31.** Use your OR Courseware to solve Prob. 17.4-22 when unsatisfied demand is not backlogged.

A* **17.4-32.** Use your OR Courseware to solve Prob. 17.4-23 when unsatisfied demand is not backlogged.

A* **17.4-33.** Use your OR Courseware to solve Prob. 17.4-24 when unsatisfied demand is not backlogged.

18

Forecasting

Chapter 17, ''Inventory Theory,'' was concerned with finding optimal inventory policies. These policies are, in part, dependent upon some forecast of sales or use of the items of interest. Forecasting is an essential component of any successful inventory system.

However, forecasting need not be associated solely with problems of inventory control. Other areas where forecasting plays an important role include marketing, financial planning, and production planning. Indeed, managerial decisions seldom are made in the absence of some form of forecasting. Thus a forecast is a basic tool to aid managerial decision making.

Forecasts can be obtained by using qualitative or quantitative techniques. In the former case, a forecast is usually the result of an expression of one or more experts' personal judgments or opinions, and the approach is called a **judgmental technique**. For example, a major research university calls in its leading economists every September to obtain their judgment on what to expect as an inflation rate for the next academic year—a number crucial to the budgeting process. This number is generally arrived at by consensus after prolonged discussion by the economists.

Two distinct quantitative techniques based on conventional statistical methods are used in forecasting, namely, time series analysis and regression analysis. A statistical **time series** is a series of numerical values that a random variable takes on over a period of time. For example, the daily market closing prices of a particular stock over 1 year constitute a time series. Time series analysis exploits techniques that utilize these data for forecasting the values of the variable of interest in a future period. For example, businesses commonly make quarterly sales forecasts based on an analysis of sales during previous quarters. Forecasting, in general, is concerned with an analysis of past time series data in order to estimate one or more future values of the time series. The forecast depends upon a model of behavior of the time series.

In **regression analysis**, the variable to be forecast (the dependent variable) is expressed as a mathematical function of other (independent) variables. For example, the forecast of the total sales of a textbook in a given period may be functionally related to the mail order sales of that book during the same period. Data on mail order sales and total sales over previous periods may be used to forecast total sales in a future period given the mail order sales for that period.

These forecasting methods may be used in conjunction with one another. Indeed, the judgmental technique is often used together with either time series analysis or regression analysis.

The first section briefly describes judgmental techniques. Sections 18.2 to 18.7 focus on forecasting based on time series analysis. Section 18.8 then turns to the use of regression analysis.

18.1 Judgmental Techniques

Judgmental techniques are, by their very nature, subjective, and they may involve such qualities as intuition, expert opinion, and experience. They generally lead to forecasts that are based upon qualitative criteria.

One commonly used technique is to bring together a group of experts who interact with each other and produce a consensus forecast. The above example of a group of university economists forecasting the inflation rate is an example of the **expert group technique**.

Perhaps the most important judgmental technique is called the **Delphi method**. Like the expert group technique, the Delphi method utilizes a group of experts, but not in a meeting setting. In addition to this group, there are one or more decision makers who ultimately are responsible for making the forecast. Finally, there is the staff who perform the duties associated with the method. These duties include the preparation of questionnaires and the analysis of their results.

The Delphi method begins with the panel of experts filling out a questionnaire. Based upon the results, a second questionnaire is developed and sent to the same panel of experts, together with the results of the first questionnaire. This second questionnaire is then completed by the panel of experts and returned for analysis. Based upon the results of the two questionnaires, and using their own expertise, the decision makers ultimately come forth with a forecast. The key to the Delphi method is the feedback of the information contained in the first questionnaire to the panel of experts. Thus each member of the panel has access to information that she or he may have lacked originally, so that all members of the panel have the same information when completing the second questionnaire.

Of course, the success of the Delphi method hinges on the quality of the design of the questionnaires. Occasionally, more than two rounds of questionnaires may be deemed desirable. This situation occurs when there is considerable disparity between the results of the first two questionnaires. A third round may then be used in the hope that the feedback from the results of the second round will lead to more convergence of the results on the third.

A recent survey of 500 U.S. corporations (see Selected Reference 8) indicates that managers rely heavily on judgmental forecasting methods.

18.2 Time Series

A **time series** can be viewed as the representation of the outcomes of a *random variable* of concern, usually taken at equally spaced intervals, over a fixed period. For example, the quarterly unemployment rate from January 1980 to July 1984 comprises a time series, as illustrated in Fig. 18.1. The quarterly sales of a particular product over the last three quarters also comprise a time series. The behavior of a time series can be displayed in graphical form, bar graphical form, or tabular form, where the first method is generally most descriptive of the pattern of behavior of the series.

Because a time series is a description of the past, a logical procedure for forecasting the future is to make use of these historical data. If the past data are indicative of what we can expect in the future, we can postulate an underlying mathematical model that is representative of the process. The model can then be used to generate forecasts.

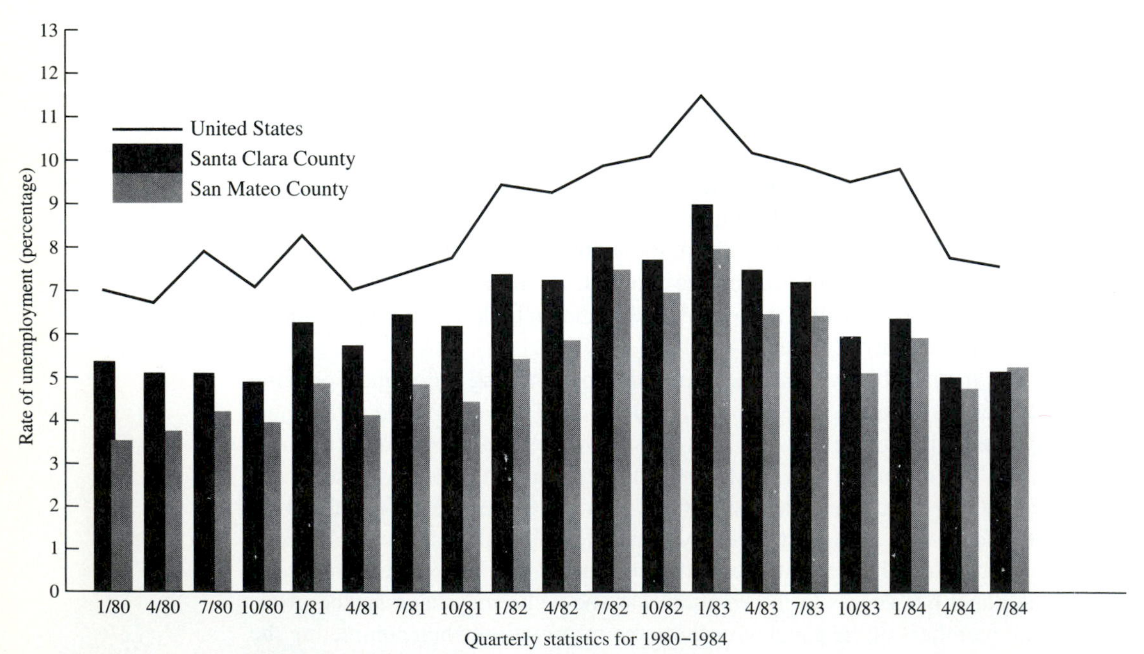

Figure 18.1 The evolution of the quarterly unemployment rate (not seasonally adjusted) for each of three geographical areas illustrates three time series.

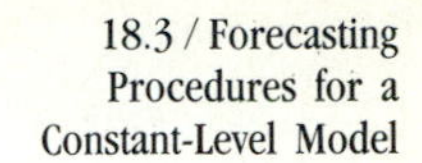

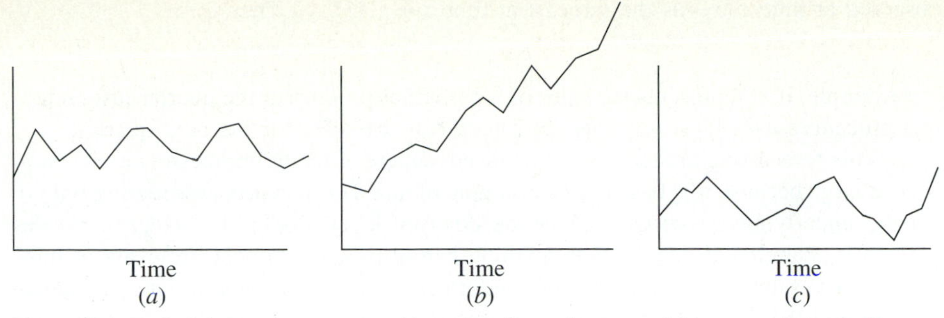

Figure 18.2 Typical time series patterns, with random fluctuations around (*a*) a constant level, (*b*) a linear trend, and (*c*) a constant level plus seasonal effects.

In most realistic situations, we do not have complete knowledge of the exact form of the model that generates the time series, so an approximate model must be chosen. Frequently, the choice is made by observing the outcomes of the time series over time. Several typical time series patterns are shown in Fig. 18.2. Figure 18.2*a* shows a time series that might be observed if the generating process were represented by a **constant level** superimposed with random fluctuations. Figure 18.2*b* shows a time series that might be observed if the generating process were represented by a **linear trend** superimposed with random fluctuations. Finally, Fig. 18.2*c* shows a time series that might be observed if the generating process were represented by a constant level superimposed with a **seasonal effect** together with random fluctuations. There are many other plausible representations, but these three are very useful in practice and so are considered in this chapter.

Once the form of the model is chosen, a mathematical representation of the generating process of the time series can be given. For example, suppose that the generating process is identified as a **constant-level model** superimposed with random fluctuations, as illustrated in Fig. 18.2*a*. Such a representation can be given by

$$X_t = A + e_t,$$

where X_t is the random variable observed at time t, A is the constant level of the model, and e_t is the random error occurring at time t (assumed to have expected value equal to zero and constant variance). Let F_{t+1} denote the forecast of the value of the time series at time $t + 1$. It is reasonable to expect that F_{t+1} will be a function of some of, or all, the observed values of the time series prior to time $t + 1$.

18.3 Forecasting Procedures for a Constant-Level Model

We now present four alternative forecasting procedures for the constant-level model introduced in the preceding paragraph.

Last-Value Forecasting Procedure

Denote by x_t the value that the random variable X_t takes on. By interpreting t as the *current time,* the last-value forecasting procedure uses the value of the time series observed at time t (x_t) as the forecast at time $t + 1$ (F_{t+1}). Therefore,

$$F_{t+1} = x_t.$$

For example, if x_t represents the sales of a particular product in the quarter just ended, this procedure uses these sales as the forecast of the sales for the next quarter.

This forecasting procedure has the disadvantage of being imprecise; i.e., its variance is large because it is based upon a sample of size 1. It is worth considering only if (1) the underlying assumption about the constant-level model is "shaky" and the process is changing so rapidly that anything before time t is almost irrelevant or misleading or (2) the assumption that the random error e_t has constant variance is unreasonable and the conditional variance at time t actually is very small.

This simple technique is frequently compared with the results of using a more sophisticated, but more cumbersome, technique (such as the three described below) to assess whether the sophisticated technique is worthwhile.

Average Forecasting Procedure

Rather than discarding all but the most recent observation, this procedure averages *all* the observations as the forecast for the next period, i.e.,

$$F_{t+1} = \sum_{i=1}^{t} \frac{x_i}{t}.$$

This estimate is an excellent one if the process is entirely stable, i.e., if the assumptions about the underlying model are correct. However, frequently there exists skepticism about the persistence of the underlying model over an extended time. Conditions inevitably change eventually. Because of a natural reluctance to use very old data, this procedure generally is limited to young processes.

Moving-Average Forecasting Procedure

Rather than using very old data that may no longer be relevant, this procedure averages the data for only the last n periods as the forecast for the next period, i.e.,

$$F_{t+1} = \sum_{i=t-n+1}^{t} \frac{x_i}{n}.$$

Note that this forecast is easily updated from period to period. All that is needed each time is to lop off the first observation and add the last one.

The *moving-average* estimator combines the advantages of the *last value* and *average* estimators in that it uses only recent history *and* it uses multiple observations. A disadvantage of this procedure is that it places as much weight on x_{t-n+1} as on x_t. Intuitively, one would expect a good procedure to place more weight on the most recent observation than on older observations that may be less representative of current conditions. Our next procedure does just this.

Exponential Smoothing Forecasting Procedure

This procedure uses the formula

$$F_{t+1} = \alpha x_t + (1 - \alpha)F_t,$$

where α ($0 < \alpha < 1$) is called the **smoothing constant**. (The choice of α is discussed later.) Thus the forecast is just a weighted sum of the last observation x_t and the preceding forecast F_t for the period just ended. Because of this recursive relationship between F_{t+1} and F_t, alternatively F_{t+1} can be expressed as

$$F_{t+1} = \alpha x_t + \alpha(1 - \alpha)x_{t-1} + \alpha(1 - \alpha)^2 x_{t-2} + \cdots.$$

In this form, it becomes evident that exponential smoothing gives the most weight to x_t and decreasing weights to earlier observations. Furthermore, the first form reveals that the forecast is simple to calculate because the data prior to period t need not be retained; all that is required is x_t and the previous forecast F_t.

Another alternative form for the exponential smoothing technique is given by

$$F_{t+1} = F_t + \alpha(x_t - F_t),$$

which gives a heuristic justification for this procedure. In particular, the forecast of the time series at time $t + 1$ is just the preceding forecast at time t plus the *product* of the forecasting error at time t and a discount factor α. This alternative form is often simpler to use.

A measure of effectiveness of exponential smoothing can be obtained under the assumption that the process is completely stable, so that $X_1, X_2, \ldots$ are independent, identically distributed random variables with variance σ^2. It then follows that (for large t)

$$\text{var}[F_{t+1}] \approx \frac{\alpha\sigma^2}{2 - \alpha} = \frac{\sigma^2}{(2 - \alpha)/\alpha},$$

so that the variance is statistically equivalent to a moving average with $(2 - \alpha)/\alpha$ observations. If α is chosen equal to 0.1, then $(2 - \alpha)/\alpha = 19$. Thus the exponential smoothing technique with this value of α is *equivalent* to a moving-average procedure that uses 19 observations. However, if a change in the process does occur (e.g., if the mean starts increasing), exponential smoothing will react more quickly with better tracking of the change than the moving-average procedure.

An important drawback of exponential smoothing is that it lags behind a continuing trend; i.e., if the constant-level model is incorrect and the mean is increasing steadily, then the forecast will be several periods behind. However, the procedure can be easily adjusted for trend (and even seasonally adjusted).

Another disadvantage of exponential smoothing is that it is difficult to choose an appropriate smoothing constant α. Exponential smoothing can be viewed as a statistical filter that inputs raw data from a stochastic process and outputs smoothed estimates of a mean that varies with time. If α is chosen to be small, response to change is slow, with resultant smooth estimators. On the other hand, if α is chosen to be large, response to change is fast, with resultant large variability in the output. Hence, there is a need to compromise, depending upon the degree of stability of the process. It has been suggested that α should not exceed 0.3 and that a reasonable choice for α is approximately 0.1. This value can be increased temporarily if a change in the process is expected or when one is just starting the forecasting. At the start, a reasonable approach is to choose the forecast for period 2 according to

$$F_2 = \alpha x_1 + (1 - \alpha)(\text{initial estimate}),$$

where some initial estimate of the constant level A must be obtained. If past data are available, such an estimate may be the average of these data.

Your OR Courseware includes a routine for applying the exponential smoothing forecasting procedure.

18.4 A Forecasting Procedure for a Linear Trend Model

All the forecasting procedures just presented for the constant-level model will lag behind the process if it actually has a continuing trend. Suppose that the generating process of the observed time series can be represented by a *linear trend* superimposed with *random fluctuations,* as illustrated in Fig. 18.2*b*. Denote the slope of the linear trend by B, where the slope is called the **trend factor**. The model is represented by

$$X_t = A + Bt + e_t,$$

where X_t is the random variable that is observed at time t, A is a constant, B is the trend factor, and e_t is the random error occurring at time t (assumed to have expected value equal to zero and constant variance). In the constant-level model, the forecast for period $t + 1$ (based upon data through period t) also provides the best current forecast for periods $t + 1 + m$, for $m = 1, 2, \ldots$. For linear trend models, such a statement no longer holds. Hence, rather than referring to forecasts immediately, the concept of a "smoothed" level will be introduced.

If x_t is the observed value of the time series at time t, then a smoothed level S_t at time t will be a linear combination of x_t and the smoothed value at time period $t - 1$ corrected by adding the trend (slope) to indicate the passage of a unit of time, i.e.,

$$S_t = \alpha x_t + (1 - \alpha)(S_{t-1} + B),$$

with the smoothing constant α satisfying $0 < \alpha < 1$. The forecast for time $t + 1$ can now be obtained:

$$F_{t+1} = S_t + B.$$

Unfortunately, the trend (slope) B is unknown so it must be estimated, and exponential smoothing can again be used for this purpose, i.e.,

$$B_t = \beta(S_t - S_{t-1}) + (1 - \beta)B_{t-1},$$

where B_t is the smoothed value of the trend at time t and β $(0 < \beta < 1)$ is another (possibly different from α) *smoothing constant.*[1] Hence, S_t can now be expressed as

$$S_t = \alpha x_t + (1 - \alpha)(S_{t-1} + B_{t-1}),$$

and the forecast for m periods ahead $(m = 1, 2, \ldots)$ is given by

$$F_{t+m} = S_t + mB_t.$$

Summary of the Forecasting Procedure with a Linear Trend

1. Using x_t, S_{t-1}, and B_{t-1}, the smoothed level of the time series at time t is given by

   $$S_t = \alpha x_t + (1 - \alpha)(S_{t-1} + B_{t-1}).$$

2. From S_t (calculated in step 1), S_{t-1}, and B_{t-1}, the smoothed value of the trend at the end of period t is given by

   $$B_t = \beta(S_t - S_{t-1}) + (1 - \beta)B_{t-1}.$$

3. The forecast of the time series for m periods ahead $(m = 1, 2, \ldots)$ is given by

   $$F_{t+m} = S_t + mB_t.$$

[1] The previous discussion concerning the choice of α is relevant to the choice of β.

As in the case of exponential smoothing for a constant-level model, an initial value is required to start the smoothing process for the linear trend model. This initialization is frequently obtained by fitting a straight line to some past data (using methods described in Secs. 18.7 and 18.8). The fitted line can be used to obtain an initial value of the smoothed level of the time series S_0 and an initial value of the smoothed trend level B_0. Thus,

$$S_1 = \alpha x_1 + (1 - \alpha)(S_0 + B_0)$$

and

$$B_1 = \beta(S_1 - S_0) + (1 - \beta)B_0.$$

If a forecast for period 2 is desired, it can be obtained from

$$F_{1+1} = F_2 = S_1 + B_1.$$

One routine in your OR Courseware is designed for applying this forecasting procedure.

Example: Suppose that a wholesale distributor of bicycles (as in Example 2 of Sec. 17.1) wishes to forecast sales of a particular model bicycle. It is early autumn, and she wants a forecast of winter sales in order to make plans for replenishing inventory.

Sales this past winter, spring, and summer have been 2,800, 2,925, and 3,040, respectively. Based upon these three observations, we will use the above forecasting procedure with $\alpha = 0.1$ and $\beta = 0.1$ to forecast winter sales. Time period t corresponds to season t of this year, so $x_1 = 2{,}800$, $x_2 = 2{,}925$, and $x_3 = 3{,}040$, with F_5 being the desired forecast.

From sales figures prior to this year, a straight line has been fitted to these data to indicate the upward trend of sales for this model bicycle. The value on the line corresponding to last autumn is 2,750, and the slope is 100. Thus $S_0 = 2{,}750$ and $B_0 = 100$.

Using the above formulas, we see that the smoothed value of sales for this past winter is

$$S_1 = 0.1(2{,}800) + 0.9(2{,}750 + 100) = 2{,}845.$$

The smoothed value of the trend during last winter is

$$B_1 = 0.1(2{,}845 - 2{,}750) + 0.9(100) = 99.5.$$

Repeating this procedure for spring leads to

$$S_2 = 0.1(2{,}925) + 0.9(2{,}845 + 99.5) = 2{,}943$$

and

$$B_2 = 0.1(2{,}943 - 2{,}845) + 0.9(99.5) = 99.4.$$

Finally, the actual sales in the summer result in

$$S_3 = 0.1(3{,}040) + 0.9(2{,}943 + 99.4) = 3{,}042$$

and

$$B_3 = 0.1(3{,}042 - 2{,}943) + 0.9(99.4) = 99.4.$$

Therefore, the forecast of sales for next winter is

$$F_{3+2} = F_5 = 3{,}042 + 2(99.4) = 3{,}241.$$

Looking back, we see that if a forecast for spring sales had been made at the end of last winter, it would have been

$$F_2 = S_1 + B_1 = 2{,}845 + 99.5 = 2{,}945.$$

Similarly, if a forecast for summer had been made at the end of spring, it would have been

$$F_3 = S_2 + B_2 = 2{,}943 + 99.4 = 3{,}042.$$

18.5 A Forecasting Procedure for a Constant Level with Seasonal-Effects Model

In many forecasting problems, there exist *seasonal effects* that must be accounted for in the model. For example, in the above case of a wholesale distributor of bicycles, it would be reasonable to have lower sales during the bad weather of winter than later in the year, and autumn sales should be boosted by Christmas shopping. We will reexamine the bicycle example in this new light later in this section after developing a model and forecasting procedure that incorporates seasonal effects.

Suppose that the generating process of the observed time series can be represented by a *constant level* superimposed with *seasonal effects* and *random fluctuations,* as illustrated in Fig. 18.2*c*. Such a model can be represented by

$$X_t = AI_t^* + e_t,$$

where X_t is the random variable that is observed at time t, A is a constant, I_t^* is the **seasonal index** for period t, and e_t is the random error occurring at time t (assumed to have expected value equal to zero and constant variance).

Unfortunately, both A and I_t^* are unknown, and smoothed levels at time t for both of these factors are useful prior to making a forecast. Exponential smoothing can again be used for this purpose; i.e.,

$$S_t = \alpha\frac{x_t}{I_{t-p}} + (1 - \alpha)S_{t-1},$$

$$I_t = \gamma\frac{x_t}{S_t} + (1 - \gamma)I_{t-p},$$

where both α $(0 < \alpha < 1)$ and γ $(0 < \gamma < 1)$ are smoothing constants. The forecast for the next period then is given by

$$F_{t+1} = S_t I_{t-p+1},$$

where p is the number of periods in the seasonal cycle. For example, if the seasonal periods are winter, spring, summer, and autumn, then $p = 4$. Note that I_{t-p} represents the *smoothed* value of the seasonal index for period t computed for the same season p periods ago; e.g., the seasonal index for autumn 1995 is based upon autumn 1994 data.

The forecast for m periods ahead $(m = 1, 2, \ldots)$ is given by

$$F_{t+m} = S_t I_{t-p+m}.$$

If m is greater than p, then I_{t-p+m} has not been calculated. For this case, I_{t-p+m} is to be interpreted as the last *computed* value of the corresponding seasonal index. For example, if a forecast for autumn 1998 is desired and no data later than the spring 1995 are available, the seasonal index for autumn 1994 is used in place of that for autumn 1997.

1. Using x_t and I_{t-p}, the smoothed level of the time series at time t is given by

$$S_t = \alpha \frac{x_t}{I_{t-p}} + (1 - \alpha) S_{t-1}.$$

2. From S_t (calculated in step 1), S_{t-1}, and I_{t-p}, the updated smoothed level of the seasonal index for period t is given by

$$I_t = \gamma \frac{x_t}{S_t} + (1 - \gamma) I_{t-p}.$$

3. The forecast of the time series for m periods ahead ($m = 1, 2, \ldots$) is given by

$$F_{t+m} = S_t I_{t-p+m}.$$

A routine for this forecasting procedure is included in your OR Courseware.

As in the other exponential smoothing procedures, initial values are required to start the smoothing process. For models containing seasonal factors, a full cycle of seasonal data is required. A description of the initialization procedure is best given in the context of an example, as done below.

Example: Consider the bicycle example of Sec. 18.4. Upon examination of past data, the linear trend model has been found to be inappropriate and has been replaced by the seasonal model described above. The seasonal sales figures for both last year and this year (to date) are shown in Table 18.1. The time index t to be used for each season also is given in the table. (The fact that $t = 1$ for winter of this year indicates that this winter's sales provide the first data point $x_1 = 2{,}800$ for the exponential smoothing, although last year's sales figures will be used for initialization and for determining the initial seasonal indices.) The goal now is to use the above procedure to develop a forecast F_4 for autumn of this year. However, for illustrative purposes, we also will show the forecast that would have been obtained for each earlier season this year based on sales prior to that season.

A reasonable way to initially estimate the seasonal indices is to divide last year's seasonal sales by the seasonal average sales over the last year, that is, $(2{,}786 + 2{,}928 + 3{,}025 + 3{,}061)/4 = 2{,}950$. These values are shown in the rightmost column of Table 18.1. The initial estimate of the constant level A is chosen to be the average of the four seasons over the past year, that is, $S_0 = 2{,}950$. We will use $\alpha = 0.1$ and $\gamma = 0.2$.

Table 18.1 **Bicycle Sales**

Season	Last Year		This Year		Initial Seasonal Index	
Winter	($t = -3$)	2,786	($t = 1$)	2,800	($t = -3$)	0.944
Spring	($t = -2$)	2,928	($t = 2$)	2,925	($t = -2$)	0.993
Summer	($t = -1$)	3,025	($t = 3$)	3,040	($t = -1$)	1.025
Autumn	($t = 0$)	3,061			($t = 0$)	1.038
		11,800				

With these initial estimates, the forecast for this winter would have been

$$F_1 = S_0 I_{-3} = 2{,}950(0.944) = 2{,}785.$$

To obtain the forecast for this spring quarter, the smoothed level of the time series S_1 must be obtained. The smoothed level of the seasonal index for next year's winter quarter forecast I_1 is also obtained:

$$S_1 = \alpha \frac{x_1}{I_{-3}} + (1 - \alpha) S_0$$

$$= 0.1 \frac{2{,}800}{0.944} + 0.9(2{,}950) = 2{,}952$$

and

$$I_1 = \gamma \frac{x_1}{S_1} + (1 - \gamma) I_{-3}$$

$$= 0.2 \frac{2{,}800}{2{,}952} + 0.8(0.944) = 0.945.$$

The forecast for this spring then would have been

$$F_2 = S_1 I_{-2} = 2{,}952(0.993) = 2{,}931.$$

To obtain the forecast for this summer quarter F_2, we need to calculate S_2 and I_2:

$$S_2 = \alpha \frac{x_2}{I_{-2}} + (1 - \alpha) S_1$$

$$= 0.1 \frac{2{,}925}{0.993} + 0.9(2{,}952) = 2{,}951,$$

and

$$I_2 = \gamma \frac{x_2}{S_2} + (1 - \gamma) I_{-2}$$

$$= 0.2 \frac{2{,}925}{2{,}951} + 0.8(0.993) = 0.993,$$

so

$$F_3 = S_2 I_{-1} = 2951(1.025) = 3{,}025.$$

Finally, in order to obtain the forecast for this autumn quarter F_4, we calculate S_3 and I_3:

$$S_3 = \alpha \frac{x_3}{I_{-1}} + (1 - \alpha) S_2$$

$$= 0.1 \frac{3{,}040}{1.025} + 0.9(2{,}951) = 2{,}952$$

and

$$I_3 = \gamma \frac{x_3}{S_3} + (1 - \gamma) I_{-1}$$

$$= 0.2 \frac{3{,}040}{2{,}952} + 0.8(1.025) = 1.026.$$

Therefore, the desired forecast is

$$F_4 = S_3 I_0 = 2{,}952(1.038) = 3{,}064.$$

Similarly, the current forecast for next winter is

$$F_5 = S_3 I_1 = 2{,}952(0.945) = 2{,}790.$$

This example now has been considered from the viewpoint of having a linear trend (the preceding section) or having seasonal effects (here). It also is possible to construct a model that has *both* a linear trend and seasonal effects, but we will not delve further into this topic.

18.6 Forecasting Errors

Several forecasting techniques have been presented together with different underlying models of the time series. How does one compare these techniques, especially if the generating process is unknown, as is frequently the case in practice? Some measure of performance is needed.

Define the **forecast error** E_t as the difference between the observed value of the time series for period t and the forecast for period t, that is,

$$E_t = x_t - F_t.$$

The forecast error is also referred to as the **residual**. A forecasting procedure that produces ''small'' absolute values of E_t is desirable. A measure of small is the **mean square error (MSE)** associated with the forecasting technique. If there are n time periods, then

$$\text{MSE} = \frac{E_1^2 + E_2^2 + \cdots + E_n^2}{n}$$

For the bicycle example, using this year's spring and summer sales as the observations together with the appropriate forecasts for these seasons, the following are mean square errors for three forecasting procedures discussed earlier:

Last-value forecasting $\quad \text{MSE} = \dfrac{(2{,}925 - 2{,}800)^2 + (3{,}040 - 2{,}925)^2}{2} = 14{,}425,$

Linear trend forecasting $\quad \text{MSE} = \dfrac{(2{,}925 - 2{,}945)^2 + (3{,}040 - 3{,}042)^2}{2} = 202,$

Seasonal forecasting $\quad \text{MSE} = \dfrac{(2{,}925 - 2{,}931)^2 + (3{,}040 - 3{,}025)^2}{2} = 131.$

Therefore, based upon the MSE measure of performance, the seasonal forecasting procedure is the best of the three. Last-value forecasting performs much worse than the other two. However, these results are based upon only two pieces of *past data* (sales figures for two past quarters), so there is no assurance that similar performances will occur with future data unless the underlying model is appropriate for that forecasting procedure.

The bicycle example illustrates the useful concept that before any forecasting procedure is chosen, it is worthwhile to try it out on some past data. Assuming that the time series does not change, obtaining the MSE from past data should give some indication of how well the forecasting procedure will perform in the future.

18.7 Box-Jenkins Method

In practice, a forecasting procedure often is chosen without adequately checking whether the underlying model is an appropriate one for the application. The beauty of the **Box-Jenkins method** is that it carefully coordinates the model and the procedure. There is a systematic approach to identifying an appropriate model, chosen from a rich class of models. The historical data are used to test the validity of the model. The model also generates an appropriate forecasting procedure.

To accomplish all this, the Box-Jenkins method requires a great amount of past data (a minimum of 50 time periods), so it is used only for major applications. It also is a sophisticated and complex technique, so we will provide only a conceptual overview of the method. (See Selected References 2 and 5 for further details.)

The Box-Jenkins method is iterative in nature. First, a model is chosen. To choose this model, we must compute autocorrelations and partial autocorrelations and examine their patterns. An *autocorrelation* measures the correlation between time series values separated by a fixed number of periods. This fixed number of periods is called the *lag*. Therefore, the autocorrelation for a lag of two periods measures the correlation between every other observation; i.e., it is the correlation between the original time series and the same series moved forward two periods. The *partial autocorrelation* is a conditional autocorrelation between the original time series and the same series moved forward a fixed number of periods, holding the effect of the other lagged times fixed. Good estimates of both the autocorrelations and the partial autocorrelations for all lags can be obtained by using a computer to calculate the *sample* autocorrelations and the *sample* partial autocorrelations. (These are "good" estimates because we are assuming large amounts of data.)

From the autocorrelations and the partial autocorrelations, we can identify the functional form of one or more possible models because a rich class of models is characterized by these quantities. Next we must estimate the parameters associated with the model by using the historical data. Then we can compute the residuals and examine their behavior. Similarly, we can examine the behavior of the estimated parameters. If both the residuals and the estimated parameters behave as expected under the presumed model, the model appears to be validated. If they do not, then the model should be modified and the procedure repeated until a model is validated. At this point, we can obtain a forecast.

For example, suppose that the sample autocorrelations and the sample partial autocorrelations have the patterns shown in Fig. 18.3. The sample autocorrelations appear to decrease exponentially as a function of the time lags, while the same partial autocorrelations have spikes at the first and second time lags followed by values that seem to be of negligible magnitude. This behavior is characteristic of the functional form

$$X_t = B_0 + B_1 X_{t-1} + B_2 X_{t-2} + e_t,$$

Assuming this functional form, we use the time series data to estimate B_0, B_1, and B_2. Denote these estimates by b_0, b_1, and b_2, respectively. Together with the time series data, we then obtain the residuals

$$x_t - (b_0 + b_1 x_{t-1} + b_2 x_{t-2}).$$

If the assumed functional form is adequate, the residuals and the estimated parameters should behave in a predictable manner. In particular, the sample residuals should be-

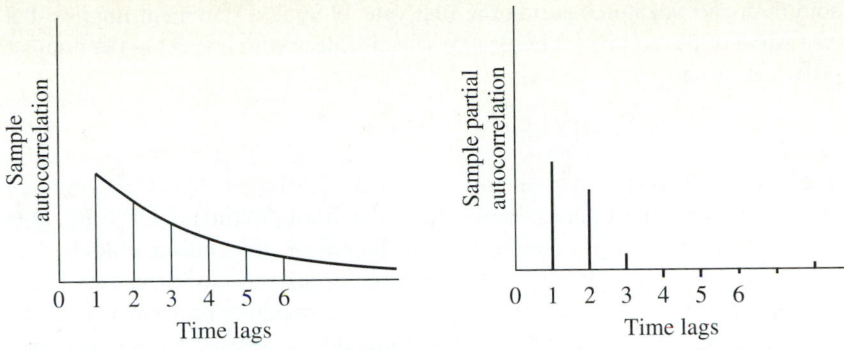

Figure 18.3 Plot of sample autocorrelation and partial autocorrelation versus time lags.

have approximately as independent, normally distributed random variables, each having mean 0 and variance σ^2 (assuming that e_t, the random error at time period t, has mean 0 and variance σ^2). The estimated parameters should be uncorrelated and significantly different from zero. Statistical tests are available for this diagnostic checking.

The Box-Jenkins procedure appears to be a complex one, and it is. Fortunately, computer software is available. The programs calculate the sample autocorrelations and the sample partial autocorrelations necessary for identifying the form of the model. They also estimate the parameters of the model and do the diagnostic checking. These programs, however, cannot accurately identify one or more models that are compatible with the autocorrelations and the partial autocorrelations. Expert human judgment is required. This expertise can be acquired, but it is beyond the scope of this text. Although the Box-Jenkins method is complicated, the resultant forecasts are extremely accurate and, when the time horizon is short, better than most other forecasting techniques. Furthermore, the procedure produces a measure of the forecast error.

18.8 Linear Regression

Statistical problems often are concerned with data where there exists a relationship between two variables. This section highlights the results when the relationship is linear. For example, suppose that a publisher of textbooks is concerned about the initial press run for her books. She sells books both through bookstores and through mail orders. This latter method uses an extensive advertising campaign through publishing media and direct mail. The advertising campaign is conducted prior to the publication of the book. The sales manager has noted that there is a rather interesting linear relationship between the number of mail orders and the number sold through bookstores during the first year. He suggests that this relationship be exploited to determine the initial press run for subsequent books.

Thus, if the number of mail order sales for a book is denoted by X and the number of bookstore sales by Y, then the random variables X and Y exhibit a *degree of association.* However there is *no functional relationship* between these two random variables; i.e., given the number of mail order sales, one does not expect to determine *exactly* the number of bookstore sales. For any given number of mail order sales, there are a range of possible bookstore sales, and vice versa.

What, then, is meant by the statement, "The sales manager has noted that there is a rather interesting linear relationship between the number of mail orders and the number sold through bookstores during the first year''? Such a statement implies that the *expected value* of the number of bookstore sales is linear with respect to the number of mail order sales, i.e.,

$$E[Y \mid X = x] = A + Bx.$$

Thus, if the number of mail order sales is x for many different books, the average number of corresponding bookstore sales would tend to be approximately $A + Bx$. This relationship between X and Y is referred to as a **degree of association model**.

Other examples of this degree of association model can easily be found. An educator may be interested in the relationship between a student's performance on the college entrance examination and subsequent performance in college. An engineer may be interested in the relationship between tensile strength and hardness of a material. An economist may wish to predict a measure of inflation as a function of the cost of living index, and so on.

The degree of association model is not the only model of interest. In some cases, there exists a **functional relationship** between two variables that may be linked linearly. In a forecasting context, one of the two variables is time, while the other is the variable of interest. In Sec. 18.4, such an example was mentioned in the context of the generating process of the time series being represented by a linear trend superimposed with random fluctuations, i.e.,

$$X_t = A + Bt + e_t,$$

where A is a constant, B is the slope, and e_t is the random error, assumed to have expected value equal to zero and constant variance. (The symbol X_t can also be read as X given t or as $X \mid t$.) It follows that

$$E(X_t) = A + Bt.$$

Note that both the degree of association model and the *exact functional relationship* model lead to the same linear relationship, and their subsequent treatment is almost identical. Hence the publishing example will be explored further to illustrate how to treat both kinds of models, although the special structure of the model

$$E(X_t) = A + Bt,$$

with t taking on integer values starting with 1, leads to certain simplified expressions. In the standard notation of regression analysis, X represents the *independent variable* and Y represents the *dependent variable* of interest. Consequently, the notational expression for this special time series model now becomes

$$Y_t = A + Bt + e_t.$$

Method of Least Squares

Suppose that bookstore sales and mail order sales are given for 15 books. These data appear in Table 18.2, and the resulting plot is given in Fig. 18.4 on p. 822.

It is evident that the points in Fig. 18.4 do not lie on a straight line. Hence it is not clear where the line should be drawn to show the linear relationship. Suppose that an arbitrary line, given by the expression $\tilde{y} = a + bx$, is drawn through the data. A

Table 18.2 **Data for the Mail-Order and Bookstore Sales Example**

Mail-Order Sales	Bookstore Sales
1,310	4,360
1,313	4,590
1,320	4,520
1,322	4,770
1,338	4,760
1,340	5,070
1,347	5,230
1,355	5,080
1,360	5,550
1,364	5,390
1,373	5,670
1,376	5,490
1,384	5,810
1,395	6,060
1,400	5,940

measure of how well this line fits the data can be obtained by computing the *sum of squares* of the vertical deviations of the actual points from the fitted line. Thus let y_i represent the bookstore sales of the ith book and x_i the corresponding mail order sales. Denote by $\tilde{y}_i$ the point on the fitted line corresponding to the mail order sales of x_i. The proposed measure of fit is then given by

$$Q = (y_1 - \tilde{y}_1)^2 + (y_2 - \tilde{y}_2)^2 + \cdots + (y_{15} - \tilde{y}_{15})^2 = \sum_{i=1}^{15} (y_i - \tilde{y}_i)^2.$$

The usual method for identifying the "best" fitted line is the **method of least squares**. This method chooses that line $a + bx$ that makes Q a minimum. Thus a and b are obtained simply by setting the partial derivatives of Q with respect to a and b equal to zero and solving the resultant equations. This method yields the solution

$$b = \frac{\sum_{i=1}^{n} (x_i - \bar{x})(y_i - \bar{y})}{\sum_{i=1}^{n} (x_i - \bar{x})^2} = \frac{\sum_{i=1}^{n} x_i y_i - \left(\sum_{i=1}^{n} x_i \sum_{i=1}^{n} y_i\right) \Big/ n}{\sum_{i=1}^{n} x_i^2 - \left(\sum_{i=1}^{n} x_i\right)^2 \Big/ n}$$

and

$$a = \bar{y} - b\bar{x},$$

where

$$\bar{x} = \sum_{i=1}^{n} \frac{x_i}{n}$$

and

$$\bar{y} = \sum_{i=1}^{n} \frac{y_i}{n}.$$

(Note that $\bar{y}$ is not the same as $\tilde{y} = a + bx$ discussed above.)

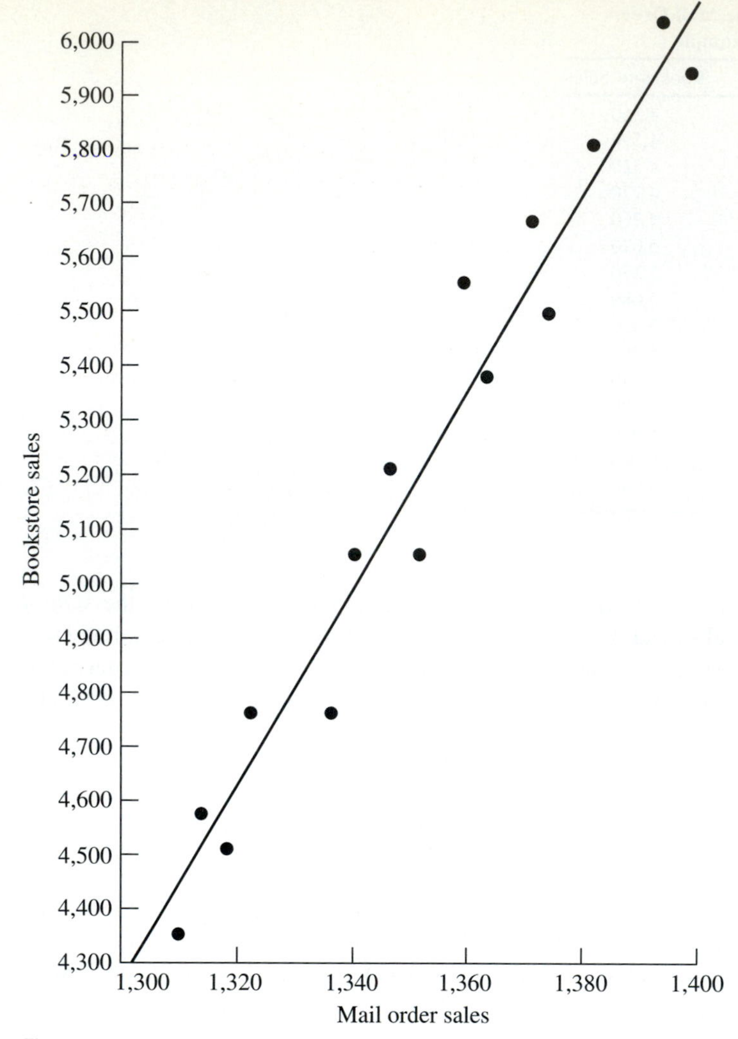

Figure 18.4 Plot of mail order sales versus bookstore sales from Table 18.2.

For the publishing example, the data in Table 18.2 and Fig. 18.4 yield

$$\bar{x} = 1{,}353.1,$$

$$\bar{y} = 5{,}219.3,$$

$$\sum_{i=1}^{15} (x_i - \bar{x})(y_i - \bar{y}) = 214{,}543.9,$$

$$\sum_{i=1}^{15} (x_i - \bar{x})^2 = 11{,}966,$$

$$a = -19{,}041.9,$$

$$b = 17.930.$$

Hence, the least-squares estimate of bookstore sales $\tilde{y}$ with mail order sales x is given by

$$\tilde{y} = -19{,}041.9 + 17.930x,$$

and this is the line drawn in Fig. 18.4. Such a line is referred to as a **regression line**.

A routine called *Method of Least Squares* is available in your OR Courseware for calculating a regression line in this way.

This regression line is useful for forecasting purposes. For a given value of x, the corresponding value of y represents the forecast.

The decision maker may be interested in some measure of uncertainty that is associated with this forecast. This measure is easily obtained provided that certain assumptions can be made. Therefore, for the remainder of this section, it is assumed that

1. A random sample of n pairs $(x_1, Y_1), (x_2, Y_2), \ldots, (x_n, Y_n)$ is to be taken.
2. The Y_i are normally distributed with mean $A + Bx_i$ and variance σ^2 (independent of i).

The assumption that Y_i is normally distributed is not a critical assumption in determining the uncertainty in the forecast, but the assumption of constant variance is crucial. Furthermore, an estimate of this variance is required.

An unbiased estimate of σ^2 is given by $s^2_{y|x}$, where

$$s^2_{y|x} = \sum_{i=1}^{n} \frac{(y_i - \tilde{y}_i)^2}{n-2}.$$

Confidence Interval Estimation of $E(Y|x = x_*)$

A very important reason for obtaining the linear relationship between two variables is to use the line for future decision making. From the regression line, it is possible to estimate $E(Y|x)$ by a *point* estimate (the forecast) and a *confidence interval* estimate (a measure of forecast uncertainty).

For example, the publisher might want to use this approach to estimate the expected number of bookstore sales corresponding to mail order sales of, say, 1,400, by both a point estimate and a confidence interval estimate for forecasting purposes.

A point estimate of $E(Y|x = x_*)$ is given by

$$\tilde{y}_* = a + bx_*.$$

The endpoints of a $(100)(1-\alpha)$ percent confidence interval for $E(Y|x = x_*)$ are given by

$$a + bx_* - t_{\alpha/2;n-2}s_{y|x}\sqrt{\frac{1}{n} + \frac{(x_* - \bar{x})^2}{\sum_{i=1}^{n}(x_i - \bar{x})^2}}$$

and

$$a + bx_* + t_{\alpha/2;n-2}s_{y|x}\sqrt{\frac{1}{n} + \frac{(x_* - \bar{x})^2}{\sum_{i=1}^{n}(x_i - \bar{x})^2}},$$

where $s^2_{y|x}$ is the estimate of σ^2, and $t_{\alpha/2;n-2}$ is the $100\alpha/2$ percentage point of the t distribution with $n - 2$ degrees of freedom (see Table A5.2 of App. 5). Note that the interval is narrowest where $x_* = \bar{x}$, and it becomes wider as x_* departs from the mean.

In the publishing example with $x_* = 1{,}400$, $s^2_{y|x}$ is computed from the data in Table 18.2 to be 17,030, so $s_{y|x} = 130.5$. If a 95 percent confidence interval is required, Table A5.2 gives $t_{0.025;13} = 2.160$. The earlier calculation of a and b yields

$$a + bx_* = -19{,}041.9 + 17.930(1{,}400) = 6{,}060$$

as the point estimate of $E(Y \mid 1{,}400)$, that is, the forecast. Hence, the confidence limits corresponding to mail order sales of 1,400 are

$$\text{Lower confidence limit} = 6{,}060 - 2.160(130.5)\sqrt{\frac{1}{15} + \frac{46.9^2}{11{,}966}}$$

$$= 5{,}919,$$

$$\text{Upper confidence limit} = 6{,}060 + 2.160(130.5)\sqrt{\frac{1}{15} + \frac{46.9^2}{11{,}966}}$$

$$= 6{,}201.$$

The fact that the confidence interval was obtained at a data point ($x = 1{,}400$) is purely coincidental.

The *Method of Least Squares* routine in your OR Courseware does most of the computational work involved in calculating these confidence limits. In addition to computing a and b (the regression line), it calculates $s^2_{y|x}$ and $\sum_{i=1}^{n} (x_i - \bar{x})^2$.

Predictions

The confidence interval statement for the expected number of bookstore sales corresponding to mail order sales of 1,400 may be useful for budgeting purposes, but it is not too useful for making decisions about the *actual* press run. Instead of obtaining bounds on the *expected* number of bookstore sales, this kind of decision requires bounds on what the *actual* bookstore sales will be, i.e., a **prediction interval** on the value that the random variable (bookstore sales) takes on. This measure is a *different* measure of forecast uncertainty.

The two endpoints of a prediction interval are given by the expressions

$$a + bx_+ - t_{\alpha/2;n-2}s_{y|x}\sqrt{1 + \frac{1}{n} + \frac{(x_+ - \bar{x})^2}{\sum_{i=1}^{n} (x_i - \bar{x})^2}}$$

and

$$a + bx_+ + t_{\alpha/2;n-2}s_{y|x}\sqrt{1 + \frac{1}{n} + \frac{(x_+ - \bar{x})^2}{\sum_{i=1}^{n} (x_i - \bar{x})^2}}$$

For any given value of x (denoted here by x_+), the probability is $1 - \alpha$ that the value of the future Y_+ associated with x_+ will fall in this interval.

Thus, in the publishing example, if x_+ is 1,400, then the corresponding 95 percent prediction interval for the number of bookstore sales is given by $6{,}060 \pm 315$, which is naturally wider than the confidence interval for the expected number of bookstore sales.

Whereas the publisher can find an interval that will contain bookstore sales corresponding to particular mail order sales with probability $1 - \alpha$, she is unable to use this type of result repeatedly and still maintain a measure for making correct statements. The reason is that these statements would all be based upon the same statistical data, so that the statements would not be statistically independent. If the statements were independent and if k future bookstore sales are predicted, with each statement being made with probability $1 - \alpha$, then the probability is $(1 - \alpha)^k$ that *all* k predictions of future bookstore sales are correct. However, if k is large or possibly unknown, even this technique based upon the (incorrect) assumption of independence would be useless.

This difficulty can be overcome by using **simultaneous tolerance intervals**. Using this technique, the publisher can take the mail order sales of any book, find an interval (based on the previously determined linear regression line) that will contain the actual bookstore sales with probability at least $1 - \alpha$, and repeat this for any number of books having the same or different mail order sales. Furthermore, the probability is P that *all* these predictions are correct. An alternative interpretation is as follows. If every publisher followed this procedure, each using his or her own linear regression line, then $100P$ percent of the publishers (on average) would find that at least $100(1 - \alpha)$ percent of their bookstore sales fell into the predicted intervals. The expression for the endpoints of each such tolerance interval is given by

$$a + bx_+ - c^{**}s_{y|x}\sqrt{\frac{1}{n} + \frac{(x_+ - \bar{x})^2}{\sum_{i=1}^{n}(x_i - \bar{x})^2}}$$

and

$$a + bx_+ + c^{**}s_{y|x}\sqrt{\frac{1}{n} + \frac{(x_+ - \bar{x})^2}{\sum_{i=1}^{n}(x_i - \bar{x})^2}},$$

where c^{**} is given in Table 18.3.

Thus the publisher can predict that the bookstore sales corresponding to known mail order sales will fall in these tolerance intervals. Such statements can be made for as many books as the publisher desires. Furthermore, the probability is P that at least $100(1 - \alpha)$ percent of bookstore sales corresponding to mail order sales will fall in these intervals. If P is chosen as 0.90 and $\alpha = 0.05$, the appropriate value of c^{**} is 11.625. Hence, the number of bookstore sales corresponding to mail order sales of 1,400 books is predicted to fall in the interval $6{,}060 \pm 759$. If another book had mail order sales of 1,353, the bookstore sales are predicted to fall in the interval $5{,}258 \pm 390$, and so on. At least 95 percent of the bookstore sales will fall into their predicted intervals, and these statements are made with confidence 0.90.

To summarize, we now have described three *measures of forecast uncertainty.* The first (in the preceding subsection) is a *confidence interval* on the *expected value* of the random variable Y (for example, bookstore sales) given the observed value x of the independent variable X (for example, mail order sales). The second is a *prediction interval* on the *actual value* that Y will take on, given x. The third is *simultaneous tolerance intervals* on a *succession* of *actual values* that Y will take on given a succession of observed values of X.

Table 18.3 **Values of c^{**}**

n	$\alpha = 0.50$	$\alpha = 0.25$	$\alpha = 0.10$	$\alpha = 0.05$	$\alpha = 0.01$	$\alpha = 0.001$
			$P = 0.90$			
4	7.471	10.160	13.069	14.953	18.663	23.003
6	5.380	7.453	9.698	11.150	14.014	17.363
8	5.037	7.082	9.292	10.722	13.543	16.837
10	4.983	7.093	9.366	10.836	13.733	17.118
12	5.023	7.221	9.586	11.112	14.121	17.634
14	5.101	7.394	9.857	11.447	14.577	18.232
16	5.197	7.586	10.150	11.803	15.057	18.856
18	5.300	7.786	10.449	12.165	15.542	19.484
20	5.408	7.987	10.747	12.526	16.023	20.140
			$P = 0.95$			
4	10.756	14.597	18.751	21.445	26.760	32.982
6	6.652	9.166	11.899	13.669	17.167	21.266
8	5.933	8.281	10.831	12.484	15.750	19.568
10	5.728	8.080	10.632	12.286	15.553	19.369
12	5.684	8.093	10.701	12.391	15.724	19.619
14	5.711	8.194	10.880	12.617	16.045	20.050
16	5.771	8.337	11.107	12.898	16.431	20.559
18	5.848	8.499	11.357	13.204	16.845	21.097
20	5.937	8.672	11.619	13.521	17.272	21.652
			$P = 0.99$			
4	24.466	33.019	42.398	48.620	60.500	74.642
6	10.444	14.285	18.483	21.215	26.606	32.920
8	8.290	11.453	14.918	17.166	21.652	26.860
10	7.567	10.539	13.796	15.911	20.097	24.997
12	7.258	10.182	13.383	15.479	19.579	24.403
14	7.127	10.063	13.267	15.355	19.485	24.316
16	7.079	10.055	13.306	15.410	19.582	24.467
18	7.074	10.111	13.404	15.552	19.794	24.746
20	7.108	10.198	13.566	15.745	20.065	25.122

Source: Reprinted by permission from G. J. Lieberman and R. G. Miller, "Simultaneous Tolerance Intervals in Regression," *Biometrika,* **50**(1 and 2): 164, 1963.

18.9 Conclusions

It is important to consider the entire forecasting system carefully. The need to obtain a forecast has to be identified at the appropriate management level. The historical data required must be compiled. By studying these data, an appropriate model can be structured. A forecasting procedure that behaves well under the model should be selected. The forecasting procedure may require choosing one or more parameters—e.g., the smoothing constant α in exponential smoothing—and the historical data may prove useful in making this choice. Finally, the forecasting process should be viewed as dynamic, with the accumulated new data compared with their associated forecasts. These data may also be used to update the parameters and the form of the underlying model, as well as the parameters of the forecasting procedure itself.

SELECTED REFERENCES

1. Armstrong, J. S., with comments by S. Makridakis, R. L. Schultz, E. S. Gardner, Jr., and B. Fischoff, "Research on Forecasting: A Quarter-Century Review, 1960–84," *Interfaces,* **16**(1): 89–109, January-February 1986.

2. Box, G. E. P., and G. M. Jenkins, *Time Series Analysis, Forecasting and Control,* Holden-Day, San Francisco, 1976.
3. Gardner, E. S., Jr., ''Exponential Smoothing: The State of the Art,'' *Journal of Forecasting,* **4**: 1–38, 1985.
4. Gilchrist, W. G., *Statistical Forecasting,* Wiley, New York, 1976.
5. Hoff, J. C., *A Practical Guide to Box-Jenkins Forecasting,* Lifetime Learning Publications, Belmont, CA, 1983.
6. Levin, R. I., D. S. Rubin, J. P. Stinson, and E. S. Gardner, Jr., *Quantitative Approaches to Management,* 8th ed., McGraw-Hill, New York, 1993.
7. Makridakis, S., S. C. Wheelwright, and V. E. McGee, *Forecasting: Methods and Applications,* 2d edition, Wiley, New York, 1983.
8. Sanders, N. R., and K. B. Manrodt, ''Forecasting Practices in U.S. Corporations: Survey Results,'' *Interfaces,* **24**(2): 92–100, March–April 1994.
9. Willis, R. E., *A Guide to Forecasting for Planners and Managers,* Prentice-Hall, Englewood Cliffs, NJ, 1987.
10. Wilson, J. H., and B. Keating, *Business Forecasting,* Irwin, Homewood, IL, 1990.

RELEVANT ROUTINES IN YOUR OR COURSEWARE

Automatic routines: *Exponential Smoothing Forecasting Procedure*
Exponential Smoothing with Linear Trend Forecasting Procedure
Exponential Smoothing with Seasonal Effects Forecasting Procedure
Method of Least Squares

To access these routines, call the ProbMod program and then choose *Forecasting* under the Area menu. See App. 1 for documentation of the software.

PROBLEMS

To the left of each of the following problems (or their parts), we have inserted an A (for automatic routine) whenever a corresponding routine listed above can be helpful. An asterisk on the A indicates that this routine definitely should be used (unless your instructor gives contrary instructions) and that the printout from this routine is all that needs to be turned in to show your work in executing the procedure. An asterisk on the problem number indicates that at least a partial answer is given in the back of the book.

18.3-1.* Suppose that the preceding forecast was 2,083, the actual value of the variable of interest for the last period was 1,975, and the oldest value of the variable of interest was 1,945. Using the moving-average technique based upon the most recent four observations, what is the new forecast for the next period?

18.3-2. Suppose that the preceding forecast was 2,083, the actual value of the variable of interest for the last period was 1,975, and $\alpha = 0.3$. Using exponential smoothing, what is the new forecast for the next period?

18.3-3.* Suppose that the preceding forecast was 782, the actual value of the variable of interest for the last period was 794, and $\alpha = 0.1$. Using exponential smoothing, what is the new forecast for the next period?

18.3-4. Suppose that the preceding forecast was 782, the actual value of the variable of interest for the last period was 794, and the oldest value of the variable of interest was 805. Using the moving-average technique based upon the most recent three observations, what is the new forecast for the next period?

18.3-5. You are the new person at a statistical forecasting service, and you have been asked to update a moving-average forecast based upon the most recent 10 observations. You know that the preceding forecast was 1,551, the actual value of the variable of interest for the last period was 1,532, and the oldest value (that is, 11 periods ago) of the variable of interest was 1,632. What is the new forecast for the next period?

18.3-6. If α is set equal to 0 or 1 in the exponential smoothing expression, what happens to the forecast?

A **18.3-7.** Refer to the data for the bicycle sales example presented in Sec. 18.4. Use simple exponential smoothing with $\alpha = 0.3$ and an initial estimate of 2,750 to obtain a forecast for the sales in the fourth period.

18.3-8. A company uses exponential smoothing with $\alpha = \frac{1}{2}$ to forecast demand for a product. For each month, the company keeps a record of the forecast demand (made at the end of the preceding month) and the actual demand. Some of the records have been lost; the remaining data appear in the table below.

	January	February	March	April	May	June
Forecast			400	380	390	380
Actual	400		360	—	—	

(*a*) Using only data in the table for March, April, May, and June, determine the actual demands in April and May.

(*b*) Suppose now that a clerical error is discovered; the actual demand in January was 432, not 400, as shown in the table. Using only the actual demands going back to January (even though the February actual demand is unknown), give the corrected forecast for June.

A* **18.4-1.** Use exponential smoothing adjusted for trend to solve the bicycle sales example presented in Sec. 18.4 when α and β are changed to $\alpha = \beta = 0.3$.

A* **18.5-1.** The U.S. unemployment rate time series shown in Fig. 18.1 can be represented in tabular form as follows:

Date	Unemployment Rate, %	Date	Unemployment Rate, %	Date	Unemployment Rate, %
1/80	6.9	1/81	8.2	1/82	9.4
4/80	6.7	4/81	7.0	4/82	9.2
7/80	7.9	7/81	7.3	7/82	9.8
10/80	7.1	10/81	7.5	10/82	9.9

Date	Unemployment Rate, %	Date	Unemployment Rate, %
1/83	11.4	1/84	8.8
4/83	10.0	4/84	7.6
7/83	9.4	7/84	7.5
10/83	8.4	10/84	—

Assume a seasonal-effects model (four periods per year) and use the data for 1980 to estimate the initial seasonal factors. Use exponential smoothing with seasonal effects to forecast the unemployment for each period of 1981. Use the final forecast (October 1981) to forecast the unemployment rate for October 1984. Let $\alpha = \gamma = 0.1$.

A* **18.5-2.** Solve Prob. 18.5-1 by using 1982 data to estimate the initial seasonal factors, and forecast the unemployment rate for each period of 1983. Use the final forecast (October 1983) to forecast the unemployment rate for October 1984.

18.6-1.* Use the unemployment data of Prob. 18.5-1.

(*a*) Starting with a forecast for October 1980, use the moving-average technique based upon the past three periods to forecast each unemployment rate through October 1984.

(*b*) Examine the residuals for October 1980 through July 1984, and compute the mean square error

$$\frac{\sum E_i^2}{16}.$$

18.6-2. Use the unemployment data of Prob. 18.5-1.

A* (*a*) Starting with a forecast for October 1980, use the exponential smoothing technique with an initial estimate of 7.2 percent and $\alpha = 0.1$, that is,

$$\text{Forecast for } 10/80 = (0.1)7.9 + (0.9)7.2,$$

to forecast each unemployment rate through October 1984.

(*b*) Examine the residuals for October 1980 through July 1984, and compute the mean square error

$$\frac{\sum E_i^2}{16}.$$

18.6-3. Reconsider Prob. 18.6-2.

(*a*) Solve this problem by using $\alpha = 0.3$.

(*b*) Which α ($\alpha = 0.1$ or 0.3) would you use to forecast the unemployment rate for October 1984?

18.6-4. Based upon the results of Probs. 18.6-1*b* and 18.6-2*b*, which forecasting procedure would you choose—moving-average or exponential smoothing?

18.6-5. Use the unemployment data of Prob. 18.5-1.

A* (*a*) Starting with a one-step forecast for October 1980, use the exponential smoothing technique adjusted for trend, with an initial estimate of 7.2 percent for unemployment and with an initial estimate of trend of 0.2 percent, to forecast each unemployment rate through October 1984. Use $\alpha = \beta = 0.1$.

(*b*) Examine the residuals for October 1980 through July 1984, and compute the mean square error

$$\frac{\sum E_i^2}{16}.$$

18.6-6. Based upon the results of Probs. 18.6-1*b*, 18.6-2*b*, and 18.6-5*b*, which forecasting procedure would you choose—moving-average, exponential smoothing, or exponential smoothing adjusted for trend?

18.6-7. To plan for a suitable labor force, you need to know the demand for a particular product. The following table presents demand over the past 11 quarters.

Quarter	Demand	Quarter	Demand	Quarter	Demand
1	546	5	647	9	736
1	528	6	594	10	724
3	530	7	665	11	813
4	508	8	630	12	—

(*a*) Starting with a forecast for quarter 5, use the moving-average technique based upon the past four quarters to forecast each demand through quarter 12.

(*b*) Examine the residuals for quarter 5 through quarter 11, and compute the mean square error

$$\frac{\sum E_i^2}{7}.$$

18.6-8. Use the quarterly demand data of Prob. 18.6-7.

A* (*a*) Starting with a forecast for quarter 3, use the exponential smoothing technique with an initial estimate of 546 and $\alpha = 0.1$, that is,

$$\text{Forecast for quarter } 3 = (0.1)528 + (0.9)546,$$

to forecast each demand through quarter 12.

(*b*) Examine the residuals for quarter 3 through quarter 11, and compute the mean square error

$$\frac{\sum E_i^2}{9}.$$

18.6-9. Reconsider Prob. 18.6-8.

(*a*) Solve this problem, using $\alpha = 0.3$.

(*b*) Which α ($\alpha = 0.1$ or 0.3) would you use to forecast the demand for quarter 12?

18.6-10. Based upon the results of Probs. 18.6-7*b* and 18.6-8*b*, which forecasting procedure would you choose—moving-average or exponential smoothing?

18.6-11. Use the quarterly demand data for Prob. 18.6-7.

A* (*a*) Starting with a one-step forecast for quarter 3, use the exponential smoothing technique adjusted for trend, with an initial estimate of 546 for demand and an initial estimate of trend of 18, to forecast each demand through quarter 12. Use $\alpha = \beta = 0.1$.

(*b*) Examine the residuals for quarter 3 through quarter 11, and compute the mean square error $\Sigma E_i^2/9$.

18.6-12. Based upon the results of Probs. 18.6-7*b*, 18.6-8*b*, and 18.6-11*b*, which forecasting procedure would you choose—moving-average, exponential smoothing, or exponential smoothing adjusted for trend?

18.8-1.* Suppose that a time series behaves as follows:

t	Y	t	Y
1	430	6	514
2	446	7	532
3	464	8	548
4	480	9	570
5	498	10	591

A* (*a*) Using the method of least squares, estimate the line $A + Bt$.
(*b*) From the line found in (*a*), forecast Y_{11}.
A* (*c*) Starting with a forecast in period 3, use the exponential smoothing technique with an initial estimate of 430 and $\alpha = 0.1$, that is,

$$\text{Forecast for period } 3 = (0.1)446 + (0.9)430,$$

to forecast each demand through period 11.
A* (*d*) Starting with a one-step forecast for period 3, use the exponential smoothing technique adjusted for trend, with an initial estimate of 430 for the variable of interest and an initial estimate of trend of 18, to forecast each demand through period 11. Use $\alpha = \beta = 0.2$.

18.8-2. The following data relate road width x and accident frequency Y. Road width (in feet) was treated as the independent variable, and values y of the random variable Y, in accidents per 10^8 vehicle miles, were observed.

Number of Observations = 7		x	y
$\sum_{i=1}^{7} x_i = 354$	$\sum_{i=1}^{7} y_i = 481$	26 30	92 85
$\sum_{i=1}^{7} x_i^2 = 19{,}956$	$\sum_{i=1}^{7} y_i^2 = 35{,}451$	44 50 62	78 81 54
$\sum_{i=1}^{7} x_i y_i = 22{,}200$		68 74	51 40

Assume that Y is normally distributed with mean $A + Bx$ and constant variance for all x and that the sample is random. Interpolate if necessary.

(*a*) Fit a least-squares line to the data, and forecast the accident frequency when the road width is 55 feet.
(*b*) Construct a 95 percent prediction interval for Y_+, a future observation of Y, corresponding to $x_+ = 55$ feet.
(*c*) Suppose that two future observations on Y, both corresponding to $x_+ = 55$ feet, are to be made. Construct prediction intervals for both of these observations so that the probability is *at least* 95 percent that *both* future values of Y will fall into them simultaneously. [*Hint:* If k predictions are to be made, such as given in part (*d*), each with probability $1 - \alpha$, then the probability is *at least* $1 - k\alpha$ that all k future observations will fall into their respective intervals.]
(*d*) Construct a simultaneous tolerance interval for the future value of Y corresponding to $x_+ = 55$ feet with $P = 0.90$ and $1 - \alpha = 0.95$.

A* **18.8-3.** The following data are observations on a dependent random variable Y taken at various levels of an independent variable x. [It is assumed that $E(Y_i | x_i) = A + Bx_i$, and the Y_i are independent normal random variables with mean 0 and variance σ^2.]

x_i	0	2	4	6	8
y_i	0	4	7	13	16

(*a*) Estimate the linear relationship by the method of least squares, and forecast the value of Y when $x = 10$.
(*b*) Find a 95 percent confidence interval for the expected value of Y at $x_* = 10$.

(*c*) Find a 95 percent prediction interval for a future observation to be taken at $x_+ = 10$.

(*d*) For $x_+ = 10$, $P = 0.90$, and $1 - \alpha = 0.95$, find a simultaneous tolerance interval for the future value of Y_+. Interpolate if necessary.

A* **18.8-4.** If a particle is dropped at time $t = 0$, physical theory indicates that the relationship between the distance traveled r and the time elapsed t is $r = gt^k$ for some positive constants g and k. A transformation to linearity can be obtained by taking logarithms:

$$\log r = \log g + k \log t.$$

By letting $y = \log r$, $A = \log g$, and $x = \log t$, this relation becomes $y = A + kx$. Due to random error in measurement, however, it can be stated only that $E(Y \mid x) = A + kx$. Assume that Y is normally distributed with mean $A + kx$ and variance σ^2.

A physicist who wishes to estimate k and g performs the following experiment: At time 0 the particle is dropped. At time t the distance r is measured. He performs this experiment five times, obtaining the following data (where all logarithms are to base 10).

$y = \log r$	$x = \log t$
−395	−2.0
−2.12	−1.0
0.08	0.0
2.20	+1.0
3.87	+2.0

(*a*) Obtain least-squares estimates for k and $\log g$, and forecast the distance traveled when $\log t = +3.0$.

(*b*) Starting with a forecast for $\log r$ when $\log t = 0$, use the exponential smoothing technique with an initial estimate of $\log r = -3.95$ and $\alpha = 0.1$, that is,

$$\text{Forecast of } \log r \text{ (when } \log t = 0) = 0.1(-2.12) + 0.9(-3.95),$$

to forecast each $\log r$ for all integer $\log t$ through $\log t = +3.0$.

(*c*) Repeat part (*b*), using the exponential smoothing technique adjusted for trend and a one-step forecast. Use an initial estimate of trend equal to the slope found in part (*a*). Let $\beta = 01$.

18.8-5. Suppose that the relation between Y and x is given by

$$E(Y \mid x) = Bx,$$

where Y is assumed to be normally distributed with mean Bx and *known* variance σ^2. Also n independent pairs of observations are taken and are denoted by $x_1, y_1; x_2, y_2; \ldots; x_n, y_n$. Find the least-squares estimate of B.

19

Markov Decision Processes

Chapter 14 introduced *Markov chains* and their analysis. Most of the chapter was devoted to *discrete time* Markov chains, i.e., Markov chains that are observed only at discrete points in time (e.g., the end of each day) rather than continuously. Each time it is observed, the Markov chain can be in any one of a number of *states.* Given the current state, a (one-state) *transition matrix* gives the probabilities for what the state will be next time. Given this transition matrix, Chap. 14 focused on *describing the behavior* of a Markov chain, e.g., finding the steady-state probabilities for what state it is in.

Many important systems (e.g., many queueing systems) can be modeled as either a discrete time or continuous time Markov chain. It is useful to describe the behavior of such a system (as we did in Chap. 15 for queueing systems) in order to evaluate its performance. However, it may be even more useful to *design the operation* of the system so as to *optimize its performance* (as we did in Chap. 16 for queueing systems).

This chapter focuses on how to design the operation of a discrete time Markov chain so as to optimize its performance. Therefore, rather than passively accepting the design of the Markov chain and the corresponding fixed transition matrix, we now are being proactive. For each possible state of the Markov chain, we make a decision about which one of several alternative actions should be taken in that state. The action chosen affects the *transition probabilities* as well as both the *immediate costs* (or rewards) and *subsequent costs* (or rewards) from operating the system. We want to choose the optimal actions for the respective states when considering both immediate and subsequent costs. The decision process for doing this is referred to as a *Markov decision process.*

The first section gives a prototype example of an application of a Markov decision process. Section 19.2 formulates the basic model for these processes. The next three sections describe how to solve them.

19.1 A Prototype Example

A manufacturer has one key machine at the core of one of its production processes. Because of heavy use, the machine deteriorates rapidly in both quality and output. Therefore, at the end of each week, a thorough inspection is done that results in classifying the condition of the machine into one of four possible states:

State	Condition
0	Good as new
1	Operable—minor deterioration
2	Operable—major deterioration
3	Inoperable—output of unacceptable quality

After historical data on these inspection results are gathered, statistical analysis is done on how the state of the machine evolves from month to month. The following matrix shows the relative frequency (probability) of each possible transition from the state in one month (a row of the matrix) to the state in the following month (a column of the matrix).

State	0	1	2	3
0	0	$\frac{7}{8}$	$\frac{1}{16}$	$\frac{1}{16}$
1	0	$\frac{3}{4}$	$\frac{1}{8}$	$\frac{1}{8}$
2	0	0	$\frac{1}{2}$	$\frac{1}{2}$
3	0	0	0	1

In addition, statistical analysis has found that these transition probabilities are unaffected by also considering what the states were in prior months. This "lack-of-memory property" is the *Markovian property* described in Chap. 14. Therefore, for the random variable X_t, which is the state of the machine at the end of month t, it has been concluded that the stochastic process $\{X_t, t = 0, 1, 2, \ldots\}$ is a *discrete time Markov chain* whose (one-step) *transition matrix* is just the above matrix.

As the last entry in this transition matrix indicates, once the machine becomes inoperable (enters state 3), it remains inoperable. In other words, state 3 is an *absorbing state.* Leaving the machine in this state would be intolerable, since this would shut down the production process; so the machine must be replaced. (Repair is not feasible in this state.) The new machine then will start off in state 0.

The replacement process takes 1 week to complete so that production is lost for this period. The cost of the lost production (lost profit) is $2,000, and the cost of replacing the machine is $4,000, so the total cost incurred whenever the current machine enters state 3 is $6,000.

Even before the machine reaches state 3, costs may be incurred from the production of defective items. The expected costs per week from this source are as follows:

State	Expected Cost due to Defective Items, $
0	0
1	1,000
2	3,000

We now have mentioned all the relevant costs associated with one particular *maintenance policy* (replace the machine when it becomes inoperable but do no maintenance otherwise). Under this policy, the evolution of the state of the *system* (the succession of machines) still is a Markov chain, but now with the following transition matrix:

State	0	1	2	3
0	0	$\frac{7}{8}$	$\frac{1}{16}$	$\frac{1}{16}$
1	0	$\frac{3}{4}$	$\frac{1}{8}$	$\frac{1}{8}$
2	0	0	$\frac{1}{2}$	$\frac{1}{2}$
3	1	0	0	0

To evaluate this maintenance policy, we should consider both the immediate costs incurred over the coming week (described above) and the subsequent costs that result from having the system evolve in this way. As introduced in Sec. 14.6, one such widely used measure of performance for Markov chains is the (long-run) **expected average cost per unit time.**[1]

To calculate this measure, we first derive the *steady-state probabilities* π_0, π_1, π_2, and π_3 for this Markov chain by solving the following steady-state equations:

$$\pi_0 = \pi_3,$$

$$\pi_1 = \tfrac{7}{8}\pi_0 + \tfrac{3}{4}\pi_1,$$

$$\pi_2 = \tfrac{1}{16}\pi_0 + \tfrac{1}{8}\pi_1 + \tfrac{1}{2}\pi_2,$$

$$\pi_3 = \tfrac{1}{16}\pi_0 + \tfrac{1}{8}\pi_1 + \tfrac{1}{2}\pi_2,$$

$$1 = \pi_0 + \pi_1 + \pi_2 + \pi_3.$$

The simultaneous solution is

$$\pi_0 = \tfrac{2}{13}, \qquad \pi_1 = \tfrac{7}{13}, \qquad \pi_2 = \tfrac{2}{13}, \qquad \pi_3 = \tfrac{2}{13}.$$

[1] The term *long-run* indicates that the average should be interpreted as being taken over an *extremely* long time so that the effect of the initial state disappears. As time goes to infinity, Sec. 14.6 discusses the fact that the *actual* average cost per unit time essentially always converges to the *expected* average cost per unit time.

Table 19.1 **Cost Data for the Prototype Example**

Decision	State	Expected Cost due to Producing Defective Items, $	Maintenance Cost, $	Cost (Lost Profit) of Lost Production, $	Total Cost per Week, $
1. Do nothing	0	0	0	0	0
	1	1,000	0	0	1,000
	2	3,000	0	0	3,000
2. Overhaul	2	0	2,000	2,000	4,000
3. Replace	1, 2, 3	0	4,000	2,000	6,000

Hence, the (long-run) expected average cost per week for this maintenance policy is

$$0\pi_0 + 1{,}000\pi_1 + 3{,}000\pi_2 + 6{,}000\pi_3 = \frac{25{,}000}{13} = \$1{,}923.08.$$

However, there also are other maintenance policies that should be considered and compared with this one. For example, perhaps the machine should be replaced before it reaches state 3. Another alternative is to *overhaul* the machine at a cost of $2,000. This option is not feasible in state 3 and does not improve the machine while in state 0 or 1, so it is of interest only in state 2. In this state, an overhaul would return the machine to state 1. A week is required, so another consequence is $2,000 in lost profit from lost production.

In summary, the possible decisions after each inspection are as follows:

Decision	Action	Relevant States
1	Do nothing	0, 1, 2
2	Overhaul (return system to state 1)	2
3	Replace (return system to state 0)	1, 2, 3

For easy reference, Table 19.1 also summarizes the relevant costs for each decision for each state where that decision could be of interest.

What is the optimal maintenance policy? We will be addressing this question to illustrate the material in the next four sections.

19.2 A Model for Markov Decision Processes

The model for the Markov decision processes considered in this chapter can be summarized as follows.

1. The state i of a discrete time Markov chain is observed after each transition ($i = 0, 1, \ldots, M$).
2. After each observation, a *decision* (action) k is chosen from a set of K possible decisions ($k = 1, 2, \ldots, K$). (Some of the K decisions may not be relevant for some of the states.)
3. If decision $d_i = k$ is made in state i, an immediate *cost* is incurred that has an expected value C_{ik}.

4. The decision $d_i = k$ in state i determines what the *transition probabilities*[1] will be for the next transition from state i. Denote these transition probabilities by $p_{ij}(k)$, for $j = 0, 1, \ldots, M$.
5. A specification of the decisions for the respective states $(d_0, d_1, \ldots, d_M)$ prescribes a *policy* for the Markov decision process.
6. The objective is to find an *optimal policy* according to some cost criterion which considers both immediate costs and subsequent costs that result from the future evolution of the process. One common criterion is to minimize the (long-run) *expected average cost per unit time.* (An alternative criterion is considered in Sec. 19.5.)

To relate this general description to the prototype example presented in Sec. 19.1, recall that the Markov chain being observed there represents the state (condition) of a particular machine. After each inspection of the machine, a choice is made between three possible decisions (do nothing, overhaul, or replace). The resulting immediate expected cost is shown in the rightmost column of Table 19.1 for each relevant combination of state and decision. Section 19.1 analyzed one particular policy $(d_0, d_1, d_2, d_3) = (1, 1, 1, 3)$, where decision 1 (do nothing) is made in states 0, 1, and 2 and decision 3 (replace) is made in state 3. The resulting transition probabilities are shown in the last transition matrix given in Sec. 19.1.

Our general model qualifies to be a Markov decision process because it possesses the Markovian property that characterizes any Markov process. In particular, given the current state and decision, any probabilistic statement about the future of the process is completely unaffected by providing any information about the history of the process. This property holds here since (1) we are dealing with a Markov chain, (2) the new transition probabilities depend on only the current state and decision, and (3) the immediate expected cost also depends on only the current state and decision.

Our description of a policy implies two convenient (but unnecessary) properties that we will assume throughout the chapter (with one exception). One property is that a policy is *stationary;* i.e., whenever the system is in state i, the rule for making the decision always is the same regardless of the value of the current time t. The second property is that a policy is *deterministic;* i.e., whenever the system is in state i, the rule for making the decision definitely chooses one particular decision. (Because of the nature of the algorithm involved, the next section considers *randomized* policies instead, where a probability distribution is used for the decision to be made.)

Using this general framework, we now return to the prototype example and find the optimal policy by enumerating and comparing all the relevant policies. In doing this, we will let R denote a specific policy and $d_i(R)$ denote the corresponding decision to be made in state i.

Solving the Prototype Example by Exhaustive Enumeration

The relevant policies for the prototype example are these:

Policy	Verbal Description	$d_0(R)$	$d_1(R)$	$d_2(R)$	$d_3(R)$
R_a	Replace in state 3	1	1	1	3
R_b	Replace in state 3, overhaul in state 2	1	1	2	3
R_c	Replace in states 2 and 3	1	1	3	3
R_d	Replace in states 1, 2, and 3	1	3	3	3

[1] The solution procedures given in the next two sections also assume that the resulting transition matrix is *irreducible.*

Each policy results in a different transition matrix, as shown below.

State	R_a			
	0	1	2	3
0	0	$\frac{7}{8}$	$\frac{1}{16}$	$\frac{1}{16}$
1	0	$\frac{3}{4}$	$\frac{1}{8}$	$\frac{1}{8}$
2	0	0	$\frac{1}{2}$	$\frac{1}{2}$
3	1	0	0	0

State	R_b			
	0	1	2	3
0	0	$\frac{7}{8}$	$\frac{1}{16}$	$\frac{1}{16}$
1	0	$\frac{3}{4}$	$\frac{1}{8}$	$\frac{1}{8}$
2	0	1	0	0
3	1	0	0	0

State	R_c			
	0	1	2	3
0	0	$\frac{7}{8}$	$\frac{1}{16}$	$\frac{1}{16}$
1	0	$\frac{3}{4}$	$\frac{1}{8}$	$\frac{1}{8}$
2	1	0	0	0
3	1	0	0	0

State	R_d			
	0	1	2	3
0	0	$\frac{7}{8}$	$\frac{1}{16}$	$\frac{1}{16}$
1	1	0	0	0
2	1	0	0	0
3	1	0	0	0

From the rightmost column of Table 19.1, the values of C_{ik} are as follows:

Decision / State	C_{ik} *(in Thousands of Dollars)*		
	1	2	3
0	0	—	—
1	1	—	6
2	3	4	6
3	—	—	6

As indicated in Sec. 14.6, the (long-run) expected average cost per unit time $E(C)$ then can be calculated from the expression

$$E(C) = \sum_{i=0}^{M} C_{ik}\pi_i,$$

where $k = d_i(R)$ for each i and $(\pi_0, \pi_1, \ldots, \pi_M)$ represents the steady-state distribution of the state of the system under the policy R being evaluated. After $(\pi_0, \pi_1, \ldots, \pi_M)$ are solved for under each of the four policies (as can be done with your OR Courseware), the calculation of $E(C)$ is as summarized here:

Policy	$(\pi_0, \pi_1, \pi_2, \pi_3)$	$E(C)$, in Thousands of Dollars
R_a	$(\frac{2}{13}, \frac{7}{13}, \frac{2}{13}, \frac{2}{13})$	$\frac{1}{13}[2(0) + 7(1) + 2(3) + 2(6)] = \frac{25}{13} = \$1{,}923$
R_b	$(\frac{2}{21}, \frac{5}{7}, \frac{2}{21}, \frac{2}{21})$	$\frac{1}{21}[2(0) + 15(1) + 2(4) + 2(6)] = \frac{35}{21} = \$1{,}667$ ← Minimum
R_c	$(\frac{2}{11}, \frac{7}{11}, \frac{1}{11}, \frac{1}{11})$	$\frac{1}{11}[2(0) + 7(1) + 1(6) + 1(6)] = \frac{19}{11} = \$1{,}727$
R_d	$(\frac{1}{2}, \frac{7}{16}, \frac{1}{32}, \frac{1}{32})$	$\frac{1}{32}[16(0) + 14(6) + 1(6) + 1(6)] = \frac{96}{32} = \$3{,}000$

Thus, the optimal policy is R_b; that is, replace the machine when it is found to be in state 3, and overhaul the machine when it is found to be in state 2. The resulting (long-run) expected average cost per week is $1,667.

Using exhaustive enumeration to find the optimal policy is appropriate for this tiny example, where there are only four relevant policies. However, many applications have so many policies that this approach would be completely infeasible. For such cases, algorithms that can efficiently find an optimal policy are needed. The next three sections consider such algorithms.

19.3 Linear Programming and Optimal Policies

Section 19.2 described the main kind of policy (called a *stationary, deterministic* policy) that is used by Markov decision processes. We saw that any such policy R can be viewed as a rule that prescribes decision $d_i(R)$ whenever the system is in state i, for each $i = 0, 1, \ldots, M$. Thus R is characterized by the values

$$\{d_0(R), d_1(R), \ldots, d_M(R)\}.$$

Equivalently, R can be characterized by assigning values $D_{ik} = 0$ or 1 in the matrix

$$\text{State} \begin{matrix} 0 \\ 1 \\ \vdots \\ M \end{matrix} \overset{\displaystyle\text{Decision } k}{\overset{\begin{matrix} 1 & 2 & \cdots & K \end{matrix}}{\begin{bmatrix} D_{01} & D_{02} & \cdots & D_{0K} \\ D_{11} & D_{12} & \cdots & D_{1K} \\ \cdots & \cdots & \cdots & \cdots \\ D_{M1} & D_{M2} & \cdots & D_{MK} \end{bmatrix}}},$$

where each D_{ik} ($i = 0, 1, \ldots, M$ and $k = 1, 2, \ldots, K$) is defined as

$$D_{ik} = \begin{cases} 1 & \text{if decision } k \text{ is to be made in state } i, \\ 0 & \text{otherwise.} \end{cases}$$

Therefore, each row in the matrix must contain a single 1 with the rest of the elements 0s. For example, the optimal policy R_b for the prototype example is characterized by the matrix

$$\text{State} \begin{matrix} 0 \\ 1 \\ 2 \\ 3 \end{matrix} \overset{\displaystyle\text{Decision } k}{\overset{\begin{matrix} 1 & 2 & 3 \end{matrix}}{\begin{bmatrix} 1 & 0 & 0 \\ 1 & 0 & 0 \\ 0 & 1 & 0 \\ 0 & 0 & 1 \end{bmatrix}}};$$

i.e., do nothing (decision 1) when the machine is in state 0 or 1, overhaul (decision 2) in state 2, and replace the machine (decision 3) when it is in state 3.

Randomized Policies

Introducing D_{ik} provides motivation for a *linear programming formulation*. It is hoped that the expected cost of a policy can be expressed as a linear function of D_{ik} or a related variable, subject to linear constraints. Unfortunately, the D_{ik} values are integers

(0 or 1), and continuous variables are required for a linear programming formulation. This requirement can be handled by expanding the interpretation of a policy. The previous definition calls for making the same decision every time the system is in state i. The new interpretation of a policy will call for determining a probability distribution for the decision to be made when the system is in state i.

With this new interpretation, the D_{ik} now need to be redefined as

$$D_{ik} = P\{\text{decision} = k \mid \text{state} = i\}.$$

In other words, given that the system is in state i, variable D_{ik} is the *probability* of choosing decision k as the decision to be made. Therefore, $(D_{i1}, D_{i2}, \ldots, D_{iK})$ is the *probability distribution* for the decision to be made in state i.

This kind of policy using probability distributions is called a *randomized policy,* whereas the policy calling for $D_{ik} = 0$ or 1 is a *deterministic policy.* Randomized policies can again be characterized by the matrix

$$\text{State} \begin{matrix} 0 \\ 1 \\ \vdots \\ M \end{matrix} \overset{\text{Decision } k}{\overset{\begin{matrix} 1 & 2 & \cdots & K \end{matrix}}{\begin{bmatrix} D_{01} & D_{02} & \cdots & D_{0K} \\ D_{11} & D_{12} & \cdots & D_{1K} \\ \cdots & \cdots & \cdots & \cdots \\ D_{M1} & D_{M2} & \cdots & D_{MK} \end{bmatrix}}},$$

where each row sums to 1, and now

$$0 \le D_{ik} \le 1.$$

To illustrate, consider a randomized policy for the prototype example given by the matrix

$$\text{State} \begin{matrix} 0 \\ 1 \\ 2 \\ 3 \end{matrix} \overset{\text{Decision } k}{\overset{\begin{matrix} 1 & 2 & 3 \end{matrix}}{\begin{bmatrix} 1 & 0 & 0 \\ \frac{1}{2} & 0 & \frac{1}{2} \\ \frac{1}{4} & \frac{1}{4} & \frac{1}{2} \\ 0 & 0 & 1 \end{bmatrix}}}.$$

This policy calls for *always* making decision 1 (do nothing) when the machine is in state 0. If it is found to be in state 1, it is left as is with probability $\frac{1}{2}$ and replaced with probability $\frac{1}{2}$, so a coin can be flipped to make the choice. If it is found to be in state 2, it is left as is with probability $\frac{1}{4}$, overhauled with probability $\frac{1}{4}$, and replaced with probability $\frac{1}{2}$. Presumably, a random device with these probabilities (possibly a table of random numbers) can be used to make the actual decision. Finally, if the machine is found to be in state 3, it always is overhauled.

By allowing randomized policies, so that the D_{ik} are continuous variables instead of integer variables, it now is possible to formulate a linear programming model for finding an optimal policy.

A Linear Programming Formulation

The convenient decision variables (denoted here by y_{ik}) for a linear programming model are defined as follows. For each $i = 0, 1, \ldots, M$ and $k = 1, 2, \ldots, K$, let y_{ik} be the steady-state unconditional probability that the system is in state *i and* decision *k* is made; i.e.,

$$y_{ik} = P\{\text{state} = i \text{ and decision} = k\}.$$

Each y_{ik} is closely related to the corresponding D_{ik} since, from the rules of conditional probability,

$$y_{ik} = \pi_i D_{ik},$$

where π_i is the steady-state probability that the Markov chain is in state i. Furthermore,

$$\pi_i = \sum_{k=1}^{K} y_{ik},$$

so that

$$D_{ik} = \frac{y_{ik}}{\pi_i} = \frac{y_{ik}}{\sum_{k=1}^{K} y_{ik}}.$$

There exist several constraints on y_{ik}:

1. $\sum_{i=1}^{M} \pi_i = 1$ so that $\sum_{i=0}^{M} \sum_{k=1}^{K} y_{ik} = 1.$

2. From results on steady-state probabilities (see Sec. 14.6),[1]

$$\pi_j = \sum_{i=0}^{M} \pi_i p_{ij}$$

so that

$$\sum_{k=1}^{K} y_{jk} = \sum_{i=0}^{M} \sum_{k=1}^{K} y_{ik} p_{ij}(k), \qquad \text{for } j = 0, 1, \ldots, M.$$

3. $y_{ik} \geq 0,$ for $i = 0, 1, \ldots, M$ and $k = 1, 2, \ldots, K.$

The long-run expected average cost per unit time is given by

$$E(C) = \sum_{i=0}^{M} \sum_{k=1}^{K} \pi_i C_{ik} D_{ik} = \sum_{i=0}^{M} \sum_{k=1}^{K} C_{ik} y_{ik}.$$

Hence, the linear programming model is to choose the y_{ik} so as to

$$\text{Minimize} \quad Z = \sum_{i=0}^{M} \sum_{k=1}^{K} C_{ik} y_{ik},$$

subject to the constraints

$$(1) \qquad \sum_{i=0}^{M} \sum_{k=1}^{K} y_{ik} = 1.$$

[1] The argument k is introduced in $p_{ij}(k)$ to indicate that the appropriate transition probability depends upon the decision k.

(2) $$\sum_{k=1}^{K} y_{jk} - \sum_{i=0}^{M} \sum_{k=1}^{K} y_{ik} p_{ij}(k) = 0, \qquad \text{for } j = 0, 1, \ldots, M.$$

(3) $$y_{ik} \geq 0, \qquad \text{for } i = 0, 1, \ldots, M; \; k = 1, 2, \ldots, K.$$

Thus, this model has $M + 2$ functional constraints and $K(M + 1)$ decision variables. [Actually, (2) provides one *redundant* constraint, so any one of these $M + 1$ constraints can be deleted.]

Assuming that the model is not too huge, it can be solved by the *simplex method.* Once the y_{ik} values are obtained, each D_{ik} is found from

$$D_{ik} = \frac{y_{ik}}{\sum_{k=1}^{K} y_{ik}}.$$

The optimal solution obtained by the simplex method has some interesting properties. It will contain $M + 1$ basic variables $y_{ik} \geq 0$. It can be shown that $y_{ik} > 0$ for at least one $k = 1, 2, \ldots, K$, for each $i = 0, 1, \ldots, M$. Therefore, it follows that $y_{ik} > 0$ for only *one* k for each $i = 0, 1, \ldots, M$. Consequently, each $D_{ik} = 0$ or 1.

The key conclusion is that the optimal policy found by the simplex method is *deterministic* rather than randomized. Thus, allowing policies to be randomized does not help at all in improving the final policy. However, it serves an extremely useful role in this formulation by converting integer variables (the D_{ik}) to continuous variables so that linear programming (LP) can be used. (The analogy in *integer programming* is to use the *LP relaxation* so that the simplex method can be applied and then to have the *integer solutions property* hold so that the optimal solution for the LP relaxation turns out to be integer anyway.)

Solving the Prototype Example by Linear Programming

Refer to the prototype example of Sec. 19.1. The first two columns of Table 19.1 give the relevant combinations of states and decisions. Therefore, the decision variables that need to be included in the model are y_{01}, y_{11}, y_{13}, y_{21}, y_{22}, y_{23}, and y_{33}. (The general expressions given above for the model include y_{ik} for *irrelevant* combinations of states and decisions here, so these $y_{ik} = 0$ in an optimal solution, and they might as well be deleted at the outset.) The rightmost column of Table 19.1 provides the coefficients of these variables in the objective function. The transition probabilities $p_{ij}(k)$ for each relevant combination of state i and decision k also are spelled out in Sec. 19.1.

The resulting linear programming model is

Minimize $$Z = 1{,}000y_{11} + 6{,}000y_{13} + 3{,}000y_{21} + 4{,}000y_{22} + 6{,}000y_{23} + 6{,}000y_{33},$$

subject to

$$y_{01} + y_{11} + y_{13} + y_{21} + y_{22} + y_{23} + y_{33} = 1$$
$$y_{01} - (y_{13} + y_{23} + y_{33}) = 0$$
$$y_{11} + y_{13} - (\tfrac{7}{8}y_{01} + \tfrac{3}{4}y_{11} + y_{22}) = 0$$
$$y_{21} + y_{22} + y_{23} - (\tfrac{1}{16}y_{01} + \tfrac{1}{8}y_{11} + \tfrac{1}{2}y_{21}) = 0$$
$$y_{33} - (\tfrac{1}{16}y_{01} + \tfrac{1}{8}y_{11} + \tfrac{1}{2}y_{21}) = 0$$

and $$\text{all } y_{ik} \geq 0.$$

Applying the simplex method, we obtain the optimal solution

$$y_{01} = \tfrac{2}{21}, \qquad (y_{11}, y_{13}) = (\tfrac{5}{7}, 0), \qquad (y_{21}, y_{22}, y_{23}) = (0, \tfrac{2}{21}, 0), \qquad y_{33} = \tfrac{2}{21},$$

so $D_{01} = 1, \qquad (D_{11}, D_{13}) = (1, 0), \qquad (D_{21}, D_{22}, D_{23}) = (0, 1, 0), \qquad D_{33} = 1.$

This policy calls for leaving the machine as is (decision 1) when it is in state 0 or 1, overhauling it (decision 2) when it is in state 2, and replacing it (decision 3) when it is in state 3. This is the same optimal policy found by exhaustive enumeration at the end of Sec. 19.2.

19.4 Policy Improvement Algorithm for Finding Optimal Policies

You now have seen two methods for deriving an optimal policy for a Markov decision process: *exhaustive enumeration* and *linear programming*. Exhaustive enumeration is useful because it is both quick and straightforward for very small problems. Linear programming can be used to solve vastly larger problems, and software packages for the simplex method are very widely available.

We now present a third popular method, namely, a *policy improvement algorithm*. The key advantage of this method is that it tends to be very efficient, because it usually reaches an optimal policy in a relatively small number of iterations (far fewer than for the simplex method with a linear programming formulation).

By following the model of Sec. 19.2 and as a joint result of the current state i of the system and the decision $d_i(R) = k$ when operating under policy R, two things occur. An (expected) cost C_{ik} is incurred that depends upon only the observed state of the system and the decision made. The system moves to state j at the next observed time period, with transition probability given by $p_{ij}(k)$. If, in fact, state j influences the cost that has been incurred, then C_{ik} is calculated as follows. Denote by $q_{ij}(k)$ the (expected) cost incurred when the system is in state i and decision k is made and then it evolves to state j at the next observed time period. Then

$$C_{ik} = \sum_{j=0}^{M} q_{ij}(k) p_{ij}(k).$$

Preliminaries

Referring to the description and notation for Markov decision processes given at the beginning of Sec. 19.2, we can show that, for any given policy R, there exist values $g(R), v_0(R), v_1(R), \ldots, v_M(R)$ that satisfy

$$g(R) + v_i(R) = C_{ik} + \sum_{j=0}^{M} p_{ij}(k)\, v_j(R), \qquad \text{for } i = 0, 1, 2, \ldots, M.$$

We now shall give a heuristic justification of these relationships and an interpretation for these values.

Denote by $v_i^n(R)$ the total expected cost of a system starting in state i (beginning the first observed time period) and evolving for n time periods. Then $v_i^n(R)$ has two components: C_{ik}, the cost incurred during the first observed time period, and $\sum_{j=0}^{M} p_{ij}(k)\, v_j^{n-1}(R)$, the total expected cost of the system evolving over the remaining $n - 1$ time periods. This gives the *recursive equation*

$$v_i^n(R) = C_{ik} + \sum_{j=0}^{M} p_{ij}(k)\, v_j^{n-1}(R), \qquad \text{for } i = 0, 1, 2, \ldots, M,$$

where $v_i^1(R) = C_{ik}$ for all i.

It will be useful to explore the behavior of $v_i^n(R)$ as n grows large. Recall that the (long-run) expected average cost per unit time following any policy R can be expressed as

$$g(R) = \sum_{i=0}^{M} \pi_i C_{ik},$$

which is independent of the starting state i. Hence, $v_i^n(R)$ behaves approximately as $ng(R)$ for large n. In fact, if we neglect small fluctuations, $v_i^n(R)$ can be expressed as the sum of two components

$$v_i^n(R) \approx ng(R) + v_i(R),$$

where the first component is independent of the initial state and the second is dependent upon the initial state. Thus, $v_i(R)$ can be interpreted as the effect on the total expected cost due to starting in state i. Consequently,

$$v_i^n(R) - v_j^n(R) \approx v_i(R) - v_j(R),$$

so that $v_i(R) - v_j(R)$ is a measure of the effect of starting in state i rather than state j.

Letting n grow large, we now can substitute $v_i^n(R) = ng(R) + v_i(R)$ and $v_j^{n-1}(R) = (n - 1)g(R) + v_j(R)$ into the *recursive equation*. This leads to the system of equations given in the opening paragraph of this subsection.

Note that this system has $M + 1$ equations with $M + 2$ unknowns, so that one of these variables may be chosen arbitrarily. By convention, $v_M(R)$ will be chosen equal to zero. Therefore, by solving the system of linear equations, we can obtain $g(R)$, the (long-run) expected average cost per unit time when policy R is followed. In principle, all policies can be enumerated, and that policy which minimizes $g(R)$ can be found. However, even for a moderate number of states and decisions, this technique is cumbersome. Fortunately, there exists an algorithm that can be used to evaluate policies and find the optimal one without complete enumeration, as described next.

The Policy Improvement Algorithm

The algorithm begins by choosing an arbitrary policy R_1. It then solves the system of equations to find the values of $g(R_1), v_0(R), v_1(R), \ldots, v_{M-1}(R)$ [with $v_M(R) = 0$]. This step is called *value determination*. A better policy, denoted by R_2, is then constructed. This step is called *policy improvement*. These two steps constitute an iteration of the algorithm. Using the new policy R_2, we perform another iteration. These iterations continue until two successive iterations lead to identical policies, which signifies that the optimal policy has been obtained. The details are outlined below.

Summary of the Policy Improvement Algorithm

Initialization: Choose an arbitrary initial trial policy R_1. Set $n = 1$.
Iteration n:

Step 1: Value determination For policy R_n, use $p_{ij}(k)$, C_{ik}, and $v_M(R_n) = 0$ to solve the system of $M + 1$ equations

$$g(R_n) = C_{ik} + \sum_{j=0}^{M} p_{ij}(k)\, v_j(R_n) - v_i(R_n), \qquad \text{for } i = 0, 1, \ldots, M,$$

for all $M + 1$ unknown values of $g(R_n), v_0(R_n), v_1(R_n), \ldots, v_{M-1}(R_n)$.

Step 2: Policy improvement Using the current values of $v_i(R_n)$ computed for policy R_n, find the alternative policy R_{n+1} such that, for each state i, $d_i(R_{n+1}) = k$ is the decision that minimizes

$$C_{ik} + \sum_{j=0}^{M} p_{ij}(k)\, v_j(R_n) - v_i(R_n)$$

i.e., for *each* state i,

$$\underset{k=1,2,\ldots,K}{\text{Minimize}} \quad \left[C_{ik} + \sum_{j=0}^{M} p_{ij}(k)\, v_j(R_n) - v_i(R_n) \right],$$

and then set $d_i(R_{n+1})$ equal to the minimizing value of k. This procedure defines a new policy R_{n+1}.

Optimality test: The current policy R_{n+1} is optimal if this policy is identical to policy R_n. If it is, stop. Otherwise, reset $n = n + 1$ and perform another iteration.

Two key properties of this algorithm are

1. $g(R_{n+1}) \le g(R_n)$, for $n = 1, 2, \ldots$.
2. The algorithm terminates with an optimal policy in a finite number of iterations.[1]

Solving the Prototype Example by the Policy Improvement Algorithm

Referring to the prototype example presented in Sec. 19.1, we outline the application of the algorithm below.

INITIALIZATION: For the initial trial policy R_1, we arbitrarily choose the policy that calls for replacement of the machine (decision 3) when it is found to be in state 3, but doing nothing (decision 1) in other states. This policy, its transition matrix, and its costs are summarized below.

Policy R_1

State	Decision
0	1
1	1
2	1
3	3

Transition Matrix

State	0	1	2	3
0	0	$\frac{7}{8}$	$\frac{1}{16}$	$\frac{1}{16}$
1	0	$\frac{3}{4}$	$\frac{1}{8}$	$\frac{1}{8}$
2	0	0	$\frac{1}{2}$	$\frac{1}{2}$
3	1	0	0	0

Costs

State	C_{ik}
0	0
1	1,000
2	3,000
3	6,000

[1] This termination is guaranteed under the assumptions of the model given in Sec. 19.2, including particularly the (implicit) assumptions of a finite number of states ($M + 1$) and a finite number of decisions (K), but not necessarily for more general models. See R. Howard, *Dynamic Programming and Markov Processes,* M.I.T. Press, Cambridge, Mass., 1960. Also see pp. 1291–1293 in A. F. Veinott, Jr., "On Finding Optimal Policies in Discrete Dynamic Programming with No Discounting," *Annals of Mathematical Statistics,* **37**: 1284–1294, 1966.

ITERATION 1: With this policy, the value determination step requires solving the following four equations simultaneously for $g(R_1)$, $v_0(R_1)$, $v_1(R_1)$, and $v_2(R_1)$ [with $v_3(R_1) = 0$].

$$g(R_1) = \qquad + \tfrac{7}{8}v_1(R_1) + \tfrac{1}{16}v_2(R_1) - v_0(R_1).$$

$$g(R_1) = 1{,}000 \qquad + \tfrac{3}{4}v_1(R_1) + \tfrac{1}{8}v_2(R_1) - v_1(R_1).$$

$$g(R_1) = 3{,}000 \qquad + \tfrac{1}{2}v_2(R_1) - v_2(R_1).$$

$$g(R_1) = 6{,}000 + v_0(R_1).$$

The simultaneous solution is

$$g(R_1) = \frac{25{,}000}{13} = 1{,}923$$

$$v_0(R_1) = -\frac{53{,}000}{13} = -4{,}077$$

$$v_1(R_1) = -\frac{34{,}000}{13} = -2{,}615$$

$$v_2(R_1) = \frac{28{,}000}{13} = 2{,}154.$$

Step 2 (policy improvement) can now be applied. We want to find an improved policy R_2 such that decision k in state i minimizes the corresponding expression below.

State 0: $\quad C_{0k} - p_{00}(k)(4{,}077) - p_{01}(k)(2{,}615) + p_{02}(k)(2{,}154) + 4{,}077$

State 1: $\quad C_{1k} - p_{10}(k)(4{,}077) - p_{11}(k)(2{,}615) + p_{12}(k)(2{,}154) + 2{,}615$

State 2: $\quad C_{2k} - p_{20}(k)(4{,}077) - p_{21}(k)(2{,}615) + p_{22}(k)(2{,}154) - 2{,}154$

State 3: $\quad C_{3k} - p_{30}(k)(4{,}077) - p_{31}(k)(2{,}615) + p_{32}(k)(2{,}154).$

Actually, in state 0, the only decision allowed is decision 1 (do nothing), so no calculations are needed. Similarly, we know that decision 3 (replace) must be made in state 3. Thus, only states 1 and 2 require calculation of the values of these expressions for alternative decisions.

For state 1, the possible decisions are 1 and 3. For each one, we show below the corresponding C_{1k}, the $p_{1j}(k)$, and the resulting value of the expression.

	State 1						
Decision	C_{1k}	$p_{10}(k)$	$p_{11}(k)$	$p_{12}(k)$	$p_{13}(k)$	Value of Expression	
1	1,000	0	$\frac{3}{4}$	$\frac{1}{8}$	$\frac{1}{8}$	1,923	←Minimum
3	6,000	1	0	0	0	4,538	

Since decision 1 minimizes the expression, it is chosen as the decision to be made in state 1 for policy R_2 (just as for policy R_1).

The corresponding results for state 2 are shown below for its three possible decisions.

	State 2					
Decision	C_{2k}	$p_{20}(k)$	$p_{21}(k)$	$p_{22}(k)$	$p_{23}(k)$	Value of Expression
1	3,000	0	0	$\frac{1}{2}$	$\frac{1}{2}$	1,923
2	4,000	0	1	0	0	−769 ←Minimum
3	6,000	1	0	0	0	−231

Therefore, decision 2 is chosen as the decision to be made in state 2 for policy R_2. Note that this is a change from policy R_1.

We summarize our new policy, its transition matrix, and its costs below.

Policy R_2

State	Decision
0	1
1	1
2	2
3	3

Transition Matrix

State	0	1	2	3
0	0	$\frac{7}{8}$	$\frac{1}{16}$	$\frac{1}{16}$
1	0	$\frac{3}{4}$	$\frac{1}{8}$	$\frac{1}{8}$
2	0	1	0	0
3	1	0	0	0

Costs

State	C_{ik}
0	0
1	1,000
2	4,000
3	6,000

Since this policy is not identical to policy R_1, the optimality test says to perform another iteration.

Iteration 2: For step 1 (value determination), the equations to be solved for this policy are shown below.

$$g(R_2) = \quad + \tfrac{7}{8}v_1(R_2) + \tfrac{1}{16}v_2(R_2) - v_0(R_2).$$

$$g(R_2) = 1{,}000 \quad + \tfrac{3}{4}v_1(R_2) + \tfrac{1}{8}v_2(R_2) - v_1(R_2).$$

$$g(R_2) = 4{,}000 \quad + \ v_1(R_2) \quad - v_2(R_2).$$

$$g(R_2) = 6{,}000 + v_0(R_2).$$

The simultaneous solution is

$$g(R_2) = \frac{5{,}000}{3} = 1{,}667$$

$$v_0(R_2) = -\frac{13{,}000}{3} = -4{,}333$$

$$v_1(R_2) = -3{,}000$$

$$v_2(R_2) = -\frac{2{,}000}{3} = -667.$$

Step 2 (policy improvement) can now be applied. For the two states with more than one possible decision, the expressions to be minimized are

State 1: $C_{1k} - p_{10}(k)(4{,}333) - p_{11}(k)(3{,}000) - p_{12}(k)(667) + 3{,}000$

State 2: $C_{2k} - p_{20}(k)(4{,}333) - p_{21}(k)(3{,}000) - p_{22}(k)(667) + 667.$

The first iteration provides the necessary data (the transition probabilities and C_{ik}) required for determining the new policy, except for the values of each of these expressions for each of the possible decisions. These values are

Decision	Value for State 1	Value for State 2
1	1,667	3,333
2	—	1,667
3	4,667	2,334

Since decision 1 minimizes the expression for state 1 and decision 2 minimizes the expression for state 2, our next trial policy R_3 is

Policy R_3

State	Decision
0	1
1	1
2	2
3	3

Note that policy R_3 is identical to policy R_2. Therefore, the optimality test indicates that this policy is optimal, so the algorithm is finished.

Another example illustrating the application of this algorithm is included under the Demo menu in your OR Courseware. The Procedure menu also includes an *interactive* routine for efficiently learning and applying the algorithm.

19.5 Discounted Cost Criterion

Throughout this chapter, we have measured policies on the basis of their (long-run) expected average cost per unit time. We now turn to an alternative measure of performance, namely, the *expected total discounted cost.*

As first introduced in Chap. 17, this measure uses a *discount factor* α, where $0 < \alpha < 1$. The discount factor α can be interpreted as equal to $1/(1 + i)$, where i is the current interest rate per period. Thus, α is the *present value* of one unit of cost one period in the future. Similarly, α^m is the *present value* of one unit of cost m periods in the future.

This *discounted cost criterion* becomes preferable to the *average cost criterion* when the time periods for the Markov chain are sufficiently long that the *time value of money* should be taken into account in adding costs in future periods to the cost in the current period. Another advantage is that the discounted cost criterion can readily be adapted to dealing with a *finite-period* Markov decision process where the Markov chain will terminate after a certain number of periods.

Both the policy improvement technique and the linear programming approach still can be applied here with relatively minor adjustments from the average cost case, as we describe next. Then we will present another technique, called the *method of successive approximations,* for quickly approximating an optimal policy.

A Policy Improvement Algorithm

To derive the expressions needed for the value determination and policy improvement steps of the algorithm, we now adopt the viewpoint of *probabilistic dynamic programming* (as described in Sec. 10.4). In particular, for each state i ($i = 0, 1, \ldots, M$) of a Markov decision process operating under policy R, let $V_i^n(R)$ be the *expected total discounted cost* when the process starts in state i (beginning the first observed time period) and evolves for n time periods. Then $V_i^n(R)$ has two components: C_{ik}, the cost incurred during the first observed time period, and $\alpha \sum_{j=0}^{M} p_{ij}(k) V_j^{n-1}(R)$, the expected total discounted cost of the process evolving over the remaining $n - 1$ time periods. For each $i = 0, 1, \ldots, M$, this yields the recursive equation

$$V_i^n(R) = C_{ik} + \alpha \sum_{j=0}^{M} p_{ij}(k) V_j^{n-1}(R),$$

with $V_i^1(R) = C_{ik}$, which closely resembles the recursive relationships of probabilistic dynamic programming found in Sec. 10.4.

As n approaches infinity, this recursive equation converges to

$$V_i(R) = C_{ik} + \alpha \sum_{j=0}^{M} p_{ij}(k) V_j(R), \qquad \text{for } i = 0, 1, \ldots, M,$$

where $V_i(R)$ can now be interpreted as the expected total discounted cost when the process starts in state i and continues indefinitely. There are $M + 1$ equations and $M + 1$ unknowns, so the simultaneous solution of this system of equations yields the $V_i(R)$.

To illustrate, consider again the prototype example of Sec. 19.1. Under the average cost criterion, we found in Secs. 19.2, 19.3, and 19.4 that the optimal policy is to do nothing in states 0 and 1, overhaul in state 2, and replace in state 3. Under the discounted cost criterion, with $\alpha = 0.9$, this same policy gives the following system of equations:

$$\begin{aligned}
V_0(R) &= + 0.9\left[\tfrac{7}{8}V_1(R) + \tfrac{1}{16}V_2(R) + \tfrac{1}{16}V_3(R)\right]\\
V_1(R) &= 1{,}000 + 0.9\left[\tfrac{3}{4}V_1(R) + \tfrac{1}{8}V_2(R) + \tfrac{1}{8}V_3(R)\right]\\
V_2(R) &= 4{,}000 + 0.9\left[V_1(R)\right]\\
V_3(R) &= 6{,}000 + 0.9\left[V_0(R)\right].
\end{aligned}$$

The simultaneous solution is

$$V_0(R) = 14{,}949$$
$$V_1(R) = 16{,}262$$
$$V_2(R) = 18{,}636$$
$$V_3(R) = 19{,}454.$$

Thus, assuming that the system starts in state 0, the expected total discounted cost is \$14,949.

This system of equations provides the expressions needed for a policy improvement algorithm. After summarizing this algorithm in general terms, we shall use it to check whether this particular policy still is optimal under the discounted cost criterion.

Summary of the Policy Improvement Algorithm (Discounted Cost Criterion)

Initialization: Choose an arbitrary initial trial policy R_1. Set $n = 1$.

Iteration n:

Step 1: Value determination For policy R_n, use $p_{ij}(k)$ and C_{ik} to solve the system of $M + 1$ equations

$$V_i(R_n) = C_{ik} + \alpha \sum_{j=0}^{M} p_{ij}(k) V_j(R_n), \qquad \text{for } i = 0, 1, \ldots, M,$$

for all $M + 1$ unknown values of $V_0(R_n), V_1(R_n), \ldots, V_M(R_n)$.

Step 2: Policy improvement Using the current values of the $V_i(R_n)$, find the alternative policy R_{n+1} such that, for each state i, $d_i(R_{n+1}) = k$ is the decision that minimizes

$$C_{ik} + \alpha \sum_{j=0}^{M} p_{ij}(k) V_j(R_n)$$

i.e., for *each* state i,

$$\underset{k=1,2,\ldots,K}{\text{Minimize}} \quad \left[C_{ik} + \alpha \sum_{j=0}^{M} p_{ij}(k) V_j(R_n) \right],$$

and then set $d_i(R_{n+1})$ equal to the minimizing value of k. This procedure defines a new policy R_{n+1}.

Optimality test: The current policy R_{n+1} is optimal if this policy is identical to policy R_n. If it is, stop. Otherwise, reset $n = n + 1$ and perform another iteration.

Three key properties of this algorithm are as follows:

1. $V_i(R_{n+1}) \le V_i(R_n), \qquad \text{for } i = 0, 1, \ldots, M \text{ and } n = 1, 2, \ldots.$
2. The algorithm terminates with an optimal policy in a finite number of iterations.
3. The algorithm is valid without the assumption (used for the average cost case) that the Markov chain associated with every transition matrix is irreducible.

Your OR Courseware includes an *interactive* routine for applying this algorithm.

Solving the Prototype Example by This Policy Improvement Algorithm: We now pick up the prototype example where we left it before summarizing the algorithm.

We already have selected the optimal policy under the average cost criterion to be our initial trial policy R_1. This policy, its transition matrix, and its costs are summarized below.

Policy R_1

State	Decision
0	1
1	1
2	2
3	3

Transition Matrix

State	0	1	2	3
0	0	$\frac{7}{8}$	$\frac{1}{16}$	$\frac{1}{16}$
1	0	$\frac{3}{4}$	$\frac{1}{8}$	$\frac{1}{8}$
2	0	1	0	0
3	1	0	0	0

Costs

State	C_{ik}
0	0
1	1,000
2	4,000
3	6,000

We also have already done step 1 (value determination) of iteration 1. This transition matrix and these costs led to the system of equations used to find $V_0(R_1) = 14{,}949$, $V_1(R_1) = 16{,}262$, $V_2(R_1) = 18{,}636$, and $V_3(R_1) = 19{,}454$.

To start step 2 (policy improvement), we only need to construct the expression to be minimized for the two states (1 and 2) with a choice of decisions.

State 1: $C_{1k} + 0.9[p_{10}(k)(14{,}949) + p_{11}(k)(16{,}262) + p_{12}(k)(18{,}636) + p_{13}(k)(19{,}454)]$

State 2: $C_{2k} + 0.9[p_{20}(k)(14{,}949) + p_{21}(k)(16{,}262) + p_{22}(k)(18{,}636) + p_{23}(k)(19{,}454)]$.

For each of these states and their possible decisions, we show below the corresponding C_{ik}, the $p_{ij}(k)$, and the resulting value of the expression.

	State 1						
Decision	C_{1k}	$p_{10}(k)$	$p_{11}(k)$	$p_{12}(k)$	$p_{13}(k)$	Value of Expression	
1	1,000	0	$\frac{3}{4}$	$\frac{1}{8}$	$\frac{1}{8}$	16,262	←Minimum
3	6,000	1	0	0	0	19,454	

	State 2						
Decision	C_{2k}	$p_{20}(k)$	$p_{21}(k)$	$p_{22}(k)$	$p_{23}(k)$	Value of Expression	
1	3,000	0	0	$\frac{1}{2}$	$\frac{1}{2}$	20,140	
2	4,000	0	1	0	0	18,636	←Minimum
3	6,000	1	0	0	0	19,454	

Since decision 1 minimizes the expression for state 1 and decision 2 minimizes the expression for state 2, our next trial policy (R_2) is as follows:

Policy R_2

State	Decision
0	1
1	1
2	2
3	3

Since this policy is identical to policy R_1, the optimality test indicates that this policy is optimal. Thus, the optimal policy under the average cost criterion also is optimal under the discounted cost criterion in this case. (This often occurs, but not always.)

Linear Programming Formulation

The linear programming formulation for the discounted cost case is similar to that for the average cost case given in Sec. 19.3. However, we no longer need the first constraint given in Sec. 19.3; but the other functional constraints do need to include the discount factor α. The other difference is that the model now contains constants β_j for $j = 0, 1, \ldots, M$. These constants must satisfy the conditions

$$\sum_{j=0}^{M} \beta_j = 1, \qquad \beta_j > 0 \qquad \text{for } j = 0, 1, \ldots, M,$$

but otherwise they can be chosen arbitrarily without affecting the optimal policy obtained from the model.

The resulting model is to choose the values of the *continuous* decision variables y_{ik} so as to

$$\text{Minimize} \qquad Z = \sum_{i=0}^{M} \sum_{k=1}^{K} C_{ik} y_{ik},$$

subject to the constraints

$$(1) \qquad \sum_{k=1}^{K} y_{jk} - \alpha \sum_{i=0}^{M} \sum_{k=1}^{K} y_{ik} p_{ij}(k) = \beta_j, \qquad \text{for } j = 0, 1, \ldots, M,$$

$$(2) \qquad y_{ik} \geq 0, \qquad \text{for } i = 0, 1, \ldots, \text{M}; \ k = 1, 2, \ldots, K.$$

Once the simplex method is used to obtain an optimal solution for this model, the corresponding optimal policy then is defined by

$$D_{ik} = P\{\text{decision} = k \text{ and state} = i\} = \frac{y_{ik}}{\sum_{k=1}^{K} y_{ik}}.$$

The y_{ik} now can be interpreted as the *discounted* expected time of being in state i and making decision k, when the probability distribution of the *initial state* (when observations begin) is $P\{X_0 = j\} = \beta_j$ for $j = 0, 1, \ldots, M$. In other words, if

$$z_{ik}^{n} = P\{\text{at time } n, \text{ state} = i \text{ and decision} = k\},$$

then

$$y_{ik} = z_{ik}^{0} + \alpha z_{ik}^{1} + \alpha^2 z_{ik}^{2} + \alpha^3 z_{ik}^{3} + \cdots.$$

With the interpretation of the β_j as *initial state probabilities* (with each probability greater than zero), Z can be interpreted as the corresponding expected total discounted cost. Thus, the choice of β_j affects the optimal value of Z (but not the resulting optimal policy).

It again can be shown that the optimal policy obtained from solving the linear programming model is deterministic; that is, $D_{ik} = 0$ or 1. Furthermore, this technique is valid without the assumption (used for the average cost case) that the Markov chain associated with every transition matrix is irreducible.

Solving the Prototype Example by Linear Programming: The linear programming model for the prototype example (with $\alpha = 0.9$) is

$$\text{Minimize} \quad Z = 1{,}000y_{11} + 6{,}000y_{13} + 3{,}000y_{21} + 4{,}000y_{22} + 6{,}000y_{23} + 6{,}000y_{33},$$

subject to

$$y_{01} - 0.9(y_{13} + y_{23} + y_{33}) = \tfrac{1}{4}$$

$$y_{11} + y_{13} - 0.9(\tfrac{7}{8}y_{01} + \tfrac{3}{4}y_{11} + y_{22}) = \tfrac{1}{4}$$

$$y_{21} + y_{22} + y_{23} - 0.9(\tfrac{1}{16}y_{01} + \tfrac{1}{8}y'_{11} + \tfrac{1}{2}y_{21}) = \tfrac{1}{4}$$

$$y_{33} - 0.9(\tfrac{1}{16}y_{01} + \tfrac{1}{8}y_{11} + \tfrac{1}{2}y_{21}) = \tfrac{1}{4}$$

and

$$\text{all } y_{ik} \geq 0,$$

where β_0, β_1, β_2, and β_3 are arbitrarily chosen to be $\frac{1}{4}$. By the simplex method, the optimal solution is

$$y_{01} = 1.210, \quad (y_{11}, y_{13}) = (6.656, 0), \quad (y_{21}, y_{22}, y_{23}) = (0, 1.067, 0), \quad y_{33} = 1.067,$$

so

$$D_{01} = 1, \quad (D_{11}, D_{13}) = (1, 0), \quad (D_{21}, D_{22}, D_{23}) = (0, 1, 0), \quad D_{33} = 1.$$

This optimal policy is the same as that obtained earlier in this section by the policy improvement algorithm.

The value of the objective function for the optimal solution is $Z = 17{,}325$. This value is closely related to the values of the $V_i(R)$ for this optimal policy that were obtained by the policy improvement algorithm. Recall that $V_i(R)$ is interpreted as the expected total discounted cost given that the system starts in state i, and we are interpreting β_i as the probability of starting in state i. Because each β_i was chosen to equal $\frac{1}{4}$,

$$17{,}325 = \tfrac{1}{4}[V_0(R) + V_1(R) + V_2(R) + V_3(R)]$$

$$= \tfrac{1}{4}(14{,}949 + 16{,}262 + 18{,}636 + 19{,}454).$$

Finite-Period Markov Decision Processes and the Method of Successive Approximations

We now turn our attention to an approach, called the *method of successive approximations,* for *quickly* finding at least an *approximation* to an optimal policy.

We have assumed that the Markov decision process will be operating indefinitely, and we have sought an optimal policy for such a process. The basic idea of the method of successive approximations is to instead find an optimal policy for the decisions to make in the first period when the process has only n time periods to go before termination, starting with $n = 1$, then $n = 2$, then $n = 3$, and so on. As n grows large, the corresponding optimal policies will converge to an optimal policy for the infinite-period problem of interest. Thus, the policies obtained for $n = 1, 2, 3, \ldots$ provide *successive approximations* that lead to the desired optimal policy.

The reason that this approach is attractive is that we already have a quick method of finding an optimal policy when the process has only n periods to go, namely, probabilistic dynamic programming as described in Sec. 10.4.

In particular, for $i = 0, 1, \ldots, M$, let

V_i^n = expected total discounted cost of following an optimal policy, given that process starts in state i and has only n periods to go.[1]

By the *principle of optimality* for dynamic programming (see Sec. 10.2), the V_i^n are obtained from the recursive relationship

$$V_i^n = \min_k \left\{ C_{ik} + \alpha \sum_{j=0}^{M} p_{ij}(k) V_j^{n-1} \right\}, \qquad \text{for } i = 0, 1, \ldots, M.$$

The minimizing value of k provides the optimal decision to make in the first period when the process starts in state i.

To get started, with $n = 1$, all the $V_i^0 = 0$ so that

$$V_i^1 = \min_k \{C_{ik}\}, \qquad \text{for } i = 0, 1, \ldots, M.$$

Although the method of successive approximations may not lead to an optimal policy for the infinite-period problem after only a few iterations, it has one distinct advantage over the policy improvement and linear programming techniques. It never requires solving a system of simultaneous equations, so each iteration can be performed simply and quickly.

Furthermore, if the Markov decision process actually does have just n periods to go, n iterations of this method definitely will lead to an optimal policy. (For an n-period problem, it is permissible to set $\alpha = 1$, that is, no discounting, in which case the objective is to minimize the expected total cost over n periods.)

Your OR Courseware includes an interactive routine to help guide you to use this method efficiently.

Solving the Prototype Example by the Method of Successive Approximations: We again use $\alpha = 0.9$. Refer to the rightmost column of Table 19.1 at the end of Sec. 19.1 for the values of C_{ik}.

For the first iteration ($n = 1$), the value obtained for each V_i^1 is shown below, along with the minimizing value of k (given in parentheses).

$$V_0^1 = \min_k \{C_{0k}\} = 0 \qquad (k = 1)$$

$$V_1^1 = \min_k \{C_{1k}\} = 1{,}000 \qquad (k = 1)$$

$$V_2^1 = \min_k \{C_{2k}\} = 3{,}000 \qquad (k = 1)$$

$$V_3^1 = \min_k \{C_{3k}\} = 6{,}000 \qquad (k = 3).$$

Thus the first approximation calls for making decision 1 (do nothing) when the system is in state 0, 1, or 2. When the system is in state 3, decision 3 (replace the machine) is made.

[1] Since we want to allow n to grow indefinitely, we are letting n be the *number of periods to go,* instead of the *number of periods from the beginning* (as in Chap. 10).

The second iteration leads to

$$V_0^2 = \min \{0 + 0.9[\tfrac{7}{8}(1{,}000) + \tfrac{1}{16}(3{,}000) + \tfrac{1}{16}(6{,}000)],\ 4{,}000 + 0.9[1(1{,}000)],\ 6{,}000 + 0.9[1(0)]\} = 1{,}294 \qquad (k = 1)$$

$$V_1^2 = \min \{1{,}000 + 0.9[\tfrac{3}{4}(1{,}000) + \tfrac{1}{8}(3{,}000) + \tfrac{1}{8}(6{,}000)],\ 4{,}000 + 0.9[1(1{,}000)],\ 6{,}000 + 0.9[1(0)]\} = 2{,}688 \qquad (k = 1)$$

$$V_2^2 = \min \{3{,}000 + 0.9[\tfrac{1}{2}(3{,}000) + \tfrac{1}{2}(6{,}000)],\ 4{,}000 + 0.9[1(1{,}000)],\ 6{,}000 + 0.9[1(0)]\} = 4{,}900 \qquad (k = 2)$$

$$V_3^2 = \qquad 6{,}000 + 0.9[1(0)] = 6{,}000 \qquad (k = 3).$$

Thus the second approximation calls for leaving the machine as is when it is in state 0 or 1, overhauling when it is in state 2, and replacing the machine when it is in state 3. Note that this policy is the optimal one for the infinite-period problem, as found earlier in this section by both the policy improvement algorithm and linear programming. However, the V_i^2 (the expected total discounted cost when starting in state i for the two-period problem) are not yet close to the V_i (the corresponding cost for the infinite-period problem).

The third iteration leads to

$$V_0^3 = \min \{0 + 0.9[\tfrac{7}{8}(2{,}688) + \tfrac{1}{16}(4{,}900) + \tfrac{1}{16}(6{,}000)],\ 4{,}000 + 0.9[1(2{,}688)],\ 6{,}000 + 0.9[1(1{,}294)]\} = 2{,}730 \qquad (k = 1)$$

$$V_1^3 = \min \{1{,}000 + 0.9[\tfrac{3}{4}(2{,}688) + \tfrac{1}{8}(4{,}900) + \tfrac{1}{8}(6{,}000)],\ 4{,}000 + 0.9[1(2{,}688)],\ 6{,}000 + 0.9[1(1{,}294)]\} = 4{,}041 \qquad (k = 1)$$

$$V_2^3 = \min \{3{,}000 + 0.9[\tfrac{1}{2}(4{,}900) + \tfrac{1}{2}(6{,}000)],\ 4{,}000 + 0.9[1(2{,}688)],\ 6{,}000 + 0.9[1(1{,}294)]\} = 6{,}419 \qquad (k = 2)$$

$$V_3^3 = \qquad 6{,}000 + 0.9[1(1{,}294)] = 7{,}165 \qquad (k = 3).$$

Again the optimal policy for the infinite-period problem is obtained, and the costs are getting closer to those for that problem. This procedure can be continued, and V_0^n, V_1^n, V_2^n, and V_3^n will converge to 14,949, 16,262, 18,636, and 19,454, respectively.

Note that termination of the method of successive approximations after the second iteration would have resulted in an optimal policy for the infinite-period problem, although there is no way to know this fact without solving the problem by other methods.

As indicated earlier, the method of successive approximations definitely obtains an optimal policy for an n-period problem after n iterations. For this example, the first, second, and third iterations have identified the optimal immediate decision for each state if the remaining number of periods is one, two, and three, respectively.

19.6 Conclusions

Markov decision processes provide a powerful tool for optimizing the performance of stochastic processes that can be modeled as a discrete time Markov chain. Applications arise in a variety of areas. Selected Reference 6 provides an interesting survey of these applications.

The two primary measures of performance used are the (long-run) *expected average cost per unit time* and the *expected total discounted cost.* The latter measure requires determination of the appropriate value of a discount factor, but this measure is useful when it is important to take into account the time value of money.

The two most important methods for deriving optimal policies for Markov decision processes are *policy improvement algorithms* and *linear programming.* Under the discounted cost criterion, the *method of successive approximations* provides a quick way of approximating an optimal policy.

SELECTED REFERENCES

1. Bertsekas, D. P., *Dynamic Programming: Deterministic and Stochastic Models,* Prentice-Hall, Englewood Cliffs, NJ, 1987.
2. Denardo, E., *Dynamic Programming Theory and Applications,* Prentice-Hall, Englewood Cliffs, NJ, 1982.
3. Derman, C., *Finite State Markovian Decision Processes,* Academic Press, New York, 1970.
4. Heyman, D., and M. Sobel, *Stochastic Models in Operations Research,* vol. 2, McGraw-Hill, New York, 1982.
5. Ross, S., *Introduction to Stochastic Dynamic Programming,* Academic Press, New York, 1983.
6. White, D. J., "Real Applications of Markov Decision Processes," *Interfaces,* **15**(6): 73–83, November-December 1985.
7. Whittle, P., *Optimization over Time: Dynamic Programming and Stochastic Control,* Wiley, New York, vol. 1, 1982; vol. 2, 1983.

RELEVANT ROUTINES IN YOUR OR COURSEWARE

A demonstration example: *Policy Improvement Algorithm—Average Cost Case*

Interactive routines:
Enter Markovian Decision Model
Interactive Policy Improvement Algorithm—Average Cost
Interactive Policy Improvement Algorithm—Discounted Cost
Interactive Method of Successive Approximations

Automatic routines:
Enter or Revise a General Linear Programming Model
Solve Automatically by the Simplex Method
Enter Transition Matrix
Steady-State Probabilities

To access these routines, call the ProbMod program and then choose *Markovian Decision Processes* under the Area menu. (For the last two automatic routines, choose *Markov Chains* under the Area menu. For the first two automatic routines, call the MathProg program and then choose *General Linear Programming* under the Area menu.) See App. 1 for documentation of the software.

PROBLEMS

To the left of each of the following problems (or their parts), we have inserted a D (for demonstration), I (for interactive routine), or A (for automatic routine) whenever a corresponding routine listed above can be helpful. An asterisk on the I or A indicates that this routine definitely should be used (unless your instructor gives contrary instructions) and that the printout from this routine is all that needs to be turned in to show your work in executing the algorithm. An asterisk on the problem number indicates that at least a partial answer is given in the back of the book.

19.2-1.* During any period, a potential customer arrives at a certain facility with probability $\frac{1}{2}$. If there are already two people at the facility (including the one being served), the potential customer leaves the facility immediately and never returns. However, if there is one person or less, he enters the facility and becomes an actual customer. The manager of the facility has two types of service configurations available. At the beginning of each period, a decision must be made on which configuration to use. If she uses her "slow" configuration at a cost of $3 and any customers are present during the period, one customer will be served and leave the facility with probability $\frac{3}{5}$. If she uses her "fast" configuration at a cost of $9 and any customers are present during the period, one customer will be served and leave the facility with probability $\frac{4}{5}$. The probability of more than one customer arriving or more than one customer being served in a period is zero. A profit of $50 is earned when a customer is served.

(*a*) Formulate the problem of choosing the service configuration period by period as a Markov decision process. Identify the states and decisions. For each combination of state and decision, find the *expected net immediate cost* (subtracting any profit from serving a customer) incurred during that period.

(*b*) Identify all the (stationary deterministic) policies. For each one, find the transition matrix and write an expression for the (long-run) expected average net cost per period in terms of the unknown steady-state probabilities ($\pi_0, \pi_1, \ldots, \pi_M$).

A* (*c*) Use your OR Courseware to find these steady-state probabilities for each policy. Then evaluate the expression obtained in part (*b*) to find the optimal policy by exhaustive enumeration.

19.2-2.* A student is concerned about her car and does not like dents. When she drives to school, she has a choice of parking it on the street in one space, parking it on the street and taking up two spaces, or parking in the lot. If she parks on the street in one space, her car gets dented with probability $\frac{1}{10}$. If she parks on the street and takes two spaces, the probability of a dent is $\frac{1}{50}$ and the probability of a $15 ticket is $\frac{3}{10}$. Parking in a lot costs $5, but the car will not get dented. If her car gets dented, she can have it repaired, in which case it is out of commission for 1 day and costs her $50 in fees and cab fares. She can also drive her car dented, but she feels that the resulting loss of value and pride is equivalent to a cost of $9 per school day. She wishes to determine the optimal policy for where to park and whether to repair the car when dented in order to minimize her (long-run) expected average cost per school day.

(*a*) Formulate this problem as a Markov decision process by identifying the states and decisions and then finding the C_{ik}.

(*b*) Identify all the (stationary deterministic) policies. For each one, find the transition matrix and write an expression for the (long-run) expected average cost per period in terms of the unknown steady-state probabilities ($\pi_0, \pi_1, \ldots, \pi_M$).

A* (*c*) Use your OR Courseware to find these steady-state probabilities for each policy. Then evaluate the expression obtained in part (*b*) to find the optimal policy by exhaustive enumeration.

19.2-3. A soap company specializes in a luxury type of bath soap. The sales of this soap fluctuate between two levels—low and high—depending upon two factors: (1) whether they advertise and (2) the advertising and marketing of new products by competitors. The second factor is out of the company's control, but it is trying to determine what its own advertising policy should be. For example, the *marketing manager's proposal* is to advertise when sales are low but not to advertise when sales are high (a particular policy). Advertising in any quarter of a year has primary impact on sales in the *following* quarter. At the beginning of each quarter, the needed information is available to forecast accurately whether sales will be low or high that quarter and to decide whether to advertise that quarter.

The cost of advertising is $1 million for each quarter of a year in which it is done. When advertising is done during a quarter, the probability of having high sales the next quarter is $\frac{1}{2}$ or $\frac{3}{4}$ depending upon whether the current quarter's sales are high or low. These probabilities go down to $\frac{1}{4}$ or $\frac{1}{2}$ when advertising is not done during the current quarter. The company's quarterly profits (excluding advertising costs) are $4 million when sales are high but only $2 million when sales are low. Management now wants to determine the advertising policy that will maximize the company's (long-run) expected average *net profit* (profit minus advertising costs) per quarter.

(*a*) Formulate this problem as a Markov decision process by identifying the states and decisions and then finding the C_{ik}.

(*b*) Identify all the (stationary deterministic) policies. For each one, find the transition matrix and write an expression for the (long-run) expected average net profit per period in terms of the unknown steady-state probabilities ($\pi_0, \pi_1, \ldots, \pi_M$).

A* (*c*) Use your OR Courseware to find these steady-state probabilities for each policy. Then evaluate the expression obtained in part (*b*) to find the optimal policy by exhaustive enumeration.

19.2-4. Every Saturday night a man plays poker at his home with the same group of friends. If he provides refreshments for the group (at an expected cost of $14) on any given Saturday night, the group will begin the following Saturday night in a good mood with probability $\frac{7}{8}$ and in a bad mood with probability $\frac{1}{8}$. However, if he fails to provide refreshments, the group will begin the following Saturday night in a good mood with probability $\frac{1}{8}$ and in a bad mood with probability $\frac{7}{8}$, regardless of their mood this Saturday. Furthermore, if the group begins the night in a bad mood and then he fails to provide refreshments, the group will gang up on him so that he incurs expected poker losses of $75. Under other circumstances, he averages no gain or loss on his poker play. The man wishes to find the policy regarding when to provide refreshments that will minimize his (long-run) expected average cost per week.

(*a*) Formulate this problem as a Markov decision process by identifying the states and decisions and then finding the C_{ik}.

(*b*) Identify all the (stationary deterministic) policies. For each one, find the transition matrix and write an expression for the (long-run) expected average cost per period in terms of the unknown steady-state probabilities ($\pi_0, \pi_1, \ldots, \pi_M$).

A* (*c*) Use your OR Courseware to find these steady-state probabilities for each policy. Then evaluate the expression obtained in part (*b*) to find the optimal policy by exhaustive enumeration.

19.2-5.* When a tennis player serves, he gets two chances to serve in bounds. If he fails to do so twice, he loses the point. If he attempts to serve an ace, he serves in bounds with probability $\frac{3}{8}$. If he serves a lob, he serves in bounds with probability $\frac{7}{8}$. If he serves an ace in bounds, he wins the point with probability $\frac{2}{3}$. With an in-bounds lob, he wins the point with probability $\frac{1}{3}$. If the cost is $+1$ for each point lost and -1 for each point won, the problem is to determine the optimal serving strategy to minimize the (long-run) expected average cost per point. (*Hint:* Let state 0 denote point over, two serves to go on next point; and let state 1 denote one serve left.)

(*a*) Formulate this problem as a Markov decision process by identifying the states and decisions and then finding the C_{ik}.

(*b*) Identify all the (stationary deterministic) policies. For each one, find the transition matrix and write an expression for the (long-run) expected average cost per point in terms of the unknown steady-state probabilities ($\pi_0, \pi_1, \ldots, \pi_M$).

A* (*c*) Use your OR Courseware to find these steady-state probabilities for each policy. Then evaluate the expression obtained in part (*b*) to find the optimal policy by exhaustive enumeration.

19.2-6. Each year Ms. Fontanez has the chance to invest in two different no-load mutual funds: the Go-Go Fund or the Go-Slow Mutual Fund. At the end of each year, Ms. Fontanez liquidates her holdings, takes her profits, and then reinvests. The yearly profits of the mutual funds are dependent upon how the market reacts each year. Recently the market has been oscillating around the 4,500 mark, according to the probabilities given in the following matrix:

$$\begin{array}{c} \\ 4{,}400 \\ 4{,}500 \\ 4{,}600 \end{array}\begin{array}{c} \begin{matrix} 4{,}400 & 4{,}500 & 4{,}600 \end{matrix} \\ \begin{bmatrix} 0.3 & 0.5 & 0.2 \\ 0.1 & 0.5 & 0.4 \\ 0.2 & 0.4 & 0.4 \end{bmatrix} \end{array}$$

Each year that the market moves up (down) 100 points, the Go-Go Fund has profits (losses) of \$20, while the Go-Slow Fund has profits (losses) of \$10. If the market moves up (down) 200 points in a year, the Go-Go Fund has profits (losses) of \$50, while the Go-Slow Fund has profits (losses) of only \$20. If the market does not change, there is no profit or loss for either fund. Ms. Fontanez wishes to determine her optimal investment policy in order to minimize her (long-run) expected average cost (loss minus profit) per year.

(*a*) Formulate this problem as a Markov decision process by identifying the states and decisions and then finding the C_{ik}.

(*b*) Identify all the (stationary deterministic) policies. For each one, find the transition matrix and write an expression for the (long-run) expected average cost per period in terms of the unknown steady-state probabilities $(\pi_0, \pi_1, \ldots, \pi_M)$.

A* (*c*) Use your OR Courseware to find these steady-state probabilities for each policy. Then evaluate the expression obtained in part (*b*) to find the optimal policy by exhaustive enumeration.

19.2-7. Buck and Bill Bogus are twin brothers who work at a gas station and have a counterfeiting business on the side. Each day a decision is made as to which brother will go to work at the gas station, and then the other will stay home and run the printing press in the basement. Each day that the machine works properly, it is estimated that 60 usable \$20 bills can be produced. However, the machine is somewhat unreliable and breaks down frequently. If the machine is not working at the beginning of the day, Buck can have it in working order by the beginning of the next day with probability 0.6. If Bill works on the machine, the probability decreases to 0.5. If Bill operates the machine when it is working, the probability is 0.6 that it will still be working at the beginning of the next day. If Buck operates the machine, it breaks down with probability 0.6. (Assume for simplicity that all breakdowns occur at the end of the day.) The brothers now wish to determine the optimal policy for when each should stay home in order to maximize their (long-run) expected average *profit* (amount of usable counterfeit money produced) per day.

(*a*) Formulate this problem as a Markov decision process by identifying the states and decisions and then finding the C_{ik}.

(*b*) Identify all the (stationary deterministic) policies. For each one, find the transition matrix and write an expression for the (long-run) expected average net profit per period in terms of the unknown steady-state probabilities $(\pi_0, \pi_1, \ldots, \pi_M)$.

A* (*c*) Use your OR Courseware to find these steady-state probabilities for each policy. Then evaluate the expression obtained in part (*b*) to find the optimal policy by exhaustive enumeration.

19.2-8. A person often finds that she is up to 1 hour late for work. If she is from 1 to 30 minutes late, \$4 is deducted from her paycheck; if she is from 31 to 60 minutes late for work, \$8 is deducted from her paycheck. If she drives to work at her normal speed (which is well under the speed limit), she can arrive in 20 minutes. However, if she exceeds the speed limit a little here and there on her way to work, she can get there in 10 minutes, but she runs the risk of getting a speeding ticket. With probability $\frac{1}{8}$ she will get caught speeding and will be fined \$20 and delayed 10 minutes, so that it takes 20 minutes to reach work.

As she leaves home, let s be the time she has to reach work before being late; that is, $s = 10$ means she has 10 minutes to get to work, and $s = -10$ means she is already 10 minutes late for work. For simplicity, she considers s to be in one of four intervals: $(20, \infty)$, $(10, 19)$, $(-10, 9)$, and $(-20, -11)$.

The transition probabilities for s tomorrow if she does not speed today are given by

	$(20, \infty)$	$(10, 19)$	$(-10, 9)$	$(-20, -11)$
$(20, \infty)$	$\frac{3}{8}$	$\frac{1}{4}$	$\frac{1}{4}$	$\frac{1}{8}$
$(10, 19)$	$\frac{1}{2}$	$\frac{1}{4}$	$\frac{1}{8}$	$\frac{1}{8}$
$(-10, 9)$	$\frac{5}{8}$	$\frac{1}{4}$	$\frac{1}{8}$	0
$(-20, -11)$	$\frac{3}{4}$	$\frac{1}{4}$	0	0

The transition probabilities for s tomorrow if she speeds to work today are given by

	$(20, \infty)$	$(10, 19)$	$(-10, 9)$	$(-20, -11)$
$(20, \infty)$				
$(10, 19)$	$\frac{3}{8}$	$\frac{1}{4}$	$\frac{1}{4}$	$\frac{1}{8}$
$(-10, 9)$				
$(-20, -11)$	$\frac{5}{8}$	$\frac{1}{4}$	$\frac{1}{8}$	0

Note that there are no transition probabilities for $(20, \infty)$ and $(-10, 9)$, because she will get to work on time and from 1 to 30 minutes late, respectively, regardless of whether she speeds. Hence, speeding when in these states would not be a logical choice.

Also note that the transition probabilities imply that the later she is for work and the more she has to rush to get there, the more likely she is to leave for work earlier the next day.

She wishes to determine when she should speed and when she should take her time getting to work in order to minimize her (long-run) expected average cost per day.

(*a*) Formulate this problem as a Markov decision process by identifying the states and decisions and then finding the C_{ik}.

(*b*) Identify all the (stationary deterministic) policies. For each one, find the transition matrix and write an expression for the (long-run) expected average cost per period in terms of the unknown steady-state probabilities $(\pi_0, \pi_1, \ldots, \pi_M)$.

A* (*c*) Use your OR Courseware to find these steady-state probabilities for each policy. Then evaluate the expression obtained in part (*b*) to find the optimal policy by exhaustive enumeration.

19.2-9. Consider an infinite-period inventory problem involving a single product where, at the beginning of each period, a decision must be made about how many items to produce during that period. The setup cost is $10, and the unit production cost is $5. The holding cost for each item not sold during the period is $4 (a *maximum* of 2 items can be stored). The demand during each period has a known probability distribution, namely, a probability of $\frac{1}{3}$ of 0, 1, and 2 items, respectively. If the demand exceeds the supply available during the period, then those sales are lost and a shortage cost (including lost revenue) is incurred, namely, $8 and $32 for a shortage of 1 and 2 items, respectively.

(*a*) Consider the policy $(s, S) = (1, 2)$, where enough items are produced to raise the current inventory level to 2 items if, and only if, the inventory level at the beginning of a period is less than 1 item (i.e., no items are present). Determine the (long-run) expected average cost per period for this policy. In finding the transition matrix for the Markov chain for this policy, let the states represent the inventory levels at the beginning of the period.

(*b*) Identify all the *feasible* (stationary deterministic) inventory policies, i.e., the policies that never lead to exceeding the storage capacity.

19.3-1. Reconsider Prob. 19.2-1.

(*a*) Formulate a linear programming model for finding an optimal policy.

A* (*b*) Use the simplex method to solve this model. Use the resulting optimal solution to identify an optimal policy.

19.3-2.* Reconsider Prob. 19.2-2.

(*a*) Formulate a linear programming model for finding an optimal policy.

A* (*b*) Use the simplex method to solve this model. Use the resulting optimal solution to identify an optimal policy.

19.3-3. Reconsider Prob. 19.2-3.

(*a*) Formulate a linear programming model for finding an optimal policy.

A* (*b*) Use the simplex method to solve this model. Use the resulting optimal solution to identify an optimal policy.

19.3-4. Reconsider Prob. 19.2-4.

(*a*) Formulate a linear programming model for finding an optimal policy.

A* (*b*) Use the simplex method to solve this model. Use the resulting optimal solution to identify an optimal policy.

19.3-5.* Reconsider Prob. 19.2-5.

(*a*) Formulate a linear programming model for finding an optimal policy.

A* (*b*) Use the simplex method to solve this model. Use the resulting optimal solution to identify an optimal policy.

19.3-6. Reconsider Prob. 19.2-6.

(*a*) Formulate a linear programming model for finding an optimal policy.

A* (*b*) Use the simplex method to solve this model. Use the resulting optimal solution to identify an optimal policy.

19.3-7. Reconsider Prob. 19.2-7.

(*a*) Formulate a linear programming model for finding an optimal policy.

A* (*b*) Use the simplex method to solve this model. Use the resulting optimal solution to identify an optimal policy.

19.3-8. Reconsider Prob. 19.2-8.

(*a*) Formulate a linear programming model for finding an optimal policy.

A* (*b*) Use the simplex method to solve this model. Use the resulting optimal solution to identify an optimal policy.

19.3-9. Reconsider Prob. 19.2-9.

(*a*) Formulate a linear programming model for finding an optimal policy.

A* (*b*) Use the simplex method to solve this model. Use the resulting optimal solution to identify an optimal policy.

D, I* **19.4-1.** Use the policy improvement algorithm to find an optimal policy for Prob. 19.2-1.

D, I* **19.4-2.*** Use the policy improvement algorithm to find an optimal policy for Prob. 19.2-2.

D, I* **19.4-3.** Use the policy improvement algorithm to find an optimal policy for Prob. 19.2-3.

D, I* **19.4-4.** Use the policy improvement algorithm to find an optimal policy for Prob. 19.2-4.

D, I* **19.4-5.** Use the policy improvement algorithm to find an optimal policy for Prob. 19.2-5.

D, I* **19.4-6.** Use the policy improvement algorithm to find an optimal policy for Prob. 19.2-6.

D, I* **19.4-7.** Use the policy improvement algorithm to find an optimal policy for Prob. 19.2-7.

D, I* **19.4-8.** Use the policy improvement algorithm to find an optimal policy for Prob. 19.2-8.

D, I* **19.4-9.** Use the policy improvement algorithm to find an optimal policy for Prob. 19.2-9.

D, I* **19.4-10.** Consider the blood-inventory problem presented in problem 14.6-5. Suppose now that the number of pints of blood delivered (on a regular delivery) can be specified at the time of delivery (instead of using the old policy of receiving 1 pint at each delivery). Thus, the number of pints delivered can be 0, 1, 2, or 3 (more than 3 pints can never be used). The cost of regular delivery is \$50 per pint, while the cost of an emergency delivery is \$100 per pint. Starting with the proposed policy given in problem 14.6-5, perform two iterations of the policy improvement algorithm.

I* **19.5-1.*** Joe wants to sell his car. He receives one offer each month and must decide immediately whether to accept the offer. Once rejected, the offer is lost. The possible offers are \$600, \$800, and \$1,000, made with probabilities $\frac{5}{8}$, $\frac{1}{4}$, and $\frac{1}{8}$, respectively (where successive offers are independent of each other). There is a maintenance cost of \$60 per month for the car. Joe is anxious to sell the car and so has chosen a discount factor of $\alpha = 0.95$.

Using the policy improvement algorithm, find a policy that minimizes the expected total discounted cost. (*Hint:* There are two actions: Accept or reject the offer. Let the state for month t be the offer in that month. Also include a state ∞, where the process goes to state ∞ whenever an offer is accepted and it remains there at a monthly cost of 0.)

19.5-2.* Reconsider Prob. 19.5-1.

(*a*) Formulate a linear programming model for finding an optimal policy.

A* (*b*) Use the simplex method to solve this model. Use the resulting optimal solution to identify an optimal policy.

I* **19.5-3.*** For Prob. 19.5-1, use three iterations of the method of successive approximations to approximate an optimal policy.

I* **19.5-4.** The price of a certain stock is fluctuating between \$10, \$20, and \$30 from month to month. Market analysts have predicted that if the stock is at \$10 during any month, it will be at \$10 or \$20 the next month, with probabilities $\frac{4}{5}$ and $\frac{1}{5}$, respectively; if the stock is at \$20, it will be at \$10, \$20, or \$30 the next month, with probabilities $\frac{1}{4}$, $\frac{1}{4}$, and $\frac{1}{2}$, respectively; and if the stock is at \$30, it will be at \$20 or \$30 the next month, with probabilities $\frac{3}{4}$ and $\frac{1}{4}$, respectively. Given a discount factor of 0.9, use the policy improvement algorithm to determine when to sell and when to hold the stock to maximize the expected total discounted profit. (*Hint:* Include a state that is reached with probability 1 when the stock is sold and with probability 0 when the stock is held.)

19.5-5. Reconsider Prob. 19.5-4.

(*a*) Formulate a linear programming model for finding an optimal policy.

A* (*b*) Use the simplex method to solve this model. Use the resulting optimal solution to identify an optimal policy.

I* **19.5-6.** For Prob. 19.5-4, use three iterations of the method of successive approximations to approximate an optimal policy.

19.5-7. A chemical company produces two chemicals, denoted by 0 and 1, and only one can be produced at a time. Each month a decision is made as to which chemical to produce that month. Because the demand for each chemical is predictable, it is known that if 1 is produced this month, there is a 70 percent chance that it will also be produced again next month. Similarly, if 0 is produced this month, there is only a 20 percent chance that it will be produced again next month.

To combat the emissions of pollutants, the chemical company has two processes, process A, which is efficient in combating the pollution from the production of 1 but not from 0, and process B, which is efficient in combating the pollution from the production of 0 but not from 1. Only one process can be used at a time. The amount of pollution from the production of each chemical under each process is

	0	1
A	100	10
B	10	30

Unfortunately, there is a time delay in setting up the pollution control processes, so that a decision as to which process to use must be made in the month prior to the production decision. Management wants to determine a policy for when to use each pollution control process that will minimize the expected total discounted amount of all future pollution with a discount factor of $\alpha = 0.5$.

(*a*) Formulate this problem as a Markov decision process by identifying the states, the decisions, and the C_{ik}. Identify all the (stationary deterministic) policies.

I* (*b*) Use the policy improvement algorithm to find an optimal policy.

19.5-8. Reconsider Prob. 19.5-7.

(*a*) Formulate a linear programming model for finding an optimal policy.

A* (*b*) Use the simplex method to solve this model. Use the resulting optimal solution to identify an optimal policy.

I* **19.5-9.** For Prob. 19.5-7, use two iterations of the method of successive approximations to approximate an optimal policy.

I* **19.5-10.** Reconsider Prob. 19.5-7. Suppose now that the company will be producing either of these chemicals for only 4 more months, so a decision on which pollution control process to use 1 month hence only needs to be made three more times. Find an optimal policy for this three-period problem.

I* **19.5-11.*** Reconsider the prototype example of Sec. 19.1. Suppose now that the production process using the machine under consideration will be used for only 4 more weeks. Using the discounted cost criterion with a discount factor of $\alpha = 0.9$, find the optimal policy for this four-period problem.

20

Decision Analysis

The last six chapters have dealt largely with the analysis of *stochastic processes*—processes that evolve in a *probabilistic manner* over time (e.g., Markov chains). Some of this analysis focused on *decision making* in the face of this uncertainty (e.g., Markov decision processes).

We now turn to decision making in the face of uncertainty in a different context. Instead of making decisions over a long time, we now are concerned with making perhaps just one decision (or at most a sequence of a few decisions) about what to do in the immediate future. However, there still are random factors outside our control that create some uncertainty about the outcome of each of the alternative courses of action.

Decision analysis provides a framework and methodology for rational decision making in this context.

Frequently, one question to be addressed is whether to make the needed decision immediately or to first do some *testing* (at some expense) to reduce the level of uncertainty about the outcome of the decision. For example, the testing might be field testing of a proposed new product to test consumer reaction before making a decision on whether to proceed with full-scale production and marketing of the product. This testing is referred to as performing *experimentation.* Therefore, decision analysis divides decision making between the cases of *without experimentation* and *with experimentation.*

The first section introduces a prototype example that will be carried throughout the chapter for illustrative purposes. Sections 20.2 and 20.3 then present the basic principles of *decision making without experimentation* and *decision making with experimentation.* Section 20.4 describes *decision trees,* a useful tool for depicting and analyzing the decision process. We then conclude the chapter by introducing *utility theory,* which provides a way of calibrating the possible outcomes of the decision to reflect the true value of these outcomes to the decision maker.

20.1 A Prototype Example

The GOFERBROKE COMPANY owns a tract of land that may contain oil. A consulting geologist has reported to management that she believes there is 1 chance in 4 of oil.

Because of this prospect, another oil company has offered to purchase the land for $90,000. However, Goferbroke is considering holding the land in order to drill for oil itself. If oil is found, the company's expected profit will be approximately $700,000. A loss of $100,000 will be incurred if the land is dry (no oil).

Table 20.1 summarizes these data. Section 20.2 discusses how to approach the decision of whether to drill or sell based just on these data.

However, another option prior to making a decision is to conduct a detailed seismic survey of the land to obtain a better estimate of the probability of finding oil. Section 20.3 discusses this case of *decision making with experimentation,* at which point the necessary additional data will be provided.

This company is operating without much capital, so a loss of $100,000 would be quite serious. In Sec. 20.5, we describe how to refine the evaluation of the consequences of the various possible outcomes.

20.2 Decision Making without Experimentation

Before seeking a solution to the Goferbroke problem, we will formulate a general framework for decision making.

In general terms, the decision maker must choose an **action** *a* from a set of possible actions. The set contains all the *feasible alternatives* under consideration for how to proceed with the problem of concern.

Table 20.1 **Prospective Profits for the Goferbroke Company**

Alternative \ Status of Land	*Payoff*	
	Oil	Dry
Drill for oil	$700,000	−$100,000
Sell the land	$90,000	$90,000
Chance of status	1 in 4	3 in 4

This choice of an action must be made in the face of uncertainty, because the outcome will be affected by random factors that are outside the control of the decision maker. These random factors determine what situation will be found at the time that the action is executed. Each of these possible situations is referred to as a possible **state of nature**, denoted by θ. Frequency, the states of nature are an enumeration of possible alternative representations of the physical phenomenon being studied.

For each combination of an action a and a state of nature θ, the decision maker knows what the resulting payoff would be. The **payoff** is a quantitative measure of the value to the decision maker of the consequences of the outcome. For example, the payoff frequently is represented by the *net monetary gain* (profit), although other measures also can be used (as described in Sec. 20.5). If the consequences of the outcome do not become completely certain even when the state of nature is given, then the payoff becomes an *expected value* (in the statistical sense) of the measure of the consequences. Let

$$p(a, \theta) = \text{payoff from taking action } a \text{ when state of nature is } \theta.$$

A **payoff table** commonly is used to provide $p(a, \theta)$ for each combination of a and θ.

If you previously studied game theory (Chap. 11), we should point out an interesting analogy between this decision analysis framework and the two-person zero-sum games described in Chap. 11. The *decision maker* and *nature* can be viewed as the *two players* of such a game. The possible *actions* and the possible *states of nature* can then be viewed as the available *strategies* for these respective players, where each combination of strategies results in some *payoff* to player 1 (the decision maker). From this viewpoint, the decision analysis framework can be summarized as follows:

1. The *decision maker* needs to choose one of the possible *actions*.
2. *Nature* then would choose one of the possible *states of nature*.
3. Each combination of an action a and state of nature θ would result in a *payoff* $p(a, \theta)$, which is given as one of the entries in a *payoff table*.
4. This payoff table should be used to find an *optimal action* for the decision maker according to an appropriate criterion.

Soon we will present three possibilities for this criterion, where the first one (the maximin payoff criterion) comes from game theory.

However, this analogy to two-person zero-sum games breaks down in one important respect. In game theory, *both* players are assumed to be *rational* and choosing their strategies to *promote their own welfare*. This description still fits the decision maker, but certainly not nature. By contrast, nature now is a passive player that chooses its strategies (states of nature) in some random fashion. This change means that the game theory criterion for how to choose an optimal strategy (action) will not appeal to many decision makers in the current context.

One additional element needs to be added to the decision analysis framework. The decision maker generally will have some information that should be taken into account about the relative likelihood of the possible states of nature. Such information can usually be translated to a probability distribution, acting as though the state of nature is a random variable, in which case this distribution is referred to as a **prior distribution**. Prior distributions are often subjective in that they may depend upon the experience or intuition of an individual. The individual probabilities for the respective states of nature are called **prior probabilities**.

Formulation of the Prototype Example in This Framework

As indicated in Table 20.1, the Goferbroke Co. has two possible actions under consideration: drill for oil or sell the land. The possible states of nature are that the land contains oil and that it does not, as designated in the column headings of Table 20.1 by *oil* and *dry*. Since the consulting geologist has estimated that there is 1 chance in 4 of oil (and so 3 chances in 4 of no oil), the prior probabilities of the two states of nature are 0.25 and 0.75, respectively. Therefore, with the payoff in units of thousands of dollars of profit, the payoff table can be obtained directly from Table 20.1, as shown in Table 20.2.

We will use this payoff table next to find the optimal action according to each of the three criteria described below.

The Maximin Payoff Criterion

If the decision maker's problem were to be viewed as a *game against nature,* then game theory would say to choose the action according to the *minimax criterion* (as described in Sec. 11.2). From the viewpoint of player 1 (the decision maker), this criterion is more aptly named the *maximin payoff criterion,* as summarized below.

Maximin payoff criterion: For each possible action, find the *minimum payoff* over all possible states of nature. Next, find the *maximum* of these minimum payoffs. Choose the action whose minimum payoff gives this maximum.

Table 20.3 shows the application of this criterion to the prototype example. Thus, since the minimum payoff for a_2 (90) is larger than that for a_1 (-100), action a_2 (sell the land) will be chosen.

The rationale for this criterion is that it provides the *best guarantee* of the payoff that will be obtained. Regardless of what the true state of nature turns out to be for the example, the payoff from a_2 cannot be less than 90.

This rationale is quite valid when one is competing against a rational and malevolent opponent. However, this criterion is not often used in games against nature because it is an extremely conservative criterion in this context. In effect, it assumes that nature is a conscious opponent that wants to inflict as much damage as possible on the decision maker. Nature is not a malevolent opponent, and the decision maker does not need to focus solely on the worst possible payoff from each action. This is especially true when the worst possible payoff from an action comes from a relatively unlikely state of nature.

Thus, this criterion normally is of interest only to a very cautious decision maker.

Table 20.2 **Payoff Table for the Decision Analysis Formulation of the Goferbroke Co. Problem**

Action \ State of Nature	θ_1 Oil	θ_2 Dry
a_1: Drill for oil	700	-100
a_2: Sell the land	90	90
Prior probability	0.25	0.75

Table 20.3 **Application of the Maximin Payoff Criterion to the Goferbroke Co. Problem**

Action \ State of Nature	θ_1 Oil	θ_2 Dry	Minimum	
a_1: Drill for oil	700	−100	−100	
a_2: Sell the land	90	90	90	← Maximin value

The Maximum Likelihood Criterion

The next criterion focuses on the *most likely* state of nature, as summarized below.

> **Maximum likelihood criterion:** Identify the most likely state of nature (the one with the largest prior probability). For this state of nature, find the action with the maximum payoff. Choose this action.

Applying this criterion to the example, Table 20.4 indicates that θ_2 has the largest prior probability. In the θ_2 column, a_2 has the maximum payoff, so the choice is to sell the land.

The appeal of this criterion is that the most important state of nature is the most likely one, so the action chosen is the best one for the most important state of nature. Furthermore, the criterion does not rely on questionable subjective estimates of the probabilities of the respective states of nature other than identifying the most likely state.

The major drawback of the criterion is that it completely ignores much relevant information. No state of nature is considered other than the most likely one. In a problem with many possible states of nature, the probability of the most likely one may be quite small, so focusing on just this one state of nature is quite unwarranted. Even in the example, where the prior probability of θ_2 is 0.75, this criterion ignores the extremely attractive payoff of 700 with a_1 and θ_1. In effect, the criterion does not permit gambling on a low-probability big payoff, no matter how attractive the gamble may be.

Bayes' Decision Rule

Our third criterion, and the one commonly chosen, is *Bayes' decision rule,* described below.

> **Bayes' decision rule:** Using the best available estimates of the probabilities of the respective states of nature (currently the prior probabilities), calculate the expected value of the payoff for each of the possible actions. Choose the action with the maximum expected payoff.

Table 20.4 **Application of the Maximum Likelihood Criterion to the Goferbroke Co. Problem**

Action \ State of Nature	θ_1 Oil	θ_2 Dry	
a_1: Drill for oil	700	−100	
a_2: Sell the land	90	90	← Maximum in this column
Prior probability	0.25	0.75	
		↑ Maximum	

For the prototype example, these expected payoffs are calculated directly from Table 20.2 as follows:

For a_1: $$E[p(a_1, \theta)] = 0.25(700) + 0.75(-100) = 100.$$

For a_2: $$E[p(a_2, \theta)] = 0.25(90) + 0.75(90) = 90.$$

Since 100 is larger than 90, the action selected is a_1 (drill for oil).

Note that this choice contrasts with the selection of a_2 (sell the land) under each of the two preceding criteria.

The big advantage of Bayes' decision rule is that it incorporates all the available information, including all the payoffs and the best available estimates of the probabilities of the respective states of nature.

It is sometimes argued that these estimates of the probabilities necessarily are largely subjective and so are too shaky to be trusted. There is no accurate way of predicting the future, including a future state of nature, even in probability terms. This argument has some validity. The reasonableness of the estimates of the probabilities should be assessed in each individual situation.

Nevertheless, under many circumstances, past experience and current evidence enable one to develop reasonable estimates of the probabilities. Using this information should provide better grounds for a sound decision than ignoring it. Furthermore, experimentation frequently can be conducted to improve these estimates, as described in the next section.

Therefore, we will be using solely Bayes' decision rule throughout the remainder of the chapter.

20.3 Decision Making with Experimentation

Frequently, additional testing (experimentation) can be done to improve the preliminary estimates of the probabilities of the respective states of nature provided by the prior probabilities. These improved estimates are called **posterior probabilities**.

We first update the Goferbroke Co. example to incorporate experimentation, then describe how to derive the posterior probabilities, and finally discuss how to decide whether it is worthwhile to conduct experimentation.

Continuing the Prototype Example

As mentioned at the end of Sec. 20.1, an available option before making a decision is to conduct a detailed seismic survey of the land to obtain a better estimate of the probability of oil. The cost is $30,000.

A seismic survey obtains seismic soundings that indicate whether the geological structure is favorable to the presence of oil. To quantify this process, we let the random variable S and its possible values be defined as follows:

S = statistic obtained from seismic survey.

$S = 0$: Unfavorable seismic soundings; oil is fairly unlikely.

$S = 1$: Favorable seismic soundings; oil is fairly likely.

Based upon past experience, if there is oil (i.e., the true state of nature is θ_1), then the probability of $S = 0$ is

$$P(S = 0 \mid \theta = \theta_1) = 0.4, \qquad \text{so} \qquad P(S = 1 \mid \theta = \theta_1) = 1 - 0.4 = 0.6.$$

Similarly, if there is no oil (i.e., the true state of nature is θ_2), then the probability of $S = 0$ is estimated to be

$$P(S = 0 \mid \theta = \theta_2) = 0.8, \qquad \text{so} \qquad P(S = 1 \mid \theta = \theta_2) = 1 - 0.8 = 0.2.$$

We soon will use these data to find the posterior probabilities of the respective states of nature *given* the observed value of S.

Posterior Probabilities

Proceeding now in general terms, we let

n = number of possible states of nature;

$P(\theta = \theta_i)$ = prior probability that true state of nature is θ_i, for $i = 1, 2, \ldots, n$;

S = statistic summarizing results of experimentation (a random variable);

s = one possible value of S;

$P(\theta = \theta_i \mid S = s)$ = posterior probability that true state of nature is θ_i, given that $S = s$, for $i = 1, 2, \ldots, n$.

The question currently being addressed is the following:

Given $P(\theta = \theta_i)$ and $P(S = s \mid \theta = \theta_i)$, for $i = 1, 2, \ldots, n$, what is $P(\theta = \theta_i \mid S = s)$?

This question is answered by combining the following standard formulas of probability theory:

$$P(\theta = \theta_i \mid S = s) = \frac{P(\theta = \theta_i, S = s)}{P(S = s)}$$

$$P(S = s) = \sum_{j=1}^{n} P(\theta = \theta_j, S = s).$$

$$P(\theta = \theta_i, S = s) = P(S = s \mid \theta = \theta_i)P(\theta = \theta_i).$$

Therefore, for each $i = 1, 2, \ldots, n$, the desired formula for the corresponding posterior probability is

$$\boxed{P(\theta = \theta_i \mid S = s) = \frac{P(S = s \mid \theta = \theta_i)P(\theta = \theta_i)}{\sum_{j=1}^{n} P(S = s \mid \theta = \theta_j)P(\theta = \theta_j)}.}$$

Now let us return to the prototype example and apply this formula. If the result of the seismic survey is that $S = 0$, then the posterior probabilities are

$$P(\theta = \theta_1 \mid S = 0) = \frac{0.4(0.25)}{0.4(0.25) + 0.8(0.75)} = \frac{1}{7},$$

$$P(\theta = \theta_2 \mid S = 0) = 1 - \frac{1}{7} = \frac{6}{7}.$$

Similarly, if the seismic survey gives $S = 1$, then

$$P(\theta = \theta_1 \mid S = 1) = \frac{0.6(0.25)}{0.6(0.25) + 0.2(0.75)} = \frac{1}{2},$$

$$P(\theta = \theta_2 \mid S = 1) = 1 - \frac{1}{2} = \frac{1}{2}.$$

When you are dealing with larger problems, the **tabular algorithm** illustrated in Table 20.5 provides a convenient way of organizing these computations. The entries in the top row of Table 20.5*a* are the prior probabilities of the state of nature, whereas the entries in the bottom two rows are the conditional probabilities of the seismic statistic given the state of nature. The entries in the first two columns of Table 20.5*b* are the corresponding entries of Table 20.5*a* *multiplied* by the appropriate prior probability [for example, 0.10 = 0.4(0.25)]. The entries in the rightmost column of Table 20.5*b* are the *sum* of the two entries in the corresponding row (for example, 0.70 = 0.10 + 0.60). The entries in Table 20.5*c* are the corresponding entries of Table 20.5*b* *divided* by the entry in the rightmost column of Table 20.5*b* (for example, $\frac{1}{7} = 0.10/0.70$). These entries in Table 20.5*c* are the desired posterior probabilities of the θ_i.

Your OR Courseware also includes an automatic routine for computing these posterior probabilities, given $P(\theta = \theta_i)$ and $P(S = s \mid \theta = \theta_i)$ for all s and $i = 1, 2, \ldots, n$.

After these computations have been completed, Bayes' decision rule can be applied just as before, with the posterior probabilities now replacing the prior probabilities. Again, by using the payoffs (in units of thousands of dollars) from Table 20.2 and subtracting the cost of the experimentation, we obtain the results shown below.

Expected Payoffs if $S = 0$:

For a_1: $E[p(a_1, \theta \mid S = 0)] = \frac{1}{7}(700) + \frac{6}{7}(-100) - 30 = 270.$

For a_2: $E[p(a_2, \theta \mid S = 0)] = \frac{1}{7}(90) + \frac{6}{7}(90) - 30 = 60.$

Table 20.5 **Application of the Tabular Algorithm for Computing the Posterior Distribution for the Goferbroke Co. Problem**

	(*a*)		(*b*)			(*c*)	
$P(\theta = \theta_i)$	0.25	0.75				Posterior Distribution of θ	
	$P(S = s \mid \theta = \theta_i)$		$P(S = s \mid \theta = \theta_i)P(\theta = \theta_i)$		$P(S = s)$ = Sum	$P(S = s \mid \theta = \theta_i)P(\theta = \theta_i)/P(S = s)$	
	θ_1	θ_2	θ_1	θ_2		θ_1	θ_2
$S = 0$	0.4	0.8	0.10	0.60	0.70	$\frac{1}{7}$	$\frac{6}{7}$
$S = 1$	0.6	0.2	0.15	0.15	0.30	$\frac{1}{2}$	$\frac{1}{2}$

Expected Payoffs if $S = 1$:

For a_1: $$E[p(a_1, \theta \mid S = 1)] = \tfrac{1}{2}(700) + \tfrac{1}{2}(-100) - 30 = 270.$$

For a_2: $$E[p(a_2, \theta \mid S = 1)] = \tfrac{1}{2}(90) + \tfrac{1}{2}(90) - 30 = 60.$$

Since the objective is to maximize the expected payoff, these results yield the optimal policy shown in Table 20.6.

However, what this analysis does not answer is whether it is worth spending $30,000 to conduct the experimentation (the seismic survey). Perhaps it would be better to forgo this major expense and just use the optimal solution without experimentation (drill for oil, with an expected payoff of $100,000). We address this issue next.

The Value of Experimentation

Before performing any experiment, we should determine its potential value. We present two complementary methods of evaluating its potential value.

The first method assumes (unrealistically) that the experiment will remove *all* uncertainty about what the true state of nature is, and then this method makes a very quick calculation of what the resulting *improvement in the expected payoff* would be (ignoring the cost of the experiment). This quantity, called the *expected value of perfect information,* provides an *upper bound* on the potential value of the experiment. Therefore, if this upper bound is less than the cost of the experiment, the experiment definitely should be forgone.

However, if this upper bound exceeds the cost of the experiment, then the second (slower) method should be used next. This method calculates the *actual* improvement in the expected payoff (ignoring the cost of the experiment) that would result from performing the experiment. Comparing this improvement with the cost indicates whether the experiment should be performed.

Expected Value of Perfect Information: Suppose now that the experiment could definitely identify what the true state of nature is, thereby providing "perfect" information. Whichever state of nature is identified, you naturally choose the action with the maximum payoff for that state. We do not know in advance which state of nature will be identified, so a calculation of the expected payoff with perfect information (ignoring the cost of the experiment) requires weighting the maximum payoff for each state of nature by the prior probability of that state of nature.

This calculation is shown in Table 20.7 for the prototype example, where the expected payoff with perfect information is 242.5. Thus, if the Goferbroke Co. could learn before choosing its action whether the land contains oil, the expected payoff as of

Table 20.6 **The Optimal Policy with Experimentation, under Bayes' Decision Rule, for the Goferbroke Co. Problem**

Result of Seismic Survey	Optimal Action	Expected Payoff Excluding Cost of Survey	Expected Payoff Including Cost of Survey
$S = 0$	a_2: Sell the land	90	60
$S = 1$	a_1: Drill for oil	300	270

Table 20.7 **Expected Payoff with Perfect Information for the Goferbroke Co. Problem**

Action \ State of Nature	θ_1 Oil	θ_2 Dry
a_1: Drill for oil	700	−100
a_2: Sell the land	90	90
Maximum payoff	700	90
Prior probability	0.25	0.75

Expected payoff with perfect information
= 0.25(700) + 0.75(90) = 242.5.

now (before acquiring this information) would be $242,500 (excluding the cost of the experiment generating the information).

To evaluate whether the experiment should be conducted, we now use this quantity to calculate the expected value of perfect information.

> The **expected value of perfect information**, abbreviated **EVPI**, is calculated as
>
> EVPI = expected payoff with perfect information − expected payoff without experimentation.[1]
>
> Thus, since experimentation usually cannot provide perfect information, EVPI provides an upper bound on the expected value of experimentation.

For the prototype example, we found in Sec. 20.2 that the expected payoff without experimentation (under Bayes' decision rule) is 100. Therefore,

$$\text{EVPI} = 242.5 - 100 = 142.5.$$

Since 142.5 far exceeds 30, the cost of experimentation (a seismic survey), it may be worthwhile to proceed with the seismic survey. To find out for sure, we now go to the second method of evaluating the potential benefit of experimentation.

Expected Value of Experimentation: Rather than just obtain an upper bound on the *expected increase in payoff* (excluding the cost of the experiment) due to performing experimentation, we now will do somewhat more work to calculate this expected increase directly. This quantity is called the *expected value of experimentation.*

Calculating this quantity requires first computing the expected payoff with experimentation (excluding the cost of the experiment). Obtaining this latter quantity requires doing all the work described earlier to find all the posterior probabilities, the resulting optimal policy with experimentation, and the corresponding expected payoff (excluding the cost of the experiment) for each possible value of S. Then each of these expected payoffs needs to be weighted by the probability of the corresponding value of S, that is,

$$\text{Expected payoff with experimentation} = \sum_s P(S = s)E[\text{payoff} \mid S = s],$$

where the summation is taken over all possible values of $S = s$.

[1] The *value of perfect information* is a random variable equal to the payoff with perfect information *minus* the payoff without experimentation. EVPI is the expected value of this random variable.

For the prototype example, we have already done all the work to obtain the terms on the right side of this equation.

$$P(S = s) = \sum_{j=1}^{n} P(S = s \mid \theta = \theta_j)P(\theta = \theta_j)$$

has been obtained in the rightmost column of Table 20.5*b* for $S = 0$ and $S = 1$. The posterior probabilities shown in Table 20.5*c* were then used to obtain the optimal policy with experimentation given in the first two columns of Table 20.6. Letting a^* denote the optimal action given $S = s$, we obtain

$$E[\text{payoff} \mid S = s] = \sum_{i=1}^{n} P(\theta = \theta_i \mid S = s)p(a^*, \theta_i),$$

as calculated in the third column of Table 20.6 for $S = 0$ and $S = 1$. With these numbers,

$$\begin{aligned}\text{Expected payoff with experimentation} &= 0.70(90) + 0.30(300)\\ &= 153.\end{aligned}$$

Now we are ready to calculate the expected value of experimentation.

The **expected value of experimentation**, abbreviated **EVE**, is calculated as

EVE = expected payoff with experimentation − expected payoff without experimentation.

Thus, EVE identifies the potential value of experimentation.

For the Goferbroke Co.,

$$\text{EVE} = 153 - 100 = 53.$$

Since this value exceeds 30, the cost of conducting a detailed seismic survey (in units of thousands of dollars), this experimentation should be done.

20.4 Decision Trees

Decision trees provide a useful way of *visually displaying* the problem and then *organizing the computational work* already described in the preceding two sections. These trees are especially helpful when a *sequence of decisions* must be made.

The prototype example involves a sequence of two decisions:

1. Should a seismic survey be conducted before an action is chosen?
2. Which action (drill for oil or sell the land) should be chosen?

The corresponding decision tree (before computations are performed) is displayed in Fig. 20.1.

The nodes of the decision tree are referred to as **forks**, and the arcs are called **branches**.

A **decision fork**, represented by a square, indicates that a decision needs to be made at that point in the process. A **chance fork**, represented by a circle, indicates that a random event occurs at that point.

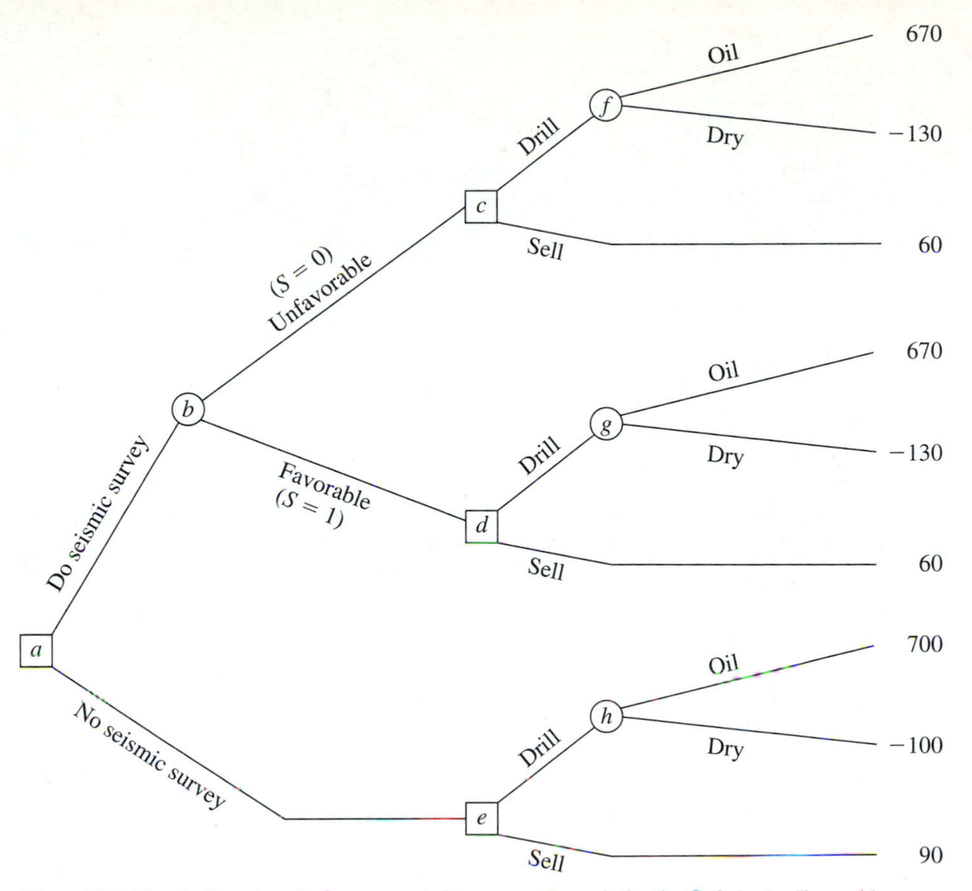

Figure 20.1 The decision tree (before computations are performed) for the Goferbroke Co. problem.

Thus, in Fig. 20.1, the first decision is represented by decision fork *a*. Fork *b* is a chance fork representing the random event of the outcome of the seismic survey. The two branches emanating from fork *b* represent the two possible outcomes of the survey. Next comes the second decision (forks, *c*, *d*, and *e*) with its two possible choices. If the decision is to drill for oil, then we come to another chance fork (forks *f*, *g*, and *h*), where its two branches correspond to the two possible states of nature. To the right of all the terminal branches are the payoffs that would be obtained at that point (including the cost of any experimentation).

Note that the path followed to reach any payoff (except the bottom one) is determined both by the decisions made and by random events that are outside the control of the decision maker. This is characteristic of problems addressed by decision analysis.

Figure 20.1 is useful for visualizing the overall decision process and the possible outcomes. In addition, it can be helpful for summarizing the computations and conclusions obtained while deriving the optimal policy. The final results of this process are shown in Fig. 20.2.

First note in Fig. 20.2 that the probability of each random event has been inserted in parentheses next to the corresponding branch emanating from a chance fork. In

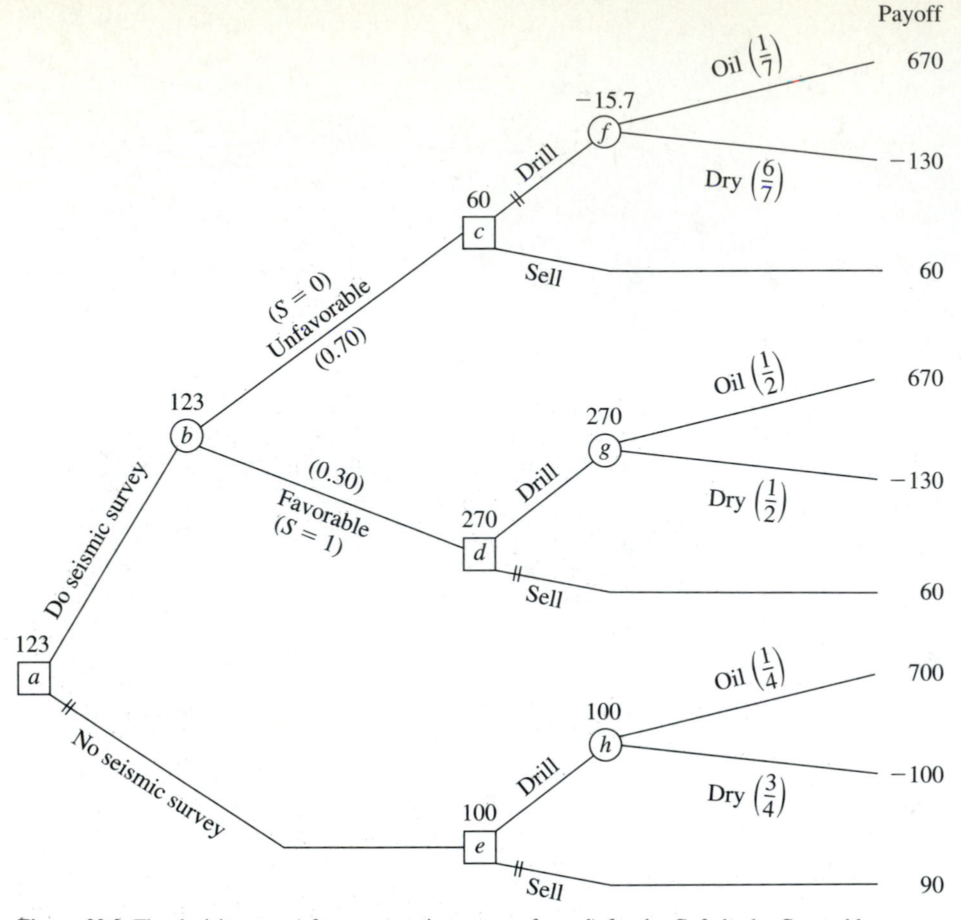

Figure 20.2 The decision tree (after computations are performed) for the Goferbroke Co. problem.

particular, for the terminal branches emanating from fork h, where no experimentation has been done, we have inserted the *prior probabilities* of these states of nature from Table 20.2. For the terminal branches emanating from forks f and g, where experimentation has been done, we have inserted the *posterior probabilities* calculated in Table 20.5c. For branches from fork b that correspond to the seismic survey resulting in $S = 0$ or $S = 1$, we have inserted $P(S = 0) = 0.70$ and $P(S = 1) = 0.30$, as calculated in the rightmost column of Table 20.5b.

Each of the other numbers that have been added to Fig. 20.2 is the *expected payoff* at that point in the process. These expected payoffs are calculated by using the given probabilities, starting the calculations on the right side and then moving left.

In particular, consider the chance forks f, g, and h. By using the given probabilities and payoffs, their expected payoffs are

$$\tfrac{1}{7}(670) + \tfrac{6}{7}(-130) = -15.7, \qquad \text{for fork } f,$$

$$\tfrac{1}{2}(670) + \tfrac{1}{2}(-130) = 270, \qquad \text{for fork } g,$$

$$\tfrac{1}{4}(700) + \tfrac{3}{4}(-100) = 100, \qquad \text{for fork } h,$$

as shown above these forks.

Now consider decision forks c, d, and e. The expected payoffs just calculated are for the action of drilling for oil at the corresponding decision fork. The alternative action of selling the land has a payoff of either 60 or 90. Since $-15.7 < 60$, but $270 > 60$ and $100 > 90$, we now conclude that the optimal action is to *sell* at fork c but *drill* at forks d and e. This conclusion is indicated by inserting a double dash as a barrier through each rejected branch. (This is done only at decision forks.) The expected payoff for the chosen branch now is inserted over the decision fork.

Now consider chance fork b. By using the given probabilities for its two branches and the expected payoffs at the end of these branches, the expected payoff at this chance fork is calculated as

$$0.70(60) + 0.30(270) = 123.$$

We now have worked our way back to the initial decision fork a. The upper branch from this fork leads to an expected payoff of 123. The lower branch leads to a smaller expected payoff of 100, so this branch is rejected. Therefore, the expected payoff at fork a is 123.

Following the open paths from left to right in Fig. 20.2 now yields the following optimal policy.

Optimal policy:
Do the seismic survey.
If the result is unfavorable ($S = 0$), sell the land.
If the result is favorable ($S = 1$), drill for oil.
The expected payoff (including the cost of experimentation) is 123.

This (unique) optimal solution naturally is the same as that obtained in the preceding section without the benefit of a decision tree. (See the optimal policy with experimentation given in Table 20.6 and the conclusion at the end of Sec. 20.3 that experimentation is worthwhile.)

For any decision tree, this **backward induction procedure** always will lead to the *optimal policy* (or policies) after the probabilities are computed for the branches emanating from a chance fork.

20.5 Utility Theory

Thus far, when applying Bayes' decision rule, we have assumed that the expected payoff in *monetary terms* is the appropriate measure of the consequences of taking an action. However, in many situations this assumption is inappropriate.

For example, suppose that an individual is offered the choice of (1) accepting a 50:50 chance of winning $100,000 or nothing or (2) receiving $40,000 with certainty. Many people would prefer the $40,000 even though the expected payoff on the 50:50 chance of winning $100,000 is $50,000. A company may be unwilling to invest a large sum of money in a new product even when the expected profit is substantial, if there is a risk of losing its investment and thereby becoming bankrupt. People buy insurance even though it is a poor investment from the viewpoint of the expected payoff.

Do these examples invalidate Bayes' decision rule? Fortunately, the answer is no, because there is a way of transforming *monetary values* to an appropriate scale that reflects the decision maker's preferences. This scale is called the *utility function for money*.

Utility Functions for Money

Figure 20.3 shows a typical utility function $u(M)$ for money M. It indicates that an individual having this utility function would value obtaining $30,000 twice as much as $10,000 and would value obtaining $100,000 twice as much as $30,000. This reflects the fact that the person's highest-priority needs would be met by the first $10,000. Having this decreasing slope of the function as the amount of money increases is referred to as having a **decreasing marginal utility for money**.

However, not all individuals have a decreasing marginal utility for money. Some people are *risk seekers* instead of *risk-averse*, and they go through life looking for the ''big score.'' The slope of their utility function *increases* as the amount of money increases, so they have an **increasing marginal utility for money**.

The fact that different people have different utility functions for money has an important implication for decision making in the face of uncertainty.

> When a *utility function for money* is incorporated into a decision analysis approach to a problem, this utility function must be constructed to fit the preferences and values of the decision maker involved. (The decision maker can be either a single individual or a group of people.)

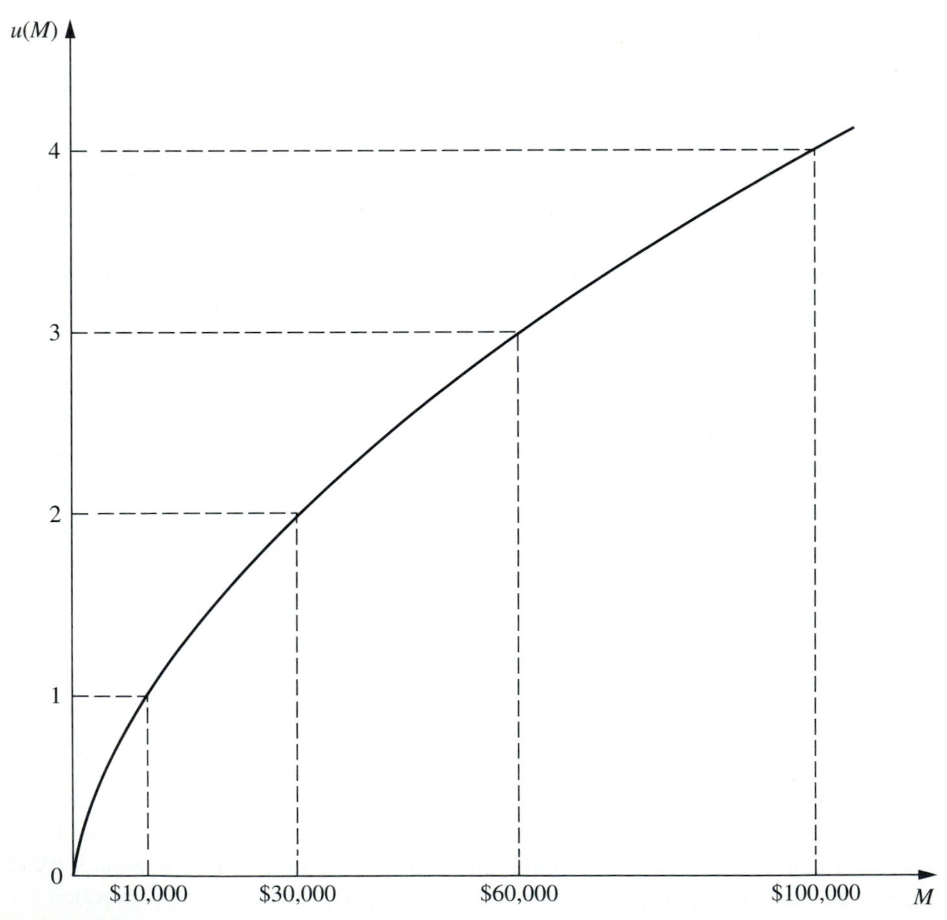

Figure 20.3 A typical utility function for money, where $u(M)$ is the utility of obtaining an amount of money M.

The key to constructing the utility function for money to fit the decision maker is the following fundamental property of utility functions.

> Under the assumptions of utility theory, the decision maker's *utility function for money* has the property that the decision maker is *indifferent* between two alternative courses of action if the two alternatives have the *same expected utility.*

To illustrate, suppose that the decision maker has the utility function shown in Fig. 20.3. Further suppose that the decision maker is offered an opportunity to obtain either \$100,000 (utility = 4) with probability p or nothing (utility = 0) with probability $1 - p$. *Thus,*

$$E(\text{utility}) = 4p, \qquad \text{for this offer.}$$

Therefore, the decision maker is indifferent between each of the following three pairs of alternatives:

1. The offer with $p = 0.25$ [E(utility) = 1] or definitely obtaining \$10,000 (utility = 1)
2. The offer with $p = 0.5$ [E(utility = 2] or definitely obtaining \$30,000 (utility = 2)
3. The offer with $p = 0.75$ [E(utility) = 3] or definitely obtaining \$60,000 (utility = 3)

This example also illustrates how the decision maker's utility function for money can be constructed in the first place. The decision maker would be made the same hypothetical offer to obtain either a large amount of money (for example, \$100,000) with probability p or nothing. Then, for each of a few smaller amounts of money (for example, \$10,000, \$30,000, and \$60,000), the decision maker would be asked to choose a value of p that would make him or her *indifferent* between the offer and definitely obtaining that amount of money.

The *scale* of the utility function (e.g., utility = 1 for \$10,000) is irrelevant. It is only the *relative values* of the utilities that matter. All the utilities can be multiplied by any positive constant without affecting which alternative course of action will have the largest expected utility.

Now we are ready to summarize the basic role of utility functions in decision analysis.

> When the decision maker's utility function for money is used to measure the relative worth of the various possible monetary outcomes, *Bayes' decision rule* replaces monetary payoffs by the corresponding utilities. Therefore, the optimal action (or series of actions) is the one which *maximizes the expected utility.*

Only utility functions *for money* have been discussed here. However, we should mention that utility functions can sometimes still be constructed when some of or all the important consequences of the alternative courses of action are *not* monetary. This is not necessarily easy, since it may require making value judgments about the relative desirability of rather intangible consequences. Nevertheless, under these circumstances, it is important to incorporate such value judgments into the decision process.

Applying Utility Theory to the Prototype Example

At the end of Sec. 20.1, we mentioned that the Goferbroke Co. was operating without much capital, so a loss of \$100,000 would be quite serious. The (primary) owner of the company already has gone heavily into debt to keep going. The worst-case scenario would be to come up with \$30,000 for a seismic survey and then still lose \$100,000 by drilling when there is no oil. This scenario would not bankrupt the company at this point, but definitely would leave it in a precarious financial position.

On the other hand, striking oil is an exciting prospect, since earning $700,000 finally would put the company in a fairly solid financial footing.

To apply the owner's (decision maker's) *utility function for money* to the problem as described in Secs. 20.1 and 20.3, it is necessary to identify the utilities for all the possible monetary payoffs. Again, in units of thousands of dollars, these possible payoffs and the corresponding utilities are given in Table 20.8. We now will discuss how these utilities were obtained.

The appropriate starting point for constructing the utility function is to consider the worst scenario and best scenario and then to address the following question.

> Suppose you have only the following two alternatives. Alternative 1 is to do nothing (payoff and utility = 0). Alternative 2 is to have a probability p of a payoff of 700 and a probability $1 - p$ of a payoff of -130 (loss of 130). What value of p makes you *indifferent* between two alternatives?
>
> The decision maker's choice: $p = \frac{1}{5}$.

If we continue to let $u(M)$ denote the utility of a monetary payoff of M, this choice of p implies that

$$\tfrac{4}{5}u(-130) + \tfrac{1}{5}u(700) = 0 \qquad \text{(utility of alternative 1).}$$

The value of either $u(-130)$ or $u(700)$ can be set arbitrarily (provided only that the first is negative and the second positive) to establish the scale of the utility function. By choosing $u(-130) = -150$, this equation then yields $u(700) = 600$.

To identify $u(-100)$, a choice of p is made that makes the decision maker indifferent between a payoff of -135 with probability p or definitely incurring a payoff of -100. The choice is $p = 0.7$, so

$$u(-100) = p\,u(-130) = 0.7(-150) = -105.$$

To obtain $u(90)$, a value of p is selected that makes the decision maker indifferent between a payoff of 700 with probability p or definitely obtaining a payoff of 90. The value chosen is $p = 0.15$, so

$$u(90) = p\,u(700) = 0.15(600) = 90.$$

At this point, a smooth curve was drawn through $u(-130)$, $u(-100)$, $u(90)$, and $u(700)$ to obtain the decision maker's *utility function for money* shown in Fig. 20.4. The values on this curve at $M = 60$ and $M = 670$ provide the corresponding utilities, $u(60) = 60$ and $u(670) = 580$, which completes the list of utilities given in the right column of Table 20.8. For contrast, the dashed line drawn at 45° in Fig. 20.4 shows the monetary value M of the amount of money M. This dashed line has provided the values of the payoffs used exclusively in the preceding sections. Note how $u(M)$ essentially equals M for small values (positive or negative) of M, and then how $u(M)$ gradually falls off M for larger values of M. This is typical for a moderately risk-averse individual.

By nature, the owner of the Goferbroke Co. is inclined to be a risk seeker. However, the difficult financial circumstances of his company, which he badly wants to keep solvent, have forced him to adopt a moderately risk-averse stance in addressing his current decisions.

Given the owner's current utility function for money, the decision-making process now is identical to that described in the preceding sections *except* for substituting *utilities* for *monetary payoffs*. Thus, our final decision tree shown in Fig. 20.5 closely resembles the one in Fig. 20.2 given in the preceding section. The forks and branches

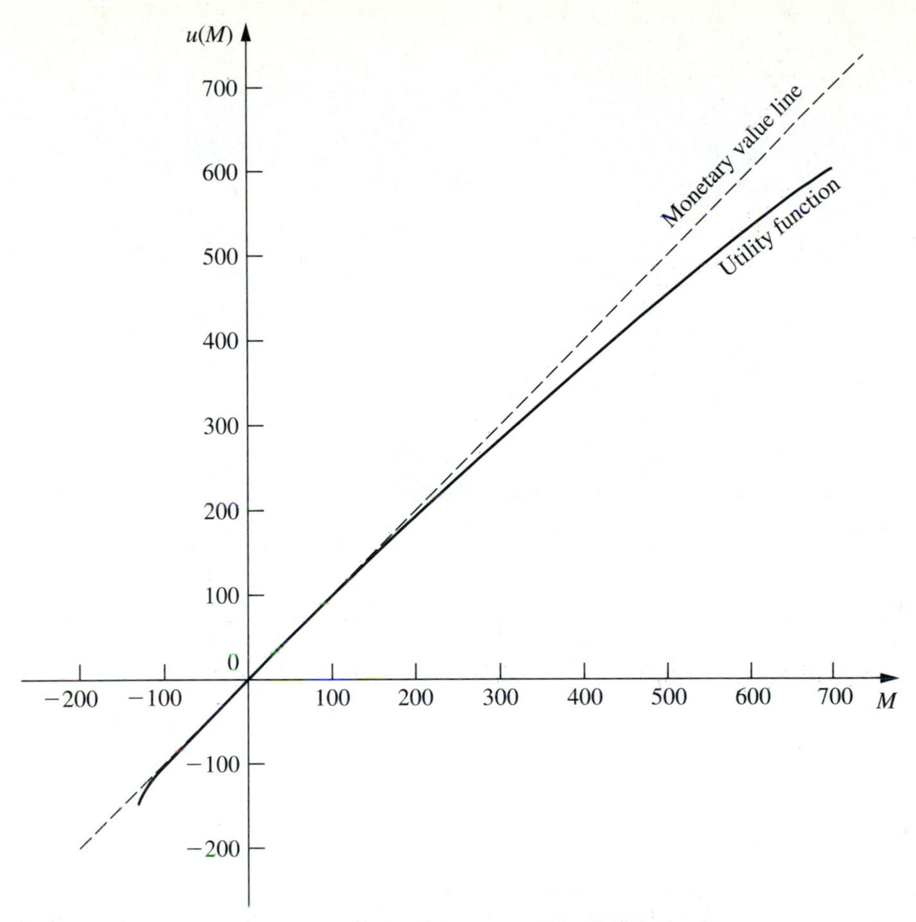

Figure 20.4 The utility function for money of the owner of the Goferbroke Co.

are exactly the same, as are the probabilities for the branches emanating from the chance forks. For informational purposes, the monetary payoffs still are given to the right of the terminal branches. However, we now have added the utilities on the right side. It is these numbers that have been used to compute the expected utilities given next to all the forks.

These expected utilities lead to the same decisions at forks *a*, *c*, and *d* as in Fig. 20.2, but the decision at fork *e* now switches to *sell* instead of *drill*. However, the backward induction procedure still leaves fork *e* on a *closed* path. Therefore, the over-

Table 20.8 **Utilities for the Goferbroke Co. Problem**

Monetary Payoff	Utility
−130	−150
−100	−105
60	60
90	90
670	580
700	600

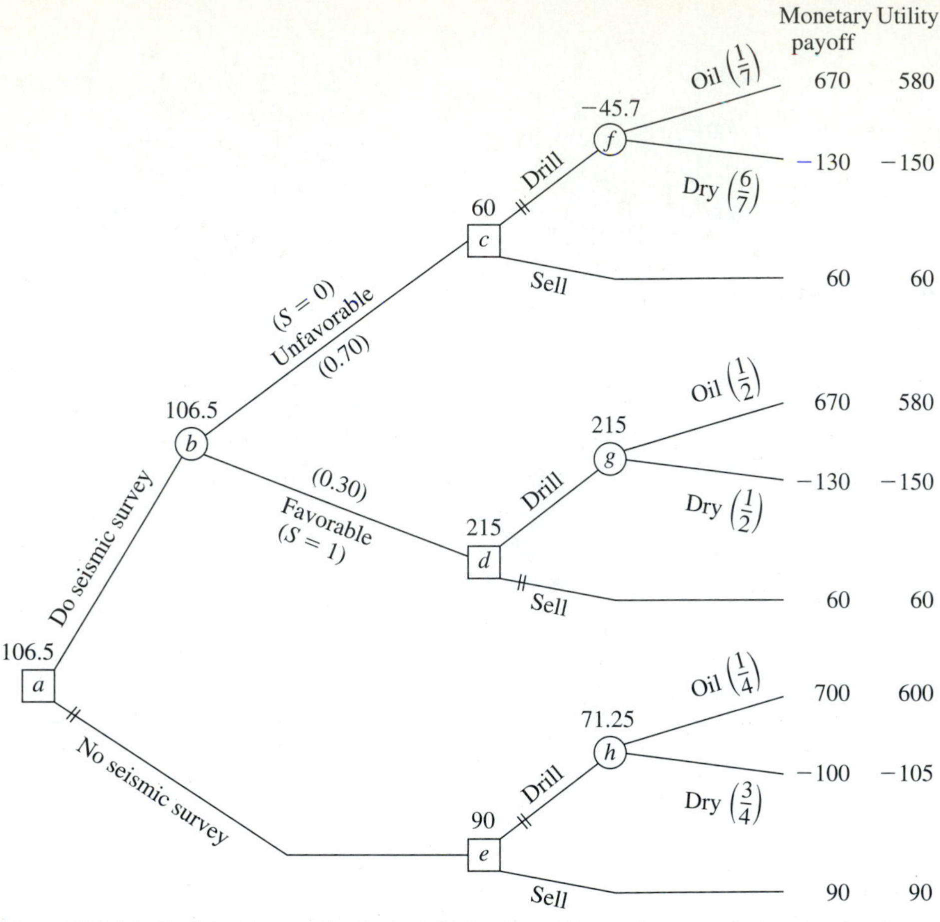

Figure 20.5 The final decision tree for the Goferbroke Co. problem, using the owner's utility function for money to maximize expected utility.

all optimal policy remains the same as given at the end of Sec. 20.4 (do the seismic survey; sell if the result is unfavorable; drill if the result is favorable).

The approach used in the preceding sections of maximizing the expected monetary payoff amounts to assuming that the decision maker is neutral toward risk, so that $u(M) = M$. By using utility theory, the optimal solution now reflects the decision maker's attitude about risk. Because the owner of the Goferbroke Co. adopted only a moderately risk-averse stance, the optimal policy did not change from before. For a somewhat more risk-averse owner, the optimal solution would switch to the more conservative approach of immediately selling the land (no seismic survey). (See Prob. 20.5-1.)

The current owner is to be commended for incorporating utility theory into a decision analysis approach to his problem. Utility theory helps to provide a rational approach to decision making in the face of uncertainty. However, many decision makers are not sufficiently comfortable with the relatively abstract notion of utilities, or with working with probabilities to construct a utility function, to be willing to use this approach. Consequently, utility theory is not yet used very widely in practice.

20.6 Conclusions

Decision analysis has become an important technique for decision making in the face of uncertainty. It is characterized by enumerating all the available courses of action, identifying the payoffs for all possible outcomes, and quantifying the subjective probabilities for all the possible random events. When these data are available, decision analysis becomes a powerful tool for determining an optimal course of action.

One option that can be readily incorporated into the analysis is to perform experimentation to obtain better estimates of the probabilities of the possible states of nature. Decision trees are a useful visual tool for analyzing this option or any series of decisions.

Utility theory provides a way of incorporating the decision maker's attitude toward risk into the analysis.

Good software is becoming widely available for performing decision analysis. (See Selected References 1 and 8 for further information.)

SELECTED REFERENCES

1. Buede, D., ''Aiding Insight II,'' *OR/MS Today,* June 1994, pp. 62–71. A survey of decision analysis software.
2. Bunn, D. W., *Applied Decision Analysis,* McGraw-Hill, New York, 1984.
3. Fishburn, P. C., ''Foundations of Decision Analysis: Along the Way,'' *Management Science,* **35:** 387–405, 1989.
4. Fishburn, P. C., *Nonlinear Preference and Utility Theory,* The Johns Hopkins Press, Baltimore, MD, 1988.
5. Goodwin, P., and G. Wright, *Decision Analysis for Management Judgment,* Wiley, New York, 1991.
6. Holloway, C. A., *Decision Making under Uncertainty,* Prentice-Hall, Englewood Cliffs, NJ, 1979.
7. Keeney, R. L., and H. Raiffa, *Decisions with Multiple Objectives,* Wiley, New York, 1976.
8. McNamee, P., and J. Celona, *Decision Analysis with SUPERTREE,* 2d ed., Scientific Press, South San Francisco, 1991.

RELEVANT ROUTINES IN YOUR OR COURSEWARE

Automatic Routines: *Enter Prior Distribution*
Solve for Posterior Distribution

To access these routines, call the ProbMod program and then choose *Decision Analysis* under the Area menu. See App. 1 for documentation of the software.

PROBLEMS[1]

To the left of each of the following problems (or their parts), we have inserted an A (for automatic routine) whenever a corresponding routine listed above can be helpful for at least

[1] Problems 20.2-2, 20.2-3, 20.2-5 to 20.2-7, 20.3-2, 20.3-3, 20.3-7 to 20.3-9, 20.3-13, 20.3-14, 20.4-2, 20.4-3, 20.4-7 to 20.4-9, 20.4-11 to 20.4-13, 20.4-15, 20.5-2 to 20.5-4, 20.5-6, and 20.5-7 have been adapted, with permission, from previous operations research examinations given by the Society of Actuaries.

checking your work. An asterisk on the A indicates that this routine definitely should be used (unless your instructor gives contrary instructions) and that the printout from this routine is all that needs to be turned in to show your work in executing the procedure. An asterisk on the problem number indicates that at least a partial answer is given in the back of the book.

20.2-1.* A company has developed a new chip that will enable it to enter the microcomputer field if it so desires. Alternatively, it can sell the rights to it for $800,000. If the company chooses to build computers, the profitability of the venture depends upon the company's ability to market the microcomputer during the first year. It has sufficient access to retail outlets that it can guarantee sales of 1,000 computers. On the other hand, if this computer catches on, the company can sell 10,000 machines. The company believes that both sales alternatives are equally likely and that all other alternatives are negligible. The cost of setting up the assembly line is $600,000. The difference between the selling price and the variable cost of each computer is $600.

(*a*) Develop a decision analysis formulation of this problem by identifying the actions, states of nature, and payoff table.
(*b*) Determine the optimal action under Bayes' decision rule.

20.2-2. A university must decide between two plans for starting a graduate program in a new academic area. The goal is to maximize the increase in student population, but it is unclear whether the interest in this new area will be high, medium, or low. Projected increases in student populations and their probabilities are shown below for each plan.

		Increase	
Interest	Probability	Plan 1	Plan 2
High	0.6	220	200
Medium	0.3	170	180
Low	0.1	110	150

(*a*) Develop a decision analysis formulation of this problem by identifying the actions, states of nature, and payoff table.
(*b*) What is the optimal action under the minimax criterion?
(*c*) What is the optimal action under the maximum likelihood criterion?
(*d*) What is the optimal action under Bayes' decision rule?

20.2-3. You sell strawberries in a market. You do not know in advance the number of cases of strawberries that will be sold tomorrow, but you must decide today how many cases to stock tomorrow in order to maximize profits. You can purchase strawberries for $3 per case and sell them for $8 per case. Because of their perishability, strawberries have no value after the first day that they are offered for sale.

An analysis of past sales has given you the following information:

Cases Sold Daily	Probability
10	0.2
11	0.4
12	0.3
13	0.1

(*a*) Develop a decision analysis formulation of this problem by identifying the actions, states of nature, and payoff table.
(*b*) What is the optimal action under the maximin payoff criterion?
(*c*) What is the optimal action under the maximum likelihood criterion?
(*d*) What is the optimal action under Bayes' decision rule?

20.2-4.* A new type of airplane is to be purchased by the Air Force, and the number of spare engines to be ordered must be determined. The Air Force must order these spare engines in batches of five, and it can choose among only 15, 20, or 25 spares. The supplier of these engines has two plants, and the Air Force must make its decision prior to knowing which plant will be used. However, the Air Force knows from past experience that two-thirds of all types of airplane engines are produced in Plant *A*, and only one-third are produced in Plant *B*. The Air Force also knows that the number of spare engines required when production takes place at Plant *A* is approximated by a Poisson distribution with mean $\theta = 21$, whereas the number of spare engines required when production takes place at Plant *B* is approximated by a Poisson distribution with mean $\theta = 24$. The cost of a spare engine purchased now is $400,000, whereas the cost of a spare engine purchased at a later date is $900,000. Spares must always be supplied if they are demanded, and unused engines will be scrapped when the airplanes become obsolete. Holding costs and interest are to be neglected. From these data, the total cost (negative payoffs) have been computed as follows:

Action \ State of Nature	θ_1	θ_2
	$\theta = 21$	$\theta = 24$
a_1: Order 15	1.155×10^7	1.414×10^7
a_2: Order 20	1.012×10^7	1.207×10^7
a_3: Order 25	1.047×10^7	1.135×10^7

Determine the optimal action under Bayes' decision rule.

20.2-5. A large bank is going to make one of three investments. The economy will have one of three possible states during the life of the investment: improve (E_1), remain stable (E_2), or worsen (E_3). The trust officer believes the respective probabilities are 0.1, 0.5, and 0.4. The estimated payoff table is as follows:

Investment	E_1	E_2	E_3
a_1	30	5	−10
a_2	40	10	−30
a_3	−10	0	15

(*a*) What is the optimal choice of investment under the maximin payoff criterion?
(*b*) What is the optimal choice of investment under the maximum likelihood criterion?
(*c*) What is the optimal choice of investment under Bayes' decision rule? What is the resulting expected payoff?
(*d*) What would be the expected payoff if the choice of investment were postponed (with no change in the payoffs) until a definitely accurate prediction could be made of the state of the economy?

20.2-6. You are deciding which one of four crops to grow on 1,000 acres of land. You are given the following estimate of crop yields and your prices per bushel under various weather conditions:

Weather	*Expected Yield, Bushels/Acre*			
	Crop 1	Crop 2	Crop 3	Crop 4
Dry	20	15	30	40
Moderate	35	20	25	40
Damp	40	30	25	40
Price per bushel, $	1.00	1.50	1.00	0.50

You also are given the following probabilities for the weather:

Dry	0.3
Moderate	0.5
Damp	0.2

(*a*) Use Bayes' decision rule to determine which crop to grow.
(*b*) Use the maximin payoff criterion to determine which crop to grow.
(*c*) Calculate the difference in expected revenues from the decisions made in parts (*a*) and (*b*).

20.2-7. An individual makes decisions according to Bayes' decision rule. For her current problem, she has constructed the following payoff table, and she now wishes to maximize the expected payoff.

Action \ State of Nature	*Payoff*		
	θ_1	θ_2	θ_3
a_1	$2x$	50	10
a_2	25	40	90
a_3	35	$3x$	30
Prior probability	0.4	0.2	0.4

The value of x currently is 50, but there is an opportunity to increase x by spending some money now.

What is the maximum amount that should be spent to increase x to 75?

20.3-1.* Reconsider Prob. 20.2-1. Suppose now that market research can be performed at a cost of $400,000 to predict which of the two levels of demand is likely to occur. Previous experience indicates that such market research is correct two-thirds of the time.

(*a*) Find EVPI.
A (*b*) Use the tabular algorithm illustrated in Table 20.5 to find the posterior probabilities of the two levels of demand for each of the two possible outcomes of the market research.
(*c*) Determine the optimal policy with experimentation.
(*d*) Find EVE. Is it worthwhile to perform the market research?

20.3-2. You are given the following payoff table.

Action \ State of Nature	Payoff θ_1	θ_2
a_1	400	-100
a_2	0	100
Prior probability	0.4	0.6

You have the option of paying 100 to have research done to better predict which state of nature will occur. The probabilities of an accurate prediction are $P(\text{predict } \theta_1 \mid \theta_1) = 0.6$ and $P(\text{predict } \theta_2 \mid \theta_2) = 0.8$.

A* (*a*) Given that research is done, calculate by hand the posterior probabilities of the states of nature. Then use your OR Courseware to check your answer.

(*b*) Determine the optimal policy regarding whether to do research and which action to take. What is the resulting expected payoff?

20.3-3. You are given the opportunity to guess whether a coin is fair or two-headed, where the prior probabilities are 0.5 for each of these possibilities. If you are correct, you win $5; otherwise, you lose $5. You are also given the option of seeing a demonstration flip of the coin before making your guess. You wish to use Bayes' decision rule to maximize expected profit.

(*a*) Develop a decision analysis formulation of this problem by identifying the actions, states of nature, and payoff table.

(*b*) What is the optimal action, given that you decline the option of seeing a demonstration flip?

(*c*) Find EVPI.

(*d*) Use the procedure presented in Sec. 20.3 to calculate the posterior distribution if the demonstration flip is a tail. Do the same if the flip is a head.

A* (*e*) Use your OR Courseware to confirm your results in part (*c*).

(*f*) Determine your optimal policy.

(*g*) Now suppose that you must pay to see the demonstration flip. What is the most that you should be willing to pay?

20.3-4.* Reconsider Prob. 20.2-4. Suppose now that the Air Force knows that a similar type of engine was produced for an earlier version of the type of airplane currently under consideration. The order size for this earlier version was the same as for the current type. Furthermore, the probability distribution of the number of spare engines required, given the plant where production takes place, is believed to be the same for this earlier airplane model and the current one. The engine for the current order will be produced in the same plant as the previous model, although the Air Force does not know which of the two plants this is. The Air Force does have access to the data on the number of spares actually required for the older version, but the supplier has not revealed the production location.

(*a*) How much money is it worthwhile to pay for perfect information on which plant will produce these engines?

A (*b*) Assume that the cost of the data on the old airplane model is free and that 30 spares were required. You are given that the probability of 30 spares, given a Poisson distribution with mean θ, is 0.013 for $\theta = 21$ and 0.036 for $\theta = 24$. Find the optimal action under Bayes' decision rule.

20.3-5. A large mill is faced with the question of whether to extend $100,000 credit to a new customer, a dress manufacturer. The mill classifies typical companies into the following categories: poor risk, average risk, and good risk. Experience indicates that 20 percent of similar companies are poor risks, 50 percent are average risks, and 30 percent are good risks. If credit is extended, the expected profit for poor risks is −$15,000, for average risks $10,000, and for good risks $20,000. If credit is not extended, the dress manufacturer will turn to another mill. The mill is able to consult a credit-rating organization for a fee of $5,000 per company evaluated. The mill's experience with this credit-rating organization's evaluation of companies whose actual credit record with the mill turns out to fall into each of the three categories is as follows:

	Actual Credit Record, %		
Credit Evaluation	Poor	Average	Good
Poor	50	40	20
Average	40	50	40
Good	10	10	40

(*a*) Develop a decision analysis formulation of this problem by identifying the possible actions and states of nature and then constructing the payoff table.
(*b*) What is the optimal action under Bayes' decision rule, assuming the credit-rating organization is not used?
(*c*) Find EVPI.
A* (*d*) Assume now that the credit-rating organization is used to evaluate this dress manufacturer. Use your OR Courseware to find the posterior distribution of the state of nature for each possible evaluation of the dress manufacturer.
(*e*) Determine the optimal policy with experimentation.
(*f*) Find EVE. Is it worthwhile to use the credit-rating organization in this case?

20.3-6. A manufacturer produces items that have a probability p of being defective. These items are formed in lots of 150. Past experience indicates that p is either 0.05 or 0.25. Furthermore, in 80 percent of the lots produced, p equals 0.05 (so p equals 0.25 in 20 percent of the lots). These items are then used in an assembly, and ultimately their quality is determined before the final assembly leaves the plant. Initially the manufacturer can *either* screen each item in a lot at a cost of $10 per item and replace defective items *or* use the items directly without screening. If the latter action is chosen, the cost of rework is ultimately $100 per defective item. From these data, the expected costs per lot can be calculated as follows:

	$p = 0.05$	$p = 0.25$
Screen	1,500	1,500
Do not screen	750	3,750

Because screening requires scheduling of inspectors and equipment, the decision to screen or not screen must be made 2 days before the screening is to take place. However, one item can be taken from the lot and sent to a laboratory, and its quality (defective or nondefective) can be reported before the screen/no-screen decision must be made. The cost of this initial inspection is $125.
(*a*) What is the optimal action under Bayes' decision rule without looking at the single item?
(*b*) Find EVPI.

A (*c*) Assume now that the single item is inspected. For each of the two possible outcomes, find the posterior probabilities for $p = 0.05$ and $p = 0.25$ *without* using your OR Courseware (except perhaps to check your answer).

(*d*) Determine the optimal policy with experimentation.

(*e*) Find EVE. Is inspecting the single item worthwhile?

20.3-7. You are given the following payoff table.

Action \ State of Nature	Payoff θ_1	θ_2	θ_3
a_1	4	0	0
a_2	0	2	0
a_3	3	0	1
Prior probability	0.2	0.5	0.3

(*a*) Determine the optimal action under Bayes' decision rule.

(*b*) Find EVPI.

20.3-8. An individual makes decisions according to Bayes' decision rule. For her current problem, she has constructed the following payoff table.

Action \ State of Nature	Payoff θ_1	θ_2	θ_3
a_1	50	100	−100
a_2	0	10	−10
a_3	20	40	−40
Prior probability	0.5	0.3	0.2

(*a*) Determine the optimal action.

(*b*) Find EVPI.

20.3-9. Using Bayes' decision rule, consider the decision analysis problem having the sates of nature, probabilities, actions, and payoffs shown below.

Action \ State of Nature	Payoff θ_1	θ_2	θ_3
a_1	−100	10	100
a_2	−10	20	50
a_3	10	10	60
Prior probability	0.2	0.3	0.5

(*a*) What are the optimal action and the resulting expected payoff?
(*b*) You are offered the opportunity to obtain information which will tell you with certainty whether θ_1 will occur. What is the maximum amount you should pay for the information? Assuming you will obtain the information, what are the optimal policy and the resulting expected payoff (excluding the payment)?
(*c*) Now repeat part (*b*) if the information offered concerns θ_2 instead of θ_1.
(*d*) Now repeat part (*b*) if the information offered concerns θ_3 instead of θ_1.
(*e*) Now suppose that the opportunity is offered to provide information which will tell you with certainty which state of nature will occur (perfect information). What is the maximum amount you should pay for the information? Assuming you will obtain the information, what are the optimal policy and the resulting expected payoff (excluding the payment)?
(*f*) If you have the opportunity to do some testing that will give you partial additional information (not perfect information) about the state of nature, what is the maximum amount you should consider paying for this information?

20.3-10.* Consider two weighted coins. Coin 1 has a probability of 0.3 of turning up heads, and coin 2 has a probability of 0.6 of turning up heads. A coin is tossed once; the probability that coin 1 is tossed is 0.6, and the probability that coin 2 is tossed is 0.4. The decision maker uses Bayes' decision rule to decide which coin is tossed. The payoff table is as follows:

	Coin 1 Tossed	Coin 2 Tossed
a_1: Say coin 1 tossed	0	−1
a_2: Say coin 2 tossed	−1	0

(*a*) What is the optimal action (a_1 or a_2) before the coin is tossed?
A (*b*) What is the optimal action after the coin is tossed if the outcome is heads? If it is tails?

20.3-11. A new type of photographic film has been developed. It is packaged in sets of five sheets, where each sheet provides an instantaneous snapshot. Because this process is new, the manufacturer has attached an additional sheet to the package, so that the store may test one sheet before it sells the package of five. In promoting the film, the manufacturer offers to refund the entire purchase price of the film if one of the five is defective. This refund must be paid by the camera store, and the selling price has been fixed at \$2 if this guarantee is to be valid. The camera store may sell the film for \$1 if the preceding guarantee is replaced by one that pays \$0.20 for each defective sheet. The cost of the film to the camera store is \$0.40, and the film is not returnable. The store may take three actions:

a_1: Scrap the film.
a_2: Sell the film for \$2.
a_3: Sell the film for \$1.

(*a*) If the six states of nature correspond to 0, 1, 2, 3, 4, and 5 defective sheets in the package, complete the following payoff table:

a \ θ	0	1	2	3	4	5
a_1	−0.40					
a_2	1.60		−0.40			
a_3	0.60	0.40		0.00		

(*b*) The store has accumulated the following information on sales of 60 such packages:

Quality of Attached Sheet	*Defectives in Package*					
	0	1	2	3	4	5
Good	10	8	6	4	2	0
Bad	0	2	4	6	8	10
Total	10	10	10	10	10	10

These data indicate that each state of nature is equally likely, so that this prior distribution can be assumed. What is the optimal action under Bayes' decision rule (before the attached sheet is tested) for a package of film?

A (*c*) Now assume that the attached sheet is tested. Use the tabular algorithm illustrated in Table 20.5 to find the posterior probabilities of the state of nature for each of the two possible outcomes of this testing.

(*d*) What is the optimal expected payoff for a package of film if the attached sheet is tested? What is the optimal action if the sheet is good? If it is bad?

20.3-12. There are two biased coins with probabilities of landing heads of 0.8 and 0.4, respectively. One coin is chosen at random (each with probability $\frac{1}{2}$) to be tossed twice. You are to receive $100 if you correctly predict how many heads will occur in two tosses.

(*a*) Using Bayes' decision rule, what is the optimal prediction, and what is the corresponding expected payoff?

A* (*b*) Suppose now that you may observe a practice toss of the chosen coin before predicting. Use your OR Courseware to find the posterior probabilities for which coin is being tossed.

(*c*) Determine your optimal policy for predicting after observing the practice toss. What is the resulting expected payoff?

(*d*) Find EVE for observing the practice toss.

20.3-13. A company is considering developing and marketing a new product. It is estimated to be twice as likely that the product would prove to be successful as unsuccessful. If it were successful, the expected profit would be $1,500,000. If unsuccessful, the expected loss would be $1,800,000. A marketing survey can be conducted at a cost of $300,000 to predict whether the product would be successful. Past experience with such surveys indicates that successful products have been predicted to be successful 80 percent of the time, whereas unsuccessful products have been predicted to be unsuccessful 70 percent of the time.

(*a*) What is the optimal action under Bayes' decision rule, assuming that the marketing survey is not conducted?

(*b*) Find EVPI.

(*c*) Use the procedure presented in Sec. 20.3 to calculate the posterior distribution of whether the product would be successful if the marketing survey predicts success. Do the same if the survey predicts that the product would be unsuccessful.

A* (*d*) Use your OR Courseware to confirm your results in part (*c*).

(*e*) Find the optimal policy regarding whether to conduct the market survey and whether to develop and market this new product.

A **20.3-14.** An athletic league is considering testing its athletes for drugs, 10 percent of whom use drugs. The test, however, is only 95 percent reliable. That is, a drug user will test positive with probability 0.95 and negative with probability 0.05, and a nonuser will test negative with probability 0.95 and positive with probability 0.05. Whenever anyone tests positive, further investigation is done to determine whether the individual actually is a drug user.

The league fines every drug user who tests positive. The league must pay $1 to each athlete who tests negative and $100 to each nonuser who falsely tests positive.

Determine the amount of the fine such that the expected value of fines equals the expected value of payments to athletes.

20.4-1. Use the scenario given in Prob. 20.3-1.

(*a*) Draw and properly label the decision tree. Include the payoff for each of the terminal branches.

A* (*b*) Find the probabilities for the branches emanating from the chance forks.

(*c*) Apply the backward induction procedure, and identify the resulting optimal policy.

20.4-2. You are given the following decision tree, where the numbers in parentheses are probabilities and the numbers on the right are payoffs at these terminal points.

Apply the backward induction procedure, and identify the resulting optimal policy.

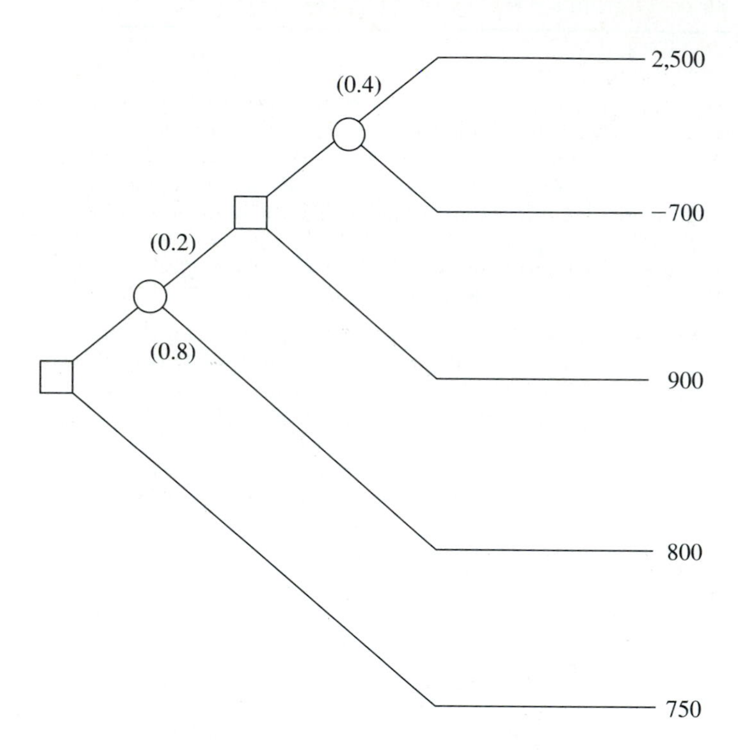

20.4-3. Use the scenario given in Prob. 20.3-3.

(*a*) Draw and properly label the decision tree. Include the payoff for each of the terminal branches.

A* (*b*) Find the probabilities for the branches emanating from the chance forks.

(*c*) Apply the backward induction procedure, and identify the resulting optimal policy.

20.4-4.* A private university is considering whether to hold an extensive centennial campaign next fall to raise funds for a new athletic field. The response to the campaign depends heavily upon the success of the varsity baseball team this spring. If the baseball team has a winning season (W), then many of the alumnae and alumni will contribute and the campaign will raise \$3 million. If the team has a losing season (L), few will contribute and the campaign will lose \$2 million. If no campaign is undertaken, no costs are incurred.

(*a*) In the past, the baseball team has had winning seasons 60 percent of the time. How much should the university be willing to pay for perfect information about whether the baseball team will have a winning season?

A (*b*) Now there is the possibility of hiring a professional talent scout to look at the baseball team before February 10, the date when both the baseball season begins and the decision must be made of whether to hold the campaign. For \$100,000, this talent scout predicts what kind of season, W or L, a team will have, and is correct three-team's season, given each of the two possible predictions by the scout.

(*c*) Use a decision tree to determine the optimal policy for what the university should do. What is the expected payoff for this optimal policy?

20.4-5. Use the scenario given in Prob. 20.3-5.

(*a*) Draw and properly label the decision tree. Include the payoff for each of the terminal branches.

A* (*b*) Find the probabilities for the branches emanating from the chance forks.

(*c*) Apply the backward induction procedure, and identify the resulting optimal policy.

20.4-6. Use the scenario given in Prob. 20.3-6.

(*a*) Draw and properly label the decision tree. Include the payoff for each of the terminal branches.

A* (*b*) Find the probabilities for the branches emanating from the chance forks.

(*c*) Apply the backward induction procedure, and identify the resulting optimal policy.

20.4-7. You are given the following decision tree, with the probabilities at chance forks shown in parentheses and with the payoffs at terminal points shown on the right.

(*a*) Apply the backward induction procedure to determine the optimal policy.

(*b*) Adapt the procedure for evaluating this decision tree to determine the optimal policy under the maximin payoff criterion.

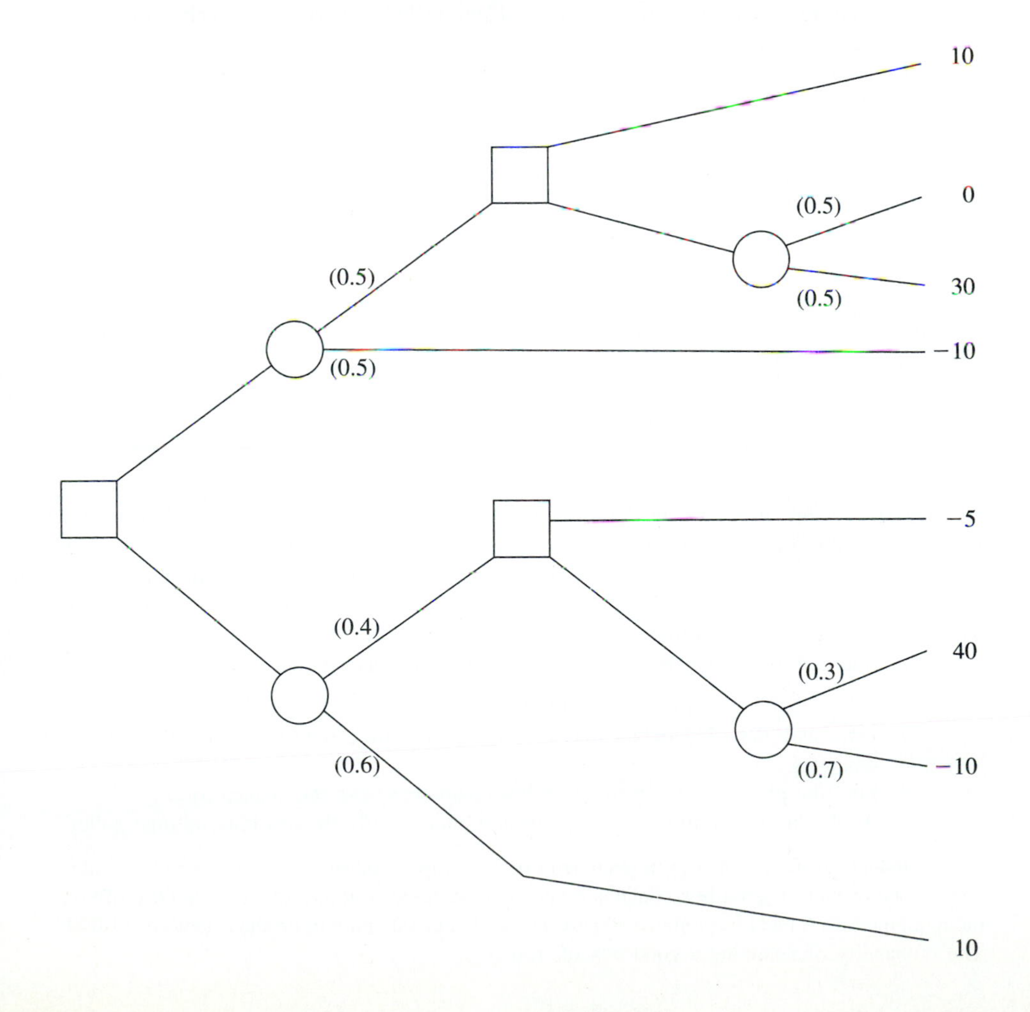

20.4-8. The comptroller of a major corporation has $100 million of excess funds to invest. She has been instructed to invest the entire amount for 1 year in either stocks or bonds (but not both) and then to reinvest the entire fund in either stocks or bonds (but not both) for 1 year more. The objective is to maximize the expected monetary value of the fund at the end of the second year.

The annual rates of return on these investments depend on the economic environment, as shown in the following table:

	Rate of Return, Percent	
Economic Environment	Stocks	Bonds
Growth	20	5
Recession	−10	10
Depression	−50	20

The probabilities of growth, recession, and depression for the first year are 0.7, 0.3, and 0.0, respectively. If growth occurs in the first year, these probabilities remain the same for the second year. However, if a recession occurs in the first year, these probabilities change to 0.2, 0.7 and 0.1, respectively, for the second year.

(*a*) Construct the decision tree for this problem. Include the payoff for each of the terminal branches.

A* (*b*) Find the probabilities for the branches emanating from the chance forks.

(*c*) Apply the backward induction procedure, and identify the resulting optimal policy.

20.4-9. On Monday, a certain stock closed at 10 per share. On Tuesday, you expect the stock to close at 9, 10, or 11 per share, with respective probabilities 0.3, 0.3, and 0.4. On Wednesday, you expect the stock to close 10 percent lower, unchanged, or 10 percent higher than Tuesday's close, with the following probabilities:

Tuesday's Close	10 Percent Lower	Unchanged	10 Percent Higher
9	0.4	0.3	0.3
10	0.2	0.2	0.6
11	0.1	0.2	0.7

On Tuesday, you are directed to buy 100 shares of the stock before Thursday. All purchases are made at the end of the day, at the known closing price for that day, so your only options are to buy at the end of Tuesday or at the end of Wednesday. You wish to determine the optimal strategy for whether to buy on Tuesday or defer the purchase until Wednesday, given the Tuesday closing price, in order to minimize the expected purchased price.

Develop and evaluate a decision tree for determining the optimal strategy.

20.4-10. Use the scenario given in Prob. 20.3-10.

(*a*) Draw and properly label the decision tree. Include the payoff for each of the terminal branches.

A* (*b*) Find the probabilities for the branches emanating from the chance forks.

(*c*) Apply the backward induction procedure, and identify the resulting optimal policy.

20.4-11. A merchant buys boxes of fruit from a grower and sells them. Each box of fruit is either good or bad. A good box contains 80 percent excellent fruit and will earn $200 profit on the retail market. A bad box contains 30 percent excellent fruit and will produce a loss of $1,000. The probability of receiving a good box of fruit is 0.9.

Before the merchant decides to put the box on the market, he has the option of sampling one piece of fruit to test whether it is excellent. Based on that sample, he then has the option of rejecting the box without paying for it.

(*a*) Construct the decision tree for this problem. Include the payoff for each of the terminal branches.

A* (*b*) Find the probabilities for the branches emanating from the chance forks.

(*c*) Apply the backward induction procedure, and identify the resulting optimal policy.

(*d*) Find EVE.

(*e*) Find EVPI.

(*f*) How much money would it be worth to have perfect information when this is compared to using the sampling option?

20.4-12. A company is considering the introduction of a new product that is believed to have a probability of 0.5 of being successful and a probability of 0.5 of being unsuccessful. One option is to try out the product in a test market, at a cost of $5 million, before making the introduction decision. Past experience shows that ultimately successful products are approved in the test market 80 percent of the time, whereas ultimately unsuccessful products are approved in the test market only 25 percent of the time. If the product is successful, the net profit to the company will be $40 million; if unsuccessful, the net loss will be $15 million.

(*a*) Discarding the option of trying out the product in a test market, develop a decision analysis formulation of the problem by identifying the actions, states of nature, and payoff table. Finally, apply Bayes' decision rule to determine the optimal action.

(*b*) Find EVPI.

(*c*) Now including the option of trying out the product in a test market, construct a decision tree for the problem. Include the payoff for each of the terminal branches.

A* (*d*) Find the probabilities for the branches emanating from the chance forks.

(*e*) Apply the backward induction procedure, and identify the resulting optimal policy.

20.4-13. Use the scenario given in Prob. 20.3-13.

(*a*) Draw and properly label the decision tree. Include the payoff for each of the terminal branches.

A* (*b*) Find the probabilities for the branches emanating from the chance forks.

(*c*) Apply the backward induction procedure, and identify the resulting optimal policy.

20.4-14. An emerging Presidential candidate is considering whether to run in the high-stakes Super Tuesday primaries. If she enters the Super Tuesday (S.T.) primaries, she and her advisers believe that she will either do well (finish first or second) or do poorly(finish third or worse) with probabilities 0.4 and 0.6, respectively. Doing well on Super Tuesday will net the candidate's campaign approximately $4 million, whereas a poor showing will mean a loss of $2.5 million after numerous TV ads are paid for. Alternatively, she may choose not to run at all on Super Tuesday and incur no costs.

The candidate's advisers realize that her chances of success on Super Tuesday may be affected by the outcome of the smaller New Hampshire (N.H.) primary occurring 3 weeks before Super Tuesday. Political analysts feel that the results of New Hampshire's primary are correct two-thirds of the time in predicting the results of the Super Tuesday primaries. Among the advisers is a decision analysis expert who uses this information to calculate the following probabilities:

$$P\{\text{candidate does well in S.T. primaries, does well in N.H.}\} = \frac{4}{7}$$

$$P\{\text{candidate does well in S.T. primaries, does poorly in N.H.}\} = \frac{1}{4}$$

$$P\{\text{candidate does well in N.H. primary}\} = \frac{7}{15}$$

The cost of entering and campaigning in the New Hampshire primary is estimated to be $400,000.

(*a*) Draw and properly label the decision tree, including all the payoffs and probabilities.
(*a*) Apply the backward induction procedure, and identify the resulting optimal policy.

20.4-15. A company is considering a candidate for a senior systems position with responsibility for development of a new management information system. The president of the company believes that the candidate has a probability of 0.7 of designing the system successfully. If he is successful, the company will realize a profit of $500,000 (net of the candidate's salary, training, recruiting costs, and expenses). If he is not successful, the company will realize a net loss of $100,000.

A consulting firm has developed an aptitude test that is 90 percent reliable in determining the candidate's potential for success; i.e., a candidate who would successfully design the system will pass the test with probability 0.9, and a candidate who would not successfully design the system will fail the test with probability 0.9. If the company decides to have the consulting firm administer this aptitude test to the candidate, the cost will be $5,000.

(*a*) Construct the decision tree for this problem. Include the payoff for each of the terminal branches.
A* (*b*) Find the probabilities for the branches emanating from the chance forks.
(*c*) Apply the backward induction procedure, and identify the resulting optimal policy.
(*d*) Now suppose that the cost to the company of having the aptitude test administered to the candidate is negotiable. What is the maximum amount that the company should pay?

20.5-1. Reconsider the prototype example introduced in Sec. 20.1 and developed through Sec. 20.5. Suppose now that the company has been sold to a much more risk-averse individual than the owner described in Sec. 20.5. Therefore, instead of the utility values shown in Table 20.8, the utility function for money of the new owner has the values $u(-130) = -200$, $u(-100) = -130$, $u(60) = 60$, $u(90) = 90$, $u(670) = 440$, and $u(700) = 450$. Apply the backward induction procedure to the decision tree corresponding to Fig. 20.5 in order to obtain the optimal policy for this new owner.

20.5-2. You live in an area that has a possibility of incurring a massive earthquake, so you are considering buying earthquake insurance on your home at an annual cost of $180. The probability of an earthquake's damaging your home during 1 year is 0.001. If this happens, you estimate that the cost of the damage (fully covered by earthquake insurance) will be $160,000. Your total assets (including your home) are worth $250,000.

(*a*) Apply Bayes' decision rule to determine which alternative (take the insurance or not) minimizes your expected loss.
(*b*) Suppose that you value your total assets at an amount x ($x \geq 0$) according to the utility function $u(x) = \sqrt{x}$. Compare the utility of reducing your total assets next year by the cost of the earthquake insurance with the expected utility next year of not taking the earthquake insurance. Should you take the insurance?

20.5-3. You have been given the choice of definitely receiving $19 or participating in a gamble with the following two possible outcomes:

Outcome	Probability
Receive $10	0.3
Receive $30	0.7

Your utility for receiving M dollars is given by the utility function $u(M) = \sqrt{M + 6}$. Which choice should you make in order to maximize expected utility?

20.5-4. You are given the following payoff table:

Action \ State of Nature	Payoff θ_1	θ_2
a_1	25	36
a_2	100	0
a_3	0	49
Prior probability	p	$1 - p$

Your utility function for the payoffs is $u(x) = \sqrt{x}$. Determine the largest value of p for which choosing action a_1 maximizes the expected utility.

20.5-5. A doctor has a seriously ill patient but has had trouble diagnosing the specific cause of the illness. The doctor now has narrowed the cause to two alternatives: disease A or disease B. Based on the evidence so far, she feels that the two alternatives are equally likely.

Beyond the testing already done, there is no test available to determine if the cause is disease B. One test is available for disease A, but it has two major problems. First, it is very expensive. Second, it is somewhat unreliable, giving an accurate result only 80 percent of the time. Thus, it will give a positive result (indicating disease A) for only 80 percent of patients who have disease A, whereas it will give a positive result for 20 percent of patients who actually have disease B instead.

Disease B is a very serious disease with no known treatment. It is sometimes fatal, and those who survive remain in poor health with a poor quality of life thereafter. The prognosis is similar for victims of disease A if it is left untreated. However, there is a fairly expensive treatment available that eliminates the danger for those with disease A, and it may return them to good health. Unfortunately, it is a relatively radical treatment that always leads to death if the patient actually has disease B instead.

The probability distribution for the prognosis for the current patient is given for each case in the following table, where the column headings (after the first one) indicate the disease for this patient.

	Outcome Probabilities			
	No Treatment		***Receive Treatment for Disease A***	
Outcome	A	B	A	B
Die	0.2	0.5	0	1.0
Survive with poor health	0.8	0.5	0.5	0
Return to good health	0	0	0.5	0

The patient has assigned the following utilities to the possible outcomes

Outcome	Utility
Die	0
Survive with poor health	10
Return to good health	30

In addition, these utilities should be incremented by -2 if the patient incurs the cost of the test for disease A and by -1 if the patient (or the patient's estate) incurs the cost of the treatment for disease A.

Use decision analysis with a complete decision tree to determine if the patient should undergo the test for disease A and then how to proceed (receive the treatment for disease A?) in order to maximize the patient's expected utility.

20.5-6. Consider the following decision tree, where the probabilities for each chance fork are shown in parentheses.

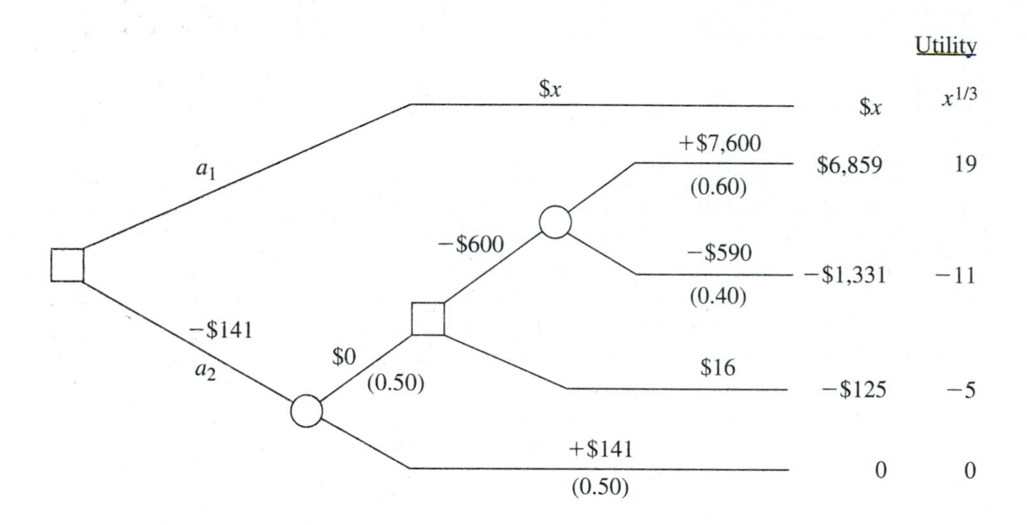

The dollar amount given next to each branch is the cash flow generated along that branch, where these intermediate cash flows add up to the total net cash flow shown to the right of each terminal branch. The decision maker has a utility function $u(y) = y^{1/3}$ where y is the total net cash flow after a terminal branch. The resulting utilities for the various terminal branches are shown to the right of the decision tree.

Use these utilities to apply the backward induction procedure to the decision tree. Then determine the value of x for which the decision maker is indifferent between actions a_1 and a_2.

20.5-7. You want to choose between actions a_1 and a_2 in the following decision tree, but you are uncertain about the value of the probability p, so you need to perform sensitivity analysis of p as well.

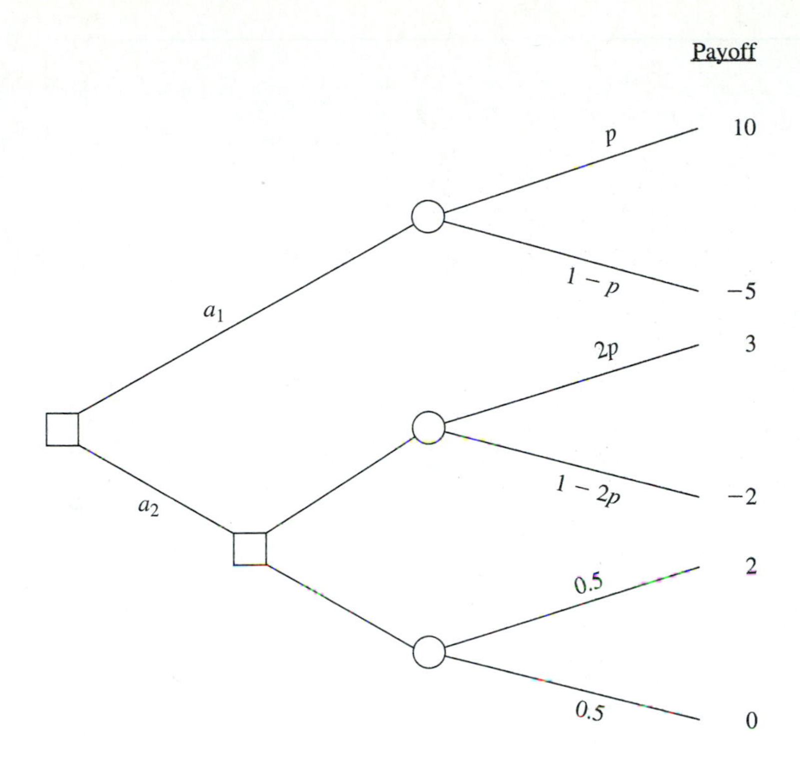

Your utility function for money (the payoff received) is

$$u(M) = \begin{cases} M^2 & \text{if } M \geq 0, \\ M & \text{if } M < 0. \end{cases}$$

(*a*) For $p = 0.25$, determine which action is optimal in the sense that it maximizes the expected utility of the payoff.

(*b*) Determine the range of values of the probability p $(0 \leq p \leq 0.5)$ for which this same action remains optimal.

21

Simulation

The technique of **simulation** has long been an important tool of the designer. For example, simulating airplane flight in a wind tunnel is standard practice when a new airplane is designed. Theoretically, the laws of physics could be used to obtain the same information about how the performance of the airplane changes as design parameters are altered, but, as a practical matter, the analysis would be too complicated to do it all. Another alternative would be to build real airplanes with alternative designs and test them in actual flight to choose the final design, but this would be far too expensive (as well as unsafe). Therefore, after some preliminary theoretical analysis is performed to develop a *rough* design, simulating flight in a wind tunnel is a vital tool for experimenting with *specific* designs. This simulation amounts to *imitating* the performance of a real airplane in a controlled environment in order to *estimate* what its actual performance would be. After a detailed design is developed in this way, a prototype model can be built and tested in actual flight to fine-tune the final design.

Simulation plays essentially this same role in many OR studies. However, rather than designing an airplane, the operations research team is concerned with developing a design (or operating policy) for some *stochastic system* (a system that evolves *probabilistically* over time). Some of these stochastic systems resemble the examples of Markov chains and queueing systems described in Chaps. 14 to 16, and others are more complicated. Rather than use a wind tunnel, the performance of the real system is *imitated* by using probability distributions to *randomly generate* various events that occur in the system. Therefore, a simulation model *synthesizes* the system by building it up component by component and event by event. Then the model *runs* the simulated system to obtain *statistical observations* of the performance of the system that result from various randomly generated events. Because the *simulation runs* typically require generating and processing a vast amount of data, these simulated statistical experiments are inevitably performed on a computer.

When simulation needs to be used as part of an OR study, commonly it is preceded and followed by the same steps as for the airplane design. In particular, some preliminary theoretical analysis is done first to develop a rough design of the system. Then simulation is used to experiment with specific designs in order to estimate the actual performance of the system. After a detailed design is developed in this way, the system is tested in actual use in order to fine-tune the final design.

OR teams typically use simulation when the stochastic system involved is too complex to be analyzed satisfactorily by the kinds of analytical models described in Chaps. 14 to 20. One of the main strengths of the analytical approach is that it abstracts the essence of the problem and reveals its underlying structure, thereby providing insight into the cause-and-effect relationships within the system. Therefore, if it is possible to construct an analytical model that is both a reasonable idealization of the problem and amenable to solution, this approach usually is superior to simulation. However, many problems are so complex that they cannot be solved analytically. Thus, even though simulation tends to be a relatively expensive procedure, it often provides the only practical approach to a problem.

In essence, then, the OR view of simulation is that it is a *controlled statistical sampling technique* for *estimating* the performance of complex stochastic systems when analytical models do not suffice. Rather than describe the overall behavior of the system directly, the *simulation model* describes the operation of the system in terms of *individual events* of the individual components of the system. In particular, the system is divided into elements whose behavior can be predicted, at least in terms of probability distributions, for each of the various possible *states of the system* and its inputs. The interrelationships among the elements also are built into the model.

Thus simulation provides a means of dividing the model-building job into smaller component parts that can be formulated more readily (e.g., a component part might be a simple queueing system) and then combining these component parts in their natural order. After constructing the model, we can activate it by using random numbers to generate simulated events over time according to the appropriate probability distributions. The result is a simulation of the actual operation of the system over time, and we can record its aggregate behavior. By repeating this process for the various alternative configurations for the design and operating policies of the system, and by comparing their performance, we can identify the most promising configurations. Because of statistical error, it is impossible to guarantee that the configuration yielding the best simulated performance is indeed the optimal one, but it should be at least near optimal if the simulated experiment was designed properly.

Thus simulation typically is nothing more or less than the technique of performing *sampling experiments* on the model of the system. The experiments are done on the model rather than on the real system itself only because the latter would be too inconvenient, expensive, and time-consuming. Otherwise, simulated experiments should be viewed as virtually indistinguishable from ordinary statistical experiments, so they also should be based upon sound statistical theory.

This chapter focuses on *discrete event simulations* (as opposed to *continuous simulations*), i.e., simulations where changes in the *state of the system* occur at random points in time (as opposed to continuously) as a result of the occurrence of discrete events. The basic building blocks of a model for a discrete event simulation are the possible states and events, a simulation clock for recording the passage of (simulated) time, a mechanism for randomly generating the different kinds of events, and a mechanism for then generating state transitions.

These building blocks are described in Sec. 21.1 in the context of illustrative examples. The second section elaborates on the formulation and implementation of simulation models. The next two sections then focus on the design and analysis of the program of statistical experimentation inherent in a simulation study.

21.1 Illustrative Examples

In this section we use some relatively simple stochastic systems to introduce and illustrate some basic concepts of simulation. The first system is so simple, in fact, that the simulation does not even need to be performed on a computer. The second system incorporates more of the normal features of a simulation, although it, too, is simple enough to be solved analytically. Following these two examples, we survey some more typical applications of simulation.

Example 1: A Coin-Flipping Game

Suppose you are offered a chance to play a game in which you repeatedly flip an unbiased coin until the *difference* between the number of heads tossed and the number of tails tossed is three. You are required to pay \$1 for each flip of the coin, but you receive \$8 at the end of each play of the game. You are not allowed to quit during a play of the game. Thus you win money if the number of flips required is fewer than eight, but you lose money if more than eight flips are required. How would you decide whether to play this game?

Many people would base this decision on simulation, although they probably would not call it by that name. (There is also an analytical solution for this game, but it is not a particularly elementary one.) In this case, simulation amounts to nothing more than playing the game alone many times until it becomes clear whether it is worthwhile to play for money. Half an hour spent in repeatedly flipping a coin and recording the earnings or losses that would have resulted might be sufficient. This is a true simulation because you are *imitating* the actual play of the game *without* actually winning or losing any money.

How would this simulated experiment be executed on a computer? Although the computer cannot flip coins, it can generate numbers. Therefore, it would generate (or be given) a sequence of random digits, each corresponding to a flip of a coin. (The generation of random numbers is discussed in Sec. 21.2.) The probability distribution for the outcome of a flip is that the probability of a head is $\frac{1}{2}$ and the probability of a tail is $\frac{1}{2}$, whereas there are 10 possible values of a random digit, each having a probability of $\frac{1}{10}$. Therefore, five of these values (say, 0, 1, 2, 3, 4) would be assigned an association with a head and the other five (say, 5, 6, 7, 8, 9) with a tail. Thus the computer would simulate the playing of the game by examining each new random digit generated and labeling it a head or a tail, according to its value. It would continue doing this, recording the outcome of each simulated play of the game, as long as desired.

To illustrate the computer approach to this simulated experiment, we suppose that the computer generated the following sequence of random digits:

8, 1, 3, 7, 2, 7, 1, 6, 5, 5, 7, 9, 0, 0, 3, 4, 3, 5, 6, 8, 5,
8, 9, 4, 8, 0, 4, 8, 6, 5, 3, 5, 9, 2, 5, 7, 9, 7, 2, 9, 3, 9,
8, 5, 8, 9, 2, 5, 7, 6, 9, 7, 6, 0, 7, 3, 9, 8, 2, 7, 1, 0, 3,
2, 6, 2, 7, 1, 3, 7, 0, 4, 4, 1, 8, 3, 2, 1, 3, 9, 5, 9, 0, 5,
0, 3, 8, 7, 8, 9, 5, 4, 0, 8, 3, 8, 0, 1.

Thus, if we denote a head by H and a tail by T, the first simulated play of the game is THHTHTHTTTT, requiring 11 simulated flips of a coin. The subsequent simulated plays of the game require 5, 5, 9, 7, 7, 5, 3, 17, 5, 5, 3, 9, and 7 simulated flips, respectively. This experiment has a sample size of 14 (14 simulated plays of the game), where the individual observations are the number of flips required for a play of the game. One useful statistic is

$$\text{Sample average} = \frac{11 + 5 + \cdots + 7}{14} = 7,$$

because the sample average provides an *estimate* of the true *mean* of the underlying probability distribution.

This sample average of 7 would seem to indicate that, on the average, you should win about \$1 each time you play the game. Therefore, if you do not have a relatively high aversion to risk, it appears that you should choose to play this game, preferably a large number of times.

However, *beware*! One common error in the use of simulation is that conclusions are based on overly small samples, because statistical analysis was inadequate or totally lacking. In this case, the *sample standard deviation* is 3.67, so that the estimated *standard deviation* of the *sample average* is $3.67/\sqrt{14} \approx 0.98$. Therefore, even if it is assumed that the probability distribution of the number of flips required for a play of the game is a *normal distribution* (which is a gross assumption because the true distribution is *skewed*), any reasonable *confidence interval* for the true *mean* of this distribution would extend far above 8. Hence a much larger sample size is required before we can draw a valid conclusion at a reasonable level of statistical significance. Unfortunately, because the standard deviation of a sample average is inversely proportional to the *square root* of the sample size, a large increase in the sample size is required to yield a relatively small increase in the precision of the estimate of the true mean. In this case, it appears that an additional 100 simulated plays of the game might be adequate.

It so happens that the true *mean* of the number of flips required for a play of this game is 9. Thus, in the long run, you actually would lose about \$1 each time you played the game.

Although formally constructing a full-fledged *simulation model* is not really necessary for this simple simulation, we do so now for illustrative purposes. The *stochastic system* being simulated is the successive flipping of the coin for a play of the game. The *simulation clock* records the number of (simulated) flips t that have occurred so far. The information about the system that defines its current status, i.e., the *state of the system,* is

$$N(t) = \text{number of heads minus number of tails after } t \text{ flips.}$$

The *events* that change the state of the system are the flipping of a head or the flipping of a tail. The *event generation mechanism* is the generation of a *random digit* where

$$0 \text{ to } 4 \Rightarrow \text{a head,}$$
$$5 \text{ to } 9 \Rightarrow \text{a tail.}$$

The *state transition mechanism* is to set

$$N(t) = \begin{cases} N(t-1) + 1 & \text{if flip } t \text{ is a head,} \\ N(t-1) - 1 & \text{if flip } t \text{ is a tail.} \end{cases}$$

The simulated game then ends at the first value of t where $N(t) = \pm 3$, where the resulting sampling *observation* for the simulated experiment is $8 - t$, the amount won (positive or negative) for that play of the game.

The next example will illustrate these building blocks of a simulation model for a prominent stochastic system from queueing theory.

Example 2: An *M/M/*1 Queueing System

Consider the *M/M/*1 queueing theory model (Poisson input, exponential service times, and single server) that was discussed at the beginning of Sec. 15.6. Although this model already has been solved analytically, it will be instructive to consider how to study it by using simulation. To be specific, suppose that the values of the *arrival rate* λ and *service rate* μ are

$$\lambda = 3 \text{ per hour,} \qquad \mu = 5 \text{ per hour.}$$

To summarize the physical operation of the system, arriving customers enter the queue, eventually are served, and then leave. Thus it is necessary for the simulation model to describe and synchronize the arrival of customers and the serving of customers.

Starting at time 0, the simulation clock records the amount of (simulated) time t that has transpired so far during the simulation run. The information about the queueing system that defines its current status, i.e., the state of the system, is

$$N(t) = \text{number of customers in system at time } t.$$

The events that change the state of the system are the *arrival* of a customer or a service completion for the customer currently in service (if any). We shall describe the event generation mechanism a little later. The state transition mechanism is to

$$\text{Reset} \qquad N(t) = \begin{cases} N(t) + 1 & \text{if arrival occurs at time } t, \\ N(t) - 1 & \text{if service completion occurs at time } t. \end{cases}$$

There are two basic methods used for advancing the simulation clock and recording the operation of the system. We did not distinguish between these methods for Example 1 because they actually coincide for that simple situation. However, we now describe and illustrate these two **time advance mechanisms** (fixed-time incrementing and next-event incrementing) in turn.

With the **fixed-time incrementing** time advance mechanism, the following two-step procedure is used repeatedly.

Summary of Fixed-Time Incrementing

1. *Advance time* by a small *fixed amount.*
2. *Update the system* by determining what events occurred during the elapsed time interval and what the resulting state of the system is. Also record desired information about the performance of the system.

For the queueing theory model under consideration, only two types of events can occur during each of these elapsed time intervals, namely, one or more *arrivals* and one or more *service completions.* Furthermore, the probability of two or more arrivals or of two or more service completions during an interval is negligible for this model if the interval is relatively short. Thus the only two possible events during such an interval that need to be investigated are the arrival of one customer and the service completion for one customer. Each of these events has a known probability.

To illustrate, let us use 0.1 hour (6 minutes) as the small fixed amount by which the clock is advanced each time. (Normally, a considerably smaller time interval would be used to render negligible the probability of multiple arrivals or multiple service completions, but this choice will create more action for illustrative purposes.) Because both interarrival times and service times have an exponential distribution, the probability P_A that a time interval of 0.1 hour will include an *arrival* is

$$P_A = 1 - e^{-3/10} = 0.259,$$

and the probability P_D that it will include a *departure* (service completion), given that a customer was being served at the beginning of the interval, is

$$P_D = 1 - e^{-5/10} = 0.393.$$

To randomly generate either kind of event according to these probabilities, the approach is similar to that in Example 1. The computer again is used to generate a random number, but this time with multiple digits rather than one. Placing a decimal point in front of the number then makes it a *uniform random number* on (0, 1), that is, a random observation from the *uniform distribution* between 0 and 1. If we denote this uniform random number by r_A,

$$r_A < 0.259 \Rightarrow \text{arrival occurred},$$

$$r_A \geq 0.259 \Rightarrow \text{arrival did not occur}.$$

Similarly, with *another* uniform random number r_D,

$$r_D < 0.393 \Rightarrow \text{departure occurred},$$

$$r_D \geq 0.393 \Rightarrow \text{departure did not occur},$$

given that a customer was being served at the beginning of the time interval. With no customer in service then (i.e., no customers in the system), it is assumed that no departure can occur during the interval even if an arrival does occur.

Table 21.1 shows the result of using this approach for 10 iterations of the *fixed-time incrementing* procedure, starting with no customers in the system and using time units of minutes.

Table 21.1 **Fixed-Time Incrementing Applied to Example 2**

t, time (min)	$N(t)$	r_A	Arrival in Interval?	r_D	Departure in Interval?
0	0				
6	1	0.096	Yes	—	
12	1	0.569	No	0.665	No
18	1	0.764	No	0.842	No
24	0	0.492	No	0.224	Yes
30	0	0.950	No	—	
36	0	0.610	No	—	
42	1	0.145	Yes	—	
48	1	0.484	No	0.552	No
54	1	0.350	No	0.590	No
60	0	0.430	No	0.041	Yes

Step 2 of the procedure (updating the system) includes recording the desired measures of performance about the aggregate behavior of the system during this time interval. For example, it could record the *number of customers* in the queueing system and the *waiting time* of any customer who just completed his or her wait. If it is sufficient to estimate only the mean rather than the probability distribution of each of these random variables, the computer will merely add the value (if any) at the end of the current time interval to a cumulative sum. The sample averages will be obtained after the simulation run is completed by dividing these sums by the sample sizes involved, namely, the total number of time intervals and the total number of customers, respectively.

Next-event incrementing differs from fixed-time incrementing in that the simulation clock is incremented by a *variable* amount rather than by a fixed amount each time. This variable amount is the time from the event that has just occurred until the *next event* of any kind occurs; i.e., the clock jumps from event to event. A summary follows.

Summary of Next-Event Incrementing

1. *Advance time* to the time of the *next event* of any kind.
2. *Update the system* by determining its new state that results from this event and by randomly generating the time until the next occurrence of any event type that can occur from this state (if not previously generated). Also record desired information about the performance of the system.

For this example the computer needs to keep track of two future events, namely, the next arrival and the next service completion (if a customer currently is being served). These times are obtained by taking a random observation from the probability distribution of interarrival and service times, respectively. As before, the computer takes such a random observation by generating and using a random number. (This technique will be discussed in Sec. 21.2.) Thus, each time an arrival or service completion occurs, the computer determines how long it will be until the next time this event will occur, adds this time to the current clock time, and then stores this sum in a computer file. (If the service completion leaves no customers in the system, then the generation of the time until the next service completion is postponed until the next arrival occurs.) To determine which event will occur next, the computer finds the minimum of the clock times stored in the file. To expedite the bookkeeping involved, simulation programming languages provide a "timing routine" that determines the occurrence time and type of the next event, advances time, and transfers control to the appropriate subprogram for the event type.

Table 21.2 shows the result of applying this approach through five iterations of the next-event incrementing procedure, starting with no customers in the system and using time units of minutes. For later reference, we include the *uniform random numbers* r_A and r_D used to generate the interarrival times and service times, respectively, by the method to be described in Sec. 21.2. These r_A and r_D are the same as those used in Table 21.1 in order to provide a truer comparison between the two time advance mechanisms.

The next-event incrementing procedure is considerably better suited for this example and similar stochastic systems than the fixed-time incrementing procedure. Next-event incrementing requires fewer iterations to cover the same amount of simulated time, and it generates a precise schedule for the evolution of the system rather than a rough approximation.

The next-event incrementing procedure will be illustrated again in Sec. 21.4 (see Table 21.12) in the context of a full statistical experiment for estimating certain measures of performance for another queueing system.

Several pertinent questions about how to conduct a simulation study of this type still remain to be answered. These answers are presented in a broader context in subsequent sections.

More Examples in Your OR Courseware

Simulation examples are easier to understand when they can be *observed in action,* rather than just talked about on a printed page. Therefore, the simulation area of your OR Courseware includes two *demonstration examples* under the Demo menu that should be viewed at this time.

Both examples involve a bank that plans to open up a new branch office. The questions address how many teller windows to provide and then how many tellers to have on duty at the outset. Therefore, the system being studied is a *queueing system.* However, in contrast to the $M/M/1$ queueing system considered in Example 2 above, this queueing system is too complicated to be solved analytically. This system has multiple servers (tellers), and the probability distributions of interarrival times and service times do not fit the standard models of queueing theory. Furthermore, in the second demonstration, it has been decided that one class of customers (merchants) needs to be given nonpreemptive priority over other customers, but the probability

Table 21.2 **Next-Event Incrementing Applied to Example 2**

t, time (min)	$N(t)$	r_A	Next Interarrival Time	r_D	Next Service Time	Next Arrival	Next Departure	Next Event
0	0	0.096	2.019	—	—	2.019	—	Arrival
2.019	1	0.569	16.833	0.665	13.123	18.852	15.142	Departure
15.142	0	—	—	—	—	18.852	—	Arrival
18.852	1	0.764	28.878	0.842	22.142	47.730	40.994	Departure
40.994	0	—	—	—	—	47.730	—	Arrival
47.730	1							

distributions for this class are different from those for other customers. These complications are typical of those that can be readily incorporated into a simulation study.

In both demonstrations, you will be able to see customers arrive and served customers leave as well as the next-event incrementing procedure being applied simultaneously to the simulation run.

The demonstrations also introduce you to two other routines that you should find very helpful in dealing with some of the problems at the end of this chapter. One is an *interactive routine* and the other is an *automatic routine* for simulating queueing systems with the same kinds of complications as in the demonstrations.

Typical Applications

During the early 1980s, a survey was made of many leading U.S. firms to learn more about their use of simulation, as reported by David Christie and Hugh Watson in Selected Reference 3. One major finding was the identification of the functional areas of the company where simulation was being applied. The results are shown in Table 21.3, where production (manufacturing) leads the list, closely followed by corporate planning, engineering, finance, and research and development. As the percentages indicate, many of the companies are applying simulation in most of these areas.

This survey also found that the development of simulation models had spread far beyond centralized OR (or management science) departments. In fact, in 54 percent of the companies, simulation models were being created in functional area departments, and corporate planning departments were doing so in 30 percent of the companies. The clear majority of both kinds of departments, as well as OR departments, reported that they perceived the results of their simulation applications to be good, as opposed to fair or poor.

More recent reports indicate that the use of simulation continues to grow rapidly in the production (manufacturing) as well as the corporate planning and finance areas. Growth in the latter areas has been aided by the development of specialized simulation programming languages for financial planning.

Another important recent development has been the increasing use of *computer graphics* to generate animated displays of the movement of entities through the simulated system. For example, in the simulation of manufacturing systems, the moving entities can represent components being manufactured and various materials-handling devices such as automated guided vehicles. The computer graphics provide greater insight into the performance of the system for any given design, and they also add credibility to the results of the simulation study.

Table 21.3 **Percentage of Surveyed Companies Using Simulation in Certain Functional Areas**

Functional Area	Percentage
Production	59
Corporate planning	53
Engineering	46
Finance	41
Research and development	37
Marketing	24
Data processing	16
Personnel	10

There have been numerous applications of simulation in a wide variety of contexts. Some examples are listed here to illustrate the great versatility of this technique:

1. Simulation of the operations at a large airport by an airline company to test changes in company policies and practices (e.g., amounts of maintenance capacity, berthing facilities, spare aircraft, and so on)
2. Simulation of the passage of traffic across a junction with time-sequenced traffic lights to determine the best time sequences
3. Simulation of a maintenance operation to determine the optimal size of repair crews
4. Simulation of the flux of uncharged particles through a radiation shield to determine the intensity of the radiation that penetrates the shield
5. Simulation of steel-making operations to evaluate changes in both operating practices and the capacity and configuration of the facilities
6. Simulation of the U.S. economy to predict the effect of economic policy decisions
7. Simulation of large-scale military battles to evaluate defensive and offensive weapons systems
8. Simulation of large-scale distribution and inventory control systems to improve the design of these systems
9. Simulation of the overall operation of an entire business firm to evaluate broad changes in the policies and operation of the firm as well as to provide a business game for training executives
10. Simulation of a telephone communications system to determine the capacity of the respective components required to provide satisfactory service at the most economical level
11. Simulation of the operation of a developed river basin to determine the best configuration of dams, power plants, and irrigation works to provide the desired level of flood control and water resource development
12. Simulation of the operation of a production line to determine the amount of in-process storage space that should be provided

21.2 Formulating and Implementing a Simulation Model

Constructing a Model

The first step in a simulation study is to develop a model representing the system to be investigated. This step requires the analyst to become thoroughly familiar with the operating realities of the system and the objectives of the study. Given this requirement, the analyst probably will attempt to reduce the real system to a logical flow diagram. The system is thereby broken down into a set of components linked together by a master flow diagram, where the components themselves may be broken down into subcomponents, and so on. Ultimately the system is decomposed into a set of elements for which operating rules may be given. These operating rules predict the events that will be generated by the corresponding elements, perhaps in terms of probability distributions. After specifying these elements, rules, and logical linkages, the analyst needs to test the model thoroughly piece by piece. This testing can be done partially by performing a gross version of the simulation on a calculator and checking whether each input is received from the appropriate source and whether each output is acceptable to the next submodel. However, the individual components of the model also should be tested alone to verify that their internal performance is reasonably consistent with reality.

It should be emphasized that, like any OR model, the simulation model need not be a completely realistic representation of the real system. In fact, it appears that most simulation models err on the side of being overly realistic rather than overly idealized. With the former approach, the model easily degenerates into a mass of trivia and meandering details, so that a great deal of programming and computer time is required to obtain a small amount of information. Furthermore, failing to strip away trivial factors to get down to the core of the system may obscure the significance of those results that are obtained.

If the behavior of an element cannot be predicted exactly, given the state of the system, it is better to take random observations from the probability distributions involved than to use averages to simulate the performance of this element. This statement is true even when one is interested only in the average aggregate performance of the system, because combining average performances for the individual elements may result in something far from average for the overall system.

One question that may arise when one is choosing probability distributions for the model is whether to use frequency distributions of historical data or to seek the theoretical probability distribution that best fits these data. The latter alternative usually is preferable because it avoids reproducing the idiosyncrasies of a certain period in the past.

When the building blocks of a simulation model described and illustrated in the preceding section are constructed, one key step is the definition of the *state of the system.* The state must include the relevant information about the current status of the system so that generating the simulated evolution of the system based upon the state provides an accurate representation of the behavior of the real system. The state must also allow measuring and combining quantities that yield meaningful estimates for measures of performance of the system. Frequently, the state must be represented by a *vector* (a set of state variables) rather than a single variable. For complex stochastic systems, there sometimes are alternative reasonable definitions for the state of the system.

Generating Random Numbers

As the examples in Sec. 21.1 demonstrated, implementing a simulation model requires random numbers to obtain random observations from probability distributions. One method for generating such random numbers is to use a physical device such as a spinning disk or an electronic randomizer. Several tables of random numbers have been generated in this way, including one containing 1 million random digits, published by the Rand Corporation. An excerpt from the Rand table is given in Table 21.4.

When a simulation is performed on a computer, the needed random numbers normally are generated directly by the computer by using a random number generator. A **random number generator** is an algorithm that produces sequences of numbers that follow a specified probability distribution and possess the appearance of randomness. The reference to *sequences of numbers* means that the algorithm produces many random numbers in a serial manner. Although an individual user may need only a few of the numbers, generally the algorithm must be capable of producing many numbers. *Probability distribution* implies that a probability statement can be associated with the occurrence of each number produced by the algorithm.

Table 21.4 **Table of Random Digits**

09656	96657	64842	49222	49506	10145	48455	23505	90430	04180
24712	55799	60857	73479	33581	17360	30406	05842	72044	90764
07202	96341	23699	76171	79126	04512	15426	15980	88898	06358
84575	46820	54083	43918	46989	05379	70682	43081	66171	38942
38144	87037	46626	70529	27918	34191	98668	33482	43998	75733
48048	56349	01986	29814	69800	91609	65374	22928	09704	59343
41936	58566	31276	19952	01352	18834	99596	09302	20087	19063
73391	94006	03822	81845	76158	41352	40596	14325	27020	17546
57580	08954	73554	28698	29022	11568	35668	59906	39557	27217
92646	41113	91411	56215	69302	86419	61224	41936	56939	27816
07118	12707	35622	81485	73354	49800	60805	05648	28898	60933
57842	57831	24130	75408	83784	64307	91620	40810	06539	70387
65078	44981	81009	33697	98324	46928	34198	96032	98426	77488
04294	96120	67629	55265	26248	40602	25566	12520	89785	93932
48381	06807	43775	09708	73199	53406	02910	83292	59249	18597
00459	62045	19249	67095	22752	24636	16965	91836	00582	46721
38824	81681	33323	64086	55970	04849	24819	20749	51711	86173
91465	22232	02907	01050	07121	53536	71070	26916	47620	01619
50874	00807	77751	73952	03073	69063	16894	85570	81746	07568
26644	75871	15618	50310	72610	66205	82640	86205	73453	90232

Source: Reproduced with permission from The Rand Corporation, *A Million Random Digits with 100,000 Normal Deviates.* Copyright, The Free Press, Glencoe, IL, 1955, top of p. 182.

We shall reserve the term **random number** to mean a random observation from some form of a *uniform distribution,* so that all possible numbers are *equally likely.* When we are interested in some other probability distribution (as in the next subsection), we shall refer to *random observations* from that distribution.

Random numbers can be divided into two main categories, random integer numbers and uniform random numbers, defined as follows:

A **random integer number** is a random observation from a *discretized uniform distribution* over some range $\underline{n}, \underline{n} + 1, \ldots, \bar{n}$. The probabilities for this distribution are

$$P(\underline{n}) = P(\underline{n} + 1) = \cdots = P(\bar{n}) = \frac{1}{\bar{n} - \underline{n} + 1}.$$

Usually, $\underline{n} = 0$ or 1, and these are convenient values for most applications. (If $\underline{n}$ has another value, then subtracting either $\underline{n}$ or $\underline{n} - 1$ from the random integer number changes the lower end of the range to either 0 or 1.)

A **uniform random number** is a random observation from a (continuous) *uniform distribution* over some interval $[a, b]$. The probability density function of this uniform distribution is

$$f(x) = \begin{cases} \dfrac{1}{b - a} & \text{if } a \le x \le b, \\ 0 & \text{otherwise.} \end{cases}$$

When a and b are not specified, they are assumed to be $a = 0$ and $b = 1$.

The random numbers initially generated by a computer usually are random integer numbers. However, if desired, these numbers can immediately be converted to a uniform random number as follows:

For a given *random integer number* in the range 0 to $\bar{n}$, dividing this number by $\bar{n}$ yields (approximately) a *uniform random number.* (If $\bar{n}$ is small, this approximation should be improved by adding $\frac{1}{2}$ to the random integer number and then dividing by $\bar{n} + 1$ instead.)

This is the usual method used for generating uniform random numbers. With the huge values of $\bar{n}$ commonly used, it is an essentially exact method.

Strictly speaking, the numbers generated by the computer should not be called random numbers because they are predictable and reproducible (which sometimes is advantageous), given the random number generator being used. Therefore, they are sometimes given the name **pseudo-random numbers**. However, the important point is that they satisfactorily play the role of random numbers in the simulation if the method used to generate them is valid.

Various relatively sophisticated statistical procedures have been proposed for testing whether a generated sequence of numbers has an acceptable appearance of randomness. Basically the requirements are that each successive number in the sequence have an equal probability of taking on any one of the possible values and that it be statistically independent of the other numbers in the sequence.

There are a number of random number generators available, of which the most popular are the *congruential methods* (additive, multiplicative, and mixed). The mixed congruential method includes features of the other two, so we shall discuss it first.

The **mixed congruential method** generates a *sequence* of random integer numbers over the range from 0 to $m - 1$. The method always calculates the next random number from the last one obtained, given an initial random number x_0, called the **seed**, which may be obtained from some published source such as the Rand table. In particular, it calculates the $(n + 1)$st random number x_{n+1} from the nth random number x_n by using the recurrence relation

$$x_{n+1} \equiv (ax_n + c)(\text{modulo } m),$$

where a, c, and m are positive integers ($a < m, c < m$). This mathematical notation signifies that x_{n+1} is the *remainder* when $ax_n + c$ is divided by m. Thus the *possible* values of x_{n+1} are $0, 1, \ldots, m - 1$, so that m represents the desired number of *different* values that could be generated for the random numbers.

To illustrate, suppose that $m = 8$, $a = 5$, $c = 7$, and $x_0 = 4$. The resulting sequence of random numbers is calculated in Table 21.5. (The sequence cannot be continued further because it would just begin repeating the numbers in the same order.) Note that this sequence includes each of the eight possible numbers exactly once. This property is a necessary one for a sequence of *random* integer numbers, but it does not

Table 21.5 **Illustration of the Mixed Congruential Method**

n	x_n	$5x_n + 7$	$(5x_n + 7)/8$	x_{n+1}
0	4	27	$3 + \frac{3}{8}$	3
1	3	22	$2 + \frac{6}{8}$	6
2	6	37	$4 + \frac{5}{8}$	5
3	5	32	$4 + \frac{0}{8}$	0
4	0	7	$0 + \frac{7}{8}$	7
5	7	42	$5 + \frac{2}{8}$	2
6	2	17	$2 + \frac{1}{8}$	1
7	1	12	$1 + \frac{4}{8}$	4

occur with some choices of a and c. (Try $a = 4$, $c = 7$, and $x_0 = 3$.) Fortunately, there are rules available for choosing values of a and c that will guarantee this property. (There are no restrictions on the seed x_0 because it affects only where the sequence begins and not the progression of numbers.)

The number of consecutive numbers in a sequence before it begins repeating itself is referred to as the **cycle length**. Thus, the cycle length in the example is 8. The *maximum* cycle length is m, so the only values of a and c considered are those that yield this maximum cycle length.

Table 21.6 illustrates the conversion of random integer numbers to uniform random numbers. The left column gives the random integer numbers obtained in the rightmost column of Table 21.5. The right column gives the corresponding uniform random numbers from the formula

$$\text{Uniform random number} = \frac{\text{random integer number} + \frac{1}{2}}{m}.$$

Note that each of these uniform random numbers lies at the midpoint of one of the eight equal-sized intervals 0 to 0.125, 0.125 to 0.25, . . . , 0.875 to 1. The small value of $m = 8$ does not enable us to obtain other values over the interval [0, 1], so we are obtaining fairly rough approximations of real uniform random numbers. In practice, far larger values of m generally are used.

For a binary computer with a word size of b bits, the usual choice for m is $m = 2^b$; this is the total number of nonnegative integers that can be expressed within the capacity of the word size. (Any undesired integers that arise in the sequence of random numbers are just not used.) With this choice of m, we can ensure that each possible number occurs exactly once before any number is repeated by selecting any of the values $a = 1, 5, 9, 13, \ldots$ and $c = 1, 3, 5, 7, \ldots$. For a decimal computer with a word size of d digits, the usual choice for m is $m = 10^d$, and the same property is ensured by selecting any of the values $a = 1, 21, 41, 61, \ldots$ and $c = 1, 3, 7, 9, 11, 13, 17, 19, \ldots$ (that is, all positive *odd* integers *except* those ending with the digit 5). The specific selection can be made on the basis of the *serial correlation* between successively generated numbers, which differs considerably among these alternatives.[1]

Occasionally, random integer numbers with only a relatively small number of digits are desired. For example, suppose that only three digits are desired, so that the

[1] See R. R. Coveyou, ''Serial Correlation in the Generation of Pseudo-Random Numbers,'' *Journal of the Association of Computing Machinery,* **7:** 72–74, 1960.

Table 21.6 **Converting Random Integer Numbers to Uniform Random Numbers**

Random Integer Number	Uniform Random Number
3	0.4375
6	0.8125
5	0.6875
0	0.0625
7	0.9375
2	0.3125
1	0.1875
4	0.5625

possible values can be expressed as 000, 001, . . . , 999. In such a case, the usual procedure still is to use $m = 2^b$ or $m = 10^d$, so that an extremely large number of random integer numbers can be generated before the sequence starts repeating itself. However, except for purposes of calculating the next random integer number in this sequence, all but three digits of each number generated would be discarded to obtain the desired three-digit random integer number. One convention is to take the *last* three digits (i.e., the three trailing digits).

The **multiplicative congruential method** is just the special case of the mixed congruential method where $c = 0$. The **additive congruential method** also is similar, but it sets $a = 1$ and replaces c by some random number preceding x_n in the sequence, for example, x_{n-1} (so that more than one seed is required to start calculating the sequence).

Among the possible random number generators (choices of a and m) based on the multiplicative congruential method, perhaps the most widely used is the *Learmouth-Lewis generator*

$$x_{n+1} \equiv 7^5 x_n (\text{modulo } 2^{31} - 1).$$

This generator has been tested extensively, and the results of the statistical tests indicate that it is very satisfactory. Versions of this generator are used, e.g., in IBM versions of APL, in the International Mathematics and Statistics Library (IMSL) package, and in the random number generator package LLRANDOM. Tables of suitable seeds also are available.

Generating Random Observations from a Probability Distribution

Given a sequence of random integer numbers, how can one generate a sequence of random observations from a given probability distribution?

For simple discrete distributions, one answer is quite evident, as demonstrated by Example 1 in Sec. 21.1. Merely allocate the possible values of a random number to the various numbers in the probability distribution in direct proportion to the respective probabilities of those numbers.

For example, consider the probability distribution of the outcome of a throw of two dice. It is known that the probability of throwing a 2 is $\frac{1}{36}$ (as is the probability of throwing a 12), the probability of throwing a 3 is $\frac{2}{36}$, and so on. Therefore, $\frac{1}{36}$ of the possible values of a random integer number should be associated with throwing a 2, $\frac{2}{36}$ of the values with throwing a 3, and so forth. Thus, if two-digit random integer numbers are being used, 72 of the 100 values will be selected for consideration, so that a random integer number will be rejected if it takes on any one of the other 28 values. Then 2 of the 72 possible values (say, 00 and 01) will be assigned an association with throwing a 2, four of them (say 02, 03, 04, and 05) will be assigned an association with throwing a 3, and so on.

For more complicated distributions, whether discrete or continuous, a generalization of this approach called the **inverse transformation method** can sometimes be used to generate random observations. Letting X be the random variable involved, we denote the cumulative distribution function by

$$F(x) = P\{X \leq x\}.$$

Generating each observation then requires the following two steps.

Summary of Inverse Transformation Method

1. Generate a *uniform random number* r between 0 and 1.
2. Set $F(x) = r$ and solve for x, which then is the desired random observation from the probability distribution.

This procedure is illustrated in Fig. 21.1 for the case where $F(x)$ is plotted graphically and the uniform random number r happens to be 0.5269.

Although the graphical procedure illustrated by Fig. 21.1 is convenient if the simulation is done manually, the computer must revert to some alternative approach. For discrete distributions, a *table lookup approach* can be taken by constructing a table that gives a ''range'' (jump) in the value of $F(x)$ for each possible value of $X = x$. For certain *continuous* distributions, $F(x) = r$ can be solved *analytically* for x, as will now be illustrated.

Consider the **exponential distribution** (see Sec. 15.4) that has the cumulative distribution function

$$F(x) = 1 - e^{-\alpha x}, \qquad \text{for } x \geq 0,$$

where $1/\alpha$ is the mean of the distribution. Setting $F(x) = r$ thereby yields

$$1 - e^{-\alpha x} = r,$$

so that

$$e^{-\alpha x} = 1 - r.$$

Therefore, taking the natural logarithm of both sides gives

$$\ln e^{-\alpha x} = \ln (1 - r),$$

so that

$$-\alpha x = \ln (1 - r),$$

which yields

$$x = \frac{\ln (1 - r)}{-\alpha}$$

as the desired random observation from the exponential distribution. (Note that other, more complicated techniques have also been developed for the exponential distribution[1] that are faster for a computer than calculating a logarithm.)

[1] For example, see J. H. Ahrens and V. Dieter, ''Efficient Table-Free Sampling Methods for Exponential, Cauchy, and Normal Distributions,'' *Communications of the ACM.* **31:** 1330–1337, 1988.

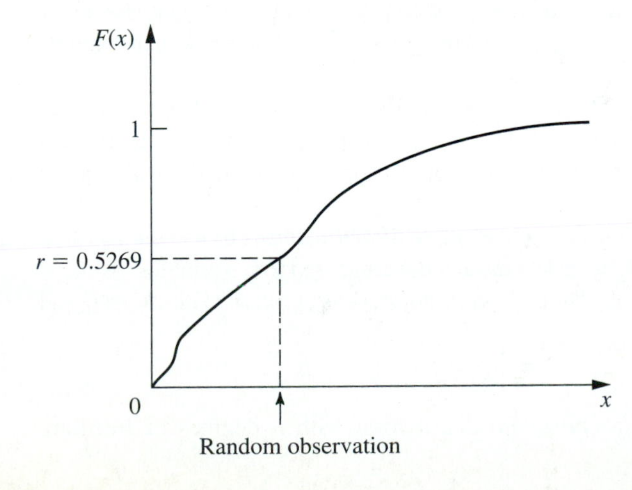

Figure 21.1 Illustration of the inverse transformation method for obtaining a random observation from a given probability distribution.

Note that $1 - r$ is itself a uniform random number. Therefore, to save a subtraction, it is common in practice simply to use the *original* uniform random number r directly in place of $1 - r$.

A natural extension of this procedure for the exponential distribution also can be used to generate a random observation from an **Erlang** (gamma) **distribution** (see Sec. 15.7). The sum of k independent exponential random variables, each with mean $1/(k\alpha)$, has the Erlang distribution with shape parameter k and mean $1/\alpha$. Therefore, given a sequence of k random numbers between 0 and 1, say, $r_1, r_2, \ldots, r_k$, the desired random observation from the Erlang distribution is

$$x = \sum_{i=1}^{k} \frac{\ln (1 - r_i)}{-k\alpha},$$

which reduces to

$$x = -\frac{1}{k\alpha} \ln \left[\prod_{i=1}^{k} (1 - r_i) \right],$$

where Π denotes multiplication. Once again, the subtractions may be eliminated simply by using r_i directly in place of $1 - r_i$.

A particularly simple technique for generating a random observation from a **normal distribution** is obtained by applying the central limit theorem. Because a uniform random number has a *uniform distribution* from 0 to 1, it has mean $\frac{1}{2}$ and standard deviation $1/\sqrt{12}$. Therefore, this theorem implies that the sum of n uniform random numbers has approximately a normal distribution with mean $n/2$ and standard deviation $\sqrt{n/12}$. Thus, if $r_1, r_2, \ldots, r_n$ are a sample of uniform random numbers, then

$$x = \frac{\sigma}{\sqrt{n/12}} \sum_{i=1}^{n} r_i + \mu - \frac{n}{2} \frac{\sigma}{\sqrt{n/12}}$$

is a random observation from an approximately normal distribution with mean μ and standard deviation σ. This approximation is an excellent one (except in the tails of the distribution), even with small values of n. Thus values of n from 5 to 10 may be adequate; $n = 12$ also is a convenient value, because it eliminates the square root terms from the preceding expression.

Since tables of the normal distribution are widely available (e.g., see App. 5), another simple method to generate a close approximation of a random observation is to use such a table to implement the inverse transformation method directly. This is fairly convenient when you are generating a few random observations by hand, but less so for computer implementation since it requires storing a large table and then using a table lookup.

Various *exact* techniques for generating random observations from a normal distribution have also been developed.[1] These exact techniques are sufficiently fast that, in practice, they generally are used instead of the approximate methods described above.

A simple method for handling the **chi-square distribution** is to use the fact that it is obtained by summing squares of standardized normal random variables. Thus, if $y_1, y_2, \ldots, y_n$ are n random observations from a normal distribution with mean 0 and standard deviation 1, then

$$x = \sum_{i=1}^{n} y_i^2$$

is a random observation from a chi-square distribution with n degrees of freedom.

[1] Ibid.

For many continuous distributions, it is not feasible to apply the inverse transformation method because $x = F^{-1}(r)$ cannot be computed (or at least computed efficiently). Therefore, several other types of methods have been developed to generate random observations from such distributions. Frequently, these methods are considerably faster than the inverse transformation method even when the latter method can be used. To provide some notion of the approach for these alternative methods, we now illustrate one called the **acceptance-rejection method** on a simple example.

Consider the *triangular distribution* having the probability density function

$$f(x) = \begin{cases} x & \text{if } 0 \le x \le 1, \\ 1 - (x - 1) & \text{if } 1 \le x \le 2, \\ 0 & \text{otherwise.} \end{cases}$$

The acceptance-rejection method uses the following two steps (perhaps repeatedly) to generate a random observation.

1. Generate a uniform random number r_1 between 0 and 1, and set $x = 2r_1$ (so that the range of possible values of x is 0 to 2).
2. Accept x with

$$\text{Probability} = \begin{cases} x & \text{if } 0 \le x \le 1, \\ 1 - (x - 1) & \text{if } 1 \le x \le 2, \end{cases}$$

   to be the desired random observation [since this probability equals $f(x)$]. *Otherwise, reject* x and repeat the two steps.

To randomly generate the event of accepting (or rejecting) x according to this probability, the method implements step 2 as follows:

2. Generate a uniform random number r_2 between 0 and 1.

$$\text{Accept } x \quad \text{if } r_2 \le f(x).$$

$$\text{Reject } x \quad \text{if } r_2 > f(x).$$

   If x is rejected, repeat the two steps.

Because $x = 2r_1$ is being accepted with a probability $= f(x)$, the probability distribution of *accepted values* has $f(x)$ as its density function, so accepted values are valid *random observations* from $f(x)$.

We were fortunate in this example that the *largest* value of $f(x)$ for any x was exactly 1. If this largest value were $L \neq 1$ instead, then r_2 would be multiplied by L in step 2. With this adjustment, the method is easily extended to other probability density functions over a finite interval, and similar concepts can be used over an infinite interval as well.

Preparing a Simulation Program

A number of detailed decisions confront the person or team who must write the computer program for executing a simulation. Although an extensive discussion of these issues is beyond the scope of this book, we shall mention several major considerations.

The basic purpose of most simulation studies is to compare alternatives. Therefore, the simulation program must be flexible enough to accommodate readily the alternatives that will be considered. Because it often is impossible to predict exactly what interesting alternatives will be uncovered during the course of the study, it is essential that flexibility and provision for rapid, simple modifications be built into the program.

Most of the instructions in a simulation program are logical operations, whereas the relatively little actual arithmetic work required is usually of a very simple type. This consideration should be reflected in the choice of computer equipment and programming language to be used.

The considerations just mentioned actually provided part of the motivation for the development of general **simulation programming languages**. For example, GPSS and SIMSCRIPT are two such languages that are widely used. These languages are designed especially to expedite the type of programming (and reprogramming) unique to simulation.

Simulation programming languages have several specific purposes. One is to provide a convenient means of describing the elements that commonly appear in simulation models. Another is to expedite changing the design and operating policies of the system being simulated, so that a large number of configurations (including some suggested during the course of the study) can be considered easily. Another service provided by the simulation languages is some type of internal timing and control mechanism, with related commands, to assist in the kind of bookkeeping that is required in executing a simulation run. These languages also are designed to obtain data and statistics conveniently on the aggregate behavior of the system being simulated. Finally, these languages provide simple operational procedures, such as introducing changes into the simulation model, initializing the state of the model, altering the kind of output data to be generated, and stacking a series of simulation runs.

For all these reasons, a simulation program often should be written in one of these simulation languages rather than in a general programming language. The tremendous savings in programming time ordinarily provided by the simulation languages usually compensate for any slight loss in computer running time.

Finally, it should be emphasized that the strategy of the simulation study should be planned carefully before the simulation program is finished. Merely letting the computer compile masses of data in a blind search for attractive alternatives is far from adequate. Simulation basically is a means for conducting an experimental investigation. Therefore, just as with a physical experiment, careful attention should be given to the construction of a theory of formal hypotheses to be tested and to the skillful design of a statistical experiment that will yield valid conclusions. This subject is discussed in Secs. 21.3 and 21.4.

Validating the Model

The typical simulation model consists of a high number of elements, rules, and logical linkages. Therefore, even when the individual components have been carefully tested, numerous small approximations can still cumulate into gross distortions in the output of the overall model. Consequently, after the writing and debugging of the computer program, it is important to test the *validity* of the model for reasonably predicting the aggregate behavior of the system being simulated.

When some form of the real system has already been in operation, its performance data should be compared with the corresponding output data from the model. Standard statistical tests can sometimes be used to determine whether the differences in the means, variances, and probability distributions generating the two sets of data are statistically significant. The time-dependent behavior of the data might also be compared statistically. If the data are not amenable to statistical analysis, another approach is to ask personnel familiar with the behavior of the real system if they can discriminate between the two sets of data.

If the model is intended to simulate alternative design configurations or operating policies for a proposed system for which no actual data are available, it may be worthwhile to conduct a *field test* to collect some real data to compare with the output of the model. Conducting such a test might involve constructing a small prototype of some version of the proposed system and placing it into operation. Another possibility might be to temporarily alter an existing system to correspond to one of the proposals.

However, field tests frequently are too expensive and time-consuming to be used. Without any real data as a standard of comparison, the only way to validate the overall model is to have knowledgeable people carefully check the credibility of output data for a variety of situations. Even when no basis exists for checking the reasonableness of the data for a *single* situation, some conclusions usually can be drawn about how the *relative* performance of the system should change as various parameters are changed. It is especially important to convince the decision makers of the credibility of the model, so they will be willing to use it at least to aid their decisions. If the model may be used again in the future, careful records of its prediction and of actual results should be kept to continue the validation process.

21.3 Experimental Design for Simulation

Selecting a Statistical Procedure[1]

The underlying statistical theory applicable to *simulated* experimentation is essentially indistinguishable from that for *physical* experimentation. Thus the design of a simulated experiment should be based upon the large body of knowledge comprising the science of statistics.

There are, however, differences between physical and simulated experimentation regarding the emphasis placed on using the various types of statistical procedures. Physical experiments frequently involve testing hypotheses about the value of a population parameter or about the equality of several population means. Simulated experiments typically place more emphasis on *optimization*. It probably is taken for granted that alternative design configurations have different population means for the measure of performance of the system. Instead, the objective of the simulation study often is to find the alternative yielding the largest mean for the measure of performance.[2] Hence *multiple decision tests* and complete or partial *ordering procedures* frequently are appropriate for simulated experiments. Furthermore, *sequential procedures* tend to be useful, both because the evolution of the experiments may be difficult to predict and because a simulated experiment often can be resumed relatively easily.

[1] This subsection assumes some knowledge of statistical procedures.

[2] In some cases, however, the objective is just to *describe* the performance of proposed systems or policies for management's evaluation and decision making, so *point estimates* and *confidence intervals* probably would be obtained. Simulated experiments also are occasionally conducted to determine which factors significantly influence the performance of the system (perhaps to guide subsequent experimentation), in which case *analysis of variance* probably would be used.

Another difference between these two types of experiments is the degree to which the experimental conditions can be held constant when alternatives are compared. Only simulated experiments can control the variability in the behavior of the elements of the system during the course of the experiment. By reproducing the same sequence of random numbers for each alternative simulated, often it is possible to reproduce an identical sequence of events. This reproduction sharpens the contrast between alternatives by reducing the residual variation in the differences in the aggregate performance of the system, so that much smaller sample sizes are required to detect statistically significant differences. Therefore, this approach usually is far superior to generating new random numbers for each alternative.

The fact that reproducing the same random numbers does not yield statistically independent results should not be of great concern. The correct procedure for comparing only two alternatives is to pair the results regarding the aggregate performance of the system that were produced by the same events. Because these pairs of results are obtained under the same experimental conditions, the *differences* between them become the relevant sample observations. This sample would be used to test the hypothesis that the mean of these differences is zero and to obtain a confidence interval for this mean. This result would thereby indicate whether there is a statistically significant difference between the means of the measure of performance for the two alternatives. If more than two alternatives need to be compared, the *Bonferroni inequality* can be used to construct *simultaneous confidence intervals* on the means of the differences for the various pairs of alternatives.

Often it is possible to express the alternatives in terms of the values of one or more continuous design variables. In these cases, there actually are an infinite number of alternatives (although the differences among some of them are minute). Because it would be impossible to simulate all of them, it is necessary to take a selective sample of these alternatives and then estimate the value of the design variables that will maximize some expected measure of performance for the system. There exists considerable literature that gives efficient procedures for experimentally determining the maximum of a mathematical function to within a specified accuracy.

Variance-Reducing Techniques

Because considerable computer time usually is required for simulation runs, it is important to obtain as much and as precise information as possible from the amount of simulation that can be done. Unfortunately, there has been a tendency in practice to apply simulation uncritically without giving adequate thought to the efficiency of the experimental design. This tendency has occurred despite the fact that considerable progress has been made in developing special techniques for increasing the precision (i.e., decreasing the variance) of sample estimators.

These variance-reducing techniques often are called **Monte Carlo techniques** (a term sometimes applied to simulation in general). Because they tend to be rather sophisticated, it is not possible to explore them deeply here. However, we shall attempt to impart the flavor of these techniques and the great increase in precision they sometimes provide by presenting two in the following example.

Consider the exponential distribution whose parameter has a value of 1. Thus its probability density function is $f(x) = e^{-x}$, as shown in Fig. 21.2, and its cumulative distribution function is $F(x) = 1 - e^{-x}$. It is known that the mean of this distribution is 1. However, suppose that this mean is not known and that we want to estimate this mean by using simulation.

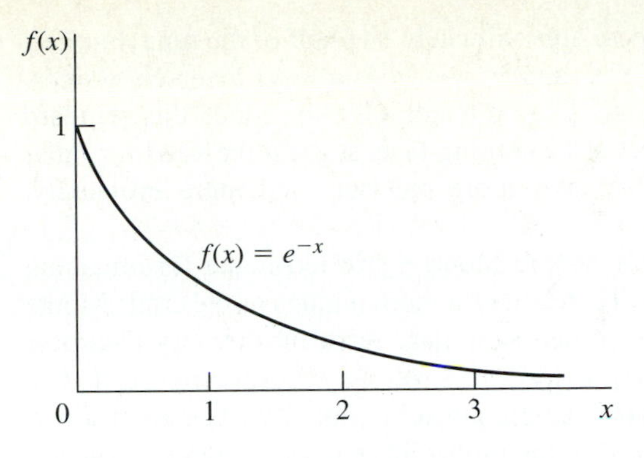

Figure 21.2 Probability density function for the example.

To provide a standard of comparison of the two variance-reducing techniques, we consider first the straightforward simulation approach, sometimes called the **crude Monte Carlo technique**. This approach involves generating some *random observations* from the exponential distribution under consideration and then using the *average* of these observations to estimate the mean. As described in Sec. 21.2, these random observations would be

$$x_i = -\ln(1 - r_i), \qquad \text{for } i = 1, 2, \ldots, n,$$

where $r_1, r_2, \ldots, r_n$ are uniform random numbers between 0 and 1. We use the first three digits in the fifth column of Table 21.4 to obtain 10 such uniform random numbers; the resulting random observations are shown in Table 21.7. (These same random numbers also are used to illustrate the variance-reducing techniques to sharpen the comparison.)

Notice that the sample average in Table 21.7 is 0.779, as opposed to the true mean of 1.000. However, because the standard deviation of the sample average happens to be $1/\sqrt{n}$, or $1/\sqrt{10}$ in this case (as could be estimated from the sample), an

Table 21.7 **Example for the Crude Monte Carlo Technique**

i	Random Number* r_i	Random Observation $x_i = -\ln(1 - r_i)$
1	0.495	0.684
2	0.335	0.408
3	0.791	1.568
4	0.469	0.633
5	0.279	0.328
6	0.698	1.199
7	0.013	0.014
8	0.761	1.433
9	0.290	0.343
10	0.693	1.183
		Total = 7.793
		Estimate of mean = 0.779

* Actually, 0.0005 was added to the indicated value for each of the r_i, so that the range of their possible values would be from 0.0005 to 0.9995 rather than from 0.000 to 0.999.

error of this amount or larger would occur approximately one-half of the time. Furthermore, because the standard deviation of a sample average is always inversely proportional to $\sqrt{n}$, this sample size would need to be quadrupled to reduce this standard deviation by one-half. These somewhat disheartening facts suggest the need for other techniques that would obtain such estimates more precisely and more efficiently.

STRATIFIED SAMPLING: A relatively simple Monte Carlo technique for obtaining better estimates is **stratified sampling**. There are two shortcomings of the crude Monte Carlo approach that are rectified by stratified sampling. First, by the very nature of randomness, a random sample may not provide a particularly uniform cross section of the distribution. For example, the random sample given in Table 21.7 has no observations between 0.014 and 0.328, even though the probability that a random observation will fall inside this interval is greater than $\frac{1}{4}$. Second, certain portions of a distribution may be more critical than others for obtaining a precise estimate, but random sampling gives no special priority to obtaining observations from these portions. For example, the tail of an exponential distribution is especially critical in determining its mean. However, the random sample in Table 21.7 includes no observations larger than 1.568, even though there is at least a small probability of *much* larger values. This explanation is the basic one for why this particular sample average is far below the true mean. Stratified sampling circumvents these difficulties by dividing the distribution into portions called *strata,* where each stratum would be sampled individually with disproportionately heavy sampling of the more critical strata.

To illustrate, suppose that the distribution is divided into three strata in the manner shown in Table 21.8. These strata were chosen to correspond to observations approximately from 0 to 1, from 1 to 3, and from 3 to ∞, respectively. To ensure that the random observations generated for each stratum actually lie in that portion of the distribution, the uniform random numbers must be converted to the indicated range for $F(x)$, as shown in the third column of Table 21.8. The number of observations to be generated from each stratum is given in the fourth column.[1] The rightmost column then shows the resulting *sampling weight* for each stratum, i.e., the *ratio* of the *sampling proportion* (the fraction of the total sample to be drawn from the stratum) to the *distribution proportion* (the probability of a random observation falling inside the stratum). These sampling weights roughly reflect the relative importance of the respective strata in determining the mean.

[1] These sample sizes are roughly based on a recommended guideline that they be proportional to the *product* of the *probability* of a random observation's falling inside the corresponding stratum *times* the *standard deviation* within this stratum.

Table 21.8 **Formulation of the Stratified Sampling Example**

Stratum	Portion of Distribution	Stratum Random No.	Sample Size	Sampling Weight
1	$0 \leq F(x) \leq 0.64$	$r_i' = 0 + 0.64r_i$	4	$w_i = \dfrac{4/10}{0.64} = \dfrac{5}{8}$
2	$0.64 \leq F(x) \leq 0.96$	$r_i' = 0.64 + 0.32r_i$	4	$w_i = \dfrac{4/10}{0.32} = \dfrac{5}{4}$
3	$0.96 \leq F(x) \leq 1$	$r_i' = 0.96 + 0.04r_i$	2	$w_i = \dfrac{2/10}{0.04} = 5$

Given the formulation of the stratified sampling approach shown in Table 21.8, the same uniform random numbers used in Table 21.7 yield the observations given in the fifth column in Table 21.9. However, it would not be correct to use the unweighted average of these observations to estimate the mean, because certain portions of the distribution have been sampled more than others. Therefore, before we take the average, we divide the observations from each stratum by the sampling weight for that stratum to give proportionate weightings to the different portions of the distribution, as shown in the rightmost column of Table 21.9. The resulting *weighted* average of 0.948 provides the desired estimate of the mean.

Method of Complementary Random Numbers: The second variance-reducing technique we shall mention is the method of *complementary random numbers.*[1] The motivation for this method is that the "luck of the draw" on the uniform random numbers generated may cause the average of the resulting random observations to be substantially on one side of the true mean, whereas the *complements* of those uniform random numbers (which are themselves uniform random numbers) would have tended to yield a nearly opposite result. (For example, the uniform random numbers in Table 21.7 average less than 0.5, and none are as large as 0.8, which led to an estimate substantially below the true mean.) Therefore, using *both* the original uniform random numbers *and* their complements to generate random observations and then calculating the *combined* sample average should provide a more precise estimator of the mean. This approach is illustrated in Table 21.10,[2] where the first three columns come from Table 21.7 and the two rightmost columns use the complementary uniform random numbers, which results in a combined sample average of 0.920.

[1] This method is a special case of the method of *antithetic variates,* which attempts to generate *pairs* of random observations having a high *negative* correlation, so that the combined average will tend to be closer to the mean.

[2] Note that 20 calculations of a logarithm were required in this case, in contrast to the 10 that were required by each of the preceding techniques.

Table 21.9 **Example for Stratified Sampling**

Stratum	i	Random Number r_i	Stratum Random No. r_i'	Stratum Random Observation $x_i' = -\ln(1 - r_i')$	Sampling Weight w_i	x_i'/w_i
1	1	0.495	0.317	0.381	$\frac{5}{8}$	0.610
	2	0.335	0.215	0.242	$\frac{5}{8}$	0.387
	3	0.791	0.507	0.707	$\frac{5}{8}$	1.131
	4	0.469	0.300	0.357	$\frac{5}{8}$	0.571
2	5	0.279	0.729	1.306	$\frac{5}{4}$	1.045
	6	0.698	0.864	1.995	$\frac{5}{4}$	1.596
	7	0.013	0.644	1.033	$\frac{5}{4}$	0.826
	8	0.761	0.884	2.154	$\frac{5}{4}$	1.723
3	9	0.290	0.9716	3.561	5	0.712
	10	0.693	0.9877	4.398	5	0.880

Total = 9.481
Estimate of mean = 0.948

Table 21.10 **Example for Method of Complementary Random Numbers**

i	Random Number r_i	Random Observation $x_i = -\ln(1 - r_i)$	Complementary Random Number $r_i' = 1 - r_i$	Random Observation $x_i' = -\ln(1 - r_i')$
1	0.495	0.684	0.505	0.702
2	0.335	0.408	0.665	1.092
3	0.791	1.568	0.209	0.234
4	0.469	0.633	0.531	0.756
5	0.279	0.328	0.721	1.275
6	0.698	1.199	0.302	0.359
7	0.013	0.014	0.987	4.305
8	0.761	1.433	0.239	0.272
9	0.290	0.343	0.710	1.236
10	0.693	1.183	0.307	0.366
		Total = 7.793		Total = 10.597
			Estimate of mean = $\frac{1}{2}(0.779 + 1.060)$ =	0.920

This example has suggested that the variance-reducing techniques provide a much more precise estimator of the mean than does straightforward simulation (the crude Monte Carlo technique). These results definitely were not a coincidence, as a derivation of the variance of the estimators would show. In comparison with straightforward simulation, these techniques (including several more complicated ones not presented here) do indeed provide a much more precise estimator with the same amount of computer time, or they provide an equally precise estimator with much less computer time. Despite the fact that additional analysis may be required to incorporate one or more of these techniques into the simulation study, the rewards should not be forgone readily.

Although this example was particularly simple, it is often possible, though more difficult, to apply these techniques to much more complex problems. For example, suppose that the objective of the simulation study is to estimate the mean waiting time of customers in a queueing system (such as those described in Sec. 16.1). Because both the probability distribution of interarrival times and the probability distribution of service times are involved, and because consecutive waiting times are not statistically independent, this problem may appear to be beyond the capabilities of the variance-reducing techniques. However, as has been described in detail elsewhere,[1] these techniques and others can indeed be applied to this type of problem very advantageously. For example, the method of *complementary random numbers* can be applied simply by repeating the original simulation run, substituting the complements of the original uniform random numbers to generate the corresponding random observations.

Tactical Problems

There are several special *tactical* issues that arise in connection with gathering the data from simulated experiments. We shall briefly describe these here and then subsequently elaborate on certain ways of dealing with them.

[1] S. Ehrenfeld and S. Ben-Tuvia, "The Efficiency of Statistical Simulation Procedures," *Technometrics*, **4**(2): 257–275, 1962. Also see Chap. 11 of Selected Reference 8.

Many (although not all) simulation studies are concerned with investigating systems that operate continually in a steady-state condition. Unfortunately, a simulation model cannot be operated this way; it must be started and stopped. Because of the artificiality introduced by the abrupt beginning of operation, the performance of the simulated system does not become representative of the corresponding real-world system until it, too, has essentially reached a steady-state condition (i.e., until the probability distribution of the state of the simulated system has essentially reached a limiting *equilibrium* distribution). Thus one tactical problem is how to obtain data that are relevant for predicting the *steady-state* behavior of the real system.

The traditional way of dealing with this problem is to run the simulation model for some time without collecting data until it is believed that the simulated system has essentially reached a steady-state condition. Unfortunately, it is difficult to estimate just how long this *stabilization period* needs to be. Furthermore, available analytical results suggest that a surprisingly long period is required, so that a great deal of unproductive computer time must be expended. Section 21.4 presents a statistical approach that eliminates these difficulties.

A related tactical issue is the selection of the *starting conditions* for the simulated system. The traditional recommendation is that the simulated system be started in a state as representative of steady-state conditions as possible to minimize the required length of the stabilization period. However, the underlying objective of the simulated experiment is to *estimate* these conditions, so little advance information may be available to guide the selection in this way. The procedure in Sec. 21.4 also eliminates this difficulty.

Most statistical sampling procedures assume that the experimental output data are in the form of a collection of distinct and statistically independent random observations from some underlying probability distribution. By contrast, because of the nature of the problems for which simulation is used, the observations from a simulated experiment are likely to be highly correlated. For example, there is a high correlation between the waiting times of consecutive customers in a queueing system. Furthermore, many measures of performance are such that the simulated experiment yields this measure continuously as a function of time rather than as a sequence of separate observations. Thus another tactical problem is how to collect the data so as to circumvent these difficulties.

One traditional method is to execute a series of completely separate and independent simulation runs of equal length and to use the average measure of performance for each run (excluding the initial stabilization period) as an individual observation. The main disadvantage is that each run requires an initial stabilization period for approaching a steady-state condition, so that much of the simulation time is unproductive. The second traditional method eliminates this disadvantage by making the runs consecutively, using the ending condition of one run as the steady-state starting condition for the next run. In other words, one continuous overall simulation run (except for the one initial stabilization period) is divided for bookkeeping purposes into a series of equal portions (runs). The average measure of performance for each portion is then treated as an individual observation. The disadvantage of this method is that it does not eliminate the correlation between observations entirely, even though it may reduce it considerably by making the portions sufficiently long.

Once again, these difficulties are eliminated by the statistical approach described in the next section.

21.4 Regenerative Method of Statistical Analysis

We have just described several difficult tactical problems in gathering data from simulated experiments and the shortcomings of traditional statistical procedures in dealing with these problems. We now present an innovative statistical approach that is especially designed to eliminate these problems.

The basic concept underlying this approach is that for many systems a simulation run can be divided into a series of **cycles** such that the evolution of the system in a cycle is a probabilistic replica of the evolution in any other cycle. Thus, if we calculate an appropriate measure of the length of the cycle along with some *statistic* to summarize the behavior of interest within each cycle, these statistics for the respective cycles constitute a series of independent and identically distributed observations that can be analyzed by standard statistical procedures. Because the system keeps going through these independent and identically distributed cycles regardless of whether it is in a steady-state condition, these observations are directly applicable from the outset for estimating the steady-state behavior of the system.

For cycles to possess these properties, they must each *begin* at the same **regeneration point**, i.e., at the point where the system probabilistically restarts, and can proceed without any knowledge of its past history. The system can be viewed as *regenerating* itself at this point in the sense that the probabilistic structure of the future behavior of the system depends upon being at this point and not on anything that happened previously. (This property is the *Markovian property* described in Sec. 14.2 for Markov chains.) A cycle *ends* when the system again reaches the regeneration point (when the next cycle begins). Thus the **length of a cycle** is the elapsed time between consecutive occurrences of the regeneration point. This elapsed time is a random variable that depends upon the evolution of the system.

When *next-event incrementing* is used, a typical regeneration point is a point at which an event has just occurred but no future events have yet been scheduled. Thus nothing needs to be known about the history of previous schedulings, and the simulation can start from scratch in scheduling future events. When *fixed-time incrementing* is used, a regeneration point is a point at which the probabilities of possible events occurring during the next unit of time do not depend upon when any past events occurred, only on the current state of the system.

Not every system possesses regeneration points, so this **regenerative method** of collecting data cannot always be used. Furthermore, even when there are regeneration points, the one chosen to define the beginning and ending points of the cycles must recur frequently enough that a substantial number of cycles will be obtained with a reasonable amount of computer time.[1] Thus some care must be taken to choose a suitable regeneration point.

Perhaps the most important application of the regenerative method to date has been to simulation of queueing systems, including queueing networks (see Sec. 15.9) such as the ones that arise in computer modeling.[2]

[1] The theoretical requirements for the method are that the expected cycle length be *finite* and that the number of cycles would go to infinity if the system continued operating indefinitely.

[2] See, e.g., D. L. Iglehart and G. S. Shedler, *Regenerative Simulation of Passage Times in Networks of Queues,* Lecture Notes in Control and Information Sciences, vol. 4, Springer-Verlag, New York, 1980. For another exposition that emphasizes applications to computer system modeling, see G. S. Shedler, *Regeneration and Networks of Queues,* Springer-Verlag, New York, 1987.

Example

Suppose that information needs to be obtained about the steady-state behavior of a system that can be formulated as a *single-server queueing system* (see Sec. 15.2). However, both the interarrival and service times have a *discrete uniform distribution* with a probability of $\frac{1}{10}$ of the values of 6, 8, . . . , 24 and the values of 1, 3, . . . , 19, respectively. Because analytical results are not available, simulation with *next-event incrementing* is to be used to obtain the desired results.

Except for the distributions involved, the general approach is the same as that described in Sec. 21.1 for Example 2. In particular, the building blocks of the simulation model are the same as specified there, including defining the state of the system as the number of customers in the system. Suppose that one-digit random integer numbers are used to generate the random observations from the distributions, as shown in Table 21.11. Beginning the simulation run with no customers in the system then yields the results summarized in Table 21.12 and Fig. 21.3, where the random numbers are obtained sequentially as needed from the tenth row of Table 21.4.[1] (Note in Table 21.12 that, at time 98, the arrival of one customer and the service completion for another customer occur simultaneously, so these canceling events are not visible in Fig. 21.3.)

For this system, one *regeneration point* is where an *arrival* occurs with *no* previous customers left. At this point, the process probabilistically restarts, so the probabilistic structure of when future arrivals and service completions will occur is completely independent of any previous history. The only relevant information is that the system has just entered the special state of having had no customers *and* having the time until the next arrival reach zero. The simulation run would not previously have scheduled any future events but would now generate *both* the next interarrival time and the service time for the customer that just arrived.

The only other regeneration points for this system are where an arrival and a service completion occur simultaneously, with a prespecified number of customers in the system. However, the regeneration point described in the preceding paragraph occurs much more frequently and thus is a better choice for defining a cycle. With this selection, the first five complete cycles of the simulation run are those shown in Fig. 21.3. (In most cases, you should have a considerably larger number of cycles in the entire simulation run in order to have sufficient precision in the statistical analysis.)

Various types of information about the steady-state behavior of the system can be obtained from this simulation run, including *point estimates* and *confidence intervals*

[1] When both an interarrival time and a service time need to be generated at the same time, the interarrival time is obtained first.

Table 21.11 **Correspondence between Random Numbers and Random Observations for the Queueing System Example**

Random Number	Interarrival Time	Service Time
0	6	1
1	8	3
⋮	⋮	⋮
9	24	19

Table 21.12 **Simulation Run for the Queueing System Example**

Time	Number of Customers	Random Number	Next Arrival	Next Service Completion
0	0	9	24	—
24	1	2, 6	34	37
34	2	4	48	37
37	1	6	48	50
48	2	4	62	50
50	1	1	62	53
53	0	—	62	—
62	1	1, 1	70	65
65	0	—	70	—
70	1	3, 9	82	89
82	2	1	90	89
89	1	4	90	98
90	2	1	98	98
98	2	1, 5	106	109
106	3	6	124	109
109	2	2	124	114
114	1	1	124	117
117	0	—	124	—
124	1	5, 6	140	137
137	0	—	140	—
140	1	9, 3	164	147
147	0	—	164	—
164	1			

for the expected number of customers in the system, the expected waiting time, and so on. In each case, it is necessary to use only the corresponding statistics from the respective cycles and the lengths of the cycles. We shall first present the general statistical expressions for the regenerative method and then apply them to this example.

Statistical Formulas

Formally speaking, the statistical problem for the regenerative method is to obtain estimates of the expected value of some random variable X of interest. This estimate is

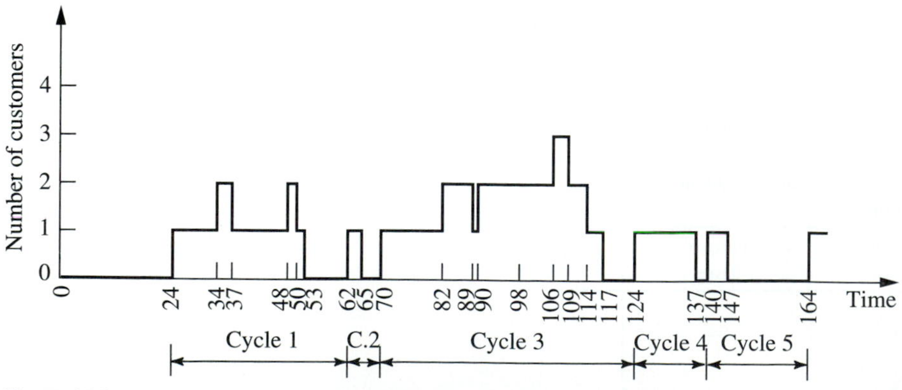

Figure 21.3 Outcome of the simulation run for the queueing system example.

to be obtained by calculating a statistic Y for each cycle and an appropriate measure Z of the *size* of the cycle such that

$$E(X) = \frac{E(Y)}{E(Z)}.$$

(The regenerative property ensures that such a *ratio formula* holds for many steady-state random variables X.) Thus, if n complete cycles are generated during the simulation run, the data gathered are $Y_1, Y_2, \ldots, Y_n$ and $Z_1, Z_2, \ldots Z_n$ for the respective cycles.

By letting $\bar{Y}$ and $\bar{Z}$, respectively, denote the sample averages for these two sets of data, the corresponding *point estimate* of $E(X)$ would be obtained from the formula

$$\text{Est}\,\{E(X)\} = \frac{\bar{Y}}{\bar{Z}}.$$

To obtain a *confidence interval* for $E(X)$, we must first calculate several quantities from the data. These quantities include the *sample variances*

$$s_{11}^2 = \frac{1}{n-1}\sum_{i=1}^{n}(Y_i - \bar{Y})^2 = \frac{1}{n-1}\sum_{i=1}^{n}Y_i^2 - \frac{1}{n(n-1)}\left(\sum_{i=1}^{n}Y_i\right)^2,$$

$$s_{22}^2 = \frac{1}{n-1}\sum_{i=1}^{n}(Z_i - \bar{Z})^2 = \frac{1}{n-1}\sum_{i=1}^{n}Z_i^2 - \frac{1}{n(n-1)}\left(\sum_{i=1}^{n}Z_i\right)^2,$$

and the combined *sample covariance*

$$s_{12}^2 = \frac{1}{n-1}\sum_{i=1}^{n}(Y_i - \bar{Y})(Z_i - \bar{Z})$$

$$= \frac{1}{n-1}\sum_{i=1}^{n}Y_iZ_i - \frac{1}{n(n-1)}\left(\sum_{i=1}^{n}Y_i\right)\left(\sum_{i=1}^{n}Z_i\right).$$

Also let

$$s^2 = s_{11}^2 - 2\frac{\bar{Y}}{\bar{Z}}s_{12}^2 + \left(\frac{\bar{Y}}{\bar{Z}}\right)^2 s_{22}^2.$$

Finally, let α be the constant such that $1 - 2\alpha$ is the desired *confidence coefficient* for the confidence interval, and look up K_α in Table A5.1 (see App. 5) for the normal distribution. If n is not too small, an *asymptotic confidence interval* for $E(X)$ is then given by

$$\frac{\bar{Y}}{\bar{Z}} - \frac{K_\alpha\, s}{\bar{Z}\sqrt{n}} \le E(X) \le \frac{\bar{Y}}{\bar{Z}} + \frac{K_\alpha\, s}{\bar{Z}\sqrt{n}};$$

i.e., the probability is approximately $1 - 2\alpha$ that the endpoints of an interval generated in this way will surround the actual value of $E(X)$.

Application of the Statistical Formulas to the Example

Consider first how to estimate the *expected waiting time* for a customer *before* beginning service (denoted by W_q in Chap. 15). Thus the random variable X now represents a customer's waiting time excluding service, so that

$$W_q = E(X).$$

The corresponding information gathered during the simulation run is the *actual* waiting time (excluding service) incurred by the respective customers. Therefore, for each cycle, the summary statistic Y is the *sum of the waiting times,* and the size of the cycle Z is the *number of customers,* so that

$$W_q = \frac{E(Y)}{E(Z)}.$$

Refer to Fig. 21.3 and Table 21.12; for cycle 1, a total of three customers are processed, so $Z_1 = 3$. The first customer incurs no waiting before beginning service, the second waits 3 units of time (from 34 to 37), and the third waits 2 units of time (from 48 to 50), so $Y_1 = 5$. We proceed similarly for the other cycles. The data for the problem are

$$\begin{aligned} Y_1 &= 5, & Z_1 &= 3 \\ Y_2 &= 0, & Z_2 &= 1 \\ Y_3 &= 34, & Z_3 &= 5 \\ Y_4 &= 0, & Z_4 &= 1 \\ Y_5 &= 0, & Z_5 &= 1 \\ \overline{Y} &= 7.8, & \overline{Z} &= 2.2. \end{aligned}$$

Therefore, the *point estimate* of W_q is

$$\text{Est}\ \{W_q\} = \frac{\overline{Y}}{\overline{Z}} = \frac{7.8}{2.2} = 3\tfrac{6}{11}.$$

To obtain a 95 percent confidence interval for W_q, the preceding formulas are first used to calculate

$$s_{11}^2 = 219.20, \qquad s_{22}^2 = 3.20, \qquad s_{12}^2 = 24.80, \qquad s = 9.14.$$

Because $1 - 2\alpha = 0.95$, then $\alpha = 0.025$, so that $K_\alpha = 1.96$ from Table A5.1. The resulting confidence interval is

$$-0.09 \leq W_q \leq 7.19;$$

or

$$W_q \leq 7.19.$$

The reason that this confidence interval is so wide (even including impossible negative values) is that the number of sample observations (cycles), $n = 5$, is so small. Note in the general formula that the width of the confidence interval is *inversely proportional* to the *square root* of n, so that, e.g., quadrupling n reduces the width by half (assuming no change in s or $\overline{Z}$). Given preliminary values of s and $\overline{Z}$ from a short preliminary simulation run (such as the run in Table 21.12), this relationship makes it possible to estimate in advance the width of the confidence interval that would result from any given choice of n for the full simulation run. The final choice of n can then be made based on the trade-off between computer time and the precision of the statistical analysis.

Now suppose that this simulation run is to be used to estimate P_0, the probability of having no customers in the system. (Because λ/μ is the utilization factor for the server in a single-server queueing system, the theoretical value is known to be $P_0 = 1 - \lambda/\mu = 1 - \frac{1}{15}/\frac{1}{10} = \frac{1}{3}$.) The corresponding information obtained during the simulation run is the fraction of time during which the system is empty. Therefore, the summary statistic Y for each cycle is the *total time* during which no customers are present, and the size Z is the *length* of the cycle, so that

$$P_0 = \frac{E(Y)}{E(Z)}.$$

The length of cycle 1 is 38 (from 24 to 62), so that $Z_1 = 38$. During this time, the system is empty from 53 to 62, so that $Y_1 = 9$. Proceeding in this manner for the other cycles, we obtain the following data for the problem:

$$\begin{aligned} Y_1 &= 9, & Z_1 &= 38 \\ Y_2 &= 5, & Z_2 &= 8 \\ Y_3 &= 7, & Z_3 &= 54 \\ Y_4 &= 3, & Z_4 &= 16 \\ Y_5 &= 17, & Z_5 &= 24 \\ \overline{Y} &= 8.2, & \overline{Z} &= 28. \end{aligned}$$

Thus the *point estimate* of P_0 is

$$\text{Est } \{P_0\} = \frac{8.2}{28} = 0.293.$$

By calculating

$$s_{11}^2 = 29.20, \qquad s_{22}^2 = 334, \qquad s_{12}^2 = 17, \qquad s = 6.92,$$

a 95 percent confidence interval for P_0 is found to be

$$0.076 \le P_0 \le 0.510.$$

(The wide range of this interval indicates that a much longer simulation run would be needed to obtain a relatively precise estimate of P_0.)

If we redefine Y appropriately, the same approach also can be used to estimate other probabilities involving the number of customers in the system. However, because this number never exceeded 3 during this simulation run, a much longer run will be needed if the probability involves larger numbers.

The other basic expected values of queueing theory defined in Sec. 15.2 (W, L_q, and L) can be estimated from the estimate of W_q by using the relationships among these four expected values given near the end of Sec. 15.2. However, the other expected values can also be estimated directly from the results of the simulation run. For example, because the expected number of customers waiting to be served is

$$L_q = \sum_{n=2}^{\infty} (n-1)P_n,$$

it can be estimated by defining

$$Y = \sum_{n=2}^{\infty} (n-1)T_n,$$

where T_n is the *total time* that exactly n customers are in the system during the cycle. (This definition of Y actually is equivalent to the definition used for estimating W_q.) In this case, Z is defined as it would be for estimating any P_n, namely, the *length* of the cycle. The resulting *point estimate* of L_q then turns out to be simply the *point estimate* of W_q *multiplied by* the actual *average arrival rate* for the complete cycles observed.

It is also possible to estimate *higher moments* of these probability distributions by redefining Y accordingly. For example, the *second moment* about the origin of the number of customers waiting to be served N_q

$$E(N_q^2) = \sum_{n=2}^{\infty} (n-1)^2 P_n$$

can be estimated by redefining

$$Y = \sum_{n=2}^{\infty} (n-1)^2 T_n.$$

This point estimate, along with the point estimate of L_q (the first moment of N_q) just described, can then be used to estimate the *variance* of N_q. Specifically, because of the general relationship between variance and moments, this variance is

$$\text{Var}(N_q) = E(N_q^2) - L_q^2.$$

Therefore, its point estimate is obtained by substituting in the point estimates of the quantities on the right-hand side of this relationship.

Finally, we should mention that it was unnecessary to generate the first *interarrival* time (24) for the simulation run summarized in Table 21.12 and Fig. 21.3, because this time played no role in the statistical analysis. It is more efficient with the regenerative method just to start the run at the regeneration point.

Selected Reference 4 provides considerably more information about the regenerative method, including how it can be applied to more complicated kinds of problems than those considered here. (Also see the references given in the second footnote at the beginning of this section.)

21.5 Conclusions

Simulation is a widely used tool for estimating the performance of complex stochastic systems if contemplated designs or operating policies are to be used.

We have focused in this chapter on the use of simulation for predicting the *steady-state* behavior of systems whose states change only at discrete points in time. However, by having a series of runs begin with the prescribed *starting conditions,* we can also use simulation to describe the *transient* behavior of a proposed system. Furthermore, if we use differential equations, simulation can be applied to systems whose states change *continuously* with time.

Simulation is indeed a very versatile tool. However, it is by no means a panacea. Simulation is inherently an imprecise technique. It provides only *statistical estimates* rather than exact results, and it *compares alternatives* rather than generating an optimal one. Furthermore, simulation is a *slow and costly* way to study a problem. It usually requires a large amount of time and expense for analysis and programming, in addition to considerable computer running time. Simulation models tend to become unwieldy, so that the number of cases that can be run and the accuracy of the results obtained often turn out to be very inadequate. Finally, simulation yields only *numerical data* about the performance of the system, so that it provides no additional insight into the cause-and-effect relationships within the system except for the clues that can be gleaned from these numbers (and from the analysis required to construct the simulation model). Therefore, it is very expensive to conduct a sensitivity analysis of the parameter values assumed by the model. The only possible way would be to conduct new series of simulation runs with different parameter values, which would tend to provide relatively little information at a relatively high cost.

Simulation provides a way of *experimenting* with proposed systems or policies without actually implementing them. Sound statistical theory should be used in designing these experiments. Surprisingly long simulation runs often are needed to obtain *statistically significant* results. However, *variance-reducing techniques* can be very helpful in reducing the length of the runs needed.

Several *tactical* problems arise when we apply traditional statistical estimation procedures to simulated experiments. These problems include prescribing appropriate *starting conditions,* determining when a *steady-state condition* has essentially been reached, and dealing with *statistically dependent* observations. These problems can be eliminated by using the *regenerative method* of statistical analysis. However, there are some restrictions on when this method can be applied.

Simulation unquestionably has an important place in the theory and practice of OR. It is an invaluable tool for use on those problems where analytical techniques are inadequate, and it is widely used.

SELECTED REFERENCES

1. Banks, J., and J. S. Carson, II, *Discrete-Event System Simulation,* Prentice-Hall, Englewood Cliffs, NJ, 1984.
2. Carrie, A., *Simulation of Manufacturing Systems,* Wiley, New York, 1988.
3. Christy, D. P., and H. J. Watson, ''The Application of Simulation: A Survey of Industry Practice,'' *Interfaces,* **13**(5): 47–52, October 1983.
4. Crane, M. A., and A. J. Lemoine, *An Introduction to the Regenerative Method for Simulation Analysis,* Springer-Verlag, Berlin, 1977.
5. Hoover, S. V., and R. F. Perry, *Simulation: A Problem Solving Approach,* Addison-Wesley, Reading, MA, 1989.
6. Kleijnen, J. P. C., *Statistical Tools for Simulation Practitioners,* Marcel Dekker, New York, 1987.
7. Law, A. M., ''Introduction to Simulation: A Powerful Tool for Complex Manufacturing Systems,'' *Industrial Engineering,* **18**(5): 46–63, May 1986.
8. Law, A. M., and W. D. Kelton, *Simulation Modeling and Analysis,* 2d ed., McGraw-Hill, New York, 1991.
9. Pegden, C. D., R. P. Sadowski, and R. E. Shannon, *Introduction to Simulation Using SIMAN,* McGraw-Hill, New York, 1991.
10. Pooch, U. W., and J. A. Wall, *Discrete Event Simulation: A Practical Approach,* CRC Press, Boca Raton, FL, 1992.

11. Ross, S. M., *A Course in Simulation,* Macmillan, New York, 1990.
12. Schriber, T. J., *An Introduction to Simulation,* Wiley, New York, 1991.
13. Taha, H. A., *Simulation Modeling and SIMNET,* Prentice-Hall, Englewood Cliffs, NJ, 1987.
14. Watson, H. J., *Computer Simulation in Business,* Wiley, New York, 1981.

RELEVANT ROUTINES IN YOUR OR COURSEWARE

Demonstration examples: *Simulating a Basic Queueing System*
Simulating a Queueing System with Priorities
Interactive routines: *Enter Queueing Problem*
Interactively Simulate Queueing Problem
An automatic routine: *Automatically Simulate Queueing Problem*

To access these routines, call the ProbMod program and then choose *Simulation* under the Area menu. See App. 1 for documentation of the software.

PROBLEMS[1]

To the left of each of the following problems (or their parts), we have inserted a D (for demonstration), I (for interactive routine), or A (for automatic routine) whenever a corresponding routine listed above can be helpful. An asterisk on the I or A indicates that this routine definitely should be used (unless your instructor gives contrary instructions) and that the printout from this routine is all that needs to be turned in to show your work in executing the procedure. An asterisk on the problem number indicates that at least a partial answer is given in the back of the book.

When an OR Courseware routine is not applicable, the random numbers needed to do these problems manually should be obtained from Table 21.4. For each part, use the digits *consecutively,* starting from the front of the top row, to form *three-digit* random integer numbers 096, 569, 665, and so on.

21.1-1.* Use the one-digit integer random numbers 5, 2, 4, 9, 7 to generate random observations for each of the following situations.
(*a*) Throwing an unbiased coin
(*b*) Throwing a die
(*c*) The color of a traffic light found by a randomly arriving car when it is green 40 percent of the time, yellow 10 percent of the time, and red 50 percent of the time

21.1-2. The weather can be considered a stochastic system, because it evolves in a probabilistic manner from one day to the next. Suppose for a certain location that this probabilistic evolution satisfies the following description:

The probability of rain tomorrow is 0.6 if it is raining today.
The probability of its being clear (no rain) tomorrow is 0.8 if it is clear today.

Simulate the evolution of the weather for 10 days, beginning the day after a clear day.

21.1-3. Select one of the typical applications of simulation listed at the end of Sec. 21.1 and develop a simulation model for this type of problem. In particular, describe the simulation clock, the state of the system, the events that change the state of the system, the state transition mechanism, and the event transition mechanism.

[1] Problems 21.2-6, 21.2-9, 21.2-12, 21.2-13, 21.2-18 to 21.2-20, 21.2-24, 21.2-25, and 21.3-3 to 21.3-7 have been adapted, with permission, from previous operations research examinations given by the Society of Actuaries.

21.1-4. Consider the $M/M/1$ queueing theory model that was discussed in Sec. 15.6 and Example 2, Sec. 21.1. Suppose that the mean arrival rate is 5 per hour, the mean service rate is 10 per hour, and you are required to estimate the expected waiting time before service begins by using simulation.

(*a*) Starting with the system empty, use next-event incrementing to perform the simulation by hand until two service completions have occurred.

(*b*) Starting with the system empty, use fixed-time incrementing (with 2 minutes as the time unit) to perform the simulation by hand until two service completions have occurred.

D,I* (*c*) Use the interactive routine in your OR Courseware (which incorporates next-event incrementing) to interactively execute a simulation run until 20 service completions have occurred.

A* (*d*) Use the automatic routine in your OR Courseware to execute a long simulation run. (Use the default run size.)

A* (*e*) Use the first routine under the Procedure menu in the queueing theory area of your OR Courseware to obtain the usual measures of performance for this queueing system. Then compare these exact results with the corresponding point estimates and 95 percent confidence intervals obtained from the simulation run in part (*d*). Identify any measure whose exact result falls outside the 95 percent confidence interval.

D,I* **21.1-5.** View the first demonstration example (*Simulating a Basic Queueing System*) under the Demo menu in the simulation area of your OR Courseware. Then enter this *same problem* into the interactive routine under the Procedure menu. Interactively execute a simulation run for 20 minutes of simulated time.

D,I* **21.1-6.** View the second demonstration example (*Simulating a Queueing System with Priorities*) under the Demo memo in the simulation area of your OR Courseware. Then enter this same problem into the interactive routine under the Procedure menu. Interactively execute a simulation run for 20 minutes of simulated time.

D,I* **21.1-7.** Cars arrive at a repair shop according to a Poisson process at a mean rate of 1 per hour. The shop has two mechanics. For each mechanic, the time required to repair a car has an exponential distribution with a mean of 90 minutes. (Only one mechanic works on any one car.)

Use the interactive routine in your OR Courseware to interactively execute a simulation run for 20 hours of simulated time.

21.1-8. A factory produces two products, A and B. A single inspector inspects all units of both products before they are released. These units are brought to the inspection station one at a time as they are completed. For product A, the interarrival time has a uniform distribution between 20 and 30 minutes, and the inspection time has an Erlang distribution with a mean of 10 minutes and shape parameter $k = 4$. For product B, the interarrival time is a constant 15 minutes, and the inspection time has an exponential distribution with a mean of 5 minutes. Product A is in high demand, so the inspector gives nonpreemptive priority to product A over product B.

D,I* (*a*) Use the interactive routine in your OR Courseware to interactively execute a simulation run for 3 hours of simulated time.

A* (*b*) Use the automatic routine in your OR Courseware to execute a long simulation run. (Use the default run size.)

21.2-1.* Use the mixed congruential method to generate the following sequences of random numbers.

(*a*) A sequence of 10 *one-digit* random integer numbers such that $x_{n+1} \equiv (x_n + 3)$ (modulo 10) and $x_0 = 2$

(*b*) A sequence of eight random integer numbers between 0 and 7 such that $x_{n+1} \equiv (5x_n + 1)$(modulo 8) and $x_0 = 1$

(*c*) A sequence of five *two-digit* random integer numbers such that $x_{n+1} \equiv (61x_n + 27)$ (modulo 100) and $x_0 = 10$

21.2-2. Reconsider Prob. 21.2-1. Suppose now that you want to convert these random integer numbers to (approximate) uniform random numbers. For each of the three parts, give a formula for this conversion that makes the approximation as close as possible.

21.2-3. Use the mixed congruential method to generate a sequence of five *two-digit* random integer numbers such that $x_{n+1} \equiv (41x_n + 33)(\text{modulo } 100)$ and $x_0 = 48$.

21.2-4. Use the mixed congruential method to generate a sequence of three *three-digit* random integer numbers such that $x_{n+1} \equiv (201x_n + 503)(\text{modulo } 1{,}000)$ and $x_0 = 485$.

21.2-5. Generate five uniform random numbers.

(*a*) Prepare to do this by using the mixed congruential method to generate a sequence of five random integer numbers between 0 and 31 such that $x_{n+1} \equiv (13x_n + 15)(\text{modulo } 32)$ and $x_0 = 14$.

(*b*) Convert these random integer numbers to uniform random numbers as closely as possible.

21.2-6. You are given the *multiplicative congruential generator* $x_0 = 1$ and $x_{n+1} \equiv 7x_n$ (modulo 13) for $n = 0, 1, 2, \ldots$.

(*a*) Calculate x_n for $n = 1, 2, \ldots, 12$.

(*b*) How often does each integer between 1 and 12 appear in the sequence generated in part (*a*)?

(*c*) Without performing additional calculations, indicate how $x_{13}, x_{14}, \ldots$ will compare with $x_1, x_2, \ldots$.

21.2-7. Generate five random observations from a uniform distribution between -10 and $+40$.

21.2-8.* Suppose that random observations are needed from the triangular distribution whose probability density function is

$$f(x) = \begin{cases} 2x & \text{if } 0 \le x \le 1, \\ 0 & \text{otherwise.} \end{cases}$$

(*a*) Derive an expression for each random observation as a function of the uniform random number r.

(*b*) Generate five random observations.

21.2-9. Consider the probability distribution whose cumulative distribution function is

$$F(x) = x^2, \qquad \text{if } 0 \le x \le 1.$$

The inverse transformation method was applied to generate the following three random observations from this distribution: 0.09, 0.64, 0.49.

Identify the three uniform random numbers that were used.

21.2-10. Generate three random observations from each of the following probability distributions.

(*a*) The uniform distribution from 25 to 75.

(*b*) The distribution whose probability density function is

$$f(x) = \begin{cases} \frac{1}{4}(x+1)^3 & \text{if } -1 \le x \le 1, \\ 0 & \text{otherwise.} \end{cases}$$

(*c*) The distribution whose probability density function is

$$f(x) = \begin{cases} \frac{1}{200}(x-40) & \text{if } 40 \le x \le 60, \\ 0 & \text{otherwise.} \end{cases}$$

21.2-11. Generate three random observations from each of the following probability distributions.

(*a*) The random variable X has $P\{X = 0\} = \frac{1}{2}$. Given $X \neq 0$, it has a uniform distribution between -5 and 15.

(*b*) The distribution whose probability density function is

$$f(x) = \begin{cases} x - 1 & \text{if } 1 \le x \le 2, \\ 3 - x & \text{if } 2 \le x \le 3. \end{cases}$$

(*c*) The geometric distribution with parameter $p = \frac{1}{3}$, so that

$$P\{X = k\} = \begin{cases} \frac{1}{3}\left(\frac{2}{3}\right)^{k-1} & \text{if } k = 1, 2, \ldots, \\ 0 & \text{otherwise.} \end{cases}$$

21.2-12. The random variable X has the cumulative distribution function $F(X)$ whose value or derivative $F'(x)$ is shown below for various values of x.

$$F(0) = 0.$$

$$F'(x) = \tfrac{1}{8}, \qquad \text{for } 0 < x < 2.$$

$$P\{X = 2\} = \tfrac{1}{2}, \qquad \text{so} \qquad F(2) = \tfrac{3}{4}.$$

$$F'(x) = \tfrac{1}{4}, \qquad \text{for } 2 < x < 3.$$

$$F(3) = 1.$$

Generate four random observations from this probability distribution by using the following uniform random numbers: $\frac{3}{4}$, $\frac{1}{2}$, $\frac{1}{4}$, $\frac{7}{8}$. Also calculate the sample average and compare it with the true mean ($\frac{15}{8}$) for this probability distribution.

21.2-13. Use the inverse transformation method and the table of the normal distribution given in App. 5 (with linear interpolation between values in the table) to generate 10 random observations (to three decimal places) from a normal distribution with mean = 1 and variance = 4. Then calculate the sample average of these random observations.

21.2-14. Generate three random observations (approximately) from a normal distribution with mean = 10 and standard deviation = 5.

(*a*) Do this by applying the central limit theorem, using three uniform random numbers to generate each random observation.

(*b*) Now do this by using the table for the normal distribution given in App. 5 and applying the inverse transformation method.

21.2-15. Generate four random observations (approximately) from a normal distribution with mean = 0 and standard deviation = 1.

(*a*) Do this by applying the central limit theorem, using three uniform random numbers to generate each random observation.

(*b*) Now do this by using the table for the normal distribution given in App. 5 and applying the inverse transformation method.

(*c*) Use your random observations from parts (*a*) and (*b*) to generate random observations from a chi-square distribution with 2 degrees of freedom.

21.2-16.* Generate two random observations from each of the following probability distributions.

(*a*) The exponential distribution with mean = 4

(*b*) The Erlang distribution with mean = 4 and shape parameter $k = 2$ (that is, standard deviation = $2\sqrt{2}$)

(*c*) The normal distribution with mean = 4 and standard deviation = $2\sqrt{2}$. (Use the central limit theorem and $n = 6$ for each observation.)

21.2-17. Generate four random observations from an exponential distribution with mean = 1. Then use these four observations to generate one random observation from an Erlang distribution with mean = 4 and shape parameter $k = 4$.

21.2-18. Generate 10 random observations from the probability distribution

$$P\{X = n\} = \begin{cases} \frac{1}{10} & \text{if } n = 0, 1, 2, \ldots, 9, \\ 0 & \text{otherwise.} \end{cases}$$

(*a*) Prepare to do this by generating 16 random integer numbers from the mixed congruential generator, $x_{n+1} \equiv (5x_n + 3)(\text{modulo } 16)$ and $x_0 = 1$.
(*b*) Use the single-digit random integer numbers from part (*a*) to generate the desired random observations.
(*c*) Note that once a particular value of X is generated in part (*b*), it can never be repeated because each of the 16 possible random integer numbers is generated exactly once in part (*a*). In which ways does this violate the desirable properties of random observations? What change would you make in what was done in parts (*a*) and (*b*) to alleviate this problem?
(*d*) Now convert the first 10 random integer numbers from part (*a*) to (approximate) uniform random numbers, and then apply the inverse transformation method to obtain the desired random observations.
(*e*) Does the procedure prescribed in part (*d*) actually give a probability of $\frac{1}{10}$ of generating each of the 10 possible values of X each time? Explain. What change would you make in what was done in parts (*a*) and (*d*) to alleviate this problem?

21.2-19. Let $r_1, r_2, \ldots, r_n$ be uniform random numbers. Define $x_i = -\ln r_i$ and $y_i = -\ln(1 - r_i)$, for $i = 1, 2, \ldots, n$, and $z = \sum_{i=1}^{n} x_i$. Label each of the following statements as true or false, and then justify your answer.

(*a*) The numbers $x_1, x_2, \ldots, x_n$ and $y_1, y_2, \ldots, y_n$ are random observations from the same exponential distribution.
(*b*) The average of $x_1, x_2, \ldots, x_n$ is equal to the average of $y_1, y_2, \ldots, y_n$.
(*c*) z is a random observation from an Erlang (gamma) distribution.

21.2-20. Consider the random variable X that is uniformly distributed (equal probabilities) on the set $\{1, 2, \ldots, 8\}$. You wish to generate a series of random observations x_i ($i = 1, 2, \ldots$) of X. The following three proposals have been made for doing this. For each one, analyze whether it is a valid method and, if not, how it can be adjusted to become a valid method.

(*a*) Proposal 1: Generate uniform random numbers r_i ($i = 1, 2, \ldots$), and then set $x_i = n$, where n is the integer satisfying $n/8 \leq r_i < (n + 1)/8$.
(*b*) Proposal 2: Generate uniform random numbers r_i ($i = 1, 2, \ldots$), and then set x_i equal to the greatest integer less than or equal to $1 + 8r_i$.
(*c*) Proposal 3: Generate x_i from the mixed congruential generator $x_{n+1} \equiv (5x_n + 7)$ (modulo 8), with starting value $x_0 = 4$.

21.2-21. Use the acceptance-rejection method to generate three random observations from the triangular distribution used to illustrate this method in Sec. 21.2.

21.2-22. Use the acceptance-rejection method to generate three random observations from the probability density function

$$f(x) = \begin{cases} \frac{1}{50}(x - 10) & \text{if } 10 \leq x \leq 20, \\ 0 & \text{otherwise.} \end{cases}$$

21.2-23. The game of craps requires the player to throw two dice one or more times until a decision has been reached as to whether he wins or loses. He wins if the first throw results in a sum of 7 or 11 or, alternatively, if the first sum is 4, 5, 6, 8, 9, or 10 and the same sum reappears before a sum of 7 has appeared. Conversely, he loses if the first throw results in a sum of 2, 3, or 12 or, alternatively, if the first sum is 4, 5, 6, 8, 9, or 10 and a sum of 7 appears before the first sum reappears.

(*a*) Simulate five plays of this game to start the process of estimating the probability of winning.

(*b*) For a large number of plays of the game, the proportion of wins has *approximately* a normal distribution with mean = 0.493 and standard deviation = $0.5\sqrt{n}$. Use this information to calculate the number of simulated plays that would be required to have a probability of at least 0.95 that the proportion of wins will be less than 0.5.

21.2-24. An insurance company insures four large risks. The number of losses for each risk is independent and identically distributed on the points {0, 1, 2} with probabilities 0.7, 0.2, and 0.1, respectively. The size of an individual loss has the following cumulative distribution function:

$$F(x) = \begin{cases} \dfrac{\sqrt{x}}{20} & \text{if } 0 \le x \le 100, \\ \dfrac{x}{200} & \text{if } 100 < x \le 200, \\ 1 & \text{if } \quad x > 200. \end{cases}$$

Perform a simulation experiment twice of the total loss generated by the four large risks.

21.2-25. A company provides its three employees with health insurance under a group plan. For each employee, the probability of incurring medical expenses during a year is 0.9, so the number of employees incurring medical expenses during a year has a binomial distribution with $p = 0.9$ and $n = 3$. Given that an employee incurs medical expenses during a year, the total amount for the year has the distribution $100 with probability 0.9 or $10,000 with probability 0.1. The company has a $5,000 deductible clause with the insurance company so that each year the insurance company pays the total medical expenses for the group in excess of $5,000. Use the uniform random numbers 0.01 and 0.02, in the order given, to generate the number of claims based on a binomial distribution for each of 2 years. Use the following uniform random numbers, in the order given, to generate the amount of each claim: 0.80, 0.95, 0.70, 0.96, 0.54, 0.01. Calculate the total amount that the insurance company pays for 2 years.

21.3-1.* Consider the probability distribution whose probability density function is

$$f(x) = \begin{cases} \dfrac{1}{x^2} & \text{if } x \ge 1, \\ 0 & \text{otherwise.} \end{cases}$$

The problem is to perform a simulated experiment, with the help of variance-reducing techniques, for estimating the mean of this distribution. To provide a standard of comparison, also derive the mean analytically.

For each of the following cases, use the same 10 uniform random numbers to generate random observations, and calculate the resulting estimate of the mean.

(*a*) Use the crude Monte Carlo technique.

(*b*) Use stratified sampling with three strata—$0 \le F(x) \le 0.6$, $0.6 < F(x) \le 0.9$, and $0.9 < F(x) \le 1$—with 3, 3, and 4 observations, respectively.

(*c*) Use the method of complementary random numbers.

21.3-2. Simulation is being used to study a system whose measure of performance X will be partially determined by the outcome of a certain external factor. This factor has three possible outcomes (unfavorable, neutral, and favorable) that will occur with equal probability ($\frac{1}{3}$). Because the favorable outcome would greatly increase the spread of possible values of X, this outcome is more critical than the others for estimating the mean and variance of X. Therefore, a stratified sampling approach has been adopted, with six random observations of the value of X generated under the favorable outcome, three generated under the neutral outcome, and one generated under the unfavorable outcome, as follows:

Outcome of External Factor	Simulated Values of X
Favorable	8, 5, 1, 6, 3, 7
Neutral	3, 5, 2
Unfavorable	2

(*a*) Develop the resulting estimate of $E(X)$.
(*b*) Develop the resulting estimate of $E(X^2)$.

21.3-3. A random X has $P\{X = 0\} = 0.9$. Given $X \neq 0$, it has a uniform distribution between 5 and 15. Thus, $E(X) = 1$. Use simulation to estimate $E(X)$.
(*a*) Estimate $E(X)$ by generating five random observations from the distribution of X and then calculating the same average. (This is the crude Monte Carlo technique.)
(*b*) Estimate $E(X)$ by using stratified sampling with two strata—$0 \leq F(x) \leq 0.9$ and $0.9 < F(x) \leq 1$—with 1 and 4 observations, respectively.

21.3-4. A company's employees receive health insurance through a group plan. During the past year, 40 percent of the employees did not file any medical claims, 40 percent filed only a small claim, and 20 percent filed a large claim. The small claims were spread uniformly, between 0 and $2,000, whereas the large claims were spread uniformly between $2,000 and $20,000.

Based on this experience, the insurance carrier now is negotiating the company's premium payment per employee for the upcoming year. You are an OR analyst for the insurance carrier, and you have been assigned the task of estimating the average cost of insurance coverage for the company's employees.
(*a*) Analytically derive the mean of the probability distribution of an employee's annual medical cost covered by this insurance, based on last year's experience.
(*b*) Estimate this mean by using the crude Monte Carlo technique with 10 random observations.
(*c*) Estimate this mean by using stratified sampling with three strata—$0 \leq F(x) \leq 0.4$, $0.4 < F(x) \leq 0.8$, and $0.8 < F(x) \leq 1$—with 1, 3, and 6 observations, respectively.

21.3-5. Consider the probability distribution whose probability density function is

$$f(x) = \begin{cases} 1 - |x| & \text{if } -1 \leq x \leq 1, \\ 0 & \text{otherwise.} \end{cases}$$

Use the method of complementary random numbers with two uniform random numbers to estimate the mean of this distribution.

21.3-6. Consider the probability distribution whose probability density function is

$$f(x) = \begin{cases} \frac{3}{2}x^2 & \text{if } -1 \leq x \leq 1, \\ 0 & \text{otherwise.} \end{cases}$$

Use the method of complementary random numbers with two uniform random numbers to estimate the mean of this distribution.

21.3-7. Consider the probability distribution whose probability density function is

$$f(x) = \begin{cases} 0.4 & \text{if } 0 \le x \le 1, \\ 0.6 & \text{if } 1 < x \le 2, \\ 0 & \text{otherwise.} \end{cases}$$

The mean of this distribution is 1.1.

(*a*) Conduct three simulated experiments to estimate the mean of this distribution where, for each experiment, you generate three random observations from the distribution and calculate the sample average to estimate the mean.

(*b*) Now repeat these three experiments [using the same uniform random numbers as in part (*a*)] with the method of complementary random numbers.

21.3-8. The probability distribution of the number of heads in 3 flips of a fair coin is the binomial distribution with $n = 3$ and $p = \frac{1}{2}$, so that

$$P\{X = k\} = \binom{3}{k}\left(\frac{1}{2}\right)^k\left(\frac{1}{2}\right)^{3-k} = \frac{3!}{k!\,(3-k)!}\left(\frac{1}{2}\right)^3 \qquad \text{for } k = 0, 1, 2, 3.$$

The mean is 1.5.

(*a*) Use the inverse transformation method to generate three random observations from this distribution, and then calculate the sample average to estimate the mean.

(*b*) Use the method of complementary random numbers [with the same uniform random numbers as in part (*a*)] to estimate the mean.

(*c*) Use uniform random numbers to simulate repeatedly flipping a coin in order to generate three random observations from this distribution, and then calculate the sample average to estimate the mean.

(*d*) Repeat part (*c*) with the method of complementary random numbers [with the same uniform random numbers as in part (*c*)] to estimate the mean.

21.3-9. For one product produced by a certain company, bushings must be drilled into a metal block and cylindrical shafts inserted into the bushings. The shafts are required to have a radius of at least 1.0000 inch, but the radius should be as little larger than this as possible. In actuality, the probability distribution of the radius of a shaft (in inches) has the probability density function

$$f_s(x) = \begin{cases} 400e^{-400(x-1.0000)} & \text{if } x \ge 1.0000, \\ 0 & \text{otherwise.} \end{cases}$$

Similarly, the probability distribution of the radius of a bushing (in inches) has the probability density function

$$f_B(x) = \begin{cases} 100 & \text{if } 1.0000 \le x \le 1.0100, \\ 0 & \text{otherwise.} \end{cases}$$

The clearance between a bushing and a shaft is the difference in their radii. Because they are selected at random, there occasionally is interference (i.e., negative clearance) between a bushing and a shaft to be mated. The objective is to determine how frequently this interference will occur under the current probability distributions.

Perform a simulated experiment for estimating the probability of interference. Notice that almost all cases of interference will occur when the radius of the bushing is much closer to 1.0000 inch than to 1.0100 inches. Therefore, it appears that an efficient experiment would generate most of the simulated bushings from this critical portion of the distribution. Take this observation into account in part *(b)*. For each of the following cases, use the same 10 pairs of uniform random numbers to generate random observations, and calculate the resulting estimate of the probability of interference.

(*a*) Use the crude Monte Carlo technique.
(*b*) Develop and apply a stratified sampling approach to this problem.
(*c*) Use the method of complementary random numbers.

21.4-1.* A certain single-server system has been simulated, with the following sequence of waiting times before service for the respective customers. Use the regenerative method to obtain a point estimate and 90 percent confidence interval for the steady-state expected waiting time before service.

(*a*) 0, 5, 4, 0, 2, 0, 3, 1, 6, 0
(*b*) 0, 3, 2, 0, 3, 1, 5, 0, 0, 2, 4, 0, 3, 5, 2, 0

21.4-2. Consider the queueing system example presented in Sec. 21.4 for the regenerative method. Explain why the point where a *service completion* occurs with *no* other customers left is *not* a regeneration point.

21.4-3. A company has been having a maintenance problem with a certain complex piece of equipment. This equipment contains four identical vacuum tubes that have been the cause of the trouble. The problem is that the tubes fail fairly frequently, thereby forcing the equipment to be shut down while a replacement is made. The current practice is to replace tubes only when they fail. However, a proposal has been made to replace all four tubes whenever any one of them fails to reduce the frequency with which the equipment must be shut down. The objective is to compare these two alternatives on a cost basis.

The pertinent data are the following. For each tube, the operating time until failure has approximately a uniform distribution from 1,000 to 2,000 hours. The equipment must be shut down for 1 hour to replace one tube or for 2 hours to replace all four tubes. The total cost associated with shutting down the equipment and replacing tubes is $100 per hour plus $20 for each new tube.

(*a*) Starting with four new tubes, simulate the operation of the two alternative policies for 5,000 hours of simulated time.
(*b*) Use the data from part (*a*) to make a preliminary comparison of the two alternatives on a cost basis.
(*c*) For the *proposed* policy, describe an appropriate regeneration point for defining cycles that will permit applying the regenerative method of statistical analysis. Explain why the regenerative method cannot be applied to the *current* policy.
(*d*) For the proposed policy, use the regenerative method to obtain a point estimate and 95 percent confidence interval for the steady-state expected cost per hour from the data obtained in part (*a*).
(*e*) Write a computer simulation program for the two alternative policies. Then repeat parts (*a*), (*b*), and (*d*) on the computer, with 100 cycles for the proposed policy and 55,000 hours of simulated time (including a stabilization period of 5,000 hours) for the current policy.

21.4-4. One of the main lessons of queueing theory (Chaps. 15 and 16) is that the amount of variability in the service times and interarrival times has a substantial impact on the measures of performance of the queueing system. Significantly decreasing variability helps considerably.

This phenomenon is well illustrated by the $M/G/1$ queueing model presented at the beginning of Sec. 15.7. For this model, the four fundamental measures of performance (L, L_q, W, and W_q) are expressed directly in terms of the *variance* of service times (σ^2), so we can see immediately what the impact of decreasing σ^2 would be.

Consider an $M/G/1$ queueing system with mean arrival rate $\lambda = 0.8$ and mean service rate $\mu = 1$, so the utilization factor is $\rho = \lambda/\mu = 0.8$.

A* (*a*) Use the automatic routine in your OR Courseware to execute a simulation run for each of the following cases: (*i*) $\sigma = 1$ (corresponds to an exponential distribution of service times), (*ii*) $\sigma = 0.5$ (corresponds to an Erlang distribution of service times with shape parameter $k = 4$), and (*iii*) $\sigma = 0$ (constant service times). (Use the default run size.) Using the point estimates of L_q obtained, calculate the ratio of L_q for case (*ii*) to L_q for case (*i*). Also calculate the ratio of L_q for case (*iii*) to L_q for case (*i*).

(*b*) For each of the three cases considered in part (*a*), use the formulas given in Sec. 15.7 to compute the exact values of L, L_q, W, and W_q. Compare these exact values to the point estimates and 95 percent confidence intervals obtained in part (*a*). Identify any exact values that fall outside the 95 percent confidence interval. Also calculate the exact values of the ratios requested in part (*a*).

A* **21.4-5.** Follow the instructions of part (*a*) of Prob. 21.4-4 for an $M/G/2$ queueing system (two servers), with $\lambda = 1.6$ and $\mu = 1$ [so $\rho = \lambda/(2\mu) = 0.8$] and with σ^2 still being the variance of service times.

21.4-6. Reconsider Prob. 21.4-4. For the single-server queueing system under consideration, suppose now that service times definitely have an exponential distribution. However, it now is possible to reduce the variability of *interarrival times,* so we want to explore the impact of doing so.

Assume now that $\lambda = 1$ and $\mu = 1.25$, so $\rho = 0.8$. Let σ^2 now denote the variance of interarrival times.

Follow the instructions of Prob. 21.4-4*a*, where the distributions for the three cases now are for interarrival times instead of service times.

21.4-7. When the waiting times for the customers of a queueing system are excessive, one way to alleviate the problem is to divide the customers into *priority classes,* as described in Sec. 15.8. This will reduce the waiting times for the more important (higher-priority) customers while increasing the waiting times for low-priority customers.

Consider a queueing system with two servers, where arrivals occur according to a Poisson process (exponential interarrival times) with a mean rate λ of 9 customers per hour. Service times have an Erlang distribution with a mean of 0.2 hour and shape parameter $k = 8$. Thus, the mean service rate for each busy server is $\mu = 5$ customers per hour, and the utilization factor is $\rho = \lambda/(2\mu) = 0.9$.

Currently, this system uses a first-come-first-served queue discipline (no priorities), so the $M/E_8/2$ queueing model applies. the bottom curve in Fig. 15.13 shows that the value of L for this model is slightly over 6 (the exact value is $L = 6.15$), so $W = L/\lambda = 0.683$ hour and $W_q = W - 1/\mu = 0.483$ hour.

These average waiting times (with some actual waiting times much larger) are considered excessive, so a proposal has been made to switch to a nonpreemptive priority queue discipline with three priority classes. The most important customers ($\frac{1}{9}$ of the total) would be placed in the top-priority class, the fairly important customers ($\frac{1}{3}$ of the total) would be placed in the second-priority class, and the rest ($\frac{5}{9}$ of the total) would be placed in the bottom-priority class. Thus, arrivals of the three priority classes occur according to Poisson processes with mean rates of 1, 3, and 5 customers per hour, respectively. Management's goal is to reduce W_q to below 0.2 hour, below 0.4 hour, and below 1 hour for the respective priority classes.

The proposed queueing system would fit the nonpreemptive priorities model presented in Sec. 15.8 *except* that service times have an *Erlang* distribution instead of the *exponential* distribution assumed in Sec. 15.8. No analytical results are available for this model with an Erlang distribution, so simulation must be used to analyze the proposal.

A* (*a*) Use the automatic routine in your OR Courseware to execute a simulation run for the current queueing system. (Use the default run size.) Compare the point estimates and 95 percent confidence intervals obtained for W and W_q with their exact values.

A* (*b*) Use the automatic routine in your OR Courseware to execute a simulation run for the proposed queueing system. (Use the default run size.) Does it appear that the proposal would meet management's goal?

CASE PROBLEM

A* **CP21-1.** A manufacturing company has two planers for cutting flat surfaces in large workpieces of two different types (A and B). The time required to perform each job varies somewhat, depending largely upon the number of passes that must be made. In particular, for workpieces of type A, the time required by a planer has an Erlang distribution with a mean of 25 minutes and shape parameter $k = 4$. For workpieces of type B, the time required has a translated exponential distribution with probability density function

$$f(x) = \begin{cases} 0 & \text{if } x < 10, \\ 0.1e^{-0.1(x-10)} & \text{if } x \geq 10, \end{cases}$$

where x is measured in minutes. Thus, the mean of this latter distribution is 20 minutes.

Workpieces of both types arrive one at a time to the planer department. For workpieces of type A, the arrivals occur according to a Poisson process (exponential interarrival times) with a mean rate of 2 per hour. For workpieces of type B, the interarrival times have a uniform distribution over the interval from 20 to 40 minutes.

Unfortunately, the planer department has had a difficult time keeping up with its workload. Frequently there are a number of workpieces waiting for a free planer. This waiting has seriously disrupted the production schedule for subsequent operations, thereby greatly increasing the cost of in-process inventory as well as the cost of idle equipment and resulting lost production. Therefore, three proposals have been made to relieve this bottleneck:

Proposal 1: Obtain one additional planer. The total incremental cost (including capital recovery cost) is estimated to be \$30 per hour. (This estimate takes into account the fact that, even with an additional planer, the total running time for all the planers will remain the same.)

Proposal 2: Eliminate the variability in the interarrival times of workpieces of type A, so that the pieces would arrive regularly, one every 30 minutes. This would require making some changes in the preceding production processes, with an incremental cost of \$40 per hour.

Proposal 3: This is the same as proposal 2, but now for the type B workpieces. The incremental cost in this case would be \$20 per hour.

These proposals are not mutually exclusive, so any combination can be adopted.

It is estimated that the total cost associated with workpieces having to wait to be processed (including processing time) is \$200 per workpiece per hour for type A and \$100 per workpiece per hour for type B. Because of this difference in costs, type A workpieces always are given nonpreemptive priority over type B. In other words, if a planer becomes free when workpieces of both types are waiting, a workpiece of type A always is chosen to be processed next.

The objective is to minimize the total cost per hour.

Use simulation to evaluate and compare all the alternatives, including the status quo and the various combinations of proposals. Then make your recommendation.

Appendixes

Appendix 1

Documentation for the OR Courseware

Special Features of the Software

The *OR Courseware* packaged in the back of most versions of this book is distinctive *tutorial software* for your personal use. It has been specially developed to accompany this book as a fully integrated *teaching supplement.* Thus, it has been designed to be your *personal tutor,* illustrating and illuminating key concepts in an interactive manner.

Your OR Courseware features three kinds of routines: demonstration examples, interactive routines, and automatic routines.

The **demonstration examples** supplement the examples in the book in ways that cannot be duplicated on the printed page. Each one vividly demonstrates one of the algorithms or concepts of OR in action. Most combine an *algebraic description* of each step with a *geometric display* of what is happening. Some of these geometric displays become quite dynamic, with moving points or moving lines, to demonstrate the evolution of the algorithm. The demonstration examples also are integrated with the book, using the same notation and terminology, with references to material in the book, etc. Students find them an enjoyable and effective learning aid.

The **interactive routines** also are a key tutorial feature of this software. Each one enables you to *interactively execute* one of the algorithms of OR. While viewing all relevant information on the computer screen, you make the decision on how the next step of the algorithm should be performed, and then the computer does all the necessary number crunching to execute that step. A Help file always is available to guide you through the computer mechanics. When uncertain about the logic of the algorithm for how to perform the next algorithm step, you can switch temporarily to reviewing the corresponding demonstration example and then switch back to the same point in the interactive routine. When a previous mistake is discovered, the routine allows you to quickly backtrack to correct the mistake. To get you started properly, the computer points out any mistake made on the first iteration (where possible). When done, you can print out all the work performed to turn in for homework.

In our judgment, these interactive routines provide the ''right'' way in this computer age for students to do homework designed to help them learn the algorithms of OR. The routines enable you to focus on concepts rather than mindless number crunching, thereby making the learning process far more efficient and effective as well as stimulating. They also point you in the right direction, including organizing the work to be done. However, the routines do not do the thinking for you. As in any good homework assignment, you are allowed to make mistakes (and to learn from those mistakes), so that hard thinking will need to be done to try to stay on the right path. We have been careful in designing the division of labor between the computer and the student to provide an efficient, complete learning process. In certain cases, the computer will take over a relatively routine task after you have demonstrated the ability to perform it correctly on the first iteration.

The software also includes a considerable number of **automatic routines** to provide number-crunching help. Many are for stochastic modeling problems involving complicated formulas. Several others are for algorithms that are not well suited to interactive execution. In a few cases (e.g., the simplex method), an algorithm will have both kinds of routines available. In these cases, the automatic routine can be used to check your answer from the interactive routine. In addition, after learning the algorithm with the help of the interactive routine, you can quickly apply the algorithm with the automatic routine whenever you formulate a model that is solved in this way.

These automatic routines are not intended to compete with the corresponding routines available in powerful commercial software packages. However, they will introduce you to what can be done with such packages. For example, the output for the automatic routine for the simplex method has been designed to emulate that in popular commercial packages.

A *selection rectangle* is used to enter data for either interactive or automatic routines. To enter a number at a certain location, the selection rectangle is moved there with the arrow keys, the number is entered, and the Enter or Return key is pressed. A number can be replaced in the same way. In interactive routines, the selection rectangle may also be used for making a choice, e.g., choosing a location where a certain operation should be done. In such cases, instructions will appear on the screen.

The OR Courseware is contained on a single floppy disk, where one disk is available for Apple Macintosh computers and another for IBM or IBM-compatible computers with a graphics card. On each disk, there are two software programs: one called MathProg and another called ProbMod. MathProg covers the subjects in the first 13 chapters on linear and mathematical programming, while ProbMod covers Chaps. 14 through 21 on probabilistic modeling.

Once the software is running on the computer, all the information necessary to use it is provided on the screen. A general introduction to the software is offered every time the software is run (or any time by choosing *New* under the File menu). Furthermore, there is a help menu which will provide context-sensitive help at any time. Instructions to get you started using the software are provided below. In addition, a Quick Reference is provided for your convenience, summarizing each of the menus available in the software.

Getting Started

The OR Courseware software can be used either directly from a floppy disk or from a hard drive. The directions for using the software directly from a floppy disk are shown below for both the Macintosh computer and IBM-compatibles. It is recommended, however, that you use the software from a hard drive, since it will operate much more quickly. To do this, *all* the files on the floppy disk will need to be copied into a folder on the hard drive. (If you do not know how to copy files from a floppy disk to the hard drive, refer to the computer's instruction manual.)

To use the OR Courseware software directly from a floppy disk on a Macintosh computer:

1. Insert the OR Courseware disk (for Macintosh) into the floppy drive.
2. Double-click on the floppy disk icon to open it.
3. Double-click on either MathProg or ProbMod to use the corresponding program.

To use the OR Courseware software directly from a floppy disk on an IBM-compatible computer:[1]

1. Insert the OR Courseware disk (for IBM-compatibles) into the floppy drive.
2. Type A: and then press the Enter key to make the floppy the active disk.
3. Type either MATHPROG or PROBMOD and then press the Enter key to use the corresponding program.

After getting started the first time, you should take the guided tour through the OR Courseware outlined at the end of Chap. 1 (see Prob. 1.4-1).

Quick Reference

The OR Courseware software is completely menu-driven. There are six menus available, and the purpose of each is summarized below.

File

The File menu provides you with a choice of basic filing operations. These include New (to clear the program's memory and start fresh), Open (to load a previously saved file), Save As . . . (to save your work), Print (to print out your work), and Quit (to quit the program).

Area

The Area menu provides you with a choice of several areas of either mathematical programming (in MathProg) or probabilistic modeling (in ProbMod). Each area corresponds to one or more of the chapters in this book.

Procedure

Once you have chosen your general area, the Procedure menu provides you with a choice of procedures, such as enter a model, solve the model with a particular algorithm, and so on. Once a procedure is chosen, you may choose *Introduction to Current Procedure* from the Help menu at any time. You should do this immediately the first time you use a procedure.

[1] These instructions assume that you are using DOS. If your computer is running another operating system (e.g., Windows or OS/2), follow the instructions of that operating system for running a program contained on a floppy disk.

Option

Once you have chosen your general area and procedure, the Option menu provides you with specific choices for how to display or carry out that procedure. Whenever you choose a procedure for the first time, you should check the Option menu to see which (if any) options are available beyond the default option.

Help

If at *any* time you do not know how to execute the next computer step, choose *Specific Help on Current Step* from the Help menu. This menu choice displays specific instructions on the screen. Since only brief instructions (if any) are displayed on procedure screens, you probably will need to make frequent use of this menu option the first time through a procedure. To receive general information about the current procedure, choose the *Introduction to Current Procedure* option.

Demo

In many of the areas, one or more demonstration examples are available by selecting the Demo menu. We highly recommend that you go through these enlightening examples before you use the corresponding procedures. You proceed at your own pace, pressing the Return key whenever you want to go to the next step. As you subsequently work problems, you can temporarily switch back to the demonstration example at any time with this menu option, in order to clarify how to proceed further with your problem. Then you return to your current spot in the problem whenever you are ready by pressing the \ or Esc key. When viewing a demonstration example, you can backtrack to review a previous screen of the demonstration by pressing b.

Appendix 2

Convexity

As introduced in Chap. 13, the concept of *convexity* is frequently used in OR work, especially in the area of nonlinear programming. Therefore, we further introduce the properties of convex or concave functions and convex sets here.

Convex or Concave Functions of a Single Variable

We begin with definitions.

> **Definitions**: A *function* of a single variable $f(x)$ is a **convex function** if, for *each* pair of values of x, say, x' and x'' ($x' < x''$),
>
> $$f[\lambda x'' + (1 - \lambda)x'] \leq \lambda f(x'') + (1 - \lambda)f(x')$$
>
> for all values of λ such that $0 < \lambda < 1$. It is a **strictly convex function** if $\leq$ can be replaced by $<$. It is a **concave function** (or a **strictly concave function**) if this statement holds when $\leq$ is replaced by $\geq$ (or by $>$).

This definition of a convex function has an enlightening geometric interpretation. Consider the graph of the function $f(x)$ drawn as a function of x, as illustrated in Fig. A2.1 for a function $f(x)$ that decreases for $x < 1$, is constant for $1 \leq x \leq 2$, and increases for $x > 2$. Then $[x', f(x')]$ and $[x'', f(x'')]$ are two points on the graph of $f(x)$, and $[\lambda x'' + (1 - \lambda)x', \lambda f(x'') +$

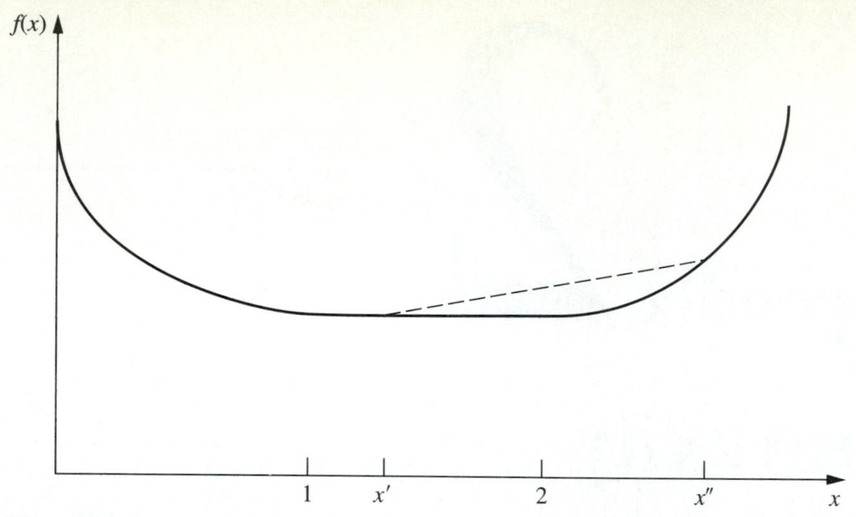

Figure A2.1 A convex function.

$(1 - \lambda)f(x')]$ represents the various points on the line segment between these two points (but excluding these endpoints) when $0 < \lambda < 1$. Thus the $\leq$ inequality in the definition indicates that this line segment lies entirely above or on the graph of the function, as in Fig. A2.1. Therefore, $f(x)$ is *convex* if, for *each* pair of points on the graph of $f(x)$, the line segment joining these two points lies entirely above or on the graph of $f(x)$.

For example, the particular choice of x' and x'' shown in Fig. A2.1 results in the entire line segment (except the two endpoints) lying *above* the graph of $f(x)$. This also occurs for other choices of x' and x'' where either $x' < 1$ or $x'' > 2$ (or both). If $1 \leq x' < x'' \leq 2$, then the entire line segment lies *on* the graph of $f(x)$. Therefore, this $f(x)$ is convex.

This geometric interpretation indicates that $f(x)$ is convex if it only "bends upward" whenever it bends at all. (This condition is sometimes referred to as *concave upward*, as opposed to *concave downward* for a concave function.) To be more precise, if $f(x)$ possesses a second derivative everywhere, then $f(x)$ is convex if and only if $d^2f(x)/dx^2 \geq 0$ for all possible values of x.

The definitions of a *strictly convex function*, a *concave function*, and a *strictly concave function* also have analogous geometric interpretations. These interpretations are summarized below in terms of the second derivative of the function, which provides a convenient test of the status of the function.

Convexity test for a function of a single variable: Consider any function of a single variable $f(x)$ that possesses a second derivative at all possible values of x. Then $f(x)$ is

1. *Convex* if and only if $\dfrac{d^2f(x)}{dx^2} \geq 0$ for all possible values of x
2. *Strictly convex* if and only if $\dfrac{d^2f(x)}{dx^2} > 0$ for all possible values of x
3. *Concave* if and only if $\dfrac{d^2f(x)}{dx^2} \leq 0$ for all possible values of x
4. *Strictly concave* if and only if $\dfrac{d^2f(x)}{dx^2} < 0$ for all possible values of x

Note that a strictly convex function also is convex, but a convex function is *not* strictly convex if the second derivative equals zero for some values of x. Similarly, a strictly concave function is concave, but the reverse need not be true.

Figures A2.1 to A2.6 show examples that illustrate these definitions and this convexity test.

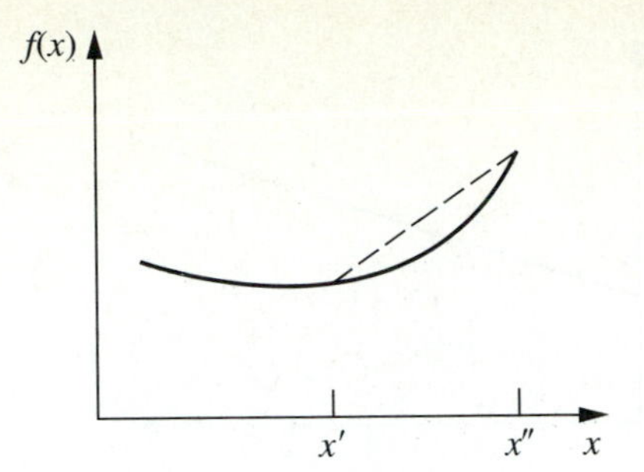

Figure A2.2 A strictly convex function.

Applying this test to the function in Fig. A2.1, we see that as x is increased, the slope (first derivative) either increases (for $0 \leq x < 1$ and $x > 2$) or remains constant (for $1 \leq x_1 \leq 2$). Therefore, the second derivative always is nonnegative, which verifies that the function is convex. However, it is *not* strictly convex because the second derivative equals zero for $1 \leq x \leq 2$.

However, the function in Fig. A2.2 is strictly convex because its slope always is increasing so its second derivative always is greater than zero.

The piecewise linear function shown in Fig. A2.3 changes its slope at $x = 1$. Consequently, it does not possess a first or second derivative at this point, so the convexity test cannot be fully applied. (The fact that the second derivative equals zero for $0 \leq x < 1$ and $x > 1$ makes the function eligible to be either convex or concave, depending upon its behavior at $x = 1$.) Applying the definition of a concave function, we see that if $0 < x' < 1$ and $x'' > 1$ (as shown in Fig. A2.3), then the entire line segment joining $[x', f(x')]$ and $[x'', f(x'')]$ lies *below* the graph of $f(x)$, except for the two endpoints of the line segment. If either $0 \leq x' < x'' \leq 1$ or $1 \leq x' < x''$, then the entire line segment lies *on* the graph of $f(x)$. Therefore, $f(x)$ is concave (but *not* strictly concave).

The function in Fig. A2.4 is strictly concave because its second derivative always is less than zero.

As illustrated in Fig. A2.5, any linear function has its second derivative equal to zero everywhere and so is both convex and concave.

The function in Fig. A2.6 is *neither* convex nor concave because as x increases, the slope fluctuates between decreasing and increasing so the second derivative fluctuates between being negative and positive.

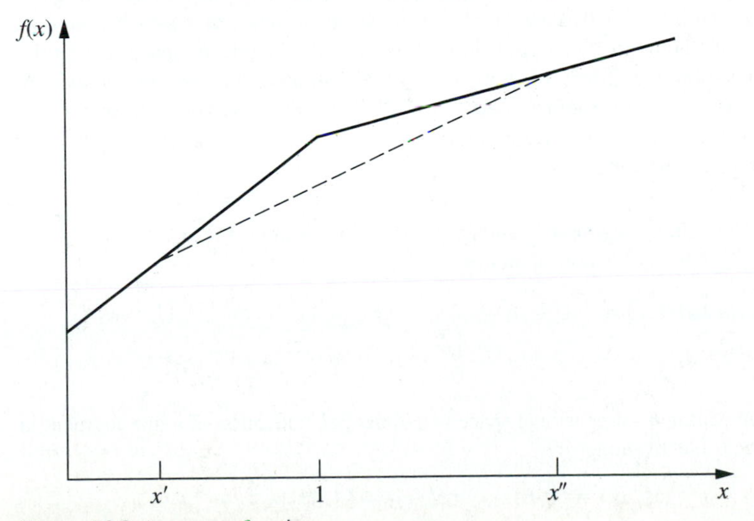

Figure A2.3 A concave function.

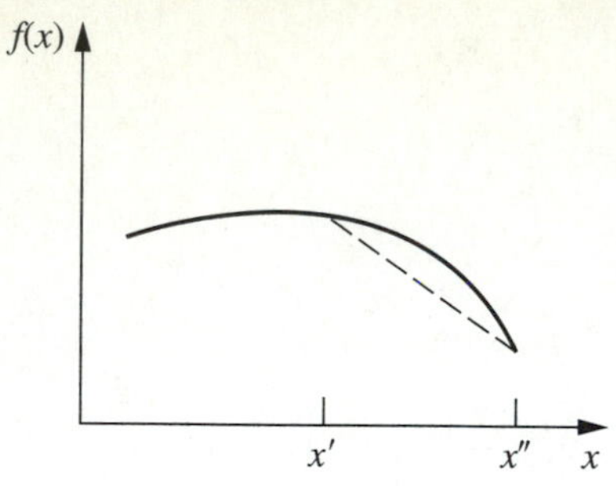

Figure A2.4 A strictly concave function.

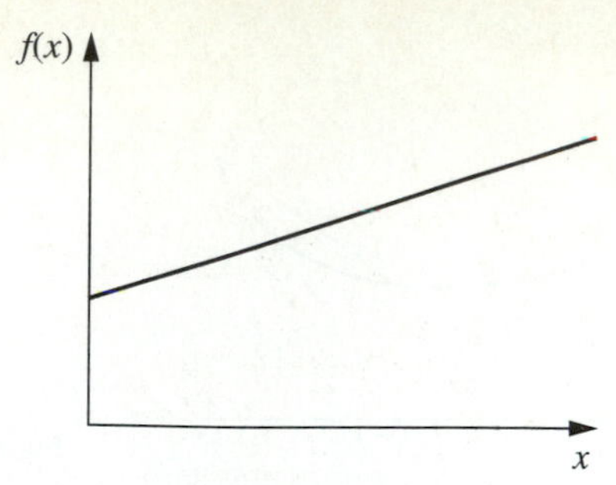

Figure A2.5 A function that is both convex and concave.

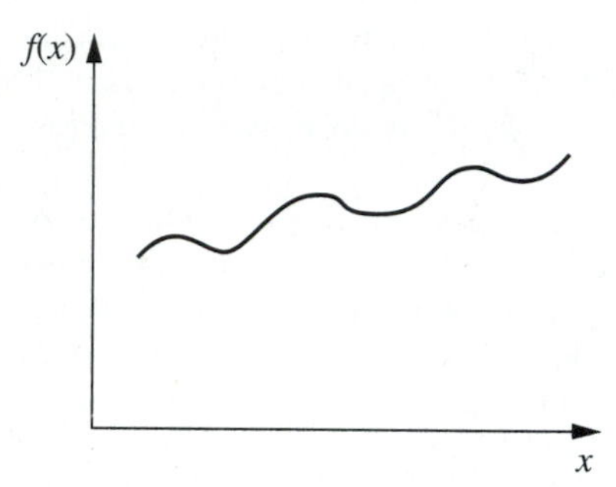

Figure A2.6 A function that is neither convex nor concave.

Convex or Concave Functions of Several Variables

The concept of a convex or concave function of a single variable also generalizes to functions of more than one variable. Thus if $f(x)$ is replaced by $f(x_1, x_2, \ldots, x_n)$, the definition still applies if x is replaced everywhere by $(x_1, x_2, \ldots, x_n)$. Similarly, the corresponding geometric interpretation is still valid after generalization of the concepts of *points* and *line segments.* Thus, just as a particular value of (x, y) is interpreted as a point in two-dimensional space, each possible value of $(x_1, x_2, \ldots, x_m)$ may be thought of as a point in m-dimensional (Euclidean) space. By letting $m = n + 1$, the points on the graph of $f(x_1, x_2, \ldots, x_n)$ become the possible values of $[x_1, x_2, \ldots, x_n, f(x_1, x_2, \ldots, x_n)]$. Another point, $(x_1, x_2, \ldots, x_n, x_{n+1})$, is said to lie above, on, or below the graph of $f(x_1, x_2, \ldots, x_n)$, according to whether x_{n+1} is larger, equal to, or smaller than $f(x_1, x_2, \ldots, x_n)$, respectively.

Definition: The **line segment** joining any two points $(x_1', x_2', \ldots, x_m')$ and $(x_1'', x_2'', \ldots, x_m'')$ is the collection of points

$$(x_1, x_2, \ldots, x_m) = [\lambda x_1'' + (1 - \lambda)x_1', \lambda x_2'' + (1 - \lambda)x_2', \ldots, \lambda x_m'' + (1 - \lambda)x_m']$$

such that $0 \leq \lambda \leq 1$.

Thus a line segment in m-dimensional space is a direct generalization of a line segment in two-dimensional space. For example, if

$$(x_1', x_2') = (2, 6), \qquad (x_1'', x_2'') = (3, 4).$$

then the line segment joining them is the collection of points

$$(x_1, x_2) = [3\lambda + 2(1-\lambda), 4\lambda + 6(1-\lambda)],$$

where $0 \le \lambda \le 1$.

Definition: $f(x_1, x_2, \ldots x_n)$ is a **convex function** if, for each pair of points on the graph of $f(x_1, x_2, \ldots, x_n)$, the line segment joining these two points lies entirely above or on the graph of $f(x_1, x_2, \ldots, x_n)$. It is a **strictly convex function** if this line segment actually lies entirely above this graph except at the endpoints of the line segment. **Concave functions** and **strictly concave functions** are defined in exactly the same way, except that *above* is replaced by *below*.

Just as the second derivative can be used (when it exists everywhere) to check whether a function of a single variable is convex, so second partial derivatives can be used to check functions of several variables, although in a more complicated way. For example, if there are two variables and all partial derivatives exist everywhere, then the convexity test assesses whether *all three quantities* in the first column of Table A2.1 satisfy the inequalities shown in the appropriate column for *all possible values* of (x_1, x_2).

When there are more than two variables, the convexity test is a generalization of the one shown in Table A2.1. For example, in mathematical terminology, $f(x_1, x_2, \ldots, x_n)$ is convex if and only if its $n \times n$ Hessian matrix is positive semidefinite for all possible values of $(x_1, x_2, \ldots, x_n)$.

To illustrate the convexity test for two variables, consider the function

$$f(x_1, x_2) = (x_1 - x_2)^2 = x_1^2 - 2x_1x_2 + x_2^2.$$

Therefore,

(1) $$\frac{\partial^2 f(x_1, x_2)}{\partial x_1^2}\frac{\partial^2 f(x_1, x_2)}{\partial x_2^2} - \left[\frac{\partial^2 f(x_1, x_2)}{\partial x_1 \partial x_2}\right]^2 = 2(2) - (-2)^2 = 0,$$

(2) $$\frac{\partial^2 f(x_1, x_2)}{\partial x_1^2} = 2 > 0,$$

(3) $$\frac{\partial^2 f(x_1, x_2)}{\partial x_2^2} = 2 > 0.$$

Since ≥ 0 holds for all three conditions, $f(x_1, x_2)$ is convex. However, it is *not* strictly convex because the first condition only gives $= 0$ rather than > 0.

Table A2.1 **Convexity Test for a Function of Two Variables**

Quantity	Convex	Strictly Convex	Concave	Strictly Concave
$\frac{\partial^2 f(x_1, x_2)}{\partial x_1^2}\frac{\partial^2 f(x_1, x_2)}{\partial x_2^2} - \left[\frac{\partial^2 f(x_1, x_2)}{\partial x_1 \partial x_2}\right]^2$	≥ 0	> 0	≥ 0	> 0
$\frac{\partial^2 f(x_1, x_2)}{\partial x_1^2}$	≥ 0	> 0	≤ 0	< 0
$\frac{\partial^2 f(x_1, x_2)}{\partial x_2^2}$	≥ 0	> 0	≤ 0	< 0
Values of (x_1, x_2)	All possible values			

Now consider the negative of this function

$$g(x_1, x_2) = -f(x_1, x_2) = -(x_1 - x_2)^2 = -x_1^2 + 2x_1x_2 - x_2^2.$$

In this case,

$$\frac{\partial^2 g(x_1, x_2)}{\partial x_1^2} \frac{\partial^2 g(x_1, x_2)}{\partial x_2^2} - \left[\frac{\partial^2 g(x_1, x_2)}{\partial x_1 \partial x_2}\right]^2 = -2(-2) - 2^2 = 0, \tag{4}$$

$$\frac{\partial^2 g(x_1, x_2)}{\partial x_1^2} = -2 < 0, \tag{5}$$

$$\frac{\partial^2 g(x_1, x_2)}{\partial x_2^2} = -2 < 0. \tag{6}$$

Because ≥ 0 holds for the first condition and ≤ 0 holds for the other two, $g(x_1, x_2)$ is a concave function. However, it is *not* strictly concave since the first condition gives $= 0$.

Thus far convexity has been treated as a general property of a function. However, many nonconvex functions do satisfy the conditions for convexity over certain intervals for the respective variables. Therefore, it is meaningful to talk about a function being convex over a certain region. For example, a function is said to be convex within a neighborhood of a specified point if its second derivative or partial derivatives satisfy the conditions for convexity at that point. This concept is useful in App. 3.

Finally, two particularly important properties of convex or concave functions should be mentioned. First, if $f(x_1, x_2, \ldots, x_n)$ is a convex function, then $g(x_1, x_2, \ldots, x_n) = -f(x_1, x_2, \ldots, x_n)$ is a concave function, and vice versa, as illustrated by the above example where $f(x_1, x_2) = (x_1 - x_2)^2$. Second, the sum of convex functions is a convex function, and the sum of concave functions is a concave function. To illustrate.

$$f_1(x_1) = x_1^4 + 2x_1^2 - 5x_1$$

and

$$f_2(x_1, x_2) = x_1^2 + 2x_1x_2 + x_2^2$$

are both convex functions, as you can verify by calculating their second derivatives. Therefore, the sum of these functions

$$f(x_1, x_2) = x_1^4 + 3x_1^2 - 5x_1 + 2x_1x_2 + x_2^2$$

is a convex function, whereas its negative

$$g(x_1, x_2) = -x_1^4 - 3x_1^2 + 5x_1 - 2x_1x_2 - x_2^2,$$

is a concave function.

Convex Sets

The concept of a convex function leads quite naturally to the related concept of a **convex set**. Thus, if $f(x_1, x_2, \ldots, x_n)$ is a convex function, then the collection of points that lie above or on the graph of $f(x_1, x_2, \ldots, x_n)$ forms a convex set. Similarly, the collection of points that lie below or on the graph of a concave function is a convex set. These cases are illustrated in Figs. A2.7 and A2.8 for the case of a single independent variable. Furthermore, convex sets have the important property that, for any given group of convex sets, the collection of points that lie in all of them (i.e., the intersection of these convex sets) is also a convex set. Therefore, the collection of points that lie both above or on a convex function and below or on a concave function is a

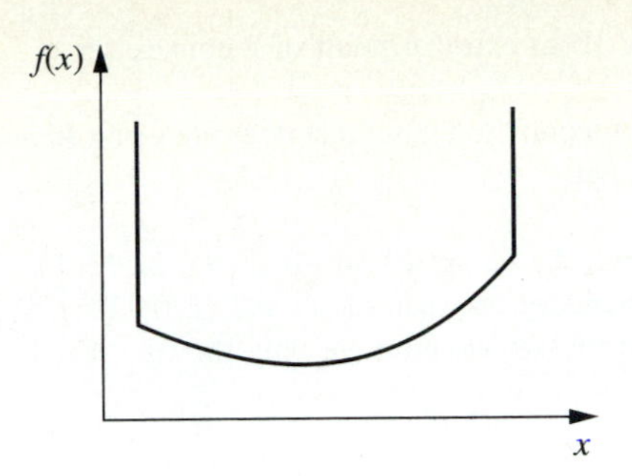

Figure A2.7 Example of a convex set determined by a convex function.

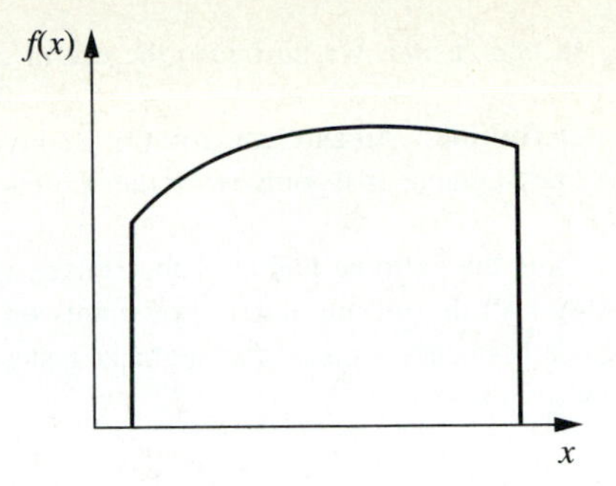

Figure A2.8 Example of a convex set determined by a concave function.

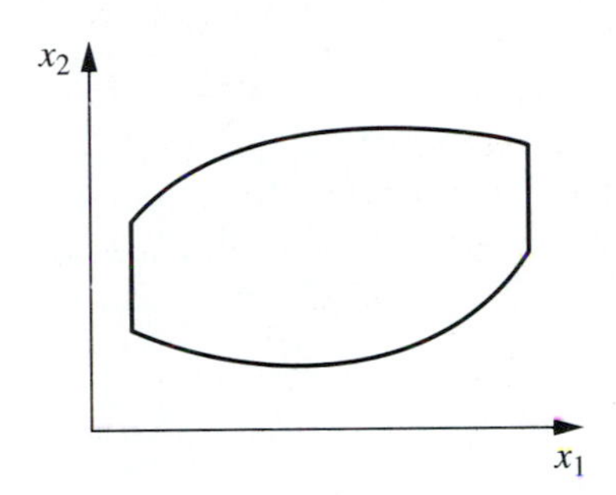

Figure A2.9 Example of a convex set determined by both convex and concave functions.

convex set, as illustrated in Fig. A2.9. Thus convex sets may be viewed intuitively as a collection of points whose bottom boundary is a convex function and whose top boundary is a concave function.

Although describing convex sets in terms of convex and concave functions may be helpful for developing intuition about their nature, their actual definition has nothing to do (directly) with such functions.

Definition: A **convex set** is a collection of points such that, for each pair of points in the collection, the entire line segment joining these two points is also in the collection.

The distinction between nonconvex sets and convex sets is illustrated in Figs. A2.10 and A2.11. Thus the set of points shown in Fig. A2.10 is not a convex set because there exist many pairs of these points, for example, (1, 2) and (2, 1), such that the line segment between them does not lie entirely within the set. This is not the case for the set in Fig. A2.11, which is convex.

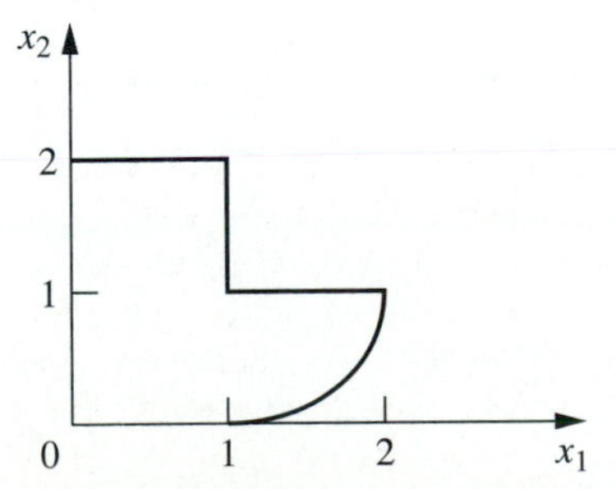

Figure A2.10 Example of a set that is not convex.

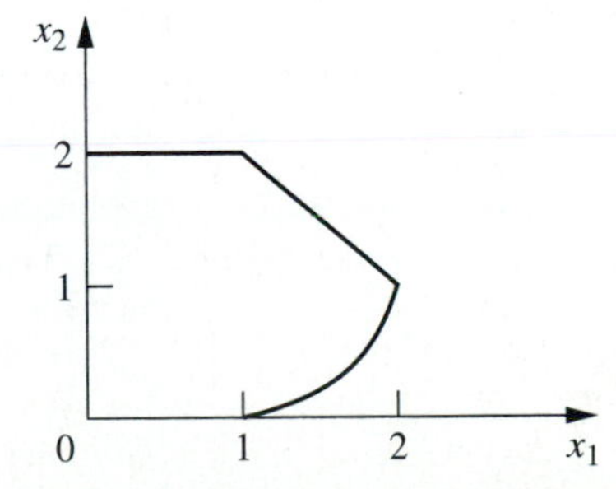

Figure A2.11 Example of a convex set.

In conclusion, we introduce the useful concept of an extreme point of a convex set.

Definition: An **extreme point** of a convex set is a point in the set that does not lie on any line segment that joins two other points in the set.

Thus the extreme points of the convex set in Fig. A2.11 are (0, 0), (0, 2), (1, 2), (2, 1), (1, 0), and all the infinite number of points on the boundary between (2, 1) and (1, 0). If this particular boundary were a line segment instead, then the set would have only the five listed extreme points.

Appendix 3

Classical Optimization Methods

This appendix reviews the classical methods of calculus for finding a solution that maximizes or minimizes (1) a function of a single variable, (2) a function of several variables, and (3) a function of several variables subject to equality constraints on the values of these variables. It is assumed that the functions considered possess continuous first and second derivatives and partial derivatives everywhere. Some of the concepts discussed next have been introduced briefly in Secs. 13.2 and 13.3.

Unconstrained Optimization of a Function of a Single Variable

Consider a function of a single variable, such as that shown in Fig. A3.1. A necessary condition for a particular solution $x = x^*$ to be either a minimum or a maximum is that

$$\frac{df(x)}{dx} = 0 \qquad \text{at } x = x^*.$$

Thus in Fig. A3.1 there are five solutions satisfying these conditions. To obtain more information about these five **critical points**, it is necessary to examine the second derivative. Thus, if

$$\frac{d^2f(x)}{dx^2} > 0 \qquad \text{at } x = x^*,$$

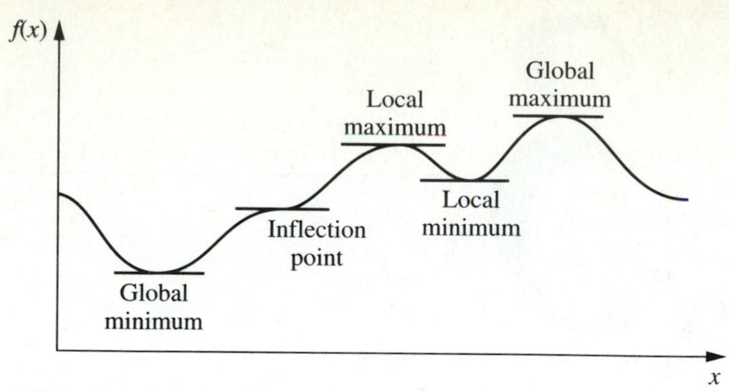

Figure A3.1 A function having several maxima and minima.

then x^* must be at least a **local minimum** [that is, $f(x^*) \leq f(x)$ for all x sufficiently close to x^*]. Using the language introduced in App. 2, we can say that x^* must be a local minimum if $f(x)$ is *strictly convex* within a neighborhood of x^*. Similarly, a sufficient condition for x^* to be a **local maximum** (given that it satisfies the necessary condition) is that $f(x)$ be *strictly concave* within a neighborhood of x^* (that is, the second derivative is *negative* at x^*). If the second derivative is zero, the issue is not resolved (the point may even be an *inflection point*), and it is necessary to examine higher derivatives.

To find a **global minimum** [i.e., a solution x^* such that $f(x^*) \leq f(x)$ for all x], it is necessary to compare the local minima and identify the one that yields the smallest value of $f(x)$. If this value is less than $f(x)$ as $x \to -\infty$ and as $x \to +\infty$ (or at the endpoints of the function, if it is defined only over a finite interval), then this point is a global mininum. Such a point is shown in Fig. A3.1, along with the **global maximum**, which is identified in an analogous way.

However, if $f(x)$ is known to be either a convex or a concave function (see App. 2 for a description of such functions), the analysis becomes much simpler. In particular, if $f(x)$ is a *convex* function, such as the one shown in Fig. A2.1, then any solution x^* such that

$$\frac{df(x)}{dx} = 0 \qquad \text{at } x = x^*$$

is known automatically to be a *global minimum*. In other words, this condition is not only a *necessary* but also a *sufficient* condition for a global minimum of a convex function. This solution need not be unique, since there could be a tie for the global minimum over a single interval where the derivative is zero. On the other hand, if $f(x)$ actually is *strictly convex,* then this solution must be the only global minimum. (However, if the function is either always decreasing or always increasing, so the derivative is nonzero for all values of x, then there will be no global minimum at a finite value of x.)

Similarly, if $f(x)$ is a *concave* function, then having

$$\frac{df(x)}{dx} = 0 \qquad \text{at } x = x^*$$

becomes both a *necessary* and *sufficient* condition for x^* to be a *global maximum.*

Unconstrained Optimization of a Function of Several Variables

The analysis for an unconstrained function of several variables $f(\mathbf{x})$, where $\mathbf{x} = (x_1, x_2, \ldots, x_n)$, is similar. Thus a *necessary* condition for a solution $\mathbf{x} = \mathbf{x}^*$ to be either a minimum or a maximum is that

$$\frac{\partial f(\mathbf{x})}{\partial x_j} = 0 \qquad \text{at } \mathbf{x} = \mathbf{x}^*, \text{ for } j = 1, 2, \ldots, n.$$

After the critical points that satisfy this condition are identified, each such point is then classified as a local minimum or maximum if the function is *strictly convex* or *strictly concave*, respectively, within a neighborhood of the point. (Additional analysis is required if the function is neither.) The *global minimum* and *maximum* would be found by comparing the local minima and maxima and then checking the value of the function as some of the variables approach $-\infty$ or $+\infty$. However, if the function is known to be *convex* or *concave*, then a critical point must be a *global minimum* or a *global maximum*, respectively.

Constrained Optimization with Equality Constraints

Now consider the problem of finding the *minimum* or *maximum* of the function $f(\mathbf{x})$, subject to the restriction that $\mathbf{x}$ must satisfy all the equations

$$\begin{aligned} g_1(\mathbf{x}) &= b_1 \\ g_2(\mathbf{x}) &= b_2 \\ &\vdots \\ g_m(\mathbf{x}) &= b_m, \end{aligned}$$

where $m < n$. For example, if $n = 2$ and $m = 1$, the problem might be

$$\text{Maximize} \qquad f(x_1, x_2) = x_1^2 + 2x_2,$$

subject to

$$g(x_1, x_2) = x_1^2 + x_2^2 = 1.$$

In this case (x_1, x_2) is restricted to be on the circle of radius 1, whose center is at the origin, so that the goal is to find the point on this circle that yields the largest value of $f(x_1, x_2)$. This example will be solved after a general approach to the problem is outlined.

A classical method of dealing with this problem is the **method of Lagrange multipliers**. This procedure begins by formulating the **Lagrangian function**

$$h(\mathbf{x}, \boldsymbol{\lambda}) = f(\mathbf{x}) - \sum_{i=1}^{m} \lambda_i [g_i(\mathbf{x}) - b_i],$$

where the new variables $\boldsymbol{\lambda} = (\lambda_1, \lambda_2, \ldots, \lambda_m)$ are called *Lagrange multipliers*. Notice the key fact that for the *feasible* values of $\mathbf{x}$.

$$g_i(\mathbf{x}) - b_i = 0, \qquad \text{for all } i,$$

so $h(\mathbf{x}, \boldsymbol{\lambda}) = f(\mathbf{x})$. Therefore, it can be shown that if $(\mathbf{x}, \boldsymbol{\lambda}) = (\mathbf{x}^*, \boldsymbol{\lambda}^*)$ is a *local* or *global minimum* or *maximum* for the unconstrained function $h(\mathbf{x}, \boldsymbol{\lambda})$, then $\mathbf{x}^*$ is a corresponding *critical point* for the original problem. As a result, the method now reduces to analyzing $h(\mathbf{x}, \boldsymbol{\lambda})$ by the procedure just described for unconstrained optimization. Thus the $n + m$ partial derivatives would be set equal to zero

$$\frac{\partial h}{\partial x_j} = \frac{\partial f}{\partial x_j} - \sum_{i=1}^{m} \lambda_i \frac{\partial g_i}{\partial x_j} = 0, \qquad \text{for } j = 1, 2, \ldots, n,$$

$$\frac{\partial h}{\partial \lambda_i} = -g_i(\mathbf{x}) + b_i = 0, \qquad \text{for } i = 1, 2, \ldots, m,$$

and then the critical points would be obtained by solving these equations for $(\mathbf{x}, \boldsymbol{\lambda})$. Notice that the last m equations are equivalent to the constraints in the original problem, so only feasible solutions are considered. After further analysis to identify the *global minimum* or *maximum* of $h(\cdot)$, the resulting value of $\mathbf{x}$ is then the desired solution to the original problem.

From a practical computational viewpoint, the method of Lagrange multipliers is not a particularly powerful procedure. It is often essentially impossible to solve the equations to obtain the critical points. Furthermore, even when the points can be obtained, the number of critical points may be so large (often infinite) that it is impractical to attempt to identify a global minimum or maximum. However, for certain types of small problems, this method can sometimes be used successfully.

To illustrate, consider the example introduced earlier. In this case,

$$h(x_1, x_2) = x_1^2 + 2x_2 - \lambda(x_1^2 + x_2^2 - 1),$$

so that

$$\frac{\partial h}{\partial x_1} = 2x_1 - 2\lambda x_1 = 0,$$

$$\frac{\partial h}{\partial x_2} = 2 - 2\lambda x_2 = 0,$$

$$\frac{\partial h}{\partial \lambda} = -(x_1^2 + x_2^2 - 1) = 0.$$

The first equation implies that either $\lambda = 1$ or $x_1 = 0$. If $\lambda = 1$, then the other two equations imply that $x_2 = 1$ and $x_1 = 0$. If $x_1 = 0$, then the third equation implies that $x_2 = \pm 1$. Therefore, the two critical points for the original problem are $(x_1, x_2) = (0, 1)$ *and* $(0, -1)$. Thus it is apparent that these points are the global maximum and minimum, respectively.

The Derivative of a Definite Integral

In presenting the classical optimization methods just described, we have assumed that you are already familiar with derivatives and how to obtain them. However, there is a special case of importance in OR work that warrants additional explanation, namely, the derivative of a definite integral. In particular, consider how to find the derivative of the function

$$F(y) = \int_{g(y)}^{h(y)} f(x, y)\, dx,$$

where $g(y)$ and $h(y)$ are the limits of integration expressed as functions of y.

To begin, suppose that these limits of integration are constants, so that $g(y) = a$ and $h(y) = b$, respectively. For this special case, it can be shown that, given the regularity conditions assumed at the beginning of this appendix, the derivative is

$$\frac{d}{dy}\int_a^b f(x, y)\, dx = \int_a^b \frac{\partial f(x, y)}{\partial y}\, dx.$$

For example, if $f(x, y) = e^{-xy}$, $a = 0$, and $b = \infty$, then

$$\frac{d}{dy}\int_0^\infty e^{-xy}\, dx = \int_0^\infty (-x)e^{-xy}\, dx = -\frac{1}{y^2}$$

at any positive value of y. Thus the intuitive procedure of interchanging the order of differentiation and integration is valid for this case.

However, finding the derivative becomes a little more complicated than this when the limits of integration are functions. In particular,

$$\frac{d}{dy}\int_{g(y)}^{h(y)} f(x, y)\, dx = \int_{g(y)}^{h(y)} \frac{\partial f(x, y)}{\partial y}\, dx + f(h(y), y)\frac{dh(y)}{dy} - f(g(y), y)\frac{dg(y)}{dy},$$

where $f(h(y), y)$ is obtained by writing out $f(x, y)$ and then replacing x by $h(y)$ wherever it appears, and similarly for $f(g(y), y)$. To illustrate, if $f(x, y) = x^2y^3$, $g(y) = y$, and $h(y) = 2y$, then

$$\frac{d}{dy}\int_y^{2y} x^2y^3\, dx = \int_y^{2y} 3x^2y^2\, dx + (2y)^2y^3(2) - y^2y^3(1) = 14y^5$$

at any positive value of y.

Appendix 4

Matrices and Matrix Operations

A **matrix** is a rectangular array of numbers. For example,

$$\mathbf{A} = \begin{bmatrix} 2 & 5 \\ 3 & 0 \\ 1 & 1 \end{bmatrix}$$

is a 3×2 matrix (where 3×2 is said "3 by 2") because it is a rectangular array of numbers with three rows and two columns. (Matrices are denoted in this book by **boldface capital letters**.) The numbers in the rectangular array are called the **elements** of the matrix. For example,

$$\mathbf{B} = \begin{bmatrix} 1 & 2.4 & 0 & \sqrt{3} \\ -4 & 2 & -1 & 15 \end{bmatrix}$$

is a 2×4 matrix whose elements are 1, 2.4, 0, $\sqrt{3}$, -4, 2, -1, and 15. Thus, in more general terms,

$$\mathbf{A} = \begin{bmatrix} a_{11} & a_{12} & \cdots & a_{1n} \\ a_{21} & a_{22} & \cdots & a_{2n} \\ \cdots & \cdots & \cdots & \cdots \\ a_{m1} & a_{m2} & \cdots & a_{mn} \end{bmatrix} = \|a_{ij}\|$$

is an $m \times n$ matrix, where $a_{11}, \ldots, a_{mn}$ represent the numbers that are the elements of this matrix; $\|a_{ij}\|$ is shorthand notation for identifying the matrix whose element in row i and column j is a_{ij} for every $i = 1, 2, \ldots, m$ and $j = 1, 2, \ldots, n$.

Because matrices do not possess a numerical value, they cannot be added, multiplied, and so on as if they were individual numbers. However, it is sometimes desirable to perform certain manipulations on arrays of numbers. Therefore, rules have been developed for performing operations on matrices that are analogous to arithmetic operations. To describe these, let $\mathbf{A} = \|a_{ij}\|$ and $\mathbf{B} = \|b_{ij}\|$ be two matrices having the same number of rows and the same number of columns. (We shall change this restriction on the size of **A** and **B** later when discussing matrix multiplication.)

Matrices **A** and **B** are said to be *equal* ($\mathbf{A} = \mathbf{B}$) if and only if *all* the corresponding elements are equal ($a_{ij} = b_{ij}$ for all i and j).

The operation of *multiplying a matrix by a number* (denote this number by k) is performed by multiplying each element of the matrix by k, so that

$$k\mathbf{A} = \|ka_{ij}\|.$$

For example,

$$3\begin{bmatrix} 1 & \frac{1}{3} & 2 \\ 5 & 0 & -3 \end{bmatrix} = \begin{bmatrix} 3 & 1 & 6 \\ 15 & 0 & -9 \end{bmatrix}.$$

To add two matrices **A** and **B**, simply add the corresponding elements, so that

$$\mathbf{A} + \mathbf{B} = \|a_{ij} + b_{ij}\|.$$

To illustrate,

$$\begin{bmatrix} 5 & 3 \\ 1 & 6 \end{bmatrix} + \begin{bmatrix} 2 & 0 \\ 3 & 1 \end{bmatrix} = \begin{bmatrix} 7 & 3 \\ 4 & 7 \end{bmatrix}.$$

Similarly, *subtraction* is done as follows:

$$\mathbf{A} - \mathbf{B} = \mathbf{A} + (-1)\mathbf{B},$$

so that

$$\mathbf{A} - \mathbf{B} = \|a_{ij} - b_{ij}\|.$$

For example,

$$\begin{bmatrix} 5 & 3 \\ 1 & 6 \end{bmatrix} - \begin{bmatrix} 2 & 0 \\ 3 & 1 \end{bmatrix} = \begin{bmatrix} 3 & 3 \\ -2 & 5 \end{bmatrix}.$$

Note that, with the exception of multiplication by a number, all the preceding operations are defined only when the two matrices involved are the same size. However, all of these operations are straightforward because they involve performing only the same comparison or arithmetic operation on the corresponding elements of the matrices.

There exists one additional elementary operation that has not been defined—**matrix multiplication**—but it is considerably more complicated. To find the element in row i, column j of the matrix resulting from multiplying matrix **A** times matrix **B**, it is necessary to multiply each element in row i of **A** by the corresponding element in column j of **B** and then to add these products. To do this element-by-element multiplication, we need the following restriction on the sizes of **A** and **B**:

> Matrix multiplication **AB** is defined if and only if the *number of columns* of **A** equals the *number of rows* of **B**.

Thus, if **A** is an $m \times n$ matrix and **B** is an $n \times s$ matrix, then their product is

$$\mathbf{AB} = \left\| \sum_{k=1}^{n} a_{ik} b_{kj} \right\|,$$

where this product is an $m \times s$ matrix. However, if **A** is an $m \times n$ matrix and **B** is an $r \times s$ matrix, where $n \neq r$, then **AB** is not defined.

To illustrate matrix multiplication,

$$\begin{bmatrix} 1 & 2 \\ 4 & 0 \\ 2 & 3 \end{bmatrix} \begin{bmatrix} 3 & 1 \\ 2 & 5 \end{bmatrix} = \begin{bmatrix} 1(3)+2(2) & 1(1)+2(5) \\ 4(3)+0(2) & 4(1)+0(5) \\ 2(3)+3(2) & 2(1)+3(5) \end{bmatrix} = \begin{bmatrix} 7 & 11 \\ 12 & 4 \\ 12 & 17 \end{bmatrix}.$$

On the other hand, if one attempts to mutliply these matrices in the reverse order, the resulting product

$$\begin{bmatrix} 3 & 1 \\ 2 & 5 \end{bmatrix} \begin{bmatrix} 1 & 2 \\ 4 & 0 \\ 2 & 3 \end{bmatrix}$$

is not even defined.

Even when both **AB** and **BA** are defined,

$$\mathbf{AB} \neq \mathbf{BA}$$

in general. Thus *matrix multiplication* should be viewed as a specially designed operation whose properties are quite different from those of *arithmetic multiplication.* To understand why this special definition was adopted, consider the following system of equations:

$$\begin{aligned} 2x_1 - x_2 + 5x_3 + x_4 &= 20 \\ x_1 + 5x_2 + 4x_3 + 5x_4 &= 30 \\ 3x_1 + x_2 - 6x_3 + 2x_4 &= 20. \end{aligned}$$

Rather than write out these equations as shown here, they can be written much more concisely in matrix form as

$$\mathbf{Ax} = \mathbf{b},$$

where

$$\mathbf{A} = \begin{bmatrix} 2 & -1 & 5 & 1 \\ 1 & 5 & 4 & 5 \\ 3 & 1 & -6 & 2 \end{bmatrix}, \quad \mathbf{x} = \begin{bmatrix} x_1 \\ x_2 \\ x_3 \\ x_4 \end{bmatrix}, \quad \mathbf{b} = \begin{bmatrix} 20 \\ 30 \\ 20 \end{bmatrix}.$$

It is this kind of multiplication for which matrix multiplication is designed.

Carefully note that *matrix division* is *not* defined.

Although the matrix operations described here do not possess certain of the properties of arithmetic operations, they do satisfy these laws

$$\mathbf{A} + \mathbf{B} = \mathbf{B} + \mathbf{A},$$

$$(\mathbf{A} + \mathbf{B}) + \mathbf{C} = \mathbf{A} + (\mathbf{B} + \mathbf{C}),$$

$$\mathbf{A}(\mathbf{B} + \mathbf{C}) = \mathbf{AB} + \mathbf{AC},$$

$$\mathbf{A}(\mathbf{BC}) = (\mathbf{AB})\mathbf{C},$$

when the relative sizes of these matrices are such that the indicated operations are defined.

Another type of matrix operation, which has no arithmetic analog, is the **transpose operation**. This operation involves nothing more than interchanging the rows and columns of the matrix, which is frequently useful for performing the multiplication operation in the desired way. Thus, for any matrix $\mathbf{A} = \|a_{ij}\|$, its transpose $\mathbf{A}^T$ is

$$\mathbf{A}^T = \|a_{ji}\|.$$

For example, if

$$\mathbf{A} = \begin{bmatrix} 2 & 5 \\ 1 & 3 \\ 4 & 0 \end{bmatrix},$$

then

$$\mathbf{A}^T = \begin{bmatrix} 2 & 1 & 4 \\ 5 & 3 & 0 \end{bmatrix}.$$

Special Kinds of Matrices

In arithmetic, 0 and 1 play a special role. There also exist special matrices that play a similar role in matrix theory. In particular, the matrix that is analogous to 1 is the **identity matrix I**, which is a square matrix whose elements are 0s except for 1s along the main diagonal. Thus

$$\mathbf{I} = \begin{bmatrix} 1 & 0 & 0 & \cdots & 0 \\ 0 & 1 & 0 & \cdots & 0 \\ 0 & 0 & 1 & \cdots & 0 \\ \cdots & \cdots & \cdots & \cdots & \cdots \\ 0 & 0 & 0 & \cdots & 1 \end{bmatrix}$$

The number of rows or columns of **I** can be specified as desired. The analogy of **I** to 1 follows from the fact that for any matrix **A**,

$$\mathbf{IA} = \mathbf{A} = \mathbf{AI},$$

where **I** is assigned the appropriate number of rows and columns in each case for the multiplication operation to be defined.

Similarly, the matrix that is analogous to 0 is the **null matrix 0**, which is a matrix of any size whose elements are *all* 0s. Thus

$$\mathbf{0} = \begin{bmatrix} 0 & 0 & \cdots & 0 \\ 0 & 0 & \cdots & 0 \\ \cdots & \cdots & \cdots & \cdots \\ 0 & 0 & \cdots & 0 \end{bmatrix}$$

Therefore, for any matrix **A**,

$$\mathbf{A} + \mathbf{0} = \mathbf{A}, \qquad \mathbf{A} - \mathbf{A} = \mathbf{0}, \qquad \text{and} \qquad \mathbf{0A} = \mathbf{0} = \mathbf{A0},$$

where **0** is the appropriate size in each case for the operations to be defined.

On certain occasions, it is useful to partition a matrix into several smaller matrices, called **submatrices**. For example, one possible way of partitioning a 3 × 4 matrix would be

$$\mathbf{A} = \left[\begin{array}{c|ccc} a_{11} & a_{12} & a_{13} & a_{14} \\ \hline a_{21} & a_{22} & a_{23} & a_{24} \\ a_{31} & a_{32} & a_{33} & a_{34} \end{array}\right] = \begin{bmatrix} a_{11} & \mathbf{A}_{12} \\ \mathbf{A}_{21} & \mathbf{A}_{22} \end{bmatrix},$$

where

$$\mathbf{A}_{12} = [a_{12}, \quad a_{13}, \quad a_{14}], \qquad \mathbf{A}_{21} = \begin{bmatrix} a_{21} \\ a_{31} \end{bmatrix}, \qquad \mathbf{A}_{22} = \begin{bmatrix} a_{22} & a_{23} & a_{24} \\ a_{32} & a_{33} & a_{34} \end{bmatrix}$$

all are submatrices. Rather than perform operations element by element on such partitioned matrices, we can do them in terms of the submatrices, provided the partitionings are such that the operations are defined. For example, if **B** is a partitioned 4×1 matrix such that

$$\mathbf{B} = \begin{bmatrix} b_1 \\ b_2 \\ b_3 \\ b_4 \end{bmatrix} = \begin{bmatrix} b_1 \\ \mathbf{B}_2 \end{bmatrix},$$

then

$$\mathbf{AB} = \left[\frac{a_{11}b_1 + \mathbf{A}_{12}\mathbf{B}_2}{\mathbf{A}_{21}b_1 + \mathbf{A}_{22}\mathbf{B}_2} \right].$$

Vectors

A special kind of matrix that plays an important role in matrix theory is the kind that has either a *single row* or a *single column.* Such matrices are often referred to as **vectors**. Thus

$$\mathbf{x} = [x_1, x_2, \ldots, x_n]$$

is a **row vector**, and

$$\mathbf{x} = \begin{bmatrix} x_1 \\ x_2 \\ \vdots \\ x_n \end{bmatrix}$$

is a **column vector**. (Vectors are denoted in this book by **boldface lowercase letters**.) These vectors also are sometimes called *n-vectors* to indicate that they have n elements. For example,

$$\mathbf{x} = [1, 4, -2, \tfrac{1}{3}, 7]$$

is a 5-vector.

A **null vector 0** is either a row vector or a column vector whose elements are *all* 0s, that is,

$$\mathbf{0} = [0, 0, \ldots, 0] \qquad \text{or} \qquad \mathbf{0} = \begin{bmatrix} 0 \\ 0 \\ \vdots \\ 0 \end{bmatrix}.$$

(Although the same symbol **0** is used for either kind of *null vector,* as well as for a *null matrix,* the context normally will identify which it is.)

One reason vectors play an important role in matrix theory is that any $m \times n$ matrix can be partitioned into either m row vectors or n column vectors, and important properties of the matrix can be analyzed in terms of these vectors. To amplify, consider a set of n-vectors $\mathbf{x}_1, \mathbf{x}_2, \ldots, \mathbf{x}_m$ of the same type (i.e., they are either all row vectors or all column vectors).

Definition: A set of vectors $\mathbf{x}_1, \mathbf{x}_2, \ldots, \mathbf{x}_m$ is said to be **linearly dependent** if there exist m numbers (denoted by $c_1, c_2, \ldots, c_m$), some of which are not zero, such that

$$c_1\mathbf{x}_1 + c_2\mathbf{x}_2 + \cdots + c_m\mathbf{x}_m = \mathbf{0}.$$

Otherwise, the set is said to be **linearly independent**.

To illustrate, if $m = 3$ and

$$\mathbf{x}_1 = [1, 1, 1], \qquad \mathbf{x}_2 = [0, 1, 1], \qquad \mathbf{x}_3 = [2, 5, 5],$$

then

$$2\mathbf{x}_1 + 3\mathbf{x}_2 - \mathbf{x}_3 = \mathbf{0},$$

so that

$$\mathbf{x}_3 = 2\mathbf{x}_1 + 3\mathbf{x}_2.$$

Thus $\mathbf{x}_1$, $\mathbf{x}_2$, $\mathbf{x}_3$ would be linearly dependent because one of them is a linear combination of the others. However, if $\mathbf{x}_3$ were changed to

$$\mathbf{x}_3 = [2, 5, 6]$$

instead, then $\mathbf{x}_1$, $\mathbf{x}_2$, $\mathbf{x}_3$ would be linearly independent.

Definition: The **rank** of a *set* of vectors is the largest number of *linearly independent vectors* that can be chosen from the set.

Continuing the preceding example, we see that the rank of the set of vectors $\mathbf{x}_1$, $\mathbf{x}_2$, $\mathbf{x}_3$ was 2, but it became 3 after $\mathbf{x}_3$ was changed.

Definition: A **basis** for a *set* of vectors is a *collection* of linearly independent vectors taken from the set such that every vector in the set is a linear combination of the vectors in the collection (i.e., every vector in the set equals the sum of certain multiples of the vectors in the collection).

To illustrate, $\mathbf{x}_1$ and $\mathbf{x}_2$ constituted a basis for $\mathbf{x}_1$, $\mathbf{x}_2$, $\mathbf{x}_3$ in the preceding example before $\mathbf{x}_3$ was changed. After $\mathbf{x}_3$ is changed, the basis becomes all three vectors.

The following theorem relates the last two definitions.

Theorem A4.1: A collection of r linearly independent vectors chosen from a set of vectors is a basis for the set if and only if the set has rank r.

Some Properties of Matrices

Given the preceding results regarding vectors, it is now possible to present certain important concepts regarding matrices.

Definition: The **row rank** of a matrix is the rank of its set of row vectors. The **column rank** of a matrix is the rank of its column vectors.

For example, if matrix **A** is

$$\mathbf{A} = \begin{bmatrix} 1 & 1 & 1 \\ 0 & 1 & 1 \\ 2 & 5 & 5 \end{bmatrix},$$

then its row rank was shown to be 2. Note that the column rank of **A** is also 2. This fact is no coincidence, as the following general theorem indicates.

Theorem A4.2: The row rank and column rank of a matrix are equal.

Thus it is only necessary to speak of the rank of a matrix.

The final concept to be discussed is the **inverse of a matrix**. For any nonzero number k, there exists a reciprocal or inverse $k^{-1} = 1/k$ such that

$$kk^{-1} = 1 = k^{-1}k.$$

Is there an analogous concept that is valid in matrix theory? In other words, for a given matrix $\mathbf{A}$ other than the null matrix, does there exist a matrix $\mathbf{A}^{-1}$ such that

$$\mathbf{AA}^{-1} = \mathbf{I} = \mathbf{A}^{-1}\mathbf{A}?$$

If $\mathbf{A}$ is not a square matrix (i.e., if the number of rows and the number of columns of $\mathbf{A}$ differ), the answer is *never,* because these matrix products would necessarily have a different number of rows for the multiplication to be defined (so that the equality operation would not be defined). However, if $\mathbf{A}$ is square, then the answer is *under certain circumstances,* as described by the following definition and Theorem A4.3.

Definition: A matrix is **nonsingular** if its rank equals both the number of rows and the number of columns. Otherwise, it is **singular**.

Thus only square matrices can be *nonsingular.* A useful way of testing for nonsingularity is provided by the fact that a square matrix is nonsingular if and only if *its determinant is nonzero.*

Theorem A4.3: (*a*) If $\mathbf{A}$ is nonsingular, there is a unique nonsingular matrix $\mathbf{A}^{-1}$, called the **inverse** of $\mathbf{A}$, such that $\mathbf{AA}^{-1} = \mathbf{I} = \mathbf{A}^{-1}\mathbf{A}$.
(*b*) If $\mathbf{A}$ is nonsingular and $\mathbf{B}$ is a matrix for which either $\mathbf{AB} = \mathbf{I}$ or $\mathbf{BA} = \mathbf{I}$, then $\mathbf{B} = \mathbf{A}^{-1}$.
(*c*) Only nonsingular matrices have inverses.

To illustrate matrix inverses, consider the matrix

$$\mathbf{A} = \begin{bmatrix} 5 & -4 \\ 1 & -1 \end{bmatrix}.$$

Notice that the rank of $\mathbf{A}$ is 2, so it is nonsingular. Therefore, $\mathbf{A}$ must have an inverse, which happens to be

$$\mathbf{A}^{-1} = \begin{bmatrix} 1 & -4 \\ 1 & -5 \end{bmatrix}.$$

Hence,

$$\mathbf{AA}^{-1} = \begin{bmatrix} 5 & -4 \\ 1 & -1 \end{bmatrix}\begin{bmatrix} 1 & -4 \\ 1 & -5 \end{bmatrix} = \begin{bmatrix} 1 & 0 \\ 0 & 1 \end{bmatrix},$$

and

$$\mathbf{A}^{-1}\mathbf{A} = \begin{bmatrix} 1 & -4 \\ 1 & -5 \end{bmatrix}\begin{bmatrix} 5 & -4 \\ 1 & -1 \end{bmatrix} = \begin{bmatrix} 1 & 0 \\ 0 & 1 \end{bmatrix}.$$

Appendix 5

Tables

Table A5.1 **Areas under the Normal Curve from K_α to ∞**

$$P\{\text{normal} \geq K_\alpha\} = \int_{K_\alpha}^{\infty} \frac{1}{\sqrt{2\pi}} e^{-x^2/2}\, dx = \alpha$$

K_α	.00	.01	.02	.03	.04	.05	.06	.07	.08	.09
0.0	.5000	.4960	.4920	.4880	.4840	.4801	.4761	.4721	.4681	.4641
0.1	.4602	.4562	.4522	.4483	.4443	.4404	.4364	.4325	.4286	.4247
0.2	.4207	.4168	.4129	.4090	.4052	.4013	.3974	.3936	.3897	.3859
0.3	.3821	.3783	.3745	.3707	.3669	.3632	.3594	.3557	.3520	.3483
0.4	.3446	.3409	.3372	.3336	.3300	.3264	.3228	.3192	.3156	.3121
0.5	.3085	.3050	.3015	.2981	.2946	.2912	.2877	.2843	.2810	.2776
0.6	.2743	.2709	.2676	.2643	.2611	.2578	.2546	.2514	.2483	.2451
0.7	.2420	.2389	.2358	.2327	.2296	.2266	.2236	.2206	.2177	.2148
0.8	.2119	.2090	.2061	.2033	.2005	.1977	.1949	.1922	.1894	.1867
0.9	.1841	.1814	.1788	.1762	.1736	.1711	.1685	.1660	.1635	.1611
1.0	.1587	.1562	.1539	.1515	.1492	.1469	.1446	.1423	.1401	.1379
1.1	.1357	.1335	.1314	.1292	.1271	.1251	.1230	.1210	.1190	.1170
1.2	.1151	.1131	.1112	.1093	.1075	.1056	.1038	.1020	.1003	.0985
1.3	.0968	.0951	.0934	.0918	.0901	.0885	.0869	.0853	.0838	.0823
1.4	.0808	.0793	.0778	.0764	.0749	.0735	.0721	.0708	.0694	.0681
1.5	.0668	.0655	.0643	.0630	.0618	.0606	.0594	.0582	.0571	.0559
1.6	.0548	.0537	.0526	.0516	.0505	.0495	.0485	.0475	.0465	.0455
1.7	.0446	.0436	.0427	.0418	.0409	.0401	.0392	.0384	.0375	.0367
1.8	.0359	.0351	.0344	.0336	.0329	.0322	.0314	.0307	.0301	.0294
1.9	.0287	.0281	.0274	.0268	.0262	.0256	.0250	.0244	.0239	.0233
2.0	.0228	.0222	.0217	.0212	.0207	.0202	.0197	.0192	.0188	.0183
2.1	.0179	.0174	.0170	.0166	.0162	.0158	.0154	.0150	.0146	.0143
2.2	.0139	.0136	.0132	.0129	.0125	.0122	.0119	.0116	.0113	.0110
2.3	.0107	.0104	.0102	.00990	.00964	.00939	.00914	.00889	.00866	.00842
2.4	.00820	.00798	.00776	.00755	.00734	.00714	.00695	.00676	.00657	.00639
2.5	.00621	.00604	.00587	.00570	.00554	.00539	.00523	.00508	.00494	.00480
2.6	.00466	.00453	.00440	.00427	.00415	.00402	.00391	.00379	.00368	.00357
2.7	.00347	.00336	.00326	.00317	.00307	.00298	.00289	.00280	.00272	.00264
2.8	.00256	.00248	.00240	.00233	.00226	.00219	.00212	.00205	.00199	.00193
2.9	.00187	.00181	.00175	.00169	.00164	.00159	.00154	.00149	.00144	.00139

K_α	.0	.1	.2	.3	.4	.5	.6	.7	.8	.9
3	.00135	$.0^3968$	$.0^3687$	$.0^3483$	$.0^3337$	$.0^3233$	$.0^3159$	$.0^3108$	$.0^4723$	$.0^4481$
4	$.0^4317$	$.0^4207$	$.0^4133$	$.0^5854$	$.0^5541$	$.0^5340$	$.0^5211$	$.0^5130$	$.0^6793$	$.0^6479$
5	$.0^6287$	$.0^6170$	$.0^7996$	$.0^7579$	$.0^7333$	$.0^7190$	$.0^7107$	$.0^8599$	$.0^8332$	$.0^8182$
6	$.0^9987$	$.0^9530$	$.0^9282$	$.0^9149$	$.0^{10}777$	$.0^{10}402$	$.0^{10}206$	$.0^{10}104$	$.0^{11}523$	$.0^{11}260$

Source: F. E. Croxton, *Tables of Areas in Two Tails and in One Tail of the Normal Curve*. Copyright 1949 by Prentice-Hall, Inc., Englewood Cliffs, NJ.

Table A5.2 **100 α Percentage Points of Student's t Distribution**
P{Student's t with ν Degrees of Freedom ≥ Tabled Value} = α

ν \ α	0.40	0.25	0.10	0.05	0.025	0.01	0.005	0.0025	0.001	0.0005
1	0.325	1.000	3.078	6.314	12.706	31.821	63.657	127.32	318.31	636.62
2	0.289	0.816	1.886	2.920	4.303	6.965	9.925	14.089	22.327	31.598
3	0.277	0.765	1.638	2.353	3.182	4.541	5.841	7.453	10.214	12.924
4	0.271	0.741	1.533	2.132	2.776	3.747	4.604	5.598	7.173	8.610
5	0.267	0.727	1.476	2.015	2.571	3.365	4.032	4.773	5.893	6.869
6	0.265	0.718	1.440	1.943	2.447	3.143	3.707	4.317	5.208	5.959
7	0.263	0.711	1.415	1.895	2.365	2.998	3.499	4.029	4.785	5.408
8	0.262	0.706	1.397	1.860	2.306	2.896	3.355	3.833	4.501	5.041
9	0.261	0.703	1.383	1.833	2.262	2.821	3.250	3.690	4.297	4.781
10	0.260	0.700	1.372	1.812	2.228	2.764	3.169	3.581	4.144	4.587
11	0.260	0.697	1.363	1.796	2.201	2.718	3.106	3.497	4.025	4.437
12	0.259	0.695	1.356	1.782	2.179	2.681	3.055	3.428	3.930	4.318
13	0.259	0.694	1.350	1.771	2.160	2.650	3.012	3.372	3.852	4.221
14	0.258	0.692	1.345	1.761	2.145	2.624	2.977	3.326	3.787	4.140
15	0.258	0.691	1.341	1.753	2.131	2.602	2.947	3.286	3.733	4.073
16	0.258	0.690	1.337	1.746	2.120	2.583	2.921	3.252	3.686	4.015
17	0.257	0.689	1.333	1.740	2.110	2.567	2.898	3.222	3.646	3.965
18	0.257	0.688	1.330	1.734	2.101	2.552	2.878	3.197	3.610	3.922
19	0.257	0.688	1.328	1.729	2.093	2.539	2.861	3.174	3.579	3.883
20	0.257	0.687	1.325	1.725	2.086	2.528	2.845	3.153	3.552	3.850
21	0.257	0.686	1.323	1.721	2.080	2.518	2.831	3.135	3.527	3.819
22	0.256	0.686	1.321	1.717	2.074	2.508	2.819	3.119	3.505	3.792
23	0.256	0.685	1.319	1.714	2.069	2.500	2.807	3.104	3.485	3.767
24	0.256	0.685	1.318	1.711	2.064	2.492	2.797	3.091	3.467	3.745
25	0.256	0.684	1.316	1.708	2.060	2.485	2.787	3.078	3.450	3.725
26	0.256	0.684	1.315	1.706	2.056	2.479	2.779	3.067	3.435	3.707
27	0.256	0.684	1.314	1.703	2.052	2.473	2.771	3.057	3.421	3.690
28	0.256	0.683	1.313	1.701	2.048	2.467	2.763	3.047	3.408	3.674
29	0.256	0.683	1.311	1.699	2.045	2.462	2.756	3.038	3.396	3.659
30	0.256	0.683	1.310	1.697	2.042	2.457	2.750	3.030	3.385	3.646
40	0.255	0.681	1.303	1.684	2.021	2.423	2.704	2.971	3.307	3.551
60	0.254	0.679	1.296	1.671	2.000	2.390	2.660	2.915	3.232	3.460
120	0.254	0.677	1.289	1.658	1.980	2.358	2.617	2.860	3.160	3.373
∞	0.253	0.674	1.282	1.645	1.960	2.326	2.576	2.807	3.090	3.291

Source: Table 12 of *Biometrika Tables for Statisticians,* vol. I, 3d ed., 1966, by permission of the Biometrika Trustees.

Answers to Selected Problems

Chapter 3

3.1-1. $(x_1, x_2) = (13, 5)$; $Z = 31$.

3.1-3. (*a*)

$$\text{Maximize} \quad Z = 50x_1 + 20x_2 + 25x_3,$$

subject to

$$\begin{aligned} 9x_1 + 3x_2 + 5x_3 &\leq 500 \\ 5x_1 + 4x_2 &\leq 350 \\ 3x_1 + 2x_3 &\leq 150 \\ x_3 &\leq 20 \end{aligned}$$

and

$$x_1 \geq 0, \quad x_2 \geq 0, \quad x_3 \geq 0.$$

(*b*) $(x_1, x_2, x_3) = (26.19, 54.76, 20)$; $Z = 2{,}904.76$.

3.2-3. (*b*)

$$\text{Maximize} \quad Z = 4{,}500x_1 + 4{,}500x_2,$$

subject to

$$\begin{aligned} x_1 \phantom{+ 4{,}000x_2} &\leq 1 \\ x_2 &\leq 1 \\ 5{,}000x_1 + 4{,}000x_2 &\leq 6{,}000 \\ 400x_1 + 500x_2 &\leq 600 \end{aligned}$$

and

$$x_1 \geq 0, \quad x_2 \geq 0.$$

3.4-1. (*a*) *Proportionality*: OK since it is implied that a fixed fraction of the radiation dosage at a given entry point is absorbed by a given area.
Additivity: OK since it is stated that the radiation absorption from multiple beams is additive.
Divisibility: OK since beam strength can be any fractional level.
Certainty: Due to the complicated analysis required to estimate the data on radiation absorption in different tissue types, there is considerable uncertainty about the data, so sensitivity analysis should be used.

3.4-3. 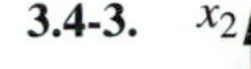

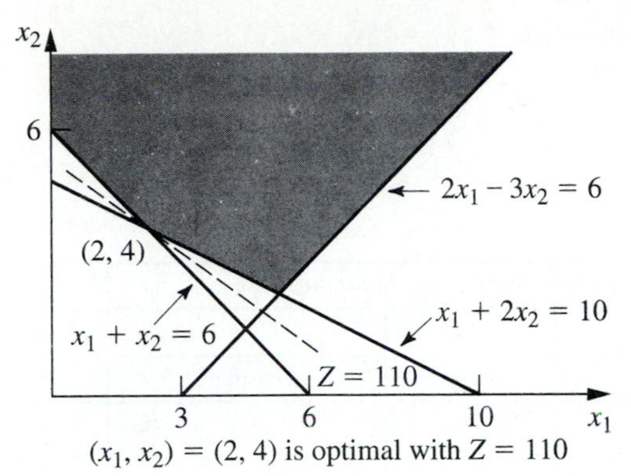

$(x_1, x_2) = (2, 4)$ is optimal with $Z = 110$

3.4-8. (*b*) $x_{1L} = 516.667$, $x_{1M} = 177.778$, $x_{1S} = 0$, $x_{2L} = 0$, $x_{2M} = 666.67$, $x_{2S} = 166.667$, $x_{3L} = 0$, $x_{3M} = 0$, $x_{3S} = 416.667$, $Z = 696{,}000$.

3.4-11. (*a*) Let R_t (for $t = 1, 2, 3, 4$) denote dollars *not* invested in year t.

Maximize $Z = 1.9C_2 + 1.7B_3 + 1.4A_4 + 1.3D_5$,

subject to

$$A_1 + B_1 + R_1 = 60{,}000$$
$$A_2 + B_2 + C_2 + R_2 = R_1$$
$$A_3 + B_3 + R_3 = 1.4A_1 + R_2$$
$$A_4 + R_4 = 1.4A_2 + 1.7B_1 + R_3$$
$$D_5 = 1.4A_3 + 1.7B_2 + R_4$$

and $A_t \geq 0$ for $t = 1, 2, 3, 4$; $B_t \geq 0$ for $t = 1, 2, 3$
$C_2 \geq 0$, $D_5 \geq 0$, $R_t \geq 0$ for $t = 1, 2, 3, 4$.

(*b*) $Z = \$152{,}880$; $A_1 = 60{,}000$; $A_3 = 84{,}000$; $D_5 = 117{,}600$. All other decision variables are 0.

Chapter 4

4.1-1. (*a*) The corner-point solutions that are *feasible* are (0, 0), (0, 1), $(\frac{1}{4}, 1)$, $(\frac{2}{3}, \frac{2}{3})$, $(1, \frac{1}{4})$, and (1, 0).

4.3-4. $(x_1, x_2, x_3) = (0, 10, 6\frac{2}{3})$; $Z = 70$.

4.6-1. $(x_1, x_2) = (2, 1)$; $Z = 7$.

4.6-4. $(x_1, x_2, x_3) = (\frac{4}{5}, \frac{9}{5}, 0)$; $Z = 7$.

4.6-8. (*a*),(*b*) $(x_1, x_2, x_3) = (0, 15, 15)$; $Z = 90$.
(*c*) For both the Big M method and the two-phase method, only the final tableau represents a feasible solution for the real problem.

4.6-13. $(x_1, x_2) = (-\frac{8}{7}, \frac{18}{7})$; $Z = \frac{80}{7}$.

4.7-6. (*a*) $(x_1, x_2, x_3) = (0, 1, 3)$; $Z = 7$.
(*b*) $y_1^* = \frac{1}{2}$, $y_2^* = \frac{5}{2}$, $y_3^* = 0$. These are the marginal values of resources 1, 2, and 3, respectively.

Chapter 5

5.1-1. (*a*) $(x_1, x_2) = (2, 2)$ is optimal. Other CPF solutions are (0, 0), (3, 0), and (0, 3).
5.1-13. $(x_1, x_2, x_3) = (0, 15, 15)$ is optimal.
5.2-2. $(x_1, x_2, x_3, x_4, x_5) = (0, 5, 0, \frac{5}{2}, 0)$; $Z = 50$.
5.3-1. (*a*) Right side is $Z = 8$, $x_2 = 14$, $x_6 = 5$, $x_3 = 11$.
(*b*) $x_1 = 0$, $2x_1 - 2x_2 + 3x_3 = 5$, $x_1 + x_2 - x_3 = 3$.

Chapter 6

6.1-2. (*a*) Minimize $y_0 = 15y_1 + 12y_2 + 45y_3$,

subject to
$$-y_1 + y_2 + 5y_3 \geq 10$$
$$2y_1 + y_2 + 3y_3 \geq 20$$
and
$$y_1 \geq 0, \quad y_2 \geq 0, \quad y_3 \geq 0.$$

6.3-1. (*c*)

Complementary Basic Solutions

Primal Problem			***Dual Problem***	
Basic Solution	Feasible?	$Z = y_0$	Feasible?	Basic Solution
(0, 0, 20, 10)	Yes	0	No	(0, 0, −6, −8)
(4, 0, 0, 6)	Yes	24	No	$(1\frac{1}{5}, 0, 0, -5\frac{3}{5})$
(0, 5, 10, 0)	Yes	40	No	(0, 4, −2, 0)
$(2\frac{1}{2}, 3\frac{3}{4}, 0, 0)$	Yes and optimal	45	Yes and optimal	$(\frac{1}{2}, 3\frac{1}{2}, 0, 0)$
(10, 0, −30, 0)	No	60	Yes	(0, 6, 0, 4)
(0, 10, 0, −10)	No	80	Yes	(4, 0, 14, 0)

6.3-7. (*c*) Basic variables are x_1 and x_2. The other variables are nonbasic.
(*e*) $x_1 + 3x_2 + 2x_3 + 3x_4 + x_5 = 6$, $4x_1 + 6x_2 + 5x_3 + 7x_4 + x_5 = 15$, $x_3 = 0$, $x_4 = 0$, $x_5 = 0$. Optimal CPF solution is $(x_1, x_2, x_3, x_4, x_5) = (\frac{3}{2}, \frac{3}{2}, 0, 0, 0)$.

6.4-3. (*a*) Maximize $y_0 = 8y_1 + 6y_2$,

subject to
$$y_1 + 3y_2 \leq 2$$
$$4y_1 + 2y_2 \leq 3$$
$$2y_1 \leq 1$$
and
$$y_1 \geq 0, \quad y_2 \geq 0.$$

6.4-8. (*a*) Minimize $y_0 = 120y_1 + 80y_2 + 100y_3$,

subject to
$$y_2 - 3y_3 = -1$$
$$3y_1 - y_2 + y_3 = 2$$
$$y_1 - 4y_2 + 2y_3 = 1$$
and
$$y_1 \geq 0, \quad y_2 \geq 0, \quad y_3 \geq 0.$$

6.6-1. (*d*) Not optimal, since $2y_1 + 3y_2 \geq 3$ is violated for $y_1^* = \frac{1}{5}$, $y_2^* = \frac{3}{5}$.
(*f*) Not optimal, since $3y_1 + 2y_2 \geq 2$ is violated for $y_1^* = \frac{1}{5}$, $y_2^* = \frac{3}{5}$.

6.7-1.

Part	New Basic Solution $(x_1, x_2, x_3, x_4, x_5)$	Feasible?	Optimal?
(*a*)	(0, 30, 0, 0, −30)	No	No
(*b*)	(0, 20, 0, 0, −10)	No	No
(*c*)	(0, 10, 0, 0, 60)	Yes	Yes
(*d*)	(0, 20, 0, 0, 10)	Yes	Yes
(*e*)	(0, 20, 0, 0, 10)	Yes	Yes
(*f*)	(0, 10, 0, 0, 40)	Yes	No
(*g*)	(0, 20, 0, 0, 10)	Yes	Yes
(*h*)	(0, 20, 0, 0, 10, $x_6 = -10$)	No	No
(*i*)	(0, 20, 0, 0, 0)	Yes	Yes

6.7-2. $-10 \le \theta \le \frac{10}{9}$

6.7-12. (*a*) $b_1 \ge 2$, $6 \le b_2 \le 18$, $12 \le b_3 \le 24$

(*b*) $0 \le c_1 \le \frac{15}{2}$, $c_2 \ge 2$

6.7-16. The current basic solution is optimal if $0 \le \theta \le \frac{3}{2}$.

Chapter 7

7.1-3. $(x_1, x_2, x_3) = (\frac{2}{3}, 2, 0)$ with $Z = \frac{22}{3}$ is optimal.

7.1-7.

Part	New Optimal Solution	Value of Z
(*a*)	$(x_1, x_2, x_3, x_4, x_5) = (0, 0, 9, 3, 0)$	117
(*b*)	$(x_1, x_2, x_3, x_4, x_5) = (0, 5, 5, 0, 0)$	90

7.2-1.

Range of θ	Optimal Solution	$Z(\theta)$
$0 \le \theta \le 2$	$(x_1, x_2) = (0, 5)$	$120 - 10\theta$
$2 \le \theta \le 8$	$(x_1, x_2) = (\frac{10}{3}, \frac{10}{3})$	$\frac{320 - 10\theta}{3}$
$8 \le \theta$	$(x_1, x_2) = (5, 0)$	$40 + 5\theta$

7.2-6.

	Optimal Solution		
Range of θ	x_1	x_2	$Z(\theta)$
$0 \le \theta \le 1$	$10 + 2\theta$	$10 + 2\theta$	$30 + 6\theta$
$1 \le \theta \le 5$	$10 + 2\theta$	$15 - 3\theta$	$35 + \theta$
$5 \le \theta \le 25$	$25 - \theta$	0	$50 - 2\theta$

7.3-3. $(x_1, x_2, x_3) = (1, 3, 1)$ with $Z = 8$ is optimal.

7.5-2. $(x_1, x_2) = (15, 0)$ is optimal.

Chapter 8

8.1-2.

		Destination			
		Today	Tomorrow	Dummy	Supply
Source	Dick	3.0	2.7	0	5
	Harry	2.9	2.8	0	4
	Demand	3	4	2	

8.2-2. (*a*) Basic variables: $x_{11} = 4$, $x_{12} = 0$, $x_{22} = 4$, $x_{23} = 2$, $x_{24} = 0$, $x_{34} = 5$, $x_{35} = 1$, $x_{45} = 0$; $Z = 53$.

(*b*) Basic variables: $x_{11} = 4$, $x_{23} = 2$, $x_{25} = 4$, $x_{31} = 0$, $x_{32} = 0$, $x_{34} = 5$, $x_{35} = 1$, $x_{42} = 4$; $Z = 45$.

(*c*) Basic variables: $x_{11} = 4$, $x_{23} = 2$, $x_{25} = 4$, $x_{32} = 0$, $x_{34} = 5$, $x_{35} = 1$, $x_{41} = 0$, $x_{42} = 4$; $Z = 45$.

8.2-8. (*a*) $x_{11} = 3$, $x_{12} = 2$, $x_{22} = 1$, $x_{23} = 1$, $x_{33} = 1$, $x_{34} = 2$; three iterations to reach optimality.

(*b*), (*c*) $x_{11} = 3$, $x_{12} = 0$, $x_{13} = 0$, $x_{14} = 2$, $x_{23} = 2$, $x_{32} = 3$; already optimal.

8.2-11. $x_{11} = 10$, $x_{12} = 15$, $x_{22} = 0$, $x_{23} = 5$, $x_{25} = 30$, $x_{33} = 20$, $x_{34} = 10$, $x_{44} = 10$; cost = \$77.30. Also have other tied optimal solutions.

8.2-12. (*b*) Let x_{ij} be the shipment from plant i to distribution center j. Then $x_{13} = 2$, $x_{14} = 10$, $x_{22} = 9$, $x_{23} = 8$, $x_{31} = 10$, $x_{32} = 1$; cost = \$20,200.

8.3-3. (*a*)

		Task				
		Backstroke	Breaststroke	Butterfly	Freestyle	Dummy
	Carl	37.7	43.4	33.3	29.2	0
	Chris	32.9	33.1	28.5	26.4	0
Assignee	David	33.8	42.2	38.9	29.6	0
	Tony	37.0	34.7	30.4	28.5	0
	Ken	35.4	41.8	33.6	31.1	0

Chapter 9

9.3-2. (*a*) $O \rightarrow A \rightarrow B \rightarrow D \rightarrow T$ or $O \rightarrow A \rightarrow B \rightarrow E \rightarrow D \rightarrow T$, with length = 16.

9.4-1. (*a*) $\{(O, A); (A, B); (B, C); (B, E); (E, D); (D, T)\}$, with length = 18.

9.5-1. (*a*)

Arc	(1, 2)	(1, 3)	(1, 4)	(2, 5)	(3, 4)	(3, 5)	(3, 6)	(4, 6)	(5, 7)	(6, 7)
Flow	4	4	1	4	1	0	3	2	4	5

9.8-1.

Event	1	2	3	4	5	6	7	8	9	10
Earliest time	0	6	3	5	10	10	11	14	13	20
Latest time	0	7	3	6	11	10	13	14	15	20
Slack	0	1	0	1	1	0	2	0	2	0

Critical path: $1 \rightarrow 3 \rightarrow 6 \rightarrow 8 \rightarrow 10$.

9.8-13. $t_e = 37$, $\sigma^2 = 9$.

Chapter 10

10.3-1.

	Store		
	1	2	3
Allocations	1	2	2
	3	2	0

10.3-8.

Phase	(*a*)	(*b*)
1	$2M$	$2.945M$
2	$1M$	$1.055M$
3	$1M$	0
Market share	6%	6.302%

10.3-14. $x_1 = -2 + \sqrt{13} \approx 1.6056$, $x_2 = 5 - \sqrt{13} \approx 1.3944$; $Z = 98.233$.

10.4-3. Produce 2 on first production run; if none acceptable, produce 2 on second run. Expected cost = \$575.

Chapter 11

11.2-2. (*a*) Player 1: strategy 2; player 2: strategy 1.

11.2-7. (*a*) Politician 1: issue 2; politician 2: issue 2.
(*b*) Politician 1: issue 1; politician 2: issue 2.
(*c*) Minimax criterion says politician 1 can use any issue, but issue 1 offers politician 1 the only chance of winning if politician 2 is not ''smart.''

11.4-3. (*a*) $(x_1, x_2) = (\frac{2}{5}, \frac{3}{5})$; $(y_1, y_2, y_3) = (\frac{1}{5}, 0, \frac{4}{5})$; $v = \frac{8}{5}$.

11.5-2. (*a*) Maximize x_4,

subject to

$$\begin{aligned} 5x_1 + 2x_2 + 3x_3 - x_4 &\geq 0 \\ 4x_2 + 2x_3 - x_4 &\geq 0 \\ 3x_1 + 3x_2 \qquad - x_4 &\geq 0 \\ x_1 + 2x_2 + 4x_3 - x_4 &\geq 0 \\ x_1 + x_2 + x_3 \qquad &= 1 \end{aligned}$$

and $x_1 \geq 0, \quad x_2 \geq 0, \quad x_3 \geq 0, \quad x_4 \geq 0.$

Chapter 12

12.1-1. (*a*) Minimize $Z = 4.5x_{em} + 7.8x_{ec} + 3.6x_{ed} + 2.9x_{el} + 4.9x_{sm} + 7.2x_{sc} + 4.3x_{sd} + 3.1x_{sl}$,

subject to

$$\begin{aligned} x_{em} + x_{ec} + x_{ed} + x_{el} &= 2 \\ x_{sm} + x_{sc} + x_{sd} + x_{sl} &= 2 \\ x_{em} + x_{sm} &= 1 \\ x_{ec} + x_{sc} &= 1 \\ x_{ed} + x_{sd} &= 1 \\ x_{el} + x_{sl} &= 1 \end{aligned}$$

and all x_{ij} are binary.

12.2-6. (*b*), (*d*) (long, medium, short) = (14, 0, 16), with profit of $47.8 million.

12.3-6. (*a*) Let $x_{ij} = \begin{cases} 1 & \text{if arc } i \rightarrow j \text{ is included in shortest path,} \\ 0 & \text{otherwise.} \end{cases}$

Mutually exclusive alternatives: For each column of arcs, exactly one arc is included in the shortest path. Contingent decisions: The shortest path leaves node i only if it enters node i.

12.4-1. (*a*) $(x_1, x_2) = (2, 3)$ is optimal.
(*b*) None of the feasible rounded solutions are optimal for the integer programming problem.

12.5-1. $(x_1, x_2, x_3, x_4, x_5) = (0, 0, 1, 1, 1)$, with $Z = 6$.

12.5-7.

Task	1	2	3	4	5
Assignee	1	3	2	4	5

12.5-9. $(x_1, x_2, x_3, x_4) = (0, 1, 1, 0)$, with $Z = 36$.

12.6-1. (*a*), (*b*) $(x_1, x_2) = (2, 1)$ is optimal.

Chapter 13

13.2-7. (*a*) Concave.

13.4-1. Approximate solution = 1.0125.

13.5-4. Exact solution is $(x_1, x_2) = (2, -2)$.

13.5-8. (*a*) Approximate solution is $(x_1, x_2) = (0.75, 1.5)$.

13.6-3.

$$\begin{aligned} -4x_1^3 - 4x_1 - 2x_2 + 2u_1 + u_2 &= 0 \quad (\text{or } \leq 0 \text{ if } x_1 = 0). \\ -2x_1 - 8x_2 + u_1 + 2u_2 &= 0 \quad (\text{or } \leq 0 \text{ if } x_2 = 0). \\ -2x_1 - x_2 + 10 &= 0 \quad (\text{or } \leq 0 \text{ if } u_1 = 0). \\ -x_1 - 2x_2 + 10 &= 0 \quad (\text{or } \leq 0 \text{ if } u_2 = 0). \end{aligned}$$

$x_1 \geq 0, \quad x_2 \geq 0, \quad u_1 \geq 0, \quad u_2 \geq 0.$

$(x_1, x_2) = (0, 10)$ cannot be optimal.

13.6-8. $(x_1, x_2) = (1, 2)$ cannot be optimal.

13.6-10. (*a*) $(x_1, x_2) = (1 - 3^{-1/2}, 3^{-1/2})$.

13.7-2. (*a*) $(x_1, x_2) = (2, 0)$ is optimal.

(*b*) Minimize $Z = z_1 + z_2$,

subject to

$$\begin{aligned} 2x_1 \quad + u_1 - y_1 \quad\quad + z_1 \quad &= 8 \\ 2x_2 + u_1 \quad - y_2 \quad\quad + z_2 &= 4 \\ x_1 + x_2 \quad\quad + v_1 \quad\quad &= 2 \end{aligned}$$

$x_1 \geq 0, \quad x_2 \geq 0, \quad u_1 \geq 0, \quad y_1 \geq 0, \quad y_2 \geq 0, \quad v_1 \geq 0, \quad z_1 \geq 0, \quad z_2 \geq 0.$

(*c*) $(x_1, x_2, u_1, y_1, y_2, v_1, z_1, z_2) = (2, 0, 4, 0, 0, 0, 0, 0)$ is optimal.

13.8-3. (*b*) Maximize $Z = 3x_{11} - 3x_{12} - 15x_{13} + 4x_{21} - 4x_{23}$,

subject to

$$\begin{aligned} x_{11} + x_{12} + x_{13} + 3x_{21} + 3x_{22} + 3x_{23} &\leq 8 \\ 5x_{11} + 5x_{12} + 5x_{13} + 2x_{21} + 2x_{22} + 2x_{23} &\leq 14 \end{aligned}$$

and $0 \leq x_{ij} \leq 1, \quad$ for $i = 1, 2, 3; j = 1, 2, 3.$

13.9-1. $(x_1, x_2) = (5, 0)$ is optimal.

13.9-10. (*a*) $(x_1, x_2) = (\frac{1}{3}, \frac{2}{3})$.

13.10-5. (*a*) $P(x; r) = -2x_1 - (x_2 - 3)^2 - r\left(\frac{1}{x_1 - 3} + \frac{1}{x_2 - 3}\right).$

(*b*) $(x_1, x_2) = \left[3 + \left(\frac{r}{2}\right)^{1/2}, 3 + \left(\frac{r}{2}\right)^{1/3}\right].$

Chapter 14

14.3-3. (*c*) $\pi_0 = \pi_1 = \pi_2 = \pi_3 = \pi_4 = \frac{1}{5}$.

14.4-1. (*a*) All states belong to the same recurrent class.

14.6-10. (*a*) $\pi_0 = 0.182, \ \pi_1 = 0.285, \ \pi_2 = 0.368, \ \pi_3 = 0.165.$

(*b*) 31.42.

Chapter 15

15.2-1. Input source: population having hair; customers: customers needing haircuts; queue: customers waiting for a barber; queue discipline: first-come-first-served; service mechanism: barber(s).

15.4-2. (*a*) 0.135

(*b*) 0.270

(*c*) 0.0527

15.5-5. (*a*) State:

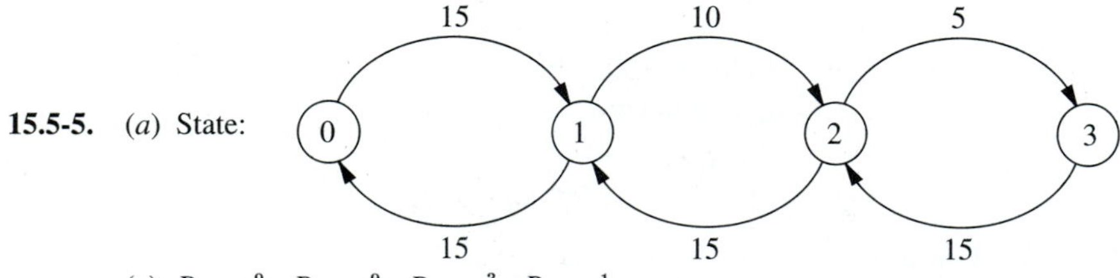

(*c*) $P_0 = \frac{9}{26}, P_1 = \frac{9}{26}, P_2 = \frac{3}{13}, P_3 = \frac{1}{13}$.

(*d*) $W = 0.11$ hour.

15.5-9. (*b*) $P_0 = \frac{2}{5}, P_n = (\frac{3}{5})(\frac{1}{2})^n$

(*c*) $L = \frac{6}{5}, L_q = \frac{3}{5}, W = \frac{1}{25}, W_q = \frac{1}{50}$

15.6-4. $\frac{31}{32}$

15.6-27. (*a*) 0.429

(*b*) 0.154

(*c*) 0.072

15.6-31. (*a*) three machines

(*b*) 0.268

15.7-1. (*a*) W_q (exponential) = $2W_q$ (constant) = $\frac{8}{5}W_q$ (Erlang).
(*b*) W_q (new) = $\frac{1}{2}W_q$ (old) and L_q (new) = L_q (old) for all distributions.

15.7-4. Current policy: $L = 1$; proposed policy: $L = \frac{13}{16}$.

15.7-8.

Service Distribution	P_0	P_1	P_2	L
Erlang	0.561	0.316	0.123	0.561
Exponential	0.571	0.286	0.143	0.571

15.8-2. (*a*) $W = \frac{1}{2}$
(*b*) $W_1 = 0.20$, $W_2 = 0.35$, $W_3 = 1.10$
(*c*) $W_1 = 0.125$, $W_2 = 0.3125$, $W_3 = 1.250$

Chapter 16

16.3-1. (*a*) $E(\text{WC}) = 16$
(*c*) $E(\text{WC}) = 26\frac{1}{2}$

16.4-2. (*a*) Model 2 with s fixed at 1
(*b*) Adopt the proposal.

16.4-7. Status quo: $E(\text{TC}) = 50$; proposal: $E(\text{TC}) = 75.75$; keep status quo.

16.4-10. (*a*) Crew size = 2
(*b*) Crew size = 3

16.4-14. $\mu = 1.15$ minimizes $E(\text{TC})$.

16.4-21. One doctor: $E(\text{TC}) = \$624.80$; two doctors: \$92.95; have two doctors.

Chapter 17

17.3-1. (*a*) $t = 1.83$, $Q = 54.77$
(*b*) $t = 1.91$, $Q = 57.45$, $S = 52.22$

17.3-4. $t = 3.26$, $Q = 26{,}046$, $S = 24{,}572$

17.3-11. Produce 7 units in period 1 and 7 units in period 3.

17.3-13. Produce 3 units in period 1 and 4 units in period 3.

17.4-3. Produce 1,857 loaves.

17.4-9. (*a*) $G(y) = \frac{3}{10}y + 70e^{-y/25} - \frac{15}{2}$
(*b*) $(k, Q) = (21, 100)$ policy

17.4-12. If $x \leq 46$, order $46 - x$ units; otherwise, do not order.

17.4-15. If $x \leq 2$, order $2 - x$ units; otherwise, do not order.

17.4-21. $s = 24$, $Q = 58$, $(s, S) = (24, 82)$

Chapter 18

18.3-1. 2,090.5

18.3-3. 783.2

18.6-1. (*a*)

10/80	7.167						
1/81	7.233	1/82	7.267	1/83	9.633	1/84	9.267
4/81	7.733	4/82	8.067	4/83	10.367	4/84	8.867
7/81	7.433	7/82	8.700	7/83	10.433	7/84	8.267
10/81	7.500	10/82	9.467	10/83	10.267	10/84	7.967

(*b*) 1.184

18.8-1. (*a*) $410.333 + 17.630t$
(*b*) 604.267
(*c*)

3	431.6	6	445.220	9	468.880
4	434.84	7	452.098	10	478.992
5	439.356	8	460.089	11	490.193

(*d*)

3	465.52	6	517.405	9	567.864
4	483.075	7	534.236	10	585.671
5	500.196	8	551.212	11	604.329

19.2-1. (*c*) Use slow service when no customers or one customer is present and fast service when two customers are present.

19.2-2. (*a*) The possible states of the car are dented and not dented.

(*c*) When the car is not dented, park it on the street in one space. When the car is dented, get it repaired.

19.2-5. (*c*) State 0: attempt ace; state 1: attempt lob.

19.3-2. Minimize $Z = 4.5y_{02} + 5y_{03} + 50y_{14} + 9y_{15}$,

subject to

$$y_{01} + y_{02} + y_{03} + y_{14} + y_{15} = 1$$
$$y_{01} + y_{02} + y_{03} - (\tfrac{9}{10}y_{01} + \tfrac{49}{50}y_{02} + y_{03} + y_{14}) = 0$$
$$y_{14} + y_{15} - (\tfrac{1}{10}y_{01} + \tfrac{1}{50}y_{02} + y_{15}) = 0$$

and all $y_{ik} \geq 0$.

19.3-5. Minimize $Z = -\tfrac{1}{8}y_{01} + \tfrac{7}{24}y_{02} + \tfrac{1}{2}y_{11} + \tfrac{5}{12}y_{12}$,

subject to

$$y_{01} + y_{02} - (\tfrac{3}{8}y_{01} + y_{11} + \tfrac{7}{8}y_{02} + y_{12}) = 0$$
$$y_{11} + y_{12} - (\tfrac{5}{8}y_{01} + \tfrac{1}{8}y_{02}) = 0$$
$$y_{01} + y_{02} + y_{11} + y_{12} = 1$$

and $y_{ik} \geq 0$ for $i = 0, 1$; $k = 1, 2$.

19.4-2. Car not dented: park it on the street in one space. Car dented: repair it.

19.4-5. State 0: attempt ace. State 1: attempt lob.

19.5-1. Reject $600 offer, accept any of the other two.

19.5-2. Minimize $Z = 60(y_{01} + y_{11} + y_{21}) - 600y_{02} - 800y_{12} - 1{,}000y_{22}$,

subject to

$$y_{01} + y_{02} - (0.95)(\tfrac{5}{8})(y_{01} + y_{11} + y_{21}) = \tfrac{5}{8}$$
$$y_{11} + y_{12} - (0.95)(\tfrac{1}{4})(y_{01} + y_{11} + y_{21}) = \tfrac{1}{4}$$
$$y_{21} + y_{22} - (0.95)(\tfrac{1}{8})(y_{01} + y_{11} + y_{21}) = \tfrac{1}{8}$$

and $y_{ik} \geq 0$ for $i = 0, 1, 2$; $k = 1, 2$.

19.5-3. After three iterations, approximation is, in fact, the optimal policy given for Prob. 19.5-1.

19.5-11. In periods 1 to 3: Do nothing when the machine is in state 0 or 1; overhaul when machine is in state 2; and replace when machine is in state 3. In period 4: Do nothing when machine is in state 0, 1, or 2; replace when machine is in state 3.

Chapter 20

20.2-1. (*b*) Produce the chip.

20.2-4. Choose a_3 (order 25).

20.3-1. (*a*) $400,000

(*b*) Performing market research is not worthwhile.

20.3-4. (*a*) Up to $230,000

(*b*) a_3

20.3-10. (*a*) Guess coin 1.

(*b*) Heads: coin 2; tails: coin 1.

20.4-4. (*a*) Up to $800,000.

(*b*)

		State of Nature	
		W	L
Prediction	*W*	0.8182	0.1818
	L	0.3333	0.6667

(*c*) Hire the scout. Hold the campaign only if the scout predicts a winning season. Expected payoff = $1,050,000.

21.1-1. (*a*) Assigning numbers 0, 1, 2, 3, 4 to heads and 5, 6, 7, 8, 9 to tails gives the sequence THHTT.

21.2-1. (*a*) 5, 8, 1, 4, 7, 0, 3, 6, 9, 2

21.2-8. (*a*) $x = \sqrt{r}$

21.2-16. (*a*) $x = -4 \ln (1 - r)$

(*b*) $x = -2 \ln [(1 - r_1)(1 - r_2)]$

(*c*) $x = 4 \sum_{i=1}^{6} r_i - 8$

21.3-1. Use the first 10 three-digit decimals from Table 21.4 and generate observations from

$$x_i = \frac{1}{1 - r_i}.$$

Method:	Analytic	Monte Carlo	Stratified sampling	Complementary random numbers
Mean:	∞	4.3969	8.7661	3.812

21.4-1. (*a*) Est. $\{W_q\} = \frac{7}{3}$ and $P\{1.572 \le W_q \le 3.094\} = 0.90$

Name Index

Subject Index